More information about this series at http://www.springer.com/series/7407

T-H. Hubert Chan · Minming Li
Lusheng Wang (Eds.)

Combinatorial Optimization and Applications

10th International Conference, COCOA 2016
Hong Kong, China, December 16–18, 2016
Proceedings

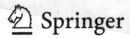 Springer

Editors
T-H. Hubert Chan
University of Hong Kong
Hong Kong
China

Lusheng Wang
City University of Hong Kong
Hong Kong
China

Minming Li
City University of Hong Kong
Hong Kong
China

ISSN 0302-9743   ISSN 1611-3349 (electronic)
Lecture Notes in Computer Science
ISBN 978-3-319-48748-9   ISBN 978-3-319-48749-6 (eBook)
DOI 10.1007/978-3-319-48749-6

Library of Congress Control Number: 2016955504

LNCS Sublibrary: SL1 – Theoretical Computer Science and General Issues

Printed on acid-free paper

This Springer imprint is published by Springer Nature
The registered company is Springer International Publishing AG
The registered company address is: Gewerbestrasse 11, 6330 Cham, Switzerland

Preface

The 10th Annual International Conference on Combinatorial Optimization and Applications (COCOA 2016) was held during December 16–18, 2016, in Hong Kong, SAR China. COCOA 2016 provided a forum for researchers working in the area of theoretical computer science and combinatorics.

The technical program of the conference included 60 contributed papers selected by the Program Committee from 122 full submissions received in response to the call for papers. All the papers were peer reviewed by Program Committee members or external reviewers. The topics cover most aspects of theoretical computer science and combinatorics related to computing, including classic combinatorial optimization, geometric optimization, complexity and data structures, graph theory, games, and miscellaneous. Some of the papers were selected for publication in special issues of *Algorithmica, Theoretical Computer Science*, and *Journal of Combinatorial Optimization*. It is expected that the journal version of the papers will appear in a more complete form.

We thank everyone who made this meeting possible: the authors for submitting papers, the Program Committee members, and external reviewers for volunteering their time to review conference papers. We would also like to extend special thanks to the publicity chairs, web chair, and financial chair for their work in making COCOA 2016 a successful event.

September 2016

T.-H. Hubert Chan
Minming Li
Lusheng Wang

Organization

General Chair

Lusheng Wang   City University of Hong Kong, Hong Kong,
SAR China

Publicity Chairs

Hongwei Du   Harbin Institute of Technology Shenzhen Graduate
School, China

Ke Yi   Hong Kong University of Science and Technology,
Hong Kong, SAR China

Financial Chair

Yingchao Zhao   Caritas Institute of Higher Education, Hong Kong,
SAR China

Web Chair

Ken C.K. Fong   City University of Hong Kong, Hong Kong,
SAR China

Program Chairs

Hubert Chan   University of Hong Kong, Hong Kong, SAR China

Minming Li   City University of Hong Kong, Hong Kong, SAR
China

Program Committee

Wolfgang Bein   University of Nevada, Las Vegas, USA

Gruia Calinescu   Illinois Institute of Technology, USA

Yixin Cao   Hong Kong Polytechnic University, Hong Kong,
SAR China

Kun-Mao Chao   National Taiwan University, Taiwan

Vincent Chau   Hong Kong Baptist University, SAR China

Jing Chen   Stony Brook University, USA

Xujin Chen   Academy of Mathematics and Systems Science, China

Zhi-Zhong Chen   Tokyo Denki University, Japan

Ovidiu Daescu   University of Texas at Dallas, USA

Additional Reviewers

Susanne Albers
James Andro-Vasko
Bogdan Armaselu
Harish Babu Arunachalam
Doina Bein
Hans-Joachim Böckenhauer
Yi-Jun Chang
Li-Hsuan Chen
Xujin Chen
Zhiyin Chen
Yongxi Cheng
Yukun Cheng
Vladimir Deineko
Ye Du
Martin Fürer
Tomas Gavenciak
Jiong Guo
Yike Guo
Xin Han
Yuya Higashikawa

Ling-Ju Hung
Johann Hurink
Hiro Ito
Taisuke Izumi
Mong-Jen Kao
Asem Kasem
Yasushi Kawase
Dennis Komm
Dimitrios Letsios
Chun-Cheng Lin
Jiaxi Liu
Tian Liu
Rashika Mishra
Tobias Mömke
Atsuki Nagao
Shin-ichi Nakano
Yoshio Okamoto
Maurizio Patrignani
Marcel Roeloffzen
Toshiki Saitoh

Feng Shan
Yongtang Shi
Suguru Tamaki
Shinichi Tanigawa
Changjun Wang
Fengmin Wang
Yinling Wang
Derek Williams
Hsiang-Yun Wu
Weiwei Wu
Mingyu Xiao
Yuan Xue
Xiaotian You
Fang Yu
Tian-Li Yu
Yong Zhang
Rong Zhou
Martin Zsifkovits

Contents

Graph Theory

On the Capture Time of Cops and Robbers Game on a Planar Graph   3
Photchchara Pisantechakool and Xuehou Tan

The Mixed Evacuation Problem   18
*Yosuke Hanawa, Yuya Higashikawa, Naoyuki Kamiyama, Naoki Katoh,
and Atsushi Takizawa*

A Comprehensive Reachability Evaluation for Airline Networks
with Multi-constraints   33
*Xiaotian You, Xiaofeng Gao, Yaru Dang, Guihai Chen,
and Xinglong Wang*

Approximation and Hardness Results for the Max k-Uncut Problem   49
Peng Zhang, Chenchen Wu, Dachuan Xu, and Xinghe Zhang

On Strong Tree-Breadth   62
Arne Leitert and Feodor F. Dragan

Computing a Tree Having a Small Vertex Cover   77
Takuro Fukunaga and Takanori Maehara

On the Approximability of PARTIAL VC DIMENSION   92
Cristina Bazgan, Florent Foucaud, and Florian Sikora

Improved Precise Fault Diagnosis Algorithm for Hypercube-Like Graphs. . . .   107
Tai-Ling Ye, Dun-Wei Cheng, and Sun-Yuan Hsieh

Finding Disjoint Paths on Edge-Colored Graphs:
A Multivariate Complexity Analysis   113
Riccardo Dondi and Florian Sikora

Total Dual Integrality of Triangle Covering   128
Xujin Chen, Zhuo Diao, Xiaodong Hu, and Zhongzheng Tang

Time-Optimal Broadcasting of Multiple Messages in 1-in Port Model   144
Petr Gregor, Riste Škrekovski, and Vida Vukašinović

Fast Searching on Complete k-partite Graphs   159
Yuan Xue, Boting Yang, Farong Zhong, and Sandra Zilles

Cliques in Regular Graphs and the Core-Periphery Problem
in Social Networks   175
    Ulrik Brandes, Eugenia Holm, and Andreas Karrenbauer

Constant Factor Approximation for the Weighted Partial Degree
Bounded Edge Packing Problem   187
    Pawan Aurora, Monalisa Jena, and Rajiv Raman

An Introduction to Coding Sequences of Graphs   202
    Shamik Ghosh, Raibatak Sen Gupta, and M.K. Sen

Minimum Eccentricity Shortest Path Problem: An Approximation
Algorithm and Relation with the k-Laminarity Problem   216
    Étienne Birmelé, Fabien de Montgolfier, and Léo Planche

On the Complexity of Extracting Subtree with Keeping Distinguishability . . .   230
    Xianmin Liu, Zhipeng Cai, Dongjing Miao, and Jianzhong Li

Safe Sets in Graphs: Graph Classes and Structural Parameters   241
    Raquel Águeda, Nathann Cohen, Shinya Fujita, Sylvain Legay,
    Yannis Manoussakis, Yasuko Matsui, Leandro Montero, Reza Naserasr,
    Yota Otachi, Tadashi Sakuma, Zsolt Tuza, and Renyu Xu

On Local Structures of Cubicity 2 Graphs   254
    Sujoy Bhore, Dibyayan Chakraborty, Sandip Das, and Sagnik Sen

Approximability of the Distance Independent Set Problem on Regular
Graphs and Planar Graphs   270
    Hiroshi Eto, Takehiro Ito, Zhilong Liu, and Eiji Miyano

Algorithmic Aspects of Disjunctive Total Domination in Graphs   285
    Chin-Fu Lin and Sheng-Lung Peng

Instance Guaranteed Ratio on Greedy Heuristic for Genome Scaffolding   294
    Clément Dallard, Mathias Weller, Annie Chateau,
    and Rodolphe Giroudeau

Geometric Optimization

Performing Multicut on Walkable Environments: Obtaining a Minimally
Connected Multi-layered Environment from a Walkable Environment   311
    Arne Hillebrand, Marjan van den Akker, Roland Geraerts,
    and Han Hoogeveen

Minimum Weight Polygon Triangulation Problem in Sub-Cubic
Time Bound   326
    Sung Eun Bae, Tong-Wook Shinn, and Tadao Takaoka

The Mixed Center Location Problem   340
  Yi Xu, Jigen Peng, and Yinfeng Xu

Constrained Light Deployment for Reducing Energy Consumption
in Buildings   350
  Huamei Tian, Kui Wu, Sue Whitesides, and Cuiying Feng

On the 2-Center Problem Under Convex Polyhedral Distance Function   365
  Sergey Bereg

Algorithms for Colourful Simplicial Depth and Medians in the Plane   378
  Olga Zasenko and Tamon Stephen

Realizability of Graphs as Triangle Cover Contact Graphs   393
  Shaheena Sultana and Md. Saidur Rahman

A Quadratic Time Exact Algorithm for Continuous Connected 2-Facility
Location Problem in Trees (Extended Abstract)   408
  Wei Ding and Ke Qiu

Complexity and Data Structure

On the (Parameterized) Complexity of Recognizing Well-Covered
(r, ℓ)-graphs   423
  Sancrey Rodrigues Alves, Konrad K. Dabrowski, Luerbio Faria,
  Sulamita Klein, Ignasi Sau, and Uéverton dos Santos Souza

Algorithmic Analysis for Ridesharing of Personal Vehicles   438
  Qian-Ping Gu, Jiajian Leo Liang, and Guochuan Zhang

On the Complexity of Bounded Deletion Propagation   453
  Dongjing Miao, Yingshu Li, Xianmin Liu, and Jianzhong Li

On Residual Approximation in Solution Extension Problems   463
  Mathias Weller, Annie Chateau, Rodolphe Giroudeau,
  Jean-Claude König, and Valentin Pollet

On the Parameterized Parallel Complexity and the Vertex Cover Problem . . .   477
  Faisal N. Abu-Khzam, Shouwei Li, Christine Markarian,
  Friedhelm Meyer auf der Heide, and Pavel Podlipyan

A Linear Potential Function for Pairing Heaps   489
  John Iacono and Mark Yagnatinsky

Amortized Efficiency of Ranking and Unranking Left-Child Sequences
in Lexicographic Order   505
  Kung-Jui Pai, Ro-Yu Wu, Jou-Ming Chang, and Shun-Chieh Chang

Combinatorial Optimization

Optimal Speed Scaling with a Solar Cell (Extended Abstract)............ 521
Neal Barcelo, Peter Kling, Michael Nugent, and Kirk Pruhs

An Approximation Algorithm for the *k*-Median Problem with Uniform
Penalties via Pseudo-Solutions 536
Chenchen Wu, Donglei Du, and Dachuan Xu

On-Line Pattern Matching on Uncertain Sequences and Applications........ 547
Carl Barton, Chang Liu, and Solon P. Pissis

Scheduling with Interjob Communication on Parallel Processors 563
*Jürgen König, Alexander Mäcker, Friedhelm Meyer auf der Heide,
and Sören Riechers*

Cost-Efficient Scheduling on Machines from the Cloud 578
*Alexander Mäcker, Manuel Malatyali, Friedhelm Meyer auf der Heide,
and Sören Riechers*

Strategic Online Facility Location 593
Maximilian Drees, Björn Feldkord, and Alexander Skopalik

An Efficient PTAS for Parallel Machine Scheduling
with Capacity Constraints..................................... 608
Lin Chen, Klaus Jansen, Wenchang Luo, and Guochuan Zhang

A Pseudo-Polynomial Time Algorithm for Solving the Knapsack Problem
in Polynomial Space 624
Noriyuki Fujimoto

Game Theory

An Incentive Mechanism for Selfish Bin Covering 641
Weian Li, Qizhi Fang, and Wenjing Liu

Congestion Games with Mixed Objectives 655
Matthias Feldotto, Lennart Leder, and Alexander Skopalik

An Optimal Strategy for Static Black-Peg Mastermind with Two Pegs...... 670
Gerold Jäger

Miscellaneous

The Incentive Ratio in Exchange Economies 685
Ido Polak

w-Centroids and Least (w, l)-Central Subtrees in Weighted Trees   693
   Erfang Shan and Liying Kang

Solving Dynamic Vehicle Routing Problem with Soft Time Window
by iLNS and hPSO   702
   Xiaohan He, Xiaoli Zeng, Liang Song, Hejiao Huang, and Hongwei Du

Convex Independence in Permutation Graphs   710
   Wing-Kai Hon, Ton Kloks, Fu-Hong Liu, and Hsiang-Hsuan Liu

The Connected p-Center Problem on Cactus Graphs   718
   Chunsong Bai, Liying Kang, and Erfang Shan

Comparison of Quadratic Convex Reformulations to Solve the Quadratic
Assignment Problem   726
   Sourour Elloumi and Amélie Lambert

Using Unified Model Checking to Verify Heaps   735
   Xu Lu, Zhenhua Duan, and Cong Tian

A Filtering Heuristic for the Computation of Minimum-Volume
Enclosing Ellipsoids   744
   Linus Källberg and Thomas Larsson

Relaxations of Discrete Sets with Semicontinuous Variables   754
   Gustavo Angulo

Unfolding the Core Structure of the Reciprocal Graph of a Massive
Online Social Network   763
   Braulio Dumba and Zhi-Li Zhang

Tackling Common Due Window Problem with a Two-Layered Approach . . .   772
   Abhishek Awasthi, Jörg Lässig, Thomas Weise, and Oliver Kramer

A Polynomial Time Solution for Permutation Scaffold Filling   782
   Nan Liu, Peng Zou, and Binhai Zhu

Author Index   791

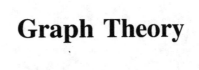

Graph Theory

On the Capture Time of Cops and Robbers Game on a Planar Graph

Photchchara Pisantechakool$^{(\boxtimes)}$ and Xuehou Tan

School of Science and Technology, School of Information Science and Technology,
Tokai University, 4-1-1 Kitakaname, Hiratsuka 259-1292, Japan
3btad008@mail.tokai-u.jp

Abstract. This paper examines the capture time of a planar graph in a pursuit-evasion games' variant called the cops and robbers game. Since any planar graph requires at most three cops, we study the capture time of a planar graph G of n vertices using three cops, which is denoted by $capt_3(G)$. We present a new capture strategy and show that $capt_3(G) \leq 2n$. This is the first result on $capt_3(G)$.

Keywords: Pursuit-evasion game · Cops and robber game · Capture time · Game length

1 Introduction

Pursuit-evasion games are turn-based ones in which one player, controlling the "evader", tries to avoid being captured by the "pursuers" controlled by another player. In each round, two players take turns to move their pieces. The games have many versions, varied through many means; such as environments, knowledge of terrain and the opponent's locations, or movements. The pursuers win if they can capture the evader, and the evader wins if he can avoid being captured indefinitely.

There are two well-known variants of pursuit-evasion games in a graph setting. In a variant where the pursuers do not know the location of the evader (see Gal [6], and Parsons [12,13]), the problem is similar to "graph searching problem". The difference is that in the pursuit-evasion game, the evader may move from an unexplored area to some areas that had been visited or cleared by the pursuers. To prevent such recontaminations, some pursuers are required to stand guard while others are searching. Finding the number of pursuers to successfully clear the graph has been the main focus of the problem. There are two approaches in finding the number of pursuers, based on searching strategy; non-monotonic search, which allows recontaminations (Kehagias et al.'s [7]), and monotonic search, which does not (such as Bienstock's [4], LaPaugh's [8] and many in Alspach's survey [2]).

This work was partially supported by JSPS KAKENHI Grant Number 15K00023.

T.-H.H. Chan et al. (Eds.): COCOA 2016, LNCS 10043, pp. 3–17, 2016.
DOI: 10.1007/978-3-319-48749-6_1

Another variant is often called a "cops and robbers game". In this variant, both players have full knowledge of the terrain and the opponent's locations. The earlier researches focused on *cop number* or how many cops are needed to win the game in a given setting (such as Nowakowski and Winkler's [11] and Quilliot's [14]). In a graph setting, Aigner and Fromme [1] proved that some instances of this game can be won by the cop player controlling one cop; such instances are called cop-win graphs. Also, they proved that for a planar graph, three cops always suffice to win. So, the cop number for a planar graph is at most three. Maurer et al. provided a report about the cops and robbers game on many different planar graphs [9].

The length of the game, or the capture time of a graph by j cops, denoted as $capt_j(G)$, has been studied recently. In 2009, Bonato et al. [5] considered the capture time of various cop-win graphs, and concluded that while the capture time of a cop-win graph of n vertices is bounded above by $n-3$, half the number of vertices is sufficient for a large class of graphs including chordal graphs. For the graphs with multiple cops required to win, the capture time can be calculated by a polynomial-time algorithm if the number of cops is fixed. They also proved that the problem of determining the minimum number of cops needed to win under the time constraint is NP-complete. In 2011, Mehrabian [10] showed that the capture time of grids, whose cop number is known to be two, is half the diameter of the graph, or $capt_2(G) \leq \lfloor \frac{m+n}{2} \rfloor - 1$ for $m \times n$ grid.

Aigner and Fromme [1] have proved that any planar graph requires at most three cops. In their proof, they introduced two concepts which can be used in actual capture strategy. One concept is the assignment of a stage i in the capture strategy to a certain subgraph $R_i \subset G$, which contains all the vertices that the robber can safely enter. The graph R_i is called the *robber territory*. Another is a concept called the *guarded path*, which is the shortest path between two vertices such that a cop can capture the robber if the robber ever enters any vertex on the path. After one cop successfully controlled a path, the robber territory changes such that $R_{i+1} \subsetneq R_i$. Their method is to repeatedly find a new guarded path that differs from previous ones, until the robber territory is eventually reduced to one vertex. However, the capture time of their strategy is not examined in [1].

In this paper, we focus on the capture time of the cops and robbers game on a planar graph in general, which has not yet been studied well. We present a new capture strategy by refining the work of Aigner and Fromme in the following two sides: (i) a new guarded path introduced at a stage shares only its end vertices with any current path, and (ii) the end vertices of a newly introduced guarded path are on or very close to some *outer cycle*, whose all vertices belong to the infinite face of the robber territory. These two refinements are involved, specially the second needs some deep observations. All guarded paths in our strategy are chosen so that they are almost distinct, excluding their end vertices and a special situation in which two new paths are simultaneously introduced at a stage. A strategy with capture time less than $2n$ can then be obtained. The capture time using our strategy is provably faster than $2n$.

In Sect. 2 of this paper, we introduce the definitions and notations for the Cops and Robbers game. In Sect. 3 we review the known results from Aigner and Fromme's research. We introduce the new concepts for our strategy in Sect. 4. Section 5 describes how to choose the guarded paths, so as to make our refinitions. The correctness and completeness of the capture strategy are provided in Sect. 6, and the capture time of our strategy is analyzed in Sect. 7. The conclusion is given in Sect. 8.

2 Preliminaries

The game of Cops and Robbers in this research is played on a planar graph, and both players know the locations of one another's pieces [1]. The cop player controls three cops, and the robber player controls one robber. The game starts by letting the cop player choose the vertices to occupy, or place her cops first, and then the robber player occupies his vertex. Such a cop-robber turn is often called a *Round*. The first cop-robber turn is considered as Round 0. For simplicity, we assume that all cops occupy the same vertex in Round 0. In one's turn or a round, a player can move his/her piece to only an adjacent vertex of that piece, and he/she can choose not to move as well. In her turn, the cop player can move up to three cops at the same time.

Let $capt_3(G)$ denote the capture time of the cops and robber game on a planar graph G. Clearly, if there is a capture strategy on G such that after a finite number k (>0) of rounds, the robber is captured (i.e., the vertex occupied by the robber is also occupied by a cop), then $capt_3(G) \leq k$.

In order to capture the robber, the cop player may employ a cop to prevent the robber from crossing a certain line, like a goalkeeper preventing a ball from entering the goal in soccer. Once she moved to the right spot on that line, she is occupied with moving along the line in reaction to the robber's movement. This kind of actions limits the robber's movement to one side of the line, and thus diminishes the area the robber can safely enter. For this purpose, we define below *stages* and *robber territories* as well.

Definition 1. *The stage i, $0 \leq i \leq t$, is the assignment of a subgraph R_i which has all the vertices the robber can still safely enter. The assignment of R_i is done after the cop player has fixed her pieces at the end of stage $i-1$, and we assume $R_0 = G$. The subgraph R_i is called the robber territory.*

We assume that stage 0 exactly coincides with Round 0, and thus $R_1 = G - \{e\}$ where e denotes the vertex initially occupied by the cops. At a stage i (>0), the cop player constructs one or two new guarded paths so that R_i is reduced to $R_{i+1} \subsetneq R_i$. Thus, stage i may consist of several rounds. The *length* of stage i is then defined as the number of rounds it takes for the cop player to fix her pieces, and the length of stage 0 is assumed to be zero. If we have a strategy that ends after t stage, the capture time of the strategy is equal to $\sum_{i=1}^{t} length$ *of stage i*. We will focus on the movements of cops because the length of a stage is actually decided by the cop whose number of movements is the largest among three cops.

A graph $G = (V, E)$ is defined as a set $V(G)$ of vertices which are connected by a set $E(G)$ of edges. We assume the graph is planar and undirected. A *cut vertex* $v \in G$ is a vertex such that when removed, the graph G has the increased number of connected components. By the nature of our problem, graph G must be connected; otherwise the cops may not be on the same connected component as the robber and thus it may not be cop-win.

Denoted by $\pi(u, v)$ a shortest path between $u \in V(G)$ and $v \in V(G)$, and $|\pi(u, v)|$ the length of $\pi(u, v)$, measured by the number of edges in $\pi(u, v)$. The distance between two vertices u and v is then defined as $|\pi(u, v)|$. A cycle, denoted by C, is a path whose start vertex and end vertex are the same. The *diameter* of a graph G is the greatest distance among all pairs of vertices in G.

For a vertex $v \in V(G)$, $N(v)$ is defined as the set of the vertices adjacent to v. For a vertex set $S \subseteq V(G)$, $G[S]$ is defined as the subgraph of G that is induced by the vertex set S, i.e., an edge $e \in E(G)$ belongs to $G[S]$ if two vertices of e belong to S.

3   Known Results

In this section, we review some known results from Aigner and Fromme's work [1].

Lemma 1 *(1-cop win [Aigner and Fromme]). Graph G is a 1-cop-win if and only if by successively removing pitfalls, G can be reduced to a single vertex.*

A pitfall is a vertex $v \in V(G)$ such that there exists a vertex $u \in N(v)$ and $N(v) \subset N(u)$. Lemma 1 can be applied to any tree, since we can simply chase down the robber from some vertex to a dead end at some leaf.

Lemma 2 *(Planar Graph Cop Number [Aigner and Fromme]). For any planar graph G, the cop number of G is at most three.*

Their proof of Lemma 2 provided a series of concepts and tools that can be used in the actual strategy. One is the concept of the robber territory, which we introduced in Definition 2. The other concept, of *guarded paths*, is devoted to diminishing the area of the robber territory.

Lemma 3 *(Guarded Shortest Path [Aigner and Fromme]). Let G be any graph, $u, v \in V(G)$, $u \neq v$ and $P = \pi(u, v)$. We assume that at least two cops are in the play. Then a single cop c on P can, after the movements no more than twice the diameter of G, prevents the robber r from entering P. That is, r will immediately be caught if he moves into P.*

It is imperative that we provide the proof of Lemma 3 as well, particularly for the new claim that "a single cop c on P can, after the movements *no more than twice the diameter of G*, prevents r from entering P".

Modified Proof of Lemma 3. Suppose the cop c is on vertex $i \in P$ and the robber r is on vertex $j \in V(G)$. Assume $\forall z \in P, |\pi(z,j)| \geq |\pi(z,i)|$; denote this as (*).

Claim A. No matter what the robber does, the cop, by moving in the appropriate direction on P, can preserve condition (*). If the r does not move, then neither does c, and (*) holds. If r moves to a new vertex k, then $\forall z \in P, |\pi(k,z)| \geq |\pi(j,z)| - 1 \geq |\pi(i,z)| - 1$. If $i' \in P$ exists with $|\pi(k,i')| \geq |\pi(i,i')| - 1$, then c, by moving on P toward i', also reduces the distance by 1 and (*) still holds. Suppose there exist vertices $x, y \in P$ such that they are on the different sides of i on the path P, and $|\pi(k,x)| = |\pi(i,x)-1|, |\pi(k,y)| \leq |\pi(i,y)|$ or $|\pi(k,x)| \leq |\pi(i,x)|$, $|\pi(k,y)| = |\pi(i,y)| - 1$. This is impossible, since by the triangle inequality and minimality of P;
$|\pi(x,y)| \leq |\pi(k,x)| + |\pi(k,y)| \leq |\pi(i,x)| + |\pi(i,y)| - 1 = |\pi(x,y)| - 1$; a contradiction.

Claim B. It takes a number of movements no more than twice the diameter of G for c to enforce (*). First, c moves to some $i \in P$, which takes at most the diameter of G. By the same argument as described above, $|\pi(j,z)| < |\pi(i,z)|$ only holds for z's on P on one side of i. By moving in the direction of z which takes at most $|P|$ or the diameter of G moves, (*) is eventually forced.   □

At a single stage, an unoccupied (free) cop c will move to position herself on some vertex of a guarded path P so as to enforce the condition (*), i.e., be on the vertex such that she takes less time than the robber to move to any other vertex on P. Until (*) is enforced by c, the robber r may cross P safely. Once (*) is enforced and constantly preserved by c, r can no longer cross P without being captured. The location satisfying (*) on P changes as the robber moves, but there always exists at least one at a given time.

Corollary 1. *If a cop c is already on P, then the number of rounds it take for c to eventually enforce (*) is bounded above by the length of P.*

Corollary 2. *When a cop successfully controls (i.e., enforce (*) on) a shortest path, all vertices of that path do not belong to the robber territory.*

From Lemma 3, capturing the robber can be done by letting the cops alternatively take the role of a free cop and guard a new shortest path within the robber territory. Once the free cop successfully guards the path, she becomes occupied and another cop, whose guarded path no longer interacts with the robber territory, becomes the free cop at the next stage. At stage i, the path guarded at stage $i-1$ by a now-free cop is called an *obsolete path*.

Remark. From Lemma 3, the length of each stage of Aigner and Fromme's strategy is bounded by twice the diameter of R_i, or loosely by $2|V(R_i)|$, as their guarded paths are usually not distinct [1]. Suppose each stage only reduces one vertex in the worst case. Then, the capture time of Aigner and Fromme's strategy can roughly be bounded by $\sum_{i=1}^{n} 2i = n(n-1)$. Their capture time may actually be much faster than $O(n^2)$, but it needs a careful calculation (such as more detailed evaluation on how paths with shared vertices interact with the movements of the cops at each stage).

4   New Concepts

In order to establish a better upper bound on $capt_3(G)$, we make two refinitions over [1]. The first one is that a new path shares only its end vertices with any current path. It is worth pointing out that in our strategy, two new paths may also be introduced at a stage. As we will see, the guarded paths in our strategy are almost distinct, excluding their starting/ending vertices. (In [1], a new path may share more than just end vertices with a current path.). Our second refinition is to choose the end vertices of the new guarded path to be on or very close to the infinite face of robber territory. To precisely define where to choose the end vertices, we need a new concept called *outer cycles*.

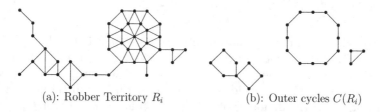

(a): Robber Territory R_i (b): Outer cycles $C(R_i)$

Fig. 1. A robber territory (a) and its outer cycles (b).

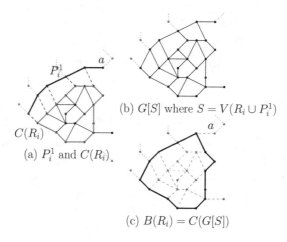

(b) $G[S]$ where $S = V(R_i \cup P_i^1)$

(a) P_i^1 and $C(R_i)$.

(c) $B(R_i) = C(G[S])$

Fig. 2. In (a) the current guarded path P_i^1 is shown in thick black line, and the outer cycle of R_i ($C(R_i)$) is shown in thick shaded cycle. (b) shows the subgraph induced on G by the vertex set $V(R_i \cup P_i^1)$, drawn in black lines and dots. The thick cycle in (c) represents the graph $B(R_i) = C(G[V(R_i \cup P_i^1)])$.

Definition 2. *The subgraph $C(R_i)$ of R_i is defined as the set of the outer cycles, whose all the vertices and edges belong to the* infinite *(exterior) face of R_i (Fig. 1). In the case of a polyhedral graph, where all of its faces can be considered as interior ones, we can choose any face as the infinite face. For graphs with multiple planar embeddings, outer cycles are made from the infinite face of the planar straight line drawing.*

Suppose that at the beginning of stage i, one or two current paths are guarded by the cops so as to prevent the robber from leaving R_i. These paths will be denoted by P_i^1 and P_i^2. Since P_i^1 and P_i^2 are assigned at the end of R_{i-1}, $P_i^1 \cap R_i = \emptyset$ and $P_i^2 \cap R_i = \emptyset$. We assume that P_i^1 always exists, i.e., R_i $(i > 0)$ can NEVER be assigned without P_i^1. The newly introduced path(s) at stage i will be denoted by P or/and Q.

For two end vertices of a new guarded path, one may consider to choose them from $C(R_i)$. But, it is difficult or even impossible in some cases for the new path to share a common vertex with P_i^1 or P_i^2. To overcome this difficulty, we will select the end vertices from the outer cycle of the subgraph induced by the vertices of the union of R_i, P_i^1 and P_i^2.

Definition 3. *Let $S = V(R_i \cup P_i^1 \cup P_i^2)$. The graph $B(R_i)$ (of enlarged outer cycles) is defined as $C(G[S])$.*

Note that $B(R_i)$ consists of only cycles, and thus not all vertices of R_i, P_i^1 and P_i^2 belong to $B(R_i)$. For instance, the vertex a of P_i^1 (Fig. 2(a)) does not belong to $B(R_i)$. See Fig. 2(c).

5   Capture Strategy

This section focuses on how to choose the guarded paths at each stage of our strategy. The analysis on the capture time of our strategy will be given in Sect. 7. We first give two propositions which we use throughout our capture strategy.

Proposition 1: At the end of each stage i, we have at least one free cop.

Proposition 2: During our strategy, any guarded path introduced at stage i shares only its end vertices with each of the current paths. For the two guarded paths introduced at the same stage, they have a common end vertex, and may share a subpath starting from that common vertex.

Before we begin, keep in mind that at some stage the robber territory may become a tree; in this case, only one cop suffices (Lemma 1). The main idea of our strategy is to let at most two cops guard two different paths, and then employ the free cop(s) to guard the new path(s), which makes one of the current guarded paths obsolete and thus reduces the robber territory.

Our capture strategy consists of two phases.

1. Initial Phase: We first find a location to place the cops, based on the structure of the graph G. This phase also establishes the very first pair of the guarded paths, and thus goes from the start to the end of stage 1. When it is over, R_1 is reduced to R_2.

2. Recursive Phase: At a stage i (≥ 2), we construct new guarded path(s) using a case-analysis method. At the end of stage i, R_i is reduced to R_{i+1}. We do this recursively until R_i becomes the tree case.

Before describing the initial and recursive phases, we first give a procedure, called *TwoCopsStart*, which is used when we have two free cops at stage i (hence P_i^2 does not exist). The procedure takes two inputs: a cycle C of $B(R_i)$ and a starting vertex $u \in C$. The procedure *TwoCopsStart* finds the other end vertex for either of the two new paths. It is done by first setting the pointers x and y such that $|\pi(u,x)|$ and $|\pi(u,y)|$ on C are at $\lfloor \frac{|C|}{3} \rfloor$. Recall that a portion of P_i^1 may be included in $C \subseteq B(R_i)$, but $P_i^1 \cap R_i = \emptyset$. Thus $P_i^1 \cap C$ is the only portion of C that is not in R_i. Since P_i^1 as well as $P_i^1 \cap C$ are shortest paths in R_{i-1}, the length of the path $C - (P_i^1 \cap C)$, which is in R_i, is at least $|P_i^1 \cap C|$. Hence, at least a half of C is in R_i, and either x or y is initially in R_i. *TwoCopsStart* then repeatedly checks whether the pointer x (or y) belongs to R_i; if *not*, x (y) is moved on C further away from u, until x (y) belongs to R_i. Finally, x and y are returned as v_1 and v_2, respectively. The outputs v_1 and v_2 will be used to construct the new paths $\pi(u,v_1)$ and $\pi(u,v_2)$ in the graph $G[V(R_i) + u]$.

Procedure: *TwoCopsStart*
Input: a cycle $C \subseteq B(R_i)$ and a vertex $u \in C$.
Output: two vertices v_1 and v_2 for the new guarded paths $\pi(u,v_1)$ and $\pi(u,v_2)$.

1. Set a pointer x at u, move x clockwise along C for $\lceil \frac{|C|}{3} \rceil$ vertices.
2. **While** x is not in R_i, **do** move x to the next vertex on C clockwise.
3. Set a pointer y at u, move y counterclockwise along C for $\lceil \frac{|C|}{3} \rceil$ vertices.
4. **While** y is not in R_i, **do** move y to the next vertex on C counterclockwise.
5. **Return** $v_1 \leftarrow x$ and $v_2 \leftarrow y$.

5.1   Initial Phase

The initial phase has the following two objectives: (i) find a vertex to place the cops at stage 0, and (ii) establish the first pair of the guarded paths at stage 1 or in R_1.

At stage 0, if $B(R_0)$ is empty, then we know that the graph is a tree, which can be easily dealt with as stated in Lemma 1. In the case that $B(R_0)$ is not empty, we choose a vertex $e_0 \in B(R_0)$ such that e_0 is not a cut vertex (for the simplicity of assigning R_1). We place all cops c_1, c_2 and c_3 at e_0, and then wait for the robber player to place his piece r on the graph.

Suppose r is now located at some vertex in $R_0 - e_0$. At stage 1, we have $R_1 = R_0 - e_0$ and $B(R_1) = B(R_0)$ (as P_1^1 is the vertex e_0 and $P_1^2 = \emptyset$). Note that $e_0 \in B(R_1)$ is on some cycle C_1 of $B(R_1)$. Using R_1, we execute *TwoCopsStart* by letting $C \leftarrow C_1$ and $u \leftarrow e_0$. After obtaining the outputs v_1 and v_2, we find the shortest paths $P = \pi(e_0, v_1)$ and $Q = \pi(e_0, v_2)$ in $G[V(R_1) + e_0]$ and send c_1 and c_2 to guard P and Q, respectively. Note that the length of stage 1 is mainly determined by the operation of moving a cop to a vertex on the shortest path and then enforcing condition (*) on that path, which can be done without concerning the movements of r.

At the end of the initial phase, if P and Q separate R_1 into two or more components, R_2 is then the connected component containing the robber r. Otherwise, $R_2 = G[V(R_1 - P - Q)]$. The reduced robber territory R_2 is either a subgraph with at least one outer cycle, or a tree. If it is the tree case, the robber can simply be captured (Lemma 1). Otherwise, we enter Recursive Phase.

5.2   Recursive Phase

In this phase, we recursively reduce the robber territory R_i into R_{i+1}, until R_i is a tree. The reduction of R_i into R_{i+1} is done by constructing and controlling one or two new guarded paths (Corollary 2).

Recall that the current paths P_i^1 and P_i^2 were given at the end of stage $i - 1$, and R_i can never be assigned without P_i^1. We distinguish the following situations.

Case (a): $B(R_i) \cap P_i^1$ has at most one vertex.

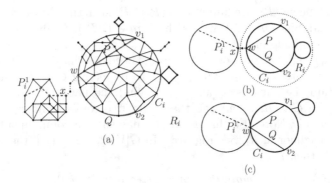

Fig. 3. An example of case (a); P_i^1, shown in thick dashed line, with the unique path $\pi(x, w)$, $x \in P_i^1$ and $w \in B(R_i)$. The new paths P and Q, with the common end vertex w, are shown in thick dark line.

Note that $P_i^2 = \emptyset$ in this case. If $B(R_i) \cap P_i^1 = \emptyset$, we find a vertex $x \in P_i^1$ and a vertex $w \in B(R_i)$ such that $|\pi(x, w)|$ is minimum among the shortest paths from a vertex of P_i^1 to the other of $B(R_i)$. Since $B(R_i) \cap P_i^1 = \emptyset$, the pair (x, w) is unique, and all vertices of $\pi(x, w)$ are cut vertices, see Fig. 3(b). We first move all three cops c_1, c_2 and c_3 to x, and then along $\pi(x, w)$ to u. In the case that $B(R_i) \cap P_i^1$ has one vertex, we let w be that vertex, see Fig. 3(c).

Let $C_i \subseteq B(R_i)$ be the cycle containing w. Since one cop can simply guard vertex w, two cops are free in this case. We execute *TwoCopsStart* procedure by letting $C \leftarrow C_i$ and $u \leftarrow w$. Note that w has some neighbors in R_i (two of them are on C_i). After obtaining the outputs v_1 and v_2 (which belong to R_i), we find the shortest paths $P = \pi(w, v_1)$ and $Q = \pi(w, v_2)$ in $G[V(R_i) + w]$, and then send the free cops, say, c_1 and c_2 to guard P and Q, respectively. In $G[V(R_i - P - Q)]$, the component containing r is then R_{i+1}.

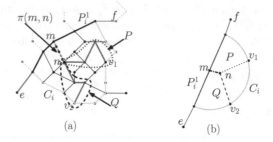

(a)   (b)

Fig. 4. An example of case (b); P_i^1 is shown in thick line, and $C_i \subseteq B(R_i)$ is shown in a combination of thick black line and thick gray lines. $P(m, v_1)$ is shown in thick dotted line, and $Q(m, v_2)$ in thick dashed line. P and Q may share some subpath $\pi(m, n)$, which is shown in thick dash dotted line.

Case (b): $B(R_i) \cap P_i^1$ has more than one vertex and $P_i^2 = \emptyset$.

Suppose $P_i^1 = \pi(e, f)$. Let $N(R_i)$ denote the set of all vertices u, where $u \in N(v)$ for some $v \in R_i$. We find $m \in N(R_i) \cap P_i^1$ such that $|\pi(m, f)| - |\pi(e, m)|$ (≥ 0) is minimal. Since $m \in P_i^1$, we also have $m \in B(R_i)$. Again, let $C_i \subseteq B(R_i)$ be the cycle containing m.

Since $B(R_i) \cap P_i^2 = \emptyset$, only one cop, say, c_1, needs to guard P_i^1. Thus two cops are free. We execute *TwoCopsStart* procedure by letting $C \leftarrow C_i$ and $u \leftarrow m$. After we obtain the outputs v_1 and v_2, we find the paths $P = \pi(m, v_1)$ and $Q = \pi(m, v_2)$ in $G[V(R_i) + m]$ and then send free cops c_2 and c_3 to guard P and Q, respectively. Note also that P and Q may share a common subpath if m has only one neighbor in R_i. See Fig. 4(a). In $G[V(R_i - P - Q)]$, the component containing r is then R_{i+1}.

Case (c): $B(R_i) \cap P_i^1$ has more than one vertex and $P_i^2 \neq \emptyset$.

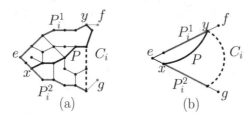

(a)   (b)

Fig. 5. An example of case (c); $P_i^1 = \pi(e, f)$, is shown in thick dark gray line, $P_i^2 = \pi(e, g)$ in thick dark line, and $P = \pi(x, y)$ in thick black line. $C_i \subseteq B(R_i)$ is shown in a combination of thick dashed lines and thick lines.

In this case, two cops have to guard P_i^1 and P_i^2, and thus we have only one free cop, say c_3. It can be deduced from Proposition 2 that P_i^1 and P_i^2 have a common vertex, say e. Let f (g) be the other end vertex of P_i^1 (P_i^2). By the denotations, $P_i^1 = \pi(e, f)$ and $P_i^2 = \pi(e, g)$. Note that f or g may not belong to $B(R_i)$ because f (g) may not be on any outer cycle of $G[V(R_i) \cup P_i^1 \cup P_i^2]$.

Next, we find two vertices $x \in N(R_i) \cap P_i^2$ and $y \in B(R_i) \cap P_i^1$ such that x and y are the vertices of P_i^2 and P_i^1, which are closest to e and f along P_i^2 and P_i^1, respectively. See Fig. 5 for example. Note that x has to be chosen from $N(R_i)$, instead of $B(R_i)$, because we want it to have some neighbor in R_i. It is also possible for $x = e$ in the case that $e \in N(R_i)$, and for $y = f$ when $f \in B(R_i)$. Finally, we find the shortest path $P = \pi(x, y)$ in $G[V(R_i) + x + y]$, and send the free cop c_3 to guard P. Again, R_{i+1} is the component containing r in $G[V(R_i - P)]$.

6 Correctness and Completeness

In this section we show that our capture strategy is correct and complete.

Theorem 1. *Proposition 1 is upheld for the whole of capture strategy.*

Proof. In case (a) and initial phase, two new paths are introduced and both start from a common vertex. Therefore, at least one cop is free at the next stage.

In case (b), as shown in Fig. 4(b), the new guarded paths P and Q may partition R_i into three components; each of them is a candidate of R_{i+1}, which is guarded by two cops at the next stage. Therefore, at least one cop is free at the next stage.

Similarly in case (c), as shown in Fig. 5(b), no matter which component becomes R_{i+1}, either P_i^1 or P_i^2 becomes obsolete and its cop is free at the next stage. □

Theorem 2. *At the end of each stage i of the robber territory, $R_{i+1} \subsetneq R_i$.*

Proof. It simply follows from the definition of the robber territory and Corollary 2. □

Theorem 3. *Proposition 2 is upheld for the whole of capture strategy.*

Proof. Using *TwoCopsStart* procedure, the two paths introduced in the initial phase, case (a) and case (b) always have a common vertex. For case (b), the newly introduced paths share exactly one vertex m with P_i^1, and two new paths may also share a subpath (e.g., $\pi(m, n)$ in Fig. 4).

In case (c), the newly introduced path $P = \pi(x, y)$ shares at most two common vertices x (when $x=e \in P_i^1$) and y with P_i^1, and exactly one common vertex x with P_i^2. □

The correctness of our strategy follows from Theorem 2, and the completeness follows from Theorems 1 and 3.

7   Capture Time of Our Strategy

In this section we describe in detail the movements of the cops during our strategy so as to get a better understanding of the capture time. We first introduce another concept, called the *active paths*.

Definition 4. *Active Path: Let $P = \pi(a, b)$ be a current guarded path at stage i, m and n the first vertex of P from a and b that has a neighbor in R_i, respectively. The subpath $P(m, n)$, from m to n, is called an active path of P.*

Sometimes a path P as whole may be an active path if both end vertices are in $N(R_i)$. If the path persists through many stages without becoming obsolete, the active portion may become smaller due to the change in robber territory. This active path can be represented as $B(R_j) \cap P$ $(j > i)$, since P is still a current path and thus the active portion of P at stage j belongs to $B(R_j)$.

Lemma 4. *Suppose path P is guarded by some cop c. Then it suffices for c to guard the active subpath of P.*

Proof. Let $P(m, n)$ be the active path of P in current stage i. By Definition 4, the robber territory R_i has no vertex adjacent to any vertices on $P - P(m, n)$. Suppose the robber r wants to travel to some vertex $u \in P - P(m, n)$. But u cannot be reached in R_i without traversing through P. The robber has to enter $P(m, n)$ first, which is a subpath of P, and by Lemma 3, he will get captured. $\square$

The guarding action of a cop requires three types of movements; (i) moving into the path, (ii) moving along the path to satisfy (*), and (iii) moving along the path while keeping to preserve (*). The length of a stage is mainly determined by the action of the free cop trying to control a new path. When a free cop successfully controls a path by enforcing (*) (i.e., be on the position that can move to any vertices on that path in smaller number of movements than the robber), the stage is then over. So (iii) movements can safely be ignored. In the following, we give a method to count the movements (i) and (ii) taken by the free cops.

In the initial phase, the cops are already on their own paths. The length of stage 1 involves only (ii) movements, which takes at most $max\{|P|, |Q|\}$. In the recursive phase, assume every stage $i > 1$ does not have any path that separates subgraph R_i into multiple components. This is the worst case because none of the vertices that do not belong to any guarded paths is removed from the robber territory.

When a path, say, $U = \pi(p, q)$, is introduced at stage i (>1), it is traversed by a cop c^i at most $|U|$ (ii) movements (Lemma 3). See Fig. 6(a). As discussed above, the introduction of U makes one of the current paths, say W, obsolete at the end of stage i. At the next stage $i + 1$, a new free cop (differing from c^i), say, c^{i+1}, must move out of her obsolete path W (made obsolete by U) to a newly introduced path, say X. It might be faster to move c^{i+1} directly from W to X. But, for simplicity, we do the following: move to the common vertex of

W and U, and then (i.1) move along the intermediate path, which is a portion of the path (U), to reach destination path (X). See Fig. 6(b). At some later stage $i+j$ $(j>1)$, U becomes obsolete by the introduction of some new path Y (at stage $i+j-1$) and the cop c^{i+j}, who was guarding U and labeled as c^i, moves to guard another new path Z. Note that Y can be X if U is obsolete at stage $i+2$. The cop c^{i+j} must first move out of the obsolete path (U) by (i.2) moving along the obsolete path to its vertex that is common with the current path (Y) at stage $i+j$, and then travels to Z using the method described above. See Fig. 6(c).

In summary, a path U is traversed by the free cops in three separate occasions: (ii) movements when being introduced at stage i, (i.1) movements when being a current guarded path at stage $i+1$, and (i.2) movements after becoming obsolete at stage $i+j$.

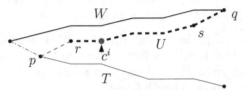

(a) Newly introduced U is traversed by c^i at stage i

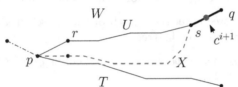

(b) Current path U is traversed by c^{i+1} $(\neq c^i)$ at stage $i+1$

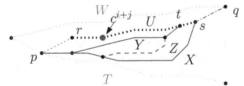

(c) Obsolete path U is traversed by c^{i+j} $(=c^i)$ at stage $i+j$

Fig. 6. A single path U is traversed by free cops on three separate occasions. At each stage, an obsolete path is shown in dotted line, a current path in normal heavy line, and a new path in dashed line. The portion of U traversed by a free cop is shown in thick line.

Let $U(r,s)$ be the active subpath of U. Following from Lemma 4, the cop c^i stops somewhere between r and s after she perform type (ii) movements. Thus, U is traversed by a free cop for at most $|U(p,s)|$ or $|U(r,q)|$ (ii) movements. In order for c^{i+1} to move from obsolete path (W) to newly introduce path (X) at stage $i+1$, at most $|U(p,r)|$ or $|U(s,q)|$ (i.1) movements are made on U. At the beginning of stage $i+j$, U is made obsolete by Y, and the cop c^{i+j} must be somewhere between r and s on U. At stage $i+j-1$, the path Y is introduced, and the active portion of the current path U is $B(R_{i+j-1}) \cap U$. Thus, an end vertex of Y coincides with an end vertex, say, t of $B(R_{i+j-1}) \cap U$ (Proposition 2). From the discussion made above, t is somewhere between r and s on U. Therefore, U is traversed by the cop c^{i+j} for at most $|U(r,t)|$ or $|U(t,s)|$ (i.2) movements to move out of obsolete path U.

In conclusion, the number of the movements (i.1), (i.2) and (ii) on U is no more than $2|U|$. Hence, we have the following result.

Lemma 5. *In our capture strategy, each guarded path is traversed no more than twice of its length.*

Theorem 4. *In our capture strategy, all the paths used in evaluating the lengths of stages are distinct, excluding their end vertices.*

Proof. Supposed the guarded paths P_i and P_{i+1} are introduced during stage i and stage $i+1$, respectively. It follows from Theorem 3 that, excluding their end vertices, P_i is distinct from P_{i+1}. When two paths are introduced at the same stage and they may share a subpath (case (b)), only one of them (i.e., the one traversed by the free cop whose number of movements is larger) is used in evaluating the length of that stage. Therefore, the theorem follows. □

For completeness, in a tree case, the length of the chase on a tree is simply bounded above by the diameter of the tree.

Theorem 5. *For the cops and robbers game on a planar graph G of n vertices with three cops, $capt_3(G) \leq 2n$.*

Proof. The theorem directly follows from Theorem 4 and Lemma 5. □

8   Conclusion

We have presented a new capture strategy for the Cops and Robbers Game on a planar graph with three cops, and shown that the capture time of our strategy is no more than $2n$. This gives the first linear result on $capt_3(G)$.

An extension for future work is to apply our new concepts to improve a capture strategy in a polygonal environment; a problem researched by Bhadauria and Isler [3]. Their strategy, which is used to prove that three cops suffice in the polygonal environment with obstacles, is also based on the concepts given by Aigner and Fromme. The upper bound on the capture time is suggested to be as large as $2nA$, where A is the area and n is the number of vertices of the environment. In a geometric setting, even if all the paths are distinct, the sum of their lengths is still bounded above by A. It is rather loose as a large portion of the given environment is removed from the robber territory at the end of a stage. This thus suggests that by applying our strategy we may obtain a smaller upper bound, say, $2A$. We are working in this direction.

References

1. Aigner, M., Fromme, M.: A game of cops and robbers. Discrete Appl. Math. **8**(1), 1–12 (1984)
2. Alspach, B.: Searching and sweeping graphs: a brief survey. Le Mathematiche **59**, 2 (2004)
3. Bhaduaria, D., Klein, K., Isler, V., Suri, S.: Capturing an evader in polygonal environments with obstacles: the full visibility case. Int. J. Robot. Res. **31**(10), 1176–1189 (2012)
4. Bienstock, D., Seymour, P.: Monotonicity in graph searching. J. Algorithm **12**, 239–245 (2011)
5. Bonato, A., Golovach, P., Hahn, G., Kratochvíl, J.: The capture time of a graph. Discrete Math. **309**, 5588–5595 (2009)
6. Gal, I.: Search Games. Addison-Wesley, Reading (1982)
7. Kehagias, A., Hollinger, G.A., Singh, S.: A graph search algorithm for indoor pursuit-evasion. Math. Comput. Model. **50**(9–10), 1305–1317 (2008)
8. LaPaugh, A.S.: Recontamination does not help to search a graph. J. ACM **40**, 224–245 (1993)
9. Maurer, A., McCauley, J., Valeva, S.: Cops and robbers on planar graphs. In: Summer 2010 Interdisciplinary Research Experience for Undergraduates. University of Minnesota (2010)
10. Mehrabian, A.: The capture time of grids. Discrete Math. **311**, 102–105 (2011)
11. Nowakowski, R., Winkler, R.P.: Vertex-to-vertex pursuit in a graph. Discrete Math. **43**, 235–239 (1983)
12. Parsons, T.D.: Pursuit-evasion in a graph. In: Alavi, Y., Lick, D.R. (eds.) Theory and Applications of Graphs. Lecture Notes in Mathematics, vol. 642, pp. 426–441. Springer, Heidelberg (1978)
13. Parsons, T.D.: The search number of a connected graph. In: Proceedings of 9th South-Eastern Conference on Combinatorics Graph Theory and Computing, pp. 549–554 (1978)
14. Quilliot, A.: Some results about pursuit games on metric spaces obtained through graph theory techniques. Eur. J. Comb. **7**(1), 55–66 (1986)

The Mixed Evacuation Problem

Yosuke Hanawa[1], Yuya Higashikawa[2,3], Naoyuki Kamiyama[4,5(✉)],
Naoki Katoh[3,6], and Atsushi Takizawa[3,7]

[1] Kyoto University, Kyoto, Japan
yosuke.hanawa@gmail.com
[2] Chuo University, Tokyo, Japan
higashikawa.874@g.chuo-u.ac.jp
[3] JST, CREST, Saitama, Japan
[4] Kyushu University, Fukuoka, Japan
kamiyama@imi.kyushu-u.ac.jp
[5] JST, PRESTO, Saitama, Japan
[6] Kwansei-Gakuin University, Hyogo, Japan
naoki.katoh@gmail.com
[7] Osaka City University, Osaka, Japan
takizawa@arch.eng.osaka-cu.ac.jp

Abstract. A dynamic network introduced by Ford and Fulkerson is a directed graph with capacities and transit times on its arcs. The quickest transshipment problem is one of the most fundamental problems in dynamic networks. In this problem, we are given sources and sinks. Then, the goal of this problem is to find a minimum time limit such that we can send exactly the right amount of flow from sources to sinks. In this paper, we introduce a variant of this problem called the mixed evacuation problem. This problem models an emergent situation in which people can evacuate on foot or by car. The goal is to organize such a mixed evacuation so that an efficient evacuation can be achieved. In this paper, we study this problem from the theoretical and practical viewpoints. In the first part, we prove the polynomial-time solvability of this problem in the case where the number of sources and sinks is not large, and also prove the polynomial-time solvability and computational hardness of its variants with integer constraints. In the second part, we apply our model to the case study of Minabe town in Wakayama prefecture, Japan.

1 Introduction

The coastal area facing the Pacific Ocean in Japan ranging from Shizuoka prefecture to Miyazaki prefecture has a high risk of a tsunami. In particular, it

This research was the result of the joint research with CSIS, the University of Tokyo (No. 573) and used the following data: Digital Road Map Database extended version 2013 provided by Sumitomo Electric Industries, Ltd and Zmap TOWN II 2008/09 Shapefile Wakayama prefecture provided by Zenrin Co. Ltd.
Y. Higashikawa, N. Katoh and A. Takizawa—This work was partially supported by JSPS Grant-in-Aid for Scientific Research(A) (25240004).
N. Kamiyama—This work was supported by JST, PRESTO.

T-H.H. Chan et al. (Eds.): COCOA 2016, LNCS 10043, pp. 18–32, 2016.
DOI: 10.1007/978-3-319-48749-6_2

is predicted that Nankai Trough Earthquake will occur with 70 % probability within thirty years, and it will trigger a tsunami of the huge size which will quickly arrive at the coast (see, e.g., [1]). Based on several assumptions and esti- mated data, Wakayama prefecture recently designated several areas in which it is difficult for all people in the area to evacuate to safety places such as tsunami evacuation buildings before a tsunami arrives when Nankai Trough Earthquake occurs. For example, it is predicted that in Kushimoto town located at the south end of the main land of Japan, a tsunami arrives at earliest within ten minutes. One of assumptions the prefecture used is that the evacuation is done only by walking. In principle, it used to be not allowed to use cars for evacuation because the usage of cars in such an emergent situation may block evacuation of pedestrians which was observed at the time of Tohoku-Pacific Ocean Earth- quake. However, if it is allowed to use cars and the smooth evacuation by car is organized, then the evacuation completion time may be shortened. The aim of this paper is to propose a mathematical model for making such a good "mixed" evacuation plan.

In this paper, we use a dynamic network flow introduced by Ford and Fulkerson [2,3] for modeling such a mixed evacuation. A dynamic network is a directed graph with capacities and transit times on its arcs. The quickest trans- shipment problem is one of the most fundamental problems in dynamic networks. In this problem, we are given a dynamic network with several sources and sinks. Furthermore, we are given a supply for each source and a demand for each sink. Then, the goal of this problem is to find the minimum time limit such that we can send exactly the right amount of flow from sources to sinks. Hoppe and Tardos [4] proved that this problem can be solved in polynomial time. In this paper, we introduce a variant of the quickest transshipment problem called the mixed evacuation problem. This problem models an emergent situation in which people can evacuate on foot or by car. The goal of this problem is to organize such a mixed evacuation so that an efficient evacuation can be achieved. In the first part of this paper, we study the mixed evacuation problem from the theoret- ical viewpoint. First we prove that if the number of sources and sinks is at most $C \log_2 n$ (n is the number of vertices) for some constant C, then mixed evacua- tion problem can be solved in polynomial time (Sect. 3). In addition, we consider variants of the mixed evacuation problem with integer constraints (Sect. 4). In the second part of this paper, we study the mixed evacuation problem from the practical viewpoint. In this part, we apply our model to the case study in Japan (Sect. 5). More precisely, we apply our model for Minabe town in Wakayama pre- fecture, which was designated as a city in which safe evacuation from a tsunami is difficult when Nankai Trough Earthquake occurs.

Let $\mathbb{R}$, $\mathbb{R}_+$, $\mathbb{Z}$, $\mathbb{Z}_+$, and $\mathbb{Z}_-$ be the sets of reals, non-negative reals, integers, non-negative integers, and non-positive integers, respectively. For each finite set U, each vector x in $\mathbb{R}^U$, and each subset W of U, we define $x(W) := \sum_{u \in W} x(u)$. Furthermore, for each finite set U and each pair of vectors x, y in $\mathbb{R}^U$, we define $\langle x, y \rangle := \sum_{u \in U} x(u)y(u)$.

2   Preliminaries

The MIXED EVACUATION problem is defined as follows. We are given a directed graph $D = (V, A)$ and two disjoint subsets S, T of V. Define $n := |V|$. The subset S (resp., T) is the set of source vertices (resp., sink vertices) in V. We assume that no arc in A enters (resp., leaves) a vertex in S (resp., T). In addition, we are given an arc capacity vector c in $\mathbb{Z}_+^A$, a supply vector b in $\mathbb{Z}_+^S$, a sink capacity vector u in $\mathbb{Z}_-^T$, transit time vectors τ_1, τ_2 in $\mathbb{Z}_+^A$, and fluid coefficients q_1, q_2 in $\mathbb{Z}_+$. In our application, τ_1 represents the speed of walking, and q_1 represents the number of people that can walk in the one unit of the arc capacity. The values τ_2, q_2 represent the information for cars. Lastly, we are given a time limit Θ in $\mathbb{Z}_+$. Define $[\Theta] := \{0, 1, \ldots, \Theta\}$.

For each integer i in $\{1, 2\}$, each function $f \colon A \times [\Theta] \to \mathbb{R}_+$, each vertex v in V, and each integer θ in $[\Theta]$, we define

$$\partial_i f(v, \theta) := \sum_{a \in \delta(v; A)} \sum_{t=0}^{\theta} f(a, t) - \sum_{a \in \varrho(v; A)} \sum_{t=0}^{\theta - \tau_i(a)} f(a, t),$$

where $\delta(v; A)$ (resp., $\varrho(v; A)$) represents the set of arcs in A leaving (resp., entering) v. A vector d in $\mathbb{R}^{S \cup T}$ is called an *allocation*, if $d(v) \geq 0$ for every vertex v in S and $d(v) \leq 0$ for every vertex v in T. For each integer i in $\{1, 2\}$, each allocation d in $\mathbb{R}^{S \cup T}$, and each vector w in $\mathbb{R}_+^A$, d is said to be (i, w)-*feasible*, if there exists a function $f \colon A \times [\Theta] \to \mathbb{R}_+$ satisfying the following conditions.

(D1) Let a and θ be an arc in A and an integer in $[\Theta]$, respectively.
     - If $\theta \leq \Theta - \tau_i(a)$, then $f(a, \theta) \leq q_i \cdot w(a)$.
     - If $\theta > \Theta - \tau_i(a)$, then $f(a, \theta) = 0$.
(D2) Let v and θ be a vertex in V and an integer in $[\Theta]$, respectively.
     - If $v \in V \backslash (S \cup T)$, then $\partial_i f(v, \theta) \leq 0$.
     - If $v \in S$ (resp., T), then $\partial_i f(v, \theta) \leq d(v)$ (resp., $\partial_i f(v, \theta) \geq d(v)$).
(D3) Let v be a vertex in V.
     - If $v \in V \backslash (S \cup T)$, then $\partial_i f(v, \Theta) = 0$.
     - If $v \in S$, then $\partial_i f(v, \Theta) = d(v)$.

For each integer i in $\{1, 2\}$ and each vector w in $\mathbb{R}_+^A$, let $\mathcal{F}_i(w)$ be the set of (i, w)-feasible allocations in $\mathbb{R}^{S \cup T}$. An *assignment* is a tuple (d_1, d_2, w_1, w_2) such that d_1, d_2 are allocations in $\mathbb{R}^{S \cup T}$ and $w_1, w_2 \in \mathbb{R}_+^A$. Furthermore, an assignment (d_1, d_2, w_1, w_2) is said to be *feasible*, if it satisfies the following conditions.

(F1) For every vertex v in S (resp., T), $d_1(v) + d_2(v) = b(v)$ (resp., $\geq u(v)$).
(F2) For every arc a in A, we have $w_1(a) + w_2(a) \leq c(a)$.
(F3) We have $d_1 \in \mathcal{F}_1(w_1)$ and $d_2 \in \mathcal{F}_2(w_2)$.

The goal of MIXED EVACUATION (ME for short) is to decide whether there exists a feasible assignment. Notice that we can straightforwardly formulate ME as a linear programming problem (in Sect. 5, we use an algorithm based on the linear programming). However, since the input size of Θ is $\log_2 \Theta$, its size is not bounded by a polynomial in the input size of ME.

3   Mixed Evacuation with Few Sources and Sinks

In this section, we prove that if $|S \cup T| \leq C \log_2 n$ for some constant C, then ME can be solved in polynomial time. Assume that we are given an integer i in $\{1, 2\}$, a vector w in $\mathbb{R}_+^A$, and a subset X of $S \cup T$. Define $\mathbf{D}_i^w(X)$ as the set of functions $f \colon A \times [\Theta] \to \mathbb{R}_+$ satisfying (D1) and the following conditions.

(D4) Let v and θ be a vertex in V and an integer in $[\Theta]$, respectively.
- If $v \in V \backslash X$, then $\partial_i f(v, \theta) \leq 0$.
- If $v \in X$, then $\partial_i f(v, \theta) \geq 0$.

(D5) For every vertex v in $V \backslash (S \cup T)$, we have $\partial_i f(v, \Theta) = 0$.

Recall that no arc in A enters (resp., leaves) a vertex in S (resp., T). Thus, for every function f in $\mathbf{D}_i^w(X)$ and every vertex v in $(S \backslash X) \cup (T \cap X)$, we have $\partial_i f(v, \Theta) = 0$. Furthermore, we define a function $o_i^w \colon 2^{S \cup T} \to \mathbb{R}_+$ by

$$o_i^w(X) := \max \left\{ \sum_{v \in X} \partial_i f(v, \Theta) \ \middle| \ f \in \mathbf{D}_i^w(X) \right\}.$$

Theorem 1 (Klinz [4, Theorem 5.1]). *Assume that we are given an integer i in $\{1, 2\}$, an allocation d in $\mathbb{R}^{S \cup T}$, and a vector w in $\mathbb{R}_+^A$. Then, $d \in \mathcal{F}_i(w)$ if and only if $d(X) \leq o_i^w(X)$ for every subset X of $S \cup T$.*

For each pair of vertices s in S and t in T, we denote by $a(t, s)$ an arc from t to s. Define $E := A \cup \{a(t, s) \mid s \in S, t \in T\}$. Furthermore, we define H as the directed graph with the vertex set V and the arc set E. Then, for each integer i in $\{1, 2\}$, each vector w in $\mathbb{R}_+^A$, each subset X of $S \cup T$, and each vector ξ in $\mathbb{R}_+^E$, ξ is called a *feasible static flow* with respect to i, w, and X, if it satisfies the following conditions (S1), (S2), and (S3).

(S1) For every arc a in A, we have $\xi(a) \leq q_i \cdot w(a)$.

(S2) For every pair of vertices s in S and t in T, if at least one of $s \in S \backslash X$ and $t \in X$ holds, then $\xi(a(t, s)) = 0$.

(S3) For every vertex v in V, we have $\xi(\delta(v; E)) = \xi(\varrho(v; E))$.

For each integer i in $\{1, 2\}$, each vector w in $\mathbb{R}_+^A$, and each subset X of $S \cup T$, we denote by $\mathbf{S}_i^w(X)$ the set of feasible static flows with respect to i, w, and X. In addition, for each integer i in $\{1, 2\}$, we define a vector k_i in $\mathbb{R}^E$ as follows. For each arc a in $E \backslash A$ (resp., A), we define $k_i(a) := \Theta + 1$ (resp., $-\tau_i(a)$).

Theorem 2 (Ford and Fulkerson [2,3]). *For every integer i in $\{1, 2\}$, every vector w in $\mathbb{R}_+^A$, and every subset X of $S \cup T$, $o_i^w(X)$ is equal to the optimal objective value of the problem of maximizing $\langle k_i, \xi \rangle$ such that $\xi \in \mathbf{S}_i^w(X)$.*

Define $\mathcal{P}$ as the set of assignments (d_1, d_2, w_1, w_2) such that it satisfies (F2) and (F3), and the following condition.

(F1$^+$) For every vertex v in S (resp., T), $d_1(v) + d_2(v) \leq b(v)$ (resp., $\geq u(v)$).

Let P1 be the problem of maximizing $d_1(S)+d_2(S)$ such that $(d_1, d_2, w_1, w_2) \in \mathcal{P}$. If the optimal objective value of P1 is equal to $b(S)$, then we can conclude that there exists a feasible assignment. Otherwise, we can conclude that there exists no feasible assignment. This observation implies that if we can formulate P1 by a linear programming problem whose size is bounded by a polynomial in the size of ME, then the polynomial-time solvability of ME follows from the polynomial-time solvability of the linear programming problem [5]. Theorems 1, 2 imply that P1 can be formulated as follows. Define a vector b° in $\mathbb{R}^{S \cup T}$ by $b^\circ(v) := b(v)$ for each vertex v in S and $b^\circ(v) := 0$ for each vertex v in T. Define a vector u° in $\mathbb{R}^{S \cup T}$ by $u^\circ(v) := 0$ for each vertex v in S and $u^\circ(v) := u(v)$ for each vertex v in T.

$$
\begin{aligned}
\text{Maximize} \quad & d_1(S) + d_2(S) \\
\text{subject to} \quad & d_1(v) \geq 0, \; d_2(v) \geq 0 \quad (v \in S) \\
& d_1(v) \leq 0, \; d_2(v) \leq 0 \quad (v \in T) \\
& u^\circ(v) \leq d_1(v) + d_2(v) \leq b^\circ(v) \quad (v \in S \cup T) \\
& \langle k_i, \xi_{i,X} \rangle \geq d_i(X) \quad (i \in \{1,2\}, X \subseteq S \cup T) \\
& \xi_{i,X} \in \mathbf{S}_i^{w_i}(X) \quad (i \in \{1,2\}, X \subseteq S \cup T) \\
& w_1(a) + w_2(a) \leq c(a) \quad (a \in A) \\
& d_1, d_2 \in \mathbb{R}^{S \cup T}, \; w_1, w_2 \in \mathbb{R}_+^A.
\end{aligned}
$$

If $|S \cup T| \leq C \log_2 n$ for some constant C, then it is not difficult to see that the size of this linear programming problem is bounded by a polynomial of the input size of ME. This completes the proof.

4   Mixed Evacuation with Integer Constraints

4.1   Integral Arc Capacities, Supplies, and Sink Capacities

Here we consider INTEGRAL MIXED EVACUATION (IME for short). This problem is a variant of ME in which a feasible assignment (d_1, d_2, w_1, w_2) must satisfy that $d_1, d_2 \in \mathbb{Z}^{S \cup T}$ and $w_1, w_2 \in \mathbb{Z}_+^A$. We prove the **NP**-completeness of IME by reduction from DISJOINT PATHS WITH DIFFERENT COSTS [6] (DPDC for short) defined as follows. In what follows, we do not distinguish a simple directed path in a directed graph and the set of arcs contained in this directed path. We are given a directed graph $G = (N, L)$, a source vertex v^+ in N, and a sink vertex v^- in N. Furthermore, we are given cost vectors ℓ_1, ℓ_2 in $\mathbb{Z}_+^L$ and a non-negative integer h in $\mathbb{Z}_+$. The goal of DPDC is to decide whether there exist arc-disjoint simple directed paths P_1, P_2 from v^+ to v^- such that $\ell_1(P_1) + \ell_2(P_2) \leq h$.

Theorem 3 (Li, McCormick, and Simchi-Levi [6]). *The problem* DPDC *is* **NP**-*complete even if $h = 0$ and $\ell_i(a) \in \{0,1\}$ for every integer i in $\{1,2\}$ and every arc a in L.*[1]

---

[1] In [6, Theorem 1], although the condition that $h = 0$ and $\ell_i(a) \in \{0,1\}$ for every integer i in $\{1,2\}$ and every arc a in L is not explicitly stated, the reduction in their proof satisfies this condition.

For proving the fact that IME is in **NP**, we need the following theorems. For each finite set U and each function $g\colon 2^U \to \mathbb{R}$, g is said to be *submodular*, if $g(X) + g(Y) \geq g(X \cup Y) + g(X \cap Y)$ for every pair of subsets X, Y of U.

Theorem 4 (E.g., [7,8]). *Assume that we are given a finite set U and a submodular function $g\colon 2^U \to \mathbb{R}$. Then, we can find a subset X of U minimizing $g(X)$ among all subsets of U in time bounded by a polynomial in $|U|$ and* EO, *where* EO *is the time required to compute $g(X)$ for a subset X of U.*

Theorem 5 (Hoppe and Tardos [4]). *For every integer i in $\{1,2\}$ and every vector w in $\mathbb{R}_+^A$, the function o_i^w is a submodular function.*

Theorem 6 (E.g., [9]). *For every integer i in $\{1,2\}$ and every subset X of $S \cup T$, we can compute $o_i^{w_i}(X)$ in polynomial time.*

Assume that we are given an integer i in $\{1,2\}$, an allocation d in $\mathbb{R}^{S \cup T}$, and a vector w in $\mathbb{R}_+^A$. Then, Theorem 1 implies that $d \in \mathcal{F}_i(w)$ if and only if $o_i^w(X) - d(X) \geq 0$ for a subset X of $S \cup T$ minimizing $o_i^w(X) - d(X)$ among all subsets of $S \cup T$. Thus, since Theorem 5 implies that $o_i^w - d$ is submodular, Theorems 4, 6 imply that we can check whether $d \in \mathcal{F}_i(w)$ in polynomial time.[2]

Theorem 7. *The problem* IME *is* **NP***-complete.*

Proof. Theorems 1, 4, 5, and 6 imply that IME is in **NP**. We prove that IME is **NP**-complete by reduction from DPDC. Assume that we are given an instance of DPDC such that $h = 0$ and $\ell_i(a) \in \{0,1\}$ for every integer i in $\{1,2\}$ and every arc a in L. Then, we construct an instance of IME as follows. Define $V := N \cup \{s^*\}$, where s^* is a new vertex. Define $A := L \cup \{a_1, a_2\}$, where a_1 and a_2 are new arcs from s^* to v^+. Define $S := \{s^*\}$ and $T := \{v^-\}$. Define $c(a) := 1$ for each arc a in A. For each arc a in A, we define

$$\tau_1(a) := \begin{cases} \ell_1(a) & \text{if } a \in L \\ 0 & \text{if } a = a_1 \\ 1 & \text{if } a = a_2 \end{cases} \qquad \tau_2(a) := \begin{cases} \ell_2(a) & \text{if } a \in L \\ 1 & \text{if } a = a_1 \\ 0 & \text{if } a = a_2. \end{cases}$$

Define $b(s^*) := 2$ and $u(v^-) := -2$. Define $q_1 := 1$, $q_2 := 1$, and $\Theta := 0$.

Assume that there are arc-disjoint simple directed paths P_1, P_2 in G from v^+ to v^- such that $\ell_1(P_1) + \ell_2(P_2) \leq 0$. Since $\ell_i(a) \geq 0$ for every integer i in $\{1,2\}$ and every arc a in L, we have $\ell_1(P_1) = \ell_2(P_2) = 0$. For each integer i in $\{1,2\}$, we define a directed path P_i^+ in D as the directed path obtained by adding a_i to P_i. Since P_1 and P_2 are arc-disjoint, P_1^+ and P_2^+ are arc-disjoint. Furthermore, for every integer i in $\{1,2\}$, since $\ell_i(P_i) = 0$ and $\tau_i(a_i) = 0$, we have $\tau_i(P_i^+) = 0$. For each integer i in $\{1,2\}$, we define $d_i(s^*) := 1$ and $d_i(v^-) := -1$. In addition, for each integer i in $\{1,2\}$, we define a vector w_i in $\mathbb{Z}_+^A$ as follows. If $a \in P_i^+$, then we define $w_i(a) := 1$. Otherwise, we define $w_i(a) := 0$. Since P_1^+ and P_2^+ are

---

[2] This proof is the same as the proof of the polynomial-time solvability of the decision version of the quickest transshipment problem in [4].

arc-disjoint, $w_1(a) + w_2(a) \leq 1 = c(a)$ for every arc a in A. Thus, (d_1, d_2, w_1, w_2) is a feasible assignment.

Next we assume that there exists a feasible assignment (d_1, d_2, w_1, w_2) such that $d_1, d_2 \in \mathbb{Z}^{S \cup T}$ and $w_1, w_2 \in \mathbb{Z}_+^A$. Since $\tau_1(a_2) = \tau_2(a_1) = 1$, $c(a_1) = c(a_2) = 1$, and $\Theta = 0$, we have $d_1(s^*) = d_2(s^*) = 1$. Since $c(a) = 1$ for every arc a in A, we have $w_1(a), w_2(a) \in \{0, 1\}$ and at most one of $w_1(a)$ and $w_2(a)$ is equal to 1 for every arc a in A. For each integer i in $\{1, 2\}$, we denote by L_i the sets of arcs a in A such that $w_i(a) = 1$. Then, L_1 and L_2 are disjoint. For every integer i in $\{1, 2\}$, since $d_i(s^*) = 1$ and $\Theta = 0$, L_i contains a simple directed path L_i' from s^* to v^- such that $\tau_i(L_i') = 0$ as a subset. Furthermore, for every integer i in $\{1, 2\}$, the definition of τ_i, we have $a_i \in L_i'$. For each integer i in $\{1, 2\}$, let P_i be the directed path obtained by removing a_i from L_i'. Then, for every integer i in $\{1, 2\}$, we have $\ell_i(P_i) = 0$. This completes the proof.   □

4.2   Integral Supplies and Sink Capacities

Here we consider problems of finding an integral allocation of supplies and sink capacities. We consider INTEGRAL MIXED EVACUATION WITH ARC CAPACITIES (IMEAC for short) defined as follows. We are given vectors $w_1, w_2 \in \mathbb{Z}_+^A$ such that $w_1(a) + w_2(a) \leq c(a)$ for every arc a in A. Then, the goal is to decide whether there exists a feasible assignment (d_1, d_2, w_1, w_2) such that $d_1, d_2 \in \mathbb{Z}^{S \cup T}$. We prove that IMEAC can be solved in polynomial time. In the rest of this section, we define $o_i := o_i^{w_i}$ for each integer i in $\{1, 2\}$. For each finite set U, each function $g \colon 2^U \to \mathbb{R}$, and each pair of subsets P_1, P_2 of $\mathbb{R}^U$, we define $P(g) := \{x \in \mathbb{R}^U \mid x(X) \leq g(X) \ (\forall X \subseteq U)\}$ and $P_1 + P_2 := \{x + y \mid x \in P_1, y \in P_2\}$. Then, Theorem 1 implies that IMEAC can be formulated as the following problem P2.

$$
\begin{aligned}
\text{Maximize} \quad & d_1(S) + d_2(S) \\
\text{subject to} \quad & d_1(v) \geq 0, \ d_2(v) \geq 0 \quad (v \in S) \\
& d_1(v) \leq 0, \ d_2(v) \leq 0 \quad (v \in T) \\
& u^\circ(v) \leq d_1(v) + d_2(v) \leq b^\circ(v) \quad (v \in S \cup T) \\
& d_1 \in P(o_1), \ d_2 \in P(o_2), \ d_1, d_2 \in \mathbb{Z}^{S \cup T}.
\end{aligned}
$$

If the optimal objective value is equal to $b(S)$, then we can conclude that there exists a desired assignment. Otherwise, there exists no such an assignment.

Assume that we are given an integer i in $\{1, 2\}$. Define $o_i^- \colon 2^{S \cup T} \to \mathbb{R}_+$ by setting $o_i^-(X)$ to be the minimum value of $o_i(Y)$ over all subsets Y of X such that $X \cap S \subseteq Y$. Let X be a subset of $S \cup T$. Define a function $o_{i,X} \colon 2^{X \setminus S} \to \mathbb{R}$ by setting $o_{i,X}(Y) := o_i(Y \cup (X \cap S)) - o_i(X \cap S)$. It is not difficult to see that $o_{i,X}$ is submodular and $o_i^-(X) = \min\{o_{i,X}(Y) \mid Y \subseteq X \setminus S\} + o_i(X \cap S)$. That is, we can evaluate $o_i^-(X)$ by evaluating the value of o_i and using the algorithm for submodular function minimization. It is known [10, Eq. (3.10)] that o_i^- is a submodular function. Furthermore, it is known [10, Theorem 3.3] that $P(o_i^-)$ is equal to the set of vectors d in $P(o_i)$ such that $d(v) \leq 0$ for every vertex v in T.

Thus, P2 is equivalent to the following problem.

$$\begin{aligned}
\text{Maximize} \quad & d_1(S) + d_2(S) \\
\text{subject to} \quad & d_1(v) \geq 0, \ d_2(v) \geq 0 \quad (v \in S) \\
& u^\circ(v) \leq d_1(v) + d_2(v) \leq b^\circ(v) \quad (v \in S \cup T) \\
& d_1 \in \mathrm{P}(o_1^-), \ d_2 \in \mathrm{P}(o_2^-), \ d_1, d_2 \in \mathbb{Z}^{S \cup T}.
\end{aligned}$$

We consider the following problem P3.

$$\begin{aligned}
\text{Maximize} \quad & d_1(S) + d_2(S) \\
\text{subject to} \quad & u^\circ(v) \leq d_1(v) + d_2(v) \leq b^\circ(v) \quad (v \in S \cup T) \\
& d_1 \in \mathrm{P}(o_1^-), \ d_2 \in \mathrm{P}(o_2^-), \ d_1, d_2 \in \mathbb{Z}^{S \cup T}.
\end{aligned}$$

Lemma 1. *The optimal objective values of* P2 *and* P3 *are the same.*

Proof. For each optimal solution (d_1, d_2) of P3, we define $\gamma(d_1, d_2)$ as the number of pairs (i, v) of an integer i in $\{1, 2\}$ and a vertex v in S such that $d_i(v) < 0$. Let (d_1, d_2) be an optimal solution of P3 minimizing $\gamma(d_1, d_2)$ among all optimal solutions of P3. If $\gamma(d_1, d_2) = 0$, then since P3 is a relaxation problem of P2, (d_1, d_2) is clearly an optimal solution of P2, and thus the proof is done. Assume that $\gamma(d_1, d_2) \geq 1$. Let (i, v) be a pair of an integer i in $\{1, 2\}$ and a vertex v in S such that $d_i(v) < 0$. We assume that $i = 1$ (we can treat the case of $i = 2$ in the same way). For proving this lemma by contradiction, we prove that there exists an optimal solution (d_1', d_2') of P3 such that $\gamma(d_1, d_2) > \gamma(d_1', d_2')$. This contradicts the definition of (d_1, d_2), and thus this completes the proof.

Define vectors d_1', d_2' in $\mathbb{Z}^{S \cup T}$ as follows. Define $d_1'(v') := d_1(v')$ and $d_2'(v') := d_2(v')$ for each vertex v' in $(S \cup T) \setminus \{v\}$. Furthermore, define $d_1'(v) := 0$ and $d_2'(v) := \min\{d_2(v), b(v)\}$. We first prove that $d_1(v) + d_2(v) \leq d_1'(v) + d_2'(v)$. If $d_2'(v) = d_2(v)$, then since $d_1(v) < 0$, this clearly holds. If $d_2'(v) = b(v)$, then since $d_1(v) + d_2(v) \leq b(v)$, this clearly holds. This implies that the objective value of (d_1', d_2') is no less than that of (d_1, d_2). Thus, what remains is to prove that (d_1', d_2') is a feasible solution of P3. The above inequality implies that (d_1', d_2') satisfies the first constraint of P3. In addition, d_2' clearly belongs to $\mathrm{P}(o_2^-)$. Thus, it suffices to prove that $d_1' \in \mathrm{P}(o_1^-)$. Assume that this does not hold. Then, there exists a subset of X of $S \cup T$ such that $v \in X$ and $o_1^-(X) - d_1(X) < -d_1(v)$. Since it is not difficult to see that $o_1(Y \setminus \{v\}) \leq o_1(Y)$ for every subset Y of X such that $X \cap S \subseteq Y$, $o_1^-(X \setminus \{v\}) \leq o_1^-(X)$. Thus, $o_1^-(X \setminus \{v\}) < d_1(X \setminus \{v\})$, which contradicts that $d_1 \in \mathrm{P}(o_1^-)$. This completes the proof. $\qquad \square$

Lemma 2. *For every integer i in $\{1, 2\}$ and every subset X of $S \cup T$, we have $o_i(X) \in \mathbb{Z}$, which implies that $o_i^-(X) \in \mathbb{Z}$.*

Proof. This lemma follows from Theorem 2 and [11, Theorem 12.8]. $\qquad \square$

Theorem 8 (E.g., [11, Corollary 46.2c]**).** *Assume that we are given a finite set U and submodular functions $\sigma, \pi: 2^U \to \mathbb{Z}$ such that $\sigma(\emptyset) = \pi(\emptyset) = 0$. Then, we have* $(\mathrm{P}(\sigma) \cap \mathbb{Z}^U) + (\mathrm{P}(\pi) \cap \mathbb{Z}^U) = (\mathrm{P}(\sigma) + \mathrm{P}(\pi)) \cap \mathbb{Z}^U.$[3]

Lemmas 1, 2 and Theorem 8 imply that P2 is equivalent to the following problem.

$$
\begin{aligned}
\text{Maximize} \quad & d(S) \\
\text{subject to} \quad & u^\circ(v) \le d(v) \le b^\circ(v) \quad (v \in S \cup T) \\
& d \in \mathrm{P}(o_1^-) + \mathrm{P}(o_2^-), \ d \in \mathbb{Z}^{S \cup T}.
\end{aligned}
$$

Theorem 9 (E.g., [11, Theorem 44.6]**).** *Assume that we are given a finite set U and submodular functions $\sigma, \pi: 2^U \to \mathbb{Z}$ such that $\sigma(\emptyset) = \pi(\emptyset) = 0$. Then, we have* $\mathrm{P}(\sigma) + \mathrm{P}(\pi) = \mathrm{P}(\sigma + \pi).$[4]

Theorem 9 implies that P2 is equivalent to the following problem.

$$
\begin{aligned}
\text{Maximize} \quad & d(S) \\
\text{subject to} \quad & u^\circ(v) \le d(v) \le b^\circ(v) \quad (v \in S \cup T) \\
& d \in \mathrm{P}(o_1^- + o_2^-), \ d \in \mathbb{Z}^{S \cup T}.
\end{aligned}
$$

We consider the following relaxation problem LP2 of P2.

$$
\begin{aligned}
\text{Maximize} \quad & d(S) \\
\text{subject to} \quad & u^\circ(v) \le d(v) \le b^\circ(v) \quad (v \in S \cup T) \\
& d \in \mathrm{P}(o_1^- + o_2^-).
\end{aligned}
$$

Lemma 3. *The optimal objective values of* P2 *and* LP2 *are the same.*

Proof. Since o_1^-, o_2^- are submodular functions, $o_1^- + o_2^-$ is a submodular function. Furthermore, Lemma 2 implies that $o_1^-(X) + o_2^-(X) \in \mathbb{Z}$ for every subset X of $S \cup T$. Thus, this lemma follows from [11, Corollary 44.3c] (i.e., the box-total dual integrality of the constraints corresponding to $\mathrm{P}(o_1^- + o_2^-)$). □

Theorem 10. *The problem* IMEAC *can be solved in polynomial time.*

Proof. In the same way as the algorithm described after Theorem 6, we can check in polynomial time whether $d \in \mathrm{P}(o_1^- + o_2^-)$ for a given vector d in $\mathbb{R}^{S \cup T}$ by minimizing the submodular function $o_1^- + o_2^- - d$. In addition, we can check in polynomial time whether a given vector d in $\mathbb{R}^{S \cup T}$ satisfies the first constraint of LP2. Thus, we can solve the separation problem for LP2 by using Theorem 4 in polynomial time (if $d \notin \mathrm{P}(o_1^- + o_2^-)$, then a separating hyperplane can be obtained from a minimizer of $o_1^- + o_2^- - d$). This implies that we can solve IMEAC in polynomial time by using the results of [12] (see also [13, Theorem 6.36]). □

---

[3] Precisely speaking, [11, Corollary 46.2c] considers $\mathrm{P}(\sigma) \cap \mathbb{R}_+^U$. However, the similar result holds for $\mathrm{P}(\sigma)$ (see the paragraph after the proof of [11, Theorem 44.6]).

[4] In [11, Theorem 44.6], the monotonicity of functions are assumed. However, even if functions are not monotone, this theorem holds. See also [10, Eq. (3.32)].

4.3   Unsplittable Supplies and Sink Capacities

Here we consider the following variant of IMEAC called UNSPLITTABLE MIXED EVACUATION WITH ARC CAPACITIES (UMEAC for short). In this problem, we are given vectors $w_1, w_2 \in \mathbb{Z}_+^A$ such that $w_1(a) + w_2(a) \le c(a)$ for every arc a in A. The goal is to decide whether there exists a feasible assignment (d_1, d_2, w_1, w_2) such that $d_1(v), d_2(v) \in \{0, b(v)\}$ for every vertex v in S and $d_1(v), d_2(v) \in \{0, u(v)\}$ for every vertex v in T. In what follows, we prove that UMEAC is **NP**-complete. Notice that if $|S \cup T| \le C \log_2 n$ for some constant C, then it follows from Theorems 1, 4, 5, and 6 that UMEAC can be solved in polynomial time by enumerating all subsets of $S \cup T$. We will prove the **NP**-completeness of UMEAC by reduction from PARTITION. In this problem, we are given a finite set I and a vector π in $\mathbb{Z}_+^I$ such that $\pi(I)$ is even. Then, the goal is to decide whether there exists a subset J of I such that $\pi(J) = \pi(I \backslash J)$. It is well known [14] that PARTITION is **NP**-complete.

Theorem 11. *The problem* UMEAC *is* **NP***-complete.*

Proof. In the same ways as the proof of Theorem 7, we can prove that UMEAC is in **NP**. We prove that the **NP**-completeness of UMEAC by reduction from PARTITION. Assume that we are given an instance of PARTITION, and then we construct an instance of UMEAC as follows. Define $V := \{v_i \mid i \in I\} \cup \{v^\circ, v^\bullet, v_1^*, v_2^*\}$ and $A := \{(v_i, v^\circ), (v_i, v^\bullet) \mid i \in I\} \cup \{(v^\circ, v_1^*), (v^\bullet, v_2^*)\}$. Define $S := \{v_i \mid i \in I\}$ and $T := \{v_1^*, v_2^*\}$. Define $c(a) := 1$, $\tau_1(a) := 0$, and $\tau_2(a) := 0$ for each arc a in A. Define $b(v_i) := \pi(i)$ for each element i in I. Define $u(v_1^*) := -\pi(I)$ and $u(v_2^*) := -\pi(I)$. Define $q_1 := 1$ and $q_2 := 1$. Define $\Theta := (\pi(I)/2) - 1$. Lastly, we define vectors w_1, w_2 in $\mathbb{Z}_+^A$ as follows. For each arc $a = (x, y)$ in A, we define

$$w_1(a) := \begin{cases} 1 & \text{if } x = v^\circ \text{ or } y = v^\circ \\ 0 & \text{otherwise} \end{cases} \qquad w_2(a) := \begin{cases} 1 & \text{if } x = v^\bullet \text{ or } y = v^\bullet \\ 0 & \text{otherwise.} \end{cases}$$

Assume that there exists a subset J of I such that $\pi(J) = \pi(I \backslash J)$. Then, we define vectors d_1, d_2 in $\mathbb{Z}^{S \cup T}$ by

$$d_1(v) \ (\text{resp., } d_2(v)) := \begin{cases} \pi(i) \ (\text{resp., } 0) & \text{if } v = v_i \text{ for some } i \in J \\ 0 \ (\text{resp., } \pi(i)) & \text{if } v = v_i \text{ for some } i \in I \backslash J \\ -\pi(I) \ (\text{resp., } 0) & \text{if } v = v_1^* \\ 0 \ (\text{resp., } -\pi(I)) & \text{if } v = v_2^*. \end{cases}$$

Since $\pi(J) = \pi(I \backslash J) = \pi(I)/2$, (d_1, d_2, w_1, w_2) is a feasible assignment.

Conversely, we assume that there exists a feasible assignment (d_1, d_2, w_1, w_2) such that $d_1(v), d_2(v) \in \{0, b(v)\}$ for every vertex v in S and $d_1(v), d_2(v) \in \{0, u(v)\}$ for every vertex v in T. Since $\Theta = (\pi(I)/2) - 1$, we have $d_1(S) = \pi(I)/2$ and $d_2(S) = \pi(I)/2$. Thus, if we define J as the set of elements i in I such that $d_1(v_i) = b(v_i)$, then $\pi(J) = \pi(I \backslash J)$. This completes the proof. $\square$

5   Case Study

Here we apply our model to the case study of Minabe town in Wakayama prefecture, which was designated as those in which safe evacuation from tsunami is difficult when Nankai Trough Earthquake occurs. The population of Minabe town is about 12000. According to the census data of 2013, the number of people living in the tsunami inundation area of this town is 4745. The left figure of Fig. 1 shows the map of this area and the expected height of the tsunami caused by Nankai Trough Earthquake. The town is surrounded by mountains of height ranging from 100 to 200 m.

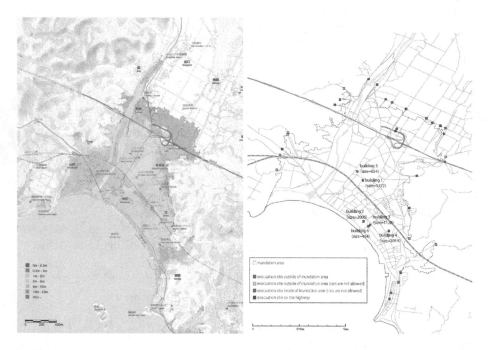

Fig. 1. (Left) The target area and its inundation depth. (Right) The road network in the target areas and evacuation sites.

It is predicted that in twelve minutes after that earthquake occurs, the first tsunami of height 1 m arrives, and then that of height 5m (and of 10 m, respectively) will arrive after 15 min (and 24 min, respectively). Since people usually start evacuation five minutes after the earthquake occurs, the actual time remaining for evacuation is from five to fifteen minutes depending on where they live. Since there are not enough evacuation buildings in the center of the town, most of the people will have to go to the outside of the tsunami inundation area, and thus some of them may not succeed to evacuate to a safety place.

Under this circumstance, we consider the following experiments. Our computational experiment aims at the inundation area of Minabe town whose population is 4745. We prepare two scenarios. The first one is that people should have to evacuate to the outside of the inundation area. The second is that people should have to evacuate to the outside of the inundation area or to tsunami evacuation buildings located inside of the inundation area. There exist six evacuation buildings inside the inundation area (numbered from 1 through 6 in the right figure of Fig. 1) whose sizes (i.e., the maximum number of evacuees that can be accommodated) are 1472, 2000, 1128, 3014, 654 and 454, respectively. We constructed a model of a dynamic network by using the GIS databases: the fundamental map information (1/2500, the Geospatial Information Authority of Japan), the population census (2010, the Ministry of Internal Affairs and Communications of Japan), and the Japan digital road map (Japan Digital Road Map Association). The road network has 860 nodes and 1,106 arcs.

We assign to a sink vertex the capacity of the evacuation site located at the vertex, i.e., the maximum number of evacuees that the site accommodates. In our experiment, the capacity of a building was computed based on the available floor space, assuming that two persons per m^2 can be accommodated. The capacity of an evacuation site which is outside the tsunami inundation area is assumed to be infinity. However, since a hill top may have an upper limit on the number of evacuees that can be accommodated, its capacity is estimated based on an aerial photograph. Evacuation by cars is only possible to the outside of the tsunami inundation area, and thus is assumed to be not allowed to tsunami evacuation buildings or hill tops. Since there are not enough tsunami evacuation buildings, the delay of evacuation is predicted. (In our experiment, we solve ME by using a linear programming solver. Thus, we can add additional constraints to our model. Furthermore, the minimum evacuation completion time can be computed by the binary search.)

5.1   Computational Results

We use Gurobi Optimizer (see http://www.gurobi.com/) as the solver to solve linear programs corresponding to our experimental data.

As seen from Table 1, in each scenario, the result for the case where cars are allowed to use is much better in the minimum evacuation completion time than the one where they are not allowed. Comparing the scenario 2 with the scenario 1, the number of evacuees who walked to the evacuation site increased since evacuation buildings located in the town center can be used in the scenario 2.

Table 1. Computational results of each scenario.

Scenario	Evacuation time	Percentage of car usage	Pedestrians only
1	9m40s	68.1 %	18m00s
2	9m05s	31.4 %	17m30s

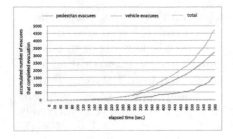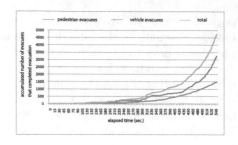

Fig. 2. (Left) The transition of the accumulated number of evacuees that completed evacuation in the scenario 1. (Right) The transition of the accumulated number of evacuees that completed evacuation in the scenario 2.

Now let us look at Fig. 2 that shows how the number of evacuees that have completed evacuation increases as time proceeds since the evacuation starts. It is observed that in the latter half for the whole time period, the number of evacuees that completed evacuation rapidly increases in both scenarios. Ideally, it is desired that the number of evacuees that completed evacuation is large in the early stage. This point should be taken into account in order to improve the current model.

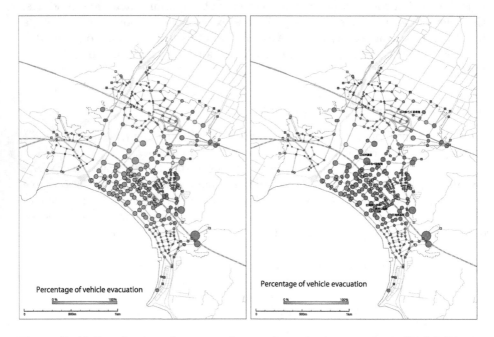

Fig. 3. (Left) Distribution of evacuees that used cars in the scenario 1. (Right) Distribution of evacuees that used cars in the scenario 2. (Color figure online)

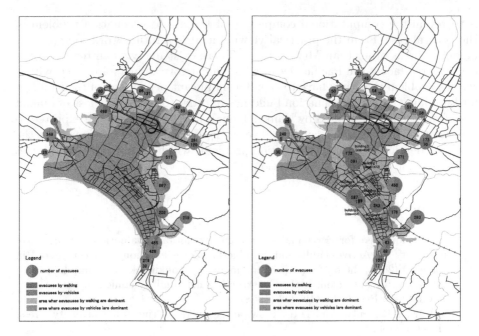

Fig. 4. (Left) The number of evacuees that arrived at each evacuation site, and the ratio of evacuees who arrived at the site by walking and those who arrived there by a car in the scenario 1 (Right) The number of evacuees that arrived at each evacuation site, and the ratio of evacuees who arrived at the site by walking and those who arrived there by a car in the scenario 2.

Let us look at the way of evacuation (by walking or a car) at each vertex. In Fig. 3, if the color at each vertex is close to blue, it means that a majority of people used cars for evacuation while on the other hand, if it is close to red, a majority of people walked for evacuation. Comparing the scenarios 1 and 2, the car usage significantly decreased in the scenario 2 near the coast since there are evacuation buildings nearby.

Figure 4 shows the number of evacuees that arrived at each evacuation site, and the ratio of evacuees who arrived at the site by walking and those who arrived there by a car. In the scenario 1, for most of evacuation sites, the number of evacuees who arrived by cars exceeds that of evacuees who arrived by walking. On the other hand in the scenario 1, many evacuees living near the town center evacuated to evacuation buildings inside the inundation area.

6 Conclusion

In this paper, we introduce the mixed evacuation problem that is motivated by making an evacuation plan in an emergent situation in which people can evacuation on foot or by car. We study this problem from the theoretical and practical viewpoints. An apparent future work from the theoretical viewpoint is to reveal the computational complexity of the mixed evacuation problem in the general case. From the practical viewpoint, it is a future work to apply our model to areas other than Minabe town. There exist many small towns on the coastal area facing the Pacific Ocean whose local governments are faced with a serious problem that they have to spend a significant percentage of their budget for building a tsunami evacuation buildings in order to reduce the loss of human lives from tsunami triggered by Nankai Trough Earthquake that are expected to occur with 70 % within the coming 30 years [1]. In this respect, we hope that the methods developed for facility location problems will help to reduce the budget to be used for such disaster prevention.

References

1. The Headquarters for Earthquake Research Promotion: Evaluation of Long-Term Probability of Active Fault and Subduction-Zone Earthquake Occurrence (in Japanese) (2016). http://www.jishin.go.jp/main/choukihyoka/ichiran.pdf
2. Ford Jr., L.R., Fulkerson, D.R.: Constructing maximal dynamic flows from staticflows. Oper. Res. **6**(3), 419–433 (1958)
3. Ford Jr., L.R., Fulkerson, D.R.: Flows in Networks. Princeton University Press, Princeton (1962)
4. Hoppe, B., Tardos, É.: The quickest transshipment problem. Math. Oper. Res. **25**(1), 36–62 (2000)
5. Khachiyan, L.G.: Polynomial algorithms in linear programming. USSR Comput. Math. Math. Phys. **20**(1), 53–72 (1980)
6. Li, C., McCormick, S.T., Simchi-Levi, D.: Finding disjoint paths with different path-costs: complexity and algorithms. Networks **22**(7), 653–667 (1992)
7. Schrijver, A.: A combinatorial algorithm minimizing submodular functions in strongly polynomial time. J. Comb. Theor. Ser. B **80**(2), 346–355 (2000)
8. Iwata, S., Fleischer, L., Fujishige, S.: A combinatorial strongly polynomial algorithm for minimizing submodular functions. J. ACM **48**(4), 761–777 (2001)
9. Orlin, J.B.: A faster strongly polynomial minimum cost flow algorithm. Oper. Res. **41**(2), 338–350 (1993)
10. Fujishige, S.: Submodular Functions and Optimization. Annals of Discrete Mathematics, vol. 58, 2nd edn. Elsevier, Amsterdam (2005)
11. Schrijver, A.: Combinatorial Optimization: Polyhedra and Efficiency, vol. 24. Springer, Heidelberg (2003)
12. Grötschel, M., Lovász, L., Schrijver, A.: Geometric Algorithms and Combinatorial Optimization, 2nd edn. Springer, Heidelberg (1993)
13. Cook, W.J., Cunningham, W.H., Pulleyblank, W.R., Schrijver, A.: Combinatorial Optimization. Wiley, Hoboken (1998)
14. Garey, M.R., Johnson, D.S.: Computers and Intractability: A Guide to the Theory of NP-Completeness. W. H Freeman and Company, New York (1979)

A Comprehensive Reachability Evaluation for Airline Networks with Multi-constraints

Xiaotian You[1], Xiaofeng Gao[1], Yaru Dang[2(✉)],
Guihai Chen[1], and Xinglong Wang[2]

[1] Shanghai Key Laboratory of Scalable Computing and Systems,
Department of Computer and Engineering, Shanghai Jiao Tong University,
Shanghai 200240, China
sjtuyxt@gmail.com, {gao-xf,gchen}@cs.sjtu.edu.cn
[2] Civil Aviation University of China, Tianjin 300300, China
yarudang@cauc.edu.cn, xinglong1979@163.com

Abstract. Airline network, including airports as network nodes and flight routes as directed network edges, has a lot of special features such as departure and arrival times, air ticket budget, flight capacity, transportation cost, etc. Thus, analyzing network behavior and service performance for such a network is much more difficult than that for many other networks. In this paper, taking China domestic airline network as a representative, we try to discuss the reachability issue for each airport respectively, which could reflect its regional connectivity level and service quality of civil aviation. More specifically, we evaluate reachability through many features including node degree, betweenness, closeness, etc. To get the values of some features, we design a fast Dijkstra-based all-pair shortest path algorithm with both time and budget requirements, then use Fenwick Tree to further improve the time efficiency. Finally, we implement Analytic Hierarchy Process (AHP) to convert the reachability feature into numerical values for all airports to measure their service qualities precisely. Our results for China domestic airline network with 210 airports and 69,160 flight routes will definitely become a guide to airline companies and civil aviation administration for their further development and management.

1 Introduction

Reachability and connectivity have been widely used as a measure to evaluate networks and for graph there are also many widely used features to benchmark. However, when it comes to the airline network, things will change. We consider an airline network as a graph including airports as nodes and flight routes as directed edges, which has a lot of differences from other networks. For instance, the graph may have hundreds or even thousands of parallel edges. Every edge in the graph has its own time window, which corresponds to the departure and arrival time of the flight route, and only in the time window can the edge be valid. What's more, every node has its own values such as ground transportation cost, which makes all nodes essentially different. Since an airline network is a

© Springer International Publishing AG 2016
T.-H.H. Chan et al. (Eds.): COCOA 2016, LNCS 10043, pp. 33–48, 2016.
DOI: 10.1007/978-3-319-48749-6_3

graph with a lot of distinctive features, it can hardly be measured by one or more explicit numerical indicators.

In this paper, we introduce a new definition of reachability, which is closely related to centrality and capacity. Centrality identifies how influential and important the node is in the graph, indicating the level of the corresponding airport. Capacity identifies the own value of the node itself regardless of the form of the graph. When it comes to centrality, we mainly consider degree centrality, betweenness centrality, and closeness centrality with time and budget requirements. For capacity, we consider flight frequency, seating capacity, and flight duration as auxiliary features. Then taking China airline network as a representative, we try to compute the reachability issue for each airport respectively through Analytic Hierarchy Process (AHP), which could indicate its rationality and superiority and be a guide to service improvement.

To figure out all the feature values, we design an algorithm to calculate the betweenness and closeness centrality, which is the most complicated part. There has been a fast algorithms for betweenness centrality [3], requiring $O(n + m)$ space and running in $O(nm + n^2 \log n)$ on weighted graph. However as is stated above, airline network is a network with many realistic features, so we should compromise to some realistic constraints. We presume the time of transfers is at most 7 and the total journey will take the passenger at most 7 days, otherwise the journey is hardly seen in the realistic world. To meet the two demands, we convert the original graph to a new three-dimensional graph, of which the three ordinates denote the airport, the time of transfers and the time of that moment respectively. After adding the corresponding edges to the new graph, we use Dijkstra algorithm with min-priority queue [3] to solve the all-pair shortest path problem. Considering some unnecessary cases, we use Fenwick Tree to reduce the size of the status, which seemingly makes the time complexity worse but actually accelerate it a lot.

To summarize, in this paper we propose a novel way to measure the reachability issue of an airport in airline networks. Taking China Airline Network as a representative, we define eight features for evaluation, and use Analytic Hierarchy Process (AHP) with expert grading matrix to quantitatively evaluate the reachability of airports. We further design a fast detection method using Dijkstra-based algorithm with min-priority queue and Fenwick tree to calculate the betweenness and closeness features in this network. Our results for China domestic airline network with 210 airports and 69,160 flight routes will denitely become a guide to airline companies and civil aviation administration.

The rest of this paper is organized as follows: Sect. 2 summarizes the related works in this area. Section 3 introduces the definitions and requirements of the problem. Section 4 design a fast algorithm to detect the betweenness and closeness centrality. Next, Sect. 5 discusses the rating procedure to evaluate the reachability issue of an airport. Finally, Sect. 6 gives the conclusion.

2   Related Work

As for airline networks, Barros [1] used DEA two-stage procedure to evaluate operational performance of European airlines, considering number of employees, number of planes and operational cost; Tsaur [13] evaluated airline service quality by fuzzy MCDM considering tangibility, reliability, responsiveness, assurance and empathy; Yu [15] used the SBM-NDEA model to assess the performance of airports considering the airside service and landside service; Bowen [2] used Airline Quality Rating (AQR) Methodology to evaluate the US airline industry.

As for centrality, Brandes [3] designed a fast algorithm for betweenness centrality requiring $O(n + m)$ space and running in $O(nm + n^2 \log n)$ time on weighted networks; Kourtellis [10] proposed a randomized algorithm for estimating betweenness centrality to identify nodes with high betweenness centrality; Lee [11] constructed tourism-management strategies for villages by evaluating spatial centrality; Guimera [9] claimed that nodes with high betweenness tend to play a more important role than those with high degree in the world-wide airport network.

3   Definitions and Features

In this paper, we take China airline network as a representative to discuss the reachability issue for each airport. First, we construct China airline network according to weekly domestic flight statistics with 210 airports and 69,160 flight routes (data are collected from July 1st, 2016 to July 7th, 2016). Figure 1 exhibits the connections between airports in mainland China with the help of JavaScript Visualization libraries like D3.js. In this figure, the blue nodes represent airports, while the grey links represent flight routes. Note that: (1) We focus on civil aviation and do not consider service aviation or cargo airline. Thus we do not plot Nanhai area. (2) We only record domestic flight routes and do not count flights from/to Taiwan, Hong Kong, and Macao. (3) Links in the figure only represents connectivity situation between airports (we omit the link direction for clarity), which do not reflect the real flight routes, since the domestic routes should be arcs within the territory of China. (4) We focus on regular flight routes, so we do not record the other irregular flight routes such as extra section flights.

Correspondingly, We construct a directed graph $G = (V, E)$. Every node in V denotes an airport while every edge in E corresponds to a flight route. For each $e \in E$, we denote the average air ticket price of this flight route as p_e, the number of seats on this flight (according to its aircraft type) as s_e, its departure time as d_e, and arrival time as a_e. For each $v \in V$, we denote its average ground transportation budget as b_v (denoting the average cost that city residents should spend to the airport) and the average ground transportation time as t_v.

To evaluate the reachability of an airport, we always need to detect the shortest path from one airport u to another v through graph G. In airline network, a path $\{u \rightarrow v \rightarrow w\}$ not only represents the connectivity property from u to w, but also denotes a valid flight transfer schedule for a passenger. Additionally,

Fig. 1. A diagrammatic sketch to show the China airline network

when considering a valid schedule, people care about the travel duration as well as the total budget. Thus, we denote the shortest path from airport u to airport v under the time and budget requirements respectively as follows.

Definition 1 (Shortest Path with Minimum Time). *A valid path from node u to v in an airline network G with Minimum time is a path $P = \{e_1, \cdots, e_n\}$, where the tail of e_i should be the head of e_{i-1} and e_1 goes out of u, e_n goes into v respectively. Additionally, $d_{e_i} \geq a_{e_{i-1}} + t$, for $i = 2, \cdots, n$ and t is a flight transfer time. The shortest path with minimum time is a valid path such that the total time duration $a_{e_n} - d_{e_1}$ is minimized.*

Let $SPT(u, v)$ denote the *shortest path with minimum time* for u and v. Easy to know, $|SPT(u, v)|$ is the path length (or cardinality of the edges) and to be realistic, $|SPT(u, v)| \leq K$ where K is the maximum transfer number (usually $K \leq 7$), and usually $t \geq 90$ min according to a regular flight transfer procedure. For convenience, we denote the time on the path $SPT(u, v)$ i.e. $a_{e_n} - d_{e_1}$ as $spt(u, v)$.

Similarly, if we consider the *shortest path with minimum budget* for u and v, instead of computing $a_{e_n} - d_{e_1}$, we want to minimize $\sum_{i=1}^{n} p_{e_i}$ with $a_{e_n} - d_{e_1} \leq T$ where T is the upper bound of total travel time duration (usually $T \leq$

7 days = 10080 min) and we use $SPB(u,v)$ to denote such a path and $spb(u,v)$ to denote the budget i.e. $\sum_{i=1}^{n} p_{e_i}$.

To analyze the reachability of a node u, we need to comprehensively consider the connectivity and importance of this airport. With the help of complex network theory, we select several features of u and evaluate the level of reachability quantitatively. These features are explained as follows.

Definition 2 (Clustering Coefficient [14]). *The clustering coefficient of a node u (CC_u) in an airline network is the portion of the pairs of connected nodes (of which the number is k_u) within its neighborhood divided by the maximal possible edges $(k_u(k_u - 1))$ between them, written as:*

$$CC_u = \frac{1}{k_u(k_u - 1)} \sum_v \sum_w I(u,v)I(u,w)I(v,w)I(w,v), \tag{1}$$

where $I(u,v)$ is presented as an indicator:

$$I(u,v) = \begin{cases} 1, & \exists e \in E, s.t. \ e \ is \ from \ u \ to \ v; \\ 0, & otherwise. \end{cases} \tag{2}$$

Degree centrality [8], betweenness centrality [7] and closeness centrality [12] are used as three major measures to evaluate a network node. Degree centrality symbolizes the importance of the node in a network, while betweenness centrality measures the extent to which a particular node lies between other nodes in a network and closeness centrality indicates the distance from all other nodes. Their definitions are represented as follows.

Definition 3 (Degree Centrality). *Degree centrality of a node u (DC_u) in an airline network is the portion of connected nodes by all the other nodes. In other words, it is the ratio of through-flight, written as:*

$$DC_u = \frac{\sum\limits_v I(u,v)}{|V| - 1}. \tag{3}$$

Definition 4 (Betweenness Centrality with Time and Budget Requirements). *Betweenness centrality of a node u (BC_u) in an airline network is defined as the ratio of all shortest paths (with time and budget requirements) passing through it and reflects its transitivity. More formally,*

$$BC_u = \frac{\sum\limits_{v \in E \wedge v \neq u} \sum\limits_{w \in E \wedge w \neq u \neq v} \left(I_t(u,v,w) + I_b(u,v,w)\right)}{2 \cdot (|V| - 1)(|V| - 2)}, \tag{4}$$

where $I_t(u,v,w)$ and $I_b(u,v,w)$ are indicators defined as Eqs. 5 and 6.

$$I_t(u,v,w) = \begin{cases} 1, & \exists e_i \in SPT(v,w), s.t. \ e_i \ is \ from \ u; \\ 0, & otherwise. \end{cases} \tag{5}$$

$$I_b(u,v,w) = \begin{cases} 1, & \exists e_i \in SPB(v,w), s.t. \ e_i \ is \ from \ u; \\ 0, & otherwise. \end{cases} \tag{6}$$

Definition 5 (Closeness Centrality of Time). *Closeness Centrality of Time of a node u (CCT_u) is the sum of the average time needed on $SPT(u,v)$ ($spt(u,v)$) for v among all the other nodes and the ground transportation time. More formally,*

$$CCT_u = \frac{\sum\limits_{v \in V \wedge v \neq u} spt(u,v)}{|V| - 1} + t_u. \tag{7}$$

Definition 6 (Closeness Centrality of Budget). *Closeness Centrality of Time of a node u (CCB_u) is the sum of the average budget needed on $SPB(u,v)$ ($spb(u,v)$) for v among all the other nodes and the ground transportation budget. More formally,*

$$CCB_u = \frac{\sum\limits_{v \in V \wedge v \neq u} spb(u,v)}{|V| - 1} + b_u. \tag{8}$$

Considering a node u, we define the *centrality* feature of u (CT_u) comprehensively in Eq. 9, combining clustering coefficient, degree centrality, betweenness centrality, and closeness centrality together.

$$CT_u = f_{CT}(CC_u, BC_u, DC_u, CCT_u, CCB_u). \tag{9}$$

Next, we consider the specific features for airline networks and give Definitions 7–9.

Definition 7 (Flight Frequency). *The Flight Frequency of a node u (FF_u) is the amount of the edges from/to u. More formally,*

$$FF_u = \sum_{e \in E} \Big(I_{in}(u,e) + I_{out}(u,e) \Big), \tag{10}$$

where $I_{in}(u,e)$ and $I_{out}(u,e)$ are defined as follows:

$$I_{in}(u,e) = \begin{cases} 1, & e \text{ is to } u; \\ 0, & \text{otherwise.} \end{cases} \qquad I_{out}(u,e) = \begin{cases} 1, & e \text{ is from } u; \\ 0, & \text{otherwise.} \end{cases} \tag{11}$$

Definition 8 (Seating Capacity). *The Seating capacity of a node u (SC_u) is the average number of the seats on the flight from the airport. More formally,*

$$SC_u = \frac{\sum\limits_{e \in E} I_{out}(u,e) s_e}{\sum\limits_{e \in E} I_{out}(u,e)}. \tag{12}$$

Definition 9 (Flight Duration). *The Flight Duration of a node u (FD_u) is the average duration of the flight routes from the airport. More formally,*

$$FD_u = \frac{\sum\limits_{e \in E} I_{out}(u,e)(a_e - d_e)}{\sum\limits_{e \in E} I_{out}(u,e)}. \tag{13}$$

For a node u, we can define the *capacity* property of u (CP_u) comprehensively as shown in Eq. 14.

$$CP_u = f_{CP}(FF_u, SC_u, FD_u). \tag{14}$$

Finally we can define the *reachability* of a node u (R_u) as follows:

$$R_u = f_R(CT_u, CP_u). \tag{15}$$

Note that functions f_{CT} in Eq. 9, f_{CP} in Eq. 14 and f_R in Eq. 15 can be customized according to different airline network architectures and requirements.

4   Fast Detection of Betweenness and Closeness

To figure out betweenness centrality and closeness centrality, we must solve the *all-pair shortest path* problem for not only the optimal $spt(u,v)$ and $spb(u,v)$ but also the trajectory of all the paths. As we mentioned above, G is a graph with multiple edges and loops and every edge has a time window so that the existing algorithms such as Dijkstra Algorithm cannot solve the problem with the transfer and duration constraints.

To implement Dijkstra-based algorithm, we need to first convert the airline network G into a simple directed graph G'. Note that a valid path links one edge to another only when the two edges are connected and their corresponding flight route can be transferred successfully. Thus the connection in the new generated graph should not only represent the connectivity issues, but also reflect the time and transfer hop information. Motivated by this observation, we introduce *duplicated graph* generated from G as follows:

Definition 10 (Duplicated Graph). *The duplicated graph of G is a simple directed graph $G' = (V', E')$ where a node in V' is denoted as (v, k, t) and an edge in E' from v'_1 to v'_2 is denoted as $\langle v'_1, v'_2 \rangle$. If there is a node $v' = (v, k, t) \in V'$, it means that at time t (take minutes as the unit) the passenger reach the airport v after the kth transfer. $\forall e \in E$ (from u to v), there is an edge $e' = \langle (u, k, t), (v, k+ 1, a_e) \rangle$ with weight $w_{e'} = p_e$ (with $k + 1 \leq$ transfer constraints and $t \leq d_e - 90$).*

In G', k reflects the transfer number, while t reflects the time information. In reality, we usually set $k \leq K$ and $t \leq T$.

Obviously every path in G' corresponds to a valid path in G and for every valid path in G, there must exist at least one corresponding path in G'. If we solve the all-pair shortest path problem on G', we can easily figure out betweenness centrality and closeness centrality in the original graph.

Figure 2 is an example to illustrate this conversion. Figure 2(a) is a simple flight network G with two nodes A and B, as well as two flight routes both departure on Monday. One is flight $A \to B$ with duration 09:00–11:00, and another is $B \to A$ during 14:00–16:00. We set $k \leq 2$ (means we only allow less than two transfers), and $0 \leq t \leq 1440$ (means the trip should be finished within one day). Figure 2(b) is the converted graph G'. Since the first flight departures at 09:00, nodes $(A, 0, 0)$ to $(A, 0, 450)$ are all valid to reach $(B, 1, 660)$ since the flight departures at 09:00, so the latest valid boarding time should be $9 \times 60 - 90 = 450$. Next, the flight takes 120 min and reaches B at 11:00, so the t value at B part should be 660. (There are totally 451 nodes if we count per minutes.) Each of the edge has weight ¥500 as ticket cost. Similarly, nodes labeled $(B, 0, 0)$ to $(B, 0, 750)$ all link to $(A, 1, 960)$ if we departure at 14×60 min $- 90$ min $= 750$ min.

Easy to see, G' has two connected components depicting all the possible transfer plans. $|V'| = 2 \times 3 \times 1441 = 8646$ and $|E'| = (451 + 751) \times 2 = 2404$ edges (but obviously there are only 4 vertices with positive in-degree, which is the key observation to reduce the algorithm complexity).

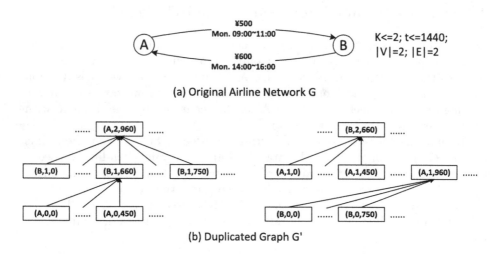

Fig. 2. An example duplicated graph G' generated from G

The next step is to solve the shortest path problem in the new graph. As a simple graph with non-negative weights (on edges as budget), we implement Dijkstra algorithm [4]. Initially we choose a $v \in V'$ and start from this specific source. At each step, we pick the closest node in the set of undiscovered nodes, update the shortest path to each undiscovered node and remove the node from the set. We keep doing it until every node is picked. Here we use array $path[u]$ to record the in-edge of the shortest path from the starting node to u, and use array $prev[u]$ to record u's previous hop along the path.

The final step is to find the corresponding shortest path in the original graph. We have got the minimum distance from the source to all other nodes with specific time and specific transfer numbers, so we can easily get the minimum duration and the minimum budget needed from the source to all the nodes. By tracing down $path[\cdot]$ and $prev[\cdot]$ arrays, we can easily get the corresponding shortest paths in original graph G.

Note that for G', there are only $O(|E|)$ nodes with positive in-degree (can be found from Fig. 2), so there are $O(|E|)$ valid nodes, i.e. $|V'| = O(|E|)$. For each node, there are at most $O(|E|)$ edges from it, so there are $O(|E|^2)$ edges i.e. $|E'| =$

$O(|E|^2)$. If we compute all-pair shortest path with Dijkstra algorithm, the total time complexity is $O(|V'|^3) = O(|E|^3)$. Apparently, for China airline network, the algorithm can hardly hit the upper bound of the time complexity. However, it still works with extremely low efficiency as a matter of fact. We choose to invoke min-priority queue [6] to improve the time efficiency. The detailed description is shown in Algorithm 1. Here $dist[u]$ is the array to record the minimum distance from the starting node to u. Actually, we need to run Algorithm 1 on G' for $|E|$ times to compute all-pair shortest path (starting from the representative of the departure point of each flight route in G).

Algorithm 1. Dijkstra Algorithm with Priority Queue

Input: the graph G', the source s.
Output: the set of minimum distance $dist[\cdot]$, the set of previous nodes $prev[\cdot]$,
        the set of previous path $path[\cdot]$.
1 Create vertex min-distance priority queue Q;
2 **foreach** *vertex v in* V' **do**
3      $dist[v] \leftarrow +\infty; prev[v] \leftarrow \emptyset; path[v] \leftarrow \emptyset$;   `// initialization`
4      $Q \leftarrow push(Q, v)$;   `// push v into queue Q`
5 $dist[s] \leftarrow 0$;
6 **while** Q *is not empty* **do**
7      $u \leftarrow pop(Q)$;   `// pop out the first vertex from Q`
8   **foreach** *edge e from u to v* **do**
9          $alt \leftarrow dist[u] + w_e$;
10   **if** $alt < dist[v]$ **then**
11             $dist[v] \leftarrow alt; prev[v] \leftarrow u; path[v] \leftarrow e$;

After adding min-priority queue, the time complexity (with Fibonacci heap) is still $O(|V'|^2 \log |V'| + |V'||E'|) = O(|E|^3)$, but later we will show how it improves the time efficiency via numerical experiments. Actually, we can further optimize the efficiency of this algorithm. Now we focus on the distribution of the shortest distances to give the following lemma:

Lemma 1. $\forall v \in V, \forall k \leq K, \forall t_2 < T$, *if* $\exists t_1 < t_2$ *s.t.* $dist[(v, k, t_1)] < dist[(v, k, t_2)]$, *then all the results (minimum time/budget path) will not change if we delete the node* (v, k, t_2).

Proof. To simplify the problem, assume now we are exploring the shortest paths from a source $(s, \cdot, \cdot)$ to all other nodes with time and budget requirements. Let $x, y \in [0, 1]$, then what we want to minimize is a linear combination of flight time and budget, shown as $x \cdot time + y \cdot budget$. Especially in this paper we choose $x = 0, y = 1$ (as minimum budget) and $x = 1, y = 0$ (as minimum time).

Now, given a t_2, if $\exists t_1 < t_2 < T$, we have $dist[(v, k, t_1)] < dist[(v, k, t_2)]$, then we can get that

$$x \cdot t_1 + y \cdot dist[(v, k, t_1)] < x \cdot t_2 + y \cdot dist[(v, k, t_2)],$$

which means the path starting from $(s, \cdot, \cdot)$ and ending up at node (v, k, t_2) can never be an answer to any requirement.

Next, let us check other paths passing through (v, k, t_2). Assume there is a shortest path starting from $(s, \cdot, \cdot)$ and passing from (v, k, t_2) to $(v', k + 1, t_3)$, denoted as $P_1 = \{(s, \cdot, \cdot) \to \cdots \to (v, k, t_2) \to (v', k + 1, t_3) \to \cdots\}$, then we can change it to $P_2 = \{(s, \cdot, \cdot) \to \cdots \to (v, k, t_1) \to (v', k + 1, t_3) \to \cdots\}$. By Definition 10, we know that there must exist an edge from (v, k, t_1) to $(v', k + 1, t_3)$ with the same weight as (v, k, t_2)'s and $dist[(v, k, t_1)] < dist[(v, k, t_2)]$. Then the length of P_2 is shorter then that of P_1, which violates the shortest property of P_1, a contradiction! Hence node (v, k, t_2) will not be involved in any shortest path starting from $(s, \cdot, \cdot)$. Additionally, since $dist[(v, k, t_1)] < dist[(v, k, t_2)]$, (v, k, t_1) can be picked earlier than (v, k, t_2) in Algorithm 1. Therefore, (v, k, t_2) can never be used to update the $dist[\cdot]$ value for all other nodes.

In conclusion, (v, k, t_2) can never be an answer to any requirement and it can never be used to update the $dist[\cdot]$ value of all other nodes, so it is safe to remove node (v, k, t_2) when exploring the shortest path from starting point $(s, \cdot, \cdot)$. □

Inspired by Lemma 1, whenever we select a node $(v, k, t) \in V'$, we can first check whether there exists another node v' with earlier t' and smaller $dist[v']$ value. To achieve this purpose, we use a Fenwick tree [5] to calculate the prefix minimum value as $F[v][k][\cdot]$, which is defined in Definition 11.

Definition 11 (Fenwick Tree [5]). *A Fenwick Tree (in our algorithm), denoted as $F[\cdot][\cdot][\cdot]$, is a data structure to calculate the prefix minimum value. More formally, $\forall v \in V$, $\forall k \leq K$, $\forall t \leq T$,*

$$F[v][k][t] = \min_{0 \leq i \leq t} dist[(v, k, i)].$$

There are two operations on the Fenwick Tree: insert a new value (alt) into the old minimum prefix (denoted as $F[v][k][t] \leftarrow alt$) and calculate the minimum prefix (just denoted as $F[v][k][t]$). Both of them occupies $O(\log T)$ time, where T is the time range (in our examples $T = 10080$).

For a node (v, k, t) in G', whenever we plan to update the $dist[\cdot]$ of all its neighbors (like Line 8 to Line 11 in Algorithm 1), we will compare $dist[(v, k, t)]$ to the prefix minimum value of $F[v][k][t]$. If $dist[(v, k, t)] > F[v][k][t]$, it means we meet a case satisfying Lemma 1, and thus (v, k, t) is redundant for the shortest path exploration. We can safely remove this node without influencing the final result, which will obviously reduce the intermediate operations for computation. Note that once we change a $dist[\cdot]$ value, we need to modify the Fenwick tree in $O(\log T)$ time. Thus we get an optimized algorithm shown in Algorithm 2.

Theorem 1. *The **if** statement in Line 9 of Algorithm 2 will not change the minimum time or minimum budget we finally calculate.*

Proof. When $dist[u] \leq F[v_1][k_1][t_1]$ is false, we can get

$$dist[(v_1, k_1, t_1)] > \min_{0 \leq i \leq t_1} dist[(v_1, k_1, i)].$$

Algorithm 2. Dijkstra Algorithm with Priority Queue and Fenwick Tree

---

Input: the graph G', the source s.
Output: the set of minimum distance $dist[\cdot]$, the set of previous node $prev[\cdot]$,
        the set of previous path $path[\cdot]$.

1 create vertex min-distance priority queue Q;
2 create min-prefix Fenwick tree $F[\cdot][\cdot][\cdot]$;
3 **foreach** *vertex v in V'* **do**
4 | $dist[v] \leftarrow +\infty$; $prev[v] \leftarrow \emptyset$; $path[v] \leftarrow \emptyset$;
5 |___ $Q \leftarrow push(Q, v)$;   // push v into queue Q

6 $dist[s] \leftarrow 0$;
7 **while** *Q is not empty* **do**
8 | $u = (v_1, k_1, t_1) \leftarrow pop(Q)$;   // pop out the first vertex from Q
9 | **if** $dist[u] \leq F[v_1][k_1][t_1]$ **then**   // Compare u with the min value in F
        and decide whether to update $dist[\cdot]$ for u's neighbors
10 | | **foreach** *edge e from u to $v = (v_2, k_2, t_2)$* **do**
11 | | | $alt \leftarrow dist[u] + w_e$;
12 | | | **if** $alt < dist[v]$ **then**
13 | | |___ $dist[v] \leftarrow alt$; $F[v_2][k_2][t_2] \leftarrow alt$; $prev[v] \leftarrow u$; $path[v] \leftarrow e$;

That means $\exists t_0 < t_1$ s.t. $dist[(v_1, k_1, t_0)] < dist[(v_1, k_1, t_1)]$ and by Lemma 1, all the results will not change if we delete u.   □

By Theorem 1, Algorithm 2 turns out the right answers. After adding Fenwick tree, the complexity of the algorithm seemingly turns even worse to $O(|E|^3 \log T)$. However, in practice we find that Fenwick tree improves the algorithm more than ten times faster, and Fig. 3 exhibits this phenomenon. We randomly choose 2 nodes as the sources and randomly choose different scales of edges, use these three algorithms to detect the shortest paths to all the other nodes, repeat three times, record the average elapsed time, and plot a chart for comparison. The x-axis denotes the scale of edges, while the y-axis is the logarithmic value of computing time. From this figure we can see clearly that the two optimization strategies improve the practical performance efficiency greatly.

5 AHP-Based Reachability Evaluation

Now, we implement Analytic Hierarchy Process (AHP) to convert the reachability feature into numerical values for all airports to measure their service qualities precisely. Figure 4 shows our AHP model with 2 layers and 8 indices (defined in Sect. 3). At first, we need to normalize these data into $[0, 1]$ range. The normalization methods are summarized in Table 1. Table 2 shows some sample data triples after the normalization.

We then survey and get expert grading matrix for these indices. Table 3 is an example matrix for centrality with 5 indices. Next, we use the matrix to

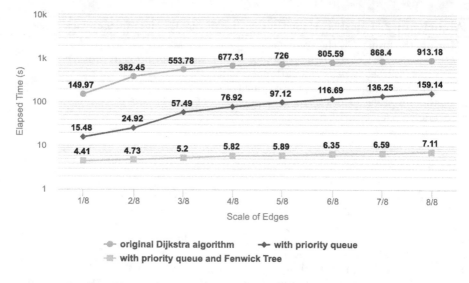

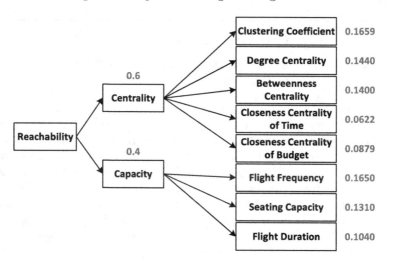

Fig. 3. A comparison among three algorithms

Fig. 4. AHP properties for reachability evaluation (Color figure online)

grade and calculate the weights of these features, with properties values and Consistency Ratio (CR). Since here $CR = CI/RI = 0.0469 < 0.1$, meaning the value of Consistency Ratio is smaller or equal to 10 %, the inconsistency is acceptable. Finally, we get all the properties for indices, shown in Fig. 4 as blue numbers next to each block.

Finally, we calculate the reachability value of all the airports. Table 4 is a sample of accurate results and we also provide a visualization map to show the distribution of the reachability in Fig. 5.

Table 1. Normalization methods for indices

Index	Normalization methods
Clustering Coefficient (CC)	Min-max normalization
Degree Centrality (DC)	Min-max normalization
Betweenness Centrality (BC)	Min-max normalization
Closeness Centrality of Time (CCT)	Reciprocal + min-max normalization
Closeness Centrality of Budget (CCB)	Reciprocal + min-max normalization
Flight Frequency (FF)	Log of the minimum + min-max normalization
Seating Capacity (SC)	Min-max normalization
Flight Duration (FD)	Reciprocal + min-max normalization

Table 2. Data after normalization

IATA	City	CC	DC	BC	CCT	CCB	FF	SC	FD
XIY	Xi'an	0.298	0.811	1.000	0.920	0.996	0.932	0.781	0.175
PEK	Beijing	0.240	1.000	0.632	1.000	0.792	1.000	1.000	0.063
CTU	Chengdu	0.312	0.803	0.527	0.992	0.821	0.928	0.861	0.122
CKG	Chongqing	0.338	0.724	0.544	0.923	0.979	0.919	0.767	0.164
KMG	Kunming	0.364	0.717	0.557	0.856	0.814	0.945	0.716	0.177
⋮	⋮	⋮	⋮	⋮	⋮	⋮	⋮	⋮	⋮
SHF	Shihezi	0.000	0.016	0.004	0.172	0.419	0.095	0.498	0.000
EJN	Ejina	0.000	0.008	0.000	0.216	0.165	0.116	0.081	0.221
RHT	Alxa Youqi	0.000	0.008	0.000	0.000	0.000	0.000	0.000	0.606

Table 3. Pairwise comparison matrix for centrality

	CC	DC	BC	CCT	CCB	Properties	CR
CC	1	1	2	3	1	0.2765	
DC	1	1	1	2	2	0.2400	
BC	1/2	1	1	3	2	0.2333	0.0469
CCT	1/3	1/2	1/2	1	1	0.1037	
CCB	1	1/2	1/2	1	1	0.1465	

Table 4. Reachability results

IATA	City	CC	DC	BC	CCT	CCB	FF	SC	FD	**R**
XIY	Xi'an	0.298	0.811	1.000	0.920	0.996	0.932	0.781	0.175	0.725
PEK	Beijing	0.240	1.000	0.632	1.000	0.792	1.000	1.000	0.063	0.707
CTU	Chengdu	0.312	0.803	0.527	0.992	0.821	0.928	0.861	0.122	0.654
CKG	Chongqing	0.338	0.724	0.544	0.923	0.979	0.919	0.767	0.164	0.649
KMG	Kunming	0.364	0.717	0.557	0.856	0.814	0.945	0.716	0.177	0.635
⋮	⋮	⋮	⋮	⋮	⋮	⋮	⋮	⋮	⋮	
NGQ	Gunsa	0.000	0.016	0.000	0.207	0.114	0.150	0.540	0.343	0.156
RKZ	Shigatse	0.000	0.008	0.000	0.270	0.205	0.116	0.535	0.115	0.137
SHF	Shihezi	0.000	0.016	0.004	0.172	0.419	0.095	0.498	0.000	0.131
EJN	Ejina	0.000	0.008	0.000	0.216	0.165	0.116	0.081	0.221	0.082
RHT	Alxa Youqi	0.000	0.008	0.000	0.000	0.000	0.000	0.000	0.606	0.064

Fig. 5. Visualization results for reachability of airports in China

Obviously, in Fig. 5 a lighter and bigger circle represents an airport with higher reachability. It is apparently that the top two airports are Xi'an and Beijing. It is seemingly surprising that Xi'an airport is with higher reachability than Beijing, but noting that what we discuss about is reachability. Considering the location of Xi'an in China and the fact that almost 15 % shortest paths passing through it, the conclusion is meaningful to some extent.

6   Conclusion

In this paper, we take China airline network as a representative and try to discuss the reachability issue for each airport respectively, which could reflect its regional connectivity level and service quality of civil aviation. More specifically, we evaluate reachability through many features including node degree, closeness, betweenness, etc. To get some feature values, we design a fast Dijkstra-based all-pair shortest path algorithm with both time and budget requirements, using priority Queue and Fenwick Tree to improve the time efficiency. Finally, we implement Analytic Hierarchy Process (AHP) to convert the reachability feature as numerical values for all airports to measure their service qualities precisely. We not only give a algorithm design and an AHP evaluation, but also prove theoretically the correctness of our method. Such methodology can be easily extended to other airline networks or any arbitrary network with transfer and duration constraints. Our results for China domestic airline network with 210 nodes and 69,160 flight routes will definitely become guide and reference to airline companies and civil aviation administration for their further development and management.

Acknowledgement. This work has been supported in part by the China 973 project (2014CB340303), the National Natural Science Foundation of China (No. 61571441, 61672353, 61472252, 61133006), and the Opening Project of Key Lab of Information Network Security of Ministry of Public Security (The Third Research Institute of Ministry of Public Security) Grant number C15602.

References

1. Barros, C.P., Peypoch, N.: An evaluation of European airlines's operational performance. Int. J. Prod. Econ. **122**(2), 525–533 (2009)
2. Bowen, B.D., Headley, D.E.: Evaluation of the US airline industry: the airline quality rating 2012 (2013)
3. Brandes, U.: A faster algorithm for betweenness centrality. J. Math. Sociol. **25**(2), 163–177 (2001)
4. Dijkstra, E.W.: A note on two problems in connexion with graphs. Numerische Mathematik **1**(1), 269–271 (1959)
5. Fenwick, P.M.: A new data structure for cumulative frequency tables. Softw. Pract. Experience **24**(3), 327–336 (1994)
6. Fredman, M.L., Tarjan, R.E.: Fibonacci heaps and their uses in improved network optimization algorithms. J. ACM (JACM) **34**(3), 596–615 (1987)

7. Freeman, L.C.: A set of measures of centrality based on betweenness. Sociometry **40**, 35–41 (1977)
8. Freeman, L.C.: Centrality in social networks conceptual clarification. Soc. Netw. **1**(3), 215–239 (1978)
9. Guimera, R., Amaral, L.A.N.: Modeling the world-wide airport network. Eur. Phys. J. B-Condens. Matter Complex Syst. **38**(2), 381–385 (2004)
10. Kourtellis, N., Alahakoon, T., Simha, R., Iamnitchi, A., Tripathi, R.: Identifying high betweenness centrality nodes in large social networks. Soc. Netw. Anal. Min. **3**(4), 899–914 (2013)
11. Lee, S.H., Choi, J.Y., Yoo, S.H., Oh, Y.G.: Evaluating spatial centrality for integrated tourism management in rural areas using GIS and network analysis. Tourism Manag. **34**, 14–24 (2013)
12. Sabidussi, G.: The centrality index of a graph. Psychometrika **31**(4), 581–603 (1966)
13. Tsaur, S.H., Chang, T.Y., Yen, C.H.: The evaluation of airline service quality by fuzzy MCDM. Tourism Manag. **23**(2), 107–115 (2002)
14. Watts, D.J., Strogatz, S.H.: Collective dynamics of small-world networks. Nature **393**(6684), 440–442 (1998)
15. Yu, M.M.: Assessment of airport performance using the SBM-NDEA model. Omega **38**(6), 440–452 (2010)

Approximation and Hardness Results for the Max k-Uncut Problem

Peng Zhang[1], Chenchen Wu[2], Dachuan Xu[3]([✉]), and Xinghe Zhang[4]

[1] School of Computer Science and Technology,
Shandong University, Jinan 250101, China
`algzhang@sdu.edu.cn`
[2] College of Science, Tianjin University of Technology, Tianjin 300384, China
[3] Department of Information and Operations Research,
College of Applied Sciences, Beijing University of Technology,
Beijing 100124, China
`xudc@bjut.edu.cn`
[4] Shandong Experimental High School (East Campus), Jinan 250109, China

Abstract. In this paper, we propose the Max k-Uncut problem. Given an n-vertex undirected graph $G = (V, E)$ with nonnegative weights $\{w_e \mid e \in E\}$ defined on edges, and a positive integer k, the Max k-Uncut problem asks to find a partition $\{V_1, V_2, \cdots, V_k\}$ of V such that the total weight of edges that are *not* cut is maximized. This problem is just the complement of the classic Min k-Cut problem. We get this problem from the study of complex networks. For Max k-Uncut, we present a randomized $(1 - \frac{k}{n})^2$-approximation algorithm, a greedy $(1 - \frac{2(k-1)}{n})$-approximation algorithm, and an $\Omega(\frac{1}{2}\alpha)$-approximation algorithm by reducing it to Densest k-Subgraph, where α is the approximation ratio for the Densest k-Subgraph problem. More importantly, we show that Max k-Uncut and Densest k-Subgraph are in fact equivalent in approximability up to a factor of 2. We also prove a weak approximation hardness result for Max k-Uncut under the assumption P $\neq$ NP.

1 Introduction

In this paper, we investigate the Max k-Uncut problem, which is obtained from the study of the *homophyly* law [8, Chap. 4] of large scale networks. Being one of the basic laws governing the structures of large scale networks, the homophyly law states that edges in a network tend to connect nodes with the same or similar attributes, just as an old proverb says, "birds of a feather flock together". For example, in a paper citation network, papers are more likely to cite papers with which they have the same or similar keywords.

While it is common to list keywords in a paper by its authors, in a paper citation network there are still many papers whose keywords are not explicitly given. Consequently, it is natural to predict keywords for these papers using

P. Zhang—Part of the author's work was done while he was visiting at the University of California - Riverside, USA, and Beijing University of Technology, China.

© Springer International Publishing AG 2016
T.-H.H. Chan et al. (Eds.): COCOA 2016, LNCS 10043, pp. 49–61, 2016.
DOI: 10.1007/978-3-319-48749-6_4

the homophyly law. Inspired by this observation, Zhang (the first author of this paper) and Li [21] proposed the Maximum Happy Edges (MHE) problem. In the MHE problem, we are given an undirected graph $G = (V, E)$ and a color set $C = \{1, 2, \cdots, k\}$. Only part of vertices are given colors in C. An edge is *happy* if its two endpoints share the same color. The goal of MHE is to color all the uncolored vertices such that the number of happy edges is maximized. Here, vertices correspond to papers, edges correspond to citations (neglecting directions), and colors correspond to keywords.

A natural variant of MHE is that in the input graph all vertices are uncolored and the problem just asks to color them in k colors such that the number (or total weight) of happy edges is maximized. This suggests the Max k-Uncut problem we investigate in this paper.

Definition 1. The Max k-Uncut Problem

(Instance). *We are given an undirected graph $G = (V, E)$ with nonnegative edge weights $\{w_e \mid e \in E\}$, and a positive integer k.*

(Goal). *The problem asks to find a partition $\{V_1, V_2, \cdots, V_k\}$ of V (i.e., to find a k-coloring of vertices) such that the total weight of happy edges is maximized.*

In the definition of Max k-Uncut, by *k-coloring* we mean a coloring scheme that uses *exactly* k colors, which results in a *k-partition* $\{V_1, V_2, \cdots, V_k\}$, where V_i is the set of vertices whose color is i. In the paper we will interchangeably use k-coloring and k-partition. Note that the requirement of exactly k colors is necessary, otherwise we can color all vertices in one color and all edges are happy.

In the Max k-Uncut problem, if $k = 1$ or $k = n$, the problem becomes trivial. The optimum would be respectively the number of all edges and 0 in these two cases. So, throughout the paper we always assume $2 \leq k \leq n - 1$ for the Max k-Uncut problem.

Note that Max k-Uncut is *not* a special case of MHE. In MHE, if all vertices are un-colored, then the problem becomes trivial: Just color all vertices in one color, then all edges will become happy. In contrast, if all vertices in Max k-Uncut are uncolored, we cannot color them in one color. In Max k-Uncut, we must figure out a k-coloring.

Two problems that are closely related to Max k-Uncut have already appeared in literature. Choudhurya et al. [7] proposed the capacitated Max k-Uncut problem. Given an undirected graph G and k integers $s_1, s_2, \cdots, s_k$, this problem is to partition $V(G)$ into k subsets of sizes $s_1, s_2, \cdots, s_k$ respectively, such that the total weight of happy edges is maximized. Very recently, Wu et al. [18] studied the balanced Max 3-Uncut problem, in which an input graph is partitioned into 3 equal-sized parts so that the total weight of happy edges is maximized.

Notations and Terms. Some common notations and terms are listed here. Given a graph G, let n be the number of its vertices. Given an optimization problem, let OPT denote the value of its optimal solution. By r-clique for some integer r, we mean a clique (i.e., a complete subgraph) that contains exactly r vertices.

1.1   Related Work

To the best of our knowledge, the general Max k-Uncut problem is new and has not been studied in literature. Though it is new, Max k-Uncut has rich connection to the classic and existing problems.

Max k-Uncut is just the complement of the classic Min k-Cut problem. The Min k-Cut problem asks for a k-partition such that the total weight of cut edges is minimized. The Min k-Cut problem is strongly NP-hard [12], so is the Max k-Uncut problem. The best approximation ratio for Min k-Cut is 2 [17]. When k is a constant, the Min k-Cut problem can be optimally solved in polynomial time [12]. Obviously, Max k-Uncut with constant k is also polynomial time solvable. In a word, Max k-Uncut is strongly NP-hard (when k is given in the input), and is polynomial time solvable when k is a constant.

Previously we have pointed out two closely related variants of Max k-Uncut, i.e., the capacitated Max k-Uncut problem and the balanced Max 3-Uncut problem. Using the heuristic of local search, Choudhurya et al. [7] gave a $\frac{1}{d(k-1)+1}$-approximation algorithm for capacitated Max k-Uncut, where d is the ratio of the largest size and the smallest size in the partition. This ratio is somewhat poor and cannot extend to the Max k-Uncut problem studied in this paper. Using the semidefinite programming technique, Wu et al. [18] gave a 0.3456-approximation algorithm for the balanced Max 3-Uncut problem.

The cut problems are classic and rich. They play an important role in the study of approximation algorithms and operations research. In literature, the "uncut" problems are also been studied. Besides Max k-Uncut, three examples are Min Uncut [1], Multiway Uncut [15,20], and the complement of Min Bisection [19]. Min Uncut is the complement of the classic Max Cut problem. Agarwal et al. [1] gave an $O(\sqrt{\log n})$-approximation algorithm for Min Uncut, where n is the number of vertices in the input graph. Multiway Uncut is the complement of the classic Multiway Cut problem [5,6]. Langberg et al. [15] proposed the Multiway Uncut problem. The current best approximation ratio for Multiway Uncut is $\frac{1}{2} + \frac{\sqrt{2}}{4} f(k) \geq 0.8535$ [20], where $f(k) \geq 1$ is a function of k. Ye and Zhang [19] gave a 0.602-approximation algorithm for the complement of the Min Bisection problem.

Due to the close relation of Max k-Uncut to Multiway Uncut, we have to say more about Multiway Uncut. Given a graph $G = (V, E)$ with edge weights and a terminal set $\{s_1, s_2, \cdots, s_k\} \subseteq V$, the Multiway Uncut problem asks a partition of V that separates the k terminals from each other and maximizes the total weight of happy edges. (Multiway Uncut is a special case of MHE.) Both Max k-Uncut and Multiway Uncut ask for a k-partition. The only difference is that in Max k-Uncut, there is no terminal, and in Multiway Uncut, there are terminals.

Another closely related problem is Max k-Cut, which is the problem to find a k-partition such that the total weight of cut edges is maximized. When $k = 2$, Max k-Cut (namely, Max Cut) is already NP-hard. The current best approximation ratio for Max Cut is 0.87856, given by Goemans and Williamson [11] using the semidefinite programming technique. Frieze and Jerrum [10] extended Goemans-Williamson's technique to the Max k-Cut problem, obtained the approximation ratio $\alpha_k = 1 - \frac{1}{k} + (1 + \epsilon(k))\frac{2\ln k}{k^2}$ for Max k-Cut, where $\epsilon(k)$ is a function of k which tends to zero as $k \to \infty$. When $k = 3, 4, 5$, α_k is no less than 0.800217, 0.850304, and 0.874243, respectively.

1.2   Our Results

In this paper, we give three approximation algorithms for the Max k-Uncut problem and prove a (weak) approximation hardness result of Max k-Uncut. These three algorithms share the same idea, which is simple but powerful: To find a k-partition with many happy edges, one may just find a dense subgraph as large as possible. The subgraph is used as one part of the k-partition. The larger and denser the subgraph is, the more happy edges we will get. Along this line, we finally find that Max k-Uncut is in fact equivalent to the Densest k-Subgraph problem in approximability (up to a factor of 2). Note that Densest k-Subgraph is one of the current hot topics in approximation algorithms. This may be our most important find in this paper.

The first algorithm is a randomized algorithm (Algorithm 2.1) whose approximation ratio is $(1 - \frac{k}{n})^2$. This algorithm can be derandomized in polynomial time. The second algorithm is a greedy algorithm (Algorithm 2.2) whose approximation ratio is $1 - \frac{2(k-1)}{n}$. While the ratios of these two algorithms are very close, they are still incomparable. Specifically, when $k < \sqrt{2n}$, the ratio $1 - \frac{2(k-1)}{n}$ is better than the ratio $(1 - \frac{k}{n})^2$. Otherwise (when $k > \sqrt{2n}$), the latter is better than the former.

The ratio $\rho = \max\{(1 - \frac{k}{n})^2, 1 - \frac{2(k-1)}{n}\}$ for Max k-Uncut we obtain so far is already good when k is not too large. For example, if $k \leq n/2$, then $\rho \geq 1/4$. However, when k approaches $n-1$, ρ becomes worse and worse, and equals to $\frac{1}{n^2}$ finally. This observation suggests that the most difficult case of approximating Max k-Uncut should be the case when k is close to n, say, $k = n - O(\log n)$. And in this case (i.e., when k is large), we may make use of the connection to Densest k-Subgraph.

Therefore, in the third algorithm (Algorithm 2.3), we reduce Max k-Uncut to Densest $\bar{k}$-Subgraph (for some suitable $\bar{k}$) by exploring the structure of optimal solutions to Max k-Uncut. It is convenient to define the Densest k-Subgraph problem here.

Definition 2. The Densest k-Subgraph Problem

(Instance). *We are given an undirected graph $G = (V, E)$ with nonnegative edge weights $\{w_e \mid e \in E\}$, and a positive integer k.*

(Goal). *The problem asks to find a k-vertex subgraph G' such that the total weight of edges in $E(G')$ is maximized.*

The reduction used in Algorithm 2.3 is nontrivial. Let α be the approximation ratio for Densest k-Subgraph. Then Algorithm 2.3 approximates Max k-Uncut within $\frac{1}{2}\alpha$ in polynomial time. The current best value of α is $\Omega(1/n^{\frac{1}{4}+\epsilon})$ [4].

Consequently, Algorithm 2.3 repairs the deficiencies of Algorithms 2.1 and 2.2. Now, the approximation ratio we obtain for Max k-Uncut is $\max\{\rho, \frac{1}{2}\alpha\}$.

Surprisingly and interestingly, our technique in the analysis of Algorithm 2.3 also implies that if Max k-Uncut can be approximated within a factor of β, then Densest k-Subgraph can be approximated within $\frac{1}{2}\beta$. Therefore, Max k-Uncut and Densest k-Subgraph are equivalent in approximability up to a factor of 2. This reveals the strong connection between Max k-Uncut and Densest k-Subgraph, and may open a new viewpoint in tackling the Densest k-Subgraph problem, since this problem is known as a notorious hard problem in approximation algorithms. (There is a wide gap between its best approximation factor and its best hardness factor).

Next, we prove an approximation hardness result for Max k-Uncut: For any small constant $\epsilon > 0$, Max k-Uncut cannot be approximated within $1 - \frac{1}{2n^\epsilon}$ in polynomial time, where n is the number of vertices in the input graph. This is proved via a gap-preserving reduction from the hardness result of the Max Clique problem [3,13]. As a result, the hardness $1 - \frac{1}{2n^\epsilon}$ for any small constant $\epsilon > 0$ implies that Max k-Uncut does not admit FPTAS.

Honestly speaking, this hardness result is weak since Max k-Uncut is indeed strongly NP-hard, and the strong NP-hardness already rules out FPTAS. However, we make twofold contribution in proving the approximation hardness of Max k-Uncut. First, we give an explicit expression of the approximation hardness factor of Max k-Uncut, instead of just speaking that it is strongly NP-hard. Second, we prove a technical lemma (Lemma 2), which gives an upper bound of the number of happy edges that can be produced by any k-partition on a graph with no $(r + 1)$-clique. The technical lemma is of independent interest and may find more applications in related problems. In fact, the upper bound is obtained by a special k-partition which consists of $k - 1$ singletons and one subset of size $n - (k - 1)$. This again hints the connection of Max k-Uncut to Densest k-Subgraph and Max Clique.

2   Approximation Algorithms

2.1   A Randomized Algorithm

A straightforward idea for Max k-Uncut is to color vertices randomly. However, if we color *every* vertex randomly, we may not get an approximation algorithm with good ratio. (We can prove that an algorithm of this type has approximation ratio $\frac{1}{k} - \frac{k-1}{n(n-1)}$. The details are omitted here.)

In graphs with only unit weight on edges, to maximize the total weight of happy edges is equivalent to leave as many as possible edges uncut. So, a clever randomized strategy is to randomly color $k - 1$ vertices only, making the remaining vertices as many as possible. Intuitively, these many vertices would induce many happy edges. Algorithm $\mathcal{R}$ below is a randomized algorithm for Max k-Uncut of this idea.

Let W_{tot} be the total weight of edges in graph G.

Algorithm 2.1. (Algorithm $\mathcal{R}$ for Max k-Uncut)

1 Pick randomly $k-1$ vertices from V, and color them respectively in colors 1 to $k-1$.
2 Color all the remaining vertices in color k.

Theorem 1. *Algorithm $\mathcal{R}$ is a randomized $\left(1-\frac{k}{n}\right)^2$-approximation algorithm for the* Max k-Uncut *problem.*

Proof. First note that Algorithm $\mathcal{R}$ runs in polynomial time. Let V_i be the set of vertices of color i. Take any edge $e=(u,v)$. Then, e is happy (uncut) if and only if both u and v are not chosen in the first $k-1$ random choices (step 1). This means that

$$\Pr[\text{edge } e \text{ is happy}] = \frac{\binom{n-2}{k-1}}{\binom{n}{k-1}} = \frac{(n-k+1)(n-k)}{n(n-1)} > \frac{(n-k)^2}{n^2} = \left(1-\frac{k}{n}\right)^2.$$

Let SOL be the solution value obtained by Algorithm $\mathcal{R}$. Therefore, we have

$$\mathrm{E}[SOL] = \sum_{e \in E} w_e \cdot \Pr[\text{edge } e \text{ is happy}] \geq \left(1-\frac{k}{n}\right)^2 W_{tot}.$$

On the other hand, the optimum OPT is obviously at most W_{tot}. So, the approximation ratio of Algorithm $\mathcal{R}$ is at least $\left(1-\frac{k}{n}\right)^2$. □

Algorithm $\mathcal{R}$ can be derandomized by the conditional expectation method in polynomial time. This is sketched as follows in rounds. In the first round we determine the first vertex to be removed. We remove each vertex v_i ($1 \leq i \leq n$) from G to obtain G_i. That is, $\forall 1 \leq i \leq n$, $G_i = G \setminus v_i$. For each G_i, we compute the expected solution value a_i of Algorithm $\mathcal{R}$ for the Max$(k-1)$-Uncut problem. We find the largest expected value in this round, say a_j. Then, v_j is the first vertex we pick and is colored in color 1. The next round begins from G with v_j removed. Repeating the above procedure for $k-1$ rounds, we obtain a solution whose value is at least as better as the expected value of Algorithm $\mathcal{R}$.

2.2   A Greedy Algorithm

The idea in Sect. 2.1 can be restated as finding a subgraph of size $n-k+1$ as dense as possible, where by dense subgraph we mean a subgraph whose total weight of edges is as much as possible. This leads to a greedy algorithm for Max k-Uncut, shown as Algorithm $\mathcal{G}$ below. For the sake of description, we define the weighted degree $d_w(v)$ of a vertex v as the sum of weights of edges incident to v. By definition, the weight of vertex v is equal to the capacity of the cut $(\{v\}, V \setminus \{v\})$. Obviously, when each edge in the graph has unit weight, the weighted degree of a vertex is simply its degree.

Algorithm 2.2. (Algorithm $\mathcal{G}$ for Max k-Uncut)

1 Pick vertices from V with the first $k-1$ smallest weighted degrees, and color them in colors 1 to $k-1$, respectively.
2 Color all the remaining vertices in color k.

Theorem 2. *Algorithm $\mathcal{G}$ is a* $\left(1 - \frac{2(k-1)}{n}\right)$*-approximation algorithm for the* Max k-Uncut *problem.*

Proof. Algorithm $\mathcal{G}$ obviously runs in polynomial time. Let $v_1, \cdots, v_{k-1}$ be the vertices picked in the first step of Algorithm $\mathcal{G}$. By the algorithm, only edges incident to vertices in $\{v_1, \cdots, v_{k-1}\}$ would be unhappy. So, the total weight of unhappy edges is at most

$$\sum_{i=1}^{k-1} d_w(v_i) \leq \frac{k-1}{n} \sum_v d_w(v) = \frac{2(k-1)}{n} W_{tot}.$$

Therefore, the total weight of happy edges is at least

$$W_{tot} - \frac{2(k-1)}{n} W_{tot}.$$

Since OPT $\leq W_{tot}$, this means the approximation ratio of Algorithm $\mathcal{G}$ is at least $1 - \frac{2(k-1)}{n}$. □

The approximation ratios $(1 - \frac{k}{n})^2$ and $1 - \frac{2(k-1)}{n}$ behave well when k is not too large. For example, $(1 - \frac{k}{n})^2 \geq \frac{1}{4}$ when $k \leq \frac{n}{2}$. However, when k is large enough, say, $k = n - O(\log n)$, the approximation ratio $\max\{(1 - \frac{k}{n})^2, 1 - \frac{2(k-1)}{n}\}$ we obtained so far becomes bad. To remedy this deficiency, we design another approximation algorithm for Max k-Uncut, that is, Algorithm $\mathcal{T}$ in Sect. 2.3. Actually, our subsequent study on Max k-Uncut in this paper makes us realize that the hard core of Max k-Uncut just lies in the case when k is large.

2.3   Reduces to Densest k-Subgraph

In this section, we reduce Max k-Uncut to Densest $\bar{k}$-Subgraph for some suitable $\bar{k}$. For clarity, when the instance of Densest k-Subgraph is given as, e.g., $(G, w, \bar{k})$, we call it the instance of the Densest $\bar{k}$-Subgraph problem. The reader should be aware of that Densest k-Subgraph and Densest $\bar{k}$-Subgraph are the same problem. This usage also happens to the Max k-Uncut problem.

Given a vertex subset S of an edge-weighted graph G, let $w(S)$ denote the total weight of happy edges induced by S. Given a k-partition $\mathcal{P} = \{V_1, V_2, \cdots, V_k\}$ of $V(G)$, let $w(\mathcal{P})$ denote the total weight of happy edges induced by $\mathcal{P}$, i.e., $w(\mathcal{P}) = \sum_i w(V_i)$.

First we prove a technical lemma.

Lemma 1. *Let $\mathcal{P} = \{V_1, V_2, \cdots, V_k\}$ be a k-partition of graph G with weights defined on edges. Then in polynomial time (in terms of $|V(G)|$) we can construct a k-partition $\mathcal{P}' = \{V_1', V_2', \cdots, V_k'\}$ which satisfies*

(i) $|V_1'| = \cdots = |V_{k-1}'| = 1$, $|V_k'| = n - k + 1$, and
(ii) $w(\mathcal{P}') \geq \frac{1}{2}w(\mathcal{P})$.

Proof. We renumber the vertex subsets in $\mathcal{P}$ according to the non-decreasing order of their $w(\cdot)$ values, and rewrite $\mathcal{P}$ as $\{R_1, R_2, \cdots, R_a, S_1, S_2, \cdots, S_b\}$, where we assume that in $\mathcal{P}$ there are a singletons $R_1, \cdots, R_a$, and b non-singletons $S_1, \cdots, S_b$ (that is, each S_i has size at least two). So, we have $a+b = k$,

$$w(R_1) \leq \cdots \leq w(R_a) \leq w(S_1) \leq \cdots \leq w(S_b),$$

and

$$w(\mathcal{P}) = w(S_1) + \cdots + w(S_b).$$

Note that a may be zero.

If $b = 1$, then the theorem is proved by just letting $\mathcal{P}' = \mathcal{P}$. So, in the following we assume that $b \geq 2$. We shall convert $S_1, \cdots, S_{b-1}, S_b$ to $S_1', \cdots, S_{b-1}', S_b'$ such that the first $b - 1$ S_i''s are singletons. This is done as follows.

We pick the unique $\ell \in \left[1, \lceil \frac{b-1}{2} \rceil\right]$ such that

$$|S_1| + |S_2| + \cdots + |S_\ell| \geq b - 1$$

and

$$|S_1| + |S_2| + \cdots + |S_{\ell-1}| < b - 1. \tag{1}$$

Note that it may be the case that $\ell = 1$, and in this case we do not need the condition (1). Also note that since $b \geq 2$ and $\forall 1 \leq i \leq b$, $|S_i| \geq 2$, ℓ must be at most $\lceil \frac{b-1}{2} \rceil$.

Initially S_b' is empty. We merge all vertices in $S_{\ell+1}, \cdots, S_b$ into S_b'. Then we pick arbitrarily $b - 1$ vertices from $S_1, \cdots, S_\ell$ to make $b - 1$ singletons S_1', $S_2', \cdots, S_{b-1}'$. If there are still remaining vertices in $S_1, \cdots, S_\ell$ (in case that $|S_1| + |S_2| + \cdots + |S_\ell| > b - 1$), we then move all of them to S_b'. This finishes the construction of $S_1', \cdots, S_{b-1}', S_b'$.

Since $\ell \leq \lceil \frac{b-1}{2} \rceil \leq \frac{1}{2}b$, the number of subsets $S_{\ell+1}, \cdots, S_b$ is at least half of b. In the above construction, all the happy edges in these subsets are kept in S_b'. Since $S_1, \cdots, S_b$ are in the non-decreasing order of the total weights of happy edges they contain, we know that

$$w(S_b') \geq \frac{1}{2}(w(S_1) + \cdots + w(S_b))$$

The desired k-partition $\mathcal{P}'$ is just $\{R_1, \cdots, R_a, S_1', \cdots, S_b'\}$.   □

Algorithm $\mathcal{T}$ is the algorithm reducing Max k-Uncut to Densest $\bar{k}$-Subgraph. Since the subgraph G' found in step 2 contains $\bar{k}$ vertices, there are exactly $n - \bar{k} = k - 1$ vertices in $V(G) \setminus V(G')$. So, in step 4 we can color them in colors $1, \cdots, k - 1$, respectively.

Algorithm 2.3. (Algorithm $\mathcal{T}$ for Max k-Uncut)

Input: An instance (G, w, k) of Max k-Uncut.
Output: A k-partition of $V(G)$.
  1 $\bar{k} \leftarrow n - (k - 1)$.
  2 Find a subgraph G' of G by an approximation algorithm for Densest $\bar{k}$-Subgraph on instance $(G, w, \bar{k})$.
  3 Color all vertices in $V(G')$ in color k.
  4 Color all vertices in $V(G) \setminus V(G')$ in colors $1, \cdots, k - 1$, respectively.

Theorem 3. *Let α be the approximation ratio of* Densest k-Subgraph. *Then Algorithm $\mathcal{T}$ is a $\frac{\alpha}{2}$-approximation algorithm for the* Max k-Uncut *problem.*

By [4], α can be $\Omega(1/n^{1/4+\epsilon})$ for every small constant $\epsilon > 0$. (The ratio in [4] is for unweighted Densest k-Subgraph. Using the technique in [9], this ratio can be extended to weighted Densest k-Subgraph.) This means that Max k-Uncut can be approximated within $\frac{1}{2}\alpha = \Omega(1/n^{1/4+\epsilon})$ in polynomial time.

Proof (of Theorem 3). Let $\mathrm{OPT}_{\mathrm{M}k\mathrm{U}}$ be the optimal value of Max k-Uncut on instance (G, w, k). Let $\mathcal{P}^* = \{V_1, V_2, \cdots, V_k\}$ be the corresponding optimal solution. By Lemma 1, we can build a k-partition $\mathcal{P}' = \{V_1', V_2', \cdots, V_k'\}$ from $\mathcal{P}^*$ such that $V_1', \cdots, V_{k-1}'$ are singletons. This means that $|V_k'| = n - (k - 1)$. So, V_k' is a feasible solution to Densest $\bar{k}$-Subgraph on instance $(G, w, \bar{k})$ satisfying

$$w(V_k') = w(\mathcal{P}') \geq \frac{1}{2}w(\mathcal{P}^*) = \frac{1}{2}\mathrm{OPT}_{\mathrm{M}k\mathrm{U}}, \tag{2}$$

where the inequality is by Lemma 1.

Note that G' is the subgraph found in step 2 by the approximation algorithm for Densest $\bar{k}$-Subgraph. There are $k - 1$ vertices in $V(G) \setminus V(G')$. Then, step 4 builds $k - 1$ singletons using the vertices in $V(G) \setminus V(G')$. These singletons, together with $V(G')$, constitute a k-partition, denoted by $\mathcal{P}$, which is a feasible solution to Max k-Uncut. We have

$$w(\mathcal{P}) = w(V(G')) \geq \alpha \cdot \mathrm{OPT}_{\mathrm{D}\bar{k}\mathrm{S}} \geq \alpha \cdot w(V_k') \underset{(2)}{\geq} \frac{\alpha}{2} \cdot \mathrm{OPT}_{\mathrm{M}k\mathrm{U}},$$

where the first inequality holds since G' is an α-approximate solution to the Densest $\bar{k}$-Subgraph instance $(G, w, \bar{k})$, and the second inequality holds since V_k' is a feasible solution to $(G, w, \bar{k})$. The theorem is proved. $\qquad\square$

Note that the running time of Algorithm $\mathcal{T}$ does not depend on the construction time of the k-partition in Lemma 1. Lemma 1 is only used in the analysis of Algorithm $\mathcal{T}$. (This construction time is useful in the following Algorithm $\mathcal{C}$.)

Interestingly and somewhat surprisingly, Lemma 1 also implies the converse of Theorem 3: Densest k-Subgraph reduces to Max k-Uncut. This is shown in Algorithm $\mathcal{C}$ and Theorem 4.

Algorithm 2.4. (Algorithm $\mathcal{C}$ for Densest k-Subgraph)

Input: An instance (G, w, k) of Densest k-Subgraph.
Output: A vertex subset $V' \subseteq V(G)$ containing exactly k vertices.
 1 $\bar{k} \leftarrow n - k + 1$.
 2 Find a $\bar{k}$-partition $\mathcal{P} = \{V_1, V_2, \cdots, V_{\bar{k}}\}$ of $V(G)$ by an approximation algorithm for Max $\bar{k}$-Uncut on instance $(G, w, \bar{k})$.
 3 Convert $\mathcal{P}$ to a $\bar{k}$-partition $\mathcal{P}' = \{V_1', V_2', \cdots, V_{\bar{k}}'\}$ by Lemma 1, where $|V_{\bar{k}}'| = n - (\bar{k} - 1) = k$.
 4 **return** $V' \leftarrow V_{\bar{k}}'$.

Theorem 4. *If* Max k-Uncut *can be approximated within a factor of* α, *then* Densest k-Subgraph *can be approximated within a factor of* $\alpha/2$.

Proof. We design Algorithm $\mathcal{C}$ as the approximation algorithm for Densest k-Subgraph. Step 2 calls the supposed α-approximation algorithm for Max $\bar{k}$-Uncut. By Lemma 1, step 3 can be finished in polynomial time. Therefore, the overall running time of Algorithm $\mathcal{C}$ is polynomial.

Let V^* be an optimal solution to the Densest k-Subgraph instance (G, w, k), whose value is denoted by $\mathrm{OPT}_{\mathrm{D}k\mathrm{S}}$. Note that in Algorithm $\mathcal{C}$ we have $\bar{k} = n - k + 1$. By viewing each vertex in $V(G) \setminus V^*$ as a singleton, we can build a $\bar{k}$-partition $\mathcal{P}^\circ = \{V_1^\circ, V_2^\circ, \cdots, V_{\bar{k}}^\circ\}$, where $V_{\bar{k}}^\circ = V^*$. Obviously we have $w(\mathcal{P}^\circ) = \mathrm{OPT}_{\mathrm{D}k\mathrm{S}}$. A crucial observation is that $\mathcal{P}^\circ$ is a feasible solution to the Max $\bar{k}$-Uncut instance $(G, w, \bar{k})$. This helps us get the connection

$$\mathrm{OPT}_{\mathrm{M}\bar{k}\mathrm{U}} \geq \mathrm{OPT}_{\mathrm{D}k\mathrm{S}}, \tag{3}$$

where $\mathrm{OPT}_{\mathrm{M}\bar{k}\mathrm{U}}$ is the optimal value of the instance $(G, w, \bar{k})$ of Max $\bar{k}$-Uncut.

For the two $\bar{k}$-partitions $\mathcal{P}$ and $\mathcal{P}'$, we have $w(\mathcal{P}') \geq \frac{1}{2} w(\mathcal{P})$ by Lemma 1. Since Max $\bar{k}$-Uncut can be approximated within α, we have $w(\mathcal{P}) \geq \alpha \cdot \mathrm{OPT}_{\mathrm{M}\bar{k}\mathrm{U}}$. These facts, together with (3), conclude the theorem. $\qquad\square$

Theorems 3 and 4 show that Max k-Uncut and Densest k-Subgraph are in fact equivalent in approximability up to a factor of two.

3   Approximation Hardness

3.1   Ruling Out Constant Factor Approximation

The approximability equivalence (up to a factor of two) of Max k-Uncut and Densest k-Subgraph naturally suggests that the approximation hardness results of Densest k-Subgraph may extend to Max k-Uncut. In particular, the following conditional hardness result holds.

Corollary 1. *If* Densest k-Subgraph *cannot be approximated within any constant factor, then so do* Max k-Uncut.

Under some appropriate complexity assumptions, people indeed proved that Densest k-Subgraph cannot be approximated within any constant factor. Raghavendra and Steurer [16] proved that assuming that the Unique Games with Small Set Expansion conjecture is true, it is NP-hard to approximate the Densest k-Subgraph problem within any constant factor. Alon $et\ al.$ [2] also ruled out constant approximation factor for Densest k-Subgraph, under an average case hardness assumption. For the exact meaning of these complexity assumptions, we refer the reader to [2,16].

Khot [14] proved that assuming NP $\not\subseteq \cap_{\epsilon>0}$BPTIME$(2^{n^\epsilon})$, Densest k-Subgraph has no PTAS. However, this result cannot be extended to Max k-Uncut directly, since the approximability equivalence of Max k-Uncut and Densest k-Subgraph proved above omits a constant factor 2.

3.2 An Explicit Hardness Factor

The approximation hardness results for Densest k-Subgraph mentioned above all use stronger complexity assumptions than the general assumption P $\neq$ NP. In the following, we shall prove an approximation hardness result for Max k-Uncut, assuming that P $\neq$ NP. The proved hardness factor is $1 - \frac{1}{2n^\epsilon}$, where $\epsilon > 0$ is an arbitrarily small constant, and n is the vertex number of the input graph. This result implies that, if P $\neq$ NP, Max k-Uncut does not admit FPTAS.

The hardness result $1 - \frac{1}{2n^\epsilon}$ for Max k-Uncut is rather weak since Max k-Uncut is strongly NP-hard, and this (the strong NP-hardness) already rules out FPTAS. However, we make twofold contribution in proving such a result. First, we give an explicit expression of the approximation hardness factor of Max k-Uncut, instead of just speaking that it is strongly NP-hard. Second, we prove a technical lemma (Lemma 2), which gives an upper bound of the number of happy edges that can be produced by any k-partition on a graph with no $(r + 1)$-clique. The technical lemma is of independent interest and may find more applications in related problems.

Lemma 2. *Any k-partition on an n-vertex undirected graph with no $(r + 1)$-clique can produce at most $\frac{1}{2}(1 - \frac{1}{r})u^2$ happy edges, where $u = n - (k - 1)$.*

Håstad [13] proved the following remarkable approximation hardness result for the Max Clique problem: For any $\epsilon > 0$, unless P $=$ NP, there is no polynomial time algorithm that approximates Max Clique within a factor of $n^{1/2-\epsilon}$, where n is the vertex number of the input graph. By this result and Lemma 2, we can prove that

Theorem 5. *For any $\epsilon > 0$, unless $P = NP$, there is no polynomial time algorithm that approximates Max k-Uncut within a factor of $1 - \frac{n^{1/2-\epsilon}-1}{n^{1/2}-1}$, where n is the vertex number of the input graph. The hardness factor is $\leq 1 - \frac{1}{2n^\epsilon}$ for sufficiently large n.*

The proof of Lemma 2 is rather complicated. Due to space limitation, the proofs of Lemma 2 and Theorem 5 are omitted here and will be given in the journal version of the paper.

Håstad [13] also proved that assuming ZPP $\neq$ NP, Max Clique cannot be approximated within $n^{1-\epsilon}$ for any small constant $\epsilon > 0$. However, for technical reasons, this stronger hardness factor cannot improve the result of Theorem 5 accordingly.

A corollary of Theorem 5 is that Max k-Uncut has no FPTAS, if P $\neq$ NP.

Corollary 2. Max k-Uncut *does not admit FPTAS, if $P \neq NP$.*

Proof. Suppose for contradiction that there is an FPTAS for Max k-Uncut which for any small $\epsilon' > 0$, gets a $(1-\epsilon')$-approximation to Max k-Uncut instance I, with running time poly($\frac{1}{\epsilon'}, |I|$), where $|I|$ denotes the length of instance I, and poly() denotes some polynomial. Given any small constant $\epsilon > 0$, if we set $\epsilon' = \frac{1}{2n^\epsilon}$ and run the FPTAS, where n is the number of vertices in the input graph, then we can get a $(1-\frac{1}{2n^\epsilon})$-approximation to instance I in time poly($2n^\epsilon, |I|$) = poly($|I|$), contradicting Theorem 5. □

Acknowledgements. We thank Marek Chrobak for the initial helpful discussion on this topic. Peng Zhang is supported by the National Natural Science Foundation of China (61672323), the State Scholarship Fund of China, the Natural Science Foundation of Shandong Province (ZR2013FM030 and ZR2015FM008), and the Fundamental Research Funds of Shandong University (2015JC006). Chenchen Wu is supported by the National Natural Science Foundation of China (11501412). Dachuan Xu is supported by the National Natural Science Foundation of China (11371001 and 11531014) and Collaborative Innovation Center on Beijing Society-Building and Social Governance.

References

1. Agarwal, A., Charikar, M., Makarychev, K., Makarychev, Y.: $O(\sqrt{\log n})$ approximation algorithms for min uncut, min 2CNF deletion, and directed cut problems. In: Proceedings of the 37th Annual ACM Symposium on Theory of Computing (STOC), pp. 573–581 (2005)
2. Alon, N., Arora, S., Manokaran, R., Moshkovitz, D., Weinstein, O.: Inapproximability of densest κ-subgraph from average case hardness. Manuscript (2011)
3. Bellare, M., Goldreich, O., Sudan, M.: Free bits, PCPs and non-approximability – towards tight results. SIAM J. Comput. **27**(3), 804–915 (1998)
4. Bhaskara, A., Charikar, M., Chlamtac, E., Feige, U., Vijayaraghavan, A.: Detecting high log-densities: an $O(n^{1/4})$ approximation for densest k-subgraph. In: Proceedings of the 42nd Annual ACM Symposium on Theory of Computing (STOC), pp. 201–210 (2010)
5. Buchbinder, N., Naor, J., Schwartz, R.: Simplex partitioning via exponential clocks and the multiway cut problem. In: Proceedings of the Annual ACM Symposium on Theory of Computing (STOC), pp. 535–544 (2013)
6. Calinescu, G., Karloff, H., Rabani, Y.: An improved approximation algorithm for multiway cut. J. Comput. Syst. Sci. **60**(3), 564–574 (2000)
7. Choudhurya, S., Gaurb, D.R., Krishnamurtic, R.: An approximation algorithm for max k-uncut with capacity constraints. Optimization **61**(2), 143–150 (2012)
8. Easley, D., Kleinberg, J.: Networks, Crowds, and Markets: Reasoning About a Highly Connected World. Cambridge University Press, Cambridge (2010)

9. Feige, U., Kortsarz, G., Peleg, D.: The dense k-subgraph problem. Algorithmica **29**, 410–421 (2001)
10. Frieze, A., Jerrum, M.: Improved approximation algorithms for max k-cut and max bisection. Algorithmica **18**, 67–81 (1997)
11. Goemans, M.X., Williamson, D.P.: Improved approximation algorithms for maximum cut and satisfiability problems using semidefinite programming. J. ACM **42**(6), 1115–1145 (1995)
12. Goldschmidt, O., Hochbaum, D.: A polynomial algorithm for the k-cut problem for fixed k. Math. Oper. Res. **19**(1), 24–37 (1994)
13. Håstad, J.: Clique is hard to approximate within $n^{1-\epsilon}$. Acta Math. **182**, 105–142 (1999)
14. Khot, S.: Ruling out PTAS for graph min-bisection, densest subgraph and bipartite clique. In: Proceedings of the 44th Annual IEEE Symposium on the Foundations of Computer Science (FOCS), pp. 136–145 (2004)
15. Langberg, M., Rabani, Y., Swamy, C.: Approximation algorithms for graph homomorphism problems. In: Díaz, J., Jansen, K., Rolim, J.D.P., Zwick, U. (eds.) APPROX/RANDOM -2006. LNCS, vol. 4110, pp. 176–187. Springer, Heidelberg (2006). doi:10.1007/11830924_18
16. Raghavendra, P., Steurer, D.: Graph expansion and the unique games conjecture. In: Proceedings of the 42nd ACM Symposium on Theory of Computing (STOC), pp. 755–764 (2010)
17. Saran, H., Vazirani, V.: Finding k-cuts within twice the optimal. SIAM J. Comput. **24**, 101–108 (1995)
18. Wu, C., Xu, D., Du, D., Xu, W.: A complex semidefinite programming rounding approximation algorithm for the balanced max-3-uncut problem. In: Cai, Z., Zelikovsky, A., Bourgeois, A. (eds.) COCOON 2014. LNCS, vol. 8591, pp. 324–335. Springer, Heidelberg (2014). doi:10.1007/978-3-319-08783-2_28
19. Ye, Y., Zhang, J.: Approximation of dense-$n/2$-subgraph and the complement of min-bisection. J. Global Optim. **25**(1), 55–73 (2003)
20. Zhang, P., Jiang, T., Li, A.: Improved approximation algorithms for the maximum happy vertices and edges problems. In: Xu, D., Du, D., Du, D. (eds.) COCOON 2015. LNCS, vol. 9198, pp. 159–170. Springer, Heidelberg (2015). doi:10.1007/978-3-319-21398-9_13
21. Zhang, P., Li, A.: Algorithmic aspects of homophyly of networks. Theoret. Comput. Sci. **593**, 117–131 (2015)

On Strong Tree-Breadth

Arne Leitert[✉] and Feodor F. Dragan

Department of Computer Science, Kent State University, Kent, OH, USA
{aleitert,dragan}@cs.kent.edu

Abstract. In this paper, we introduce and investigate a new notion of *strong tree-breadth*. We say that a graph G has *strong tree-breadth* ρ if there is a tree-decomposition T for G such that each bag B of T is equal to the complete ρ-neighbourhood of some vertex v in G, i.e., $B = N_G^\rho[v]$. We show that

- it is NP-complete to determine if a given graph has strong tree-breadth ρ, even for $\rho = 1$;
- if a graph G has strong tree-breadth ρ, then we can find a tree-decomposition for G with tree-breadth ρ in $\mathcal{O}(n^2 m)$ time;
- with some additional restrictions, a tree-decomposition with strong breadth ρ can be found in polynomial time;
- some graph classes including distance-hereditary graphs have strong tree-breadth 1.

1 Introduction

Decomposing a graph into a tree is an old concept. It was introduced already by Halin [14]. However, a more popular introduction was given by Robertson and Seymour [15,16]. The idea is to decompose a graph into multiple induced subgraphs, usually called *bags*, where each vertex can be in multiple bags. These bags are combined to a tree in such a way that the following requirements are fulfilled: Each vertex is in at least one bag, each edge is in at least one bag, and, for each vertex, the bags containing it induces a subtree. We will give formal definitions in the next section.

For a given graph, there can be up to exponentially many different tree-decompositions. The easiest is to have only one bag containing the whole graph. To make the concept more interesting, it is necessary to add additional restrictions. The most known is called *tree-width*. A decomposition has *width* ω if each bag contains at most $\omega + 1$ vertices. Then, a graph G has tree-width ω if there is a tree-decomposition for G which has width ω.

In the last years, a new perspective on tree-decompositions was invested. Instead of limiting the number of vertices in each bag, the distance between vertices inside a bag is limited [8,9]. In this paper, we are interested in a variant called *tree-breadth*. It was introduced by Dragan and Köhler in [9]. The *breadth* of a tree-decomposition is ρ, if, for each bag B, there is a vertex v such that each vertex in B has distance at most ρ to v. Accordingly, we say the *tree-breadth* of a graph G is ρ (written as $\mathrm{tb}(G) = \rho$) if there is a tree-decomposition

© Springer International Publishing AG 2016
T.-H.H. Chan et al. (Eds.): COCOA 2016, LNCS 10043, pp. 62–76, 2016.
DOI: 10.1007/978-3-319-48749-6_5

for G with breadth ρ and there is no tree-decomposition with smaller breadth. This new concept of tree-breadth played a crucial role in designing an efficient and best to the date approximation algorithm for the well-known tree t-spanner problem (see [9] for details). Recently, Ducoffe et al. [13] have shown that it is NP-complete to determine if a graph has tree-breadth ρ for all $\rho \geq 1$. On the other hand, for a given graph G, a tree-decomposition of breadth at most $3\,\text{tb}(G)$ can be computed in linear time [1].

By definition, a tree-decomposition has breadth ρ if each bag B is the subset of the ρ-neighbourhood of some vertex v, i.e., the set of bags is the set of *subsets* of the ρ-neighbourhoods of *some* vertices. Tree-breadth 1 graphs contain the class of *dually chordal* graphs which can be defines as follows: A graph G is dually chordal if it admits a tree-decomposition T such that, for each vertex v in G, T contains a bag $B = N_G[v]$ [4]. That is, the set of bags in T is the set of *complete* neighbourhoods of *all* vertices.

In this paper, we investigate the case which lays between dually chordal graphs and general tree-breadth ρ graphs. In particular, tree-decompositions are considered where the set of bags are the *complete* ρ-neighbourhoods of *some* vertices. We call this *strong tree-breadth*. The *strong breadth* of a tree-decomposition is ρ, if, for each bag B, there is a vertex v such that $B = N_G^\rho[v]$. Accordingly, a graph G has strong tree-breadth smaller than or equal to ρ (written as $\text{stb}(G) \leq \rho$) if there is a tree-decomposition for G with strong breadth at most ρ.

Dually chordal graphs and their powers are exactly the graphs admitting a tree-decomposition where the set of bags is equal to the set of *complete* neighbourhoods (complete ρ-neighbourhoods) of *all* vertices. It is a known fact that the dually chordal graphs (the powers of dually chordal graphs) can be recognised in linear time (respectively, polynomial time) [4]. General tree-breadth ρ graphs cannot be recognised in polynomial time unless P = NP [13]. It remained an interesting open question if the graphs with strong tree-breadth ρ can be recognised in polynomial time.

In this paper we show that it is NP-complete to determine if a given graph has strong tree-breadth ρ, even for $\rho = 1$. Furthermore, we demonstrate that: if a graph G has strong tree-breadth ρ, then we can find a tree-decomposition for G with tree-breadth ρ in $\mathcal{O}(n^2 m)$ time; with some additional restrictions, a tree-decomposition with strong breadth ρ can be found in polynomial time; some graph classes including distance-hereditary graphs have strong tree-breadth 1. Our future research plans are to investigate algorithmic implications of the existence for a graph of a tree-decomposition with small strong tree-breadth. Can some algorithmic problems that remain NP-complete on general tree-breadth ρ graphs be solved/approximated efficiently on the graphs with strong tree-breadth ρ? Recall that, for example, greedy routing with aid of a spanning tree [12], (connected) r-domination [3], Steiner tree [3], and (weighted) efficient domination [5,6] can be efficiently solved on dually chordal graphs and their powers.

2   Preliminaries

All graphs occurring in this paper are (if not stated or constructed otherwise) connected, finite, unweighted, undirected, without loops, and without multiple edges. For a graph $G = (V, E)$, we use $n = |V|$ and $m = |E|$ to denote the cardinality of the vertex set and the edge set of G. The *length* of a path from a vertex v to a vertex u is the number of edges in the path. The *distance* $d_G(u, v)$ of two vertices u and v is the length of a shortest path connecting u and v. The distance between a vertex v and a set $S \subseteq V$ is defined as $d_G(v, S) = \min_{u \in S} d_G(u, v)$.

For a vertex v of G, $N_G(v) = \{ u \in V \mid uv \in E \}$ is called the *open neighborhood* of v. Similarly, for a set $S \subseteq V$, we define $N_G(S) = \{ u \in V \mid d_G(u, S) = 1 \}$. The r-*neighbourhood* of a vertex v in G is $N_G^r[v] = \{ u \mid d_G(u, v) \leq r \}$; if r is not specified, then $r = 1$. Two vertices u and v are *true twins* if $N_G[u] = N_G[v]$ and are *false twins* if they are non-adjacent and $N_G(u) = N_G(v)$.

For a vertex set S, let $G[S]$ denote the subgraph of G induced by S. With $G - S$, we denote the graph $G[V \backslash S]$. A vertex set S is a *separator* for two vertices u and v in G if each path from u to v contains a vertex $s \in S$; in this case we say S *separates* u from v. If a separator S contains only one vertex s, i.e., $S = \{s\}$, then s is an *articulation point*. A *block* is a maximal subgraph without articulation points.

A *chord* in a cycle is an edge connecting two non-consecutive vertices of the cycle. A cycle is called *induced* if it has no chords. For each $k \geq 3$, an induced cycle of length k is called as C_k. A subgraph is called *clique* if all its vertices are pairwise adjacent. A *maximal clique* is a clique that cannot be extended by including any additional vertex.

A *tree-decomposition* of a graph $G = (V, E)$ is a tree T with the vertex set $\mathcal{B}$ where each vertex of T, called bag, is a subset of V such that: (i) $V = \bigcup_{B \in \mathcal{B}} B$, (ii) for each edge $uv \in E$, there is a bag $B \in \mathcal{B}$ with $u, v \in B$, and (iii) for each vertex $v \in V$, the bags containing v induce a subtree of T. A tree-decomposition T of G has *breadth* ρ if, for each bag B of T, there is a vertex v in G with $B \subseteq N_G^\rho[v]$. The *tree-breadth* of a graph G is ρ, written as $\mathrm{tb}(G) = \rho$, if ρ is the minimal breadth of all tree-decomposition for G. Similarly, a tree-decomposition T of G has *strong breadth* ρ if, for each bag B of T, there is a vertex v in G with $B = N_G^\rho[v]$. The *strong tree-breadth* of a graph G is the minimal ρ for which G admits a tree-decomposition with strong breadth ρ. This is written as $\mathrm{stb}(G) = \rho$.

3   NP-Completeness

In this section, we will show that it is NP-complete to determine if a given graph has strong tree-breadth ρ even if $\rho = 1$. To do so, we will first show that, for some small graphs, the choice of possible centers is restricted. Then, we will use these small graphs to construct a reduction.

Lemma 1. *Let $C = \{v_1, v_2, v_3, v_4\}$ be an induced C_4 in a graph G with the edge set $\{v_1v_2, v_2v_3, v_3v_4, v_4v_1\}$. If there is no vertex $w \notin C$ with $N_G[w] \supseteq C$, then $N_G[v_1]$ and $N_G[v_2]$ cannot both be bags in the same tree-decomposition with strong breadth 1.*

Proof. Assume that there is a decomposition T with strong breadth 1 containing the bags $B_1 = N_G[v_1]$ and $B_2 = N_G[v_2]$. Because v_3 and v_4 are adjacent, there is a bag $B_3 \supseteq \{v_3, v_4\}$. Consider the subtrees T_1, T_2, T_3, and T_4 of T induced by v_1, v_2, v_3, and v_4, respectively. These subtrees pairwise intersect in the bags B_1, B_2, and B_3. Because pairwise intersecting subtrees of a tree have a common vertex, T contains a bag $N_G[w] \supseteq C$. Note that there is no $v_i \in C$ with $N_G[v_i] \supseteq C$. Thus, $w \notin C$. This contradicts with the condition that there is no vertex $w \notin C$ with $N_G[w] \supseteq C$. □

Let $C = \{v_1, \ldots, v_5\}$ be a C_5 with the edges $E_5 = \{v_1v_2, v_2v_3, \ldots, v_5v_1\}$. We call the graph $H = (C \cup \{u\}, E_5 \cup \{uv_1, uv_3, uv_4\})$, with $u \notin C$, an *extended C_5 of degree 1* and refer to the vertices u, v_1, v_2, and v_5 as *middle*, *top*, *right*, and *left* vertex of H, respectively. Based on $H = (V_H, E_H)$, we construct an *extended C_5 of degree ρ* (with $\rho > 1$) as follows. First, replace each edge $xy \in E_H$ by a path of length ρ. Second, for each vertex w on the shortest path from v_3 to v_4, connect u with w using a path of length ρ. Figure 1 gives an illustration.

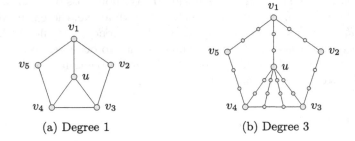

(a) Degree 1   (b) Degree 3

Fig. 1. Two *extended C_5* of degree 1 and degree 3. We refer to the vertices u, v_1, v_2, and v_5 as middle, top, right, and left vertex, respectively.

Lemma 2. *Let B be a bag of a tree-decomposition T for a graph G and let C be a connected component in $G - B$. Then, T contains a bag B_C with $B_C \supseteq N_G(C)$ and $B_C \cap C \neq \emptyset$.*

Proof. Let B_C be the bag in T for which $B_C \cap C \neq \emptyset$ and the distance between B and B_C in T is minimal. Additionally, let B' be the bag in T adjacent to B_C which is closest to B and let $S = B_C \cap B'$. Note that $S \cap C = \emptyset$ and, by properties of tree-decompositions, S separates C from all vertices in $B \backslash S$. Assume that there is a vertex $u \in N_G(C) \backslash S$. Because $u \in N_G(C)$, there is a vertex $v \in C$ which is adjacent to u. This contradicts with S being a separator for u and v. Therefore, $N_G(C) \subseteq S \subseteq B_C$. □

Lemma 3. *Let H be an extended C_5 of degree ρ in a graph G as defined above. Additionally, let H be a block of G and its top vertex v_1 be the only articulation point of G in H. Then, there is no vertex w in G with $d_G(w, v_1) < \rho$ which is the center of a bag in a tree-decomposition for G with strong breadth ρ.*

Proof. Let T be a tree-decomposition for G with strong breadth ρ. Assume that T contains a bag $B_w = N_G^\rho[w]$ with $d_G(w, v_1) < \rho$. Note that the distance from v_1 to any vertex on the shortest path from v_3 to v_4 is 2ρ. Hence, $G - B_w$ has a connected component C containing the vertices v_3 and v_4. Then, by Lemma 2, there has to be a vertex $w' \neq w$ in G and a bag $B'_w = N_G^\rho[w']$ in T such that (i) $B'_w \supseteq N_G(C)$ and (ii) $B'_w \cap C \neq \emptyset$. Thus, if we can show, for a given w, that there is no such w', then w cannot be center of a bag.

First, consider the case that w is in H. We will construct a set $X = \{x, y\} \subseteq N_G(C)$ such that there is a unique shortest path from x to y in G passing w. If $w = v_1$, let $x = v_2$ and $y = v_5$. If w is on the shortest path from v_1 to u, let x and y be on the shortest path from v_1 to v_2 and from v_4 to u, respectively. If w is on the shortest path from v_1 to v_2, let x and y be on the shortest path from v_1 to v_5 and from v_2 to v_3, respectively. In each case, there is a unique shortest path from x to y passing w. Note that, for all three cases, $d_G(v_1, y) \geq \rho$. Thus, each w' with $d_G(w', y) \leq \rho$ is in H. Therefore, w is the only vertex in G with $X \subseteq N_G^\rho[w]$, i.e., there is no vertex $w' \neq w$ satisfying condition (i). This implies that w cannot be center of a bag in T.

Next, consider the case that w is not in H. Without loss of generality, let w be a center for which $d_G(v_1, w)$ is minimal. As shown above, there is no vertex w' in H with $d_G(v_1, w') < \rho$ which is center of a bag. Hence, w' is not in H either. However, because v_1 is an articulation point, w' has to be closer to v_1 than w to satisfy condition (ii). This contradicts with $d_G(v_1, w)$ being minimal. Therefore, there is no vertex w' satisfying condition (ii) and w cannot be center of a bag in T. $\square$

Theorem 1. *It is NP-complete to decide, for a given graph G, if $\mathrm{stb}(G) = 1$.*

Proof. Clearly, the problem is in NP: Select non-deterministically a set S of vertices such that their neighbourhoods cover each vertex and each edge. Then, check deterministically if the neighbourhoods of the vertices in S give a valid tree-decomposition. This can be done in linear time [18]. The algorithm in [18] also creates the corresponding tree.

To show that the problem is NP-hard, we will make a reduction from 1-in-3-SAT [17]. That is, you are given a boolean formula in CNF with at most three literals per clause; find a satisfying assignment such that, in each clause, only one literal becomes true.

Let $\mathcal{I}$ be an instance of 1-in-3-SAT with the literals $\mathcal{L} = \{p_1, \ldots, p_n\}$, the clauses $\mathcal{C} = \{c_1, \ldots, c_m\}$, and, for each $c \in \mathcal{C}$, $c \subseteq \mathcal{L}$. We create a graph $G = (V, E)$ as follows. Create a vertex for each literal $p \in \mathcal{L}$ and, for all literals p_i and p_j with $p_i \equiv \neg p_j$, create an induced $C_4 = \{p_i, p_j, q_i, q_j\}$ with the edges $p_i p_j$, $q_i q_j$, $p_i q_i$, and $p_j q_j$. For each clause $c \in \mathcal{C}$ with $c = \{p_i, p_j, p_k\}$, create an extended C_5 with c as top vertex, connect c with an edge to all literals it contains, and make all literals in c pairwise adjacent, i.e., the vertex set $\{c, p_i, p_j, p_k\}$ induces a maximal clique in G. Additionally, create a vertex v and make v adjacent to all literals. Figure 2a gives an illustration for the construction so far.

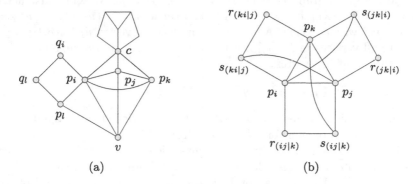

(a)   (b)

Fig. 2. Illustration to the proof of Theorem 1. The graphs shown are subgraphs of G as created by a clause $c = \{p_i, p_j, p_k\}$ and a literal p_l with $p_i \equiv \neg p_l$.

Next, for each clause $\{p_i, p_j, p_k\}$ and for each $(xy|z) \in \{(ij|k), (jk|i), (ki|j)\}$, create the vertices $r_{(xy|z)}$ and $s_{(xy|z)}$, make $r_{(xy|z)}$ adjacent to $s_{(xy|z)}$ and p_x, and make $s_{(xy|z)}$ adjacent to p_y and p_z. See Fig. 2b for an illustration. Note that $r_{(ij|k)}$ and $s_{(ij|k)}$ are specific for the clause $\{p_i, p_j, p_k\}$. Thus, if p_i and p_j are additionally in a clause with p_l, then we also create the vertices $r_{(ij|l)}$ and $s_{(ij|l)}$. For the case that a clause only contains two literals p_i and p_j, create the vertices $r_{(ij)}$ and $s_{(ij)}$, make $r_{(ij)}$ adjacent to p_i and $s_{(ij)}$, and make $s_{(ij)}$ adjacent to p_j, i.e., $\{p_i, p_j, r_{(ij)}, s_{(ij)}\}$ induces a C_4 in G.

For the reduction, first, consider the case that $\mathcal{I}$ is a *yes*-instance for 1-in-3-SAT. Let $f\colon \mathcal{P} \to \{T, F\}$ be a satisfying assignment such that each clause contains only one literal p_i with $f(p_i) = T$. Select the following vertices as centers of bags: v, the middle, left and right vertex of each extended C_5, p_i if $f(p_i) = T$, and q_j if $f(p_j) = F$. Additionally, for each clause $\{p_i, p_j, p_k\}$ with $f(p_i) = T$, select the vertices $s_{(ij|k)}$, $r_{(jk|i)}$, and $r_{(ki|j)}$. The neighbourhoods of the selected vertices give a valid tree-decomposition for G. Therefore, stb$(G) = 1$.

Next, assume that stb$(G) = 1$. Recall that, for a clause $c = \{p_i, p_j, p_k\}$, the vertex set $\{c, p_i, p_j, p_k\}$ induces a maximal clique in G. By Lemma 3, c cannot be center of a bag because it is top of an extended C_5. Therefore, at least one vertex in $\{p_i, p_j, p_k\}$ must be center of a bag. Without loss of generality, let p_i be a center of a bag. By construction, p_i is adjacent to all $p \in \{p_j, p_k, p_l\}$, where $p_l \equiv \neg p_i$. Additionally, p and p_i are vertices in an induced C_4, say C, and there is no vertex w in G with $N_G[w] \supseteq C$. Thus, by Lemma 1, at most one vertex in $\{p_i, p_j, p_k\}$ can be center of a bag. Therefore, the function $f\colon \mathcal{L} \to \{T, F\}$ defined as

$$f(p_i) = \begin{cases} T & \text{if } p_i \text{ is center of a bag,} \\ F & \text{else} \end{cases}$$

is a satisfying assignment for $\mathcal{I}$.   □

In [13], Ducoffe et al. have shown how to construct a graph G'_ρ based on a given graph G such that $\mathrm{tb}(G'_\rho) = 1$ if and only if $\mathrm{tb}(G) \leq \rho$. We will slightly extend their construction to achieve a similar result for strong tree-breadth.

Consider a given graph $G = (V, E)$ with $\mathrm{stb}(G) = \rho$. We will construct G'_ρ as follows. Let $V = \{v_1, v_2, \ldots, v_n\}$. Add the vertices $U = \{u_1, u_2, \ldots, u_n\}$ and make them pairwise adjacent. Additionally, make each vertex u_i, with $1 \leq i \leq n$, adjacent to all vertices in $N_G^\rho[v_i]$. Last, for each $v_i \in V$, add an extended C_5 of degree 1 with v_i as top vertex.

Lemma 4. $\mathrm{stb}(G) \leq \rho$ if and only if $\mathrm{stb}(G'_\rho) = 1$.

Proof. First, consider a tree-decomposition T for G with strong breadth ρ. Let T'_ρ be a tree-decomposition for G'_ρ created from T by adding all vertices in U into each bag of T and by making the center, left, and right vertices of each extended C_5 centers of bags. Because the set U induces a clique in G'_ρ and $N_G^\rho[v_i] = N_{G'_\rho}[u_i] \cap V$, each bag of T'_ρ is the complete neighbourhood of some vertex.

Next, consider a tree-decomposition T'_ρ for G'_ρ with strong breadth 1. Note that each vertex v_i is top vertex of some extended C_5. Thus, v_i cannot be center of a bag. Therefore, each edge $v_i v_j$ is in a bag $B_k = N_{G'_\rho}[u_k]$. By construction of G'_ρ, $B_k \cap V = N_G^\rho[v_k]$. Thus, we can construct a tree-decomposition T for G with strong breadth ρ by creating a bag $B_i = N_G^\rho[v_i]$ for each bag $N_{G'_\rho}[u_i]$ of T'_ρ.   □

Next, consider a given graph $G = (V, E)$ with $V = \{v_1, v_2, \ldots, v_n\}$ and $\mathrm{stb}(G) = 1$. For a given $\rho > 1$, we obtain the graph G_ρ^+ by doing the following for each $v_i \in V$:

- Add the vertices $u_{i,1}, \ldots, u_{i,5}$, x_i, and y_i.
- Add an extended C_5 of degree ρ with the top vertex z_i.
- Connect
  - $u_{i,1}$ and x_i with a path of length $\lfloor \rho/2 \rfloor - 1$,
  - $u_{i,2}$ and y_i with a path of length $\lfloor \rho/2 \rfloor$,
  - $u_{i,3}$ and v_i with a path of length $\lceil \rho/2 \rceil - 1$,
  - $u_{i,4}$ and v_i with a path of length $\lfloor \rho/2 \rfloor$, and
  - $u_{i,4}$ and z_i with a path of length $\lceil \rho/2 \rceil - 1$.
- Add the edges $u_{i,1}u_{i,2}$, $u_{i,1}u_{i,3}$, $u_{i,2}u_{i,3}$, $u_{i,2}u_{i,4}$, and $u_{i,3}u_{i,4}$.

Note that, for small ρ, it can happen that $v_i = u_{i,4}$, $x_i = u_{i,1}$, $y_i = u_{i,2}$, or $z_i = u_{i,5}$. Figure 3 gives an illustration.

Lemma 5. $\mathrm{stb}(G) = 1$ if and only if $\mathrm{stb}(G_\rho^+) = \rho$.

Proof. First, assume that $\mathrm{stb}(G) = 1$. Then, there is a tree-decomposition T for G with strong breadth 1. We will construct for G_ρ^+ a tree-decomposition T_ρ^+ with strong breadth ρ. Make the middle, left, and right vertex of each extended C_5 center of a bag. For each $v_i \in V$, if v_i is center of a bag of T, make x_i a center of a bag of T_ρ^+. Otherwise, make y_i center of a bag of T_ρ^+. The distance in G_ρ^+

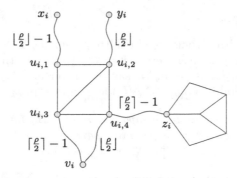

Fig. 3. Illustration for the graph G_ρ^+. The graph shown is a subgraph of G_ρ^+ as constructed for each v_i in G.

from v_i to x_i is $\rho - 1$. The distances from v_i to y_i, from x_i to z_i, and from y_i to z_i are ρ. Thus, $N_{G_\rho^+}^\rho[x_i] \cap V = N_G[v_i]$, $N_{G_\rho^+}^\rho[y_i] \cap V = \{v_i\}$, and there is no conflict with Lemma 3. Therefore, the constructed T_ρ^+ is a valid tree-decomposition with strong breadth ρ for G_ρ^+.

Next, assume that $\mathrm{stb}(G_\rho^+) = \rho$ and there is a tree-decomposition T_ρ^+ with strong breadth ρ for G_ρ^+. By Lemma 3, no vertex in distance less than ρ to any z_i can be a center of a bag in T_ρ^+. Therefore, because the distance from v_i to z_i in G_ρ^+ is $\rho - 1$, no $v_i \in V$ can be a center of a bag in T_ρ^+. The only vertices with a large enough distance to z_i to be a center of a bag are x_i and y_i. Therefore, either x_i or y_i is selected as center. To construct a tree-decomposition T with strong breadth 1 for G, select v_i as center if and only if x_i is a center of a bag in T_ρ^+. Because $N_{G_\rho^+}^\rho[x_i] \cap V = N_G[v_i]$ and $N_{G_\rho^+}^\rho[y_i] \cap V = \{v_i\}$, the constructed T is a valid tree-decomposition with strong breadth 1 for G. $\square$

Constructing G_ρ' can be done in $\mathcal{O}(n^2)$ time and constructing G_ρ^+ can be done in $\mathcal{O}(\rho \cdot n + m)$ time. Thus, combining Lemmas 4 and 5 allows us, for a given graph G, some given ρ, and some given ρ', to construct a graph H in $\mathcal{O}(\rho \cdot n^2)$ time such that $\mathrm{stb}(G) \leq \rho$ if and only if $\mathrm{stb}(H) \leq \rho'$. Additionally, by combining Theorem 1 and Lemma 3, we get:

Theorem 2. *It is NP-complete to decide, for a graph G and a given ρ, if* $\mathrm{stb}(G) = \rho$.

4 Polynomial Time Cases

In the previous section, we have shown that, in general, it is NP-complete to determine the strong tree-breadth of a graph. In this section, we will investigate cases for which a decomposition can be found in polynomial time.

4.1   General Graphs

Let G be a graph with strong tree-breadth ρ and let T be a corresponding tree-decomposition. For a given vertex u in G, we denote the set of connected components in $G - N_G^\rho[u]$ as $\mathcal{C}_G[u]$. We say that a vertex v is a *potential partner* of u for some $C \in \mathcal{C}_G[u]$ if $N_G^\rho[v] \supseteq N_G(C)$ and $N_G^\rho[v] \cap C \neq \emptyset$.

Lemma 6. *Let C be a connected component in $G - B_u$ for some $B_u \subseteq N_G^\rho[u]$. Also, let $C \in \mathcal{C}_G[u]$ and v be a potential partner of u for C. Then, for all connected components C_v in $G[C] - N_G^\rho[v]$, $C_v \in \mathcal{C}_G[v]$.*

Proof. Consider a connected component C_v in $G[C] - N_G^\rho[v]$. Clearly, $C_v \subseteq C$ and there is a connected component $C' \in \mathcal{C}[v]$ such that $C' \supseteq C_v$.

Let x be an arbitrary vertex in C'. Then, there is a path $P \subseteq C'$ from x to C_v. Because $N_G(C) \subseteq B_u$ and v is a potential partner of u for C, $N_G(C) \subseteq B_u \cap N_G^\rho[v]$. Also, $N_G(C)$ separates all vertices in C from all other vertices in G. Therefore, $x \in C$ and $C' \subseteq C$; otherwise, P would intersect $N_G^\rho[v]$. It follows that each vertex in P is in the same connected component of $G[C] - N_G^\rho[v]$ and, thus, $C_v = C'$.   □

From Lemma 2, it directly follows:

Corollary 1. *If $N_G^\rho[u]$ is a bag in T, then T contains a bag $N_G^\rho[v]$ for each $C \in \mathcal{C}_G[u]$ such that v is a potential partner of u for C.*

Because of Corollary 1, there is a vertex set U such that each $u \in U$ has a potential partner $v \in U$ for each connected component $C \in \mathcal{C}_G[u]$. With such a set, we can construct a tree-decomposition for G with the following approach: Pick a vertex $u \in U$ and make it center of a bag B_u. For each connected component $C \in \mathcal{C}_G[u]$, u has a potential partner v. $N_G^\rho[v]$ splits C in more connected components and, because $v \in U$, v has a potential partner $w \in U$ for each of these components. Hence, create a bag $B_v = N_G^\rho[v] \cap (B_u \cup C)$ and continue this until the whole graph is covered. Algorithm 1 will determine such a set of vertices with their potential partners (represented as a graph H) and then construct a decomposition as described above.

Theorem 3. *Algorithm 1 constructs, for a given graph G with strong tree-breadth ρ, a tree-decomposition T with breadth ρ in $\mathcal{O}(n^2 m)$ time.*

Proof (Correctness). The Algorithm 1 works in two parts. First, it creates a graph H with potential centers (line 1 to line 6). Second, it uses H to create a tree-decomposition for G (line 9 to line 15). To show the correctness of the algorithm, we will, first, show that centers of a tree-decomposition for G are vertices in H and, then, show that a tree-decomposition created based on H is a valid tree-decomposition for G.

A vertex u is added to H (line 4) if, for *at least one* connected component $C \in \mathcal{C}_G[u]$, u has a potential partner v. Later, u is kept in H (line 5 and 6) if it has a potential partner v *for all* connected components in $C \in \mathcal{C}_G[u]$. By Corollary 1,

---

Algorithm 1. Constructs, for a given graph $G = (V, E)$ with strong tree-breadth ρ, a tree-decomposition T with breadth ρ.

---

1 Create an empty directed graph $H = (V_H, E_H)$. Let ϕ be a function that maps each edge $(u, v) \in E_H$ to a connected component $C \in \mathcal{C}_G[u]$.

2 **foreach** $u, v \in V$ *and all* $C \in \mathcal{C}_G[u]$ **do**

3   **if** v *is a potential parter of* u *for* C **then**

4   Add the directed edge (u, v) to H and set $\phi(u, v) := C$. (Add u and v to H if necessary.)

5 **while** *there is a vertex* $u \in V_H$ *and some* $C \in \mathcal{C}_G[u]$ *such that there is no* $(u, v) \in E_H$ *with* $\phi(u, v) = C$ **do**

6   Remove u from H.

7 **if** H *is empty* **then**

8   **Stop.** stb$(G) > \rho$.

9 Create an empty tree-decomposition T.

10 Let $G - T$ be the subgraph of G that is not covered by T and let ψ be a function that maps each connected component in $G - T$ to a bag $B_u \subseteq N_G^\rho[u]$.

11 Pick an arbitrary vertex $u \in V_H$, add $B_u = N_G^\rho[u]$ as bag to T, and set $\psi(C) := B_u$ for each connect component C in $G - T$.

12 **while** $G - T$ *is non-empty* **do**

13   Pick a connected component C in $G - T$, determine the bag $B_v := \psi(C)$ and find an edge $(v, w) \in E_H$ with $\phi(v, w) = C$.

14   Add $B_w = N_G^\rho[w] \cap (B_v \cup C)$ to T, and make B_v and B_w adjacent in T.

15   For each new connected component C' in $G - T$ with $C' \subseteq C$, set $\psi(C_w) := B_w$.

16 Output T.

---

each center of a bag in a tree-decomposition T with strong breadth ρ satisfies these conditions. Therefore, after line 6, H contains all centers of bags in T, i.e., if G has strong tree-breadth ρ, H is non-empty.

Next, we show that T created in the second part of the algorithm (line 9 to line 15) is a valid tree-decomposition for G with breadth ρ. To do so, we will show the following invariant for the loop starting in line 12: (i) T is a valid tree-decomposition with breadth ρ for the subgraph covered by T and (ii) for each connected component C in $G - T$, the bag $B_v = \psi(C)$ is in T, $N_G(C) \subseteq B_v$, and $C \in \mathcal{C}_G[v]$. After line 11, the invariant clearly holds. Assume by induction that the invariant holds each time line 12 is checked. If T covers the whole graph, the check fails and the algorithm outputs T. If T does not cover G completely, there is a connected component C in $G - T$. By condition (ii), the bag $B_v = \psi(C)$ is in T, $N_G(C) \subseteq B_v$, and $C \in \mathcal{C}_G[v]$. Because of the way H is constructed and $C \in \mathcal{C}_G[v]$, there is an edge $(v, w) \in E_H$ with $\phi(v, w) = C$, i.e., w is a potential partner of v for C. Thus, line 13 is successful and the algorithm adds a new bag $B_w = N_G^\rho[w] \cap (B_v \cup C)$ (line 14). Because w is a potential partner of v for C, i.e., $N_G(C) \subseteq N_G^\rho[w]$, and $N_G(C) \subseteq B_v$, $B_w \supseteq N_G(C)$. Therefore, after adding B_w to T, T still satisfies condition (i). Additionally, B_w splits C in a set $\mathcal{C}'$ of connected components such that, for each $C' \in \mathcal{C}'$, $N_G(C') \subseteq B_w$ and, by Lemma 6, $C' \in \mathcal{C}_G[w]$. Thus, condition (ii) is also satisfied. □

Proof (Complexity). First, determine the pairwise distance of all vertices. This can be done in $\mathcal{O}(nm)$ time and allows to check the distance between vertices in constant time.

For a vertex u, let $\mathcal{N}[u] = \{N_G(C) \mid C \in \mathcal{C}_G[u]\}$. Note that, for some $C \in \mathcal{C}_G[u]$ and each vertex $x \in N_G(C)$, there is an edge xy with $y \in C$. Therefore, $|\mathcal{N}[u]| := \sum_{C \in \mathcal{C}_G[u]} |N_G(C)| \leq m$. To determine, for some vertex u, all its potential partners v, first, compute $\mathcal{N}[u]$. This can be done in $\mathcal{O}(m)$ time. Then, check, for each vertex v and each $N_G(C) \in \mathcal{N}[u]$, if $N_G(C) \subseteq N_G^\rho[v]$ and add the edge (u, v) to H if successful. For a single vertex v this requires $\mathcal{O}(m)$ time because $|\mathcal{N}[u]| \leq m$ and distances can be determined in constant time. Therefore, the total runtime for creating H (line 1 to line 4) is $\mathcal{O}(n(m + nm)) = \mathcal{O}(n^2 m)$.

Assume that, for each $\phi(u, v) = C$, C is represented buy two values: (i) a characteristic vertex $x \in C$ (for example the vertex with the lowest index) and (ii) the index of C in $\mathcal{C}_G[u]$. While creating H, count and store, for each vertex u and each connected component $C \in \mathcal{C}_G[u]$, the number of edges $(u, v) \in E_H$ with $\phi(u, v) = C$. Note that there is a different counter for each $C \in \mathcal{C}_G[u]$. With this information, we can implement line 5 and 6 as follows. First check, for every vertex v in H, if one of its counters is 0. In this case, remove v from H and update the counters for all vertices u with $(u, v) \in E_H$ using value (ii) of $\phi(u, v)$. If this sets a counter for u to 0, add u to a queue Q of vertices to process. Continue this until each vertex is checked. Then, for each vertex u in Q, remove u form H and add its neighbours into Q if necessary until Q is empty. This way, a vertex is processed at most twice. A single iteration runs in at most $\mathcal{O}(n)$ time. Therefore, line 5 and 6 can be implemented in $\mathcal{O}(n^2)$ time.

Assume that ψ uses the characteristic vertex x to represent a connected component, i. e., value (i) of ϕ. Then, finding an edge $(v, w) \in E_H$ (line 13) can be done in $\mathcal{O}(m)$ time. Creating B_w (line 14), splitting C into new connected components C', finding their characteristic vertex, and setting $\psi(C')$ (line 15) takes $\mathcal{O}(m)$ time, too. In each iteration, at least one more vertex of G is covered by T. Hence, there are at most n iterations and, thus, the loop starting in line 12 runs in $\mathcal{O}(mn)$ time.

Therefore, Algorithm 1 runs in total $\mathcal{O}(n^2 m)$ time. □

Algorithm 1 creates for each graph G with $\mathrm{stb}(G) \leq \rho$ a tree-decomposition T with breadth ρ. Next, we will invest a case where we can construct a tree-decomposition for G with strong breadth ρ.

We say that two vertices u and v are *perfect partners* if (i) u and v are potential partner of each other for some $C_u \in \mathcal{C}_G[u]$ and some $C_v \in \mathcal{C}_G[v]$, (ii) C_u is the only connected component in $\mathcal{C}_G[u]$ which is intersected by $N_G^\rho[v]$, and (iii) C_v is the only connected component in $\mathcal{C}_G[v]$ which is intersected by $N_G^\rho[u]$. Accordingly, we say that a tree-decomposition T has *perfect strong breadth ρ* if it has strong breadth ρ and, for each center u of some bag and each connected component $C \in \mathcal{C}_G[u]$, there is a center v such that v is a perfect partner of u for C.

Theorem 4. *A tree-decomposition with perfect strong breadth ρ can be constructed in polynomial time.*

Proof. To construct such a tree-decomposition, we can modify Algorithm 1. Instead of checking if u has a potential partner v (line 3), check if u and v are perfect partners.

Assume by induction that, for each bag B_v in T, $B_v = N_G^\rho[v]$. By definition of perfect partners v and w, $N_G^\rho[w]$ intersects only one $C \in \mathcal{C}_G[v]$, i.e., $N_G^\rho[w] \subseteq N_G^\rho[v] \cup C$. Thus, when creating the bag B_w (line 14), $B_w = N_G^\rho[w] \cap (B_v \cup C) = N_G^\rho[w] \cap (N_G^\rho[v] \cup C) = N_G^\rho[w]$. Therefore, the created tree-decomposition T has perfect strong tree-breadth ρ. □

We conjecture that there are weaker cases than perfect strong breadth which allow to construct a tree-decomposition with strong-breadth ρ. For example, if the centers of two adjacent bags are perfect partners, but a center u does not need to have a perfect partner for each $C \in \mathcal{C}_G[u]$. However, when using a similar approach as in Algorithm 1, this would require a more complex way of constructing H.

4.2 Special Graph Classes

A graph G is *distance-hereditary* if, in any connected induced subgraph, the distances are the same as in G.

Theorem 5. *Distance-hereditary graphs have strong tree-breadth 1. An according decomposition can be computed in linear time.*

Proof. Let $\sigma = \langle v_1, v_2, \ldots, v_n \rangle$ be an ordering for the vertices of a graph G, $V_i = \{v_1, v_2, \ldots, v_i\}$, and G_i denote the graph $G[V_i]$. An ordering σ is called a *pruning sequence* for G if, for $1 < i \leq n$, each v_i satisfies one of the following conditions in G_i:

  (i) v_i is a pendant vertex,
 (ii) v_i is a true twin of some vertex v_j, or
(iii) v_i is a false twin of some vertex v_j.

A graph G is distance-hereditary if and only if there is a pruning sequence for G [2].

Assume that we are given such a pruning sequence. Additionally, assume by induction over i that G_i has a tree-decomposition T_i with strong breadth 1. Then, there are three cases:

  (i) v_{i+1} *is a pendant vertex in* G_{i+1}. If the neighbour u of v_{i+1} is a center of a bag B_u, add v_{i+1} to B_u. Thus, T_{i+1} is a valid decomposition for G_{i+1}. Otherwise, if u is not a center, make v_{i+1} center of a bag. Because u is an articulation point, $T_{i+1} = T_i + N_G[v]$ is a valid decomposition for G_{i+1}.
 (ii) v_{i+1} *is a true twin of a vertex* u *in* G_{i+1}. Simply add v_{i+1} into any bag containing u. The resulting decomposition is a valid decomposition for G_{i+1}.

(iii) v_{i+1} *is a false twin of a vertex* u *in* G_{i+1}. If u is not center of a bag, add v_{i+1} into any bag u is in. Otherwise, make a new bag $B_{i+1} = N_G[v_{i+1}]$ and make it adjacent to the bag $N_G[u]$. Because no vertex in $N_G(u)$ is center of a bag, the resulting decomposition is a valid decomposition for G_{i+1}.

Therefore, distance-hereditary graphs have strong tree-breadth 1.

Next, we will show how to compute an according tree-decomposition in linear time. The argument above already gives an algorithmic approach. First, we compute a pruning sequence for G. This can be done in linear time with an algorithm by Damiand et al. [7]. Then, we determine which vertex becomes a center of a bag. Note that we can simplify the three cases above with the following rule: If v_i has no neighbour in G_i which is center of a bag, make v_i center of a bag. Otherwise, proceed with v_{i+1}. This can be easily implemented in linear time with a binary flag for each vertex.   □

Algorithm 2 formalizes the method described in the proof of Theorem 5.

Algorithm 2. Computes, for a given distance-hereditary graph G, a tree-decomposition T with strong breadth 1.

---

1 Compute a pruning sequence $\langle v_1, v_2, \ldots, v_n \rangle$ (see [7]).
2 Create a set $C := \emptyset$.
3 **for** $i := 1$ **to** n **do**
4 ⎡ **if** $N_G[v_i] \cap V_i \cap C = \emptyset$ **then**
5 ⎣ ⎣ Add v_i to C.

6 Create a tree-decomposition T with the vertices in C as centers of its bags.

A bipartite graph is *chordal bipartite* if each cycle of length at least 6 has a chord. In [11], it was shown that any chordal bipartite graph $G = (X, Y, E)$ admits a tree-decomposition with the set of bags $\mathcal{B} = \{B_1, B_2, \ldots, B_{|X|}\}$, where $B_i = N_G[x_i]$, $x_i \in X$. As far as we can tell, there is no linear time algorithm known to recognise chordal bipartite graphs. However, we can still compute a tree-decomposition in linear time with three steps. First, compute a 2-colouring. Second, select a colour and make the neighbourhood of all vertices with this colour bags. Third, use the algorithm in [18] to check if the selected bags give a valid tree-decomposition.

Theorem 6 [11]. *Each chordal bipartite graph has strong tree-breadth* 1. *An according tree-decomposition can be found in linear time.*

Consider two parallel lines (upper and lower) in the plane. Assume that each line contains n points, labelled 1 to n. Each two points with the same label define a segment with that label. The intersection graph of such a set of segments between two parallel lines is called a *permutation graph*. In [10], an algorithm was presented that finds, for a given permutation graph, a path-decomposition with strong breadth 1 in linear time.

Theorem 7 [10]. *Permutation graphs have strong tree-breadth 1. An according tree-decomposition can be found in linear time.*

5 Conclusion

We have shown that, in general, it is NP-complete to determine if a given graph G admits a tree-decomposition with strong breadth ρ for all $\rho \geq 1$. Consider the case that a vertex v is center of a bag. Part of the hardness of finding a decomposition, even for $\rho = 1$, lays in determining which connected component $C \in \mathcal{C}_G[v]$ will be covered by which neighbouring bag $N_G[u]$. If, for two vertices u and w, $N_G[u]$ and $N_G[w]$ intersect C and are bags in the same decompositions T, both cannot be separated in T by $N_G[v]$. Additionally, if u is adjacent to v, it might happen that $N_G[u]$ intersects multiple connected components. This leads to a potentially exponential number of combinations.

A *path-decomposition* of graph is a tree-decomposition with the restriction that the bags form a path instead of a tree with multiple branches. Accordingly, a graph has *(strong) path-breadth* ρ if it admits a path-decomposition with (strong) breadth ρ. In [10], it was shown that, for graphs with bounded path-breadth, a constant factor approximation for the bandwidth problem and the line-distortion problem can be found in polynomial time.

Now, consider the case that we want to compute if a given graph admits a path-decomposition P with strong breadth 1. In this case, there can be at most two bags adjacent to a bag $N_G[v]$ in P. Hence, for each v, there is at most a quadratic number of combinations. This leads to the following conjecture.

Conjecture. *The strong path-breadth of a graph can be computed in polynomial time.*

Another question is if a bounded strong tree-breadth leads to a lower bound for the tree-breadth of a graph. That is, is there a constant c such that, for any graph G, $\mathrm{stb}(G) \leq c \cdot \mathrm{tb}(G)$. Using Algorithm 1, a small constant might lead to a new approach for approximating the tree-breadth of a graph.

References

1. Abu-Ata, M., Dragan, F.F.: Metric tree-like structures in real-life networks: an empirical study. Networks **67**(1), 49–68 (2016)
2. Bandelt, H.-J., Mulder, H.M.: Distance-hereditary graphs. J. Comb. Theory Ser. B **41**, 182–208 (1986)
3. Brandstädt, A., Chepoi, V.D., Dragan, F.F.: The algorithmic use of hypertree structure and maximum neighborhood orderings. Discret. Appl. Math. **82**, 43–77 (1998)
4. Brandstädt, A., Dragan, F.F., Chepoi, V.D., Voloshin, V.: Dually chordal graphs. SIAM J. Discret. Math. **11**(3), 437–455 (1998)
5. Brandstädt, A., Fičur, P., Leitert, A., Milanič, M.: Polynomial-time algorithms for weighted efficient domination problems in AT-free graphs and dually chordal graphs. Inf. Process. Lett. **115**(2), 256–262 (2015)

6. Brandstädt, A., Leitert, A., Rautenbach, D.: Efficient dominating and edge dominating sets for graphs and hypergraphs. In: Chao, K.-M., Hsu, T., Lee, D.-T. (eds.) ISAAC 2012. LNCS, vol. 7676, pp. 267–277. Springer, Heidelberg (2012)
7. Damiand, G., Habib, M., Paul, C.: A simple paradigm for graph recognition: application to cographs and distance hereditary graphs. Theoret. Comput. Sci. **263**(1–2), 99–111 (2001)
8. Dourisboure, Y., Gavoille, C.: Tree-decompositions with bags of small diameter. Discret. Math. **307**(16), 2008–2029 (2007)
9. Dragan, F.F., Köhler, E.: An approximation algorithm for the tree t-spanner problem on unweighted graphs via generalized chordal graphs. Algorithmica **69**, 884–905 (2014)
10. Dragan, F.F., Köhler, E., Leitert, A.: Line-distortion, bandwidth and path-length of a graph. Algorithmica (in print)
11. Dragan, F.F., Lomonosov, I.: On compact and efficient routing in certain graph classes. Discret. Appl. Math. **155**, 1458–1470 (2007)
12. Dragan, F.F., Matamala, M.: Navigating in a graph by aid of its spanning tree. SIAM J. Discret. Math. **25**(1), 306–332 (2011)
13. Ducoffe, G., Legay, S., Nisse, N.: On computing tree and path decompositions with metric constraints on the bags. CoRR abs/1601.01958 (2016)
14. Halin, R.: S-functions for graphs. J. Geom. **8**(1–2), 171–186 (1976)
15. Robertson, N., Seymour, P.D.: Graph minors. I. Excluding a forest. J. Comb. Theory Ser. B **35**(1), 39–61 (1983)
16. Robertson, N., Seymour, P.D.: Graph minors. III. Planar tree-width. J. Comb. Theory Ser. B **36**(1), 49–64 (1984)
17. Schaefer, T.J.: The complexity of satisfiability problems. In: Proceedings of the Tenth Annual ACM Symposium on Theory of Computing (STOC 1978), pp. 216–226 (1978)
18. Tarjan, R.E., Yannakakis, M.: Simple linear-time algorithms to test chordality of graphs, test acyclicity of hypergraphs, and selectively reduce acyclic hypergraphs. SIAM J. Comput. **13**(3), 566–579 (1984)

Computing a Tree Having a Small Vertex Cover

Takuro Fukunaga[1,2](✉) and Takanori Maehara[3]

[1] National Institute of Informatics, Tokyo, Japan
takuro@nii.ac.jp
[2] JST, ERATO, Kawarabayashi Large Graph Project, Tokyo, Japan
[3] Department of Mathematical and Systems Engineering,
Shizuoka University, Shizuoka, Japan
maehara.takanori@shizuoka.ac.jp

Abstract. In this paper, we consider a new Steiner tree problem. This problem defines the weight of a Steiner tree as the minimum weight of vertex covers in the tree, and seeks a minimum-weight Steiner tree in a given vertex-weighted undirected graph. Since it is included by the Steiner tree activation problem, the problem admits an $O(\log n)$-approximation algorithm in general graphs with n vertices. This approximation factor is tight because it is known to be NP-hard to achieve an $o(\log n)$-approximation for the problem with general graphs. In this paper, we present constant-factor approximation algorithms for the problem with unit disk graphs and with graphs excluding a fixed minor.

1 Introduction

The problem of finding a minimum-weight tree in a graph has been extensively studied in the field of combinatorial optimization. A typical example is the Steiner tree problem in edge-weighted graphs; it has a long history of approximation algorithms, culminating in the currently best approximation factor of 1.39 [1,2]. The Steiner tree problem has also been studied in vertex-weighted graphs, where the weight of a Steiner tree is defined as the total weight of the vertices spanned by the tree. We call this problem the *vertex-weighted Steiner tree problem* while the problem in the edge-weighted graphs is called the *edge-weighted Steiner tree problem*. There is an $O(\log k)$-approximation algorithm for the vertex-weighted Steiner tree problem with k terminals, and it is NP-hard to improve this factor because the problem includes the set cover problem [3].

In this paper, we present a new variation of the Steiner tree problem. Our problem is motivated by the following situation in communication networks. We assume that messages are exchanged along a tree in a network; this is the case in many popular routing protocols such as the spanning tree protocol [4]. We consider locating devices that will monitor the traffic in the tree. If a device is located at a vertex, it can monitor all the traffic that passes through links incident to that vertex. How many devices do we need for monitoring all of the traffic in the tree? Obviously, it depends on the topology of the tree. If the tree is a star, it suffices to locate one device at the center. If the tree is a path

T.-H.H. Chan et al. (Eds.): COCOA 2016, LNCS 10043, pp. 77–91, 2016.
DOI: 10.1007/978-3-319-48749-6_6

on n vertices, then it requires $\lceil n/2 \rceil$ devices, because any vertex cover of the path consists of at least $\lceil n/2 \rceil$ vertices. Our problem is to compute a tree that minimizes the number (or, more generally, the weight) of devices required to monitor all of the traffic.

More formally, our problem is defined as follows. Let $G = (V, E)$ be an undirected graph associated with nonnegative vertex weights $w \in \mathbb{R}_+^V$. Throughout this paper, we will denote $|V|$ by n. Let $T \subseteq V$ be a set of vertices called *terminals*. The problem seeks a pair comprising a tree F and a vertex set $U \subseteq V(F)$ such that (i) F is a Steiner tree with regard to the terminal set T (i.e., $T \subseteq V(F)$), and (ii) U is a vertex cover of F (i.e., each edge in F is incident to at least one vertex in U). The objective is to find such a pair (F, U) that minimizes the weight $w(U) := \sum_{v \in U} w(v)$ of the vertex cover. We call this the *vertex-cover-weighted (VC-weighted) Steiner tree problem*. We call the special case in which $V = T$ the *vertex-cover-weighted (VC-weighted) spanning tree problem*. The aim of this paper is to investigate these fundamental problems.

Besides the motivation from the communication networks, there is another reason for the importance of the VC-weighted Steiner tree problem. The VC-weighted Steiner tree problem is a special case of the *Steiner tree activation problem*, which was formulated by Panigrahi [5]. In the Steiner tree activation problem, we are given a set W of nonnegative real numbers, and each edge uv in the graph is associated with an activation function $f_{uv} : W \times W \to \{\top, \bot\}$, where $\top$ indicates that an edge uv is activated, and $\bot$ indicates that it is not. A solution for the problem is defined as a $|V|$-dimensional vector $x \in W^V$. We say that a solution x *activates* an edge uv if $f_{uv}(x(u), x(v)) = \top$. The problem seeks a solution x that minimizes $x(V) := \sum_{v \in V} x(v)$ subject to the constraint that the edges activated by x include a Steiner tree. In previous studies of this problem, an algorithm is allowed to run in polynomial time of $|W|$, and it is assumed that the activation function is monotone (i.e., if $f_{uv}(i, j) = \top$, $i \leq i'$, and $j \leq j'$, then $f_{uv}(i', j') = \top$). The Steiner tree activation problem models various natural settings in design of wireless networks [5]. To see that the Steiner tree activation problem includes the VC-weighted Steiner tree problem, define W as $\{w(v): v \in V\}$, and let $f_{uv}(i, j) = \top$ if and only if $i \geq w(u)$ or $j \geq w(v)$ for each edge uv. Under this setting, if x is a minimal vector that activates an edge set F, the objective $x(V)$ is equal to the minimum weight of vertex covers of the subgraph induced by F. Hence the Steiner tree activation problem under this setting is equivalent to the VC-weighted Steiner tree problem.

The Steiner tree activation problem also contains the vertex-weighted Steiner tree problem. Indeed, vertex-weighted Steiner tree problem corresponds to the activation function f_{uv} such that $f_{uv}(i, j) = \top$ if and only if $i \geq w(u)$ and $j \geq w(v)$ for each edge uv. Notice that the similarity of the activation functions for the VC-weighted and the vertex-weighted Steiner tree problems. Thus the VC-weighted Steiner tree problem is an interesting variant of the vertex-weighted Steiner tree and the Steiner tree activation problems, which are studied actively in the literature.

It is known that the Steiner tree activation problems admits an $O(\log k)$-approximation algorithm when $|T| = k$. Indeed, there is an approximation-preserving reduction from the problem to the vertex-weighted Steiner tree problem, and hence the $O(\log k)$-approximation algorithm for the latter problem implies that for the former problem. This approximation factor is proven to be tight even in the spanning tree variant of the problem [5].

Since the VC-weighted Steiner tree problem is included by the Steiner tree activation problem, the $O(\log k)$-approximation algorithm can also be applied to the VC-weighted problem. Moreover, Angel et al. [6] presented a reduction from the dominating set problem to the VC-weighted spanning tree problem with uniform vertex-weights. This reduction implies the it is NP-hard to approximate the VC-weighted spanning tree problem within a factor of $o(\log n)$ even if the given vertex weights are uniform.

1.1   Our Contributions

Because of the hardness of the VC-weighted spanning tree problem on general graphs, we will consider restricted graph classes. We show that the VC-weighted Steiner tree problem admits constant-factor approximation algorithms for unit disk graphs (Corollary 3) and graphs excluding a fixed minor (Theorem 5). Note that the later graph class contains planar graphs. For these graphs, it is known that the vertex-weighted Steiner tree problem admits constant-factor approximation algorithms [7–9]. Hence it is natural to ask whether the VC-weighted Steiner tree problem in these graph classes admits constant-factor approximation algorithms. Moreover, unit disk graphs are regarded as a reasonable model of wireless networks, and the vertex-weighted Steiner tree problem in unit disk graphs has been actively studied in this context (see, e.g., [8–12]). Since our problem is motivated by an application in communication networks, it is reasonable to investigate the problem in unit disk graphs.

Our algorithm for unit disk graphs is based on a novel reduction to another optimization problem. The problem used in the reduction is similar to the connected facility location problem studied in [13,14], but it is slightly different. In the connected facility location problem, we are given sets $C, D \subseteq V$ of clients and facilities with an edge-weighted undirected graph $G = (V, E)$. If a facility $f \in D$ is opened by paying an associated opening cost, any client $i \in C$ can be allocated to f by paying the allocation cost, which is defined as the shortest path length from i to f multiplied by the demand of i. The opened facilities must be spanned by a Steiner tree, which incurs a connection cost defined as the edge weight of the tree. The objective is to find a set of opened facilities and a Steiner tree connecting them, that minimizes the sum of the opening cost, the allocation cost, and the connection cost. Our problem differs from the connected facility location problem in the fact that each client i can be allocated to an opened facility f only when i is adjacent to f in G, and there is no cost for the allocation. It can be regarded as a combination of the dominating set and the edge-weighted Steiner tree problems. Hence we call this the *connected dominating set problem*, although in the literature, this name is usually reserved for the case where the connection cost is defined by vertex weights and all vertices in the graph are clients. From a geometric property of unit disk graphs, we show that our reduction preserves the approximation guarantee up to a constant factor if the graph is a unit disk graph (Theorem 1). To solve the connected dominating set problem, we present a linear programming (LP) rounding algorithm. This algorithm relies on an idea presented by Huang et al. [11], who considered a variant of the connected dominating set problem in unit disk graphs. Although their algorithm is only for minimizing the number of vertices in a solution, we prove that it can be extended to our problem.

For graphs excluding a fixed minor, we solve the VC-weighted Steiner tree problem by presenting a constant-factor approximation algorithm for the Steiner tree activation problem. Our algorithm simply combines the reduction to the vertex-weighted Steiner tree problem and the algorithm of Demaine et al. [7] for the vertex-weighted Steiner tree problem in graphs excluding a fixed minor. However, analyzing it is not straightforward, because the reduction does not preserve the minor-freeness of the input graphs. Nevertheless, we show that the algorithm of Demaine et al. achieves a constant-factor approximation for the graphs constructed by the reduction (Sect. 4).

1.2   Organization

Section 2 introduces the notation and preliminary facts used throughout the paper. Sections 3 and 4 provide constant-factor approximation algorithms for unit disk graphs and for graphs excluding a fixed minor, respectively. Section 5 concludes the paper.

2   Preliminaries

We first define the notation used in this paper. Let $G = (V, E)$ be a graph with the vertex set V and the edge set E. We sometimes identify the graph G with its edge set E and by $V(G)$ denote the vertex set of G. When G is a tree, $L(G)$ denotes the set of leaves of G.

Let U be a subset of V. Then $G - U$ denotes the subgraph of G obtained by removing all vertices in U and all edges incident to them. $G[U]$ denotes the subgraph of G induced by U.

We denote a singleton vertex set $\{v\}$ by v. An edge joining vertices u and v is denoted by uv. For a vertex v, $N_G(v)$ denotes the set of neighbors of v in a graph G, i.e., $N_G(v) = \{u \in V : uv \in E\}$. $N_G[v]$ indicates $N_G(v) \cup v$. We let $d_G(v)$ denote $|N_G(v)|$. For a set U of vertices, $N_G(U)$ denotes $\left(\bigcup_{v \in U} N_G(v)\right) \setminus U$. When the graph G is clear from the context, we may remove the subscripts from our notation. We say that a vertex set U dominates a vertex v if $v \in U$, or U contains a vertex u that is adjacent to v. If a vertex set U dominates each vertex v in another vertex set W, then we say that U dominates W.

A graph G is a unit disk graph when there is an embedding of the vertex set into the Euclidean plane such that two vertices u and v are joined by an edge if and only if their Euclidean distance is at most 1. If G is a unit disk graph, we call such an embedding a *geometric representation* of G.

Let G and H be undirected graphs. We say that H is a *minor* of G if H is obtained from G by deleting edges and vertices and by contracting edges. If H is not a minor of G, G is called H-*minor-free*. By Kuratowski's theorem, a graph is planar if and only if it is K_5-minor-free and $K_{3,3}$-minor-free.

As mentioned in Sect. 1, the Steiner tree activation problem contains both the VC-weighted and the vertex-weighted Steiner tree problems. In addition, the Steiner tree activation problem can be reduced to the vertex-weighted Steiner tree problem, as summarized in the following theorem.

Theorem 1. *There is an approximation-preserving reduction from the Steiner tree activation problem to the vertex-weighted Steiner tree problem. Hence, if the latter problem admits an α-approximation algorithm, the former problem does also.*

Proof. Recall that an instance I of the Steiner tree activation problem consists of an undirected graph $G = (V, E)$, a terminal set T, a range $W \subseteq \mathbb{R}_+$, and an activation function $f_{uv} \colon W \times W \to \{\top, \bot\}$ for each $uv \in E$. We define a copy v_i of a vertex v for each $v \in V$ and $i \in W$, and associate v_i with the weight $w(v_i) := i$. We join u_i and v_j by an edge if $uv \in E$ and $f_{uv}(i, j) = \top$. In addition, we join each terminal $t \in T$ with its copies t_i, $i \in W$. The weight $w(t)$ of t is defined to be 0. Let G' be the obtained graph on the vertex set $T \cup \{v_i : v \in V, i \in W\}$. Let I' be the instance of the vertex weighted Steiner tree problem that consists of the graph G', the vertex weights w, and the terminal set T. From an inclusion-wise minimal Steiner tree F feasible to I', define a vector $x \in W^V$ by $x(v) = \max\{i \in W : v_i \in V(F)\}$ for each $v \in V$. Then x activates a Steiner tree in the original instance I, and $x(V)$ is equal to the vertex weight of F. Hence there is a one-to-one correspondence between a minimal Steiner tree in I' and a feasible solution in I, and they have the same objective values in their own problems. Hence the above reduction is an approximation-preserving reduction from the Steiner tree activation problem to the vertex-weighted Steiner tree problem.   □

We do not claim the originality of Theorem 1; we believe that this reduction has been already known although we are aware of no previous study describing this reduction explicitly.

We note that the reduction claimed in Theorem 1 transforms the input graph, and hence it may not be closed in a graph class. In fact, we can observe that the reduction is not closed in unit disk graphs or planar graphs.

3   VC-weighted Steiner Tree Problem in Unit Disk Graphs

This section presents a constant-factor approximation algorithm for the VC-weighted Steiner tree problem in unit disk graphs. Our algorithm consists of two steps. In the first step, we reduce the VC-weighted Steiner tree problem to another optimization problem, which is called the connected dominating set problem. Combined with a constant-factor approximation algorithm for the connected dominating set problem, the reduction gives the required algorithm for the VC-weighted Steiner tree problem.

Let us discuss the reduction. As noted in Theorem 1, the Steiner tree activation problem can be reduced to the vertex-weighted Steiner tree problem. Since the VC-weighted Steiner tree problem is included in the Steiner tree activation problem, the reduction also applies to the VC-weighted Steiner tree problem. Since there is a constant-factor approximation algorithm for the vertex-weighted Steiner tree problem in unit disk graphs, this reduction gives a constant-factor approximation for the VC-weighted problem if the graph constructed by the reduction is a unit disk graph. However, the constructed graph may not be a unit disk graph even if the original graph is a unit disk graph.

Our idea is to reduce the VC-weighted Steiner tree problem to another optimization problem. This is inspired by a constant-factor approximation algorithm for the vertex-weighted Steiner tree problem on a unit disk graph [8,9]. This algorithm is based on a reduction from the vertex-weighted to the edge-weighted Steiner tree problems. The reduction is possible because the former problem always admits an optimal Steiner tree in which the maximum degree is a constant if the graph is a unit disk graph. Even in the VC-weighted Steiner tree problem, if there is an optimal solution (F, U) such that the maximum degree of vertices in the vertex cover U is a constant in the Steiner tree F, then we can reduce the problem to the edge-weighted Steiner tree problem. However, there is an instance of the VC-weighted Steiner tree problem that admits no such optimal solution. For example, if the vertex weights are uniform, and the graph includes a star in which all of the terminals are its leaves, then the star is the Steiner tree in the optimal solution, and its minimum vertex cover consists of only the center of the star. The degree of the center of the star is not bounded by a constant. Hence it seems that it would be difficult to reduce the VC-weighted Steiner tree problem to the edge-weighted problem.

We reduce the VC-weighted Steiner tree problem to a problem similar to the connected facility location problem. The reduction is based on a geometric property of unit disk graphs, and we will begin by proving this property. The following lemma gives a basic claim about geometry. For two points i and j on the plane, we denote their Euclidean distance by l_{ij}.

Lemma 1. *Let i be a point on the Euclidean plane, and let $\alpha \in (1/2, 3/4]$. Let P be a set of points on the plane such that $\alpha < l_{ik}/l_{ij} \leq 1/\alpha$ holds for all $j, k \in P$. If $|P| > 2\pi/\arccos(\alpha/2 + 3/(8\alpha))$, then there exist $j, k \in P$ such that $l_{jk} < \max\{l_{ij}, l_{ik}\}/2$.*

Proof. Since $|P| > 2\pi/\arccos(\alpha/2 + 3/(8\alpha))$, there exist $j, k \in P$ such that $\theta := \angle jik < \arccos(\alpha/2 + 3/(8\alpha))$. We note that $l_{jk}^2 = l_{ij}^2 + l_{ik}^2 - 2l_{ij}l_{ik}\cos\theta$. Without loss of generality, we assume $l_{ij} \geq l_{ik}$. Then, $(\max\{l_{ij}, l_{ik}\})^2 = l_{ij}^2$. Hence it suffices to show that $-4l_{ik}^2 - 3l_{ij}^2 + 8l_{ij}l_{ik}\cos\theta > 0$.

Let $\beta := l_{ik}/l_{ij}$. Then, $\alpha < \beta \le 1$. $\max_{\alpha < \beta \le 1} 4\beta + 3/\beta = 4\alpha + 3/\alpha$ holds, where the maximum is attained by $\beta = \alpha$. Hence the required inequality is verified by

$$-4l_{ik}^2 - 3l_{ij}^2 + 8l_{ij}l_{ik}\cos\theta = l_{ij}l_{ik}\left(-4\beta - \frac{3}{\beta} + 8\cos\theta\right)$$

$$\ge l_{ij}l_{ik}\left(-4\alpha - \frac{3}{\alpha} + 8\cos\theta\right)$$

$$= 0. \qquad \square$$

Our reduction requires the assumption that there is an optimal solution (F, U) for the VC-weighted Steiner tree problem such that the degree of each vertex $v \in U$ is bounded by a constant α in the tree $F - (L(F) \setminus U)$. The following lemma proves that the unit disk graph satisfies this assumption with $\alpha = 29$.

Lemma 2. *If the input graph $G = (V, E)$ is a unit disk graph, the VC-weighted Steiner tree problem admits an optimal solution consisting of a Steiner tree F and a vertex cover U of F such that the degree of each vertex in U is at most 29 in $F - (L(F) \setminus U)$.*

Proof. For two vertices $u, v \in V$, let l_{uv} denote the Euclidean distance between u and v in the geometric representation of G. Let (F, U) be an optimal solution for the VC-weighted Steiner tree problem. Without loss of generality, we can assume that (F, U) satisfies the following conditions:

(a) (F, U) maximizes $|L(F) \setminus U|$ over all optimal solutions;
(b) F minimizes $\sum_{e \in F} l_e$ over all optimal solutions subject to (a);
(c) (F, U) minimizes the number of vertices $v \in U$ such that $|\{u \in U : uv \in F\}| \ge 6$ over all optimal solutions subject to (a) and (b).

Let $v \in U$. Let $M_v := \{u \in U : uv \in F\}$ and $M'_v := \{u \in V \setminus (U \cup L(F)) : uv \in F\}$. The lemma can be proven by showing that $|M_v| \le 5$ and $|M'_v| \le 24$.

Let us prove $|M_v| \le 5$. We first prove $|M_v| \le 6$, and then prove $|M_v| \ne 6$. Suppose that there are two distinct vertices $i, j \in M_v$ such that $l_{ij} < \max\{l_{vi}, l_{vj}\}$. Let $l_{vi} = \max\{l_{vi}, l_{vj}\}$, and denote $F \setminus \{vi\} \cup \{ij\}$ by F'. Then F' is a Steiner tree, U is a vertex cover of F', $L(F) \setminus U = L(F') \setminus U$, and $\sum_{e \in F'} l_e < \sum_{e \in F} l_e$. Since the existence of such an F' contradicts the condition (b), M_v contains no such vertices i and j. If $|M_v| \ge 7$, there must be two vertices $i, j \in M_v$ such that $\angle ivj < \pi/3$, and $l_{ij} < \max\{l_{vi}, l_{vj}\}$ holds for these vertices. Hence $|M_v| \le 6$ holds.

Suppose that $|M_v| = 6$. In this case, $M_v = \{u_1, \ldots, u_6\}$, and $l_{vu_k} = l_{vu_{k+1}} = l_{u_k u_{k+1}}$ holds for all $k = 1, \ldots, 6$, where u_7 denotes u_1 for notational convenience. If $|M_{u_1}| \le 4$, we define F' as $F \setminus \{vu_2\} \cup \{u_1 u_2\}$. Then, F' is a Steiner tree, U is a vertex cover of F', $L(F') \setminus U = L(F) \setminus U$, and $\sum_{e \in F'} l_e = \sum_{e \in F} l_e$. Replacing F by F' decreases the number of vertices $v \in U$ such that $|M_v| \ge 6$, which contradicts the condition (c). If $|M_{u_1}| \ge 5$, then (i) there exist $i, j \in M_{u_1} \setminus v$ such that $l_{ij} < \max\{l_{u_1 i}, l_{u_1 j}\}$, or (ii) there exist $i \in M_{u_1} \setminus v$ and $j \in \{v, u_2, u_6\}$

such that $l_{ij} < \max\{l_{u_1 i}, l_{u_1 j}\}$. Case (i) contradicts the condition (b) as observed above. In case (ii), we define F' as $F \setminus \{u_1 i\} \cup \{ij\}$ if $\max\{l_{u_1 i}, l_{u_1 j}\} = l_{u_1 i}$, and as $F \setminus \{u_1 v\} \cup \{ij\}$ if $\max\{l_{u_1 i}, l_{u_1 j}\} = l_{u_1 j}$. In either case, F' is a Steiner tree, U is a vertex cover of F', $L(F') \setminus U = L(F) \setminus U$, and $\sum_{e \in F} l_e > \sum_{e \in F'} l_e$, which contradicts the condition (b). Hence $|M_v| \le 5$ holds.

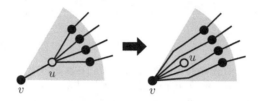

Fig. 1. Transformation of F when $l_{vu'} \le 1$ for all $u' \in A_u$

We now prove $|M'_v| \le 24$. Let $u \in M'_v$. Since u is not a leaf, u has a neighbor other than v. We denote by A_u the set of neighbors of u other than v. Since $u \notin U$, each vertex in A_u is included in U. If $l_{vu'} \le 1$ holds for all vertices $u' \in A_u$, consider $F' := F \setminus (vu \cup \{uu' : u' \in A_u\}) \cup \{vu' : u' \in A_u\}$. Then, F' is a Steiner tree, U is a vertex cover of F', and $L(F') \setminus U = (L(F) \setminus U) \cup \{u\}$ (Fig. 1). Since the existence of such an F' contradicts condition (a), there is at least one vertex $u' \in A_u$ with $l_{vu'} > 1$. We choose one of these vertices for each $u \in M'_v$, and let B denote the set of those chosen vertices (hence B includes exactly one vertex in A_u for each $u \in M'_v$).

Suppose there exist two vertices $i, j \in B$ such that $l_{ij} < \max\{l_{vi}, l_{vj}\}/2$. Let $l_{vi} = \max\{l_{vi}, l_{vj}\}$. Let u denote the common neighbor of v and i. Then l_{vu} or l_{ui} is at least $l_{vi}/2$. If $l_{vu} \ge l_{vi}/2$, then replace edge vu by ij in F (see Fig. 2(a)). Otherwise, replace edge ui by ij in F (see Fig. 2(b)). Let F' denote the tree obtained by this replacement. F' is a Steiner tree, U is a vertex cover of F', $L(F) \setminus U \subseteq L(F') \setminus U$, and $\sum_{e \in F'} l_e < \sum_{e \in F} l_e$ hold. Since this contradicts condition (a) or (b), there exists no such pair of vertices $i, j \in B'$.

We divide B into $B' := \{i \in B \mid l_{vi} \le 1.41\}$ and $B'' := \{i \in B \mid 1.41 \le l_{vi}\}$. Notice that $1/1.41 \le l_{vi}/l_{vj} \le 1.41$ holds for any $i, j \in B'$. Hence, by Lemma 1, $|B'| \le \lfloor 2\pi/\arccos(1/2.82 + 4.23/8) \rfloor = 12$. Moreover, $3/5 \le l_{vi}/l_{vj} \le 5/3$ holds for any $i, j \in B''$. Hence, by Lemma 1, $|B''| \le \lfloor 2\pi/\arccos(3/10 + 5/8) \rfloor = 12$. Since $|M'_v| \le |B| = |B'| + |B''| \le 24$, this proves the lemma. □

In the remainder of this section, we assume that G is not necessarily a unit disk graph, but there is an optimal solution (F, U) for the VC-weighted Steiner tree problem such that the degree of each vertex $v \in U$ is at most a constant α in the tree $F - (L(F) \setminus U)$. Based on this assumption, we reduce the VC-weighted Steiner tree problem to another optimization problem. First, let us define the problem used in the reduction.

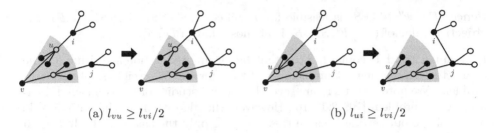

(a) $l_{vu} \geq l_{vi}/2$ (b) $l_{ui} \geq l_{vi}/2$

Fig. 2. Transformation of a tree F when $l_{ij} < \max\{l_{vi}, l_{vj}\}/2$

Definition 1 (Connected Dominating Set Problem). *Let $G = (V, E)$ be an undirected graph, and let $T \subseteq V$ be a set of terminals. Each edge e is associated with the length $l(e) \in \mathbb{R}_+$, each vertex v is associated with the weight $w(v) \in \mathbb{R}_+$, and $l(e) \leq \min\{w(u), w(v)\}$ holds for each edge $e = uv \in E$. The problem seeks a pair of a tree $F \subseteq E$ and a vertex set $S \subseteq V$ such that S dominates T and F spans S. Let $l(F)$ denote $\sum_{e \in F} l(e)$. The objective is to minimize $w(S) + l(F)$.*

Theorem 2. *Suppose that the VC-weighted Steiner tree problem with input graph G admits an optimal solution (F, U) such that the degree of each vertex in U is at most α in the tree $F - (L(F) \setminus U)$. If there is a β-approximation algorithm for the connected dominating set problem in G, then there is an $(\alpha + 1)\beta$-approximation algorithm for the VC-weighted Steiner tree problem with G.*

Proof. Suppose that an instance I of the VC-weighted Steiner tree problem consists of an undirected graph $G = (V, E)$, a terminal set $T \subseteq V$, and vertex weights $w \in \mathbb{R}_+^V$. We define the edge length $l(e)$ as $\min\{w(u), w(v)\}$ for each $e = uv \in E$, and define an instance I' of the connected dominating set problem from G, T, w, and l. We show that the optimal objective value of I' is at most $\alpha + 1$ times that of I, and a feasible solution for I can be constructed from the one for I' without increasing the objective value.

First, we prove that the optimal objective value of I' is at most $\alpha + 1$ times that of I. Let (F, U) be an optimal solution for I. Then, the optimal objective value of I is $w(U)$. Since F spans T and U is a vertex cover of F, U dominates T. Define $F' := F - (L(F) \setminus U)$. Since F' is a tree spanning U, (F', U) is a feasible solution for I'. If $e = uv \in F'$, then u or v is included in U, and $l(e)$ is at most $w(u)$ and $w(v)$. Hence $l(F') \leq \sum_{v \in U} w(v) d_{F'}(v)$. By assumption, $d_{F'}(v) \leq \alpha$ holds for each $v \in U$. Hence $l(F') \leq \alpha w(U)$. Since the objective value of (F', U) in I' is $l(F') + w(U)$, the optimal objective value of I' is at most $(\alpha + 1)w(U)$.

Next, we prove that a feasible solution (F, S) for I' provides a feasible solution for I, and its objective value is at most that of (F, S). Since S dominates T, if a terminal $t \in T$ is not spanned by F, there is a vertex $v \in S$ with $tv \in E$. We let F' be the set of such edges tv. Notice that $F \cup F'$ is a Steiner tree of the terminal set T. For each edge $e \in F$, choose an end vertex v of e such that $l(e) = w(v)$. Let S' denote this set of chosen vertices. Then, $S' \cup S$ is a vertex cover of $F \cup F'$.

Hence $(F \cup F', S' \cup S)$ is feasible for I. Since $w(S' \cup S) \leq w(S) + l(F)$, the objective value of $(F \cup F', S' \cup S)$ is at most that of (F, S).   □

By Lemma 2 and Theorem 2, a constant-factor approximation algorithm for the connected dominating set problem gives that for the VC-weighted Steiner tree problem. We note that there are several previous studies of the connected dominating set problem [10,12,15,16]. However, the algorithms in those studies do not apply to our setting because they consider only the case $T = V$. Indeed, if $T = V$, the connected dominating set problem can be solved by a simple algorithm as follows; compute an approximate solution S for the minimum weight dominating set of the graph, and then compute a tree spanning S by an approximation algorithm for the edge-weighted Steiner tree problem. We note that the dominating set problem admits a constant-factor approximation algorithm if the graph is a unit disk graph [12,19]. This achieves a constant-factor approximation for the connected dominating set problem with $T = V$. However, this simple algorithm does not work for the case of $T \subset V$.

We can observe through an example that computing a dominating set and a Steiner tree for connecting it separately does not give a good approximate solution for the connected dominating set problem. Motivated by this observation, we consider an LP rounding algorithm. Our key idea is to use an LP relaxation for coordinating a dominating set and a Steiner tree. The same idea was previously given by Huang et al. [11] for a problem, which is a special case of the connected dominating set problem. We can prove that their algorithm can be extended to the connected dominating set problem. In the present paper, we omit its detail due to the space limitation. We will present it in the full version.

Theorem 3. *The VC-weighted Steiner tree problem admits a constant-factor approximation algorithm in unit disk graphs.*

4   Steiner Tree Activation Problem in Graphs Excluding a Fixed Minor

In this section, we present a constant-factor approximation algorithm for the Steiner tree activation problem in graphs excluding a fixed minor. In particular, our algorithms is a 11-approximation for planar graphs.

Our algorithm is based on the reduction mentioned in Theorem 1. We reduce the problem to the vertex-weighted Steiner tree problem by using that reduction, and we solve the obtained instance by using the constant-factor approximation algorithm proposed by Demaine et al. [7] for the vertex-weighted Steiner tree problem in graphs excluding a fixed minor. We prove that this achieves a constant-factor approximation for the Steiner tree activation problem when the input graph is H-minor-free for some graph H such that $|V(H)|$ is a constant. This seems to be an easy corollary to Demaine et al., but it is not because the reduction does not preserve the H-minor-freeness of the input graph. In spite of this, we can prove that the approximation guarantee given by Demaine et al. extends to the graphs constructed from a H-minor-free graph by the reduction.

We recall that the reduction constructs a graph G' on the vertex set $T \cup \{v_i \colon v \in V, i \in W\}$ from the input graph $G = (V, E)$ and the monotone activation functions $f_{uv} \colon W \times W \to \{\top, \bot\}$, $uv \in E$. We denote the vertex set $\{v_i \colon i \in W\}$ defined from an original vertex $v \in V$ by U_v. Let U denote $\bigcup_{v \in V} U_v$.

First, let us illustrate how the algorithm of Demaine et al. behaves for G'. The algorithm maintains a vertex set $X \subseteq T \cup U$, where X is initialized to T at the beginning. Let $\mathcal{A}(X) \subseteq 2^X$ denote the family of connected components that include some terminals in the subgraph of $G'[X]$. We call each member of $\mathcal{A}(X)$ an *active set*. The algorithm consists of two phases, called the increase phase and the reverse-deletion phase. In the increase phase, the algorithm iteratively adds vertices to X until $|\mathcal{A}(X)|$ is equal to one. This implies that, when the increase phase terminates, the subgraph induced by X connects all of the terminals. In the reverse-deletion phase, X is transformed into an inclusion-wise minimal vertex set that induces a Steiner tree. This is done by repeatedly removing vertices from X in the reverse of the order in which they were added.

Let $\bar{X}$ be the vertex set X when the algorithm terminates, and let X be the vertex set at some point during the increase phase. We denote $\bar{X} \setminus X$ by $\bar{X}'$. Note that $\bar{X}'$ is a minimal augmentation of X such that $X \cup \bar{X}'$ induces a Steiner tree. Each $Y \in \mathcal{A}(X)$ is disjoint from $\bar{X}'$, because $Y \subseteq X$. Demaine et al. showed the following analysis of their algorithm.

Theorem 4 ([7]). *Let X be a vertex set maintained at some moment in the increase phase, and let $\bar{X}'$ be a minimal augmentation of X so that $X \cup \bar{X}'$ induces a Steiner tree. If there is a number γ such that $\sum_{Y \in \mathcal{A}(X)} |\bar{X}' \cap N(Y)| \le \gamma |\mathcal{A}(X)|$ holds for any X and $\bar{X}'$, the algorithm of Demaine et al. achieves an approximation factor γ.*

In $G'[\bar{X}' \cup (\bigcup_{Y \in \mathcal{A}(X)} Y)]$, contract each $Y \in \mathcal{A}(X)$ into a single vertex, discard all edges induced by $\bar{X}'$ and all isolated vertices in $\bar{X}'$, and replace multiple edges by single edges. This gives us a simple bipartite graph with the bipartition $\{A, B\}$ of the vertex set, where each vertex in A corresponds to an active set, and B is a subset of $\bar{X}'$. Let D denote this graph. This construction of D is illustrated in Fig. 3. We note that $\sum_{Y \in \mathcal{A}(X)} |\bar{X}' \cap N(Y)|$ is equal to the number of edges in D. Hence, by Theorem 4, if the number of edges is at most a constant factor of $|A|$, the algorithm achieves a constant-factor approximation.

Demaine et al. proved that $|B| \le 2|A|$, and D is H-minor-free if G is H-minor-free. By [17, 18], these two facts imply that the number of edges in D is $O(|A||V(H)|\sqrt{\log |V(H)|})$. When G is planar, together with Euler's formula and the fact that D is bipartite, they imply that the number of edges in D is at most $6|A|$.

The proof of Demaine et al. for $|B| \le 2|A|$ can be carried to our case. However, D is not necessarily H-minor-free even if G is H-minor-free. Nevertheless, we can bound the number of edges in D, as follows.

Lemma 3. *Suppose that the given activation function is monotone. If G is H-minor-free, the number of edges in D is $O(|A||V(H)|\sqrt{\log |V(H)|})$. If G is planar, the number of edges in D is at most $11|A|$.*

The following theorem is immediate from Theorem 4 and Lemma 3.

Theorem 5. *If an input graph is H-minor-free for some graph H, then the Steiner tree activation problem with a monotone activation function admits an $O(|V(H)|\sqrt{\log |V(H)|})$-approximation algorithm. In particular, if the input graph is planar, then the problem admits a 11-approximation algorithm.*

In the rest of this section, we prove Lemma 3. We first provide several preparatory lemmas.

Lemma 4. *If G' includes an edge $u_i v_j$ for some $u, v \in V$ and $i, j \in W$, then G' also includes an edge $u_{i'} v_{j'}$ for any $i', j' \in W$ with $i' \geq i$ and $j' \geq j$.*

Proof. The lemma is immediate from the construction of G' and the assumption that each edge in G is associated with a monotone activation function.   □

Lemma 5. *$\bar{X}$ does not contain any two distinct copies of an original vertex.*

Proof. For the sake of a contradiction, suppose that $v_i, v_j \in \bar{X}$ for some $v \in V$ and $i, j \in W$ with $i < j$. If an edge $u_k v_i$ exists in G', then another edge $u_k v_j$ also exists by Lemma 4. This means that $\bar{X} \setminus v_i$ induces a Steiner tree in G', which contradicts the minimality of $\bar{X}$.   □

Lemma 6. *Let $Y, Y' \in \mathcal{A}(X)$ with $Y \neq Y'$. If $Y \cap U_v \neq \emptyset$ for some $v \in V$, then $Y' \cap U_v = \emptyset$.*

Proof. Suppose that $Y \cap U_v \neq \emptyset \neq Y' \cap U_v$. Let $v_i \in Y$ and $v_j \in Y'$ with $i < j$. A vertex adjacent to v_i is also adjacent to v_j in G' by Lemma 4. By the definition, Y induces a connected component of $G'[X]$ that includes a terminal t. Hence v_i has at least one neighbor in Y. This implies that v_i and v_j are connected in $G'[X]$. This contradicts the fact that Y and Y' are different connected components of $G'[X]$.   □

Consider $Y \in \mathcal{A}(X)$ and $v \in V$ such that $Y \cap U_v \neq \emptyset$. Let v_i be the vertex that has the largest subscript in $Y \cap U_v$ (i.e., $i = \max\{i' \in W : v_{i'} \in Y \cap U_v\}$). Then, from Y, we remove all vertices in $Y \cap U_v$ but v_i. Moreover, if a copy v_j of v is included in B, we replace v_i by v_j. Notice that $j > i$ holds in this case by Lemma 4, and B does not include more than one copy of v because of Lemma 5. Let $\bar{Y}$ denote the vertex set obtained from Y by doing these operations for each $v \in V$ with $Y \cap U_v \neq \emptyset$. $\bar{Y}$ induces a connected subgraph of G' because of Lemma 4.

We let V_B denote $\{v \in V : B \cap U_v \neq \emptyset\}$, and let $V_{B,Y}$ denote $\{v \in V_B : \bar{Y} \cap U_v \neq \emptyset\}$ for each $Y \in \mathcal{A}(X)$. Moreover, let B' denote $B \cap \{U_v : v \in \bigcup_{Y \in \mathcal{A}(X)} V_{B,Y}\}$, and B'' denote $B \setminus B'$. In other words, each vertex $v_j \in B$ belongs to B' if and only if some copy v_i of the same original vertex $v \in V$ is contained by an active set in $\mathcal{A}(X)$.

If $k := |V_{B,Y}| \geq 2$, we divide $\bar{Y}$ into k subsets such that the copies of the vertices in $V_{B,Y}$ belong to different subsets, and each subset induces a connected subgraph of G'. Let $\mathcal{A}'(X)$ denote the family of vertex sets obtained by doing

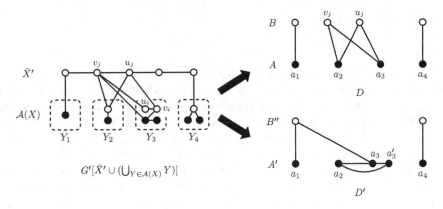

Fig. 3. An example of $G'[\bar{X}' \cup (\bigcup_{Y \in \mathcal{A}(X)} Y)]$, D, and D'; in construction of D', $\bar{Y}_3$ is divided into two subsets, one of which contains u_j and the other contains v_j; the former is shrunken into a_3 and the later is shrunken into a'_3

these operations to all active sets in $\mathcal{A}(X)$. Notice that $|\mathcal{A}'(X)| = |\mathcal{A}(X)| + \sum_{Y \in \mathcal{A}(X)} \max\{0, |V_{B,Y}| - 1\}$. Lemma 6 indicates that, if a vertex $v \in V_B$ belongs to $V_{B,Y}$ for some $Y \in \mathcal{A}(X)$, then it does not belong to $V_{B,Y'}$ for any $Y' \in \mathcal{A}(X) \setminus \{Y\}$. Thus, $\sum_{Y \in \mathcal{A}} |V_{B,Y}| \leq |B'|$, and hence $|\mathcal{A}'(X)| \leq |\mathcal{A}(X)| + |B'|$.

We shrink each $Z \in \mathcal{A}'(X)$ into a single vertex in the induced subgraph $G'[B'' \cup (\bigcup_{Z \in \mathcal{A}'(X)} Z)]$ of G', and convert the obtained graph into a simple graph by removing all self-loops and by replacing multiple edges with single edges. Let A' denote the set of vertices obtained by shrinking vertex sets in $\mathcal{A}'(X)$, and let D' denote the obtained graph (with the vertex set $A' \cup B''$). See Fig. 3 for an illustration of this construction. We observe that D' is H-minor-free in the following lemma.

Lemma 7. *If G is H-minor-free, then D' is H-minor-free.*

Proof. By Lemma 5 and the construction of $\mathcal{A}'(X)$, each vertex in V has at most one copy in $B'' \cup (\bigcup_{Z \in \mathcal{A}'(X)} Z)$. If $G'[B'' \cup (\bigcup_{Z \in \mathcal{A}'(X)} Z)]$ includes an edge $u_i v_j$ for $u_i \in U_u$ and $v_j \in U_v$, then G also includes an edge uv. Thus $G'[B'' \cup (\bigcup_{Z \in \mathcal{A}'(X)} Z)]$ is isomorphic to a subgraph of G. Since each $Z \in \mathcal{A}'(X)$ induces a connected subgraph of G', the graph D' (constructed from $G'[B'' \cup (\bigcup_{Z \in \mathcal{A}'(X)} Z)]$ by shrinking each $Z \in \mathcal{A}'(X)$) is a minor of G. Hence if G is H-minor-free, D' is also H-minor-free. ∎

The following lemma gives a relationship between D and D'.

Lemma 8. *If l is the number of edges in D', then D contains at most $l + |B'|$ edges.*

Proof. Let av_i be an edge in D that joins vertices $a \in A$ and $v_j \in B$. Suppose that a is a vertex obtained by shrinking $Y \in \mathcal{A}(X)$, and v_j is a copy of $v \in V$. Remember that v_j belongs to either B' or B''. If $v_j \in B'$, it is contained by a vertex set in $\mathcal{A}'(X)$, denoted by Z_v. We consider the following three cases:

1. $v_j \in B'$ and $Z_v \subseteq \bar{Y}$
2. $v_j \in B'$ and $Z_v \not\subseteq \bar{Y}$
3. $v_j \in B''$

In the second case, an edge in D' joins vertices obtained by shrinking Z_v and a subset of $\bar{Y}$. In the third case, v_j exists in D', and D' includes an edge that joins v_j and the vertex obtained by shrinking a subset of $\bar{Y}$. Thus D' includes an edge corresponding to av_j in these two cases. We can also observe that no edge in D' corresponds to more than two such edges av_j. This is because $Z_u \neq Z_v$ for any distinct vertices u_i and v_j in B' by the construction of $\mathcal{A}'(X)$.

In the first case, D' may not contain an edge corresponding to av_j. However, the number of such edges is at most $|B'|$ in total because $\bar{Y}$ are uniquely determined from v_j in this case. Therefore, the number of edges in D is at most $l + |B'|$. □

We now prove Lemma 3.

Proof (Lemma 3). The number of vertices in D' is at most $|\mathcal{A}'(X)| + |B''| \leq |\mathcal{A}(X)| + |B'| + |B''| = |A| + |B|$. As we mentioned, we can prove $|B| \leq 2|A|$ similar to Demaine et al. [7]. Hence D' contains at most $3|A|$ vertices. By Lemma 7, D' is H-minor-free. It is known [17,18] that the number of edges in an H-minor-free graph with n vertices is $O(n|V(H)|\sqrt{\log|V(H)|})$. Therefore, the number of edges in D' is $O(|A||V(H)|\sqrt{\log|V(H)|})$. By Lemma 8, this implies that the number of edges in D is $|B'| + O(|A||V(H)|\sqrt{\log|V(H)|}) = O(|A||V(H)|\sqrt{\log|V(H)|})$.

If G is planar, by Euler's formula, the number of edges in D' is at most $3(|A| + |B|)$. Hence, by Lemma 8, the number of edges in D is at most $3(|A| + |B|) + |B'| \leq 3|A| + 4|B| \leq 11|A|$. □

5   Conclusion

In this paper, we formulate the VC-weighted Steiner tree problem, a new variant of the vertex-weighted Steiner tree and the Steiner tree activation problems. We proved that it is NP-hard to achieve an $o(\log n)$-approximation for the VC-weighted spanning tree problem in general graphs, and we presented constant-factor approximation algorithms for the VC-weighted Steiner tree problem with unit disk graphs and for the Steiner tree activation problem with graphs excluding a fixed minor. Finding a constant-factor approximation algorithm for the Steiner tree activation problem with unit disk graphs remains an open problem.

References

1. Byrka, J., Grandoni, F., Rothvoß, T., Sanità, L.: Steiner tree approximation via iterative randomized rounding. J. ACM **60**, 6 (2013)
2. Goemans, M.X., Olver, N., Rothvoß, T., Zenklusen, R.: Matroids and integrality gaps for hypergraphic steiner tree relaxations. In: STOC, pp. 1161–1176 (2012)

3. Klein, P.N., Ravi, R.: A nearly best-possible approximation algorithm for node-weighted Steiner trees. In: IPCO, pp. 323–332 (1993)

4. Perlman, R.J.: An algorithm for distributed computation of a spanningtree in an extended LAN. In: SIGCOMM, pp. 44–53 (1985)

5. Panigrahi, D.: Survivable network design problems in wireless networks. In: SODA, pp. 1014–1027 (2011)

6. Angel, E., Bampis, E., Chau, V., Kononov, A.: Min-power covering problems. In: Elbassioni, K., Makino, K. (eds.) ISAAC 2015. LNCS, vol. 9472, pp. 367–377. Springer, Heidelberg (2015). doi:10.1007/978-3-662-48971-0_32

7. Demaine, E.D., Hajiaghayi, M., Klein, P.N.: Node-weighted Steiner tree and group Steiner tree in planar graphs. ACM Trans. Algorithms 10, 13 (2014)

8. Zou, F., Li, X., Gao, S., Wu, W.: Node-weighted Steiner tree approximation in unit disk graphs. J. Comb. Opt. 18, 342–349 (2009)

9. Zou, F., Li, X., Kim, D., Wu, W.: Two constant approximation algorithms for node-weighted steiner tree in unit disk graphs. In: Yang, B., Du, D.-Z., Wang, C.A. (eds.) COCOA 2008. LNCS, vol. 5165, pp. 278–285. Springer, Heidelberg (2008). doi:10.1007/978-3-540-85097-7_26

10. Ambühl, C., Erlebach, T., Mihalák, M., Nunkesser, M.: Constant-factor approximation for minimum-weight (connected) dominating sets in unit disk graphs. In: Díaz, J., Jansen, K., Rolim, J.D.P., Zwick, U. (eds.) APPROX/RANDOM -2006. LNCS, vol. 4110, pp. 3–14. Springer, Heidelberg (2006). doi:10.1007/11830924_3

11. Huang, L., Li, J., Shi, Q.: Approximation algorithms for the connected sensor cover problem. In: Xu, D., Du, D., Du, D. (eds.) COCOON 2015. LNCS, vol. 9198, pp. 183–196. Springer, Heidelberg (2015). doi:10.1007/978-3-319-21398-9_15

12. Zou, F., Wang, Y., Xu, X., Li, X., Du, H., Wan, P., Wu, W.: New approximations for minimum-weighted dominating sets and minimum-weighted connected dominating sets on unit disk graphs. Theor. Comp. Sci. 412, 198–208 (2011)

13. Eisenbrand, F., Grandoni, F., Rothvoß, T., Schäfer, G.: Connected facility location via random facility sampling and core detouring. J. Comp. Sys. Sci. 76, 709–726 (2010)

14. Swamy, C., Kumar, A.: Primal-dual algorithms for connected facility location problems. Algorithmica 40, 245–269 (2004)

15. Cheng, X., Huang, X., Li, D., Wu, W., Du, D.: A polynomial-time approximation scheme for the minimum-connected dominating set in ad hoc wireless networks. Networks 42, 202–208 (2003)

16. Guha, S., Khuller, S.: Approximation algorithms for connected dominating sets. Algorithmica 20, 374–387 (1998)

17. Kostochka, A.V.: Lower bound of the Hadwiger number of graphs by their average degree. Combinatorica 4, 307–316 (1984)

18. Thomason, A.: The extremal function for complete minors. J. Comb. Theory Ser. B 81, 318–338 (2001)

19. Fonseca, G.D., Sá, V.G.P., Figueiredo, C.M.H.: Linear-time approximation algorithms for unit disk graphs. In: Bampis, E., Svensson, O. (eds.) WAOA 2014. LNCS, vol. 8952, pp. 132–143. Springer, Heidelberg (2015). doi:10.1007/978-3-319-18263-6_12

On the Approximability of Partial VC Dimension

Cristina Bazgan[1], Florent Foucaud[2], and Florian Sikora[1]([⊠])

[1] Université Paris-Dauphine, PSL Research University, CNRS, LAMSADE, Paris, France
{cristina.bazgan,florian.sikora}@dauphine.fr
[2] Université Blaise Pascal - CNRS UMR 6158 -LIMOS, Clermont-Ferrand, France
florent.foucaud@gmail.com

Abstract. We introduce the problem Partial VC Dimension that asks, given a hypergraph $H = (X, E)$ and integers k and ℓ, whether one can select a set $C \subseteq X$ of k vertices of H such that the set $\{e \cap C, e \in E\}$ of distinct hyperedge-intersections with C has size at least ℓ. The sets $e \cap C$ define equivalence classes over E. Partial VC Dimension is a generalization of VC Dimension, which corresponds to the case $\ell = 2^k$, and of Distinguishing Transversal, which corresponds to the case $\ell = |E|$ (the latter is also known as Test Cover in the dual hypergraph). We also introduce the associated fixed-cardinality maximization problem Max Partial VC Dimension that aims at maximizing the number of equivalence classes induced by a solution set of k vertices. We study the approximation complexity of Max Partial VC Dimension on general hypergraphs and on more restricted instances, in particular, neighborhood hypergraphs of graphs.

1 Introduction

We study identification problems in discrete structures. Consider a hypergraph (or set system) $H = (X, E)$, where X is the vertex set and E is a collection of hyperedges, that is, subsets of X. Given a subset $C \subseteq X$ of vertices, we say that two hyperedges of E are *distinguished* (or *separated*) by C if some element in C belongs to exactly one of the two hyperedges. In this setting, one can tell apart the two distinguished hyperedges simply by comparing their intersections with C. Following this viewpoint, one may say that two hyperedges are related if they have the same intersection with C. This is clearly an equivalence relation, and one may determine the collection of equivalence classes induced by C: each such class corresponds to its own subset of C. Any two hyperedges belonging to distinct equivalence classes are then distinguished by C. We call these classes *neighborhood equivalence classes*. In general, one naturally seeks to distinguish as many pairs of hyperedges as possible, using a small set C.

It is a well-studied setting to ask for a maximum-size set C such that C induces all possible $2^{|C|}$ equivalence classes. In this case, C is said to be *shattered*. The maximum size of a shattered set in a hypergraph H is called its

---

C. Bazgan—Institut Universitaire de France.

© Springer International Publishing AG 2016
T.-H.H. Chan et al. (Eds.): COCOA 2016, LNCS 10043, pp. 92–106, 2016.
DOI: 10.1007/978-3-319-48749-6_7

Vapnis-Červonenkis dimension (VC dimension for short). This notion, introduced by Vapnis and Červonenkis [40] arose in the context of statistical learning theory as a measure of the structural complexity of the data. It has since been widely used in discrete mathematics; see the references in the thesis [9] for more references. We have the following associated decision problem.

VC Dimension
Input: A hypergraph $H = (X, E)$, and an integer k.
Question: Is there a shattered set $C \subseteq X$ of size at least k in H?

The complexity of VC Dimension was studied in e.g. [17,21,34]; it is a complete problem for the complexity class LOGNP defined in [34] (it is therefore a good candidate for an NP-intermediate problem). VC Dimension remains LOGNP-complete for *neighborhood hypergraphs of graphs* [30] (the *neighborhood hypergraph* of G has $V(G)$ as its vertex set, and the set of closed neighborhoods of vertices of G as its hyperedge set).

In another setting, one wishes to distinguish *all* pairs of hyperedges (in other words, each equivalence class must have size 1) while minimizing the size of the solution set C. Following [26], we call the associated decision problem, Distinguishing Transversal.

Distinguishing Transversal
Input: A hypergraph $H = (X, E)$, and an integer k.
Question: Is there a set $C \subseteq X$ of size at most k that induces $|E|$ distinct equivalence classes?

There exists a rich literature about Distinguishing Transversal. It was studied under different names, such as Test Set in Garey and Johnson's book [25, SP6]; other names include Test Cover [18–20], Discriminating Code [16] or Separating System [7,37].[1] A celebrated theorem of Bondy [8] also implicitly studies this notion. A version of Distinguishing Transversal called Identifying Code was defined for graphs instead of hypergraphs [23,27]. Similarly as for the well-known relation between the classic graph problem Dominating Set and the hypergraph problem Hitting Set, it is easy to check that an identifying code in graph G is the same as a distinguishing transversal in the neighborhood hypergraph of G.

The goal of this paper is to introduce and study the problem Partial VC Dimension, that generalizes both Distinguishing Transversal and VC Dimension, and defined as follows.

---

[1] Technically speaking, in Test Set, Test Cover and Separating System, the goal is to distinguish the *vertices* of a hypergraph using a set C of *hyperedges*, and in Discriminating Code the input is presented as a bipartite graph. Nevertheless, these formulations are equivalent to Distinguishing Transversal by considering either the dual hypergraph of the input hypergraph $H = (X, E)$ (with vertex set E and hyperedge set X, and hyperedge x contains vertex e in the dual if hyperedge e contains vertex x in H), or the bipartite incidence graph (defined over vertex set $X \cup E$, and where x and e are adjacent if they were incident in H).

PARTIAL VC DIMENSION
Input: A hypergraph $H = (X, E)$, and two integers k and ℓ.
Question: Is there a set $C \subseteq X$ of size k that induces at least ℓ distinct equivalence classses?

PARTIAL VC DIMENSION belongs to the category of *partial* versions of common decision problems, in which, instead of satisfying the problem's constraint task for all elements (here, all 2^k equivalence classes), we ask whether we can satisfy a certain number, ℓ, of these constraints. See for example the papers [22,29] that study some partial versions of standard decision problems, such as SET COVER or DOMINATING SET.

When $\ell = |E|$, PARTIAL VC DIMENSION is precisely the problem DISTINGUISHING TRANSVERSAL. When $\ell = 2^k$, we have the problem VC DIMENSION. Hence, PARTIAL VC DIMENSION is NP-hard, even on many restricted classes. Indeed, DISTINGUISHING TRANSVERSAL is NP-hard [25], even on hypergraphs where each vertex belongs to at most two hyperedges [20], or on neighborhood hypergraphs of graphs that are either: unit disk graphs [32], planar bipartite subcubic [23], graphs that are interval and permutation [24], split graphs [23]. MIN DISTINGUISHING TRANSVERSAL cannot be approximated within a factor of $o(\log n)$ on hypergraphs of order n [20], even on hypergraphs without 4-cycles [10], and on neighborhood hypergraphs of bipartite, co-bipartite or split graphs [23].

When $\ell = 2^k$, PARTIAL VC DIMENSION is equivalent to VC DIMENSION and unlikely to be NP-hard (unless all problems in NP can be solved in quasi-polynomial time), since $|X| \leqslant 2^k$ and a simple brute-force algorithm has quasi-polynomial running time. Moreover, VC DIMENSION (and hence PARTIAL VC DIMENSION) is W[1]-complete when parameterized by k [21].

Recently, the authors in [12] introduced the notion of (α, β)-*set systems*, that is, hypergraphs where, for any set S of vertices with $|S| \leqslant \alpha$, S induces at most β equivalence classes. Using this terminology, if a given hypergraph H is an (α, β)-set system, (H, k, ℓ) with $k = \alpha$ is a YES-instance of PARTIAL VC DIMENSION if and only if $\ell \leqslant \beta$.

We will also study the approximation complexity of the following fixed-cardinality maximization problem associated to PARTIAL VC DIMENSION.

MAX PARTIAL VC DIMENSION
Input: A hypergraph $H = (X, E)$, and an integer k.
Output: A set $C \subseteq X$ of size k that maximizes the number of equivalence classes induced by C.

Similar *fixed-cardinality* versions of classic optimization problems such as SET COVER, DOMINATING SET or VERTEX COVER, derived from the "partial" counterparts of the corresponding decision problems, have gained some attention in the recent years, see for example [13,14,29].

MAX PARTIAL VC DIMENSION is clearly NP-hard since PARTIAL VC DIMENSION is NP-complete; other than that, its approximation complexity is completely unknown since it cannot be directly related to the one of approximating MIN DISTINGUISHING TRANSVERSAL or MAX VC DIMENSION (the minimization and maximization versions of DISTINGUISHING TRANSVERSAL and VC DIMENSION, respectively).

Our Results. Our focus is on the approximation complexity of MAX PARTIAL VC DIMENSION. We give positive results in Sect. 3. We first provide polynomial-time approximation algorithms using the VC-dimension for the maximum degree or for the maximum edge-size of the input hypergraph. We apply these to obtain approximation ratios of the form n^δ (for $\delta < 1$ a constant) in certain special cases, as well as a better approximation ratio but with exponential time. For neighbourhood hypergraphs of planar graphs, MAX PARTIAL VC DIMENSION admits a PTAS (this is also shown for MIN DISTINGUISHING TRANSVERSAL). In Sect. 4, we give hardness results. We show that any 2-approximation algorithm for MAX PARTIAL VC DIMENSION implies a 2-approximation algorithm for MAX VC DIMENSION. Finally, we show that MAX PARTIAL VC DIMENSION is APX-hard, even for graphs of maximum degree at most 7.

2   Preliminaries

Twin-free Hypergraphs. In a hypergraph H, we call two equal hyperedges *twin hyperedges*. Similarly, two vertices belonging to the same set of hyperedges are *twin vertices*.

Clearly, two twin hyperedges will always belong to the same neighborhood equivalence classes. Similarly, for any set T of mutually twin vertices, there is no advantage in selecting more than one of the vertices in T when building a solution set C.

Observation 1. *Let $H = (X, E)$ be a hypergraph and let $H' = (X', E')$ be the hypergraph obtained from H by deleting all but one of the hyperedges or vertices from each set of mutual twins. Then, for any set $C \subseteq X$, the equivalence classes induced by C in H are the same as those induced by $C \cap X'$ in H'.*

Therefore, since it is easy to detect twin hyperedges and vertices in an input hypergraph, in what follows, we will always restrict ourselves to hypergraphs without twins. We call such hypergraphs *twin-free*.

Degree conditions. In a hypergraph H, the *degree* of a vertex x is the number of hyperedges it belongs to. The *maximum degree* of H is the maximum value of the degree of a vertex of H; we denote it by $\Delta(H)$.

The next theorem gives an upper bound on the number of neighborhood equivalence classes that can be induced when the degrees are bounded.

Theorem 2 ([18,20,27]). *Let $H = (X, E)$ be a hypergraph with maximum degree Δ and let C be a subset of X of size k. Then, C cannot induce more than $\frac{k(\Delta+1)}{2} + 1$ neighborhood equivalence classes.*

The Sauer-Shelah lemma. The following theorem is known as the Sauer-Shelah Lemma [38,39] (it is also credited to Perles in [39] and a weaker form was stated by Vapnik and Červonenkis [40]). It is a fundamental tool in the study of the VC dimension.

Theorem 3 (Sauer-Shelah Lemma [38,39]). *Let $H = (X, E)$ be a hypergraph with strictly more than $\sum_{i=0}^{d-1} \binom{|X|}{i}$ distinct hyperedges. Then, S has VC-dimension at least d.*

Theorem 3 is known to be tight. Indeed, the system that consists of considering all subsets of $\{1, \ldots, n\}$ of cardinality at most $d - 1$ has VC-dimension equal to $d - 1$. Though the original proofs of Theorem 3 were non-constructive, Ajtai [1] gave a constructive proof that yields a (randomized) polynomial-time algorithm, and an easier proof of this type can be found in Miccianio [31].

The following direct corollary of Theorem 3 is observed for example in [10].

Corollary 4. *Let $S = (X, E)$ be a hypergraph with VC dimension at most d. Then, for any subset $X' \subseteq X$, there are at most $\sum_{i=0}^{d} \binom{|X'|}{i} \leqslant |X'|^d + 1$ equivalence classes induced by X'.*

Approximation. An algorithm for an optimization problem is a c-approximation algorithm if it returns a solution whose value is always at most a factor of c away from the optimum. The class APX contains all optimization problems that admit a polynomial-time c-approximation algorithm for some fixed constant c. A *polynomial-time approximation scheme* (PTAS for short) for an optimization problem is an algorithm that, given any fixed constant $\epsilon > 0$, returns in polynomial time (in terms of the instance and for fixed ϵ) a solution that is a factor of $1 + \epsilon$ away from the optimum. An optimization problem is APX-*hard* if it admits no PTAS (unless P=NP).

Given an optimization problem P, an instance I of P, we denote by $opt_P(I)$ (or $opt(I)$ if there is no ambiguity) the value of an optimal solution for I.

Definition 5 (*L*-reduction [33]). *Let A and B be two optimization problems. Then A is said to be L-reducible to B if there are two constants $\alpha, \beta > 0$ and two polynomial time computable functions f, g such that: (i) f maps an instance I of A into an instance I' of B such that $opt_B(I') \leqslant \alpha \cdot opt_A(I)$, (ii) g maps each solution S' of I' into a solution S of I such that $||S| - opt_A(I)| \leqslant \beta \cdot ||S'| - opt_B(I')|$.*

L-reductions are useful in order to apply the following theorem.

Theorem 6 [33]. *Let A and B be two optimization problems. If A is APX-hard and L-reducible to B, then B is APX-hard.*

3   Positive Approximation Results for Max Partial VC Dimension

We start with a greedy polynomial-time procedure that always returns (if it exists), a set $|X'|$ that induces at least $|X'| + 1$ equivalence classes.

Lemma 7. *Let $H = (X, E)$ be a twin-free hypergraph and let $k \leqslant |X| - 1$ be an integer. One can construct, in time $O(k(|X| + |E|))$, a set $C \subseteq X$ of size k that produces at least $\min\{|E|, k + 1\}$ neighborhood equivalence classes.*

Proof. We produce C in an inductive way. First, let $C_1 = \{x\}$ for an arbitrary vertex x of X for which there exists at least one hyperedge of E with $x \notin E$ (if such hyperedge does not exist, then all edges are twin edges; since H is twin-free, $|E| \leqslant 1$ and we are done). Then, for each i with $2 \leqslant i \leqslant k$, we build C_i from C_{i-1} as follows: select vertex x_i as a vertex in $X \setminus C_{i-1}$ such that $C_{i-1} \cup \{x\}$ maximizes the number of equivalence classes.

We claim that either we already have at least $|E|$ equivalence classes, or C_i induces at least one more equivalence class than C_{i-1}. Assume for a contradiction that we have strictly less than $|E|$ equivalence classes, but C_i has the same number of equivalence classes as C_{i-1}. Since we have strictly less than $|E|$ classes, there is an equivalence class consisting of at least two edges, say e_1 and e_2. But then, since H is twin-free, there is a vertex x that belongs to exactly one of e_1 and e_2. But $C_{i-1} \cup \{x\}$ would have strictly more equivalence classes than C_i, a contradiction since C_i was maximizing the number of equivalence classes.

Hence, setting $C = C_k$ finishes the proof. $\square$

Proposition 8. MAX PARTIAL VC DIMENSION *is $\frac{\min\{2^k, |E|\}}{k+1}$-approximable in polynomial time. For hypergraphs with VC dimension at most d,* MAX PARTIAL VC DIMENSION *is k^{d-1}-approximable. For hypergraphs with maximum degree Δ,* MAX PARTIAL VC DIMENSION *is $\frac{\Delta+1}{2}$-approximable.*

Proof. By Lemma 7, we can always compute in polynomial time, a solution with at least $k + 1$ neighborhood equivalence classes (if it exists; otherwise, we solve the problem exactly). Since there are at most $\min\{2^k, |E|\}$ possible classes, the first part of the statement follows. Similarly, by Corollary 4, if the hypergraph has VC dimension at most d, there are at most $k^d + 1$ equivalence classes, and $\frac{k^d+1}{k+1} \leqslant k^{d-1}$. Finally, if the maximum degree is at most Δ, by Theorem 2 there are at most $\frac{k(\Delta+1)+2}{2}$ possible classes (and when $\Delta \geqslant 1$, $\frac{k(\Delta+1)+2}{2(k+1)} \leqslant \frac{\Delta+1}{2}$). $\square$

Corollary 9. *For hypergraphs of VC dimension at most d,* MAX PARTIAL VC DIMENSION *is $|E|^{(d-1)/d}$-approximable.*

Proof. By Proposition 8, we have a $\min\{k^{d-1}, |E|/k\}$-approximation. If $k^{d-1} < |E|^{(d-1)/d}$ we are done. Otherwise, we have $k^{d-1} \geqslant |E|^{(d-1)/d}$ and hence $k \geqslant |E|^{1/d}$, which implies that $\frac{|E|}{k} \leqslant |E|^{(d-1)/d}$. $\square$

For examples of concrete applications of Corollary 9, hypergraphs with no 4-cycles in its bipartite incidence graph[2] have VC-dimension at most 3 and hence we have an $|E|^{2/3}$-approximation for this class. Hypergraphs with maximum edge-size d also have VC dimension at most d. Other examples, arising from graphs, are neighborhood hypergraphs of: K_{d+1}-minor-free graphs (that have VC dimension at most d [11]); graphs of rankwidth at most r (VC dimension at most $2^{2^{O(r)}}$ [11]); interval graphs (VC dimension at most 2 [10]); permutation graphs (VC dimension at most 3 [10]); line graphs (VC dimension at most 3); unit disk graphs (VC dimension at most 3) [10]; C_4-free graphs (VC dimension at most 2); chordal bipartite graphs (VC dimension at most 3 [10]); undirected path graphs (VC dimension at most 3 [10]). Typical graph classes with unbounded VC dimension are bipartite graphs and their complements, or split graphs.

In the case of hypergraphs with no 4-cycles in its bipartite incidence graph (for which Max Partial VC Dimension has an $|E|^{2/3}$-approximation algorithm by Corollary 9), we can also relate Max Partial VC Dimension to Max Partial Double Hitting Set, defined as follows.

Max Partial Double Hitting Set
Input: A hypergraph $H = (X, E)$, an integer k.
Output: A subset $C \subseteq X$ of size k maximizing the number of hyperedges containing at least two elements of C.

Theorem 10. *Any α-approximation algorithm for* Max Partial Double Hitting Set *on hypergraphs without 4-cycles in its bipartite incidence graph can be used to obtain a 4α-approximation algorithm for* Max Partial VC Dimension *on hypergraphs without 4-cycles in its bipartite incidence graph.*

Proof. Let $H = (X, E)$ be a hypergraph without 4-cycles in its bipartite incidence graph, and let $C \subseteq X$ be a subset of vertices. Since H has no 4-cycles in its bipartite incidence graph, note that if some hyperedge contains two vertices of X, then no other hyperedge contains these two vertices. Therefore, the number of equivalence classes induced by C is equal to the number of hyperedges containing at least two elements of C, plus the number of equivalence classes corresponding to a single (or no) element of C. Therefore, the maximum number $opt(H)$ of equivalence classes for a set of size k is at most $opt_{2HS}(H) + k + 1$, where $opt_{2HS}(H)$ is the value of an optimal solution for Max Partial Double Hitting Set on H. Observing that $opt_{2HS}(H) \geqslant \frac{k}{2}$ (since one may always iteratively select pairs of vertices covering a same hyperedge to obtain a valid double hitting set of H), we get that $opt(H) \leqslant 3opt_{2HS}(H) + 1 \leqslant 4opt_{2HS}(H)$. Moreover, in polynomial time we can apply the approximation algorithm of Max Partial Double Hitting Set to H to obtain a set C inducing at least $\frac{opt_{2HS}(H)}{\alpha}$ neighborhood equivalence classes. Thus, C induces at least $\frac{opt(H)}{4\alpha}$ neighborhood equivalence classes. ☐

---

[2] In the dual hypergraph, this corresponds to the property that each pair of hyperedges have at most one common element, see for example [2].

Unfortunately, the complexity of approximating MAX PARTIAL DOUBLE HITTING SET seems not to be well-known, even when restricted to hypergraphs with no 4-cycles in its bipartite incidence graph. In fact, the problem MAX DENSEST SUBGRAPH (which, given an input graph, consists of maximizing the number of edges of a subgraph of order k) is precisely MAX PARTIAL DOUBLE HITTING SET restricted to hypergraphs where each hyperedge has size at most 2 (that is, to graphs), that can be assumed to contain no 4-cycles in its bipartite incidence graph (a 4-cycle would imply the existence of two twin hyperedges). Although MAX DENSEST SUBGRAPH (and hence MAX PARTIAL DOUBLE HITTING SET for hypergraphs with no 4-cycles in its bipartite incidence graph) is only known to admit no PTAS [28], the best known approximation ratio for it is $O(|E|^{1/4})$ [5].[3] We deduce from this result, the following corollary of Theorem 10 for hypergraphs of hyperedge-size bounded by 2. This improves on the $O(|E|^{1/2})$-approximation algorithm given by Corollary 9 for this case.

Corollary 11. *Let α be the best approximation ratio in polynomial time for* MAX DENSEST SUBGRAPH. *Then,* MAX PARTIAL VC DIMENSION *can be 3α-approximated in polynomial time on hypergraphs with hyperedges of size at most 2. In particular, there is a polynomial-time $O(|E|^{1/4})$-approximation algorithm for this case.*

We will now apply the following result from [4].

Lemma 12 ([4]). *If an optimization problem is $r_1(k)$-approximable in fpt-time with respect to parameter k for some strictly increasing function r_1 depending solely on k, then it is also $r_2(n)$-approximable in fpt-time w.r.t. parameter k for any strictly increasing function r_2 depending solely on the instance size n.*

Using Proposition 8 showing that MAX PARTIAL VC DIMENSION is $\frac{2^k}{k+1}$-approximable and Lemma 12, we directly obtain the following.

Corollary 13. *For any strictly increasing function r,* MAX PARTIAL VC DIMENSION *parameterized by k is $r(n)$-approximable in FPT-time.*

In the following we establish polynomial time approximation schemes for MIN DISTINGUISHING TRANSVERSAL and MAX PARTIAL VC DIMENSION on planar graphs using the layer decomposition technique introduced by Baker [3].

Given a planar embedding of an input graph, we call the vertices which are on the external face *level 1 vertices*. By induction, we define *level t vertices* as the set of vertices which are on the external face after removing the vertices of levels smaller than t [3]. A planar embedding is *t-level* if it has no vertices of level greater than t. If a planar graph is t-level, it has a t-outerplanar embedding.

Theorem 14. MAX PARTIAL VC DIMENSION *on neighborhood hypergraphs of planar graphs admits a PTAS.*

---

[3] Formally, it is stated in [5] as an $O(|V|^{1/4})$-approximation algorithm, but we may assume that the input graph is connected, and hence $|V| = O(|E|)$.

Proof. Let G be a planar graph with a t-level planar embedding for some integer t. We aim to achieve an approximation ratio of $1 + \varepsilon$. Let $\lambda = \lceil \frac{1}{\varepsilon} \rceil - 1$.

Let G_i $(0 \leqslant i \leqslant \lambda)$ be the graph obtained from G by removing the vertices on levels $i \bmod (\lambda+1)$. Thus, graph G_i is the disjoint union of several subgraphs G_{ij} $(0 \leqslant j \leqslant p$ with $p = \lceil \frac{t+i}{\lambda+1} \rceil)$ where G_{i0} is induced by the vertices on levels $0, \ldots, i-1$ (note that G_{00} is empty) and G_{ij} with $j \geqslant 1$ is induced by the vertices on levels $(j-1)(\lambda+1) + i + 1, \ldots j(\lambda+1) + i - 1$. In other words, each subgraph G_{ij} is the union of at most λ consecutive levels and is thus λ-outerplanar. Hence, G_i is also λ-outerplanar and it has treewidth at most $3\lambda - 1$ [6]. Using Courcelle's theorem[4], for any integer t and any subgraph G_{ij}, we can efficiently determine an optimal set S_{ij}^t of t vertices of G_{ij} that maximizes the number of (nonempty) induced equivalence classes in G_{ij}. We then use dynamic programming to construct a solution for G_i. Denote by $S_i(q, y)$ a solution corresponding to the maximum feasible number of equivalence classes induced by a set of y vertices of G_i $(0 \leqslant y \leqslant k)$ among the first q subgraphs $G_{i1}, \ldots, G_{iq}$ $(1 \leqslant q \leqslant p)$. We have $S_i(q, y) = \max_{0 \leqslant x \leqslant y}(S_{iq}^x + S_i(q-1, y-x))$. Let $S_i = S_i(p, k)$.

Among $S_0, \ldots, S_\lambda$, we choose the best solution, that we denote by S. We now prove that S is an $(1+\varepsilon)$-approximation of the optimal value $opt(G)$ for MAX PARTIAL VC DIMENSION on G. Let S_{opt} be an optimal solution of G. Then, there is at least one integer r such that at most $1/(\lambda+1)$ of the equivalent classes induced by S_{opt} in G are lost when we remove vertices on the levels congruent to $r \bmod (\lambda+1)$.

Thus, $val(S) \geqslant val(S_r) \geqslant opt(G) - \frac{opt(G)}{\lambda+1} = \frac{\lambda}{\lambda+1} opt(G) \geqslant (1 - \frac{1}{\varepsilon}) opt(G)$, which completes the proof.

The overall running time of the algorithm is λ times the running time for graphs of treewidth at most $3\lambda - 1$, that is, $O(\lambda n)$.   □

As a side result, using the same technique, we provide the following theorem about MIN DISTINGUISHING TRANSVERSAL, which is an improvement over the 7-approximation algorithm that follows from [36] (in which it is proved that any YES-instance satisfies $\ell \leqslant 7k$) and solves an open problem from [23]. Due to space constraints, its proof is omitted.

Theorem 15. MIN DISTINGUISHING TRANSVERSAL *on neighborhood hypergraphs of planar graphs (equivalently,* MIN IDENTIFYING CODE *on planar graphs) admits a PTAS.*

4   Hardness of Approximation Results for Max Partial VC Dimension

We define MAX VC DIMENSION as the maximization version of VC DIMENSION.

---

[4] We can indeed encode the decision version of our problem in MSOL as follows:
$$\exists x_1, \ldots, x_k, y_1, \ldots, y_l, s_1^1, s_1^2, \ldots, s_{\ell-1}^\ell, s_i^j = \bigvee_{q=1}^k x_q, \bigvee_{i,j=0}^{\binom{\ell}{2}} (s_i^j \in y_i \wedge s_i^j \notin y_j) \vee (s_i^j \notin y_i \wedge s_i^j \in y_j).$$

MAX VC DIMENSION

Input: A hypergraph $H = (X, E)$.

Output: A maximum-size shattered subset $C \subseteq X$ of vertices.

Not much is known about the complexity of MAX VC DIMENSION: it is trivially $\log_2 |E|$-approximable by returning a single vertex; a lower bound on the running time of a potential PTAS has been proved [17]. It is mentioned as an outstanding open problem in [15]. In the following we establish a connection between the approximability of MAX VC DIMENSION and MAX PARTIAL VC DIMENSION.

Theorem 16. *Any 2-approximation algorithm for* MAX PARTIAL VC DIMEN-SION *can be transformed into a randomized 2-approximation algorithm for* MAX VC DIMENSION *with polynomial overhead in the running time.*

Proof. Let H be a hypergraph on n vertices that is an instance for MAX VC DIMENSION, and suppose we have a c-approximation algorithm $\mathscr{A}$ for MAX PARTIAL VC DIMENSION.

We run $\mathscr{A}$ with $k = 1, \ldots, \log_2 |X|$, and let k_0 be the largest value of k such that the algorithm outputs a solution with at least $\frac{2^k}{c}$ neighborhood equivalence classes. Since $\mathscr{A}$ is a c-approximation algorithm, we know that the optimum for MAX PARTIAL VC DIMENSION for any $k > k_0$ is strictly less than 2^k. This implies that the VC-dimension of S is at most k_0.

Now, let X be the solution set of size k_0 computed by $\mathscr{A}$, and let H_X be the sub-hypergraph of H induced by X. By our assumption, this hypergraph has at least $\frac{2^{k_0}}{c}$ distinct edges. We can now apply the Sauer-Shelah Lemma (Theorem 3).

We have $c = 2$, and we apply the lemma with $|X| = k_0$ and $d = \frac{k_0}{2} + 1$; it follows that the VC dimension of H_X (and hence, of H) is at least $\frac{k_0}{2} + 1$. By the constructive proof of Theorem 3, a shattered set Y of this size can be computed in (randomized) polynomial time [1,31]. Set Y is a 2-approximation, since we saw in the previous paragraph that the VC dimension of H is at most k_0. $\square$

We note that the previous proof does not seem to apply for any other constant than 2, because the Sauer-Shelah Lemma would not apply. Though the approximation complexity of MAX VC DIMENSION is not known, our result shows that MAX PARTIAL VC DIMENSION is at least as hard to approximate.

Before proving our next result, we first need an intermediate result for MAX PARTIAL VERTEX COVER (also known as MAX k-VERTEX COVER [14]), which is defined as follows.

MAX PARTIAL VERTEX COVER

Input: A graph $G = (V, E)$, an integer k.

Output: A subset $S \subseteq V$ of size k covering the maximum number of edges.

Proposition 17 ([35]). MAX PARTIAL VERTEX COVER *is* APX-*hard, even for cubic graphs.*

Theorem 18. MAX PARTIAL VC DIMENSION *is* APX-*hard, even for graphs of maximum degree* 7.

Proof. We will give an L-reduction from MAX PARTIAL VERTEX COVER (which is APX-hard, by Proposition 17) to MAX PARTIAL VC DIMENSION. The result will then follow from Theorem 6. Given an instance $I = (G, k)$ of MAX PARTIAL VERTEX COVER with $G = (V, E)$ a cubic graph, we construct an instance $I' = (G', k')$ of MAX PARTIAL VC DIMENSION with $G' = (V', E')$ of maximum degree 7 in the following way. For each vertex $v \in V$, we create a gadget P_v with twelve vertices where four among these twelve vertices are special: they form the set $F_v = \{f_v^1, f_v^2, f_v^3, f_v^4\}$. The other vertices are adjacent to the subsets $\{f_v^4\}$, $\{f_v^2, f_v^3\}$, $\{f_v^1, f_v^3\}$, $\{f_v^2, f_v^4\}$, $\{f_v^1, f_v^4\}$, $\{f_v^1, f_v^3, f_v^4\}$, $\{f_v^1, f_v^2, f_v^4\}, \{f_v^1, f_v^2, f_v^3, f_v^4\}$, respectively. We also add edges between f_v^1 and f_v^2, between f_v^2 and f_v^3 and between f_v^3 and f_v^4. Since G is cubic, for each vertex v of G, there are three edges e_1, e_2 and e_3 incident with v. For each edge e_i ($1 \leqslant i \leqslant 3$), the endpoint v is replaced by f_v^i. Moreover, each of these original edges of G is replaced in G' by two edges by subdividing it once (see Fig. 1 for an illustration). We call the vertices resulting from the subdivision process, *edge-vertices*. Finally, we set $k' = 4k$.

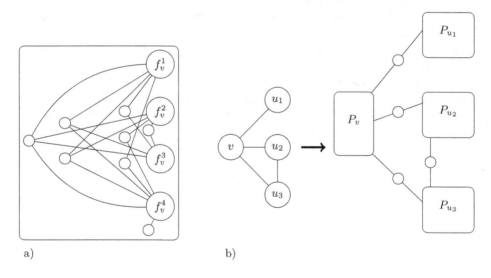

a)   b)

Fig. 1. a) Vertex-gadget P_v and b) illustration of the reduction.

From any optimal solution S with $|S| = k$ covering $opt(I)$ edges of G, we construct a set $C = \{f_v^j : 1 \leqslant j \leqslant 4, v \in S\}$ of size $4k$. By construction, C induces 12 equivalence classes in each vertex gadget. Moreover, for each covered edge $e = xy$ in G, the corresponding edge-vertex v_e in G' forms a class of size 1 (which corresponds to one or two neighbor vertices f_x^i and f_y^j of v_e in C). Finally, all vertices in G' corresponding to edges not covered by S in G, as well as all vertices in vertex gadgets corresponding to vertices not in S, belong to the same equivalence class (corresponding to the empty set). Thus, C induces in G' $12k + opt(I) + 1$ equivalence classes, and hence we have

$$opt(I') \geqslant 12k + opt(I) + 1. \tag{1}$$

Conversely, given a solution C' of I' with $|C'| = 4k$, we transform it into a solution for I as follows. First, we show that C' can be transformed into another solution C'' such that (1) C'' only contains vertices of the form f_v^i, (2) each vertex-gadget contains either zero or four vertices of C'', and (3) C'' does not induce less equivalence classes than C'. To prove this, we proceed step by step by locally altering C' whenever (1) and (2) are not satisfied, while ensuring (3).

Suppose first that some vertex-gadget P_v of G' contains at least four vertices of C'. Then, the number of equivalence classes involving some vertex of $V(P_v) \cap C'$ is at most twelve within P_v (since there are only twelve vertices in P_v), and at most three outside P_v (since there are only three vertices not in P_v adjacent to vertices in P_v). Therefore, we can replace $V(P_v) \cap C'$ by the four special vertices of the set F_v in P_v; this choice also induces twelve equivalence classes within P_v, and does not decrease the number of induced classes.

Next, we show that it is always best to select the four special vertices of F_v from some vertex-gadget (rather than having several vertex-gadgets containing less than four solution vertices each). To the contrary, assume that there are two vertex-gadgets P_u and P_v containing respectively a and b vertices of C', where $1 \leqslant b \leqslant a \leqslant 3$. Then, we remove an arbitrary vertex from $C' \cap V(P_v)$; moreover we replace $C' \cap V(P_u)$ with the subset $\{f_u^i, 1 \leqslant i \leqslant a + 1\}$, and similarly we replace $C' \cap V(P_v)$ with the subset $\{f_v^i, 1 \leqslant i \leqslant b - 1\}$. Before this alteration, the solution vertices within $V(P_u) \cup V(P_v)$ could contribute to at most $2^a + 2^b - 2$ equivalence classes. After the modification, one can check that this quantity is at least $2^{a+1} + 2^{b-1} - 2$ classes. Observing that $2^{a+1} + 2^{b-1} - 2 \geqslant 2^a + 2^b - 2$ since $2^a - 2^{b-1} \geqslant 0$ yields our claim. Hence, by this argument, we conclude that all vertex-gadgets (except possibly at most one) contain either zero or four vertices from the solution set C'.

Suppose that there exists one vertex-gadget P_v with i solution vertices, $1 \leqslant i \leqslant 3$. We show that we may add $4 - i$ solution vertices to it so that $C' \cap V(P_v) = F_v$. Consider the set of edge-vertices belonging to C'. Since we had $|C'| = 4k$ and all but one vertex-gadget contain exactly four solution vertices, there are at least $4 - i$ edge-vertices in the current solution set. Then, we remove an arbitrary set of $4 - i$ edge-vertices from C' and instead, we replace the set $V(P_v) \cap C'$ by the set F_v of special vertices of P_v. We now claim that this does not decrease the number of classes induced by C'. Indeed, any edge-vertex, since it has degree 2, may contribute to at most three equivalence classes, and the i solution vertices in P_v can contribute to at most 2^i classes. Summing up, in the old solution set, these four vertices contribute to at most $3(4 - i) + 2^i$ classes, which is less than

12 since $1 \leqslant i \leqslant 3$. In the new solution, these four vertices contribute to at least 12 classes, which proves our above claim.

We now know that there are $4i$ edge-vertices in C', for some $i \leqslant k$. All other solution vertices are special vertices in some vertex-gadgets. By similar arguments as in the previous paragraph, we may select any four of them and replace them with some set F_v of special vertices of some vertex-gadget P_v. Before this modification, these four solution vertices may have contributed to at most $3 \cdot 4 = 12$ classes, while the new four solution vertices now contribute to at least 12 classes.

Applying the above arguments, we have proved the existence of the required set C'' that satisfies conditions (1)–(3).

Therefore, we may now assume that the solution C'' contains no edge-vertices, and for each vertex-gadget P_v, $C'' \cap V(P_v) \in \{\emptyset, F_v\}$. We define as solution S for I the set of vertices v of G for which P_v contains four vertices of C''. Then, $val(S) = val(C') - 12k - 1$. Considering an optimal solution C' for I', we have $opt(I) \geqslant opt(I') - 12k - 1$. Using (1), we conclude that $opt(I') = opt(I) + 12k + 1 \leqslant opt(I) + 24opt(I) + 1$ since $k \leqslant 2opt(I)$ and thus $opt(I') \leqslant 26opt(I)$.

Moreover, we have $opt(I) - val(S) = opt(I') - 12k - 1 - (val(C') - 12k - 1) = opt(I') - val(C')$.

Thus, our reduction is an L-reduction with $\alpha = 26$ and $\beta = 1$.   □

Proposition 8 and Theorem 18 give the following corollary:

Corollary 19. MAX PARTIAL VC DIMENSION *is* APX-*complete for bounded degree graphs.*

5   Conclusion

In this paper, we defined and studied generalization of DISTINGUISHING TRANSVERSAL and VC DIMENSION. The probably most intriguing open question seems to be the approximation complexity of MAX PARTIAL VC DIMENSION. In particular, does the problem admit a constant-factor approximation algorithm? As a first step, one could determine whether such an approximation algorithm exists in superpolynomial time, or on special subclasses such as neighbourhood hypergraphs of specific graphs. We have seen that there exist polynomial-time approximation algorithms with a sublinear ratio for special cases; does one exist in the general case?

References

1. Ajtai, M.: The shortest vector problem in L2 is NP-hard for randomized reductions. In: Proceedings of the 30th Annual ACM Symposium on Theory of Computing (STOC 1998), pp. 10–19 (1998)

2. Kumar, V.S.A., Arya, S., Ramesh, H.: Hardness of set cover with intersection 1. In: Welzl, E., Montanari, U., Rolim, J.D.P. (eds.) ICALP 2000. LNCS, vol. 1853, pp. 624–635. Springer, Heidelberg (2000)
3. Baker, B.S.: Approximation algorithms for NP-complete problems on planar graphs. J. ACM 41(1), 153–180 (1994)
4. Bazgan, C., Chopin, M., Nichterlein, A., Sikora, F.: Parameterized approximability of maximizing the spread of influence in networks. J. Discrete Algorithms 27, 54–65 (2014)
5. Bhaskara, A., Charikar, M., Chlamtac, E., Feige, U., Vijayaraghavan, A.: Detecting high log-densities: an $O(n^{1/4})$ approximation for densest k-subgraph. In: Proceedings of the 42nd Annual ACM Symposium on Theory of Computing (STOC 2010), pp. 201–210 (2010)
6. Bodlaender, H.: A partial k-arboretum of graphs with bounded treewidth. Theoret. Comput. Sci. 209(12), 1–45 (1998)
7. Bollobás, B., Scott, A.D.: On separating systems. Eur. J. Comb. 28, 1068–1071 (2007)
8. Bondy, J.A.: Induced subsets. J. Comb. Theory, Ser. B 12(2), 201–202 (1972)
9. Bousquet, N.: Hitting sets: VC-dimension and multicut. Ph.D. thesis, Université Montepellier II, France (2013). http://tel.archives-ouvertes.fr/tel-01012106/
10. Bousquet, N., Lagoutte, A., Li, Z., Parreau, A., Thomassé, S.: Identifying codes in hereditary classes of graphs and VC-dimension. SIAM J. Discrete Math. 29(4), 2047–2064 (2015)
11. Bousquet, N., Thomassé, S.: VC-dimension and Erdős-Pósa property of graphs. Discrete Math. 338(12), 2302–2317 (2015)
12. Bringmann, K., Kozma, L., Moran, S., Narayanaswamy, N.S.: Hitting Set in hypergraphs of low VC-dimension. Manuscript (2015). http://arxiv.org/abs/1512.00481
13. Bruglieri, M., Ehrgott, M., Hamacher, H.W., Maffioli, F.: An annotated bibliography of combinatorial optimization problems with fixed cardinality constraints. Discrete Appl. Math. 154(9), 1344–1357 (2006)
14. Cai, L.: Parameterized complexity of cardinality constrained optimization problems. Comput. J. 51(1), 102–121 (2007)
15. Cai, L., Juedes, D., Kanj, I.: The inapproximability of non-NP-hard optimization problems. Theoret. Comput. Sci. 289(1), 553–571 (2002)
16. Charbit, E., Charon, I., Cohen, G., Hudry, O., Lobstein, A.: Discriminating codes in bipartite graphs: bounds, extremal cardinalities, complexity. Adv. Math. Commun. 2(4), 403–420 (2008)
17. Chen, J., Huang, X., Kanj, I.A., Xia, G.: On the computational hardness based on linear FPT-reductions. J. Comb. Optim. 11(2), 231–247 (2006)
18. Crowston, R., Gutin, G., Jones, M., Muciaccia, G., Yeo, A.: Parameterizations of test cover with bounded test sizes. Algorithmica 74(1), 367–384 (2016)
19. Crowston, R., Gutin, G., Jones, M., Saurabh, S., Yeo, A.: Parameterized study of the test cover problem. In: Rovan, B., Sassone, V., Widmayer, P. (eds.) MFCS 2012. LNCS, vol. 7464, pp. 283–295. Springer, Heidelberg (2012)
20. De Bontridder, K.M.J., Halldórsson, B.V., Halldórsson, M.M., Hurkens, C.A.J., Lenstra, J.K., Ravi, R., Stougie, L.: Approximation algorithms for the test cover problem. Math. Program. Ser. B 98, 477–491 (2003)
21. Downey, R.G., Evans, P.A., Fellows, M.R.: Parameterized learning complexity. In: Proceedings of the 6th Annual Conference on Computational Learning Theory (COLT 1993), pp. 51–57 (1993)
22. Fomin, F.V., Lokshtanov, D., Raman, V., Saurabh, S.: Subexponential algorithms for partial cover problems. Inf. Process. Lett. 111(16), 814–818 (2011)

23. Foucaud, F.: Decision and approximation complexity for identifying codes and locating-dominating sets in restricted graph classes. J. Discrete Algorithms **31**, 48–68 (2015)
24. Foucaud, F., Mertzios, G.B., Naserasr, R., Parreau, A., Valicov, P.: Algorithms and complexity for metric dimension and location-domination on interval and permutation graphs. In: Mayr, E.W. (ed.) WG 2015. LNCS, vol. 9224, pp. 456–471. Springer, Heidelberg (2016). doi:10.1007/978-3-662-53174-7_32
25. Garey, M.R., Johnson, D.S.: Computers and Intractability: A Guide to the Theory of NP-completeness. W. H. Freeman, New York (1979)
26. Henning, M.A., Yeo, A.: Distinguishing-transversal in hypergraphs and identifying open codes in cubic graphs. Graphs Comb. **30**(4), 909–932 (2014)
27. Karpovsky, M.G., Chakrabarty, K., Levitin, L.B.: On a new class of codes for identifying vertices in graphs. IEEE Trans. Inf. Theory **44**, 599–611 (1998)
28. Khot, S.: Ruling out PTAS for graph min-bisection, dense k-subgraph, and bipartite clique. SIAM J. Comput. **36**(4), 1025–1071 (2006)
29. Kneis, J., Mölle, D., Rossmanith, P.: Partial vs. complete domination: t-dominating set. In: Proceedings of the 33rd Conference on Current Trends in Theory and Practice of Computer Science (SOFSEM 2007), pp. 367–376 (2007)
30. Kranakis, E., Krizanc, D., Ruf, B., Urrutia, J., Woeginger, G.J.: The VC-dimension of set systems defined by graphs. Discrete Appl. Math. **77**(3), 237–257 (1997)
31. Micciancio, D.: The shortest vector in a lattice is hard to approximate to within some constant. SIAM J. Comput. **30**(6), 2008–2035 (2001)
32. Müller, T., Sereni, J.-S.: Identifying and locating-dominating codes in (random) geometric networks. Comb. Probab. Comput. **18**(6), 925–952 (2009)
33. Papadimitriou, C.H., Yannakakis, M.: Optimization, approximation, and complexity classes. J. Comput. Syst. Sci. **43**(3), 425–440 (1991)
34. Papadimitriou, C.H., Yannakakis, M.: On limited nondeterminism and the complexity of the V-C dimension. J. Comput. Syst. Sci. **53**(2), 161–170 (1996)
35. Petrank, E.: The hardness of approximation: gap location. Comput. Complex. **4**, 133–157 (1994)
36. Slater, P.J., Rall, D.F.: On location-domination numbers for certain classes of graphs. Congr. Numerantium **45**, 97–106 (1984)
37. Rényi, A.: On random generating elements of a finite Boolean algebra. Acta Sci. Math. Szeged **22**, 75–81 (1961)
38. Sauer, N.: On the density of families of sets. J. Comb. Theory Ser. A **13**, 145–147 (1972)
39. Shelah, S.: A combinatorial problem; stability and order for models and theories in infinitary languages. Pac. J. Math. **41**, 247–261 (1972)
40. Vapnik, V.N., Červonenkis, A.J.: The uniform convergence of frequencies of the appearance of events to their probabilities. Akademija Nauk SSSR **16**, 264–279 (1971)

Improved Precise Fault Diagnosis Algorithm for Hypercube-Like Graphs

Tai-Ling Ye[1(✉)], Dun-Wei Cheng[1], and Sun-Yuan Hsieh[1,2]

[1] Department of Computer Science and Information Engineering,
National Cheng Kung University, Tainan 701, Taiwan
{p78991016,hsiehsy}@mail.ncku.edu.tw, dunwei.ncku@gmail.com
[2] Institute of Medical Informatics,
National Cheng Kung University, Tainan 701, Taiwan

Abstract. The system reliability is an important issue for multiprocessor systems. The fault diagnosis has become crucial for achieving high reliability in multiprocessor systems. In the comparison-based model, it allows a processor to perform diagnosis by contrasting the responses from a pair of neighboring processors through sending the identical assignment. Recently, Ye and Hsieh devised an precise fault diagnosis algorithm to detect all faulty processors for hypercube-like networks by using the MM* model with $O(N(\log_2 N)^2)$ time complexity, where N is the cardinality of processor set in multiprocessor systems. On the basis of Hamiltonian cycle properties, we improve the aforementioned results by presenting an $O(N)$-time precise fault diagnosis algorithm to detect all faulty processors for hypercube-like networks by using the MM* model.

1 Introduction

In the semiconductor technology, the system reliability is crucial for multiprocessor systems. To maintain high reliability, multiprocessor systems should differentiate between fault-free and faulty processors, that is, each faulty processor in this system ought to be substituted by a fault-free processor. Determining all faulty processors is known as *fault diagnosis*. When all faulty processors can be evaluated precisely, and t is the upper bound of the faulty processors, we called the multiprocessor system as *t-diagnosable*. The largest cardinality of the faulty set is named as the *diagnosability* of this multiprocessor system. Numerous fault diagnosis algorithms have been presented for different topologies [2,5,6,12,13,17,18].

In self-diagnosable systems, there are several approaches which have been proposed to detect faulty processors. The PMC model [11] allows each processor to perform diagnosis by testing the neighboring processors and observing their responses. In the PMC model, a *test syndrome* collects all test results. The MM model [7] is a comparison-based model and it allows a processor to perform diagnosis by contrasting the responses from a pair of neighboring processors through sending the identical assignment. In the MM model, a *comparison syndrome* collects all comparison results. In [13], Sengupta and Dahbura modified

© Springer International Publishing AG 2016
T.-H.H. Chan et al. (Eds.): COCOA 2016, LNCS 10043, pp. 107–112, 2016.
DOI: 10.1007/978-3-319-48749-6_8

the comparison-based model and put forward the MM* model. Under this model, any processor w (comparator) detects two processors x and y if w has direct communication links to them. In the MM* model, provided that w is a fault-free processor and x, y are two neighbors of w; an *agreement* of the comparison results means that processors x and y are fault-free processors. Conversely, a *disagreement* of the comparison results means that at least one of x, y, and w must be faulty.

In self-diagnosable systems, there are two basic strategies for fault diagnosis. The first one is named as the *precise diagnositic strategy*, which detects all processors correctly [7,11]. The second one is named as the *pessimistic diagnositic strategy*, which isolated all faulty processors in a faulty set such that no more than two processors which are faulty-free processors [4].

In a multiprocessor system, using a simple graph $G = (V_G, E_G)$ depicts the topology, where every vertex in V_G depicts a processor and every edge in E_G depicts a communication link between two processors. In [14], Vaidya *et al.* proposed a class of *hypercube-like networks*.

Recently, Ye and Hsieh [15] devised an $O(N(\log_2 N)^2)$-time precise fault diagnosis algorithm to detect all faulty processors for hypercube-like networks on the basis of the MM* model, where N is the cardinality of the vertex set in this system.

In order to reduce the time complexity, we present an $O(N)$-time precise fault diagnosis algorithm to detect all faulty processors for hypercube-like networks based on the MM* model, where $N = 2^n$ is the cardinality of the vertex set in this topology. Therefore, our precise fault diagnosis algorithm is better than the Ye-Hsieh algorithm [15] for hypercube-like networks.

Next, we introduce the basic concept in self-diagnosable systems through Sect. 2. In Sect. 3, we investigate several definitions and lemmas, then propose the precise fault diagnosis algorithm to detect all faulty processors for hypercube-like networks. Finally, we provide some concluding remarks in Sect. 4.

2   Preliminaries

We first present some fundamental definitions and notations and introduce the background of fault diagnosis.

2.1   Definitions and Notations from Graph Theory

Use a *simple graph* $G = (V_G, E_G)$ to describe the link situation for a simple multiprocessor system. The processors in this system are denoted by an *vertex set* V_G and the links between each pair of processors are denoted by an *edge set* E_G. When there exists an edge between two processors (x and y) such that the processor x is *adjacent to* the processor y, x is a *neighbor* of y. Use $N_G(x)$ = $\{y \in V_G|\ y$ is adjacent to $x\}$ to denote the neighbors of x. The *degree* of a processor x is the cardinality of the edge set incident to x in a simple graph G. Use $deg_G(x)$ to denote the degree of a processor x, that is, $deg_G(x) = |N_G(x)|$.

We say a simple graph is k-*regular* when each processor has exactly k neighbors. In this research, we consider the diagnosability of the hypercube-like network, and the structure of hypercube-like network contains the property of k-regular.

In a simple graph, a *path* is constructed by a sequence of distinct processors. We use $P[x_0, x_k] = \langle x_0, x_1, \ldots, x_k \rangle$ to represent a path with k distinct processors, such that the processor x_i and the processor x_{i+1} are adjacent to each other for $0 \leq i \leq k - 1$. The processors x_0 and x_k are the two *end processors* of this path. A *subpath* is denoted by $P[x_i, x_j]$, which is a sequence of processors within the original path $P[x_0, x_k] = \langle x_0, x_1, \ldots, x_{i-1}, P[x_i, x_j], x_{j+1}, \ldots, x_k \rangle$. A path in a simple graph that visits each processor of G exactly once is named as a *Hamiltonian path*. We also denote a *cycle* by a sequence of processors in a simple graph of the form $\langle x_0, x_1, \ldots, x_t \rangle$ for $t \geq 2$, where $x_0, x_1, \ldots, x_t$ are all distinct processors such that any two consecutive processors are adjacent and the two processors x_0 and x_t are also adjacent to each other. A cycle in a simple graph that visits each processor of G exactly once is named as a *Hamiltonian cycle*. When a graph contains a Hamiltonian cycle, it is called a *Hamiltonian*.

A *matching* in a simple graph is an edge subset $M_G \subseteq E_G$, which contains the non-loop edges without any common processor. If a matching M_G saturates a processor $x \in V_G$, then x is said to be M_G-saturated; otherwise, x is M_G-unsaturated. When a matching saturates every processor of the graph, we say this matching is a *perfect matching*.

Definition 1. The class of n -*dimensional hypercube-like networks*, denoted by HL_n, can be recursively defined as follows:

1. HL_0 is a trivial graph that contains only one vertex.
2. For $i \geq 1$, $HL_i = \{G | G = PM(G_0, G_1)$, where $G_0, G_1 \in HL_{i-1}$, and $PM(G_0, G_1)$ is a graph which is connected by a perfect matching PM between V_{G_0} and V_{G_1}, and every edge in this perfect matching is labeled by $i - 1$. For a graph $G \in HL_n$, $V_G = V_{G_0} \bigcup V_{G_1}$ and $E_G = E_{G_0} \bigcup E_{G_1} \bigcup PM$.

An n-dimensional hypercube-like graph is an n-regular graph with 2^n vertices and $n2^{n-1}$ edges [14], denoted by HL_n.

2.2   System Diagnostic Models

Observing those responses from the two neighbors x and y, the set of comparison results can be used to construct a comparison multigraph $M_G = (V_G, L_G)$. In a comparison multigraph, the processor set V_G is the set of all processors and L_G is the set of labeled edges that denote which processor is used to compare the two processors of this edge. A labeled edge $(x, y)_w \in L_G$ means that the processor w compares processor x and processor y.

In the MM* model, for the pair of processors x and y such that $(w, x) \in E_G$ and $(w, y) \in E_G$, we say that w is a comparator. Let $(x, y)_w \in L_G$; the output is denoted by $\sigma((x, y)_w)$. We use $\sigma((x, y)_w) = 1$ to denote a disagreement of the output result and $\sigma((x, y)_w) = 0$ to denote an agreement of the output result. The MM* model collects all the comparison results as a set and names it as a comparison syndrome of the self-diagnosable systems. The comparison result $\sigma((x, y)_w) = 0$ indicates that the compared processors x and y are fault-free processors provided that the comparator processor w is a fault-free vertex. The comparison result $\sigma((x, y)_w) = 1$ means that at least one of x, y, and w must be faulty.

3   A Precise Fault Diagnosis Algorithm in HL_n

In this section, we use a Hamiltonian cycle in HL_n to detect all faulty vertices.

3.1   Comparison Syndrome Properties in HL_n

Definition 2. Assume that σ be a comparison syndrome of $G \in HL_n$. If $\sigma((x, y)_w) = 0$, a vertex w is named as a σ -zero vertex, where x and y are adjacent to w. If $\sigma((x, y)_w) = 1)$, a vertex w is named as a σ -one vertex, where x and y are adjacent to w.

Definition 3. Assume that σ be a comparison syndrome of $G \in HL_n$. A path $P[x_0, x_k] = \langle x_0, x_1, \ldots, x_k \rangle$ in G is a σ-zero path if $\sigma((x_{i-1}, x_{i+1})_{x_i}) = 0$ for any three consecutive vertices with $1 \leq i \leq k - 1$.

According to previous definitions, we can use the following Lemma to detect all the vertices in a σ-zero path.

Lemma 1 [16]. *All vertices in a σ-zero path must be in the same state. The vertices in this path are either all fault-free vertices or all faulty vertices.*

Lemma 2 [3]. *Provided that there exists at most l faulty vertices in a path P and σ is a syndrome on this path. Let P^* denotes a σ-zero subpath of P and the length of this subpath is k. Then,*

1. *When $k > l$, we assume that all vertices in P^* must be fault-free. Meanwhile, the σ-one neighbors of the two end vertices in P^* are also fault-free.*
2. *When $k \leq l$, we assume that the σ-one neighbors of the two end vertices and all vertices in P^* are fault-free. Otherwise, we assume that all vertices in P^* are faulty.*

Lemma 3 [10]. *Every graph in HL_n for $n \geq 2$ is Hamiltonian.*

Since the diagnosability of an n-dimensional hypercube-like network is bounded by n [1] on the basis of the MM* model for $n \geq 5$, we assume that there are no more than t faulty vertices in an n-dimensional hypercube-like network with $t = n$. According to Lemma 2, a σ-zero subpath with more than t vertices can be located if the cardinality of the faulty set existing in an N-vertex cycle does not exceed t. Let $P[v_m, v_n] = \langle v_m, v_{m+1}, \ldots, v_{n-1}, v_n \rangle$ be a σ-zero subpath with $n - m + 1$ vertices and $n - m + 1 > t$. Then, according to Lemmas 1 and 2, we can identify vertices from v_{m-1} to v_{n+1} as fault-free vertices, and v_{m-2} and v_{n+2} as faulty vertices. If v_{m-2} or v_{n+2} is adjacent to another σ-zero subpath, then identify all vertices in this σ-zero subpath as faulty vertices.

3.2   A Precise Fault Diagnosis Algorithm

A precise fault diagnosis algorithm is described in this subsection.

Algorithm 1. Algorithm_HL

Input   : An integer $n \geq 5$, and a graph $G \in HL_n$, where 2^n is the cardinality of the vertex set in G.

Output: The state of all vertices are diagnosed as either fault-free or faulty.

Step 1: Construct a Hamiltonian cycle in HL_n.

Step 2: Obtain its cycle syndrome by $\sigma((x_{i-1}, x_{i+1})_{x_i})$ for $0 \leq i \leq 2^n - 1$.

Step 3: Let $P[v_m, v_n] = \langle v_m, v_{m+1}, \ldots, v_{n-1}, v_n \rangle$ be a σ-zero subpath with $n - m + 1$ vertices and $n - m + 1 > t$.

Step 4: Identify all vertices in $P[v_m, v_n]$ as fault-free, and v_{m-1}, v_{n+1} as fault-free and v_{m-2}, v_{n+2} as faulty.

Step 5: For each unknown σ-zero subpath, if it has a vertex x that has an identified fault-free neighbor w that in turn has an identified fault-free neighbor y, then diagnose $(x, y)_w$; if this σ-zero vertex x is identified as faulty, then identify other vertices in this σ-zero subpath as faulty.

---

4   Conclusion

For hypercube-like networks, when the cardinality of the vertex set is $N = 2^n$ for $n \geq 5$, we have designed an $O(N)$-time precise fault diagnosis algorithm on the basis of the MM* model. Moreover, our previous result presented in [15] an $N(\log_2 N)^2$-time algorithm for the hypercube-like networks for $n \geq 11$. Therefore, the proposed precise fault diagnosis algorithm can run in $O(N)$ time complexity for $n \geq 5$, which is superior to the previous algorithms for the class of hypercube-like networks.

References

1. Chiang, C.F., Tan, J.M.: A novel approach to comparison-based diagnosis for hypercube-like systems. J. Inf. Sci. Eng. **24**(1), 1–9 (2008)
2. Duarte Jr., E.P., Ziwich, R.P., Albini, L.C.: A survey of comparison-based system-level diagnosis. ACM Comput. Surv. **43**(22), 1–56 (2011)
3. Hsu, H.C., Lai, P.L.: Adaptive diagnosis for grids under the comparison model, 26th International Conference on Advanced Information Networking and Applications Workshops, pp. 152–159 (1987)
4. Kavianpour, A., Friedman, A.D.: Efficient design of easily diagnosable system. In: Proceedings of the 3th IEEE Computer Society's USA-Japan Computer Conference, pp. 173–181 (1978)
5. Khanna, S., Fuchs, W.K.: A linear time algorithm for sequential diagnosis in hypercubes. J. Parallel Distrib. Comput. **26**, 48–53 (1995)
6. Khanna, S., Fuchs, W.K.: A graph partitioning approach to sequential diagnosis. IEEE Trans. Comput. **46**(1), 39–47 (1997)

7. Maeng, J., Malek, M.: A Comparison connection assignment for self-diagnosis of multiprocessor systems. In: Proc. 11th International Symposium Fault-Tolerant Computing, pp. 173–175 (1981)
8. Park, J.H.: Panconnectivity and edge-pancyclicity of faulty recursive circulant $G(2^m, 4)$. Theoret. Comput. Sci. **390**(1), 70–80 (2008)
9. Park, J.H., Chwa, K.Y.: Recursive circulant: a new topology for multicomputer networks. In: Proceedings of Internation Symposium Parallel Architectures. Algorithms and Networks (ISPAN 1994), pp. 73–80. IEEE press, New York (1994)
10. Park, J.H., Kim, H.C., Lim, H.S.: Fault-hamiltonicity of hypercube-like interconnection networks. In: Proceedings of the 19th IEEE International Parallel and Distributed Prcessing Symposium (IPDPS 2005), Denver, CA, USA, IEEE Computer Society, April 2005
11. Preparata, F.P., Metze, G., Chien, R.T.: On the connection assignment problem of diagnosable systems. IEEE Trans. Comput. **16**(12), 448–454 (1967)
12. Sullivan, G.: An $O(t^3 + |E|)$ fault identification algorithm for diagnosable systems. IEEE Trans. Comput. **37**(4), 388–397 (1988)
13. Senqupta, A., Dahbura, A.T.: On self-diagnosable multiprocessor systems: diagnosis by the comparison approach. IEEE Trans. Comput. **41**(11), 1386–1396 (1992)
14. Vaidya, A.S., Rao, P.S.N., Shankar, S.R.: A class of hypercube-like networks. In: Proceedings of Fifth IEEE symposium Parallel and Distributed Processing(SPDP), pp. 800–803 (1993)
15. Ye, T.L., Hsieh, S.Y.: A scalable comparison-based diagnosis algorithm for hypercube-like networks. IEEE Trans. Reliab. **62**(4), 789–799 (2013)
16. Yang, X.F., Megson, G.M., Evans, D.J.: A comparison-based diagnosis algorithm tailored for crossed cube multiprocessor systems. Microprocess. Microsyst. **29**(4), 169–175 (2005)
17. Yang, C.L., Masson, G.M., Leonetti, R.A.: On fault isolation and identification in t_1/t_1-diagnosable systems. IEEE Trans. Comput. **35**(6), 639–643 (1986)
18. Zhao, J., Meyer, F.J., Park, N., Lombardi, F.: Sequential diagnosis of processor array systems. IEEE Trans. Reliab. **53**(4), 487–498 (2004)

Finding Disjoint Paths on Edge-Colored Graphs: A Multivariate Complexity Analysis

Riccardo Dondi[1] and Florian Sikora[2(✉)]

[1] Dipartimento di Lettere e Comunicazione,
Università degli Studi di Bergamo, Bergamo, Italy
riccardo.dondi@unibg.it
[2] Université Paris-Dauphine, PSL Research University,
CNRS, LAMSADE, Paris, France
florian.sikora@dauphine.fr

Abstract. The problem of finding the maximum number of vertex-disjoint uni-color paths in an edge-colored graph (MAXCDP) has been recently introduced in literature, motivated by applications in social network analysis. In this paper we investigate how the complexity of the problem depends on graph parameters (distance from disjoint paths and size of vertex cover), and that is not FPT-approximable. Moreover, we introduce a new variant of the problem, called MAXCDDP, whose goal is to find the maximum number of vertex-disjoint and color-disjoint uni-color paths. We extend some of the results of MAXCDP to this new variant, and we prove that unlike MAXCDP, MAXCDDP is already hard on graphs at distance two from disjoint paths.

1 Introduction

The analysis of social networks and social media has introduced several interesting problems from an algorithmic point of view. Social networks are usually viewed as graphs, where vertices represent the elements of the network, and edges represent a binary relation between the represented elements. One of the most relevant properties to analyze social network is the vertex connectivity of two given vertices. Indeed, a relevant property of social networks is how information flows from one vertex to the other, and vertex connectivity is considered as a measure of the information flow. Furthermore, two relevant structural properties of a social network, group cohesiveness and centrality, can be identified via vertex connectivity [8,12]. Vertex connectivity has been widely investigated in graph theory; Menger's theorem shows that vertex connectivity is equivalent to the maximum number of disjoint paths between two given vertices.

Usually social networks analyses focus on a single type of relation. However, due to the availability of several social networks, a natural goal is to integrate the information into a single network. Wu [13] introduced a model to consider multi-relational social networks, where different kinds of relations are considered. In the proposed model colors are associated with edges of the graph to distinguish different kinds of relations. Given such an edge-colored graph, a natural

© Springer International Publishing AG 2016
T.-H.H. Chan et al. (Eds.): COCOA 2016, LNCS 10043, pp. 113–127, 2016.
DOI: 10.1007/978-3-319-48749-6_9

combinatorial problem to compute vertex connectivity introduced in [13], called *Maximum Colored Disjoint Paths* (MAXCDP), asks for the maximum number of vertex-disjoint paths consisting of edges of the same colors (also called uni-color paths) in the input graph.

The complexity of MAXCDP has been investigated in [3,13]. MAXCDP is polynomial time solvable when the input graph contains exactly one color, while it is NP-hard when the edges of the graph are associated with at least two colors. On general instance, MAXCDP is shown to be not approximable with factor $O(n^d)$, where n is the number of vertices of the input graph, for any constant $0 < d < 1$, and W[1]-hard if the parameter if the number of paths in the solution [3]. Moreover, MAXCDP is approximable within factor q, where q is the number of colors of the edges of the input graph, but not approximable within factor $2 - \epsilon$, for any $\epsilon > 0$, when q is a fixed constant. Moreover, in [13] it is considered a variant of the problem where the length of the paths in the solution are (upper) bounded by an integer $l \geqslant 1$, as in many real social networks the diameter of the graph is bounded by a constant and we are interesting in short paths connecting two vertices. When $l \geqslant 4$ MAXCDP is NP-hard, while it is polynomial time solvable for $l < 4$ [13]. The bounded length variant of the problem is approximable within factor $(l - 1)/2 + \epsilon$ [13], and it is fixed-parameter tractable for the combined parameter number of paths in the solution and l [3]. Moreover, this variant does not admit a polynomial kernel unless $NP \subseteq coNP/Poly$, as it follows from the results in [7].

In this paper, we further investigate the complexity of MAXCDP and of a related problem that we introduce, called MAXCDDP, whose goal is to find the maximum number of vertex-disjoint and color-disjoint (that is having different colors) uni-color paths. The color disjointness of paths can be interesting to characterize how different relations in a network connects two vertices. In this case, we are not interested to have more paths of a single color, but rather to compute the maximum number of color-disjoint paths between two vertices. We study how the complexity of MAXCDP and MAXCDDP depends on several parameters, in the spirit of a multivariate complexity analysis [9]. As described in the previous paragraph, it has already been studied how the complexity of MAXCDP depends on different constraints (number of colors of each edge, maximum length of a path). We believe that it is interesting to take into account the structure of the input graph when studying the complexity of these two problems, since real-life networks exhibit properties that leads to graphs with a specific structure. For example, it is widely believed that social networks have a "small-world phenomenon" property, and thus information on the structure of the corresponding graphs can be derived. Moreover, such a study is also of theoretical interest, since it helps to better understand the complexity of the two problems.

First, we investigate how the complexity of the two problems depends on two graph parameters: distance from disjoint paths and size of vertex cover. In Sect. 3 we show that on graphs at distance bounded by a constant from disjoint paths MAXCDP admits a polynomial-time algorithm, whereas MAXCDDP is

NP-hard. Then, in Sect. 4 we show that MaxCDP is fixed-parameter tractable when parameterized by the size of the vertex cover of the input graph. In Sect. 5 we consider the parameterized complexity of the bounded length version of MaxCDDP, where the parameters are the number of vertex and color-disjoint paths of a solution and the maximum length of a path, and we extend the FPT algorithm for MaxCDP to MaxCDDP. Finally, we show in Sect. 6 that both problems are not ρ-approximable in FPT time, for any function ρ.

2 Definitions

In this section we present some definitions as well as the formal definition of the two combinatorial problems we are interested in. First, notice that in this paper, we will consider undirected graphs. Given a graph $G = (V, E)$ and a vertex $v \in V$, we denote by $N(v)$ the vertex adjacent to v in G. Consider a set of colors $C = \{c_1, \ldots, c_q\}$, where q represents cardinality of C. A C-edge-colored graph (or simply an edge-colored graph when the set of colors is clear from the context) is defined as $G = (V, E, f_C)$, where V denotes the set of vertices of G and E denotes the set of edges, and $f_C : E \rightarrow 2^C$ is a coloring of each edge with a set of colors in $C = \{c_1, \ldots, c_q\}$. In the paper, we denote by n the size of V and by m the size of E.

A path π in G is said to be colored by $c_j \in C$ if all the edges of π are colored by c_j. A path π in G is called a *uni-color* path if there is a color $c_j \in C$ such that all the edges of π are colored by c_j.

Given two vertices $x, y \in V$, an xy-path is a path between vertices x and y. Two paths π' and π'' are *internally disjoint* (or, simply, *disjoint*) if they do not share any internal vertex, while a set P of paths is internally disjoint if the paths in P are pairwise internally disjoint. Two uni-color paths π' and π'' are *color disjoint* if they are disjoint and they have different colors.

Next, we introduce the formal definitions of the problems we deal with.

Max Colored Disjoint Path (MaxCDP)

- **Input**: a set C of colors, a C-edge-colored graph $G = (V, E, f_C)$ and two vertices $s, t \in V$.
- **Output**: the maximum number of disjoint uni-color st-paths.

Max Colored Doubly Disjoint Path (MaxCDDP)

- **Input**: a set C of colors, a C-edge-colored graph $G = (V, E, f_C)$, and two vertices $s, t \in V$.
- **Output**: the maximum number of color disjoint uni-color st-paths.

We will consider a variant of the two problems where the length of the paths in the solution is (upper) bounded by an integer $l \geqslant 1$, that is we are interested only in paths bounded by l. These variants will be denoted by l-MaxCDP and l-MaxCDDP.

Notice that in the versions of these problems parameterized by the natural parameter, we are also given an integer $k > 0$ and we look whether there exists at least k (color) disjoint uni-color st-paths.

A parameterized problem (I, k) is said *fixed-parameter tractable* (or in the class FPT) with respect to a parameter k if it can be solved in $f(k) \cdot |I|^c$ time (in *fpt-time*), where f is any computable function and c is a constant. The class XP contains problems solvable in time $|I|^{f(k)}$, where f is an unrestricted function. We defer the reader to the recent monographs of Downey and Fellows or Cygan et al. for additional information around parameterized complexity [5,6].

The natural notion of *parameterized approximation* was introduced quite recently (see the survey of Marx for an overview [10]). Informally, it aims at giving more time than polynomiality to achieve better approximation ratio. We give the definition of fpt cost ρ-approximation algorithm, as in Sect. 6 we will rule out the existence of such an algorithm for MAXCDP and MAXCDDP. This is a weaker notion than fpt-approximation, but notice that we will prove negative result (which will thus be stronger).

An NP-optimization problem Q is a tuple $(\mathcal{I}, Sol, val, goal)$, where $\mathcal{I}$ is the set of instances, $Sol(I)$ is the set of feasible solutions for instance I, $val(I, S)$ is the value of a feasible solution S of I, and $goal$ is either max or min.

Definition 1 (fpt cost ρ-approximation algorithm, Chen et al. [4]). *Let Q be an optimization problem and $\rho\colon \mathbb{N} \to \mathbb{R}$ be a function such that $\rho(k) \geqslant 1$ for every $k \geqslant 1$ and $k \cdot \rho(k)$ is nondecreasing (when goal = min) or $\frac{k}{\rho(k)}$ is unbounded and nondecreasing (when goal = max). A decision algorithm $\mathcal{A}$ is an fpt cost ρ-approximation algorithm for Q (when ρ satisfies the previous conditions) if for every instance I of Q and integer k, with $Sol(I) \neq \emptyset$, its output satisfies the following conditions:*

1. *If $opt(I) > k$ (when goal = min) or $opt(I) < k$ (when goal = max), then $\mathcal{A}$ rejects (I, k).*
2. *If $k \geqslant opt(I) \cdot \rho(opt(I))$ (when goal = min) or $k \leqslant \frac{opt(I)}{\rho(opt(I))}$ (when goal = max), then $\mathcal{A}$ accepts (I, k).*

Moreover the running time of $\mathcal{A}$ on input (I, k) is $f(k) \cdot |I|^{O(1)}$. If such a decision algorithm $\mathcal{A}$ exists then Q is called fpt cost ρ-approximable.

3   Complexity of MaxCDDP and MaxCDP on Graphs at Bounded Distance from Disjoint Paths

In this section we consider the complexity of MAXCDP and MAXCDDP on graphs having distance bounded by a constant from disjoint paths. The distance to disjoint paths is the minimum number of vertices to remove to make the graph a set of disjoint paths.

We show that MAXCDDP is NP-hard for graphs at distance two from disjoint paths, while MAXCDDP is polynomial time solvable when the input graph has distance bounded by a constant from disjoint paths.

3.1 Complexity of MaxCDDP on Graphs at Distance Two from Disjoint Paths

In this section, we show that if the input graph G has distance two from a set of disjoint path, then MaxCDDP is NP-hard.

We give a reduction from Maximum Independent Set on Cubic graphs (MaxISC)[1]. We recall the definition of MaxISC:

MAXIMUM INDEPENDENT SET ON CUBIC GRAPHS (MaxISC)

- **Input**: a cubic graph $G_I = (V_I, E_I)$.
- **Output**: a subset $V_I' \subseteq V_I$ of maximum cardinality, such that for each $v_x, v_y \in V_I'$ it holds $\{v_x, v_y\} \notin E$

We build a graph $G = (V, E, f_C)$ from $G_I = (V_I, E_I)$ by defining a gadget GV_i for each vertex $v_i \in V_I$, and connecting the gadget to vertices s and t.

Given $v_i \in V_I$, define a gadget GV_i consisting of a set V_i of 4 vertices (see Fig. 1): $V_i = \{v_i', v_{i,j}' : v_i \in V_I, 1 \leqslant j \leqslant 3\}$.

Moreover, define the set C of colors as follows: $C = \{c_i : v_i \in V_I\} \cup \{c_{i,j} : \{v_i, v_j\} \in E_I\}$.

We assume that, given a vertex v_i, the vertices adjacent to v_i (that is the vertices in $N(v_i)$) are ordered, i.e. if $v_j, v_h, v_z \in N(v_i)$ with $1 \leqslant j \leqslant h \leqslant z$, then v_j is the first vertex adjacent to v_i, v_h is the second and v_z is the third.

We define the edges of G be means of the following paths:

- a path colored c_i that consists of s, v_i', $v_{i,1}'$, $v_{i,2}'$, $v_{i,3}'$, t, with $1 \leqslant i \leqslant |V_I|$
- if, according to the ordering, v_j is the p-th vertex incident on v_i, $1 \leqslant p \leqslant 3$, then there exists a path colored $c_{i,j}$ that passes through s, $v_{i,p}'$, t

First, we prove that the graph G has distance two from disjoint paths.

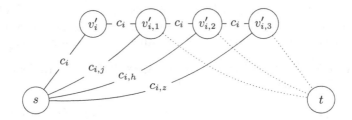

Fig. 1. Gadget GV_i associated with vertex $v_i \in V_I$. Vertices v_j, v_h, v_z are three vertices of V_I, with $N(v_i) = \{v_j, v_h, v_z\}$ and $j < h < z$. v_j is the first vertex adjacent to v_i in G_I and thus there exists a path in G colored by $c_{i,j}$ that passes through s, $v_{i,1}'$, t; v_h is the second vertex adjacent to v_i in G_I (hence there exists a path in G colored by $c_{i,h}$ that passes through s, $v_{i,2}'$, t); v_z is the third vertex adjacent to v_i in G_I (hence there exists a path in G colored by $c_{i,z}$ that passes through s, $v_{i,3}'$, t).

---

[1] A graph is cubic when each of its vertices has degree 3.

Lemma 2. *Given a cubic graph G_I, let G be the corresponding graph input of* MaxCDDP. *Then G has distance two from disjoint paths.*

Proof. After the removal of s and t, the paths left in the resulting graph are the paths colored by c_i, with $1 \leqslant i \leqslant |V_I|$, that pass through v'_i, $v'_{i,1}$, $v'_{i,2}$, $v'_{i,3}$. Since these paths are pairwise vertex disjoint, the lemma holds.   □

Next, we prove the main results of the reduction.

Lemma 3. *Let G_I be a cubic graph and G be the corresponding graph input of* MaxCDDP. *Given an independent set V'_I of G_I, then we can compute in polynomial time $|E| + |V'_I|$ disjoint uni-color color paths in G.*

Proof. Consider an independent set $V'_I \subseteq V_I$ of G_I, we define a set $\mathcal{P}$ of uni-color disjoint paths as follows. $\mathcal{P}$ contains a path $s, v'_i, v'_{i,1}, v'_{i,2}, v'_{i,3}, t$ colored by c_i, for each $v_i \in V'_I$. Moreover, for each $\{v_i, v_j\} \in E_I$, assume w.l.o.g. that $v_i \in V_I \setminus V'_I$ and that v_j is the h-th vertex, $1 \leqslant h \leqslant 3$, adjacent to v_i. Then $\mathcal{P}$ contains the path $s, v'_{i,h}, t$ colored by $c_{i,j}$. Notice that these paths, since V'_I is an independent set, are by construction color disjoint.   □

Lemma 4. *Let G_I be a cubic graph and G be the corresponding graph input of* MaxCDDP. *Given $|E| + t$ color disjoint uni-color paths in G, we can compute in polynomial time an independent set of size t for G_I.*

Proof. Consider a solution $\mathcal{P}$ of the instance of MaxCDDP consisting of $|E| + t$ color disjoint uni-color paths. First, we can assume that $\mathcal{P}$ contains, for each color $c_{i,j}$, a path colored by $c_{i,j}$. Assume this is not the case. Then, we can replace a path colored by c_i or a path colored by c_j with a path p' colored by $c_{i,j}$ that passes through the vertices of gadget VG_i or VG_j, without decreasing the number of path in $\mathcal{P}$. Moreover, notice that by replacing a path color c_i with p', the set $\mathcal{P}$ still contains color disjoint uni-color paths.

Now, starting from $\mathcal{P}$, we can compute an independent set V'_I as follows. If $\mathcal{P}$ contains a path $s, v'_i, v'_{i,1}, v'_{i,2}, v'_{i,3}, t$ colored by c_i, then $v_i \in V'_I$. Notice that V'_I is an independent set, since, if v_i, v_j, with $\{v_i, v_j\} \in E$, are both in V'_I, this implies that there is no $c_{i,j}$-path in $\mathcal{P}$, contradicting our assumption.   □

Hence, we can prove the NP-hardness of MaxCDDP on graphs at distance two from disjoint paths.

Theorem 5. MaxCDDP *is NP-hard, even if the graph G has distance two from disjoint paths.*

Proof. MaxISC is NP-hard [1]. Hence Lemmas 2, 3 and 4 imply that MaxCDDP is NP-hard, even if the graph G has distance two from disjoint paths.   □

The previous result implies that MaxCDDP cannot be solved in $n^{f(k)}$ time unless $P = NP$ (it is not in the class XP), where k is the distance to disjoint paths of G, but also for "stronger" parameters like pathwidth or treewidth [9]

3.2 A Polynomial-Time Algorithm for MaxCDP on Graphs at Constant Distance from Disjoint Paths

In this section, we show that, contrary to MaxCDDP, MaxCDP is polynomial-time solvable when the input graph G has distance bounded by a constant d from a set $\mathcal{P}$ of disjoint paths (that is, it is in the class XP for the parameter distance to disjoint paths).

Next, we present the algorithm. Notice that we assume that a set $X \subseteq V$ is given, such that after the removal of $X \cup \{s, t\}$ the resulting graph consists of a set $\mathcal{P}$ of disjoint paths[2]. We assume that X and $\mathcal{P}$ are defined so that $s, t \notin X$ and that no path in $\mathcal{P}$ contains s and t. Denote by $V(\mathcal{P})$ the set of vertices that belong to a path of $\mathcal{P}$, it holds $V = V(\mathcal{P}) \cup X \cup \{s, t\}$.

Since G has distance d, where $d > 0$ is constant, from the set of disjoint paths $\mathcal{P}$, it follows that $|X| \leqslant d$. Let $\mathcal{P}' = \{p_1, \ldots, p_b\}$, with $1 \leqslant i \leqslant b \leqslant d$, such that $V(\mathcal{P}') \subseteq V$ is the set of paths of an optimal solution of MaxCDP such that p_i contains a non-empty subset of X.

The algorithm computes each p_i, with $1 \leqslant i \leqslant b$, by iterating through sub-paths of size at most d in $\mathcal{P}$ and a subset of X. More precisely, p_i is computed as follows. Each path p_i contains at most $d + 1$ disjoint subpaths that belong to paths in $\mathcal{P}$, that are connected through a subset of at most d vertices of X. In time $O(n^{2(d+1)})$, we compute the at most $d + 1$ disjoint subpaths $p_x[j_1, j_2]$ of $P_x \in \mathcal{P}$ that belong to p_i; in time $O(2^d)$ we compute the subset $X_i \subseteq X$ that belong to each p_i. Let $V_i = V(p_i) \cup X_i$, that is the set of vertices that belong to p_i and to subset X_i. Notice that the subsets V_i, with $1 \leqslant i \leqslant b$, are computed so that they are pairwise disjoint.

The algorithm computes in polynomial time if there exists a uni-color path from s to t that passes through the vertices V_i. If for each i with $1 \leqslant i \leqslant b$ such a path exists, then the algorithm computes the maximum number of uni-color disjoint paths in the subgraph G' of G induced by $V' = V \setminus \bigcup_{i=1}^{b} V_i$. Notice that, since $V' \cap X = \emptyset$, it follows that, if we remove s and t from V', G' consists of a set of disjoint paths $\{p'_1, \ldots, p'_r\}$. The maximum number of uni-color disjoint paths in the subgraph G' can be computed in polynomial time, as shown in the following lemma.

Lemma 6. *Let $G = (V, E, f_c)$ be an edge colored graph such that $V^* = V \setminus \{s, t\}$ induces a set of disjoint paths. Then MaxCDP on G can be solved in polynomial time.*

Proof. Let $P = \{p'_1, \ldots, p'_r\}$ be the set of disjoint paths induced by V^*. Since there is no st-path in G containing a vertex of p'_i and a vertex of p'_j, with $i \neq j$, we compute a solution of MaxCDP independently on each path p'_i. Let $\mathcal{P}_i$ be the set of uni-color st-paths that contains only vertices of p'_i. For each i with $1 \leqslant i \leqslant r$, we compute a shortest uni-color st-path p that contains only vertices of p'_i, we add it to $\mathcal{P}_i$, and we remove the vertices of p from p'_i. We iterate this procedure, until there exists no st-path that contains only vertices of p'_i.

---

[2] Notice that, since $|X| \leqslant d$, X can be computed in time $O(n^d)$.

We claim that $\mathcal{P}_i$ is a set of uni-color st-paths of maximum size. Consider a shortest path p added to $\mathcal{P}_i$. Let x be the vertex of p adjacent to s and y be the vertex of p adjacent to t. Notice that each vertex in p, except for x and y, is not connected to s or t, otherwise p would not be a shortest path between s and t. Now, assume that there is an optimal solution $\mathcal{Q}$ of MaxCDP that does not contain p and that, moreover, contains an st-path that passes through some vertex of p, otherwise we can add p to $\mathcal{Q}$ and $\mathcal{Q}$ would not be optimal. Then by construction, since P is a set of disjoint paths, $\mathcal{Q}$ must contain a path p' that contains p as a subpath. But then we can replace p' with p in $\mathcal{Q}$, without decreasing the size of the optimal solution. □

Now, we give the main result of this section.

Theorem 7. MaxCDP *is in XP when the distance to disjoint paths is bounded by a constant.*

Proof. Notice that for each i with $1 \leqslant i \leqslant b \leqslant d$, we compute the set V_i in time $O(2^d n^{2(d+1)})$; hence the d disjoint sets $V_1, \ldots, V_b$ are computed in time $O(2^{d^2} n^{2d(d+1)})$. Since the existence of a uni-color path that passes through the vertices V_i can be computed in polynomial time and since by Lemma 6 we compute in polynomial time the maximum number of uni-color disjoint paths in the subgraph G', the theorem holds. □

4   FPT Algorithm Parameterized by Vertex Cover for MaxCDP

In this section, we will show that MaxCDP is FPT when parameterized by the size of the vertex cover of the input graph.

Theorem 8. MaxCDP *is in FPT when parameterized by the size of the vertex cover of the input graph.*

Proof. First, consider uni-color paths of length three s, v, t, for some $v \in V$. Uni-color path of length three are greedily added to a solution of MaxCDP. Since any solution of MaxCDP contains at most one uni-color path that passes through v, it follows that there exists an optimal solution of MaxCDP that contains path s, v, t. Hence, we add such path to our solution $\mathcal{P}$, and we remove vertex v from G.

Let V' be a vertex cover of the resulting graph $G = (V, E, f_C)$, $|V'| = k$ (which can be computed in FPT-time). Since in G there is no uni-color path of length three connecting s and t, the following property holds. Consider a uni-color path p of G, then p either consists of vertices in V' or each vertex of $V \setminus V'$ that belongs to p is adjacent in p to vertices of $V' \cup \{s\} \cup \{t\}$. This is true since V' is a vertex cover (and thus $V \setminus V'$ is an independent set in G).

A consequence of this property is that each uni-color path has length at most $2k$. Moreover, there can be at most k uni-color paths in a solution (since each path must contain a vertex of V' and $|V'| \leqslant k$). Since both the number of paths and the length of paths are bounded by k and MaxCDP is known to be in FPT w.r.t. the combination of these two parameters [3], the claimed result follows. $\square$

This algorithm does not easily extend to MaxCDDP. The main difference between MaxCDDP and MaxCDP, when considering as parameter the vertex cover of the input graph, is that in the latter we can safely add a uni-color path s, v, t of length three to a solution, while in the former we are not allowed to do it. Consider for example the uni-color path s, v, t of length three colored by c; if this path belongs to a solution of MaxCDDP, it prevents any other uni-color path p' that passes through v (colored by some color c'), but also any path p'' colored by c (that does not pass through v) to be part of the solution. So, by adding the path s, v, t to the solution we are computing, we may get a suboptimal solution, since by removing p and by adding p' and p'', we possibly compute a larger set of disjoint color uni-color paths.

5   A Fixed-Parameter Algorithm for l-MaxCDDP

In this section, we give a fixed-parameter algorithm for l-MaxCDDP, the length-bounded version of MaxCDDP, parameterized by the number k of uni-color color disjoint st-paths of a solution. Notice that MaxCDDP is W[1]-hard when parameterized by k, as the reduction that prove the W[1]-hardness of MaxCDP parameterized by k consists of paths having distinct colors [3].

Next, we present a parameterized algorithm based on the *color coding* technique [2]. The algorithm is inspired by the one for MaxCDP [3]. However, in this case we must combine two different labelings, one to label the vertices that belong to a uni-color path, one to label the color associated with a uni-color path of MaxCDDP.

First, we introduce the definition of perfect hash function on which our algorithm is based. A family F of hash functions from a set U (the vertex set in the traditional applications of color coding) to the set $\{l_1, \ldots, l_k\}$ of labels is *k-perfect* if, for each subset U' of U with $|U'| = k$, there exists a hash function $f \in F$ such that f assigns a distinct label to each element of U'. Function f is called a *labelling function*.

Let $f_v \in F_V$ be a labelling function that assigns to each vertex $v \in V \setminus \{s, t\}$ a label $f_v(v) \in L_v = \{1_v, \ldots, h_v\}$, where $h_v = |L_v| \leqslant lk$.

Consider a second labelling function $f_c \in F_C$ that assigns to each color $c \in C$ a label $f_c(v) \in L_c = \{1_c, \ldots, h_c\}$, where $h_c = |L_c| \leqslant k$.

By the property of perfect hash functions, we assume that each vertex that belongs to a solution of MaxCDDP is assigned a distinct label by f_v and that each color d, such that there exists a uni-color path of MaxCDDP colored by d, is associated with distinct label f_c.

A simple path p in G is *perfect* for a set L_v of labels assigned to V if and only if for each vertex v of p, with $v \notin \{s, t\}$, $f_v(v) \in L_v$, and for each pair of distinct vertices u, v of p, $f_v(u) \neq f_v(v)$. A set $\{p_1, \ldots, p_k\}$ of uni-color paths is *perfect* for the set L_v and L_c of labels if and only if: (1) there exists a partition $\{L_{v,1}, \ldots, L_{v,k}\}$ of L_v such that each p_i is perfect for L_i; (2) each path p_i, with $1 \leqslant i \leqslant k$, is colored by $c \in C$ associated with a distinct label in L_c. We combine two dynamic-programming recurrences to compute, given the labelling functions f_v and f_c, whether there exists a set of perfect uni-color paths in G.

First, consider the function $S[L'_v, u, \lambda]$, with $L'_v \subseteq L_v$, $u \in V$ and $\lambda \in C$. $S[L'_v, u, \lambda] = 1$ if and only if there exists a path from s to vertex $u \neq t$, such that the path is perfect for L'_v and p is colored by λ.

We consider a second function $\Pi[L'_v, M, z]$, with $L'_v \subseteq L_v$ and $M \subseteq L_c$, $0 \leqslant z \leqslant k$. $\Pi[L'_v, M, z] = 1$ if and only if there exist a set of labels $L'_v \subseteq L_v$ and a set of labels $M \subseteq L_c$, such that there exists a set of z uni-color paths perfect for L'_v and M.

$S[L'_v, u, \lambda]$ is defined as follows (we recall that $\uplus$ represents the disjoint union operator). In the base case, when $u = s$, $S[L'_v, u, \lambda] = 1$ if $L'_v = \emptyset$, otherwise (when $L'_v \neq \emptyset$), $S[L'_v, u, \lambda] = 0$.

When $u \neq s$:

$$S[L'_v, u, \lambda] = \max_{w \in N(u)} \left\{ S[L''_v, w, \lambda] \mid L'_v = L''_v \uplus \{f_v(u)\} \wedge \{w, u\} \text{ is colored by } \lambda \right\}$$

Next, we give the recurrence, $\Pi[L'_v, M, z]$. In the base case, that is when $z = 0$, then $\Pi[L'_v, M, 0] = 1$ if $L'_v = \emptyset$ and $M = \emptyset$, else $\Pi[L'_v, M, 0] = 0$. Recall that l is the bound on the length of each path, $L'_v \subseteq L_v$, $M \subseteq C$ and $0 \leqslant z \leqslant k$, $\Pi[L'_v, M, z]$ is defined as follows:

$$\Pi[L'_v, M, z] = \begin{cases} \max \left\{ \Pi[L''_v, M \setminus \{f_c(\lambda)\}, z-1] \ \wedge \ S[L^*_v, u, \lambda] \mid \right. \\ \qquad L'_v = L''_v \uplus L^*_v \wedge \ |L^*| \leqslant l-1 \ \wedge \ \lambda \in C \wedge f_c(\lambda) \in M \ \wedge \\ \qquad \left. \{u, t\} \in E \text{ is colored by } \lambda \right\} \end{cases}$$

$$(1)$$

Next, we prove the correctness of the two recurrences.

Lemma 9. *Given a labelling f_v of the vertices of G, a color $\lambda \in C$, a vertex u and a set $L'_v \subseteq L_v$, there exists a simple path p in G from s to u perfect for L'_v if and only if $S[L'_v, u, \lambda] = 1$.*

Proof. We prove the lemma by induction on the length of the path p. First, we consider the base case, that is $u = s$. Since s is not associated with a label in L_v, it holds $S[L'_v, u, \lambda] = 1$ if and only if $L'_v = \emptyset$.

Consider now the general case and assume that there exists a path p in G perfect for L'_v such that p is colored by λ. Consider the last vertex u of p, and let w be the vertex adjacent to u in p. By induction hypothesis, it follows that $S[L''_v, w, \lambda] = 1$, where $L'_v = L''_v \uplus \{f_v(u)\}$. By the definition of the recurrence, then $S[L'_v, u, \lambda] = 1$.

Assume that $S[L'_v, u, \lambda] = 1$. By the definition of the recurrence it holds that $S[L''_v, w, \lambda] = 1$, where $L'_v = L''_v \uplus \{f_v(u)\}$ and there is an edge $\{u, w\} \in E$

colored by λ. By induction hypothesis, since $S[L''_v, w, \lambda] = 1$, there exists a path p' from s to w perfect for L''_v such that p' is colored by λ. Since $\{u, w\} \in E$ is colored by λ, it follows that there exists a simple path p in G from s to u perfect for L'_v. $\qquad\square$

Lemma 10. *Given a labelling f_v of the vertices of G and a labelling f_c of the set C of colors, a set $L'_v \subseteq L_v$, a set $M \subseteq L_c$, and integer z with $0 \leqslant z \leqslant k$, there exists a set $\{p_1, \ldots, p_z\}$ of uni-color paths which is perfect for L'_v and M if and only if $\Pi[L'_v, M, z] = 1$.*

Proof. We prove the lemma by induction on the number of uni-color paths. First, we consider the base case, that is $z = 0$. Then there is no uni-color path perfect for $L'_v = \emptyset$ and $M = \emptyset$ if and only if $\Pi[\emptyset, \emptyset, 0] = 1$.

Consider now that there exist z disjoint color uni-color paths. Consider one of such paths, denoted by p, which is colored by λ and whose vertices are associated with set of labels L^*_v and such that the vertex of p adjacent to t is u, hence $\{u, t\} \in E$ is colored by λ. Then, by Lemma 9 $S[L^*_v, u, \lambda] = 1$. Moreover, by induction hypothesis it holds $\Pi[L''_v, M \setminus \{f_c(\lambda)\}, z - 1] = 1$, where $L'_v = L^*_v \uplus L''_v$ and $f_c(\lambda) \in M$. Hence, by the definition of the recurrence for Π, it holds $\Pi[L'_v, M, z] = 1$.

Consider the case that $\Pi[L'_v, M, z] = 1$. By the definition of function Π, it follows that there exists a color $\lambda \in C$, with $f_c(\lambda) \in M$, and a set of labels $L^*_v \subseteq L_v$, such that $\Pi[L''_v, M \setminus \{f_c(\lambda)\}, z - 1] = 1$, where $L'_v = L''_v \uplus L^*_v$, and $S[L^*_v, u, \lambda] = 1$. By induction hypothesis, since $\Pi[L''_v, M \setminus \{f_c(\lambda)\}, z - 1] = 1$, there exists a set P' of $z - 1$ paths perfect for the sets L''_V and $M \setminus \{f_c(\lambda)\}$. By Lemma 9 there exists a path p' from s to u colorful for L''_v and that has color λ. Moreover, since $\Pi[L'_v, M, z] = 1$, $\{u, t\} \in E$ is colored by λ. By the property of labelling f_c, no path of P' has label $f_c(\lambda)$, hence $P' \cup p$ is perfect for L'_v and M. $\qquad\square$

We can now state the main result.

Theorem 11. *l-MaxCDDP can be solved in time $2^{O(lk)}poly(n)$.*

Proof. An optimal solution of l-MaxCDDP consisting of k color disjoint paths exists if and only if $\Pi[L_v, M, k] = 1$. The correctness of the recurrence to compute Π follows from Lemma 10. Now, we discuss the time complexity to compute $\Pi[L'_v, M, z]$ and $S[L'_v, u, \lambda]$. First, consider $S[L'_v, u, \lambda]$. It consists of $2^{lk}nq$ entries and each entry can be computed in time $O(n)$, as we consider each vertex $w \in N(u)$.

Now, consider $\Pi[L'_v, M, z]$. It consists of $2^{k(l+1)}k$ entries. In order to compute $\Pi[L'_v, M, z]$, at most $2^{kl}k$ entries must be considered, since $\Pi[L^*_v, M \setminus \{f_c(v)\}, z - 1]$ is considered, where we have 2^{kl} subsets $L^*_v \subseteq L'_v$ and k labels $f_c(v)$. Given two labelling functions f_v and f_c, the time complexity to compute the entries $\Pi[L'_v, M, z]$ is $O(2^{k(2l+1)}kn)$. By the property of color-coding [2], a function $f_v \in F_v$ and a function $f_v \in F_v$ can be computed in time $2^{O(lk)}poly(n)$ and $2^{O(k)}poly(n)$, respectively, hence in time $2^{O(lk)}poly(n)$. $\qquad\square$

6   FPT Inapproximation

Since MAXCDP and MAXCDDP are hard to approximate in poly-time and do not admit fixed-parameterized algorithm for parameter number of paths, it is worth to investigate approximation in FPT time, i.e. find approximate solution with additional time. Unfortunately, in this section, we show that both MAXCDP and MAXCDDP do not admit an FPT cost ρ-approximation, for any function ρ of the optimum, unless FPT = W[1]. We will show the result by giving a reduction from the THRESHOLD SET problem. Marx [11] showed that the THRESHOLD SET problem does not admit a fpt cost ρ-approximation, for any function ρ of the optimum, unless FPT = W[1].

First, we introduce the definition of the THRESHOLD SET problem.

THRESHOLD SET

• **Input**:   a set U of elements, a collection $\mathcal{S} = \{S_1, \ldots, S_q\}$ of subsets of U and a positive integer weight $w(S_i)$ for each $S_i \in \mathcal{S}$, with $1 \leqslant i \leqslant q$.
• **Output**: a set $T \subseteq U$ of maximum cardinality such that $|T \cap S_i| \leqslant w(S_i)$ for every $S_i \in \mathcal{S}$.

The cost of a solution of THRESHOLD SET is denoted by $|T|$. Notice that this problem can be seen as a generalization of the INDEPENDENT SET problem; indeed, for a graph $G = (V, E)$, we can define $U = V$, $\mathcal{S} = E$ and $w(S) = 1$ for every set $S \in \mathcal{S}$.

We will reduce THRESHOLD SET to MAXCDP in polynomial time such that there is a "one-to-one" correspondence between the solutions of the two problem, therefore the inapproximability result transfers to MAXCDP, and then to MAX-CDDP. The reduction is inspired by the one in [3], that shows inapproximability in polynomial time and W[1]-hardness of MAXCDP.

First, we design the reduction for the MAXCDP problem. Notice that we assume that we are given an ordering over the sets in $\mathcal{S}$ (i.e. $S_i < S_j, i < j$). Consider an instance $(U, \mathcal{S}, w)$ of THRESHOLD SET, we define a corresponding instance $(G = (V, E, f_C), s, t)$ of MAXCDP. The set V of vertices is defined as follows:

$$V = \{s, t\} \cup \{s_i | i \in [|U|]\} \cup \{S_i^j | i \in [|\mathcal{S}|], 1 \leqslant j \leqslant w(S_i)\}$$

The set of colors C is defined as follows: $C = \{c_i : i \in U\}$.

Now, we define the set E of edges.

– for all $i \in [|U|]$, define an edge $\{s, s_i\}$ colored by c_i and an edge $\{s_i, S_q^j\}$ colored by c_i, for all $1 \leqslant j \leqslant w(S_q)$, where q is the smallest index of a set $S_q \in \mathcal{S}$ such that $i \in S_q$,
– from each S_q^j, define an edge $\{S_q^j, S_{q'}^{j'}\}$ colored by c_i, for all $1 \leqslant j' \leqslant w(S_{q'})$, such that $i \in S_q$, $i \in S_{q'}$, $q' > q$ and, for each $q < l < q'$, it holds $i \notin S_l$,
– from each S_q^j, define an edge $\{S_q^j, t\}$ colored by c_i, where $i \in S_q$ and for each $q' > q$ with $S_{q'} \in \mathcal{S}$, $i \notin S_{q'}$.

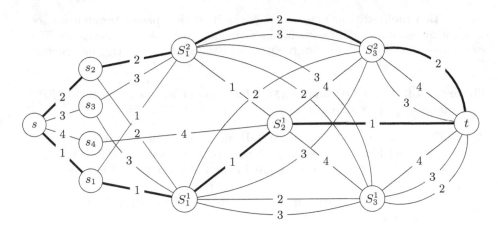

Fig. 2. Sample construction of an instance of MAXCDP from an instance of THRESH-OLD SET with $\mathcal{S} = \{\{1,2,3\},\{1,4\},\{2,3,4\}\}, w(S_1) = 2, w(S_2) = 1, w(S_3) = 2$. A solution for this instance of THRESHOLD SET could be $T' = \{1,2\}$, and we drawn with dark edges the corresponding disjoint paths for MAXCDP.

See Fig. 2 for an example. Now, we prove the main properties of the reduction.

Lemma 12. *Given an instance $(U, \mathcal{S}, w)$ of* THRESHOLD SET, *let $(G = (V, E, f_C), s, t)$ be the corresponding instance of* MAXCDP. *Then, given a solution T' of* THRESHOLD SET *on instance $(U, \mathcal{S}, w)$, we can compute in polynomial time a set of $|T'|$ disjoint uni-color paths in $(G = (V, E, f_C), s, t)$.*

Proof. Consider a solution T' of THRESHOLD SET on instance $(U, \mathcal{S}, w)$, and define a set P of $|T'|$ disjoint uni-color paths in $(G = (V, E, f_C), s, t)$ as follows. For each $i \in T'$, define a uni-color path p colored by c_i that starts in s, passes through s_i, and for each $S_q \in \mathcal{S}$, if i is the j-th element of T' in S_q, $1 \leqslant j \leqslant w(S_q)$, passes through vertex S_q^j. It follows that the path defined are disjoints, as at most one element can be the j-th element of T' in S_q and $|T' \cap S_q| \leqslant w(S_q)$. $\square$

Lemma 13. *Given an instance $(U, \mathcal{S}, w)$ of* THRESHOLD SET, *let $(G = (V, E, f_C), s, t)$ be the corresponding instance of* MAXCDP. *Then, given a set of q disjoint uni-color paths in $(G = (V, E, f_C), s, t)$, we can compute in polynomial time a solution of size q of* THRESHOLD SET *on instance $(U, \mathcal{S}, w)$.*

Proof. Consider a set P of disjoint uni-color paths in $(G = (V, E, f_C), s, t)$. First, we claim that each path in P has a distinct color. Indeed, the paths in P must be disjoint and, by construction, for each color c_i each path must pass trough vertex s_i.

Now, starting from P, we define a solution T' of of THRESHOLD SET on instance $(U, \mathcal{S}, w)$. For each path $p \in P$ colored by c_i, elements u_i belongs to T'. We show that T' is a a solution of THRESHOLD SET on instance $(U, \mathcal{S}, w)$.

Consider a set $S_i \in \mathcal{S}$, then there exists a most $w(S_i)$ elements in T'. Indeed, notice that, by construction, there exists at most $w(S_i)$ vertices S_i^j, hence by construction there exist at most $w(S_i)$ paths in P that passes through vertices of S_i^j, hence at most $w(S_i)$ elements in T' belong to S_i. As a consequence T' is a feasible solution of THRESHOLD SET on instance (U, S, w). By construction, $|T'| = q$.   □

Theorem 14. MAXCDP *and* MAXCDDP *cannot be approximated in FPT-time within any function ρ of the optimum, unless FPT = W[1].*

Proof. The theorem holds for MAXCDP since THRESHOLD SET cannot be approximated within any function ρ of the optimum, unless FPT = W[1] [11], and from the properties of the polynomial time reduction proved in Lemmas 12 and 13.

For MAXCDDP, it holds from the fact that in the described reduction all the paths have a distinct color.   □

7   Conclusion

In this paper, we continued the complexity analysis of MAXCDP and deepen the hardness analysis according to the structure of the input graph. We also introduced a new variant, called MAXCDDP, asking for a solution with vertex and disjoint colors.

In the future, we would like to further deepen the analysis on the the structural complexity of MAXCDP and MAXCDDP. For example is MAXCDP in XP when the parameter if the size of the Feedback Vertex Set of the input graph? Is MAXCDP FPT when the parameter if the distance to disjoint paths of the input graph? We would also like to improve the running time of our algorithms and to match them with some lower bounds under widely believed assumptions in order to have a fine-grained complexity analysis of these problems.

References

1. Alimonti, P., Kann, V.: Some APX-completeness results for cubic graphs. Theor. Comput. Sci. **237**(1–2), 123–134 (2000)
2. Alon, N., Yuster, R., Zwick, U.: Color-coding. J. ACM **42**(4), 844–856 (1995)
3. Bonizzoni, P., Dondi, R., Pirola, Y.: Maximum disjoint paths on edge-colored graphs: approximability and tractability. Algorithms **6**(1), 1–11 (2013)
4. Chen, Y., Grohe, M., Grüber, M.: On parameterized approximability. In: Bodlaender, H.L., Langston, M.A. (eds.) IWPEC 2006. LNCS, vol. 4169, pp. 109–120. Springer, Heidelberg (2006). doi:10.1007/11847250_10
5. Cygan, M., Fomin, F.V., Kowalik, L., Lokshtanov, D., Marx, D., Pilipczuk, M., Pilipczuk, M., Saurabh, S.: Parameterized Algorithms. Springer, Cham (2015)
6. Downey, R.G., Fellows, M.R.: Fundamentals of Parameterized Complexity. Springer, London (2013)
7. Golovach, P.A., Thilikos, D.M.: Paths of bounded length and their cuts: parameterized complexity and algorithms. Discrete Optim. **8**(1), 72–86 (2011)

8. Hanneman, R., Riddle, M.: Introduction to social network methods. In: Scott, J., Carrington, P.J. (eds.) The SAGE Handbook of Social Network Analysis, pp. 340–369. SAGE Publications Ltd, Thousand Oaks (2011)
9. Komusiewicz, C., Niedermeier, R.: New races in parameterized algorithmics. In: Rovan, B., Sassone, V., Widmayer, P. (eds.) MFCS 2012. LNCS, vol. 7464, pp. 19–30. Springer, Heidelberg (2012). doi:10.1007/978-3-642-32589-2_2
10. Marx, D.: Parameterized complexity and approximation algorithms. Comput. J. **51**(1), 60–78 (2008)
11. Marx, D.: Completely inapproximable monotone and antimonotone parameterized problems. J. Comput. Syst. Sci. **79**(1), 144–151 (2013)
12. Wasserman, S., Faust, K.: Social Network Analysis: Methods and Applications. Structural Analysis in the Social Sciences. Cambridge University Press, Cambridge (1994)
13. Wu, B.Y.: On the maximum disjoint paths problem on edge-colored graphs. Discrete Optim. **9**(1), 50–57 (2012)

Total Dual Integrality of Triangle Covering

Xujin Chen$^{(\boxtimes)}$, Zhuo Diao, Xiaodong Hu, and Zhongzheng Tang

Institute of Applied Mathematics, AMSS,
Chinese Academy of Sciences, Beijing 100190, China
{xchen,diaozhuo,xdhu,tangzhongzheng}@amss.ac.cn

Abstract. This paper concerns weighted triangle covering in undirected graph $G = (V, E)$, where a nonnegative integral vector $\mathbf{w} = (w(e) : e \in E)^T$ gives weights of edges. A subset S of E is a *triangle cover* in G if S intersects every triangle of G. The weight of a triangle cover is the sum of $w(e)$ over all edges e in it. The characteristic vector $\mathbf{x}$ of each triangle cover in G is an integral solution of the linear system

$$\pi : A\mathbf{x} \geq \mathbf{1}, \mathbf{x} \geq \mathbf{0},$$

where A is the triangle-edge incidence matrix of G. System π is *totally dual integral* if $\max\{\mathbf{1}^T\mathbf{y} : A^T\mathbf{y} \leq \mathbf{w}, \mathbf{y} \geq \mathbf{0}\}$ has an integral optimum solution $\mathbf{y}$ for each integral vector $\mathbf{w} \in \mathbb{Z}_+^E$ for which the maximum is finite. The total dual integrality of π implies the nice combinatorial min-max relation that the minimum weight of a triangle cover equals the maximize size of a triangle packing, i.e., a collection of triangles in G (repetitions allowed) such that each edge e is contained in at most $w(e)$ of them. In this paper, we obtain graphical properties that are necessary for the total dual integrality of system π, as well as those for the (stronger) total unimodularity of matrix A and the (weaker) integrality of polyhedron $\{\mathbf{x} : A\mathbf{x} \geq \mathbf{1}, \mathbf{x} \geq \mathbf{0}\}$. These necessary conditions are shown to be sufficient when restricted to planar graphs. We prove that the three notions of integrality coincide, and are commonly characterized by excluding odd pseudo-wheels from the planar graphs.

Keywords: Triangle packing and covering · Totally dual integral system · Totally unimodular matrix · Integral polyhedron · Planar graph · Hypergraph

1   Introduction

Covering and packing triangles in graphs has been extensively studied for decades in graph theory [6,7,14] and optimization theory [2,9]. In this paper, we study the problem from both a polyhedral perspective and a graphical persective – characterizing polyhedral integralities of triangle covering and packing with graphical structures.

Research supported in part by NNSF of China under Grant No. 11531014 and 11222109.

T.-H.H. Chan et al. (Eds.): COCOA 2016, LNCS 10043, pp. 128–143, 2016.
DOI: 10.1007/978-3-319-48749-6_10

Graphs considered in this paper are undirected, simple and finite. A weighted graph $(G, \mathbf{w})$ consists of a graph G (with vertex set $V(G)$ and edge set $E(G)$) and an edge weight (function) $\mathbf{w} \in \mathbb{Z}_+^{E(G)}$. The weight of any edge subset S is $w(S) = \sum_{e \in S} w(e)$. By a *triangle cover* of G we mean an edge subset S $(\subseteq E(G))$ whose removal from G leaves a triangle-free graph. Let $\tau_w(G)$ denote the minimum weight of a triangle cover of $(G, \mathbf{w})$. By a *triangle packing* of $(G, \mathbf{w})$ we mean a collection of triangles in G (repetition allowed) such that each edge $e \in E(G)$ is contained in at most $w(e)$ of them. Let $\nu_w(G)$ denote the maximum size of a triangle packing of $(G, \mathbf{w})$. In case of $\mathbf{w} = \mathbf{1}$, we write $\tau_w(G)$ and $\nu_w(G)$ as $\tau(G)$ and $\nu(G)$, respectively.

Tuza's Conjecture and Variants. A vast literature on triangle covering and packing concerns Tuza's conjecture [14] that $\tau(G) \le 2\nu(G)$ for all graphs G and its weighted version [2] that $\tau_w(G) \le 2\nu_w(G)$ for all graphs G and all $\mathbf{w} \in \mathbb{Z}_+^{E(G)}$. Both conjectures remain wide open. The best known general results $\tau(G) \le 2.87\nu(G)$ and $\tau_w(G) \le 2.92\nu_w(G)$ are due to Haxell [7] and Chapuy et al. [2], respectively. Many researchers have pursued the conjectures by showing the conjectured inequalities hold for certain special class of graphs. In particular, Tuza [15] and Chapuy et al. [2] confirmed their own conjectures for planar graphs. Haxell et al. [6] proved the stronger inequality $\tau(G) \le 1.5\nu(G)$ if G is planar and K_4-free, where K_4 denotes the complete graph on 4 vertices.

Along a different line, Lakshmanan et al. [10] proved that the equation $\tau(G) = \nu(G)$ holds whenever G is (K_4, gem)-free or G's triangle graph is odd-hole-free. A natural question arises for the weighted version: When does $\tau_w(G) = \nu_w(G)$ hold? This question is closely related to the notion of total dual integrality from the theory of polyhedral combinatorics.

Total Dual Integrality. A rational system $\{A\mathbf{x} \ge \mathbf{b}, \mathbf{x} \ge \mathbf{0}\}$ is called *totally dual integral* (TDI) if the maximum in the LP duality equation

$$\min\{\mathbf{c}^T\mathbf{x} : A\mathbf{x} \ge \mathbf{b}, \mathbf{x} \ge \mathbf{0}\} = \max\{\mathbf{b}^T\mathbf{y} : A^T\mathbf{y} \le \mathbf{c}, \mathbf{y} \ge \mathbf{0}\}$$

has an integral optimum solution $\mathbf{y}$ for each integral vector $\mathbf{c}$ for which the maximum is finite. The model of TDI systems introduced by Edmonds and Galies [5] plays a crucial role in combinatorial optimization and serves as a general framework for establishing many important combinatorial min-max relations [3,4,11,12]. Schrijver and Seymour [13] derived the following useful tool for proving total dual integrality.

Theorem 1 [13]. *The rational system $A\mathbf{x} \ge \mathbf{b}, \mathbf{x} \ge \mathbf{0}$ is TDI, if and only if*

$$\max\{\mathbf{b}^T\mathbf{y} : A^T\mathbf{y} \le \mathbf{c}, \mathbf{y} \ge \mathbf{0}, 2\mathbf{y} \text{ is integral}\}$$

has an integral optimum solution $\mathbf{y}$ for each integral vector $\mathbf{c}$ for which the maximum is finite.

Edmonds and Giles [5] showed that total dual integrality implies primal integrality as specified by the following theorem.

Theorem 2 [5]. *If rational system $Ax \geq b, x \geq 0$ is TDI and b is integral, then the polyhedron $\{x : Ax \geq b, x \geq 0\}$ is integral, i.e., $\min\{c^T x : Ax \geq b, x \geq 0\}$ is attained by an integral vector for each integral vector c for which the minimum is finite.*

Given a weighted graph (G, w), let $\Lambda(G)$ denote the set of triangles in G. To see the relation between the equation $\tau_w(G) = \nu_w(G)$ and TDI systems, let us consider the hypergraph $\mathcal{H}_G = (E(G), \Lambda(G))$ of triangles in G. We assume $\Lambda(G) \neq \emptyset$ to avoid triviality. The edge-vertex incidence matrix A_G of $\mathcal{H}_G$ is exactly the triangle-edge incidence matrix of G, whose rows and columns are indexed by triangles and edges of G, respectively, such that for any $\triangle \in \Lambda(G)$ and $e \in E(G)$, $A_{\triangle, e} = 1$ if $e \in \triangle$ and $A_{\triangle, e} = 0$ otherwise. In standard terminologies from the theory of packing and covering [4, 12], we write

$$\tau_w(\mathcal{H}_G) = \min\{w^T x : A_G x \geq 1, x \in \mathbb{Z}_+^{E(G)}\}, \tag{1.1}$$

$$\nu_w(\mathcal{H}_G) = \max\{1^T y : A_G^T y \leq w, y \in \mathbb{Z}_+^{\Lambda(G)}\}, \tag{1.2}$$

$$\tau_w^*(\mathcal{H}_G) = \min\{w^T x : A_G x \geq 1, x \geq 0\}, \tag{1.3}$$

$$\nu_w^*(\mathcal{H}_G) = \max\{1^T y : A_G^T y \leq w, y \geq 0\}. \tag{1.4}$$

Combinatorially, each feasible 0–1 solution x of (1.1) is the characteristic vector of a triangle cover of G, and vice versa. Thus such an x is also referred to as a triangle cover (or an integral triangle cover to emphasis the integrality) of G. Moreover, the minimality of $\tau_w(\mathcal{H}_G)$ implies that

$$\tau_w(\mathcal{H}_G) = \tau_w(G).$$

Similarly, each feasible solution y of (1.2) is regarded as a triangle packing (or an integral triangle packing) which contains, for each $\triangle \in \Lambda(G)$, exactly $y(\triangle)$ copies of $\triangle$. In particular,

$$\nu_w(\mathcal{H}_G) = \nu_w(G).$$

Usually, feasible solutions of (1.3) and (1.4) are called *fractional triangle covers* and *fractional triangle packings* of G, respectively. Writing $\tau_w^*(G) = \tau_w^*(\mathcal{H}_G)$ and $\nu_w^*(G) = \nu_w^*(\mathcal{H}_G)$, the LP-duality theorem gives

$$\tau_w(G) \geq \tau_w^*(G) = \nu_w^*(G) \geq \nu_w(G).$$

It is well known (see e.g., page 1397 of [12]) that

$$\tau_w(G) = \nu_w(G) \text{ holds for each } w \in \mathbb{Z}_+^{E(G)} \text{ if and only if } A_G x \geq 1, x \geq 0 \text{ is TDI.}$$

Total Unimodularity. A matrix A is *totally unimodular* (TUM) if each subdeterminant of A is 0, 1 or -1. Total unimodular matrices often imply stronger integrality than TDI systems (see e.g., [8]).

Theorem 3. *An integral matrix A is totally unimodular if and only if the system $Ax \geq b, x \geq 0$ is TDI for each vector b.*

The 0–1 TUM matrices are connected to balanced hypergraphs. Let $\mathcal{H} = (\mathcal{V}, \mathcal{E})$ be a hypergraph with vertex set $\mathcal{V}$ and edge set $\mathcal{E}$. Let $k \geq 2$ be an integer. In $\mathcal{H}$, a *cycle of length* k is a sequence $v_1 e_1 v_2 e_2 \ldots v_k e_k v_1$ such that $v_1, \ldots, v_k \in \mathcal{V}$ are distinct, $e_1, \ldots, e_k \in \mathcal{E}$ are distinct, and $\{v_i, v_{i+1}\} \subseteq e_i$ for each $i = 1, \ldots, k$, where $v_{k+1} = v_1$. Hypergraph $\mathcal{H}$ is called *balanced* if every odd cycle, i.e., cycle of odd length, has an edge that contains at least three vertices of the cycle.

Theorem 4 (Berge [1]). *Let $\mathcal{H}$ be a hypergraph such that every edge consists of at most three vertices. Then the vertex-edge incidence matrix of $\mathcal{H}$ is TUM if and only if $\mathcal{H}$ is balanced.*

Our Results. Let $\mathfrak{B}$, $\mathfrak{M}$, and $\mathfrak{I}$ be the sets of graphs G such that the triangle-edge incidence matrices A_G are TUM, systems $A_G \mathbf{x} \geq \mathbf{1}, \mathbf{x} \geq \mathbf{0}$ are TDI, and polyhedra $\{\mathbf{x} | A_G \mathbf{x} \geq \mathbf{1}, \mathbf{x} \geq \mathbf{0}\}$ are integral, respectively. In terminologies of hypergraph theory (see e.g., Part VIII of [12]),

$G \in \mathfrak{B} \Leftrightarrow \mathcal{H}_G$ is balanced (by Theorem 4 because $\mathcal{H}_G$ is 3-uniform).

$G \in \mathfrak{M} \Leftrightarrow \mathcal{H}_G$ is Mengerian, i.e., $\tau_w(G) = \nu_w(G)$ holds for each $\mathbf{w} \in \mathbb{Z}_+^{E(G)}$.

$G \in \mathfrak{I} \Leftrightarrow \mathcal{H}_G$ is ideal, i.e., $\tau_w(G) = \tau_w^*(G)$ holds for each $\mathbf{w} \in \mathbb{Z}_+^{E(G)}$.

Recalling Theorems 2 and 3, given any graph G, the total modularity (balancedness): $G \in \mathfrak{B}$ implies the total dual integrality (Mengerian property): $G \in \mathfrak{M}$, while $G \in \mathfrak{M}$ implies primal integrality: $G \in \mathfrak{I}$. It follows that

$$\mathfrak{B} \subseteq \mathfrak{M} \subseteq \mathfrak{I}. \tag{1.5}$$

In Sect. 2, first we strengthen (1.5) to $\mathfrak{B} \subsetneq \mathfrak{M} \subsetneq \mathfrak{I}$ (Theorem 5). Then we obtain necessary conditions for a graph to be a member of $\mathfrak{I}$ (Lemma 4) or a minimal graph outside $\mathfrak{B}$ (Theorem 6 and its corollaries) in terms of the pattern of the so-called odd triangle-cycles (Definition 1). Building on these conditions, we establish in Sect. 3 the following characterization for total dual integrality of covering triangle in planar graphs G (Theorem 9):

$G \in \mathfrak{M} \Leftrightarrow G \in \mathfrak{B} \Leftrightarrow G \in \mathfrak{I}$ is K_4-free $\Leftrightarrow$ G is K_4-free & odd pseudo-wheel-free,

where odd pseudo-wheels correspond to odd induced cycles in the triangle graph of G (Definition 2). We conclude in Sect. 4 with remarks on characterizing general graphs $G \in \mathfrak{M}$ and general graphs $G \in \mathfrak{I}$. For easy reference, Appendix gives a list of mathematical symbols used in the paper.

2   General Graphs

In this section, we study TUM, TDI and integral properties for covering and packing triangle in general graphs. We often identify a graph G with its edge set $E(G)$. The following definition is crucial to our discussions.

Definition 1. A *triangle-cycle* in G is a sequence $C = e_1\triangle_1 e_2 \cdots e_k\triangle_k e_1$ with $k \geq 3$ such that $e_1, \cdots, e_k$ are distinct edges, $\triangle_1, \cdots, \triangle_k$ are distinct triangles, and $\{e_i, e_{i+1}\} \subseteq \triangle_i$ for each $i \in \{1, 2, \cdots, k\}$, where $e_{k+1} = e_1$. In $\cup_{i=1}^{k}\triangle_i$, the edges $e_1, e_2, \ldots, e_k$ are *join edges* and other edges are *non-join edges*.

Let $C = e_1\triangle_1 e_2 \cdots e_k\triangle_k e_1$ be a triangle-cycle. We call C *odd* if its *length* k is odd. By abusing notations, we identify C with the graph $\cup_{i=1}^{k}\triangle_i$, whose edge set we denote as $E(C)$. We write $J_C = \{e_1, \cdots, e_k\}$ for the set of join edges, and $N_C = E(C)\backslash J_C$ for the set of non-join edges. Let $\mathscr{T}_C$ denote the set of triangles in C. A triangle in $\mathscr{T}_C$ is *basic* if it belongs to $\mathscr{B}_C = \{\triangle_1, \cdots, \triangle_k\}$. Two basic triangles $\triangle_i$ and $\triangle_j$ are *consecutive* if $|i - j| \in \{1, k - 1\}$. Triangles in $\mathscr{T}_C$ can be classified into four categories:

$$\mathscr{T}_{C,i} = \{\triangle \in \mathscr{T}_C : |\triangle \cap J_C| = i\}, \quad i = 0, 1, 2, 3.$$

It is clear from Definition 1 that $\mathscr{B}_C \subseteq \mathscr{T}_{C,2} \cup \mathscr{T}_{C,3}$. We will establish a strengthening $\mathscr{B} \subsetneq \mathfrak{M} \subsetneq \mathfrak{I}$ of the inclusion relations (1.5). The proof needs the following equivalence implied by hypergraph theory.

Lemma 1. *Let G be a graph. Then $G \in \mathscr{B}$ if and only if every odd triangle-cycle C in G (if any) contains a basic triangle that belongs to $\mathscr{T}_{C,3}$;*

Proof. Recall that $G \in \mathscr{B}$ if and only if hypergraph $\mathcal{H}_G = (E(G), \Lambda(G))$ is balanced. By definition, the balance condition amounts to saying that every odd triangle-cycle C in G (if any) has a triangle $\triangle$ which contains at least 3 joins. It must be the case that $\triangle$ is formed by exactly 3 joins, giving $\triangle \in \mathscr{T}_{C,3}$. $\square$

Observe that the balanced, Mengerian, and integral properties are all closed under taking subgraphs (see, e.g., Theorems 78.2 and 79.1 of [12]).

Lemma 2. *Let G be a graph and H a subgraph of G. If $G \in \mathfrak{X}$ for some $\mathfrak{X} \in \{\mathscr{B}, \mathfrak{M}, \mathfrak{I}\}$, then $H \in \mathfrak{X}$.* $\square$

Lemma 3. $K_4 \in \mathfrak{I} \setminus \mathfrak{M}$.

Proof. Note that $K_4 \notin \mathfrak{M}$ follows from the fact that $\tau(K_4) = 2$ and $\nu(K_4) = 1$. To see $K_4 = (V, E) \in \mathfrak{I}$, for any $\mathbf{x} \in \mathbb{Q}^E$, let $F(\mathbf{x}) = \{e \in E : 0 < x(e) < 1\}$ consist of "fractional" edges w.r.t $\mathbf{x}$. Taking arbitrary $\mathbf{w} \in \mathbb{Z}_+^E$, we consider an optimal fractional triangle cover $\mathbf{x}^*$ for $(K_4, \mathbf{w})$ such that

$$F(\mathbf{x}^*) \text{ is as small as possible.}$$

We are done by showing that $\mathbf{x}^*$ is integral. Suppose it were not the case. The optimality says that $\mathbf{w}^T\mathbf{x}^* = \tau_w^*(K_4)$ and $\mathbf{x}^* \leq \mathbf{1}$. Thus $F(\mathbf{x}^*) \neq \emptyset$.

If $x^*(e) = 1$ for some $e \in E$, then $\mathbf{x}^*|_{E\backslash\{e\}}$ is a fractional triangle cover for $K_4 \backslash e$ such that $(\mathbf{w}|_{E\backslash\{e\}})^T\mathbf{x}^*|_{E\backslash\{e\}} = \tau_w^*(K_4) - w(e)$. Since $K_4 \backslash e \in \mathscr{B} \subseteq \mathfrak{I}$, there is a triangle cover S of $K_4 \backslash e$ with minimum weight $w(S) \leq \tau_w^*(K_4) - w(e)$. So $S \cup \{e\}$ is a triangle cover of K_4 with weight $w(S) + w(e) \leq \tau_w^*(K_4)$, and hence the incidence vector $\mathbf{x} \in \{0, 1\}^E$ of $S \cup \{e\}$ is an optimal fractional triangle cover for $(K_4, \mathbf{w})$ with $F(\mathbf{x}) = \emptyset \subsetneq F(\mathbf{x}^*)$ contradicting the minimality of $F(\mathbf{x}^*)$.

Therefore $x^*(e) < 1$ for all $e \in E$, and $A_{K_4}\mathbf{x}^* \geq \mathbf{1}$ enforces that every triangle of K_4 intersects $F(\mathbf{x}^*)$ with at least 2 edges. Thus $F(\mathbf{x}^*)$ contains four edges e_1, e_2, e_3, e_4 that induce a cycle of K_4, where $\{e_1, e_3\}$ and $\{e_2, e_4\}$ are two matchings of K_4. Without loss of generality we may assume that $x^*(e_1) = \min_{i=1}^4 x^*(e_i)$. Let $\mathbf{x} \in \mathbb{Q}_+^E$ be defined by $x(e_i) = x^*(e_i) + (-1)^i x^*(e_1)$ for $i = 1, 2, 3, 4$ and $x(e) = x^*(e)$ for $e \in E \setminus \{e_1, e_2, e_3, e_4\}$. It is straightforward that

$$\mathbf{w}^T \mathbf{x} = \mathbf{w}^T \mathbf{x}^* \text{ and } F(\mathbf{x}) \subseteq F(\mathbf{x}^*) \setminus \{e_1\}.$$

Since every triangle of K_4 intersects each of $\{e_1, e_3\}$ and $\{e_2, e_4\}$ with exactly one edge, we have $A_{K_4}\mathbf{x} = A_{K_4}\mathbf{x}^* \geq \mathbf{1}$, which along with $\mathbf{w}^T \mathbf{x} = \mathbf{w}^T \mathbf{x}^*$ says that $\mathbf{x} \in \{0,1\}^E$ is an optimal fractional triangle cover for $(K_4, \mathbf{w})$. However, $F(\mathbf{x}) \subsetneq F(\mathbf{x}^*)$ gives a contradiction. $\qquad\square$

Theorem 5. $\mathfrak{B} \subsetneq \mathfrak{M} \subsetneq \mathfrak{J}$.

Proof. In view of Lemma 3, it suffices to show that the graph $G = (V, E)$ depicted in Fig. 1 belongs to $\mathfrak{M} \setminus \mathfrak{B}$. Note that $G = e_1 \triangle_1 e_2 \cdots e_7 \triangle_7 e_1$ is an odd triangle-cycle of length 7, where $\mathscr{B}_G = \{\triangle_1, \triangle_2, \ldots, \triangle_7\}$ and $\Lambda = \Lambda(G) = \mathscr{T}_G = \{\triangle_1, \ldots, \triangle_7, \triangle_8\}$.

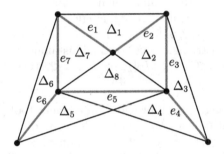

Fig. 1. Graph $G \in \mathfrak{M} \setminus \mathfrak{B}$.

It is routine to check that none of G's basic triangles $\triangle_1, \triangle_2, \ldots, \triangle_7$ belongs to $\mathscr{T}_{G,3}$. Hence Lemma 1 asserts that $G \notin \mathfrak{B}$. To prove $G \in \mathfrak{M}$, by Theorem 1, it suffices to prove that, for any $\mathbf{w} \in \mathbb{Z}_+^E$ and an optimal solution $\mathbf{y}^*$ of $\max\{\mathbf{1}^T\mathbf{y} : A_G^T\mathbf{y} \leq \mathbf{w}, \mathbf{y} \geq \mathbf{0}, 2\mathbf{y} \in \mathbb{Z}_+^\Lambda\}$, there is an integral triangle packing $\mathbf{z} \in \mathbb{Z}_+^\Lambda$ of $(G, \mathbf{w})$ such that $\mathbf{1}^T\mathbf{z} \geq \mathbf{1}^T\mathbf{y}^*$.

Let $\mathbf{y}' \in \{0, 1/2\}^\Lambda$ be defined by $y'(\triangle) = y^*(\triangle) - \lfloor y^*(\triangle) \rfloor$ for each $\triangle \in \Lambda$, and let $\mathbf{w}' \in \mathbb{Z}_+^E$ be defined by $w'(e) = w(e) - \sum_{\triangle \in \Lambda : e \in \triangle} \lfloor y(\triangle) \rfloor$ for each $e \in E$. Then $\mathbf{y}'$ is a fractional triangle packing of $(G, \mathbf{w}')$ such that

$$\mathbf{1}^T\mathbf{y}' = \mathbf{1}^T\mathbf{y}^* - \sum_{\triangle \in \Lambda} y^*(\triangle).$$

If there is an integral packing $\mathbf{z}'$ of $(G, \mathbf{w}')$ such that $\mathbf{1}^T\mathbf{z}' \geq \mathbf{1}^T\mathbf{y}'$, then $\mathbf{z}$ with $z(\triangle) = \lfloor y^*(\triangle) \rfloor + z'(\triangle)$ for each $\triangle \in \Lambda$ is an integral packing of $(G, \mathbf{w})$ satisfying $\mathbf{1}^T\mathbf{z} \geq \sum_{\triangle \in \Lambda} y^*(\triangle) + \mathbf{1}^T\mathbf{y}' = \mathbf{1}^T\mathbf{y}^*$ as desired. We next show such a $\mathbf{z}'$ does exist by distinguishing two cases for integral weight $\mathbf{w}'$.

In case of $w'(e) \geq 1$ for each $e \in E$, we observe that $\mathbf{z}'$ with $z'(\triangle_i) = 1$ for $i = 1, 3, 6, 8$ and $z'(\triangle_i) = 0$ for $i = 2, 4, 5, 7$ is a triangle packing of $(G, \mathbf{w}')$ with $\mathbf{1}^T \mathbf{z}' = 4 = |\Lambda|/2 \geq \mathbf{1}^T \mathbf{y}'$.

In case of $w'(e) = 0$ for some $e \in G$, the restriction $\mathbf{y}''$ of $\mathbf{y}'$ to $\Lambda(G \setminus e)$ is a fractional triangle packing of $(G \setminus e, \mathbf{w}'|_{E \setminus e})$ with $\mathbf{1}^T \mathbf{y}'' = \mathbf{1}^T \mathbf{y}'$. Using Lemma 1, it is routine to check that $G \setminus e \in \mathfrak{B}$, which along with $\mathfrak{B} \subseteq \mathfrak{M}$ gives an integral triangle packing $\mathbf{z}''$ of $(G \setminus e, \mathbf{w}'|_{E \setminus e})$ with $\mathbf{1}^T \mathbf{z}'' \geq \mathbf{1}^T \mathbf{y}''$. For each triangle $\triangle \in \Lambda$, set $z'(\triangle)$ to 0 if $e \in \triangle$ and to $z''(\triangle)$ otherwise. It follows that $\mathbf{z}' \in \mathbb{Z}_+^{\Lambda}$ is an integral triangle packing of $(G, \mathbf{w}')$ with $\mathbf{1}^T \mathbf{z}' = \mathbf{1}^T \mathbf{z}'' \geq \mathbf{1}^T \mathbf{y}'$ as desired.   □

Lemma 4. *If C is an odd triangle-cycle of graph $G \in \mathfrak{J}$, then C contains either a basic triangle belonging to $\mathfrak{T}_{C,3}$ or a non-basic triangle belonging to $\mathfrak{T}_{C,0} \cup \mathfrak{T}_{C,1}$.*

Proof. By contradiction, suppose that graph $G \in \mathfrak{J}$ and its odd triangle-cycle C of length $2k + 1$ form a counterexample, i.e., $\mathfrak{B}_C \subseteq \mathfrak{T}_{C,2}$ and $\mathfrak{T}_C \setminus \mathfrak{B}_C \subseteq \mathfrak{T}_{C,2} \cup \mathfrak{T}_{C,3}$. By Observation 2, we have $C \in \mathfrak{J}$. Let $\mathbf{w} \in \{1, \infty\}^{E(C)}$ be defined by $w(e) = 1$ for all $e \in J_C$ and $w(e) = \infty$ for all $e \in N_C$. On one hand, $\mathfrak{B}_C \subseteq \mathfrak{T}_{C,2}$ implies that each join edge of C exactly belongs to two basic triangles. To break all $2k + 1$ basic triangles, we have to delete at least $k + 1$ join edges unless we use some non-join edge (with infinity weight). Thus $\tau_w(C) \geq k + 1$.

On the other hand, note that every triangle of C contains at least two join edges in J_C. Thus $\mathbf{x} \in \{1/2, 0\}^{E(C)}$ with $x(e) = 1/2$ if $e \in J_C$ and $x(e) = 0$ otherwise is a fractional triangle cover of C. This along with $|J_C| = 2k + 1$ and $\mathbf{w}|_{J_C} = \mathbf{1}$ shows that $\tau_w^*(C) \leq |J_C|/2 = k + 1/2$. However, $\tau_w(C) > \tau_w^*(C)$ contradicts $C \in \mathfrak{J}$.   □

The concept of triangle graph provides an efficient tool for studying triangle covering. Suppose that G is a graph with at least a triangle. Its *triangle graph*, denoted as $T(G)$, is a graph whose vertices are named as triangles of G such that $\triangle_i \triangle_j$ is an edge in $T(G)$ if and only if $\triangle_i$ and $\triangle_j$ are distinct triangles in G which share a common edge. For example, the graph G in Fig. 1 has its triangle graph as depicted in Fig. 2.

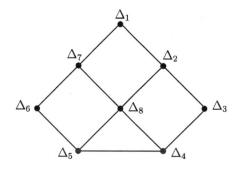

Fig. 2. The triangle graph $T(G)$ of G in Fig. 1.

A graph $G \notin \mathfrak{B}$ is called *minimal* if every proper subgraph H of G belongs to $\mathfrak{B}$. Let $\mathfrak{N}$ denote the set of these minimal graphs.

Theorem 6. *If $G \in \mathfrak{N}$, then G is either K_4 or an odd triangle-cycle with length at least 5 such that $\mathscr{B}_G \subseteq \mathscr{T}_{G,2}$ and $\mathscr{T}_G \setminus \mathscr{B}_G \subseteq \mathscr{T}_{G,1} \cup \mathscr{T}_{G,3}$.*

Proof. Clearly, $K_4 \in \mathfrak{N}$. So we consider $G \neq K_4$. Since $G \notin \mathfrak{B}$ is minimal, G is K_4-free, and by Lemma 1, $G = e_1 \triangle_1 e_2 \cdots e_k \triangle_k e_1$ is an odd triangle-cycle such that $\mathscr{B}_G \subseteq \mathscr{T}_{C,2}$, where $k \geq 5$ is odd. Observe that triangle-cycle G corresponds to a cycle $\tilde{C} = \tilde{e}_1 \triangle_1 \tilde{e}_2 \cdots \tilde{e}_k \triangle_k \tilde{e}_1$ in triangle graph $T(G)$. We first present a series of useful properties.

Property 1. If $\triangle_i \triangle_j$ is a chord of $\tilde{C}$, then the common edge of $\triangle_i$ and $\triangle_j$ is an non-join edge.

Since $\{\triangle_i, \triangle_j\} \subseteq \mathscr{T}_{G,2}$ and they are not consecutive in G, $\triangle_i \cap J_G$ and $\triangle_j \cap J_G$ are disjoint. ∎

Property 2. If both $\triangle_i \triangle_j$ and $\triangle_j \triangle_k$ are chords of $\tilde{C}$, then $\triangle_i, \triangle_j, \triangle_k$ share the same non-join edge in G, and $\triangle_i \triangle_k$ is a chord of $\tilde{C}$.

It follows from Property 1 that each of $\triangle_i, \triangle_j, \triangle_k$ has only one non-join edge. ∎

Property 3. If $\triangle_{i_1}, \triangle_{i_2}, \ldots, \triangle_{i_t}$ are all basic triangles in $\mathscr{B}_G$ that contain $e \in N_G$, where $t \geq 2$ and $i_1 < i_2 < \cdots < i_t$, then for each $j = 1, 2, \ldots, t$, $|\{\triangle_{i_j}, \triangle_{i_j+1} \ldots, \triangle_{i_{j+1}-1}, \triangle_{i_{j+1}}\}|$ is even (where $i_{t+1} = i_1$ in case of $j = t$).

Otherwise, $C_j = e \triangle_{i_j} e_{i_j+1} \triangle_{i_j+1} \cdots \triangle_{i_{j+1}-1} e_{i_{j+1}} \triangle_{i_{j+1}} e$ is an odd triangle-cycle of G for some $1 \leq j \leq t$. Observe that every basic triangle of C_j belongs to $\mathscr{T}_{C_j,2}$. Thus Lemma 1 says that $C_j \notin \mathfrak{B}$, which along with the minimality of $G \in \mathfrak{N}$ enforces that $C_j = G$. However this is absurd because C_j does not contain the join edge $e_{i_{j+2}} \in J_G$ of G. ∎

Property 4. For each $e \in N_G$, there are exactly an odd number of basic triangles in $\mathscr{B}_G$ that contain e.

Since G is the union of its basic triangles, e is contained by some basic triangle of G. The property is instant from Property 3 and the odd length k of the triangle-cycle G. ∎

We now proceed to prove $\mathscr{T}_G \setminus \mathscr{B}_G \subseteq \mathscr{T}_{G,1} \cup \mathscr{T}_{G,3}$. Suppose for a contradiction that there exists $\triangle \in \mathscr{T}_G \setminus \mathscr{B}_G$ with $\triangle \in \mathscr{T}_{G,0}$. Then $\triangle$ consists of three non-join edges $p, q, r \in N_G$. Let

$$\mathscr{B}_p = \{\triangle \in \mathscr{B}_G : p \in \triangle\}, \mathscr{B}_q = \{\triangle \in \mathscr{B}_G : q \in \triangle\}, \mathscr{B}_r = \{\triangle \in \mathscr{B}_G : r \in \triangle\}$$

denote the sets of basic triangles (of G) that contain p, q, r, respectively. Notice from Property 4 that

$$|\mathscr{B}_p|, |\mathscr{B}_q| \text{ and } |\mathscr{B}_r| \text{ are odd numbers.}$$

We distinguish between two cases depending on whether all of $\mathscr{B}_p, \mathscr{B}_q, \mathscr{B}_r$ are singletons or not.

Case 1. $|\mathscr{B}_p| = |\mathscr{B}_q| = |\mathscr{B}_r| = 1$. We may assume without loss of generality that $\mathscr{B}_j = \{\triangle_{i_j}\}$ for $j \in \{p, q, r\}$ and $i_p < i_q < i_r$. Note that

$$C_{pq} = p\triangle_{i_p} e_{i_p+1}\triangle_{i_p+1} \cdots e_{i_q}\triangle_{i_q} q\triangle p,$$
$$C_{qr} = q\triangle_{i_q} e_{i_q+1}\triangle_{i_q+1} \cdots e_{i_r}\triangle_{i_r} r\triangle q,$$
$$C_{rp} = r\triangle_{i_r} e_{i_r+1}\triangle_{i_r+1} \cdots e_{i_p}\triangle_{i_p} p\triangle r$$

are triangle-cycles of G whose basic triangles each contain exactly two join edges. Observe that the sum of lengths of C_{pq}, C_{qr}, C_{rp} equals $k + 6$, which is odd. So at least one of C_{pq}, C_{qr}, C_{rp}, say C_{pq}, has an odd length. It follows from $\mathscr{B}_{C_{pq}} \subseteq \mathscr{T}_{C_{pq},2}$ and Lemma 1 that $C_{pq} \notin \mathfrak{B}$. Now the minimality of $G \in \mathfrak{N}$ enforces $C_{pq} = G$. Hence the join edge $e_{i_q+2} \in J_G$ must be one of $e_{i_p}, e_{i_p+1}, \ldots, e_{i_q-1}$, from which we deduce that $e_{i_q+2} = e_{i_p}$ (and $i_q + 1 = i_r$). As e_{i_q+2} has a common vertex with e_{i_q}, it follows that e_{i_p}, e_{i_q+1} and r form a triangle, and $p, q, r, e_{i_p}, e_{i_p+1}, e_{i_q+1}$ induce a K_4, contradicting the fact that G is K_4-free.

Case 2. $\max\{|\mathscr{B}_p|, |\mathscr{B}_q|, |\mathscr{B}_r|\} \geq 3$. Suppose without loss of generality that $\mathscr{B}_p = \{\triangle_{i_1}, \cdots, \triangle_{i_t}\}$ where $t \geq 3$ and $i_1 < i_2 \cdots < i_t$. Setting $i_{t+1} = i_1$, since $\mathscr{B}_p \cap \mathscr{B}_q = \emptyset$, we have $|\mathscr{B}_q| = \sum_{j=1}^{t} |\{\triangle_{i_j}, \triangle_{i_j+1} \cdots, \triangle_{i_{j+1}}\} \cap \mathscr{B}_q|$. Recall that $|\mathscr{B}_q|$ is odd. So there exists $j \in \{1, \ldots, t\}$ such that $\{\triangle_{i_j}, \triangle_{i_j+1} \cdots, \triangle_{i_{j+1}}\} \cap \mathscr{B}_q$ consists of

an odd number s of basic triangles $\triangle_{h_1}, \ldots, \triangle_{h_s}$,

where $i_j < h_1 < \cdots < h_s < i_{j+1}$. By Property 3, $|\{\triangle_{i_j}, \triangle_{i_j+1} \ldots, \triangle_{i_{j+1}}\}|$ is even, and $|\{\triangle_{h_\ell}, \triangle_{h_\ell+1}, \ldots, \triangle_{h_{\ell+1}}\}|$ is even for each $\ell \in \{1, \ldots, s-1\}$. Note that

$$|\{\triangle_{i_j}, \triangle_{i_j+1} \ldots, \triangle_{i_{j+1}}\}|$$
$$= |\{\triangle_{i_j}, \triangle_{i_j+1} \ldots, \triangle_{h_1}\}| + \left(\sum_{\ell=1}^{s-1} |\{\triangle_{h_\ell}, \triangle_{h_\ell+1}, \ldots, \triangle_{h_{\ell+1}}\}|\right)$$
$$+ |\{\triangle_{h_s}, \triangle_{h_s+1} \ldots, \triangle_{i_{j+1}}\}| - s$$
$$\equiv (h_1 - i_j) + (i_{j+1} - h_s) - s \pmod 2$$

Since s is odd, either $h_1 - i_j$ or $i_{j+1} - h_s$ is odd. Suppose by symmetry that $h_1 - i_j$ is odd. It follows that $C = p\triangle_{i_j} e_{i_j+1}\triangle_{i_j+1} \cdots e_{h_1}\triangle_{h_1} q\triangle p$ is a triangle-cycle of G such that $\mathscr{B}_C \subseteq \mathscr{T}_{C,2}$. As the length $h_1 - i_j + 2$ is odd, we deduce from Lemma 1 that $C \notin \mathfrak{B}$. In turn $G \in \mathfrak{N}$ enforces $C = G$. Similar to Case 1, $e_{h_1+2} \in J_G \subseteq C$ implies that $\triangle_{h_1}, \triangle_{i_{j+1}}, \triangle$ form a K_4, a contradiction to the K_4-freeness of G. The contradiction shows that $(\mathscr{T}_G \setminus \mathscr{B}_G) \cap \mathscr{T}_{G,0} = \emptyset$.

It remains to prove $(\mathscr{T}_G \setminus \mathscr{B}_G) \cap \mathscr{T}_{G,2} = \emptyset$. Suppose on the contrary that there exists $\triangle \in \mathscr{T}_G \setminus \mathscr{B}_G$ which consists of two join edges $p, q \in J_G$ and one non-join edge $r \in N_G$. Again we set $\mathscr{B}_p = \{\triangle \in \mathscr{B}_G : p \in \triangle\}$, $\mathscr{B}_q = \{\triangle \in \mathscr{B}_G : q \in \triangle\}$ and $\mathscr{B}_r = \{\triangle \in \mathscr{B}_G : r \in \triangle\}$. Recalling $\mathscr{B}_G \subseteq \mathscr{T}_{G,2}$, we derive

$|\mathscr{B}_p| = |\mathscr{B}_q| = 2$. Suppose without loss of generality that $\mathscr{B}_p = \{\triangle_{i_p}, \triangle_{i_p+1}\}$, $\mathscr{B}_q = \{\triangle_{i_q}, \triangle_{i_q+1}\}$ and $i_p < i_p + 1 < i_q < i_q + 1$ (note $p = e_{i_p+1}, q = e_{i_q+1}$). Recall from Property 4 that $|\mathscr{B}_r|$ is an odd number. Observe that both $C = p\triangle_{i_p+1}e_{i_p+2}\triangle_{i_p+2}\cdots e_{i_q}\triangle_{i_q}q\triangle p$ and $C' = q\triangle_{i_q+1}e_{i_q+2}\triangle_{i_q+2}\cdots e_{i_p}\triangle_{i_p}p\triangle q$ are triangle-cycles whose basic triangles each contain exactly 2 join edges. Because the length of G is odd, exactly one of C and C', say C, whose length is odd. By Lemma 1(i), $C \notin \mathscr{B}$. In turn $G \in \mathfrak{N}$ gives $C = G$. Since neither $\triangle_{i_p}$ nor $\triangle_{i_q+1}$ is a basic triangle of C and $\triangle_{i_p} \neq \triangle_{i_q+1}$, we derive that $e_{i_q+2} \in G \setminus C$, a contradiction to $C = G$. This completes the proof of Theorem 6.   □

Let $\mathfrak{X} \in \{\mathfrak{M}, \mathfrak{I}\}$. If graph $G \in \mathfrak{X} \setminus \mathfrak{B}$ is *minimal* in the sense that every proper subgraph H of G is outside $\mathfrak{X} \setminus \mathfrak{B}$, then $H \in \mathfrak{X}$ (by Lemma 2) enforces $H \in \mathfrak{B}$. Hence $G \in \mathfrak{N}$. Conversely, if $G \in \mathfrak{X} \cap \mathfrak{N}$, then every subgraph H of G satisfies $H \in \mathfrak{B} \subseteq \mathfrak{X}$, giving $H \notin \mathfrak{X} \setminus \mathfrak{B}$. Thus the set of *minimal graphs* in $\mathfrak{X} \setminus \mathfrak{B}$ is

$$\{G \in \mathfrak{X} \setminus \mathfrak{B} : H \notin \mathfrak{X} \setminus \mathfrak{B} \text{ for every } H \subsetneq G\} = \mathfrak{N} \cap \mathfrak{X}, \text{ where } \mathfrak{X} \in \{\mathfrak{M}, \mathfrak{I}\}. \quad (2.1)$$

Corollary 1. *If $G \in \mathfrak{N} \cap \mathfrak{I}$ (i.e., $G \in \mathfrak{I} \setminus \mathfrak{B}$ is minimal) , then G is either K_4 or an odd triangle-cycle such that $\mathscr{B}_G \subseteq \mathscr{T}_{G,2}$, $\mathscr{T}_G \setminus \mathscr{B}_G \subseteq \mathscr{T}_{G,1} \cup \mathscr{T}_{G,3}$, and $\mathscr{T}_{G,1} = \mathscr{T}_{G,1} \setminus \mathscr{B}_G \neq \emptyset$.*

Proof. In view of Theorem 6, it suffices to consider G being an odd triangle-cycle such that $\mathscr{B}_G \subseteq \mathscr{T}_{G,2}$ and $\mathscr{T}_G \setminus \mathscr{B}_G \subseteq \mathscr{T}_{G,1} \cup \mathscr{T}_{G,3}$. In turn, Lemma 4 implies the existence of at least a non-basic triangle of G that belongs to $\mathscr{T}_{G,1}$.   □

Corollary 2. *If $G \in \mathfrak{N} \cap \mathfrak{M}$ (i.e., $G \subseteq \mathfrak{M} \setminus \mathfrak{B}$ is minimal) , then G is an odd triangle-cycle such that $\mathscr{B}_G \subseteq \mathscr{T}_{G,2}$, $\mathscr{T}_G \setminus \mathscr{B}_G \subseteq \mathscr{T}_{G,1} \cup \mathscr{T}_{G,3}$, and $\mathscr{T}_{G,1} = \mathscr{T}_{G,1} \setminus \mathscr{B}_G \neq \emptyset$.*

Proof. Note from $G \in \mathfrak{M}$ that $G \neq K_4$. As $\mathfrak{M} \subseteq \mathfrak{I}$, the conclusion is immediate from Corollary 1.   □

3   Planar Graphs

In this section, we study the planar case more closely, and characterize planar graphs in $\mathfrak{M}$ by excluding pseudo-wheels defined as follows.

Definition 2. A *triangle-cycle* C is a *pseudo-wheel* if it has length at least 4, $\mathscr{T}_C = \mathscr{B}_C$ and each pair of non-consecutive basic triangles of C is edge-disjoint.

It is easy to see that a triangle-cycle C is a pseudo-wheel if and only if its triangle graph $T(C)$ is an induced cycle with length at least 4. Thus every wheel other than K_4 is a pseudo-wheel. Two pseudo-wheels that are not wheels are shown in Fig. 3.

Lemma 5. *If C is an odd pseudo-wheel, then $C \notin \mathfrak{I}$.*

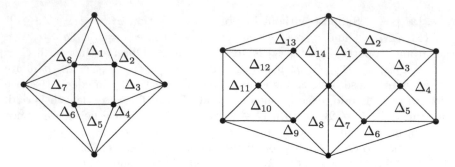

Fig. 3. Examples of pseudo-wheels.

Proof. Suppose the length of C is $2k+1$. Let $\mathbf{w} \in \mathbb{Z}_+^{E}(C)$ be defined by $w(e) = 1$ for all $e \in J_C$ and $w(e) = \infty$ for all $e \in N_C$. Then $\tau_w(C) = k + 1$. On the other hand $\mathbf{x} \in \{0, 1/2\}^{E(C)}$ with $x(e) = 1/2$ for all $e \in J_C$ and $x(e) = 0$ for all $e \in N_C$ is a fractional triangle cover of C, showing $\tau_w^*(C) \leq \mathbf{w}^T\mathbf{x} = k + 1/2 < \tau_w(C)$. $\square$

If $\triangle^i, \triangle^o, \triangle$ are distinct triangles of plane graph G such that $\triangle^i$ is inside $\triangle$ and $\triangle^o$ is outside $\triangle$, then we say that $\triangle$ is a *separating triangle* of $\triangle^i$ and $\triangle^o$, or $\triangle$ *separates* $\triangle^i$ from $\triangle^o$.

A *triangle-path* in graph G is a sequence $P = \triangle_1 e_1 \cdots e_k \triangle_{k+1}$ with $k \geq 1$ such that $e_1, \cdots, e_k$ are distinct edges, $\triangle_1, \cdots, \triangle_{k+1}$ are distinct triangles of G, and $\{e_1\} \subseteq \triangle_1, \{e_k\} \subseteq \triangle_{k+1}, \{e_i, e_{i+1}\} \subseteq \triangle_{i+1}$ for each $i \in [k-1]$. In $\cup_{i=1}^{k+1}\triangle_i$, the edges $e_1, e_2, \ldots, e_k$ are called *join edges* and other edges are called *non-join edges*. Let J_P denote the set of join edges of P. The length of P is defined as k. We often say that P is a triangle-path from $\triangle_1$ to $\triangle_{k+1}$.

Lemma 6. *Let G be a plane graph in which $\triangle$ is a separating triangle of triangles $\triangle^i$ and $\triangle^o$. Then $\triangle$ contains at least one join edge of every triangle-path from $\triangle^i$ to $\triangle^o$ in G.*

Proof. Consider an arbitrary triangle-path $P = \triangle_1 e_1 \cdots e_k \triangle_{k+1}$ in G from $\triangle_1 = \triangle^i$ to $\triangle_{k+1} = \triangle^o$. We prove $\triangle \cap \{e_1, \ldots, e_k\} \neq \emptyset$ by induction on k. The basic case of $k = 1$ is trivial. We consider $k \geq 2$ and assuming that the lemma holds when triangle-path involved has length at most $k - 1$. If $\triangle_2 = \triangle$, then we are done. If $\triangle_2 \neq \triangle$, then either $\triangle$ separates $\triangle_1$ from $\triangle_2$ or separates $\triangle_2$ and $\triangle_{k+1}$. Observe that $\triangle_1 e_1 \triangle_2$ is a triangle-path of length $1 < k$, and $\triangle_2 e_2 \ldots, e_k \triangle_{k+1}$ is a triangle-path of length $k - 1$. From the induction hypothesis, we derive $e_1 \in \triangle$ in the former case, and $e_j \in \triangle$ for some $j = 2, \ldots, k$ in the latter case. $\square$

Lemma 7. *Let $C = e_1 \triangle_1 e_2 \cdots e_k \triangle_k e_1$ with $k \geq 3$ be a triangle-cycle. If C is plane and $\mathscr{B}_C \subseteq \mathscr{T}_{C,2}$, then $\triangle_h$ does not separate $\triangle_i$ from $\triangle_j$ for any distinct $h, i, j \in \{1, \ldots, k\}$.*

Proof. Note that C contains a triangle-path P from $\triangle_i$ and $\triangle_j$ with $J_P \subseteq J_C \setminus \triangle_h$. The triangle-path P along with Lemma 6 implies the result. $\square$

Theorem 7. *If C is a planar triangle-cycle such that $\mathscr{B}_C \subseteq \mathscr{T}_{C,2}$, then $\mathscr{T}_C \subseteq \mathscr{T}_{C,0} \cup \mathscr{T}_{C,2}$.*

Proof. Suppose that $C = e_1 \triangle_1 e_2 \cdots e_k \triangle_k e_1$ with $k \geq 3$ is plane, and there exists $\triangle \in \mathscr{T}_C$ with $\triangle \in \mathscr{T}_{C,1} \cup \mathscr{T}_{C,3}$.

Case 1. $\triangle \in \mathscr{T}_{C,3}$ consists of three join edges e_h, e_i, e_j, where $1 \leq h < i < j \leq k$. The structure of the triangle graph $T(C)$ is illustrated in the left part of Fig. 4.

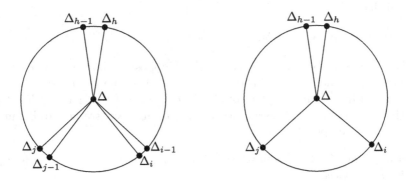

Fig. 4. The triangle graph $T(C)$ in the two cases of the proof for Theorem 7.

For each pair $(s,t) \in \{(h, i-1), (i, j-1), (j, h-1)\}$, there is a triangle-path in C from $\triangle_s$ to $\triangle_t$ whose set of join edges is disjoint from $\{e_h, e_i, e_j\} = \triangle$. It follows from Lemma 6 that

$$\triangle \text{ does not separate } \triangle_s \text{ from } \triangle_t \text{ for each} \\ (s,t) \in \{(h, i-1), (i, j-1), (j, h-1)\}. \tag{3.1}$$

Suppose that $\triangle$ separates $\triangle_{h-1}$ from $\triangle_h$, and separates $\triangle_{i-1}$ from $\triangle_i$. Without loss of generality let $\triangle_{h-1}$ and $\triangle_h$ sit inside and outside $\triangle$, respectively. Then (3.1) implies that $\triangle_j$ and $\triangle_{i-1}$ are inside and outside $\triangle$, respectively. In turn, i is inside $\triangle$, and (3.1) says that $\triangle_{j-1}$ is inside $\triangle$. Now $\triangle_{j-1}$ and $\triangle_j$ are both inside $\triangle$, i.e., $\triangle$ does not separate $\triangle_{j-1}$ from $\triangle_j$. Hence, by symmetry we

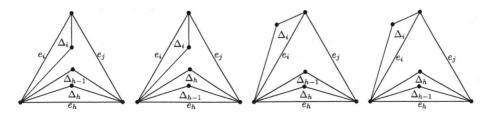

Fig. 5. $\triangle_{h-1}$ and $\triangle_h$ are both inside $\triangle$.

may assume that $\triangle$ does not separate $\triangle_{h-1}$ from $\triangle_h$, and further that $\triangle_{h-1}$ and $\triangle_h$ are both inside $\triangle$. as illustrated in Fig. 5.

As $e_h \in \triangle_{h-1} \cap \triangle_h$, it is easy to see that either $\triangle_{h-1}$ separates $\triangle_h$ from $\triangle_i$ or $\triangle_h$ separates $\triangle_{h-1}$ from $\triangle_i$. The contradiction to Lemma 7 finishes our discussion on Case 1.

Case 2. $\triangle \in \mathscr{T}_{C,1}$ consist of join edge e_h of C (shared with $\triangle_{h-1}, \triangle_h$), non-join edge f (shared with $\triangle_i$) and non-join edge g (shared with $\triangle_j$), where h, i, j are distinct. See the right part of Fig. 4. Similar to Case 1, it can be derived from Lemma 6 that

$$\triangle \text{ does not separate} \triangle_s \text{ from } \triangle_t \text{ for each} (s,t) \in \{(h,i),(i,j),(j,h-1)\}.$$

Therefore $\triangle$ does not separate $\triangle_{h-1}$ and $\triangle_h$. Suppose without loss of generality that both $\triangle_{h-1}$ and $\triangle_h$ are inside $\triangle$. Then C has one of the structures as illustrated in Fig. 5 with f in place of e_i and g in place of e_j. Again, either $\triangle_{h-1}$ separating $\triangle_h$ from $\triangle_i$ or $\triangle_h$ separating $\triangle_{h-1}$ from $\triangle_i$ contradicts to Lemma 7. This completes the proof. $\square$

Theorem 8. *Let G be a planar graph. Then $G \in \mathfrak{N}$ if and only if G is K_4 or an odd pseudo-wheel.*

Proof. Sufficiency: Clearly $K_4 \in \mathfrak{N}$. If G is an odd pseudo-wheel C, then G is an odd triangle-cycle such that $\mathscr{B}_G \subseteq \mathscr{T}_{G,2}$. By Lemma 1, $G \notin \mathfrak{B}$. Since the triangle graph $T(C)$ is an induced cycle, every proper subgraph of C is triangle-cycle-free, and hence belongs to $\mathfrak{B}$, giving $G \in \mathfrak{N}$.

Necessity: Suppose that $G \in \mathfrak{N}$ and $G \neq K_4$. By Theorem 6, an odd triangle-cycle with length at least 5 such that $\mathscr{B}_G \subseteq \mathscr{T}_{G,2}$ and $\mathscr{T}_G \setminus \mathscr{B}_G \subseteq \mathscr{T}_{G,1} \cup \mathscr{T}_{G,3}$. In turn, Theorem 7 enforces

$$\mathscr{T}_C = \mathscr{B}_C.$$

Suppose for a contradiction that there exists non-consecutive triangles $\triangle_i, \triangle_j \in \mathscr{B}_G$ that share a common non-join edge e, where $i < j-1$. Then G contains two triangle-cycles $C_1 = e\triangle_i e_{i+1} \triangle_{i+1} \cdots e_j \triangle_j e$ and $C_2 = e\triangle_j e_{j+1}\triangle_{j+1}\cdots e_i \triangle_i e$. Because G is odd, one of C_1 and C_2, say C_1, is odd. As C_1 is a proper subgraph of $G \in \mathfrak{N}$, we have $C_1 \in \mathfrak{B}$. By Lemma 1, there exists a basic triangle $\triangle_h$ in $\mathscr{B}_{C_1} \cap \mathscr{T}_{C_1,3}$. Because $\triangle_h \in \mathscr{T}_{G,2}$, it must be the case that $e \in \triangle_h$. Thus $\triangle_i, \triangle_j, \triangle_h$ share a common non-join edge e of G. However in any planar embedding for G, there is one triangle in $\{\triangle_i, \triangle_j, \triangle_h\}$, which is a separating triangle of the other two. This is a contradiction to Lemma 7. Thus each pair of non-consecutive basic triangles of G is edge-disjoint, and G is an odd pseudo-wheel. $\square$

Theorem 9. *Let G be a planar graph, then the following are equivalent:*

(i) $G \in \mathfrak{B}$;
(ii) $G \in \mathfrak{M}$;

(iii) $G \in \mathfrak{I}$ *is* K_4*-free; and*
(iv) G *is* K_4*-free and odd pseudo-wheel free.*

Proof. Recalling (1.5) and Lemma 3, $\mathfrak{B} \subseteq \mathfrak{M} \subseteq \mathfrak{I}$ and $K_4 \in \mathfrak{I} \setminus \mathfrak{M}$ imply the relation $(i) \Rightarrow (ii) \Rightarrow (iii)$. If G contains an odd pseudo-wheel H, then $H \notin \mathfrak{I}$ by Lemma 5, which along with Lemma 2 would give $G \notin \mathfrak{I}$. So we have $(iii) \Rightarrow (iv)$.

It remains to prove $(iv) \Rightarrow (i)$. If $G \notin \mathfrak{B}$, we take $H \subseteq G$ to be minimal, i.e., $H \in \mathfrak{N}$. Theorem 8 says that H is K_4 or an odd pseudo-wheel, i.e., G is not K_4-free and G is not odd pseudo-wheel free. $\qquad\square$

4   Remarks

Lemma 4 provides us a necessary condition for $G \in \mathfrak{I}$ as follows:

$$(\mathscr{B}_C \cap \mathscr{T}_{C,3}) \cup ((\mathscr{T}_{C,0} \cup \mathscr{T}_{C,1}) \setminus \mathscr{B}_C) \neq \emptyset \text{ for any odd triangle-cycle } C \text{ of } G. \quad (4.1)$$

It would be interesting to see if the condition is sufficient for $G \in \mathfrak{I}$. A supporting evidence is the following.

Remark 1. Condition (4.1) is a necessary and sufficient condition for K_4-free planar graph G to be a member of $\mathfrak{I}$.

Proof. By Theorem 9, a K_4-free planar graph $G \in \mathfrak{I}$ implies $G \in \mathfrak{B}$, and thus $\mathscr{B}_C \cap \mathscr{T}_{C,3} \neq \emptyset$ for every odd triangle-cycle C in G. On the other hand, given a K_4-free planar graph G satisfying (4.1), we see from Definition 2 that G does not contain any odd pseudo-wheel. It follows from Theorem 8 that G does not contain any subgraph in $\mathfrak{N}$, which implies $G \in \mathfrak{B} \subseteq \mathfrak{I}$. $\qquad\square$

As $\mathfrak{M} \subseteq \mathfrak{I}$, condition (4.1) is also necessary for $G \in \mathfrak{M}$, but it is not sufficient for the total dual integrality. This can be seen from $K_4 \notin \mathfrak{M}$, which satisfies (4.1): K_4 has four odd triangle-cycles with length 3 each containing a triangle without any join edge, and for each odd triangle-cycle C, there is a triangle in $\mathscr{T}_C \setminus \mathscr{B}_C$ that belongs to $\mathscr{T}_{C,0}$. This motivates us to ask about the necessity and sufficiency of the following conditions for $G \in \mathfrak{M}$:

$$(\mathscr{B}_C \cap \mathscr{T}_{C,3}) \cup (\mathscr{T}_{C,1} \setminus \mathscr{B}_C) \neq \emptyset \text{ for any odd triangle-cycle } C \text{ of } G. \quad (4.2)$$

Note that condition (4.2) implies G contains neither K_4 nor odd pseudo-wheels. Similar to Remark 1, Theorems 8 and 9 provide the following fact.

Remark 2. Condition (4.2) is a necessary and sufficient condition for planar graph G to be a member of $\mathfrak{M}$. $\qquad\square$

Appendix: A List of Mathematical Symbols

$(G, \mathbf{w})$	Weighted graph $G = (V(G), E(G))$ with $\mathbf{w} \in \mathbb{Z}_+^{E(G)}$
$\tau_w(G)$	The minimum weight of an integral triangle cover in $(G, \mathbf{w})$
$\nu_w(G)$	The maximum size of an integral triangle packing in $(G, \mathbf{w})$
$\tau_w^*(G)$	The minimum weight of a fractional triangle cover in $(G, \mathbf{w})$
$\nu_w^*(G)$	The maximum size of a fractional triangle packing in $(G, \mathbf{w})$
$\tau(G)$	$\tau_w(G)$ when $\mathbf{w} = \mathbf{1}$
$\nu(G)$	$\nu_w(G)$ when $\mathbf{w} = \mathbf{1}$
A_G	The triangle-edge incidence matrix of graph G
$\Lambda(G)$	The set of triangles in graph G
$\mathfrak{B}$	The set of graphs G such that A_G are TUM
$\mathfrak{M}$	The set of graphs G such that systems $A_G x \geq 1, x \geq 0$ are TDI
$\mathfrak{J}$	The set of graphs G such that $\{\mathbf{x} : A_G \mathbf{x} \geq 1, \mathbf{x} \geq \mathbf{0}\}$ are intergal
$\mathfrak{N}$	The set of minimal graphs not belonging to $\mathfrak{B}$
$\mathscr{T}_C$	The set of triangles in triangle-cycle $C = e_1 \triangle_1 e_2 \cdots e_k \triangle_k e_1 = \cup_{i=1}^{k} \triangle_i$
$\mathscr{B}_C$	The set of basic triangles in triangle-cycle C, i.e., $\{\triangle_1, \cdots, \triangle_k\}$
J_C	The set of join edges in triangle-cycle C, i.e., $\{e_1, \cdots, e_k\}$
N_C	The set of nonjoin edges in triangle-cycle C, i.e., $E(C) \backslash J_C$
$\mathscr{T}_{C,i}$	$\{\triangle \in \mathscr{T}_C : |\triangle \cap J_C| = i\}$, $i = 0, 1, 2, 3$

References

1. Berge, C.: Hypergraphs: Combinatorics of Finite Sets. Elsevier, Amsterdam (1989)
2. Chapuy, G., DeVos, M., McDonald, J., Mohar, B., Scheide, D.: Packing triangles in weighted graphs. SIAM J. Discret. Math. **28**(1), 226–239 (2014)
3. Chen, X., Hu, X., Zang, W.: Dual integrality in combinatorial optimization. In: Pardalos, P.M., Du, D.-Z., Graham, R.L. (eds.) Handbook of Combinatorial Optimization, pp. 995–1063. Springer, New York (2013)
4. Cornuéjols, G.: Combinatorial optimization: packing and covering. In: Society for Industrial and Applied Mathematics (2001)
5. Edmonds, J., Giles, R.: A min-max relation for submodular functions on graphs. Ann. Discret. Math. **1**, 185–204 (1977)
6. Haxell, P., Kostochka, A., Thomassé, S.: Packing and covering triangles in K_4-free planar graphs. Graphs Comb. **28**(5), 653–662 (2012)
7. Haxell, P.E.: Packing and covering triangles in graphs. Discret. Math. **195**(1), 251–254 (1999)
8. Korte, B., Vygen, J.: Combinatorial Optimization: Theory and Algorithms. Springer, Berlin (2012)
9. Krivelevich, M.: On a conjecture of Tuza about packing and covering of triangles. Discret. Math. **142**(1), 281–286 (1995)

10. Lakshmanan, S.A., Bujtás, C., Tuza, Z.: Small edge sets meeting all triangles of a graph. Graphs Comb. **28**(3), 381–392 (2012)
11. Schrijver, A.: Theory of Linear and Integer Programming. Wiley, New York (1986)
12. Schrijver, A.: Combinatorial Optimization: Polyhedra and Efficiency. Springer, Berlin (2003)
13. Schrijver, A., Seymour, P.: A proof of total dual integrality of matching polyhedra. Stichting Mathematisch Centrum. Zuivere Wiskunde (ZN 79/77), pp. 1–12 (1977)
14. Tuza, Z.: Conjecture in: finite and infinite sets. In: Proceedings of Colloque Mathematical Society Jnos Bolyai, p. 888, Eger, Hungary, North-Holland (1981)
15. Tuza, Z.: A conjecture on triangles of graphs. Graphs Comb. **6**(4), 373–380 (1990)

Time-Optimal Broadcasting of Multiple Messages in 1-in Port Model

Petr Gregor[1]([✉]), Riste Škrekovski[2,3], and Vida Vukašinović[4]

[1] Department of Theoretical Computer Science and Mathematical Logic,
Charles University, Malostranské nám. 25, 11800 Prague, Czech Republic
gregor@ktiml.mff.cuni.cz
[2] Department of Mathematics, University of Ljubljana,
Jadranska 19, 1000 Ljubljana, Slovenia
[3] Faculty of Information Studies, Ljubljanska cesta 31A, 8000 Novo Mesto, Slovenia
skrekovski@gmail.com
[4] Computer Systems Department, Jožef Stefan Institute,
Jamova 39, 1000 Ljubljana, Slovenia
vida.vukasinovic@ijs.si

Abstract. In the 1-in port model, every vertex of a synchronous network can receive each time unit at most one message. We consider simultaneous broadcasting of multiple messages from the same source in such networks with an additional restriction that every received message can be sent out to neighbors only in the next time unit and never to already informed vertex. We use a general concept of level-disjoint partitions developed for this scenario. Here we introduce a subgraph extension technique for efficient spreading information within this concept. Surprisingly, this approach with so called biwheels leads to simultaneous broadcasting of optimal number of messages on a wide class of graphs in optimal time. In particular, we provide tight results for bipartite tori, meshes, hypercubes. Several problems and conjectures are proposed.

Keywords: Simultaneous broadcasting · Multiple message broadcasting · Level-disjoint partitions · Torus · Mesh · Hypercube

1 Introduction

A massive amount of traffic in communication networks that flows from providers of large data (such as video streaming services) to many clients at once leads to various optimization problems for broadcasting of multiple messages. Similar types of problems arise in master/workers parallel computations on specific networks when multiple tasks are simultaneously distributed from one node (master) to all other nodes (workers). This has been subject of research for many years. For surveys on broadcasting and other communication protocols in various kinds of networks see e.g. [8,9,12–14].

This research was supported by the Czech Science Foundation grant GA14-10799S, ARRS Program P1-0383, and by ARTEMIS-JU project "333020 ACCUS".

© Springer International Publishing AG 2016
T.-H.H. Chan et al. (Eds.): COCOA 2016, LNCS 10043, pp. 144–158, 2016.
DOI: 10.1007/978-3-319-48749-6_11

We restrict ourselves to synchronous networks, where at each time unit messages can be sent from nodes to all their neighbors in one unit of time. A network is modeled by a graph. As an example we consider namely tori, meshes, and hypercubes, perhaps the most popular and extensively studied networks [15], but our approach is more general.

Since networks have limited capacity of links, any larger data to be broadcast needs to be split into multiple messages and sent individually. This leads to a more general variant of *broadcasting* in which several different messages need to be simultaneously transmitted from one source node, called the *originator*. The problem of multiple broadcasting was first defined in [5] and previously studied under several different models in [1,2,10]. The minimal overall time needed for simultaneous broadcasting and the maximal number of messages that can be simultaneously broadcast were considered in [1,6,10,16–18], respectively.

Here we consider a scenario when each message (or task) needs to be handled (or processed) at each node in a time unit before it is sent out further to other selected neighbors. It is reasonable to demand that each node has to handle at each time unit only a single message (task). Equivalently, each node receives at most one message in each time unit. We call this restriction a *1-in-port model*. Furthermore, every received message is send out only in the next time unit and no message is sent to already informed vertex. In other words, nodes have no buffers to store messages for delayed transmission. This simplification is motivated by memory or security restrictions, or a need for uninterrupted data flow. As usual, we also assume full-duplex mode.

For this scenario, the concept of level-disjoint partitions was developed in [6] to study how many messages and in what time they can be simultaneously broadcast from a given originator vertex in a given graph, see the definitions in the next section. The same concept was further developed in [7] where results on existence of optimal number of level-disjoint partitions in general graphs were obtained. It was also shown in [7] that the problem of simultaneous broadcasting in a graph G can be solved locally on a suitable subgraph H of G and then extended to a solution for the whole graph G (c.f. Proposition 2), but without guarantee of optimality.

In this paper, the latter result is improved in the terms of optimality by showing that if H satisfies additional properties, namely if H contains all neighbours of the originator vertex v and preserves distances to v, then simultaneous broadcasting from v on H with optimal time for each destination vertex can be extended to simultaneous broadcasting from v on G again with optimal time for each destination vertex (Theorem 1).

Furthermore, we identify particular subgraphs, namely wheels and biwheels, that play a key role for simultaneous broadcasting. We show (Theorems 2–4) that they can be used for simultaneous broadcasting (of optimal number) of messages in optimal time for a wide class of graphs.

In particular, since biwheels naturally occur in Cartesian products (Propositions 3 and 4), we obtain tight results for bipartite tori, meshes, and hypercubes. For these graphs we also provide an explicit description how optimal simultaneous broadcasting can be realized (Sect. 4). We also answer affirmatively a conjecture from [6] asserting that the n-dimensional hypercube admits simultaneous broadcasting of n messages in optimal time $3n - 2$. We conclude with summary of open problems and conjectures (Sect. 5).

2   Concept of Level-Disjoint Partitions

In this paper we use the concept of level-disjoint partitions, introduced in [6], to capture broadcasting under the considered communication model. We use standard graph terminology and notation. An open neighborhood of a vertex u in a graph G is denoted by $N_G(u)$, the degree of u by $\deg_G(u)$, the distance between vertices u and v by $d_G(u, v)$. The *eccentricity* of a vertex u, i.e. the maximal distance from u to other vertices, is denoted by $\mathrm{ecc}_G(u)$. The subscript G is omitted whenever the graph is clear from context.

A *level partition* of a graph G is a partition $\mathcal{S} = (S_0, \ldots, S_h)$ of $V(G)$ into a tuple of sets, called *levels*, such that $S_i \subseteq N(S_{i-1})$ for every $1 \leq i \leq h$; that is, every vertex has a neighbor from previous level. The number $h = h(\mathcal{S}) = |\mathcal{S}| - 1$ is called the *height* of $\mathcal{S}$. The broadcasting starts at all vertices from the level S_0, at each time unit the same message is sent from all vertices of the current level to all vertices in the next level through edges of the graph, till the hth time unit, when the message is spread to all vertices of G. Note that we do not care which particular edges are used. In the case when the starting level S_0 is a singleton, say $S_0 = \{v\}$, we say that the level partition is *rooted* at v (or v-*rooted*) and the vertex v is called the *root* of $\mathcal{S}$.

A level partition $(S_0, \ldots, S_h)$ of G with $S_i = \{u \in V(G) \mid d_G(u, S_0) = i\}$ for every $0 \leq i \leq h$ is called a *distance level partition*. Clearly, a distance level partition is determined by the choice of the starting level S_0 and it has minimal height among all level partitions with the same starting level. If, moreover, it is rooted at a vertex v, it corresponds to the breadth-first-search tree from v (up to the choice of edges).

Two level partitions $\mathcal{S} = (S_0, \ldots, S_{h(\mathcal{S})})$ and $\mathcal{T} = (T_0, \ldots, T_{h(\mathcal{T})})$ are said to be *level-disjoint* if $S_i \cap T_i = \emptyset$ for every $1 \leq i \leq \min(h(\mathcal{S}), h(\mathcal{T}))$. Note that we allow $S_0 \cap T_0 \neq \emptyset$ since we consider the case when different messages have the same originator. Level partitions $\mathcal{S}^1, \ldots, \mathcal{S}^k$ are said to be (mutually) level-disjoint if each two partitions are level-disjoint. Then we say that $\mathcal{S}^1, \ldots, \mathcal{S}^k$ are *level-disjoint partitions*, shortly *LDPs*. If every partition is rooted in the same vertex v and they are level-disjoint (up to the starting level $\{v\}$), we say that $\mathcal{S}^1, \ldots, \mathcal{S}^k$ are *level-disjoint partitions with the same root v*, shortly v-*rooted LDPs*. For an example of four v-rooted LDPs of a circulant graph, see Fig. 2. Note that the 4-tuple at a vertex denotes its levels in each partition.

Let $\mathcal{S}^1, \ldots, \mathcal{S}^k$ be level partitions of G, not necessarily level-disjoint. The set of levels $\{l \mid u \in S_l^i \text{ for some } 1 \leq i \leq k\}$ in which a given vertex u occurs is called the *range* of u *with respect to* $\mathcal{S}^1, \ldots, \mathcal{S}^k$, denoted by $R(u)$.

The number of level-disjoint partitions determines how many messages can be broadcast simultaneously while their maximal height determines the overall time of the broadcasting. Hence a general aim is to construct for a given graph

- as many as possible (mutually) level-disjoint partitions; and
- with as small maximal height as possible.

In [7] some necessary conditions on the number of v-rooted LDPs as well as on their maximal height were given. Assume that $\mathcal{S}^1, \ldots, \mathcal{S}^k$ are v-rooted LDPs of G. Clearly, for every vertex u except v, $\max(R(u)) \geq d(u, v) + k - 1$ since u cannot appear in a level smaller than the distance to the root v and $|R(u)| = k$. If equality holds, we say that u has *perfect range*; that is,

$$R(u) = \{d(u, v), d(u, v) + 1, \ldots, d(u, v) + k - 1\}.$$

This means that all k messages will be delivered to the vertex u in the best time possible for this vertex. If all vertices (up to the root v) have perfect range, we say that level-disjoint partitions $\mathcal{S}^1, \ldots, \mathcal{S}^k$ are *perfect*.

Furthermore, the above definition is adjusted for bipartite graphs. If G is bipartite, then for any same-rooted LDPs of G, the range of each vertex contains elements of the same parity. It follows that no vertex can have perfect range as defined above (except the trivial case of a single partition). So the concept of perfect range is relaxed for bipartite graphs as follows. In a bipartite graph G, for every vertex u except v, $\max(R(u)) \geq d(u, v) + 2k - 2$. If equality holds, we say that u has *biperfect range*; that is,

$$R(u) = \{d(u, v), d(u, v) + 2, \ldots, d(u, v) + 2k - 2\}.$$

If all vertices (up to the root v) have biperfect range, we say that level-disjoint partitions $\mathcal{S}^1, \ldots, \mathcal{S}^k$ are *biperfect*. Further, the following necessary conditions on same-rooted LDPs were proven in [7].

Proposition 1 ([7]). *Let $\mathcal{S}^1, \ldots, \mathcal{S}^k$ be level-disjoint partitions of a graph G with the same root v. Then,*

$$k \leq \deg(v) \tag{1}$$

$$\max_{1 \leq i \leq k} h(\mathcal{S}^i) \geq \begin{cases} \mathrm{ecc}(v) + k - 1 & \text{if } G \text{ is not bipartite,} \\ \mathrm{ecc}(v) + 2k - 2 & \text{if } G \text{ is bipartite.} \end{cases} \tag{2}$$

2.1 Subgraph Extension Technique

In [7] it was shown that it suffices to find v-rooted LDPs on some suitable subgraph H of G and then extend them to v-rooted LDPs of the whole graph G as stated by the following proposition. Let $G - v$ denote the graph obtained by removing a vertex v and all incident edges from a G.

Proposition 2 ([7]). *Let v be a vertex of a graph G and H be a subgraph of G containing v and some vertex from each component of $G - v$. Then any k v-rooted level-disjoint partitions of H can be extended to k v-rooted level-disjoint partitions of G.*

Our first result in this paper extends Proposition 2 in terms of preserving (bi)perfectness. It shows that it suffices to find (bi)perfect LDPs locally on a subgraph H of G that covers all neighbors of the root v and preserves distances to v. Then they can be extended to (bi)perfect, respectively, LDPs with the same root to the whole graph G. We say that a subgraph H of a graph G *preserves distances* to a vertex $v \in V(H)$ if $d_H(u,v) = d_G(u,v)$ for every $u \in V(H)$. If a subgraph $H \subseteq G$ does not contain a vertex v of G, we denote by $H + v$ the subgraph of G obtained by adding v and all incident edges from G to H.

Theorem 1. *Let v be a vertex of a graph G and H be a subgraph of G containing $N(v) \cup \{v\}$ and preserving distances to v. Then any k (bi)perfect v-rooted level-disjoint partitions of H can be extended to k (bi)perfect, respectively, v-rooted level-disjoint partitions of G.*

Proof. Let $\mathcal{S}^1, \ldots, \mathcal{S}^k$ be (bi)perfect level-disjoint partitions of H rooted in v and assume $V(H) \subsetneq V(G)$; for otherwise we are done. We show that they can be extended to (bi)perfect, respectively, v-rooted level-disjoint partitions of $H' = H + u$ for some vertex u of G uncovered by H such that H' preserves distances to v. Then, by incremental extension until no uncovered vertex remains, we obtain (bi)perfect v-rooted level-disjoint partitions of G.

Let u be a vertex of G that is not in H but has a neighbor w in H distinct from v such that w belongs to some shortest path in G between u and v. Note that such u exist since $N(v) \subseteq V(H)$. Since H preserves distances to v, for $H' = H + u$ we have

$$d_{H'}(u,v) = d_H(w,v) + 1 = d_G(w,v) + 1 = d_G(u,v),$$

and thus H' preserves distances to v as well.

Let us denote by l_i the level of w in $\mathcal{S}^i$; that is, $w \in S^i_{l_i}$ for every $1 \le i \le k$. Then, we extend $\mathcal{S}^1, \ldots, \mathcal{S}^k$ to H' by adding u to the $(l_i + 1)$-th level of $\mathcal{S}^i$ for every $1 \le i \le k$. Clearly, such extended partitions are level partitions of H'. Moreover, they are level-disjoint since u was added into distinct levels of level-disjoint partitions $\mathcal{S}^1, \ldots, \mathcal{S}^k$. Finally, if $\mathcal{S}^1, \ldots, \mathcal{S}^k$ are perfect, then

$$R(w) = \{l_i \mid 1 \le i \le k\} = \{d_H(w,v), d_H(w,v) + 1, \ldots, d_H(w,v) + k - 1\},$$

and therefore the vertex u has perfect range as well:

$$R(u) = \{l_i + 1 \mid 1 \le i \le k\} = \{d_{H'}(u,v), d_{H'}(u,v) + 1, \ldots, d_{H'}(u,v) + k - 1\}.$$

Similarly, if $\mathcal{S}^1, \ldots, \mathcal{S}^k$ are biperfect, then

$$R(w) = \{l_i \mid 1 \le i \le k\} = \{d_H(w,v), d_H(w,v) + 2, \ldots, d_H(w,v) + 2k - 2\},$$

and therefore the vertex u has biperfect range as well:

$$R(u) = \{l_i + 1 \mid 1 \le i \le k\} = \{d_{H'}(u,v), d_{H'}(u,v) + 2, \ldots, d_{H'}(u,v) + 2k - 2\}.$$

$\square$

The above theorem is applied in the next section to obtain (bi)perfect level-disjoint partitions of a wide class of graphs, including particular Cartesian products.

3 Simultaneous Broadcasting in Cartesian Products

A Cartesian product of graphs G and H is the graph $G \,\square\, H$ with the vertex set $V(G \,\square\, H) = V(G) \times V(H)$ and the edge set $E(G \,\square\, H) = \{(u,v)(u',v) \mid uu' \in E(G), v \in V(H)\} \cup \{(u,v)(u,v') \mid u \in V(G), vv' \in E(H)\}$. As an example, consider the hypercube. The n-dimensional hypercube Q_n is the graph on vertices $V(Q_n) = \{0,1\}^n$ and edges between vertices that differ in precisely one coordinate. Observe that Q_n can be viewed as the n-fold Cartesian product of K_2.

In [6] we developed a concept of composing level-disjoint partitions with certain properties of graphs G and H into level-disjoint partitions of $G \,\square\, H$. Here we present a different approach of so called *biwheels*. We define biwheels in the next Subsect. 3.1 and show that they naturally occur in Cartesian products. Then in the Subsect. 3.2 we show how they can be used for construction of optimal number of same-rooted level-disjoint partitions of optimal height in Cartesian products. Finally, in the Subsect. 4 we consider particular Cartesian products: meshes, tori, hypercubes and we present optimal constructions for them also explicitly.

3.1 Biwheels in Cartesian Products

First we formally define wheels and biwheels. A k-*wheel* W_k for $k \geq 0$ *centered at* a vertex v is the graph on vertices $v, w_1, \ldots, w_k$ with edges joining v to all w_i's and edges joining w_i and w_{i+1} for every $1 \leq i \leq k$ where w_{k+1} is identified as w_1. Note that for technical reasons we allow $k \leq 2$. A k-wheel for $k \geq 3$ is a join of a k-cycle and a vertex.

A k-*biwheel* $\widehat{W}_k$ for $k \geq 0$ *centered at* a vertex v is the subdivision of W_k centered at v obtained by inserting a new vertex x_i to the edge between w_i and w_{i+1} for every $1 \leq i \leq k$. Clearly, k-biwheel is bipartite for every k whereas k-wheel is bipartite only for $k = 0, 1$. See Fig. 1 for an illustration of small wheels and biwheels.

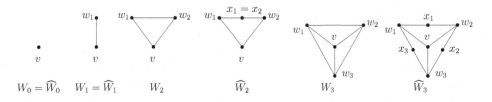

Fig. 1. k-wheels and k-biwheels centered at v for $k = 0, 1, 2, 3$.

Biwheels naturally occur in Cartesian products of graphs as stated by the following proposition.

Proposition 3. *Let u, v be vertices in graphs G, H respectively. Then $G \square H$ has a 2k-biwheel centered at (u, v) for any $0 \leq k \leq \min(\deg_G(u), \deg_H(v))$.*

Proof. It suffices to prove the statement for $k = \deg_G(u) = \deg_H(v)$ for otherwise we may use subgraphs G' of G and H' of H such that $k = \deg_{G'}(u) = \deg_{H'}(v)$ as $G' \square H'$ is a subgraph of $G \square H$. For $1 \leq i \leq k$ let us denote the i-th neighbor of u, v by u_i, v_i, respectively. We define the vertices of a 2k-biwheel in $G \square H$ as follows:

$$w_{2i-1} = (u_i, v), \qquad\qquad x_{2i-1} = (u_i, v_i),$$
$$w_{2i} = (u, v_i), \qquad\qquad x_{2i} = (u_{i+1}, v_i)$$

for every $1 \leq i \leq k$ where u_{k+1} is identified as u_1.

Note that all vertices $w_{2i-1}, x_{2i-1}, w_{2i}, x_{2i}$ are distinct, vertices w_{2i-1}, w_{2i} are adjacent to (u, v), and $w_{2i-1}x_{2i-1}, x_{2i-1}w_{2i}, w_{2i}x_{2i}, x_{2i}w_{2i+1}$ are edges in $G \square H$ for every $1 \leq i \leq k$. Hence these vertices form a 2k-biwheel in $G \square H$ centered at (u, v). ∎

In particular, if $\deg_G(u) = \deg_H(v)$ then $G \square H$ has a $\deg_{G \square H}((u, v))$-biwheel centered at (u, v); that is, the largest possible biwheel at (u, v). For example, $P_2 \square P_2$ contains a 2-biwheel (with center in any vertex) or $P_3 \square P_3$ contains a 4-biwheel centered in the degree-4 vertex. In fact, $P_2 \square P_2 \simeq \widehat{W}_2$ and $P_3 \square P_3 \simeq \widehat{W}_4$.

For another example, by recursive applications we obtain an n-biwheel in the hypercube Q_n for every $n = 2^m$ where m is an integer since $Q_{2^m} \simeq Q_{2^{m-1}} \square Q_{2^{m-1}}$. However, we would like to have n-biwheel in Q_n for any n. For this purpose we need a more general result as follows.

Proposition 4. *Let u, v be vertices in graphs G, H respectively, with $\deg_G(u) \geq \deg_H(v) \geq 1$ and $l = \max(2, \deg_G(u) - \deg_H(v) + 1)$. If G has at least l-biwheel centered at u, then $G \square H$ has a k-biwheel centered at (u, v) for any $0 \leq k \leq \deg_G(u) + \deg_H(v) = \deg_{G \square H}((u, v))$.*

Proof. Let $k = 2k' + l'$ where k' is the maximal integer such that $k' \leq \deg_H(v)$ and $l' \geq 0$. It follows that $l' \leq l - 1$. Indeed, if $k' < \deg_H(v)$ then $l' = 0$ or $l' = 1$, and if $k' = \deg_H(v)$ then

$$l' \leq \deg_G(u) + \deg_H(v) - 2k' = \deg_G(u) - \deg_H(v) \leq l - 1.$$

(The first inequality holds since $k = 2k' + l' \leq \deg_G(u) + \deg_H(v)$ and the last inequality holds since $\deg_G(u) - \deg_H(v) + 1 \leq l$.)

Let us denote by u_i for $1 \leq i \leq \deg_G(u)$ the i-th neighbor of u, and by v_j for $1 \leq j \leq \deg_H(v)$ the j-th neighbor of v. Furthermore, let $(u_1, y_1, \ldots, u_{l'}, y_{l'}, u_{l'+1})$ be a subpath of the at least l-biwheel of G centered at u where $d_G(u, y_i) = 2$ for all $1 \leq i \leq l'$. We define the vertices of a k-biwheel in $G \square H$ as follows:

$$
\begin{array}{llll}
w_i = (u_i, v), & x_i = (y_i, v) & \text{for all } i = 1, \ldots, l', \\
w_{l'+2j-1} = (u_{l'+j}, v), & x_{l'+2j-1} = (u_{l'+j}, v_j) & \text{for all } j = 1, \ldots, k', \\
w_{l'+2j} = (u, v_j), & x_{l'+2j} = (u_{l'+j+1}, v_j) & \text{for all } j = 1, \ldots, k'-1, \\
w_{l'+2j} = (u, v_j), & x_{l'+2j} = (u_1, v_j) & \text{for } j = k'.
\end{array}
$$

Note that all vertices w_i, x_i are distinct, vertices w_i are adjacent to (u, v), and $w_i x_i$, $x_i w_{i+1}$ are edges in $G \square H$ for every $1 \leq i \leq k = 2k' + l'$. Hence these vertices form a k-biwheel in $G \square H$ centered at (u, v). □

3.2   (Perfect) Level-Disjoint Partitions from Wheels and Biwheels

Both k-wheels and k-biwheels (except for $k = 2$) have obvious k perfect, respectively biperfect, level-disjoint partitions rooted in their centers. Indeed, let the i-th level partition $\mathcal{S}^i$ for $1 \leq i \leq k$ of W_k where $k \geq 1$ be

$$\mathcal{S}^i = (\{v\}, \{w_i\}, \{w_{i+1}\}, \dots, \{w_{i+k-1}\})$$

with the indices taken cyclically; that is, modulo $(k + 1)$ plus 1. Clearly, $\mathcal{S}^i$ and $\mathcal{S}^j$ are level-disjoint up to the root v for any distinct i, j. Similarly for a k-biwheel where $k \geq 3$, let the i-th level partition $\mathcal{T}^i$ for $1 \leq i \leq k$ of $\widehat{W}_k$ be

$$\mathcal{T}^i = (\{v\}, \{w_i\}, \{x_i\}, \{w_{i+1}\}, \{x_{i+1}\}, \dots, \{w_{i+k-1}\}, \{x_{i+k-1}\})$$

with the indices taken cyclically; that is, modulo $(k + 1)$ plus 1. Clearly, $\mathcal{T}^i$ and $\mathcal{T}^j$ are level-disjoint up to the root v for any distinct i, j.

Note that the above level-disjoint partitions of W_k and $\widehat{W}_k$ are perfect, respectively biperfect. Hence their maximal height is optimal in the sense of Proposition 1. Also their number is optimal. Indeed, $k = \deg_{W_k}(v) = \deg_{\widehat{W}_k}(v)$, W_k is non-bipartite, $\widehat{W}_k$ is bipartite, and the maximal heights are

$$\max_{1 \leq i \leq k} h(\mathcal{S}^i) = k = \mathrm{ecc}_{W_k}(v) + k - 1, \qquad \max_{1 \leq i \leq k} h(\mathcal{T}^i) = 2k = \mathrm{ecc}_{\widehat{W}_k}(v) + 2k - 2.$$

The above partitions together with Proposition 2 lead to following sufficient conditions on existence of k level-disjoint partitions with the same root. A vertex v is a *cut-vertex* in a graph G if $G - v$ is disconnected.

Theorem 2. *If a graph G has a k-wheel for $k \geq 1$ or k-biwheel for $k \geq 3$ centered at a vertex v and v is not a cut-vertex, then G has k level-disjoint partitions rooted at v.*

Note that the above theorem can be easily generalized for a vertex v that is not a cut-vertex and is adjacent to k vertices on an arbitrarily large cycle in G. Theorem 2 together with Propositions 3 and 4 applies in particular for Cartesian products of (nontrivial) connected graphs as they are 2-connected.

Furthermore, applying Theorem 1 we obtain a sufficient condition on existence of optimal number of (bi)perfect level-disjoint partitions with the same root.

Theorem 3. *Let v be a vertex of degree $k \geq 1$ in a graph G. If G has a k-wheel centered at v, then G has k perfect level-disjoint partitions rooted at v. If G is bipartite, $k \geq 3$, and G has a k-biwheel centered at v, then G has k biperfect level-disjoint partitions rooted at v.*

Proof. Let H denote the k-wheel resp. k-biwheel centered at v. All neighbors of v in G are also in H. In addition, distances to v from G are preserved in H. (Note that in the case of k-biwheel we have $d_G(v, x_i) = d_H(v, x_i) = 2$ for every $1 \leq i \leq k$ since G is bipartite.) Hence we may apply Theorem 1 to extend the above (bi)perfect level-disjoint partitions of H to G.   □

Theorem 3 can be applied to obtain perfect or biperfect level-disjoint partitions for various graphs. For an example, see the four biperfect level-disjoint partitions of the circulant graph in Fig. 2. Further examples are provided in the next subsection.

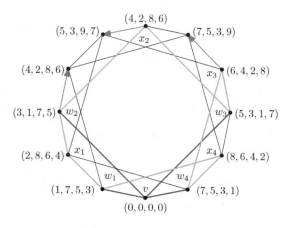

Fig. 2. Four perfect level-disjoint partitions of a circulant graph rooted at v obtained from a 4-biwheel.

Next we generalize Theorem 3. If l divides k, then the above k (bi)perfect level-disjoint partitions of W_k or $\widehat{W}_k$ can be compressed into l (bi)perfect level-disjoint partitions. Let $k = pl$ for some integers l, p and let the i-th level partition of W_k for $1 \leq i \leq l$ be $\mathcal{U}^i = (U_0^i = \{v\}, U_1^i, \ldots, U_l^i)$ where

$$U_j^i = \{w_{i+j-1+l}, w_{i+j-1+2l}, \ldots, w_{i+j-1+pl}\}$$

for $1 \leq j \leq l$ and the indices are taken cyclically; that is, modulo $(k+1)$ plus 1. Clearly, $\mathcal{U}^i$ and $\mathcal{U}^j$ are level-disjoint for any distinct $1 \leq i, j \leq l$. Similarly for a k-biwheel with $k = pl$, let the i-th level partition of $\widehat{W}_k$ for $1 \leq i \leq l$ be $\mathcal{V}^i = (\{v\}, U_1^i, X_1^i, \ldots, U_l^i, X_l^i)$ where U_j^i is as above and

$$X_j^i = \{x_{i+j-1+l}, x_{i+j-1+2l}, \ldots, x_{i+j-1+pl}\}$$

for $1 \leq j \leq l$ and the indices are taken cyclically; that is, modulo $(k+1)$ plus 1. Clearly, $\mathcal{V}^i$ and $\mathcal{V}^j$ are level-disjoint for any distinct $1 \leq i, j \leq l$.

These partitions lead to generalization of Theorem 3 as follows. Additional properties of these partitions, called partitions *modulo p*, have been studied in [6].

Theorem 4. *Let v be a vertex in a graph G of degree $k \geq 1$ divisible by an integer $l \geq 1$. If G has a k-wheel centered at v, then G has l perfect level-disjoint partitions rooted at v. If G is bipartite, $k \geq 3$, and G has a k-biwheel centered at v, then G has l biperfect level-disjoint partitions rooted at v.*

4   Particular Networks

In this section we consider particular examples of Cartesian products and provide explicit constructions for them. We also propose several problems and conjectures.

4.1   Torus $C_{2n} \,\square\, C_{2m}$

First we consider a bipartite 2-dimensional torus; that is, the graph $C_{2n} \,\square\, C_{2m}$ where $n, m \geq 2$. By Proposition 3, it has a 4-biwheel centered at any vertex r. Hence by Theorem 3, it has four level-disjoint partitions rooted at r of (optimal) height

$$ecc_{C_{2n}\square C_{2m}}(r) + 6 = n + m + 6.$$

Explicitly, let us denote the vertices of cycles C_{2n}, C_{2m} by $C_{2n} = (u_1, \ldots, u_{2n})$, $C_{2m} = (v_1, \ldots, v_{2m})$ and assume $r = (u_1, v_1)$. We define a function $f(i, j, k)$ for $1 \leq i \leq 2n$, $1 \leq j \leq 2m$, $1 \leq k \leq 4$ determining the level of each vertex (u_i, v_j) in the k-th level partition by

$$f(i,j,k) = d(i,j) + \begin{cases} 2((k-1) \bmod 4) & \text{if } 1 < i \leq n \text{ and } 1 \leq j \leq m, \\ 2(k \bmod 4) & \text{if } i = 1 \text{ or } n < i \leq 2n, \text{ and } 2 \leq j \leq m, \\ 2((k+1) \bmod 4) & \text{if } n < i \leq 2n, \text{ and } j = 1 \text{ or } m < j \leq 2m, \\ 2((k+2) \bmod 4) & \text{if } 1 \leq i \leq n \text{ and } m < j \leq 2m \end{cases}$$

and $f(1, 1, k) = 0$, where $d(i, j) = d_{C_{2n}\square C_{2m}}((u_i, v_j), (u_1, v_1))$ is

$$d(i,j) = \begin{cases} i + j - 2 & \text{if } 1 \leq i \leq n \text{ and } 1 \leq j \leq m, \\ 2n - i + j & \text{if } n < i \leq 2n \text{ and } 1 \leq j \leq m, \\ 2n - i + 2m - j + 2 & \text{if } n < i \leq 2n \text{ and } m < j \leq 2m, \\ i + 2m - j & \text{if } 1 \leq i \leq n \text{ and } m < j \leq 2m. \end{cases}$$

Then it is easy to verify that $\mathcal{S}^k = (S_0^k, \ldots, S_h^k)$ for $k = 1, \ldots, 4$ where $h = n + m + 6$ and

$$S_l^k = \{(u_i, v_j) \mid f(i, j, k) = l, 1 \leq i \leq 2n, 1 \leq j \leq 2m\}$$

for every $0 \leq l \leq h$ are biperfect level-disjoint partitions of $C_{2n} \,\square\, C_{2m}$. For example see Fig. 3.

Furthermore, by applying Theorem 4 we obtain two same-rooted level-disjoint partitions of $C_{2n} \,\square\, C_{2m}$ of (optimal) height $n + m + 2$. Trivially, a single distance partition from the root has (optimal) height $n + m$ as well. Hence it remains a question whether $C_{2n} \,\square\, C_{2m}$ has three same-rooted level-disjoint partitions of (optimal) height $n + m + 4$, which is perhaps easy to resolve. More interestingly, this can be generalized for higher dimensions as follows.

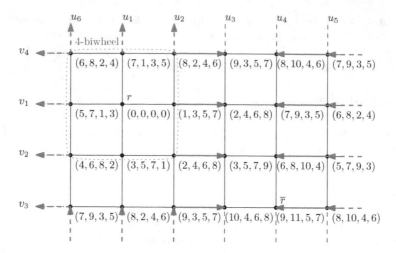

Fig. 3. Four biperfect level-disjoint partitions of $C_6 \,\square\, C_4$ rooted at $r = (u_1, v_1)$ of maximal height 11. The eccentric vertex to r, denoted by $\bar{r}$, is in the (last) 11th level of the second partition. The arrows denote how these LDPs are constructed from LDPs of the 4-biwheel centered at r.

Conjecture 1. The (bipartite) generalized torus $C_{2n_1} \,\square\, C_{2n_2} \,\square\, \cdots \,\square\, C_{2n_d}$ where $d \geq 2$ and $n_1, \ldots, n_d \geq 2$ has l same-rooted level-disjoint partitions of optimal height for every $1 \leq l \leq 2d$.

We only know from Theorem 4 that Conjecture 1 holds if l divides $2d$. Note that by (2), an optimal height of four r-rooted level-disjoint partitions of a non-bipartite torus is $ecc(r) + 3$ instead of $ecc(r) + 6$ for bipartite case. This leads us to pose the following problem. Clearly, if Conjecture 1 holds, this problem reduces only to non-bipartite cases.

Problem 1. Which of generalized tori $C_{m_1} \,\square\, C_{m_2} \,\square\, \cdots \,\square\, C_{m_d}$ admit $2d$ same-rooted level-disjoint partitions of optimal height?

4.2   Mesh $P_n \,\square\, P_m$

For 2-dimensional meshes $P_n \,\square\, P_m$ where $n, m \geq 3$ we obtain similar results as for tori, up to choice of the root. Let us denote the vertices of paths P_n, P_m by $P_n = (u_1, \ldots, u_n)$, $P_m = (v_1, \ldots, v_m)$. A vertex (u_i, v_j) of $P_n \,\square\, P_m$ is an *inner vertex* if $1 < i < n$ and $1 < j < m$; and a *border vertex* otherwise.

By Proposition 3, the mesh $P_n \,\square\, P_m$ has a 4-biwheel centered at any inner vertex. Hence by Theorem 3 it has four level-disjoint partitions rooted at the same inner vertex $r = (u_i, v_j)$ of (optimal) height $ecc_{P_n \square P_m}(r) + 6$ where

$$ecc_{P_n \square P_m}((u_i, u_j)) = \max(i + j - 2, i + m - j - 1, n - i + m - j, n - i + j - 1).$$

For example, see Fig. 4.

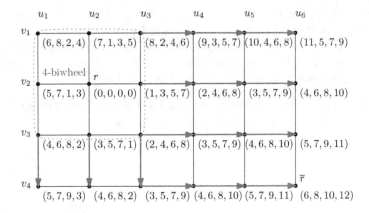

Fig. 4. Four biperfect level-disjoint partitions of $P_6 \square P_4$ rooted at $r = (u_2, v_2)$ of maximal height 12. The eccentric vertex to r, denoted by $\overline{r}$, is in the (last) 12th level of the fourth partition. The arrows denote how these LDPs are constructed from LDPs of the 4-biwheel centered at r.

Furthermore, by applying Theorem 4 we obtain two level-disjoint partitions rooted at the same inner vertex r of (optimal) height $ecc(r) + 2$. Explicit constructions of such level-disjoint partitions can easily be derived in a similar way as for torus. We leave them out as they are merely technical. Similarly as for bipartite tori, we propose the following conjecture.

Conjecture 2. The generalized mesh $P_{m_1} \square P_{m_2} \square \cdots \square P_{m_d}$ where $d \geq 2$ and $m_1, \ldots, m_d \geq 3$ has l r-rooted level-disjoint partitions of optimal height for every $1 \leq l \leq 2d$ and every inner vertex r.

Note that Conjecture 2 implies Conjecture 1 since a bipartite torus contains a mesh with the same parameters $2n_1, \ldots, 2n_d$ as a spanning subgraph, and the mesh has an inner vertex with eccentricity equal to the eccentricity of any vertex of the torus. We only know from Theorem 4 that Conjecture 2 holds if l divides $2d$. Note that we considered only inner vertices as roots since for border vertices in 2-dimensional meshes there are no k-biwheels with $k \geq 3$.

Problem 2. Determine the maximal number of r-rooted level-disjoint partitions of optimal height in the generalized mesh $P_{m_1} \square P_{m_2} \square \cdots \square P_{m_d}$ for all vertices r and all parameters $d \geq 2$, $m_1, \ldots, m_d \geq 2$.

4.3   Hypercube Q_n

We view the n-dimensional hypercube Q_n for $n \geq 3$ as the Cartesian product of $C_4 \simeq \widehat{W}_2$ and the $(n-2)$-fold Cartesian product of K_2; that is, $Q_n \simeq C_4 \square (K_2)^{n-2}$. By recursive application of Proposition 4, we obtain that Q_n for any $n \geq 3$ has an n-biwheel centered at any vertex v. Explicitly, let us assume that $v = 0^n = (0, \ldots, 0)$. Then an n-biwheel centered at v is formed (for example) by vertices $w_i = e_i$ for $i = 1, \ldots, n$, $x_i = e_i \oplus e_{i+1}$ for $i = 1, \ldots, n-1$, and $x_n = e_1 \oplus e_n$, where e_i denotes the vector with 1 exactly in the ith coordinate.

Hence by Theorem 3 we obtain the following result, answering affirmatively a conjecture from [6] where only the case when $n = 3 \cdot 2^i$ or $n = 4 \cdot 2^i$ for some integer $i \geq 0$ was shown. See examples for $n = 3$ and $n = 4$ on Fig. 5.

Corollary 1. *For every $n \geq 3$ there exist n level-disjoint partitions of Q_n with the same root and with the maximal height $3n - 2$.*

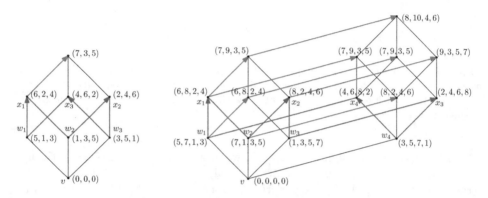

Fig. 5. (a) Three biperfect level-disjoint partitions of Q_3 rooted at v of maximal height 7. (b) Four biperfect level-disjoint partitions of Q_4 rooted at v of maximal height 10.

Explicitly, we define a function $f(u, k)$ for $u \in V(Q_n)$, $1 \leq k \leq n$ determining the level of each vertex u in the k-th level partition as

$$
f(u, k) = \begin{cases} 0 & \text{if } u = v \ (= 0^n), \\ 2((n+k) \bmod n) + 2 & \text{if } u = x_n \ (= e_1 \oplus e_n), \\ 2((i+k) \bmod n) + j & \text{otherwise} \end{cases}
$$

where i is the position of the leftmost 1 in u and j is the number of 1's in u. Then it is easy to verify that $\mathcal{S}^k = (S_0^k, \ldots, S_h^k)$ for $k = 1, \ldots, n$ where $h = 3n - 2$ and $S_l^k = \{u \in V(Q_n) \mid f(u, k) = l\}$ for every $0 \leq l \leq h$ are biperfect level-disjoint partitions of Q_n.

Note that the above definition of $f(u, k)$ is based on the fact that each vertex u except v or x_n has a shortest path to the root v that goes through $w_i = e_i$ and avoids x_n, where i is the position of the leftmost 1 in u. Indeed, from u by consecutively changing the rightmost 1 to 0 we obtain all vertices of such a path. Furthermore, from x_i we go to $w_i = e_i$ along these paths for each $i = 1, \ldots, n-1$ which agrees with the partition of the n-biwheel $\widehat{W}_n$. Therefore, we may extend the canonical biperfect level-disjoint partitions of $\widehat{W}_n$ along these paths to biperfect level-disjoint partitions of Q_n by applying Theorem 1, which corresponds to the above prescription for $f(u, k)$.

Furthermore, from Theorem 4 we obtain that Q_n for any $n \geq 3$ has k biperfect level-disjoint partitions rooted at the same vertex (of maximal height $n + 2k - 2$) if k divides n. We propose that it holds for any $k \geq 1$.

Conjecture 3. For any $1 \leq k \leq n$, $n \geq 3$, the hypercube Q_n has k same-rooted level-disjoint partitions of optimal height.

5   Conclusions

In this work the concept of level-disjoint partitions which was originally introduced in [6] is employed to describe simultaneous broadcasting of multiple messages from the same originator in the considered communication model.

It is shown that a local solution on a suitable subgraph can be extended to the whole graph without loss of optimality. In this paper we use specifically wheels and biwheels as local subgraphs. This could be further generalized for other subgraphs such as subdivisions of wheels.

This approach leads to simultaneous broadcasting in optimal time on particular Cartesian products of graphs. However, it can be applied for a much larger class of graphs. For example, for some circulant graphs or Knödel graphs that have been previously studied in the context of broadcasting [3,4,11].

For bipartite tori, meshes, and hypercubes we provided tight results based on construction of optimal number of biperfect level-disjoint partitions from biwheels. We believe that simultaneous broadcasting can be achieved in optimal time for any number of messages on generalized bipartite tori, generalized meshes, and hypercubes (Conjectures 1–3). The problem of simultaneous broadcasting in optimal time remains open for general tori (Problem 1) and meshes with border originator vertices (Problem 2).

References

1. Bar-Noy, A., Kionis, S., Schieber, B.: Optimal multiple message broadcasting in telephone-like communication systems. Discrete Appl. Math. **100**, 1–15 (2000)
2. Bruck, J., Cypher, R., Ho, C.T.: Multiple message broadcasting with generalized Fibonacci trees. In: Proceedings of the 4th Symposium on Parallel and Distributed Processing, pp. 424–431 (1992)
3. Chang, F.-H., Chen, Y.-M., Chia, M.-L., Kuo, D., Yu, M.-F.: All-to-all broadcast problem of some classes of graphs under the half duplex all-port model. Discrete Appl. Math. **173**, 28–34 (2014)
4. Fertin, G., Raspaud, A.: A survey on Knödel graphs. Discrete Appl. Math. **137**, 276–289 (2013)
5. Farley, A.: Broadcast time in communication networks. SIAM J. Appl. Math. **39**, 385–390 (1980)
6. Gregor, P., Škrekovski, R., Vukašinović, V.: Rooted level-disjoint partitions of Cartesian products. Appl. Math. Comput. **266**, 244–258 (2015)
7. Gregor, P., Škrekovski, R., Vukašinović, V.: Modelling simultaneous broadcasting by level-disjoint partitions. Preprint arXiv:1609.01116

8. Grigoryan, H.: Problems related to broadcasting in graphs. Ph.D. thesis, Concordia University, Montreal, Quebec, Canada (2013)
9. Grigoryan, H., Harutyunyan, H.A.: Diametral broadcast graphs. Discrete Appl. Math. **171**, 53–59 (2014)
10. Harutyunyan, H.A.: Minimum multiple message broadcast graphs. Networks **47**, 218–224 (2006)
11. Harutyunyan, H.A.: Multiple message broadcasting in modified Knödel graphs. In: Proceedings of the 7th International Colloquium on Structural Information and Communication Complexity, pp. 157–165 (2000)
12. Hedetniemi, S.M., Hedetniemi, S.T., Liestman, A.L.: A survey of gossiping and broadcasting in communication networks. Networks **18**, 319–349 (1988)
13. Hromkovič, J., Klasing, R., Monien, B., Piene, R., Du, D.-Z., Hsu, D.F.: Dissemination of information in communication networks (broadcasting and gossiping). Combinatorial Network Theory. Applied Optimization, vol. 1, pp. 125–212. Springer, New York (1996)
14. Hromkovič, J., Klasing, R., Pelc, A., Ružička, P., Unger, W.: Dissemination of Information in Communication Networks: Broadcasting, Gossiping, Leader Election, and Fault-Tolerance. Texts in Theoretical Computer Science. Springer, Berlin (2005)
15. Leighton, F.T.: Introduction to Parallel Algorithms and Architectures: Arrays, Trees, Hypercubes. Morgan Kaufmann, San Mateo (1992)
16. Sun, C.M., Lin, C.K., Huang, H.M., Hsu, L.H.: Mutually independent Hamiltonian cycles in hypercubes. In: Proceedings of 8th Symposium on Parallel Architectures, Algorithms and Networks (2005)
17. Vukašinović, V., Gregor, P., Škrekovski, R.: On the mutually independent Hamiltonian cycles in faulty hypercubes. Inform. Sci. **236**, 224–235 (2013)
18. Wu, K.-S., Juan, JS.-T.: Mutually independent Hamiltonian cycles of $C_m \times C_n$ when m, n are odd. In: Proceedings of 29th Workshop on Combinatorial Mathematics and Computation Theory, pp. 165–170 (2012)

Fast Searching on Complete k-partite Graphs

Yuan Xue[1](✉), Boting Yang[1], Farong Zhong[2], and Sandra Zilles[1]

[1] Department of Computer Science, University of Regina, Regina, Canada
{xue228,boting,zilles}@cs.uregina.ca
[2] College of Math, Physics and Information Technology,
Zhejiang Normal University, Jinhua, China
zfr@zjnu.edu.cn

Abstract. Research on graph searching has recently gained interest in computer science, mathematics, and physics. This paper studies *fast searching* of a fugitive in a graph, a model that was introduced by Dyer, Yang and Yaşar in 2008. We provide lower bounds and upper bounds on the fast search number (i.e., the minimum number of searchers required for capturing the fugitive) of complete k-partite graphs. We also investigate some special classes of complete k-partite graphs, such as complete bipartite graphs and complete split graphs. We solve the open problem of determining the fast search number of complete bipartite graphs, and present upper and lower bounds on the fast search number of complete split graphs.

1 Introduction

Graph searching, also called Cops and Robbers games or pursuit-evasion problems, has many models, such as edge searching, node searching, mixed searching, fast searching, etc. [1,3,4,7–10]. Let G denote an undirected graph. In the fast search model, a fugitive hides either on vertices or on edges of G. The fugitive can move at a great speed at any time from one vertex to another along a path that contains no searchers. We call an edge *contaminated* if it may contain the fugitive, and we call an edge *cleared* if we are certain that it does not contain the fugitive. In order to capture the fugitive, one launches a set of searchers on some vertices of the graph; these searchers then clear the graph edge by edge while at the same time guarding the already cleared parts of the graph. This idea is modelled by rules that describe the searchers' allowed moves, as explained in Sect. 2. A *fast search strategy* of a graph is a sequence of actions of searchers that clear all contaminated edges of the graph. The *fast search number* of G, denoted by fs(G), is the smallest number of searchers needed to capture the fugitive in G.

Stanley and Yang [11] presented a linear time algorithm for computing the fast search number of Harlin graphs and their extensions, as well as a quadratic time algorithm for computing the fast search number of cubic graphs. Yang [13] proved that the problem of finding the fast search number of a graph is NP-complete; and it remains NP-complete for Eulerian graphs. He also proved that the problem of determining whether the fast search number of G equals to a

© Springer International Publishing AG 2016
T.-H.H. Chan et al. (Eds.): COCOA 2016, LNCS 10043, pp. 159–174, 2016.
DOI: 10.1007/978-3-319-48749-6_12

half of the number of odd vertices in G is NP-complete for planar graphs with maximum degree 4. Dereniowski et al. [5] gave characterizations of graphs for which 2 or 3 searchers are sufficient in the fast search model. Xue and Yang [12] investigated Cartesian products of graphs, and proved an explicit formula for computing the fast search number of the Cartesian product of an Eulerian graph and a path. They also presented upper and lower bounds on the fast search number of hypercubes.

The fast search problem has a close relationship with the edge search problem [6]. Alspach et al. [2] presented a formula for the edge search number of complete k-partite graphs. Dyer et al. [6] proved the fast search number of complete bipartite graphs $K_{m,n}$ when m is even. They also presented lower and upper bounds respectively on the fast search number of $K_{m,n}$ when m is odd. However, the gap between the lower and upper bounds can be arbitrarily large, and this open problem remains unsolved for eight years.

In this paper, we provide lower and upper bounds on the fast search number of complete k-partite graphs. Further, we investigate some special classes of k-partite graphs, such as complete bipartite graphs and complete split graphs. We solve the open problem of determining the fast search number of complete bipartite graphs. We also present lower and upper bounds on the fast search number of complete split graphs.

2   Preliminaries

Throughout this paper, we only consider finite undirected graphs that have no loops or multiple edges. Let $G = (V, E)$ denote a graph with vertex set V and edge set E. We also use $V(G)$ and $E(G)$ to denote the vertex set and edge set of G respectively. Let uv be an edge with two endpoints u and v. For a vertex $v \in V$, the *degree* of v is the number of edges incident on v, denoted by $\deg_G(v)$. We say a vertex is *odd* if its degree is odd, and we say a vertex is *even* if its degree is even. An *odd graph* is a graph in which all vertices are odd. An *even graph* is a graph in which all vertices are even. Define $V_{\mathrm{odd}}(G) = \{v \in V : v \text{ is odd}\}$.

For a subset $V' \subseteq V$, we use $G[V']$ to denote the subgraph induced by V', which consists of all vertices of V' and all the edges of G between vertices in V'. We use $G - V'$ to denote the induced subgraph $G[V \setminus V']$. For a subset $E' \subseteq E$, we use $G - E'$ to denote the subgraph $(V, E \setminus E')$. Let $G_1 = (V_1, E_1)$ and $G_2 = (V_2, E_2)$ be two subgraphs of G. The *union* of two graphs G_1 and G_2 is the graph $G_1 \cup G_2 = (V_1 \cup V_2, E_1 \cup E_2)$. We use $G_1 + V_2$ to denote the induced subgraph $G[V_1 \cup V_2]$.

A *walk* is a list $v_0, e_1, v_1, \ldots, e_k, v_k$ of vertices and edges such that each edge e_i, $1 \leq i \leq k$, has endpoints v_{i-1} and v_i. A *path* is a walk that does not contain the same vertex twice, except that its first vertex might be the same as its last vertex. We use $v_0 v_1 \ldots v_k$ to denote a path with ends v_0 and v_k. A *trail* is a walk in which no edge occurs multiple times. For a connected subgraph G' with at least one edge, an *Eulerian trail* of G' is a trail that traverses every edge of G' exactly once. A *circuit* is a trail whose first vertex is the same as its last.

An *Eulerian circuit* is an Eulerian trail that begins and ends on the same vertex. A graph is called *Eulerian* if it contains an Eulerian circuit that traverses all its edges. Note that we only consider finite graphs with no loops or multiple edges in this paper. So, throughout this paper, we assume that an Eulerian circuit or Eulerian subgraph contains at least three edges.

In the fast search model, initially every vertex in V and every edge in E is considered *contaminated*. We call a vertex $v \in V$ *cleared* if all edges incident on v are cleared, and we call v *partially cleared* if v has both contaminated and cleared incident edges. A fast search strategy proceeds as follows. First, it places some number of searchers on some vertices in V. Then, it performs sliding actions along contaminated edges until either every edge in E is cleared or no more sliding actions are possible. A searcher on vertex u can slide along the edge $e = uv$ if e is contaminated and (1) u contains one additional searcher or (2) e is the only contaminated edge incident on u. After sliding along e, the searcher then resides on v and e is cleared. Intuitively, the sliding rules ensure that the searchers guard the already cleared parts of the graph, so that the fugitive cannot hide there. The following lemmas give two known lower bounds on the fast search number.

Lemma 1 [6]. *For any connected graph G, $\mathrm{fs}(G) \geq \frac{1}{2}|V_{\mathrm{odd}}(G)|$.*

Lemma 2 [11]. *For any connected graph G with no leaves, $\mathrm{fs}(G) \geq \frac{1}{2}|V_{\mathrm{odd}}(G)| + 2$.*

Let $K_{n_1,\ldots,n_k} = (V_1, \ldots, V_k, E)$ denote a complete k-partite graph, where $V_1, \ldots, V_k$ are disjoint independent sets, $|V_i| = n_i$ and $n_i \leq n_{i+1}$ for all $1 \leq i \leq k - 1$. Each vertex in V_i is adjacent to all the vertices in $V(K_{n_1,\ldots,n_k}) \setminus V_i$. We use $K_{m,n} = (V_1, V_2, E)$ to denote a complete bipartite graph, where $|V_1| = m$, $|V_2| = n$ and $1 \leq m \leq n$. We use $S_{m,n} = (V_1, V_2, E)$ to denote a complete split graph, where V_1 and V_2 are disjoint sets, V_1 induces a clique with m vertices and V_2 is an independent set with n vertices. In $S_{m,n}$, each vertex in V_1 is adjacent to all the other vertices in $V_1 \cup V_2$.

Note that for any connected graph G, the fast search number of G is always at least the edge search number of G. From Theorem 2 in [2], we have the next lemma.

Lemma 3. *For any connected graph G that contains a clique K_m of order m, where $m \geq 4$, we have $\mathrm{fs}(G) \geq m$.*

3   Complete k-partite Graphs

In the following, we give lower bounds and upper bounds on the fast search number of complete k-partite graphs. Throughout this section, in order to better describe our proof ideas, we assume that placing actions of searchers can be inserted after sliding actions of searchers in a fast search strategy. If we want all placing actions to happen before all sliding actions in a fast search strategy, then we can simply move all placing actions before all sliding actions in that fast search strategy.

Lemma 4. *For a complete k-partite graph $K_{n_1,\ldots,n_k}$, where $k \geq 2$ and $n_1 \leq \cdots \leq n_k$, we have $\mathrm{fs}(K_{n_1,\ldots,n_k}) \geq \sum_{i=1}^{k-1} n_i$.*

Lemma 5. *For a complete k-partite graph $K_{n_1,\ldots,n_k}$, where $k \geq 3$ and $n_1 \leq \cdots \leq n_k$, if $\sum_{i=1}^{k-1} n_i \geq 3$ and $n_k \geq 3$, then $\mathrm{fs}(K_{n_1,\ldots,n_k}) \geq 2 + \sum_{i=1}^{k-1} n_i$.*

Proof. For any graph G, $\mathrm{fs}(G)$ is greater than or equal to the edge search number of G. Thus, it follows from Theorem 6 in [2] that $\mathrm{fs}(K_{n_1,\ldots,n_k}) \geq 2 + \sum_{i=1}^{k-1} n_i$.

Theorem 1. *For a complete k-partite graph $K_{n_1,\ldots,n_k}$, where $k \geq 3$, $n_1 \leq \cdots \leq n_k$ and $\sum_{i=1}^{k} n_i = n$, if $\sum_{i=1}^{k-1} n_i \geq n_k = 3$, then $\mathrm{fs}(K_{n_1,\ldots,n_k}) = n - 1$.*

Proof. From Lemma 5, we have $\mathrm{fs}(K_{n_1,\ldots,n_k}) \geq n - n_k + 2 = n - 1$. We will show that $n - 1$ searchers can clear the graph. Let $V_k = \{v_1, v_2, v_3\}$ and $X = K_{n_1,\ldots,n_k} - V_k$. Place $n - 3$ searchers on v_1 and slide them to each vertex of X. Since $k \geq 3$, X is connected. We have three cases for the graph X.

   Case 1. X is Eulerian. The following fast search strategy can clear all edges of the graph $K_{n_1,\ldots,n_k} - \{v_1\}$ using $n - 1$ searchers.

1. Place a searcher on a vertex u of X.
2. Slide one of the two searchers on u along the Eulerian circuit of X to clear all its edges.
3. Slide the two searchers on u to v_2 and v_3 respectively.
4. Place a searcher on v_2. Let Y be the graph formed by all the remaining contaminated edges of $K_{n_1,\ldots,n_k}$.
   (a) If $\deg_Y(v_2)$ is even (Y is Eulerian in this case), then slide one of the two searchers on v_2 along the Eulerian circuit of Y to clear all its edges.
   (b) If $\deg_Y(v_2)$ is odd (Y has an Eulerian trail in this case), then slide one of the two searchers on v_2 to v_3 along the Eulerian trail of Y to clear all its edges.

   Case 2. X is odd. So $X + \{v_2\}$ is Eulerian. We first place two searchers on v_2. Then slide one of the two searchers on v_2 along the Eulerian circuit of $X + \{v_2\}$ to clear all its edges. Finally, slide all searchers on X to v_3 to clear all the remaining contaminated edges of $K_{n_1,\ldots,n_k}$.

   Case 3. X has both even and odd vertices. Suppose that X has $2h$ odd vertices. Let a_1 and b_1 be two odd vertices of X such that there is a path P_1 between them which does not contain any vertex in $V_{\mathrm{odd}}(X)$ as an internal vertex. Let $H_1 = X - E(P_1)$. For $i = 2, \ldots, h$, let a_i and b_i be two odd vertices of H_{i-1} such that there is a path P_i between them which does not contain any vertex in $V_{\mathrm{odd}}(H_{i-1})$ as an internal vertex. Let $H_i = H_{i-1} - E(P_i)$. It is easy to see that H_h contains no odd vertices. In particular, we select P_i in the following manner:

   (1) If X contains at least two even vertices, say u and u', then for $i = 1, \ldots, h$, let $P_i = a_i u' b_i$.

   (2) If X contains only one even vertex, say u, then we first show that $V_1 = \{u\}$. Note that all vertices in V_j, $1 \leq j \leq k - 1$, have the same degree in X.

Therefore, we know $|V_1| = 1$ and u is the only vertex in V_1. Further, if there is a vertex set V_j, $2 \le j \le k - 1$, which contains three vertices, then each of the three vertices is even in X. This is a contradiction. Hence, $|V_j| = 2$ for all $2 \le j \le k - 1$. We have two subcases for k.

(2.1) If $k > 3$, then we can find a matching for all odd vertices of X. Note that there are $2k - 4$ odd vertices on X. Let $V_2 = \{a_1, b_{k-2}\}$ and $V_j = \{a_{j-1}, b_{j-2}\}$, $3 \le j \le k - 1$. For $1 \le i \le k - 2$, it is easy to see that a_i is adjacent to b_i. Hence, we can let $P_i = a_i b_i$. Clearly, u is not included in P_i.

(2.2) If $k = 3$, then we have $|V_1| = 1$, $|V_2| = 2$ and $|V_3| = 3$. Further, a_1 and b_1 are the only two odd vertices of X. Let $V(X) = \{u, a_1, b_1\}$ and $P_1 = a_1 u b_1$.

If X contains at least two even vertices or X contains only one even vertex and $k > 3$, then similar to Case 1, we clear all edges of the graph $K_{n_1,\ldots,n_k} - \{v_1\}$ using the following fast search strategy. Let U be a connected component in H_h that contains u.

1. Place a searcher on the vertex u.
2. Slide one of the two searchers on u along the Eulerian circuit of U to clear all its edges. Note that all edges of X incident on u are cleared after this step.
3. Slide the two searchers on u to v_2 and v_3 respectively.
4. Place a searcher on v_2. Let H be the graph formed by all the remaining contaminated edges of $K_{n_1,\ldots,n_k}$ except edges in $\cup_{i=1}^{h} E(P_i)$.
   (a) If $\deg_H(v_2)$ is even (so H is Eulerian), then slide one of the two searchers on v_2 along the Eulerian circuit of H to clear all its edges.
   (b) If $\deg_H(v_2)$ is odd (so H has an Eulerian trail), then slide one of the two searchers on v_2 from v_2 to v_3 along the Eulerian trail of H to clear all its edges.
5. Let G_P be the graph formed by the paths $P_1, \ldots, P_h$ ($E(G_P)$ is the set of all the remaining contaminated edges of $K_{n_1,\ldots,n_k}$). Note that a_h and b_h are two vertices of degree one on G_P. Slide the searcher on a_h along P_h to b_h. Then a_{h-1} and b_{h-1} are two vertices of degree one on $G_P - E(P_h)$. Slide the searcher on a_{h-1} along P_{h-1} to b_{h-1}. Continuing like this we see that all edges of G_P can be cleared.

If X contains only one even vertex and $k = 3$, then similar to Case 1, we clear all edges of the graph $K_{1,2,3} - \{v_1\}$ using the following fast search strategy. Place a searcher on a_1 and v_2 respectively. Slide one of the two searchers on a_1 along P_1 to b_1. Slide the two searchers on b_1 to v_2 and v_3 respectively. Note that the graph formed by all the remaining contaminated edges of $K_{1,2,3}$ is Eulerian. Slide one of the searchers on v_2 along the path $v_2 u v_3 a_1 v_2$ to clear all its edges. Then, $K_{1,2,3}$ is cleared.

Theorem 2. *For a complete k-partite graph $K_{n_1,\ldots,n_k}$, if there is an n_j, $1 \le j \le k$, such that $\sum_{i=1}^{k} n_i - n_j \ge 4$ and $\sum_{i=1}^{k} n_i - n_j$ is even, then $\mathrm{fs}(K_{n_1,\ldots,n_k}) \le \sum_{i=1}^{k} n_i - n_j + 3$.*

Proof. If $n_j \le 3$, from Theorem 5.1 in [13], we see that the claim holds. If $k = 2$ and $\sum_{i=1}^{k} n_i - n_j \ge 6$, from Lemma 5 in [6], we know that the claim holds.

If $k = 2$ and $\sum_{i=1}^{k} n_i - n_j = 4$, similar to Lemma 5 in [6], we can show that the claim also holds. So we assume that $n_j \geq 4$ and $k \geq 3$ in the rest of the proof. Let $V_j = \{v_1, v_2, \ldots, v_{n_j}\}$ and $X = K_{n_1,\ldots,n_k} - V_j$. Let $\sum_{i=1}^{k} n_i - n_j = m$ and $V(X) = \{u_1, u_2, \ldots, u_m\}$. If n_j is odd, then place m searchers on v_{n_j} and slide them to each vertex of X. If n_j is even, then place m searchers on each vertex of X. Without loss of generality, we assume that n_j is even. Place three additional searchers on u_1, u_2 and u_3 respectively.

Since $k \geq 3$, we know that X is a complete $(k-1)$-partite graph. So X is connected. If X is Eulerian, then slide a searcher from u_1 along the Eulerian circuit of X to clear all its edges. Without loss of generality, we assume that X is not Eulerian. Suppose that X has $2h$ odd vertices. Let $H_0 = X$. Similar to Case 3 in the proof of Theorem 1, let a_i and b_i be two odd vertices of H_{i-1} such that there is a path P_i between them which does not contain any vertex in $V_{odd}(H_{i-1})$ as an internal vertex. Let $H_i = H_{i-1} - E(P_i)$, $1 \leq i \leq h$. We now describe a fast search strategy that can clear all edges of $K_{n_1,\ldots,n_k}$ using $m + 3$ searchers.

1. In the following procedure, at any moment when a vertex u_i $(1 \leq i \leq m)$ contains two searchers, if H_h has a connected component that contains u_i and no edges of the component are cleared, then slide a searcher from u_i along the Eulerian circuit of the component to clear all its edges.
2. Slide a searcher from u_1 to v_1 along $u_1 v_1$, slide a searcher from u_2 to v_1 along $u_2 v_1$ and slide a searcher from u_3 to v_2 along $u_3 v_2$.
3. Note that the subgraph induced by all the edges across $\{u_4, \ldots, u_m\}$ and $\{v_1, v_2\}$ has an Eulerian trail (since m is even). Slide a searcher from v_1 to v_2 along the Eulerian trail to clear all its edges.
4. Slide a searcher from v_1 to u_3 along $v_1 u_3$, slide a searcher from v_2 to u_1 along $v_2 u_1$ and slide a searcher from v_2 to u_2 along $v_2 u_2$. After this step, v_1 and v_2 are cleared.
5. Similar to Steps 2, 3 and 4, we can clear v_3 and v_4, and then clear v_5 and v_6 (if they exist), and so on, until v_{n-1} and v_n are cleared.
6. Let G_P be the graph formed by the paths $P_1, \ldots, P_h$ ($E(G_P)$ is the set of all the remaining contaminated edges of $K_{n_1,\ldots,n_k}$). Similar to Step 5 in Case 3 of the proof of Theorem 1, we can clear all edges of G_P.

Theorem 3. *For a complete k-partite graph $K_{n_1,\ldots,n_k}$, if there is an n_j, $1 \leq j \leq k$, such that $\sum_{i=1}^{k} n_i - n_j \geq 3$ and $\sum_{i=1}^{k} n_i - n_j$ is odd, then $\mathrm{fs}(K_{n_1,\ldots,n_k}) \leq \sum_{i=1}^{k} n_i - \lfloor \frac{n_j}{2} \rfloor$.*

Proof. If $n_j \leq 3$, similar to Theorem 5.1 in [13], we can prove the claim. If $k = 2$, from Lemma 7 in [6], we see that the claim holds. So we assume that $n_j \geq 4$ and $k \geq 3$ in the remainder of the proof. Let $V_j = \{v_1, v_2, \ldots, v_{n_j}\}$ and $X = K_{n_1,\ldots,n_k} - V_j$. Let $\sum_{i=1}^{k} n_i - n_j = m$ and $V(X) = \{u_1, u_2, \ldots, u_m\}$. Note that X is connected since $k \geq 3$. Suppose that X has $2h$ odd vertices. Similar to Case 3 in the proof of Theorem 1, we can define a_i, b_i, P_i and H_i for $1 \leq i \leq h$.

Case 1. $n_j = 4\ell + 1$. Place m searchers on v_1, place one searcher on each of u_1, v_2 and v_3. Place one searcher on each of v_{4i+2} and v_{4i+3} for $i = 1, \ldots, \ell - 1$

(i.e., we place two searchers for every four vertices in $V_j \setminus \{v_1\}$). In total we use $m + 1 + \frac{n_j - 1}{2}$ searchers.

1. In the following process, at any moment when a vertex u_i ($1 \leq i \leq m$) contains two searchers, if H_h has a connected component that contains u_i and no edges of the component are cleared, then slide a searcher from u_i along the Eulerian circuit of the component to clear all its edges.
2. Slide m searchers from v_1 to each vertex of X. Slide one of the two searchers on u_1 along the Eulerian circuit induced by all the edges across $\{u_1, u_2, \ldots, u_{m-1}\}$ and $\{v_2, v_3\}$ to clear all its edges.
3. Slide a searcher from v_2 to v_4 along $v_2 u_m v_4$ and slide a searcher from v_3 to u_5 along $v_3 u_m v_5$ to clear v_2 and v_3. Slide a searcher on u_1 along the Eulerian circuit induced by all the edges across $\{u_1, u_2, \ldots, u_{m-1}\}$ and $\{v_4, v_5\}$ to clear all its edges.
4. Repeat the above step for all of v_{4i+2} and v_{4i+3} where $i = 1, \ldots, \ell - 1$. First clear the Eulerian circuit induced by all the edges across $\{u_1, u_2, \ldots, u_{m-1}\}$ and $\{v_{4i+2}, v_{4i+3}\}$ with a searcher on u_1. Slide the searcher on v_{4i+2} along $v_{4i+2} u_m v_{4i+4}$ and the searcher on v_{4i+3} along $v_{4i+3} u_m v_{4i+5}$. Then clear the Eulerian circuit induced by all the edges across $\{u_1, u_2, \ldots, u_{m-1}\}$ and $\{v_{4i+4}, v_{4i+5}\}$ with a searcher on u_1.
5. Let G_P be the graph formed by the paths $P_1, \ldots, P_h$. Similar to Step 5 in Case 3 of the proof of Theorem 1, we can clear all edges of G_P.

Case 2. $n_j = 4\ell + 2$. Place the searchers as in Case 1. So $m + 1 + \frac{n_j - 2}{2} = m + \frac{n_j}{2}$ searchers are placed on the graph. Clear all vertices in $V_j \setminus \{v_{n_j}\}$ with the same strategy used in Steps 1–4 in Case 1. Note that the only contaminated edges are the ones incident on v_{n_j} and the edges of G_P. We can arrange the vertices of X before placing actions such that $u_1 = a_h$, which is a vertex of degree one on G_P. Since m is odd, there is at least one vertex u such that $\deg_X(u)$ is even. For each vertex $u \in V(X)$ whose $\deg_X(u)$ is even, if $u \notin V(G_P)$, then slide a searcher on u to v_{n_j} along $u v_{n_j}$. Slide a searcher from u_1 to v_{n_j} along $u_1 v_{n_j}$; slide the other searcher on u_1 (i.e., a_h) along P_h to b_h, during which, when a vertex u_i of P_h has only one contaminated edge (i.e., $u_i v_{n_j}$), incident on it, slide a searcher on u_i along $u_i v_{n_j}$ to v_{n_j}. Then a_{h-1} and b_{h-1} are two vertices of degree one on $G_P - E(P_h)$. Slide a searcher from v_{n_j} to a_{h-1} along $v_{n_j} a_{h-1}$, and slide this searcher along P_{h-1} to b_{h-1}, during which, when a vertex u_i of P_{h-1} has only one contaminated edge incident on it, slide a searcher on u_i along $u_i v_{n_j}$ to v_{n_j}. Continuing like this we can clear all edges of G_P and all edges incident on v_{n_j}.

Case 3. $n_j = 4\ell + 3$. Place the searchers as in Case 1. Place another searcher on u_m. Hence we use $m + 1 + \frac{n_j - 3}{2} + 1 = m + \frac{n_j + 1}{2}$ searchers. Use the same strategy as in Steps 1–4 in Case 1 to clear every vertex in $V_j \setminus \{v_{n_j - 1}, v_{n_j}\}$. Now there is one searcher on every vertex of X except u_1 and u_m on which there are two searchers. We can arrange the vertices of X before placing actions such that $u_m = a_h$. Slide one of the two searchers on u_m along P_h to b_h to clear all its edges. Then, b_h contains two searchers. Slide a searcher on b_h along $b_h v_{n_j - 1}$ and $b_h v_{n_j}$ respectively. Slide a searcher on u_1 to clear the Eulerian circuit induced by all the edges across $V(X) \setminus \{b_1\}$ and $\{v_{n_j-1}, v_{n_j}\}$. Finally, similar to Step 5 in Case 1, we can clear all edges of $G_P - E(P_h)$.

Case 4. $n_j = 4\ell$. Place a searcher on every vertex in $\{u_1, u_2, \ldots, u_{m-1}, v_1, v_2, \ldots, v_{2\ell}\}$ and place a second searcher on u_1. Hence we use $m + \frac{n_j}{2}$ searchers. We can arrange the vertices of X before placing actions such that $\deg_X(u_m)$ is even and $u_1 = a_h$. Let $P_i = a_i u_m b_i$, $1 \leq i \leq h$.

1. Slide the searcher from u_1 along the Eulerian circuit induced by all the edges across $\{u_1, u_2, \ldots, u_{m-1}\}$ and $\{v_1, v_2, \ldots, v_{2\ell}\}$. Then slide each searcher on $v_i \in \{v_1, v_2, \ldots, v_{2\ell}\}$ along $v_i u_m$ to clear $\{v_1, v_2, \ldots, v_{2\ell}\}$.
2. Slide a searcher on u_m to each vertex in $\{v_{2\ell+1}, v_{2\ell+2}, \ldots, v_{4\ell-2}\}$. Slide a searcher on u_1 to clear the Eulerian circuit induced by all the edges across $\{u_1, u_2, \ldots, u_{m-1}\}$ and $\{v_{2\ell+1}, v_{2\ell+2}, \ldots, v_{4\ell-2}\}$. Slide a searcher on u_1 to b_h along P_h.
3. In the following process, at any moment when a vertex u_i $(1 \leq i \leq m)$ contains two searchers, if H_h has a connected component that contains u_i and no edges of the component are cleared, then slide a searcher from u_i along the Eulerian circuit of the component to clear all its edges.
4. Slide a searcher on b_h along $b_h v_{4\ell-1}$ and $b_h v_{4\ell}$ respectively and b_h is cleared. Then, slide a searcher on u_m to clear the Eulerian circuit induced by all the edges across $V(X) \setminus \{b_h\}$ and $\{v_{4\ell-1}, v_{4\ell}\}$.
5. Finally, similar to Step 5 in Case 1, we can clear all edges of $G_P - E(P_h)$.

Corollary 1. *For a complete k-partite graph $K_{n_1, \ldots, n_k}$, define α_j, $1 \leq j \leq k$, as*

$$
\alpha_j = \begin{cases}
\sum_{i=1}^{k} n_i - n_j + 3, & \text{if } \sum_{i=1}^{k} n_i - n_j \text{ is even and } \sum_{i=1}^{k} n_i - n_j \geq 4, \\
\sum_{i=1}^{k} n_i - \left\lfloor \frac{n_j}{2} \right\rfloor, & \text{if } \sum_{i=1}^{k} n_i - n_j \text{ is odd and } \sum_{i=1}^{k} n_i - n_j \geq 3, \\
\sum_{i=1}^{k} n_i, & \text{else.}
\end{cases}
$$

Then $\mathrm{fs}(K_{n_1, \ldots, n_k}) \leq \min_{1 \leq j \leq k} \alpha_j$.

4   Complete Bipartite Graphs

In Sects. 4 and 5, we focus on some special classes of complete k-partite graphs. When $k = 2$, $K_{n_1, \ldots, n_k}$ is a complete bipartite graph. Dyer et al. [6] proved several results on the fast search number of $K_{m,n}$. The fast search problem on $K_{m,n}$ has been solved when m is even. However, the fast search problem remains open when m is odd, and they only gave lower and upper bounds on $\mathrm{fs}(K_{m,n})$ in [6]:

- When m is odd, n is even and $3 \leq m \leq n$, we have $\max\{m + 2, \frac{n}{2}\} \leq$ fs$(K_{m,n}) \leq \min\{n + 3, m + \frac{n}{2}\}$.
- When m and n are odd and $3 \leq m \leq n$, we have $\max\{m + 2, \frac{m+n}{2}\} \leq$ fs$(K_{m,n}) \leq m + \frac{n+1}{2}$.

In the following, we will prove that for a complete bipartite graph $K_{m,n}$ with $3 \leq m \leq n$, if m is odd, then fs$(K_{m,n})$ equals to the upper bounds given above. Let $\mathcal{S}_{K_{m,n}}$ denote an optimal fast search strategy for $K_{m,n}$, which uses the minimum number of sliding actions to clear the first cleared vertex of $K_{m,n}$ among all optimal fast search strategies for $K_{m,n}$. We use w_1 to denote the first cleared vertex of $K_{m,n}$. Let t_1 denote the moment at which w_1 is cleared (see Fig. 1(1)). Note that vertices of $K_{m,n}$ are partitioned into two vertex sets V_1 and V_2. We use w_2 to denote the first cleared vertex in another vertex set of $K_{m,n}$ which does not contain w_1. That is, if $w_1 \in V_1$, then $w_2 \in V_2$; if $w_1 \in V_2$, then $w_2 \in V_1$. Let t_2 denote the moment after which the next sliding action clears w_2 (see Fig. 1(2)). Without loss of generality, we first assume that $w_1 \in V_2$. In a similar way, we can prove the lower bound on fs$(K_{m,n})$ when $w_1 \in V_1$.

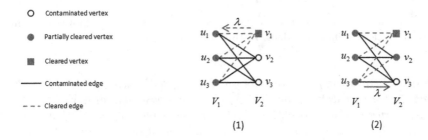

Fig. 1. (1) After searcher λ slides from v_1 to u_1, v_1 becomes the first cleared vertex of $K_{3,3}$. Let this moment be denoted by t_1, and we have $w_1 = v_1$. (2) Searcher λ will slide from u_3 to v_3 in the next step. After that, u_3 becomes the first cleared vertex in V_1. Let t_2 denote this moment, and we have $w_2 = u_3$.

Throughout this section, we assume m is odd. We use A_1 to denote the set of all vertices in $V_2 \setminus \{w_1\}$ which contain a searcher at t_1 and have cleared incident edges at t_2. We use A_2 to denote the set of all vertices in $V_2 \setminus \{w_1\}$ which contain a searcher and have cleared incident edges at t_2. Let $a_1 = |A_1|$ and $a_2 = |A_2|$, it is easy to see that $a_1 + a_2 \geq |A_1 \cup A_2|$. Figures 2 and 3 illustrate A_1 and A_2.

Note that at the moment t_1, all vertices in $A_2 \setminus \{A_1 \cap A_2\}$ are contaminated and contain no searchers, and hence contain no searchers at the beginning of $\mathcal{S}_{K_{m,n}}$ either. Since m is odd, we know all vertices in A_2 are odd. Therefore, each vertex in $A_2 \setminus \{A_1 \cap A_2\}$ must contain a searcher at the end of $\mathcal{S}_{K_{m,n}}$.

Lemma 6. *For a complete bipartite graph $K_{m,n}$ with $m, n \geq 3$, let $\mathcal{S}_{K_{m,n}}$ be an optimal fast search strategy for clearing $K_{m,n}$. Suppose that $w_1 \in V_2$ in $\mathcal{S}_{K_{m,n}}$, then we have $a_1 + a_2 \geq |A_1 \cup A_2| \geq n - 2$.*

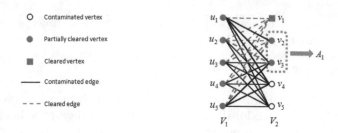

Fig. 2. At the moment t_1, each vertex in A_1 contains a searcher. Further, each vertex in A_1 has cleared incident edges at t_2 (see Fig. 3). In this case, $A_1 = \{v_2, v_3\}$.

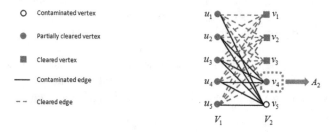

Fig. 3. At the moment t_2, each vertex in A_2 contains a searcher, and all vertices in A_1 and A_2 have cleared incident edges. In this case, $A_2 = \{v_4\}$.

Lemma 7. *For a complete bipartite graph $K_{m,n}$ with $m, n \geq 3$, let $\mathcal{S}_{K_{m,n}}$ be an optimal fast search strategy for clearing $K_{m,n}$. Suppose that $w_1 \in V_2$ in $\mathcal{S}_{K_{m,n}}$. If (1) each vertex in $V_1 \cup A_1$ contains exactly one searcher at t_1, and (2) w_1 contains no searchers at t_1, then each vertex in A_1 has at least two contaminated incident edges at t_1.*

4.1 Both m and n Are Odd

Lemma 8. *For a complete bipartite graph $K_{m,n}$ with $3 \leq m \leq n$, suppose that both m and n are odd. If $w_1 \in V_2$, then $\mathrm{fs}(K_{m,n}) \geq m + \frac{n+1}{2}$.*

Proof. If $3 = m \leq n$, then it follows from Lemma 2 that $\mathrm{fs}(K_{m,n}) \geq \frac{m+n}{2} + 2 = \frac{n+1}{2} + 3 = m + \frac{n+1}{2}$. So we only need to consider $5 \leq m \leq n$ in the following. Since $w_1 \in V_2$ and w_1 is cleared at t_1, we know each vertex in V_1 must be guarded by a searcher at the moment t_1. If $\max\{a_1, a_2\} \geq \frac{n+1}{2}$, then $\mathrm{fs}(K_{m,n}) \geq m + \frac{n+1}{2}$. Suppose that $\max\{a_1, a_2\} \leq \frac{n-1}{2}$. Note that $a_1 + a_2 \geq n - 2$ and both m and n are odd. We know $\min\{a_1, a_2\} \geq \frac{n-3}{2}$. Further, a_1 and a_2 cannot both equal to $\frac{n-3}{2}$; otherwise, $a_1 + a_2 = n - 3 < n - 2$. Hence, there are two cases.

Case 1. $a_1 = \frac{n-1}{2}$. If w_1 contains a searcher at t_1, then $\mathrm{fs}(K_{m,n}) \geq |V_1| + |A_1| + 1 = m + a_1 + 1 = m + \frac{n+1}{2}$. If w_1 contains no searchers at t_1, then for the sake of contradiction, we assume that $m + \frac{n-1}{2}$ searchers can clear $K_{m,n}$. Since $|V_1 \cup A_1| = m + \frac{n-1}{2}$, we know each vertex in $V_1 \cup A_1$ contains exactly one searcher at t_1, and no searchers are located on other vertices. Consider the moment

t_1. From Lemma 7, we know each vertex in A_1 has at least two contaminated incident edges at t_1. Further, since $|V_2 \setminus \{A_1 \cup \{w_1\}\}| = n - \frac{n-1}{2} - 1 \geq 2$, there are at least two vertices in V_2 which have no cleared incident edges. Therefore, each vertex in V_1 has at least two contaminated incident edges. Observe that every vertex in $V_1 \cup A_1$ contains exactly one searcher and has at least two contaminated incident edges. Therefore, all searchers get stuck at t_1, which contradicts that $m + \frac{n-1}{2}$ searchers can clear $K_{m,n}$. Hence, $\mathrm{fs}(K_{m,n}) \geq m + \frac{n+1}{2}$.

Case 2. $a_1 = \frac{n-3}{2}$. Since $\max\{a_1, a_2\} \leq \frac{n-1}{2}$ and $a_1 + a_2 \geq n - 2$, we know $a_2 = \frac{n-1}{2}$. Further, since $a_1 + a_2 = n - 2$, we know $A_1 \cap A_2 = \emptyset$, and hence each vertex in A_2 should always contain a searcher after t_2. For the sake of contradiction, assume that $m + \frac{n-1}{2}$ searchers can clear $K_{m,n}$. Recall that at the moment t_2, each vertex in $A_2 \cup V_1$ is occupied by a searcher and $|A_2 \cup V_1| = m + \frac{n-1}{2}$, we know each vertex in $A_2 \cup V_1$ is occupied by exactly one searcher at t_2. Let $x_1 x_2$ denote the last cleared edge before t_2, which is cleared by sliding a searcher from x_1 to x_2. Note that each vertex in V_1 is occupied by a searcher between t_1 and t_2. We know x_2 must be in A_2, and x_2 contains no searchers before $x_1 x_2$ is cleared. Thus, $x_1 x_2$ is the only cleared edge incident on x_2 at t_2. Recall that $a_1 + a_2 = n - 2$, it is easy to see that there is still a vertex in V_2, say x_3, which has no cleared incident edges at t_2. Hence, $w_2 x_3$ must be cleared by the next sliding action after t_2. When w_2 is cleared, we know both of x_2 and x_3 have exactly one cleared incident edge, and the two edges must be $w_2 x_2$ and $w_2 x_3$. Therefore, when w_2 is cleared, each vertex in V_1 except w_2 has at least two contaminated incident edges. Note that each vertex in A_2 should be guarded by a searcher after t_2. Hence, every searcher gets stuck after w_2 is cleared. This contradicts that $m + \frac{n-1}{2}$ searchers can clear $K_{m,n}$. Therefore, $\mathrm{fs}(K_{m,n}) \geq m + \frac{n+1}{2}$.

Corollary 2. *For a complete bipartite graph $K_{m,n}$ with $3 \leq m \leq n$, suppose that both m and n are odd. If $w_1 \in V_1$, then $\mathrm{fs}(K_{m,n}) \geq m + \frac{n+1}{2}$ when $m = 3$, and $\mathrm{fs}(K_{m,n}) \geq n + \frac{m+1}{2}$ when $m \geq 5$.*

From Lemma 8 and Corollary 2, we are ready to present the lower bound on $\mathrm{fs}(K_{m,n})$ when both m and n are odd. Note that since $m \leq n$, $\min\{m + \frac{n+1}{2}, n + \frac{m+1}{2}\} = m + \frac{n+1}{2}$.

Theorem 4. *Given a complete bipartite graph $K_{m,n}$ with $3 \leq m \leq n$, if both m and n are odd, then $\mathrm{fs}(K_{m,n}) \geq m + \frac{n+1}{2}$.*

4.2 m is Odd and n is Even

Lemma 9. *For a complete bipartite graph $K_{m,n}$ with $3 \leq m < n$, suppose that m is odd and n is even. If $w_1 \in V_2$, then $\mathrm{fs}(K_{m,n}) \geq m + \frac{n}{2}$.*

Proof. If $\max\{a_1, a_2\} \geq \frac{n}{2}$, then it is easy to see that $\mathrm{fs}(K_{m,n}) \geq m + \frac{n}{2}$. Suppose that $\max\{a_1, a_2\} < \frac{n}{2}$. Since $a_1 + a_2 \geq n - 2$ and n is even, we know $a_1 = a_2 = \frac{n-2}{2}$ and $A_1 \cap A_2 = \emptyset$. Consider the moment t_1. We know each vertex in $V_1 \cup A_1$ contains a searcher. For the sake of contradiction, we assume that

$m + \frac{n-2}{2}$ searchers can clear $K_{m,n}$. Then each vertex in $V_1 \cup A_1$ contains exactly one searcher at t_1. From Lemma 7, we know each vertex in A_1 has at least two contaminated incident edges. Further, since $A_1 \cap A_2 = \emptyset$ and $|V_2 \setminus \{A_1 \cup \{w_1\}\}| = n - \frac{n-2}{2} - 1 \geq 2$, we know there are at least two vertices in V_2 which have no cleared incident edges at t_1. Thus, each vertex in V_1 has at least two contaminated incident edges at t_1, and hence, all searchers get stuck at t_1. This contradicts that $m + \frac{n-2}{2}$ searchers can clear $K_{m,n}$. Therefore, $\mathrm{fs}(K_{m,n}) \geq m + \frac{n}{2}$.

In the following, we consider the case when $w_1 \in V_1$.

Lemma 10. *For a complete bipartite graph $K_{m,n}$ with $3 \leq m < n$, suppose that m is odd and n is even. If $w_1 \in V_1$, then $\mathrm{fs}(K_{m,n}) \geq n + 1$ when $m = 3$, and $\mathrm{fs}(K_{m,n}) \geq n + 3$ when $m \geq 5$.*

Proof. If $w_1 \in V_1$, then $w_2 \in V_2$. At the moment t_1, since w_1 is the first cleared vertex, each vertex in V_2 is occupied by a searcher. Let w_3 denote the second cleared vertex of $K_{m,n}$. If $w_3 \in V_2$, then we know each vertex of $K_{m,n}$ except w_1 and w_3 must be occupied by a searcher before w_3 is cleared. Hence, $\mathrm{fs}(K_{m,n}) \geq m + n - 2$. If $w_3 \in V_1$, then we have two cases:

Case 1. $m = 3$. Assume that n searchers can clear $K_{m,n}$. Consider the moment t_1. Note that $|V_2| = n$ and each vertex in V_2 is occupied by a searcher at t_1. Hence, each vertex in V_2 contains exactly one searcher at t_1 and no searchers are located on other vertices. Since there are still two vertices in V_1 which have no cleared incident edges, then each vertex in V_2 has two contaminated incident edges. Thus, it is impossible to move any of the searchers located on V_2 after t_1. This contradicts our assumption that n searchers can clear $K_{m,n}$. Therefore, $\mathrm{fs}(K_{m,n}) \geq n + 1$ when $m = 3$.

Case 2. $m \geq 5$. For the sake of contradiction, we assume that $n + 2$ searchers are sufficient to clear $K_{m,n}$. We have three subcases:

Case 2.1. w_3 contains no searchers after it is cleared. Then the last two cleared edges incident on w_3 are both cleared by sliding a searcher from w_3 to V_2. After w_3 is cleared, all searchers will get stuck within five steps. This contradicts the assumption that $n + 2$ searchers are sufficient to clear $K_{m,n}$. Therefore, $\mathrm{fs}(K_{m,n}) \geq n + 3$.

Case 2.2. w_3 contains exactly one searcher after it is cleared. Note that w_3 has degree at least 6, we know the last cleared edge incident on w_3 has to be cleared by sliding a searcher from w_3 to V_2. Consider the moment when w_3 is cleared. Note that each vertex in V_2 is occupied by a searcher between t_1 and t_2, and there are at least $m - 2 \geq 3$ vertices in V_1 which contain no searchers and have no cleared incident edges. Since we assume that $n + 2$ searchers are sufficient to clear $K_{m,n}$, hence, there is only one vertex in V_2 which contains two searchers. It is easy to see that all searchers get stuck within one step after w_3 is cleared, which is a contradiction. Therefore, $\mathrm{fs}(K_{m,n}) \geq n + 3$.

Case 2.3. w_3 contains exactly two searchers after it is cleared. Consider the moment at which w_3 is cleared. Note that there are still at least $m - 2 \geq 3$ vertices in V_1 which contain no searchers and have no cleared incident edges.

Further, each vertex in V_2 is occupied by exactly one searcher. Hence, it is easy to see that all searchers get stuck after w_3 is cleared. Therefore, $\mathrm{fs}(K_{m,n}) \geq n+3$.

From the above cases, if $w_1 \in V_1$, then $\mathrm{fs}(K_{m,n}) \geq \min\{m+n-2, n+1\} = n+1$ when $m = 3$, and $\mathrm{fs}(K_{m,n}) \geq \min\{m+n-2, n+3\} = n+3$ when $m \geq 5$.

From Lemmas 9 and 10, we know: (1) when $m = 3$, $\mathrm{fs}(K_{m,n}) \geq \min\{m + \frac{n}{2}, n+1\} = m + \frac{n}{2}$; (2) when $m \geq 5$, $\mathrm{fs}(K_{m,n}) \geq \min\{m + \frac{n}{2}, n+3\}$. Hence, we are now ready to give the lower bound on $\mathrm{fs}(K_{m,n})$ when m is odd, n is even and $3 \leq m \leq n$.

Theorem 5. *For a complete bipartite graph $K_{m,n}$ with $3 \leq m < n$, if m is odd and n is even, then $\mathrm{fs}(K_{m,n}) \geq \min\{n+3, m + \frac{n}{2}\}$.*

From Theorems 4 and 5 above, in combination with Lemma 4 and Theorem 4 in [6], we have a complete solution to $\mathrm{fs}(K_{m,n})$.

Theorem 6. *For a complete bipartite graph $K_{m,n}$ with $3 \leq m \leq n$,*

$$\mathrm{fs}(K_{m,n}) = \begin{cases} \left\lceil \dfrac{n}{2} \right\rceil, & m = 1, \\[2mm] 2, & m = n = 2, \\[2mm] 3, & m = 2 \text{ and } n \geq 3, \\[2mm] m + \dfrac{n+1}{2}, & 3 \leq m \leq n, \text{ both } m \text{ and } n \text{ are odd}, \\[2mm] \min\{n+3, m + \dfrac{n}{2}\}, & 3 \leq m < n, m \text{ is odd and } n \text{ is even}, \\[2mm] 6, & m = 4 \text{ and } n \geq 4, \\[2mm] m + 3, & 6 \leq m \leq n \text{ and } m \text{ is even.} \end{cases}$$

5 Complete Split Graphs

In this section, we consider complete split graphs $S_{m,n}$ with $m, n \geq 1$, which also form a special class of k-partite graphs $K_{n_1,\ldots,n_k}$ when $1 = n_1 = \cdots = n_{k-1} \leq n_k$. We start with some initial cases.

Lemma 11. *For a complete split graph $S_{m,n}$, if $n = 1$, then*

$$\mathrm{fs}(S_{m,1}) = \begin{cases} 1, & m = 1, \\ 2, & m = 2, \\ m+1, & m \geq 3. \end{cases}$$

In the following, we consider the fast search number of $S_{m,n}$ when $n \geq 2$. Let $\mathcal{S}_{S_{m,n}}$ denote an optimal fast search strategy for clearing $S_{m,n}$. Let w_1' denote the first cleared vertex in $\mathcal{S}_{S_{m,n}}$, and let t_1' denote the moment at which w_1' is cleared.

5.1   m is Odd and $n \geq 2$

When $m = 1$ and $n \geq 2$, $S_{m,n}$ is a star with n leaves. It is easy to see that $S_{1,n}$ can be cleared with $\lceil \frac{n}{2} \rceil$ searchers. Further, it follows from Lemma 1 that $\mathrm{fs}(S_{1,n}) \geq \frac{1}{2}|V_{\mathrm{odd}}(S_{1,n})| = \lceil \frac{n}{2} \rceil$. Hence, we have the next lemma.

Lemma 12. *For a complete split graph with $m = 1$, if $n \geq 2$, then $\mathrm{fs}(S_{1,n}) = \lceil \frac{n}{2} \rceil$.*

Lemma 13. *For a complete split graph $S_{m,n}$ with $m \geq 3$ and $n \geq 2$, if m is odd, then $\mathrm{fs}(S_{m,n}) = m + \lceil \frac{n}{2} \rceil$.*

Proof. If $w_1' \in V_1$, then each vertex of $S_{m,n}$ except w_1' should be guarded by a searcher at the moment t_1'. Hence, $\mathrm{fs}(S_{m.n}) \geq m - 1 + n$. If $w_1' \in V_2$, then we have two cases:

Case 1. n is even. If $n = 2$, then it follows from Lemma 3 that $\mathrm{fs}(S_{m,n}) \geq m + 1 = m + \frac{n}{2}$. If $n \geq 4$, then similar to the proof of Lemma 9, we can show that $\mathrm{fs}(S_{m,n}) \geq m + \frac{n}{2}$.

Case 2. n is odd. If $n = 3$, then it follows from Lemma 5 that $\mathrm{fs}(S_{m,n}) \geq 2 + m = m + \frac{n+1}{2}$. If $n = 5$, then similar to the proof of Lemma 8 when $n \geq 5$, we can show that $\mathrm{fs}(S_{m,n}) \geq m + \frac{n+1}{2}$.

From the above cases, when $m \geq 3$ and $n \geq 2$, $\mathrm{fs}(S_{m,n}) \geq \min\{m-1+n, m+ \lceil \frac{n}{2} \rceil\} = m + \lceil \frac{n}{2} \rceil$. In combination with Theorem 3, we have $\mathrm{fs}(S_{m,n}) = m + \lceil \frac{n}{2} \rceil$, when $m \geq 3$ and $n \geq 2$.

From Lemmas 12 and 13, we are ready to give the fast search number of $S_{m,n}$ when m is odd and $n \geq 2$.

Theorem 7. *For a complete split graph $S_{m,n}$, if m is odd, then*

$$
\mathrm{fs}(S_{m,n}) =
\begin{cases}
\left\lceil \dfrac{n}{2} \right\rceil, & m = 1,\ n \geq 2, \\[2mm]
m + \left\lceil \dfrac{n}{2} \right\rceil, & m \geq 3,\ n \geq 2.
\end{cases}
$$

5.2   m is Even and $n \geq 2$

Now we consider the complete split graph $S_{m,n}$ where m is even and $n \geq 2$. We first give the following upper bound on $\mathrm{fs}(S_{m,n})$.

Lemma 14. *For a complete split graph $S_{m,n}$ with $m = 2$ and $n \geq 2$, we have $\mathrm{fs}(S_{2,n}) \leq 3$.*

Proof. Let $V_1 = \{u_1, u_2\}$ and $V_2 = \{v_1, v_2, \ldots, v_n\}$. Place a searcher on u_1 and u_2 respectively. Place a second searcher, say λ, on u_1. Hence we use 3 searchers. Let λ clear v_1 by sliding along the path $u_1 v_1 u_2$. Next let λ clear v_2 by sliding along the path $u_2 v_2 u_1$. Repeat this process to clear all the other vertices of $S_{m,n}$.

Lemma 15. *For a complete split graph $S_{m,n}$ with $m = 4$ and $n \geq 3$, we have $\mathrm{fs}(S_{4,n}) \leq 6$.*

Lemma 16. *For a complete split graph $S_{m,n}$ with $m \geq 4$ and $n = 2$, we have $\mathrm{fs}(S_{m,2}) \leq m + 1$.*

Theorem 8. *For a complete graph $S_{m,n}$,*

$$
\mathrm{fs}(S_{m,n}) = \begin{cases}
3, & m = 2,\ n \geq 2, \\
6, & m = 4,\ n \geq 3, \\
m + 1, & m \geq 4,\ n = 2, \\
m + 2, & m \geq 6,\ n = 3.
\end{cases}
$$

Proof.

(1) $m = 2$ and $n \geq 2$. If $w_1' \in V_1$, then $\mathrm{fs}(S_{2,n}) \geq |V_1 \cup V_2| - 1 = 2 + n - 1 \geq 3$. If $w_1' \in V_2$, then let $w_1' x_1$ denote the last sliding action at t_1'. Suppose that two searchers are sufficient to clear $S_{m,n}$. When w_1' is cleared, each vertex in V_1 should be occupied by a searcher. Therefore, at the moment t_1', each vertex in V_1 is occupied by exactly one searcher and no searchers are located on other vertices. Hence, x_1 has no cleared incident edges before $w_1' x_1$ is cleared. Further, the only edge between two vertices in V_1 is contaminated when $w_1' x_1$ is cleared. Since there is at least one vertex in V_2 which has no cleared incident edges, we know each vertex in V_1 has at least two contaminated incident edges. Therefore, no searchers can move after w_1' is cleared. This is a contradiction. Thus, when $m = 2$ and $n \geq 2$, $\mathrm{fs}(S_{2,n}) \geq 3$.

(2) $m = 4$ and $n \geq 3$. It follows from Lemmas 5 and 15 that $\mathrm{fs}(S_{4,n}) = m + 2 = 6$.

(3) $m \geq 4$ and $n = 2$. Clearly, $S_{m,2}$ contains a clique K_{m+1}. From Lemmas 3 and 16, we have $\mathrm{fs}(S_{m,2}) = m + 1$.

(4) $m \geq 6$ and $n = 3$. It follows from Theorem 1 that $\mathrm{fs}(S_{m,3}) = m + n - 1 = m + 2$.

From Lemma 5 and Theorem 2, we give a lower bound and an upper bound on $\mathrm{fs}(S_{m,n})$ when $m \geq 6$ and $n \geq 4$.

Theorem 9. *For a complete split graph $S_{m,n}$ with $m \geq 6$ and $n \geq 4$, if m is even, then $m + 2 \leq \mathrm{fs}(S_{m,n}) \leq m + 3$.*

6   Conclusion and Open Problems

We established both lower bounds and upper bounds on the fast search number of complete k-partite graphs. For $k = 2$, in combination with existing upper bounds, we completely resolved the open question of determining the fast search number of complete bipartite graphs. In addition, we presented some new and nontrivial bounds on the fast search number of complete split graphs.

State-of-the-art knowledge and intuition about the fast search model is not developed as well as for most other search models. Our lower bounds required new proof approaches compared to the existing results in the literature; thus our results shed light on the general problem of finding optimal fast search strategies.

The following problems are left open which we consider worth to investigate:

(1) For complete split graphs $S_{m,n}$ with $m \geq 6$ and $n \geq 4$, resolve the gap of 1 between the upper bound and lower bound on the fast search number when m is even.

(2) Determine the fast search number of $K_{n_1,\ldots,n_k}$ for general values of $n_1, \ldots, n_k$. We conjecture that in Corollary 1, if $\sum_{i=1}^{k} n_i - n_j$ is odd and $\sum_{i=1}^{k} n_i - n_j \geq 3$, then $\mathrm{fs}(K_{n_1,\ldots,n_k}) = \min_{1 \leq j \leq k} \alpha_j$, where $\alpha_j = \sum_{i=1}^{k} n_i - \lfloor \frac{n_j}{2} \rfloor$.

References

1. Alspach, B.: Sweeping and searching in graphs: a brief survey. Matematiche **59**, 5–37 (2006)
2. Alspach, B., Dyer, D., Hanson, D., Yang, B.: Lower bounds on edge searching. In: Chen, B., Paterson, M., Zhang, G. (eds.) ESCAPE 2007. LNCS, vol. 4614, pp. 516–527. Springer, Heidelberg (2007)
3. Bienstock, D.: Graph searching, path-width, tree-width and related problems (a survey). DIMACS Ser. Discrete Math. Theoret. Comput. Sci. **5**, 33–49 (1991)
4. Bonato, A., Nowakowski, R.J.: The Game of Cops and Robbers on Graphs. American Mathematical Soc., Philadelphia (2011)
5. Dereniowski, D., Diner, Ö., Dyer, D.: Three-fast-searchable graphs. Discrete Appl. Math. **161**(13), 1950–1958 (2013)
6. Dyer, D., Yang, B., Yaşar, Ö.: On the fast searching problem. In: Fleischer, R., Xu, J. (eds.) AAIM 2008. LNCS, vol. 5034, pp. 143–154. Springer, Heidelberg (2008)
7. Fomin, F.V., Thilikos, D.M.: An annotated bibliography on guaranteed graph searching. Theoret. Comput. Sci. **399**(3), 236–245 (2008)
8. Hahn, G.: Cops, robbers and graphs. Tatra Mt. Math. Publ. **36**(163), 163–176 (2007)
9. Makedon, F.S., Papadimitriou, C.H., Sudborough, I.H.: Topological bandwidth. SIAM J. Algebraic Discrete Methods **6**(3), 418–444 (1985)
10. Megiddo, N., Hakimi, S.L., Garey, M.R., Johnson, D.S., Papadimitriou, C.H.: The complexity of searching a graph. J. ACM **35**(1), 18–44 (1988)
11. Stanley, D., Yang, B.: Fast searching games on graphs. J. Comb. Optim. **22**(4), 763–777 (2011)
12. Xue, Y., Yang, B.: Fast searching on cartesian products of graphs. In: The 14th Annual Conference on Theory and Applications of Models of Computation (2017, accepted)
13. Yang, B.: Fast edge searching and fast searching on graphs. Theoret. Comput. Sci. **412**(12), 1208–1219 (2011)

Cliques in Regular Graphs
and the Core-Periphery Problem in Social
Networks

Ulrik Brandes[1], Eugenia Holm[1], and Andreas Karrenbauer[2(✉)]

[1] Department of Computer & Information Science,
University of Konstanz, Konstanz, Germany
{ulrik.brandes,eugenia.holm}@uni-konstanz.de
[2] Max Planck Institute for Informatics, Saarbrücken, Germany
andreas.karrenbauer@mpi-inf.mpg.de

Abstract. The existence of a densely knit core surrounded by a loosely connected periphery is a common macro-structural feature of social networks. Formally, the CorePeriphery problem is to partition the nodes of an undirected graph $G = (V, E)$ such that a subset $X \subset V$, the *core*, induces a dense subgraph, and its complement $V \setminus X$, the *periphery*, induces a sparse subgraph. Split graphs represent the ideal case in which the core induces a clique and the periphery forms an independent set. The number of missing and superfluous edges in the core and the periphery, respectively, can be minimized in linear time via edit distance to the closest split graph.

We show that the CorePeriphery becomes intractable for standard notions of density other than the absolute number of misclassified pairs. Our main tool is a regularization procedure that transforms a given graph with maximum degree d into a d-regular graph with the same clique number by adding at most $d \cdot n$ new nodes. This is of independent interest because it implies that finding a maximum clique in a regular graph is NP-hard to approximate to within a factor of $n^{1/2-\varepsilon}$ for all $\varepsilon > 0$.

1 Introduction

In the CorePeriphery problem, we are given a graph $G = (V, E)$ and our goal is to find a bipartition of V into a tightly knit core and a loosely connected periphery. To formalize the CorePeriphery problem, we compare the given graph with the class of *split graphs*, i.e., graphs that admit a bipartition into a complete induced subgraph and a set of mutually non-adjacent vertices. Our aim is now to minimize the error, i.e., the deviation from the ideal case.

To this end, we want to simultaneously maximize the density in the core and minimize the density in the periphery.

We gratefully acknowledge financial support from Deutsche Forschungsgemeinschaft (DFG) under grants Br 2158/6-1 and Ka 3042/3-1. This work is partially supported by the Zukunftskolleg of the University of Konstanz, and the Max Planck Center for Visual Computing and Communication (www.mpc-vcc.org).

© Springer International Publishing AG 2016
T.-H.H. Chan et al. (Eds.): COCOA 2016, LNCS 10043, pp. 175–186, 2016.
DOI: 10.1007/978-3-319-48749-6_13

The CorePeriphery problem is highly relevant for the analysis of social networks [2,21] in various domains [1,5,6,8,9,15,16,18–20,22–25]. The most common formalizations are due to Borgatti and Everett [2] and heuristic algorithms are used [3,4] to separate a core from its periphery. Rombach et al. [21] present a method for identifying multiple cores in a network. Holme [12] introduced a core-periphery-coefficient to measure if the network can be bisected in core and periphery. Zhang et al. [26] developed a statistically principled method, where they use a maximum-likelihood fit, for detecting a core-periphery-decomposition.

The CorePeriphery is closely related to other problems in graph theory. For example, if we omit the condition of a bipartition, we get the problem of finding a densest subgraph in a given graph. Goldberg [10] shows that this problem can be solved in polynomial time for the linear density by using an algorithm based on a network flow computation. The problem to find a vertex partition with maximal sum of the densities of the subsets is known to be NP-hard [7]. Khuller and Saha [14] give 2-approximation algorithms for computing a densest subgraph with at least k vertices for a given k.

The problems of finding large cliques and independent sets are notoriously difficult: it is NP-hard to approximate the size of the largest clique/independent set within a factor of $n^{1-\varepsilon}$ for all $\varepsilon > 0$ [13,27]. Mathieson and Szeider [17] showed that the clique problem remains $W[1]$-hard even in regular graphs. To this end, they proposed a gadget to regularize a given graph, which yields $n^{1/3-\varepsilon}$-hardness for approximating the size of the largest clique in regular graphs. To the best of our knowledge, this was the best previously known lower bound for the hardness of approximation of cliques in regular graphs.

1.1   Our Contribution

We propose a novel regularization procedure that transforms a given graph with n nodes and maximum degree d into a d-regular graph by adding $O(d \cdot n)$ nodes and $O(d^2 n)$ edges without increasing the size of the largest clique provided that the given graph was not already triangle-free. This improves the construction in [17], which uses $O(d^2 n)$ extra nodes. Furthermore, we show that this implies that it is NP-hard to approximate the size of the largest clique in a regular graph within a factor of $O(n^{1/2-\varepsilon})$ for all $\varepsilon > 0$. Finally we show by a new proof technique that the CorePeriphery problem is NP-hard for linear and quadratic densities.

1.2   Preliminaries

We start by a formal treatment of our problems. A graph $G = (V, E)$ is d-regular if all nodes in V have exactly d neighbors, i.e., all nodes in V have the same degree $deg(v) = d$.

We say that a graph $G' = (V', E')$ is an *induced subgraph* of the graph G, $G' \subseteq G$, if $V' \subseteq V$ and E' consists of all edges in E, which have both endnodes in V'. We also write $G' = G[V']$. If we consider the number of adjacent nodes of $v \in V'$ in a subgraph $G' = (V', E')$, we express this by $deg_{G'}(v)$.

The *complement edge set* $\bar{E} := \{\{v, w\} \subseteq V \; : \; \{v, w\} \notin E, v \neq w\}$ of $G = (V, E)$ consists of all edges which are not included in E. The *complement graph* $\bar{G} = (V, \bar{E})$ of G is a graph with the same set of vertices and the complement edge set of G.

An *ℓ-clique K_ℓ* is an induced *complete* subgraph of G with ℓ nodes, i.e., every pair of nodes in the node set of K_ℓ is connected by an edge. A famous problem in the graph theory is the MAXCLIQUE problem. The goal of it is to find a maximum clique in the given graph G, i.e., a clique with the biggest number of nodes. We call this number the *clique number* $\omega(G)$ of the graph G. If G is a regular graph, we call the problem of computing $\omega(G)$ REGULARCLIQUE.

A graph $G = (V, E)$ is called *bipartite* if the node set V can be partitioned into two not empty subsets V_1 and V_2 so that every edge in E has one endnode in V_1 and one endnode in V_2. A bipartite subgraph $K_{a,b} = (V_1' \cup V_2', E')$ in G with $|V_1'| = a$ and $|V_2'| = b$ is called *biclique*, if every node of V_1' is adjacent to every node of V_2'. A set of edges in the graph G that are mutually disjoint is called *matching* in G. If every node in G is incident to one edge of the matching then it is a *perfect* matching. A complete bipartite graph with $2n$ nodes from which the edges of a perfect matching have been removed is called *crown graph*.

2   RegularClique is NP-Hard to Approximate

To prove the hardness result, we first describe a polynomial-time algorithm to regularize a given graph $G = (V, E)$ with $|V| = n$ and $|E| = m$ and maximum degree d.

2.1   Regularization Procedure

Our goal is to augment a graph $G = (V, E)$ with maximum degree d by additional nodes and edges to obtain a d-regular graph $G_d = (V_d, E_d)$ such that $G \subseteq G_d$ and $\omega(G) = \omega(G_d)$. We assume w.l.o.g. that $|V| = 2x$ with $x \in \mathbb{Z}^+$: If the number of nodes is odd, we add an isolated node to V. This does not change $\omega(G)$ and increases n to $n + 1$, which does not harm the asymptotic statements later on.

Since we will not remove any edges, we must fill up the degree of each node until it reaches d. Then the regularized graph $G_d = (V_d, E_d)$ will contain exactly $y := d \cdot n - 2m$ new edges, each with exactly one incident node in V. We choose $r, s \in \mathbb{Z}_{\geq 0}$ so that $m = d \cdot s - r$ with $0 \leq r < d$. Thus,

$$y = 2d \cdot x - 2d \cdot s + 2r = 2d \cdot (x - s) + 2r.$$

Now we consider a crown graph with $2d$ nodes, i.e., biclique $K_{d,d}$ without a perfect matching. Every node of it has degree equal to $d - 1$. We add $x - s$ such graphs to G and connect every node of them with one of the nodes in V with degree less than d. Thereby, we may connect one node in V with one or several nodes of these auxiliary graphs until its degree is equal to d.

For the $2r$ remaining required new edges we add a further biclique $K_{d,d}$ without a matching with r edges. This auxiliary graph contains $2r$ nodes with degree $d - 1$. We connect those nodes with the nodes in V whose degree is still smaller than d (Fig. 1).

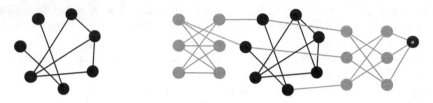

Fig. 1. A graph G with $n = 7$, $m = 8$ and $d = 3$. We add an auxiliary node a, set $x = 4$, $s = 3$ and $r = 1$. Then we construct G_d by adding one $K_{3,3}$ without a perfect matching (on the right side of the figure) and one $K_{3,3}$ without the matching with $r = 1$ edges. Finally we create new edges according to the above description.

Theorem 1. *The graph G contains a k-clique with $k \geq 4$, if and only if its regularized version G_d contains a k-clique.*

Proof. Let $k \geq 4$. Assume that G contains a k-clique. As the regularization procedure only adds nodes and edges and does not remove any of the original nodes or edges, G is an induced subgraph of G_d and thus G_d contains the same k-clique.

On the other hand, if G_d contains a k-clique K with $k \geq 4$, at most two of its nodes may be contained in $V_d \backslash V$ because the added bipartite graphs are triangle-free. But if K contains exactly two nodes from $V_d \backslash V$, then these two nodes can have at most one common neighbor in K because each of them is only incident to exactly one node in V, a contradiction to the assumption that K is a clique with at least 4 nodes. Similarly, if exactly one node from $V_d \backslash V$ is contained in K, then it is incident to exactly one other node in K, again a contradiction. Thus, K is completely contained in V, which proves the claim. $\square$

2.2   Hardness of Approximation

The REGULARCLIQUE problem is formally defined as follows.

*Problem 1 (*REGULARCLIQUE*).* Given a regular graph G and an integer k, decide whether G contains a clique of size k.

This problem is not only NP-hard, but also very hard to approximate. To prove this, we recall the situation for MAXCLIQUE in general graphs. Zuckerman [27] derandomized a construction of Håstad [13] to obtain the following theorem.

Theorem 2 ([13, 27]). *Let $\varepsilon > 0$. Given a graph $G = (V, E)$, it is NP-hard to approximate* MAXCLIQUE *to within a factor of $|V(G)|^{1-\varepsilon}$.*

From this, we derive a similar hardness result, where we lose a $\sqrt{n}$-factor due to the blow-up of $O(dn)$ in the number of nodes with our regularization procedure. We restrict ourselves to the case $k \geq 4$ because we can decide whether there is a K_3 by enumerating all triples of nodes in $O(n^3)$ time and because the cases for $k \in \{1, 2\}$ are trivial.

Theorem 3. *It is* NP-*hard to approximate* REGULARCLIQUE *within a factor of $n^{\frac{1}{2}-\varepsilon}$ for all $\varepsilon > 0$.*

Proof. Let $G = (V, E)$ be a given undirected graph with $n := |V|$ and $m := |E|$, and let $G_d = (V_d, E_d)$ denote its regularized version. For our considerations, we can assume that the graph G is a connected graph and therefore $m \geq n - 1 \Leftrightarrow 2m \geq 2n - 2$ and hence $2m \geq n$ for all $n \geq 2$. By the regularizing construction we get

$$N := |V_d| = n + 2d \cdot (x - s) + 2d = n + d \cdot n - 2m - 2r + 2d \leq d \cdot n \leq n^2.$$

Suppose that there exists an approximation algorithm $\mathcal{A}_{RegClique}$ for regular clique within a factor of $N^{\frac{1-\varepsilon}{2}}$ for an $\varepsilon > 0$. Then we can find a k-clique K in G_d with $k \geq \omega(G_d)/N^{\frac{1-\varepsilon}{2}}$ nodes. According to the Theorem 1, K is contained in G and $\omega(G_d) = \omega(G)$. Thus, $k \geq \omega(G)/N^{\frac{1-\varepsilon}{2}} \geq \omega(G)/n^{1-\varepsilon}$. Thus, we would have an $n^{1-\varepsilon}$-approximation for MAXCLIQUE. Theorem 2 proves the statement above. $\qquad\square$

For the sake of presentation, we further restrict the range for k to $\{4, \ldots, d\}$. This is w.l.o.g. because d-regular graphs cannot have a clique with more than $d+1$ nodes, the cases for $k \in \{1, 2, 3\}$ can be decided in polynomial time, as well as the case for $k = d + 1$ as the following Lemma shows.

Lemma 1. *A d-regular graph $G = (V, E)$ contains a clique K_{d+1}, if and only if G contains a connected component with $d + 1$ nodes.*

Proof. Let $K \subseteq V$ be a clique in G with $d + 1$ nodes. Since $deg_K(v) = d$ for all $v \in K$ every node in K is adjacent to all other nodes in K. Because G is a d-regular graph, there cannot be an edge $\{v, w\}$ with $v \in K$ and $w \notin K$.

On the other hand, let G contain a connected component $V' \subseteq V$ with $d+1$ nodes. Because G is a d-regular graph, every node in V' is adjacent to d nodes. This means that every node in V' is adjacent to all nodes in V'. Thus, $G[V']$ is a clique with $d + 1$ nodes. $\qquad\square$

3   Application to CorePeriphery

We apply the results from the previous section to prove NP-hardness of two versions of the COREPERIPHERY problem. Generally speaking our aim is to decompose a given graph $G = (V, E)$ into a core, i.e., nodes that are tightly connected, and a periphery, i.e., vertices that are loosely connected. The ideal case is a so called *split graph*. This is a graph for which there exists a bipartition of its vertices into a clique and an *independent set*, i.e., a set of nodes that not induce any edge. Hammer and Simeone [11] showed that the split graphs can be recognized in linear time.

Theorem 4 (Hammer and Simeone, 1981 [11]). *Let $G = (V, E)$ be an undirected graph with $|V| = n$ and the degree sequence $d_1 \geq \cdots \geq d_n$. Define $k := \max\{i : d_i \geq i - 1\}$. Then, G is a split graph if and only if the splittance*

$$\frac{1}{2}\left(k(k-1) - \sum_{i=1}^{k} d_i + \sum_{j=k+1}^{n} d_j\right) = 0.$$

Furthermore, if this is the case, then k is the clique number of G.

For all other graphs we try to minimize the deviation from this ideal case, i.e., to minimize the splittance. Observe that $\frac{1}{2}\left(k(k-1) - \sum_{i=1}^{k} d_i + \sum_{j=d+1}^{n} d_j\right)$ edges have to be added/deleted to make G a split graph. That is, the splittance is the number of edges that have to be added or to be deleted to obtain a bipartition of a given graph into a clique as a core and the independent set as periphery. Furthermore, Hammer and Simeone showed in [11] that the splittance of any graph can be determined in linear time.

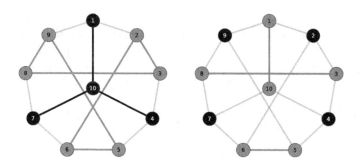

Fig. 2. Here we see two different cores (black nodes) in the Petersen graph. The split-tance of the graph is always equal to 9 if the core consists of 4 nodes, although the core induces no edges.

The fact that the splittance of a d-regular graph is equal for all cores with $d + 1$ nodes, illustrates that the splittance is not able to discriminate certain situations. For example, Fig. 2 shows two cores in the Petersen graphs that are both optimal w.r.t. the splittance, but one of them even induces an independent set — the opposite of a clique. So it is natural to ask for the size normalized deviation like how many edges on average per node must be added in the core and be deleted in the periphery to obtain a split graph. Analogously, we can consider the problem so that we ask which fraction of edges of the clique is to be added to the core and which fraction of potential edges in periphery is to be deleted to obtain the ideal case.

To facilitate the comparison of further objective functions, we introduce the following notations. Let $G = (V, E)$ be a graph and $X \subseteq V$ a candidate for the core of G. The number of edges and non-edges in the core X are denoted by

$$c_1(X) := |\{e \in E : e \subseteq X\}|, \quad c_0(X) := |\{\bar{e} \in \bar{E} : \bar{e} \subseteq X\}|$$

and similarly in the periphery $V \setminus X$:

$$p_1(X) := |\{e \in E : e \cap X = \emptyset\}|, \quad p_0(X) := |\{\bar{e} \in \bar{E} : \bar{e} \cap X = \emptyset\}|.$$

We write c_1, c_0, p_1, or p_0 if X is clear from the context. Using our notion the splittance is equal to $c_0 + p_1$.

Our aim is to decompose the nodes of the graph into the core X and periphery $V \setminus X$ such that the density of the subgraph induced by X is maximal and the density of the subgraph induced by $V \setminus X$ is minimal. To combine these criteria in a single objective function that mimics the splittance, we minimize $sparsity_G(X) + density_G(V \setminus X)$, where $sparsity_G(X) := density_{\bar{G}}(X)$, i.e., the sparsity is defined as density in the complement graph.

Popular density functions are the linear and the quadratic density. The *linear density* $d_1(X)$ of $X \subseteq V$ in a graph $G = (V, E)$ is defined as the average degree in the subgraph induced by X. That is,

$$d_1(X) = \frac{1}{|X|} \sum_{v \in X} deg_X(v) = \frac{2c_1}{|X|}.$$

The *quadratic density* $d_2(X)$ of $X \subseteq V$ in a graph $G = (V, E)$ is the ratio of existent edges to the number of all possible edges in the subgraph induced by X. That is,

$$d_2(G) := \frac{c_1}{\frac{|X|(|X|-1)}{2}} = \frac{2c_1}{|X|(|X| - 1)}.$$

To facilitate the discussion, we split the contribution of the sparsity of the core X and the density of the periphery $V \setminus X$ into two functions $f(X)$ and $g(X)$, respectively, such that their sum defines the objective function $h(X)$. This is summarized in the Table 1.

The function $f(X)$ counts non-edges in X, in relation to the size of X and $g(X)$ counts edges having both incident nodes in the set $V \setminus X$, in relation to the size of $V \setminus X$. These quantities yield the deviation of the core X and the periphery $V \setminus X$ to a perfect core-periphery structure, i.e., the average number of edges per node to be added to the core or to be deleted from the periphery in the case of linear density, and the percentage of missing edges in the core or surplus edges in the periphery, respectively, to make X a clique and $V \setminus X$ an independent set.

Table 1. The decomposition of the objective function into contributions of the core and the periphery for splittance, linear, and quadratic normalization.

Deviation	Absolute	Linear	Quadratic
Core X	$c_0(X)$	$f(X) := \frac{2c_0(X)}{|X|}$	$f(X) := \frac{2c_0(X)}{|X||X-1|}$
Periphery $V \setminus X$	$p_1(X)$	$g(X) := \frac{2p_1(X)}{|V|-|X|}$	$g(X) := \frac{2p_1(X)}{(|V|-|X|)(|V|-|X|-1)}$
Total deviation	$c_0(X) + p_1(X)$	$h(X) := f(X) + g(X)$	

We will show in the following that the CorePeriphery problem is NP-hard for both of these densities. For this we will use the hardness result for RegularClique from the previous section. The main idea of our proof is that we augment the graph by isolated nodes such that any reasonably good solution will take a clique as the core. That is, all solutions that have an incomplete subgraph as a core will have a worse objective value than taking the two endpoints of any edge as the core. It is important for our argument that the input graph is regular and therefore we will make use of our regularization procedure.

3.1 Linear Density

First we show that if the core-candidate X is not a clique, then the value of the density of X never falls below a certain value.

Lemma 2. *If a non-empty set $X \subseteq V$ does not induce a clique, then $f(X) \geq \frac{2}{d+1}$ for any $d \geq 2$.*

Proof. We first consider the case $|X| \leq d+1$. Since X does not induce a clique, the induced subgraph misses at least one edge from being complete, i.e., $c_0 \geq 1$. Thus, we have

$$f(X) = \frac{2c_0}{|X|} \geq \frac{2}{|X|} \geq \frac{2}{d+1}.$$

If $|X| \geq d+2$, we have $|X| - 1 \geq d+1$ and so X must miss more than

$$\frac{|X|(|X|-1)}{2} - \frac{d|X|}{2} \geq \frac{|X|(d+1) - d|X|}{2} = \frac{|X|}{2}$$

edges, i.e., $c_0 \geq \frac{|X|}{2}$. Thus, $f(X) = \frac{2c_0}{|X|} \geq 1 \geq 2/(d+1)$ for all $d \geq 2$.   □

The idea for proving NP-hardness is to augment a given graph by isolated nodes such that $h(X)$ is at most $1/d < 2/(d+1)$ whenever X induces a clique (even if it induces single edge) in a graph with $d \geq 2$.

Lemma 3. *A d-regular graph G contains a clique of size $k \leq d$, if and only if, $G' = (V', E)$ with the node set V' consisting of the n nodes of V and q additional isolated nodes (i.e., $|V'| = n + q$) contains a core $X \subseteq V'$ with*

$$h(X) \leq \frac{nd - 2dk + k(k-1)}{n + q - k}$$

for all $q \geq d^2 n$.

Proof. Let X be a k-clique in G and thus also in G'. Note that $f(X) = 0$. Moreover, the total number of edges is $\frac{nd}{2}$ due to the regularity and the number of edges incident to nodes in X is given by $kd - \binom{k}{2}$. Hence,

$$h(X) = g(X) = 2\frac{\frac{nd}{2} - (dk - \binom{k}{2})}{n + q - k} = \underbrace{\frac{nd - 2dk + k(k-1)}{n + q - k}}_{=:g(k)}.$$

To prove the converse direction, we show that $g(k)$ is decreasing in the range of $1 \le k \le d$. To this end, we consider the first derivative of $g(k)$, i.e.,

$$\begin{aligned}
g'(k) &= \frac{(2k - 2d - 1)q + 2kn - dn - n - k^2}{(n + q - k)^2} \\
&\le -\frac{q - dn + n + 1}{(n + q - k)^2} \\
&\le -\frac{(d^2 - d + 1)n + 1}{(n + q - k)^2} < 0
\end{aligned}$$

Thus, $g(k) \le g(1) = (dn - 2d)/(n + q - 1) \le (dn - 2d)/(n + d^2 n - 1) \le 1/d < 2/(d + 1)$ for all $d \ge 2$. Hence, G contains a k-clique if and only if there is an $X \subset V'$ with

$$h(X) \le \frac{nd - 2dk + k(k-1)}{q + n - k}.$$

$\square$

As a consequence, we obtain a reduction to prove NP-hardness of linear CorePeriphery.

Theorem 5. *Solving the problem* CorePeriphery *with linear density is* NP-*complete.*

Proof. Given a d-regular graph $G = (V, E)$ and an integer $k \in \{1, \ldots, d\}$, we wish to decide whether G contains a k-clique (by Lemma 1 it is sufficient to consider k-cliques with $k \le d$). We add $q = nd^2$ isolated nodes to G and thereby obtain $G' = (V', E')$ for which we compute the CorePeriphery problem. Combining Lemmas 2 and 3, the reported core will be a clique. $\square$

3.2 Quadratic Density

Lemma 4. *If $X \subseteq V$ does not induce a clique, then $f(X) \ge \frac{1}{d^2}$ for $d \ge 2$.*

Proof. We again consider the case $|X| \le d + 1$ first. Since X does not induce a clique, it contains at least one non-edge in $\overline{E}$ and thereby

$$f(X) \ge \frac{2}{|X|(|X| - 1)} \ge \frac{2}{(d + 1)d} \ge \frac{1}{d^2}.$$

If $|X| = d + \ell$ with $2 \leq \ell \in \mathbb{Z}$, than there are $(d+\ell)(d+\ell-1)/2 - d(d+\ell)/2$ non-edges and thus

$$
\begin{aligned}
f(X) &\geq \frac{(d+\ell)(d+\ell-1) - d(d+\ell)}{(d+\ell)(d+\ell-1)} \\
&= \frac{d+\ell-1-d}{d+\ell-1} = 1 - \frac{d}{d+\ell-1} \\
&\geq 1 - \frac{d}{d+1} = \frac{1}{d+1}.
\end{aligned}
$$

$\square$

For proving NP-hardness for the quadratic density, we augment a given graph by q isolated nodes analogous to the case of linear density such that $h(X)$ is strictly less than $1/d^2$ whenever X induces a clique (even if it induces single edge).

Lemma 5. *A d-regular graph G contains a clique of size $k \leq d$, if and only if, $G' = (V', E)$ with the node set V' consisting of the n nodes of V and q additional isolated nodes (i.e., $|V'| = n + q$) contains a core $X \subseteq V'$ with*

$$
h(X) \leq \frac{nd - 2dk + k(k-1)}{(n+q-k)(n+q-k-1)}
$$

for all $q \geq dn$.

Proof. At first we assume that X is a k-clique in G and also in G'. Analogous to the proof for the linear density we obtain

$$
\begin{aligned}
h(X) = g(X) &= \underbrace{\frac{nd - 2dk + k(k-1)}{(n+q-k)(n+q-k-1)}}_{=:g(k)} \\
&= \frac{nd - k(2d - k + 1)}{q^2 + k^2 + n(n - 2k - 1) + q(2n - 2k - 1) + k} < \frac{nd}{q^2}.
\end{aligned}
$$

For $q \geq dn$ we get $g(X) < \frac{1}{dn}$ and therefore $h(X) < \frac{1}{dn} < \frac{1}{d^2}$. Recall that $h(X) \geq \frac{1}{d^2}$ if X is not a clique by Lemma 4.

Now we have to prove that a larger clique is preferred instead of a smaller one. To this end, we show that the function $g(k)$ decreases for increasing k by considering the difference

$$
g(k-1) - g(k) = \frac{-2(1 - dk - k - n - q + d - dq + kn + kq)}{(n+q-k+1)(n+q-k)(n+q-k-1)}
$$

It is easy to verify that the denominator of this difference is positive. So we have to investigate the numerator only.

$$
-2(1 - dk - k - n - q + d - dq + kn + kq) = 2(q(d - k + 1) + dk + k - kn - d + n - 1)
$$

We show that for $q \geq dn$ this term is positive.

$$2(q(d - k + 1) + dk + k - kn - d + n - 1)$$
$$\geq 2(dn(d - k + 1) + dk + k - kn - d + n - 1)$$
$$\geq 2(dn(d - k + 1) + d + 1 - kn - d + n - 1)$$
$$\geq 2(dn(d - d + 1) + d + 1 - dn - d + n - 1)$$
$$= 2n > 0$$

As a consequence a minimizer for the CorePeriphery problem for quadratic density is a clique $X \subseteq V$.   □

Theorem 6. *Solving the problem* CorePeriphery *with quadratic density is* NP- *complete.*

Proof. Given a d-regular graph $G = (V, E)$ and an integer $k \in \{1, \ldots, d\}$. The question is again whether G contains a clique with k nodes. We construct a graph $G' = (V', E')$ by adding $q = nd$ isolated nodes to G analogous to the case of linear density. Then we solve the CorePeriphery problem on G'. Combining Lemmas 4 and 5, we obtain a clique as the core.   □

References

1. Alba, R.D., Moore, G.: Elite social circles. Sociol. Methods Res. **7**(2), 167–188 (1978)
2. Borgatti, S.P., Everett, M.G.: Models of core/periphery structures. Soc. Netw. **21**(4), 375–395 (2000)
3. Borgatti, S.P., Everett, M.G., Freeman, L.C.: Ucinet for windows: software for social network analysis (2002)
4. Boyd, J.P., Fitzgerald, W.J., Beck, R.J.: Computing core/periphery structures and permutation tests for social relations data. Soc. Netw. **28**(2), 165–178 (2006)
5. Chase-Dunn, C.K.: Global Formation: Structures of the World-Economy. Rowman & Littlefield, Lanham (1998)
6. Corradino, C.: Proximity structure in a captive colony of Japanese monkeys (macaca fuscata fuscata): an application of multidimensional scaling. Primates **31**(3), 351–362 (1990)
7. Darlay, J., Brauner, N., Moncel, J.: Dense and sparse graph partition. Discrete Appl. Math. **160**(16), 2389–2396 (2012)
8. Doreian, P.: Structural equivalence in a psychology journal network. J. Am. Soc. Inf. Sci. **36**(6), 411–417 (1985)
9. Faulkner, R.R.: Music on Demand. Transaction Publishers, Piscataway (1983)
10. Goldberg, A.V.: Finding a maximum density subgraph. University of California Berkeley, CA (1984)
11. Hammer, P.L., Simeone, B.: The splittance of a graph. Combinatorica **1**(3), 275–284 (1981)
12. Holme, P.: Core-periphery organization of complex networks. Phys. Rev. E **72**(4), 046111 (2005)
13. Håstad, J.: Clique is hard to approximate within $n^{1-\epsilon}$. Acta Math. **182**, 105–142 (1999)

14. Khuller, S., Saha, B.: On finding dense subgraphs. In: Albers, S., Marchetti-Spaccamela, A., Matias, Y., Nikoletseas, S., Thomas, W. (eds.) ICALP 2009, Part I. LNCS, vol. 5555, pp. 597–608. Springer, Heidelberg (2009)
15. Krugman, P.: The self-organizing economy. Number 338.9 KRU 1996. CIMMYT (1996)
16. Laumann, E.O., Pappi, U.: Networks of collective actions, New York (1976)
17. Mathieson, L., Szeider, S.: The parameterized complexity of regular subgraph problems andgeneralizations. In: Proceedings of the Fourteenth Symposium on Computing: The Australasian Theory - Volume 77, CATS 2008, pp. 79–86, Darlinghurst, Australia. Australian Computer Society Inc. (2008)
18. Mintz, B., Schwartz, M.: Interlocking directorates and interest group formation. Am. Sociol. Rev. **46**, 851–869 (1981)
19. Mullins, N.C., Hargens, L.L., Hecht, P.K., Kick, E.L.: The group structure of cocitation clusters: a comparative study. Am. Sociol. Rev. **42**, 552–562 (1977)
20. Nemeth, R.J., Smith, D.A.: International trade and world-system structure: a multiple network analysis. Rev. (Fernand Braudel Center) **8**(4), 517–560 (1985)
21. Rombach, M.P., Porter, M.A., Fowler, J.H., Mucha, P.J.: Core-periphery structure in networks. SIAM J. Appl. Math. **74**(1), 167–190 (2014)
22. Smith, D.A., White, D.R.: Structure and dynamics of the global economy: network analysis of international trade 1965–1980. Soc. Forces **70**(4), 857–893 (1992)
23. Snyder, D., Kick, E.L.: Structural position in the world system, economic growth, 1955–1970: a multiple-network analysis of transnational interactions. Am. J. Sociol. **84**, 1096–1126 (1979)
24. Steiber, S.R.: The world system and world trade: an empirical exploration of conceptual conflicts. Sociol. Q. **20**(1), 23–36 (1979)
25. Wallerstein, I.: The Modern World-System I: Capitalist Agriculture and the Origins of the European World-Economy in the Sixteenth Century, with a New Prologue, vol. 1. University of California Press, Berkeley (2011)
26. Zhang, X., Martin, T., Newman, M.E.: Identification of core-periphery structure in networks. Phys. Rev. E **91**(3), 032803 (2015)
27. Zuckerman, D.: Linear degree extractors and the inapproximability of max clique and chromatic number. In: Proceedings of the Thirty-Eighth Annual ACM Symposium on Theory of Computing, STOC 2006, pp. 681–690. ACM, New York (2006)

Constant Factor Approximation for the Weighted Partial Degree Bounded Edge Packing Problem

Pawan Aurora[1(✉)], Monalisa Jena[2], and Rajiv Raman[2,3]

[1] IISER, Bhopal, India
paurora@iiserb.ac.in
[2] IIIT-Delhi, Delhi, India
{monalisaj,rajiv}@iiitd.ac.in
[3] NYU, Abu Dhabi, UAE

Abstract. In the partial degree bounded edge packing problem (PDBEP), the input is an undirected graph $G = (V, E)$ with capacity $c_v \in \mathbb{N}$ on each vertex. The objective is to find a *feasible* subgraph $G' = (V, E')$ maximizing $|E'|$, where G' is said to be feasible if for each $e = \{u, v\} \in E'$, $\deg_{G'}(u) \leq c_u$ or $\deg_{G'}(v) \leq c_v$. In the weighted version of the problem, additionally each edge $e \in E$ has a weight $w(e)$ and we want to find a feasible subgraph $G' = (V, E')$ maximizing $\sum_{e \in E'} w(e)$. The problem is already NP-hard if $c_v = 1$ for all $v \in V$ [Zhang, FAW-AAIM 2012].

In this paper, we introduce a generalization of the PDBEP problem. We let the edges have weights as well as demands, and we present the first constant-factor approximation algorithms for this problem. Our results imply the first constant-factor approximation algorithm for the weighted PDBEP problem, improving the result of Aurora et al. [FAW-AAIM 2013] who presented an $O(\log n)$-approximation for the weighted case.

We also present a PTAS for H-minor free graphs, if the demands on the edges are bounded above by a constant, and we show that the problem is APX-hard even for cubic graphs and bounded degree bipartite graphs with $c_v = 1$, $\forall v \in V$.

1 Introduction

Packing problems are central objects of study in the theory of algorithms. Quintessential examples of such problems are the Independent Set problem [4] in graphs, Maximum Matchings in graphs [10,12], and the Knapsack Problem [16]. Due to their fundamental nature, and wide applicability, these problems and variants thereof have been studied intensively over several decades. In this paper, we study a variant of the matching problem that is called the Partial Degree Bounded Edge Packing problem (PDBEP for short).

In the most basic version of this problem, the input is an undirected graph $G = (V, E)$, with unit capacities on the vertices, and unit weight on the edges.

M. Jena—The author is supported by a TCS Scholarship.

T.-H.H. Chan et al. (Eds.): COCOA 2016, LNCS 10043, pp. 187–201, 2016.
DOI: 10.1007/978-3-319-48749-6_14

The goal is to pack a maximum cardinality set $E' \subseteq E$ of edges such that in the resulting sub-graph $G' = (V, E')$, each edge $e = \{u, v\} \in E'$ is *satisfied*, where an edge is said to be satisfied if either of its end-points has degree at most 1 in the sub-graph G', i.e., $\deg_{G'}(u) \leq 1$ *or* $\deg_{G'}(v) \leq 1$.

The Maximum Matching problem, phrased as above would be to find a sub-graph $G' = (V, E')$ with maximum number of edges E' such that each edge is *satisfied*, where an edge is said to be satisfied if both of its end-points have degree at most 1 in the sub-graph G', i.e., $\deg_{G'}(u) \leq 1$ *and* $\deg_{G'}(v) \leq 1$.

The difference between the Maximum Matching problem and the PDBEP problem is only in the definition of when an edge is satisfied. While in the case of Maximum Matching, we require that the degree condition at both end-points be satisfied, we only require a weaker condition to be satisfied for the PDBEP problem, namely that for each edge, the degree condition be satisfied at at least one end-point. One can immediately observe that despite the seeming similarity with the maximum matching problem, the solutions to the two problems can be vastly different. For example, consider a star $K_{1,n}$. The maximum matching problem has a solution of size 1, whereas the PDBEP problem on the same instance has a solution of size n. In fact, while the Maximum Matching problem admits a polynomial time algorithm [10], our problem is NP-hard [19] even in the case of unit capacities.

Motivated by an application in computing on binary strings by Bu et al. [6], Zhang introduced the PEPD problem in [19]. The problem he studied is called the Maximum Expressive Independent Set (MEIS for short) problem where the objective is to find a subset X of maximum cardinality from a set of binary strings W such that no string $t \in X$ is *expressible* from $X \setminus \{t\}$, where a binary string s is expressible from a set of binary strings S, if it can be obtained by combining the strings in S using bitwise AND and OR operators. He studied a restricted version of this problem where each string is 2-regular which means that it has exactly two ones. This he posed as a graph problem where the graph has a vertex for every bit position and an edge $\{i, j\}$ corresponds to a string that has ones at positions i, j. Now a solution to the PDBEP problem with uniform $c_u = 2$ corresponds to a subset of strings such that for any string in the subset with ones at positions i, j, at most one other string can have a 1 at one of these two positions which means that the subset of edges gives a solution to the MEIS problem (this follows from Lemma 2.4 in [19]).

Another natural application of the PDBEP problem is in resource allocation. Here, we are given $|V|$ types of resources and $|E|$ jobs. Each job needs two types of resources. A job u can be accomplished if either one of its necessary resources is shared by no more than c_u other jobs. The problem then asks to finish as many jobs as possible.

The rest of the paper is organized as follows. In Sect. 2, we present the notation used, give a formal definition of the problems studied, and present preliminary results. Section 3 describes related work. We study the PDBEP problem with unit capacities in Sect. 4, and then present results for the general setting in Sect. 5. We give a PTAS for the weighted PDBEP problem on H-minor free graphs in Sect. 6. In Sect. 7, we prove that the PDBEP problem is APX-hard.

2   Preliminaries

Let $G = (V, E)$ denote an undirected graph. In our setting, the graphs come equipped with a weight function $w : E \to \mathbb{N}$ and a demand function $d : E \to \mathbb{N}$ on the edges, and a capacity function $c : V \to \mathbb{N}$ on the vertices. We also consider the special case when all vertices have unit capacity. In this setting, we assume that the demand of each edge is also 1, and that the graphs are simple. When we consider the general problem, the graph is no longer assumed to be simple.

We also consider directed graphs, denoted $D = (V, A)$. Each edge $(u, v) \in A$ with head v and tail u is said to be *entering* v, and *exiting* u. We use $in(v), out(v)$ respectively to denote the edges entering and exiting v.

For a vertex v, we let $N_e(v) = \{e = \{u, v\} \in E\}$ denote the set of edges incident on v. In the weighted setting, we let $N_{\max}(v) = \{e \in N_e(v) : w(e) \geq w(f) \; \forall f \in N_e(v)\}$. Thus, $N_{\max}(v)$ is the set of heaviest weighted edges incident on v. For a set $F \subseteq E$ of edges, we let $w(F) = \sum_{e \in F} w(e)$, and $d(F) = \sum_{e \in F} d_e$.

We now formally define the Weighted Partial Degree Bounded Edge Packing Problem with Demands, denoted PEPD: The input is an undirected (multi-)graph, with $w, d : E \to \mathbb{N}$, the weights and demands respectively, on the edges, and $c : V \to \mathbb{N}$ the capacity on the vertices. We want to compute a sub-graph $G' = (V, E')$ such that $w(E')$ is maximized, and each edge is *satisfied*. We say that an edge is satisfied if at least one of its end-points is not overloaded. Thus, we want for each $e = \{u, v\} \in E'$, $d(N_e(u)) \leq c_u$ or $d(N_e(v)) \leq c_v$. For an edge $e = \{u, v\}$ an end-point that is not overloaded, we call a *good end-point* of e. If neither end-point is overloaded, we pick a good end-point arbitrarily. We also consider the unit-capacity case, i.e., $c_v = 1$, $\forall v \in V$. In this case, we assume $d_e = 1$, $\forall e \in E$. We use PEP to denote the unit-capacity problem.

We also consider in this paper, two graph orientation problems. The Maximum Degree-Bounded Orientation Problem with Demands (ORD) is defined as follows: The input is identical to the PEPD problem, namely an undirected graph $G = (V, E)$, $w, d : E \to \mathbb{N}$ and $c : V \to \mathbb{N}$. The goal is to select a maximum weight subset of edges $E' \subseteq E$, and compute an orientation $\overrightarrow{E'}$ of the edges in E' such that the total demand of $in(v)$ is at most its capacity for each $v \in V$, i.e., $\sum_{e \in in(v)} d_e \leq c_v$ for all $v \in V$. When all vertex capacities are 1, we assume $d_e = 1$, $\forall e \in E$, and use OR to denote this problem with unit capacity and demands.

A solution to PEPD yields a solution to the ORD problem on the same instance. To see this, each edge e in a feasible PEPD solution has a good end-point, and orienting e towards its good end-point is a feasible solution to ORD of the same weight. It would be tempting to hope that the reverse might be true; and if so, this would be cause for cheer as we will show that the ORD problem is tractable. Unfortunately, this is not the case even in the unit-capacity case. Consider for example, a triangle with unit capacity on the vertices, and unit weight on the edges. Any feasible solution to the PEP problem consists of at most 2 edges, but orienting the three edges in a cycle is feasible for OR. In fact, it is known that the PEP problem is NP-hard [19].

Our approximation algorithms for the PEPD problem nevertheless use a solution to the ORD problem as a starting point, and in fact, any solution to the ORD problem can be transformed into one for the PEPD problem on the same instance at the loss of a small constant factor. The relation between the two problems is useful, and we capture this in the following proposition.

Proposition 1. *For any instance I, $OPT_{\text{PEPD}}(I) \leq OPT_{\text{ORD}}(I)$.*

We show that OR problem can be reduced in polynomial time to the b-matching problem in bipartite graphs, which can be solved in polynomial time [17]. Hence, OR can be solved in polynomial time. Similarly, the ORD problem with demands can be reduced to the demand matching problem in bipartite graphs [18], and hence, 2-approximation follows.

Lemma 1. *The OR problem can be solved in polynomial time.*

Proof. We prove this by giving a polynomial time reduction from OR to b-matching in bipartite graphs. The reduction is as follows: Given an instance $G = (V, E)$, $c : V \to \mathbb{N}$, $w : E \to \mathbb{N}$ of the OR problem, we construct a bipartite graph $H = (E \cup V, F)$, with capacities $b_e = 1$ for all $e \in E$ and $b_v = c_v$ for all $v \in V$. The edges F are defined as follows: For each edge $e = \{u, v\}$, add two edges $\{e, u\}, \{e, v\}$ to F, each of weight $w(e)$.

Suppose $E' \subseteq E$ is a feasible solution to OR, where $\overrightarrow{E'}$ denotes a feasible orientation of E'. We claim that E' yields a feasible matching M in H of the same weight. Corresponding to each $e = (u, v) \in \overrightarrow{E'}$, pick $\{e, v\}$ in M. Then, exactly one edge is chosen in M for each $e \in E'$, and for each v, at most c_v edges in M are incident on it. Thus, M is feasible and has weight $w(M) = w(E')$.

Let M be a maximum weight b-matching in H. Let $E' \subseteq E$ be the set of edges of G covered by M. We claim that E' is a feasible solution to OR. To see this, since $b_e = 1$ for all $e \in E$, for each $e = \{u, v\}$, at most one of $\{e, u\}$ or $\{e, v\}$ is in M. This defines an orientation of e in the graph G. If $\{e, u\} \in M$, let $e \in E'$ and $\overrightarrow{e} = (v, u)$. Else, if $\{e, v\} \in M$, let $e \in E'$ and let $\overrightarrow{e} = (u, v)$. Since edges $\{e, v\}$ and $\{e, u\}$ have the same weight as that of e, it follows that $w(E') = w(M)$.

Since b-matching in bipartite graphs admits a polynomial time algorithm [17], it follows that OR can be solved in polynomial time. $\qquad\square$

A similar reduction, implies that ORD has a 2-approximation algorithm.

Lemma 2. *The ORD problem has a 2-approximation algorithm.*

Proof. We use a reduction similar to that in Lemma 1. The only difference is that the edges in F inherit both the weight as well as demand of the corresponding edge, and in the bipartite graph H, we set $b_e = d_e$ for $e \in E$, and $b_v = c_v$ for each $v \in V$. Since demand-matching on bipartite graphs has a 2-approximation algorithm [15], the lemma follows. $\qquad\square$

3   Related Work

The PDBEP problem was introduced by Zhang [19] motivated by a problem of resource allocation as well as a problem of finding large independent sets. This is the unit demand and unit weight version of our problem, namely PEPD with the additional constraint that $w(e) = d_e = 1$ for all $e \in E$. As mentioned earlier, Zhang proved that unit capacity version of PDBEP i.e., PEP with unit edge weights is NP-hard. This result follows from the fact that for a graph $G = (V, E)$, a solution of size k for PEP implies a dominating set of size $|V| - k$. Since the Dominating Set problem is NP-hard [11], this implies PEP is NP-hard. Zhang also presented a 2-approximation algorithm for PEP, with unit edge weights, and a 32/11-approximation algorithm, again with unit edge-weights and a uniform capacity of 2 for all vertices. Dehne et al. [8] studied a parameterized version of the PDBEP problem (with vertex capacity 1 and edge demands 1), and obtained algorithms that are exponential in the size of the PEP solution.

A related problem is the problem of packing vertex disjoint T-stars where a T-star is a complete bipartite graph $K_{1,t}$ for some $1 \leq t \leq T$. In [3] the authors gave a $\frac{9}{4}\frac{T}{T+1}$-approximation for this problem. When $T \geq |V| - 1$, the T-star packing in an edge weighted graph where the objective is to maximize the total weight of the edges in the stars is exactly the PEP problem.

Aurora et al. [2] presented a simple 2-approximation algorithm for PEP with arbitrary vertex capacity, but unit demands and unit weights on the edges. In the setting with weighted edges, but unit demands on the edges, they presented only an $O(\log n)$-approximation algorithm. We introduce the version of the PEP problem with demands on the edges, and present the first constant-factor approximation for this general case.

The PEP problem, as stated earlier is similar to the Maximum weight matching problem, for which several polynomial time algorithms are known [10,13]. The demand version of the problem, PEPD is similar to the *demand matching* problem introduced by Shepherd and Vetta [18]. For the demand matching problem, Shepherd and Vetta presented a 3.264-approximation for general graphs and a 2.764-approximation for bipartite graphs. These results have since been improved using the technique of iterative rounding to a factor 3 for general graphs, and a factor of 2 for bipartite graphs [14,15].

The degree-bounded orienting problem OR is a classic combinatorial optimization problem. However, it has mostly received attention in terms of maintaining connectivity. See [17] for more details.

4   A 2-Approximation Algorithm for Unit Capacity Instances

In this section, we present a 2-approximation algorithm for PEP. In this setting, recall that $c_v = 1$ for all $v \in V$. Earlier, a 2-approximation algorithm was known only for the unweighted case by Zhang [19], i.e., when $w(e) = 1$ for all $e \in E$.

Our algorithm is combinatorial. We show that we can carefully select a subset of edges such that an upper bound on OPT_{PEP} can be constructed from this subset. Recall that for a vertex $v \in V$, $N_{\max}(v)$ denotes the set of edges of maximum weight incident on v, and we use $e_{\max}(v)$ to denote an edge in $N_{\max}(v)$.

The set of edges $E_{\max} \subseteq E$ is constructed as follows: Let $v_1, \ldots, v_n$ be an arbitrary ordering of the vertices. Starting with $E_{\max}^1 = E_{\max}^2 = \emptyset$, for each i from $1, \ldots, n$, if an edge from $N_{\max}(v_i)$ has not been chosen, pick an arbitrary edge $e = \{v_i, v_j\}$ from $N_{\max}(v_i)$. If $e \in N_{\max}(v_j)$ and no edges from $N_{\max}(v_j)$ have been chosen yet we add e to $E_{\max}^2$, otherwise add e to $E_{\max}^1$. We denote the set $E_{\max} = E_{\max}^1 \cup E_{\max}^2$. Observe that by the way we choose the edges in $E_{\max}^2$ at most one edge from $N_{\max}(v)$ is chosen for each vertex. This is encoded in the Proposition below.

Proposition 2. *For each $v \in V$, $|E_{\max}^2 \cap N_{\max}(v)| \leq 1$*

Lemma 3. $\text{OPT}_{\text{PEP}} \leq \sum_{v \in V} w(e_{\max}(v))$

Proof. We show that $\sum_{v \in V} w(e_{\max}(v))$ is an upper-bound on the Max-orientation problem OR on the same instance. Then, the lemma follows from Proposition 1. Let $F \subseteq E$ be a feasible solution to the OR problem in G. Since the vertices have unit capacity, for any vertex $v \in V$, there is at most one edge of F in-coming to v. Let $w(in(v))$ denote the weight of this edge if any, and zero otherwise. Then, $\sum_{(u,v) \in F} w(u,v) = \sum_{v \in V} w(in(v)) \leq \sum_{v \in V} w(e_{\max}(v))$. $\square$

Note that in order to obtain an upper bound, we require that we sum up $w(e_{\max}(v))$ over all vertices as $w(E_{\max})$ by itself does not constitute an upper bound. For example consider the graph $G = (\{a, b, c, d\}, \{\{a, b\}, \{b, c\}, \{c, d\}, \{a, d\}, \{b, d\}\})$. Let the weights on the edges be $\{5, 3, 4, 1, 2\}$ in the same order. In this example, $\sum_{e \in E_{\max}} w(e) = w(a,b) + w(d,c) = 9$. However, $\text{OPT}_{\text{PEP}} = w(a,b) + w(b,c) + w(b,d) = 10$.

In order to obtain our claimed approximation, we construct an orientation of the edges in $E_{\max}$ such that $\overrightarrow{G}_{\max} = (V, \overrightarrow{E}_{\max})$ is acyclic, and show that each connected component of $\overrightarrow{G}_{\max}$ is an out-tree.

Lemma 4. *There exists an orientation of the edges in $E_{\max}$ such that each connected component of $\overrightarrow{G}_{\max} = (V, \overrightarrow{E}_{\max})$ is an out-tree.*

Proof. We suggest a natural orientation of the edges in $E_{\max}$. Suppose the vertices are considered in the order of $v_1, \ldots, v_n$ during the construction of $E_{\max}$. In iteration i of the construction, if an edge is added to either $E_{\max}^1$ or $E_{\max}^2$, then orient that edge towards v_i. Since in each iteration at most one edge can be added to $E_{\max}$, our orientation ensures that for any vertex $v_i \in V$, at most one edge in $\overrightarrow{E}_{\max}$ is oriented towards v_i, which implies the in-degree of any vertex is at most 1. Note that this shows the set $E_{\max}$ is in fact a feasible solution to the OR problem.

To show that each connected component of $\overrightarrow{G}_{\max}$ is an out-tree, we also require that the graph $\overrightarrow{G}_{\max}$ is acyclic. In contrary, let $C = v_0, v_1, \ldots, v_k, v_0$ be a directed cycle in $\overrightarrow{G}_{\max}$. Our orientation ensures that if an edge e is oriented towards v, then $e \in N_{\max}(v)$. Thus, the weight of an in-coming edge into a vertex v has weight at least as large as any out-going edge. Thus, following the cycle from v_0, the weight of the edges can not increase. Therefore, the only possibility is that all edges in C have equal weight. However, we add an edge e incident on a vertex v into $E_{\max}$ only if no edge in $N_{max}(v)$ is present in $E_{\max}$ when v is processed, and we then orient e towards v. If, without loss of generality, the vertex v_0 was the first to be processed during the construction of $E_{\max}$, then the edge $\{v_k, v_0\}$ is already present in the solution when we process v_k. Since $\{v_k, v_0\} \in N_{\max}(v_k)$, no edge is added during the processing of v_k. This implies no edge is oriented towards v_k, which is a contradiction. Hence, $\overrightarrow{G}_{max}$ is acyclic.

Thus, we have $\overrightarrow{G}_{\max}$ is acyclic and each vertex has in-degree at most 1. This ensures each connected component of $G_{\max}$ is an out-tree. $\qquad \square$

The example discussed above shows that $w(E_{\max})$ is not an upper bound on $\mathrm{OPT}_{\mathrm{PEP}}$. However, we can re-write the upper bound by distinguishing the contribution from the edges of $E_{\max}^1$ and $E_{\max}^2$.

Lemma 5. $\mathrm{OPT}_{\mathrm{PEP}} \leq w(E_{\max}^1) + 2w(E_{\max}^2)$

Proof. Consider the above oriented graph $\overrightarrow{G}_{\max} = (V, \overrightarrow{E}_{\max})$. We claim that for any vertex v_i, if no edge is oriented towards v_i, then there is exactly one edge in $E_{\max}^2$ incident on v_i. To see this, note that if no edge is oriented towards v_i, then before iteration i, we have already added at least one edge from $N_{\max}(v_i)$ to $E_{\max}$. Let j be the minimal iteration such that an edge from the set $N_{\max}(v_i)$ is added to $E_{\max}$. Then according to our construction, the edge $\{v_j, v_i\}$ is the only edge added to the set $E_{\max}^2$ among all the edges incident on v_i in $E_{\max}$.

Let v be any vertex towards which no edge is oriented in $\overrightarrow{G}_{\max}$. Then there is exactly one edge in $E_{\max}^2$ incident on v. Let $e = \{u, v\}$ be that edge. We add another edge of weight $w(e)$ between u and v to $\overrightarrow{G}_{\max}$, and orient this edge towards v. Let the resulting multi-graph be $\overrightarrow{G}'_{\max} = (V, \overrightarrow{E}'_{\max})$. In $\overrightarrow{G}'_{\max}$, exactly one edge from $N_{\max}(v)$ is oriented towards v for each vertex v. Thus, $\sum_{v \in V} w(e_{\max}(v)) = \sum_{e \in \overrightarrow{E}'_{\max}} w(e) = \sum_{e \in E_{\max}^1} w(e) + 2\sum_{e \in E_{\max}^2} w(e)$. Using Lemma 3, we get $\mathrm{OPT}_{\mathrm{PEP}} \leq w(E_{\max}^1) + 2w(E_{\max}^2)$. $\qquad \square$

Theorem 1. *There exists a 2-approximation algorithm for* PEP.

Proof. Since each connected component of the graph $\overrightarrow{G}_{\max}$ is an out-tree, we can partition the vertices into two sets such that all the edges in $\overrightarrow{E}_{\max}$ crosses the partition. To see this, for any tree T in $\overrightarrow{G}_{\max}$, label the vertices with distance from the root. Since T is a tree, a vertex with odd label is only adjacent to vertices with even label, and vice-versa. Therefore, we can partition V into two sets X and Y, where X consists of odd-labeled vertices, and Y consists of even-labeled vertices. The set of edges between X and Y consists of all the edges.

Now consider the cut (X, Y) in $G_{\max}$, and orient the edges in E_{max} as follows: Orient all the edges in $E_{\max}^1$ in the same way it is oriented in $\overrightarrow{G}_{\max}$, and orient each edge $e \in E_{\max}^2$, towards both the end-points. Note that $\sum_{e \in \overrightarrow{\delta}(X)} w(e) + \sum_{e \in \overrightarrow{\delta}(Y)} w(e) = w(E_{\max}^1) + 2w(E_{\max}^2)$, since each edge in $E_{\max}^2$ is present in both $\overrightarrow{\delta}(X)$ and $\overrightarrow{\delta}(Y)$, where for any Z, $\overrightarrow{\delta}(Z)$ denotes the set of out-going edges from Z. Therefore, $\max\{\overrightarrow{\delta}(X), \overrightarrow{\delta}(Y)\} \geq (w(E_{\max}^1) + 2w(E_{\max}^2))/2$. Thus, by returning the maximum among $\overrightarrow{\delta}(X)$, and $\overrightarrow{\delta}(Y)$, we guarantee a solution of weight at least $\text{OPT}_{\text{PEP}}/2$ (Using Lemma 5).

Now it remains to prove the feasibility of $\overrightarrow{\delta}(X)$, and $\overrightarrow{\delta}(Y)$. Note that the in-degree of any vertex is at most 1 in the oriented graph, which ensures both $\overrightarrow{\delta}(X)$ and $\overrightarrow{\delta}(Y)$ are individually feasible for PEP.   □

Note that our 2-approximation algorithm for PEP improves the result of Babenko and Gusakov [3] for the special case of T-star packing when $T = |V| - 1$. They proposed a $\frac{9}{4}\frac{T}{T+1}$-approximation algorithm for this problem.

5   A Constant-Factor Approximation Algorithm

In this section, we obtain a $(4+\epsilon)$-approximation algorithm for any $\epsilon > 0$ for the PEPD problem on general graphs, and a $(2+\epsilon)$-approximation algorithm for the PEPD problem on bipartite graphs. Our algorithm holds for a slightly more general problem. Instead of demands on the edges, we let each edge $e = \{u, v\}$ have possibly different demands $d(e, u)$, $d(e, v)$ at its end-points. It is possible that $d(e, u)$ exceeds the capacity c_u of vertex u, and yet, e could be in our solution since its other end-point, namely v could be its good end-point.

Given a graph $G = (V, E)$, our algorithm finds an oriented multi-graph $\overrightarrow{G}' = (V, \overrightarrow{E}')$ having $w(E')$ almost equal to the optimal PEPD solution such that for all the vertices, in-degree constraint is satisfied. Next, by finding a directed cut of weight at least $w(E')/4$ in $\overrightarrow{G}'$, we guarantee a $(4 + \epsilon)$-approximate solution for PEPD in G.

Lemma 6. *Given a graph $G = (V, E)$, weights $w : E \to \mathbb{N}$, demands $d : E \to \mathbb{N}$ on the edges, and capacities $c : V \to \mathbb{N}$ on the vertices, there exists a directed multi-graph $\overrightarrow{G}' = (V, \overrightarrow{E}')$ with $w(E')$ at least $(1 - \epsilon)\text{OPT}_{\text{PEPD}}$ such that $\sum_{e \in in(v)} d_e \leq c_v$, for all $v \in V$.*

Proof. Let OPT_{PEPD} denote an optimal solution for the PEPD problem, and $F \subseteq E$ be the edges picked in this solution. For any vertex v_i, if the degree condition is satisfied in F, then we set $\text{OPT}_{\text{PEPD}}^i$ to be the total weight of the edges incident on v_i in F, otherwise we set $\text{OPT}_{\text{PEPD}}^i$ to be 0. Since for any edge in F, the degree condition is satisfied at at least one end-point, we have $\sum_{i=1}^{n} \text{OPT}_{\text{PEPD}}^i \geq \text{OPT}_{\text{PEPD}}$.

For all $v_i \in V$, we consider the problem of picking a maximum weight sub-set of edges from $N_e(v_i)$ such that the degree condition is satisfied. Observe that at each vertex, this amounts to solving an independent Knapsack problem. Since Knapsack admits an FPTAS [16], we obtain a solution $\mathcal{A}_i$ of weight at least

$(1 - \epsilon)\text{OPT}_i$, where OPT_i is the optimal solution to this problem w.r.t. vertex v_i. Therefore, we have

$$\sum_{i=1}^{n} w(\mathcal{A}_i) \geq (1 - \epsilon) \sum_{i=1}^{n} \text{OPT}_i$$

$$\geq (1 - \epsilon) \sum_{i=1}^{n} \text{OPT}_{\text{PEPD}}^{i}$$

$$\geq (1 - \epsilon)\text{OPT}_{\text{PEPD}}.$$

The second inequality is true because $\text{OPT}_{\text{PEPD}}^{i}$ is a feasible solution to the knapsack problem w.r.t. v_i.

Observe that in a similar way, we can show that $\sum_{i=1}^{n} w(\mathcal{A}_i)$ is at least $(1 - \epsilon)\text{OPT}_{\text{ORD}}$. The set of edges in $\cup_{i=1}^{n}\mathcal{A}_i$ is in fact a feasible solution to the ORD problem. To see this, for each vertex v_i, orient the edges in $\mathcal{A}_i \setminus \cup_{j=1}^{i-1}\mathcal{A}_j$ towards the vertex v_i. This ensures that for any vertex, the total demand of the incoming edges is at most its capacity. Since, each edge can appear at most twice in the sum $\sum_{i=1}^{n} w(\mathcal{A}_i)$, we have a $(2 + \epsilon)$-approximation algorithm for the ORD problem.

However, the set of edges in $\cup_{i=1}^{n}\mathcal{A}_i$ may not be a feasible solution for the PEPD problem. In order to obtain the $(4 + \epsilon)$-approximation, we construct a directed multi-graph $\overrightarrow{G}' = (V, \overrightarrow{E}')$ as follows: Pick an arbitrary ordering of the vertices, say $v_1, \ldots, v_n$. Starting with $E' = \emptyset$, for each i from $1, \ldots n$, add the edges in $\mathcal{A}_i$ to the multi-set of edges E' and orient the edges in $\mathcal{A}_i$ towards v_i. By doing this, we ensure that $\sum_{e \in E'} w(e) = \sum_{i=1}^{n} w(\mathcal{A}_i) \geq (1 - \epsilon)\text{OPT}_{\text{PEPD}}$.

Theorem 2. *For any $\epsilon > 0$, there exists a $(2 + \epsilon)$-approximation algorithm for PEPD on bipartite graphs.*

Proof. Given a bipartite graph $G = (U \cup V, E)$, using Lemma 6, we can find the directed multi-graph $\overrightarrow{G}' = (U \cup V, \overrightarrow{E}')$ with $w(E') \geq (1 - \epsilon')\text{OPT}_{\text{PEPD}}$, for any $\epsilon' > 0$ such that total demand of the in-coming edges to any vertex is at most its capacity. So, the set of incoming edges to U and the set of in-coming edges to V in $\overrightarrow{G}'$ are separately feasible for the PEPD problem on G, and the maximum of both has weight at least $(1 - \epsilon')\text{OPT}_{\text{PEPD}}/2$. Choosing $\epsilon' = \epsilon/(2 + \epsilon)$, we obtain a solution of weight at least $\text{OPT}_{\text{PEPD}}/(2 + \epsilon)$ for the PEPD problem on bipartite graphs.

In order to get the desired approximation ratio for general graphs, we find a directed cut (DICUT) of weight at least $w(E')/4$ in $\overrightarrow{G}'$. Given a directed multi-graph $\overrightarrow{G}_m$ and an edge weight function $w : E(G_m) \rightarrow \mathbb{N}$, a DICUT is defined to be the set of out-going edges from some vertex subset X (we denote it by $\overrightarrow{\delta}(X)$). Note that any directed cut in $\overrightarrow{G}'$ is a feasible PEPD solution. Lemma 7 captures this.

Lemma 7. *Any directed cut of the directed multi-graph $\overrightarrow{G}'$ is a feasible PEPD solution of G.*

Proof. Let $\overrightarrow{\delta}(X)$ be a DICUT of $\overrightarrow{G}'$. This implies for all $v \in V \setminus X$, $out(v) = \emptyset$, and for any edge (u, v) in $\overrightarrow{\delta}(X)$ directed towards v, the degree condition of v is satisfied in G. This ensures $\overrightarrow{\delta}(X)$ is a feasible PEPD solution.

Lemma 8. *Given a directed multi-graph $\overrightarrow{G}' = (V, \overrightarrow{E}')$, weights $w : E' \to \mathbb{N}$, there exists a directed cut of size at least $w(E')/4$.*

Proof. Consider the trivial randomized algorithm which adds any vertex in V to the set X with probability $1/2$. Any directed edge $e = (u, v)$ is a cut if $u \in X$ and $v \in V \setminus X$. This happens with probability $1/4$. So, the expected weight of the DICUT is

$$\mathbb{E}\left(\sum_{e \in \overrightarrow{\delta}(X)} w(e)\right) = \sum_{e \in \overrightarrow{E}'} w(e) \cdot \Pr\left(e \in \overrightarrow{\delta}(X)\right) = \sum_{e \in \overrightarrow{E}'} w(e) \cdot \frac{1}{4} = \frac{w(E')}{4}.$$

This ensures, there exists a DICUT of weight at least $w(E')/4$. To find it, de-randomize this by using the method of conditional expectations.

Armed with Lemmas 6, 7, and 8 we can now complete the proof.

Theorem 3. *There exists a $(4+\epsilon)$-approximation algorithm for PEPD, for any $\epsilon > 0$.*

Proof. Given an instance of PEPD, let OPT_{PEPD} be an optimal solution to PEPD. Lemma 6 shows that we can obtain an oriented graph $\overrightarrow{G}' = (V, \overrightarrow{E}')$ having $w(E') \geq (1 - \epsilon')\text{OPT}_{\text{PEPD}}$, for any $\epsilon' > 0$ such that $\sum_{e \in in(v)} d_e \leq c_v$, for all $v \in V$. Combining this with Lemma 7 and Lemma 8 we obtain a PEPD solution in G of weight at least $w(E')/4$ which is at least $(1 - \epsilon')\text{OPT}_{\text{PEPD}}/4$. Using $\epsilon' = \epsilon/(4+\epsilon)$, we obtain a PEPD solution of weight at least $\text{OPT}_{\text{PEPD}}/(4 + \epsilon)$.

6 PTAS for PEpD on Minor-free Graphs

In this section, we obtain a PTAS for PEPD in H-minor-free graphs. Our result follows the standard procedure for proving a PTAS for such graphs. We present a polynomial time algorithm for graphs of bounded-treewidth. However, the algorithm only work in the setting where the demands on the edges are bounded by a constant. Then, a PTAS for H-minor-free graphs then follows from the results of Demaine et al. [9]. For ease of exposition, we only describe the polynomial time algorithm for bounded tree-width graphs with $d_e = 1$ for all $e \in E$. However, the extension to arbitrary, but constant demands is straight forward.

6.1 A Polynomial Time Algorithm for Bounded-Treewidth Graphs

For the sake of completeness, we give a definition of a tree-decomposition. See [5] for a description and results on tree-decompositions.

Definition 1. *A tree decomposition of a graph $G = (V, E)$ is a pair (T, X), where $T = (I, F)$ is a tree and $X = \{X_i | \ i \in I\}$ is a set with $X_i \subseteq V$ satisfying*

- $\bigcup_{i \in I} X_i = V$.
- *for any edge $e = (u, v) \in E$, there exists an $i \in I$ with $u \in X_i$ and $v \in X_i$.*
- *for all $v \in V$, the set of nodes $\{i \in I | \ v \in X_i\}$ forms a connected subtree of T.*

We refer to the vertices of T as nodes and the corresponding X_i's as bags in order to distinguish them from the vertices of G. The width of any tree decomposition $T = (I, F)$ is $\max_{i \in I} |X_i| - 1$ and the tree-width of a graph G, denoted as $tw(G)$ is the minimum width among all possible tree decompositions of G. Let G be a graph with $tw(G) = t - 1$, for constant $t > 0$, and let (T, X) with $T = (I, F)$ and $X = \{X_i | \ i \in I\}$ be a tree decomposition of G of width $t - 1$. It is also well known that without loss of generality we can assume that T is a rooted binary tree [5].

Define for all $i \in I$, $Y_i = \{v \in X_j | \ j$ is a descendant of $i\}$. Let $G[X_i] = (X_i, E_i)$, and $G[Y_i] = (Y_i, F_i)$ denote the vertex induced subgraphs of G with vertices in X_i and Y_i respectively. Let $G' = (V, E')$ be an optimal solution for PEPD, and $F_i' = F_i \cap E'$, $E_i' = E_i \cap E'$. For any bag X_i, let $\mathbf{d_i}$ be the degree sequence of the vertices in X_i in the subgraph $G_i' = (V, F_i' \cup E_i')$. Suppose $\mathbf{f}$ be a vector representing whether $\deg_{G'}(v) \leq c_v$, or $\deg_{G'}(v) > c_v$. For any $v \in V$, we set $\mathbf{f}(v) = c_v$, if $\deg_{G'}(v) \leq c_v$, and $\mathbf{f}(v) = \infty$, otherwise. The vector $\mathbf{f_i}$ denotes the vector $\mathbf{f}$ with restriction to the vertices in X_i.

We now describe our dynamic program. The dynamic program works bottom-up. Each DP cell $C(i, E_i', \mathbf{d_i}, \mathbf{f_i})$ represents the subproblem of choosing a set of edges $F_i' \subseteq F_i$ with maximum total weight, such that $F_i' \cup E_i'$ are feasible assuming the degrees on vertices of X_i in the subgraphs (V, E') and $(V, F_i' \cup E_i')$ are bounded above by vectors $\mathbf{f_i}$ and $\mathbf{d_i}$ respectively.

We calculate each DP cell as follows: for each node i, we guess the set of edges E_i', and the vectors $\mathbf{f_i}$, $\mathbf{d_i} \preceq \mathbf{f_i}$, where $\mathbf{d_i} \preceq \mathbf{f_i}$ denotes that the vector $\mathbf{d_i}$ is component-wise less than or equal to the vector $\mathbf{f_i}$. Since $\mathbf{f}$ provides an upper bound on the degree of any vertex in G', edges in any subgraph of G' must satisfy the feasibility constraint assuming degree of vertex v can be at most $\mathbf{f}(v)$. Let $W(i, E_i', \mathbf{f_i}, \mathbf{d_i})$ be the weight of the DP cell $C(i, E_i', \mathbf{f_i}, \mathbf{d_i})$. For any leaf node i, we compute $W(i, E_i', \mathbf{f_i}, \mathbf{d_i})$ as follows:

$$W(i, E_i', \mathbf{f_i}, \mathbf{d_i}) = \begin{cases} w(E_i'), & \text{if } \forall \{u, v\} \in E_i', \ \mathbf{f_i}(u) \leq c_u \text{ or } \mathbf{f_i}(v) \leq c_v \\ -\infty, & \text{otherwise.} \end{cases}$$

For any internal node i, with children j and k for which we have already computed the DP cells, we can compute the DP cell as follows:

$$W(i, E_i', \mathbf{f_i}, \mathbf{d_i}) = \begin{cases} \mathcal{A}\big(W(i, E_i', \mathbf{f_i}, \mathbf{d_i})\big), & \text{if } \forall \{u, v\} \in E_i', \ \mathbf{f_i}(u) \leq c_u \text{ or } \mathbf{f_i}(v) \leq c_v \\ -\infty, & \text{otherwise.} \end{cases}$$

Where, $\mathcal{A}\big(\mathrm{W}(i, E_i', \mathbf{f_i}, \mathbf{d_i})\big)$ can be computed as follows:

$$\mathcal{A}\big(\mathrm{W}(i, E_i', \mathbf{f_i}, \mathbf{d_i})\big) = \max_{\substack{E_j', \mathbf{f_j}, \mathbf{d_j} \preceq \mathbf{f_j}, \\ E_k', \mathbf{f_k}, \mathbf{d_k} \preceq \mathbf{f_k}}} \Bigg\{ \mathrm{W}(j, E_j', \mathbf{f_j}, \mathbf{d_j}) + \mathrm{W}(k, E_k', \mathbf{f_k}, \mathbf{d_k})$$

$$+ w(E_i' \setminus (E_j' \cup E_k')) - w(E_j' \cap E_k') \Bigg|$$

$$\forall v \in \{X_i \cap X_j \cap X_k\}, \mathbf{f_i}(v) = \mathbf{f_j}(v) = \mathbf{f_k}(v),$$
$$\forall v \in \{X_i \cap X_j\}, \mathbf{f_i}(v) = \mathbf{f_j}(v),$$
$$\forall v \in \{X_i \cap X_k\}, \mathbf{f_i}(v) = \mathbf{f_k}(v),$$
$$\forall v \in X_i, \mathbf{d_i}(v) \geq \mathbf{d_j}(v) + \mathbf{d_k}(v) +$$

$$\deg_{E_i' \setminus (E_j' \cup E_k')}(v) - \deg_{E_j' \cap E_k'}(v) \Bigg\},$$

Where, $\deg_{E_i' \setminus (E_j' \cup E_k')}(v)$ and $\deg_{E_j' \cap E_k'}(v)$ denote the degrees of v in the subgraphs $(V, E_i' \setminus (E_j' \cup E_k'))$ and $(V, E_j' \cap E_k')$ respectively. Note that if $v \notin X_j$, then $\mathbf{d_j}(v) = 0$ and if $v \notin X_k$, then $\mathbf{d_k}(v) = 0$.

The optimal solution is the $\max\{\mathrm{W}(r, E_r', \mathbf{f_r}, \mathbf{f_r})\}$, where r is the root node of the tree T.

The number of nodes in the tree-decomposition T of G is at most $O(nt)$ [5]. For any node $i \in I$, $|X_i| \leq t$, so E_i' can take at most 2^{t^2} values, and f_i can take at most 2^t values. For any vertex $v \in X_i$, $\mathbf{d_i}(v)$ can take at most n values. $\mathbf{d_i}(v)$ can take at most n values. So, total number of cells in the DP can be at most $O(nt) \cdot 2^{t^2} \cdot 2^t \cdot n^t = O(n^t)$. Each DP cell takes $O(n^{2t})$ computation time. So the running time of the DP is bounded by $O(n^{3t})$.

Theorem 4. *The PEPD problem, on graphs with bounded tree-width can be solved in polynomial time if the demand of any edge is bounded above by a constant.*

6.2   Partition into Bounded Treewidth Graphs

We use the following result of Demaine et al. [9] on the structure of H-minor-free graphs.

Lemma 9 [9]. *For a fixed graph H, there is a constant c_H such that, for any integer $k \geq 1$ and for every H-minor-free graph G, the vertices of G (or the edges of G) can be partitioned into $k + 1$ sets such that any k of the sets induce a graph of tree-width at most $c_H k$. Furthermore, such a partition can be found in polynomial time.*

Theorem 5. *In H-minor free graphs, there is a PTAS for the PEPD problem if the demand of any edge is bounded above by a constant.*

Proof. Let $G = (V, E)$ be any H-minor-free graph. We apply lemma 9 with $k = 1/\epsilon$ to partition E into sets $E_1, E_2, \cdots E_{1+1/\epsilon}$. Let E' be the edges in the optimal solution, and $E'_1 = E' \cap E_1, E'_2 = E' \cap E_2, \cdots, E'_{1+1/\epsilon} = E' \cap E_{1+1/\epsilon}$. Let E'_m be the set with minimum weight among $\{E'_1, E'_2, \cdots, E'_{1+1/\epsilon}\}$. Since,

$w(E'_m) \leq \frac{w(E')}{k+1}$. this implies $w(E' \setminus E'_m) \geq (1 - 1/k + 1)w(E')$.

Let G_i be the subgraph of G with edge set $\overline{E}_i = \cup_{j \neq i} E_j$. Each G_i has tree-width bounded by $c_H k$ for which we can get the optimal solution OPT_i by using theorem 4. Since the set $E' \setminus E'_i$ is a feasible solution to G_i, we have $OPT_i \geq w(E' \setminus E'_i)$.

$$\max\{OPT_1, \cdots, OPT_{k+1}\} \geq OPT_m \geq w(E' \setminus E'_m)$$

$$\geq \left(1 - \frac{1}{k+1}\right) w(E') \geq (1 - \epsilon)w(E').$$

Hence, the maximum weighted solution among the solutions for $G_1, G_2, \cdots G_{k+1}$ gives a PTAS.

7   APX-Hardness

In this section, we prove that PEp problem is APX-hard even for unweighted cubic graphs, and unweighted bipartite graphs of bounded degree. Earlier, only NP-hardness was known. This was proved by Zhang [19] by showing that for a graph $G = (V, E)$, a solution to unweighted PEp of size k implies a Dominating Set in G of size $|V| - k$. Our result follows from the following facts: Cubic graphs and bipartite graphs of bounded degree have large dominating set, and the fact that the Dominating set problem is APX-hard on cubic graphs and bounded degree bipartite graphs. We show that a PTAS for the unweighted PEp would imply a PTAS for the Dominating set problem.

Proposition 3. *Let $G = (U \cup V, E)$ be bipartite graph with degree bounded by B. Then, any dominating set in G has size at least $|U \cup V|/(1 + B)$.*

Proof. Let $|U \cup V| = n$ and suppose there is a dominating set $S \subset U \cup V$ of size $< n/(1+B)$. Since each $v \in S$ can dominate at most B vertices in $\{U \cup V\} \setminus S$, all vertices in S together can dominate $< nB/(1 + B)$ vertices. Since any vertex in G either belongs to S or is dominated by a vertex in S, we have the total number of vertices in $G < n/(1 + B) + nB/(1 + B) = n$ contradicting our assumption that $|U \cup V| = n$.

We use the following result of Zhang [19] on the relation between Dominating Sets and PEp on unweighted graphs. Following which, our result on bipartite graphs follows by using the result of Chlebík and Chlebíková [7] that Dominating Set on bounded degree bipartite graphs is APX-hard.

Lemma 10 [19]. *Let $G = (V, E)$ be a graph without isolated vertices having $w(e) = 1$, for all $e \in E$. In G, there is a maximal solution of size k to the PEp if and only if there is a solution of size $|V| - k$ to the Dominating Set problem.*

Theorem 6 [7]. *It is NP-hard to approximate the Dominating Set problem in bipartite graphs of degree at most $B \geq 3$ within a factor of $\ln B - C \ln \ln B$ for some absolute constant C.*

Theorem 7. *The PEP problem is APX-hard for unweighted bipartite graphs having degree at most $B \geq 3$.*

Proof. We prove that an existence of a PTAS for the PEP problem on bounded degree bipartite graphs implies a PTAS for the Dominating Set problem on the same class of graphs, contradicting Theorem 6.

Let $G = (U \cup V, E)$ be a bipartite graph with degree bounded by B and $|U \cup V| = n$. By Proposition 3, $\text{OPT}_{DS}(G) \geq n/(1 + B)$. Lemma 10 implies that $\text{OPT}_{\text{PEP}}(G) = n - \text{OPT}_{DS}(G)$. If there exists a PTAS for PEP, this implies that for every $\epsilon > 0$, we can find a sub-graph $G' = (V, E')$ such that $|E'| \geq (1 - \epsilon)(n - \text{OPT}_{DS}(G))$. We can assume that G' is a collection of disjoint stars with no isolated vertices as a PEP solution with isolated vertices can always be transformed into one without isolated vertices. Therefore G' is a collection of stars spanning V and the set C of central vertices form a dominating set (Lemma 10). $|C| = n - |E'|$. Therefore,

$$
\begin{aligned}
|C| = n - |E'| &\leq n - (1 - \epsilon)(n - \text{OPT}_{DS}(G)) \\
&\leq (1 + B)\epsilon OPT_{DS}(G) + (1 - \epsilon)OPT_{DS}(G) \\
&\leq (1 + B\epsilon)OPT_{DS}(G).
\end{aligned}
$$

Therefore, PEP is APX-hard, even on unweighted bipartite graphs of bounded degree.

Remark: The APX-hardness result on cubic graphs can be obtained by using a similar argument and the result of Alimonti and Kann [1] that Dominating Set on cubic graphs is APX-hard.

8   Conclusion

To obtain better than 2-approximation for PEP, $(2+\epsilon)$-approximation for PEPD on bipartite graphs, and $(4 + \epsilon)$-approximation for PEPD on general graphs, we need to find better upper bounds on OPT_{PEP}, OPT_{PEPD} on bipartite graphs, and OPT_{PEPD} on general graphs respectively. For example consider the graph C_4 (cycle on 4 vertices) with unit weight on all the edges. In this case $\text{OPT}_{\text{PEP}} = 2$, whereas the upper bound is $\sum_{v \in V} w(e_{\max}(v)) = 4$. So we cannot expect to get a better than 2-approximation with this upper bound.

References

1. Alimonti, P., Kann, V.: Hardness of approximating problems on cubic graphs. In: Bongiovanni, G., Bovet, D.P., Di Battista, G. (eds.) CIAC 1997. LNCS, vol. 1203, pp. 288–298. Springer, Heidelberg (1997)

2. Aurora, P., Singh, S., Mehta, S.K.: Partial degree bounded edge packing problem with arbitrary bounds. In: Tan, X., Zhu, B., Fellows, M. (eds.) FAW-AAIM 2013. LNCS, vol. 7924, pp. 24–35. Springer, Heidelberg (2013)
3. Babenko, M.A., Gusakov, A.: New exact and approximation algorithms for the star packing problem in undirected graphs. In: 28th International Symposium on Theoretical Aspects of Computer Science, STACS 2011, 10–12 March 2011, Dortmund, Germany, pp. 519–530 (2011)
4. Baker, B.S.: Approximation algorithms for NP-complete problems on planar graphs. J. ACM **41**(1), 153–180 (1994)
5. Bodlaender, H.L.: A partial k-arboretum of graphs with bounded treewidth. Theor. Comput. Sci. **209**(1–2), 1–45 (1998)
6. Bu, T.-M., Yuan, C., Zhang, P.: Computing on binary strings. Theoret. Comput. Sci. **562**, 122–128 (2015)
7. Chlebík, M., Chlebíková, J.: Approximation hardness of dominating set problems in bounded degree graphs. Inf. Comput. **206**(11), 1264–1275 (2008)
8. Dehne, F., Fellows, M.R., Fernau, H., Prieto, E., Rosamond, F.A.: NONBLOCKER: parameterized algorithmics for MINIMUM DOMINATING SET. In: Wiedermann, J., Tel, G., Pokorný, J., Bieliková, M., Štuller, J. (eds.) SOFSEM 2006. LNCS, vol. 3831, pp. 237–245. Springer, Heidelberg (2006)
9. Demaine, E.D., Hajiaghayi, M.T., Kawarabayashi, K.-i.: Algorithmic graph minor theory: decomposition, approximation, and coloring. In: Proceedings of the 46th Annual IEEE Symposium on Foundations of Computer Science (FOCS 2005), 23–25 October 2005, Pittsburgh, PA, USA, pp. 637–646 (2005)
10. Edmonds, J.: Paths, trees, and flowers. Can. J. Math. **17**(3), 449–467 (1965)
11. Karp, R.M.: Reducibility among combinatorial problems. In: Miller, R.E., Thatcher, J.W., Bohlinger, J.D., et al. (eds.) Complexity of Computer Computations. The IBM Research Symposia Series, pp. 85–103. Springer, Heidelberg (1972)
12. Lovász, L., Plummer, M.D.: Matching Theory, vol. 367. American Mathematical Soc., Providence (2009)
13. Micali, S., Vazirani, V.V.: An $O(\sqrt{|V|}|E|)$ algoithm for finding maximum matching in general graphs. In: 21st Annual Symposium on Foundations of Computer Science, pp. 17–27. IEEE (1980)
14. Parekh, O.: Iterative packing for demand and hypergraph matching. In: Günlük, O., Woeginger, G.J. (eds.) IPCO 2011. LNCS, vol. 6655, pp. 349–361. Springer, Heidelberg (2011)
15. Parekh, O., Pritchard, D.: Generalized hypergraph matching via iterated packing and local ratio. In: Bampis, E., Svensson, O. (eds.) WAOA 2014. LNCS, vol. 8952, pp. 207–223. Springer, Heidelberg (2015)
16. Sahni, S.K.: On the knapsack and other computationally related problems (1973)
17. Schrijver, A.: Combinatorial Optimization: Polyhedra and Efficiency, vol. 24. Springer Science & Business Media, Heidelberg (2003)
18. Shepherd, B., Vetta, A.: The demand matching problem. In: Cook, W.J., Schulz, A.S. (eds.) IPCO 2002. LNCS, vol. 2337, pp. 457–474. Springer, Heidelberg (2002)
19. Zhang, P.: Partial degree bounded edge packing problem. In: Snoeyink, J., Lu, P., Su, K., Wang, L. (eds.) AAIM 2012 and FAW 2012. LNCS, vol. 7285, pp. 359–367. Springer, Heidelberg (2012)

An Introduction to Coding Sequences of Graphs

Shamik Ghosh[1]([✉]), Raibatak Sen Gupta[1], and M.K. Sen[2]

[1] Department of Mathematics, Jadavpur University,
Kolkata 700032, India
sghosh@math.jdvu.ac.in, raibatak2010@gmail.com
[2] Department of Pure Mathematics, University of Calcutta,
Kolkata 700019, India
senmk6@yahoo.com

Abstract. In this paper, we introduce a new representation of simple undirected graphs in terms of set of vectors in finite dimensional vector spaces over $\mathbb{Z}_2$ which satisfy consecutive 1's property, called a coding sequence of a graph G. Among all coding sequences we identify the one which is unique for a class of isomorphic graphs, called the code of a graph. We characterize several classes of graphs in terms of coding sequences. It is shown that a graph G with n vertices is a tree if and only if any coding sequence of G is a basis of the vector space $\mathbb{Z}_2^{n-1}$ over $\mathbb{Z}_2$.

Moreover, considering coding sequences as binary matroids, we obtain a characterization for simple graphic matroids. Introducing concepts of segment binary matroid and strong isomorphisms we show that two simple undirected graphs are isomorphic if and only if their canonical sequences are strongly isomorphic simple segment binary matroids.

AMS Subject Classifications: 05C62, 05C50, 05B35.

Keywords: Simple undirected graph · Graph representation · Graph isomorphism · Incidence matrix · Consecutive 1's property · Binary matroid · Graphic matroid

1 Introduction

There are various representations of simple undirected graphs in terms of adjacency matrices, adjacency lists, incidence matrices, unordered pairs etc. In this paper, we introduce another representation of a simple undirected graph with n vertices in terms of certain vectors in the vector space $\mathbb{Z}_2^{n-1}$ over $\mathbb{Z}_2$. We call the set of vectors representing a graph G as a coding sequence of G and denote it by $\beta(G, n)$. Among all such coding sequences we identify the one which is unique for a class of isomorphic graphs. We call it the code of the graph. We find characterizations of graphs which are connected, acyclic, bipartite, Eulerian or Hamiltonian in terms of $\beta(G, n)$. We prove that a graph G with n vertices is a tree if and only if any coding sequence of G is a basis of the vector space $\mathbb{Z}_2^{n-1}$ over $\mathbb{Z}_2$.

© Springer International Publishing AG 2016
T.-H.H. Chan et al. (Eds.): COCOA 2016, LNCS 10043, pp. 202–215, 2016.
DOI: 10.1007/978-3-319-48749-6_15

In his pioneering paper [9] on matroids in 1935, Whitney left the problem of characterizing graphic matroid open by making the following comment: "The problem of characterizing linear graphs from this point of view is the same as that of characterizing matroids which correspond to matrices (mod 2) with exactly two ones in each column." In 1959, Tutte obtained a characterization of graphic matroids in terms of forbidden minors [6]. But it is clear that Whitney indicated about incidence matrices of simple undirected graphs. In this paper, we use a variation of incidence matrix for the same characterization.

In Sect. 3, we introduce the concept of a segment binary matroid which corresponds to matrices over $\mathbb{Z}_2$ that has the consecutive 1's property (i.e., 1's are consecutive) for columns and a characterization of graphic matroids is obtained by considering $\beta(G, n)$ as a segment binary matroid. We introduce the concept of a strong isomorphism for segment binary matroids and show that two simple graphs G and H (with n vertices each) are isomorphic if and only if $\beta(G, n)$ and $\beta(H, n)$ are strongly isomorphic segment binary matroids.

For graph theoretic concepts, see [8] and for matroid related terminologies, one may consult [5].

2   Coding Sequences

Definition 1. *Let $G = (V, E)$ be a simple undirected graph with n vertices and m edges. Let $V = \{v_0, v_1, \ldots, v_{n-1}\}$. Define a map $f : V \longrightarrow \mathbb{N}$ by $f(v_i) = 10^i$ and another map $f^* : E \longrightarrow \mathbb{N}$ by $f^*(v_i v_j) = |f(v_i) - f(v_j)|$, if $E \neq \emptyset$. Let $\sigma(G, n)$ be the sequence $\{f^*(e) \mid e \in E\}$ sorted in the increasing order. If $E = \emptyset$, then $\sigma(G, n) = \emptyset$. It is worth noticing that for $e = v_i v_j \in E$, $f^*(c) = |10^i - 10^j|$ uniquely determines the pair (i, j) as it is a natural number with i digits, starting with $i - j$ number of 9's and followed by j number of 0's, when $i > j$. Thus the m entries of $\sigma(G, n)$ are all distinct.*

*Now for $E \neq \emptyset$, we define a map $f^\# : E \longrightarrow \mathbb{Z}_2^{n-1}$ by $f^\#(e) = (x_1, x_2, \ldots, x_{n-1})$, where $x_i = 1$, if the $(n - i)^{th}$ digit of $f^*(e)$ from the right is 9, otherwise $x_i = 0$ for $i = 1, 2, \ldots, n - 1$. For convenience we write the field $\mathbb{Z}_2$ as $\{0, 1\}$ instead of $\{\bar{0}, \bar{1}\}$. Let $\beta(G, n)$ be the sequence $\{f^\#(e) \mid e \in E\}$ sorted in the same order as in $\sigma(G, n)$. If $E = \emptyset$, then $\beta(G, n) = \emptyset$. The sequence $\beta(G, n)$ is called a **coding sequence** of the graph G.*

*Naturally, $\beta(G, n)$ is not unique for a graph G as it depends on the labeling f of vertices. Now there are $n!$ such labellings and consequently we have at most $n!$ different $\sigma(G, n)$ for a graph G. Among which we choose the one, say, $\sigma_c(G, n)$ which is the minimum in the lexicographic ordering of $\mathbb{N}^m$. The corresponding $\beta(G, n)$ is called the **code** of the graph G and is denoted by $\beta_c(G, n)$. Clearly $\beta_c(G, n)$ is unique for a class of isomorphic graphs with n vertices, though it is not always easy to determine generally.*

Example 1. Consider the graph G in Fig. 1 (left). We have

$$\sigma(G, 4) = (9, 90, 900, 990) \quad \text{and} \quad \beta(G, 4) = \{(0, 0, 1), (0, 1, 0), (1, 0, 0), (1, 1, 0)\}$$

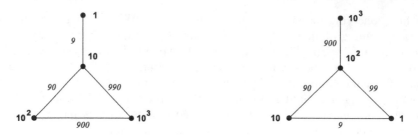

Fig. 1. The graph G in Example 1 with different labellings

according to the labeling of vertices given in Fig. 1 (left). One may verify that

$$\sigma_c(G, 4) = (9, 90, 99, 900) \text{ and } \beta_c(G, 4) = \{(0, 0, 1), (0, 1, 0), (0, 1, 1), (1, 0, 0)\}$$

according to the labeling of vertices shown in Fig. 1 (right).

Remark 1. An incidence matrix of a (simple undirected) graph $G = (V, E)$ is obtained by placing its vertices in rows and edges in columns and an entry in a row corresponding to a vertex v and in a column corresponding to an edge e of the matrix is 1 if and only if v is an end point of e, otherwise it is 0. It is important to note that a coding sequence of a graph has a similarity with the incidence matrix of the graph. In fact, given a coding sequence of a graph G, one can easily obtain the incidence matrix of G and vice-versa. Also cut-set and circuit subspaces of a vector space of dimension $|E|$ over $\mathbb{Z}_2$ constructed from edges of G are well known [1,3]. For studying various matrix representations of graphs, one may see [2,4,8]. Now, as we mentioned in the introduction, Whitney expected the characterization of graphic matroids would be obtained from the incidence matrix. Here we consider a variation of it with a consecutive 1's representation as it helps us to build a very natural interplay between graph theory, matroids and linear algebra which is evident from Theorem 1, Corollary 8 and Theorem 4.

Throughout this section by a graph we mean a simple undirected graph.

Definition 2. *A non-null vector $e = (x_1, x_2, \ldots, x_{n-1}) \in \mathbb{Z}_2^{n-1}$ is said to satisfy the* **consecutive 1's property** *if 1's appear consecutively in the sequence of coordinates of e. Let*

$$C(n - 1) = \{v \in \mathbb{Z}_2^{n-1} \mid v \text{ satisfies the consecutive 1's property}\}.$$

Clearly, $|C(n - 1)| = \binom{n}{2} = \frac{n(n-1)}{2}$ and G is a complete graph with n vertices if and only if $\beta(G, n) = C(n-1)$. In fact, for every $S \subseteq C(n-1)$, there is a unique graph $G(S)$ of n vertices such that $\beta(G, n) = S$. If $S = \emptyset$, then G is the null graph with n vertices and no edges. If $S \neq \emptyset$, each member e of S represents an edge of $G(S) = (V, E)$ with end points 10^{n-i} and 10^{n-j-1}, where the consecutive *stretch of 1's in e starts from the ith entry and ends at the jth entry from the left and $V = \{1, 10, 10^2, \ldots, 10^{n-1}\}$. Also it is clear that $C(n-1) \smallsetminus \beta(G, n)$ is a coding sequence of the complement $\bar{G}$ of a graph G with n vertices.*

Let $\emptyset \neq S \subseteq C(n-1)$. Let $\tilde{G}(S)$ be the subgraph of $G(S)$ obtained by removing all isolated vertices (if there is any) from $G(S)$. Then $\tilde{G}(S)$ is the subgraph of the complete graph of n vertices induced by the edges represented by the vectors in S.

We denote the null vector in the vector space $\mathbb{Z}_2^{n-1}$ by $\mathbf{0}$ for any $n \in \mathbb{N}$ and write $\mathbb{Z}_2^0$ for the zero-dimensional space $\{\mathbf{0}\}$. Let $S = \{e_1, e_2, \ldots, e_k\} \subseteq \mathbb{Z}_2^{n-1} \smallsetminus \{\mathbf{0}\}$, $(k \in \mathbb{N}, k \leqslant 2^{n-1})$. As e_i's are distinct, we have $e_i + e_j \neq \mathbf{0}$ for all $i \neq j$, $i, j \in \{1, 2, \ldots, k\}$. Thus S is a set of linearly dependent vectors in $\mathbb{Z}_2^{n-1}$ over $\mathbb{Z}_2$ if and only if there exists $A \subseteq S$, $|A| \geqslant 3$ such that $\sum_{e \in A} e = \mathbf{0}$. In other words, $S \subseteq \mathbb{Z}_2^{n-1}$ is linearly independent over $\mathbb{Z}_2$ if and only if $S = \emptyset$ or $S = \{e_1, e_2, \ldots, e_k\}$ for some $k \in \mathbb{N}, k \leqslant 2^{n-1}$ and $\sum_{e \in A} e \neq \mathbf{0}$ for all $\emptyset \neq A \subseteq S$. We denote the linear span (over $\mathbb{Z}_2$) of a subset S of $\mathbb{Z}_2^{n-1}$ by $\text{Sp}(S)$, i.e., $\text{Sp}(S)$ is the smallest subspace of $\mathbb{Z}_2^{n-1}$ containing S.

Proposition 1. *Let $S = \{e_1, e_2, e_3\} \subseteq C(n-1)$ for some $n \in \mathbb{N}, n \geqslant 3$. Then $\tilde{G}(S)$ is a 3-cycle if and only if $e_1 + e_2 + e_3 = \mathbf{0}$.*

Proof. First suppose that $\tilde{G}(S)$ is the 3-cycle shown in Fig. 2, where $\alpha, \beta, \gamma \in \{0, 1, \ldots, n-1\}$. Without loss of generality we assume $\alpha > \beta > \gamma$. Then

$$e_1 = (\underbrace{0, 0, \ldots, 0}_{n-\alpha-1}, \underbrace{1, 1, \ldots, 1}_{\alpha-\beta}, \underbrace{0, 0, \ldots, 0}_{\beta}, 0, 0, \ldots, 0)$$

$$e_2 = (\underbrace{0, 0, \ldots, 0}_{n-\alpha-1}, \underbrace{1, 1, \ldots, 1}_{\alpha-\gamma}, 1, 1, 1, \ldots, \underbrace{1, 0, 0, \ldots, 0}_{\gamma})$$

$$e_3 = (\underbrace{0, 0, \ldots, 0, 0, 0}_{n-\beta-1}, \ldots, \underbrace{0, 1, 1, \ldots, 1}_{\beta-\gamma}, \underbrace{0, 0, \ldots, 0}_{\gamma})$$

Clearly $e_1 + e_2 + e_3 = \mathbf{0}$.

Conversely, let $e_1 + e_2 + e_3 = \mathbf{0}$. Consider the matrix $M = \begin{pmatrix} e_1 \\ e_2 \\ e_3 \end{pmatrix}$, where we represent each e_i as a row matrix consisting of $n-1$ columns for $i = 1, 2, 3$. Since $e_1 + e_2 + e_3 = \mathbf{0}$, in each column where 1's appear, they appear exactly in two rows. Let i be the least column number of M that contains 1. Without loss of generality we assume that 1's appear in the first two rows in the ith column (otherwise we rearrange rows of M). Also suppose that the number of zeros after the stretch of 1's in the first row, say, β is more than that of the second, say, γ (otherwise again we rearrange rows of M). Let $\alpha = n - i$. Then the end points of the edge of $\tilde{G}(S)$ corresponding to e_1 are 10^α and 10^β and those of the edge corresponding to e_2 are 10^α and 10^γ. Since $\beta > \gamma$ and $e_1 + e_2 + e_3 = \mathbf{0}$, we have

$$e_3 = e_1 + e_2 = (\underbrace{0, 0, \ldots, 0}_{n-\beta-1}, \underbrace{1, 1, \ldots, 1}_{\beta-\gamma}, \underbrace{0, 0, \ldots, 0}_{\gamma}).$$

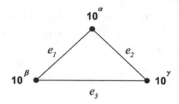

Fig. 2. A 3-cycle

Thus the end points of the edge of $\tilde{G}(S)$ corresponding to e_3 are 10^β and 10^γ. So the vertices labeled by $10^\alpha, 10^\beta$ and 10^γ form a 3-cycle with edges corresponding to e_1, e_2, e_3, as required.

Definition 3. *A set $S \neq \emptyset$ of non-null vectors in $\mathbb{Z}_2^{n-1}$ is called* **reduced** *if $\sum\limits_{e \in A} e \neq \mathbf{0}$ for all $\emptyset \neq A \subsetneq S$.*

Lemma 1. *Let $S = \{e_1, e_2, \ldots, e_k\} \subseteq C(n-1)$ for some $k, n \in \mathbb{N}$, $3 \leqslant k \leqslant n$. Then $\tilde{G}(S)$ is a k-cycle if and only if S is reduced and $e_1 + e_2 + \cdots + e_k = \mathbf{0}$.*

Proof. We prove by induction on k. By Proposition 1, the result is true for $k = 3$. Suppose the result is true for $k = r - 1 \geqslant 3$. Let $S = \{e_1, e_2, \ldots, e_r\} \subseteq C(n-1)$ for some $r, n \in \mathbb{N}$, $3 < r \leqslant n$ form the r-cycle shown in Fig. 3 (we renumber e_i's, if necessary). Consider the chord e so that e_1, e_2 and e form a triangle. Then $e_1 + e_2 + e = \mathbf{0}$ by Proposition 1. So $e = e_1 + e_2$. Also $\{e, e_3, e_4, \ldots, e_r\}$ forms an $(r-1)$-cycle. So by induction hypothesis $e + e_3 + e_4 + \cdots + e_r = \mathbf{0}$ which implies $e_1 + e_2 + e_3 + e_4 + \cdots + e_r = \mathbf{0}$. Moreover, since S is a cycle, no proper subset of S forms a cycle. Thus S is reduced by induction hypothesis.

Conversely, let $S = \{e_1, e_2, \ldots, e_r\} \subseteq C(n-1)$ for some $r, n \in \mathbb{N}$, $3 < r \leqslant n$ be reduced and $e_1 + e_2 + \cdots + e_r = \mathbf{0}$. Consider the matrix $M = \begin{pmatrix} e_1 \\ e_2 \\ \cdots \\ e_r \end{pmatrix}$, where we

represent each e_i as a row matrix consisting of $n-1$ columns for $i = 1, 2, \ldots, r$. Let i be the least column number of M that contains 1. Since the row sum of M is zero (the null vector), the ith column contains even number of 1's. So there are at least two rows with 1 in the ith column. Rearrange rows of M such that e_1 and e_2 be two such rows. Since both of these rows begin with 1 in the ith column, the edges corresponding to them have a common end point with label 10^{n-i}. Join the other end points of e_1 and e_2 by an edge, say, e to form a triangle with edges e_1, e_2, e. Then $e_1 + e_2 + e = \mathbf{0}$ which implies $e = e_1 + e_2$. So $e + e_3 + e_4 + \cdots + e_r = \mathbf{0}$.

We claim that $S_1 = \{e, e_3, e_4, \ldots, e_r\}$ is reduced. Suppose $A \subsetneq S_1$, $|A| \geqslant 3$ be such that $a = \sum\limits_{x \in A} x = \mathbf{0}$. If $e \notin A$, then $a \neq \mathbf{0}$ as S is reduced. So $e \in A$. Then replacing e by $e_1 + e_2$ in a would again contradict the fact that S is reduced. So S_1 is reduced. Hence by induction hypothesis, S_1 form an $(r-1)$-cycle. Now

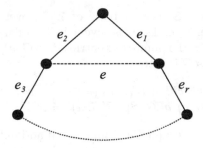

Fig. 3. An r-cycle

replacing the edge e by the path consisting of edges e_1 and e_2 gives us an r-cycle formed by S.

The following two corollaries follow immediately from the above lemma.

Corollary 1. *A graph G with $n \geqslant 3$ vertices $(n \in \mathbb{N})$ is Hamiltonian if and only if for any coding sequence $\beta(G, n)$ of G, there exists $S = \{e_1, e_2, \ldots, e_n\} \subseteq \beta(G, n)$ such that S is reduced and $e_1 + e_2 + \cdots + e_n = \mathbf{0}$.*

Corollary 2. *A graph G with n vertices $(n \in \mathbb{N})$ is acyclic if and only if any coding sequence $\beta(G, n)$ of G is linearly independent over $\mathbb{Z}_2$.*

Corollary 3. *A graph G with at most one non-trivial component and with n vertices $(n \in \mathbb{N})$ is Eulerian if and only if $\displaystyle\sum_{e \in \beta(G,n)} e = \mathbf{0}$ for any coding sequence $\beta(G, n)$ of G.*

Proof. Follows from Lemma 1 and the fact that a circuit in a graph can be decomposed into edge-disjoint cycles.

Corollary 4. *A graph G with $n \geqslant 2$ vertices $(n \in \mathbb{N})$ is bipartite if and only if for any coding sequence $\beta(G, n)$ of G, $\displaystyle\sum_{e \in S} e \neq \mathbf{0}$ for every $S \subseteq \beta(G, n)$ where $|S|$ is odd.*

Proof. Follows from Lemma 1 and the fact that a graph is bipartite if and only if it does not contain any odd cycle.

Theorem 1. *A graph G with n vertices $(n \in \mathbb{N})$ is a tree if and only if any coding sequence $\beta(G, n)$ of G is a basis of the vector space $\mathbb{Z}_2^{n-1}$ over the field $\mathbb{Z}_2$.*

Proof. Suppose G is a tree. Then G is acyclic which implies $\beta(G, n)$ is linearly independent over $\mathbb{Z}_2$ by Corollary 2. Again since G is a tree, the number of entries in $\beta(G, n)$ is $n - 1$, we have $n - 1$ linearly independent vectors in $\mathbb{Z}_2^{n-1}$ over $\mathbb{Z}_2$. Thus $\beta(G, n)$ is a basis of $\mathbb{Z}_2^{n-1}$ over $\mathbb{Z}_2$.

Conversely, suppose $\beta(G,n)$ is a basis of $\mathbb{Z}_2^{n-1}$ over $\mathbb{Z}_2$. Then $\beta(G,n)$ is linearly independent over $\mathbb{Z}_2$ and so G is acyclic by Corollary 2. Also since $\beta(G,n)$ is a basis of $\mathbb{Z}_2^{n-1}$ over $\mathbb{Z}_2$, the number of entries in $\beta(G,n)$ is $n-1$ which implies G has $n-1$ edges. Thus G is a tree.

Corollary 5. *A graph G with n vertices $(n \in \mathbb{N})$ is connected if and only if for any coding sequence $\beta(G,n)$ of G, $\mathrm{Sp}\,(\beta(G,n)) = \mathbb{Z}_2^{n-1}$.*

Proof. We first note that a graph G is connected if and only if G has a spanning tree. Suppose $G = (V,E)$ has a spanning tree $H = (V, E_1)$. Then $\beta(H,n) \subseteq \beta(G,n)$ with the same vertex labeling. But $\beta(H,n)$ is a basis of $\mathbb{Z}_2^{n-1}$ over $\mathbb{Z}_2$ by Theorem 1. Thus $\mathrm{Sp}\,(\beta(G,n)) \supseteq \mathrm{Sp}\,(\beta(H,n)) = \mathbb{Z}_2^{n-1}$. So $\mathrm{Sp}\,(\beta(G,n)) = \mathbb{Z}_2^{n-1}$.

Conversely, if $\mathrm{Sp}\,(\beta(G,n)) = \mathbb{Z}_2^{n-1}$, then $\beta(G,n)$ contains a basis, say B of $\mathbb{Z}_2^{n-1}$ over $\mathbb{Z}_2$. Then $G(B)$ is a spanning tree of G by Theorem 1 as $B = \beta(G(B), n)$. Thus G is connected.

Corollary 6. *Let G be a connected graph with n vertices and $S \subseteq \beta(G,n)$. Then $G(S)$ is a spanning tree of G if and only if S is a basis of the vector space $\mathbb{Z}_2^{n-1}$ over the field $\mathbb{Z}_2$.*

3   Matroid Representation

Whitney introduced the concept of a matroid in [9]. There are several ways of defining matroids. We take the one that will serve our purpose. A *matroid* M is an ordered pair $(E, \mathscr{B})$ consisting of a finite set E of *elements* and a nonempty collection $\mathscr{B}$ of subsets of E, called *bases* which satisfies the properties: (i) no proper subset of a base is a base and (ii) if $B_1, B_2 \in \mathscr{B}$ and $e \in B_1 \smallsetminus B_2$, then there exists $f \in B_2 \smallsetminus B_1$ such that $(B_1 \smallsetminus \{e\}) \cup \{f\} \in \mathscr{B}$. *Independent sets* of M are subsets of bases and minimal dependent sets are *circuits*. The *cycle matroid* $M[G]$ of graph G is the matroid whose elements are edges of G and circuits are cycles of G. Independent sets and bases of $M[G]$ are forests and maximal forests of G respectively. A matroid is *graphic (simple graphic)* if it is a cycle matroid of a graph (respectively, simple graph).

Let E be the set of column labels of an $n \times m$ matrix A over a field F, and $\mathscr{B}$ be the set of maximal subsets X of E for which the multiset of columns labeled by X is linearly independent in the vector space F^m over F. Then the pair $(E, \mathscr{B})$ is the *column (vector) matroid* of A and is denoted by $M[A]$. In particular, if $F = \mathbb{Z}_2$, then $M[A]$ is a *binary matroid*. A binary matroid $M[A]$ is *simple* if A does not contain zero columns and no two columns of A are identical (i.e., columns of A are non-zero and distinct).

Definition 4. *A binary matroid $M[A]$ is called a* **segment binary matroid** *if A satisfies the consecutive 1's property for columns. Moreover, if it is simple, then we call it a* **simple segment binary matroid**. *For any $\emptyset \neq S \subseteq \mathbb{Z}_2^{n-1}$, $M[S]$ denotes the column (vector) matroid of the matrix whose columns are precisely the elements of S. Clearly, $M[S]$ is a binary matroid.*

Remark 2. In particular, when $\emptyset \neq S \subseteq C(n-1)$, $M[S]$ becomes a simple segment binary matroid. So for any simple graph G with n vertices, $M[\beta(G,n)]$ is a simple segment binary matroid. Conversely, every simple segment binary matroid $M[A]$ with $n-1$ rows is same as $M[S]$, where S is the set of column vectors of A over $\mathbb{Z}_2$. Also in this case $S \subseteq C(n-1)$.

Two matroids $M_1 = (E_1, \mathscr{B}_1)$ and $M_2 = (E_2, \mathscr{B}_2)$ are *isomorphic* if there is a bijection ψ from E_1 onto E_2 such that for all $X \subseteq E_1$, X is independent in M_1 if and only if $\psi(X)$ is independent in M_2 (or, equivalently, X is a circuit in M_1 if and only if $\psi(X)$ is a circuit in M_2). In this case, we denote by $M_1 \cong M_2$. Also abusing notations we sometimes identify elements of $M[A]$ with its corresponding column vector representation. Thus a simple binary matroid $M[A]$ may be considered as the set of column vectors of A. The following theorem characterizes isomorphisms of simple binary matroids in terms of linear transformations.

Theorem 2. *Let $M[A]$ and $M[A_1]$ be two simple binary matroids such that both A and A_1 are of same order $n \times m$, $(m,n \in \mathbb{N})$. Then $M[A] \cong M[A_1]$ if and only if there exists a bijective linear operator T on $\mathbb{Z}_2^n$ such that T restricted on $M[A]$ is a bijective map from $M[A]$ onto $M[A_1]$.*

Proof. Let ψ be an isomorphism from $M[A]$ onto $M[A_1]$. Let B be a base in $M[A]$. Then B is linearly independent over $\mathbb{Z}_2$. We extend B to a basis B_1 (say) of $\mathbb{Z}_2^n$ over $\mathbb{Z}_2$. Now since ψ is an isomorphism, $\psi(B)$ is also a base in $M[A_1]$ and $|\psi(B)| = |B|$. We also extend $\psi(B)$ to B_2, a basis of $\mathbb{Z}_2^n$ over $\mathbb{Z}_2$. Then $|B_1 \setminus B| = |B_2 \setminus \psi(B)| = n - |B|$. Let f be a bijection from $B_1 \setminus B$ onto $B_2 \setminus \psi(B)$. Now define a map $T_1 : B_1 \longrightarrow B_2$ by

$$T_1(e) = \begin{cases} \psi(e), e \in B \\ f(e), e \in B_1 \setminus B \end{cases}$$

We next verify that ψ is 'linear' on $M[A]$, i.e., if e_1, e_2 are columns of A such that $e_1 + e_2$ is also a column of A, then $\psi(e_1 + e_2) = \psi(e_1) + \psi(e_2)$. Let $e = e_1 + e_2$. Then $e + e_1 + e_2 = \mathbf{0}$ which implies that $\{e, e_1, e_2\}$ is a circuit of $M[A]$. Again since ψ is an isomorphism, $\{\psi(e), \psi(e_1), \psi(e_2)\}$ is also a circuit in $M[A_1]$. Hence $\psi(e) + \psi(e_1) + \psi(e_2) = \mathbf{0}$, i.e., $\psi(e_1 + e_2) = \psi(e) = \psi(e_1) + \psi(e_2)$. This completes the verification. We extend T_1 linearly to obtain a linear operator T on $\mathbb{Z}_2^n$ over $\mathbb{Z}_2$. Then T is bijective as T_1 maps a basis bijectively to another basis of $\mathbb{Z}_2^n$ over $\mathbb{Z}_2$ and the restriction of T on $M[A]$ is ψ due to the above verification.

Conversely, let T be a bijective linear operator on $\mathbb{Z}_2^n$ such that the map ψ, the restriction of T on $M[A]$ is a bijective map from $M[A]$ onto $M[A_1]$. Let X be a circuit in $M[A]$. Then $\sum_{e \in X} e = \mathbf{0}$ and $\sum_{e \in A} e \neq \mathbf{0}$ for all $\emptyset \neq A \subsetneq X$. Now since T is bijective and linear, we have $\sum_{e \in A} e = \mathbf{0}$ if and only if $\sum_{e \in A} T(e) = \mathbf{0}$ for all $\emptyset \neq A \subseteq X$. Thus X is a circuit in $M[A]$ if and only if $\psi(X)$ is a circuit in $M[A_1]$. Hence ψ is an isomorphism from $M[A]$ onto $M[A_1]$.

Corollary 7. *Let $M[A]$ and $M[A_1]$ be two simple binary matroids such that both A and A_1 are of same order $n \times m$, $(m, n \in \mathbb{N})$. Then $M[A] \cong M[A_1]$ if and only if there exist a non-singular matrix P of order $n \times n$ and a permutation matrix Q of order $m \times m$ such that $PAQ = A_1$.*

Proof. If $M[A] \cong M[A_1]$, then following the proof of the direct part of the above theorem, consider two bases B_1 and B_2 of $\mathbb{Z}_2^n$ over $\mathbb{Z}_2$ and the bijective linear operator T that maps B_1 onto B_2. Let P be the matrix representation of T with respect to these bases. Then P is a non-singular matrix and $PA = A_2$ where A_2 is obtained from A_1 by rearranging columns such that i^{th} column of A_2 is the image of the ith column of A under T. Thus $PAQ = A_1$ for some permutation matrix Q.

Conversely, let $A_1 = PAQ$ for some non-singular matrix P and some permutation matrix Q. Let $A_2 = A_1 Q^{-1}$. Then $PA = A_2$. Since P is non-singular, it corresponds to a bijective linear operator T on $\mathbb{Z}_2^n$ (over $\mathbb{Z}_2$) defined by $T(e) = Pe$ (considering elements of $\mathbb{Z}_2^n$ as column matrices) such that the restriction of T on $M[A]$ is a bijective map from $M[A]$ onto $M[A_2]$. Then $M[A] \cong M[A_2]$ by the above theorem. Since $M[A_1] = M[A_2]$, we have $M[A] \cong M[A_1]$.

Now we proceed to characterize simple graphic matroids.

Lemma 2. *Let G be a simple graph with n vertices. Then $M[G] \cong M[\beta(G, n)]$ for any coding sequence $\beta(G, n)$ of G.*

Proof. It follows from Lemma 1 that cycles of G are precisely the circuits of the matroid $M[\beta(G, n)]$. So $M[G] \cong M[\beta(G, n)]$ as matroids.

Theorem 3. *A matroid is simple graphic if and only if it is isomorphic to a simple segment binary matroid.*

Proof. Let M be a simple graphic matroid. Then $M \cong M[G]$ for a simple graph G. By Lemma 2, we have $M[G] \cong M[\beta(G, n)]$ where n is the number of vertices of G. Thus M is isomorphic to a simple segment binary matroid by Remark 2.

Conversely, let $M[A]$ be a simple segment binary matroid. Then we may consider $M[A]$ as $M[S]$ where S is the set of columns of A. By Remark 2, we have $S \subseteq C(n-1)$, where $A \in M_{n-1, m}(\mathbb{Z}_2)$. Then by Definition 2, there is a unique simple graph G such that $S = \beta(G, n)$. Therefore, by Lemma 2, $M[S] = M[\beta(G, n)] \cong M[G]$. Thus, $M[A]$ is a simple graphic matroid.

Corollary 8. *A simple binary matroid $M[A]$ (where A is of order $(n-1) \times m$) is simple graphic if and only if $m \leqslant \binom{n}{2}$ and there exists a non-singular matrix P such that PA satisfies the consecutive 1's property for columns.*

Proof. Follows from Theorem 3 and Corollary 7.

Remark 3. Since any non-singular matrix is obtained from identity matrix by finite number of elementary row operations, a simple binary matroid $M[A]$ is simple graphic if and only if the consecutive 1's property for columns can be obtained from A by finite number of elementary row operations.

It is well known [5] that an ordinary matroid isomorphism does not guarantee the corresponding graph isomorphism for graphic matroids. We now introduce the concept of a strong isomorphism of simple segment binary matroids.

Definition 5. *Two simple segment binary matroids $M[A_1]$ and $M[A_2]$ are called* **strongly isomorphic** *if*

(1) $A_1, A_2 \in M_{n-1,m}(\mathbb{Z}_2)$ for some $m, n \in \mathbb{N}$, $n \geqslant 2$.
(2) There exists a bijective linear operator T on $\mathbb{Z}_2^{n-1}$ such that:
   (i) T restricted on $C(n-1)$ is a bijection onto itself.
   (ii) T restricted on $M[A_1]$ is a bijective map from $M[A_1]$ onto $M[A_2]$.

We write $M[A_1] \cong_s M[A_2]$ to denote that $M[A_1]$ is strongly isomorphic to $M[A_2]$.

Note that, if $M[A_1] \cong_s M[A_2]$, then the restriction of T on $M[A_1]$ is a matroid isomorphism onto $M[A_2]$ and the restriction of T on $C(n-1)$ is a matroid automorphism. These follow from the fact that T is linear and injective, as then for any subset X of the set of columns of A_1, $\sum_{e \in X} e = \mathbf{0}$ if and only if $\sum_{e \in X} T(e) = \mathbf{0}$. In the sequel, we show that strong isomorphism of simple segment binary matroids would guarantee the corresponding graph isomorphism.

Let $G = (V, E)$ be a (simple undirected) graph with $|V| = n$. Then for any $e \in E$, we use the symbol $p \sim_n |10^x - 10^y|$ if $f^*(e) = |10^x - 10^y|$ and $p = f^\#(e) \in \beta(G, n)$.

Lemma 3. *Let p_1, p_2 be distinct elements in $\beta(G, n)$ for any coding sequence $\beta(G, n)$ of a graph G with n vertices. If $p_1 \sim_n |10^x - 10^y|$ and $p_2 \sim_n |10^x - 10^z|$, then $p_1 + p_2 \sim_n |10^y - 10^z|$.*

Proof. Let $p_3 = p_1 + p_2$. So $p_1 + p_2 + p_3 = 0$. From the converse part of the proof of Proposition 1, we have that p_3 corresponds to the end points 10^y and 10^z. Thus, $p_1 + p_2 \sim_n |10^y - 10^z|$.

Lemma 4. *Let $S = \{e_1, e_2, \ldots, e_k\} \subseteq \beta(G, n)$ for any coding sequence $\beta(G, n)$ of a graph G with n vertices. Then $\tilde{G}(S)$ induces a path in G if and only if $\sum_{j=1}^{k} e_j \in C(n-1)$ and S is reduced.*

Proof. If $e_1, e_2, \ldots, e_k$ induce a path (in that order) then it is easy to see that we have $e_i \sim_n |10^{x_{i+1}} - 10^{x_i}|$ for some distinct $x_1, x_2, \ldots, x_{k+1} \in \mathbb{N} \cup \{0\}$. By applying Lemma 3 repetitively, we have $\sum_{j=1}^{k} e_j \sim_n |10^{x_{k+1}} - 10^{x_1}|$. Thus $\sum_{j=1}^{k} e_j \in C(n-1)$. Let $e = \sum_{j=1}^{k} e_j$. Then elements of $S \cup \{e\}$ form a cycle. Then by Lemma 1, $S \cup \{e\}$ is reduced and so S is reduced.

Conversely, let $\sum_{j=1}^{k} e_j \in C(n-1)$ and S is reduced. Let $\sum_{j=1}^{k} e_j = e$. So $e +$

$\sum_{j=1}^{k} e_j = \mathbf{0}$ and since S is reduced, $S \cup \{e\}$ is also reduced. Now consider the graph G' such that $\beta(G', n) = \beta(G, n) \cup \{e\}$. Clearly, $S \cup \{e\}$ induces a cycle in $\beta(G', n)$ by Lemma 1. One edge of that cycle corresponds to e, all the other edges correspond precisely to the members of S. Hence, $\tilde{G}(S)$ induces a path in G.

Corollary 9. *Suppose* $p \sim_n |10^i - 10^j|, q \sim_n |10^r - 10^s|$ *where* $p \neq q$ *and* $i, j, r, s \in \{0, 1, 2, \ldots, n - 1\}$. *Then we have* $p + q \in C(n-1)$ *if and only if* $|\{i, j\} \cap \{r, s\}| = 1$. *Moreover,* $p + q \sim_n |10^x - 10^y|$, *where* $x \in \{i, j\}$, $y \in \{r, s\}$ *and* $x, y \notin \{i, j\} \cap \{r, s\}$.

Proof. Consider a graph G with n vertices such that $p, q \in \beta(G, n)$ for some coding sequence $\beta(G, n)$ of G. Clearly, p corresponds to end-points 10^i and 10^j, and q corresponds to end-points 10^r and 10^s. First, let $p + q \in C(n-1)$. So by Lemma 4, it follows that $\tilde{G}(\{p, q\})$ induces a path in G. This implies that the two edges corresponding to p and q have a common vertex. Since $p \neq q$, this gives that $|\{i, j\} \cap \{r, s\}| = 1$.

Conversely, let $|\{i, j\} \cap \{r, s\}| = 1$. Suppose $j = r$, without loss of generality. Then by Lemma 3, $p + q \sim_n |10^i - 10^s|$ (which proves the next part also). Thus, $p + q \in C(n-1)$.

Lemma 5. *Suppose* $p_1 \sim_n |10^i - 10^j|, p_2 \sim_n |10^k - 10^l|, p_3 \sim_n |10^r - 10^s|$, *where* $i, j, k, l, r, s \in \{0, 1, 2, \ldots, n-1\}$ *and* p_1, p_2, p_3 *are distinct. If* $p_1 + p_2, p_2 + p_3, p_1 + p_3 \in C(n-1)$, *then either* $p_1 + p_2 + p_3 = \mathbf{0}$ *or* $\{i, j\} \cap \{k, l\} = \{i, j\} \cap \{r, s\} = \{k, l\} \cap \{r, s\}$.

Proof. Consider a graph G with n vertices such that $p_1, p_2, p_3 \in \beta(G, n)$ for some coding sequence $\beta(G, n)$ of G. Clearly, p_1 corrsponds to end-points 10^i and 10^j, p_2 corresponds to end-points 10^k and 10^l and p_3 corresponds to end-points 10^r and 10^s. Now since $p_1 + p_2 \in C(n-1)$, by Corollary 9 we have $|\{i, j\} \cap \{k, l\}| = 1$. Without loss of generality, let $j = k$. Then 10^j is the common end-point between edges corresponding to p_1 and p_2. Since we also have that $p_1 + p_3 \in C(n-1)$ and $p_2 + p_3 \in C(n-1)$, it follows that the edge corresponding to p_3 has a common end-point with the edge corresponding to p_1 and a common end-point with the edge correpponding to p_2. If the common end-point in both the cases is $10^j (= 10^k)$ then we have $\{i, j\} \cap \{k, l\} = \{i, j\} \cap \{r, s\} = \{k, l\} \cap \{r, s\}$. Otherwise, the common end-point between p_1 and p_3 must be 10^i and the common end-point between p_2 and p_3 must be 10^l. Thus, edges corresponding to p_1, p_2, p_3 form a cycle involving the vertices $10^i, 10^j, 10^l$. So by Proposition 1, we have $p_1 + p_2 + p_3 = \mathbf{0}$.

Lemma 6. *Let* G, H *be two simple graphs with* n *vertices each and suppose* $M[\beta(G, n)] \cong_s M[\beta(H, n)]$. *Let* T *be any bijective linear operator on* $\mathbb{Z}_2^{n-1}$ *giving a strong isomorphism between* $M[\beta(G, n)]$ *and* $M[\beta(H, n)]$. *Then for* $e_1, e_2 \in C(n-1)$, *we have* $e_1 + e_2 \in C(n-1)$ *if and only if* $T(e_1) + T(e_2) \in C(n-1)$.

Proof. Let $e_1, e_2 \in \beta(G, n)$. First, let $e_1 + e_2 \in C(n-1)$. Now $T(e_1) + T(e_2) = T(e_1 + e_2) \in T(C(n-1)) = C(n-1)$ as restriction of T maps $C(n-1)$ onto itself. Conversely, let $T(e_1) + T(e_2) \in C(n-1)$. So $T(e_1 + e_2) \in C(n-1)$. Again since restriction of T maps $C(n-1)$ onto itself, there exists some e in $C(n-1)$ such that $T(e) = T(e_1 + e_2)$. Finally, since T is injective, we have $e = e_1 + e_2$. So $e_1 + e_2 \in C(n-1)$.

Now we prove the theorem which gives a necessary and sufficient condition for two simple graphs to be isomorphic.

Theorem 4. *Let G and H be two simple graphs of n vertices each. Then $G \cong H$ if and only if $M[\beta(G, n)] \cong_s M[\beta(H, n)]$ for any coding sequences $\beta(G, n)$ and $\beta(H, n)$ of G and H, respectively.*

Proof. We consider vertices of both G and H are labeled by $1, 10, 10^2, \ldots, 10^{n-1}$. First, let $G \cong H$. So there exists a permutation g on the set $\{0, 1, 2, \ldots, n-1\}$ such that for any $r, s \in \{0, 1, 2, \ldots, n-1\}$, we have 10^r and 10^s are adjacent in G if and only if $10^{g(r)}$ and $10^{g(s)}$ are adjacent in H. We consider the graphs $K^{(1)}, K^{(2)}$, where $K^{(1)} = G \cup \bar{G}$ and $K^{(2)} = H \cup \bar{H}$, where $\bar{G}$ and $\bar{H}$ are complements of graphs G and H respectively. Now for each $e \in C(n-1)$, if $e \sim_n |10^i - 10^j|$, we define $T(e) \sim_n |10^{g(i)} - 10^{g(j)}|$. Clearly, T is a well-defined mapping from $C(n-1)$ into itself, since $|10^p - 10^q|$ uniquely determines the pair $\{p, q\}$ for any p, q.

Now g, being a permutation, is a bijection. Suppose e_1, e_2 are distinct elements of $C(n-1)$. Let $e_1 \sim_n |10^a - 10^b|$ and $e_2 \sim_n |10^c - 10^d|$. Clearly, $\{a, b\} \neq \{c, d\}$. Bijectiveness of g implies that $\{g(a), g(b)\} \neq \{g(c), g(d)\}$. This shows that $T(e_1) \neq T(e_2)$. So we have that T is one-to-one. Since T is a mapping from a finite set into itself, injectiveness of T implies that T is a bijective mapping from $C(n-1)$ onto itself. Again, since 10^r and 10^s are adjacent in G if and only if $10^{g(r)}$ and $10^{g(s)}$ are adjacent in H for distinct $r, s \in \{0, 1, 2, \ldots, n-1\}$, we have that T restricted on $\beta(G, n)$ is a mapping from $\beta(G, n)$ into $\beta(H, n)$. Injectiveness of T ensures the injectiveness of T restricted to $\beta(G, n)$. Since $\beta(G, n)$ and $\beta(H, n)$ are finite sets with equal cardinality, we have that T restricted to $\beta(G, n)$ is a bijection from $\beta(G, n)$ onto $\beta(H, n)$.

Next, we observe that T is 'linear' on $C(n-1)$, i.e., if $p, q \in C(n-1)$ such that $p + q \in C(n-1)$, then $T(p) + T(q) = T(p + q)$. Let $p \sim_n |10^i - 10^j|, q \sim_n |10^r - 10^s|$ such that $p + q \in C(n-1)$. Then by Corollary 9, we have $|\{i, j\} \cap \{r, s\}| = 1$. Without loss of generality, we assume $j = r$. So $g(j) = g(r)$ and $p + q \sim_n |10^i - 10^s|$. Now $T(p) \sim_n |10^{g(i)} - 10^{g(j)}|$ and $T(q) \sim_n |10^{g(r)} - 10^{g(s)}|$. Since $g(j) = g(r)$, we have $T(p) + T(q) \in C(n-1)$ and $T(p) + T(q) \sim_n |10^{g(i)} - 10^{g(s)}|$. Since $T(p + q) \sim_n |10^{g(i)} - 10^{g(s)}|$, we have $T(p + q) = T(p) + T(q)$.

Now let $e_i \sim_n |10^i - 10^{i-1}|$ for all $i = 1, 2, \ldots, n$. We know that $B = \{e_i \mid i = 1, 2, \ldots, n\}$ is a basis of $\mathbb{Z}_2^{n-1}$. Since T is defined on each e_i as the latter is in $C(n-1)$ (in fact, $T(e_i) \sim_n |10^{g(i)} - 10^{g(i-1)}|$), we can extend T linearly to a linear operator T_1 on $\mathbb{Z}_2^{n-1}$. Bijectiveness of T_1 follows from injectiveness of T on $C(n-1)$ (which ensures distinct images under T for distinct elements of B, thus ensuring injectiveness of T_1) and finiteness of $\mathbb{Z}_2^{n-1}$. As $M[\beta(G,n)]$ and $M[\beta(H,n)]$ are also of the same order, we have $M[\beta(G,n)] \cong_s M[\beta(H,n)]$.

Conversely, let $M[\beta(G,n)] \cong_s M[\beta(H,n)]$. So there exists a bijective linear operator T satisfying the properties mentioned in the Definition 5. We find a permutation g on the set $\{0,1,2,\ldots,n-1\}$ such that for any distinct $r,s \in \{0,1,2,\ldots,n-1\}$, 10^r and 10^s are adjacent in G if and only if $10^{g(r)}$ and $10^{g(s)}$ are adjacent in H. Clearly, such a g acts as an isomorphism between G and H. Now for $i = 1,2,\ldots,n$, define e_i as the element in $C(n-1)$ which has 1 in its ith co-ordinate from the right $((n-i)^{\text{th}}$ co-ordinate from the left) and 0 in remaining coordinates. Then $e_i \sim_n |10^i - 10^{i-1}|$. Now for $i = 1,2,\ldots,n-1$, we have $e_i + e_{i+1} \in C(n-1)$. So from Lemma 6, we have $T(e_i) + T(e_{i+1}) \in C(n-1)$. From Corollary 9, the edges corresponding to $T(e_i)$ and $T(e_{i+1})$ have a (unique) common point, say 10^x. We define $g(i-1) = x$. This defines g as a mapping from $\{0,1,2,\ldots,n-2\}$ into $\{0,1,2,\ldots,n-1\}$. Now we show that g is one-to-one.

If possible, let $g(i-1) = g(j-1)$ for some $i \neq j$, $i,j \in \{1,2,\ldots,n-1\}$. By the above definition of g, $g(i-1)$ and $g(j-1)$ are one of the common end points of $T(e_i), T(e_{i+1})$ and $T(e_j), T(e_{j+1})$ respectively. So the edges corresponding to $T(e_i)$ and $T(e_j)$ have a common end-point $g(i-1)$ $(= g(j-1))$. So from Corollary 9, we have $T(e_i) + T(e_j) \in C(n-1)$. Linearity of T implies that $T(e_i + e_j) \in C(n-1)$. Since T restricted to $C(n-1)$ is a bijection from T onto itself, we have $e_i + e_j \in C(n-1)$. Again, $e_j + e_{i+1} \in C(n-1)$ for the same reason as $g(i-1)$ is a common end-point between edges corresponding to $T(e_j)$ and $T(e_{i+1})$. We also have $e_i + e_{i+1} \in C(n-1)$. So $e_i + e_j, e_{i+1} + e_i, e_j + e_{i+1} \in C(n-1)$. If possible, let $e_i + e_j + e_{i+1} = \mathbf{0}$. However, then $e_j = e_i + e_{i+1}$, which is impossible by the definition of e_i's. Thus, from Lemma 5, 10^i is either 10^j or 10^{j-1}, i.e., $i = j$ or $j-1$. Similar argument on e_i, e_j, e_{j+1} gives us $j = i$ or $i-1$. But $i = j-1$ and $j = i-1$ both cannot be true simultaneously. So we have $i = j$ which is a contradiction. Thus g is injective. So we define $g(n-1) = x$, where $x \in \{0,1,2,\ldots,n-1\} \setminus \{g(0),g(1),\ldots,g(n-2)\}$. Then g is defined on $\{0,1,2,\ldots,n-1\}$ into itself. Moreover since g is injective, it is a permutation on the set $\{0,1,2,\ldots,n-1\}$.

Acknowledgements. Authors are thankful to learned referees for their kind comments and suggestions. The second author is grateful to the University Grants Commission, Government of India, for providing research support (Grant no. F. 17-76/2008 (SA-I).

References

1. Chen, W.K.: On vector spaces associated with a graph. SIAM J. Appl. Math. **20**, 526–529 (1971)
2. Deo, N.: Graph Theory with Applications to Engineering and Computer Science. Prentice Hall of India Pvt. Ltd., New Delhi (1997)
3. Gould, R.: Graphs and vector spaces. J. Math. Phys. **37**, 193–214 (1958)
4. Gross, J.L., Yellen, J.: Graph Theory and Its Applications, 2nd edn. CRC Press, New York (2006)

5. Oxley, J.G.: Matroid Theory. Oxford University Press, Oxford (1992)
6. Tutte, W.T.: Matroids and graphs. Trans. Am. Math. Soc. **90**, 527–552 (1959)
7. Tutte, W.T.: An algorithm for determining whether a given binary matroid is graphic. Proc. Am. Math. Soc. **11**, 905–917 (1960)
8. West, D.B.: Introduction to Graph Theory. Prentice-Hall of India Pvt. Ltd., New Delhi (2003)
9. Whitney, H.: On the abstract properties of linear dependence. Am. J. Math. **57**, 509–533 (1935)

Minimum Eccentricity Shortest Path Problem: An Approximation Algorithm and Relation with the k-Laminarity Problem

Étienne Birmelé[1], Fabien de Montgolfier[2(✉)], and Léo Planche[1,2]

[1] MAP5, UMR CNRS 8145, Univ. Sorbonne Paris Cité, Paris, France
etienne.birmele@parisdescartes.fr
[2] IRIF, UMR CRNS 8243, Univ. Sorbonne Paris Cité, Paris, France
{fm,leo_planche}@liafa.univ-paris-diderot.fr

Abstract. The Minimum Eccentricity Shortest Path (MESP) Problem consists in determining a shortest path (a path whose length is the distance between its extremities) of minimum eccentricity in a graph. It was introduced by Dragan and Leitert [9] who described a linear-time algorithm which is an 8-approximation of the problem. In this paper, we study deeper the double-BFS procedure used in that algorithm and extend it to obtain a linear-time 3-approximation algorithm. We moreover study the link between the MESP problem and the notion of laminarity, introduced by Völkel *et al.* [12], corresponding to its restriction to a diameter (*i.e.* a shortest path of maximum length), and show tight bounds between MESP and laminarity parameters.

Keywords: Graph search · Graph theory · Eccentricity · Diameter · BFS · Approximation algorithms · k-Laminar graph

1 Introduction

For both graph classification purposes and applications, it is an important issue to determine to which extent a graph can be summarized by a path. Different path constructions and metrics to characterize how far the graph is from the constructed path can be used, for example path-decompositions and path-width [11] or path-distance-decompositions and path-distance-width [13]. Another approach, on which we focus in this article, is to characterize the graph by a spine defined by one of its paths.

This problem was first studied in terms of domination, that is finding a path such that every vertex in the graph belongs to or has a neighbor in the path. Several graphs classes were defined in terms of dominating paths. [7] studies the graphs for which the dominating path is a diameter. [8] introduces dominating pairs, that is vertices such that every path linking them is dominating. Graphs such that short dominating paths are present in all induced subgraphs are characterized in [2]. Linear-time algorithms to find dominating paths or dominating vertex pairs were also developed for AT-free graphs [4,6].

© Springer International Publishing AG 2016
T.-H.H. Chan et al. (Eds.): COCOA 2016, LNCS 10043, pp. 216–229, 2016.
DOI: 10.1007/978-3-319-48749-6_16

Dominating paths do not exist however in every graph and have no associated metric to measure the distance from the graph to the path. A natural extension of the notion of domination is the notion of k-coverage for a given integer k, defined by the fact that a path k-covers the graphs if every vertex is at distance at most k from the path. The smallest k such that a path k-covers the graph is then a metric as desired.

In the present paper, we study the latter problem in which the covering path is required to be a shortest path between its end-vertices. It was introduced in [9] as the Minimum Eccentricity Shortest Path Problem, and shown to be linked to the minimum line distortion problem [14].

The MESP problem is also closely related to the notion of k-laminar graphs introduced in [12], in which the covering path is required to be a diameter.

The MESP problem, as well as determining if a graph is k-laminar for a given k, are NP-hard [9,12]. However, Dragan and Leitert [9] develop a 2-approximation algorithm for MESP of time complexity $\mathcal{O}(n^3)$, a 3-approximation algorithm in $\mathcal{O}(nm)$ and a linear 8-approximation. The latter is extremely simple as it consists in a double-BFS procedure.

Roadmap. In this paper, we introduce a different analysis of the double-BFS procedure and prove that it is in fact a 5-approximation algorithm, and that the bound is tight. We then develop the idea of this algorithm and reach a 3-approximation, which still runs in linear time. Finally, we establish bounds relating the MESP problem and the notion of laminarity.

Definitions and Notations. Through this paper $G = (V, E)$ denotes a finite connected undirected graph. A *shortest path* between two vertices u and v is a path whose length is minimal among all u, v-paths. This length (counting edges) is the *distance* $d(u, v)$. Depending on the context, we consider a path either as a sequence, or as a set of vertices. The *distance* $d(v, S)$ between a vertex v and a set S is smallest distance between v and a vertex from S.

The *eccentricity* $ecc(S)$ of a set S is the largest distance between S and any vertex of G.

The maximal eccentricity of any singleton $\{v\}$, or equivalently the largest distance between two vertices, denoted here $diam(G)$, is often called the diameter of the graph, but for clarity in this paper a *diameter* is always a shortest path of maximum length, *i.e.* a shortest path of length $diam(G)$, and not its length.

2 Double-BFS Is a 5-Approximation Algorithm

Let us define the problem we are interested in:

Definition 1 (Minimum Eccentricity Shortest Path Problem (MESP)). *Given a graph G, find a shortest path P such that, for every shortest path Q, $ecc(P) \leq ecc(Q)$.*

$k(G)$ denotes the eccentricity of a MESP of G.

Theorem 1 (Dragan and Leitert [9]**).** *Computing $k(G)$ or finding a MESP are NP-complete problems.*

It is therefore worth using polynomial-time approximation algorithms. We say that an algorithm is an α-approximation of the MESP if every path output by this algorithm is a shortest path of eccentricity at most $\alpha k(G)$.

Double-BFS is a widely used tool for approximating $diam(G)$ [3]. It simply consists in the following procedure:

1. Pick an arbitrary vertex r
2. Perform a BFS (Breadth-First Search) starting at r and ending at x. x is thus one of the furthest vertices from r.
3. Perform a BFS (Breadth-First Search) starting at x and ending at y.

The output of the algorithm is the path from x to y, called a *spread path*, while its extremities (x, y) are called a *spread pair*. A folklore result is that the distance between x and y 2-approximates the diameter of G. As noted by Dragan and Leitert, Double-BFS may also be used for approximating MESP: they have shown in [9] that any spread path is an 8-approximation of the MESP problem.

The first result of the present paper is that any spread path is in fact a 5-approximation of the MESP problem and that the bound is tight. But before we prove this result (Theorem 2), let us give the key lemma used for proving our three theorems:

Lemma 1. *Let G be a graph having a shortest path $v_0, v_1 \ldots v_t$ of eccentricity k.*
Let $P = x_0, x_1, \ldots x_s$ be a shortest path of G.
Let i_{min}^P (resp. i_{max}^P) be the smallest (resp. largest) integer such that $v_{i_{min}^P}$ (resp. $v_{i_{max}^P}$) is at distance at most k of P.
For every integer i such that $i_{min}^P \leq i \leq i_{max}^P$, v_i is then at distance at most $2k$ from P.
Subsequently, every vertex v of G at distance at most k from the subpath between $v_{i_{min}^P}$ and $v_{i_{max}^P}$ is at distance at most $3k$ of P.

One may think, at first glance, that this lemma looks similar to the following:

Lemma 2 (from Dragan et al. [9]**).** *If G has a shortest path of eccentricity at most k from s to t, then every path Q with s in Q and $d(s, t) \leq max_{v \in Q} d(s, v)$ has eccentricity at most $3k$.*

The difference lies in the fact that the k in Lemma 2 is specific to the given couple of vertices (s, t) while the k in Lemma 1 is global. On the other hand, Lemma 2 gives a bound on the eccentricity of a path with respect to the whole graph, while Lemma 1 only guarantees an eccentricity for a defined subgraph.

Proof (of Lemma 1). The second assertion of the lemma is straightforward given the first one. To prove the latter, we define, for all l between 0 and s, the subpath $P_l = x_0, x_1 \ldots x_l$.

Let us show by induction on l that for all i between $i_{min}^{P_l}$ and $i_{max}^{P_l}$, v_i is at distance at most $2k$ of P_l.

- $l = 0$, $P_0 = x_0$.

Using the triangle inequality:

$$d(v_{i_{min}^{P_0}}, v_{i_{max}^{P_0}}) \leq d(v_{i_{min}^{P_0}}, x_0) + d(x_0, v_{i_{max}^{P_0}}) \leq 2k \tag{1}$$

Hence, for all i between $i_{min}^{P_0}$ and $i_{max}^{P_0}$,

$$d(v_{i_{min}^{P_0}}, v_i) \leq k \text{ or } d(v_{i_{max}^{P_0}}, v_i) \leq k \tag{2}$$

The result is thus verified for $l = 0$.

- Let l in $(1 \ldots s)$ such that the property if verified for $l - 1$.

For all i between $i_{min}^{P_{l-1}}$ and $i_{max}^{P_{l-1}}$, v_i is at distance at most $2k$ of P_{l-1} by the induction hypothesis. Hence, v_i is at distance at most $2k$ of P_l.

Moreover,

$$d(v_{i_{max}^{P_{l-1}}}, v_{i_{max}^{P_l}}) \leq d(v_{i_{max}^{x_{l-1}}}, v_{i_{max}^{x_l}}) \tag{3}$$

and by the triangle inequality:

$$d(v_{i_{max}^{x_{l-1}}}, v_{i_{max}^{x_l}}) \leq d(v_{i_{max}^{x_{l-1}}}, x_{l-1}) + d(x_{l-1}, x_l) + d(x_l, v_{i_{max}^{x_l}}) \leq 2k + 1 \tag{4}$$

As the sub-path of P between $v_{i_{max}^{P_{l-1}}}$ and $v_{i_{max}^{P_l}}$ is a shortest path, it follows that for all i between $i_{max}^{P_{l-1}}$ and $i_{max}^{P_l}$,

$$d(v_{i_{max}^{P_{l-1}}}, v_i) \leq k \text{ or } d(v_{i_{max}^{P_l}}, v_i) \leq k, \tag{5}$$

meaning that v_i is at distance at most $2k$ of P_{l-1} or of x_l.

A similar proof shows that for all i between $i_{min}^{P_l}$ and $i_{min}^{P_{l-1}}$, v_i is at distance at most $2k$ from P_{l-1} or from x_l.

The property is verified by induction, and the lemma follows for $l = s$.

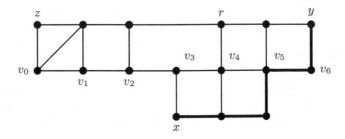

Fig. 1. The bound shown in Theorem 2 is tight. Indeed the graph is such that $v_0, v_1 \ldots v_6$ is a shortest path of eccentricity 1. The vertex z is at distance 5 from the shortest path (shown by thick edges) between x and y computed by double-BFS starting at r.

Theorem 2. *A double-BFS is a linear-time 5-approximation algorithm for the MESP problem.*

Before we prove it, notice that Fig. 1 shows that this bound is tight.

Proof. Let k be $k(G)$, $P = v_0, v_1 \ldots v_t$ be a MESP (its eccentricity is thus k), and $Q = x, \ldots, y$ be the result of a double-BFS starting at some arbitrary vertex r, then reaching x, then reaching y. We shall prove that Q is a $5k$-dominating path of G.

Let i (resp. j) be such that v_i (resp. v_j) is at distance at most k of r (resp. x). The following inequalities are verified:

$$d(r, x) \geq d(r, v_t) \geq d(v_i, v_t) - d(r, v_i) \geq d(v_i, v_t) - k \tag{6}$$

$$d(r, x) \leq d(r, v_i) + d(v_i, v_j) + d(v_j, x) \leq d(v_i, v_j) + 2k \tag{7}$$

Combining those inequalities,

$$d(v_i, v_t) - 3k \leq d(v_i, v_j) \tag{8}$$

Similarly:

$$d(v_i, v_0) - 3k \leq d(v_i, v_j) \tag{9}$$

Therefore v_j is at distance at most $3k$ of v_0 or v_t. Without loss of generality, assume that v_j is at distance at most $3k$ of v_0.

Let l be such that v_l is at distance at most k of y. We distinguish two cases:

(i) $l \leq j$:

Then y is at distance at most $5k$ of x. As y is a vertex most distant from x, x is a $5k$-dominating vertex of the graph. The lemma is then verified.

(ii) $l > j$:

Applying to (x, y) the inequalities established at the beginning of the proof:

$$d(v_j, v_t) - 3k \leq d(v_j, v_l) \tag{10}$$

As $l > j$, it follows that:

$$d(v_l, v_t) \leq 3k \tag{11}$$

Figure 2 shows the configuration of the graph in that case. The vertices at distance at most k of a vertex v_s such that $s \leq j$ (resp. $s \geq l$) are at distance at most $5k$ of x (resp. y).

According to Lemma 1, every vertex v of G at distance at most k of a vertex v_s such that s is between j and l is at distance at most $3k$ of any shortest path between x and y. The lemma is thus verified.

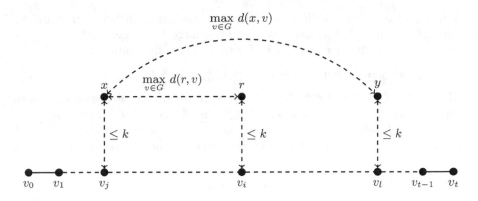

Fig. 2. Notations used in the proof of Theorem 2

3   A 3-Approximation Algorithm

We show now that by using more BFS runs we may obtain a $3k$-approximation of MESP, still in linear time.

Let *bestPath* and *bestEcc* be global variables used as return values for the path and its eccentricity. *bestPath* stores a path and is uninitialized, and *bestEcc* is an integer initialized with $|V(G)|$.

Data: G graph, x,y vertices of G, *step* integer
1 Compute a shortest path Q between x and y;
2 Select a vertex z of G most distant from Q;
3 **if** $d(Q,z) < bestEcc$ **then**
4 $\quad$ $bestPath \leftarrow Q$;
5 $\quad$ $bestEcc \leftarrow d(Q,z)$;
6 **end**
7 **if** $step < 8$ **then**
8 $\quad$ Algorithm3k(G,x,z,$step+1$);
9 $\quad$ Algorithm3k(G,y,z,$step+1$);
10 **end**

Algorithm 1. Algorithm3k

Theorem 3. *A 3-approximation of the MESP Problem can be computed in linear time by considering a spread pair (s,l) of G and running Algorithm3k(G,s,l,0).*

Proof (Correctness). Let G be a graph admitting a shortest path $P = v_0, v_1 \ldots v_t$ of eccentricity k.

Let x and y be any vertices of G, $Q_{x,y}$ a shortest path between x and y. Define $i_{min}^{x,y}$ (resp. $i_{max}^{x,y}$) as the smallest (resp. largest) integer such that $v_{i_{min}^{x,y}}$ (resp. $v_{i_{max}^{x,y}}$) is at distance at most k of x or y. Then, by Lemma 1,

$$\text{For all } j \text{ such that } i_{min}^{x,y} - k \leq j \leq i_{max}^{x,y} + k, \quad d(Q_{x,y}, v_j) \leq 2k \tag{12}$$

Hence, if $i_{min}^{x,y} \leq k$ and $i_{max}^{x,y} \geq t - k$, every vertex of P is at distance at most $2k$ of $Q_{x,y}$ and, as P is of eccentricity k, $Q_{x,y}$ is of eccentricity at most $3k$.

Algorithm3k uses this implication to exhibit a pair x, y such that $Q_{x,y}$ is of eccentricity at most $3k$. Indeed, in each recursive call, one of the following cases holds:

1. the vertex z selected at line 3 is at distance at most $3k$ from $Q_{x,y}$. In that case, $bestPath$ will be set to $Q_{x,y}$ unless it already contains a path of even better eccentricity. In any case, the result of the algorithm is a path of eccentricity at most $3k$.
2. the vertex z is at a distance greater than $3k$ of $Q_{x,y}$. Let i_z be such that v_{i_z} is at distance at most k of z. Then, according to Eq. (12),

$$i_z \leq i_{min}^{x,y} - k \ or \ i_z \geq i_{max}^{x,y} + k \tag{13}$$

(a) Suppose that $i_z \geq i_{max}^{x,y} + k$. Then, in the case $d(v_{i_{min}^{x,y}}, x) = k$, we get $i_{min}^{x,z} \leq i_{min}^{x,y}$ and $i_{max}^{x,z} \geq i_{max}^{x,y} + k$. And in the case $d(v_{i_{min}^{x,y}}, y) = k$ we get $i_{min}^{x,z} \leq i_{min}^{x,y} - k$ and $i_{max}^{x,z} \geq i_{max}^{x,y}$.
(b) A similar reasoning can be applied if $i_z \leq i_{min}^{x,y} - k$, also yielding to $i_{min}^{x,z} \leq i_{min}^{x,y}$ and $i_{max}^{x,z} \geq i_{max}^{x,y} + k$ or $i_{min}^{x,z} \leq i_{min}^{x,y} - k$ and $i_{max}^{x,z} \geq i_{max}^{x,y}$.

Therefore, either the algorithm already found a path of eccentricity at most $3k$, or it makes one of its two new calls with a couple (x', y') such that the interval $[i_{min}^{x',y'}, i_{max}^{x',y'}]$ contains $[i_{min}^{x,y}, i_{max}^{x,y}]$ but has length increased by at least k.

Consider now a spread pair (s, l) for which Algorithm3k$(G, s, l, 0)$ is run. It follows from case (i) and (ii) of the proof of Theorem 2 that

$$i_{min}^{s,l} \leq 5k \text{ and } i_{max}^{s,l} \geq t - 5k \tag{14}$$

At each of the recursive calls, if no path of eccentricity at most $3k$ has already been discovered, one of the new calls expands the interval $[i_{min}^{x,y}, i_{max}^{x,y}]$ length by at least k, while containing the previous interval. As the recursive calls are made until $step = 8$, it follows that either a path of eccentricity $3k$ has been discovered, or one of the explored possibilities corresponds to eight extensions of size at least k starting from $[i_{min}^{s,l}, i_{max}^{s,l}]$.

In the latter case, Eq. (14) implies that the final couple of vertices (x, y) fulfills $i_{min}^{x,y} \leq k$ and $i_{max}^{x,y} \geq t - k$. Every vertex of P is then of distance at most $2k$ of $Q_{x,y}$ and thus $Q_{x,y}$ is of eccentricity at most $3k$.

Proof (Complexity). The algorithm computes two BFS trees at line 1 and 2, taking $\mathcal{O}(n + m)$ time. The rest of the operations is computed in constant time.

The recursivity width is 2 and, since it is first called with $step = 0$, the recursivity length is 8. The algorithm is thus called 255 times. Therefore the total runtime of the algorithm is $\mathcal{O}(n + m)$.

Proof (Tightness of the approximation). Fig. 3 shows a graph for which the algorithm may produce a path of eccentricity $3k(G)$ (see caption).

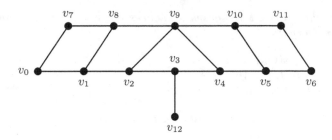

Fig. 3. Tightness of the bound shown in Theorem 3. The algorithm may indeed loop between the following couples of vertices: $(v_0, v_6), (v_0, v_{12}), (v_6, v_{12}), (v_0, v_{11}), (v_{11}, v_{12})$, $(v_6, v_7), (v_7, v_{12}), (v_{11}, v_7)$. Each time, it may choose a shortest path of eccentricity 3 (passing through v_8 v_9 and v_{10} whenever v_{12} is not an endvertex of the path) while $v_0..v_3..v_6$ has eccentricity 1.

4   Bounds Between MESP and Laminarity

In this section, we investigate the link between the MESP problem and the notion of laminarity introduced by Völkel *et al.* in [12]. The study of the k-laminar graph class finds motivation both from a theoretical and practical point of view. On the theoretical side, AT-free graphs form a well known graph class introduced half a century ago by Lekkerkerker and Boland [10], which contains many graph classes like co-comparability graphs. An AT-free graph admits a diameter all other vertices are adjacent with [5]. It is then natural to extend this notion of dominating diameter. On the practical side, some large graphs constructed from reads similarity networks of genomic or metagenomic data appear to have a very long diameter and all vertices at short distance from it [12], and exhibiting the "best" diameter allows to better understand their structure.

Definition 2 (laminarity). *A graph G is*

– *l-laminar if G has a diameter of eccentricity at most l.*
– *s-strongly laminar if every diameter has eccentricity at most s.*

$l(G)$ and $s(G)$ denote the minimal values of l and s such that G is respectively l-laminar and s-strongly laminar.

A natural question about laminarity and MESP is to ask what link exists between them.

Theorem 4. *For every graph* G,

$$k(G) \leq l(G) \leq 4k(G) - 2$$
$$k(G) \leq s(G) \leq 4k(G)$$

Moreover, there exist three graph sequences $(G_k)_{k \geq 1}$, $(H_k)_{k \geq 1}$ *and* $(J_k)_{k \geq 1}$ *such that, for every* k,

- $k(G_k) = l(G_k) = s(G_k) = k$;
- $k(H_k) = k$ *and* $l(H_k) = 4k - 2$;
- $k(J_k) = k$ *and* $s(J_k) = 4k$;

The bounds given by the inequalities are therefore tight.

Proof $(k(G) \leq l(G)$ *and* $k(G) \leq s(G))$. Those inequalities are straightforward as every diameter is by definition a shortest path. The eccentricity of every diameter is therefore always greater than $k(G)$.

Proof $(s(G) \leq 4k(G))$. Let $D = x_0, x_1, \ldots x_s$ be a diameter of G and $P = v_0, v_1 \ldots v_t$ a shortest path of eccentricity k. We shall show $ecc(D) \leq 4k$. Let z be any vertex of G. Since $ecc(P) = k$ there exists a vertex v_i of P such that $d(z, v_i) \leq k$. Let us distinguish three cases:

- Case 1: there exists vertices x_a, x_b of D and v_a, v_b of P such that $a \leq i \leq b$ and $d(v_a, x_a) \leq k$ and $d(v_b, x_b) \leq k$. Then by Lemma 1, z is at distance at most $3k$ from any shortest path between x_a and x_b, and thus at distance at most $3k$ of D.
- Case 2: there exists no vertex v_a of P with $a \leq i$ and $d(v_a, D) \leq k$
- Case 3: there exists no vertex v_a of P with $i \leq a$ and $d(v_a, D) \leq k$.

Without loss of generality we focus on Case 2 (illustrated in Fig. 4), which is symmetric with Case 3. Let l (resp. m) be such that v_l (resp. v_m) is at distance at most k of x_0 (resp. x_s), assume $l \leq m$:

$$d(v_l, v_m) \geq d(x_0, x_s) - 2k \tag{15}$$

D being a diameter,

$$d(x_0, x_s) \geq d(v_0, v_t) \tag{16}$$

By combining those inequalities,

$$d(v_l, v_m) \geq d(v_0, v_t) - 2k \tag{17}$$

$$d(v_l, v_m) \geq d(v_0, v_i) + d(v_i, v_l) + d(v_l, v_m) + d(v_m, v_t) - 2k \tag{18}$$

$$2k \geq d(v_i, v_l) \tag{19}$$

It follows that z is at distance at most $4k$ of x_0.

Proof ($l(G) \leq 4k(G) - 2$). Let $D = x_0, x_1, \ldots x_s$ be a diameter of G and $P = v_0, v_1 \ldots v_t$ a shortest path of eccentricity k. We shall show that either $ecc(D) \leq 4k - 2$ or G contains a diameter D' of eccentricity $3k$. If P is a diameter we are done. Let us suppose from now it is of length at most $|D| - 1$.

Let z be any vertex of G and v_i a vertex of P such that $d(z, v_i) \leq k$. Let us distinguish the same three cases than in the proof that $s(G) \leq 4k(G)$. The first case also leads to $d(z, D) \leq 3k$. The second and third being symmetric, let us suppose there exists no vertex v_j of P at distance at most k of D such that $j \leq i$.

Let v_l (resp. v_m) be a vertex of P at distance at most k from x_0 (resp. x_s), clearly,

$$d(v_l, v_m) \geq |D| - 2k. \tag{20}$$

Let us distinguish two subcases:

- Case 2.1: $d(v_l, v_m) > |D| - 2k$,

$$d(v_i, v_l) \leq d(v_0, v_t) - d(v_l, v_m) \leq (|D| - 1) - (|D| - 2k + 1) \leq 2k - 2 \tag{21}$$

It follows that z is at distance at most $4k - 2$ of D.
- Case 2.2: $d(v_l, v_m) = |D| - 2k$
In this case, a path $D' = x_0, ..v_l, v_{l+1}, ..v_m, ..x_s$ is a diameter. Assuming $l \leq m$, Eq. 19 in previous proof shows that:

$$d(v_i, v_l) \leq 2k \tag{22}$$

and with a symmetrical reasoning,

$$d(v_m, v_t) \leq 2k \tag{23}$$

It follows that any vertex v of G at distance at most k of a vertex v_a with $a \leq l$ (resp. $a \geq m$) is at distance at most $3k$ of v_l (resp. v_m). Hence at distance at most $3k$ of D'. $v_l, v_{l+1}, ..v_m$ being a subpath of D', any vertex v of G at distance at most k of a vertex v_a with a between m and t is at distance at most k of D'. Finally, any vertex of G is at distance at most $3k$ of D'.

Proof (Tightness of the bounds). Consider the graph G_k reduced to a path P of length $4k$ to which a second path of length k is attached in the middle. P is then simultaneously the only diameter and the MESP, and it k-covers G_k but doesn't $(k - 1)$-cover it. Hence the inequalities $k(G) \leq l(G)$ and $k(G) \leq s(G)$ are tight.

Figure 5 shows how to build the graph sequence $(J_k)_{k \geq 1}$ (only J_1 and J_6 are drawn). J_k is a graph with a shortest path of eccentricity k and a diameter of eccentricity $4k$. The inequality $s(G) \leq 4k(G)$ is thus tight.

Figure 6 shows how to build the graph sequence $(H_k)_{k \geq 1}$ (only H_1, H_2 and and H_6 are drawn). H_k is a graph with a shortest path of eccentricity k, while the unique diameter has eccentricity $4k - 2$ (H_1 is a special case with two diameters). The inequality $l(G) \leq 4k(G) - 2$ is therefore tight.

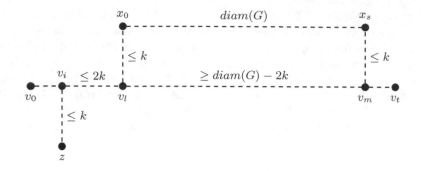

Fig. 4. Notations used in Case 2 of the proof of Theorem 4

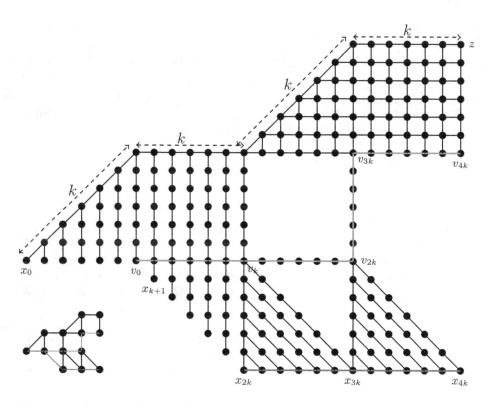

Fig. 5. Proof that $s(G) \geq 4k(G)$. The red path $x_0, x_1, \ldots x_{4k}$ is a diameter of length $4k$ and at distance $4k$ of z; while the green path $v_0, v_1, \ldots v_{4k}$ is a shortest path (another diameter indeed) of eccentricity k. The large graph is J_6 (using the graph sequence $(J_k)_k$ from Theorem 4) and the small one on the bottom left is J_1. The other members of the sequence car easily be derived. (Color figure online)

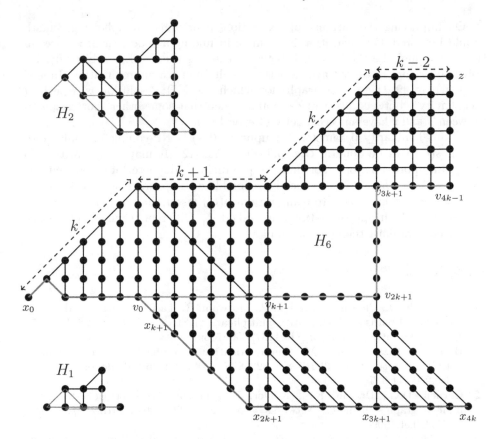

Fig. 6. Proof that $l(G) \geq 4k(G) - 2$. It is a graph sequence $(H_k)_k$, using the notation from Theorem 4. For $k \geq 2$, the red path $x_0, x_1, \ldots x_{4k}$ is the unique diameter. Its length is $4k$ and it is at distance $4k - 2$ of z. The green path $v_0, v_1, \ldots v_{4k-1}$ is a shortest path of length $4k - 1$ and of eccentricity k. Graphs H_2 and H_6 are drawn but all graphs H_k, $k \geq 2$ can be derived from the pattern of H_6. The small graph on the bottom left is the special case H_1 who do not follow this pattern. It admits exactly two diameters, both of eccentricity 2 (red), and a shortest path of eccentricity 1 (green). (Color figure online)

5   Conclusion

We have investigated the Minimum Eccentricity Shortest Path problem for general graphs and proposed a linear time algorithm computing a 3-approximation. The algorithm is a 2-recursive function with constant recursivity depth, launching two BFSs each time, thus taking linear time. Additionally, we've established some tight bounds linking the MESP parameter $k(G)$ and the k-laminarity parameters $s(G)$ and $l(G)$.

On improving the current approximation algorithms, the following remark should be noted. Our algorithm is confined in finding a good pair of vertices in the graph, and the shortest path between them is then picked arbitrarily. By doing so, we are unlikely to get a better result than a 3-approximation. Indeed as shown by [9] there exist graphs for which the MESP solution is a path of eccentricity k between two vertices s and t such that some other shortest paths between s and t have an eccentricity of exactly $3k$.

About laminarity parameters, computing $l(G)$ is NP-complete, while computing $s(G)$ can be done in $O(n^2 m \log n)$ time [12]. It may be interesting to design an approximation algorithm, $i.e$ producing a diameter of eccentricity at most $\alpha s(G)$ or $\beta l(G)$. Linear-time algorithms like BFS cannot be used however, since we do not know how to compute $diam(G)$ faster than a matrix product, and even surlinear approximation are studied [1]. Different techniques than the ones used here must therefore be employed.

References

1. Aingworth, D., Chekuri, C., Indyk, P., Motwani, R.: Fast estimation of diameter and shortest paths (without matrix multiplication). SIAM J. Comput. **28**(4), 1167–1181 (1999). http://dx.doi.org/10.1137/S0097539796303421
2. Bacsó, G., Tuza, Z., Voigt, M.: Characterization of graphs dominated by induced paths. Discret. Math. **307**(7–8), 822–826 (2007). http://dx.doi.org/10.1016/j.disc.2005.11.035
3. Corneil, D.G., Dragan, F.F., Köhler, E.: On the power of BFS to determine a graph's diameter. Networks **42**(4), 209–222 (2003). http://dx.doi.org/10.1002/net.10098
4. Corneil, D.G., Olariu, S., Stewart, L.: A linear time algorithm to compute a dominating path in an at-free graph. Inf. Process. Lett. **54**(5), 253–257 (1995). http://dx.doi.org/10.1016/0020-0190(95)00021-4
5. Corneil, D.G., Olariu, S., Stewart, L.: Asteroidal triple-free graphs. SIAM J. Discret. Math. **10**(3), 399–430 (1997). http://dx.doi.org/10.1137/S0895480193250125
6. Corneil, D.G., Olariu, S., Stewart, L.: Linear time algorithms for dominating pairs in asteroidal triple-free graphs. SIAM J. Comput. **28**(4), 1284–1297 (1999). http://dx.doi.org/10.1137/S0097539795282377
7. Deogun, J.S., Kratsch, D.: Diametral path graphs. In: Nagl, M. (ed.) WG 1995. LNCS, vol. 1017, pp. 344–357. Springer, Heidelberg (1995). doi:10.1007/3-540-60618-1_87
8. Deogun, J.S., Kratsch, D.: Dominating pair graphs. SIAM J. Discret. Math. **15**(3), 353–366 (2002). http://dx.doi.org/10.1137/S0895480100367111
9. Dragan, F.F., Leitert, A.: On the minimum eccentricity shortest path problem. In: Dehne, F., Sack, J.-R., Stege, U. (eds.) WADS 2015. LNCS, vol. 9214, pp. 276–288. Springer, Heidelberg (2015). doi:10.1007/978-3-319-21840-3_23
10. Lekkerkerker, C., Boland, J.: Representation of a finite graph by a set of intervals on the real line. Fund. Math. **51**, 45–64 (1962)
11. Robertson, N., Seymour, P.D.: Graph minors. I. Excluding a forest. J. Comb. Theory Ser. B **35**(1), 39–61 (1983). http://dx.doi.org/10.1016/0095-8956(83)90079-5

12. Völkel, F., Bapteste, E., Habib, M., Lopez, P., Vigliotti, C.: Read networks and
    k-laminar graphs. CoRR abs/1603.01179 (2016). arXiv:1603.01179
13. Yamazaki, K., Bodlaender, H.L., Fluiter, B., Thilikos, D.M.: Isomorphism for
    graphs of bounded distance width. In: Bongiovanni, G., Bovet, D.P., Battista,
    G. (eds.) CIAC 1997. LNCS, vol. 1203, pp. 276–287. Springer, Heidelberg (1997).
    doi:10.1007/3-540-62592-5_79
14. Yan, S., Xu, D., Zhang, B., Zhang, H., Yang, Q., Lin, S.: Graph embed-
    ding and extensions: a general framework for dimensionality reduction. IEEE
    Trans. Pattern Anal. Mach. Intell. **29**(1), 40–51 (2007). http://dx.doi.org/10.1109/
    TPAMI.2007.250598

On the Complexity of Extracting Subtree with Keeping Distinguishability

Xianmin Liu[1]([⊠]), Zhipeng Cai[2], Dongjing Miao[1], and Jianzhong Li[1]

[1] Harbin Institute of Technology, Harbin, China
{liuxianmin,miaodongjing,lijzh}@hit.edu.cn
[2] Georgia State University, Atlanta, USA
zcai@gsu.edu

Abstract. We consider the problem of subtree extraction with guarantee of preserving *distinguishability*. Given a query q and a tree T, evaluating q on T will output $q(T)$ which is a set of nodes of T. For two nodes a and b in T, they can be *distinguished* by some query q in T, iff exactly one of them belongs to $q(T)$. Then, given a tree T, a query class $\mathcal{L}$, and two disjoint node sets A and B of T, a subtree T' of T is called *preserving distinguishability* of T, iff (1) T' contains all nodes in $A \cup B$, (2) for any node pair $(a,b) \in A \times B$, if a and b can be *distinguished* by some query in $\mathcal{L}$ in T, they can also be *distinguished* by some query (not necessarily the same one) in $\mathcal{L}$ in T', and (3) for any node pair $(a,b) \in A \times B$ and a query $q \in \mathcal{L}$, if a and b can be *distinguished* by q in T', they can also be *distinguished* by q in T. The subtree extraction problem considered by this paper is to determine whether there is a small enough subtree T' of T, such that for query class $\mathcal{L}$ and node sets A and B, T' preserves the *distinguishability* of T. In this paper, as an initial attempt of investigating this problem, fixing $\mathcal{L}$ to be a specific part of *tree pattern queries* (introduced later), the subtree extraction problem is shown to be NP-*complete*.

Keywords: Subtree extraction · Distinguishability · Computational complexity

1 Introduction

Recently, computation on large-scale data has increased lots of research interests. An important strategy for large data is to preprocess data before the computation, which aims to reduce the data size by only keeping the *necessary* information related to the computation. This paper studies that idea on the platform of tree-structured data and tree queries.

Tree-structured data (*e.g.* XML) and its query language have been investigated very much [1,2]. For a query q and data tree T, let $q(T)$ be the result node set outputted by evaluating q on T, then two nodes u and v in T can be *distinguished* by q, iff exactly one of them belongs to $q(T)$. They can be distinguished by query class $\mathcal{L}$, iff they can be distinguished by some query $q \in \mathcal{L}$.

T.-H.H. Chan et al. (Eds.): COCOA 2016, LNCS 10043, pp. 230–240, 2016.
DOI: 10.1007/978-3-319-48749-6_17

In this paper, the Distinguishability Preserving Subtree Extraction problem (dpSE for short) is considered which can be described informally as follows. Given data tree T and an integer $k > 0$, dpSE determines whether or not T has a subtree T' satisfying $|T| - |T'| \geq k$ such that for specific query class $\mathcal{L}$ and two node sets A, B of T, the following two conditions are satisfied.

- Preserving Node Distinguishability. Each node pair in $A \times B$ which can be distinguished in T can still be distinguished in T'.
- Preserving Query Distinguishability. For each node pair $(u, v) \in A \times B$, if query $q \in \mathcal{L}$ can distinguish them in T', it can still distinguish them in T.

There are lots of useful applications of the dpSE problem in many areas such as *Mass Customization, Category Management* and so on. We are not aware of any previous works focusing on the same problem. As an initial attempt, in this paper, the query class is fixed to be a specific part of tree pattern queries, and, in that case, dpSE is shown to be NP-*complete*.

1.1 Related Work

Some related works are in the area of XML data, a popular and very studied tree-structured data. Finding concise representations for both syntactical and semantic information of XML data is useful for many applications, such as schema inference [3,4], information extraction [5,6], query learning [6–8] and so on. Schema inference usually uses regular expressions or its fragments as representations [3,4,9], and extracts schemas from given valid data. Information extraction usually utilizes semantic rules or queries [5,6,10] to extract information interested. Query learning uses query languages defined over XML [6–8], and searches proper queries accepting positive examples and rejecting negative examples. The problem dpSE can be extended to those applications. Aso, this paper is motivated by the work of learning XML queries by [6,7,11], which shows that, even for very simple queries, it is NP-*hard* to find a query consistent with given positive and negative examples. The problem considered here investigates the possibility of preprocessing techniques for reducing input data size. The tree-map used in this paper is related to the mappings studied by [12–14], which mainly focus on the relations between mappings and query containment or query evaluation.

2 Preliminary

A tree is referred to the *rooted, directed* and *labeled* tree here, formally defined as below.

Definition 2.1 (Tree Structure Data). A data tree T can be represented by a 5-tuple $(V_T, E_T, r_T, \Omega, l_T)$, where V_T and E_T are the node and edge sets respectively, r_T is the unique *root* node in V_T, Ω is an alphabet for labeling the nodes, and l_T is a *label* function which is defined as a mapping from V_T to Ω. $\qquad\square$

For two trees T_1 and T_2, it is denoted by $T_1 \subseteq T_2$ if T_1 is a subtree of T_2. For a tree T and a node $u \in V_T$, we use T_u to represent the whole subtree rooted at u in T.

Tree pattern query is a popular tree querying language [12], and this paper considers one *simple* and *common used* fragment of tree pattern query.

Definition 2.2 (Simple Tree Pattern Query). A simple tree pattern query q is a binary tuple (T_q, t_q), where T_q is a tree represented by $(V_q, E_q, r_q, \Omega, l_q)$ and $t_q \in V_q$ is the unique *target* node of q. Additionally, the set of all simple tree pattern queries is denoted by $\mathcal{Q}'$. □

Also, for convenience, two other representations for query are also used. Given a query q, a graphical *tree definition* of q can be built by first drawing the embedded tree of q and then marking the *target* node in the tree T. The *string definition* can be built as follows. First, string s is built by concatenating the label of the nodes on the path p between root and target node of q using '/' orderly. Then, the strings of each subtree of node u on p is wrapped by '[ ]' and inserted after the label of u in s.

The semantics of $\mathcal{Q}'$ will be given after introducing a kind of map functions on trees.

Definition 2.3 (Tree-map). A *tree-map* is a map defined between two trees. For given trees T and T', a function $f : V_T \mapsto V_{T'}$ is a *tree-map* from V_T to $V_{T'}$, if

(1) $f(r_T) = r_{T'}$,
(2) for node $n \in V_T$, $l_T(n) = l_{T'}(f(n))$, and
(3) for edge $(x, y) \in E_T$, $(f(x), f(y)) \in E_{T'}$. □

Definition 2.4 (Semantic of $\mathcal{Q}'$). For a tree T and $q = (T_q, t_q) \in \mathcal{Q}'$, T *matches* (or *satisfies*) q, denoted by $T \models q$, iff there exists a tree-map f from V_q to V_T. For special f, it is also denoted by $T \models_f q$, if f just defines such a tree-map. Additionally, a node $u \in V_T$ is *selected* by q on T, denoted by $u \in q(T)$, if there exists a tree-map f such that $T \models_f q$ and $f(t_q) = u$. □

There are some properties of $\mathcal{Q}'$ useful in this paper.

Lemma 2.1: $\mathcal{Q}'$ is in PTIME. That is, given a tree T, a query $q \in \mathcal{Q}'$ and a node $u \in V_T$, it can be determined in PTIME whether or not $u \in q(T)$. □

Lemma 2.2: Given trees T_1 and T_2, it can be determined in PTIME whether there is a tree-map from T_1 to T_2. □

Lemma 2.3: Given trees T_1, T_2 and T_3, if f_1 is a tree-map from T_1 to T_2 and f_2 is a tree-map from T_2 to T_3, then the function f_3 defined as $f_3(v) = f_2(f_1(v))$ is a tree-map from T_1 to T_3. □

Given a tree T, two nodes $n_1, n_2 \in V_T$ are called to be *distinguished* by query q, denoted by $n_1 \neq_{T,q} n_2$, if exactly one of them is selected by q. Otherwise, they can not be *distinguished* by q, denoted by $n_1 \equiv_{T,q} n_2$. For a query class $\mathcal{L}$ and tree T, two nodes $n_1, n_2 \in V_T$ are called to be *distinguishable*, denoted by $n_1 \neq_{T,\mathcal{L}} n_2$, if there is a $q \in \mathcal{L}$ such that $n_1 \neq_{T,q} n_2$. Otherwise, they are not *distinguishable*, denoted by $n_1 \equiv_{T,\mathcal{L}} n_2$. If the query class is clear from the context, we only use the representations $n_1 \neq_T n_2$ and $n_1 \equiv_T n_2$.

For query class $\mathcal{Q}'$, the Distinguishability Preserving Subtree Extraction problem, dpSE for short, can be formally defined as follows.

Definition 2.5 (Problem $\langle \mathsf{dpSE}, \mathcal{Q}' \rangle$). Given a tree T, node sets $A, B \subseteq V_T$, and an integer k, the dpSE problem is to determine whether there is a subtree T' of T such that $|V_T| - |V_{T'}| \geq k$, $A \cup B \subseteq V_{T'}$, and the following condition is satisfied. For each pair of nodes $(u, v) \in A \times B$,
- Node Distinguishability Preserving. If $u \neq_{T,\mathcal{Q}'} v$, then $u \neq_{T',\mathcal{Q}'} v$, and
- Query Distinguishability Preserving. For a query $q \in \mathcal{Q}'$, if $u \neq_{T',q} v$, then $u \neq_{T,q} v$. $\qquad\square$

3   The Hardness Result

The main result of this paper can be described by the following theorem.

Theorem 3.1: $\langle \mathsf{dpSE}, \mathcal{Q}' \rangle$ is NP-*complete*. $\qquad\square$

Before describing the proof, we introduce some useful notations first. For two paths p_1 and p_2, let $\bowtie(p_1, p_2)$ be the set of node pairs with the same depths. For nodes n_1 and n_2, let $\mathsf{path}(n_1, n_2)$ be the path from n_1 to n_2. For two trees $T_1 = (V_1, E_1, r_1, \Omega_1, l_1)$ and $T_2 = (V_2, E_2, r_2, \Omega_2, l_2)$ and a node $u \in V_1$, w.l.o.g. assuming $V_1 \cap V_2 = \emptyset$, a operator Link can be defined such that $\mathsf{Link}(T_1, u, T_2)$ is the tree $\big(V_1 \cup V_2, E_1 \cup E_2 \cup (u, r_{T_2}), r_{T_1}, \Omega_1 \cup \Omega_2, l_1 \cup l_2\big)$. Intuitively, $\mathsf{Link}(T_1, u, T_2)$ is obtained by linking T_2 to the node u and letting *root* of T_1 be the new *root*.

3.1   The Upper Bound

The upper bound for $\langle \mathsf{dpSE}, \mathcal{Q}' \rangle$ problem can be described by the following lemma.

Lemma 3.1: $\langle \mathsf{dpSE}, \mathcal{Q}' \rangle$ is in NP. $\qquad\square$

To prove the Lemma 3.1, based on two functions gc and lc, a NP algorithm for the problem $\langle \mathsf{dpSE}, \mathcal{Q}' \rangle$ is given in Lemma 3.2.

Definition 3.1 (gc and lcfunction). Given trees T_1 and T_2 satisfying $T_2 \subseteq T_1$ and two nodes $u, v \in V_{T_1} \cap V_{T_2}$,
- gc $(u, v, T_1, T_2) = \textbf{true}$ iff $u \neq_{T_1} v$ implies $u \neq_{T_2} v$.
- lc $(u, v, T_1, T_2) = \textbf{true}$ iff $u \neq_{T_2,q} v$ implies $u \neq_{T_1,q} v$ for any query $q \in \mathcal{Q}'$. $\qquad\square$

In fact, it is easy to find that the conditions for gc and lc are same as preserving the node and query distinguishability.

Lemma 3.2: If gc and lc can be determined in PTIME, there is an NP algorithm for the $\langle \mathsf{dpSE}, \mathcal{Q}' \rangle$ problem.   □

Proof: It is easy to verify by considering the definitions directly.   □

Lemma 3.3: Function gc can be evaluated in PTIME.   □

Proof: Since gc (u, v, T_1, T_2) is equivalent to $(u \equiv_{T_1} v) \vee (u \not\equiv_{T_2} v)$, a PTIME algorithm for determining the relation $\equiv$ will simply a PTIME algorithm for gc (u, v, T_1, T_2) immediately. Next, a PTIME algorithm ifEquiv will be given such that ifEquiv (u, v, T) returns **true** iff $u \equiv_T v$.

A function ifMap is utilized here, which takes two trees T' and T'' as input, and returns **true** iff there is a tree-map from T' to T''. The existence of ifMap can be known from Lemma 2.2.

Given inputs T and $u, v \in V_T$, the ifEquiv algorithm works as follows. Let $p_u = \mathsf{path}(r_T, u)$, $p_v = \mathsf{path}(r_T, v)$.

- If $|p_u| \neq |p_v|$ return **false**, otherwise, continue.
- If there is a pair (u_i, v_i) in $\bowtie (p_u, p_v)$ such that $l_T(u_i) \neq l_T(v_i)$, return **false**, otherwise continue.
- Return $res=$

$$\bigwedge_{(u_i, v_i) \in \bowtie(p_u, p_v)} \left(\mathsf{ifMap}(T_{u_i}, T_{v_i}) \wedge \mathsf{ifMap}(T_{v_i}, T_{u_i}) \right)$$

Obviously, the ifEquiv algorithm is in PTIME. The correctness of ifEquiv algorithm can be verified by considering constructing maps according to the outputs of functions.   □

Lemma 3.4: Function lc can be evaluated in PTIME.   □

Proof: To prove the lemma, a PTIME algorithm ifRight is given in Fig. 1.

An important observation is introduced first. Let p_u represent $\mathsf{path}(r_{T_1}, u)$ or $\mathsf{path}(r_{T_2}, u)$ (*they are same since* $T_2 \subseteq T_1$). Similarly, we can define p_v. The observation is that if p_u and p_v have different lengths or different node labels one some position, lc(u, v, T_1, T_2) must be *true*. To prove this, let us consider an arbitrary query $q \in \mathcal{Q}'$, and assume $u \not\equiv_{T_2, q} v$. Moreover, *w.l.o.g.*, we can assume $u \in q(T_2)$ and $v \notin q(T_2)$. (1) First, we have $v \notin q(T_1)$. If not, both u and v are in $q(T_1)$ will imply p_u and p_v have same lengths and labels, which is a contradiction. (2) Second, we have $u \in q(T_1)$ since T_2 is a part of T_1. Combining them, we have $u \not\equiv_{T_1, q} v$. Then, according to the definition of lc, lc(u, v, T_1, T_2) is *true*.

The first step of Algorithm ifRight is to return **true** for the case that the conditions for the above observation are satisfied (line 1–4). The output of Join has two properties.

+ Property P1. There are tree-maps from Join (T', T'') to T' and T''.

Algorithm ifRight

Input: Trees T_1 and T_2, *s.t.* $T_2 \subseteq T_1$,

   nodes $u, v \in V_{T_1} \cap V_{T_2}$

Output: The boolean value *ans* of lc

/ $* * *$ *The first step* $* * *$ /

1. Let $p_u = \mathsf{path}(r_{T_2}, u)$, $p_v = \mathsf{path}(r_{T_2}, v)$;
2. **if** $|p_u| \neq |p_v|$ **then return true**;
3. **for** $(u_i, v_i) \in \bowtie(p_u, p_v)$ **do**
4.   **if** $l_{T_2}(u_i) \neq l_{T_2}(v_i)$ **then return true**;

/ $* * *$ *The second step* $* * *$ /

5. **for** $(u_i, v_i) \in \bowtie(p_u, p_v)$ **do**
6. $C_{i1} = \mathsf{Join}(T_{2,u_i}, T_{1,v_i})$;
7. $C_{i2} = \mathsf{Join}(T_{2,v_i}, T_{1,u_i})$;
8. $q_{i1} = \mathsf{Link}(p_u, u_i, C_{i1})$;
9. $q_{i2} = \mathsf{Link}(p_u, u_i, C_{i2})$;
10.   **if** $u \not\equiv_{T_2, q_{i1}} v \vee u \not\equiv_{T_2, q_{i2}} v$ **then return false**;
11. **return true**;

Procedure Join

Input: Two trees T' and T''

Output: The tree $T = \mathsf{Join}(T', T'')$

1. **if** $T_1 = \varnothing$ or $T_2 = \varnothing$ **then return** $\varnothing$;
2. **if** $l_{T'}(r_{T'}) \neq l_{T''}(r_{T''})$ **then return** $\varnothing$;
3. Initialize T to be node r_T, *s.t.* $l_T(r_T) = l_{T'}(r_{T'})$;
4. **for** each node x *s.t.* $(r_{T'}, x) \in E_{T'}$ **do**
5.   **for** each node y *s.t.* $(r_{T''}, y) \in E_{T''}$ **do**
6. $T_{xy} = \mathsf{Join}(T'_x, T''_y)$;
7.   **if** $T_{xy} \neq \varnothing$ **then**
8. $T = \mathsf{Link}(T, r_T, T_{xy})$;
9. **return** T;

Fig. 1. Algorithm for lc

+Property P2. Any query q having tree-maps from T_q to T' and T'' also has a tree-map from T_q to Join (T', T'').

The second step of Algorithm ifRight builds q_{i1} and q_{i2} based on Join procedure for each node pair $(u_i, v_i) \in \bowtie(p_u, p_v)$ first (line 5–9), and then returns **false** if there is a query q_{ij} ($i \in [1, |p_u|]$, $j \in [1,2]$) *s.t.* $u \not\equiv_{T_2, q_{ij}} v$ (line 10).

The proof of correctness of the second step can be divided into two directions.

$\Rightarrow$: If there is a query q_{ij} satisfying $u \not\equiv_{T_2, q_{ij}} v$, we will prove that there is a query q satisfying both $u \not\equiv_{T_2, q} v$ and $u \equiv_{T_1, q} v$. In fact, q_{ij} can just the role of q. Obviously, we only need to prove $u \equiv_{T_1, q_{ij}} v$. There are two cases.

For $j = 1$, we have $q_{ij} = q_{i1} = \mathsf{Link}(p_u, u_i, C_{i1})$ and $C_{i1} = \mathsf{Join}(T_{2,u_i}, T_{1,v_i})$. Then, we know $u \in q_{i1}(T_2)$, because there is a tree-map from C_{i1} to T_{2,u_i} (Property P1), and it can be extended to a tree-map from q_{i1} to T_2 such that

t_q is mapped to u. Similarly, $v \in q_{i1}(T_1)$. We also have $u \in q_{i1}(T_1)$, since $T_2 \subseteq T_1$. Therefore, $u \equiv_{T_1,q_{i1}} v$.

For $j = 2$, we have similar proof.

$\Leftarrow$: If there is a query q such that $u \not\equiv_{T_2,q} v$ and $u \equiv_{T_1,q} v$, we will prove that there is a query q_{ij} such that $u \not\equiv_{T_2,q_{ij}} v$.

Since $u \not\equiv_{T_2,q} v$, w.l.o.g., it is assumed that $u \in q(T_2)$ and $v \notin q(T_2)$. Then, because $T_2 \subseteq T_1$ and $u \equiv_{T_1,q} v$, we also have $u \in q(T_1)$ and $v \in q(T_1)$. Since p_u and p_v are totally same after the first step of ifRight, $\text{path}(r_{T_q}, t_q)$ must be also same as p_u and p_v. Then, we will prove this part by assuming each q_{ij} satisfies $u \equiv_{T_2,q_{ij}} v$ and making the contradiction that $v \in q(T_2)$.

Let $p_q = (z_1, \ldots, z_{|p_q|})$ be the $\text{path}(r_{T_q}, t_q)$, and T_{z_i} be the subtree rooted at z_i obtained after deleting all edges on p_q from T_q. (1) First, because $v \in q(T_1)$ and $u \in q(T_2)$, for each i, there must be a tree-map from T_{z_i} to T_{1,v_i} and a tree-map from T_{z_i} to T_{2,u_i}. According to Property P2, there must be a tree-map f_i from T_{z_i} to $\text{Join}(T_{2,u_i}, T_{1,v_i}) = C_{i1}$. (2) Second, consider C_{i1}, according to Property P1, there must be a tree-map from C_{i1} to T_{2,u_i}, which implies $u \in q_{i1}(T_2)$. Also, according to the assumption, q_{i1} satisfies $u \equiv_{T_2,q_{i1}} v$, we have $v \in q_{i1}(T_2)$. Thus, there must be a tree-map g_i from q_{i1} to T_2 such that $g_i(t_{q_{i1}}) = v$. Obviously, g_i is also a tree-map from C_{i1} to T_{2,v_i}. (3) Then, according to Lemma 2.3, a tree-map h_i from T_{z_i} to T_{2,v_i} can be built by composing f_i and g_i. (4) Finally, a tree-map h from q to T_2 satisfying $h(t_q) = v$ can be built by taking a union of all h_is, which will make the contradiction that $v \in q(T_2)$.

Finally, the construction of the contradiction proves this direction.

In conclusion, lc can be evaluated by algorithm ifRight in PTIME. $\square$

3.2   The Lower Bound

The lower bound of dpSE on $\mathcal{Q}'$ can be described by the following lemma.

Lemma 3.5: $\langle \text{dpSE}, \mathcal{Q}' \rangle$ is NP-*hard*. $\square$

Proof: We show $\langle \text{dpSE}, \mathcal{Q}' \rangle$ is NP-*hard* by reduction from 3SAT problem, which is known to be NP-*complete* very much [15]. For given variable set $X = \{x_1, x_2, \ldots, x_n\}$, a 3SAT instance can be represented as $\varphi = C_1 \wedge C_2 \wedge \cdots \wedge C_m$, and each C_j is in the form $l_{j1} \vee l_{j2} \vee l_{j3}$ where each l_{jk} is x_i or $\bar{x}_i$ for $i \in [1, n]$. Without loss of generality, for any variable $z \in X$, it is assumed that $\{z, \bar{z}\}$ can not appear in the same clause.

Given an instance φ of 3SAT, we construct an instance $S_\varphi = \langle T, A, B, k \rangle$ of $\langle \text{dpSE}, \mathcal{Q}' \rangle$ such that the answer of S_φ is 'yes' iff φ is satisfied. S_φ can be defined as follows.

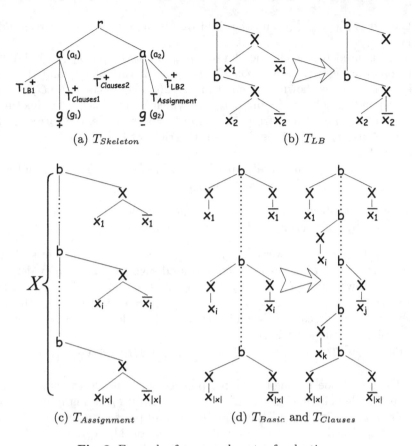

Fig. 2. Examples for several parts of reduction

- First, $T_{Skeleton}$ shown in Fig. 2(a) gives an overview of the structure of T. The *root* node r_T is labelled r uniquely, and in the followings, node r also means the node r_T. There are two paths (a_1, g_1) and (a_2, g_2) under r, where a_1 and a_2 are labelled a, and g_1 and g_2 are labelled g. Under a_1, there are two subtree sets T_{LB1} and $T_{Clause1}$, and under a_2, there are three subtree sets T_{LB2}, $T_{Clause2}$ and $T_{Assignment}$. T_{LB1} and T_{LB2} have totally same structures, and so do $T_{Clause1}$ and $T_{Clause2}$. Therefore, in the followings, T_{LB} presents both T_{LB1} and T_{LB2}, and T_{Clause} represents both $T_{Clause1}$ and $T_{Clause2}$.
- As shown in Fig. 2(c), there is only one tree in $T_{Assignment}$, which is obtained by first constructing a tree $/b/X[/x_i][/\overline{x}_i]$ for each variable x_i, and then concatenating them orderly. In the followings, $T_{Assignment}$ also represents the only subtree in it.
- T_{LB} is obtained based on $T_{Assignment}$. In detail, for each x_i, there is a corresponding tree T_{x_i} in T_{LB} which is constructed by deleting x_i and $\overline{x}_i$ from $T_{Assignment}$. For example, as shown in Fig. 2(b), when $X = \{x_1, x_2\}$, the tree

on the left is $T_{Assignment}$, and the right one is the tree in T_{LB} corresponding to x_1.

- $T_{Clauses}$ is used to encode the clauses in φ. First, a tree T_{Basic} is built by first constructing a tree $/b[/X/x_i]/X/\overline{x}_i$ for each variable x_i, and then concatenating them as shown on the left of Fig. 2(d). For each clause C_j, there is a corresponding subtree T_{C_j} in $T_{Clauses}$, which is constructed by deleting x_i (resp. $\overline{x}_i$) from T_{Basic} if x_i (resp. $\overline{x}_i$) is in C_j. For example, the corresponding tree for clause $\overline{x}_i \wedge x_j \wedge \overline{x}_k$ is shown on the right of Fig. 2(d).
- Let $k = |X|$.
- A is the defined to contain all leaf nodes in T_{Clause} and T_{LB}, g_1, and all X nodes in $T_{Assignment}$. B only contains g_2.

Obviously, the above reduction is in PTIME. The correctness of this reduction can be shown by considering the following two directions.

$\Rightarrow$: If φ is satisfied, the answer of S_φ is 'yes'. Suppose the assignment satisfying φ is ν_φ, a subtree T' of T will be constructed such that T' is an evidence for the 'yes' answer of S_φ. T' is got by deleting $|X|$ leaf nodes in $T_{Assignment}$ from T. In detail, for each variable $x_i \in X$, if $\nu_\varphi(x_i) =$**true**, delete $\overline{x}_i$, otherwise, delete x_i. Then, T' can be shown to be an evidence for the 'yes' answer of S_φ by considering the followings.

(1) Obviously, T' is a subtree of T containing $A \cup B$, and $|V_T| - |V_{T'}| = |X| = k$.

(2) To show Node Distinguishability preserving, it needs to show that, for each node pair $(u, v) \in A \times B$, if $u \not\equiv_T v$, then $u \not\equiv_{T'} v$. Let $\bar{q}$ be the query $/r/a/g$. Obviously, in T', except (g_1, g_2), all pairs (u, v) in $A \times B$ can be distinguished by $\bar{q}$. Thus, we only need to consider (g_1, g_2). (i) In T, g_1 and g_2 can be distinguished $q = \left(\mathsf{Link}(T_{\bar{q}}, a, T_{Assignment}), g\right)$, where, to be convenient, a and g represent the nodes labelled a and g in $T_{\bar{q}}$ respectively. Obviously, $g_2 \in q(T)$. Also, we have $g_1 \notin q(T)$, because, for g_1's parent node a_1, there does not exist a tree-map from $T_{Assignment}$ to T_{a_1}. (ii) In T', g_1 and g_2 can be distinguished $q' = \left(\mathsf{Link}(T_{\bar{q}}, a, T'_{Assignment}), g\right)$. Obviously, $g_2 \in q'(T')$. Also, we have $g_1 \notin q'(T')$ because of following observations. First, there is no tree-map from $T'_{Assignment}$ to trees in T_{LB}, because each X node in $T'_{Assignment}$ has at least one child, but there is always one X has no child in each tree of T_{LB}. Second, there is no tree-map from $T'_{Assignment}$ to trees in $T_{Clauses}$. For each $T_{C_j} \in T_{Clauses}$, because ν_φ satisfies C_j, according to the definitions of T_{C_j} and $T'_{Assignment}$, there is a variable x_i such that x_i (resp. $\overline{x}_i$) exists in $T'_{Assignment}$ but not in T_{C_j}, which means that a tree-map from $T'_{Assignment}$ to T_{C_j} is impossible. (iii) Combining (i) and (ii), we have $g_1 \not\equiv_T g_2$ and $g_1 \not\equiv_{T'} g_2$. Finally, Node Distinguishability is preserved in T'.

(3) To show Query Distinguishability preserving, it needs to show, for each node pair (u, v), if a query q satisfies $u \not\equiv_{T', q} v$, it also has $u \not\equiv_{T, q} v$. Similar with the (2) part, we only need to consider (g_1, g_2). Consider an arbitrary query q such that $g_1 \not\equiv_{T', q} g_2$. Because, in T', $T'_{a_1} \subseteq T'_{a_2}$, there

must be $g_2 \in q(T')$ and $g_1 \notin q(T')$. Since $T' \subseteq T$, obviously, $g_2 \in q(T)$. Because T_{a_1} and T'_{a_1} are totally same, we also have $g_1 \notin q(T)$. Therefore, $g_1 \not\equiv_{T,q} g_2$.

Combing them, the answer of S_φ is 'yes'.

$\Leftarrow$: If the answer of S_φ is 'yes', φ is satisfied. Suppose the subtree of T satisfying conditions in dpSE is T', an assignment ν_φ satisfying φ will be constructed. Before defining ν_φ, following observations are explained. First, the only difference between T and T' is between $T_{Assignment}$ and $T'_{Assignment}$. Second, in $T_{Assignment}$, at most $2|X| = 2n = 2k$ nodes can be deleted, which are $\{x_1, \overline{x}_1, \ldots, x_k, \overline{x}_k\}$. Third, at least k nodes are deleted. Forth, for each X, at most one child can be deleted. Otherwise, there will be a tree-map from $T'_{Assignment}$ to some tree in T_{LB}, and g_1 and g_2 can not be distinguished in T'. Combined with $g_1 \not\equiv_T g_2$ (by query $(\mathsf{Link}(T_{\tilde{q}}, a, T_{Assignment}), g))$, it contradicts that T' is a solution of S_φ. Fifth, taking the *third* and *forth* observations, there is, in $T'_{Assignment}$, each X selects exactly one child from x_i and $\overline{x}_i$.

Then, based on $T'_{Assignment}$, ν_φ is built as follows. For each variable $x_i \in X$, if x_i appears in $T'_{Assignment}$, let $\nu_\varphi(x_i) =$ **true**, otherwise let $\nu_\varphi(x_i) =$ **false**.

Finally, it will be checked that ν_φ indeed satisfies φ. First, because g_1 and g_2 can be distinguished in T as discussed above, it is known that $g_1 \not\equiv_{T'} g_2$. Then, we have, for each T_{C_j}, there is no tree-map from $T'_{Assignment}$ to T_{C_j}. If not, we can build a tree-map from T'_{a_1} to T'_{a_2} (*because* $T'_{a_1} \subseteq T'_{a_2}$) and a tree-map from T'_{a_2} to T'_{a_1} (*extending the tree-map from* $T'_{Assignment}$ *to* T_{C_j} *naturally*). Then, there must be a x_i such that, x_i (or $\overline{x}_i$) is in $T'_{Assignment}$, but not in T_{C_j}. Assume x_i is in $T'_{Assignment}$ and $\overline{x}_i$ is in T_{C_j}. According to the definition of T_{C_j}, there is $x_i \in C_j$. Since x_i is in $T'_{Assignment}$, according to the construction of ν_φ, we have $\nu_\varphi(x_i) =$ **true**, that is C_j can be satisfied by ν_φ. For the case that $\overline{x}_i$ is in $T'_{Assignment}$ and x_i is in T_{C_j}, a similar proof can show C_j is satisfied by ν_φ. At last, it is known that φ is satisfied by ν_φ.

In conclusion, after showing the correctness of the reduction, it is shown that 3SAT problem can be reduced to the problem $\langle \mathsf{dpSE}, \mathcal{Q}' \rangle$ in PTIME. Therefore, $\langle \mathsf{dpSE}, \mathcal{Q}' \rangle$ is NP-*hard*. $\square$

4   Conclusion

We have proposed dpSE problem in this paper, whose question is whether the size of a given tree can be reduced while preserving two kinds of distinguishability. Considering the query class $\mathcal{Q}'$, dpSE is proved to be NP-*complete*.

Acknowledgement. This work was supported in part by the National Grand Fundamental Research 973 Program of China under grant 2012CB316200, the General Program of the National Natural Science Foundation of China under grant 61502121, 61402130, the China Postdoctoral Science Foundation under grant 2016M590284, the Fundamental Research Funds for the Central Universities (Grant No. HIT.NSRIF.201649), and Heilongjiang Postdoctoral Foundation (Grant No. LBH-Z15094).

References

1. Neven, F., Schwentick, T.: Expressive, efficient pattern languages for tree-structured data. In: PODS (2000)
2. Gou, G., Chirkova, R.: Efficiently querying large XML data repositories: a survey. IEEE Trans. Knowl. Data Eng. **19**(10), 1381–1403 (2007)
3. Min, J.-K., Ahn, J.-Y., Chung, C.-W.: Efficient extraction of schemas for XML documents. Inf. Process. Lett. **85**, 7–12 (2003)
4. Bex, G.J., Neven, F., Schwentick, T., Vansummeren, S.: Inference of concise regular expressions and DTDs. ACM Trans. Database Syst. (TODS) **35**(2), 11:1–11:47 (2010)
5. Sarawagi, S.: Information extraction. Found. Trends Databases **1**(3), 261–344 (2008)
6. Raeymaekers, S., Bruynooghe, M., Bussche, J.: Learning (k, l)-contextual tree languages for information extraction. Mach. Learn. **71**(2–3), 155–183 (2008)
7. Staworko, S., Wieczorek, P.: Learning twig, path queries. In: ICDT (2012)
8. Carme, J., Gilleron, R., Lemay, A., Niehren, J.: Interactive learning of node selecting tree transducer. Mach. Learn. **66**(1), 33–67 (2007)
9. Bex, G.J., Neven, F., Vansummeren, S.: Learning deterministic regular expressions for the inference of schemas from xml data. ACM Trans. Web **4**(4), 1–32 (2010)
10. Liu, L., Pu, C., Han, W.: XWRAP: an XML-enabled wrapper construction system for web information sources. In: ICDE, pp. 611–621 (2000)
11. Suzuki, Y., Shoudai, T., Uchida, T., Miyahara, T.: Ordered term tree languages which are polynomial time inductively inferable from positive data. Theoret. Comput. Sci. **350**(1), 63–90 (2006)
12. Amer-Yahia, S., Cho, S., Lakshmanan, L.V.S., Srivastava, D.: Tree pattern query minimization. VLDB J. **11**(4), 315–331 (2002)
13. Miklau, G., Suciu, D.: Containment and equivalence for a fragment of XPath. J. ACM **51**(1), 2–45 (2004)
14. Benedikt, M., Koch, C.: XPath leashed. ACM Comput. Surv. (CSUR) **41**(1), 1–54 (2009)
15. Papadimitriou, C.H.: Computational Complexity. Addison-Wesley, Boston (1994)

Safe Sets in Graphs: Graph Classes and Structural Parameters

Raquel Águeda[1], Nathann Cohen[2], Shinya Fujita[3], Sylvain Legay[2], Yannis Manoussakis[2], Yasuko Matsui[4], Leandro Montero[2], Reza Naserasr[5], Yota Otachi[6(✉)], Tadashi Sakuma[7], Zsolt Tuza[8,9], and Renyu Xu[10]

[1] Universidad de Castilla-La Mancha, Ciudad Real, Spain
[2] LRI, University Paris-Sud, Orsay, France
[3] Yokohama City University, Yokohama, Japan
[4] Tokai University, Tokyo, Japan
[5] LIAFA, University Paris-Diderot, Paris, France
[6] Japan Advanced Institute of Science and Technology, Nomi, Japan
otachi@jaist.ac.jp
[7] Yamagata University, Yamagata, Japan
[8] MTA Rényi Institute, Budapest, Hungary
[9] University of Pannonia, Veszprém, Hungary
[10] Shandong University, Jinan, China

Abstract. A *safe set* of a graph $G = (V, E)$ is a non-empty subset S of V such that for every component A of $G[S]$ and every component B of $G[V \setminus S]$, we have $|A| \geq |B|$ whenever there exists an edge of G between A and B. In this paper, we show that a minimum safe set can be found in polynomial time for trees. We then further extend the result and present polynomial-time algorithms for graphs of bounded treewidth, and also for interval graphs. We also study the parameterized complexity of the problem. We show that the problem is fixed-parameter tractable when parameterized by the solution size. Furthermore, we show that this parameter lies between tree-depth and vertex cover number.

Keywords: Graph algorithm · Safe set · Treewidth · Interval graph · Fixed-parameter tractability

1 Introduction

In this paper, we only consider finite and simple graphs. The subgraph of a graph G induced by $S \subseteq V(G)$ is denoted by $G[S]$. A *component* of G is a connected induced subgraph of G with an inclusionwise maximal vertex set. For vertex-disjoint subgraphs A and B of G, if there is an edge between A and B, then A and B are *adjacent*.

In a graph $G = (V, E)$, a non-empty set $S \subseteq V$ of vertices is a *safe set* if, for every component A of $G[S]$ and every component B of $G[V \setminus S]$ adjacent to A, it holds that $|A| \geq |B|$. If a safe set induces a connected subgraph, then it is a *connected safe set*. The *safe number* $s(G)$ of G is the size of a minimum safe

© Springer International Publishing AG 2016
T.-H.H. Chan et al. (Eds.): COCOA 2016, LNCS 10043, pp. 241–253, 2016.
DOI: 10.1007/978-3-319-48749-6_18

set of G, and the *connected safe number* $\mathsf{cs}(G)$ of G is the size of a minimum connected safe set of G. It is known that $\mathsf{s}(G) \leq \mathsf{cs}(G) \leq 2 \cdot \mathsf{s}(G) - 1$ [10].

The concept of (connected) safe number was introduced by Fujita et al. [10]. Their motivation came from a variant of facility location problems, where the goal is to find a "safe" subset of nodes in a network to place facilities. They showed that the problems of finding a minimum safe set and a minimum connected safe set are NP-hard in general. They also showed that a minimum connected safe set in a tree can be found in linear time.

The main contribution of this paper is to give polynomial-time algorithms for finding a minimum safe set on trees, graphs of bounded treewidth, and interval graphs. We also show that the problems are fixed-parameter tractable when parameterized by the solution size.

The rest of the paper is organized as follows. In Sect. 2, we present an $O(n^5)$-time algorithm for finding a minimum safe set on trees. In Sect. 3, we generalize the algorithm to make it work on graphs of bounded treewidth. In Sect. 4, we show that the problem can be solved in $O(n^8)$ time for interval graphs. In Sect. 5, we show the fixed-parameter tractability of the problem when the parameter is the solution size. We also discuss the relationship of safe number to other important and well-studied graph parameters. In the final section, we conclude the paper with a few open problems.

2   Safe Sets in Trees

Recall that a tree is a connected graph with no cycles. In this section, we prove the following theorem.

Theorem 2.1. *For an n-vertex tree, a safe set of the minimum size can be found in time $O(n^5)$.*

We only show that the size of a minimum safe set can be computed in $O(n^5)$ time. It is straightforward to modify the dynamic program below for computing an actual safe set in the same running time.

In the following, we assume that a tree $T = (V, E)$ has a root and that the children of each vertex are ordered. For a vertex $u \in V$, we denote the set of children of u by $C_T(u)$. By V_u we denote the vertex set that consists of u and its descendants. We define some subtrees induced by special sets of vertices as follows (see Fig. 1):

- For a vertex $u \in V$, let $T(u) = T[V_u]$.
- For an edge $\{u, v\} \in E$ where v is the parent of u, let $T(u \rightarrow v) = T[\{v\} \cup V_u]$.
- For $u \in V$ with children $w_1, \ldots, w_d$, let $T(u, i) = T\left[\{u\} \cup \bigcup_{1 \leq j \leq i} V_{w_j}\right]$.

Note that $T(u, 1) = T(w_1 \rightarrow u)$ if w_1 is the first child of u, $T(u) = T(u, |C_T(u)|)$ if u is not a leaf, and $T = T(\rho)$ if ρ is the root of T.

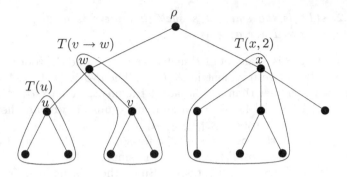

Fig. 1. Subtrees $T(u)$, $T(v \rightarrow w)$, and $T(x, 2)$.

Fragments: For a subtree T' of T and $S \subseteq V(T')$, a *fragment in T' with respect to S* is the vertex set of a component in $T'[S]$ or $T'[V(T') \setminus S]$. We denote the set of fragments in T' with respect to S by $\mathcal{F}(T', S)$. The fragment that contains the root of T' is *active*, and the other fragments are *inactive*. Two fragments in $\mathcal{F}(T', S)$ are *adjacent* if there is an edge of T' between them. A fragment $F \in \mathcal{F}(T', S)$ is *bad* if it is inactive, $F \subseteq S$, and there is another inactive fragment $F' \in \mathcal{F}(T', S)$ adjacent to F with $|F| < |F'|$.

$(T', \mathsf{b}, s, a) - feasiblesets$: For $\mathsf{b} \in \{\mathsf{t}, \mathsf{f}\}$, $s \in \{0, \dots, n\}$, and $a \in \{1, \dots, n\}$, we say $S \subseteq V(T')$ is (T', b, s, a)-*feasible* if $|S| = s$, the size of the active fragment in $\mathcal{F}(T', S)$ is a, there is no bad fragment in $\mathcal{F}(T', S)$, and $\mathsf{b} = \mathsf{t}$ if and only if the root of T' is in S.

Intuitively, a (T', b, s, a)-feasible set S is "almost safe." If A is the active fragment in $\mathcal{F}(T', S)$, then $S \setminus A$ is a safe set of $T'[V(T') \setminus A]$.

For $S \subseteq V(T')$, we set $\partial_{T'}^{\max}(S)$ and $\partial_{T'}^{\min}(S)$ to be the sizes of maximum and minimum fragments, respectively, adjacent to the active fragment in $\mathcal{F}(T', S)$. If there is no adjacent fragment, then we set $\partial_{T'}^{\max}(S) = -\infty$ and $\partial_{T'}^{\min}(S) = +\infty$.

DP Table: We construct a table with values $\mathsf{ps}(T', \mathsf{b}, s, a) \in \{0, \dots, n\} \cup \{+\infty, -\infty\}$ for storing information of partial solutions, where $\mathsf{b} \in \{\mathsf{t}, \mathsf{f}\}$, $s \in \{0, \dots, n\}$, and $a \in \{1, \dots, n\}$, and T' is a subtree of T such that either $T' = T(u)$ for some $u \in V$, $T' = T(u \rightarrow v)$ for some $\{u, v\} \in E$, or $T' = T(u, i)$ for some $u \in V$ and $1 \leq i \leq |C_T(u)|$. The table entries will have the following values:

$$\mathsf{ps}(T', \mathsf{t}, s, a) = \begin{cases} +\infty & \text{if no } (T', \mathsf{t}, s, a)\text{-feasible set exists,} \\ \min_{(T', \mathsf{t}, s, a)\text{-feasible } S} \partial_{T'}^{\max}(S) & \text{otherwise,} \end{cases}$$

$$\mathsf{ps}(T', \mathsf{f}, s, a) = \begin{cases} -\infty & \text{if no } (T', \mathsf{f}, s, a)\text{-feasible set exists,} \\ \max_{(T', \mathsf{f}, s, a)\text{-feasible } S} \partial_{T'}^{\min}(S) & \text{otherwise.} \end{cases}$$

The definition of the table ps implies the following fact.

Lemma 2.2. $s(T)$ *is the smallest* s *such that there is* $a \in \{1, \ldots, n\}$ *with* $\mathsf{ps}(T, \mathsf{t}, s, a) \le a$ *or* $\mathsf{ps}(T, \mathsf{f}, s, a) \ge a$.

Proof. Assume that S is a safe set of T such that $|S| = s$ and the root is contained in S. Let A be the active fragment in $\mathcal{F}(T, S)$. Then, S is $(T, \mathsf{t}, s, |A|)$-feasible. Since A cannot be smaller than any adjacent fragment, we have $\partial_T^{\max}(S) \le |A|$. Hence $\mathsf{ps}(T, \mathsf{t}, s, |A|) \le |A|$ holds. By a similar argument, we can show that if the root is not in S, then $\mathsf{ps}(T, \mathsf{f}, s, |A|) \ge |A|$.

Conversely, assume that $\mathsf{ps}(T, \mathsf{t}, s, a) \le a$ for some $a \in \{1, \ldots, n\}$. (The proof for the other case, where $\mathsf{ps}(T, \mathsf{f}, s, a) \ge a$, is similar.) Let S be a (T, t, s, a)-feasible set with $\partial_T^{\max}(S) = \mathsf{ps}(T, \mathsf{t}, s, a)$. Since there is no bad fragment in $\mathcal{F}(T, S)$ and the active fragment (of size a) is not smaller than the adjacent fragments (of size at most $\partial_T^{\max}(S) = \mathsf{ps}(T, \mathsf{t}, s, a) \le a$), all fragments included in S are not smaller than their adjacent fragments. This implies that S is a safe set of size s. □

By Lemma 2.2, after computing all entries $\mathsf{ps}(T', \mathsf{b}, s, a)$, we can compute $s(T)$ in time $O(n^2)$. There are $O(n^3)$ tuples (T', b, s, a), and thus to prove the theorem, it suffices to show that each entry $\mathsf{ps}(T', \mathsf{b}, s, a)$ can be computed in time $O(n^2)$ assuming that the entries for all subtrees of T' are already computed.

We compute all entries $\mathsf{ps}(T', \mathsf{b}, s, a)$ in a bottom-up manner: We first compute the entries for $T(u)$ for each leaf u. We then repeat the following steps until none of them can be applied. (1) For each u such that the entries for $T(u)$ are already computed, we compute the entries for $T(u \to v)$, where v is the parent of u. (2) For each u such that the entries for $T(u, i-1)$ and $T(w_i \to u)$ are already computed, where w_i is the ith child of u, we compute the entries for $T(u, i)$.

Lemma 2.3. *For a leaf u of T, each table entry $\mathsf{ps}(T(u), \mathsf{b}, s, a)$ can be computed in constant time.*

Proof. The set $\{u\}$ is the unique $(T(u), \mathsf{t}, 1, 1)$-feasible set. Since $\mathcal{F}(T(u), \{u\})$ contains no inactive fragment, we set $\mathsf{ps}(T(u), \mathsf{t}, 1, 1) = -\infty$. Similarly the empty set is the unique $(T(u), \mathsf{f}, 0, 1)$-feasible set. We set $\mathsf{ps}(T(u), \mathsf{f}, 0, 1) = +\infty$. For the other tuples, there are no feasible sets. We set the values accordingly for them. Clearly, each entry can be computed in constant time. □

Lemma 2.4. *For a vertex u and its parent v in T, each table entry $\mathsf{ps}(T(u \to v), \mathsf{b}, s, a)$ can be computed in $O(n)$ time, using the table entries for the subtree $T(u)$.*

Proof. We separate the proof into two cases: $a \ge 2$ and $a = 1$. If $a \ge 2$, then we can compute the table entry in constant time. If $a = 1$, we need $O(n)$ time.

Case 1: $a \ge 2$. In this case, for every $(T(u \to v), \mathsf{b}, s, a)$-feasible set S, u and v are in the active fragment of $\mathcal{F}(T(u \to v), S)$ since the root v of $T(u \to v)$ has the unique neighbor u.

Case 1-1: $\mathsf{b} = \mathsf{t}$. Let S be a $(T(u \to v), \mathsf{t}, s, a)$-feasible set that minimizes $\partial_{T(u \to v)}^{\max}(S)$. Observe that $S \setminus \{v\}$ is $(T(u), \mathsf{t}, s-1, a-1)$-feasible and that

$\partial_{T(u \to v)}^{\max}(S) = \partial_{T(u)}^{\max}(S \setminus \{v\})$. We claim that $\partial_{T(u)}^{\max}(S \setminus \{v\}) = \mathsf{ps}(T(u), \mathsf{t}, s - 1, a - 1)$, and thus

$$\mathsf{ps}(T(u \to v), \mathsf{t}, s, a) = \mathsf{ps}(T(u), \mathsf{t}, s - 1, a - 1).$$

Suppose that some $(T(u), \mathsf{t}, s - 1, a - 1)$-feasible set Q satisfies $\partial_{T(u)}^{\max}(Q) < \partial_{T(u)}^{\max}(S \setminus \{v\})$. Now $Q \cup \{v\}$ is $(T(u \to v), \mathsf{t}, s, a)$-feasible. However, it holds that

$$\partial_{T(u \to v)}^{\max}(Q \cup \{v\}) = \partial_{T(u)}^{\max}(Q) < \partial_{T(u)}^{\max}(S \setminus \{v\}) = \partial_{T(u \to v)}^{\max}(S).$$

This contradicts the optimality of S.

Case 1-2: b = f. Let S be a $(T(u \to v), \mathsf{f}, s, a)$-feasible set that maximizes $\partial_{T(u \to v)}^{\min}(S)$. The set S is also $(T(u), \mathsf{f}, s, a-1)$-feasible and satisfies $\partial_{T(u \to v)}^{\min}(S) = \partial_{T(u)}^{\min}(S)$. We claim that $\partial_{T(u)}^{\min}(S) = \mathsf{ps}(T(u), \mathsf{t}, s, a - 1)$, and thus

$$\mathsf{ps}(T(u \to v), \mathsf{f}, s, a) = \mathsf{ps}(T(u), \mathsf{f}, s, a - 1).$$

Suppose that there is a $(T(u), \mathsf{t}, s, a-1)$-feasible set Q with $\partial_{T(u)}^{\min}(Q) > \partial_{T(u)}^{\min}(S)$. Since Q is also $(T(u \to v), \mathsf{f}, s, a)$-feasible, it holds that

$$\partial_{T(u \to v)}^{\min}(Q) = \partial_{T(u)}^{\min}(Q) > \partial_{T(u)}^{\min}(S) = \partial_{T(u \to v)}^{\min}(S).$$

This contradicts the optimality of S.

Case 2: a = 1. For every $(T(u \to v), \mathsf{b}, s, 1)$-feasible set S, the set $\{v\}$ is the active fragment, and the vertex u is in the unique fragment adjacent to the active fragment.

Case 2-1: b = t. Let S be a $(T(u \to v), \mathsf{t}, s, 1)$-feasible set. Then $S \setminus \{v\}$ is a $(T(u), \mathsf{f}, s - 1, a')$-feasible set for some a'. Moreover, since $\mathcal{F}(T(u \to v), S)$ does not contain any bad fragment, $\partial_{T(u)}^{\min}(S \setminus \{v\}) \geq a'$. Thus we can set $\mathsf{ps}(T(u \to v), \mathsf{t}, s, 1)$ as follows:

$$\mathsf{ps}(T(u \to v), \mathsf{t}, s, 1) = \begin{cases} \min\{a' \mid \mathsf{ps}(T(u), \mathsf{f}, s - 1, a') \geq a'\} & \text{if such } a' \text{ exists,} \\ +\infty & \text{otherwise.} \end{cases}$$

Case 2-2: b = f. Let S be a $(T(u \to v), \mathsf{f}, s, 1)$-feasible set. The set S is a $(T(u), \mathsf{t}, s, a')$-feasible set for some a'. Since $\mathcal{F}(T(u \to v), S)$ does not contain any bad fragment, $\partial_{T(u)}^{\max}(S) \leq a'$. Thus we can set $\mathsf{ps}(T(u \to v), \mathsf{f}, s, 1)$ as follows:

$$\mathsf{ps}(T(u \to v), \mathsf{f}, s, 1) = \begin{cases} \max\{a' \mid \mathsf{ps}(T(u), \mathsf{t}, s, a') \leq a'\} & \text{if there is such an } a', \\ -\infty & \text{otherwise.} \end{cases}$$

In both Cases 2-1 and 2-2, we can compute the entry $\mathsf{ps}(T(u \to v), \mathsf{b}, s, 1)$ in $O(n)$ time by looking up at most n table entries for the subtree $T(u)$. □

Lemma 2.5. *For a non-leaf vertex u with the children $w_1, \ldots, w_d$ and an integer i with $2 \leq i \leq d$, each table entry $\mathsf{ps}(T(u,i), \mathsf{b}, s, a)$ can be computed in $O(n^2)$ time, using the table entries for the subtrees $T(u, i-1)$ and $T(w_i \to u)$.*

Proof. For the sake of simplicity, let $T_1 = T(u, i-1)$ and $T_2 = T(w_i \to u)$. Let S be a $(T(u,i), \mathsf{b}, s, a)$-feasible set and A be the active fragment in $\mathcal{F}(T(u,i), S)$. For $j \in \{1, 2\}$, let $S_j = S \cap V(T_j)$ and $A_j = A \cap V(T_j)$. Observe that S_j is a $(T_j, \mathsf{b}, |S_j|, |A_j|)$-feasible set. If $\mathsf{b} = \mathsf{t}$, then $S_1 \cap S_2 = \{u\}$; otherwise $S_1 \cap S_2 = \emptyset$. Thus $|S_1| + |S_2| = |S| + 1$ if $\mathsf{b} = \mathsf{t}$, and $|S_1| + |S_2| = |S|$ otherwise. Similarly, since $A_1 \cap A_2 = \{u\}$, it holds that $|A_1| + |A_2| = |A| + 1$. Therefore, we can set the table entries as follows:

$$\mathsf{ps}(T(u,i), \mathsf{t}, s, a) = \min_{\substack{s_1 + s_2 = s+1 \\ a_1 + a_2 = a+1}} \max\{\mathsf{ps}(T_1, \mathsf{t}, s_1, a_1), \mathsf{ps}(T_2, \mathsf{t}, s_2, a_2)\},$$

$$\mathsf{ps}(T(u,i), \mathsf{f}, s, a) = \max_{\substack{s_1 + s_2 = s \\ a_1 + a_2 = a+1}} \min\{\mathsf{ps}(T_1, \mathsf{f}, s_1, a_1), \mathsf{ps}(T_2, \mathsf{f}, s_2, a_2)\}.$$

In both cases, we can compute the entry $\mathsf{ps}(T(u,i), \mathsf{b}, s, a)$ in $O(n^2)$ time since there are $O(n)$ possibilities for each (s_1, s_2) and (a_1, a_2). □

A graph is *unicyclic* if it can be obtained by adding an edge to a tree. Using the algorithm for weighted paths presented in [2] as a subroutine, we can extend the algorithm in this section to find a minimum safe set and a minimum connected safe set of a unicyclic graph in the same running time.

3   Safe Sets in Graphs of Bounded Treewidth

In this section, we show that for any fixed constant k, a minimum safe set and a minimum connected safe set of a graph of treewidth at most k can be found in $O(n^{5k+8})$ time.

Basically, the algorithm in this section is a generalization of the one in the previous section. The most crucial difference is that here we may have many active fragments, and each active fragment may have many vertices adjacent to the "outside." This makes the algorithm much more complicated and slow.

A *tree decomposition* of a graph $G = (V, E)$ is a pair $(\{X_p \mid p \in I\}, T)$ such that each X_p, called a *bag*, is a subset of V, and T is a tree with $V(T) = I$ such that

– for each $v \in V$, there is $p \in I$ with $v \in X_p$;
– for each $\{u, v\} \in E$, there is $p \in I$ with $u, v \in X_p$;
– for $p, q, r \in I$, if q is on the p–r path in T, then $X_p \cap X_r \subseteq X_q$.

The *width* of a tree decomposition is the size of a largest bag minus 1. The *treewidth* of a graph, denoted by $\mathsf{tw}(G)$, is the minimum width over all tree decompositions of G.

A tree decomposition $(\{X_p \mid p \in I\}, T)$ is *nice* if

- T is a rooted tree in which every node has at most two children;
- if a node p has two children q, r, then $X_p = X_q = X_r$ (such a node p is a *join node*);
- if a node p has only one child q, then either
  - $X_p = X_q \cup \{v\}$ for some $v \notin X_q$ (p is a *introduce node*), or
  - $X_p = X_q \setminus \{v\}$ for some $v \in X_q$ (q is a *forget node*);
- if a node p is a leaf, then $X_p = \{v\}$ for some $v \in V$.

Theorem 3.1. *Let k be a fixed constant. For an n-vertex graph of treewidth at most k, a (connected) safe set of minimum size can be found in time $O(n^{5k+8})$.*

Proof. We only show that $\mathsf{s}(G)$ and $\mathsf{cs}(G)$ can be computed in the claimed running time. It is straightforward to modify the dynamic program below for computing an actual set in the same running time.

Let $G = (V, E)$ be a graph of treewidth at most k. We compute a nice tree decomposition $(\{X_p \mid p \in I\}, T)$ with at most $4n$ nodes. It can be done in $O(n)$ time [4,12]. For each $p \in I$, let $V_p = X_p \cup \bigcup_q X_q$, where q runs through all descendants of p in T.

Fragments: For a node p and a vertex set $S \subseteq V_p$, a *fragment* is a component in $G[S]$ or $G[V_p \setminus S]$. We denote the set of fragments with respect to p and S by $\mathcal{F}(p, S)$. A fragment $F \in \mathcal{F}(p, S)$ is *active* if $F \cap X_p \neq \emptyset$, and it is *inactive* otherwise. Two fragments in $\mathcal{F}(p, S)$ are *adjacent* if there is an edge of $G[V_p]$ between them. A fragment F is *bad* if it is inactive, $F \subseteq S$, and there is another inactive fragment F' adjacent to F with $|F| < |F'|$.

DP Table: For storing information of partial solutions, we construct a table with values $\mathsf{ps}(p, s, \mathcal{A}, \beta, \gamma, \phi, \psi) \in \{\mathsf{t}, \mathsf{f}\}$ with indices $p \in I$, $s \in \{0, \dots, n\}$, a partition $\mathcal{A}$ of X_p, $\beta \colon \mathcal{A} \to \{1, \dots, n\}$, $\gamma \colon \mathcal{A} \to \{1, \dots, n\} \cup \{\pm\infty\}$, $\phi \colon \mathcal{A} \to \{\mathsf{t}, \mathsf{f}\}$, and $\psi \colon \binom{\mathcal{A}}{2} \to \{\mathsf{t}, \mathsf{f}\}$. We set

$$\mathsf{ps}(p, s, \mathcal{A}, \beta, \gamma, \phi, \psi) = \mathsf{t}$$

if and only if there exists a set $S \subseteq V_p$ of size s with the following conditions:

- there is no bad fragment in $\mathcal{F}(p, S)$,
- for each active fragment F in $\mathcal{F}(p, S)$,
  - there is a unique element $A_F \in \mathcal{A}$ such that $A_F = F \cap X_p$,
  - $\beta(A_F) = |F|$,
  - $\phi(A_F) = \mathsf{t}$ if and only if $F \subseteq S$,
  - if $F \subseteq S$, then $\gamma(A_F)$ is the size of a *maximum* inactive fragment adjacent to F (if no such fragment exists, we set $\gamma(A_i) = -\infty$),
  - if $F \nsubseteq S$, then $\gamma(A_F)$ is the size of a *minimum* inactive fragment adjacent to F if no such fragment exists, we set $\gamma(A_i) = +\infty$),
- for two active fragments F, F' in $\mathcal{F}(p, S)$, $\psi(\{A_F, A_{F'}\}) = \mathsf{t}$ if and only if F and F' are adjacent, where $A_F = F \cap X_p$ and $A_{F'} = F' \cap X_p$.[1]

---

[1] In the following, we (ab)use simpler notation $\psi(A_F, A_{F'})$ instead of $\psi(\{A_F, A_{F'}\})$.

Let ρ be the root of T. The definition of the table ps implies the following fact.

Observation 3.2. $\mathsf{s}(G)$ *is the smallest s with* $\mathsf{ps}(\rho, s, \mathcal{A}, \beta, \gamma, \phi, \psi) = \mathsf{t}$ *for some* $\mathcal{A}, \beta, \gamma, \phi,$ *and ψ such that $\beta(A) \geq \gamma(A)$ for each $A \in \mathcal{A}$ with $\phi(A) = \mathsf{t}$, $\beta(A) \leq \gamma(A)$ for each $A \in \mathcal{A}$ with $\phi(A) = \mathsf{f}$, and $\beta(A) \geq \beta(A')$ for any $A, A' \in \mathcal{A}$ with $\phi(A) = \mathsf{t}$ and $\psi(A, A') = \mathsf{t}$.*

For computing $\mathsf{cs}(G)$, we need to compute additional information for each tuple $(p, s, \mathcal{A}, \beta, \gamma, \phi, \psi)$. For $A \in \mathcal{A}$, let $\beta'(A)$ be the size of the fragment in $\mathcal{F}(\rho, S)$ that is a superset of the fragment $F_A \supseteq A$ in $\mathcal{F}(p, S)$. If $A \subseteq S$, then $\beta'(A) = \beta(A)$; otherwise $\beta'(A) = |C_A \setminus X_p| + \sum_{A' \in \mathcal{A}, \, A' \subseteq C_A} \beta(A')$, where C_A is the component in $G[(V \setminus V_p) \cup (X_p \setminus S)]$ that includes A. We can compute $\beta'(A)$ for all $A \in \mathcal{A}$ in time $O(n)$ by running a breadth-first search from $X_p \setminus S$.

Observation 3.3. $\mathsf{cs}(G)$ *is the smallest s with* $\mathsf{ps}(p, s, \mathcal{A}, \beta, \gamma, \phi, \psi) = \mathsf{t}$ *for some* $p, \mathcal{A}, \beta, \gamma, \phi,$ *and ψ such that $\beta(A) \geq \gamma(A)$ for each $A \in \mathcal{A}$ with $\phi(A) = \mathsf{t}$, $\beta(A) \leq \gamma(A)$ for each $A \in \mathcal{A}$ with $\phi(A) = \mathsf{f}$, and $\beta(A) \geq \beta'(A')$ for any $A, A' \in \mathcal{A}$ with $\phi(A) = \mathsf{t}$ and $\psi(A, A') = \mathsf{t}$.*

By Observations 3.2 and 3.3, provided that all entries $\mathsf{ps}(p, s, \mathcal{A}, \beta, \gamma, \phi, \psi)$ are computed in advance, we can compute $\mathsf{s}(G)$ and $\mathsf{cs}(G)$ by spending time $O(1)$ and $O(n)$, respectively, for each tuple. We compute all entries $\mathsf{ps}(p, s, \mathcal{A}, \beta, \gamma, \phi, \psi)$ by a bottom-up dynamic program. Due to the space limitation, we omit this part in the conference version.   □
For a vertex-weighted graph $G = (V, E)$ with a weight function $w \colon V \to \mathbb{Z}^+$, a set $S \subseteq V$ is a *weighted safe set* of weight $\sum_{s \in S} w(s)$ if for each component C of $G[S]$ and each component D of $G[V \setminus S]$ with an edge between C and D, it holds that $w(C) \geq w(D)$. Bapat et al. [2] show that finding a minimum (connected) weighted safe set is weakly NP-hard even for stars. Let $W = \sum_{v \in V} w(v)$. Our dynamic program above works for the weighted version if we extend the ranges of parameters s, β, and γ by including $\{1, \ldots, W\}$. The running time becomes polynomial in W.

Theorem 3.4. *For a vertex weighted graph of bounded treewidth, a weighted (connected) safe set of the minimum weight can be found in pseudo-polynomial time.*

4   Safe Sets in Interval Graphs

In this section, we present a polynomial-time algorithm for finding a minimum safe set and a minimum connected safe set in an interval graph.

A graph is an *interval graph* if it can be represented as the intersection graph of intervals on a line. Given a graph, one can determine in linear time whether the graph is an interval graph, and if so, find a corresponding interval representation in the same running time [6].

Theorem 4.1. *For an n-vertex interval graph, a minimum safe set and a minimum connected safe set can be found in time $O(n^8)$.*

Proof. Let G be a given interval graph. As we can deal with each component of G separately, we assume that G is connected. The algorithm is a dynamic programming on an interval representation of G. We assume that its vertices (i.e. intervals) $v_1, \ldots, v_n$ are ordered increasingly according to their left ends, and write $X_i = \{v_1, \ldots, v_i\}$.

At each step i of the algorithm, we want to store all subsets $S \subseteq X_i$ which can potentially be completed (with vertices from $G \setminus X_i$) into a safe set. The number of such sets can be exponential: we thus define a notion of *signature*, and store the signatures of the sets instead of storing the sets themselves. The cost of this storage is bounded by the number of possible signatures, which is polynomial in n.

We will then prove that all possible signatures of sets at step i can be deduced from the set of signatures at step $i-1$. The cardinality of a minimum safe set (and a minimum connected safe set) can finally be deduced from the set of signatures stored during the last step. We can easily modify the algorithm so that it also outputs a minimum set.

We define the *signature of S at step i* as the 8-tuple that consists of the following items (see Fig. 2):

1. The size of S.
2. The vertex v_r^S of S with the most neighbors in $G \setminus X_i$.
3. The vertex $v_r^{\bar{S}}$ of $\bar{S} := X_i \setminus S$ with the most neighbors in $G \setminus X_i$.
4. The size of S_r (the rightmost component of S).
5. The size of $\bar{S}_r$ (the rightmost component of $\bar{S}$).
6. The largest size of a component of $\bar{S} \setminus \bar{S}_r$ adjacent with S_r.
7. The smallest size of a component of $S \setminus S_r$ adjacent with $\bar{S}_r$.
8. A boolean value indicating whether S is connected.

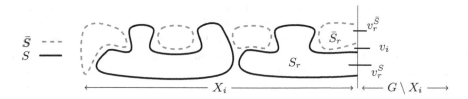

Fig. 2. The dynamic programming on an interval graph.

Assuming that we know the signature of a set S at step i, we show how to obtain the signature at step $i + 1$ of (a) $S' = S$ and (b) $S' = S \cup \{v_{i+1}\}$. With this procedure, all signatures of step $i + 1$ can be obtained from all signatures at step i.

1. The size of S' (at step i: $|S|$).
   (a) $|S|$.
   (b) $|S| + 1$.

2. The vertex of S' with the most neighbors in $G \setminus X_{i+1}$ (at step i: v_r^S).
   (a) v_r^S.
   (b) The one of v_r^S and v_{i+1} which has the most neighbors in $G \setminus X_{i+1}$.
3. The vertex of $\bar{S}' := X_{i+1} \setminus S'$ with the most neighbors in $G \setminus X_{i+1}$ (at step i: $v_r^{\bar{S}}$).
   (a) The one of $v_r^{\bar{S}}$ and v_{i+1} which has the most neighbors in $G \setminus X_{i+1}$.
   (b) $v_r^{\bar{S}}$.
4. The size of the rightmost component S_r' of S' (at step i: $|S_r|$).
   (a) $|S_r|$.
   (b) $|S_r| + 1$ if v_{i+1} and v_r^S are adjacent, and 1 otherwise (new component). In the latter case, we discard the signature if $|S_r|$ is strictly smaller than the largest size of a component of $\bar{S} \setminus \bar{S}_r$ adjacent with S_r at step i.
5. The size of the rightmost component $\bar{S}_r'$ of $\bar{S}'$ (at step i: $|\bar{S}_r|$).
   (a) 1 if v_{i+1} and $v_r^{\bar{S}}$ are not adjacent (new component), and $|\bar{S}_r|+1$ otherwise. In the latter case, we discard the signature if $|\bar{S}_r|$ is strictly larger than the smallest size of a component of $S \setminus S_r$ adjacent with $\bar{S}_r$ at step i.
   (b) $|\bar{S}_r|$.
6. The largest size of a component of $\bar{S}' \setminus \bar{S}_r'$ adjacent with S_r' (at step i: c).
   (a) c if no new component of $\bar{S}'$ was created (see 5.), and $\max\{c, |\bar{S}_r'|\}$ otherwise.
   (b) c if no new component of S' was created (see 4.), and $-\infty$ otherwise.
7. The smallest size of a component of $S' \setminus S_r'$ adjacent with $\bar{S}_r'$ (at step i: c).
   (a) c if no new component of $\bar{S}'$ was created (see 5.), and $+\infty$ otherwise.
   (b) c if no new component of S' was created (see 4.), and $\min\{c, |S_r'|\}$ otherwise.
8. A boolean variable indicating whether S' is connected (at step i: b).
   (a) b
   (b) t if $|S| = 0$, b if v_{i+1} and v_r^S are adjacent, and f otherwise.

When all signatures at step n have been computed, we use the additional information that S and $\bar{S}$ cannot be further extended to discard the remaining signatures corresponding to non-safe sets. That is, we discard a signature if $|S_r| < |\bar{S}_r|$, or $|S_r|$ is strictly smaller than the largest size of a component of $\bar{S} \setminus \bar{S}_r$ adjacent to it, or $|\bar{S}_r| + 1$ is strictly larger than the smallest size of a component of $S \setminus S_r$ adjacent to it.

The minimum sizes of a safe set and a connected safe set can be obtained from the remaining signatures. For each step i, there are $O(n^7)$ signatures. From a signature for step i, we can compute the corresponding signature for step $i+1$ in $O(1)$ time. Therefore, the total running time is $O(n^8)$. □

5   Fixed-Parameter Tractability

In this section, we show that the problems of finding a safe set and a connected safe set of size at most s is fixed-parameter tractable when the solution size s is the parameter. For the standard concepts in parameterized complexity, see the recent textbook [8].

We first show that graphs with small safe sets have small treewidth. We then show that for any fixed constants s the property of having a (connected) safe set of size at most s can be expressed in the monadic second-order logic on graphs. Then we use the well-known theorems by Bodlaender [4] and Courcelle [7] to obtain an FPT algorithm that depends only linearly on the input size.

Lemma 5.1. *Let $G = (V, E)$ be a connected graph. If $\mathsf{tw}(G) \geq s^2 - 1$, then $\mathsf{s}(G) \geq s$.*

Proof. It is known that every graph G has a path of $\mathsf{tw}(G) + 1$ vertices as a subgraph [3]. Thus $\mathsf{tw}(G) \geq s^2 - 1$ implies that G has a path of s^2 vertices as a subgraph.

Let P be a path of s^2 vertices in G, and let $S \subseteq V$ be an arbitrary set of size less than s. By the pigeon-hole principle, there is a subpath Q of P such that $|Q| \geq s$ and $S \cap V(Q) = \emptyset$. Hence there is a component B of $G[V \setminus S]$ with $V(Q) \subseteq B$. Since G is connected there is a component A of $G[S]$ adjacent to B. Now we have $|A| \leq |S| < s \leq |Q| \leq |B|$, which implies that S is not a safe set.   □

The syntax of the *monadic second-order logic of graphs* (MS_2) includes (i) the logical connectives $\vee, \wedge, \neg, \Leftrightarrow, \Rightarrow$, (ii) variables for vertices, edges, vertex sets, and edge sets, (iii) the quantifiers $\forall$ and $\exists$ applicable to these variables, and (iv) the following binary relations:

- $v \in U$ for a vertex variable v and a vertex set variable U;
- $e \in F$ for an edge variable e and an edge set variable F;
- $\mathsf{inc}(e, v)$ for an edge variable e and a vertex variable v, where the interpretation is that e is incident with v;
- equality of variables.

Now we can show the following.

Lemma 5.2. *For a fixed constant s, the property of having a safe set of size at most s can be expressed in MS_2.*

Corollary 5.3. *For a fixed constant s, the property of having a connected safe set of size at most s can be expressed in MS_2.*   □

Theorem 5.4. *The problems of finding a safe set and a connected safe set of size at most s is fixed-parameter tractable when the solution size s is the parameter. Furthermore, the running time depends only linearly on the input size.*

Proof. Let G be a given graph. Since we can handle the components separately, we assume that G is connected. We first check whether $\mathsf{tw}(G) < (s+1)^2 - 1$ in $O(n)$ time by Bodlaender's algorithm [4]. If not, Lemma 5.1 implies that $\mathsf{s}(G) \geq s+1$. Otherwise, Bodlaender's algorithm gives us a tree decomposition of G with width less than $(s+1)^2 - 1$. Courcelle's theorem [7] says that it can be checked in linear time whether a graph satisfies a fixed MS_2 formula if the graph is given with a tree decomposition of constant width (see also [1]). Therefore, Lemma 5.2 and Corollary 5.3 imply the theorem.   □

5.1 Relationship to Other Structural Graph Parameters

As we showed in Lemma 5.1, the treewidth of a graph is bounded by a constant if it has constant safe number. Here we further discuss the relationship to other well-studied graph parameters: tree-depth and vertex cover number. As bounding these parameters is more restricted than bounding treewidth, more problems can be solved efficiently when the problems are parameterized by tree-depth or vertex cover number (see [9,11]). In the following, we show that safe number *lies between* these two parameters. This implies that parameterizing a problem by safe number may give a finer understanding of the parameterized complexity of the problem.

Tree-Depth. The *tree-depth* [13] (also known as *elimination tree height* [14] and *vertex ranking number* [5]) of a connected graph G is the minimum depth of a rooted tree T such that T^* contains G as a subgraph, where T^* is the supergraph of T with the additional edges connecting all comparable pairs in T. We can easily see that the tree-depth of a graph is at least its treewidth. It is known that a graph has constant tree-depth if and only if it has a constant upper bound on the length of paths in it [13]. Hence the proof of Lemma 5.1 implies the following relation.

Lemma 5.5. *The tree-depth of a connected graph is bounded by a constant if it has constant safe number.*

The converse of the statement above is not true in general. The complete k-ary tree of depth 2 has tree-depth 2 and safe number k.

Vertex Cover Number. A set $C \subseteq V(G)$ is a *vertex cover* of a graph G if each edge in G has at least one end in C. The *vertex cover number* of a graph is the size of a minimum vertex cover in the graph. We can see that C is a vertex cover if and only if each component of $G \setminus C$ has size 1. Thus a vertex cover is a safe set, and the following relation follows.

Lemma 5.6. *The safe number of a graph is at most its vertex cover number.*

Again the converse is not true. Consider the graph obtained from the star graph $K_{1,k}$ by subdividing each edge. It has a (connected) safe set of size 2, while its vertex cover number is k.

Note that Lemma 5.6 and Theorem 5.4 together imply that the problem of finding a (connected) safe set is fixed-parameter tractable when parameterized by vertex cover number.

6 Concluding Remarks

A graph is *chordal* if it has no induced cycle of length 4 or more. Trees and interval graphs form the most well-known subclasses of the class of chordal graphs. A natural question would be the complexity of the problems on chordal graphs.

Another question is about planar graphs. As the original motivation of the problem was from a facility location problem, it would be natural and important to study the problem on planar graphs.[2]

Our algorithm for graphs of treewidth at most k runs in $n^{O(k)}$ time. Such an algorithm is called an XP algorithm, and an FPT algorithm with running time $f(k) \cdot n^c$ is more preferable, where f is an arbitrary computable function and c is a fixed constant. It would be interesting if one can show that such an algorithm exists (or does not exist under some complexity assumption).

References

1. Arnborg, S., Lagergren, J., Seese, D.: Easy problems for tree-decomposable graphs. J. Algorithms **12**, 308–340 (1991)
2. Bapat, R.B., Fujita, S., Legay, S., Manoussakis, Y., Matsui, Y., Sakuma, T., Tuza, Z.: Network majority on tree topological network (2016). http://www2u.biglobe. ne.jp/~sfujita/fullpaper.pdf
3. Bodlaender, H.L.: On linear time minor tests with depth-first search. J. Algorithms **14**, 1–23 (1993)
4. Bodlaender, H.L.: A linear-time algorithm for finding tree-decompositions of small treewidth. SIAM J. Comput. **25**, 1305–1317 (1996)
5. Bodlaender, H.L., Deogun, J.S., Jansen, K., Kloks, T., Kratsch, D., Müller, H., Tuza, Z.: Rankings of graphs. SIAM J. Discrete Math. **11**, 168–181 (1998)
6. Booth, K.S., Lueker, G.S.: Testing for the consecutive ones property, interval graphs, and graph planarity using PQ-tree algorithms. J. Comput. Syst. Sci. **13**, 335–379 (1976)
7. Courcelle, B.: The monadic second-order logic of graphs III: tree-decompositions, minor and complexity issues. Theor. Inform. Appl. **26**, 257–286 (1992)
8. Cygan, M., Fomin, F.V., Kowalik, L., Lokshtanov, D., Marx, D., Pilipczuk, M., Pilipczuk, M., Saurabh, S.: Parameterized Algorithms. Springer, Heidelberg (2015)
9. Fellows, M.R., Lokshtanov, D., Misra, N., Rosamond, F.A., Saurabh, S.: Graph layout problems parameterized by vertex cover. In: Hong, S.-H., Nagamochi, H., Fukunaga, T. (eds.) ISAAC 2008. LNCS, vol. 5369, pp. 294–305. Springer, Heidelberg (2008)
10. Fujita, S., MacGillivray, G., Sakuma, T.: Safe set problem on graphs. Discrete Appl. Math. **215**, 106–111 (2016)
11. Gutin, G., Jones, M., Wahlström, M.: Structural parameterizations of the mixed chinese postman problem. In: Bansal, N., Finocchi, I. (eds.) ESA 2015. LNCS, vol. 9294, pp. 668–679. Springer, Heidelberg (2015)
12. Kloks, T. (ed.): Treewidth: Computations and Approximations. LNCS, vol. 842. Springer, Heidelberg (1994)
13. Nešetřil, J., de Mendez, P.O.: Sparsity: Graphs, Structures, and Algorithms. Algorithms and combinatorics, vol. 28. Springer, Heidelberg (2012)
14. Pothen, A.: The complexity of optimal elimination trees. Technical report CS-88-13. Pennsylvania State University (1988)

---

[2] After the submission of the conference version, together with Hirotaka Ono, we found that the problem is NP-hard for these graph classes.

On Local Structures of Cubicity 2 Graphs

Sujoy Bhore[2], Dibyayan Chakraborty[1(✉)], Sandip Das[1], and Sagnik Sen[3]

[1] Indian Statistical Institute, Kolkata, India
dibyayancg@gmail.com
[2] Ben-Gurion University, Beersheba, Israel
[3] Indian Statistical Institute, Bangalore, India

Abstract. A 2-stab unit interval graph (2SUIG) is an axes-parallel unit square intersection graph where the unit squares intersect either of the two fixed lines parallel to the X-axis, distance $1 + \epsilon$ ($0 < \epsilon < 1$) apart. This family of graphs allow us to study local structures of unit square intersection graphs, that is, graphs with cubicity 2. The complexity of determining whether a tree has cubicity 2 is unknown while the graph recognition problem for unit square intersection graph is known to be NP-hard. We present a linear time algorithm for recognizing trees that admit a 2SUIG representation.

1 Introduction

We know that geometric intersection graphs have been studied for over 50 years [9,10] and interval graphs [9] are probably the most studied such family. To begin with, let us present a few definitions. Cubicity, $cub(G)$ of a graph G is the minimum d such that G is representable as a geometric intersection graph of d-dimensional (axes-parallel) cubes [10]. The notion of cubicity is a special case of boxicity [10]. Boxicity $box(G)$ of a graph G is the minimum d such that G is representable as a geometric intersection graph of d-dimensional (axes-parallel) hyper-rectangles. The family of graphs with boxicity 1 and the family of graphs with cubicity 1 are the families of interval and unit interval graphs, respectively.

It is curious that several decision problems such as graph recognition, k-coloring, finding minimum dominating set etc. are polynomial time solvable for interval graphs and unit interval graphs but are NP-hard for their 2-dimensional counterparts, the families of graphs with boxicity 2 and cubicity 2, respectively [3,4,6,8]. There seems to be a jump in the difficulty level of the study of these families while going from dimension 1 to 2. To be precise, interval graphs and unit interval graphs are intersection graphs of geometric objects embedded in $\mathbb{R}^1$ while boxicity 2 and cubicity 2 graphs are intersection graphs of geometric objects embedded in $\mathbb{R}^2$.

Our goal is to understand the reason of this jump and study what lies "in between". A number of efforts [5,12] in this direction has been made yielding different graphs families, each of which are generalization of interval graphs. For each such family, determining the complexity of the graph recognition problem has been one of the most difficult and important algorithmic question. The answer

© Springer International Publishing AG 2016
T.-H.H. Chan et al. (Eds.): COCOA 2016, LNCS 10043, pp. 254–269, 2016.
DOI: 10.1007/978-3-319-48749-6_19

to this question is either extremely difficult or had remained unsolved unless the graph family in question are a family of perfect graphs. In general, such graph families are of interest and solving problems in these set ups are challenging.

In a recent work [2] we introduced and studied new families of geometric intersection graphs which satisfies the properties prescribed by Scheinerman [11]. One such graph family is the focus of our study here in order to get insight towards understanding the structure of graphs with cubicity 2.

Let $y = 1$ be the lower stab line and $y = 2 + \epsilon$ be the upper stab line where $\epsilon \in (0, 1)$ is a constant. Now consider unit squares that intersects one of the stab lines. A 2-stab unit interval graph (2SUIG) is a graph G that can be represented as an intersection graph of such unit squares. Such a representation R of G is called a 2SUIG representation (for example, see Fig. 1).

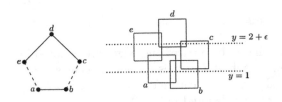

Fig. 1. A representation (right) of a *2SUIG* graph (left).

Note that this family is highly relevant in the study of cubicity 2 graphs as they, informally speaking, capture the local structures of these graphs. Note that the complexity of the graph recognition problem for 2SUIG is not known and seems to be a difficult problem. Whereas, recognition problem of trees is challenging, which we address in this article. In general, recognizing tree for graph families are either polynomial time solvable, such as in the case of interval graphs, unit interval graphs, boxicity 2 graphs or NP-hard, such as in the case of the family of induced subgraphs of the 2-dimensional infinite grid graph [1]. Trees with boxicity 1 are the caterpiller graphs while all trees have boxicity 2. On the contrary, determining cubicity of a tree seems to be a more difficult problem. It is easy to note that trees with cubicity 1 are paths. For higher dimensions, Babu et al. [7] presented a randomized algorithm that runs in polynomial time and computes cube representations of trees, of dimension within a constant factor of the optimum. The complexity of determining the cubicity of a tree is unknown [7]. In our case, the problem is linear time solvable but the solution is highly non-trivial.

In this article, we characterize all trees that admit a 2SUIG representation using forbidden structures. We prove the following:

Theorem 1. *Determining whether a given tree $T = (V, E)$ is a 2SUIG can be done in $O(|V|)$ time.*

We propose an algorithm that outputs a 2SUIG representation of the trees which are 2SUIG. Otherwise, our algorithm finds a forbidden structure responsible for the tree not having a 2SUIG representation.

One interesting aspect of the family of 2-stab unit interval graphs is that there is a scope of naturally generalizing the concept by defining a family of "k-stab unit interval graphs or kSUIG". Intuitively, for $k = 1$, this family will give us the family of unit interval graphs and for $k = \infty$ it will give us the family of cubicity 2 graphs. So our definition, in this sense, manages to go "in between". Varying the number of stab-lines, k, we obtain an infinite chain of graph families starting with the unit interval graphs and tending towards graphs with cubicity 2. Thus the study of 2SUIG graphs maybe the beginning of solving the mystery of why studying 2 dimensional geometric graphs are more difficult than studying one dimensional graphs.

In Sect. 2, we present some definitions. In Sect. 3, we present the proof of Theorem 1. Finally in Sect. 4, conclusions are drawn.

2   Preliminaries

Let G be a graph. The set of vertices and edges are denoted by $V(G)$ and $E(G)$, respectively. A vertex subset I of G is an *independent set* if all the vertices of I are pairwise non-adjacent. The cardinality of the largest independent set of G is its *independence number*, denoted by $\alpha(G)$.

Let G be a unit square intersection graph with a fixed representation R. We denote the unit square in R corresponding to the vertex v of G by s_v. In this article, by a unit square we will always mean a closed unit square. The coordinates of the left lower-corner of s_u are denoted by (x_u, y_u). Given a graph G with a 2SUIG representation R and two vertices $u, v \in V(G)$ we say $s_u <_x s_v$ if $x_u < x_v$ and $s_u <_y s_v$ if $y_u < y_v$ (see Fig. 2). Let H be a connected subgraph of G. Consider the union of intervals obtained from the projection of unit squares corresponding to the vertices of H on $X-$axis. This so obtained interval $span(H)$ is called the *span* of H in R.

Fig. 2. In the above picture $s_u <_x s_v$ and $s_u <_y s_v$.

A *leaf* is a vertex with degree 1. A *caterpillar* is a tree where every leaf vertex is adjacent to a vertex of a fixed path. A *branch vertex* is a vertex having degree more than 2. A *branch edge* is an edge incident to a branch vertex. A *claw* is the complete bipartite graph $K_{1,3}$. Given a 2SUIG representation R of a graph G, an edge uv is a *bridge edge* if s_u and s_v intersect different stab lines. Given a 2SUIG representation of a graph G, the vertices corresponding to the unit squares intersecting the upper stab line are called *upper* vertices and the vertices corresponding to the unit squares intersecting the lower stab line are called *lower* vertices. If all the vertices of a set intersect the same stab line then we say they are in the *same stab*.

Let $P = v_1 v_2 ... v_k$ be a path with a 2SUIG representation R. The path P is a *monotone path* if either $s_{v_1} <_x s_{v_2} <_x ... <_x s_{v_k}$ or $s_{v_k} <_x s_{v_{k-1}} <_x ... <_x s_{v_1}$. Let $P = v_1 v_2 ... v_k$ be a monotone path with $s_{v_1} <_x s_{v_2} <_x ... <_x s_{v_k}$ and all v_i's are in the same stab. Observe that $\alpha(P) = \lceil \frac{k}{2} \rceil < span(P) \leq k$ are tight bounds for $span(P)$. Fix some constant $c \in (0, .5)$. A monotone representation of P is *stretched* if $span(P) = k$ and is *shrinked* if $span(P) = \lceil \frac{k}{2} \rceil + c$ (see Fig. 3). The value of c within that range will not affect our proof. If all the vertices of P have distinct lower left corners and are in the same stab, then P must be monotone. Such monotone path can be classified as follows.

1. *right monotone*: all the vertices of the path are $s_{v_1} <_x s_{v_2} <_x ... <_x s_{v_k}$;
2. *left monotone*: all the vertices of the path are $s_{v_k} <_x s_{v_{k-1}} <_x ... <_x s_{v_1}$;

Lower right monotone path is right monotone path with all lower vertices. *Upper right monotone* path is right monotone path with all upper vertices. Similalry, we have *lower left monotone* path and *upper left monotone* path.

A path $P = v_1 v_2 ... v_k$ is called a *folded path* if it has a degree two vertex u such that either $s_u <_x s_v$ for all $v \in V(P) \backslash \{u\}$ or $s_v <_x s_u$ for all $v \in V(P) \backslash \{u\}$. Note that all vertices of a folded path cannot be in the same stab.

A *red edge* of a tree T is an edge e such that each component of $T \backslash \{e\}$ contains a claw. A *red path* is a path induced by red edges. A *maximal red path* is a red path that is not properly contained in another red path. Let $P = v_1 v_2 ... v_k$ be a maximal red path in T. The vertices v_1 and v_k are *endpoints* of P.

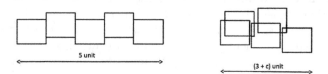

Fig. 3. A streached (left) and shrinked (right) representation of a path on five vertices.

3   Proof of Theorem 1

Given a tree T, our objective is to determine whether T has a 2SUIG representation. Let $T = (V, E)$ be a fixed tree. We will now prove several necessary conditions for T being a 2SUIG. On the other hand, we will show that these conditions together are also sufficient and can be verified in $O(|V|)$ time.

3.1   Structural Properties

We will start by proving some structural properties of T assuming it is a 2SUIG.

Lemma 1. *If T has a 2SUIG representation, then its vertices have degree at most four.*

A tree T that admits a 2SUIG representation does not necessarily have a red edge (defined in Sect. 2). But if T has at least one red edge then the red edges of T must induce a path.

Lemma 2. *If T has a 2SUIG representation, then either T has no red edge or the set of red edges of T induces a connected path.*

Proof. Let T has at least one red edge and T' be the graph induced by all red edges. First we will show that T' is connected. Thus, assume that T' has at least two components T_1 and T_2. Then there is a path P in T connecting T_1 and T_2. Note that removing an edge e of P creates two components of T each of which contains a claw. Thus, e should be a red edge. Therefore, T' is connected.

Now we will show that T' is a path. Assume that v is a vertex of T' with degree at least 3. Also, let v_1, v_2 and v_3 be three neighbors of v in T'. In any 2SUIG representation of T, at least three corners of s_v must be intersected by s_{v_1}, s_{v_2} and s_{v_3}. Without loss of generality, we assume that s_{v_1} intersects the upper-left corner of s_v, s_{v_2} intersect the upper-right corner of s_v, and s_{v_3} intersects the left lower-corner of s_v. This implies that s_{v_1}, s_{v_2} intersect the upper stab line while s_{v_3} intersects the lower stab line.

Note that a claw has a 2SUIG representation. Any such representation of a claw will have squares intersecting the upper stab line and squares intersecting the lower stab line. As each component of $T \setminus \{vv_1\}$ has a claw, there must be a path of the form $v_1v_{11}v_{12}...v_{1k}$ in T such that $s_{v_{1i}} <_x s_{v_1}$ for all $i \in \{1, 2, ..., k\}$ where $s_{v_{1k}}$ intersects the lower stab line. Similarly, as each component of $T \setminus \{vv_2\}$ has a claw, there must be a path of the form $v_2v_{21}v_{22}...v_{2k'}$ in T such that $s_{v_2} <_x s_{v_{2i}}$ for all $i \in \{1, 2, ..., k'\}$ where $s_{v_{2k'}}$ intersects the lower stab line. Moreover, as each component of $T \setminus \{vv_3\}$ has a claw, there must be a path of the form $v_3v_{31}v_{32}...v_{3k''}$ in T where $s_{v_{3k''}}$ intersects the upper stab line. This will force a cycle in the representation of T, a contradiction. Thus, T' must be a path. ∎

The above result leads us to two cases: when T has a red path and when T does not have any red edge. If the red edges of T induces a path P, then construct the *extended red path* $A = a_1a_2...a_k$ by including the edge(s), that are not red, incident to the endpoint(s) of P that have degree two in T. In particular, if both the end points of P are branch vertices, the extended red path $A = P$. On the other hand, if T has no red edges, then distance between any two branch vertices is at most 2. Thus, there exist a vertex v in T whose closed neighborhood $N[v]$ contains all the branch vertices of T. Choose (if not found to be unique) one such special vertex v. If v has degree two then consider the path uvw induced by the closed neighborhood of v and call it the extended red path of T. If v does not have degree two, then the extended red path of T is the singleton vertex v. In any case, rename the vertices of the extended red path $A = a_1a_2...a_k$ so that we can speak about it in an uniform framework along with the case T having red edges. We fix such an extended red path $A = a_1a_2...a_k$ for the rest of this article. The vertices $V_A = \{a_1, a_2, ..., a_k\}$ of this extended red path A are called the *red vertices*. The following lemmas will provide intuitions about how a tree having a 2SUIG representation looks like.

Lemma 3. *If T has a 2SUIG representation and does not have any red edge, then the number of branch vertex in T is at most 5.*

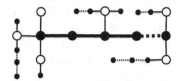

Fig. 4. The nomenclature – red edges = thick lines; new edge(s) added to the red path for obtaining the extended red path = thick dotted line(s); red vertices = big solid circles, agents = big hollow circles, tails = thin dotted lines; vertices of the tails = small solid circles.

The above follows directly from the fact that the vertices of T has degree at most four.

Lemma 4. *If T has at most one branch vertex with degree at most four, then T has a 2SUIG representation.*

Proof. If T has no branch vertices, then T is a path which admits a unit interval representation. On the other hand, T with one branch vertices of degree at most four is a subdivision of $K_{1,3}$ or $K_{1,4}$. These graphs clearly admit 2SUIG representation.

Lemma 5. *A branch vertex of a tree T is either a red vertex or is adjacent to a red vertex.*

Proof. If T has no red edges, then there exists a red vertex v in T such that all the branch vertices are in the closed neighborhood $N[v]$ of v.

Thus suppose that the red edges of T induces a path. Let a red vertex v and a non-red branch vertex u be connected by a path with no red edges of length at least 2. Clearly after deleting this path, the component containing v contatins a claw. Thus if we delete the edge $\{e\}$ of the path incident to v, then both the components of $T \setminus \{e\}$ contains a claw, a contradiction.

Note that if T is a 2SUIG tree with at least two branch vertices, then the endpoints a_1 and a_k of the extended path A must be branch vertices of T. Assume that $A_b = \{a_1 = a_{i_1}, a_{i_2}, ..., a_{i_{k'}} = a_k\}$ be the branch vertices of A where $1 = i_1 < i_2 < ... < i_{k'} = k$. The neighbors of red vertices that are not red are called *agents*. An agent v is adjacent to exactly one red vertex, say a_j, of T. We call v is an agent of a_j in this case.

If we delete all the red vertices and agents from a 2SUIG tree T then by Lemma 5 we will be left with some disjoint paths. Each such path actually starts from (that is, is one endpoint of) one of the agents. Let $P = v_1v_2...v_l$ be a path where v_1 is an agent and the other vertices are neither agent nor red vertices. Also $v_2, v_3, ..., v_{l-1}$ are degree 2 vertices and v_l is a leaf. Then the path $P' = v_2v_3...v_{l-1}v_l$ is called a *tail* of agent v_1. Let v_1 be an agent of the red vertex a_j. Sometimes we will also use the term "tail P' of the red vertex a_j". Deleting the tail P' is to delete all the vertices of P'. The red vertex a_j, all its agents and tails are together called a_j *and its associates* for each $j \in \{1, 2, ...k\}$. Note that an agent has exactly two tails by allowing tails with zero vertices. Let us set the following conventions: the tails of an agent z are the *long tail* $lt(z)$ and the *short tail* $st(z)$ such that $|lt(z)| \geq |st(z)|$ where $|lt(z)|$ and $|st(z)|$ denotes the number of vertices in the respective tails. Now we have enough nomenclatures (see Fig. 4) to present the rest of the proof.

3.2 Partial Description of the Canonical Representation

In the following, we will show that there is a canonical way to represent a 2SUIG tree. First we will describe the representation of the extended red path followed by representation of the agents and their tails.

Lemma 6. *If T is a 2SUIG with at least one red edge, then there exists a 2SUIG representation where the extended red path of T is monotone and stretched.*

Proof. Let A be the extended red path of T with a representation R. If A is not monotone then one of the following is true. (i) A is a folded path. (ii) There are three vertices $\{a_i, a_{i+1}, a_{i+2}\} \in V(A)$ such that $s_{a_{i+1}} <_x s_{a_i}$ and $s_{a_{i+1}} <_x s_{a_{i+2}}$ where $i > 1$ and for all $j < i$ we have $s_{a_j} <_x s_{a_{j+1}}$.

(i) There is a vertex $u \in V(A)$ with $s_v <_x s_u$ for all $v \in V(A) \setminus \{u\}$. Then there will be two claws C_1, C_2 in two different components of $T \setminus \{u\}$ with $s_v <_x s_u$ for all $v \in V(C_1) \cup V(C_2)$. But as T is a 2SUIG, this configuration will force a cycle in it. This is a contradiction.

(ii) Without loss of generality assume s_{a_i} intersects the lower stab line. Then $s_{a_{i+1}}$ and $s_{a_{i+2}}$ must intersect the upper stab line. Let T_{i+1} be the component of T obtained by deleting the edge a_ia_{i+1} and contains a_{i+1}. There is a claw C_3 in T_{i+1} with $s_{a_{i+1}} <_x s_w$ for all $w \in V(C_3)$. Thus, an agent z of a_i with $s_{a_i} <_x s_z$ is not a branch vertex as otherwise this will force a cycle. Hence its tail can be presented by a lower-right monotone representation. Therefore, we can translate (rigid motion) the component T_{i+1} to the right to obtain a 2SUIG representation of T where the extended red path A is monotone.

To prove that A can be stretched let $e = a_ia_{i+1}$ be an edge of the extended red path A with $x_{a_i} < x_{a_{i+1}}$ in R. Let T_i and T_{i+1} be the components of $T \setminus \{e\}$ containing a_i and a_{i+1}, respectively. Now translate (rigid motion) the component T_{i+1} to the right obtaining a 2SUIG representation with $x_{a_{i+1}} = x_{a_i} + 1$. We are done by performing this operation on every edge of A.

Similarly it can be shown that if T is a 2SUIG with no red edge, still it admits a representation where the extended red path A is monotone and streched. We turn our focus on the bridge edges of the extended red path.

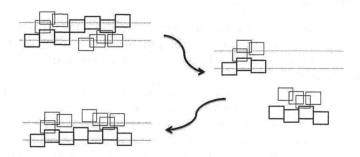

Fig. 5. Explaining the reflection and translation described in the proof of Lemma 7 with an example.

Lemma 7. *If T admits a 2SUIG representation R with a stretched monotone extended red path, then there exists a 2SUIG representation where every red bridge vertex is a branch vertex.*

Proof. Let $A = a_1 a_2 ... a_k$ be a right monotone extended red path of T with respect to R. Let $e = a_i a_{i+1}$ be a bridge edge of the extended red path A where a_{i+1} is not a branch vertex. Also assume that a bridge vertex a_j is a branch vertex for all $j < i$.

Let $S = \{v \in V(T) | s_{a_i} <_x s_v\}$. Let T' be the graph induced by S. Note that S contains the vertex a_{i+1}. Now consider the reflection of the 2SUIG representation of T' induced by R with respect to the X-axis (Fig. 5). This will give us a picture where every unit square corresponding to the vertices of T' lies under the X-axis. This is a particular unit square representation R' of T' which in fact is also a 2SUIG representation of it if we consider the stab lines to be $y = -1$ and $y = -2 - \epsilon$. In this 2SUIG representation R', the lower vertices with respect to R of T' became upper vertices and vice versa. Now translate the unit square representation R' upwards until all the upper vertices (with respect to R') of T' intersects $y = 2 + \epsilon$, all the lower vertices intersect $y = 1$. Note that a_i can have at most one degree 2 agent in T' and thus, that agent can have at most one tail. After what we did above, we can adjust the Y-co-ordinates of that agent and its tail, if needed, to obtain a 2SUIG representation of T.

We will be done by induction after handling one more case. The case where a_i is the first bridge vertex of A which is not a branch vertex while a_{i+1} is a branch vertex. Let $S' = \{v \in V(T) | s_v <_x s_{a_{i+1}}\}$ and let T'' be the graph induced by S'. To achieve our goal, we do the exact same thing with T'' that we did with T'. This will provide us a 2SUIG representation of T where each bridge vertex a_j is a branch vertex for all $j \leq i + 1$. Hence we are done by induction.

Now we will describe a way to represent the tails in the following lemma.

Lemma 8. *If T admits a 2SUIG representation R, then there exists a 2SUIG representation where each tail is a shrinked monotone path and all its vertices are in the same stab.*

Proof. Let $P = v_2 v_3 ... v_{l-1}$ be a tail of agent v_1. Note that if all vertices of the tail P are in the same stab then P must be monotone. Furthermore, if P is not shrinked in R then we can shrink it to obtain a new representation of T without changing anything else of R.

Therefore, to complete the proof, let us assume that not all vertices of P are in the same stab. Then at least one edge e of P is a bridge edge. Any brige edge divides the stab lines into two parts, left and right. Assume, without loss of generality, that s_{v_1} is in the left part. Thus, as there are no branch vertices in the tail, we do not have any vertex $w \in V(T) \setminus V(P)$ with s_w lying in the right part. Thus, we can modify the representation R by placing all the vertices of the tail P in the same stab making use of the empty right part.

3.3 Properties of the Canonical Representation

Assume that R is a 2SUIG representation of T such that extended red path is a monotone stretched path. The other vertices of T are the agents and the vertices of the tails. Note that the endpoints a_1 and a_k can have at most 3 agents and 6 tails while the other red vertices can have at most 2 agents and 4 tails.

Lemma 9. *Let T be a 2SUIG tree with a representation R where the extended red path A is a stretched right monotone path and R satisfies the conditions of Lemmas 7 and 8.*

(a) If each right monotone tail of T is such that it is not possible to make the tail left monotone and obtain a 2SUIG representation of T from R without changing anything else, then any red vertex other than a_k has at most one right monotone tail having at least two vertices.

(b) If each left monotone tail of T is such that it is not possible to make the tail right monotone and obtain a 2SUIG representation of T from R without changing anything else, then any red vertex other than a_1 has at most one left monotone tail having at least two vertices.

Proof. (a) Let a_j $(j < k)$ be a red vertex of T with at least two monotone tails. Without loss of generality, assume that a_j is a lower vertex. Note that $s_{a_{j+1}}$ contains either the upper-right corner or the lower-right corner of s_{a_j}. Let v be an agent of a_j.

Case 1. If s_v contains the lower-left corner of s_{a_j}, then v cannot have a right monotone tail for any $j \geq 2$ as $s_{a_{j-1}}$ contains the upper-left corner of s_{a_j}. If $j = 1$, then a right monotone tail $P = v_1 v_2 ... v_l$ of v must be upper-right monotone. This means $v v_1$ is a bridge edge. This will mean, there is no upper vertex z in T with $s_z <_x s_{v_1}$. Thus, if we make the tail P an upper-left monotone tail instead (we keep the position of s_{v_1} as before but change the positions of the other vertices of P), then we obtain a 2SUIG representation of T. But this should not be possible according to the assumptions. Hence v cannot have a right-monotone tail for any red vertex a_j $(j \neq k)$.

Case 2. If s_v contains the upper-left corner of s_{a_j}, then v can indeed have an upper-right monotone tail. Note that, in this case, v can have at most one right monotone path, an upper-right monotone that is. If some other neighbor of a_j contains the upper-right corner of s_{a_j}, then the right monotone tail of v can have at most one vertex in order to avoid cycles in T.

Case 3. If s_v contains the upper-right corner of s_{a_j}, then v can have at most one right monotone tail, an upper-right monotone tail to be specific, as this situation implies that $s_{a_{j+1}}$ contains the lower-right corner of s_{a_j}.

Therefore, only the agents containing upper-left corner or upper-right corner of s_{a_j} can have at most one right monotone tail each. But if both types of agents are present, then the right monotone tail of the agent containing upper-left corner of s_{a_j} can have at most one vertex.

(b) This proof can be done similarly like (a).

From the above we can infer the following:

Lemma 10. *Let T be a 2SUIG tree with a representation R where a lower (or upper) red vertex has two upper (or lower) neighbors.*

1. *If the left neighbor is an agent then it can have at most one right monotone tail having at most one vertex.*
2. *If the right neighbor is an agent then it can have at most one left monotone tail having at most one vertex.*
3. *If both the neighbors are agents then the left neighbor can have a right monotone tail and the right neighbor can have a left monotone tail having one vertex each.*

In the above lemma, the third case is to say that the worst case scenarios of the first two cases can take place simultaniously. Now we will discuss the length of the tails that can be accommodated between two red branch vertices in the following lemma.

Lemma 11. *Let T be a 2SUIG tree with a representation R with a stretched right monotone extended red path A. Let a_l, a_{l+m} ($m = 0$ is possible) be two red branch vertices such that a_l has a right monotone tail P and a_{l+m} has a left monotone tail P' in the same stab with no vertex v of T satisfying $s_p <_x s_v <_x s_{p'}$ where p, p' are the leaf vertices of the tails P, P', respectively. Then, depending on the positions of the corresponding agents v of P and v' of P', R must satisfy one of the following conditions:*

1. *If $s_{a_l} <_x s_v$ and $s_{v'} <_x s_{a_{l+m}}$, then $\alpha(P_v) + \alpha(P'_{v'}) \leq m$;*
2. *If $s_v <_x s_{a_l}$ and $s_{v'} <_x s_{a_{l+m}}$, then $\alpha(P_v) + \alpha(P'_{v'}) - 1 \leq m$;*

3. If $s_v <_x s_{a_l}$ and $s_{a_{l+m}} <_x s_{v'}$, then $\alpha(P_v) + \alpha(P'_{v'}) - 2 \leq m$;

4. If $s_{a_l} <_x s_v$ and $s_{a_{l+m}} <_x s_{v'}$, then $\alpha(P_v) + \alpha(P'_{v'}) - 1 \leq m$;

where P_v is the path induced by $V(P) \cup \{v\}$ and $P'_{v'}$ is the path induced by $V(P') \cup \{v'\}$.

Proof. Let $A' = a_l a_{l+1}...a_{l+m}$ and $s_v <_x s_{a_l}$ and $s_{a_{l+m}} <_x s_{v'}$. As A' is stretched and P_v and $P'_{v'}$ are both shrinked we have $span(A') + 2 \geq span(P_v) + span(P'_{v'})$. This implies $m + 1 + 2 = m + 3 \geq \lceil \alpha(P_v) + \alpha(P'_{v'}) + 2c \rceil = \alpha(P_v) + \alpha(P'_{v'}) + 1$ and hence condition 3. The other conditions can be proved similarly.

As we discuss in the following lemma, in our prescribed representation of T, assuming it is a 2SUIG, bridge edges of the extended red path are induced by red branch vertices.

Lemma 12. *If two adjacent red vertices both have degree 4, then they must be in different stabs.*

Proof. Without loss of generality assume that $a_l a_{l+1}$ is such an edge where a_l, a_{l+1} are both degree 4 lower vertices with $s_{a_l} <_x s_{a_{l+1}}$. Then either the upper-right corner of s_{a_l} is contained in $s_{a_{l+1}}$ or the upper-left corner of $s_{a_{l+1}}$ is contained in s_{a_l}. If the upper-right corner of s_{a_l} is contained in $s_{a_{l+1}}$, then a_l cannot have more than three neighbors as no neighbor other than $s_{a_{l+1}}$ can contain one of the right corners of s_{a_l}. We can argue similarly for the other case as well.

3.4   The Canonical Representation

In this section suppose that T is a tree with maximum degree 4 such that either there is no red edge or the red edges induces a path. We will try to obtain a 2SUIG representation of T and if our process fails to obtain such a presentation, then we will conclude that T is not a 2SUIG. Also assume that the extended red path of T is $A = a_1 a_2 ... a_k$. Due to Lemma 6 we can assume that A is a stretched right monotone path and a_1 is a lower vertex.

Our strategy is to first represent a_1 and its associates and then to represent a_i and its associates one by one in ascending order of indices where $i \in \{2, 3, ..., k\}$. In each step our strategy is to represent a_i and its associates in such a way that the maximum value of x_v is minimized where v is a vertex from a_i and its associates. Note that the main difficulty is to represent a_i and its associates when $d(a_i) \geq 3$ as otherwise a_i do not have any agents or tails. We begin with the following lemma.

Lemma 13. *There exists a 2SUIG representation, satisfying all the properties of the canonical representation proved till now, with a_1 and a_2 in the same stab if and only if either $d(a_1) \leq 3$ or $d(a_2) \leq 3$.*

Proof. The "only if" part follows from Lemma 12. The "if" part can be proved similarly like the proof of Lemma 7.

Representation of a_1 and Its Associates When $k \neq 1$: First we will handle the case $k \neq 1$. Now we are going to list out the way to obtain the canonical representation of a_1 and its associates and the conditions for it to be valid through case analysis. Also in any representation the agents intersecting the lower-left corner, the upper-left corner, the upper-right corner and the lower-right corner of s_{a_1} are renamed as z_1, z_2, z_3 and z_4, respectively. The conditions below are simple conditions for avoiding cycles in the graph.

Case 1: $d(a_1) = 4, d(a_2) = 4$. In this case a_2 is an upper vertex by Lemma 12, s_{a_2} intersects the upper-right corner of s_{a_1} and the three agents of a_1 are z_1, z_2, z_4.
  (1) $lt(z_1)$ is shrinked lower-left monotone and $st(z_1)$ is shrinked upper-left monotone.
  (2) $lt(z_2)$ is shrinked upper-left monotone and $st(z_2)$ is shrinked upper-right monotone.
  (3) $|st(z_2)| \leq 1$ and if $|st(z_1)| > 0$, then $|lt(z_2)| \leq 1$.
  (4) $|lt(z_4)| = 0$.
Case 2: $d(a_1) = 3, d(a_2) = 4$. In this case a_2 is a lower vertex by Lemma 13, s_{a_2} intersects the upper-right corner of s_{a_1} and the two agents of a_1 are z_1, z_2.
  (1) conditions (1)–(3) of Case 1.
Case 3: $d(a_1) = 4, d(a_2) = 3$. In this case a_2 is a lower vertex by Lemma 13, s_{a_2} intersects the lower-right corner of s_{a_1} and the three agents of a_1 are z_1, z_2, z_3.
  (1) conditions (1)–(3) from Case 1.
  (2) $|lt(z_3)| \leq 1$, $lt(z_3)$ is shrinked upper-left monotone and $st(z_3)$ is shrinked upper-right monotone.
Case 4: $d(a_1) = 3, d(a_2) = 3$. In this case a_2 is a lower vertex by Lemma 13, s_{a_2} intersects the lower-right corner of s_{a_1} and the two agents of a_1 are z_1, z_2.
  (1) $lt(z_1)$ is shrinked lower-left monotone and $st(z_1)$ is shrinked upper-left monotone.
  (2) if $|st(z_1)| > 0$, then $lt(z_2)$ is shrinked upper-right monotone with $|lt(z_2)| \leq 3$ and $st(z_2)$ is shrinked upper-left monotone $|st(z_2)| \leq 1$.
  (3) if $|st(z_1)| = 0$, then $lt(z_2)$ is shrinked upper-left monotone and $st(z_2)$ is shrinked upper-right monotone $|st(z_2)| \leq 3$.
Case 5: $d(a_1) = 4, d(a_2) = 2$. In this case a_2 is a lower vertex by Lemma 13, s_{a_2} intersects the lower-right corner of s_{a_1} and the three agents of a_1 are z_1, z_2, z_3.
  (1) condition (1)–(3) from Case 1.
  (2) if $|lt(z_3)| \leq 1$, then $lt(z_3)$ is shrinked upper-left monotone and $st(z_3)$ is shrinked upper-right monotone.
  (3) if $|lt(z_3)| > 1$, then $|st(z_3)| \leq 1$ and $lt(z_3)$ is shrinked upper-right monotone and $st(z_3)$ is shrinked upper-left monotone.
Case 6: $d(a_1) = 3, d(a_2) = 2$. In this case a_2 is a lower vertex by Lemma 13, s_{a_2} intersects the lower-right corner of s_{a_1} and the two agents of a_1 are from $\{z_1, z_2, z_3\}$.
  (1) $lt(z_1)$ is shrinked lower-left monotone and $st(z_1)$ is shrinked upper-left monotone.

(2) if both z_1, z_2 exists and $|st(z_1)| = 0$, then $lt(z_2)$ is shrinked upper-left monotone and $st(z_2)$ is shrinked upper-right monotone.

(3) if both z_1, z_2 exists, $|st(z_1)| > 0$ and $|lt(z_2)| \leq 1$, then $lt(z_2)$ is shrinked upper-left monotone and $st(z_2)$ is shrinked upper-right monotone.

(4) if both z_1, z_2 exists, $|st(z_1)| > 0$ and $|lt(z_2)| > 1$, then $st(z_2)$ is shrinked upper-left monotone and $lt(z_2)$ is shrinked upper-right monotone with $|st(z_1)| \leq 1$.

(5) there is no case when both z_2, z_3 exists as we can always modify this representation by making the agent playing the role of z_2 play the role of z_1 instead.

(6) there is no case when both z_1, z_3 exists with $|st(z_1)| = 0$ as we can always modify this representation by making the agent playing the role of z_3 play the role of z_2 instead.

(7) there is no case when both z_1, z_3 exists with $|st(z_3)| \leq 1$ and $|lt(z_3)| \leq 3$ as we can always modify this representation by making the agent playing the role of z_3 play the role of z_2 instead.

(8) if both z_1, z_3 exists, $|st(z_1)| > 0$ and $|lt(z_3)| \leq 3$, then $lt(z_3)$ is shrinked upper-left monotone and $st(z_2)$ is shrinked upper-right monotone.

(9) if both z_1, z_3 exists, $|st(z_1)| > 0$ and $|lt(z_3)| > 3$, then $st(z_3)$ is shrinked upper-left monotone with $|st(z_1)| \leq 3$ and $lt(z_2)$ is shrinked upper-right monotone.

The square s_{a_2} must intersect one of the right corners of s_{a_1}. In each of the cases listed above, there can be at most $3! = 6$ possible ways of in which the agents of a_1 can play the role of z_1, z_2, z_3, z_4. Among all possible ways those which satisfies the above conditions, we choose the one for which the leaf of the right-monotone tail of z_2 (only when z_3, z_4 does not exist) or z_3 (z_4 cannot exist) or z_4 (z_3 cannot exist) is minimized with respect to $<_x$. As there are at most a constant number of probes to be made, this is achieveable in constant time. Moreover, such a representation, if found, will be called the *optimized representation of a_1 and its associates*. Otherwise, T is not a 2SUIG.

Representation of a_1 and Its Associates When $k = 1$: The canonical representation of a_k and its associates is similar as above.

Representation of a_i and Its Associates for All $1 < i < k$: We describe the canonical representation of a_i and its associates given the canonical representation of a_j and its associates for all $j < i$. Throughout the case analysis we will assume without loss of generality that a_{i-1} is an upper vertex. Also assume that $a_{i'}$ be the maximum $i' < i$ such that $d(a_{i'}) \geq 3$ ($i - 1 = i'$ is possible). Moreover, in any representation the agents intersecting the lower-left corner, the upper-right corner and the lower-right corner of s_{a_i} are renamed as z_1, z_3 and z_4, respectively. The conditions below are simple conditions for avoiding cycles in the graph.

(1) If z_1 exists and a_i is a lower vertex, then $|st(z_1)| = 0$ and $lt(z_1)$ is a lower-left shrinked monotone path satisfying conditions of Lemma 11.

(2) If z_1 exists, a_i is an upper vertex and either z_3 or z_4 exists, then $st(z_1)$ is a lower-right shrinked monotone path with $|st(z_1)| \leq 1$ and $lt(z_1)$ is a lower-left shrinked monotone path satisfying conditions of Lemma 11.

(3) If z_1 exists, a_i is an upper vertex and neither z_3 nor z_4 exists, P_{11} is shrinked lower-left monotone satisfying conditions of Lemma 11 and P_{12} is shrinked lower-right monotone where $\{P_{11}, P_{12}\} = \{st(z_1), lt(z_1)\}$.

(4) If z_3 exists with a_i being an upper vertex, then $|st(z_3)| = 0$ and $lt(z_3)$ is an upper-right shrinked monotone path.

(5) If z_3 exists with a_i being a lower vertex, then z_3 has an upper-left monotone tail P_1 satisfying $|P_1| \leq 1$ and an upper-right monotone tail P_2 with some $\{P_1, P_2\} = \{st(z_3), lt(z_3)\}$.

(6) If z_4 exists with a_i being an upper vertex and z_1 also exists, then z_4 has a lower-left monotone tail P_1 satisfying $|P_1| \leq 1$ and a lower-right monotone tail P_2 with some $\{P_1, P_2\} = \{st(z_4), lt(z_4)\}$.

(7) If z_4 exists with a_i being an upper vertex and z_1 does not exist, then z_4 has a lower-left monotone tail P_1 conditions of Lemma 11 and a lower-right monotone tail P_2 with some $\{P_1, P_2\} = \{st(z_4), lt(z_4)\}$.

(8) If z_4 exists with a_i being a lower vertex, then $|st(z_4)| = 0$ and $lt(z_4)$ is a lower-right shrinked monotone path.

To check the above conditions, as there are at most a constant number of probes to be made, this is achievable in constant time. Moreover, such a representation, if found, will be called the *optimized representation of a_i and its associates*. Otherwise, T is not a 2SUIG.

Representation of a_k and Its Associates: The canonical representation of a_k and its associates is similar as above.

3.5   Algorithm

Finally we will describe the algorithm for recongnizing if a given tree T is a 2SUIG. Whenever our algorithm concludes that the given tree T is not a 2SUIG, there is a configuration responsible for it. These configurations are forbidden configurations for 2SUIG trees.

(1) Check if maximum degree of T is at most 4. If not, then T is not a 2SUIG (Lemma 1).

(2) Check if there at most one branch vertex in T. If yes, then T is a 2SUIG by Lemma 4.

(3) Find out the graph induced by the red edges of the tree. If that graph has at least one edge but not a path, then T is not a 2SUIG by Lemma 2.

(4) Find out a (but for some trivial cases it is unique) extended red path $A = a_1 a_2 ... a_k$. Assign $x_{a_i} = i$ for all $i \in \{1, 2, ..., k\}$. Moreover, put s_{a_1} in the lower stab.

(5) For $i = 1$ to k find out the optimized representation of a_i and its associates. If we fail to find such a representation for some $i \in \{1, 2, ..., k\}$, then T is not a 2SUIG.

Correctness of the algorithm implies from the previous results and discussions. Given a tree it is possible to find out its set of red edges in linear time using post-order traversal. For the other steps we need to probe at most a constant number cases for each red vertex. Thus, it is possible to run the algorithm in $O(|V|)$ time.

4 Conclusions

In this paper we consider the problem of recognizing 2SUIG trees. While doing that we proved a number of structural properties and provided insights regarding how a canonical 2SUIG representation of a tree can be obtained. Recall our discussion on red edges and red vertices of a tree. Observe that, if the red vertices induce a path, then the tree has a unit square intersection representation. Hence, we hope our work can be extended for "k-stab unit interval graphs" and will help solving the tree recognition problem for cubicity two graphs. Even though the recognition of trees turns out to be solvable in linear time for 2SUIG, the natural and probably, the more important question is the following:

Question 1. Given a graph G what is the complexity for determining if G is a 2SUIG?

References

1. Bhatt, S.N., Cosmadakis, S.S.: The complexity of minimizing wire lengths in VLSI layouts. Inf. Process. Lett. **25**(4), 263–267 (1987)
2. Bhore, S.K., Chakraborty, D., Das, S., Sen, S.: On a special class of boxicity 2 graphs. In: Ganguly, S., Krishnamurti, R. (eds.) CALDAM 2015. LNCS, vol. 8959, pp. 157–168. Springer, Heidelberg (2015)
3. Breu, H.: Algorithmic aspects of constrained unit disk graphs. Ph.D. thesis, University of British Columbia (1996)
4. Clark, B.N., Colbourn, C.J., Johnson, D.S.: Unit disk graphs. Discrete Math. **86**(1–3), 165–177 (1990)
5. Correa, J., Feuilloley, L., Pérez-Lantero, P., Soto, J.A.: Independent and hitting sets of rectangles intersecting a diagonal line: algorithms and complexity. Discrete Comput. Geom. **53**(2), 344–365 (2015)
6. Imai, H., Asano, T.: Finding the connected components and a maximum clique of an intersection graph of rectangles in the plane. J. Algorithms **4**(4), 310–323 (1983)
7. Babu, J., Basavaraju, M., Chandran, L.S., Rajendraprasad, D., Sivadasan, N.: Approximating the cubicity of trees. CoRR, abs/1402.6310 (2014)
8. Kratochvíl, J.: A special planar satisfiability problem and a consequence of its NP-completeness. Discrete Appl. Math. **52**(3), 233–252 (1994)

9. Lekkeikerker, C., Boland, J.: Representation of a finite graph by a set of intervals on the real line. Fundamenta Math. **51**(1), 45–64 (1962)

10. Roberts, F.S.: On the boxicity and cubicity of a graph. In: Recent Progresses in Combinatorics, pp. 301–310 (1969)

11. Scheinerman, E.R.: Characterizing intersection classes of graphs. Discrete Math. **55**(2), 185–193 (1985)

12. West, D.B., Shmoys, D.B.: Recognizing graphs with fixed interval number is NP-complete. Discrete Appl. Math. **8**(3), 295–305 (1984)

Approximability of the Distance Independent Set Problem on Regular Graphs and Planar Graphs

Hiroshi Eto[1], Takehiro Ito[2,3], Zhilong Liu[1(✉)], and Eiji Miyano[1]

[1] Kyushu Institute of Technology, Fukuoka 820-8502, Japan
{eto,liu}@theory.ces.kyutech.ac.jp, miyano@ces.kyutech.ac.jp
[2] Tohoku University, Sendai 980-8579, Japan
takehiro@ecei.tohoku.ac.jp
[3] CREST, JST, 4-1-8 Honcho, Kawaguchi 332-0012, Japan

Abstract. This paper studies generalized variants of the MAXIMUM INDEPENDENT SET problem, called the MAXIMUM DISTANCE-d INDEPENDENT SET problem (MaxDdIS for short). For an integer $d \geq 2$, a distance-d independent set of an unweighted graph $G = (V, E)$ is a subset $S \subseteq V$ of vertices such that for any pair of vertices $u, v \in S$, the number of edges in any path between u and v is at least d in G. Given an unweighted graph G, the goal of MaxDdIS is to find a maximum-cardinality distance-d independent set of G. In this paper, we analyze the (in)approximability of the problem on r-regular graphs ($r \geq 3$) and planar graphs, as follows: (1) For every fixed integers $d \geq 3$ and $r \geq 3$, MaxDdIS on r-regular graphs is APX-hard. (2) We design polynomial-time $O(r^{d-1})$-approximation and $O(r^{d-2}/d)$-approximation algorithms for MaxDdIS on r-regular graphs. (3) We sharpen the above $O(r^{d-2}/d)$-approximation algorithms when restricted to $d = r = 3$, and give a polynomial-time 2-approximation algorithm for MaxD3IS on cubic graphs. (4) Finally, we show that MaxDdIS admits a polynomial-time approximation scheme (PTAS) for planar graphs.

1 Introduction

Let G be an unweighted graph; we denote by $V(G)$ and $E(G)$ the sets of vertices and edges, respectively, and let $n = |V(G)|$. An *independent set* (or *stable set*) of G is a subset $S \subseteq V(G)$ of vertices such that $\{u, v\} \notin E$ holds for all $u, v \in S$. In theoretical computer science and combinatorial optimization, one of the most important and most investigated computational problems is the MAXIMUM INDEPENDENT SET problem (MaxIS for short): Given a graph G, the goal of MaxIS is to find an independent set S of maximum cardinality in G. There are a huge number of its applications in diverse fields, such as scheduling, computer vision, pattern recognition, coding theory, map labeling, and computational biology; many different problems have been modeled using independent sets.

This work is partially supported by JSPS KAKENHI Grant Numbers JP15J05484, JP15H00849, JP16K00004, and JP26330017.

© Springer International Publishing AG 2016
T.-H.H. Chan et al. (Eds.): COCOA 2016, LNCS 10043, pp. 270–284, 2016.
DOI: 10.1007/978-3-319-48749-6_20

1.1   Our Problems

In this paper, we consider a generalization of MaxIS, named the MAXIMUM DISTANCE-d INDEPENDENT SET problem (MaxDdIS for short). For an integer $d \geq 2$, a *distance-d independent set* of an unweighted graph G is a subset $S \subseteq V(G)$ of vertices such that for any pair of vertices $u, v \in S$, the distance (i.e., the number of edges) of any path between u and v is at least d in G. For an integer $d \geq 2$, MaxDdIS is formulated as the following class of problems [1,8]:

Maximum Distance-d Independent Set (MaxDdIS)
Input: An unweighted graph G
Output: A distance-d independent set of G with the maximum cardinality

---

When $d = 2$, MaxDdIS (i.e., MaxD2IS) is equivalent to the original MaxIS. Zuckerman [18] proved that MaxD2IS cannot be approximated in polynomial time, unless P = NP, within a factor of $n^{1-\varepsilon}$ for any $\varepsilon > 0$. Moreover, MaxD2IS remains NP-hard even if the input graph is a cubic planar graph, a triangle-free graph, or a graph with large girth. Fortunately, however, it is well known that MaxD2IS can be solved in polynomial time when restricted to, for example, bipartite graphs [14], chordal graphs [9], circular-arc graphs [10], comparability graphs [11], and many other classes [4,16,17].

For every fixed integer $d \geq 3$, Eto et al. [8] proved that MaxDdIS is NP-hard even for planar bipartite graphs of maximum degree three. Furthermore, they showed that it is NP-hard to approximate MaxDdIS on bipartite graphs and chordal graphs within a factor of $n^{1/2-\varepsilon}$ $(\varepsilon > 0)$ for every fixed integer $d \geq 3$ and every fixed *odd* integer $d \geq 3$, respectively. On the other hand, interestingly, they showed that MaxDdIS on chordal graphs is solvable in polynomial time for every fixed *even* integer $d \geq 2$. As the other positive results, Agnarsson et al. [1] showed the tractability of MaxDdIS on interval graphs, trapezoid graphs, and circular-arc graphs.

1.2   Known Results for Regular Graphs and Planar Graphs

In this paper, we focus on the (in)approximability of MaxDdIS on regular graphs and planar graphs. As far as we know, this is the first paper which studies the problem on those graphs for general $d \geq 2$. Thus, known results exist only for MaxD2IS (MaxIS) on those graphs. Recall that the problem is NP-hard even for cubic (i.e., 3-regular) planar graphs.

Chlebík and Chlebíková [6] proved the 1.0107, 1.0216, 1.0225, and 1.0236-inapproximability for MaxD2IS on 3-regular, 4-regular, 5-regular, and r-regular $(r \geq 6)$ graphs, respectively. On the other hand, we can obtain polynomial-time 1.2, 1.4, and 1.6-approximation algorithms for MaxD2IS on 3-regular, 4-regular, and 5-regular graphs, respectively, by applying the $\frac{\Delta+3}{5}$-approximation algorithm proposed by Berman and Fujito [3] for the problem on general graphs of maximum degree $\Delta \leq 613$. We note that, for a larger maximum degree Δ

(and hence general r), Halldórsson and Radhakrishnan developed polynomial-time approximation algorithms within factors of $\frac{\Delta+2}{3}$ [12] and $O(\frac{\Delta}{\log\log\Delta})$ [13].

For planar graphs, it is well known that the Baker's shifting technique [2] for NP-hard optimization problems can be applied to MaxD2lS on planar graphs; it yields a polynomial-time approximation scheme (PTAS). Thus, MaxD2lS can be approximated within an arbitrarily small factor for planar graphs.

1.3   Our Contribution

In this paper, we study the (in)approximability of MaxDdlS on regular graphs and planar graphs for a fixed integer $d \geq 3$. Our main results are summarized as follows:

(i) For every fixed integers $d \geq 3$ and $r \geq 3$, MaxDdlS on r-regular graphs is APX-hard. In particular, when restricted to $d = r = 3$, we show that it is NP-hard to approximate MaxD3lS on 3-regular graphs within 1.00105.

(ii) We then design polynomial-time $O(r^{d-1})$-approximation and $O(r^{d-2}/d)$-approximation algorithms for MaxDdlS on r-regular graphs. (The approximation ratio of each algorithm will be analyzed precisely.) Note that the running time of each algorithm is independent from r and d.

(iii) We consider the problem when restricted to $d = r = 3$, and give a polynomial-time 2-approximation algorithm for MaxD3lS on 3-regular graphs. We note that the simple applications of the above $O(r^{d-2}/d)$-approximation algorithm yields an approximation ratio strictly greater than two. To improve the ratio to two, we sharpen and precisely analyze the approximation algorithm.

(iv) Finally, by employing the Baker's shifting technique [2], we show that MaxDdlS on planar graphs admits a PTAS for every fixed constant $d \geq 3$.

Due to the page limitation, we omit some proofs from this extended abstract.

2   Notation

Let $G = (V, E)$ be an unweighted graph, where V and E denote the set of vertices and the set of edges, respectively. $V(G)$ and $E(G)$ also denote the vertex set and the edge set of G, respectively. We denote an edge with endpoints u and v by $\{u, v\}$. For a pair of vertices u and v, the length of a shortest path from u to v, i.e., the distance between u and v is denoted by $dist_G(u, v)$, and the diameter G is defined as $diam(G) = \max_{u,v \in V} dist_G(u, v)$.

For a graph G and its vertex v, we denote the (open) neighborhood of v in G by $D_1(v) = \{u \in V(G) \mid \{v, u\} \in E(G)\}$, i.e., for any $u \in D_1(v)$, $dist_G(v, u) = 1$ holds. More generally, for $d \geq 1$, let $D_d(v) = \{w \in V(G) \mid dist_G(v, w) = d\}$ be the subset of vertices that are distance-d away from v. Similarly, let $D_1(S)$ be the open neighborhood of a subset S of vertices, $D_2(S)$ be the open neighborhood of $D_1(S)$, and so on. The degree of v is denoted by $deg(v) = |D_1(v)|$. A graph is

r-*regular* if the degree $deg(v)$ of every vertex v is exactly $r \geq 0$, and a 3-regular graph is often called *cubic* graph.

A graph G_S is a subgraph of a graph G if $V(G_S) \subseteq V(G)$ and $E(G_S) \subseteq E(G)$. For a subset of vertices $U \subseteq V$, let $G[U]$ be the subgraph induced by U. For a positive integer $d \geq 1$ and a graph G, the dth power of G, denoted by $G^d = (V(G), E^d)$, is the graph formed from $V(G)$, where all pairs of vertices $u, v \in G$ such that $dist_G(u, v) \leq d$ are connected by edges $\{u, v\}$'s. Note that $E(G) \subseteq E^d$, i.e., the original edges in $E(G)$ are retained.

We say that an algorithm ALG is a σ-approximation algorithm for MaxDdIS or that ALG's approximation ratio is at most σ if $|OPT(G)| \leq \sigma \cdot |ALG(G)|$ holds for any input G, where $ALG(G)$ is a distance-d independent set returned by ALG and $OPT(G)$ is an optimal distance-d independent set on input G.

3   APX-Hardness of MaxDdIS on Regular Graphs

In this section we analyze the inapproximability of MaxDdIS on r-regular graphs.

3.1   MaxDdIS for Cubic Graphs

First, we prove the following inapproximability for MaxD3IS on cubic (i.e., 3-regular) graphs:

Theorem 1. *There exists no σ-approximation algorithm for MaxD3IS on cubic graphs for constant $\sigma < 1.00105 < \frac{950}{949}$.*

Proof. The hardness of approximation of MaxD3IS on cubic graphs is shown by a *gap-preserving reduction* from MaxD2IS on cubic graphs. It is known [6] that there exists no σ'-approximation algorithm for the latter problem for constant $\sigma' < \frac{95}{94}$. Consider an input cubic graph $G_0 = (V_0, E_0)$ with n-vertices and m edges of MaxD2IS. Then, we construct another cubic graph $G = (V, E)$ as an instance of MaxD3IS on cubic graphs from G_0.

Let $\#OPT_2(G_0)$ (and $\#OPT_3(G)$, resp.) denote the number of vertices of an optimal distance-2 independent set in G_0 (and one of an optimal distance-3 independent set in G, resp.). Let $V_0 = \{v_1, v_2, \cdots, v_n\}$ and $E_0 = \{e_1, e_2, \cdots, e_m\}$ be vertex and edge sets of G_0, respectively. Also, let $g(n)$ be a parameter function of the instance G_0, meaning a solution size. Then, we provide the gap preserving reduction such that (C1) if $\#OPT_2(G_0) \geq g(n)$, then $\#OPT_3(G) \geq g(n) + 2m$, and (C2) if $\#OPT_2(G_0) < \frac{g(n)}{\gamma'}$ for a constant $\gamma' > 1$, then $\#OPT_3(G) < \frac{g(n)}{\gamma'} + 2m$.

From G_0, we construct the cubic graph G which consists of (i) n vertices, u^1 through u^n, which are associated with n vertices in V_0, v_1 through v_n, respectively, and (ii) m subgraphs, G_1 through G_m, which are associated with m edges in E_0, e_1 through e_m, respectively. We often call those subgraphs *edge-gadgets* in the following. See Fig. 1(a). For every p, $1 \leq p \leq m$, the pth diamond-shape gadget G_p contains ten vertices $V(G_p) = \{u_1^p, u_2^p, u_3^p, u_4^p\} \cup \{\alpha_1^p, \alpha_2^p\} \cup \{\beta_1^p, \beta_2^p, \beta_3^p, \beta_4^p\}$,

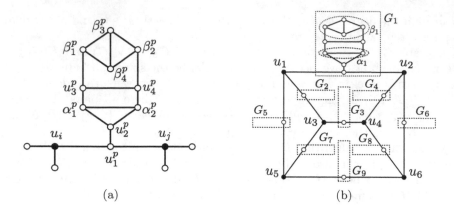

Fig. 1. (a) two vertices u_i, u_j and edge-gadget $G_p^{5,3}$ and (b) reduced graph G

and the pth edge set $E(G_p)$ has 14 edges as illustrated in Fig. 1(a). (iii) If $e_i = \{v_i, v_j\} \in E_0$, then we introduce two edges $\{u_1^p, u^i\}$ and $\{u_1^p, u^j\}$. As shown in Fig. 1(b), all the edges are replaced with edge-gadgets. This completes the reduction. One can see that the constructed graph G is cubic. Also, the above construction can be accomplished in polynomial time.

For the above construction of G, we show that G has a distance-3 independent set S such that $|S| \geq g(n) + 2m$ if and only if G_0 has a distance-2 independent set S_0 such that $|S_0| \geq g(n)$. Suppose that the graph G_0 of MaxD2IS has the distance-2 independent set $S_0 = \{v_{1^*}, v_{2^*}, \cdots, v_{g(n)^*}\}$ in G_0, where $\{1^*, 2^*, \cdots, g(n)^*\} \subseteq \{1, 2, \cdots, n\}$. Then, we select a subset of vertices $S' = \{u_{1^*}, u_{2^*}, \cdots, u_{g(n)^*}\}$ and two vertices in each edge-gadget, arbitrary one of the four pairs $\{\alpha_1^p, \beta_3^p\}$, $\{\alpha_1^p, \beta_4^p\}$, $\{\alpha_2^p, \beta_3^p\}$, and $\{\alpha_2^p, \beta_4^p\}$. Let S'' be the set of vertices in edge-gadgets. Hence $|S'| = g(n)$ and $|S''| = 2m$. One can see that $S = S' \cup S''$ is a distance-3 independent set in G since the pairwise distance in S' is at least four, the pairwise distance in S'' is at least six, and the distance between α_1^p (or α_2^p) in S'' and every vertex in S' is at least three for each p.

Conversely, suppose that the graph G has the distance-3 independent set S such that $|S| \geq g(n) + 2m$. Then, one can see that for each subgraph $G[V(G_p) \cup \{u_i, u_j\}]$, the diameter $diam(G_p)$ of G_p is five, and $diam(G[V(G_p) \cup \{u_i\}])$ (or $diam(G[V(G_p) \cup \{u_j\}])$) is six. Therefore, we can select at most three vertices as the distance-3 independent set from the graph in Fig. 1(a). If we select the other vertices in each subgraph G_p, then we can select at most two vertices of the distance-3 independent set in $G[V(G_p) \cup \{u_i, u_j\}]$. Thus, the maximum size of the distance-3 independent set in $V(G_1) \cup V(G_2) \cup \cdots \cup V(G_m)$ is at most $2m$, which means that $|S \cap \{u_1, u_2, \ldots, u_n\}| \geq g(n)$. Let $\{u_{1^*}, u_{2^*}, \cdots, u_{g(n)^*}\}$ be a subset of $g(n)$ vertices in $S \cap \{u_1, u_2, \cdots, u_n\}$. Then, the pairwise distance in the corresponding subset of vertices $\{v_{1^*}, v_{2^*}, \cdots, v_{g(n)^*}\}$ of G_0 is surely at least 2, i.e., G_0 has a distance-2 independent set S_0 such that $|S_0| \geq g(n)$. Hence, the reduction satisfies the conditions (C1) and (C2). This implies that MaxD3IS on cubic graphs cannot be approximated within $\gamma = (g(n) + 2m)/(\frac{g(n)}{\gamma'} + 2m)$.

In the remaining we obtain the value of γ: Note that a cubic graph has $m = \frac{3n}{2}$ edges. Thus, $(g(n) + 2m)/(\frac{g(n)}{\gamma'} + 2m) = (g(n) + 3n)/(\frac{g(n)}{\gamma'} + 3n)$. It is important to note that any optimal solution of MaxD2IS on a cubic graph with $n \geq 5$ is at least $\frac{n}{3}$ since Brooks' theorem says [5] that such a graph has a (proper) coloring using three colors, and hence has an independent set of cardinality at least $\frac{n}{3}$. Thus, $g(n) \geq \frac{n}{3}$ and $\gamma = (g(n) + 3n)/(\frac{g(n)}{\gamma'} + 3n) \geq \frac{10\gamma'}{9\gamma'+1}$ since $\gamma' > 1$. By setting $\gamma' = \sigma' = \frac{95}{94}$, we obtain $\gamma \geq \frac{950}{949} > 1.00105$, i.e., the approximation gap remains at least 1.00105. This completes the proof of this theorem. □

3.2   MaxDdIS for r-Regular Graphs

Next, we give the inapproximability for MaxDdIS on r-regular graphs:

Theorem 2. *There exists no σ-approximation algorithm* (i) *for* MaxDdIS *on r-regular graphs for $r \geq 5$ and constant $\sigma < \frac{575\lceil d/2 \rceil - 529}{575\lceil d/2 \rceil - 530}$. For small d and r,* (ii) *if $d = 3$ and $r = 4$, then $\sigma < 1.00122$;* (iii) *if $d = 4$ and $r = 3$, then $\sigma < 1.00191$; and* (iv) *if $d = 4$ and $r = 4$, then $\sigma < 1.00122$.*

Proof. Omitted in this abstract. □

4   Approximation Algorithms for MaxDdIS on Regular Graphs

In this section we design two approximation algorithms for MaxDdIS on regular graphs. The first one finds a (distance-2) independent set from the $(d-1)$th power of an input graph by using the previously known approximation algorithm for MaxIS. The second one iteratively executes the following: (i) Picks one vertex v into a solution and (ii) removes all vertices whose distance from the "center" vertex v is less than d. Then, we show that, from the point of view of the approximation ratio, the latter is better than the former for sufficiently large d and/or r.

4.1   Power-Graph-Based Algorithms

In this section we design an $\frac{r(r-1)^{d-1}+2r-6}{5(r-2)}$-approximation algorithm for MaxDdIS on r-regular graphs, which uses the following approximation algorithm for MaxIS, i.e., MaxD2IS as a subroutine:

Proposition 1 [3]. *There exists a polynomial-time $\frac{\Delta+3}{5}$-approximation algorithm for* MaxD2IS *on graphs with the maximum degree Δ.*

Let ALG$_2$ be such a $\frac{\Delta+3}{5}$-approximation algorithm for MaxD2IS on graphs with the maximum degree Δ. The above proposition immediately suggests the following simple algorithm: First, construct the $(d-1)$th power G^{d-1} of an input graph G, and then obtain a distance-2 independent set of G^{d-1}. The following is a description of the algorithm POWER$_d$.

Algorithm POWER$_d$

Input: r-regular graph $G = (V(G), E(G))$
Output: Distance-d independent set $DdIS(G)$ in G
Step 1. Obtain the $(d-1)$th power G^{d-1} of G by the following:
   (1-1) Compute $dist_G(u, v)$ for any pair $u, v \in V$.
   (1-2) Add an edge $\{u, v\}$ if $dist_G(u, v) \leq d - 1$.
Step 2. Apply ALG$_2$ to G^{d-1}, and then obtain a distance-2 independent
   set $ALG_2(G^{d-1})$ in G^{d-1}.
Step 3. Output $DdIS(G) = ALG_2(G^{d-1})$ as a solution.

Theorem 3. *The algorithm POWER$_d$ runs in polynomial time, and achieves an* $\frac{r(r-1)^{d-1}+2r-6}{5(r-2)}$-*approximation ratio for* MaxDd*IS* *on* r-*regular graphs.*

Proof. First, we must verify that the output $DdIS(G) = ALG_2(G^{d-1})$ of POWER$_d$ is a feasible solution for MaxDd*IS*, i.e., the distance-2 independent set in G^{d-1} is a distance-d independent set in G. Suppose for contradiction that there is a pair of vertices $u, v \in ALG_2(G^{d-1})$ (i.e., $dist_{G^{d-1}}(u, v) \geq 2$) such that $dist_G(u, v) \leq d-1$. Since $dist_G(u, v) \leq d-1$, in **Step 1** of POWER$_d$, an edge $\{u, v\}$ must be added between u and v. That is, $dist_{G^{d-1}}(u, v) = 1$ holds, which is a contradiction. Therefore, the output of POWER$_d$ is always feasible.

Next, we show the approximation ratio of POWER$_d$ by estimating the maximum degree of the $(d-1)$th power graph G^{d-1}. Now consider a vertex $v \in V(G)$. Since G is an r-regular graph, v has r neighbor vertices, i.e., $|D_1(v)| = r$. Also, $|D_2(v)| \leq r(r-1)$ holds since each neighbor vertex $u \in D_1(v)$ has at most $r-1$ neighbors, each of which is not v. That is, $|D_i(v)| \leq r(r-1)^{i-1}$ holds for each $1 \leq i \leq d-1$. Therefore, the maximum degree Δ of G^{d-1} is at most:

$$\Delta \leq r + r(r-1) + r(r-1)^2 + \cdots + r(r-1)^{d-2}$$
$$= \frac{r}{r-2}\{(r-1)^{d-1} - 1\}.$$

Since POWER$_d$ applies the $\frac{\Delta+3}{5}$-approximation algorithm ALG$_2$ for G^{d-1}, the approximation ratio of POWER$_d$ is as follows:

$$\frac{r(r-1)^{d-1} + 2r - 6}{5(r-2)}.$$

The algorithm clearly runs in polynomial time and hence this completes the proof of this theorem. □

Roughly, the approximation ratio of POWER$_d$ is $O(r^{d-1})$.

4.2 Iterative-Pick-One Algorithms

Next, we consider a naive algorithm for MaxDd*IS* on r-regular graphs, which iteratively picks a vertex v into the distance-d independent set and eliminates all the vertices in $D_1(v) \cup D_2(v) \cup \cdots \cup D_{d-1}(v)$ from candidates of the solution. Then we show its approximation ratio. Here is a description of the "pick-one" algorithm, where $DdIS(G)$ stores vertices in the distance-d independent set, B does vertices which are determined to be *not* candidates of the solution, W does vertices which can be picked in the next iteration, and V does vertices which are not processed yet:

Algorithm PICK_ONE$_d$

Input: r-regular graph $G = (V(G), E(G))$
Output Distance-d independent set $DdIS(G)$
Step 1. Set $DdIS(G) = \emptyset$, $B = \emptyset$, $W = V(G)$, and $V = V(G)$.
Step 2. If $V \neq \emptyset$, then repeat the following; else goto **Step 3**:
   Select one arbitrary vertex v from W. Then, let $B_i = \{v\} \cup \bigcup_{1 \le i \le d-1} D_i(v)$ for the ith iteration of this step, update $DdIS(G) = DdIS(G) \cup \{v\}$, $V = V \setminus B_i$, $B = B \cup B_i$, and set $W = D_1(B) \setminus B$.
Step 3. Terminate and output $DdIS(G)$ as a solution.

In order to prove the approximation ratio of the above algorithm PICK_ONE$_d$, we now provide an upper bound of the maximum number of vertices in the distance-d independent set in an input graph G with n vertices:

Lemma 1. *Consider an r-regular graph $G = (V, E)$ with $|V| = n$ vertices. Then, if $r \ge 3$ and $d \ge 4$, then the size $\#OPT_d(G)$ of optimal solutions of MaxDdIS satisfies the following inequality:*

$$
\#OPT_d(G) \le
\begin{cases}
\dfrac{3n}{r(d-2)} & d \text{ is even,} \\[2ex]
\dfrac{3n}{r(d-1)} & \text{otherwise.}
\end{cases}
$$

Proof. Given an r-regular graph G, let $OPT_d(G) = \{v_1^*, v_2^*, \cdots, v_L^*\}$ be an optimal solution of MaxDdIS and let $\#OPT_d(G) = L$. Then, if d is even, then, for every $1 \le i \le L$, consider a ball $Ball(v_i^*) = D_1(v_i^*) \cup D_2(v_i^*) \cup \cdots \cup D_{(d-2)/2}(v_i^*)$, where the center of the ball is v_i^* and its radius is $(d-2)/2$ (or, equivalently, its diameter is $(d-2)$). If d is odd, then we consider a ball $Ball(v_i^*) = D_1(v_i^*) \cup D_2(v_i^*) \cup \cdots \cup D_{(d-1)/2}(v_i^*)$ of diameter $(d-1)$. Since, for every pair of i and j ($i \neq j$), $dist_G(v_i^*, v_j^*) \ge d$ holds from the feasibility of the solution, $Ball(v_i^*) \cap Ball(v_j^*) = \emptyset$ is surely satisfied for every pair i and j. It follows that $\sum_{i=1}^{L} |Ball(v_i^*)| \le n$.

Now, we estimate the value of $\sum_{i=1}^{L} |Ball(v_i^*)|$ by considering the "smallest" r-regular graph of diameter $diam$, that is, a lower bound of the size of $|Ball(v_i^*)|$. Recently, Knor has proven [15] that the minimum number of vertices in an r-regular graph of diameter $diam$ is at least $\frac{r \cdot diam}{3}$ if $r \ge 3$ and $diam \ge 4$. As a result, the following inequality holds:

$$\sum_{i=1}^{L} |Ball(v_i^*)| \geq \frac{r \cdot diam}{3} \times L.$$

Then, we have

$$\#OPT_d(G) = L \leq \frac{3n}{r \cdot diam},$$

where $diam = d-2$ if d is even and $diam = d-1$ if d is odd as mentioned above. This completes the proof of this lemma. □

Now we calculate the number $\#ALG_d(G)$ of vertices in $DdIS(G)$ output by PICK_ONE$_d$, and obtain the following lemma:

Lemma 2. *Assume that* PICK_ONE$_d$ *finds a solution of size* $\#ALG_d(G)$, *give an r-regular graph with n vertices. Then, the following is satisfied:*

$$\#ALG(G) \geq \begin{cases} \dfrac{n(r-2)-r(r-1)^{\frac{d}{2}-1}+2}{r(r-1)^{d-1}-r(r-1)^{\frac{d}{2}-1}} & d \text{ is even,} \\[2ex] \dfrac{n(r-2)-2(r-1)^{\frac{d-1}{2}}+2r-2}{r(r-1)^{d-1}-2(r-1)^{\frac{d-1}{2}}+2r-4} & \text{otherwise.} \end{cases}$$

Proof. Omitted in this abstract. □

Theorem 4. *The approximation ratio σ of* PICK_ONE$_d$ *is as follows:*

$$\sigma = \begin{cases} \dfrac{3(r-1)^{d-1}-3(r-1)^{\frac{d}{2}-1}}{(r-2)(d-2)} + O(\frac{1}{n}) & d \text{ is even,} \\[2ex] \dfrac{3r(r-1)^{d-1}-6(r-1)^{\frac{d-1}{2}}+6r-12}{r(r-2)(d-1)} + O(\frac{1}{n}) & \text{otherwise.} \end{cases}$$

Proof. The approximation ratio σ is bounded by $\#OPT_d(G)/\#ALG_d(G)$. From the upperbound of $\#OPT_d(G)$ and the lowerbound of $\#ALG_d(G)$ shown in Lemmas 1 and 2, respectively, we can obtain this theorem. □

That is, the approximation ratio of PICK_ONE$_d$ is $O(r^{d-2}/d)$, while the approximation ratio of POWER$_d$ is $O(r^{d-1})$.

5 2-Approximation Algorithm for MaxD3IS on Cubic Graphs

In this section, as a special case, we study the approximability of MaxD3IS on cubic graphs, i.e., $d = 3$ and $r = 3$ and show the approximation ratios of POWER$_3$ and PICK_ONE$_3$. Furthermore, by a slight modification, we obtain a 2-approximation algorithm for MaxD3IS on cubic graphs.

5.1   Power-Graph-Based Algorithm

First, as an immediate consequence of Theorem 3, we have the following corollary:

Corollary 1. *The algorithm* POWER$_3$ *achieves a 2.4-approximation ratio for* MaxD3IS *on cubic graphs.*

Proof. Since the maximum degree of the second power G^2 of an input 3-regular graph G is nine, the approximation ratio is $12/5 = 2.4$.   □

5.2   Iterative-Pick-One Algorithm

In this section, we prove that PICK_ONE$_3$ achieves $2 + O(1/n)$-approximation ratio, and furthermore, the ratio can be improved into exactly 2 by a slight modification of PICK_ONE$_3$ and careful observations.

Recall that the upperbound of optimal solutions of MaxDdIS on r-regular graphs provided in Lemma 1 holds only for the case where $d \geq 4$. Then, we give an estimation of the upperbound of the maximum number of vertices in an optimal solution for the case where $r = 3$ and $d = 3$:

Lemma 3. *Consider a cubic graph* $G = (V, E)$ *with* $|V| = n$ *vertices. Then, the size* $\#OPT_3(G)$ *of every optimal solution of* MaxD3IS *satisfies the following inequality:*

$$\#OPT_3(G) \leq \frac{n}{4}.$$

Proof. Given a 3-regular graph G of n vertices, let $OPT_3(G) = \{v_1^*, v_2^*, \cdots, v_L^*\}$ be an optimal solution of MaxD3IS and let $\#OPT_3(G) = L$. Also, let $\overline{OPT_3(G)}$ be the set of vertices not in $OPT_3(G)$, i.e., $\overline{OPT_3(G)} = V(G) \setminus OPT_3(G)$. Then, three edges, say, $\{\{v_i^*, u_{i,1}\}, \{v_i^*, u_{i,2}\}, \{v_i^*, u_{i,3}\}\}$, are incident to every vertex $v_i^* \in OPT_3(G)$ for $1 \leq i \leq L$, and $u_{i,1}, u_{i,2}, u_{i,3} \in \overline{OPT_3(G)}$. Therefore, $|\overline{OPT_3(G)}| \geq 3L$. From the definition, $|\overline{OPT_3(G)}| = n - L$ holds. As a result, the following inequality is obtained:

$$\#OPT_3(G) = L \leq \frac{n}{4}.$$

This completes the proof of this lemma.   □

Consider a graph $D_2 = (\{v_1, v_2, v_3, v_4, v_5, v_6, v_7, v_8\}, \{\{v_1, v_2\}, \{v_1, v_3\}, \{v_2, v_3\}, \{v_2, v_4\}, \{v_3, v_4\}, \{v_5, v_6\}, \{v_5, v_7\}, \{v_6, v_7\}, \{v_6, v_8\}, \{v_7, v_8\}, \{v_4, v_5\}, \{v_8, v_1\}\})$ of eight vertices, which consists of two diamond graphs and two edges. One can verify that D_2 is cubic and $|OPT_3(D_2)| = 2 = 8/4$. Similarly, by circularly joining diamond graphs, we can obtain an infinite family of tight examples for Lemma 3; for a graph D_ℓ having ℓ diamond graphs (4ℓ vertices), $|OPT_3(D_\ell)| = \ell$.

Theorem 5. *The algorithm* PICK_ONE$_3$ *achieves a* $\left(2 + \frac{4}{n-2}\right)$*-approximation ratio for* MaxD3IS *on cubic graphs.*

Proof. Let $D3IS(G) = \{s_1, s_2, \cdots, s_\ell\}$ be an output of PICK_ONE$_3$, and without loss of generality, assume that PICK_ONE$_3$ picks those ℓ vertices into $D3IS(G)$ in this order, i.e., first s_1, next s_2, and so on.

(i) In the first iteration of **Step 2** of PICK_ONE$_3$, the first vertex s_1 is selected into $D3IS(G)$, then $B_1 = \{s_1\} \cup D_1(s_1) \cup D_2(s_1)$ are removed from $V(G)$, and set $V = V(G) \setminus B_1$. One can see that the number of vertices in B_1 is at most 10 since s_1 has at most three neighbors, i.e., $|D_1(s_1)| \le 3$, and each vertex in $D_1(s_1)$ has at most two other vertices, i.e., $|D_2(s_1)| \le 6$.

(ii) In the second iteration, the second vertex s_2 is selected from neighbor vertices of B_1 into $D3IS(G)$, and then $B_2 = \{s_2\} \cup D_1(s_2) \cup D_2(s_2)$ are removed from V updated in **Step 2**. The number of vertices in B_2 is again at most 10, but $|B_1 \cap B_2| \ge 2$ because there must exist at least two vertices between s_1 and s_2 from the fact $dist_G(s_1, s_2) \ge 3$. That is, $|B_2 \setminus B_1| \le 8$ and thus at most eight vertices currently in V are removed from V in the second iteration. Similarly, when s_i for $3 \le i \le \ell$ are selected into $D3IS(G)$, at most eight vertices in V are removed from V. Therefore,

$$|B_1| + |B_2 \setminus B_1| + \cdots + |B_\ell \setminus (\bigcup_{1 \le i \le \ell-1} B_i)| \le 10 + 8(\ell - 1).$$

At the time when PICK_ONE$_3$ terminates, $V = \emptyset$ and thus the following inequality holds since the value of the left-hand side of the above inequality is equal to n:

$$10 + 8(\ell - 1) \ge n.$$

Namely,

$$\ell \ge \frac{n-2}{8}.$$

Since $\#OPT_3(G) \le \frac{n}{4}$, the approximation ratio of PICK_ONE$_3$ is as follows:

$$\frac{\#OPT_3(G)}{\ell} \le 2 + \frac{4}{n-2}. \qquad \square$$

To improve the above ratio of $2 + \varepsilon$ ($\varepsilon > 0$) to 2, we slightly modify **Step 2** of PICK_ONE$_3$, and get the following algorithm, called REV_PICK_ONE$_3$:

Algorithm REV_PICK_ONE$_3$:

Input: 3-regular graph $G = (V(G), E(G))$
Output: Distance-3 independent set $D3IS(G)$
Step 1. Set $D3IS(G) = \emptyset$, $B = \emptyset$, $W = V(G)$, and $V = V(G)$.
Step 2. If $V \ne \emptyset$, then repeat the following; else goto **Step 3**:
    Select one vertex v from W such that $|(D_1(v) \cup D_2(v)) \setminus B|$ is minimum among all vertices in W. Then, let $B_i = \{v\} \cup D_1(v) \cup D_2(v)$ in the ith iteration of this step, update $D3IS(G) = D3IS(G) \cup \{v\}$, $V = V \setminus B_i$, $B = B \cup B_i$, and set $W = D_1(B) \setminus B$.
Step 3. Terminate and output $D3IS$ as a solution.

Recall that PICK_ONE$_3$ selects an arbitrary vertex v in each iteration in **Step 2**. On the other hand, REV_PICK_ONE$_3$ selects a vertex v such that $|(D_1(v) \cup D_2(v)) \setminus B|$ is minimum among all vertices in W in each iteration, only which is the difference between PICK_ONE$_3$ and REV_PICK_ONE$_3$.

Theorem 6. *The algorithm REV_PICK_ONE$_3$ runs in polynomial time, and achieves a 2-approximation ratio for MaxD3IS on cubic graphs.*

Proof. Again, let $D3IS(G) = \{s_1, s_2, \cdots, s_\ell\}$ be an output of REV_PICK_ONE$_3$, and assume that REV_PICK_ONE$_3$ picks those ℓ vertices into $D3IS(G)$ in this order. That is, in the first iteration, REV_PICK_ONE$_3$ picks s_1 such that $|(D_1(s_1) \cup D_2(s_1))|$ is minimum among all vertices in $V(G)$ since $B = \emptyset$. Then, we update $B = B_1 = \{s_1\} \cup D_1(s_1) \cup D_2(s_1)$. We have the following three cases according to the size of $|B_1|$: (i) $|B_1| \leq 8$, (ii) $|B_1| = 9$, and (iii) $|B_1| = 10$.

(i) First consider the case where $|B_1| \leq 8$. Similarly to the proof of Theorem 5, in the second iteration of **Step 2**, the second vertex s_2 is selected from neighbor vertices of B_1 into $D3IS(G)$, and then $B_2 = \{s_2\} \cup D_1(s_2) \cup D_2(s_2)$ are removed from V updated in **Step 2**. Recall that $|B_2 \setminus B_1| \leq 8$. Similarly, when s_i for $3 \leq i \leq \ell$ are selected into $D3IS(G)$, $|B_i \setminus (\bigcup_{1 \leq j \leq i-1} B_j)| \leq 8$ holds. Therefore,

$$|B_1| + |B_2 \setminus B_1| + \cdots + |B_\ell \setminus (\bigcup_{1 \leq i \leq \ell-1} B_i)| \leq 8\ell. \tag{1}$$

Namely,

$$\ell \geq \frac{n}{8}.$$

Since $\#OPT_3(G) \leq \frac{n}{4}$, the approximation ratio of REV_PICK_ONE$_3$ is as follows:

$$\frac{\#OPT_3(G)}{\ell} \leq 2.$$

(ii) Next suppose that $|B_1| = 9$. Similarly, again $|B_i \setminus (\bigcup_{1 \leq j \leq i-1} B_j)| \leq 8$ holds for the ith iteration, $2 \leq i \leq \ell$. It is now important to note that the number n of vertices in the cubic graph G must be even since the degree r is odd. Thus, actually, at least one of $|B_i \setminus (\bigcup_{1 \leq j \leq i-1} B_j)|$ for $2 \leq i \leq \ell$ must be at most seven. Therefore, the left-hand side of the inequality (1) is at most $9 + 7 + 8(\ell - 2) = 8\ell$. As a result, the inequality (1) holds again, which means that the approximation ratio is two.

(iii) Finally, suppose that $|B_1| = 10$, which implies that $|\{s_i\} \cup D_1(s_i) \cup D_2(s_i)| = 10$ for every vertex s_i since $|\{s_1\} \cup D_1(s_1) \cup D_2(s_1)|$ is minimum. Indeed, for example, $|\{v\} \cup D_1(v) \cup D_2(v)| = 10$ holds for any vertex v in a C_4-*free cubic graph* (i.e., the graph including no induced cycles of length 3 and 4). Fortunately, if at least one, say, $|B_i \setminus (\bigcup_{1 \leq j \leq i-1} B_j)|$ is seven, then there must exist at least one iteration, say, $i' (\neq i)$ such that $|B_{i'} \setminus (\bigcup_{1 \leq j \leq i'-1} B_j)| \leq 7$ holds since n is even. That is, the inequality (1) is true as well. Unfortunately, however, if $|B_i \setminus (\bigcup_{1 \leq j \leq i-1} B_j)| = 8$ holds for every $2 \leq i \leq \ell$, then the ratio of REV_PICK_ONE$_3$ is $2 + 4/(n-2)$ similarly to PICK_ONE$_3$. Now, as the worst case, we suppose that in the second through the $(\ell - 1)$th iterations, s_2 through $s_{\ell-1}$ are selected and $|B_2 \setminus B_1|$ through $|B_{\ell-1} \setminus (\bigcup_{1 \leq j \leq \ell-2} B_j)|$ are all eight. Then, we take a look at the last iteration in detail. (iii-1) If the current V has at least nine vertices, then we can get further two vertices in the distance-3 independent set since $|B_\ell \setminus (\bigcup_{1 \leq j \leq \ell-1} B_j)| \leq 8$, which is a contradiction from the assumption of $|D3IS(G)| = \ell$. Thus, (iii-2) we can assume that the number of the remaining vertices in V is at most eight after the $(\ell - 1)$th iteration. Then, one can see that if those eight vertices are connected, then we again get two vertices in the distance-3 independent set, which is another contradiction. (iii-3) Now suppose that the remaining graph $G[V]$ has at least two connected components. Then, there must exist a vertex s_ℓ such that $|B_\ell \setminus (\bigcup_{1 \leq j \leq \ell-1} B_j)| \leq 5$. As a result, again we can obtain the inequality (1), which follows that the approximation ratio is two. This completes the proof of this theorem.   □

6   Approximation Scheme for Planar Graphs

An *outerplanar* graph (often called a 1-*outerplanar* graph) is a graph that can be drawn in the plane without any edge-crossing such that all vertices lie on the unbounded face. A planar graph G is said to be k-*outerplanar* for $k \geq 2$ if it has a plane-embedding such that by removing the vertices on the unbounded face, we obtain a $(k-1)$-outerplanar graph; the deleted vertices form the kth *layer* of G. Note that every planar graph G can be regarded as a k-outerplanar graph for some integer k, although k can be $\Omega(\sqrt{|V(G)|})$. Also note that the treewidth of a k-outerplanar graph is at most $3k + 1$. The outerplanar factor k plays an important role in many polynomial-time approximation schemes based on the *Baker's shifting technique* for NP-hard optimization problems on planar graphs [2]. The Baker's shifting technique can be applied to MaxDdIS on planar graphs, as follows:

Algorithm SHIFTING$_d$

Input: D-outerplanar graph G
Output: Distance-d independent set $DdIS(G)$ of G
Step 1. For each $i \in \{1, 2, \ldots, k\}$, repeat the following:
    (1-1) Delete all vertices in layers i through $i + (d-2)$, $k + i + (d-2)$ through $k + i + 2(d-2)$, $2k + i + 2(d-2)$ through $2k + i + 3(d-2)$, and so on. Let G_i be the resulting graph.
       – Note that each connected component of G_i is a $(k-1)$-outerplanar graph, and hence its treewidth is at most $3k - 2$.
    (1-2) Solve MaxDdIS for each connected component of G_i, and obtain an optimal distance-d independent set S_i^* of G_i.
Step 2. Output the best S^* among the k obtained distance-d independent sets S_1^* through S_k^* as the solution $DdIS(G)$.

Theorem 7. *For a fixed constant $d \geq 2$, MaxDdIS admits a polynomial-time approximation scheme for planar graphs.*

Proof. As a seminal result of Courcelle [7], it is known that every problem definable in monadic second-order logic can be solved for graphs with bounded treewidth in time linear in the number of vertices of the graph. By a simple extension of the independent set problem (i.e., MaxD2IS), MaxDdIS can be also defined in monadic second order logic. Therefore, MaxDdIS can be solved in linear time (although its running time depends exponentially on the treewidth and the distance d). Thus, the algorithm SHIFTING$_d$ runs in time polynomial in n, which is the number of vertices.

Let S be any optimal distance-d independent set in a given planar graph. Let S_i be the distance-d independent set obtained from S by deleting all vertices in layers i through $i+(d-2)$, $k+i+(d-2)$ through $k+i+2(d-2)$, $2k+i+2(d-2)$ through $2k + i + 3(d - 2)$, and so on. Let S^* be the output of the algorithm SHIFTING$_d$, and S_i^* be the distance-d independent set of G_i (and hence of G) obtained by Step 1–2. From the definitions of these sets, both $|S_i| \leq |S_i^*|$ and $|S_i^*| \leq |S^*|$ hold for every $i \in \{1, 2, \ldots, k\}$. Then, since $|S_i| \leq |S_i^*|$ for every $i \in \{1, 2, \ldots, k\}$, we have

$$|S_1| + |S_2| + \cdots + |S_k| \leq |S_1^*| + |S_2^*| + \cdots + |S_k^*|.$$

Next, since G_i (or S_i) does not include any vertices in layers i through $i+(d-2)$, $k+i+(d-2)$ through $k+i+2(d-2)$, $2k+i+2(d-2)$ through $2k+i+3(d-2)$, and so on, the following inequality holds:

$$|S_1| + |S_2| + \cdots + |S_k| = (k - (d - 1))|S|.$$

Since $|S^*| = \max\{|S_i^*| : 1 \leq i \leq k\}$, we have

$$|S_1^*| + |S_2^*| + \cdots + |S_k^*| \leq k|S^*|.$$

Therefore, the following holds:

$$(k - (d - 1))|S| \leq k|S^*|,$$

that is,

$$\frac{|S|}{|S^*|} \leq 1 + \frac{d - 1}{k - (d - 1)}.$$

Thus, by setting $\varepsilon = \frac{d-1}{k-(d-1)}$, we can conclude that SHIFTING$_d$ is a $(1 + \varepsilon)$-approximation algorithm, that is, it is a polynomial-time approximation scheme for MaxDdIS on planar graphs. This completes the proof.

References

1. Agnarsson, G., Damaschke, P., Halldórsson, M.H.: Powers of geometric intersection graphs and dispersion algorithms. Discret. Appl. Math. **132**, 3–16 (2004)
2. Baker, B.S.: Approximation algorithms for NP-complete problems on planar graphs. J. ACM **41**(1), 153–180 (1994)
3. Berman, P., Fujito, T.: On approximation properties of the independent set problem for low degree graphs. Theory Comput. Syst. **32**(2), 115–132 (1999)
4. Brandstädt, A., Giakoumakis, V.: Maximum weight independent sets in hole- and co-chair-free graphs. Inf. Process. Lett. **112**, 67–71 (2012)
5. Brooks, R.L.: On colouring the nodes of a network. In: Proceedings of Cambridge Philosophical Society, Math. Phys. Sci. **37**, 194–197 (1941)
6. Chlebík, M., Chlebíková, J.: Complexity of approximating bounded variants of optimization problems. Theoret. Comput. Sci. **354**, 320–338 (2006)
7. Courcelle, B.: The monadic second-order logic of graphs. I. Recognizable sets of finite graphs. Inf. Comput. **85**(1), 12–75 (1990)
8. Eto, H., Guo, F., Miyano, E.: Distance-d independent set problems for bipartite and chordal graphs. J. Comb. Optim. **27**(1), 88–99 (2014)
9. Gavril, F.: Algorithms for minimum coloring, maximum clique, minimum covering by cliques, and maximum independent set of chordal graph. SIAM J. Comput. **1**, 180–187 (1972)
10. Gavril, F.: Algorithms on circular-arc graphs. Networks **4**, 357–369 (1974)
11. Golumbic, M.C.: The complexity of comparability graph recognition and coloring. Computing **18**, 199–208 (1977)
12. Halldórsson, M.M., Radhakrishnan, J.: Greed is good: approximating independent sets in sparse and bounded-degree graphs. Algorithmica **18**(1), 145–163 (1997)
13. Halldórsson, M.M., Radhakrishnan, J.: Improved approximations of independent sets in bounded-degree graphs. In: Schmidt, E.M., Skyum, S. (eds.) SWAT 1994. LNCS, vol. 824, pp. 195–206. Springer, Heidelberg (1994). doi:10.1007/3-540-58218-5_18
14. Harary, F.: Graph Theory. Addison-Wesley, Boston (1969)
15. Knor, M.: Smallest regular graphs of given degree and diameter. Discuss. Math. Graph Theory **34**, 187–191 (2014)
16. Lozin, V.V., Milanič, M.: A polynomial algorithm to find an independent set of maximum weight in a fork-free graph. J. Discret. Algorithms **6**, 595–604 (2008)
17. Minty, G.J.: On maximal independent sets of vertices in claw-free graphs. J. Combin. Theory Ser. B **28**, 284–304 (1980)
18. Zuckerman, D.: Linear degree extractors and the inapproximability of max clique and chromatic number. Theory Comput. **3**(1), 103–128 (2007)

Algorithmic Aspects of Disjunctive Total Domination in Graphs

Chin-Fu Lin and Sheng-Lung Peng[✉]

Department of Computer Science and Information Engineering,
National Dong Hwa University, Hualien 97401, Taiwan
j813811987@gmail.com, slpeng@mail.ndhu.edu.tw

Abstract. For a graph $G = (V, E)$, $D \subseteq V$ is a dominating set if every vertex in $V \setminus D$ has a neighbor in D. If every vertex in V has to be adjacent to a vertex of D, then D is called a total dominating set of G. The (total) domination problem on G is to find a (total) dominating set D of the minimum cardinality. The (total) domination problem is well-studied. Recently, the following variant is proposed. Vertex subset D is a disjunctive total dominating set if every vertex of V is adjacent to a vertex of D or has at least two vertices in D at distance 2 from it. The disjunctive total domination problem on G is to find a disjunctive total dominating set D of the minimum cardinality. For the complexity issue, the only known result is that the disjunctive total domination problem is NP-hard on general graphs. In this paper, by using a minimum-cost flow algorithm as a subroutine, we show that the disjunctive total domination problem on trees can be solved in polynomial time. This is the first polynomial-time algorithm for the problem on a special class of graphs. Besides, we show that the problem remains NP-hard on bipartite graphs and planar graphs.

Keywords: Trees · Total domination · Disjunctive total domination · Minimum-cost flow algorithm

1 Introduction

Let $G = (V, E)$ be a simple and undirected graph with vertex set V and edge set E. For two vertices $u, v \in V$, the distance between u and v is denoted as $d(u, v)$ which is the length of the shortest path from u to v. The *open neighborhood* of a vertex v is denoted by $N(v) = \{u \in V \mid (u, v) \in E\}$. The *close neighborhood* of vertex v is denoted by $N[v] = N(v) \cup \{v\}$. The concept of open (close) neighborhood can be extended to a vertex subset. For a vertex set S, we let $N(S) = \{u \in V \mid (u, v) \in E$ and $v \in S\}$ and $N[S] = N(S) \cup S$.

A vertex subset $D \subseteq V$ is called a *dominating set* if every vertex in $V \setminus D$ has a neighbor in D, *i.e.*, $V \setminus D \subseteq N(D)$. If every vertex in V has to be adjacent to a vertex of D, then D is called a *total dominating set* of G, *i.e.*, $V = N(D)$ for a total dominating set D. The (*total*) *domination problem* on G is to find a (total) dominating set D of the minimum cardinality. The (total) domination

© Springer International Publishing AG 2016
T.-H.H. Chan et al. (Eds.): COCOA 2016, LNCS 10043, pp. 285–293, 2016.
DOI: 10.1007/978-3-319-48749-6_21

problem is well-studied. For some theoretical and algorithmic results, please refer [1,2,5,6,12,13,15,16,21,22].

Recently, a variant of domination problem called disjunctive domination problem is proposed. A vertex subset D is a *disjunctive dominating set* if every vertex of $V\backslash D$ is adjacent to a vertex of D or has at least two vertices in D at distance 2 from it. The *disjunctive domination problem* on G is to find a disjunctive dominating set D of the minimum cardinality [4,7,8]. Some algorithmic results are studied in [19]. Similarly, if every vertex of V is adjacent to a vertex of D or has at least two vertices in D at distance 2 from it, then D is called a *disjunctive total dominating set* of G. The *disjunctive total domination problem* (DTDP) on G is to find the minimum cardinality of a disjunctive total dominating set [11]. Some theoretical results are studied in [9,10].

For the disjunctive total domination problem, there are few results about the issue of algorithmic complexity. In [7], the authors proposed a linear-time algorithm for solving the disjunctive domination problem on trees based on the dynamic programming approach. In this paper, we propose the first polynomial-time algorithm for the DTDP on trees. Our algorithm runs from leaves to an internal vertex. For each phase we determine a partial dominating set that can dominate all the undominated leaves. This partial dominating set is determined by using a minimum-cost maximum-flow algorithm [14,18]. Then, we delete all the dominated vertices (including some non-leaf vertices) and the next phase is initialized and processed again until all the vertices are disjunctively dominated. For the remaining of this paper, we propose our polynomial-time algorithm for DTDP on trees in Sect. 2. The NP-hardness results are presented in Sect. 3. Finally, we give a concluding remarks in the last section.

2   Polynomial Algorithm on Trees

Trees are a class of graphs that are connected and without any cycle. In this section, the graph we considered is a tree denoted as $T = (V, E)$. Assume that $L(T)$ is the set of all leaves of T. Many tree algorithms using dynamic programming usually take a tree as a rooted tree. Then, a bottom-up process from leaves to root can work for solving many combinatorial problems on trees. However, for DTDP, the behavior of our algorithm runs like an algorithm for finding the center of a tree. A center-finding algorithm works as follows. At first, we delete all the leaves from the input tree. We call it a deletion process. If the resulting tree is neither a vertex nor an edge, then we do the deletion process on the resulting tree. The algorithm runs deletion process many times until the final tree becomes a vertex or an edge. Vertex (vertices) in the final tree is called the center of the tree. This is also a standard approach for many tree algorithms.

Labeling is another useful tool for many domination-like algorithms. In our algorithm, we use numbers in $\{0, 1, 2\}$ to denote the status of a vertex. For convenience, for each vertex v, we define $N_1(v) = N(v)$ and $N_2(v) = \{u \mid d(u, v) = 2\}$. With respect to a partial dominating set D, the label of vertex v is defined as follows.

Definition 1. *The label lab(v) for a vertex v is defined as follows:*

1. $lab(v) = 0$ *if* $|N_1(v) \cap D| \geq 1$ *or* $|N_2(v) \cap D| \geq 2$
2. $lab(v) = 1$ *if* $|N_2(v) \cap D| = 1$
3. $lab(v) = 2$ *otherwise*

Initially, the label of every vertex in V is 2 since the partial dominating set D is empty. For each phase (or iteration), we only consider how to make the label of each leaf become 0, *i.e.*, disjunctively total dominated. To do this, we transform the tree into an instance of network flow problem. Then, we determine which vertex should be included in D. Once we select a vertex u into D. We then update the label of each vertex in $N_1(u) \cup N_2(u)$ according to the definitions of labels. Let us consider the following tree T as an example.

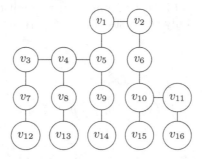

Fig. 1. A tree T with 16 vertices.

Similar to most domination problems on trees, the following lemma is not hard to be proved.

Lemma 1. *For any tree T, there is a disjunctive total dominating set D such that D does not contain any leaf vertices.*

By Lemma 1, our first intension is to find a partial dominating set to dominate all the leaves of T. To find such a set, we transform the tree into an instance of a flow network. For each leaf v, we only need to consider vertices in $N_1(v) \cup N_2(v)$ as candidates for disjunctively total dominating v. Thus we obtain two vertex subsets X and Y such that $X = \bigcup_{v \in L(T)} N_1(v) \cup N_2(v)$ and $Y = \{v \mid v \in L(T)\}$. For constructing a flow network $\mathcal{N}$, we need two extra vertices, namely, source vertex s and sink vertex t. Note that edges in $\mathcal{N}$ are directed and weighted. Thus, the direction of an edge (u, v) in $\mathcal{N}$ is from u to v. Finally, the flow network $\mathcal{N} = (s, t, X \cup Y, E')$ can be constructed as follows.

Definition 2. *In $\mathcal{N} = (s, t, X \cup Y, E')$, E' is defined as follows:*

1. $(s, x) \in E'$ *for every vertex $x \in X$.*
2. $(x, y) \in E'$ *if $x \in X$, $y \in Y$, and $d(x, y) \leq 2$.*
3. $(y, t) \in E'$ *for every vertex $y \in Y$.*

Since we consider the minimum-cost maximum-flow problem, edges in $\mathcal{N}$ are directed and weighted. For convenience, let $\deg(u) = |\{v \mid (u, v) \in E'\}|$.

Definition 3. *In* $\mathcal{N} = (s, t, X \cup Y, E')$, *for each edge* $(u, v) \in E'$, *we define its capacity and cost as follows:*

1. $u \in X$ *and* $v \in Y$: *capacity* $c(u, v) = 3 - d(u, v)$ *and cost* $w(u, v) = 1$
2. $u = s$ *and* $v \in X$: *capacity* $c(u, v) = \sum_{y \in Y} c(v, y)$ *and cost* $w(u, v) = \frac{1}{\deg(u)}$
3. $u \in Y$ *and* $v = t$: *capacity* $c(u, v) = lab(u)$ *and cost* $w(u, v) = 1$

Note that in $\mathcal{N}$, the capacity of each edge contributes the dominating power. However, the cost of each edge lets the flow algorithm prefer vertices with more dominating power that will minimize the cardinality of the selected disjunctive total dominating set. Figure 2 is the resulting flow network that is constructed from the tree T depicted in Fig. 1. For simplicity, we omit the cost for each edge. It is not hard to see that only the costs of v_{10} and v_{11} are $\frac{1}{2}$; the others are 1.

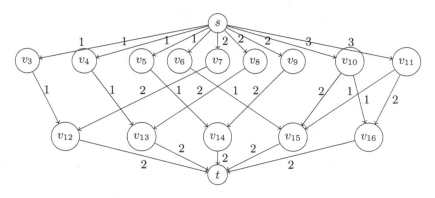

Fig. 2. The flow network instance transformed from tree T with capacity.

For a tree $T = (V, E)$ and a vertex subset $U \subseteq V$, let $T[U]$ be the forest induced by U. We have the following lemma.

Lemma 2. *Let* $\mathcal{N} = (s, t, X \cup Y, E')$ *be the flow network constructed by* T. *Let* f *be the minimum-cost maximum-flow of* $\mathcal{N}$. *Then,* $\{x \mid x \in X$ *and* x *is in any nonzero flow path from* s *to* $t\}$ *is an optimal disjunctive total dominating set of* $T[X \cup Y]$.

By considering the flow network in Fig. 2, we run a minimum-cost maximum-flow algorithm [14,18] and obtain a partial dominating set $D = \{v_7, v_8, v_9, v_{10}, v_{11}\}$. Then we update vertex labels and delete each leaf whose label is 0 until there is no label-0 leaf. For convenience, we list the algorithm for clearing label-0 leaves as follows.

Algorithm 1. Algorithm for removing label-0 leaves

Input: A tree $T = (V, E)$
Output: A tree without label-0 leaf
1: initially, a queue $Q = \emptyset$;
2: put every label-0 leaf into Q;
3: **for** $Q \neq \emptyset$ **do**
4:     $v = Dequeue(Q)$;
5:     let $(v, u) \in E$;
6:     delete v from V;
7:     **if** u becomes a leaf and $label(u) = 0$ **then**
8:         $Enqueue(Q, u)$;
9:     **end if**
10: **end for**
11: **return** the resulting tree T

Consider our example again. The resulting tree is depicted in Fig. 3. Note that, in Fig. 3, the labels of v_1 and v_2 are 1. However, the labels of v_3, v_4, and v_5 are 0 and the labels of v_7, v_8, and v_9 are 2.

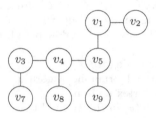

Fig. 3. The resulting tree for deleting label-0 leaves after a partial dominating set is determined.

For completeness, we construct the flow network of the remaining tree in Fig. 4. It is not hard to check that $\{v_3, v_5\}$ is a disjunctive total dominating set. Thus, by combining the previous partial dominating set, we obtain that $\{v_3, v_5, v_7, v_8, v_9, v_{10}, v_{11}\}$ is a disjunctive total dominating set of the tree T (depicted in Fig. 1).

Note that during the computation of disjunctive total dominating set, for each phase, we have to construct a flow network $\mathcal{N}$ from the current tree T. Since our $\mathcal{N}$ is four layers, namely, s, t, X, and Y (also known as bipartite, i.e., X and Y). Thus, if we cannot find X and Y from T, then the network-flow algorithm has to be terminated. In this final phase, T is possible to be empty, a vertex, or all the vertices of T are leaves, i.e., an edge (u, v). If T contains only one edge (u, v), then $\{u, v\}$ is the disjunctive total dominating set of T. The detailed algorithm is listed as follows.

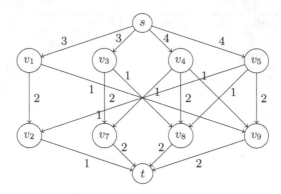

Fig. 4. The flow network instance transformed from the tree in Fig. 3.

Algorithm 2. DTDP-algorithm

Input: A tree $T = (V, E)$
Output: A disjunctive total dominating set D
 1: initially, $lab(v) = 2$ for each vertex v in V and $D = \emptyset$;
 2: **for** T is not empty **do**
 3:     construct flow network $\mathcal{N} = (s, t, X \cup Y, E')$ from T;
 4:     run a minimum-cost maximum-flow algorithm on $\mathcal{N}$;
 5:     let $D' = \{x \in X \mid x$ is in any nonzero flow path from s to $t\}$;
 6:     $D = D \cup D'$;
 7:     update vertex labels according to D';
 8:     let T be the resulting tree by running Algorithm 1;
 9:     **if** T contains only one vertex u **then**
10:         v be any vertex in $N(u)$ in the original tree;
11:         $D = D \cup \{v\}$;
12:         let $lab(u) = 0$ and $T = \emptyset$
13:     **end if**
14:     **if** T contains only one edge (u, v) **then**
15:         $D = D \cup \{u, v\}$;
16:         let $lab(u) = lab(v) = 0$ and $T = \emptyset$
17:     **end if**
18: **end for**
19: **return** D

Theorem 1. *For any tree T, Algorithm 2 computes an optimal disjunctive total dominating set of T.*

Proof. The correctness can be proved by induction on the number of iterations in Algorithm 2. We omit the detail in this conference version. □

Now we consider the time complexity issue for Algorithm 2. For any network $\mathcal{N} = (s, t, X \cup Y, E')$, let $n = |\{s, t\} \cup X \cup Y|$ and $m = |E'|$. Then the minimum-cost maximum-flow problem on $\mathcal{N}$ can be solved in $O(m \log n(m + n \log n))$

time [18]. However, in Algorithm 2, the flow network is constructed from $T = (V, E)$. Thus, $n = O(|V|)$. Since T is a tree, each leaf has only one vertex at the distance 2. Therefore, $m = O(|V|)$. In the worst case, the number of iterations of Algorithm 2 will be at most r, the radius of T. Since $r = O(|V|)$, Algorithm 2 can be implemented in $O(n^3 \log^2 n)$ time. Therefore, we have the following theorem.

Theorem 2. *The disjunctive total domination problem on trees can be solved in polynomial time.*

Thought the disjunctive domination problem on tree can be solved in linear time [4], we find that in Algorithm 2 if the label of every selected vertex in D' is set to 0 at each iteration, then the final set D will be a disjunctive dominating set of T. That is, our proposed algorithm is a unified approach for these two variants of domination problem. Thus we have the following corollary.

Corollary 1. *The disjunctive domination problem on trees can be solved in polynomial time.*

3   NP-Hardness Results

Recall that for a graph $G = (V, E)$, a vertex subset $D \subseteq V$ is called a total dominating set if every vertex in V is adjacent to a vertex of D, i.e., $V = N(D)$. The *total domination problem* (TDP for short) on G is to find a total dominating set D of the minimum cardinality. It is known that TDP is NP-hard on bipartite graphs [20] and planar graphs [3]. By reducing the TDP to DTDP, we establish the hardness result of DTDP.

Given a graph $G = (V, E)$, we construct the graph $H_G = (V', E')$ as follows. For each edge $(u, v) \in E$, we add a vertex w in (u, v). The resulting graph G' is called the subdivision of G. We call these new vertices *subdivision vertices*. Then for each original (respectively, subdivision) vertex, we attach one (respectively, two) P_4, a path with 4 vertices. H_G is the final graph. Let $n = |V|$ and $m = |E|$. It is not hard to check that $|V'| = n + m + 4n + 8m = 5n + 9m$ and $|E'| = 2m + 4n + 8m = 4n + 10m$. Figure 5 shows an example of a graph $G = P_4$ and the corresponding graph H_G.

By this reduction, we can show that if there is a total dominating set of size k for G, then there is a disjunctive total dominating set of size $k + 2(n + 2m)$ for graph H_G. The darken vertices in Fig. 5 belong to the desired dominating set that shows this correspondence. Note that it is not hard to check that if G is a bipartite (planar) graph, then H_G is a bipartite (planar) graph too. Thus by the results of [3, 20], we have the following theorem.

Theorem 3. *The disjunctive total domination problem is NP-hard on bipartite graphs and planar graphs.*

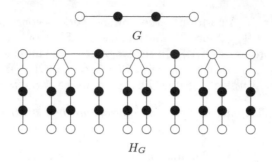

Fig. 5. Example of the reduction.

4   Conclusion

In this paper, we show that the disjunctive total domination problem can be solved in polynomial time on trees. It is interesting that our approach can be extended to some other domination problems. Though, the time complexity of our algorithm depends on the running time of a minimum-cost maximum-flow algorithm, the idea of our algorithm is easy to understand. On the other hand, we show that the disjunctive total domination problem is NP-hard on bipartite graphs and planar graphs. In fact, this idea is similar to the one in [4] for showing that the hardness of the disjunctive domination problem. Thus, by relaxing the number of dominating vertices at distance 2, we can have a more general result of hardness. In [19], the authors proposed a polynomial-time algorithm for the disjunctive domination problem on proper interval graphs. Thus, it is interesting whether the disjunctive total domination problem can be solved in polynomial time on proper interval graphs or other special classes of graphs.

Acknowledgement. This work was partially supported by the Ministry of Science and Technology of Taiwan, under Contract No. MOST 105-2221-E-259-018.

References

1. Booth, K.S., Johnson, J.H.: Dominating sets in chordal graphs. SIAM J. Comput. **11**, 191–199 (1982)
2. Chang, G.J.: Labeling algorithms for domination problems in sun-free chordal graphs. Discrete Appl. Math. **22**, 21–34 (1988)
3. Garey, M.R., Johnson, D.S.: Computers and Intractability: A Guide to the Theory of NP-Completeness. W.H. Freeman, San Francisco (1979)
4. Goddard, W., Henning, M.A., McPillan, C.A.: The disjunctive domination number of a graph. Quaestiones Math. **37**, 547–561 (2014)
5. Henning, M.A.: Graphs with large total domination number. J. Graph Theor. **35**, 21–45 (2000)
6. Henning, M.A.: A survey of selected recent results on total domination in graphs. Discrete Math. **309**, 32–63 (2009)

7. Henning, M.A., Marcon, S.A.: Domination versus disjunctive domination in trees. Discrete Appl. Math. **184**, 171–177 (2015)
8. Henning, M.A., Marcon, S.A.: Domination versus disjunctive domination in graphs. Quaestiones Math. **39**, 261–273 (2016)
9. Henning, M.A., Naicker, V.: Graphs with large disjunctive total domination number. Discrete Math. Theor. Comput. Sci. **17**, 255–282 (2015)
10. Henning, M.A., Naicker, V.: Bounds on the disjunctive total domination number of a tree. Discussiones Math. Graph Theor. **36**, 153–171 (2016)
11. Henning, M.A., Naicker, V.: Disjunctive total domination in graphs. J. Comb. Optim. **31**, 1090–1110 (2016)
12. Henning, M.A., Yeo, A.: Total Domination in Graphs. Springer, New York (2013). ISBN:978-1-4614-6525-6
13. Keil, J.M.: The complexity of domination problems in circle graphs. Discrete Appl. Math. **42**, 51–63 (1993)
14. Kovács, P.: Minimum-cost flow algorithms: an experimental evaluation. Optim. Methods Softw. **30**, 94–127 (2015)
15. Kratsch, D., Stewart, L.: Total domination and transformation. Inf. Process. Lett. **63**, 167–170 (1997)
16. Laskar, R., Pfaff, J., Hedetniemi, S.M., Hedetneimi, S.T.: On the algorithmic complexity of total domination. SIAM. J. Algebraic Discrete Methods **5**, 420–425 (1984)
17. Müller, H., Brandstädt, A.: The NP-completeness of steiner tree and dominating set for chordal bipartite graphs. Theor. Comput. Sci. **53**, 257–265 (1987)
18. Orlin, J.B.: A faster strongly polynomial minimum cost flow algorithm. Oper. Res. **41**, 338–350 (1993)
19. Panda, B.S., Pandey, A., Paul, S.: Algorithmic aspects of disjunctive domination in graphs. In: Xu, D., Du, D., Du, D. (eds.) COCOON 2015. LNCS, vol. 9198, pp. 325–336. Springer, Heidelberg (2015). doi:10.1007/978-3-319-21398-9_26
20. Pfaff, J., Laskar, R.C., Hedetniemi, S.T.: NP-completeness of total and connected domination and irredundance for bipartite graphs. Technical report 428, Clemson University. Dept. Math. Sciences (1983)
21. Ramalingam, G., Pandu Rangan, C.: Total domination in interval graphs revisited. Inf. Process. Lett. **27**, 17–21 (1988)
22. Telle, J.A.: Complexity of domination-type problems in graphs. Nord. J. Comput. **1**, 157–171 (1994)

Instance Guaranteed Ratio on Greedy Heuristic for Genome Scaffolding

Clément Dallard[1], Mathias Weller[1(✉)], Annie Chateau[1,2(✉)], and Rodolphe Giroudeau[1]

[1] LIRMM - CNRS UMR 5506, Montpellier, France
{clement.dallard,mathias.weller,annie.chateau,
rodolphe.giroudeau}@lirmm.fr
[2] IBC, Montpellier, France

Abstract. The SCAFFOLDING problem in bioinformatics, aims to complete the contig assembly process by determining the relative position and orientation of these contigs. Modeled as a combinatorial optimization problem in a graph named *scaffold graph*, this problem is $\mathcal{NP}$-hard and its exact resolution is generally impossible on large instances. Hence, heuristics like polynomial-time approximation algorithms remain the only possibility to propose a solution. In general, even in the case where we know a constant guaranteed approximation ratio, it is impossible to know if the solution proposed by the algorithm is close to the optimal, or close to the bound defined by this ratio. In this paper we present a measure, associated to a greedy algorithm, determining an upper bound on the score of the optimal solution. This measure, depending on the instance, guarantees a – non constant – ratio for the greedy algorithm on this instance. We prove that this measure is a fine upper bound on optimal score, we perform experiments on real instances and show that the greedy algorithm yields near from optimal solutions.

Keywords: Genome scaffolding · Greedy heuristic · Approximation ratio

1 Introduction

Genomic studies, especially comparative genomic and genome rearrangement inference, necessitate the production of high-quality whole sequences. Such sequences are hard to obtain from High-Throughput Sequencing data, consisting of sometimes billions of small DNA fragments. Genome scaffolding, which is our concern in this paper, consists of orienting and ordering a collection of pre-assembled DNA fragments, called *contigs*. This can be done using NGS information that locally links endpoints of contigs, that is usually available for NGS data (paired-end reads). The scaffolding problem, expressible as a combinatorial optimization problem in a graph, is $\mathcal{NP}$-hard [2,11] and represents a barrier in genome sequence production. When it comes to treat very large data like

© Springer International Publishing AG 2016
T.-H.H. Chan et al. (Eds.): COCOA 2016, LNCS 10043, pp. 294–308, 2016.
DOI: 10.1007/978-3-319-48749-6_22

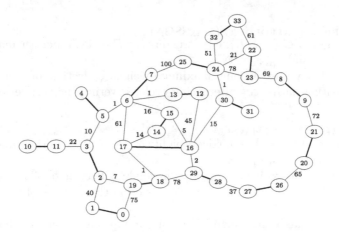

Fig. 1. A scaffold graph with 17 contigs (bold edges) and 26 (weighted) links between them, corresponding to the genome of virus *Ebola*.

eukaryotic genomes (especially plants like rice), exact methods find their limits and we have to consider heuristics.

Description of the Problem. We consider an undirected graph $G = (V, E)$ with an even number $2n$ of vertices and without self loops and a perfect matching (that is, a pairing of vertices such that each vertex is paired with *exactly one* other vertex) $M^* \subseteq E$ in G. We call a path $(v_0, v_1, \ldots, v_t)$ in G *closed* (or a cycle) if $v_0 = v_t$. Slightly abusing notation, we sometimes consider (possibly closed) paths in G as sets of edges occurring in the path. In the following, we call a (possibly closed) path p in G *alternating* with respect to M^* if each vertex on p is incident with an edge in $p \cap M^*$. Note that an alternating path has an even number ≥ 2 of vertices (≥ 4 if it is closed).

In the bioinformatic context, M^* represents contigs and, thus, all vertices are extremities of contigs and each contig has two extremities. Edges of $E \setminus M^*$ represent ways to link the contigs together. These edges are weighted by a weight function $w : E \to \mathbb{N}$ measuring the support of each of these links (e.g. the number of pairs of reads with one end in each of the two contigs). Figure 1 shows an example of such a scaffold graph. The goal is then to compute a most probable genomic structure consisting of a fixed number σ_p of linear chromosomes (paths) and σ_c of circular chromosomes (cycles). Thus, we are interested in covering G with a given number of vertex-disjoint alternating paths and cycles, maximizing the total support of the solution. Two variants of the problem are considered: in the first one, qualified "strict", the solution must exactly satisfy the number of paths and cycles; the second one allows to find less of these paths and cycles.

MAX STRICT SCAFFOLDING (MaxSSCA)
Input: a graph $G = (V, E)$, edge weights $\omega : E \to \mathbb{N}$, a perfect matching
  M^* of G, and $\sigma_\mathrm{p}, \sigma_\mathrm{c} \in \mathbb{N}$
Task: Find a vertex-disjoint maximum-weight collection of exactly σ_p
  alternating paths $\&\sigma_\mathrm{c}$ alternating cycles covering the vertices of G.

MAX SCAFFOLDING (MaxSCA)
Input: a graph $G = (V, E)$, edge weights $\omega : E \to \mathbb{N}$, a perfect matching
  M^* of G, and $\sigma_\mathrm{p}, \sigma_\mathrm{c} \in \mathbb{N}$
Task: Find a vertex-disjoint maximum-weight collection of $\leq \sigma_\mathrm{p}$ alternating paths $\&\leq \sigma_\mathrm{c}$ alternating cycles covering the vertices of G.

Exchanging "maximum-weight" for "minimum-weight" in the above definition yields corresponding minimization problems called MIN STRICT SCAFFOLD-
ING and MIN SCAFFOLDING. The decision versions of the maximization problems
are called STRICT SCAFFOLDING and SCAFFOLDING.

Related Work. For efficiency reasons, previous work on scaffolding is essentially
based on heuristics and exact methods on simplified instances of variants of the
scaffold graph. The literature offers a wide range of methods [5,6,8,9,12,15,16],
as well as a comparison thereof [10]. The problem received much attention in
the framework of complexity and approximation [2,17–19] and it is known to be
$\mathcal{NP}$-hard, even in constrained special cases. While neither MAX SCAFFOLDING
nor MIN SCAFFOLDING are constant-factor approximable [19], some polynomial-
time approximation algorithms are known for the case that G is a clique [2] or
a complete bipartite graph [19]. The parameterized complexity of MAX SCAF-
FOLDING has been considered [17,18], showing that, while an optimal solution
can be calculated in polynomial time on graphs of low treewidth, there is no
polynomial kernel for this parameter. Further, for the special case where removing the edges of M^* kills all cycles in G, MAX SCAFFOLDING remains $\mathcal{NP}$-hard
to approximate, but can be solved optimally in $O(n^{2\sigma_p+1})$ time [19].

Some of our previous work on the problem is dedicated to approximation
algorithms. We described algorithms with a ratio of three and two in special
cases [2,17,19], in a first attempt to control the extent to which a heuristic
solution diverges from the optimal. This result was improved by randomization
techniques [3]. However, since this ratio in obtained for complete graphs, or
graphs incompletely describing real instances, it is not completely satisfying.
When applying this approach to real instances, one needs to complete the graph
with zero-weighted edges that have are not biologically meaningful and void any
computational advantage derived from the sparsity of real-world instances.

Our Contribution. In this paper, we focus on the greedy heuristic described by
Chateau and Giroudeau [2]. This algorithm was proven to guarantee a constant
approximation ratio of three, which is tight[1], on complete graphs. However, this

---

[1] This means that instances for which the computed solution has a third of the optimal
weight exist. It does not exclude better approximation algorithms.

bound is of pure theoretical nature and we suspected the solutions to behave much better in practice. Hence, we propose here a measure of quality computed for each instance, instead of improving the guaranteed constant ratio on special instances. The main idea is to exploit the greedy process and, at each step, analyze to what extent the greedy choice diverges from potential optimality. We define an upper bound on the optimal score for each given instance, computed at no additional cost during the greedy algorithm. We implemented and tested this bound on a set of real instances and show that the greedy algorithm is really performant on those instances.

Organization of the Paper. In the following, Sect. 2 recalls the principles of the greedy algorithm, Sect. 3 gives a description of the dynamic upper bound and its proof. Finally, Sect. 4 presents and discusses the experimental results.

2   Greedy Algorithm

In this work, we modify the existing 3-approximation algorithm [2] which is described on complete scaffold graphs. To apply this algorithm to any instance, we may have to complete the input graph first. However, it is possible to stop the algorithm at the first edge with weight zero, and arbitrarily and quickly complete the solution.

Notation. Let G be the input graph and let $n := |V|/2 = |M^*|$. In a partial cover of G with alternating paths and cycles, we refer to the number of paths and cycles by n_p and n_c, respectively. The number of paths of length 1 (*i.e.* isolated contigs) within the partial solution is p_1, while the number of paths containing at least 3 edges is $p_{\geq 3}$. At the beginning of each iteration of the algorithm we have $n_p = p_1 + p_{\geq 3}$. We also maintain the following data structures:

- for each path p in the partial solution, we maintain its length $L(p)$,
- for each vertex v, we store the unique path $p(v)$ containing v,
- for each endpoint v of a path p in the partial solution, $o(v)$ refers to the other endpoint of p.

Note that, at the beginning of the algorithm, for each $uv \in M^*$, we have $p(u) = p(v) = uv$, $o(u) = v$, $o(v) = u$ and, since all cycles should contain at least four edges, we also have $n \geq \sigma_p + 2\sigma_c$. If, during the execution of Algorithm 1, we have $\sigma_p = 0$ and $n_c = \sigma_c > 0$ and $n_p \geq 1$, then we call any alternating path that is not part of an alternating cycle an *orphan path*.

Description. Algorithm 1 is a greedy algorithm which consecutively considers edges of $E \backslash M^*$ by decreasing order of weight. When an edge is added to the partial solution, the algorithm removes all edges adjacent to the added edge, forcing every vertex to have degree at most two in the partial solution (see Procedure Take). Algorithm 1 has been proven 3-approximate for MIN STRICT SCAFFOLD-ING on complete graphs satisfying $n \geq 2(\sigma_p + 2\sigma_c)$, with a time complexity of $\mathcal{O}(m \log_2 n)$ [2].

Algorithm 1. 3-approximated greedy algorithm.

Data: an edge-weighted clique $G = (V, E)$, a perfect matching M^* of G,
    integers σ_p and σ_c.
Result: a set $E' \subset E$ inducing a collection of exactly σ_p and σ_c alternating
    paths and cycles, respectively, in G

1 sort the edges of E by decreasing order of weight;
2 $(E, E') \leftarrow (E \setminus M^*, M^*)$; initialize $p(v)$, $o(v)$ for each $v \in V$;
3 $(n_p, n_c, p_1, p_{\geq 3}) \leftarrow (n, 0, n, 0)$;
4 **while** $n_p \neq \sigma_p$ *or* $n_c \neq \sigma_c$ **do**
5       $e = uv \leftarrow$ the first element of the ordered-list E;
6       $E \leftarrow E \setminus \{e\}$;
7   **if** $p(u) \neq p(v)$ *and* $n_p + n_c > \sigma_p + \sigma_c$ *and* `Permit(e)` **then**
8   `Take(E,E',e);` // merge two paths
9   update $L(p(u))$, $L(p(v))$, p_1, $p_{\geq 3}$, $o(o(u))$, and $o(o(v))$;
10   **else**
11   **if** $p(u) = p(v)$ *and* $n_c < \sigma_c$ *and no orphan path will remain* **then**
12   `Take(E,E',e);` // complete a cycle
13   update $L(p(u))$, n_p, n_c, and $p_{\geq 3}$;
14   **return** E'

Function Permit maintains the condition $\sigma_p + 2(\sigma_c - n_c) \leq p_1 + 2p_{\geq 3}$, which is neccessary for constructing a solution as it permits to build at least $\sigma_c - n_c$ cycles and σ_p paths from $n_p = p_1 + p_{\geq 3}$ paths. The function returns $True$ if and only if this condition will hold after adding the edge e to the partial solution.

Function. Permit(e).

Data: edge $uv \in E \setminus M^*$
Result: $True$ iff a solution can still be constructed after adding e to E'
1 **if** $L(p(u)) = L(p(v)) = 1$ **then**
2   **return** $True$;
3 **else**
4   **if** $L(p(u)) = 1$ *or* $L(p(v)) = 1$ **then**
5   **return** $(p_1 - 1 \geq 2(\sigma_c - n_c - p_{\geq 3}) + \sigma_p)$;
6   **else**
7   **return** $(p_1 \geq 2(\sigma_c - n_c - (p_{\geq 3} - 1)) + \sigma_p)$;

---

3   Dynamic Upper Bound

Experiments on real datasets point out the near optimality of the greedy algorithm [2]. Indeed, despite the fact that the algorithm has a fixed ratio of

---

Procedure. Take(E,E',e).

Data: $e \in E \backslash M^*$
1   $E' \leftarrow E' \cup \{e\}$;
2   remove all non-matching edges adjacent to e from E;

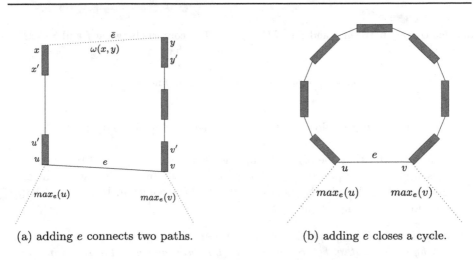

(a) adding e connects two paths.   (b) adding e closes a cycle.

Fig. 2. Calculation of $\theta(e)$ (dashed) for a partial solution (solid) with M^* (bold).

three, the real ratio of the algorithm on several real instances is close to one. If we know the weight of an optimal solution, it is easy to compute the real ratio of the weight of the approximate solution. However, if we do not have the weight of an optimal solution (for example if the instance is too big to be solved by an exact algorithm) one cannot guess the real ratio of the algorithm on the instance. Our main goal is to calculate an upper bound on the weight of an optimal solution and then guarantee an upper bound on the effective ratio of the greedy algorithm on a specific instance.

3.1   Notation and Definitions

Let $G = (V, E, M^*)$ be a scaffold graph with weight function $\omega : E \to \mathbb{N}$. For any $S \subseteq E$, we abbreviate $\omega(S) := \sum_{e \in S} \omega(e)$. Let $e = uv \notin M^*$. We say an edge f is *adjacent* to e if $e \cap f \neq \varnothing$. The set of edges of $S \subseteq E$ that are adjacent to e is denoted by $e \sqcap S := \{f \in S \mid |e \cap f| = 1\}$. We let $S_A = \{e_1, \ldots, e_{n-\sigma_p}\} \subseteq E \backslash M^*$ denote the set of edges chosen by Algorithm 1 *in that order*. We also denote by $S_{OPT} = \{f_1, \ldots, f_{n-\sigma_p}\} \subseteq E \backslash M^*$ the set of edges of an optimal solution, considered in that order during Algorithm 1. When running the algorithm, we could consider two points of view:

1. we can consider that a solution is progressively built by adding edges to M^*, and thus we denote by S_k the set of edges $M^* \cup \{e_1, \ldots, e_k\}$ corresponding

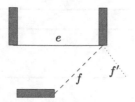
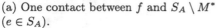
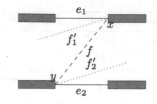

(a) One contact between f and $S_A \setminus M^*$ ($e \in S_A$).

(b) Two contacts between f and $S_A \setminus M^*$ ($e_1, e_2 \in S_A$).

Fig. 3. Contacts between $f \in S_{OPT}$ and S_A.

to a partial solution. Note that (V, S_k) consists of at least σ_p paths and at most σ_c cycles;

2. we can consider that the original clique is progressively updated by removing edges, during the update steps. Then, we denote by E_k the set E in the algorithm after the edge e_k has been added to the solution. In the same way, we denote by G_k the graph $(V, S_k \cup E_k)$.

Observation 1. *The graph G_k does not contain any edge which has been rejected by the algorithm before the edge e_k has been added to the solution.*

Definition 1. *Let $(G = (V, E), M^*, \sigma_p, \sigma_c)$ be an instance of* MAX STRICT SCAFFOLDING, *and $e_i \in S_A$ an edge of the solution given by Algorithm 1. For $u \in e_i$, we define $\max_{e_i}(u)$ to be any edge that, among all edges $f \notin M^*$ of G_{i-1} with $f \cap e_i = \{u\}$ has maximal weight.*

Definition 2. *Let $(G = (V, E), M^*, \sigma_p, \sigma_c)$ be an instance of* MAX STRICT SCAFFOLDING, *and let $e_i = uv \in S_A$ be an edge of the solution given by Algorithm 1. Let p_u and p_v be maximal-length paths of $(V, S_k \cup M^*)$ such that $p_u = (u, \dots, x)$ and $p_v = (v, \dots, y)$. Then,*

$$\theta(e_i) := \{\max{}_e(u), \max{}_e(v)\} \cup (\{xy\} \setminus \{uv\})$$

For convenience, we use $\bar{e}_i$ as shorthand for the edge xy, i.e. the edge which would close the path $p_u \cup \{e_i\} \cup p_v$ in a cycle. See Fig. 2 for an example.

Notice that the definition of θ depends on the partial solution S_i.

3.2   Computation of the Upper Bound

Lemma 1. *Let $(G = (V, E), M^*, \sigma_p, \sigma_c)$ be an instance of* MAX STRICT SCAFFOLDING, *let S_A be a solution produced by Algorithm 1 and let S_{OPT} be an optimal solution. Then,*

$$\omega(S_{OPT}) \leq \sum_{e \in S_A} \max(\omega(\theta(e)), \omega(e) + \omega(\bar{e})) \tag{1}$$

Proof. We call any function $\Gamma_i : S_{OPT} \to E$ *good* if Γ_i is injective and $\omega(f) \leq \omega(\Gamma_i(f))$ for all f for which Γ_i is defined. To prove (1), we inductively construct a good function $\Gamma : S_{OPT} \to E$ that assigns each $f \in S_{OPT}$ such that $\Gamma(f)$ is in $\{e\} \cup \theta(e)$ for some $e \in S_A$, and $\Gamma(S_{OPT})$ does not contain both e and $\theta(e) \setminus \{\bar{e}\}$ for any $e \in S_A$. For all k, we denote by S_{OPT}^k the set of edges of S_{OPT} for which Γ_k is defined.

$\underline{k = 0:}$ Then, we set $\Gamma_0(f) = f$ for all $f \in M^*$ and observe that Γ_0 is good.

$\underline{k \Rightarrow k+1:}$ we suppose that Γ_k is good and extend it to Γ_{k+1}. Let $e = e_{k+1} = xy$ be the last edge added to S_{k+1} when running Algorithm 1.

For all $f \in S_{OPT} \setminus S_{OPT}^{k-1}$, *i.e.* $\Gamma_k(f)$ is not defined:

- if $|f \cap e| = 2$ then $f = e$ and we set $\Gamma_{k+1}(f) = f$. Since e is the newly added edge to S_{k+1}, Γ_{k+1} is still injective on S_{OPT}^{k+1} and $\omega(f) \leq \omega(\Gamma(f))$;
- if $|f \cap e| = 1$ then, without loss of generality, let $x \in f$ and we set $\Gamma_{k+1}(f) = \max_e(x)$. As $\max_e(x)$ is in G_k and no two edges of $S_{OPT} \setminus M^*$ can share an endpoint, there is no edge $f' \in S_{OPT} \setminus \{f\}$ with $\Gamma_{k+1}(f') = \max_e(x)$. Hence, Γ is still injective on S_{OPT}^{k+1} and $\omega(f) \leq \omega(\Gamma(f))$;
- if $f = \bar{e}$ then we set $\Gamma_{k+1}(f) = f$. By similar argument, since two edges of $S_{OPT} \setminus M^*$ cannot share an endpoint, there is no edge $f' \in S_{OPT}$ such that $\Gamma_{k+1}(f') = \bar{e}$. Thus, Γ_{k+1} is still injective on $S_{OPT}^k + 1$ on and $\omega(f) \leq \omega(\Gamma(f))$.

Otherwise, leave $\Gamma_{k+1}(f)$ undefined.

If $\Gamma_{k+1} = \Gamma_k$, then we call e *free*. We remark that all edges rejected by the algorithm belongs to $\theta(e')$ for some $e' \subset S_A$.

At the end of the construction, $\Gamma_{n-\sigma_p}$ is good. Though, some edges of S_{OPT} may still not have an image in $\Gamma_{n-\sigma_p}$, namely *unassigned* edges. However, since $|S_A| = |S_{OPT}| = n - \sigma_p$ and $\Gamma_{n-\sigma_p}$ is injective on $S_{OPT}^{n-\sigma_p}$, there are at least as many free edges in S_A than unassigned edges in S_{OPT}. Moreover, we can assign each unassigned edge of S_{OPT} to a free edge in S_A. Indeed, because f is unassigned, either f has not been rejected by the algorithm and $\omega(f) \leq \omega(e), \forall e \in S_A$, or f was rejected and then necessarily already belongs to $\theta(e)$ for some $e \in S_A$. Hence, the association of unassigned edges is trivial. Let Γ denote the result of adding these assignments to $\Gamma_{n-\sigma_p}$.

The relation Γ is then injective on S_{OPT}, and $\forall f \in S_{OPT}$, we have $\omega(f) \leq \omega(\Gamma(f))$. In addition, for any $e \in S_A$ that is not free, $\bigcup_{f \in S_{OPT}} \Gamma(f)$ intersects either $\{e\}$ or $\{\max_e(x), \max_e(y)\}$, but not both. Actually, for $f \in S_{OPT} \setminus M^*$ and $e \in S_A$, if $\Gamma(f) = \{e\}$ then either $f = e$ or e was free before it was assigned. Moreover, if $\Gamma(f) = \max_e(x)$ (or $\max_e(y)$) then $|e \cap f| = 1$ and since neither $S_{OPT} \setminus M^*$ nor $S_A \setminus M^*$ contains a pair of adjacent edges, no $f' \in S_{OPT} \setminus M^*$ is mapped to e by Γ. Hence,

$$\sum_{f \in S_{OPT}} \omega(f) \leq \sum_{f \in S_{OPT}} \Gamma(f) \leq \sum_{e \in S_A} \max\{\omega(\theta(e)), \omega(e) + \omega(\bar{e})\}. \qquad \square$$

Lemma 2. *Let G be a clique, let $(G, M^*, \sigma_p, \sigma_c)$ be an instance of* MAX STRICT SCAFFOLDING *and let S_A be a solution computed by Algorithm 1. Then,*

$$\sum_{e \in S_A} \max(\omega(\theta(e)), \omega(e) + \omega(\bar{e})) \leq 3\omega(S_A)$$

Proof. Since, for any $e_k \in S_A, \theta(e_k)$ is defined considering the graph G_{k-1} and, by Observation 1, the edge e_k have maximum weight among E_{k-1}. Indeed it is not possible that $\bar{e}_k$ has been rejected before considering e_k, since $\bar{e}_k$ is counted in $\theta(e_k)$ when $\bar{e}_k$ closes a cycle e_k belongs to. If it is structurally possible to join both involved paths by e_k, it would also have been possible to join them by $\bar{e}_k$ and this edge would not have been rejected then. Thus $\omega(\bar{e}_k) \leq \omega(e_k)$, and more generally, for any $e' \in E_{k-1}$, we have $\omega(e') \leq \omega(e)$. Then, for each $e \in S_A$, we have $\omega(\theta(e)) \leq 3\omega(e)$ and, thus, $\sum_{e \in S_A} \max(\omega(\theta(e)), \omega(e)) \leq \sum_{e \in S_A} 3\omega(e) = 3\omega(S_A)$. □

Theorem 1 follows straightforwardly from previous Lemmas 1 and 2:

Theorem 1. *Let G be a clique, let $(G, M^*, \sigma_p, \sigma_c)$ be an instance of* MAX STRICT SCAFFOLDING *and let S_A be a solution computed by Algorithm 1. Then,*

$$\omega(S_{OPT}) \leq \sum_{e \in S_A} \max(\omega(\theta(e)), \omega(e)) \leq 3\omega(S_A)$$

3.3 Derived Results

In the following, let $G = (V, E)$ be a clique, let $I := (G, M^*, \omega, \sigma_p, \sigma_c)$ be an instance of MAX STRICT SCAFFOLDING. Let $S \subseteq E$ be a subset of (or equal to) a feasible solution for I. Then, we call S *extensible*. For any vertex $v \in V$, we define $d_S(v)$ as the degree of the vertex v in the graph (V, S). Further, any edge $e \in E \setminus S$ such that $S \cup \{e\}$ is extensible is called *valid* with respect to S.

Lemma 3. *Let G be a clique, let $I := (G, M^*, \omega, \sigma_p, \sigma_c)$ be an instance of* MAX STRICT SCAFFOLDING, *and let $S_p \supseteq M^*$ be a subset of an optimal solution for I. Let $e = xy$ and e^+ be edges in $E \setminus S_p$ that are valid for S_p, with $e \neq e^+$, $\omega(e) \geq \omega(e^+)$, and no other edge that is valid for S_p is heavier than e^+. Let*

$$\omega(\max{}_e(x)) + \omega(\max{}_e(y)) + \omega(e^+) \leq \omega(e). \tag{2}$$

Then, there is an optimal solution S' for I, such that $S_p \cup \{e\} \subseteq S'$.

Proof. Let S^* be an optimal solution for I with $S_p \subseteq S^*$ and consider the set $S = S^* \cup \{e\}$. In the following, we construct a solution S' with $S_p \cup \{e\} \subseteq S'$ and $\omega(S') \geq \omega(S^*)$. Then, we conclude that S' is an optimal solution for I. First, let $S' = S$. Then, we exhaustively apply the following rules to modify S'.

Rule 1: $\forall w \in V, d_{S'}(w) = 3$, remove from $S' \setminus S_p$ the unique edge $wz \neq e$;
Rule 2: if σ_c is strictly smaller than the number of cycles in (V, S'), then remove from S' any edge $f \in S' \setminus S_p, f \neq e$ that is on a cycle;

Rule 3: if σ_p is strictly greater than the number of paths in (V, S'), remove any
       edge $f \in S' \setminus S_p, f \neq e$ that is on a path;
Rule 4: connect path-endpoints in (V, S') to restore σ_p paths and σ_c cycles.

Note that the result S' is a feasible solution for I. Next, we prove $w(S') \geq w(S)$.

If Rule 1 is triggered, then Rule 1 removes at most two edges f_1, f_2 such that
    $f_1 \cap e = \{x\}$ and $f_2 \cap e = \{y\}$ and, as e is valid for S_p, we have $f_1, f_2 \notin S_p$.
    Since $w(f_1) \leq w(\max_e(x))$ and $w(f_2) \leq w(\max_e(y))$, (2) implies $w(f_1) +$
    $w(f_2) \leq w(e)$. Thus, $w(S' \setminus \{f_1, f_2\}) \geq w(S^*)$.
If Rules 1 and 2 are triggered, then Rule 1 removes at most two edges f_1, f_2 with
    $w(f_1) + w(f_2) \leq w(\max_e(x)) + w(\max_e(y))$ and Rule 2 one edge f. Further,
    $w(f) \leq w(e^+)$ by choice of e^+. Thus, (2) implies $w(S' \setminus \{f_1, f_2, f\}) \geq w(S^*)$.
If Rule 3 is triggered, then the edge $f \in S' \setminus S_p, f \neq e$ exists as otherwise, $S_p \cup \{e\}$
    is not extensible, contradicting the choice of e. Moreover, $w(f) \leq w(e^+)$ by
    choice of e^+ and, thus, (2) implies $w(S' \setminus \{f\}) \geq w(S^*)$.
Rule 4 only adds edges and, thus, may only increase the weight of S'.   □

An edge e like the one considered in Lemma 3 will be called *optimal edge*, since we
showed that there is an optimal solution containing e. The direct application of
Lemma 3 is the conception of an exact algorithm for the SCAFFOLDING problem.
Moreover, since our greedy algorithm considers only maximal weight edges, only
a slight modification of the algorithm is necessary to take such edges into account.

4   Experimental Results

Theorem 1 provides a dynamic upper bound on the optimal score, allowing us
to estimate the quality of Algorithm 1 on different kinds of instances. In order
to prove the near optimality of Algorithm 1 on real instances, we have developed a variant of the algorithm in which the calculation of the dynamic upper
bound of Sect. 3 is implemented. This calculation neither affects the behavior
of the algorithm, nor its time complexity of $\mathcal{O}(m \log n)$, nor its worst-case ratio
of 3. Through experimental runs, we give a direct application of the bound as
a measure of how realistic the generated scaffold graphs are and we answer the
following questions:

1. Can the greedy algorithm be used on very large scaffold graphs, and what is
   its associated computation time?
2. Is the upper bound on the ratio computed with θ closer to 1 or to the theoretical guaranteed ratio 3?
3. Is the upper bound useful to measure the efficiency of simulation?

    We tested our algorithm on real data from assembled paired-end reads and
from assembled simulated paired-end reads.
Real Dataset. A first dataset, called *real instances*, has been built using the following pipeline, except for *azucena* which comes from a personal communication
with authors of a paired-end reads library:

Table 1. Real dataset.

Species	Size (bp)	Type	Accession
Anopheles Gambiae str. PEST (anopheles)	41,963,435	Chromosome 3L	NT_078267.5
Bacillus anthracis str. Sterne (anthrax)	5,228,663	Chromosome	NC_005945.1
Arabidopsis Thaliana (arabido)	119,667,750	Complete genome	TAIR10
Zaire ebolavirus (ebola)	18,959	Complete genome	NC_002549.1
Gloeobacter violaceus PCC 7421 (gloeobacter)	4,659,019	Chromosome	NC_005125.1
Lactobacillus acidophilus NCFM (lactobacillus)	1,993,560	Chromosome	NC_006814.3
Danaus plexippus (monarch)	15,314	Mitochondrion	NC_021452.1
Pandoravirus salinus (pandora)	2,473,870	Complete genome	NC_022098.1
Pseudomonas aeruginosa PAO1 (pseudomonas)	6,264,404	Chromosome	NC_002516.2
Oryza sativa Japonica (rice)	134,525	Chloroplast	X15901.1
Saccharomyces cerevisiae (sacchr3)	316,613	Chromosome 3	X59720.2
Saccharomyces cerevisiae (sacchr12)	1,078,177	Chromosome 12	NC_001144.5

1. Choice of a reference genome, for instance on the nucleotide database from NCBI[2]. Table 1 shows selected genomes used to perform our experiments, chosen because of their various origins and sizes.
2. Simulation of paired-end reads, using `wgsim` [14]. The chosen parameters are an insert size of 500bp and a read length L of 100bp.
3. Assembly using the *de novo* assembly tool `minia` [4] with k-mer size $k = 30$.
4. Mapping of reads on contigs, using `bwa` [13]. This mapping tool was chosen according to results obtained by Hunt [10], a survey on scaffolding tools.
5. Generation of scaffold graph from the mapping file. Statistics on the numbers of vertices and edges in produced scaffold graphs can be viewed in Table 4.

Simulated Reads. The main advantage of simulated instances is that we can easily control the graph size and its density. However, it seems quite difficult to relevantly identify structures of real scaffold graphs and then simulated data are generally far from reality. We used two kinds of simulation tools:

1. A naive random generator, which uniformly generates a graph with given density and number of vertices. This generator is called *Uniform*;
2. A more complex generator, issued from the Model Driven Engineering world and especially designed to generate realistic instances: *Grimm*[3] [7].

Instances are generated by both *Grimm* and *Uniform*, either using real parameters on the number of contigs and edges, namely *semi-simulated*, or with artificial parameters sometimes far from those observed in real scaffold graphs, namely *pure simulated*. Thus, we have four simulated datasets to play with.

4.1   Computation Time Scaling

Experiments were run on a personal computer with four processors i7 3.2GHz and 16GB RAM. Memory usage was very light, except for the biggest graph

---

[2] http://www.ncbi.nlm.nih.gov/.
[3] http://www.lirmm.fr/~ferdjoukh/english/research.html.

Table 2. Computation time (in ms) for real instances.

Instance	Time	Instance	Time
monarch	<0.1	*anthrax*	34
ebola	0.1	*gloeobacter*	36
rice	0.27	*pseudomonas*	37
sacchr3	1	*anopheles*	462
sacchr12	10	*arabido*	1,500
lactobacillus	17	*azucena*	19,215
pandora	19		

Table 3. Computation time for two instances on usual scaffolders. Note that SOPRA ran out of memory on the azucena data set.

Scaffolder	*anopheles*	*azucena*
BESST [15]	102.1s	727.5s
SCARPA [6]	146.9s	>48h
SOPRA [5]	56.1s	–
Greedy	116.0s	889.4s

(*azucena*) which took up to 5GB. However, we think this amount of memory usage may be significantly lowered by a better handling of graph I/O. Table 2 shows computation time of the greedy part. Even for the biggest graph with nearly one million contigs, computation time is less than 20s. Note that this time only concerns the greedy computation time, once the graph is generated. Generation of the graph from a mapping file may take more time, up to 15 min in the case of *azucena*. The most expensive step in the production of the graph remains the mapping of reads on contigs, which takes several hours for *azucena*, but is common to all *de novo* scaffolding tools. These experiments let us answer our first question: Algorithm 1 is capable of treating huge instances, using very little computational ressources. Table 3 shows a comparison of the computation time of several scaffolders on the instances *anopheles* and *azucena*, including precomputation time (after the mapping step). We notice that our method is competitive relatively to other heuristics, in addition to providing a bound on the quality of the solution.

4.2   Exploring the Bounds on Ratio

If S_A is the solution given by Algorithm 1, and S_{OPT} a corresponding optimal solution, we denote their weights by ω_A and ω_{OPT}, respectively. For the bound B proven in Theorem 1, we have $\omega_A \leq \omega_{OPT} \leq B \leq 3\omega_A$. Thus, the ratio $\omega_{OPT}/w_A \geq 1$ is not only smaller than 3, but also than B/ω_A. Table 4 shows the results we have on the upper bound. Indeed, we are able to guarantee a ratio very close to 1 for all of the real data since the gap between the weight of the approximated solution and the upper bound is small. The solution for *ebola* is even optimal. Moreover, for *azucena*, the largest instance we have in our possession, we prove that the solution returned by the algorithm is at worst 1 % smaller than the upper bound we have calculated. As a conclusion to this experiment, we can remark that, although it is only a heuristic, the greedy algorithm provides very good solutions on real instances, and since it is very quick, has a very practical interest.

Table 4. Guaranteed ratio on real data for $\sigma_p = 1$ and $\sigma_c = 0$.

graph	#contigs	#edges	ω_A	B	$B/\omega_A - 1$
monarch	14	19	506	507	0.20 %
ebola	**17**	**26**	**776**	**776**	**0.00 %**
rice	84	139	4,293	4,321	0.65 %
sacchr3	296	527	14,524	14,629	0.72 %
sacchr12	889	1,522	46,027	46,345	0.69 %
lactobacillus	1,898	3,335	95,538	96,194	0.69 %
pandora	2,451	4,271	119,575	120,334	0.64 %
anthrax	4,055	6,958	226,709	227,925	0.54 %
gloeobacter	4,517	7,885	218,524	220,091	0.72 %
pseudomonas	5,248	9,086	279,611	280,865	0.45 %
anopheles	42,045	71,452	1,707,463	1,720,269	0.75 %
arabido	1,571,094	1,561,540	1,561,540	1,571,094	0.60 %
azucena	**956,902**	**2,425,349**	**4,692,840**	**4,740,633**	**1.06 %**

Table 5. Guaranteed ratio on simulated data.

Data type	#graphs	mean ratio	median ratio
Pure simulated, *Grimm*	375	1.433	1.320
Pure simulated, *Uniform*	520	1.729	1.814
Semi-simulated, *Grimm*	330	1.086	1.084
Semi-simulated, *Uniform*	330	1.151	1.118

4.3   Simulated Data

Because of the near optimal results we obtained on real data, we wanted to test our algorithm on simulated instances. We then compared obtained average ratios between these four types of data to point out the gap of efficiency according to the generation method.

Table 5 reveals the differences of gaps between the approximated solution weight and the calculated upper bound, depending on the generation method. Note that this gap is always smaller when simulating with *Grimm* than with *Uniform* data, which was expected since real-world instances (which *Grimm* tries to mimic) are expected to be the result of overlaying a strong signal with little noise, which makes detecting this signal much easier than looking for a signal in uniform random graphs. Furthermore, semi-simulated data obtain the smallest gap, which means our algorithm is more effective on real data like instances than on pure simulated instances. We noticed that degree distribution has an impact on the efficiency of the greedy method. However, we observe that simulated data cannot generally guarantee a ratio as good as real data, meaning that simulation process should be refined to produce fully exploitable benchmarks.

5   Conclusion

In Sect. 4 we pointed out the efficiency of our upper bound calculation previously proven in Sect. 3. By demonstrating that the worst case scenario never actually occurs in real or simulated data, these results encourage the use of Algorithm 1. In addition, the gap between real data and simulated data seems to indicate that additional effort in designing more realistic scaffold graph generators is warranted.

Perspectives of this work go into both practical and theoretical directions. On the latter one, short term perspectives include exploring possibilities to develop dynamic upper bounds on approximation ratios of other algorithms. As far as we know, this kind of technique for computing upper bounds has not been considered, and it extends straightforwardly to several $\mathcal{NP}$-complete problems such as the maximization variant of *Travelling Salesman Problem* [1]. Finally, the proof of Lemma 3 may lead to an exact algorithm which would use optimal edges to minimize branching on the search-tree. On the practical side, we are interested in testing the current implementation of Algorithm 1 in comparison with existing scaffolding tools. This may also include a careful examination of the quality of solutions, not only from a combinatorial point of view, but also from a biological point of view. As an extension of this work, we are also interested in the generalization of this greedy algorithm to a version of the problem allowing to use a same contig several times, modeling the case where there are repeats in the genomes, which is what happens in reality.

Acknowledgments. This work was partially founded by the "Projet Investissement d'Avenir" Institut de Biologie Computationnelle. We also like to thank Anne Dievart and Julien Frouin from CIRAD, for their interest to our work and the Azucena Rice illumina reads library.

References

1. Barvinok, A., Gimadi, E.K., Serdyukov, A.I.: The maximum TSP. In: Gutin, G., Punnen, A.P. (eds.) The Traveling Salesman Problem and Its Variations. Combinatorial Optimization, pp. 585–607. Springer, Heidelberg (2007)
2. Chateau, A., Giroudeau, R.: A complexity and approximation framework for the maximization scaffolding problem. Theor. Comput. Sci. **595**, 92–106 (2015). http://dx.doi.org/10.1016/j.tcs.2015.06.023
3. Chen, Z.-Z., Harada, Y., Machida, E., Guo, F., Wang, L.: Better approximation algorithms for scaffolding problems. In: Zhu, D., Bereg, S. (eds.) FAW 2016. LNCS, vol. 9711, pp. 17–28. Springer, Heidelberg (2016). http://dx.doi.org/10.1007/978-3-319-39817-4_3
4. Chikhi, R., Rizk, G.: Space-efficient and exact de bruijn graph representation based on a bloom filter. In: Raphael, B., Tang, J. (eds.) WABI 2012. LNCS, vol. 7534, pp. 236–248. Springer, Heidelberg (2012)
5. Dayarian, A., Michael, T.P., Sengupta, A.M.: SOPRA: scaffolding algorithm for paired reads via statistical optimization. BMC Bioinform. **11**(1), 1–21 (2010)

6. Donmez, N., Brudno, M.: SCARPA: scaffolding reads with practical algorithms. Bioinformatics **29**(4), 428–434 (2013)
7. Ferdjoukh, A., Bourreau, E., Chateau, A., Nebut, C.: A model-driven approach to generate relevant and realistic datasets. In: SEKE, pp. 105–109. KSI Research Inc. and Knowledge Systems Institute Graduate School (2016)
8. Gao, S., Sung, W.-K., Nagarajan, N.: Opera: reconstructing optimal genomic scaffolds with high-throughput paired-end sequences. J. Comput. Biol. **18**(11), 1681–1691 (2011)
9. Gritsenko, A.A., Nijkamp, J.F., Reinders, M.J., de Ridder, D.: GRASS: a generic algorithm for scaffolding next-generation sequencing assemblies. Bioinformatics **28**(11), 1429–1437 (2012)
10. Hunt, M., Newbold, C., Berriman, M., Otto, T.: A comprehensive evaluation of assembly scaffolding tools. Genome Biol. **15**(3), 1–15 (2014). doi:10.1186/gb-2014-15-3-r42. http://dx.doi.org/10.1186/gb-2014-15-3-r42
11. Huson, D.H., Reinert, K., Myers, E.W.: The greedy path-merging algorithm for contig scaffolding. J. ACM (JACM) **49**(5), 603–615 (2002)
12. Koren, S., Treangen, T.J., Pop, M.: Bambus 2: scaffolding metagenomes. Bioinformatics **27**(21), 2964–2971 (2011)
13. Li, H., Durbin, R.: Fast and accurate long-read alignment with Burrows-Wheeler transform. Bioinformatics **26**(5), 589–595 (2010). doi:10.1093/bioinformatics/btp698. http://dx.doi.org/10.1093/bioinformatics/btp698
14. Li, H., Handsaker, B., Wysoker, A., Fennell, T., Ruan, J., Homer, N., Marth, G.T., Abecasis, G.R., Durbin, R.: The sequence alignment/map format and SAMtools. Bioinformatics **25**(16), 2078–2079 (2009)
15. Sahlin, K., Vezzi, F., Nystedt, B., Lundeberg, J., Arvestad, L.: BESST - efficient scaffolding of large fragmented assemblies. BMC Bioinform. **15**(1), 281 (2014). ISSN 1471–2105
16. Salmela, L., Mäkinen, V., Välimäki, N., Ylinen, J., Ukkonen, E.: Fast scaffolding with small independent mixed integer programs. Bioinformatics **27**(23), 3259–3265 (2011)
17. Weller, M., Chateau, A., Dallard, C., Giroudeau, R.: Scaffolding problems revisited: complexity, approximation and fixed parameter tractable algorithms, and some special cases. In: (2016, revision)
18. Weller, M., Chateau, A., Giroudeau, R.: Exact approaches for scaffolding. BMC Bioinform. **16**(14), S2 (2015). ISSN 1471–2105
19. Weller, M., Chateau, A., Giroudeau, R.: On the complexity of scaffolding problems: from cliques to sparse graphs. In: Lu, Z., Kim, D., Wu, W., Li, W., Du, D.-Z. (eds.) COCOA 2015. LNCS, vol. 9486, pp. 409–423. Springer, Heidelberg (2015)

Geometric Optimization

Performing Multicut on Walkable Environments

Obtaining a Minimally Connected Multi-layered Environment from a Walkable Environment

Arne Hillebrand$^{(\boxtimes)}$, Marjan van den Akker,
Roland Geraerts, and Han Hoogeveen

Institute of Information and Computing Sciences,
Utrecht University, 3508 TA Utrecht, The Netherlands
{A.Hillebrand,J.M.vandenAkker,R.J.Geraerts,J.A.Hoogeveen}@uu.nl

Abstract. A multi-layered environment is a representation of the walkable space in a 3D virtual environment that comprises a set of two-dimensional layers together with the locations where the different layers touch, which are called connections. This representation can be used for crowd simulations, e.g. to determine evacuation times in complex buildings. Since the execution times of many algorithms depend on the number of connections, we will study multi-layered environments with a minimal number of connections. We show how finding a minimally connected multi-layered environment can be formulated as an instance of the multicut problem. We will prove that finding a minimally connected multi-layered environment is an NP-Hard problem. Lastly, we will present techniques which shrink the size of the underlying graph by removing redundant information. These techniques decrease the input size for algorithms that use this representation for finding multi-layered environments.

1 Introduction

Evacuation planning and crowd simulations for safety purposes are becoming more and more important in modern society. To perform such simulations in a soccer stadium for example, we need its underlying *polygonal environment* (PE), which is a common format used by architects [17] and 3D modelling tools. A PE is a collection of polygons in $\mathbb{R}^3$ which can be processed through a pipeline to mould it in an appropriate format; an example of such a pipeline is shown in Fig. 1. As is shown in Fig. 1(a), such a PE usually contains unnecessary details for simulations. We only need a filtered version of the PE that contains the polygons that are traversable. Examples of polygons that are not needed in the *walkable environment* (WE) are polygons that are too steep, too close to a ceiling or polygons for which there is not enough clearance for a character to be positioned on the polygon. We assume, without loss of generality, that all polygons $P \in$ WE are convex. Furthermore, we assume that the WE is clean, that is, there is no intersecting or degenerate geometry. Both these properties can be guaranteed when extracting the WE from the PE. An example of a WE

T.-H.H. Chan et al. (Eds.): COCOA 2016, LNCS 10043, pp. 311–325, 2016.
DOI: 10.1007/978-3-319-48749-6_23

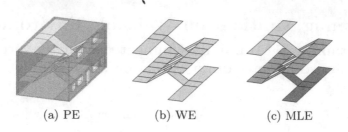

(a) PE   (b) WE   (c) MLE

Fig. 1. (a): A polygonal environment. (b): The walkable environment of this model. (c): A multi-layered environment for this walkable environment. Polygons with the same shade of grey are in the same layer. The red edges are connections. (Color figure online)

is shown in Fig. 1(b). Polygons in a WE can either overlap or be connected. Two polygons *overlap* when they share at least one point when projected on the ground plane that is not on an edge of both polygons. The polygons P and Q are *connected* when they do not overlap and share exactly one edge $e_{P,Q}$. When P and Q are connected, it is possible for a virtual point character to move from P to Q and vice versa.

On a WE we want to perform operations such as constructing visibility graphs [10], which are used for finding shortest paths in the environment, or for creating navigation meshes [15], which enable fast path planning queries that are used for crowd simulations. However, many operations currently are limited to two-dimensional environments. These operations can be extended to layered two-dimensional environments by using a *multi-layered environment* (MLE). An MLE is a decomposition of the WE in layers, such that every layer can be embedded in the plane. When two polygons P and Q share an edge $e_{P,Q}$ and do not overlap but are in different layers, $e_{P,Q}$ connects the two layers. This type of edge is called a *connection*. This set of edges is also stored for an MLE.

Definition 1 (Multi-layered Environment, Van Toll et al. [16]). An MLE for a given WE consists of a set $\mathbf{L} = \{L_1, ..., L_l\}$ of two-dimensional layers and a set $\mathcal{C}$ of connections, such that:

- No layer L_i ($i = 1, \ldots, l$) contains overlapping polygons;
- Every polygon in the WE is assigned to exactly one layer L_i ($i = 1, \ldots, l$);
- When two polygons P and Q are connected, but are in different layers, the connection between P and Q is part of the set $\mathcal{C}$;
- Every layer L_i ($i = 1, \ldots, l$) forms a single connected component, i.e. for any two polygons P, Q in L_i we can walk from P to Q without leaving L_i.

When we use this definition of an MLE, it is rarely the case that there exists only one possible MLE for a WE. Take for example the MLE given in Fig. 1(c), consisting of three layers and two connections. Another possible MLE for this WE has only two layers, but needs four different connections. We call an MLE with the minimal number of connections a *minimally connected multi-layered environment* (MICLE). In this paper we will focus on MICLEs. This is the most useful type of MLE because subsequent operations are dependent on the number of connections. For example, van Toll et al. [16] show that constructing a navigation mesh for an MLE can be done in $O(k \times n \log n)$ time, where k is the number of connections in an MLE and n is the number of obstacle points used to describe the boundaries of the individual layers of the MLE.

1.1 Related Work

There are several applications and algorithms that are already using some form of an MLE. However, the MLEs that they use are often of poor quality. They do not cover all of the walkable space or contain a high number of connections. Van Toll et al. [16] use an MLE to create a multi-layered navigation mesh, which allows for fast path planning queries. Their MLE is a decomposition of the walkable environment into layers. One strength of this type of MLE is that the representation of the corresponding WE is exact. However, they do not describe any methods to find an MLE with a low number of connections.

Deusdado et al. [3] use a discretized height map to automatically extract walkable surfaces from a PE. The locations of the connections are determined by comparing the height information of different walkable surfaces. Oliva and Pelechano [11] overlay the environment with a three-dimensional grid and mark each grid cell positive when it contains walkable geometry. These grid cells are grouped into layers which are then used to create a multi-layered navigation mesh. Pettré et al. [12] create multiple elevation maps of a PE by using the graphics card. From these elevation maps, slopes that are too steep are filtered and the elevation maps are joined, resulting in a WE. The downside of these techniques is that they are not general enough and that the resulting MLE is only an approximation of the WE and does not cover all of the WE.

Instead of extracting an MLE from a PE or WE, Jiang et al. [9] propose a method which models the environment in simple blocks that can be described in two dimensions. The blocks are linked together in a floor plan-like fashion. When an environment is described this way, an MLE is easy to extract since the layers and connections are explicitly defined.

1.2 Our Contribution

In Sect. 2, we will show that the search for a MICLE can be solved in theory by using *multi-commodity minimal-cut* (MULTICUT). For this we will encode the WE as a graph. MULTICUT is a problem in the class of NP-Hard problems [1], for which fixed parameter tractable (FPT) algorithms exist [5]. We will prove that finding a MICLE is also in the class of NP-Hard problems. In Sect. 3 we will identify situations in which edges, vertices and overlaps can be removed from the graph, without influencing the size of the cut needed for finding a valid MICLE. We have also implemented algorithms to handle these situations and we experimentally evaluate them in Sect. 4. The graph reduction algorithms have varying results on real-world environments.

2   Finding a MICLE

We first convert the WE into a graph $G = (V, E, O)$. In this graph, a vertex is added to V for every polygon in the WE. An undirected edge (v, w) is added to E whenever the polygons corresponding to the vertices v and w are connected in the WE. When two polygons with corresponding vertices v' and w' overlap, the unordered pair (v', w') is added to O. This process is illustrated in Fig. 2. The resulting graph is called the *walkable environment graph* (WEG). Sometimes, we want to assign different weights to edges in the WEG. One such situation will be described in Sect. 3. For a weighted WEG we have $G = (V, E, O, w)$. Here w is a function that maps every edge $e \in E$ to a real number. The weight of the cut set $\mathcal{C}$ is defined as $|\mathcal{C}| = w(\mathcal{C}) = \sum_{e \in \mathcal{C}} w(e)$. Using the WEG, the problem of finding a MICLE can now be formulated as follows:

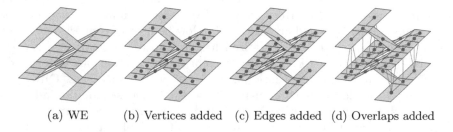

(a) WE   (b) Vertices added   (c) Edges added (d) Overlaps added

Fig. 2. Constructing the graph representation of a walkable environment. (a): The walkable environment. (b): Vertices are added for every polygon in the walkable environment. (c): Undirected edges are added for connected polygons. (d): Overlapping vertices are annotated (represented by red edges). (Color figure online)

Problem 1 (Finding a MICLE). Given WEG $G = (V, E, O)$ of WE W. Finding a MICLE is now the same as finding the set $\mathcal{C} \subseteq E$ for which:

- $\forall v, w$ such that $(v, w) \in O$: v and w are in different graph components for the graph $G' = (V, E \backslash \mathcal{C})$;
- $|\mathcal{C}|$ is minimal.

From the different connected components in the graph $G' = (V, E \backslash \mathcal{C})$, we can construct the different layers in the MLE. When $|O| = 1$, this problem is the same as the *s-t* cut problem, which can be solved using the max-flow min-cut theorem [4]. If $|O| > 1$, we have an instance of the MULTICUT problem [14]. In the MULTICUT framework, multiple source-sink pairs (s_i, t_i) exist, where each pair has to be separated. By using MULTICUT it is possible to find an MLE with a minimal number of connections for a WEG by using the overlapping pairs $(v, w) \in O$ as the terminal pair set. Unfortunately, MULTICUT has been proven to be NP-Hard when $|O| \geq 3$ [1]. For $|O| = 2$, polynomial time algorithms exist to solve MULTICUT [8]. However, since finding a MICLE is a special case of MULTICUT, we still need to prove that finding a MICLE for a WEG is indeed an NP-Hard problem. We will show this below.

2.1   Preliminaries

The proof that finding a MICLE is in the class of NP-Hard problems is heavily based on the work done by Dahlhaus et al. [2] on the MULTITERMINAL-CUT (MTC) problem. The decision version of this problem is defined below:

Definition 2 (MULTITERMINAL-CUT, **Dahlhaus et al.** [2]). *Given a graph $G = (V, E, w)$, a terminal set $T \subseteq V$ and a positive bound B. The decision version of MTC is now the same as finding a set $E' \subseteq E$ for which:*

- *$\forall v, w \in T$ such that $v \neq w$: v and w are in different components for the graph $G' = (V, E \backslash E')$;*
- *$\sum_{e \in E'} w(e) \leq B$.*

To prove that the MTC problem is NP-Hard, Dahlhaus et al. first prove a restricted version of PLANAR 3-SAT (P3R) to be NP-Complete.

Definition 3 (P3R, **Dahlhaus et al.** [2]). Given a set of variables $X = \{x_1, \ldots x_n\}$ and set of clauses $C = \{c_1, \ldots, c_m\}$, where:

- $\forall c_j \in C$: $c_j = (x_k \vee x_l)$ or $c_j = (x_k \vee x_l \vee x_m)$ for some $x_k, x_l, x_m \in X$;
- $\forall x_i \in X$: x_i occurs exactly once in three different clauses, and;
- $\forall x_i \in X$: Both the literals x_i and $\bar{x}_i$ occur at least once.

Solving P3R for a formula $F = c_1 \wedge \ldots \wedge c_m$ is now equal to finding a suitable truth assignment T for all the variables $x_i \in X$ such that F equals true.

Dahlhaus et al. show how an instance of P3R can be reduced to the decision version of MTC. Some of the widgets that they use are given in Fig. 3. In this figure, the circles represent the terminals of the MTC problem. Every variable x_i with two x_i literals is replaced with the widget shown in Fig. 3(a). The clauses of size three are replaced with the widget from Fig. 3(c). Similar widgets exist for the clauses of size two and for the variables with two $\bar{x}_i$ literals.

Next, the widgets representing the variables and the clauses are connected using weight two *link edges*. The link edges are attached to the vertices labelled $l_{i,x}$, $\bar{l}_{i,1}$ and $\hat{l}_{j,x}$ ($x = 1, 2, 3$). For details on how these widgets are connected using the link edges, we refer the reader to reference [2]. The component induced by the vertices of the link edge and the neighbours of these vertices is called the *link structure*. While proving that the decision version of the MTC problem is NP-Complete, Dahlhaus et al. also proved the following two lemmas:

Lemma 1 (Dahlhaus et al. [2]). Given are a P3R instance (X, C) with X the set of variables and C the set of clauses. Let $G_{X,C}$ be the bipartite graph representing this instance and $G'_{X,C}$ the transformed graph of $G_{X,C}$. There exists a set of edges E' with total weight $B = 10|X| + 4|C|$ separating the terminals in $G'_{X,C}$ if and only if the P3R instance (X, C) has a satisfying truth assignment.

Lemma 2 (Dahlhaus et al. [2]). When a cut set E' for MTC is found where the separation cost $\sum_{e \in E'} w(e) \leq B$, E' only contains edges from the connector triangles or link edges. Furthermore, for each link structure it will only contain one of the connector triangles or the link edge.

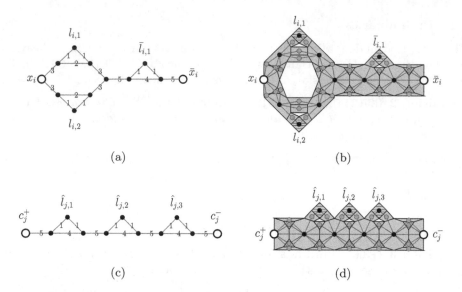

Fig. 3. The widgets used by Dahlhaus et al. [2] and their polygonal counterparts. The terminals are depicted by circles and the vertices by discs. The widget in (a) is used for a variable x_i with two x_i literals. The widget in (c) is used for clauses with three variables. For the polygonal counterparts (Figs.(b) and (d)), the newly created vertices are coloured orange. (Color figure online)

These two lemmas are of key importance for the proof that finding a MICLE is an NP-Hard problem.

2.2   Finding a MICLE Is NP-Hard in the Strong Sense

First, we will show that the transformed graph $G'_{X,C}$ can exist as a WEG. Next, we will prove that it is possible to encode the terminals of MTC in a WE. Finally, we will show that this can be done in polynomial time.

Since $G'_{X,C}$ is planar, the components from this graph can easily be represented using the polygonal structures given in Figs. 3(b) and (d). The link edges connecting the link vertices $l_{i,x}$ or $\bar{l}_{i,x}$ to $\hat{l}_{j,x}$ ($x = 1, 2, 3$) can be represented using a series of polygons. Since the link edges have weight two, two of such series of polygons are required. These series are attached to the only two free sides of the tetragons $l_{i,x}$ or $\bar{l}_{i,x}$ and $\hat{l}_{j,y}$. The tetragons that represent these vertices are hatched in Figs. 3(b) and (d). Every link edge can be represented by using at most five polygons, since there exists a straight line drawing for every planar graph, and it can be found in $O(|V|)$ time [6]. It is easy to see that the resulting WEG has the same properties as the components given by Dahlhaus et al., that is, separating any pair of the vertices from the WEG that directly corresponds to vertices from the original widgets will result in a cut of exactly the same weight.

All the terminals can be encoded by one big set T_{tower} of polygons. The required number of polygons in this set is $2|X| + 2|C|$, one for each terminal in $G'_{X,C}$. These polygons should all be centred around one point c such that, when all the polygons are projected on the ground plane, they all contain this point. As a result, all these polygons will be placed in different components when searching for any MLE.

The next step is connecting the polygons from T_{tower} to the polygons where the terminals of the MTC problem are located. These are labelled x_i, $\bar{x}_i$, c_j^- and c_j^+ in Figs. 3(b) and (d). This should be done in such a way that Lemma 1 remains valid and the final polygonal environment is a WE. The validity of Lemma 1 remains when we connect each terminal in T_{tower} to one of the terminals of MTC using a single weight 7 edge. Cutting such an edge will always increase the weight of the cut set E' to a value higher than B. Edges of weight 7 are sufficient, since the edges of weight 3 and weight 5 connected to the terminals of MTC are guaranteed not to be part of the cut set E' if a satisfying truth assignment exists (Lemma 2).

When we represent these edges of weight 7 by 7 polygonal paths, the polygons can overlap parts of clause and variable structures, generating new overlaps that have to be separated. Fortunately, we can create paths that guarantee that Lemmas 1 and 2 will still hold, even when these paths are added to the WEG.

Lemma 3. *We are given a former terminal t, a path p connecting t to the new terminal $t' \in T_{tower}$ and the polygonal structure S that contains t, the polygonal widget it belongs to and its link structures. If p does not overlap S or itself, Lemmas 1 and 2 still hold.*

Proof. Since Lemmas 1 and 2 do not state anything about the scenario where no satisfying truth assignment for P3R exists, we do not need to consider this scenario. For this reason, we will assume that there exists a satisfying truth assignment for P3R.

Note that the path p must be one of the 7 polygonal paths that connects t to t'. Therefore, cutting only path p will increase the weight of the cut set but not make t' and t disconnected in the WEG. Therefore, cutting p will also mean cutting the other 6 paths connecting t and t'. We already know that these 7 paths will not be cut when separating all terminals in T_{tower}, unless there is no possible truth assignment for P3R. Furthermore, we know from Lemma 2 that the edges from the link structures separate all the terminals. This also means that all paths connecting vertices in a variable or clause structure will be separated from any other variable or clause structure. Since p connects the former terminal t and the new terminal t', we know that cutting the correct link structures will separate t' from all other terminals in T_{tower}. The terminals that are created when p overlaps parts of the WE are also cut, since we know that these overlaps are connected through link structures, and all link structures are cut. Therefore, both Lemmas 1 and 2 must hold.   □

It is easy to see that these paths exists. We know that the various clause and variable structures can be connected using straight paths [6], not considering the polygons from T_{tower} or the paths connecting these structures to T_{tower}. This follows from the fact that we started from an instance of P3R, which can be embedded in the plane. Furthermore, the only non-planar structures are T_{tower} and the paths that lead to it. We also know that the paths connecting the former terminals to the new terminals in T_{tower} may overlap. Therefore, it is sufficient to have these paths skim the border of the polygonal structure S, until a straight path to T_{tower} is possible. This results in the following theorem:

Theorem 1. *The decision version of finding a MICLE is NP-Complete and finding a MICLE is NP-Hard.*

Proof. First we use the graph $G'_{X,C}$ of Dahlhaus et al. [2]. This graph can be represented using 45 polygons for every variable, 20 polygons for every clause with 2 variables and 32 polygons for every clause of three variables. Every link edge can be represented using 5 polygons, and there are $2|X|$ link edges in total. The number of added polygons to create the link structures and components is linear in $|X| + |C|$; the same holds for the needed number of polygons for T_{tower}. The polygons will also encode the $2(|X| + |C|)$ terminals from the MTC problem. For creating the paths to T_{tower}, at most another number of polygons linear in $|X| + |C|$ is needed.

To complete the proof, bounds for the terminal-pairs have to be given. The number of added terminal-pairs while creating the paths to T_{tower} is at most $O((|X| + |C|)^2)$. As a result, the decision version of finding a MICLE is NP-Complete, and, therefore the optimization version of finding a MICLE is NP-Hard.   □

3   Reducing the Size of the WEG

In this section we describe situations in which vertices, edges and overlaps can be removed from a WEG without changing the optimal solution. In Sect. 3.1, we will present techniques to reduce the number of edges in the WEG, and in Sect. 3.2, we will discuss the removal of overlaps that are already enforced by other overlaps of the WEG. In this section we will use a weighted WEG, as defined in Sect. 2. We start with $\forall e \in E : w(e) = 1$. We will use $N_v = \{w | (v, w) \in E\}$ for the set of neighbours of a vertex v in a weighted WEG $G = (V, E, O, w)$. Similarly, $O_v = \{w | (v, w) \in O\}$ is the set of vertices that v overlaps. A simple path between the vertices v and w, will be denoted as $[v \rightarrow w]$.

3.1   Reducing the Number of Edges

The first reduction we will discuss is called 1-CONTRACT. It is only applicable for a vertex v with $|O_v| = 0$ and $|N_v| = 1$. Such a vertex can easily be ignored. There is no reason to add this vertex to any other layer than the layer of its only neighbour.

The second graph simplification applies to a vertex with $|N_v| = 2$ and $|O_v| = 0$. Assume the two neighbours of v are u and w. Since this vertex v

(a)   (b)   (c)   (d)

Fig. 4. An example of the application of the contraction operations. In (a) the WEG with which we start is given. In (b), (c) and (d) the WEG is shown after the application of 1-CONTRACT, 2-CONTRACT and E-REDUCE, respectively. The thick edge in (d) is of weight 2.

has no vertices it overlaps with, merging v and u or v and w will not force any new overlaps. The newly created edge (u, w) does not create any new connections between u and w, nor does the minimal cost of separating u and w change. This operation is called 2-CONTRACT.

The third graph simplification is only useful in combination with the previous simplification. The operation applies to two vertices connected by multiple edges. This situation can not exist in the original WEG because of the convexity constraint for the polygons in a WE. In this situation it would be nice if we could apply the 2-CONTRACT operation, but this is not possible since the degree of v is three, not two. A simple solution solving this problem is merging all the double edges and combining their respective weights. This method is called E-REDUCE. An example of a EEG, and the resulting WEGs after we have applied each of these operations, can be seen in Fig. 4.

3.2   Reducing the Number of Overlaps

A logical next step is removing overlaps from the WEG. This situation will be subdivided into two categories. First, we have the trivial cases, where there is only one possible cut to separate two vertices. Second, there is the case of vertices with degree 2 that overlap. In this scenario several overlaps can be removed under specific circumstances.

Trivial Cases. A case is considered trivial when there exists only one vertex-disjoint path between v and w, $(v, w) \in E$, and v and w are overlapping. When this happens, there is only one possible way to separate v and w, which is done by cutting the edge (v, w). Since this action guarantees that v and w are separated, the overlap (v, w) can be removed from O. When there are other vertex-disjoint paths between v and w not containing the edge (v, w), cutting the edge (v, w) will not separate these paths. For this reason, the overlap (v, w) cannot be removed from O in this situation.

Another trivial case is when a vertex v has degree 1 and $(v, w) \in O$. In this scenario the overlap can be removed when all the neighbours of w that are on a vertex-disjoint path $[v \rightarrow w]$ are overlapped by the single neighbour of v. In this situation, every path connecting v and w has a subpath connecting a neighbour

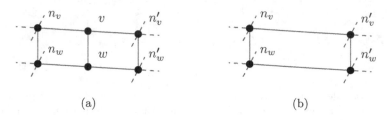

Fig. 5. An example in which the overlap (v, w) can be removed as long as the vertices have degree 2.

of v to a neighbour of w. This subpath has to be cut, since the neighbour of v is assumed to overlap all the neighbours of w. We call this operation 1-REMOVE. For a vertex v, we can check if an overlap can be removed in $O(|N_w|)$ time if all members of N_w are overlapped by the single member of N_v, say n_v. If n_v does not overlap all members of N_w, this check is more expensive. We need to check if the members of N_w that are not overlapped by n_v are in a different graph component after the temporary removal of w. This can be done using a Breadth-first search (BFS), resulting in an expensive $O(|E|)$ check to see if this operation is applicable. Performing this step on an entire WEG can take $O(|V||O||E|)$ time. Since $|O|$ can be $O(|V|^2)$, we end up with an $O(|V|^3|E|)$ algorithm, which is a costly operation for larger environments. However, we think that this worst-case scenario rarely happens in practice, because $|O|$ is only $O(|V|^2)$ when almost all polygons overlap all other polygons.

Removing Overlaps from Vertices of Degree 2. When considering a vertex v with degree 2 and its overlaps, some of these overlaps might be removed. Assuming $(v, w) \in O$, this overlap can be removed when, on each path connecting v and w, there is a pair of vertices $(x, y) \in O$. That is, for every possible path connecting v to w, there is a pair of vertices $(x, y) \in O$ and both x and y are on this path. An example of such a situation is given in Fig. 5(a). We will call such a structure a stack, and the operation that removes overlaps from this stack we will call STACK-REMOVE. In this situation, all edges have weight one and both v and w have degree two. Furthermore, the neighbours of v and w do also overlap. A formal proof of the validity of STACK-REMOVE is given in Appendix A.

Overlaps in similar cases can still be removed from a WEG whenever the edges (n_w, w) and (w, n'_w) have the same weights a, and the edges (n_v, v) and (v, n'_v) also have the same weights b. However, if edges (n_w, w) and (w, n'_w) each have different weights, this is not longer true, since these edges cannot be easily replaced by edges from the original graph.

General Overlap Removal. In the general case, overlaps can be removed whenever the separation of overlapping vertices is already forced by the surrounding environment, just as with STACK-REMOVE. Unfortunately, checking this requires an exponential amount of time in worst case scenarios. A simple BFS or DFS will not suffice since not all paths connecting v and w are traversed.

Since such a worst-case runtime is not usable in practice (especially when you consider running that algorithm for every vertex that has an overlap), it is prudent to limit the search depth at the cost of the number of overlaps that will be removed. Such an algorithm is d-REMOVE. The parameter d is a bound on the path length considered when searching for overlaps that can be removed.

For each simple path of length d originating in v, we temporarily remove the vertices that are overlapped by vertices on the simple path from the WEG. After doing this, a BFS is started from the last vertex on the simple path. If we encounter a vertex o that overlaps v during this BFS, we remember it. This process is repeated for all the simple paths. For overlaps that were not encountered during this process, there must be vertices on the simple paths of length d that guarantee the separation. Therefore, we can safely remove the overlaps that were not encountered during this process from the WEG.

The running time for this algorithm is $O(x^d \times (|E| + |V| - 2d) + x^d \times y)$, where x is the maximal branching factor and y is the maximal number of overlaps associated with a vertex. In this algorithm, x^d calls are made to a BFS algorithm. Since we already traversed d vertices and edges, the BFS does not have to visit them any more. Registering the encountered overlaps on the simple paths accounts for the remaining $x^d \times y$ time.

4   Experiments and Results

We have implemented the WEG reductions described in Sect. 3 in C++ and tested them on a number of environments. The details of the used environments are given in Table 1. The environments As_oilrig, Library and Parking lot were taken from Saaltink [13], Station 1 and Station 2 were provided by Movares, an engineering and consultancy company. The environments Halo, Cliffsides and Hexagon were taken from the Google Sketchup warehouse[1]. The Tower environment was created by the authors, based on a flat in Utrecht, the Netherlands.

Table 1. The different environments we have tested. Column **T.** gives the type of environment. V stands for "real" virtual environment and R stands for real world environment. A ✓ in column **Tri.** means that the environment is triangulated.

Environ.	T.	Tri.	$|V|$	$|E|$	$|O|$
As_oilrig	V	✓	2077	2399	10717
Halo	V	✓	179	184	346
Cliffsides	V	✓	748	764	162
Hexagon	V	✓	2368	2419	20207

Environ.	T.	Tri.	$|V|$	$|E|$	$|O|$
Library	R	✗	298	420	775
Tower	R	✗	5932	8033	116983
Station 1	R	✓	206	209	1026
Station 2	R	✓	82	86	115

Besides the size of these environments (which we can see in column $|V|$), there are two other important aspects of the environments. The first one is

---

[1] https://3dwarehouse.sketchup.com/model.html?id={13c3078fa52d14554b9e177bc9-ee06a9, 2ac949d235d65acb46697ff0ff0b9b2c, 33b2c337108275568c09573a9753f4fd}.

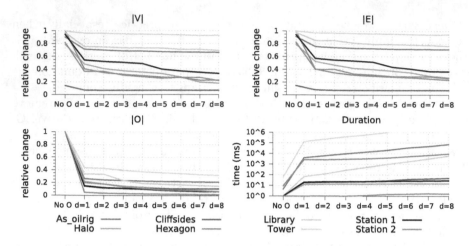

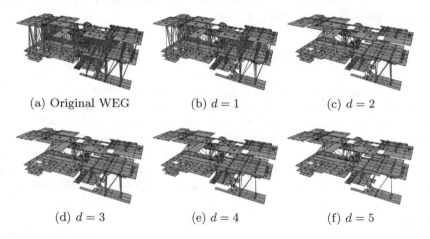

Fig. 6. Plots showing the relative changes of $|V|$, $|E|$, $|O|$ and the running time for the different values of d. For the results in column 'No O' only 1-CONTRACT, 2-CONTRACT and E-REDUCE were used.

(a) Original WEG   (b) $d = 1$ (c) $d = 2$

(d) $d = 3$ (e) $d = 4$ (f) $d = 5$

Fig. 7. Different WEGs for the Library environment. The black circles are vertices, the blue segments are edges and the red segments are overlaps. In (a) the unmodified WEG is shown. (b) through (e) show the reduced WEGs for different values of d. (Color figure online)

the ratio $\frac{|O|}{|V|}$, which is an indication of how layered the environment is. The second aspect is what types of geometric primitives were used to model these environments. If an environment consists solely of triangles, it might be easier to reduce the underlying graph. We tested our algorithms on models of real buildings (type R) and on game levels (type V).

All our experiments were performed on a machine with an Intel i5-4670 clocked at 3.4 GHz with 16 GB of DDR3 RAM. They only used a single thread and were repeated 20 times. The OS and compiler that we used are Linux Mint 13.2 (64 bit) and g++ version 5.3.0, respectively.

For each environment, we first ran experiments that only used the edge reduction algorithms described in Sect. 3.1. Next we added the overlap reduction algorithms from Sect. 3.2. We did not include the results of experiments that only used the algorithms in Sect. 3.2, because of space limitations. We tested for d from d-REMOVE in the range 1 through 8. The results for all environments can be seen in Fig. 6. This figure shows the relative changes for $|V|$, $|E|$ and $|O|$. Here, a value of 1 means that no graph reduction was performed. A value of 0 means that all the vertices (or edges or overlaps) were removed.

The first thing that we notice is that the Cliffsides environment benefits greatly from the edge-reduction algorithms. We believe this to be because of the relatively low ratio of $\frac{|O|}{|V|}$ for this environment. This ratio is a measure of how layered the environment is. Another important factor is the fact that the environment is triangulated, which is also illustrated in Fig. 7. In this figure, we see the Library environment and the corresponding WEG for different values of d. When $d = 5$, there are many vertices that do not have any overlap, but cannot be removed because the edge degree is too high. The edge degree would have been lower, if the environment had been triangulated beforehand.

We can also see that all environments benefit from the use of the overlap-reductions algorithms. There is a steep decline in the number of vertices and edges when it is first used.

5   Conclusion

We have given a definition of MLEs that can be used as input for already existing 2D algorithms. A special type of MLE, the MICLE, can be useful for solving problems in the multi-layered problem domain. Since a MICLE has the lowest number of connections possible, cross-layer operations will occur less frequently. Finding such a MICLE for a given WE is an NP-Hard problem. It is a version of the well studied MULTICUT, which is also in the class of NP-Hard problems.

Furthermore, we have described some algorithms that can reduce the size of the WEG. This is accomplished by merging vertices and removing overlaps in such a way, when searching for a MICLE, that we find a solution that is also optimal for the unmodified WEG, or one that can be adjusted to an optimal solution for the unmodified WEG. These algorithms have been implemented and tested for different environments.

Working with MICLEs can increase the efficiency of operations performed on this MLE. Examples of such operations include, but are not limited to, performing simulations of large crowds or constructing a visibility graphs for finding shortest routes. However, as we have proven, an MLE with the smallest number of connections is hard to find.

Currently, we are investigating techniques that can generate a WE from a PE and strategies that can construct an MLE with a small number of connections from a WE. Our first results on extracting MLEs from WEs can be found in reference [7].

A   Proof of STACK-REMOVE

Theorem 2. *Given a weighted WEG $G = (V, E, O, w)$, vertex v with $N_v = \{n_v, n'_v\}$, vertex w with $N_w = \{n_w, n'_w\}$, $\{(v, w), (n_v, n_w), (n'_v, n'_w)\} \subseteq O$ and $w(e) = 1$ for all $e \in E$. The optimal cut set C' for $G' = (V, E, O', w)$ with $O' = O\backslash(v, w)$, either is also optimal for G, or can be adjusted to an optimal cut set C for G.*

Proof. Assume we have the WEG $G' = (V, E, O', w)$, with $O' = O \backslash (v, w)$. Furthermore, we will also assume that we have found a MICLE for G' with the cut set C'. Are all paths $[v \rightarrow w]$ cut by C'? If not, how can we change C' without increasing the weight so that these paths are cut? This situation is also illustrated in Fig. 5.

First, we observe that all vertex-disjoint paths $[v \rightarrow w]$ can be split into four categories, namely $[v, n_v \rightarrow n_w, w]$, $[v, n'_v \rightarrow n'_w, w]$, $[v, n_v \rightarrow n'_w, w]$ and $[v, n'_v \rightarrow n_w, w]$. Categories $[v, n_v \rightarrow n_w, w]$ and $[v, n'_v \rightarrow n'_w, w]$ will be cut by C', because the vertex-disjoint paths $[n_v \rightarrow n_w]$ and $[n'_v \rightarrow n'_w]$ need to be cut for any valid MICLE of G'. But what about the paths $[v, n_v \rightarrow n'_w, w]$ and $[v, n'_v \rightarrow n_w, w]$? We know the following groups of paths will be cut using edges of C': (a) $[n_v \rightarrow n'_w, w, n_w]$; (b) $[n'_v \rightarrow n_w, w, n'_w]$; (c) $[n_v, v, n'_v \rightarrow n_w]$; (d) $[n'_v, v, n_v \rightarrow n'_w]$.

This fact does not guarantee that the vertex-disjoint paths $[v, n_v \rightarrow n'_w, w]$ and $[v, n'_v \rightarrow n_w, w]$ are cut by C'. The paths $[v, n_v \rightarrow n'_w, w]$ and $[v, n'_v \rightarrow n_w, w]$ are definitely cut when C' contains at least one edge of all vertex-disjoint paths $[n_v \rightarrow n'_w]$ and $[n'_v \rightarrow n_w]$. If this is not the case, we have one of the following three situations:

1. The paths $[n_v \rightarrow n'_w]$ are cut by C', but the paths $[n'_v \rightarrow n_w]$ are not;
2. The paths $[n'_v \rightarrow n_w]$ are not cut by C', but the paths $[n_w \rightarrow n_v]$ are;
3. Both the paths from $[n_v \rightarrow n'_w]$ and $[n'_v \rightarrow n_w]$ are not cut by C'.

For these three remaining situations we can replace edges from C' to also cut all paths $[v \rightarrow w]$ without increasing the weight of C' and cutting all paths from O and thus obtaining the cut set C for G.

For the first situation, we know that the edges (n_v, v) and (n'_w, w) need to be in C' to cut the paths of type (b) and (c). Instead of adding the edges (n_v, v) and (n'_w, w) to C', we can add the edges (n'_v, v) and (n_w, w) to C' without increasing the weight of C'. When we do this the overlapping vertices of G' will still be separated, but we also separated v from w without increasing the weight of C'. The second situation can be handled analogously.

When we have the third situation, we know that one of the edges $\{(n_w, w), (w, n'_w)\}$ needs to be in C' to cut the paths of types (a) and (b), and one of the edges $\{(n_v, v), (v, n'_v)\}$ to cut the paths of type (c) and (d). When we just pick the edges (v, n_v) and (w, n_w), we will once again not change the size of C' and still separate all overlaps of O', but also all overlaps of O. □

The same trick can be applied to prove that overlaps can also be removed in a stack of structures. This can be proven using exactly the same steps as before and can only be applied under the same restrictions.

References

1. Calinescu, G., Fernandes, C.G., Reed, B.: Multicuts in unweighted graphs with bounded degree and bounded tree-width. In: Bixby, R.E., Boyd, E.A., Ríos-Mercado, R.Z. (eds.) IPCO 1998. LNCS, vol. 1412, p. 137. Springer, Heidelberg (1998)
2. Dahlhaus, E., Johnson, D., Papadimitriou, C., Seymour, P., Yannakakis, M.: The complexity of multiterminal cuts. SIAM J. Comput. **23**(4), 864–894 (1994)
3. Deusdado, L., Fernandes, A.R., Belo, O.: Path planning for complex 3D multilevel environments. In: Proceedings of 24th Spring Conference on Computer Graphics, pp. 187–194 (2008)
4. Ford, L., Fulkerson, D.: Solving the transportation problem. Manag. Sci. **3**(1), 24–32 (1956)
5. Guo, J., Hüffner, F., Kenar, E., Niedermeier, R., Uhlmann, J.: Complexity and exact algorithms for multicut. In: Software Seminar, pp. 303–312 (2006)
6. Harel, D., Sardas, M.: An algorithm for straight-line drawing of planar graphs. Algorithmica **20**, 119–135 (1998)
7. Hillebrand, A., van den Akker, M., Geraerts, R., Hoogeveen, H.: Separating a walkable environment into layers. In: 9th International ACM SIGGRAPH Conference on Motion in Games (2016, to appear)
8. Itai, A.: Two-commodity flow. J. ACM **25**, 596–611 (1978)
9. Jiang, H., Xu, W., Mao, T., Li, C., Xia, S., Wang, Z.: A semantic environment model for crowd simulation in multilayered complex environment. ACM Symp. Virtual Reality Softw. Technol. **2015**, 191–198 (2009)
10. Lozano-Pérez, T., Wesley, M.A.: An algorithm for planning collision-free paths among polyhedral obstacles. Commun. ACM **22**(10), 560–570 (1979)
11. Oliva, R., Pelechano, N.: NEOGEN: near optimal generator of navigation meshes for 3D multi-layered environments. Comput. Graph. **37**(5), 403–412 (2013)
12. Pettré, J., Laumond, J.P., Thalmann, D.: A navigation graph for real-time crowd animation on multilayered and uneven terrain. First Int. Workshop Crowd Simul. **47**(2), 81–90 (2005)
13. Saaltink, W.: Partitioning polygonal environments into multi-layered environments. Master's thesis, Utrecht University (2011)
14. Schrijver, A.: Combinatorial Optimization - Polyhedra and Efficiency. Algorithms and Combinatorics, vol. 24. Springer, Heidelberg (2003)
15. Snook, G.: Simplified 3D movement and pathfinding using navigation meshes. In: DeLoura, M. (ed.) Game Programming Gems, pp. 288–304. Charles River Media, Newton Centre (2000)
16. van Toll, W., Cook IV, A., Geraerts, R.: Navigation meshes for realistic multi-layered environments. In: 2011 IEEE/RSJ International Conference on Intelligent Robots and Systems, pp. 3526–3532 (2011)
17. Whyte, J., Bouchlaghem, N., Thorpe, A., McCaffer, R.: From cad to virtual reality: modelling approaches, data exchange and interactive 3D building design tools. Autom. Constr. **10**(1), 43–55 (2000)

Minimum Weight Polygon Triangulation Problem in Sub-Cubic Time Bound

Sung Eun Bae, Tong-Wook Shinn$^{(\boxtimes)}$, and Tadao Takaoka

Algorithm Research Institute, Christchurch, New Zealand
{sung.bae,tongwook.shinn,tadao.takaoka}@ari.org.nz

Abstract. We break the long standing cubic time bound of $O(n^3)$ for the Minimum Weight Polygon Triangulation problem by showing that the well known dynamic programming algorithm, reported independently by Gilbert and Klincsek, can be optimized with a faster algorithm for the $(min, +)$-product using look-up tables. In doing so, we also show that the well known Floyd-Warshall algorithm can be optimized in a similar manner to achieve a sub-cubic time bound for the All Pairs Shortest Paths problem without having to resort to recursion in the semi-ring theory.

1 Introduction

Given a polygon, there are many different ways to divide up the polygon into triangles. Such an operation is commonly known as triangulation. It is well known that the total number of possible triangulations on a polygon with n vertices is the $(n-2)^{th}$ Catalan number[1] that grows exponentially as n increases. If the edge costs between vertices of the polygon are defined, the Minimum Weight Polygon Triangulation (MWPT) problem is to find the triangulation of the polygon such that the total edge cost is minimal. We assume the polygon is convex for simplicity.

There exists a well known dynamic programming algorithm that solves the MWPT problem in $O(n^3)$ time bound reported independently by Gilbert [8] and Klincsek [12]. We refer to this algorithm as the GK algorithm.

If we consider the input polygon as a graph with edge costs, then another well known problem is the All Pairs Shortest Paths (APSP) problem, which is to find the path that gives the minimal total edge cost for all possible pairs of vertices. Floyd-Warshall (FW) algorithm [6] is a well known dynamic programming algorithm for the APSP problem, also with $O(n^3)$ time bound.

The GK algorithm for solving the MWPT problem that we will review later in this paper has a striking resemblance to the FW algorithm for solving the APSP problem. The similarity may suggest that perhaps sub-cubic time bound algorithms for the MWPT problem would also be possible, as there has been much research in sub-cubic algorithms for the APSP problem, as shown in Table 1. For the APSP problem, even algorithms with deeply sub-cubic time bounds

---

[1] Named after the Belgian mathematician Eugene Charles Catalan (1814–1894).

© Springer International Publishing AG 2016
T.-H.H. Chan et al. (Eds.): COCOA 2016, LNCS 10043, pp. 326–339, 2016.
DOI: 10.1007/978-3-319-48749-6_24

Table 1. Sub-cubic algorithms for the APSP problem.

Year	Time complexity	Author(s)
1976	$O(n^3(\log\log n/\log n)^{1/3})$	Fredman [7]
1990	$O(n^3/\sqrt{\log n})$	Dobosiewicz [5]
1992	$O(n^3\sqrt{\log\log n/\log n})$	Takaoka [14]
2004	$O(n^3(\log\log n/\log n)^{5/7})$	Han [9]
2005	$O(n^3\log\log n/\log n)$	Takaoka [15]
2006	$O(n^3\sqrt{\log\log n}/\log n)$	Zwick [17]
2008	$O(n^3/\log n)$	Chan [4]
2008	$O(n^3(\log\log n/\log n)^{5/4})$	Han [10]
2012	$O(n^3\log\log n/\log^2 n)$	Han and Takaoka [11]

are known for graphs with small integer edge costs that utilize faster matrix multiplication over a ring [3,13,16].

All algorithms provided in Table 1 are based on speeding up the $(min,+)$ matrix multiplication in a semi-ring using pre-computed look-up tables. It is widely known that computing the closure of the $(min,+)$ matrix semi-ring is equivalent to solving the APSP problem, and that the closure of a matrix semi-ring can be computed in the same time complexity as multiplying two matrices in the semi-ring [1].

The MWPT problem on general polygons are known to be P-complete [2]. If the problem on convex polygons is also P-complete, it is quite unlikely that the aforementioned approach based on the semi-ring theory would be able to be applied to the MWPT problem, since the semi-ring based approach would likely mean that the algorithm is easily parallelizable, which contradicts the problem being P-complete.

In this paper, we firstly show that we do not have to resort to the semi-ring theory in order to provide a sub-cubic time bound for the APSP problem. We still rely on optimizing the $(min,+)$ matrix multiplication with look-up tables, but we embed this optimization directly within the FW algorithm rather than relying on recursion in the semi-ring theory. While we are unable to provide a faster algorithm for solving the APSP problem with this new approach, we show that we can apply the same set of principles to the GK algorithm to break the long standing cubic time barrier of the MWPT problem, which we present as the main contribution of this paper. We have tried to appeal to the reader's intuition by using analogies such as *acceleration, cruising* and *braking* in the description of our algorithms.

2   Preliminaries

Let $G = (V, E)$ be the input polygon (graph), where V is the set of vertices and E is the set of edges with costs. Let $n = |V|$. We assume that the vertices are numbered from 1 to n in the clockwise direction as shown in Fig. 1.

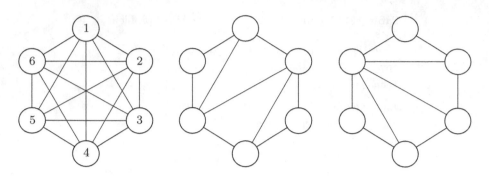

Fig. 1. Two example triangulations of a Hexagon

Let $D = \{d_{ij}\}$ be an n-by-n matrix such that $d_{ij} = D[i][j]$ is the distance
(cost) from vertex i to vertex j. We refer to D as a distance matrix. If no edge
exists between i to j, we let $d_{ij} = \infty$. $d_{ii} = 0$ for all $1 \leq i \leq n$. We assume
that enough edges exist in the input graph for a successful triangulation. We
assume that the edges are undirected for the MWPT problem, although no such
restrictions are required for the APSP problem.

If A, B and C are all distance matrices, then we can define the $(min, +)$
matrix multiplication, $C = A * B$, which we refer to as the $(min, +)$-product, as
performing the following operations for all possible pairs of i and j:

$$c_{ij} \leftarrow \infty, c_{ij} \leftarrow \min_{k=1}^{n}\{c_{ij}, a_{ik} + b_{kj}\}$$

If we were to compute the $(min, +)$-product $D * D$, intuitively, the meaning of the
operation $min\{d_{ij}, d_{ik} + d_{kj}\}$ can be understood to be the comparison between
the currently known distance from i to j against the total distance from i to
j going *via* vertex k. Vertex k in this instance is also commonly referred to as
the *witness* vertex. In other words, vertex k proves that the given distance from
i to j is possible by going *via* k. We refer to the above min operation as the
triple operation. In both the FW algorithm for solving the APSP problem and
the GK algorithm for solving the MWPT problem, we repeatedly perform this
triple operation to pick the best vertex k that gives the lowest cost/distance.

For the set of distance matrices, if we define the above $(min, +)$-product as
the matrix multiplication and define the component-wise min operation as the
matrix addition, then we can formulate a distance matrix semi-ring, such that
computing the closure of the semi-ring is equivalent to solving the APSP problem [1]. A straightforward implementation of computing the $(min, +)$-product
would take $O(n^3)$ time. Since the closure of a matrix semi-ring can be computed
within the same time bound as computing the product in the semi-ring [1], the
APSP problem can be solved with the semi-ring theory in $O(n^3)$ time, which
incidentally is also the time bound for both the FW and GK algorithms.

As briefly mentioned in Sect. 1, time bounds provided in Table 1 are achieved
by optimizing the computation of the $(min, +)$-product with look-up tables.

The key idea that allows for such optimization was first given by Fredman [7] as follows:

$$a_{ir} + b_{rj} \leq a_{is} + b_{sj}$$

If the above equation holds when computing $A * B$, then r is a better *via* vertex than s since it provides a lower cost from i to j. The equation can be rearranged as:

$$a_{ir} - a_{is} \leq b_{sj} - b_{rj}$$

Fredman showed that we can pre-compute both sides separately from each other for all possible r and s pairs within a given set of *via* vertices, and then sort and merge the results to retrieve the relative rankings for the r and s pairs, such that we can then use the rankings from both sides to index into a pre-computed look-up table to retrieve which *via* vertex gives the lowest cost from i to j in $O(1)$ time. We refer to the sorted results for r and s the sorted lists of differences. Since the size of the look-up table grows exponentially, the amount of speed up provided by this method is likely to remain within the polylog factor.

3   Sub-Cubic Floyd-Warshall Algorithm

We start with a review of the classical FW algorithm, as provided in Algorithm 1. A visualization of the algorithm is given in Fig. 2.

Algorithm 1. Original Floyd-Warshall Algorithm

1: **for** $k \leftarrow 1$ to n **do**
2:     **for** $i \leftarrow 1$ to n **do**
3:         **for** $j \leftarrow 1$ to n **do**
4:             $D[i][j] \leftarrow \min\{D[i][j], D[i][k] + D[k][j]\}$

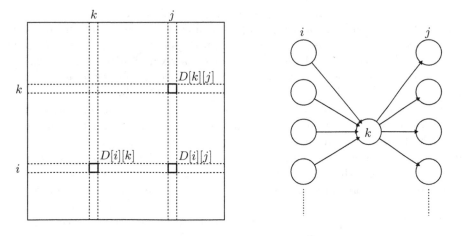

Fig. 2. Visualization of Floyd algorithm

As we can see from the right hand side diagram in Fig. 2, in the FW algorithm, for each *via* vertex k, we perform the *triple* operation for all pairs of i and j to check whether going via vertex k provides a better distance than the currently known distance from i to j.

Close inspection of the FW algorithm reveals that neither the order in which we iterate through the different values for k, or the order in which we iterate through the possible (i, j) pairs, affect the correctness of the algorithm. This observation can be easily derived from the fact that vertices in G can be numbered from 1 to n in any order for the APSP problem. This leads to Algorithm 2, which we refer to as the generalized FW algorithm. Algorithm 2 clearly shows that there is room for flexibility in implementing the FW algorithm.

Algorithm 2. Generalized Floyd-Warshall Algorithm

1: **for** each $k \in \{1...n\}$ in any order **do**
2:     **for all** (i, j) pair in any order **do**
3:         $D[i][j] \leftarrow \min\{D[i][j], D[i][k] + D[k][j]\}$

With Algorithm 2, we make another key observation that the values for k must iterate outside of the iteration for (i, j) pairs, but this restriction only exists because $D[1...n][k]$ or $D[k][1...n]$ must be used for subsequent computations of other vertex pairs that do not have k as an index. In other words, if, for example the *triple* operation was performed for $D[i][j]$ for *via* vertices k_1 and k_2 but the resulting value for $D[i][j]$ is not used for subsequent *triple* operations on other (i, j) pairs with *via* vertices k_1 and k_2, then clearly performing the two triple operations for $D[i][j]$ out of order does not affect the overall computation of the FW algorithm. From this key observation, we can derive the coarse-grained FW algorithm given by Algorithm 3, which effectively goes through the different *via* vertices in bulk in each iteration.

The visualization of the coarse-grained FW algorithm is given in Fig. 3. The main point of this new algorithm is in performing the triple operations for $O(m)$ *via* vertices in a single iteration, rather than stepping through each *via* vertex one by one, by dividing up the n-by-n distance matrix into m-by-m sub-matrices. Each iteration of Algorithm 3 is divided into two distinct phases to achieve this, which we refer to as the Acceleration phase and the Cruising phase.

Let $N = n/m$ such that D is divided into a total of N^2 square matrices, where each sub matrix is m-by-m. The term "coarse-grained" comes from iterating through the m-by-m sub-matrices in contrast to the "fine-grained" method of iterating through each column/row of D one-by-one. For clarity we use capital letters in our algorithms for coarse-grained iterators. $D[I][J]$ represents the m-by-m sub-matrix on coarse-grained row I and coarse-grained column J, where $1 \leq I, J \leq N$.

The Acceleration phase of an iteration performs "fine-grained" triple operations for a set of (i, j) pairs for a total of m *via* vertices, similarly to the original FW algorithm, iterating over the m *via* vertices one-by-one. The shaded area

Algorithm 3. Coarse-Grained Floyd-Warshall Algorithm

1: $N \leftarrow n/m$
2: **for** $K \leftarrow 1$ **to** N **do**
3:     $k_m \leftarrow (K-1)m$
4:     /* Acceleration Phase */
5:     **for** $k \leftarrow k_m + 1$ **to** $k_m + m$ **do**
6:         **for** $i \leftarrow k_m + 1$ **to** $k_m + m$ **do**
7:             **for** $j \leftarrow 1$ **to** n **do**
8:                 $D[i][j] \leftarrow min(D[i][j], D[i][k] + D[k][j])$
9:         **for** $i \leftarrow 1$ **to** n **do**
10:           **for** $j \leftarrow k_m + 1$ **to** $k_m + m$ **do**
11:              $D[i][j] \leftarrow min(D[i][j], D[i][k] + D[k][j])$
12:     /* Cruising Phase */
13:     **for** $I \leftarrow 1$ **to** N **do**
14:         **for** $J \leftarrow 1$ **to** N **do**
15:            Compute $D[I][J] \leftarrow D[I][K] * D[K][J]$

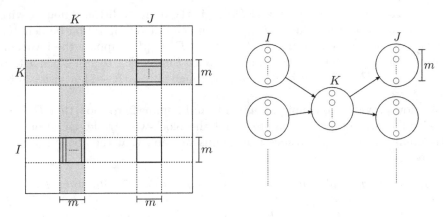

Fig. 3. Visualization of coarse-grained Floyd-Warshall algorithm

in Fig. 3 represents the subset of (i, j) pairs that finish computation for m *via* vertices in the Acceleration phase of an iteration. We refer to this shaded area as the thick row and the thick column.

We then move onto the Cruising phase, where we use the computed sub-matrices from the Acceleration phase, namely thick column $D[1...N][K]$ and thick row $D[K][1...J]$, to finish off all remaining m-by-m sub-matrices for the same set of m *via* vertices in the iteration. It is important to note that the general $(min, +)$-product can be used in the cruising phase to compute $D[I][J] = D[I][K] * D[K][J]$ that assumes no order in the computation of the *via* vertices since the sub-matrices sit outside of the shaded area in Fig. 3. In other words, for any of the given m *via* vertices, $D[i][j]$ that does not sit in the shaded area are not used in the computation of any other (i, j) pairs.

The Acceleration phase takes $O(m^2n)$ time and the Cruising phase takes $O(mn^2)$ time, assuming $O(m^3)$ time bound for computing the $(min, +)$-product of two m-by-m sub-matrices. Since we iterate $N = n/m$ times, the total time bound is $O(mn^2 + n^3)$, but since $m < n$, we have $O(n^3)$ for the coarse-grained FW algorithm given by Algorithm 3.

Theorem 1. *Algorithm 3 correctly solves the APSP problem.*

Let block I be the vertex set $\{(I-1)m+1, (I-1)m+2, ..., (I-1)m+m = Im\}$, that is, the Ith block of the sorted vertex set $V = \{1, 2, ..., n\}$. Let block J and block K be similarly defined. V can be expressed as $S(1) \cup S(2) \cup ... \cup S(N)$ where $S(K) = \{(K-1)m+1, (K-1)m+2, ..., (K-1)m+m = Km\}$. We prove the correctness of Algorithm 3 by induction.

Invariant. At the end of the Kth iteration $D[i][j]$ is the distance of the shortest path from vertex i in any block I to vertex j in any block J that goes through the set of *via* vertices $V(K) = \{1, 2, ..., K * m\}$.

Basis. At $K = 0$, $V(0)$ is empty and $D[i][j]$ is the original distance matrix with no *via* vertices. Let the value of $D[i][j]$ at the end of Kth iteration be denoted by $D^K[i][j]$ and that of sub-matrix $D[I][J]$ be $D^K[I][J]$. Suppose the invariant is correct for $K - 1$ and take up a shortest path i in block I to j in block J that goes through $V(K)$. The path either goes through only $V(K-1)$ or visits $S(K)$ on the way.

We now provide two lemmas, Lemma 1, and Lemma 2, to show that $D[I][K]$ and $D[K][J]$ are correctly computed, and the block-wise triple operation correctly computes $D^K[I][J]$, respectively. Thus we show that the above invariant is correct for K.

Lemma 1. *At the end of the accelerating phase of the Kth iteration, $D[i][j]$ is the distance of a shortest path from i in block K to j in any block $1 \leq J \leq N$ that goes through $V(K)$. Similarly $D[i][j]$ is the distance of a shortest path from i in any block $1 \leq I \leq N$ to j in block K that goes through $V(K)$.*

Proof. We prove this by induction on k. *Invariant.* At the end of the kth iteration of the update for the thick row in the global K^{th} iteration, $D[i][j]$ is the distance of the shortest path from i in block K to j in any block J that goes through $\{1, 2, ..., k\}$. *Basis.* $k = Km$. This is guaranteed by the global induction assumption. Suppose invariant is true for up to $k - 1$. The triple operation $D[i][j] = min\{D[i][j], D[i][k] + D[k][j]\}$ makes invariant true for k. The proof for the thick column is similar.

Lemma 2. *Let $D[I][K]$ be the shortest distances of paths from block I to block K that go through $V(K)$. $D[K][J]$ is similar. Then the statement $D[I][J] = min\{D[I][J], D[I][K] * D[K][J]\}$ correctly computes $D[I][J]$ for the distances of the shortest paths that go from block I to block J through $V(K)$.*

Proof. Let the two dimensional index block, block $I \times$ block J, be denoted by (I, J). Suppose $I \neq K$ and $J \neq K$. Then block (I, K) and block (K, J)

are disjoint. In the $(min, +)$-product $D[I][K] * D[K][J]$ all possibilities for the shortest paths that go from block I to block J via block K are represented. As mentioned earlier we can compute the $(min, +)$-product without any cumulative effect since the blocks are disjoint. If $I = K$ or $J = K$, we note that $D[I][K] * D[K][J] = D[I][J]$. By taking the minima with $D[I][J]$ the case of the paths from block I to block J that do not go through block K are taken care of. The possibility of a shortest path going though block K is taken care of by $D[I][K]$ or $D[K][J]$ at the end of the accelerating phase.

Theorem 2. *The FW algorithm can be optimized to have a sub-cubic time bound.*

Proof. In each iteration of 'Algorithm 3, after the Acceleration phase completes and before the Cruising phase begins, we go through each m-by-m sub matrix that was computed in the Acceleration phase (i.e. the shaded region in Fig. 3), and further divide each sub-matrix in column K as m-by-l rectangular sub-matrices, and divide each sub-matrix in row K as l-by-m rectangular sub-matrices. An example of this further break down into rectangular sub-matrices is shown in Fig. 3. Then we can pre-compute the rankings from these rectangular sub-matrices as briefly explained in Sect. 2 and use the sorted rankings to index into a pre-computed look-up table such that the $(min, +)$-product of an m-by-l matrix with an l-by-m matrix can be computed in $O(l^2 m)$ time bound, which is a much faster time bound compared to the straightforward time bound of $O(lm^2)$ [14]. Since there are m/l l-by-m or m-by-l rectangular sub-matrices in an m-by-m sub-matrix, computing the $(min, +)$-product of two m-by-m sub-matrices using the look up table method given by Takaoka [14] takes $O(l^2 m) * m/l = O(lm^2)$ for the $(min, +)$-product and $O(m^3/l)$ for the component-wise min operation to combine the m/l separate products, resulting in $O(lm^2 + m^3/l)$, which is balanced to $O(m^{2.5})$ by setting $l = \sqrt{m}$. Thus in the Cruising phase of Algorithm 3, we can compute the $(min, +)$-product of all $O(n^2/m^2)$ m-by-m sub-matrices in the total time bound of $O(\sqrt{m}n^2)$.

Since we iterate $N = n/m$ times, for the whole algorithm, we spend $O(mn^2)$ in the Acceleration phase, and $O(n^3/\sqrt{m})$ in the Cruising phase. By letting $m = O(\log n/\log \log n)$ the total time taken for pre-computation and sorting for indexing into the look-up table, as well as the computation of the whole look-up table itself, can be shown to stay within the time bound of $O(n^3/\sqrt{m})$ [14]. Since the time complexity of the Cruising phase dominates, for the whole algorithm, we have the sub-cubic time complexity of $O(n^3 \sqrt{\log \log n/\log n})$.

4   Sub-Cubic Minimum Weight Polygon Triangulation Algorithm

Similarly to the APSP problem in Sect. 3, we start with a review of the GK algorithm based on dynamic programming given by Algorithm 4. As noted earlier, we assume that the vertices in the polygon are numbered from 1 to n in the clockwise direction as shown in Fig. 1. In Algorithm 4, $t + 1$ is the total number of vertices

Algorithm 4. Original Gilbert-Klincse Algorithm

1: **for** $t \leftarrow 2$ to $n - 1$ **do**
2:     **for** $i \leftarrow 1$ to $n - t - 1$ **do**
3:         $j \leftarrow i + t$
4:         **for** $k \leftarrow i + 1$ to $j - 1$ **do**
5:             $C[i][j] \leftarrow min(C[i][j], C[i][k] + C[k][j] + D[i][j])$

in the sub-polygon for which the MWPT problem will be solved. Starting from polygons of size 3 ($t = 2$), we solve the MWPT problem for the sub-polygons in the clockwise direction (i.e. i iterates from 1 in increasing order), storing the minimum triangulation costs in the matrix $C = \{c_{ij}\}$, where $c_{ij} = C[i][j]$ is the minimum possible triangulation cost for the sub-polygon that includes the vertices $\{i, i + 1, i + 2, ..., j\}$. We assume that $C[i][j]$ is firstly initialized to ∞ for all $1 \leq i, j \leq n$, then we let $C[i][i + 1] = D[i][i + 1]$ for all $1 \leq i < n$.

We note that the main operation of Algorithm 4 to compute $C[i][j]$ in each iteration looks very similar to the *triple* operation. In fact, we observe that the addition of $D[i][j]$ can trivially be moved outside of the inner most for loop to give $C[i][j] \leftarrow min\{C[i][j], C[i][k] + C[k][j]\}$, which is exactly the *triple* operation as given in Sect. 2. For the MWPT problem the *triple* operation is performed on the triangulation cost matrix rather than the distance matrix for the APSP problem. Intuitively, the meaning of the *triple* operation performed on C is that given a sub-polygon with vertices $\{i, i + 1, i + 2, ..., j\}$, we wish to take the minimum of the currently known triangulation cost $C[i][j]$, and the triangulation cost given by the triangle i, j, k in addition to the already known minimum triangulation cost $C[i][k]$ and $C[k][j]$. In the context of the MWPT problem, we refer to k as the *dividing* vertex.

A visualization of Algorithm 4 is given by Fig. 4. The shaded region on the left hand side shows the part of the cost matrix C that is never used because j

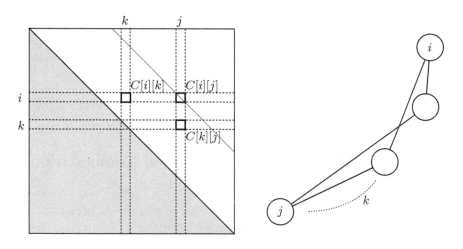

Fig. 4. Visualization of GK algorithm

is always greater than i. The diagram on the right hand side shows the iteration over k to determine which triangulation provides the minimum cost.

We can visualize the sequence of the GK algorithm by picturing the diagonal, given by the set of (i, j) such that $j - i = t$ in Fig. 4, shifting closer and closer to the top-right corner of C in each iteration. The cost matrix elements on this diagonal contains the sub-polygon triangulation costs that have just been computed in the iteration. Thus in each iteration, the diagonal shifts one element closer to the top-right corner. Once the diagonal reaches $C[1][n]$, the MWPT problem is solved and the total triangulation cost will be contained in $C[1][n]$. We refer to this moving diagonal line as the update line.

A closer inspection of the original GK algorithm reveals that iterating through the different values of t is very sequential in nature because to compute the minimum weight triangulation for a sub-polygon with $t + 1$ vertices, we must have finished the minimum weight triangulation computation for all sub-polygons with t vertices. On the other hand, given a sub-polygon starting with vertex i and ending with vertex j, we can clearly iterate through the dividing vertices k in any order to determine which triangulation option gives the minimum total cost. Together with the aforementioned possible relocation of the summation of $D[i][j]$ from the inner-most for loop, we can provide a more generalized version of the GK algorithm as given by Algorithm 5.

Algorithm 5. Generalized Gilbert-Klincse Algorithm

1: **for** $t \leftarrow 2$ to $n - 1$ **do**
2:     **for** $i \in \{1, 2, ..., n - t - 1\}$ in any order **do**
3:         $j \leftarrow i + t$
4:         **for** $k \in \{i + 1...j - 1\}$ in any order **do**
5:             $C[i][j] \leftarrow \min\{C[i][j], C[i][k] + C[k][j]\}$
6:         $C[i][j] \leftarrow C[i][j] + D[i][j]$

Finally we extend the generalized GK Algorithm to provide the coarse-grained GK algorithm, given by Algorithm 6 and visualized in Fig. 5. Algorithm 6 consists of three phases. The Acceleration Phase sits outside of the main loop, for solving the MWPT problem for all sub-polygons of size $O(m)$ where $m < n$. In other words, the update line is moved towards the top-right corner by $O(m)$. Then in the main for loop, the aim is to shift the update line closer to the top-right corner by $O(m)$ in a single iteration in a "coarse-grained" manner, instead of the "fine-grained" manner of shifting the update line by one element in each iteration. This is achieved by the Cruising Phase of each iteration performing bulk of the work by computing a series of $(min, +)$-products of m-by-m sub-matrices, followed by a necessary tidy-up in the Braking Phase due to the sequential nature of the original GK algorithm. Similarly to the coarse-grained FW algorithm described in Sect. 3, we use upper case letters as coarse-grained iterators, and $C[I][J]$ denotes an m-by-m sub-matrix such that $1 \leq I, J \leq N$.

Algorithm 6. Coarse-Grained Gilbert-Klincse Algorithm

1: /* Acceleration Phase */
2: Perform original GK Algorithm for $1 \leq t < m$
3: $N \leftarrow n/m$
4: **for** $T \leftarrow 1$ to $N - 1$ **do**
5:     **for** $I \leftarrow 1$ to $N - T - 1$ **do**
6:         /* Cruising Phase */
7:         $J \leftarrow I + T$
8:         **for** $K \leftarrow I + 1$ to $J - 1$ **do**
9:             $C[I][J] \leftarrow min\{C[I][J], C[I][K] * C[K][J]\}$
10:        $C[I][J] = C[I][J] + D[I][J]$
11:        /* Braking Phase */
12:        $t_m \leftarrow (T - 1)m; i_m \leftarrow (I - 1)m; j_m \leftarrow (J - 1)m$
13:        **for** $t \leftarrow 1$ to $2m - 1$ **do**
14:            **for** $i \leftarrow i_m + 1$ to $i_m + m$ **do**
15:                $j \leftarrow i + t_m + t$
16:                **if** $j_m + 1 \leq j \leq j_m + m$ **then**
17:                    **for** $k \leftarrow i + 1$ to $i_m + m$ **do**
18:                        $C[i][j] \leftarrow min\{C[i][j], C[i][k] + C[k][j] + D[i][j]\}$
19:                    **for** $k \leftarrow j_m + 1$ to $j - 1$ **do**
20:                        $C[i][j] \leftarrow min\{C[i][j], C[i][k] + C[k][j] + D[i][j]\}$

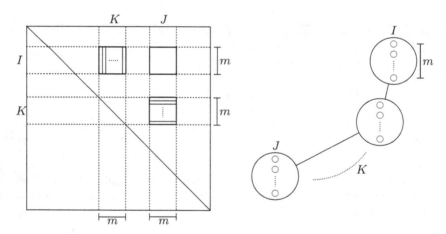

Fig. 5. Visualization of coarse grained GK algorithm

Theorem 3. *Algorithm 6 correctly solves the MWPT problem.*

Proof. Formally the proof can be given based on induction on T and also induction on t, which we refer to as the global induction and the local induction, respectively. The Acceleration Phase corresponds to the basis for the global induction and the Cruising Phase is the general iteration by T. The basis for the local induction is the fact that the bottom left corner of matrix $C[I][J]$ such that $J - I = T$ is already computed as well as $C[I][J]$ such that $J - I < T$ at

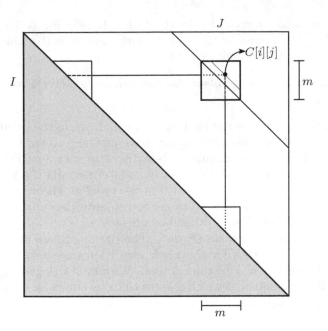

Fig. 6. Computing all values of k

the end of the current cruising Phase. The general process in the Braking Phase is to finalize the value of $C[i][j]$ on the update line in increasing order of t. The Braking Phase is essentially the GK algorithm executed on the sub-matrices $C[I][J], C[I][K]$ and $C[K][J]$, as shown in Fig. 6. Note that for $T = 1$, the Cruising Phase is skipped entirely, and the final values for $C[I][J]$ is computed entirely in the Braking Phase.

For the induction process it is sufficient to show that for all elements c_{ij} in $C[I][J]$ sub-matrices that have been computed in a given iteration, all *dividing* vertices k such that $i < k < j$ have been considered. All *dividing* vertices k such that $i_m + m < k \le j_m$ are checked in the Cruising Phase. These values of k are represented by the horizontal and vertical solid lines in Fig. 6. As noted earlier, k can be iterated in any order, thus $(min, +)$-product can be used without any restrictions in the order of computation. This leaves the two disjoint sets of *dividing* vertices $i < k \le i_m + m$ and $j_m < k < j$.

The reason we cannot simply compute $C[I][I] * C[I][J]$ and $C[I][J] * C[J][J]$ in the Cruising Phase to compute the *dividing* vertices $i < k \le i_m + m$ and $j_m < k < j$, respectively is because the sub-matrix $C[I][J]$ does not get fully computed until the end of the iteration. More specifically, due to the sequential nature of the GK algorithm, for any i and j, $C[i][j]$ can only be computed if both $C[i][j-1]$ and $C[i+1][j]$ have already been computed.

Thus the remaining *dividing* vertices are covered off in a "fine-grained" manner in the Braking Phase, which is in effect no different from the original GK algorithm. The first set of $i < k \le i + m$ is covered by the for loop with Line 18 as the main operation, represented by the dashed lines in Fig. 6. The second set of $j - m \leq k < j$ is covered by the for loop with Line 20 as the main operation, represented by the dotted lines in Fig. 6.

Theorem 4. *There exists an algorithm that solves the MWPT problem in sub-cubic time bound.*

Proof. In Algorithm 6, after the Braking Phase and before the iteration with T ends, divide the m-by-m sub-matrices that have just been solved in the iteration into m-by-l and l-by-m rectangular sub-matrices as shown in Fig. 5. We can then generate the sorted lists of differences to later perform the merge process and rank computation [14]. As explained in the proof of Theorem 2, with those lists and ranks, we can use look-up tables to compute the $(min, +)$-product of m-by-m sub-matrices in $O(m^{2.5})$ time bound [14].

The Acceleration Phase takes $O(nm^2)$. The Braking Phase of each iteration takes $O(m^3)$ time, and thus for the whole algorithm takes $O(n^2 m)$ time. The Cruising Phase takes $O(n^3 \sqrt{m})$ time in total. With $m = O(\log n / \log \log n)$ [14], the Cruising Phase dominates, which gives us the total sub-cubic time complexity of $O(n^3 \sqrt{\log \log n / \log n})$.

5   Concluding Remarks

We have made use of the sub-cubic algorithm for the $(min, +)$-product given by Takaoka [14] to optimize the FW algorithm in Sect. 3, and more importantly, the GK algorithm in Sect. 4 to break the cubic time barrier for the MWPT problem. We chose the aforementioned $(min, +)$-product algorithm not because it was the fastest algorithm that we could apply, but because it was the simplest that we could use to achieve the goal of this paper, which is to show that an algorithm with sub-cubic time bounds for solving the MWPT problem is indeed possible.

In fact, it seems straightforward to use a faster $(min, +)$-product [15] in Algorithm 6 to achieve a faster time bound, although applying the currently fastest known algorithm for $(min, +)$-product [11] seems challenging due to the inherent sequential nature of the GK problem. We conclude the paper with the obvious open question: How close can we take the time bounds of solving the MWPT problem to the time bounds of the APSP problem?

References

1. Aho, A.V., Hopcroft, J.E., Ullman, J.D.: The Design and Analysis of Computer Algorithms. Addison-Wesley, Boston (1974)
2. Atallah, M.J., Callahan, P.B., Goodrich, M.T.: P-complete geometric problems. Int. J. Comput. Geom. Appl. **3**, 443–462 (1993)
3. Alon, N., Galil, Z., Margalit, O.: On the exponent of the all pairs shortest path problem. In: FOCS, pp. 569–575 (1991)
4. Chan, T.M.: All-pairs shortest paths with real weights in $O(n^3 / \log n)$ time. Algorithmica **2**, 236–243 (2008)

5. Dobosiewicz, W.: A more efficient algorithm for the min-plus multiplication. Int. J. Comput. Math. **32**, 49–60 (1990)
6. Floyd, R.W.: Algorithm 97: shortest path. Commun. ACM **5**(6), 345 (1962)
7. Fredman, M.L.: New bounds on the complexity of the shortest path problem. SIAM J. Comput. **5**, 83–89 (1976)
8. Gilbert, P.D.: New results on planar triangulations. Report R-850, Coordinated Science Laboratory, University of Illinois, Urbana, Illinois (1979)
9. Han, Y.: Improved algorithm for all pairs shortest paths. Inf. Process. Lett. **91**, 245–250 (2004)
10. Han, Y.: An $O(n^3(\log\log n/\log n)^{5/4})$ time algorithm for all pairs shortest path. Algorithmica **51**, 428–434 (2008)
11. Han, Y., Takaoka, T.: An $O(n^3 \log\log n/\log^2 n)$ time algorithm for all pairs shortest paths. In: Fomin, F.V., Kaski, P. (eds.) SWAT 2012. LNCS, vol. 7357, pp. 131–141. Springer, Heidelberg (2012)
12. Klincsek, G.T.: Minimal triangulations of polygonal domains. Ann. Discret. Math. **9**, 121–123 (1980)
13. Seidel, R.: On the all-pairs-shortest-path problem in unweighted undirected graphs. J. Comput. Syst. Sci. **51**, 400–403 (1995)
14. Takaoka, T.: A new upper bound on the complexity of the all pairs shortest path problem. Inf. Process. Lett. **43**, 195–199 (1992)
15. Takaoka, T.: An $O(n^3 \log\log n/\log n)$ time algorithm for the all-pairs shortest path problem. Inf. Process. Lett. **96**, 155–161 (2005)
16. Zwick, U.: All pairs shortest paths using bridging sets and rectangular matrix multiplication. J. ACM **49**, 289–317 (2002)
17. Zwick, U.: A slightly improved sub-cubic algorithm for the all pairs shortest paths problem with real edge lengths. Algorithmica **46**, 181–192 (2006)

The Mixed Center Location Problem

Yi Xu[1]([✉]), Jigen Peng[1,2], and Yinfeng Xu[3,4]

[1] School of Mathematics and Statistics,
Xi'an Jiaotong University, Xi'an 710049, People's Republic of China
`xy.clark@stu.xjtu.edu.cn`
[2] Beijing Center for Mathematics and Information Interdisciplinary Sciences,
Beijing 100048, People's Republic of China
[3] School of Management, Xi'an Jiaotong University,
Xi'an 710049, People's Republic of China
[4] The State Key Lab for Manufacturing Systems Engineering,
Xi'an 710049, People's Republic of China

Abstract. This paper studies a new version of the location problem called the *mixed center location* problem. Let P be a set of n points in the plane. We first consider the mixed 2-center problem where one of the centers must be in P and solve it in $O(n^2 \log n)$ time. Next we consider the mixed k-center problem where m of the centers are in P. Motivated by two practical constraints, we propose two variations of the problem. We present an exact algorithm, a 2-approximation algorithm and a heuristic algorithm solving the mixed k-center problem. The time complexity of the exact algorithm is $O(n^{m+O(\sqrt{k-m})})$.

Keywords: k-center problem · Facility location problem · Voronoi diagram · Computational geometry

1 Introduction

The facility location problem is to choose the locations of facilities to minimize the cost of satisfying the demand for certain commodity. Sometimes the location problem is associated with the costs for building the facilities, as well as the transportation costs for distributing the commodities. In this paper, we consider a new facility location problem called the *mixed center location* problem.

Related work. Let P be a set of n points in the plane. The k-center location problem is to find k centers such that the maximum of the distances from the stations to the nearest centers is minimized. When $k = 2$, the 2-center problem ($S2CP$) has been extensively studied. Jaromczyk and Kowaluk first gave a deterministic algorithm with running time $O(n^2 \log n)$ [1]. Eppstein gave an improvement with a randomized algorithm running in $O(n \log^2 n)$ expected time [2]. In a major breakthrough, Sharir showed that the planar 2-center problem can be actually solved in near-linear time and the time bound is $O(n \log^9 n)$ time [3]. Finally the algorithm was further improved by Chan in $O(n \log^2 n \log^2 \log n)$ time [4].

© Springer International Publishing AG 2016
T.-H.H. Chan et al. (Eds.): COCOA 2016, LNCS 10043, pp. 340–349, 2016.
DOI: 10.1007/978-3-319-48749-6_25

In [3,4], $S2CP$ was divided into two parts: the fixed-size problem and the problem of computing the smallest radius. The fixed-size problem is to determine whether there exist two congruent disks with radius r whose union covers P for a given radius r. Hershberger solved the fixed-size problem for $S2CP$ in $O(n^2)$ time [5] and Sharir improved the bound to $O(n \log^3 n)$ time [3]. The discrete 2-center problem ($D2CP$) is defined as follows: covering P by the union of two congruent closed disks whose radius is as small as possible, and centers are two points of P. The first near-quadratic algorithm was proposed in [6] and finally improved to $O(n^{\frac{4}{3}} \log^5 n)$ time in [7].

The k-center problem ($SkCP$) has also been widely considered. When k is part of the input, the problem is known to be NP-complete [8]. Drezner proposed an algorithm with $O(n^{2k+1} \log n)$ time to solve $SkCP$ [9]. By combining the result of 1-center problem proposed by Megiddo [10], $SkCP$ can be revised to $O(n^{2k-1} \log n)$ time. Finally, Hwang et al. improved the time complexity to $O(n^{O(\sqrt{k})})$ using the slab dividing method [11]. Some approximation algorithms for $SkCP$ have also been considered. Under general metrics, Hochbaum and Shmoys [12] and Gonzalez [13] provided the 2-approximation algorithms in $O(n^2 \log n)$ and $O(nk)$ time respectively. Feder and Greene [14] gave a 2-approximation algorithm in $O(n \log k)$ time for the Euclidean standard k-center problem.

Problem statement. In this paper, we consider the *mixed center location* problem. First we propose a variation of the 2-center problem called the *mixed 2-center* problem ($M2CP$). $M2CP$ is to cover P by two closed disks whose maximum radius is minimized and one of the two centers is in P. We simply solve $M2CP$ in $O(n^2 \log n)$ time. See Fig. 1.

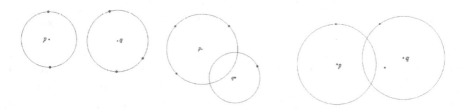

Fig. 1. (i) An example of $S2CP$ (ii) An example of $M2CP$ (iii) An example of $D2CP$

Then we consider the *mixed k-center* problem. For the k-center problem, if all the centers are in P, we call the problem the discrete k-center problem ($DkCP$), if m ($m < k$) of the centers are in P, we call the problem the mixed k-center problem ($M(k,m)CP$), otherwise we call the problem the standard k-center problem ($SkCP$). Let $\{r_{p_1}(P), r_{p_2}(P), ..., r_{p_k}(P)\}$ be the radii of k disks centered at $\{p_1, p_2, ..., p_k\}$. The three problems are listed as follows:

(1) The discrete k-center problem $(DkCP)$: find k centers $\{p_1, p_2, ..., p_k\} \in P$ such that $max\{r_{p_1}(P), r_{p_2}(P), ..., r_{p_k}(P)\}$ is minimized.

(2) The mixed k-center problem $(M(k, m)CP)$: find k centers $\{p_1, p_2, ..., p_k\}$ and m of them are in P such that $max\{r_{p_1}(P), r_{p_2}(P), ..., r_{p_k}(P)\}$ is minimized.

(3) The standard k-center problem $(SkCP)$: find k centers $\{p_1, p_2, ..., p_k\}$ such that $max\{r_{p_1}(P), r_{p_2}(P), ..., r_{p_k}(P)\}$ is minimized.

Organization. The remainder of this paper is organized as follows. In Sect. 2, we consider the $M2CP$ which can be solved in $O(n^2 \log n)$ time. In Sect. 3, we consider the $M(k, m)CP$. In Sect. 3.1 and 3.2, we consider the $M(k, m)CP$ under two constraints. One is to consider $M(k, m)CP$ under the budget constraint, and the other is that the m centers in P must be lying in a given region. We show an algorithm that determines $\{k, m\}$ and the optimal solution under the budget constraint. In Sect. 3.3, we give an exact algorithm that solves $M(k, m)CP$ in $O(n^{m+O(\sqrt{k-m})})$ time. In Sect. 3.4, we give a 2-approximation algorithm. In Sect. 3.5, we present a heuristic algorithm based on the Voronoi diagram and finally we give a conclusion in Sect. 4.

2   The Mixed 2-Center Problem

For a planar point set P, it is obvious that when one center of the optimum solution for $S2CP$ is in P, the optimum solution of $M2CP$ is the same as $S2CP$'s. We consider the case where the optimum solutions of $S2CP$ and $M2CP$ are different. Letting the radius of the standard (resp. discrete) minimum enclosing disk of a planar point set Q be r_Q^S (resp. r_Q^D), we have:

Observation 1 *Suppose* $q \notin Q$, *then* $r_Q^S \leq r_{Q \cup \{q\}}^S$.

However, r_Q^D may not be less than $r_{Q \cup \{q\}}^D$. Here we show a counter-example: Suppose Q is a straight line point set, the leftmost point is q_1, the rightmost point is q_n and no point in Q is in the middle of the segment l_{q_1,q_n} ($l_{a,b}$ denotes the line segment with two nodes a, b). Let q be the middle point of l_{q_1,q_n}, we have $r_Q^D > r_{Q \cup \{q\}}^D$. Conversely, if q is on the left of q_1, $r_Q^D \leq r_{Q \cup \{q\}}^D$. Suppose $\{m_1, m_2, ..., m_k\} \notin Q$, from Observation 1, we know that the radii of the standard minimum enclosing disks of Q, $Q \cup \{m_1\}$, $Q \cup \{m_1, m_2\}$, ... and $Q \cup \{m_1, m_2, ..., m_k\}$ are non-decreasing. Whereas, for discrete minimum enclosing disk, the radii of Q, $Q \cup \{m_1\}$, $Q \cup \{m_1, m_2\}$, ... and $Q \cup \{m_1, m_2, ..., m_k\}$ have no monotone property.

2.1   The Overall Algorithm

For the fixed point p, we go through all the points in P, and sort $P \backslash \{p\}$ in order of non-decreasing distance from p. Let $\{p_1, p_2, ..., p_{n-1}\}$ be the resulting order, i.e., $d(p, p_1) \leq d(p, p_2) \leq ... \leq d(p, p_{n-1})$, where $d(a, b)$ denotes the distance between a and b. For the disk with center p and radius $d(p, p_i)$, we compute the standard minimum enclosing disk of $P \backslash \{p_1, p_2, ..., p_i\}$, which can be done in $O(n)$ time [10]. From Observation 1, by binary search, the $M2CP$ can be solved in $O(n \log n)$ time. Finally the overall running time of the algorithm is $O(n^2 \log n)$ time using $O(n)$ storage. (See Algorithm 1.)

Algorithm 1. Algorithm for the mixed 2-center problem

1: **for** each $p \in P$ **do**
2: Sort the distance among p and all the other points in non-decreasing order, i.e., $d(p, p_1) \leq d(p, p_2) \leq ... \leq d(p, p_{n-1})$.
3:     **for** $p_i \in P$ **do**
4:         By using linear programming [10], compute the radius r_{p_i} of the standard minimum enclosing disk of $P \setminus \{p, p_1, ..., p_i\}$. $r_p = max\{d(p, p_i), r_{p_i}\}$.
5:     **end for**
6: **end for**
7: $r_{opt} \leftarrow min(r_p)$.

Theorem 1. *Let P be a planar point set of n points, the mixed 2-center problem can be solved in $O(n^2 \log n)$ time using $O(n)$ storage.*

3 The Mixed k-Center Problem

We consider the *mixed k-center* problem $(M(k, m)CP)$ in two cases, called the budget constraint problem and the region constraint problem. The budget constraint problem is to consider the $M(k, m)CP$ under the budget constraint. The region constraint problem is to consider the $M(k, m)CP$ with m centers lying in the given region.

3.1 The Budget Constraint Problem

First we consider the $M(k, m)CP$ under the budget constraint. In practice, the facility location problem is not only to find a good location for serving some stations or facilities, but also to consider the budget constraint and the cost difference among different places. In general, the point set P, which needs to be covered, is given. It means that there are some facilities or buildings at some points in P while none is located at the left points. Thus the construction costs for setting up a station in or not in P is different. Under the budget constraint, we need to find the number of the stations to be constructed and hope to serve the consumers efficiently.

From the definitions of $SkCP$, $DkCP$ and $M(k, m)CP$, let the optimal radii of these problems be r_{DkCP}, r_{SkCP} and $r_{M(k,m)CP}$ respectively, we have the following fact.

Fact 1. *Let P be a planar point set of n points. r_{DkCP}, r_{SkCP} and $r_{M(k,m)CP}$ are defined above. We have $r_{SkCP} \leq r_{M(k,m)CP} \leq r_{DkCP}$.*

Consider the center location problem under budget constraint. Suppose building a new center costs M_1 and rebuilding (changing one facility into a center) a center costs M_2 ($M_2 < M_1$). Under the budget M where $mM_2 + (k-m)M_1 \leq M$, we will try to choose the proper centers. From the definitions of $SkCP$, $DkCP$ and $M(k,m)CP$, if the center is in P, the cost can be seen as M_2, otherwise the cost is M_1. Our goal is to serve the consumers efficiently under the budget constraint; i.e., to minimize the distance of the optimal solution of the mixed k-center location problem.

Fact 2. *Let P be a planar point set of n points. For fixed k, if $m_2 < m_1$, we have $r_{M(k,m_2)CP} \leq r_{M(k,m_1)CP}$.*

Let $k_1^* = \lfloor \frac{M}{M_1} \rfloor$ and $k_2^* = \lfloor \frac{M}{M_2} \rfloor$, in order to obtain the minimum radius, we check all the possible combinations with $0 \leq m \leq k \leq k_2^*$. From Fact 2, the radius of the optimal solution is increasing when m is growing larger for each k under the budget M. From Fact 1, we need to compute the minimum $r_{M(k,m)CP}$ where $i + m = k$, $0 \leq i \leq k_1^*$ and $m = \lfloor \frac{M - iM_1}{M_2} \rfloor$. See Algorithm 2. Suppose $M(k,m)CP$ can be solved in $O(A)$ time, we have Theorem 2.

Algorithm 2. Algorithm of k-center location problem under budget constraint

1: Let M be the budget constraint, building a new facility costs M_1 and rebuilding a facility costs M_2, $k_1^* = \lfloor \frac{M}{M_1} \rfloor$.
2: **for** $i = 0$ to k_1^* **do**
3:     Consider the mixed k-center problem where $m = \lfloor \frac{M - iM_1}{M_2} \rfloor$ and $k = i + m$ (See Sect. 3.3). Let the optimal radius be r_i.
4: **end for**
5: Return $r_{opt} \leftarrow min(r_i)$.

Theorem 2. *Let P be a planar point set of n points, building a new facility cost M_1 and rebuilding a facility cost M_2. We can find the optimal solution of k-center location problem in $O(k_1^* A)$ time under the budget M where $k_1^* = \lfloor \frac{M}{M_1} \rfloor$.*

3.2 The Region Constraint Problem

In this subsection, we consider the case where the centers are located in two bounded areas. In many developing countries, the new urban area of a city is usually nearby the old urban area. This means that the two areas can be divided. Intuitively, we suggest two areas are separated by a bounded area, like a line or a convex polygon. Take Shanghai in China as an example, the economy of the urban area in the east of the Huangpu River (the new urban area) develops faster than the urban area in the west (the old urban area). We hope to build k shopping malls, where m are in the old urban area and $k - m$ are in the open area. In the new urban area, we choose to build the shopping malls in the newly freed up area. On the contrary, there aren't enough areas in the old urban area and we need to reconstruct some old buildings into shopping malls. Thus the planar has been partitioned into two regions R_1 and R_2. For a given planar point set and a partition, the $M(k, m)CP$ is to find k centers, such that the maximum distance from the clients to the nearest shopping mall is minimized, and m of the centers are in R_1 (or R_2). Figure 2 shows two examples. Figure 2(i) illustrates two regions separated by a line and Fig. 2(ii) illustrates two regions separated by a convex polygon. This problem is very similar to the k-supplier problem. In [15], Nagarajan et al. considered the k-supplier problem which is given a set of clients C and a set of facilities F, along with a bound k, to find a subset of k facilities so as to minimize the maximum distance of a client to the open facility, i.e., $min_{S \in F:|S|=k} max_{v \in C} d(v, S)$ where $d(v, S) = min_{u \in S} d(u, v)$. They gave a $(1+\sqrt{3})$-approximation algorithm for the k-supplier problem under Euclidean metrics.

Let the constraint region be R_1, the points in R_1 denoted as P_1 and $P_2 := \{P \backslash P_1\}$. Since R_1 and R_2 have already been given, we go through all the combinations of m centers in R_1 and solve the $M(k, m)CP$. Let $|P_1| = l$ and $l > m$, $S(k - m)CP$ can be solved in $O(n^{O(\sqrt{k-m})})$ time for fixed m centers. See the details in Algorithm 2. We have:

Theorem 3. *The region constraint problem can be solved in* $O(l^m n^{O(\sqrt{k-m})})$ *time.*

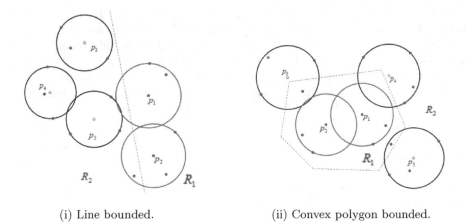

(i) Line bounded.   (ii) Convex polygon bounded.

Fig. 2. $k = 5$, $m = 2$ and p_1, p_2 are in P

In the following, we consider the $M(k, m)CP$. We give an exact algorithm, a 2-approximation algorithm and a heuristic algorithm solving the $M(k, m)CP$.

3.3   Exact Algorithm

In this section, we present an exact algorithm for $M(k, m)CP$. The main idea of the algorithm is based on the following lemma. Our algorithm combines the algorithms of Drezner [9] and Hwang et al. [11]. Let $D(o, r)$ be the disk with center o and radius r and $S1CP$ (resp. $D1CP$) be the standard (resp. discrete) minimum enclosing disk problem of a point set.

Lemma 1. *For fixed m points $\{p_1, p_2, ..., p_m\}$ in P as the centers of $M(k, m)CP$ and a given radius r, let subset P^T of P consist of the elements out of $D(p_1, r) \cup D(p_2, r) \cup, ..., \cup D(p_m, r)$ and radius r^T be the radius of the solution of the standard $(k - m)$-center problem of P^T, we have: if $r_1 > r_2$, $r_1^T \leq r_2^T$.*

Proof. As $r_1 > r_2$, we have $P_1^T \subseteq P_2^T$, then $r_1^T \leq r_2^T$. $\qquad\qquad\square$

The optimal radius of $M(k, m)CP$ is either the distance between two points in P or the radius of the minimum enclosing disk of a subset of P. If the optimal radius is the distance between two points in P, it is clear that the optimal radius of $M(k, m)CP$ is the radius of a discrete minimum enclosing disk. Otherwise, the optimal radius of $M(k, m)CP$ is the radius of a standard minimum enclosing disk. If the optimal radius of $M(k, m)CP$ is the radius of a discrete minimum enclosing disk, there are at most $n(n - 1)/2$ distances between two points in P. Otherwise the optimal radius of $M(k, m)CP$ is of a standard minimum enclosing disk, and there are at most $O(n^3)$ distances. We check all these distances and set all m possible points as the centers $\{p_1, p_2, ..., p_m\}$ in P. For m points in P, there are at most $O(n^m)$ combinations. For each possible combination and radius r, we obtain the disk $D(p_i, r)$ $(i = 1, 2, ..., m)$ and compute P^T. It can be done in $O(m \log n)$ time after computing the distances among all the points and sorting them. According to the algorithm mentioned in [4], we solve the standard $(k - m)$-center problem of P^T. Note that we can sightly modify to reduce the bound of the complexity from Lemma 1. The method described above of reducing the value of r can be replaced by binary search. Finding all the possible r and sorting them can be done in $O(n^3 \log n)$ time. By binary search, we only need $O(\log n)$ iterations in the **for** loop in Algorithm 3, rather than $O(n^3)$. Therefore the bound of the complexity of the modified algorithm is $O(n^{m+O(\sqrt{k-m})} \log n) = O(n^{m+O(\sqrt{k-m})})$ time. See Algorithm 3 for details. From the above analysis, we have the following theorem:

Theorem 4. *Let P be a planar point set of n points, the mixed k-center problem can be solved in $O(n^{m+O(\sqrt{k-m})})$ time.*

3.4   2-Approximation Algorithm

There are many 2-approximation algorithms for $SkCP$ [12–14]. Feder and Greene showed that under Euclidean metrics, $SkCP$ can not be approximated to within a factor of $\sqrt{3}$ unless P = NP [14]. In this section, we also give a 2-approximation algorithm for $M(k, m)CP$. We solve the $SkCP$ of P by the exact algorithm mentioned in [11] and sort the radii in non-decreasing order. Then solve the $D1CP$ of these m subsets consisting of the points in m disks whose radii are the m minimum. We have Lemma 2.

Algorithm 3. Exact algorithm for the mixed k-center problem for fixed m centers

1: Initial $i = 1, r = \infty$, $\{p_1, p_2, ..., p_m\}$ be the fixed m centers;
2: **for** each i **do**
3:     Compute P_i^T for fixed radius r_i and the standard $(k-m)$-center problem for P_i^T. Let the radius and the centers be r_i^T and $\{c_{i_1}, ..., c_{i_{k-m}}\}$ respectively;
4:     **if** $r > max(r_i, r_i^T)$, **then**
5:         $\{c_{i-1_1}, ..., c_{i-1_{k-m}}\} \leftarrow \{c_{i_1}, ..., c_{i_{k-m}}\}$, $r \leftarrow max(r_i, r_i^T)$.
6:     **end if**
7: **end for**
8: Return r, Centers: $\{p_1, p_2, ..., p_m, c_{i_1}, ..., c_{i_{k-m}}\}$.

Lemma 2. *Let S be a planar point set, the minimum enclosing disk and the discrete minimum enclosing disk be $D(s, r_S)$ and $D(d, r_D)$ respectively, we have $1 \le \frac{r_D}{r_S} \le 2$.*

Proof. We have $d \in S$. Let the farthest point of d be f_d. Both d and f_d are in $D(s, r_S)$, i.e., $r_D \le 2r_S$, and we have $\frac{r_D}{r_S} \le 2$. On the other hand, it is obvious that when $r_D \ge r_S$, we have $1 \le \frac{r_D}{r_S} \le 2$. $\qquad \square$

In the following, we give the algorithm and show the approximation factor is 2. We solve the $SkCP$ and sort the k radii in non-deceasing order. We choose m minimum radii and solve the $D1CP$ of these m corresponding subsets. See Algorithm 4.

Algorithm 4. Approximation algorithm

1: Solve the $SkCP$ and sort the k radii in non-deceasing order.
2: Choose m minimum radii and solve the $D1CP$ of these m corresponding subsets.
3: Return the optimal solution.

Theorem 5. *Algorithm 4 yields a 2-approximation algorithm.*

Proof. Let the optimal radius of $SkCP$ be r_{opt}^S, the disk with the maximum radius of the m discrete minimum enclosing disks in Algorithm 4 be $D(o_m, r_m)$ and r_s be the radius of the standard minimum enclosing disk of the points in $D(o_m, r_m)$. The optimal radius of $MkCP$ be r_{opt}^M. If $r_{opt}^S \ge r_m$, from Fact 1, we have $r_{opt}^M = r_{opt}^S$. Otherwise, we have $r_{opt}^M \ge r_{opt}^S$ and $r_s \le r_{opt}^S < r_m$. From Lemma 2, we have $\frac{r_m}{r_{opt}^M} \le \frac{r_m}{r_{opt}^S} \le \frac{r_m}{r_s} \le 2$. $\qquad \square$

3.5   Heuristic Algorithm

In this section, we present a heuristic algorithm for $M(k, m)CP$. Suppose there are m centers in P, the heuristic algorithm is similar to Drezner's [9]. The main idea of the heuristic algorithm is to partition P into k subsets and solve $S1CP$ of each subset. Initially, we pick a set Q of k arbitrary points $p_1, p_2, ..., p_k$ and compute the Voronoi diagram [16]. Let the Voronoi cell of each point p_i be V_i. Compute the standard minimum enclosing disk of all the Voronoi cells and let $Q' = \{p'_i | p'_i$ is the center of the standard minimum enclosing disk$\}$. We repeat this process by setting $Q = Q'$. When the iteration is terminated, go through all the radii in non-decreasing order and choose the m minimum radii. Solve the $D1CP$ of these m corresponding subsets. If any point belongs to more than one set, assign it to an arbitrary set. See Algorithm 5 for details.

Algorithm 5. Heuristic algorithm for the mixed k-center problem

---

1: Let j be the iteration number, choose k arbitrary points as the initial k centers, denoted as $\{p_1^0, p_2^0, ..., p_k^0\}$.
2: **for** $i = 0$ to $j - 1$ **do**
3:     Compute the Voronoi diagram of $\{p_1^i, p_2^i, ..., p_k^i\}$. Let each Voronoi region be $\{R_1^i, R_2^i, ..., R_k^i\}$. Solve the $S1CP$ of $\{R_1^i, R_2^i, ..., R_k^i\}$. Let the centers of the optimal solution of $S1CP$ of each Voronoi region be $\{p_1^{i+1}, p_2^{i+1}, ..., p_k^{i+1}\}$.
4: **end for**
5: Let the radii of the $S1CP$ of $\{R_1^j, R_2^j, ..., R_k^j\}$ be $\{r_1^j, r_2^j, ..., r_k^j\}$. Sort them in non-deceasing order and denote them as $r_1 \leq r_2 \leq ... \leq r_k$. For each r_l ($l = 1, 2, ..., k$), the corresponding region and center are R_l and c_l respectively. Choose $\{R_1, R_2, ..., R_m\}$ and solve $D1CP$ of the points in $\{R_1, R_2, ..., R_m\}$, let the centers of these disks be $\{q_1, q_2, ..., q_m\}$ and the maximal radius of these disks be r_{max}.
6: Return $r_{heuristics} \leftarrow max(r_{max}, r_{k-m+1})$, centers: $\{q_1, q_2, ..., q_m, c_{k-m+1}, ..., c_k\}$.

Theorem 6. *The time complexity of every iteration in Algorithm 5 is* $O(n \log k)$.

Proof. In the **for** loop, it takes $O(k \log k)$ time to compute the Voronoi diagram for a set of k points. All the points in each Voronoi region can be computed in $O(n \log k)$ time using planar point location. For all points in Voronoi region, solving the $S1CP$ takes $O(n)$ time. Thus, the overall time complexity of a single iteration is $O(n \log k)$.   □

4   Conclusion

In this paper, we consider two variations of the k center location problem. We consider the mixed k-center problem. When $k = 2$, we give an $O(n^2 \log n)$ time algorithm. When $k > 2$, we show two motivations considering such a problem and give an exact algorithm, a 2-approximation algorithm and a heuristic algorithm solving the mixed k-center problem. The time complexity of the exact algorithm is $O(n^{m+O(\sqrt{k-m})})$.

However, we have not found any geometric properties for the mixed k-center problem under region constraint to improve the time complexity. So far, we have not found any algorithms considering the mixed k-center problem for a convex polygon with constraint that m of the centers are the vertexes. All these are interesting problems for the further research.

References

1. Jaromczyk, J., Kowaluk, M.: An efficient algorithm for the Euclidean two-center problem. In: Proceedings 10th ACM Symposium on Computational Geometry, pp. 303–311 (1994)
2. Eppstein, D.: Faster construction of planar two-centers. In: Proceedings of the 8th ACM-SIAM Symposium on Discrete Algorithms, pp. 131–138 (1997)
3. Sharir, M.: A near-linear algorithm for the planar 2-center problem. Discret. Comput. Geom. **18**, 125–134 (1997)
4. Chan, T.M.: More planar two-center algorithms. Comput. Geom. Theory Appl. **13**, 189–198 (1999)
5. Hershberger, J.: A faster algorithm for the two-center decision problem. Inform. Process. Lett. **47**, 23–29 (1993)
6. Hershberger, J., Suri, S.: Finding tailored partitions. J. Algorithms **12**, 431–463 (1991)
7. Agarwal, P.K., Sharir, M., Welzl, E.: The discrete 2-center problem. Discret. Comput. Geom. **20**, 287–305 (2000)
8. Megiddo, N., Supowit, K.: On the complexity of some common geometric location problems. SIAM J. Comput. **13**, 1182–1196 (1984)
9. Drezener, Z.: The p-center problem-heuristics and optimal algorithms. J. Oper. Res. Soc. **35**, 741–748 (1984)
10. Megiddo, N.: Linear-time algorithms for the linear programming in R^3 and related problems. SIAM J. Comput. **12**, 759–776 (1983)
11. Hwang, R.Z., Lee, R.C.T., Chang, R.C.: The slab dividing approach to solve the Euclidean p-center problem. Algorithmica **9**, 1–22 (1993)
12. Hochbaum, D.S., Shmoys, D.B.: A best possible heuristic for the k-center problem. Math. Oper. Res. **10**(2), 180–184 (1985)
13. Gonzalez, T.F.: Clustering to minimize the maximum intercluster distance. Theor. Comput. Sci. **38**, 293–306 (1985)
14. Feder, T., Greene, D.: Optimal algorithms for approximate clustering. In: Proceedings of the 20th ACM Symposium on Theory of Computing, pp. 434–444 (1988)
15. Nagarajan, V., Schieber, B., Shachnai, H.: The Euclidean k-supplier problem. In: Goemans, M., Correa, J. (eds.) IPCO 2013. LNCS, vol. 7801, pp. 290–301. Springer, Heidelberg (2013)
16. Aurenhammer, F., Klein, R., Lee, D.T.: Voronoi Diagrams and Delaunay Triangulations. World Scientific Publishing Company, Singapore (2013)

Constrained Light Deployment for Reducing Energy Consumption in Buildings

Huamei Tian, Kui Wu[(⊠)], Sue Whitesides, and Cuiying Feng

Department of Computer Science, University of Victoria, Victoria, Canada
wkui@uvic.ca

Abstract. Lighting systems account for a major part of the energy consumed by large commercial buildings. This paper aims at reducing this energy consumption by defining the Contrained Light Deployment Problem. This new problem is related to the classical Art Gallery Problem (AGP) in computational geometry. In contrast to AGP, which asks for the minimum number of guards to monitor a polygonal area, our problem, CLDP, poses a new challenging requirement: not only must each point p have an unobstructed line-of-sight to a light source, but also, the combined illuminance at p from all light sources must exceed some given threshold value. We provide evidence that our new problem is NP-hard, based on known results for AGP. Then we propose fast heuristics for floor plans shaped like orthogonal polygons, with and without holes. Our problem formulation allows lights to be placed internally, not only at vertices. Our algorithm, which combines ideas from computational geometry, clustering and binary search, computes a set of light placements that satisfies the illumination requirement. The algorithm seeks a light set of minimum size by an iterative binary search procedure that progressively tightens upper and lower bounds.

1 Introduction[1]

Lighting systems, critical to our daily life in modern society, consume a tremendous and costly amount of energy resources. The U.S. Energy Information Administration (EIA) discloses that in 2015, about 404 billion kilowatthours (kWh) of electricity were used for lighting by the residential sector and the commercial sector in the United States, which amounted to about 10 % of total U.S. electricity consumption [6]. Therefore, reducing energy consumed by lighting systems could have a significant impact on the sustainable development of our society.

Reducing the energy consumption of lighting systems raises challenging trade-off issues. On the one hand, reducing energy implies lowering tota lilluminance[2]. On the other hand, lighting systems should keep the building occupants comfortable and safe, which implies maintaining a suitable illuminance level everywhere.

---

[1] This research was supported in part by NSERC and the University of Victoria.
[2] Illuminance is the amount of luminous flux per unit area.

© Springer International Publishing AG 2016
T.-H.H. Chan et al. (Eds.): COCOA 2016, LNCS 10043, pp. 350–364, 2016.
DOI: 10.1007/978-3-319-48749-6_26

The challenging problem we face is: without making major changes to the building, such as by installing skylights, can we reduce the energy used by the lighting system while still meeting the needs of the occupants?

As a start to tackling this general problem, we consider the following scenario: occupants may add their own task lighting at their own expense; the building owner meets a given uniform threshold lighting requirement at each point in the floorplan by deploying a set of light fixtures ("lights") of the same type, modeled as a discrete point set in the floorplan; and the floorplan is modelled by an orthogonal polygon, either with or without holes (see Sect. 3 for definitions). Thus we seek the minimum number of lights that will meet the given threshold level of illuminance throughout the building. We call this problem the Constrained Light Deployment Problem (CLDP) (Sect. 3 elaborates). This problem is similar to the well-known Art Gallery Problem (AGP), which asks for the minimum number of guards to monitor an art gallery with a polygonal floor plan. While a light is analogous to a guard, our problem differs from AGP in two crucial respects: (1) the illuminance at a point p provided by a light source at g decreases nonlinearly over distance, and (2) the combined illuminance at any point p must exceed a given value. Like AGP, the CLDP proposed above has many variations. It is relevant both to the design of new buildings and to the retrofitting of existing ones [13]. Our main contributions are:

1. Formulation of the Constrained Light Deployment Problem CLDP. Unlike AGP, the CLDP specifies an aggregated threshold illuminance level to be met at each point, not just visibility by a guard.
2. Evidence that CLDP is NP-hard, together with a polynomial time strategy for solving it.
3. Algorithms to determine upper bounds on the number of lights needed for floor plans modeled as orthogonal polygons with, and without, holes.

2   Related Work

Strategies for energy savings in the design of lighting systems for buildings have been extensively considered [13]. Most existing efforts, however, focus on using energy-efficient lights or developing intelligent lighting systems that turn lights on/off based on sensing data about tenant behaviour [14,16]. To the best of our knowledge, the CLDP problem we formulate is the first to consider the optimal deployment of light fixtures from an algorithmic, geometric viewpoint, for the purpose of meeting a threshold illuminance requirement. Our problem does, however, belong to a broader class of problems seeking an optimal selection of points to meet a given requirement, e.g., fully covering a monitored area [15] or minimizing the uncertainty of the monitored phenomenon [12].

The Art Gallery Problem (AGP) [15], i.e., guarding an art gallery with the minimum number of guards who together see the whole gallery, has an extensive literature that considers various shapes for the art gallery and various types of guards and candidate positions for them. The algorithms that have been proposed to solve AGP fall mainly into these groups: (1) linear programming based algorithms [3,4], (2) algorithms based on the SET COVER problem [7,9], (3) algorithms based on the HITTING SET problem [10,11], (4) genetic algorithms [1], and (5) algorithms using simulated annealing [2].

Since the illuminance at a point p is determined by all lights that have an unobstructed line-of-sight to p, the requirement that the illuminance at each point in the building should exceed a given threshold significantly complicates algorithm design for CLDP. For this reason, none of the algorithms in the extensive AGP literature can be applied directly to solve CLDP.

3 Problem Formulation

3.1 Floor Model

Definition 1. *An **orthogonal polygon** is one whose edges each align with one of a pair of given orthogonal coordinate axes.*

We consider buildings whose floor plans are orthogonal polygons in 2D. For industrial buildings, such polygons are not uncommon. Moveover, orthogonal polygons have been widely used to approximate general polygons. We consider two types of orthogonal polygons: orthogonal polygons without holes inside, called *Simple Orthogonal Polygons* (SOP), and *Orthogonal Polygons with Holes*, denoted as OPH. Note that the holes are inside the polygon and must themselves have boundaries that are orthogonal polygons. In our context, holes represent regions that block light.

3.2 Illumination Model

Definition 2. *The **unobstructed distance** between two points $p = (p_x, p_y)$ and $g = (g_x, g_y)$, inside or on the boundary of an orthogonal polygon, is:*

$$d_{pg} = \mathcal{I}\sqrt{(p_x - g_x)^2 + (p_y - g_y)^2},\tag{1}$$

where $\mathcal{I}$ is the indicator function:

$$\mathcal{I} = \begin{cases} 1, & \text{if } p \text{ and } g \text{ have an unblocked line-of-sight;} \\ \infty, & \text{otherwise.} \end{cases}\tag{2}$$

Here, no part of the line-of-sight can lie outside the polygon.

We adopt a simple illumination model that is a special case of the generic model proposed in [5]. Assume that the illuminance at a light source g is I_0. Then the *illuminance* I_{pg} at p due to light g is:

$$I_{pg} = \frac{I_0}{1 + d_{pg}^2}.\tag{3}$$

To keep the model simple, we disregard the refraction, reflection and coherence of light waves, all of which are hard to model as they depend on, for example, the wall and floor coverings and the furniture. Thus we assume that the total illuminance at a point p is the sum, over all the lights, of the illuminance due to that light. Thus, given a light set $G = \{g_1, g_2, \ldots, g_m\}$ in a polygon P, the total illuminance at a point p in P is:

$$I_p = \sum_{i=1}^{m} \frac{I_0}{1 + d_{pg_i}^2}, \tag{4}$$

where m is the number of lights and d_{pg_i} is the unobstructed distance between the point p and the light g_i. Note that while we assume homogeneous light sources by using the same I_0 for all lights, the illumination model can be easily extended to handle heterogeneous lights, where the value of I_0 may be different for different lights.

Definition 3. *The **effective illumination area** of a light g is the circular area centered at g such that at each point in that area, the illuminance due to light g is strictly greater than the given illuminance threshold δ.*

Let R denote the radius of the effective illumination area. It is easy to calculate that

$$R = \sqrt{\frac{I_0}{\delta} - 1}. \tag{5}$$

Note that we require $\frac{I_0}{\delta} > 1$ and set $R = \infty$ if $\delta = 0$.

3.3   CLDP and the Hardness of CLDP

Definition 4. *The **Constrained Light Deployment Problem** takes as input the coordinates of the vertices of an orthogonal polygon P, listed in cyclic order, together with a rational threshold value $\delta \geq 0$, and asks to find the minimum possible cardinality of a set of lights located inside or on the boundary of the polygon P, such that the illuminance at each point inside P due to the set of lights is strictly greater than δ. Here, the illuminance is determined by Eq. (4).*

The version of AGP that assumes point guards and SOP (or OPH) corresponds to the special case of CLDP for $\delta = 0$: each point is visible to at least one guard (light) and thus receives some illuminance greater than 0. Because this version of AGP is NP-hard [17], CLDP is also NP-hard. While modifying the definition of CLDP to exclude the possibility of $\delta = 0$ might better model the practical problem, nevertheless, one would expect CLDP to remain NP-hard.

Note also that modifying CLDP to ask for the deployment of the lights as well as their minimum possible number does not affect the NP-hardness. In fact our algorithms output the placement of the lights, not just the number of them.

We denote CLDP with SOP as CLDP-SOP, and we denote CLDP with OPH as CLDP-OPH. Next we describe our general approach, then adapt it to these problems in later sections.

4   Solving CLDP with Clustering and Binary Search

4.1   Main Idea

Since CLDP is NP-hard, we present a heuristic algorithm to solve CLDP. The main idea is to compute lower and upper bounds, denoted as n_{lb} and n_{ub}, respectively, for the number $m = n_s$ of lights required in solution sets of the type we will seek, and then to use iterative binary search to find a solution with the best n_s value. See Fig. 1.

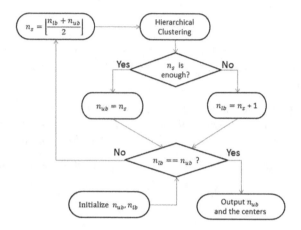

Fig. 1. Search for the smallest number n_s of lights

While the basic idea is simple, to fully explain the algorithm, we must provide answers to the questions below:

1. What is clustered in the hierarchical clustering?
2. Where are the n_s lights placed?
3. How is it determined whether n_s lights are sufficient to meet the illumination requirement?
4. How is the initial value of n_{lb} set?
5. How is the initial value of n_{ub} set?

In the rest of this section, we answer the first four questions. We answer the last question in Sects. 5 and 6.

4.2   Hierarchical Clustering of Observation Points

Given that there is an infinitude of points inside a polygon, we cannot check that the illumination requirement is met point by point. Furthermore, locations of lights might have irrational coordinates. To address these issues, we place a unit square grid on the 2D plane and make use of the vertices of the grid falling inside the orthogonal polygon; we call such grid vertices *observation points*.

Here the scale of the coordinate system is chosen so that $R \geq 1$, where R is the radius of the effective illumination area.

We first calculate the unobstructed distance between any pair of observation points. We cluster the observation points using the distance values calculated. Many clustering algorithms, e.g., K-means and hierarchical clustering [8], may be applied here. Since some clustering algorithms, e.g., K-means, are sensitive to initial selection of means, we use hierarchical clustering to avoid the difficulty in selecting initial cluster means in our context. Our implementation of hierarchical clustering uses the bottom-up approach: each observation point starts in its own cluster, and pairs of clusters are merged as one moves up the hierarchy. The process continues until the number n_s of clusters, as specified by the binary search, is reached.

4.3 Placement of Lights

After we cluster the observation points into n_s clusters, we need to calculate the location of each cluster centre. In our case, the average coordinates of observation points in each cluster may not fall within the polygon. To handle this issue, we consider what we call *cluster centrality*, denoted by d_c^p and defined to be the sum of the unobstructed distances between an observation point p and the other observation points q in the same cluster c, i.e.,

$$d_c^p = \sum_{q \in c} d_{pq}, \qquad (6)$$

where p and q belong to the same cluster c. For each cluster, we select the observation point whose cluster centrality value is the smallest in the cluster as the cluster centre. This is the place where we deploy a light.

After the above steps, an observation point may have a shorter unobstructed distance to the centre of another cluster than to its own cluster centre. We then use the following iterative procedure to further adjust the cluster centres: For an observation point p, if the unobstructed distance to its cluster centre g is greater than the unobstructed distance to another cluster centre h, point p is reassigned to the cluster centred at h. After that, we calculate the new centres of the adjusted clusters. The above procedure repeats until convergence[3].

4.4 Are n_s Lights Enough?

After we select the locations of n_s lights, we need to check whether the illuminance level at each point inside the polygon (including points that are not observation points), calculated with Eq. (4), is above the given threshold δ. Note

---

[3] While we do not prove convergence, our experiments (matlab package available upon request) suggest that adjustment is required only for a relatively small number of observation points near the boundary of two clusters, and convergence is reached after a few rounds.

that satisfying the illuminance requirement at each observation point does not immediately imply that the whole polygon area meets the requirement.

To handle this difficulty, we put a stronger illuminance requirement on observation points: each observation point must fall within the effective illumination area of at least one light. This stronger requirement brings two benefits. First, the calculation is much simpler. Since the shortest unobstructed distance from an observation point to a light is that from the observation point to its cluster centre, we only need to check whether or not the observation point falls within the effective illumination area of its cluster centre. Second, because of the next theorem, we can easily check whether or not the whole polygon area meets the illuminance requirement.

Theorem 1. *Consider a square whose side length is no greater than the effective illumination radius R of each light. If each of the four corners of the square falls within the effective illumination area of at least one light, then the whole square area satisfies the illumination requirement defined in CLDP.*

Refer to the Appendix for the proof and for all other proofs in the paper. Note that in the above theorem, the four corners of the square may fall within the effective illumination area of different lights. From Theorem 1, we conclude that if every observation point falls within the effective illumination area of its cluster centre, then the whole polygon satisfies the illuminance requirement.

4.5   The Initial Lower Bound n_{lb}

We denote the area of a given orthogonal polygon P by S_{op}. In terms of the effective illumination radius R of a light, the effective illumination area of that light is πR^2.

According to previous subsection, we now require that each observation point falls within the effective illumination area of its cluster centre. Taking the area of a region as an approximate count of the number of observation points in that region, a reasonable initial lower bound on the number n_s of lights (placed at the cluster centres) needed to cover all the observation points is:

$$n_{lb} = \lfloor \frac{S_{op}}{\pi R^2} \rfloor. \tag{7}$$

While this value for n_{lb} is easy to compute, finding a suitable initial value for n_{ub} is nontrivial. To complete the description of our algorithms, and thereby answer the last question raised in Sect. 4.1, we explain how this can be done for CLDP-SOP and for CLDP-OPH in the next two sections.

5   An Upper Bound for CLDP-SOP

In this section, we no longer use the grid and the observation points of the previous section. Our idea to obtain n_{ub} is as follows: we modify the original

CLDP problem to require that each of the infinitude of points in the polygon must fall within the effective illumination area of at least one light. Thus a solution to the modified CLDP gives an upper bound for the original CLDP as more lights may be needed to satisfy the stronger constraint.

We solve the modified CLDP in two steps: Step 1 obtains an initial solution to the modified CLDP by assuming that the radius R of the effective illumination area of a light is infinite; Step 2 then puts back the finite value for the radius R and adds more lights to the set of lights found in Step 1.

Step 1 is the same as the classical AGP. There are both heuristic algorithms and approximation algorithms to solve the classical AGP, without or with holes. Below we give a heuristic algorithm for Step 1.

5.1   Step 1

We first explain several basic concepts we use.

- **Reflex Vertex**: If the internal angle of a vertex is greater than π, this vertex is called a reflex vertex.
- **Edge Extension**: The two edges incident to a reflex vertex can be extended until they intersect the boundary of the orthogonal polygon.
- **Rectangular Block**: With edge extensions, the whole orthogonal polygon can be divided into rectangular blocks.
- **Block Number**: We assign a unique integer label, called a *block number*, to each rectangular block.
- **Coverage Sets**: The coverage set $CS(g)$ of a light g is the set of all the block numbers of the rectangular blocks that are visible to light g. The coverage set $CS(G)$ of a set G of lights is the union of the coverage sets of the lights in G. We say that a set G of lights **covers** a polygon P if all the block numbers of P belong to $CS(G)$.

We have the following propositions:

Proposition 1. *Every rectangular block is covered by at least one reflex vertex.*

From Proposition 1, we easily obtain the following propositions.

Proposition 2. *If a light is placed at each reflex vertex of a simple orthogonal polygon P, the set of these lights at the reflex vertices covers P.*

Proposition 3. *If every element in the coverage set of a light A is also an element in the coverage set of another light B, then light A is redundant and can be removed from any light set containing B. Similarly, if the coverage set of a union of lights is included in the coverage set of the union of another set of lights, then the lights generating the included coverage set are redundant and can be removed from any light set containing the lights that generate the including coverage set.*

We need a method to determine whether or not a reflex vertex covers a given rectangular block. For this, we have the following proposition.

Proposition 4. *If all four vertices of one block are visible to a light g, then the entire block is visible to the light g.*

A Heuristic Algorithm. According to Proposition 2, if a light is placed at each reflex vertex, the light set covers the whole polygon and the number of lights is $r = \frac{n-4}{2}$, where n is the number of vertices of the orthogonal polygon. To reduce the number of lights below r, we remove redundant reflex vertices from the light set, based on Proposition 3. In other words, we initialize the candidate light set with the reflex vertex set, and then remove redundant lights.

Reducing the redundancy of the coverage sets is a special case of the Set Covering Problem, one of Karp's original 21 NP-complete problems in 1972. We propose Algorithm 1 for finding an approximate solution.

Algorithm 1. Redundancy Removal Algorithm

Input: The coverage sets $C = C(1), C(2), ..., C(r)$ of a simple orthogonal polygon.
Output: A reduced light set
1: Construct a bipartite graph $G = < V_L \bigcup V_R, E >$, where each left-side vertex in $V_L = \{1, 2, \ldots, r\}$ represents a reflex vertex in the orthogonal polygon, each right-side vertex in $V_R = \{1, 2, \ldots, m\}$ represents a rectangular block, and there is an edge between $v_l \in V_L$ and $v_r \in V_R$ if the set $C(v_l)$ includes v_r.
2: Mark all left-side vertices as unlabelled and all right-side vertices as unlabelled.
3: Label all right-side vertices whose degree is 1. /*A right-side vertex has degree 1 if it is covered by only one reflex vertex. */
4: Label all left-side vertices that have an edge to a labelled right-side vertex.
5: Remove all labelled right-side vertices and the corresponding edges linked to those vertices.
6: **while** Remaining right-side vertex set is not empty **do**
7:     Label the left-side vertex of the highest degree, if it is unlabelled, in the remaining bipartite graph.
8:     Label all the right-side vertices that have an edge to a labelled left-side vertex.
9:     Remove all labelled right-side vertices and the corresponding edges linked to those vertices.
10: **end while**
11: **return** All labelled left-side vertices

5.2   Step 2

In Step 1, we find an initial value for n_{ub} for simple orthogonal polygons without considering the illuminance threshold value. In this section, we put back the stronger illumination requirement, as defined in the unmodified CLDP, i.e., we put back the illumination threshold $\delta > 0$.

Fortunately, we can reuse the solution in Step 1 after the following changes. If the diagonal of one rectangular block is greater than the radius R (calculated with Eq. (5)) of the coverage area of one light, we split this rectangular block with horizontal and vertical lines into smaller blocks until all rectangular blocks can be covered by one light. When we split one rectangular block, we introduce another important set: the **secondary-point** set. A secondary point is a vertex shared by two new adjacent blocks generated by the splitting process, and it is a candidate location for placing a light.

Combining the reflex vertex set and the secondary-point set as the candidate light set, we apply the solution in Step 1 to obtain the new n_{ub} value.

5.3   Worst Case Time Complexity Analysis

Denoting the number of vertices of the simple orthogonal polygon by n, and the number of its reflex vertices by r, we have $n = 2r + 4$. The time complexity for determining the reflex set is $O(n)$; constructing the edge extension costs $O(r)$ time; determining the intersections of edge extensions costs $O(r^2)$; labeling the blocks costs $O(n^2)$; calculating the coverage set costs $O(n^3)$; and redundancy removal costs $O(nr)$. Thus finding n_{ub} is $O(n^3)$.

6   An Upper Bound for CLDP-OPH

We use the idea in the previous section to obtain the initial upper bound n_{ub} for CLDP-OPH, namely, we modify CLDP to require that each point in the polygon fall within the effective illuminance area of at least one light. We also need to modify the algorithm to address the problems caused by the special topology of OPH. As such, we next analyse the different conditions caused by holes and give the modifications to the solution for SOP to address these conditions.

Propositions 1, 2 and 3 hold for OPH because (1) the proof of Proposition 1 applies to OPH and (2) Propositions 2 and 3 follow directly from Proposition 1. However, we cannot use Proposition 4 to determine whether or not a reflex vertex covers a rectangular block in OPH. Figure 2 shows a special case of Proposition 4 in which the orthogonal polygon has one hole. In the figure, the outer boundary of the orthogonal polygon and the hole are shown with a black line; the blue dotted lines are the extended edges. Since the four points A, B, C, D are visible to the reflex vertex g, by Proposition 1, the rectangular block $ABCD$ should be visible to g. But this is not the case here because of the hole.

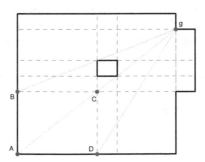

Fig. 2. Special case in OPH (Color figure online)

Fortunately, we only need to revise Proposition 4 by adding the following constraint, so that the revised proposition holds for OPHs.

Proposition 5. *If all four vertices of one block are visible to light g outside the block, and there is no other vertex inside the convex polygon determined by light g and the vertices of that block, then the entire block is visible to light g.*

7 Conclusion

We have formulated and studied the Constrained Light Deployment Problem (CLDP), which aims to deploy the minimum possible number of lights to satisfy illuminance requirements. Given the hardness of the problem, we proposed a heuristic algorithm that uses hierarchical clustering and binary search. To set proper initial values for the binary search, we proposed methods to obtain upper and lower bounds on the number of lights needed. We implemented and evaluated the algorithms in matlab. The package is available upon request.

Our formulation of CLDP leads naturally to the study of many variations, including 3D versions and versions in which light fixtures have different properties and types. The hardness of these variations can be considered from a complexity theory viewpoint, as well as the hardness of CLDP for $\delta > 0$.

Appendix

Proof of Theorem 1

Proof. Proving Theorem 1 is equivalent to solving the following problem shown in Fig. 3: Assuming that the side length of the square is R and O is an arbitrary point in the square, we need to prove that

$$\frac{1}{1 + |OA'|^2} + \frac{1}{1 + |OB'|^2} + \frac{1}{1 + |OC'|^2} + \frac{1}{1 + |OD'|^2} \geq \frac{1}{1 + R^2}. \tag{8}$$

The equivalence is based on the fact that Fig. 3 illustrates the worst case[4] scenario: each corner of the square is covered by a different light, and the locations of the lights (i.e., A', B', C', D') lead to the lowest illuminance value at O. As the assumption in Theorem 1, the right-hand side of (8), $\frac{1}{1+R^2}$, is larger than the given threshold.

First, by the inequality of arithmetic and geometric means

$$\frac{x_1 + x_2 + \cdots + x_n}{n} \geq \sqrt[n]{x_1 \cdot x_2 \cdots x_n}, \quad x_1, \cdots, x_n > 0, \tag{9}$$

---

[4] It is easy to verify Theorem 1 for other simple cases: (One-light case:) all four corners are covered by the same light. (Two-light cases:) one corner is covered by one light and the other three corners are covered by another light; two corner are covered by one light and the other two corners by another light. (Three-light cases:) two corners are covered by one light and the other two corners each by the other two lights, respectively.

we have

$$\frac{1}{1+|OA'|^2} + \frac{1}{1+|OB'|^2} + \frac{1}{1+|OC'|^2} + \frac{1}{1+|OD'|^2}$$

$$\geq \frac{4}{\sqrt[4]{(1+|OA'|^2)(1+|OB'|^2)(1+|OC'|^2)(1+|OD'|^2)}}$$

$$\geq 4 \times \frac{1}{\frac{(1+|OA'|^2)+(1+|OB'|^2)+(1+|OC'|^2)+(1+|OD'|^2)}{4}}$$

$$= \frac{16}{4+(|OA'|^2+|OB'|^2+|OC'|^2+|OD'|^2)} \tag{10}$$

Observing that $|OA'|^2 = (R+|OA|)^2 = R^2+|OA|^2+2R|OA| \leq 2(R^2+|OA|^2)$, the relation of which holds also for $|OB'|, |OC'|, |OD'|$, we have

$$\frac{16}{4+(|OA'|^2+|OB'|^2+|OC'|^2+|OD'|^2)}$$

$$\geq \frac{16}{4+8R^2+2[|OA|^2+|OB|^2+|OC|^2+|OD|^2]} \tag{11}$$

Now, we indicate the arbitrary O by (x,y) where $x,y \in [-\frac{R}{2}, \frac{R}{2}]$ in the Euclidean coordinate, and $A = (\frac{R}{2}, \frac{R}{2})$, $B = (-\frac{R}{2}, \frac{R}{2})$, $C = (-\frac{R}{2}, -\frac{R}{2})$, $D = (\frac{R}{2}, -\frac{R}{2})$. Thus we have

$$|OA|^2 = (x - \frac{R}{2})^2 + (y - \frac{R}{2})^2, |OB|^2 = (x + \frac{R}{2})^2 + (y - \frac{R}{2})^2,$$

$$|OC|^2 = (x + \frac{R}{2})^2 + (y + \frac{R}{2})^2, |OD|^2 = (x - \frac{R}{2})^2 + (y + \frac{R}{2})^2 \tag{12}$$

Substituting Eq. (12) into (11) and using the fact that $x^2+y^2 \leq (\frac{R}{2})^2+(\frac{R}{2})^2 = \frac{R^2}{2}$, we obtain

$$\frac{16}{4+8R^2+2[|OA|^2+|OB|^2+|OC|^2+|OD|^2]}$$

$$= \frac{16}{4+8R^2+2[4(x^2+y^2)+2R^2]}$$

$$\geq \frac{16}{4+8R^2+2[2R^2+2R^2]} \geq \frac{1}{1+R^2} \qquad \Box \tag{13}$$

Proof of Proposition 1

Proof. There are four kinds of blocks generated by edge extensions:

1. The block has only one edge on an extended edge.
2. The block has two edges on the extended edges.

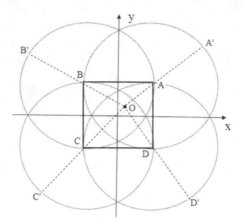

Fig. 3. Illustration of Theorem 1

3. The block has three edges on the extended edges.
4. The block has four edges on the extended edges.

For the first case, the block has two possible shapes, which are shown by (a) and (b) in Fig. 4. The solid edges are line segments on the polygon's boundary and the dotted edge is on the extended edge. Since the extended edge is a ray starting from a reflex vertex, at least one endpoint of the dotted edge is reflex vertex and the whole block is visible to this reflex vertex.

For the second case, the block has five kinds of shapes, shown by (c) to (g) in Fig. 4. In (c), although no vertex of the block is reflex, the block must be visible to the reflex vertices that emit the block's dotted edges, because in a simple orthogonal polygon, if some other points obstruct the visibility between the reflex vertex and the block, there must be another reflex vertex that divides the block into smaller ones. In (d) to (k), at least one vertex of the polygon is reflex, and hence, the whole block is visible to the vertex.

For the third case, (h) and (i) show the two possible shapes. For the shape in (h), the block must be visible to three reflex vertices and the reason is similar to that for (c). In (i), the block is visible to at least two reflex vertices, which are its own vertices.

For the fourth case, similar to the shape in (c) and (h), the block is entirely visible to at least four reflex vertices.   □

Proof of Proposition 4

Proof. We can prove this proposition by contradiction. Suppose there is a block that is not entirely visible to the light g, while all its four vertices are visible to the light g. Then there must be some other points or edges that obstruct the visibility from the light g. Since it is a simple orthogonal polygon, the obstacle

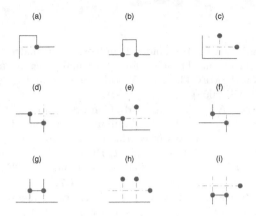

Fig. 4. Block visibility

must be part of the boundary of the orthogonal polygon. Hence, the boundary must obstruct the visibility from the light g to at least one vertex of the block, which contradicts to the condition. □

Proof of Proposition 5

Proof. This proposition can be proved by contradiction. Suppose block $ABCD$ is a small block generated by the edge extensions of an orthogonal polygon P, and all four vertices of block $ABCD$ (Fig. 5) are visible to light G_1 but the block is not entirely visible to light G_1. The convex polygon shaped by light G_1 and some vertices of block $ABCD$ is G_1ABCD. Since block $ABCD$ is not entirely visible to light G_1, there must be some vertices of the polygon P inside the polygon G_1ABCD blocking the light from light G_1. This contradicts with the condition "there is no other vertex inside the convex polygon". □

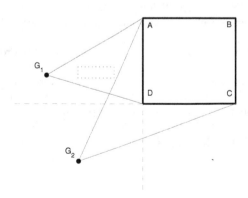

Fig. 5. Figure for the proof of Proposition 5

References

1. Bajuelos Domínguez, A.L., Hernández Peñalver, G., Canales Cano, S., Martins, A.M.: Minimum vertex guard problem for orthogonal polygons: a genetic approach. In: Proceedings of MAMECTIS 2008. World Scientific and Engineering Academy and Society, WSEAS (2008)
2. Bajuelos Domínguez, A.L., Martins, A.M., Canales Cano, S., Hernández Peñalver, G.: Metaheuristic approaches for the minimum vertex guard problem. In: Third International Conference on Advanced Engineering Computing and Applications in Sciences, ADVCOMP 2009, pp. 77–82. IEEE (2009)
3. Baumgartner, T., Fekete, S.P., Kröller, A., Schmidt, C.: Exact solutions and bounds for general art gallery problems. Science **407**, 14–1 (2011)
4. Couto, M.C., de Rezende, P.J., de Souza, C.C.: An exact algorithm for minimizing vertex guards on art galleries. Int. Trans. Oper. Res. **18**(4), 425–448 (2011)
5. csep10.phys.utk.edu. Intensity: the inverse square law (2015). http://csep.10.phys.utk.edu/astr162/lect/light/intensity.html
6. EIA. How much electricity is used for lighting in the united states? (2016). https://www.eia.gov/tools/faqs/faq.cfm?id=99&t=3
7. Ghosh, S.K.: Approximation algorithms for art gallery problems in polygons. Discrete Appl. Math. **158**(6), 718–722 (2010)
8. Jain, A.K.: Data clustering: 50 years beyond k-means. Pattern Recogn. Lett. **31**(8), 651–666 (2010)
9. Jang, D.-S., Kwon, S.-I.: Fast approximation algorithms for art gallery problems in simple polygons. arXiv preprint arXiv:1101.1346 (2011)
10. King, J.: Fast vertex guarding for polygons with and without holes. Comput. Geom. **46**(3), 219–231 (2013)
11. King, J., Kirkpatrick, D.: Improved approximation for guarding simple galleries from the perimeter. Discrete Comput. Geom. **46**(2), 252–269 (2011)
12. Krause, A., Guestrin, C.: Near-optimal observation selection using submodular functions. In: AAAI, vol. 7, pp. 1650–1654 (2007)
13. Muhamad, W.N.W., Zain, M.Y.M., Wahab, N., Aziz, N.H.A., Kadir, R.A.: Energy efficient lighting system design for building. In: 2010 International Conference on Intelligent Systems, Modelling and Simulation (ISMS), pp. 282–286. IEEE (2010)
14. Nagy, Z., Yong, F.Y., Frei, M., Schlueter, A.: Occupant centered lighting control for comfort and energy efficient building operation. Energy Build. **94**, 100–108 (2015)
15. O'rourke, J.: Art Gallery Theorems and Algorithms. Oxford University Press, Oxford (1987)
16. Schaeper, A., Palazuelos, C., Denteneer, D., Garcia-Morchon, O.: Intelligent lighting control using sensor networks. In: 2013 10th IEEE International Conference on Networking, Sensing and Control (ICNSC), pp. 170–175. IEEE (2013)
17. Schuchardt, D., Hecker, H.-D.: Two NP-hard art-gallery problems for ortho-polygons. Math. Logic Q. **41**(2), 261–267 (1995)

On the 2-Center Problem Under Convex Polyhedral Distance Function

Sergey Bereg$^{(\boxtimes)}$

Department of Computer Science,
University of Texas at Dallas, Richardson, TX, USA
besp@utdallas.edu

Abstract. Metrics or distance functions generalizing the Euclidean metric (and even L_p-metrics) have been used in Computational Geometry long ago; for example, convex distance functions appeared in the first Symposium on Computational geometry [2]. Very recently Das *et al.* [5] studied the 1-center problem under a convex polyhedral distance function where the unit ball of the distance function is a convex polytope. In this paper we develop algorithms for the 2-center problem under a convex polyhedral distance function in $\mathbb{R}^d, d = 2, 3$. We show that the 2-center can be computed in $O(n \log m)$ time for the plane and in $O(nm \log n)$ time for $d = 3$. We show that the problem of for computing the 2-center in $\mathbb{R}^3$ has an $\Omega(n \log n)$ lower bound.

1 Introduction

The Euclidean metric or L_p-metrics in general are common metrics in Computational Geometry. Chew and Drysdale [2] studied Voronoi diagrams based on convex distance functions which can be viewed as a generalization of L_p-metrics. "If a pebble is dropped into a still pond, circular waves move out from the point of impact." Then the Voronoi diagram under the Euclidean distance function is just the set of points where circular waves collide, if n pebbles are dropped simultaneously. Chew and Drysdale observed that the circular waves can be changed to waves of a different convex shape. Then the waves collide forming a new Voronoi diagram for the same set of n points using a new distance function. If P is a convex bounded region containing the origin O in the interior of P, then the distance from the origin to a point p is the stretch factor of P such that the boundary of the stretched P contains p. In other words, the distance from the origin to p is $|p|/|p_P|$ where p_P is the intersection point of the ray emanating from the origin to p and the boundary of P, see Fig. 1. The distance from a point a to a point b is defined as the distance from the origin to $b - a$. This distance is well defined since the region P is star-shaped. Note that the distance function may not be a metric since the distance from a to b may be not equal to the distance from b to a. Clearly, this definition of *convex distance function* can be extended to higher dimensions.

© Springer International Publishing AG 2016
T.-H.H. Chan et al. (Eds.): COCOA 2016, LNCS 10043, pp. 365–377, 2016.
DOI: 10.1007/978-3-319-48749-6_27

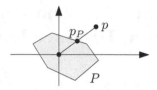

Fig. 1. The new distance from the origin to p is $|p|/|p_P|$.

Polyhedral Convex Distance Functions. Chew *et al.* [3] studied Voronoi diagrams of lines in the three-dimensional space under polyhedral convex distance functions which can be defined as follows. For $i = 1, 2, \ldots, m$, let $a_i \in \mathbb{R}^d$ and $b_i \in \mathbb{R}_+$. Consider a convex polytope with m facets

$$P = \{x \in \mathbb{R}^d \mid a_i x \leq b_i, i = 1, 2, \ldots, m\}.$$

Then the distance $\delta_P(x, y)$ (the polyhedral convex distance function) from a point $x \in \mathbb{R}^d$ to a point $y \in \mathbb{R}^d$ is defined as

$$\min r$$
$$\text{s.t. } a_i(y - x) \leq rb_i, \ \forall i = 1, 2, \ldots, m.$$

Icking *et al.* [8] studied bisectors of two points in the plane with respect to convex distance functions where the unit balls are convex polygons. They proved that the bisector consists of polylines and polygons (bounded and unbounded). It has $O(n + m)$ complexity where n and m are the sizes of convex polygons for the two given points. Icking *et al.* [7] also studied convex distance functions in $\mathbb{R}^3$.

Recently Das *et al.* [5] studied the 1-center problem under a convex polyhedral distance function where the convex region of the distance function is a convex polytope.

1-center problem. Given points $p_1, p_2, \ldots, p_n \in \mathbb{R}^d$, find a point $q \in \mathbb{R}^d$ such that $\max_i \delta_P(q, p_i)$ is minimized. We assume that
(i) $m = O(n)$,[1]
(ii) no two points $p_i, p_j, i \neq j$ are aligned with a facet of P.[2]

Das *et al.* [5] found a linear time algorithm for computing the 1-center problem under a convex polyhedral distance function. The man ingredient is the algorithm for the restricted 1-center problem where the 1-center is restricted to a line. They show that the constrained 1-center problem can be solved in linear time. Furthermore in linear time it can be decided whether the unconstrained 1-center lies on the given line or not. If the unconstrained 1-center does not lie on the given line, the side of the line contains the 1-center is computed also.

The algorithm by Das *et al.* [5] for computing the 1-center in the plane employs prune and search technique using weighted bisectors. The points are

---

[1] To simplify the analysis.
[2] This condition can be removed by a perturbation.

paired and a bisector for each pair is computed. A weight is assigned to each bisector. In one iteration of the algorithm bisectors corresponding to a quarter of total weight are pruned. The running time of the entire algorithm is $O(nm \log^2 m)$ where m is the complexity of the unit ball (the number of vertices in the polygon).

In this paper we study the 2-center problem under a convex polyhedral distance function.

2-center problem. Given points $p_1, p_2, \ldots, p_n \in \mathbb{R}^d$, find two points $q_1, q_2 \in \mathbb{R}^d$ such that $\max_i \min_j \delta_P(q_j, p_i)$ is minimized. The value of $\max_i \min_j \delta_P(q_j, p_i)$ is called *2-center radius*. We assume that
(i) $m = O(n)$,
(ii) no two points $p_i, p_j, i \neq j$ are aligned with a facet of P.

We develop algorithms for the 2-center problem under a convex polyhedral distance function in the plane and in $\mathbb{R}^3$. We also show that the problem of computing the 2-center in $\mathbb{R}^3$ has an $\Omega(n \log n)$ lower bound.

2   The 1-Center in the Plane

First, we consider the problem of computing the 1-center in the plane under a convex polyhedral distance function. We design a new algorithm different from the algorithm from by Das *et al.* [5] and use its ingredients to develop later an algorithm for computing the 2-center. Our algorithm for computing the 1-center in the plane does not use the bisectors. The unit ball of the distance function is a convex polygon P with m vertices. We consider the m directions corresponding to the sides of the polygon P, see Fig. 2.

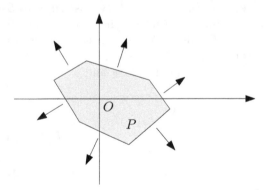

Fig. 2. Polygon P with $m = 6$ vertices and 6 extreme directions.

Algorithm 1-Center-in-2D.

1. Compute m extreme points in S using directions normal to the edges of P.
2. Compute the minimum $s > 0$ such that the points computed in Step 1 can be covered by a polygon $sP + t$ for some $t \in \mathbb{R}^2$.
3. Return 1-center t.

2.1 Extreme Points

In this section we show that the extreme points in the first step of our algorithm can be computed in $O(n \log m)$ time. We consider two parts of the polygon P: the upper part P_{up} and the lower part P_{low}. They can be viewed as two polylines between p_{left} and p_{right}, the vertices of P with the smallest and largest x-coordinates, respectively. Without loss of generality we assume that the vertices of P with minimum and maximum x-coordinate are unique. It suffices to show that the extreme points corresponding to P_{up} can be computed in $O(n \log m)$ time (the extreme points corresponding to P_{low} can be computed in $O(n \log m)$ time as well). Let $d_1, d_2, \ldots, d_u$ be the directions corresponding to P_{up} sorted by the slope. Note that the value of u is at most m.

Next, we split the directions into two lists

$$D_1 = \{d_1, d_2, \ldots, d_{\lfloor u/2 \rfloor}\} \text{ and } D_2 = \{d_{\lfloor u/2 \rfloor+1}, d_{\lfloor u/2 \rfloor+2}, \ldots, d_u\}.$$

We will split the points in S into two lists S_1 and S_2 such that, for each $i = 1, 2$ and all directions in D_i, the extreme points in P_{up} must be in the set S_i. First, we find a direction d_{split} between $d_{\lfloor u/2 \rfloor}$ and $d_{\lfloor u/2 \rfloor+1}$, for example $d_{split} = (d_{\lfloor u/2 \rfloor} + d_{\lfloor u/2 \rfloor+1})/2$. Consider pairs of points $(p_1, p_2), (p_2, p_3), (p_3, p_4), \ldots$ from the set S. For each pair of points (p_i, p_{i+1}), we draw the line l passing through p_i and p_{i+1}. Let $l_\perp$ be the line perpendicular to line l and passing through the origin, see Fig. 3(c) for an example. If line $l_\perp$ coincides with the x-axis, then one of the points p_i and p_{i+1} (which is below the other) cannot be in either S_1 or S_2, see Fig. 3(a). We delete this point. Suppose that line $l_\perp$ has the same slope as direction d_{split}. Without loss of generality we can assume that the slope of vector $p_i p_{i+1}$ is $d_{split} - \pi/2$ as shown in Fig. 3(b). Then p_i cannot be in either S_1 and p_{i+1} cannot be in S_2. In other words, we assign p_i to S_2 and p_{i+1} to S_1.

Consider a non-degenerate case where the slope of line $l_\perp$ is, say between 0 and d_{split} as shown in Fig. 3(c). Without loss of generality we can assume that the slope of vector $p_i p_{i+1}$ is greater than the slope of line $l_\perp$ by $\pi/2$ (as in Fig. 3(b)). Then point p_i cannot be in set S_2 and we assign it to set S_1.

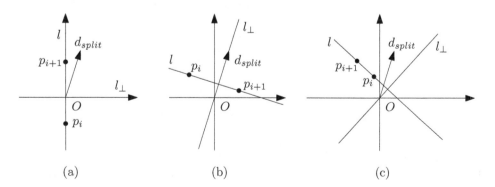

(a)   (b)   (c)

Fig. 3. (a) p_i can be deleted as it is not in S_1 nor in S_2. (b) Pair of points p_i and p_{i+1} such that the point p_{i+1} cannot be extreme for any direction in D_2.

Clearly, the number of points in S not assigned to either S_1 or S_2 will be at most $n/2$ after all pairs are processed. We repeat this algorithm until none or one unassigned point left (which can be assigned to both S_1 and S_2 or, alternatively, it can be compared with extreme points computed later). Let $T(n, m)$ be the running time of this algorithm. Let $n_i, i = 1, 2$ be the size of the computed set S_i. Note that $n_1 + n_2$ could be equal to n or $n + 1$. The recurrence for $T(n, m)$ is then

$$T(n, m) = T(n_1, m/2) + T(n_2, m/2) + O(n).$$

The solution the recurrence is $T(n, m) = O(n \log m)$.

Lemma 1. *For any set S of n points and a convex polygon P of m points, the extreme points from S in directions normal to the edges of P can be computed in $O(n \log m)$ time.*

2.2 Step 2

Our task in Step 2 is to compute the minimum $s > 0$ such that the points computed in Step 1 can be covered by a polygon $sP + t$ for some $t \in \mathbb{R}^2$. These points are assigned to the directions $d_i, i \in \{1, 2, \ldots, u\}$. Note that a group of consecutive directions $d_i, d_{i+1}, \ldots, d_j$ can be assigned to a single point in S. We show that Step 2 can be done in $O(m)$ time.

Let $p_{k_i}, i = 1, 2, \ldots, m$ be the extreme point in direction d_i computed in Step 1. Consider the following problem:

$$\begin{aligned} \min \ & s \\ \text{s.t. } & a_i(p_{k_i} - t) \le sb_i, \ \forall i = 1, 2, \ldots, m. \end{aligned} \tag{1}$$

It can be viewed as a linear programming problem in three dimensions (the unknowns are t_x, t_y and s). It can be solved in linear time [4, 6, 10]. Since both s and t are computed, we can return t in Step 3 (Fig. 4).

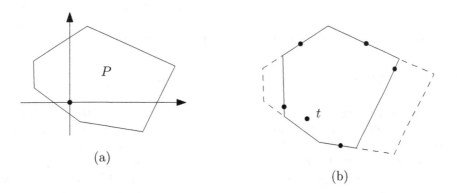

(a)

(b)

Fig. 4. (a) Polygon P. (b) Extreme points found in Step 1 and the optimal covering polygon sP.

2.3   Correctness and Running Time

Since the polygon P contains the origin it can be viewed as

$$P = \{v \in \mathbb{R}^2 \mid a_i v \leq b_i, i = 1, 2, \ldots, m\}.$$

Then the distance $\delta_P(v, u)$ from a point v to a point u is defined as

$$\min r$$
$$\text{s.t. } a_i(u - v) \leq r b_i, \ \forall i = 1, 2, \ldots, m.$$

Recall the 1-center problem: Given points $p_1, p_2, \ldots, p_n$ in the plane, find a point q such that $\max_i \delta_P(q, p_i)$ is minimized. Let q be the 1-center and r be the 1-center radius, i.e.

$$r = \max_i \delta_P(q, p_i).$$

Then

$$a_i(p_j - q) \leq r b_i, \ \forall i = 1, 2, \ldots, m \text{ and } \forall j = 1, 2, \ldots, n$$

Fix any i. The inequalities

$$a_i(p_j - q) \leq r b_i, \ \forall j = 1, 2, \ldots, n$$

hold if

$$a_i(p_{j^*} - q) \leq r b_i$$

where p_{j^*} is a point $p_j, j \in \{1, 2, \ldots, n\}$ that maximizes $a_i p_j$. The algorithm is correct since p_{j^*} is the extreme point in the ith direction.

The Running Time. By Lemma 1, Step 1 takes $O(n \log m)$ time. Step 2 takes $O(m)$ time. We conclude the following result.

Theorem 1. *Let S be a set of n points in the plane and let P be a convex polygon in the plane containing the origin. The 1-center of S under the distance function $\delta_P()$ can be computed in $O(n \log m)$ time where $m = |P|$.*

3   The 1-Center in Higher Dimensions

In this section we give a straightforward reduction of the 1-center problem in $\mathbb{R}^d$ under the convex polyhedral distance function to linear programming. We generalize the approach from Sect. 2.

Recall that in the 1-center problem we are given points $p_1, p_2, \ldots, p_n \in \mathbb{R}^d$ and we want to find a point $q \in \mathbb{R}^d$ such that $\max_i \delta_P(q, p_i)$ is minimized. For two points x and $y \in \mathbb{R}^d$, the polyhedral convex distance function $\delta_P(x, y)$ is

$$\min r$$
$$\text{s.t. } a_i(y - x) \leq r b_i, \ \forall i = 1, 2, \ldots, m.$$

Then the 1-center radius can be expressed as

$$\min r$$
$$\text{s.t. } a_i(q - p_j) \le rb_i, \ \forall i = 1, 2, \ldots, m, \ \forall j = 1, 2, \ldots, n.$$

Clearly this is a linear program with unknowns q and r. Since it is a $(d+1)$-dimensional linear program with nm constraints which can be solved in linear time in fixed dimension [4,6,10], we conclude the following.

Theorem 2. *Let S be a set of n points in $\mathbb{R}^d$ and let P be a convex polytope in $\mathbb{R}^d$ containing the origin. The 1-center of S under the distance function $\delta_P()$ can be computed in $O(nm)$ time where m is the number of facets of P.*

4 The 2-Center in the Plane

In this section we show that the 2-center in the plane under a convex polyhedral distance function can be computed efficiently. We use the ingredients of the algorithm from Sect. 2.

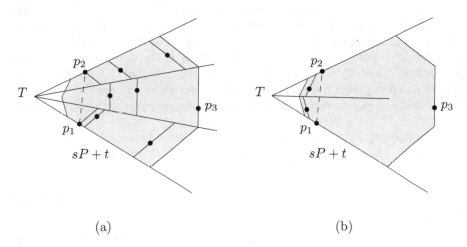

(a)   (b)

Fig. 5. A polygon $sP + t$ containing all the given points. (a) Points lying outside the triangle Tp_1p_2 and corresponding sectors. (b) Points lying inside the triangle Tp_1p_2 and corresponding sectors.

Algorithm 2-Center-in-2D.

1. Compute m extreme points in S using directions normal to the edges of P.
2. Compute the minimum $s > 0$ such that the points computed in Step 1 can be covered by a polygon $sP + t$ for some $t \in \mathbb{R}^2$. Let $p_1, p_2, \ldots, p_k, k \le m$ be at most m points from S on the boundary of $sP + t$ (at most one from each side of $sP + t$).

3. If $k \geq 5$ then s is the 2-center radius and two copies of $sP + t$ cover S (just one is sufficient).
4. Suppose that $k \leq 4$. For each pair p_i and p_j, $1 \leq i < j \leq k$ such that the directions of p_i and p_j are not opposite, compute $s_{i,j}, a_{i,j}, b_{i,j}$ such that
   (a) $S \subset s_{i,j}P + a_{i,j} \cup s_{i,j}P + b_{i,j}$,
   (b) p_i, p_j are on the boundary of $s_{i,j}P + a_{i,j}$, and
   (c) $s_{i,j}$ is minimized.
5. Return 2-center $a_{i,j}, b_{i,j}$ minimizing $s_{i,j}$.

4.1   Step 4

We show that, for any pair i and j with $1 \leq i < j \leq k$, the number $s_{i,j}$ and the points $a_{i,j}, b_{i,j}$ can be computed in linear time.

Consider the polygon $sP + t$ containing all the given points, see Fig. 5. First, compute the intersection of two rays extending the sides that contain points p_i and p_j ($i = 1$ and $j = 2$ in Fig. 5). Let T be this point. Draw rays emanating from T through the vertices of the polygon $sP+t$ that not visible (assuming that $sP+t$ is an obstacle) as shown in Fig. 5. For each point in S find the sector between consecutive rays containing the point. This can be done in $O(\log m)$ time for each point using a binary search. For each point $p_l, l \in \{1, 2, \dots, n\}, l \neq i, j$ find the stretch factor s_l such that the polygon $sP + t$ scaled by s_l about T contains p_l on its boundary. Compute the median of values s_l and test whether S can be covered by two polygons P_1 and P_2 such that P_1 is the polygon $sP + t$ scaled by s_l about T and P_2 is a translated copy of the polygon $ss_l P$. In $O(n \log m)$ time we can find points from S covered by P_1. In $O(n \log m)$ time we can decide if the remaining points can be covered by a polygon P_2 using algorithms from Sect. 2.

If the polygon P_2 exists, then we search next time for $s_{l'} > s_l$; otherwise we search for $s_{l'} < s_l$. The entire algorithm will take $O(n \log n \log m)$ time.

Improvement. We show that Step 4 can be performed in linear time (in n). Consider the triangle $Tp_i p_j$ denoted by Δ. We maintain four sets A, B, C, and D where

- set A stores points from S that must be in polygon P_1 (for example, p_i, p_j),
- set B stores points from S that must be in polygon P_2,
- set C stores points from $S \cap \Delta$ that could be in either P_1 or P_2, and
- set D stores points from $S \cap \mathbb{R}^2 \backslash \Delta$ that could be in either P_1 or P_2.

Note that we can store at most m points in both A and B as in the algorithm from Sect. 2. We use the median scale of either set C or D whichever is larger (ties may be broken arbitrarily). At least half of points in the selected set will be moved to either A or B. We update the set keeping at most m points. The algorithm stops when $C = D = \emptyset$. Then $s_{i,j}$ is computed as the maximum of the size of P_1 or P_2 relative to the size of P. Let $t(|C| + |D|, |A| + |B|)$ denote the running time of the algorithm. Then the recurrence for $t()$ is

$$t(n, m) = t(3n/4, m) + O(n \log m)$$

which is solved as $t(n, m) = O(n \log m)$.

Theorem 3. *Let S be a set of n points in the plane and let P be a convex polygon in the plane containing the origin. The 2-center of S under the distance function $\delta_P()$ can be computed in $O(n \log m)$ time where $m = |P|$.*

5   The 2-Center in Three Dimensions

In this section we show that the 2-center in $\mathbb{R}^3$ under a convex polyhedral distance function can be computed efficiently.

Algorithm 2-Center-in-3D.

1. Compute m extreme points in S using directions normal to the facets of P.
2. Compute the minimum $s > 0$ such that the points computed in step 1 can be covered by a polytope $sP + t$ for some $t \in \mathbb{R}^3$. Let $p_1, p_2, \ldots, p_k, k \leq m$ be at most m points from S on the boundary of $sP + t$ (at most one from each side of $sP + t$).
3. If $k \geq 5$ then pick 3 points p_i, p_j and p_l. Similar to the scaling in the algorithm 2-Center-in-2D, use the binary search in the cone bounded by 3 planes through the facets corresponding to p_i, p_j, p_l. This step can be done in linear time (in n) but a simple realization with $O(n \log n \log m)$ time suffices.
4. Suppose that $k = 4$. For each pair p_i and p_j, $1 \leq i < j \leq k$ such that the directions of p_i and p_j are not opposite, compute $s_{i,j}, a_{i,j}, b_{i,j}$ such that
   (a) $S \subset s_{i,j}P + a_{i,j} \cup s_{i,j}P + b_{i,j}$,
   (b) p_i, p_j are on the boundary of $s_{i,j}P + a_{i,j}$, and
   (c) $s_{i,j}$ is minimized.
5. Return 2-center $a_{i,j}, b_{i,j}$ minimizing $s_{i,j}$.

5.1   Step 4

We show that, for any pair i and j with $1 \leq i < j \leq k$, the number $s_{i,j}$ and the points $a_{i,j}, b_{i,j}$ can be computed in $O(mn \log n)$ time. Pick a direction normal to an lth facet of the polytope P that is different from the directions of p_i and p_j and is not opposite to any of them. Create a list L_l of scale factors of P such that a translated copy of the scaled polytope has p_i, p_j on its corresponding facets and a point from S on its lth facet. This can be done by (pre)sorting S in the direction normal to the lth facet of P. The list L_l is actually a truncated sublist of this sorted list. The algorithm apply a binary search using the lists L_l.

To test a particular scale factor s we slide the polytope sP such that p_i and p_j are both on its corresponding facets. The points covered by the sliding polytope undergo insertions and deletions of points from S. These points are deleted/inserted into the set S_2 that is tested for covering by a translated copy of sP. This polytope is again sliding since its two facets are defined by the other two extreme points computed in Step 2.

Note that each point is at S_2 at the beginning of the sliding of the first copy of sP. Also each point either is in S_2 all the time or is deleted and then inserted again. We compute these times for all the points in $O(n \log m)$ time by (i) projecting S and the first copy of sP onto a plane perpendicular to the sliding direction, and (ii) locating each point in two projected facets of the copy of sP. Similarly, we find two facets of the second copy of sP for each point in S. Now, for each instant of time, we compute two points from S that are extreme in two (opposite) sliding directions. We explain how to do it for one direction, say d. Each point is already assigned to one facet in direction d. We partition S into corresponding lists. These lists are sublists of the corresponding sorted lists that can be precomputed in $O(nm \log n)$ time. The sublists can be extracted in $O(nm)$ time. In $O(nm)$ time the sublists can be merged into the required list for d. Similarly, we crate the list for the opposite direction. Finally, the test can be done by traversing the two lists.

Theorem 4. *Let S be a set of n points in the plane and let P be a convex polygon in $\mathbb{R}^3$ containing the origin. The 2-center of S under the distance function $\delta_P()$ can be computed in $O(mn \log m)$ time where $m = |P|$.*

6   Lower Bound

In this section we prove a lower bound for computing the 2-center in $\mathbb{R}^3$ under a convex polyhedral distance function by a reduction from the MAX-GAP problem.

MAX-GAP problem. Given a set $A = \{a_1, a_2, \ldots, a_n\}$ of n real numbers, compute the *maximum gap* of A, which is the largest absolute value of the difference between two elements a_i and a_j that are consecutive in the sorted sequence.

Note that the value of the maximum gap is easily computed after sorting S. The MAX-GAP problem has an $\Omega(n \log n)$ lower bound in the algebraic decision tree model, as proved by Lee and Wu [9]. We will reduce the MAX-GAP problem to the 2-center in $\mathbb{R}^3$ under a convex polyhedral distance function. In fact, we use the decision problem where, given a positive real number γ, we want to determine whether or not the maximum gap is greater than or equal to γ. It has an $\Omega(n \log n)$ lower bound in the algebraic computation tree model [1].

Consider a tetrahedron T_1 with four vertices in $\mathbb{R}^3$: $(0, 0, -1), (0, 0, 1), (1, 1, 0)$ and $(-1, 1, 0)$, see Fig. 6. Let T_2 be a copy of T_1 shifted by some number $y_0 - 1 \in (0, 1)$ along Y-axis. We actually assume that $4/3 < y_0 < 2$ for a reason that will be explained later. The top facet of T_1 is the triangle with vertices $(0, 0, 1), (1, 1, 0)$ and $(-1, 1, 0)$. The equation of the plane π_1 passing through these points is then $y + z = 1$. The facet seen from X-axis (or from $(\infty, 0, 0)$) is the triangle with vertices $(0, 0, -1), (0, 0, 1), (1, 1, 0)$. The equation of the plane π_2 passing through these points is $x - y = 0$, see Fig. 6. We place two points on the two facets of T_1 close to the origin, say points $p_1 = (\sigma, \sigma, 0)$ and $p_2 = (-\sigma, \sigma, 0)$ for some small $\sigma > 0$. Similarly, place two points on the two facets of T_2 close to the point $(0, y_0, 0)$, say points $p_3 = (0, y_0 - \sigma, \sigma)$ and $p_4 = (0, y_0 - \sigma, -\sigma)$.

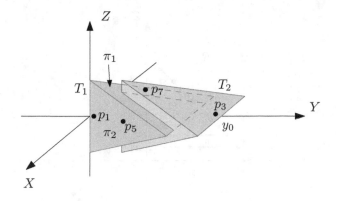

Fig. 6. Two tetrahedra covering n given points.

Polytope P. We define the unit ball of the convex polyhedral distance function as a translated copy of tetrahedron T_1 such that the origin is in its interior. For example, tetrahedron $T_1 - v$ where $v = (0, 1/2, 0)$ can be used as the polytope P. Its vertices are $(0, -1/2, -1), (0, -1/2, 1), (1, 1/2, 0)$ and $(-1, 1/2, 0)$.

Consider the plane π_3 with equation $y = y_0/2$. We place the remaining points in π_3 as follows. The intersection of plane π_3 and tetrahedron T_1 is a rectangle R_1. Its sides are defined by the intersection of plane π_3 and

- the intersection of planes π_1 and π_3, which is $\{(x, y, z)|\ y = y_0/2, z = 1 - y_0\}$, and
- the intersection of planes π_2 and π_3, which is $\{(x, y, z)|\ x = y_0/2, y = y_0/2\}$.

Similarly, the intersection of plane π_3 and tetrahedron T_2 is a $(2 - y_0) \times y_0$ rectangle R_2.

We create a set A_1 of n points corresponding to the numbers $a_1, a_2, \ldots, a_n$. Let s be the *spread* of A, i.e. $s = \max_i a_i - \min_i a_i$. Let t be the number $\gamma(2 - y_0)/(s + \gamma)$. We map a number $a_i, i = 1, 2 \ldots, n$ to a point $(x_i, y_0/2, z_i)$ where $x_i = t + 2 - y_0 - z_i, y_0/2, z_i), z_i = t + (a_i - \min_j a_j)r$, and $r = (2 - y_0 - t)/s$. Note that the points of A_1 lie on the line $x + z = t + 2 - y_0, y = y_0/2$, see Fig. 7. We create a translated copy of A_1, the set $A_2 = A_1 - (2t + 2 - y_0, 2t + 2 - y_0)$. Finally, we place two points $p_9 = (2 - y_0 + \varepsilon, t)$ and $p_{10} = (t, 2 - y_0 + \varepsilon)$ for sufficiently small ε. Thus, we constructed a set $S = \{p_i \mid i = 1, 2, \ldots, 10\} \cup A_1, A_2$ of $2n + 10$ points.

Lemma 2. *The maximum gap of A is greater than or equal to γ if and only if the 2-center radius of S is smaller than or equal to one.*

Proof. First, we show that if the set S can be covered by two copies of P under translation then the maximum gap of A is greater than or equal to γ. Suppose that two polytopes $P_1 = P + v_1$ and $P_2 = P + v_2$ cover S, i.e. $S \subset P_1 \cap P_2$. It is easy to see that polytope P cannot be translated to cover one point from $\{p_1, p_2\}$ and one point from $\{p_3, p_4\}$. Therefore, we can assume that $p_1, p_2 \in P_1$

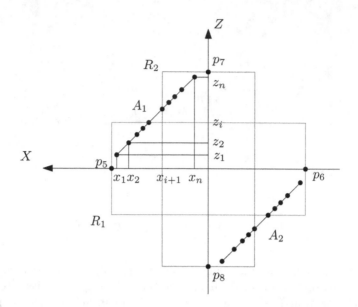

Fig. 7. The sets A_1 and A_2 in case of $a_1 \leq a_2 \leq \ldots a_n$.

and $p_3, p_4 \in P_2$. The intersection of P_1 and π_3 is a $l_1 \times w_1$ rectangle where $l_1 \leq y_0$ and $w_2 \leq 2 - y_0$. Note that polytope P_1 cannot contain a point from $\{p_5, p_6\}$ and a point from $\{p_7, p_8\}$, since $w_2/2 < z(p_7) = y_0/2$. The last inequality follows from $y_0/2 > (2 - y_0)/2$ or $y_0 > 1$.

We can choose y_0 such that P_1 cannot contain a point from $\{p_5, p_6\}$ and a point from $\{p_7, p_8\}$; for example, by restricting $w_2 < z(p_7) = y_0/2$. The last inequality follows from $2 - y_0 < y_0/2$ or $y_0 > 4/3$.

Similarly, P_2 does not contain a point from $\{p_5, p_6\}$ and a point from $\{p_7, p_8\}$. Also P_1 cannot contain p_7 or p_8 and P_2 cannot contain p_5 or p_6. Thus, P_1 must contain p_5 and p_6 and P_2 must contain p_7 and p_8. Furthermore, $p_5, p_6 \notin P_2$ and $p_7, p_8 \notin P_1$. This implies that the tetrahedron P_1 is a translated copy of T_1 along the z-axis and the tetrahedron P_2 is a translated copy of T_2 along the x-axis.

Since rectangle R_1 cannot contain point p_{10}, it should be covered by R_2. Similarly, point p_9 lies in R_1. Since the intersection of R_1 and R_2 is a square of side $2 - y_0$, $R_1 \cup R_2$ cannot cover the segment of A_2. The uncovered part of A_2 must be within a gap between two consecutive points in A_2. The claim follows if the uncovered segment has x-length (if projected onto the x-axis) at least γr. Suppose the uncovered segment has x-length smaller than γr. The x-shift between lines containing A_1 and A_2 is $2t + t - y_0$. It can be verified that $t = \gamma r$. Therefore, the uncovered segment of A_1 has x-length greater than γr and the claim follows.

Clearly, if the 2-center radius of S is smaller than or equal to one, then the maximum gap of A is greater than or equal to γ (since the set S can be covered by two copies of P under translation). This conclude "if" part of the lemma.

Now, we show that if the maximum gap of A is equal to γ then the set S can be covered by two copies of P under translation. Without loss of generality we assume that the numbers in A are sorted, i.e. $a_1 \leq a_2 \leq \ldots a_n$. Suppose that the maximum gap is between a_i and a_{i+1}. Place R_1 such that its top side is at z_i and place R_2 such that its left side is at x_{i+1} as shown in Fig. 7. Then the uncovered segment of A_2 is exactly between two points corresponding a_i and a_{i+1} since

- $x_{i+1} - x_i = \gamma r$ and the width of R_2 is $2 - y_0$, and
- $z_{i+1} - z_i = \gamma r$ and the width of R_1 is $2 - y_0$.

The positions of rectangles $R + i, i = 1, 2$ can be used to find the positions of tetrahedra $T_i, i = 1, 2$. Therefore, the "only if" part of the lemma follows.

By Lemma 2 and an $\Omega(n \log n)$ bound [1] for the problem of deciding whether the maximum gap is greater than or equal to γ, the main result of this section follows.

Theorem 5. *The problem of computing the 2-center in $\mathbb{R}^3$ under a convex polyhedral distance has an $\Omega(n \log n)$ lower bound in the algebraic decision tree model.*

References

1. Arkin, E.M., Hurtado, F., Mitchell, J.S.B., Seara, C., Skiena, S.: Some lower bounds on geometric separability problems. Int. J. Comput. Geom. Appl. **16**(1), 1–26 (2006)
2. Chew, L.P., Drysdale III, R.L.: Voronoi diagrams based on convex distance functions. In: Proceedings of the 1st Annual ACM Symposium Computational Geometry, pp. 235–244 (1985)
3. Chew, L.P., Kedem, K., Sharir, M., Tagansky, B., Welzl, E.: Voronoi diagrams of lines in 3-space under polyhedral convex distance functions. J. Algorithms **29**(2), 238–255 (1998)
4. Clarkson, K.L.: Linear programming in $O(n3^{d^2})$ time. Inform. Process. Lett. **22**, 21–24 (1986)
5. Das, S., Nandy, A., Sarvottamananda, S.: Linear time algorithm for 1-center in $\mathfrak{R}^d$ under convex polyhedral distance function. In: Zhu, D., Bereg, S. (eds.) FAW 2016. LNCS, vol. 9711, pp. 41–52. Springer, Heidelberg (2016). doi:10.1007/978-3-319-39817-4_5
6. Dyer, M.E.: On a multidimensional search technique and its application to the Euclidean one-centre problem. SIAM J. Comput. **15**, 725–738 (1986)
7. Icking, C., Klein, R., Lê, N., Ma, L.: Convex distance functions in 3-space are different. Fundam. Inform. **22**(4), 331–352 (1995)
8. Icking, C., Klein, R., Ma, L., Nickel, S., Weißler, A.: On bisectors for different distance functions. Discret. Appl. Math. **109**, 139–161 (2001)
9. Lee, D.T., Wu, Y.F.: Geometric complexity of some location problems. Algorithmica **1**, 193–211 (1986)
10. Megiddo, N.: Linear programming in linear time when the dimension is fixed. J. ACM **31**, 114–127 (1984)

Algorithms for Colourful Simplicial Depth and Medians in the Plane

Olga Zasenko and Tamon Stephen[(✉)]

Simon Fraser University, British Columbia, Canada
{ozasenko,tamon}@sfu.ca

Abstract. The *colourful simplicial depth* (CSD) of a point $x \in \mathbb{R}^2$ relative to a configuration $P = (P^1, P^2, \ldots, P^k)$ of n points in k colour classes is exactly the number of closed simplices (triangles) with vertices from 3 different colour classes that contain x in their convex hull. We consider the problems of efficiently computing the colourful simplicial depth of a point x, and of finding a point in $\mathbb{R}^2$, called a *median*, that maximizes colourful simplicial depth.

For computing the colourful simplicial depth of x, our algorithm runs in time $O(n \log n + kn)$ in general, and $O(kn)$ if the points are sorted around x. For finding the colourful median, we get a time of $O(n^4)$. For comparison, the running times of the best known algorithm for the monochrome version of these problems are $O(n \log n)$ in general, improving to $O(n)$ if the points are sorted around x for monochrome depth, and $O(n^4)$ for finding a monochrome median.

1 Introduction

The *simplicial depth* of a point $x \in \mathbb{R}^2$ relative to a set P of n data points is exactly the number of simplices (triangles) formed with the points from P that contain x in their convex hull. A *simplicial median* of the set P is any point in $\mathbb{R}^2$ which is contained in the most triangles formed by elements of P, i.e. has maximum simplicial depth with respect to P. Here we consider a set P that consists of k colour classes $P^1, \ldots, P^k$. The *colourful simplicial depth* of x with respect to configuration P is the number of triangles with vertices from 3 different colour classes that contain x. A *colourful simplicial median* of a configuration $P = (P^1, P^2, \ldots, P^k)$ is any point in the convex hull of P with maximum colourful simplicial depth.

The monochrome simplicial depth was introduced by Liu [16]. Up to a constant, it can be interpreted as the probability that x is in the convex hull of a random simplex generated by P. The colourful version, see [7], generalizes this to selecting points from k distributions. Then medians are central points which are in some sense most representative of the distribution(s). Our objective is find efficient algorithms for finding both the colourful simplicial depth of a given point x with respect to a configuration, and a colourful simplicial median of a configuration.

© Springer International Publishing AG 2016
T.-H.H. Chan et al. (Eds.): COCOA 2016, LNCS 10043, pp. 378–392, 2016.
DOI: 10.1007/978-3-319-48749-6_28

1.1   Background

Both monochrome and colourful simplicial depth extend to $\mathbb{R}^d$ and are natural objects of study in discrete geometry. For more background on simplicial depth and competing measures of data depth, see [2,11]. Monochrome depth has seen a flurry of activity in the past few years, most notably relating to the *First Selection Lemma*, which is a lower bound for the depth of the median, see e.g. [17]. Among the recent work on colourful depth are proofs of the lower [20] and upper [1] bounds conjectured by Deza et al., with the latter result showing beautiful connections to Minkowski sums of polytopes.

The monochrome simplicial depth can be computed by enumerating simplices, but in general dimension, it is quite challenging to compute it more efficiently [2,5,11]. Several authors have considered the two-dimensional version of the problem, including Khuller and Mitchell [14], Gil et al. [13] and Rousseeuw and Ruts [19]. Each of these groups produced an algorithm that computes the monochrome depth in $O(n \log n)$ time, with sorting the input as the bottleneck. If the input points are sorted, these algorithms take linear time.

We consider a *simplicial median* to be *any* point $x \in \mathbb{R}^2$ maximizing the simplicial depth. Aloupis et al. [3] considered this question, and found an $O(n^4)$ algorithm to do this. This is arguably as good as should be expected, following the observation of Lemma 2 in Sect. 3.1 that shows that there are in some sense $\Theta(n^4)$ candidate points for the location of the colourful median.

1.2   Organization and Main Results

In Sect. 2, we develop an algorithm for computing colourful simplicial depth that runs in $O(n \log n + kn)$ time. This retains the $O(n \log n)$ asymptotics of the monochrome algorithms when k is fixed. As in the monochrome case, sorting the initial input is a bottleneck, and the time drops to $O(kn)$ if the input is sorted around x. In this case, for fixed k, it is a linear time algorithm.

In Sect. 3, we turn our attention to computing a colourful simplicial median. We develop an algorithm that does this in $O(n^4)$ time using a topological sweep. This is independent of k and matches the running time from the monotone case. Section 4 contains conclusions and discussion about future directions.

2   Computing Colourful Simplicial Depth

2.1   Preliminaries

We consider a family of sets P^1, P^2, ..., $P^k \subseteq \mathbb{R}^2$, $k \geq 3$, where each P^i consists of the points of some particular colour i. Refer to the jth element of P^i as P_j^i. We generally use superscripts for colour classes, while subscripts indicate the position in the array. We will sometime perform arithmetic operations on the subscripts, in which case the indices are taken modulo the size of the array i.e. (mod n_i).

We denote the union of all colour sets by P: $P = \bigcup\limits_{i=1}^{k} P^i$. The total number of points is n, where $|P^i| = n_i$, $\sum\limits_{i=1}^{k} n_i = n$. We assume that points of $P \bigcup \{x\}$ are in general position to avoid technicalities. Without loss of generality, we can take $x = \mathbf{0}$, the zero vector.

Definition 1. *A colourful triangle is a triangle with one vertex of each colour, i.e. it is a triangle whose vertices v_1, v_2, v_3 are chosen from distinct sets P^{i_1}, P^{i_2}, P^{i_3}, where $i_i \neq i_2, i_3; i_2 \neq i_3$.*

Definition 2. *The* colourful simplicial depth *$\hat{D}(x, P)$ of a point x relative to the set P in $\mathbb{R}^2$ is the number of colourful triangles containing x in their convex hull. We reserve $D(x, P)$ for the (monochrome) simplicial depth, which counts all triangles from P regardless of the colours of their vertices.*

Remark 1. We are checking containment in *closed* triangles. With our general position assumption, this will not affect the value of $\hat{D}(x, P)$. It is more natural to consider closed triangles than open triangles in defining colourful medians; the open triangles version of this question may also be interesting.

Throughout the paper we work with polar angles θ_j^i formed by the data points P_j^i and a fixed ray from x. We remark that simplicial containment does not change as points are moved on rays from x, see for example [23]. Thus we can ignore the moduli of the P_j^i, and work entirely with the θ_j^i, which lie on the unit circle $\mathcal{C}$ with x as its origin. We will at times abuse notation, and not distinguish between P_j^i and θ_j^i.

Note that the ray taken to have angle 0 is arbitrary, and may be chosen based on an underlying coordinate system if available, or set to the direction of the first data point P_1. We can sort the input by polar angle, in other words, we can order the points around x. (Perhaps it is naturally presented this way.) We reduce the θ_j^i to lie in the range $[0, 2\pi)$.

The *antipode* of some point α on the unit circle is $\bar{\alpha} = (\alpha + \pi) \bmod 2\pi$. A key fact in computing CSD is that a triangle $\triangle abc$ does *not* contain x if and only if the corresponding polar angles of points a, b and c lie on a circular arc of less than π radians. This is illustrated in Fig. 1, and is equivalent to the following lemma, stated by Gil, Steiger and Wigderson [13]:

Lemma 1. *Given points a, b, c on the unit circle $\mathcal{C}$ centred at x, let $\bar{a}$ be antipodal to a. Then $\triangle abc$ contains x if and only if $\bar{a}$ is contained in the minor arc (i.e. of at most π radians) with endpoints b and c.*

2.2   Outline of Strategy

Recall that we denote the ordinary and colourful simplicial depth by $D(x, P)$ and $\hat{D}(x, P)$ respectively. We can compute $\hat{D}(x, P)$ by first computing $D(x, P)$ and then removing all triangles that contain less than three distinct colours.

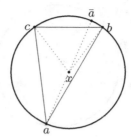

Fig. 1. Antipode $\bar{a}$ falls in the minor arc between b and c and, therefore, the triangle $\triangle abc$ contains x.

To this end, we denote the number of triangles with at least two vertices of colour i as $D^i(x, P)$. When x and P are clear from the context, we will abbreviate these to D, $\hat{D}$ and D^i.

Since we can compute $D(x, P)$ efficiently using the algorithms mentioned in the introduction [13,14,19], the challenge is to compute $D^i(x, P)$ for each $i = 1, 2, \ldots, k$. Then we conclude $\hat{D}(x, P) = D(x, P) - \sum_{i=1}^{k} D^i(x, P)$. To compute D^i efficiently for each colour i, we walk around the unit circle tracking the minor arcs between pairs of points of colour i, and the number of antipodes between them. We do this in linear time in n by moving the front and back of the interval once around the circle, and adjusting the number of relevant antipodes with each move. This builds on the approach of Gil et al. [13] for monochrome depth.

Remark 2. When computing D^i, we count antipodes of all k colours; the triangles with three vertices of colour i will be counted three times: $\triangle abc$, $\triangle bca$ and $\triangle cab$. Thus the quantity obtained by this count is in fact $D^i_* :=$ $D^i + 2 \sum_{i=1}^{k} D(x, P^i)$. We separately compute $\sum_{i=1}^{k} D(x, P^i)$, allowing us to correct for the overcounting at the end.

2.3   Data Structures and Preprocessing

We begin with the arrays θ^i of polar angles, which we sort if necessary. All elements in $\bigcup_{i=1}^{k} \theta^i$ are distinct due to the general position requirement. By construction we have:

$$0 \leq \theta^i_0 < \theta^i_1 < \ldots < \theta^i_{n_i - 1} < 2\pi, \quad \text{for all } 1 \leq i \leq k. \tag{1}$$

Let $\bar{\theta}^i$ be the array of antipodes of θ^i, also sorted in ascending order. We generate $\bar{\theta}^i$ by finding the first $\theta^i_j \geq \pi$, moving the part of the array that begins with that element to the front, and hence the front of the original array to the back; π is subtracted from the elements moved to the front and added to those moved to the back. This takes linear time.

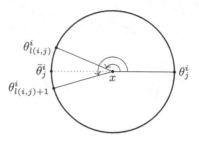

Fig. 2. Index $l(i,j)$ and index $(l(i,j)+1)$

We merge all $\bar{\theta}^i$ into a common sorted array denoted by A. Now we have all antipodes ordered as if we were scanning them in counter-clockwise order around the circle $\mathcal{C}$ with origin x. Let us index the n elements of A starting from 0. Then, for each colour $i = 1, \ldots, k$, we merge A and θ^i into a sorted array A^i. Once again, this corresponds to a counter-clockwise ordering of data points around $\mathcal{C}$.

While building A^i, we associate pointers from the elements of array θ^i to the corresponding position (index) in A^i. This is done by updating the pointers whenever a swap occurs during the process of merging the arrays. Denote the index of some θ_j^i in A^i by $p(\theta_j^i)$. Then the number of the antipodes that fall in the minor arc between two consecutive points θ_j^i and θ_{j+1}^i on $\mathcal{C}$ is $\left(p\left(\theta_{j+1}^i\right) - p\left(\theta_j^i\right) - 1\right)$, if $p\left(\theta_j^i\right) < p\left(\theta_{j+1}^i\right)$, or $\left(n + n_i - p\left(\theta_j^i\right) + p\left(\theta_{j+1}^i\right) - 1\right)$, if $p\left(\theta_j^i\right) > p\left(\theta_{j+1}^i\right)$. Note that $p\left(\theta_j^i\right)$ is never equal to $p\left(\theta_{j+1}^i\right)$.

Now, for each point θ_j^i, we find the index $l(i,j)$ in the corresponding array θ^i such that $\angle \theta_j^i, x, \theta_{l(i,j)}^i < \pi$ and $\angle \theta_j^i, x, \theta_{l(i,j)+1}^i > \pi$ (Fig. 2). Thus the sequence of points $\theta_j^i, \theta_{j+1}^i, \ldots, \theta_{l(i,j)}^i$ is maximal on an arc shorter than π. Viewing the minor arc between two points as an interval, the intervals with left endpoint θ_j^i and right end point from this sequence overlap and can be split into small disjoint intervals as follows:

$$[\theta_j^i, \theta_t^i) = \bigcup_{h=j+1}^{t} [\theta_{h-1}^i, \theta_h^i), \quad \text{where } t = j+1, \ldots, l(i,j). \tag{2}$$

2.4   Computing D_*^i

Let us denote the count of the antipodes within the minor arc between a and b by $c(a,b)$. Then D_*^i can be written as follows:

$$D_*^i = \sum_{j=0}^{n_i-1} \sum_{t=j+1}^{l(i,j)} c\left(\theta_j^i, \theta_t^i\right). \tag{3}$$

Note that index t is taken modulo n_i. From (2) we have:

$$c\left(\theta_j^i, \theta_t^i\right) = \sum_{h=j+1}^{t} c\left(\theta_{h-1}^i, \theta_h^i\right), \quad \text{for } t = j+1, \ldots, l(i,j). \tag{4}$$

Due to (3) and (4), we have:

$$D_*^i = \sum_{j=0}^{n_i-1} \sum_{t=j+1}^{l(i,j)} \sum_{h=j+1}^{t} c\left(\theta_{h-1}^i, \theta_h^i\right).$$ (5)

Let $C_h^i = c\left(\theta_{h-1}^i, \theta_h^i\right)$, $|C^i| = n_i$. Then (5) can be rewritten as:

$$D_*^i = \sum_{j=0}^{n_i-1} \sum_{t=j+1}^{l(i,j)} \sum_{h=j+1}^{t} C_h^i.$$ (6)

Let us create an array of prefix sums: S^i, where $S_t^i = \sum_{h=0}^{t} C_h^i$, $|S^i| = n_i$. This array can be filled in $O(n_i)$ time and proves to be very useful when we need to calculate a sum of the elements of C^i between two certain indices. In fact, such sum can be obtained in constant time using the elements of array S^i:

$$\sum_{h=j+1}^{t} C_h^i = \begin{cases} S_t^i - S_j^i, & \text{if } t \geq j+1, j \neq n_i-1, \\ S_{n_i-1}^i + S_t^i - S_j^i, & \text{if } t < j+1, j \neq n_i-1, \\ S_t^i, & \text{if } j = n_i-1. \end{cases}$$ (7)

Combining (6) and (7), we get:

$$D_*^i = \sum_{j=0}^{n_i-1} \sum_{t=j+1}^{l(i,j)} S_t^i - \sum_{j=0}^{n_i-1} ((l(i,j)-j) \bmod n_i) \cdot S_j^i + \begin{cases} 0, & \text{if } t \geq j+1 \text{ or } j = n_i-1, \\ \sum_{j=0}^{n_i-1} \sum_{t=j+1}^{l(i,j)} S_{n_i-1}^i, & \text{if } t < j+1. \end{cases}$$ (8)

Let us create another array of prefix sums T^i, where $T_j^i = \sum_{t=0}^{j} S_t^i$, $|T^i| = n_i$. This array is used to retrieve the sum of elements of S^i between the indices $j+1$ and $l(i,j)$ in $O(1)$ time:

$$\sum_{t=j+1}^{l(i,j)} S_t^i = \begin{cases} T_{l(i,j)}^i - T_j^i, & \text{if } l(i,j) \geq j+1, j \neq n_i-1, \\ T_{n_i-1}^i + T_{l(i,j)}^i - T_j^i, & \text{if } l(i,j) < j+1, j \neq n_i-1, \\ T_{l(i,j)}^i, & \text{if } j = n_i-1. \end{cases}$$ (9)

Also note that the index t runs from $j+1$ to $l(i,j)$. So $t < j+1$ in (8) is only possible if initially $j+1 > l(i,j)$ and we wrapped around the array. In other words, $t < j+1$ is equivalent to $j+1 > l(i,j)$ and $t = 0, \ldots, l(i,j)$.

After simplifying, we obtain:

$$D_*^i = \sum_{j=0}^{n_i-1} \left(T_{l(i,j)}^i - T_j^i - ((l(i,j)-j) \bmod n_i) \cdot S_j^i\right)$$

$$+ \begin{cases} n_i \cdot \left(T_{n_i-1}^i + ((l(i,j)+1) \bmod n_i) \cdot S_{n_i-1}^i\right), & \text{if } l(i,j) < j+1, \\ 0, & \text{otherwise.} \end{cases}$$ (10)

2.5   Algorithm and Analysis

First, we find all polar angles and their antipodes, which takes $O(n)$ in total. Second, we sort the arrays of polar angles θ^i and their corresponding antipodal elements $\bar{\theta}^i$, which gives us $O\left(\sum_{i=1}^{k} n_i \log n_i\right)$. Third, we need to rotate $\bar{\theta}^i$, so that they are in ascending order. This will take $O(n)$ time. Then we compute for each i the number of triangles with all three vertices of colour i that contain x using the algorithm of Rousseeuw and Ruts [19] for sorted data. This will run in $O(n_i)$, for each i, or $O(n)$ in total. Hence lines 2–10 of the Algorithm 1 take

---

Algorithm 1. CSD(x, P)

Input: $x, P = (P^1, \ldots, P^k)$. Output: $\hat{D}(x, P)$.

```
 1: Sum1 ← 0, Sum2 ← 0;
 2: for i ← 1, k do
 3:     for j ← 0, nᵢ − 1 do
 4:         θⁱⱼ ← polar angle of (Pⁱⱼ − x) mod 2π;
 5:         θ̄ⁱⱼ ← (θⁱⱼ + π) mod 2π;
 6:     end for
 7:     Sort(θ¹);                               ▷ while permuting θ̄ⁱ
 8:     Restore the order in θ̄ⁱ;
 9:     Sum1 ← Sum1 + D(x, θ¹);                 ▷ use the algorithm from [19]
10: end for
11: A ← Merge(θ̄¹, ..., θ̄ᵏ);                     ▷ A is sorted
12: D ← D(x, A);                                ▷ use the algorithm from [19]
13: for i ← 1, k do
14:     B ← Merge(A, θⁱ);                       ▷ update p(θⁱⱼ) the pointers of θⁱⱼ,
15:                                             ▷ B stands for Aⁱ
16:     for j ← 1, nᵢ do                        ▷ j = j mod nᵢ
17:         if p(θⁱⱼ₋₁) < p(θⁱⱼ) then
18:             Cⱼ ← p(θⁱⱼ) − p(θⁱⱼ₋₁) − 1;    ▷ C = Cⁱ - array of antipodal counts
19:         else
20:             Cⱼ ← n + nᵢ − p(θⁱⱼ₋₁) + p(θⁱⱼ) − 1;
21:         end if
22:     end for
23:     Find l(i, 0) using binary search in θⁱ;
24:     S₀ ← C₀; T₀ ← S₀;                       ▷ S = Sⁱ, T = Tⁱ - prefix sum arrays
25:     for j ← 1, nᵢ − 1 do
26:         Find l(i, j);
27:         Sⱼ ← Sⱼ₋₁ + Cⱼ;
28:         Tⱼ ← Tⱼ₋₁ + Sⱼ;
29:     end for
30:     Sum2 ← Sum2 + Dⁱ*(x, P) obtained from the formula (10);
31:     delete B, C, S, T;
32: end for
33: return D̂(x, P) = D − (Sum2 − 2 * Sum1);    ▷ Sum1 = Σᵢ₌₁ᵏ D(x, Pⁱ)
```

$$O\left(\sum_{i=1}^{k} n_i + \sum_{i=1}^{k} n_i \log n_i\right) = O(n \log n)$$ time to complete. This follows from the

facts that $\sum_{i=1}^{k} n_i = n$ and $n \log n$ is convex.

To generate the sorted array A of antipodes, we merge the k single-coloured arrays using a heap (following e.g. [6]) in $O(n \log k)$ time. We need to compute the monochrome depth $D(x, P)$ of x with respect to all points in P, regardless of colour. For this we can use the sorted array of antipodes rather than sorting the original array. Thus we again use the linear time monochrome algorithm [19] with x and A. Note that working with the antipodes is equivalent due to the fact that the simplicial depth of x does not change if we rotate the system of data points around the centre x.

After that, we execute a cycle of k iterations – one for each colour. It starts with merging two sorted arrays A and θ^i, which is linear in the size of arrays we are merging and takes $O\left(\sum_{i=1}^{k} (n + n_i)\right) = O(kn)$ in total. Filling the arrays C is linear. Since the $l(i, j)$ appear in sequence in the array θ^i, we find the first one $l(i, 0)$ using a binary search that takes $O(\log n_i)$, and $O(k \log n)$ in total. Then we find the rest of $l(i, j)$ in $O(n)$ time for each i by scanning through the array starting from the element $\theta^i_{l(i,0)}$. The remaining operations take constant time to execute. Therefore, total running time of Algorithm 1 is $O(n + n \log n + n + n \log k + kn + k \log n + kn) = O(n \log n + kn)$. The $n \log n$ term corresponds to the initial sorting of the data points, if they are presented in sorted order, the running time drops to $O(kn)$.

As for space, arrays θ^i, $\bar{\theta}^i$ and A take $O(3n) = O(n)$ space in total. Note that merging k sorted arrays into A can be done in place [12]. At each iteration i, we create B of size $O(n + n_i)$, and C, S, T of size $O(n_i)$ each. Fortunately, we only need these arrays within the ith iteration, so we can delete them in the end (line 31 of the Algorithm 1) and reuse the space freed. To store the indices $l(i, j)$, we need $O(n)$ space, which again can be reallocated when i changes. Thus the amount of space used by our algorithm is $O(n)$.

An implementation of this algorithm is available on-line [22].

Remark 3. In Sect. 3, we will want to compute the colourful simplicial depth of the data points themselves. This can be done by computing $\hat{D}(x, P \setminus \{x\})$ and counting colourful simplices which have x as a vertex. This is the number of pairs of vertices of some other colour, and can be computed in linear time.

3   Computing Colourful Simplicial Medians

3.1   Preliminaries

Consider a family of sets $P^1, P^2, \ldots, P^k \in \mathbb{R}^2$, $k \geq 3$, where each P^i consists of the points of some particular colour i. Define $n_i = |P^i|$, for $i = 1, \ldots, k$. Let P

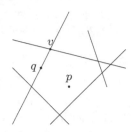

Fig. 3. An example of a cell

be the union of all colour sets: $P = \bigcup\limits_{i=1}^{k} P^i$. Recall that we denote the CSD of a point $x \in \mathbb{R}^2$ relative to P by $\hat{D}(x, P)$.

Our objective is to find a point x inside the convex hull of P, denoted $conv(P)$, maximizing $\hat{D}(x, P)$. Call the depth of such a point $\hat{\mu}(P)$. Let S be the set of line segments formed by all possible pairs of points (A, B), where $A \in P^i$, $B \in P^j$, $i < j$. The following lemma (from [3]) is here adapted to a colourful setting:

Lemma 2. *To find a point with maximum colourful simplicial depth it suffices to consider the intersection points of the colourful segments in S.*

Proof. The segments of S partition $conv(P)$ into cells[1] of dimension 2, 1, 0 of constant colourful simplicial depth [7]. Consider a 2-dimensional cell. Let p be a point in the interior of this cell, q a point on the interior of an edge and v a vertex, so that q and v belong to the same line segment (Fig. 3). Then the following inequality holds: $\hat{D}(p, P) \leq \hat{D}(q, P) \leq \hat{D}(v, P)$, since any colourful simplices containing p also contain q, and any containing q also contain v.

Let $col(A)$ denote the colour of a point A. We store the segments in S as pairs of points: $s = (A, B)$, $col(A) < col(B)$. It is helpful to view each segment as directed, i.e. a vector, with A as the tail and B as the head. Each segment s extends to a directed line h dividing $\mathbb{R}^2$ into two open half-spaces: s^+ and s^-, where s^+ lies to the right of the vector s, and s^- to the left (Fig. 4). We denote the set of lines generated by segments by H, so that every segment $s \in S$ corresponds to a line $h \in H$.

We call the intersection points of the segments in S *vertices*. Note that drawing the colourful segments is equivalent to generating a rectilinear drawing of the complete graph K_n with a few edges removed (the monochrome ones). Thus, unless the points are concentrated in a single colour class, the Crossing Lemma (see e.g. [18]) shows that we will have $\Theta(n^4)$ vertices. Computing the CSD of each of these points gives an $O(n^4 \log n)$ algorithm for finding a simplicial median.

To improve this, we follow Aloupis et al. [3], and compute the monochrome simplicial depth of most vertices based on values of their neighbours and information about the half-spaces of local segments.

---

[1] Some points of $conv(P)$ may fall outside any cell.

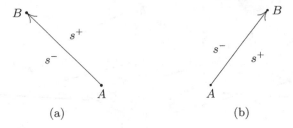

Fig. 4. s^+ and s^- of the segment $s = (A, B)$

Denote the number of points in s^+ that have colours different from the endpoints of s by $r(s)$, and those in s^- by $l(s)$. Let $r^i(s)$ and $l^i(s)$ be the number of points of a colour i in s^+ and s^- respectively. Let $\bar{r}^i(s)$ and $\bar{l}^i(s)$ be the number of points of all k colours except for the colour i in s^+ and s^- respectively. So for segment $s = (A, B)$, we have quantities as follows $\bar{r}^{col(A)}(s) = \sum_{\substack{i=1, \\ i \neq col(A)}}^{k} r^i(s)$,

$\bar{l}^{col(A)}(s) = \sum_{\substack{i=1, \\ i \neq col(A)}}^{k} l^i(s)$. Then it follows: $r(s) = \bar{r}^{col(A)}(s) - r^{col(B)}(s)$ and

$l(s) = \bar{l}^{col(A)}(s) - l^{col(B)}(s)$. The quantities $\bar{r}^{col(A)}(s)$ and $\bar{l}^{col(A)}(s)$, $r^{col(B)}(s)$ and $l^{col(B)}(s)$, can be obtained as byproducts of running an algorithm that computes half-space depth.

The *half-space depth* $HSD(x, P)$ of a point x relative to data set P is the smallest number of data points in a half-plane through the point x [21]. An algorithm to compute half-space depth is described by Rousseeuw and Ruts [19], it runs in $O(|P|)$ time when P is sorted around x. It calculates the number of points k_i in P that lie strictly to the left of each line formed by x and some point P_i, where x is the tail of the vector $\overrightarrow{xP_i}$. Then the number of points to the right $\overrightarrow{xP_i}$ is $|P| - k_i - 1$. These intermediate calculations are used in our algorithm.

The algorithm of [15] will, for each $P_i \in P$, sort $P \backslash \{P_i\}$ around P_i in $\Theta(|P|^2)$ time. In particular, it assigns every point $P_i \in P$ a list of indices that determine the order of points $P \backslash \{P_i\}$ in the clockwise ordering around P_i. Denote this by $List(P_i)$. These ideas allow us to compute $r(s)$ and $l(s)$ for every segment s. At every iteration i, we form arrays of sorted polar angles $\bar{\theta}^{col(P_i)}$ and $\theta^{i'}$. Together they take $O(2n) = O(n)$ space.

3.2   Computing a Median

To compute the CSD of all vertices, we carry out a topological sweep (see e.g. [9]). We begin by extending the segments in S to a set of lines H. The set V^* of intersection points of lines of H includes the $\Theta(n^4)$ vertices V which are on the interior of a pair of segments of S, points from P, and additional exterior intersections. We call points in $V^* \backslash V$ *phantom vertices*.

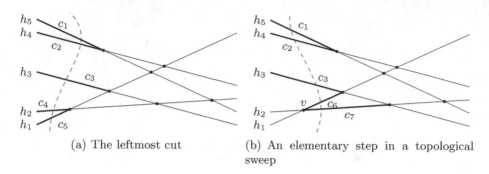

(a) The leftmost cut

(b) An elementary step in a topological sweep

Fig. 5. Topological sweep

Call a line segment of any line in H between two neighbouring vertices, or a ray from a vertex on a line that contains no further vertices an *edge*. A *topological line* is a curve in $\mathbb{R}^2$ that is topologically a line and intersects each line in H exactly once. We choose an initial topological line to be an unbounded curve that divides $\mathbb{R}^2$ into two pieces such that all the finitely many vertices in V lie on one side of the curve, by convention the right side. We call this line the *leftmost cut*. We call a *vertical cut* the list $(c_1, c_2, \ldots, c_m)$ of the $m = |H|$ edges intersecting a particular topological line. For each i, $1 \leq i \leq m - 1$, c_i and c_{i+1} share a 2-cell in the complex induced by H.

The topological sweep begins with the leftmost cut and moves across the arrangement to the right, crossing one vertex at a time. If two edges c_i and c_{i+1} of the current cut have a common right endpoint, we store the index i in the stack I. For example, in Fig. 5(a), $I = \{1, 4\}$. An *elementary step* is performed when we move to a new vertex by popping the stack I. In Fig. 5(b), we have moved past the vertex v, a common right endpoint of c_4 and c_5 which is the intersection point of h_1 and h_2. The updated stack is $I = \{1, 3\}$.

We focus on the elementary steps, because at each step we can compute the CSD of the crossed vertex. As it moves, the topological line retains the property that everything to the left of it has already been swept over. That is, if we are crossing vertex v that belongs to segment s, every vertex of the line containing s on the opposite side of the topological line prior to crossing has already been swept. For each segment $s \in S$ we store the last processed vertex and denote it by $ver(s)$, along with its CSD. Since every vertex lies at the intersection of two segments, we also store the crossing segment for s and $ver(s)$, denote it by $cross(ver(s))$. Before starting the topological sweep, for each $s \in S$ we assign $ver(s) = \emptyset$, and $cross(ver(s)) = \emptyset$. After completing an elementary step where we crossed a vertex v that lies at the intersection of s_i and s_j, we assign $ver(s_i) \leftarrow v$, $ver(s_j) \leftarrow v$, $cross(ver(s_i)) = s_j$, $cross(ver(s_j)) = s_i$.

The topological sweep skips through phantom vertices, and computes the CSD of vertices in P directly. We now explain how we process a non-phantom vertex v at an elementary step when we have an adjacent vertex already computed. Assume v is at the intersection of $s_i = \overrightarrow{AB}$ and $s_k = \overrightarrow{EF}$, see Fig. 6(a).

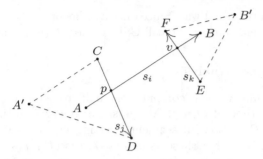

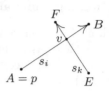

(a) Two adjacent vertices p and v and their corresponding line segments. A colourful triangle $\triangle CDA'$ contains p but not v, where $col(A') \notin \{col(C), col(D)\}$. Similarly, a colourful triangle $\triangle EFB'$ contains v but not p, where $col(B') \notin \{col(E), col(F)\}$.

(b)   Here $ver(s_i) = \emptyset$, hence $cross(ver(s_i)) = \emptyset$, and we can not run Subroutine 2.

Fig. 6. Capturing a new vertex

Without loss of generality we take $ver(s_i) = p$, where $cross(ver(s_i)) = s_j$. We view this elementary step as moving along the segment s_i from its intersection point with s_j to the one with s_k. Each intersecting segment forms a triangle with every point strictly to one side. Thus when we leave segment $s_j = (C, D)$ behind, we exit as many colourful triangles that contain p as there are points on the other side of s_j of colours different from $col(C)$ and $col(D)$. When we encounter segment $s_k = (E, F)$, we enter the colourful triangles that contain v formed by s_k and each point of a colour different from $col(E)$ and $col(F)$ on the other side of s_k. Let us denote the x and y coordinates of a point A by $A.x$ and $A.y$ respectively. Now, to compute the CSD of v knowing the CSD of p, we execute Subroutine 2.

Subroutine 2. Computing $\hat{D}(v)$ from $\hat{D}(p)$

Input: $\hat{D}(p), p, v, s_j = (C, D), s_k = (E, F)$. Output: $\hat{D}(v)$.

1: **if** $(v.x - C.x)(D.y - C.y) - (v.y - C.y)(D.x - C.x) < 0$ **then**
2:     $\hat{D}(v) \leftarrow \hat{D}(p) - r(s_j)$;
3: **else**
4:     $\hat{D}(v) \leftarrow \hat{D}(p) - l(s_j)$;
5: **end if**
6: **if** $(p.x - E.x)(F.y - E.y) - (p.y - E.y)(F.x - E.x) < 0$ **then**
7:     $\hat{D}(v) \leftarrow \hat{D}(v) + r(s_k)$;
8: **else**
9:     $\hat{D}(v) \leftarrow \hat{D}(v) + l(s_k)$;
10: **end if**

When both $ver(s_i) = \emptyset$, $ver(s_k) = \emptyset$, i.e. vertex v is the first vertex to be discovered for both segments (Fig. 6(b)), we execute $CSD(v, P)$ to find the depth, and otherwise update in the usual way. Since once a segment s has $ver(s)$ nonempty it cannot return to being empty, we call CSD at most $O(n^2)$ times.

3.3   Running Time and Space Analysis

Algorithm 3 is our main algorithm. First, it computes the half-space counts $r(s)$ and $l(s)$, which has a running time of $O(n^2)$. At the same time, we initialize the structure S that contains the colourful segments, setting $ver(s) = \emptyset$ and $cross(ver(s)) = \emptyset$ for all $s \in S$. Note that these as well as $H, List(P_i), r(s), l(s)$ require $O(n^2)$ storage.

Sorting the lines in H according to their slopes while also permuting the segments in S takes $O(n^2 \log n)$ time. We assume non-degeneracy and no vertical lines (these can use some special handling, see e.g. [10]). Computing the CSD of points where no previous vertex is available takes $O(n^2 \log n + kn^2)$ total time. The topological sweep takes linear time in the number of intersection points of

---

Algorithm 3. Computing $\underline{\beta}(P)$

Input: $P^1, \ldots, P^k, S, H, r(s), l(s)$. **Output:** $v, \underline{\beta}(P)$.

```
 1: Preprocessing: Initialize S, compute r(s),l(s);
 2: Sort H while permuting S;
 3: max ← 0;
 4: for i ← 0,n − 1 do
 5:     θ = polar angles of List(Pᵢ);
 6:     Ď(Pᵢ) ← CSD(Pᵢ, θ);
 7:     if d > max then
 8:         max ← Ď(Pᵢ);
 9:         median ← Pᵢ;
10:     end if
11: end for
12: I ← ∅;
13: Push common right endpoints of the edges of the leftmost cut onto I;
14: while I ≠ ∅ do                              ▷ Start of the topological sweep.
15:     v ← pop(I);            ▷ v lies at the intersection of sᵢ = (A,B) and sₖ = (E,F)
16:     if v lies in the interiors of sᵢ and sₖ then
17:         if ver(sᵢ) = ∅ & ver(sₖ) = ∅ then
18:             Ď(v) = CSD(v, P);
19:         else if ver(sᵢ) ≠ ∅ then
20:             Ď(v) ← Subr 2 (Ď(p),p,v,sⱼ,sₖ);      ▷ p = ver(sᵢ), sⱼ = cross(ver(sᵢ))
21:         else
22:             Ď(v) ← Subr 2 (Ď(p),p,v,sⱼ,sᵢ);      ▷ p = ver(sₖ), sⱼ = cross(ver(sₖ))
23:         end if
24:         if Ď(v) > max then
25:             max ← Ď(v);
26:             median ← v;
27:         end if
28:         ver(sᵢ) ← v, ver(sₖ) ← v, cross(ver(sᵢ)) ← sₖ, cross(ver(sₖ)) ← sᵢ;
29:     end if
30:     Push any new common right endpoints of the edges onto I;
31: end while
                                              ▷ End of the topological sweep.
32: return (median, max).
```

H, so $O(n^4)$. We do not store all the vertices, but only one per segment. Steps 15–28 (except for 18) in Algorithm 3 take O(1) time, including the calls to the Subroutine 2. As for step 18, it could happen $O(n)$ times. Therefore, the total time it will take is $O(n^2 \log n + kn^2)$. Hence overall our algorithm takes $O(n^4)$ time and needs $O(n^2)$ storage.

Algorithm 3 returns a point that has maximum colourful simplicial depth along with its CSD. It is simple to modify the algorithm to return a list of all such points if there is more than one.

4   Conclusions and Questions

Our main result is an algorithm computing the colourful simplicial depth of a point x relative to a configuration $P = (P^1, P^2, \ldots, P^k)$ of n points in $\mathbb{R}^2$ in k colour classes can be solved in $O(n \log n + kn)$ time, or in $O(kn)$ time if the input is sorted. If we assume, as seems likely, that we cannot do better without sorting the input, then for fixed k this result is optimal up to a constant factor. It is an interesting question whether we can improve the dependence on k, in particular when k is large.

Computing colourful simplicial depth in higher dimension is very challenging, in particular because there is no longer a natural (circular) order of the points. Non-trivial algorithms for monochrome depth do exist in dimension 3 [5,13], but we do not know of any non-trivial algorithms for $d \geq 4$. Algorithms for monochrome and colourful depth in higher dimension are an appealing challenge. Indeed, for $(d+1)$ colours in $\mathbb{R}^d$, it is not even clear how efficiently one can exhibit a single colourful simplex containing a given point [4,8].

Acknowledgments. This research was partially supported by an NSERC Discovery Grant to T. Stephen and by an SFU Graduate Fellowships to O. Zasenko. We thank A. Deza for comments on the presentation.

References

1. Adiprasito, K., Brinkmann, P., Padrol, A., Paták, P., Patáková, Z., Sanyal, R.: Colorful simplicial depth, Minkowski sums, and generalized Gale transforms (2016). preprint arXiv:1607.00347
2. Aloupis, G.: Geometric measures of data depth. In: Data Depth: Robust Multivariate Analysis, Computational Geometry and Applications (2006). DIMACS Ser. Discret. Math. Theoret. Comput. Sci. **72**, 147–158. Amer. Math. Soc., Providence
3. Aloupis, G., Langerman, S., Soss, M., Toussaint, G.: Algorithms for bivariate medians and a Fermat-Torricelli problem for lines. Comput. Geom. **26**(1), 69–79 (2003)
4. Bárány, I., Onn, S.: Colourful linear programming and its relatives. Math. Oper. Res. **22**(3), 550–567 (1997)
5. Cheng, A.Y., Ouyang, M.: On algorithms for simplicial depth. In: Proceedings of the 13th Canadian Conference on Computational Geometry, pp. 53–56 (2001)

6. Cormen, T.H., Leiserson, C.E., Rivest, R.L.: Introduction to Algorithms. The MIT Press and McGraw-Hill Book Company, Cambridge and New York (1989)

7. Deza, A., Huang, S., Stephen, T., Terlaky, T.: Colourful simplicial depth. Discret. Comput. Geom. **35**(4), 597–615 (2006)

8. Deza, A., Huang, S., Stephen, T., Terlaky, T.: The colourful feasibility problem. Discret. Appl. Math. **156**(11), 2166–2177 (2008)

9. Edelsbrunner, H., Guibas, L.J.: Topologically sweeping an arrangement. J. Comput. System Sci. **38**(1), 165–194 (1989). 18th Annual ACM Symposium on Theory of Computing (Berkeley, CA, 1986)

10. Edelsbrunner, H., Mücke, E.P.: Simulation of simplicity: a technique to cope with degenerate cases in geometric algorithms. In: Proceedings of the Fourth Annual Symposium on Computational Geometry (Urbana, IL, 1988), pp. 118–133. ACM, New York (1988)

11. Fukuda, K., Rosta, V.: Data depth and maximum feasible subsystems. In: Avis, D., Hertz, A., Marcotte, O. (eds.) Graph Theory and Combinatorial Optimization, vol. 8, pp. 37–67. Springer, New York (2005)

12. Geffert, V., Gajdoš, J.: Multiway in-place merging. Theoret. Comput. Sci. **411**(16–18), 1793–1808 (2010)

13. Gil, J., Steiger, W., Wigderson, A.: Geometric medians. Discret. Math. **108**(1–3), 37–51 (1992)

14. Khuller, S., Mitchell, J.S.B.: On a triangle counting problem. Inform. Process. Lett. **33**(6), 319–321 (1990)

15. Lee, D.T., Ching, Y.T.: The power of geometric duality revisited. Inform. Process. Lett. **21**(3), 117–122 (1985)

16. Liu, R.Y.: On a notion of data depth based on random simplices. Ann. Stat. **18**(1), 405–414 (1990)

17. Matousek, J., Wagner, U.: On Gromov's method of selecting heavily covered points. Discret. Comput. Geom. **52**(1), 1–33 (2014)

18. Pach, J., Radoičić, R., Tardos, G., Tóth, G.: Improving the crossing lemma by finding more crossings in sparse graphs. Discret. Comput. Geom. **36**(4), 527–552 (2006)

19. Rousseeuw, P.J., Ruts, I.: Bivariate location depth. Appl. Stat.: J. Roy. Stat. Soc. Ser. C **45**(4), 516–526 (1996)

20. Sarrabezolles, P.: The colourful simplicial depth conjecture. J. Combin. Theor. Ser. A **130**, 119–128 (2015)

21. Tukey, J.W.: Mathematics and the picturing of data. In: Proceedings of the International Congress of Mathematicians (Vancouver, B. C., 1974), vol. 2, pp. 523–531. Canad. Math. Congress, Montreal, Que. (1975)

22. Zasenko, O.: Colourful simplicial depth in the plane (2016). https://github.com/olgazasenko/ColourfulSimplicialDepthInThePlane. Accessed 13 July 2016

23. Zuo, Y., Serfling, R.: General notions of statistical depth function. Ann. Stat. **28**(2), 461–482 (2000)

Realizability of Graphs as Triangle Cover Contact Graphs

Shaheena Sultana$^{(\boxtimes)}$ and Md. Saidur Rahman

Graph Drawing and Information Visualization Laboratory,
Department of Computer Science and Engineering,
Bangladesh University of Engineering and Technology (BUET),
Dhaka 1000, Bangladesh
zareefas.sultana@gmail.com, saidurrahman@cse.buet.ac.bd

Abstract. Let $S = \{p_1, p_2, \ldots, p_n\}$ be a set of pairwise disjoint geometric objects of some type and let $C = \{c_1, c_2, \ldots, c_n\}$ be a set of closed objects of some type with the property that each element in C covers exactly one element in S and any two elements in C can intersect only on their boundaries. We call an element in S a *seed* and an element in C a *cover*. A *cover contact graph* (CCG) consists of a set of vertices and a set of edges where each of the vertex corresponds to each of the covers and each edge corresponds to a connection between two covers if and only if they touch at their boundaries. A *triangle cover contact graph* ($TCCG$) is a cover contact graph whose cover elements are triangles. In this paper, we show that every Halin graph has a realization as a $TCCG$ on a given set of collinear seeds. We introduce a new class of graphs which we call super-Halin graphs. We also show that the classes super-Halin graphs, cubic planar Hamiltonian graphs and $a \times b$ grid graphs have realizations as $TCCG$s on collinear seeds. We also show that every complete graph has a realization as a $TCCG$ on any given set of seeds. Note that only trees and cycles are known to be realizable as CCGs and outerplanar graphs are known to be realizable as $TCCG$s.

1 Introduction

Let $S = \{p_1, p_2, \ldots, p_n\}$ be a set of pairwise disjoint geometric objects of some type and $C = \{c_1, c_2, \ldots, c_n\}$ be a set of closed objects of some type with the property that each element in C covers exactly one element in S and any two elements in C can intersect only on their boundaries. We call an element in S a *seed* and an element in C a *cover*. The seeds may be points, disks or triangles and covering elements may be disks or triangles. The *cover contact graph (CCG)* consists of a set of vertices and a set of edges where each of the vertex corresponds to each of the covers and each edge corresponds to a connection between two covers if they touch at their boundaries. In other words, two vertices of a cover contact graph are adjacent if and only if the corresponding cover elements touch at their boundaries. Note that the vertices of the cover contact graph are in one-to-one correspondence to both seeds and covering objects. In a cover contact

© Springer International Publishing AG 2016
T.-H.H. Chan et al. (Eds.): COCOA 2016, LNCS 10043, pp. 393–407, 2016.
DOI: 10.1007/978-3-319-48749-6_29

graph, if disks are used as covers then it is called a *disk cover contact graph* and if triangles are used as covers then it is called a *triangle cover contact graph (TCCG)*. Figure 1(b) depicts the disk cover contact graph induced by the disk covers in Fig. 1(a), whereas Fig. 1(d) depicts the triangle cover contact graph induced by the triangle covers in Fig. 1(c).

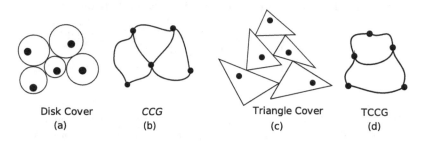

Disk Cover	CCG	Triangle Cover	TCCG
(a)	(b)	(c)	(d)

Fig. 1. Illustration for CCG and $TCCG$

There are several works [6,8,10–12] in the area of geometric optimization where the problem is how to cover geometric objects such as points by other geometric objects such as disks. The main goal is to minimize the radius of a set of k disks to cover n input points. Applications of such covering problems are found in geometric optimization problems such as facility location problems [11,12]. There are also lot of works [4,12] related to optimization of covering geometric objects. Abellanas *et al.* [1] illustrated *coin placement problem*, which is NP-complete. They tried to cover the n points using n disks (each having different radius) by placing each disk in the center position at one of the points so that no two disks overlap. Further Abellanas *et al.* [2] considered another related problem. They showed that for a given set of points in the plane, it is also NP-complete to decide whether there are disjoint disks centered at the points such that the contact graph of the disks is connected.

Atienza *et al.* [3] introduced the concept of cover contact graphs where they considered a problem which they called "realization problem." They gave some necessary conditions and then showed that it is NP-hard to decide whether a given graph can be realized as a disk cover contact graph if the correspondence between vertices and point seeds is given. They also showed that every tree and cycle have realizations as $CCGs$ on a given set of collinear point seeds. Recently, Iqbal *et al.* [7] worked on triangle cover contact graphs ($TCCGs$) where the seeds are points and the covers are triangles. First they considered the set of seeds which are in general position, i.e., no two seeds lie on a vertical line and they gave an $O(n \log n)$ algorithm to construct a 3-connected $TCCG$ of the set of seeds. They also gave an $O(n \log n)$ algorithm to construct a 4-connected $TCCG$ for a given set of six or more seeds. Addressing the realization problem, they gave an algorithm that realizes a given outerplanar graph as a $TCCG$ for a given set of seeds on a line.

In this paper, addressing the realization problem we show that every Halin graph has a realization as a $TCCG$ on a given set of collinear seeds. We introduce a new class of graphs which we call super-Halin graphs. We also show that the classes super-Halin graphs, cubic planar Hamiltonian graphs and $a \times b$ grid graphs have realizations as $TCCG$s on collinear seeds. We also show that every complete graph has a realization as a $TCCG$ on any given set of seeds.

The remaining of the paper is organized as follows. Section 2 presents some definitions and preliminary results. Section 3 gives an algorithm that realizes a given Halin graph as a $TCCG$. This section also presents a new class of graphs which we call super-Halin graphs and gives an algorithm that realizes a super-Halin graph as a $TCCG$. Two algorithms that realize a given Hamiltonian graph and an $a \times b$ grid graph as $TCCG$s are given in Sects. 4 and 5, respectively. In Sect. 6 we present an algorithm that realizes a given complete graph as a $TCCG$. Finally, Sect. 7 contains concluding remarks and directions for further research in this field.

2   Preliminaries

In this section we present some terminologies and definitions which will be used throughout the paper. For the graph theoretic definitions which have not been described here, see [5,9].

A graph is *planar* if it can be embedded in the plane without edge crossing except at the vertices where the edges are incident. A *plane graph* is a planar graph with a fixed planar embedding. A plane graph divides the plane into connected regions called *faces*. The unbounded region is called the *outer face*; the other faces are called *inner faces*. The cycle lies on the outer face is called *outer cycle*. We denote the outer cycle of G by $C_o(G)$. The edges in the outer cycle is called *outer edges*.

A graph G is *connected* if there is a path between any two distinct vertices u and v in G. A graph which is not connected is called a *disconnected graph*. Let $G = (V, E)$ be a connected simple graph with vertex set V and edge set E. A *subgraph* of a graph $G = (V, E)$ is a graph $G' = (V', E')$ such that $V' \subseteq V$ and $E' \subseteq E$. If G' contains all the edges of G that join vertices in V', then G' is called the subgraph *induced* by V'. The *connectivity* $\kappa(G)$ of a graph G is the minimum number of vertices whose removal results in a disconnected graph or a single-vertex graph. We say that G is k-connected if $\kappa(G) \geq k$. A vertex v in G is a *cut-vertex* if the removal of v results in a disconnected graph. A *biconnected component* is a maximal biconnected subgraph.

Let $S = \{p_1, p_2, \ldots, p_n\}$ be a set of pairwise disjoint geometric objects of some type and $C = \{c_1, c_2, \ldots, c_n\}$ be a set of closed objects of some type with the property that each element in C covers exactly one element in S and any two elements in C can intersect only on their boundaries. We call an element in S a *seed* and an element in C a *cover*. The seeds may be points, disks or triangles and covering elements may be disks or triangles. The *cover contact graph (CCG)* induced by C is the contact graph of the elements of C, that is, the graph $G = (C, E)$ with $E = \{\{C_i, C_j\} \subseteq C \mid C_i \neq C_j, C_i \cap C_j \neq \emptyset\}$. In other words, two vertices of a cover contact graph are adjacent if and only if the corresponding cover elements touch at their boundaries. Note that the vertices of the cover contact graph are in one-to-one correspondence to both seeds and covering objects. In a cover contact graph, if disks are used as covers, then the cover contact graph is called a *disk cover contact graph* and if triangles are used as covers, then it is called a *triangle cover contact graph (TCCG)*. Figure 1(c) shows a triangle cover of seeds and Fig. 1(d) shows the resulting *TCCG*.

A *Halin graph* is a connected planar graph that consists of an unrooted tree and a cycle connecting the end vertices of the tree. A Halin graph is constructed as follows. Let T be a tree with more than three vertices embedded in the plane. Then a Halin graph is constructed by adding to T a cycle through each of its leaves, such that the augmented graph remains planar. We call T the *core tree* of the Halin graph. Figure 2(a) shows a Halin graph.

Let T be an ordered rooted tree with root r. If T consists only of r, then r is the preorder traversal of T. Otherwise, suppose that $T_1, T_2, \ldots, T_n$ are the subtrees at r from left to right in T. The preorder traversal begins by visiting r. It continues by traversing T_1 in preorder, then T_2 in preorder, and so on, until T_n is traversed in preorder.

A *super-Halin graph* is a connected graph in which every biconnected component is a Halin graph. Figure 2(b) shows a super-Halin graph where G_1, G_2, G_3 and G_4 are Halin graphs. T_1, T_2, T_3 and T_4 are core trees of corresponding G_1, G_2, G_3 and G_4, respectively. C_1 is the cut-vertex by which G_1, G_2 and G_3 are connected and C_2 is the cut-vertex by which G_3 and G_4 are connected.

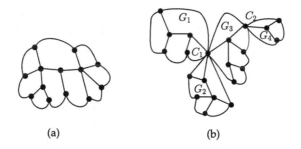

(a)   (b)

Fig. 2. (a) A Halin graph and (b) a super-Halin graph.

A path is a *Hamiltonian path* if its vertices are distinct and span V. A *cycle* is a path with at least three vertices such that its first vertex is the same as the last vertex. A cycle is a *Hamiltonian cycle* if it traverses every vertex of G exactly once. A graph is *Hamiltonian* if it has a Hamiltonian cycle. A graph is *cubic* if each of its vertex is of degree 3. A cubic planar graph which has a Hamiltonian cycle is called a *cubic planar Hamiltonian graph*. Figure 3(a) shows a cubic planar Hamiltonian graph.

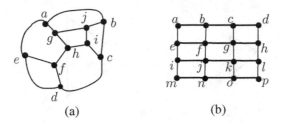

Fig. 3. (a) A cubic planar Hamiltonian graph and (b) an $a \times b$ grid graph.

A two-dimensional grid graph, also known as a *square grid graph*, is an $a \times b$ *grid graph* $G_{a,b}$ that is the graph Cartesian product $P_a \times P_b$ of path graphs on a and b vertices. Figure 3(b) shows an $a \times b$ grid graph where $a = b$.

3   Realizability of Halin Graphs and Super-Halin Graphs

In this section we show that a Halin graph has a realization as a triangle cover contact graph (TCCG) on a given set of seeds on a line as in the following theorem.

Theorem 1. *Let G be a Halin graph of n vertices. Let S be a set of n seeds aligned on a straight line. Then G is realizable on S as a TCCG in $O(n \log n)$ time.*

We give a constructive proof of Theorem 1. Here we cover each seed p by a triangle $\mathcal{T}$ such that p is the bottommost point of $\mathcal{T}$ and one side of $\mathcal{T}$ lies on the vertical line $x = x(p)$. Here $x(p)$ denotes the x-coordinate of p point. We define three corner points of $\mathcal{T}$ by α, β, γ where $\alpha = p$, β lies on the line $x = x(p)$ and γ is an arbitrary point that lies in the left half-plane of the line $x = x(p)$. Figure 4(a) illustrates an example of a triangle described above. We call this triangle the *covering triangle* of p in our algorithm. Sometimes we change covering triangle $\mathcal{T}$ by adjusting γ points arbitrarily, β points by shifting them along line $x = x(p)$ and α points by shifting them vertically downwardly on line $x = x(p)$ such that the definition of a covering triangle is not violated as shown in Fig. 4(b).

Proof. Let $G = (V, E)$ be a Halin graph with the core tree T. We make T an ordered rooted tree by taking an arbitrary leaf as the root of T and ordering the children of each vertex in a counter clockwise order which reflect the plane embedding of G as shown in Fig. 5(c). We then obtain a vertex ordering of T by preorder traversal. Let $O = v_1, v_2, \ldots, v_n$ be the vertex ordering of the vertices of T obtained by preorder traversal. Note that v_1 is the root of T. Let $p_1, p_2, \ldots, p_n$ be the seeds of S sorted according to their x-coordinates.

We cover each seed p_i by a covering triangle $\mathcal{T}_i$ corresponds to the vertex v_i of T as follows. We denote α, β and γ points of $\mathcal{T}_i$ by α_i, β_i and γ_i, respectively.

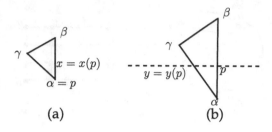

Fig. 4. Covering triangle T of seed p.

Let $x(\alpha)$ and $y(\alpha)$ be the x-coordinate and y-coordinate of α point, respectively and let $L_{\alpha\beta}$ be the line that passes through the points α and β.

We cover p_1 by T_1 corresponding to v_1 so that $x(\gamma_1) < x(\beta_1)$ and $x(\gamma_1) < x(\alpha_1)$ and $y(\gamma_1) > y(\alpha_1)$ and $y(\gamma_1) < y(\beta_1)$. We then cover p_i $(1 < i \leq n)$ by T_i according to v_i $(1 < i \leq n)$. Here γ_i of T_i corresponding to v_i which are children of T lies on $L_{\alpha_i \beta_i}$ of T_i corresponding to v_i which are parents of that children. β_i of T_i lies on $x(p_i), y(p_i) + q$ where q is sufficient small constant and α_i of T_i lies on p_i as shown in Fig. 5(d). This ensures that the triangle corresponding to a non-root vertex of T touches the triangle corresponding to its parent. Clearly this can be done safely so that T_i $(1 < i \leq n)$ does not touch or overlap with any unwanted triangles. One can observe that the resulting $TCCG$ is a tree that corresponds to the ordered rooted tree T.

We are now going to realize the edges on $C_o(G)$. Let v_p and v_q be two leaves of T such that $1 < p < q < n$ in O and $(v_p, v_q) \in E(G)$. We now realize (v_p, v_q) as follows. We have the following two cases to consider.

Case 1: *The leaves v_p and v_q have the same parent.* In this case we adjust T_p. We shift β_p on the line $L_{\alpha_q \gamma_q}$ so that T_p touches to T_q. For example, in Fig. 5(e), since v_7 and v_8 are two children of the same parent v_3, we adjust T_7 by shifting β_7 on the line $L_{\alpha_8 \gamma_8}$.

Case 2: *The leaves v_p and v_q have different parents.* We first consider the case where v_p and v_q are consecutive in O. In this case we adjust T_p. We shift β_p on the line $L_{\alpha_q \gamma_q}$ so that T_p touches to T_q. For example, in Fig. 5(e), we adjust T_1 by shifting β_1 on the line $L_{\alpha_5 \gamma_5}$. We now consider the other case, that is, v_p and v_q are not consecutive in O. Let v_k be the parent of v_q. In this case we adjust T_p and T_q. We also adjust T_k corresponding to v_k which is the parent of v_q. We put γ_q on α_p by shifting α_p vertically downward, then shift α_k on $L_{\beta_q \gamma_q}$ so that other triangles do not touch or overlap. For example, in Fig. 5(e), since v_8 and v_9 are two children of different parents i.e., v_3 is the parent of v_8 and v_2 is the parent of v_9, we put γ_9 on α_8 by shifting α_8 vertically downward, then shift α_2 on $L_{\beta_9 \gamma_9}$.

To complete the proof we now need to show the realization of two edges of $C_o(G)$ incident to v_1. Clearly (v_1, v_n) is one such edge. Let the other edge be (v_1, v_r).

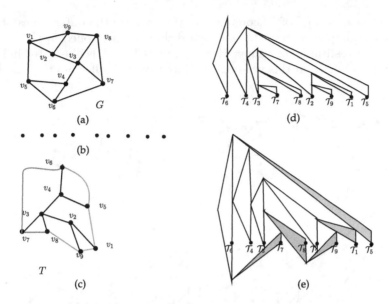

Fig. 5. (a) A Halin graph G, (b) a set of seeds S, (c) an ordered rooted tree T, (d) realization of T as $TCCG$ on S and (e) realization of G as $TCCG$ on S.

We now realize (v_1, v_r) as follows. Let v_l be the parent of v_r. In this case we adjust $\mathcal{T}_1$ and $\mathcal{T}_r$. We also adjust $\mathcal{T}_l$ corresponding to v_l which is the parent of v_r. We put γ_r on α_1 by shifting α_1 vertically downward, then shift α_l on $L_{\beta_r \gamma_r}$ so that other triangles do not touch or overlap with any unwanted triangles. For example, in Fig. 5(e), since v_6 and v_7 are adjacent in G and v_3 is the parent of v_7, we put γ_7 on α_6 by shifting α_6 vertically downward, then shift α_3 on $L_{\beta_7 \gamma_7}$.

We now realize (v_1, v_n) as follows. In this case we adjust $\mathcal{T}_n$. We put γ_n on β_1 such that shifting β_i of the previous triangles which were touched with $\mathcal{T}_n$ upward. For example, in Fig. 5(e), since v_6 and v_5 are adjacent in G, we put γ_5 on β_6 such that shifting β_4 and β_1 of $\mathcal{T}_4$ and $\mathcal{T}_1$ which were touched with $\mathcal{T}_5$ upward.

Since G is a simple graph there is no multi edge. For each edge (v_p, v_q) we ensure $\mathcal{T}_p$, $\mathcal{T}_q$ and $\mathcal{T}_k$ do not contact with any unwanted triangles. Hence G is realizable on S.

All steps of this algorithm can be done in $O(n)$ time, but for sorting we need $O(n \log n)$ time. So the overall time complexity is $O(n \log n)$. □

We now consider the realizability of super-Halin graphs. Note that a super-Halin graph is a connected graph in which every biconnected component is a Halin graph. We show that a super-Halin graph has a realization as a triangle cover contact graph (TCCG) on a given set of seeds on a line. We first construct a tree by deleting the outer edges of the super-Halin graph. We then make the tree an ordered rooted tree by taking an arbitrary cut-vertex of the super-Halin graph as the root of the tree. We realize the ordered rooted tree by using the technique mentioned in the proof of Theorem 1. We then realize the outer edges of each biconnected component by extending the technique mentioned in the proof of Theorem 1. A realization of a super-Halin graph as $TCCG$ is illustrated in Fig. 6. The detail is omitted in this version. We thus have the following theorem.

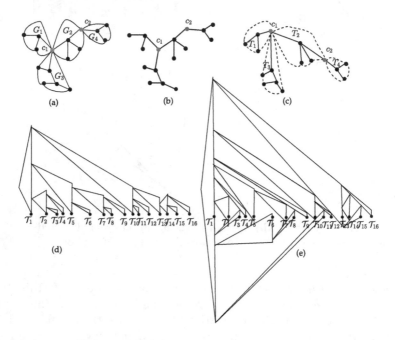

Fig. 6. (a) A super-Halin graph, (b) a tree where core trees of Halin graphs are connected by cut-vertices of the graph, (c) an ordered rooted tree with outer edges of the graph, (d) realization of the tree as $TCCG$ on S and (e) realization of the super-Halin graph.

Theorem 2. *Let G be a super-Halin graph of n vertices. Let S be a set of n seeds aligned on a straight line. Then G is realizable on S as a $TCCG$ in $O(n \log n)$ time.*

4 Realizability of Hamiltonian Graphs

In this section, we show that cubic planar Hamiltonian graphs have realizations as triangle cover contact graphs (TCCG) on a given set of seeds on a line. We have the following theorem.

Theorem 3. *Let G be a cubic planar Hamiltonian graph of n vertices. Let S be a set of n seeds aligned on a straight line. Then G is realizable on S as a $TCCG$ in $O(n \log n)$ time.*

Proof. We give a constructive proof. Let $G = (V, E)$ be a cubic planar Hamiltonian graph and Let C be a Hamiltonian cycle of G illustrated in Fig. 7(c). Let $v_1, v_2, \ldots, v_n$ be the vertices on C in clockwise order taking the starting vertex v_1 arbitrary. Let $p_1, p_2, \ldots, p_n$ be the seeds of S sorted according to their x-coordinates. We fix the vertices v_i of the cycle C on p_i of S as illustrated in Fig. 7(d). We thus find a plane embedding of G which we realize as $TCCG$. In this embedding an edge of G which is not on C is either lie inside the region bounded by C or outside the region bounded by C. We call an edge of G which is inside of the region bounded by C an *internal edge* of C and call an edge which is outside of the region an *external edge* of C.

We cover each seed p_i by a covering triangle T_i corresponding to the vertex v_i of C as follows. We denote α, β and γ points of T_i by α_i, β_i and γ_i, respectively. Let $x(\alpha)$ and $y(\alpha)$ be the x-coordinate and y-coordinate of α point, respectively and let $L_{\alpha\beta}$ be the line that passes through the points α and β.

We cover p_1 by T_1 corresponding to v_1 so that $x(\gamma_1) < x(\beta_1)$ and $x(\gamma_1) < x(\alpha_1)$ and $y(\gamma_1) > y(\alpha_1)$ and $y(\gamma_1) < y(\beta_1)$. We then cover p_i $(1 < i < n)$ by T_i where γ_i lies on $L_{\alpha_{i-1}\beta_{i-1}}$ and β_i lies on $x_{p_i}, y_{p_i} + q$ where q is a sufficient small constant. This ensures that triangle T_i touches T_{i+1} $(1 < i < n - 1)$. We now

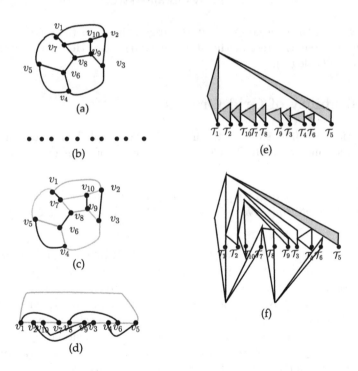

(a)

(b)

(c)

(d)

(e)

(f)

Fig. 7. (a) A cubic planar Hamiltonian graph G, (b) a set of collinear seeds S, (c) Hamiltonian cycle in G, (d) Hamiltonian cycle with external edges and internal edges in S, (e) realization of Hamiltonian cycle as $TCCG$ on S and (f) realization of G as $TCCG$ on S.

cover p_n by T_n so that it touches T_1 and T_{n-1}. The point γ_n lies on γ_1 we then shift β_{n-1} on $L_{\alpha_n\gamma_n}$. Clearly this can be done safely so that $T_2, T_3 \ldots, T_{n-2}$ does not touch or overlap line $L_{\alpha_n\gamma_n}$. One can observe that the resulting $TCCG$ is a cycle that corresponds to the Hamiltonian cycle C of G as shown in Fig. 7(e).

We are now going to realize the internal edges and external edges of C. Let v_s and v_t be two vertices of C such that $s < t$ and $(v_s, v_t) \in E(G)$. We now realize (v_s, v_t) as follows. We have the following two cases to consider.

Case 1: *The (v_s, v_t) is an internal edge.* We put γ_t on α_s by shifting α_s vertically downward. Then shift α_s of the previous triangle of T_t on $L_{\beta_t\gamma_t}$ so that other triangles do not touch or overlap with unwanted triangles. For example, We put γ_6 on α_8 by shifting α_8 vertically downward. Then shift α_4 on $L_{\beta_6\gamma_6}$ as illustrated in Fig. 7(f).

Case 2: *The (v_s, v_t) is an external edge.* We put γ_t on β_s by shifting β_s vertically upward. If T_t is previously adjusted then β_s touches on $L_{\alpha_t\gamma_t}$ by shifting β_s vertically upward. For example, We put γ_3 on β_2 by shifting β_2 vertically upward as illustrated in Fig. 7(f).

Since G is a simple graph there is no multi-edge. For each edge (v_i, v_{i+1}) we ensure $T_i, T_{i+1} \ldots, T_n$ do not touch any unwanted triangles. Hence G is realizable on S.

We now prove the time complexity. Sorting of the seeds takes $O(n \log n)$ time. One can implement the rest of the steps in $O(n)$ time. Hence the overall time complexity is $O(n \log n)$. □

5   Realizability of Grid Graphs

In this section, we now show that an $a \times b$ grid graph has a realization as a triangle cover contact graph (TCCG) on a given set of seeds on a line as in the following theorem.

Theorem 4. *Let G be an $a \times b$ grid graph of n vertices. Let S be a set of n seeds aligned on a straight line. Then G is realizable on S as a TCCG in $O(n \log n)$ time.*

Proof. We give a constructive proof. Let $G = (V, E)$ be an $a \times b$ grid graph. We find out a path on the $a \times b$ grid graph G starting from the upper-left vertex, across the top row to the upper-right vertex, then down one vertex, then across the second row to the left, then down one vertex, then across to the right again, and so on. We call this path a *zig-zag path*. This path is shown by the gray lines in Fig. 8(c). Let the vertices of the $a \times b$ grid graph be denoted $v_1, v_2, \ldots, v_n$ in the order that they encountered on the zig-zag path taking the starting vertex of this path is v_1 which is upper-left vertex of G. Let $p_1, p_2, \ldots, p_n$ be the seeds of S sorted according to their x-coordinates. We set the vertices v_i of the zig-zag path on p_i of S as illustrated in Fig. 8(d). We thus find a plane embedding of G which we realize as $TCCG$. In this embedding an edge of G which is not on the zig-zag path is either upper plane of the line or lower plane of the line. We call an edge of G which is in the upper plane of the line an *upward-edge* and call an edge which is in the lower plane of the line an *downward-edge* of G. The zig-zag path, upward-edges and downward- edges are shown in Fig. 8(d).

We cover each seed p_i by a covering triangle T_i corresponds to the vertex v_i of G as follows. We denote α, β and γ points of T_i by α_i, β_i and γ_i, respectively. Let $x(\alpha)$ and $y(\alpha)$ be the x-coordinate and y-coordinate of α point, respectively and let $L_{\alpha\beta}$ be the line that passes through the points α and β.

We cover p_1 by T_1 corresponding to v_1 so that $x(\gamma_1) < x(\beta_1)$ and $x(\gamma_1) < x(\alpha_1)$ and $y(\gamma_1) > y(\alpha_1)$ and $y(\gamma_1) < y(\beta_1)$. We then cover p_i $(1 < i \leq n)$ by T_i where γ_i lies on $L_{\alpha_{i-1}\beta_{i-1}}$ and β_i lies on $x_{p_i}, y_{p_i} + q$ where q is a sufficient small constant. This ensure that triangle T_i touches T_{i+1} $(1 < i < n - 1)$. One can observe that the resulting $TCCG$ is a path that corresponds to the zig-zag path of G as illustrated in Fig. 8(e).

We are now going to realize upward-edges and downward-edges of G. For this purpose we adjust the covering triangles by traversing vertices $v_1, v_2 \ldots, v_n$ in this order of the path. Let (v_s, v_t) be an edge such that $s < t$ which to be realized. We adjust covering triangles T_s and T_t for realizing the edge (v_s, v_t) as follows. (Initially no triangle is adjusted.) We have the following two cases to consider.

Case 1: *The edge (v_s, v_t) is an upward-edge.* Depending on whether there is a downward-edge incident to v_s or not, T_s might be preadjusted. We thus have the following two cases to consider to realize all upward-edges: (1) T_s and T_t are not adjusted and (2) T_s is adjusted downward and T_t is not adjusted. We first consider the case where T_s and T_t are not adjusted. In this case we need to adjust the triangles T_s and T_t. We put γ_t on $L_{\alpha_s\beta_s}$, then shift β_{t-1} on $L_{\alpha_t\gamma_t}$ so that $T_{s+1}, \ldots, T_{t-2}$ do not touch or overlap the line $L_{\alpha_t\gamma_t}$. For example, in Fig. 8(f), both triangles T_1 and T_8 are not adjusted. To adjust these triangles, we put γ_8 on $L_{\alpha_1\beta_1}$ and then shift β_7 on $L_{\alpha_8\gamma_8}$ so that the triangles between these two triangles do not touch or overlap the line $L_{\alpha_8\gamma_8}$. We now consider the other case, that is, T_s is adjusted downward and T_t is not adjusted. In this case we also need to adjust the triangles T_s and T_t. We put γ_t on $L_{\alpha_s\beta_s}$, then shift β_{t-1} on $L_{\alpha_t\gamma_t}$ so that $T_{s+1}, \ldots, T_{t-2}$ do not touch or overlap the line $L_{\alpha_t\gamma_t}$, then we mark T_t adjusted. For example, in Fig. 8(f), T_{10} is adjusted downward but T_{15} is not adjusted. To adjust T_{10}, we put γ_{15} on $L_{\alpha_{10}\beta_{10}}$ and then shift β_{14} on $L_{\alpha_{15}\gamma_{15}}$ so that triangles between these two triangles do not touch or overlap the line $L_{\alpha_{15}\gamma_{15}}$.

Case 2: *The edge (v_s, v_t) is a downward-edge.* Depending on whether there is an upward-edge incident to v_s or not, T_s might be preadjusted. We thus have the following two cases to consider to realize all downward-edges: (1) T_s and T_t are not adjusted and (2) T_s is adjusted upward and T_t is not adjusted. We first consider the case where T_s and T_t are not adjusted. In this case we need to adjust the triangles T_s and T_t. We put γ_t on $L_{\alpha_s\beta_s}$ by shifting α_s vertically downward, then shift β_{t-1} on $L_{\alpha_t\gamma_t}$ so that $T_{s+1}, \ldots, T_{t-2}$ do not touch or overlap the line $L_{\alpha_t\gamma_t}$. For example, in Fig. 8(f), both triangles T_5 and T_{12} are not adjusted. To adjust these triangles, we put γ_{12} on $L_{\alpha_5\beta_5}$ by shifting α_5 vertically downward and then shift β_{11} on $L_{\alpha_{12}\gamma_{12}}$ so that the triangles between these two triangles do not touch or overlap the line $L_{\alpha_{12}\gamma_{12}}$. We now consider the other case, that is, $\mathcal{T}_s$ is adjusted upward and $\mathcal{T}_t$ is not adjusted. In this case we also need to adjust the triangles $\mathcal{T}_s$ and $\mathcal{T}_t$. We put γ_t on $L_{\alpha_s\beta_s}$ by shifting α_s vertically downward, then shift β_{t-1} on $L_{\alpha_t\gamma_t}$ so that $\mathcal{T}_{s+1}, \ldots, \mathcal{T}_{t-2}$ do not touch or overlap the line $L_{\alpha_t\gamma_t}$. For example, in Fig. 8(f), triangle $\mathcal{T}_7$ is adjusted upward and $\mathcal{T}_{10}$ is not adjusted. To adjust these triangles, we put γ_{10} on $L_{\alpha_7\beta_7}$ by shifting α_7 vertically downward and then shift β_9 on $L_{\alpha_{10}\gamma_{10}}$ so that the triangles between these two triangles do not touch or overlap the line $L_{\alpha_{10}\gamma_{10}}$. Hence G is realizable on S.

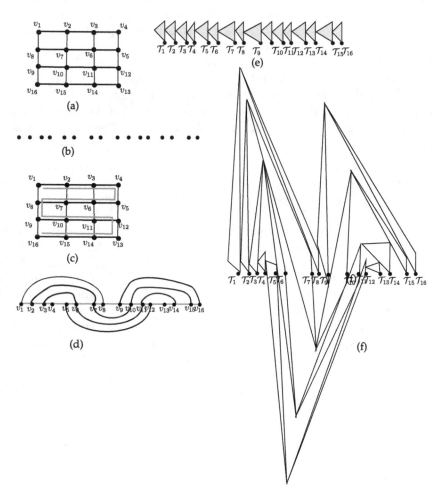

Fig. 8. (a) An $a \times b$ grid graph G, (b) a set of collinear seeds S, (c) the zig-zag path in G, (d) the zig-zag path, upward-edges and downward-edges on S, (e) realization of the zig-zag path as $TCCG$ on S and (f) realization of some upward-edges and downward-edges of G as $TCCG$ on S.

All steps of this algorithm except sorting of the seeds can be implemented in $O(n)$ time. Since sorting of the seeds takes $O(n \log n)$ time, the overall time complexity is $O(n \log n)$.   □

6 Realizability of Complete Graphs

In this section, we show that a complete graph has a realization as a triangle cover contact graph (TCCG) on any given set of seeds as in the following theorem.

Theorem 5. *Let G be a complete graph of n vertices and let S be a set of n seeds. Then G is realizable on S as a TCCG in $O(n \log n)$ time.*

Proof. We give a constructive proof. Let $G = (V, E)$ be a complete graph. Assume for the simplicity that no two seeds in S are on a horizontal or on a vertical line. If the point set does not satisfy the condition we can simply rotate the coordinate plane so that the point set satisfies the condition. Let p_1 be the left seed and p_n be the right seed among the seeds of S. Let $v_1, v_2, \ldots, v_n$ be the vertices of G. Let $p_2, p_3 \ldots, p_{n-1}$ be the seeds of S sorted according to their y-coordinates. We set the vertices v_i of G on p_i of S.

Our idea is to cover each seed p_i by a triangle $\mathcal{T}_i$ such that p_i is inside of $\mathcal{T}_i$ and one side of $\mathcal{T}_i$ lies on the vertical line $x = x(p_i) + k$. Here $x(p_i)$ denotes the x-coordinate of point p_i and k is a constant. We define three corner points of $\mathcal{T}_i$

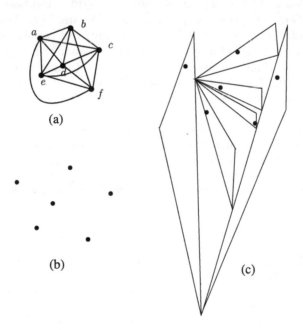

(a)

(b)

(c)

Fig. 9. (a) A complete graph G, (b) a set of seeds S and (c) realization of G as $TCCG$ on S.

by α_i, β_i, γ_i where α_i, β_i lies on the line $x = x(p_i) + k$ and γ_i is an arbitrary point.

We cover each seed p_i by a covering triangle T_i corresponds to the vertex v_i of G as follows. We denote α, β and γ points of T_i by α_i, β_i and γ_i, respectively. Let $x(\alpha)$ and $y(\alpha)$ be the x-coordinate and y-coordinate of α point, respectively and let $L_{\alpha\beta}$ be the line that passes through the points α and β. Here each triangle covers one seed.

We cover p_1 by T_1 corresponding to v_1 so that $x(\gamma_1) < x(\beta_1)$ and $x(\gamma_1) < x(\alpha_1)$ and $y(\gamma_1) > y(\alpha_1)$ and $y(\gamma_1) < y(\beta_1)$. We now cover p_n by T_n so that it touches T_1. The point γ_n lies on α_1. We then cover p_i $(1 < i < n)$ by T_i where γ_i meets at a point and lies on $L_{\alpha_1\beta_1}$ and α_i lies on $L_{\beta_n\gamma_n}$. This ensure that every triangle T_i touches with each other where $(1 \geq i \leq n)$. One can observe that the resulting $TCCG$ is a complete graph as shown in Fig. 9(c). Hence G is realizable on S.

All steps of the algorithm except sorting of the seeds can be implemented in $O(n)$ time. Since sorting of the seeds takes $O(n \log n)$ time, the overall time complexity is $O(n \log n)$. $\qquad\square$

7   Conclusion

In this paper we have shown that every Halin graph has a realization as a $TCCG$ on a given set of collinear seeds. We also have shown that every super-Halin graph, cubic planar Hamiltonian graph and $a \times b$ grid graph have a realization each as a $TCCG$ on a given set of collinear seeds. We also have shown that every complete graph has a realization as a $TCCG$ on any given set of seeds. The result is interesting since $TCCG$ of a complete graph is planar although the complete graph is non planar. This triggers a research direction to investigate which classes of non planar graphs have $TCCG$ representations. It is also interesting to know which larger classes of graphs are realizable as $TCCG$s.

Acknowledgement. This work is done in Graph Drawing and Information Visualization Laboratory, Department of Computer Science and Engineering, Bangladesh University of Engineering and Technology as a part of a Ph.D. research work. The first author is supported by ICT Fellowship of ICT division, Ministry of Posts, Telecommunications and IT, Government of the People's Republic of Bangladesh.

References

1. Abellanas, M., Bereg, S., Hurtado, F., Olaverri, A.G., Rappaport, D., Tejel, J.: Moving coins. Comput. Geom. **34**(1), 35–48 (2006)
2. Abellanas, M., Castro, N., Hernández, G., Márquez, A., Moreno-Jiménez, C.: Gear System Graphs. Manuscript (2006)
3. Atienza, N., Castro, N., Cortés, C., Garrido, M.A., Grima, C.I., Hernández, G., Márquez, A., Moreno, A., Nöllenburg, M., Portillo, J.R., Reyes, P., Valenzuela, J., Trinidad Villar, M., Wolff, A.: Cover contact graphs. J. Comput. Geom. **3**(1), 102–131 (2012)

4. De Fraysseix, H., de Mendez, P.O.: Representations by contact and intersection of segments. J. Algorithmica **47**(4), 453–463 (2007)
5. Di Battista, G., Eades, P., Tamassia, R., Tollis, I.G.: Graph Drawing: Algorithms for the Visualization of Graphs. Prentice-Hall Inc., Upper Saddle River (1999)
6. Durocher, S., Mehrabi, S., Skala, M., Wahid, M.A.: The cover contact graph of discs touching a line. In: CCCG, pp. 59–64 (2012)
7. Hossain, M.I., Sultana, S., Moon, N.N., Hashem, T., Rahman, M.S.: On triangle cover contact graphs. In: Rahman, M.S., Tomita, E. (eds.) WALCOM 2015. LNCS, vol. 8973, pp. 323–328. Springer, Heidelberg (2015). doi:10.1007/978-3-319-15612-5_29
8. Koebe, P.: Kontaktprobleme der konformen abbildung. Ber. Sächs. Akad. Wiss. Leipzig Math.-Phys. Klasse **88**(1–3), 141–164 (1936)
9. Nishizeki, T., Rahman, M.S.: Planar Graph Drawing. Lecture Notes Series on Computing. World Scientific, Singapore (2004)
10. Pach, J., Agarwal, P.K.: Combinatorial Geometry. Wiley, New York (1995)
11. Robert, J.M., Toussaint, G.T. (eds.): Computational geometry and facility location. In: Proceedings of the International Conference on Operations Research and Management Science, vol. 68 (1990)
12. Welzl, E.: Smallest enclosing disks (balls and ellipsoids). In: Maurer, I.H. (ed.) New Results and New Trends in Computer Science. LNCS, vol. 555, pp. 359–370. Springer, Heidelberg (1991)

A Quadratic Time Exact Algorithm for Continuous Connected 2-Facility Location Problem in Trees (Extended Abstract)

Wei Ding[1(✉)] and Ke Qiu[2]

[1] Zhejiang University of Water Resources and Electric Power,
Hangzhou 310018, Zhejiang, China
dingweicumt@163.com
[2] Department of Computer Science, Brock University,
St. Catharines, Canada
kqiu@brocku.ca

Abstract. This paper studies the **continuous connected 2-facility location problem (CC2FLP)** in trees. Let $T = (V, E, c, d, \ell, \mu)$ be an undirected rooted tree, where each node $v \in V$ has a weight $d(v) \geq 0$ denoting the demand amount of v as well as a weight $\ell(v) \geq 0$ denoting the cost of opening a facility at v, and each edge $e \in E$ has a weight $c(e) \geq 0$ denoting the cost on e as well as is associated with a function $\mu(e, t) \geq 0$ denoting the cost of opening a facility at a point $x(e, t)$ on e where t is a continuous variable on e. Given a subset $\mathcal{D} \subseteq V$ of clients, and a subset $\mathcal{F} \subseteq \mathcal{P}(T)$ of continuum points admitting facilities where $\mathcal{P}(T)$ is the set of all the points on edges of T, when two facilities are installed at a pair of continuum points x_1 and x_2 in $\mathcal{F}$, the total cost involved by CC2FLP includes three parts: the cost of opening two facilities at x_1 and x_2, K times the cost of connecting x_1 and x_2, and the cost of all the clients in $\mathcal{D}$ connecting to some facility. The objective is to open two facilities at a pair of continuum points in $\mathcal{F}$ to minimize the total cost, for a given input parameter $K \geq 1$. This paper considers the case of $\mathcal{D} = V$ and $\mathcal{F} = \mathcal{P}(T)$. We first study the discrete version of CC2FLP, named the **discrete connected 2-facility location problem (DC2FLP)**, where two facilities are restricted to the nodes of T, and devise a quadratic time edge-splitting algorithm for DC2FLP. Furthermore, we prove CC2FLP is almost equivalent to DC2FLP in trees, and develop a quadratic time exact algorithm based on the edge-splitting algorithm.

Keywords: Continuous connected 2-facility · Tree · Edge-splitting · Quadratic time

1 Introduction

The **connected facility location problem (CFLP)** has a wide range of applications in the design of telecommunication networks and data management problems on networks, and has attracted considerable attention both from the theoretical computer science community [5–11,15] and from the operations research community [2,12–14].

© Springer International Publishing AG 2016
T.-H.H. Chan et al. (Eds.): COCOA 2016, LNCS 10043, pp. 408–420, 2016.
DOI: 10.1007/978-3-319-48749-6_30

The CFLP can be described as follows: given an undirected graph $G = (V, E)$ with each edge $e \in E$ having a cost $c_e \geq 0$, a set $\mathcal{F} \subseteq V$ of *facilities*, a set $\mathcal{D} \subseteq V$ of *client* nodes and a parameter $K \geq 1$. Each facility $i \in \mathcal{F}$ has an opening cost $f_i \geq 0$, and each client $j \in \mathcal{D}$ has a demand amount $d_j \geq 0$. The objective is to open a subset $F \subseteq \mathcal{F}$ of facilities, assign each client $j \in \mathcal{D}$ to some facility $i_j \in F$ as well as use a *shortest path* in G to connect j and i_j, and connect all the open facilities using a Steiner tree T such that the total cost $\sum_{i \in F} f_i + K \sum_{e \in T} c_e + \sum_{j \in \mathcal{D}} d_j \rho(j, i_j)$ is minimized, where $\rho(j, i_j)$ is the j-to-i_j shortest path distance in G. The CFLP is the combination and generalization of the **uncapacitated facility location problem (UFLP)** and the **Steiner tree problem (STP)**, and is both NP-hard and APX-complete [5].

Karger and Minkoff [11] introduced CFLP and presented the first approximation algorithm of a constant factor. Gupta *et al.* [6] devised a 10.66-approximation algorithm by rounding LP of an exponential size. In [8], Gupta *et al.* obtained a 9.01-approximation algorithm by random facility sampling. Swamy and Kumar [15] developed an 8.55-approximation algorithm by integrating the primal-dual approaches for the facility location problem and the Steiner tree problem. In [7], Gupta *et al.* conjectured whether Swamy and Kumar's algorithm can be improved by a randomized sampling approach. Eisenbrand *et al.* [5] presented a 4-approximation randomized algorithm, and thus answered this question affirmatively. Later, Hasan *et al.* [9] gave an 8.29-approximation LP rounding algorithm, and Jung *et al.* [10] obtained a 6.55-approximation algorithm by improving Step 1 of Swamy and Kumar's algorithm. Moreover, a variety of heuristics have been proposed, *e.g.*, a variable neighborhood search by Ljubić [13] which combines reactive tabu search and branch-and-cut approach, and a dual-based local search by Bardossy and Raghavan [2], and etc.

A variant of CFLP with the number of open facilities being at most p, named the **connected p-facility location problem (CpFLP)**, has been studied recently. Swamy and Kumar [15] first introduced CpFLP and proposed a 15.55-approximation algorithm. Eisenbrand *et al.* [5] presented a randomized 6.85-approximation algorithm.

The previous results focused on the discrete version of CFLP, named the **discrete connected facility location problem (DCFLP)**, where all the facilities are restricted to the nodes of G. In practice, however, it commonly occurs that the facilities are installed on the edges, *e.g.*, the emergency service centers of a city are constructed on the roads of communication network of the city. Provided that the number of open facilities is at most p, we can model the scenario above as the **continuous connected p-facility location problem (CCpFLP)**. Let $\mathcal{F} \subseteq \mathcal{P}(G)$ denote the set of candidate points on edges of admitting facilities. When opening facilities at $F \subseteq \mathcal{F}$, every client $j \in \mathcal{D}$ is assigned to some facility $x_j \in F$, j connects to x_j via a *shortest path* in G, and all the open facilities in F are connected by a Steiner tree T_s. Let f_x be the cost of opening a facility at $x \in F$, $\sigma(j, x_j)$ be the j-to-x_j shortest path distance in G, and $c(T_s)$ be the cost of T_s. CCpFLP can be formulated as

$$\min_{F \in \mathcal{F}, |F| \leq p} \sum_{x \in F} f_x + Kc(T_s) + \sum_{j \in \mathcal{D}} d_j \sigma(j, x_j). \tag{1}$$

As discussed in [5], CCpFLP is both NP-hard [4] and APX-complete [1,14] since it contains STP as a special case. This paper focuses on the case of $\mathcal{D} = V$ and $\mathcal{F} = \mathcal{P}(G)$.

The CCpFLP is denoted by (C, p)-CFLP, and the case $p = 2$ of CCpFLP is denoted by CC2FLP or (C, 2)-CFLP. The paper deals with (C, 2)-CFLP in trees and presents a *nontrivial* quadratic time exact algorithm. First, we consider the discrete version of CC2FLP, that is DC2FLP and denoted by (D, 2)-CFLP, in trees. We examine an important property that an optimum to (D, 2)-CFLP in trees is the combination of two optimums to the **discrete towed 1-facility location problem (DT1FLP)** over two subtrees obtained by removing one edge of tree, respectively. We design a linear time algorithm for DT1FLP using *complementary dynamic programming (CDP)* in trees, proposed by Ding and Xue [3], which consists of a *bottom-up dynamic programming (BUDP)* and a *top-down dynamic programming (TDDP)*. The dynamic computation in trees goes on from bottom up to root in BUDP while from root down to bottom in TDDP. The combination of the CDP algorithm for DT1FLP and related properties on (D, 2)-CFLP leads to a quadratic time *edge-splitting* algorithm for (D, 2)-CFLP. Next, we study (C, 2)-CFLP. We prove that the **continuous towed 1-facility location problem (CT1FLP)** in trees is equivalent to DT1FLP, and (C, 2)-CFLP in trees is almost equivalent to (D, 2)-CFLP other than seldom exceptional cases. Based on the edge-splitting algorithm for (D, 2)-CFLP and the asymptotic equivalence, we develop a quadratic time exact algorithm for (C, 2)-CFLP in trees.

The rest of this paper is organized as follows. In Sect. 2, we define DT1FLP and DC2FLP formally, and design a linear time algorithm by using CDP for DT1FLP in trees as well as a quadratic time edge-splitting algorithm for DC2FLP in trees. In Sect. 3, we define CT1FLP and CC2FLP formally. In Sect. 4, we prove the equivalence between DT1FLP and CT1FLP in trees. In Sect. 5, we obtain the asymptotic equivalence between DC2FLP and CC2FLP in trees, and further develop a quadratic time exact algorithm for CC2FLP. In Sect. 6, we conclude this paper.

2   Optimal Discrete Connected 2-Facility in Trees

In this section, we will first design a linear time algorithm by using CDP for DT1FLP in trees, and then develop a quadratic time edge-splitting algorithm for DC2FLP in trees based on the linear time algorithm.

2.1   Problem Statements and Notations

Let $T = (V, E, c, d, \ell)$ be an undirected tree, where V is the *node set* and E is the *edge set*, each edge $e \in E$ has a weight $c(e) \geq 0$ denoting the *cost* on e, each node

$v \in V$ has one weight $\ell(v) \geq 0$ denoting the cost of *opening a facility* at v and the other weight $d(v) \geq 0$ denoting the *demand amount* of v. Let $\langle v, u \rangle$ denote a *node pair* of T, and V^2 be the set of all the node pairs. For any $\langle v, u \rangle \in V^2$, we use $\pi(v, u)$ to denote the unique path on T connecting v and u, and use $c(v, u)$ to denote the cost of $\pi(v, u)$, which is equal to the sum of costs on all the edges of $\pi(v, u)$, *i.e.*, $c(v, u) = \sum_{e \in \pi(v, u)} c(e)$. Clearly, $c(v, u) = c(u, v)$. Specifically, $c(v, v) = 0, \forall v \in V$. Suppose T is a *rooted* tree with root node r throughout the paper since an unrooted tree always can be transformed into a rooted tree by designating a node of tree as its root.

We consider that every node of V is a *client* and a *center* is fixed at node α. A facility is called a *discrete towed 1-facility (DT1F)*, if when it is installed at node $v \in V$ the total cost involved includes three parts: the cost of opening a facility at v, K times the cost of connecting v to α, and the cost of connecting all the clients to v. The cost of connecting any client $u \in V$ to v is equal to $d(u)$ times $c(v, u)$. Let $F_1(v)$ denote the cost of opening a DT1F at v. We have

$$F_1(v) = \ell(v) + Kc(v, \alpha) + \sum_{u \in V} d(u)c(v, u), \quad \forall v \in V. \tag{2}$$

Problem 1. Given $T = (V, E, c, d, \ell)$, a fixed center $\alpha \in V$ and a constant $K \geq 1$, the **discrete towed 1-facility location problem (DT1FLP)** asks to find a node of T at which the cost of opening a DT1F is minimized.

Let v^* be an optimal DT1F of T, *i.e.*,

$$F_1(v^*) = \min_{v \in V} F_1(v). \tag{3}$$

When a pair of facilities is installed at $\langle v_1, v_2 \rangle$, every client $u \in V$ gets services from the *closer* facility to itself (or say, u connects to the closer facility) in order to reduce the service cost. A pair of facilities is called a *discrete connected 2-facility (DC2F)*, if when it is installed at $\langle v_1, v_2 \rangle$ the total cost involved includes three parts: the cost of opening a pair of facilities at $\langle v_1, v_2 \rangle$, K times the cost of connecting v_1 with v_2, and the cost of connecting all clients to some facility. The cost of connecting u to some facility is equal to $d(u)$ times the minimum of $c(v_1, u)$ and $c(v_2, u)$. For any $\langle v_1, v_2 \rangle \in V^2$, we use $F_2(v_1, v_2)$ to denote the cost of opening a pair of facilities at $\langle v_1, v_2 \rangle$. We have

$$
\begin{aligned}
&F_2(v_1, v_2) \\
&= \ell(v_1) + \ell(v_2) + Kc(v_1, v_2) + \sum_{u \in V} d(u) \min\{c(v_1, u), c(v_2, u)\}.
\end{aligned} \tag{4}
$$

Problem 2. Given $T = (V, E, c, d, \ell)$ and a constant $K \geq 1$, the **discrete connected 2-facility location problem (DC2FLP)**, abbreviated to $(D, 2)$-CFLP, asks to find a node pair of T at which the cost of opening a DC2F is minimized.

Let $\langle v_1^*, v_2^* \rangle$ be an optimal DC2F of T, *i.e.*,

$$F_2(v_1^*, v_2^*) = \min_{\langle v_1, v_2 \rangle \in V^2} F_2(v_1, v_2). \tag{5}$$

2.2   A Linear Time Algorithm for DT1FLP in Trees

In this subsection, we devise a linear time exact algorithm for DT1FLP in $T = (V, E, c, d, \ell)$ by using CDP.

We claim by Eq. (3) that an optimal DT1F, v^*, of T can be derived from computing all the values of $F_1(v), v \in V$ and then finding the minimum. By Eq. (2), we let $F_1(v) = F_1^{(1)}(v) + F_1^{(2)}(v)$, where

$$F_1^{(1)}(v) = \sum_{u \in V} d(u)c(v, u), \tag{6}$$

and

$$F_1^{(2)}(v) = \ell(v) + Kc(v, \alpha). \tag{7}$$

For every $v \in V$, the value of $F_1^{(2)}(v)$ is easy to be figured out since the value of $\ell(v)$ is known as part of input of DT1CP and the value of $c(v, \alpha)$ can be obtained by using preprocessing procedure PreDT1FLP. The idea of PreDT1FLP is described roughly as follows: it uses DFS (*depth-first search*) to tour T with α as origin, and the *tree distance* $c(\alpha, v)$ is set to $c(\alpha, u) + c(u, v)$ when a new node v is visited at the first time via the arc from u to v. In fact, PreDT1FLP calls DFS once, and so PreDT1FLP produces all the tree distances with α as origin in $O(|V|)$ time.

Next, we compute all the values of $F_1^{(1)}(v), v \in V$. For any $v \in V$, we let $S(v)$ be the set of children of v and $p(v)$ be the parent of v. Let $T(v)$ be the subtree of T rooted at v. We denote by $V^+(v)$ the set of nodes in $T(v)$, and by $V^-(v)$ the set of nodes outside $T(v)$. Let $f^+(v)$ (*resp.* $f^-(v)$) denote the cost of all the nodes in $V^+(v)$ (*resp.* $V^-(v)$) connecting to v, i.e., $f^+(v) = \sum_{u \in V^+(v)} d(u)c(v, u)$ and $f^-(v) = \sum_{u \in V^-(v)} d(u)c(v, u)$. Let $d^+(v)$ (*resp.* $d^-(v)$) denote the total demand amount from $V^+(v)$ (*resp.* $V^-(v)$), i.e., $d^+(v) = \sum_{u \in V^+(v)} d(u)$ and $d^-(v) = \sum_{u \in V^-(v)} d(u)$. Given any $T = (V, E, c, d, \ell)$, the total demand amount from all the nodes in V is certainly a constant, denoted by $\mathbb{D}$. Clearly, $\mathbb{D} = d^+(r)$. In addition, we let $X \uplus Y$ denote the union of two disjoint sets X and Y. Theorem 1 shows a partition scheme of T, see Fig. 1.

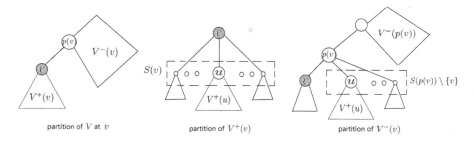

Fig. 1. The partition scheme of T at v.

Theorem 1 (see [3]). *For any $v \in V$, we have*

$$V = V^+(v) \uplus V^-(v), \tag{8}$$

where

$$V^+(v) = \{v\} \uplus \left\{ \biguplus_{u \in S(v)} V^+(u) \right\}, \tag{9}$$

and

$$V^-(v) = \{p(v)\} \uplus V^-(p(v)) \uplus \left\{ \biguplus_{u \in S(p(v)) \setminus \{v\}} V^+(u) \right\}. \tag{10}$$

Specifically, $S(v) = \emptyset$ and so $V^+(v) = \{v\}, d^+(v) = d(v)$ for each leaf v of T while $p(r) = $ null and so $V^-(r) = \emptyset$. From Eq. (9), it follows that

$$d^+(v) = d(v) + \sum_{u \in S(v)} \sum_{w \in V^+(u)} d(w) = d(v) + \sum_{u \in S(v)} d^+(u). \tag{11}$$

By Eq. (8), $d^-(v) = \mathbb{D} - d^+(v)$. Based on Eq. (11), we can use BUDP to compute all the values of $d^+(v)$ and $d^-(v)$, which can be incorporated into PreDT1FLP.

Theorem 2. *For any $v \in V$, we have*

$$F_1^{(1)}(v) = f^+(v) + f^-(v), \tag{12}$$

where

$$f^+(v) = \sum_{u \in S(v)} \left(f^+(u) + d^+(u)c(v, u) \right), \tag{13}$$

and

$$f^-(v) = f^-(p(v)) + f^+(p(v)) - f^+(v) + (d^-(v) - d^+(v))c(v, p(v)). \tag{14}$$

By Eq. (12), we can first compute the values of $f^+(v)$ and $f^-(v)$, and then get $F_1^{(1)}(v)$, for all $v \in V$. According to Eqs. (13) and (14), all the values of $d^+(v)$ and $d^-(v), v \in V$ need to be obtained in advance, which is done by PreDT1FLP. We claim from Eq. (13) that we can use BUDP on T to compute all the values of $f^+(v)$, and from Eq. (14) we can use TDDP on T to compute all the values of $f^-(v)$. This idea results in a CDP procedure, described as MainProcedure of our algorithm for DT1FLP. Therefore, the combination of PreDT1FLP and MainProcedure forms our algorithm for DT1FLP on $T = (V, E, c, d, \ell)$, called AlgDT1FLP.

In practice, we can implement AlgDT1FLP based on the search of tree. AlgDT1FLP uses BUDP twice and TDDP once. The BUDP procedure in AlgDT1FLP can be implemented by DFS as follows: tour T with r as origin. For any node $v \in V$, the values of $d^+(v)$ and $f^+(v)$ are figured out when the tour goes backward from v. The TDDP procedure in AlgDT1FLP can be implemented by BFS (*breadth-first search*) as follows: tour T with r as origin, and compute $f^-(v)$ when every $v \in V \setminus \{r\}$ is visited at the first time. The time complexity of AlgDT1FLP is shown in Theorem 3.

Theorem 3. *Given $T = (V, E, c, d, \ell)$ with n nodes, a fixed center $\alpha \in V$ and a constant $K \geq 1$, AlgDT1FLP can find an optimal DT1F of T in $O(n)$ time.*

2.3   An Edge-Splitting Algorithm for (D, 2)-CFLP in Trees

In this subsection, we present a quadratic time edge-splitting algorithm for (D, 2)-CFLP in $T = (V, E, c, d, \ell)$ based on AlgDT1FLP.

All the nodes in T are labelled from bottom up to root r and so r has the highest level. For any $e \in E$, the endpoint of e lying on a lower (*resp.* higher) level is denoted by e^+ (*resp.* e^-). So, $e = \{e^+, e^-\}$. We simplify $V^+(e^+)$ and $V^-(e^+)$ to $V^+(e)$ and $V^-(e)$, respectively. Evidently, all the nodes in $V^+(e)$ reach e^+ without passing through e while all the nodes in $V^-(e)$ reach e^+ by passing through e. The node pair of V with $v_1 \in V^+(e)$ and $v_2 \in V^-(e)$ is denoted by $\langle v_1, v_2 \rangle_e$, and all such node pairs form a set denoted by $V^2(e)$. Unless specified otherwise, we always suppose $v_1 \in V^+(e)$ and $v_2 \in V^-(e)$ for all $e \in E$. Clearly, $\langle v_1, v_2 \rangle$ is interchangeable while $\langle v_1, v_2 \rangle_e$ is not interchangeable. In addition, if one edge $\widetilde{e} \in \pi(v_1, v_2)$ satisfies that $c(v_1, \widetilde{e}^+) < c(v_2, \widetilde{e}^+)$ and $c(v_1, \widetilde{e}^-) \geq c(v_2, \widetilde{e}^-)$, it is called the *threshold edge* of $\langle v_1, v_2 \rangle$ and denoted by $t(v_1, v_2)$. All the node pairs with e as their threshold edge form a set, denoted by $V_t^2(e)$.

Lemma 1. $V^2 = \bigcup_{e \in E} V_t^2(e)$.

A tree can be partitioned into two subtrees by removing one edge. The edge removed from T is called a *splitting edge* of T. For any $\langle v_1, v_2 \rangle \in V^2$ and any $e \in \pi(v_1, v_2)$, when e is designated as the splitting edge, the objective function of DT1FLP over the subtree including v_1 and a fixed center e^+ is

$$F_1^+(v_1; e) = \ell(v_1) + Kc(v_1, e^+) + \sum_{u \in V^+(e)} d(u)c(v_1, u), \qquad (15)$$

and that over the subtree including v_2 and a fixed center e^- is

$$F_1^-(v_2; e) = \ell(v_2) + Kc(v_2, e^-) + \sum_{u \in V^-(e)} d(u)c(v_2, u). \qquad (16)$$

Below are several fundamental lemmas, which form the basis of our quadratic time edge-splitting algorithm for (D, 2)-CFLP.

Lemma 2. *Given any $\langle v_1, v_2 \rangle \in V^2$ and any $e \in \pi(v_1, v_2)$, we have*

$$\begin{aligned} &F_1^+(v_1; e) + F_1^-(v_2; e) + Kc(e) \\ &\geq F_1^+(v_1; t(v_1, v_2)) + F_1^-(v_2; t(v_1, v_2)) + Kc(t(v_1, v_2)). \end{aligned} \qquad (17)$$

Lemma 3. *Given any $\langle v_1, v_2 \rangle \in V^2$, we have*

$$F_2(v_1, v_2) = F_1^+(v_1; t(v_1, v_2)) + F_1^-(v_2; t(v_1, v_2)) + Kc(t(v_1, v_2)). \qquad (18)$$

Theorem 4. *Given $T = (V, E, c, d, \ell)$, an optimal DC2F, $\langle v_1^*, v_2^* \rangle$, of T satisfies that*

$$F_2(v_1^*, v_2^*) = \min_{e \in E} \left\{ \min_{v_1 \in V^+(e)} F_1^+(v_1; e) + \min_{v_2 \in V^-(e)} F_1^-(v_2; e) + Kc(e) \right\}. \quad (19)$$

According to Theorem 4, we need to compute the values of $F_1^+(v_1^*; e)$ and $F_1^-(v_2^*; e)$ for all $e \in E$ in order to compute the value of $F_2(v_1^*, v_2^*)$. For any $e \in E$, during the computation of $F_1^+(v_1^*; e)$, none of the nodes in $V^-(e)$ is involved while all the nodes in $V^+(e)$ are involved. Thus, we call AlgDT1FLP$(T(e^+), e^+, K)$ to compute $F_1^+(v_1^*; e)$. During the computation of $F_1^-(v_2^*; e)$, none of the nodes in $V^+(e)$ is involved while all the nodes in $V^-(e)$ are involved. So, it is certain that e is not visited during the tour on T. Thus, we call AlgDT1FLP(T, e^-, K) with e forbidden to visit to compute $F_1^-(v_2^*; e)$. The edge minimizing the sum of $F_1^+(v_1^*; e)$, $F_1^-(v_2^*; e)$ and $Kc(e)$ is an optimal splitting edge e^*. The node pair $\langle v_1^*, v_2^* \rangle$, where v_1^* and v_2^* minimize $F_1^+(v_1; e^*)$ and $F_1^-(v_2; e^*)$ respectively, is an optimal DC2F. This idea leads to our edge-splitting algorithm for $(D, 2)$-CFLP, called AlgDC2FLP. The time complexity of AlgDC2FLP is shown in Theorem 5.

Theorem 5. *Given $T = (V, E, c, d, \ell)$ with n nodes and a constant $K \geq 1$, AlgDC2FLP can find an optimal DC2F of T in $O(n^2)$ time.*

3   Optimal Continuous Connected 2-Facility in Trees

Let $T = (V, E, c, d, \ell, \mu)$ be an undirected rooted tree, where $V, E, c(\cdot)$, $d(\cdot)$ and $\ell(\cdot)$ are as defined in Sect. 2. In addition, every edge $e \in E$ has another weight denoting the cost of opening a facility on e. Let $\mathcal{P}(e)$ be the set of all the *continuum* points on e, and $\mathcal{P}(T)$ be the set of all the continuum points on edges of T. For any $e \in E$ and any point $x \in \mathcal{P}(e)$, e is partitioned into two sections at x. Suppose the cost over one section between x and e^- is t, and thus the cost over the other section between x and e^+ is $c(e) - t$. Clearly, $0 \leq t \leq c(e), \forall e \in E$. Let $x(e, t)$ be the point on e such that the cost between it and e^- is equal to t, and $\mu(e, t)$ denote the cost of opening a facility at $x(e, t)$ as follows,

$$\mu(e, t) = \left(1 - \frac{t}{c(e)} \right) \ell(e^-) + \frac{t}{c(e)} \ell(e^+), \quad \forall 0 \leq t \leq c(e). \quad (20)$$

Given a node $v \in V$ and a point $x \in \mathcal{P}(T)$, we let $\pi(x, v)$ denote the unique path on T connecting x and v, and let $c(x, v)$ denote the cost of $\pi(x, v)$. For any edge $e \in E$ and any point $x \in \mathcal{P}(e)$, $\pi(x, v)$ certainly contains one path $\pi(e^+, v)$ and the section of e between x and e^+ if v lies in $V^+(e)$ while it contains $\pi(e^-, v)$ and the section of e between x and e^- if v lies in $V^-(e)$. Thus,

$$c(x, v) = \begin{cases} c(e^-, v) - t & \text{if} \quad v \in V^+(e), \\ c(e^-, v) + t & \text{if} \quad v \in V^-(e), \end{cases} \quad \forall v \in V. \quad (21)$$

Suppose that each node of V is a client and a center is fixed at node α. A facility is called a *continuous towed 1-facility (CT1F)*, if when it is installed at point $x \in \mathcal{P}(T)$ the total cost involved includes three parts: the cost of opening a facility at x, K times the cost of connecting x to α, and the cost of connecting all the clients to x. The cost of connecting every client $v \in V$ to x is equal to $d(v)$ times $c(x, v)$. For any edge $e \in E$ and any point $x(e, t) \in \mathcal{P}(e)$, we let $\mathbb{F}_1(e, t)$ denote the cost of opening a CT1F at $x(e, t)$. We have

$$\mathbb{F}_1(e, t) = \mu(e, t) + Kc(x(e, t), \alpha) + \sum_{v \in V} d(v)c(x(e, t), v), \quad \forall 0 \le t \le c(e). \quad (22)$$

Problem 3. Given $T = (V, E, c, d, \ell, \mu)$, a fixed center $\alpha \in V$ and a constant $K \ge 1$, the **continuous towed 1-facility location problem (CT1FLP)**, asks to find a point of T at which the cost of opening a CT1F is minimized.

Let $x(e^*, t^*)$ be an optimal CT1F of T, i.e.,

$$\mathbb{F}_1(e^*, t^*) = \min_{e \in E,\ 0 \le t \le c(e)} \mathbb{F}_1(e, t). \quad (23)$$

Let $\langle x_1, x_2 \rangle$ denote a point pair of T, i.e., $x_1, x_2 \in \mathcal{P}(T)$. When a pair of facilities is installed at $\langle x_1, x_2 \rangle$, each client $v \in V$ connects to the closer facility. A pair of facilities is called a *continuous connected 2-facility (CC2F)*, if when it is installed at $\langle x_1, x_2 \rangle$ the total cost involved includes three parts: the cost of opening a pair of facilities at $\langle x_1, x_2 \rangle$, K times the cost of connecting x_1 with x_2, and the cost of connecting all the clients to some facility. For any pair of edges e_1 and e_2, given any two points $x(e_1, t_1) \in \mathcal{P}(e_1), x(e_2, t_2) \in \mathcal{P}(e_2)$, we let $\mathbb{F}_2(e_1, t_1; e_2, t_2)$ denote the cost of opening a pair of facilities at $\langle x(e_1, t_1), x(e_2, t_2) \rangle$. The objective function of Eq. (1) can rewritten to be

$$\mathbb{F}_2(e_1, t_1; e_2, t_2) = \mu(e_1, t_1) + \mu(e_2, t_2) + Kc(x(e_1, t_1), x(e_2, t_2))$$
$$+ \sum_{v \in V} d(v) \min\{c(x(e_1, t_1), v), c(x(e_2, t_2), v)\}. \quad (24)$$

Problem 4. Given $T = (V, E, c, d, \ell, \mu)$ and a constant $K \ge 1$, the **continuous connected 2-facility location problem (CC2FLP)**, abbreviated to (C, 2)-CFLP, asks to find a point pair of T at which the cost of opening a CC2F is minimized.

Let $\langle x(e_1^*, t_1^*), x(e_2^*, t_2^*) \rangle$ be an optimal CC2F of T, i.e.,

$$\mathbb{F}_2(e_1^*, t_1^*; e_2^*, t_2^*) = \min_{e_1 \in E, 0 \le t_1 \le c(e_1);\ e_2 \in E, 0 \le t_2 \le c(e_2)} \mathbb{F}_2(e_1, t_1; e_2, t_2). \quad (25)$$

In the remainder of this paper, we first study CT1FLP in trees, and then study (C, 2)-CFLP in trees. Due to page limit, the most of technical details will be omitted here.

4   Equivalence of CT1FLP in Trees

In this section, we prove the equivalence between CT1FLP and DT1FLP in trees. So, AlgDT1FLP also applies to CT1FLP in trees.

Theorem 6. *CT1FLP in* $T = (V, E, c, d, \ell, \mu)$ *is equivalent to DT1FLP in* $T = (V, E, c, d, \ell)$.

5   Algorithm for $(C, 2)$-CFLP in Trees

In this section, we develop a quadratic time exact algorithm for $(C, 2)$-CFLP in trees based on AlgDC2FLP.

Let $\langle e_1, e_2 \rangle$ be a pair of edges of T, and E^2 be the set of all the edge pairs. If e_1 and e_2 have a common node then they are called *adjacent*, and otherwise called *nonadjacent*. Let E_1^2, E_2^2 and E_3^2 be the set of all the pairs of nonadjacent edges, all the pairs of adjacent edges, and all the pairs of same edge, respectively. Clearly, $E^2 = E_1^2 \uplus E_2^2 \uplus E_3^2$. If $\langle e_1, e_2 \rangle \in E_k^2, k = 1, 2, 3$, we use $\mathbb{F}_2^{(k)}(e_1, t_1; e_2, t_2)$ to denote the objective function of $(C, 2)$-CFLP. Specifically, we have $\mathbb{F}_2^{(3)}(e, t_1; e, t_2)$ when $e_1 = e_2 = e$. Accordingly, we use $\mathbb{F}_2^{(1)}(e_{11}^*, t_{11}^*; e_{12}^*, t_{12}^*)$, $\mathbb{F}_2^{(2)}(e_{21}^*, t_{21}^*; e_{22}^*, t_{22}^*)$ and $\mathbb{F}_2^{(3)}(e_3^*, t_{31}^*; e_3^*, t_{32}^*)$ to denote the optimum, respectively.

For any point pair $\langle x_1, x_2 \rangle$ of T, there are three potential situations: they lie on a pair of nonadjacent edges, or a pair of adjacent edges or a single edge. So, we need to discuss three cases, see Fig. 2, respectively.

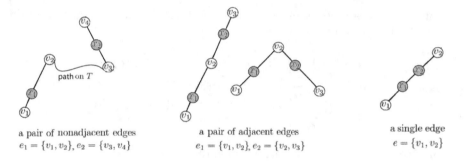

a pair of nonadjacent edges
$e_1 = \{v_1, v_2\}, e_2 = \{v_3, v_4\}$

a pair of adjacent edges
$e_1 = \{v_1, v_2\}, e_2 = \{v_2, v_3\}$

a single edge
$e = \{v_1, v_2\}$

Fig. 2. Three potential situations of a point pair $\langle x_1, x_2 \rangle$.

5.1   Case 1: A Pair of Nonadjacent Edges

In this subsection, we show a way to computing $\mathbb{F}_2^{(1)}(e_{11}^*, t_{11}^*; e_{12}^*, t_{12}^*)$. We obtain the following theorem.

Theorem 7. $\mathbb{F}_2^{(1)}(e_{11}^*, t_{11}^*; e_{12}^*, t_{12}^*) = F_2(v_1^*, v_2^*)$.

5.2   Case 2: A Pair of Adjacent Edges

In this subsection, we show a way to computing $\mathbb{F}_2^{(2)}(e_{21}^*, t_{21}^*; e_{22}^*, t_{22}^*)$. For any $\langle e_1, e_2 \rangle \in E_2^2$, we let $e_1 = \{v_1, v_2\}$ and $e_2 = \{v_2, v_3\}$. Suppose, *w.l.o.g*, that $c(e_1) \leq c(e_2)$. There are two types of the layouts of $\langle e_1, e_2 \rangle$, see Fig. 3. Let V_1, V_2, V_3 be the set of nodes which reach $x(e_1, t_1)$ and/or $x(e_2, t_2)$ by passing

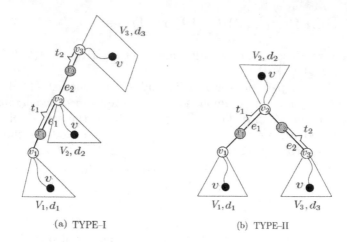

(a) TYPE–I   (b) TYPE–II

Fig. 3. A node pair $\langle x_1, x_2 \rangle$ lies on a pair of adjacent edges $\langle e_1, e_2 \rangle \in E_2^2$. There are two types of layouts of $\langle e_1, e_2 \rangle$. TYPE–I is shown in (a) and TYPE–II is shown in (b).

through v_1, v_2, v_3, and let d_1, d_2, d_3 denote the total demand amount of all the nodes in V_1, V_2, V_3, respectively. Let

$$
\begin{aligned}
\lambda_0 &= \ell(v_2) + \ell(v_3) + F_1^{(1)}(v_2) + (K - d_3)c(e_2), \\
\lambda_1 &= \frac{\ell(v_1) - \ell(v_2)}{c(e_1)} + K - d_1, \\
\lambda_2 &= \frac{\ell(v_2) - \ell(v_3)}{c(e_2)} - K + d_3.
\end{aligned}
\tag{26}
$$

For ease of presentation, we define three terms as follows:

– FALSE1 = "$\lambda_1 + d_2 < 0, \lambda_2 < 0, \lambda_1 - \lambda_2 + d_2 < 0$";
– FALSE2 = "$\lambda_1 > 0, \lambda_2 - d_2 > 0, \lambda_1 - \lambda_2 + d_2 < 0$";
– FALSE3 = "$\lambda_1 \leq 0, \lambda_2 - d_2 > 0$".

We obtain the following Lemma.

Lemma 4. *Given* $T = (V, E, c, d, \ell, \mu)$ *and any pair of adjacent edges* $\langle e_1, e_2 \rangle \in E_2^2$, *when one of FALSE1, FALSE2 and FALSE3 occurs, we infer* $\mathbb{F}_2^{(2)}(e_1, t_1; e_2, t_2)$ *gets its minimum at* $\langle e_1^+, x(e_2, c(e_2) - c(e_1)) \rangle$, *i.e.,*

$$
\mathbb{F}_2^{(2)}(e_1, t_1^*; e_2, t_2^*) = \lambda_0 + \lambda_2 c(e_2) + (\lambda_1 - \lambda_2 + d_2)c(e_1).
\tag{27}
$$

Theorem 8. *Given* $T = (V, E, c, d, \ell, \mu)$, *unless one of FALSE1, FALSE2 and FALSE3 occurs to a pair of adjacent edges of* T, *it always holds that* $\mathbb{F}_2^{(2)}(e_{21}, t_{21}^*; e_{22}, t_{22}^*) \geq F_2(v_1^*, v_2^*)$.

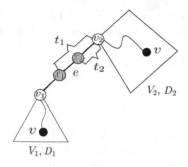

Fig. 4. A node pair lies on a single edge $e = \{v_1, v_2\}$.

5.3   Case 3: A Single Edge (see Fig. 4)

In this subsection, we show a way to computing $\mathbb{F}_2^{(3)}(e_3^*, t_{31}^*; e_3^*, t_{32}^*)$. We obtain the following theorem.

Theorem 9. $\mathbb{F}_2^{(3)}(e_3^*, t_{31}^*; e_3^*, t_{32}^*) \geq F_2(v_1^*, v_2^*)$.

5.4   A Quadratic Time Exact Algorithm

The combination of Theorems 7, 8 and 9 yields Theorem 10.

Theorem 10. $(C, 2)$-*CFLP in* $T = (V, E, c, d, \ell, \mu)$ *is equivalent to* $(D, 2)$-*CFLP in* $T = (V, E, c, d, \ell)$ *unless one of FALSE1, FALSE2 and FALSE3 occurs to a pair of adjacent edges of* $T = (V, E, c, d, \ell, \mu)$.

Let $\mathbb{F}_2^{(2)}(\overline{e}_1, \overline{t}_1^*; \overline{e}_2, \overline{t}_2^*)$ denote the minimum of all the potential values of $\mathbb{F}_2^{(2)}(e_1, t_1^*; e_2, t_2^*)$ under FALSE1, FALSE2 and FALSE3. We conclude from Theorem 10 that we can get $\mathbb{F}_2(e_1^*, t_1^*; e_2^*, t_2^*)$ and an optimal CC2F, $\langle x(e_1^*, t_1^*), x(e_2^*, t_2^*)\rangle$, by first using AlgDC2FLP to compute $F_2(v_1^*, v_2^*)$, and using Eq. (27) to compute all the potential values of $\mathbb{F}_2^{(2)}(e_1, t_1^*; e_2, t_2^*)$, and finally determining the minimum of them. If $\mathbb{F}_2(e_1^*, t_1^*; e_2^*, t_2^*)$ is equal to $F_2(v_1^*, v_2^*)$, then $\langle v_1^*, v_2^*\rangle$ is an optimal CC2F. If $\mathbb{F}_2(e_1^*, t_1^*; e_2^*, t_2^*)$ is equal to $\mathbb{F}_2^{(2)}(\overline{e}_1, \overline{t}_1^*; \overline{e}_2, \overline{t}_2^*)$, then $\langle(\overline{e}_1)^+, x(\overline{e}_2, c(\overline{e}_2) - c(\overline{e}_1))\rangle$ is an optimal CC2F by Corollary 4. This idea leads to our algorithm for $(C, 2)$-CFLP in $T = (V, E, c, d, \ell, \mu)$, called AlgCC2FLP. The time complexity of AlgCC2FLP is shown in Theorem 11.

Theorem 11. *Given* $T = (V, E, c, d, \ell, \mu)$ *with* n *nodes and a constant* $K \geq 1$, AlgCC2FLP *can find an optimal CC2F of* T *in* $O(n^2)$ *time.*

6   Conclusions

This paper first designs a quadratic time edge-splitting algorithm AlgDC2FLP for $(D, 2)$-CFLP in trees, and then develops a quadratic time exact algorithm AlgCC2FLP based on AlgDC2FLP for $(C, 2)$-CFLP in trees.

This paper deals with the case of $\mathcal{F} = \mathcal{P}(T)$ and $\mathcal{D} = V$. In essence, the approach used in this paper can be easily extended to the general case of $\mathcal{F} \subseteq \mathcal{P}(T)$ and $\mathcal{D} \subseteq V$. Furthermore, we conjecture that (C, p)-CFLP with $p > 2$ in trees admits a polynomial time exact algorithm.

References

1. Arora, S., Lund, C., Motwani, R., Sudan, M., Szegedy, M.: Proof verification and the hardness of approximation problems. J. ACM. **45**(3), 501–555 (1998)
2. Bardossy, M.G., Raghavan, S.: Dual-based local search for the connected facility location and related problems. INFORMS J. Comput. **22**(4), 584–602 (2010)
3. Ding, W., Xue, G.: A linear time algorithm for computing a most reliable source on a tree network with faulty nodes. Theor. Comput. Sci. **412**(3), 225–232 (2011)
4. Garey, M.R., Johnson, D.S.: Computers and Intractability: A Guide to the Theory of NP-Completeness. Freeman, San Francisco (1979)
5. Eisenbrand, F., Grandoni, F., Rothvoß, T., Schäfer, G.: Connected facility location via random facility sampling and core detouring. J. Comput. Syst. Sci. **76**(8), 709–726 (2010)
6. Gupta, A., Kleinberg, J., Kumar, A., Rastogi, R., Yener, B.: Provisioning a virtual private network: a network design problem for multicommodity flow. In: Proceedings of 33rd STOC, pp. 389–398 (2001)
7. Gupta, A., Kumar, A., Roughgarden, T.: Simpler and better approximation algorithms for network design. In: Proceedings of 35th STOC, pp. 365–372 (2003)
8. Gupta, A., Srinivasan, A., Tardos, É.: Cost-sharing mechanisms for network design. In: Jansen, K., Khanna, S., Rolim, J.D.P., Ron, D. (eds.) RANDOM 2004 and APPROX 2004. LNCS, vol. 3122, pp. 139–150. Springer, Heidelberg (2004)
9. Hasan, M.K., Jung, H., Chwa, K.Y.: Approximation algorithms for connected facility location problems. J. Combinatirial Optim. **16**, 155–172 (2008)
10. Jung, H., Hasan, M.K., Chwa, K.-Y.: Improved primal-dual approximation algorithm for the connected facility location problem. In: Yang, B., Du, D.-Z., Wang, C.A. (eds.) COCOA 2008. LNCS, vol. 5165, pp. 265–277. Springer, Heidelberg (2008)
11. Karger, D.R., Minkoff, M.: Building Steiner trees with incomplete global knowledge. In: Proceedings of 41st FOCS, pp. 613–623 (2000)
12. Lee, Y., Chiu, S.Y., Ryan, J.: A branch and cut algorithm for a Steiner tree-star problem. INFORMS J. Comput. **8**(3), 194–201 (1996)
13. Ljubić, I.: A hybrid VNS for connected facility location. In: Bartz-Beielstein, T., Blesa Aguilera, M.J., Blum, C., Naujoks, B., Roli, A., Rudolph, G., Sampels, M. (eds.) HCI/ICCV 2007. LNCS, vol. 4771, pp. 157–169. Springer, Heidelberg (2007)
14. Ravi, R., Sinha, A.: Approximation algorithms for problems combining facility location and network design. Oper. Res. **54**(1), 73–81 (2006)
15. Swamy, C., Kumar, A.: Primal-dual algorithms for connected facility location problems. Algorithmica **40**(4), 245–269 (2004)

Complexity and Data Structure

On the (Parameterized) Complexity
of Recognizing Well-Covered (r, ℓ)-graphs

Sancrey Rodrigues Alves[1], Konrad K. Dabrowski[2], Luerbio Faria[3],
Sulamita Klein[4], Ignasi Sau[5,6(✉)], and Uéverton dos Santos Souza[7]

[1] FAETEC, Fundação de Apoio à Escola Técn. do Estado do Rio de Janeiro,
Rio de Janeiro, Brazil
sancrey@cos.ufrj.br
[2] School of Engineering and Computing Sciences, Durham University, Durham, UK
konrad.dabrowski@durham.ac.uk
[3] UERJ, DICC, Universidade do Estado do Rio de Janeiro, Rio de Janeiro, Brazil
luerbio@cos.ufrj.br
[4] UFRJ, COPPE-Sistemas, Universidade Federal do Rio de Janeiro,
Rio de Janeiro, Brazil
sula@cos.ufrj.br
[5] CNRS, LIRMM, Université de Montpellier, Montpellier, France
ignasi.sau@lirmm.fr
[6] Departamento de Matemática, Universidade Federal do Ceará, Fortaleza, Brazil
[7] UFF, IC, Universidade Federal Fluminense, Niterói, Brazil
ueverton@ic.uff.br

Abstract. An (r, ℓ)-*partition* of a graph G is a partition of its vertex
set into r independent sets and ℓ cliques. A graph is (r, ℓ) if it admits
an (r, ℓ)-partition. A graph is *well-covered* if every maximal independent set is also maximum. A graph is (r, ℓ)-*well-covered* if it is both
(r, ℓ) and well-covered. In this paper we consider two different decision
problems. In the (r, ℓ)-WELL-COVERED GRAPH problem ((r, ℓ)WCG for
short), we are given a graph G, and the question is whether G is an
(r, ℓ)-well-covered graph. In the WELL-COVERED (r, ℓ)-GRAPH problem
(WC(r, ℓ)G for short), we are given an (r, ℓ)-graph G together with an
(r, ℓ)-partition of $V(G)$ into r independent sets and ℓ cliques, and the
question is whether G is well-covered. We classify most of these problems into P, coNP-complete, NP-complete, NP-hard, or coNP-hard. Only
the cases WC$(r, 0)$G for $r \geq 3$ remain open. In addition, we consider
the parameterized complexity of these problems for several choices of
parameters, such as the size α of a maximum independent set of the
input graph, its neighborhood diversity, or the number ℓ of cliques in an
(r, ℓ)-partition. In particular, we show that the parameterized problem
of deciding whether a general graph is well-covered parameterized by α
can be reduced to the WC$(0, \ell)$G problem parameterized by ℓ, and we
prove that this latter problem is in XP but does not admit polynomial
kernels unless coNP $\subseteq$ NP/poly.

This work was supported by FAPERJ, CNPq, CAPES Brazilian Research Agencies
and EPSRC (EP/K025090/1).

T-H.H. Chan et al. (Eds.): COCOA 2016, LNCS 10043, pp. 423–437, 2016.
DOI: 10.1007/978-3-319-48749-6_31

Keywords: Well-covered graph · (r, ℓ)-graph · coNP-completeness · FPT-algorithm · Parameterized complexity · Polynomial kernel

1   Introduction

An (r, ℓ)-*partition* of a graph $G = (V, E)$ is a partition of V into r independent sets $S^1, \ldots, S^r$ and ℓ cliques $K^1, \ldots, K^\ell$. For convenience, we allow these sets to be empty. A graph is (r, ℓ) if it admits an (r, ℓ)-partition. The P versus NP-complete dichotomy for recognizing (r, ℓ)-graphs is well known [2]: the problem is in P if $\max\{r, \ell\} \leq 2$, and NP-complete otherwise. The class of (r, ℓ)-graphs and its subclasses have been extensively studied in the literature. For instance, list partitions of (r, ℓ)-graphs were studied by Feder et al. [11]. In another paper, Feder et al. [12] proved that recognizing graphs that are both chordal and (r, ℓ) is in P.

Well-covered graphs were first introduced by Plummer [20] in 1970. Plummer defined that "a graph is said to be *well-covered* if every minimal point cover is also a minimum cover". This is equivalent to demanding that every maximal independent set has the same cardinality. The problem of recognizing a well-covered graph, which we denote by WELL-COVERED GRAPH, was proved to be coNP-complete by Chvátal and Slater [3] and independently by Sankaranarayana and Stewart [22], but is in P when the input graph is known to be claw-free [18,24].

Motivated by this latter example and by the relevance of (r, ℓ)-graphs, in this paper we are interested in recognizing graphs that are both (r, ℓ) and well-covered. We note that similar restrictions have been considered in the literature. For instance, Kolay et al. [16] recently considered the problem of removing few vertices from a perfect graph so that it additionally becomes (r, ℓ).

Let $r, \ell \geq 0$ be two fixed integers. A graph is (r, ℓ)-*well-covered* if it is both (r, ℓ) and well-covered. More precisely, in this paper we focus on the following two decision problems.

(r, ℓ)-WELL-COVERED GRAPH ((r, ℓ)WCG)
    Input: A graph G
Question: Is G (r, ℓ)-well-covered?

WELL-COVERED (r, ℓ)-GRAPH (WC(r, ℓ)G)
    Input: An (r, ℓ)-graph G, together with a partition of $V(G)$ into r
                independent sets and ℓ cliques
Question: Is G well-covered?

We establish an almost complete characterization of the complexity of the (r, ℓ)WCG and WC(r, ℓ)G problems. Our results are shown in the following tables, where r (resp. ℓ) corresponds to the rows (resp. columns) of the tables, and where coNPc stands for coNP-complete, NPh stands for NP-hard, NPc stands for NP-complete, and (co)NPh stands for both NP-hard and coNP-hard. The symbol '?' denotes that the complexity of the corresponding problem is open.

(r, ℓ)WCG	0	1	2	≥ 3
0	–	P	P	NPc
1	P	P	P	NPc
2	P	coNPc	coNPc	(co)NPh
≥ 3	NPh	(co)NPh	(co)NPh	(co)NPh

WC(r, ℓ)G	0	1	2	≥ 3
0	–	P	P	P
1	P	P	P	P
2	P	coNPc	coNPc	coNPc
≥ 3	?	coNPc	coNPc	coNPc

We note the following simple facts that we will use to fill the above tables:

Fact 1. *If* (r, ℓ)WCG *is in* P, *then* WC(r, ℓ)G *is in* P.

Fact 2. *If* WC(r, ℓ)G *is* coNP-*hard, then* (r, ℓ)WCG *is* coNP-*hard.*

Note that WC(r, ℓ)G is in coNP, since a certificate for a No-instance consists just of two maximal independent sets of different size. On the other hand, for (r, ℓ)WCG we have the following facts, which are easy to verify:

Fact 3. *For any pair of integers* (r, ℓ) *such that the problem of recognizing an* (r, ℓ)-*graph is in* P, *the* (r, ℓ)WCG *problem is in* coNP.

Fact 4. *For any pair of integers* (r, ℓ) *such that the* WC(r, ℓ)G *problem is in* P, *the* (r, ℓ)WCG *problem is in* NP.

In this paper we prove that $(0, 1)$, $(1, 0)$, $(0, 2)$, $(1, 1)$, $(2, 0)$, and $(1, 2)$WCG can be solved in polynomial time, which by Fact 1 yields that WC$(0, 1)$, $(1, 0)$, $(0, 2)$,$(1, 1)$, $(2, 0)$, and $(1, 2)$G can also be solved in polynomial time. On the other hand, we prove that WC$(2, 1)$G is coNP-complete, which by Facts 2 and 3 will yield that $(2, 1)$WCG is also coNP-complete. Furthermore, we also prove that WC$(0, \ell)$G and WC$(1, \ell)$G are polynomial, and that $(0, 3)$, $(3, 0)$, and $(1, 3)$WCG are NP-hard. Finally, we state and prove a "monotonicity" result, namely Theorem 1, stating how to extend the NP-hardness or coNP-hardness of WC(r, ℓ)G (resp. (r, ℓ)WCG) to WC$(r + 1, \ell)$G (resp. $(r + 1, \ell)$WCG), and WC$(r, \ell + 1)$G (resp. $(r, \ell + 1)$WCG).

Together, these results correspond to those shown in the above tables. Note that the only remaining open cases are WC$(r, 0)$G for $r \geq 3$. As an avenue for further research, it would be interesting to provide a complete characterization of well-covered tripartite graphs, as has been done for bipartite graphs [10,21,26]. So far, only partial characterizations exist [14,15].

In addition, we consider the parameterized complexity of these problems for several choices of the parameters, such as the size α of a maximum independent set of the input graph, its neighborhood diversity, or the number ℓ of cliques in an (r, ℓ)-partition. We obtain several positive and negative results. In particular, we show that the parameterized problem of deciding whether a general graph is well-covered parameterized by α can be reduced to the WC$(0, \ell)$G problem parameterized by ℓ, and we prove that this latter problem is in XP but does not admit polynomial kernels unless coNP $\subseteq$ NP/poly. (For an introduction to the field of Parameterized Complexity, see [4,7,13,19].)

The rest of this paper is organized as follows. In Sect. 2 we prove our results concerning the classical complexity of both problems, and in Sect. 3 we focus on their parameterized complexity.

We use standard graph-theoretic notation, and we refer the reader to [6] for any undefined notation. Throughout the paper, we let n denote the number of vertices in the input graph for the problem under consideration.

2   Classical Complexity of the Problems

We start with a monotonicity theorem that will be very helpful to fill the tables presented in Sect. 1. The remainder of this section is divided into four subsections according to whether (r, ℓ)WCG and WC(r, ℓ)G are polynomial or *hard* problems.

Theorem 1. *Let $r, \ell \geq 0$ be two fixed integers.*

(i) *If* WC(r, ℓ)G *is* coNP-*complete then* WC$(r + 1, \ell)$G *and* WC$(r, \ell + 1)$G *are* coNP-*complete.*

(ii) *If (r, ℓ)WCG is* NP-*hard (resp.* coNP-*hard) then $(r, \ell + 1)$WCG is* NP-*hard (resp.* coNP-*hard).*

(iii) *Suppose that $r \geq 1$. If (r, ℓ)WCG is* NP-*hard (resp.* coNP-*hard) then $(r + 1, \ell)$WCG is* NP-*hard (resp.* coNP-*hard).*

Proof. (i) This follows immediately from the fact that every (r, ℓ)-graph is also an $(r + 1, \ell)$-graph and an $(r, \ell + 1)$-graph.

(ii) Let G be an instance of (r, ℓ)WCG. Let H be the disjoint union of G and a clique Z with $V(Z) = \{z_1, \ldots, z_{r+1}\}$. Clearly G is well-covered if and only if H is well-covered. If G is an (r, ℓ)-graph then H is an $(r, \ell + 1)$-graph. Suppose H is an $(r, \ell + 1)$-graph, with a partition into r independent sets $S^1, \ldots, S^r$ and $\ell + 1$ cliques $K^1, \ldots, K^{\ell+1}$. Each independent set S^i can contain at most one vertex of the clique Z. Therefore, there must be a vertex z_i in some clique K^j. Assume without loss of generality that there is a vertex of Z in $K^{\ell+1}$. Then $K^{\ell+1}$ cannot contain any vertex outside of $V(Z)$, so we may assume that $K^{\ell+1}$ contains all vertices of Z. Now $S^1, \ldots, S^r, K^1, \ldots, K^\ell$ is an (r, ℓ)-partition of G, so G is an (r, ℓ)-graph. Hence, H is a YES-instance of $(r, \ell + 1)$WCG if and only if G is a YES-instance of (r, ℓ)WCG.

(iii) Let G be an instance of (r, ℓ)WCG. Let G' be the graph obtained by adding $\ell + 1$ isolated vertices to G. (This guarantees that every maximal independent set in G' contains at least $\ell + 1$ vertices.) Since $r \geq 1$, it follows that G' is an (r, ℓ)-graph if and only if G is. Clearly G' is well-covered if and only if G is.

Next, find an arbitrary maximal independent set in G' and let p be the number of vertices in this set. Note that $p \geq \ell + 1$. Let H be the join of G' and a set of p independent vertices $Z = \{z_1, \ldots, z_p\}$, i.e., $N_H(z_i) = V(G')$ for all i. Every maximal independent set of H is either Z or a maximal independent set of G' and every maximal independent set of G' is a maximal independent set of H. Therefore, H is well-covered if and only if G' is well-covered. Clearly, if G' is an (r, ℓ)-graph then H is an $(r + 1, \ell)$-graph. Suppose H is an $(r + 1, \ell)$-graph, with a partition into $r + 1$ independent sets $S^1, \ldots, S^{r+1}$ and ℓ cliques $K^1, \ldots, K^\ell$. Each clique set K^i can contain at most one vertex of Z. Therefore there must be a vertex z_i in some independent set S^j. Suppose that there is a vertex of Z in S^{r+1}. Then S^{r+1} cannot contain any vertex outside of Z. Without loss of generality, we may assume that S^{r+1} contains all vertices of Z. Now $S^1, \ldots, S^r, K^1, \ldots, K^\ell$ is an (r, ℓ)-partition of G, so G is an (r, ℓ)-graph. Thus H is a YES-instance of $(r + 1, \ell)$WCG if and only if G is a YES-instance of (r, ℓ)WCG. □

2.1   Polynomial Cases for WC(r, ℓ)G

Theorem 2. WC$(0, \ell)$G *and* WC$(1, \ell)$G *are in* P *for every integer* $\ell \geq 0$.

Proof. It is enough to prove that WC$(1, \ell)$G is polynomial. Let $V = (S, K^1, K^2, K^3, \ldots, K^\ell)$ be a $(1, \ell)$-partition for G. Then each maximal independent set I of G admits a partition $I = (I_K, S \setminus N_S(I_K))$, where I_K is an independent set of $K^1 \cup K^2 \cup K^3 \cup \cdots \cup K^\ell$.

Observe that there are at most $O(n^\ell)$ choices for an independent set I_K of $K^1 \cup K^2 \cup K^3 \cup \cdots \cup K^\ell$, which can be listed in time $O(n^\ell)$, since ℓ is constant and $(K^1, K^2, K^3, \ldots, K^\ell)$ is given. For each of them, we consider the independent set $I = I_K \cup (S \setminus N_S(I_K))$. If I is not maximal (which may happen if a vertex in $(K^1 \cup K^2 \cup K^3 \cup \cdots \cup K^\ell) \setminus I_K$ has no neighbors in I), we discard this choice of I_K. Hence, we have a polynomial number $O(n^\ell)$ of maximal independent sets to check in order to decide whether G is a well-covered graph. □

2.2   Polynomial Cases for (r, ℓ)WCG

Fact 5. *The graph induced by a clique or by an independent set is well-covered.*

The following corollary is a simple application of Fact 5.

Corollary 1. G *is a* $(0, 1)$*-well-covered graph if and only if* G *is a* $(0, 1)$*-graph. Similarly,* G *is a* $(1, 0)$*-well-covered graph if and only if* G *is a* $(1, 0)$*-graph.*

The following is an easy observation.

Theorem 3. $(0, 2)$WCG *can be solved in polynomial time.*

In the next three lemmas we give a characterization of $(1, 1)$-well-covered graphs in terms of their graph degree sequence. Note that $(1, 1)$-graphs are known in the literature as *split graphs*.

Lemma 1. *Let* $G = (V, E)$ *be a* $(1, 1)$*-well-covered graph with* $(1, 1)$*-partition* $V = (S, K)$*, where* S *is a independent set and* K *is a clique. If* $x \in K$*, then* $|N_S(x)| \leq 1$.

Proof. Suppose that G is a $(1,1)$-well-covered graph with $(1,1)$-partition $V = (S, K)$, where S is a independent set and K is a clique. Let I be a maximal independent set of G such that $x \in I \cap K$. Suppose for contradiction that $|N_S(x)| \geq 2$, and let $y, z \in N_S(x)$. Since $y, z \in S$, $N_G(y), N_G(z) \subseteq K$. Since K is a clique, vertex x is the only vertex of I in K. Hence, we have that $N_G(y) \cap (I \setminus \{x\}) = N_G(z) \cap (I \setminus \{x\}) = \emptyset$. Therefore $I' = (I \setminus \{x\}) \cup \{y, z\}$ is an independent set of G such that $|I'| = |I| + 1$. Thus, I is a maximal independent set that is not maximum, so G is not well-covered. Thus, $|N_S(x)| \leq 1$.   □

Lemma 2. *A graph G is a $(1,1)$-well-covered graph if and only if it admits a $(1,1)$-partition $V = (S, K)$ such that either for every $x \in K$, $|N_S(x)| = 0$, or for every $x \in K$, $|N_S(x)| = 1$.*

Proof. Let G be a $(1,1)$-well-covered graph. By Lemma 1 we have that, given a vertex $x \in K$, either $|N_S(x)| = 0$ or $|N_S(x)| = 1$. Suppose for contradiction that there are two vertices $x, y \in K$ such that $|N_S(x)| = 0$ and $|N_S(y)| = 1$. Let z be the vertex of S adjacent to y. Let I be a maximal independent set containing vertex y. Note that the vertex x is non-adjacent to every vertex of $I \setminus \{y\}$ since there is at most one vertex of I in K. The same applies to the vertex z. Hence, a larger independent set I', with size $|I'| = |I| + 1$, can be obtained from I by replacing vertex y with the non-adjacent vertices x, z, i.e., I is a maximal independent set of G that is not maximum, a contradiction. Thus, either for every $x \in K$, $|N_S(x)| = 0$, or for every $x \in K$, $|N_S(x)| = 1$.

Conversely, suppose that there is a $(1,1)$-partition $V = (S, K)$ of G such that either for every $x \in K$, $|N_S(x)| = 0$, or for every $x \in K$, $|N_S(x)| = 1$. If $K = \emptyset$, then G is $(1,0)$ and then G is well-covered. Hence we assume $K \neq \emptyset$. If for every $x \in K$, $|N_S(x)| = 0$, then every maximal independent set consists of all the vertices of S and exactly one vertex $v \in K$. If for every $x \in K$, $|N_S(x)| = 1$, then every maximal independent set is either $I = S$, or $I = \{x\} \cup (S \setminus N_S(x))$ for some $x \in K$. Since $|N_S(x)| = 1$ we have $|I| = 1 + |S| - 1 = |S|$, and hence G is a $(1,1)$-well-covered graph.   □

Lemma 3. *G is a $(1,1)$-well-covered graph if and only if there is a positive integer k such that G is a graph with a $(1,1)$-partition $V = (S, K)$ where $|K| = k$, with degree sequence either $(k, k, k, \ldots, k, i_1, i_2, \ldots, i_s, 0, 0, 0, \ldots, 0)$ with $\sum_{j=1}^{s}(i_j) = k$, or $(k-1, k-1, k-1, \ldots, k-1, 0, 0, 0, \ldots, 0)$, where the subsequences $k, \ldots, k$ (resp. $k-1, \ldots, k-1$) have length k.*

Proof. Let G be a $(1,1)$-well-covered graph. Then G admits a $(1,1)$-partition $V = (S, K)$ where $k := |K|, k \geq 0$. If $k = 0$, then the degree sequence is $(0, 0, 0, \ldots, 0)$. If $k \geq 1$, then by Lemma 2 either for every $x \in K$, $|N_S(x)| = 0$, or for every $x \in K$, $|N_S(x)| = 1$. If for every $x \in K$, $|N_S(x)| = 0$, then the degree sequence of G is $(k-1, k-1, k-1, \ldots, k-1, 0, 0, 0, \ldots, 0)$. If for every $x \in K$, $|N_S(x)| = 1$, then the degree sequence of G is $(k, k, k, \ldots, k, i_1, i_2, \ldots, i_s, 0, 0, 0, \ldots, 0)$, with $\sum_{j=1}^{s}(i_j) = k$.

Suppose that there is a positive integer k such that G is a graph with $(1,1)$-partition $V = (S, K)$ where $|K| = k$, with degree sequence either $(k, k, k, \ldots, k,$

$i_1, i_2, \ldots, i_s, 0, 0, 0, \ldots, 0)$, or $(k-1, k-1, k-1, \ldots, k-1, 0, 0, 0, \ldots, 0)$, such that $\sum_{j=1}^{s}(i_j) = k$. If the degree sequence of G is $(k, k, k, \ldots, k, i_1, i_2, \ldots, i_s, 0, 0, 0, \ldots, 0)$, then the vertices of K are adjacent to $k-1$ vertices of K and exactly one of S, since the vertices with degree $i_1, i_2, \ldots, i_s$, have degree at most k and the vertices with degree 0 are isolated. If the degree sequence of G is $(k-1, k-1, k-1, \ldots, k-1, 0, 0, 0, \ldots, 0)$, then the vertices of K are adjacent to $k-1$ vertices of K and none of S and the vertices with degree 0 are isolated. By Lemma 2 we have that G is a well-covered graph. $\square$

Corollary 2. $(1,1)$WCG *can be solved in polynomial time.*

Ravindra [21] gave the following characterization of $(2,0)$-well-covered graphs.

Theorem 4 (Ravindra [21]). *Let G be a connected graph. G is a $(2,0)$-well-covered graph if and only if G contains a perfect matching F such that for every edge $e = uv$ in F, $G[N(u) \cup N(v)]$ is a complete bipartite graph.*

We now prove that Theorem 4 leads to a polynomial-time algorithm.

Theorem 5. $(2,0)$WCG *can be solved in polynomial time.*

Proof. Assume that G is connnected and consider the weighted graph (G, ω) with $\omega : E(G) \to \{0, 1\}$ satisfying $\omega(uv) = 1$, if $G[N(u) \cup N(v)]$ is a complete bipartite graph, and 0 otherwise. By Theorem 4, G is well-covered if and only if (G, ω) has a weighted perfect matching with weight at least $n/2$, and this can be decided in polynomial time [9]. $\square$

Theorem 6. $(1,2)$WCG *can be solved in polynomial time.*

Proof. We can find a $(1,2)$-partition of a graph G (if any) in polynomial time [2]. After that, we use the algorithm for WC$(1, \ell)$G given by Theorem 2. $\square$

2.3   coNP-complete Cases for WC(r, ℓ)G

We note that the WELL-COVERED GRAPH instance G constructed in the reduction of Chvátal and Slater [3] is $(2,1)$, directly implying that WC$(2,1)$G is coNP-complete.

Indeed, Chvátal and Slater [3] take a 3-SAT instance $I = (U, C) = (\{u_1, u_2, u_3, \ldots, u_n\}, \{c_1, c_2, c_3, \ldots, c_m\})$, and construct a WELL-COVERED GRAPH instance $G = (V, E) = (\{u_1, u_2, u_3, \ldots, u_n, \overline{u}_1, \overline{u}_2, \overline{u}_3, \ldots, \overline{u}_n, c_1, c_2, c_3, \ldots, c_m\}, \{xc_j : x \text{ occurs in } c_j\} \cup \{u_i \overline{u}_i : 1 \leq i \leq n\} \cup \{c_i c_j : 1 \leq i < j \leq m\})$. Note that $\{c_i c_j : 1 \leq i < j \leq m\}$ is a clique, and that $\{u_1, u_2, u_3, \ldots, u_n\}$, and $\{\overline{u}_1, \overline{u}_2, \overline{u}_3, \ldots, \overline{u}_n\}$ are independent sets. Hence, G is a $(2,1)$-graph. An illustration of this construction can be found in Fig. 1. This discussion can be summarized as follows.

Theorem 7 (Chvátal and Slater [3]). WC$(2,1)$G *is coNP-complete.*

As $(2,1)$-graphs can be recognized in polynomial time [2], we have the following corollary.

Corollary 3. $(2,1)$WCG *is coNP-complete.*

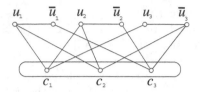

Fig. 1. Chvatal and Slater's [3] WELL-COVERED GRAPH instance $G = (V, E)$ obtained from the satisfiable 3-SAT instance $I = (U, C) = (\{u_1, u_2, u_3\}, \{(u_1, u_2, u_3), (u_1, u_2, \overline{u}_3), (\overline{u}_1, \overline{u}_2, \overline{u}_3)\})$, where $\{c_1, c_2, \ldots, c_m\}$ is a clique of G. Observe that I is satisfiable if and only if G is not well-covered, since there is a maximal independent set with size $n + 1$ (e.g. $\{c_1, \overline{u}_1, \overline{u}_2, \overline{u}_3\}$) and there is a maximal independent set of size n (e.g. $\{u_1, u_2, \overline{u}_3\}$). Note also that G is a $(2, 1)$-graph with $(2, 1)$-partition $V = (\{u_1, u_2, \ldots, u_n\}, \{\overline{u}_1, \overline{u}_2, \ldots, \overline{u}_n\}, \{c_1, c_2, \ldots, c_m\})$.

2.4   NP-hard Cases for (r, ℓ)wcg

Now we prove that $(0, 3)$ WCG is NP-complete. For this purpose, we slightly modify an NP-completeness proof of Stockmeyer [23].

Stockmeyer's [23] NP-completeness proof of 3-coloring considers a 3-SAT instance $I = (U, C) = (\{u_1, u_2, u_3, \ldots, u_n\}, \{c_1, c_2, c_3, \ldots, c_m\})$, and constructs a 3-COLORING instance $G = (V, E) = (\{u_1, u_2, u_3, \ldots, u_n, \overline{u}_1, \overline{u}_2, \overline{u}_3, \ldots, \overline{u}_n\} \cup \{v_1[j], v_2[j], v_3[j], v_4[j], v_5[j], v_6[j] : j \in \{1, 2, 3, \ldots, m\}\} \cup \{t_1, t_2\}, \{u_i\overline{u}_i : i \in \{1, 2, 3, \ldots, n\}\} \cup \{v_1[j]v_2[j], v_2[j]v_4[j], v_4[j]v_1[j], v_4[j]v_5[j], v_5[j]v_6[j], v_6[j]v_3[j], v_3[j]v_5[j] : j \in \{1, 2, 3, \ldots, m\}\} \cup \{v_1[j]x, v_2[j]y, v_3[j]z : c_j = (x, y, z)\} \cup \{t_1u_i, t_1\overline{u}_i : i \in \{1, 2, 3, \ldots, n\}\} \cup \{t_2v_6[j] : j \in \{1, 2, 3, \ldots, m\}\})$; see Fig. 2(a).

Theorem 8. $(0, 3)$wcg *is* NP-*complete.*

Proof. As by Theorem 2 the WELL-COVERED GRAPH problem can be solved in polynomial time on $(0, 3)$-graphs, by Fact 4 $(0, 3)$wcg is in NP.

Let $I = (U, C)$ be a 3-SAT instance. We produce, in polynomial time in the size of I, a $(0, 3)$wcg instance H, such that I is satisfiable if and only if H is $(0, 3)$-well-covered. Let $G = (V, E)$ be the graph of [23] obtained from I, and let G' be the graph obtained from G by adding to V a vertex x_{uv} for every edge uv of G not belonging to a triangle, and by adding to E edges ux_{uv} and vx_{uv}; see Fig. 2(b). Finally, we define $H = \overline{G'}$ as the complement of G'. Note that, by [23], I is satisfiable if and only if G is 3-colorable. Since x_{uv} is adjacent to only two different colors of G, clearly G is 3-colorable if and only if G' is 3-colorable. Hence, I is satisfiable if and only if H is a $(0, 3)$-graph. We prove next that I is satisfiable if and only if H is a $(0, 3)$-well-covered graph.

Suppose that I is satisfiable. Then, since H is a $(0, 3)$-graph, every maximal independent set of H has size 3, 2, or 1. If there is a maximal independent set I in H with size 1 or 2, then I is a maximal clique of G' of size 1 or 2. This contradicts the construction of G', since every maximal clique of G' is a triangle. Therefore, G is well-covered.

Suppose that H is $(0, 3)$-well-covered. Then G' is 3-colorable, so G is also 3-colorable. Thus, by [23], I is satisfiable. $\qquad\square$

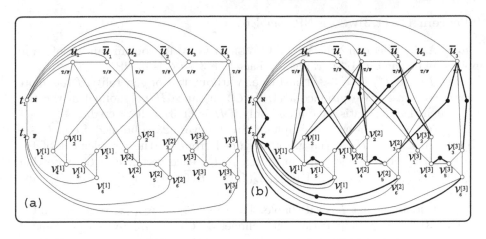

Fig. 2. (a) Stockmeyer's [23] 3-COLORING instance G obtained from the 3-SAT instance $I = (U, C) = (\{u_1, u_2, u_3\}, \{(u_1, u_2, u_3), (u_1, u_2, \overline{u}_3), (\overline{u}_1, \overline{u}_2, \overline{u}_3)\})$. (b) The graph G' obtained from G by adding a vertex x_{uv} with $N_{G'}(x_{uv}) = \{u, v\}$ for every edge uv of G not belonging to a triangle.

We next prove that $(3, 0)$WCG is NP-hard. For this, we again use the proof of Stockmeyer [23], together with the following theorem.

Theorem 9 (Topp and Volkmann [25]). *Let $G = (V, E)$ be an n-vertex graph, $V = \{v_1, v_2, v_3, \ldots, v_n\}$, and let H be obtained from G such that $V(H) = V \cup \{u_1, u_2, u_3, \ldots, u_n\}$ and $E(H) = E \cup \{v_i u_i : i \in \{1, 2, 3, \ldots, n\}\}$. Then H is a well-covered graph where every maximal independent set has size n.*

Proof. Observe that every maximal independent set I of H has a subset $I_G = I \cap V$. Let $\mathcal{U} \subseteq \{1, 2, 3, \ldots, n\}$ be the set of indices i such that $v_i \in I$. Since I is maximal, the set $\{u_i : i \in \{1, 2, 3, \ldots, n\} \setminus \mathcal{U}\}$ must be contained in I, so $|I| = n$. □

Theorem 10. $(3, 0)$WCG *is* NP-*hard.*

Proof. Let $I = (U, C)$ be a 3-SAT instance; let $G = (V, E)$ be the graph obtained from I in Stockmeyer's [23] NP-completeness proof for 3-COLORING; and let H be the graph obtained from G by the transformation described in Theorem 9. We prove that I is satisfiable if and only if H is a $(3, 0)$-well-covered graph. Suppose that I is satisfiable. Then by [23] we have that G is 3-colorable. Since a vertex $v \in V(H) \setminus V(G)$ has just one neighbor, there are 2 colors left for v to extend a 3-coloring of G, and so H is a $(3, 0)$-graph. Hence, by Theorem 9, H is a $(3, 0)$-well-covered graph. Suppose that H is a $(3, 0)$-well-covered graph. Then we have that G is a $(3, 0)$-graph. By [23], I is satisfiable. □

Note that Theorem 1 combined with Theorem 8 does not imply that $(1, 3)$WCG is NP-complete.

Theorem 11. $(1,3)$WCG *is* NP-*complete*.

Proof. As by Theorem 2 the WELL-COVERED GRAPH problem can be solved in polynomial time on $(1,3)$-graphs, by Fact 4 $(1,3)$WCG is in NP.

Let $I = (U, C)$ be a 3-SAT instance. Without loss of generality, I has more than two clauses. We produce a $(1,3)$WCG instance H polynomial in the size of I, such that I is satisfiable if and only if H is $(1,3)$-well-covered.

Let $G = (V, E)$ be the graph of Stockmeyer [23] obtained from I (see Fig. 2(a)), and let H be the graph obtained from $\overline{G}$ (the complement of the graph G) by adding one pendant vertex p_v for each vertex v of $\overline{G}$. Note that $V(H) = V(G) \cup \{p_v : v \in V(G)\}$, $E(H) = E(\overline{G}) \cup \{p_v v : v \in V(G)\}$, and $N_H(p_v) = \{v\}$.

First suppose that I is satisfiable. Then by [23], G is a $(3,0)$-graph, and $\overline{G}$ is a $(0,3)$-graph with partition into cliques $V(\overline{G}) = (K_{\overline{G}}^1, K_{\overline{G}}^2, K_{\overline{G}}^3)$. Thus it follows that $(S = \{p_v : v \in V(G)\}, K_{\overline{G}}^1, K_{\overline{G}}^2, K_{\overline{G}}^3)$ is a $(1,3)$-partition of $V(H)$. In addition, from Theorem 9 and by the construction of H, H is a well-covered graph. Hence H is $(1,3)$-well-covered.

Conversely, suppose that H is $(1,3)$-well-covered, and let $V(H) = (S, K^1, K^2, K^3)$ be a $(1,3)$-partition for H. Then we claim that no vertex $p_v \in V(H) \setminus V(G)$ belongs to $K^i, i \in \{1,2,3\}$. Indeed, suppose for contradiction that $p_v \in K^i$ for some $i \in \{1,2,3\}$. Then, $K^i \subseteq \{p_v, v\}$. Hence, $H \setminus K^i$ is a $(1,2)$-graph and $G \setminus \{v\}$ is an induced subgraph of a $(2,1)$-graph. But by construction of G, $G \setminus \{v\}$ (for any $v \in V(G)$) contains at least one $2K_3$ (that is, two vertex-disjoint copies of K_3) as an induced subgraph, which is a contradiction given that $2K_3$ is clearly a forbidden subgraph for $(2,1)$-graphs. Therefore, $\{p_v : v \in V(G)\} \subseteq S$, and since $\{p_v : v \in V(G)\}$ is a dominating set of H, $S = \{p_v : v \in V(G)\}$. Thus, $\overline{G}$ is a $(0,3)$-graph with partition $V(\overline{G}) = (K^1, K^2, K^3)$, and therefore G is a $(3,0)$-graph, i.e., a 3-colorable graph. Therefore, by [23], I is satisfiable. □

Corollary 4. *If* $(r \geq 3$ *and* $\ell = 0)$ *then* (r, ℓ)WCG *is* NP-*hard. If* $(r \in \{0,1\}$ *and* $\ell \geq 3)$ *then* (r, ℓ)WCG *is* NP-*complete.*

Proof. (r, ℓ)WCG is NP-hard in all of these cases by combining Theorems 1, 8, 10 and 11. For $(r \in \{0,1\}$ and $\ell \geq 3)$, the WELL-COVERED GRAPH problem can be solved in polynomial time on (r, ℓ)-graphs, so by Fact 4 (r, ℓ)WCG is in NP. □

3   Parameterized Complexity of the Problems

In this section we focus on the parameterized complexity of the WELL-COVERED GRAPH problem, with special emphasis on the case where the input graph is an (r, ℓ)-graph. Henceforth we let α denote the size of a maximum independent set in the input graph G for the problem under consideration. That is, G is well-covered if and only if any maximal independent set of G has size α.

Lemma 4. *The* WC(r, ℓ)G *problem can be solved in time* $2^{r \cdot \alpha} \cdot n^{O(\ell)}$. *In particular, it is* FPT *when* ℓ *is fixed and* r, α *are parameters.*

Proof. Note that each of the r independent sets $S^1, \ldots, S^r$ of the given partition of $V(G)$ must have size at most α. On the other hand, any maximal independent set of G contains at most one vertex in each of the ℓ cliques. The algorithm exhaustively constructs all maximal independent sets of G as follows: we start by guessing a subset of $\bigcup_{i=1}^{r} S^i$, and then choose at most one vertex in each clique. For each choice we just have to verify whether the constructed set is a maximal independent set, and then check that all the constructed maximal independent sets have the same size. The claimed running time follows. In fact, in the statement of the lemma, one could replace α with $\max_{1 \leq i \leq r} |S^i|$, which yields a stronger result. $\qquad\square$

The following lemma motivates the study of the WELL-COVERED $(0, \ell)$-GRAPH problem, as it shows that the WELL-COVERED GRAPH problem (on general graphs) parameterized by α can be reduced to the WC$(0, \ell)$G problem parameterized by ℓ.

Lemma 5. *The* WELL-COVERED GRAPH *problem parameterized by α, with α given as part of the input, can be fpt-reduced to the* WC$(0, \ell)$G *problem parameterized by ℓ.*

Proof. Given a general graph G and α (recall that α is the size of a maximum independent set in G), we construct a $(0, \alpha)$-graph G' as follows. For every vertex $v \in V(G)$, we add to G' a clique K_v on α vertices, which are denoted $v_1, \ldots, v_\alpha$. For every two vertices $u, v \in V(G)$ and every integer i, $1 \leq i \leq \alpha$, we add to G' the edge $u_i v_i$. Finally, for every edge $uv \in E(G)$, we add to G' a complete bipartite subgraph between $V(K_u)$ and $V(K_v)$. It is clear from the construction that G' is a $(0, \alpha)$-graph, and that G is well-covered if and only if G' is well-covered. $\qquad\square$

Lemma 6. *The* WC$(1, \ell)$G *problem can be solved in time $n^{O(\ell)}$. In other words, it is in* XP *when parameterized by ℓ.*

Proof. Let $V(G) = S^1 \cup K^1 \cup \cdots \cup K^\ell$. The algorithm chooses at most one vertex in each clique, and adds them to a potential independent set $I_k \subseteq V(G)$. If I_k is not independent, then we discard this set. Otherwise, we set $I = I_k \cup (S^1 \setminus N_{S^1}(I_k))$. This clearly defines an independent set of G, which may be maximal or not, but any maximal independent set of G can be constructed in this way. The number of sets considered by this procedure is $n^{O(\ell)}$. Finally, it just remains to check whether all the constructed independent sets, after discarding those that are not maximal, have the same size. $\qquad\square$

Note that Lemma 6 implies that the WELL-COVERED $(0, \ell)$-GRAPH problem can also be solved in time $n^{O(\ell)}$. This observation raises the following question: are the WELL-COVERED $(0, \ell)$-GRAPH and WELL-COVERED $(1, \ell)$-GRAPH problems FPT when parameterized by ℓ? Even if we still do not know the answer to this question, in the following them we prove that, in particular, the WELL-COVERED $(0, \ell)$-GRAPH and WELL-COVERED $(1, \ell)$-GRAPH problems are unlikely to admit polynomial kernels when parameterized by ℓ. We first need a definition introduced by Bodlaender et al. [1].

Definition 1 (and-composition [1]**).** *Let $\mathcal{Q} \subseteq \Sigma^* \times \mathbb{N}$ be a parameterized problem. An* AND*-composition for $\mathcal{Q}$ is an algorithm that, given t instances (x_1, k), $\ldots, (x_t, k) \in \Sigma^* \times \mathbb{N}$ of $\mathcal{Q}$, takes time polynomial in $\sum_{i=1}^{t} |x_i| + k$ and outputs an instance $(y, k') \in \Sigma^* \times \mathbb{N}$ such that:*

(i) *The parameter value k' is polynomially bounded by k.*
(ii) *The instance (y, k') is* YES *for $\mathcal{Q}$ if and only if all instances (x_i, k) are* YES *for $\mathcal{Q}$.*

The following result was formulated as the 'AND-conjecture' by Bodlaender et al. [1], and it was finally proved by Drucker [8] (see [5] for a simplified proof).

Theorem 12 (Drucker [8]**).** *If a parameterized problem admits a polynomial kernel and an* AND*-composition, then* coNP $\subseteq$ NP/poly.

We are ready the state our result.

Theorem 13. *For any fixed integer $r \geq 0$, the* WC(r, ℓ)G *problem does not admit polynomial kernels when parameterized by ℓ, unless* coNP $\subseteq$ NP/poly.

Proof. By Theorem 12, to prove the result it is enough to present an AND-composition for the WELL-COVERED $(0, \ell)$-GRAPH problem, which implies the result for the WELL-COVERED (r, ℓ)-GRAPH for any $r \geq 0$, as any $(0, \ell)$-graph is also an (r, ℓ)-graph for $r \geq 0$. Let $G_1, \ldots, G_t$ be the given $(0, \ell)$-graphs. For every $1 \leq i \leq t$, take arbitrarily a maximal independent set S_i of G_i, and let G'_i be the graph obtained from G_i by adding $\ell - |S_i|$ isolated vertices. As any independent set in G_i can have at most one vertex in each clique K^j, it follows that $|S_i| \leq \ell$, and thus G'_i is well-defined. Note also that for every $1 \leq i \leq t$, G'_i is a $(0, 2\ell)$-graph (just consider a new clique for each isolated vertex). The following two properties are easy to verify:

(i) For every $1 \leq i \leq t$, G'_i is well-covered if and only if G_i is well-covered.
(ii) For every $1 \leq i \leq t$, if G'_i is well-covered then every maximal independent set of G'_i has size ℓ.

We create a graph G by taking the disjoint union of the G_i's and then adding, for every $i, j \in \{1, \ldots, t\}$, $i \neq j$, all edges between $V(G'_i)$ and $V(G'_j)$. We set the parameter of G as $\ell' = 2\ell$. Since each G'_i is a $(0, 2\ell)$-graph, by construction G is also a $(0, 2\ell)$-graph. Note that, by construction of G, any independent set of G can intersect only one G_i, and therefore, if we let $\mathcal{S}$ (resp. $\mathcal{S}_i$) denote the set of all maximal independent sets of G (resp. G'_i), it follows that

$$\mathcal{S} = \bigcup \{\mathcal{S}_i : 1 \leq i \leq t\}. \tag{1}$$

We claim that G is well-covered if and only if G_i is well-covered for every $1 \leq i \leq t$, which will conclude the proof.

Indeed, first suppose that G is well-covered. Then, by Eq. (1), G'_i is well-covered for every $1 \leq i \leq t$, and by Property (i) this implies that G_i is also well-covered for every $1 \leq i \leq t$.

Conversely, suppose that G_i is well-covered for every $1 \leq i \leq t$. Then, by Property (i), G_i' is also well-covered for every $1 \leq i \leq t$ and, by Property (ii), for every $1 \leq i \leq t$, any maximal independent set of G_i' has size ℓ. This implies by Eq. (1) that any maximal independent set of G also has size ℓ, and therefore G is well-covered, as we wanted to prove.   □

3.1   Taking the Neighborhood Diversity as the Parameter

The vertex cover number is a classical graph parameter that has been widely used as a parameter in the multivariate complexity analysis of problems, including those that are not directly related to the vertex cover number. Neighborhood diversity is another graph parameter, defined by Lampis [17], which captures more precisely than vertex cover number the property that two vertices with the same neighborhood are "equivalent". This parameter is defined as follows.

Definition 2. *The neighborhood diversity* $\mathbf{nd}(G)$ *of a graph* $G = (V, E)$ *is the minimum t such that V can be partitioned into t sets $V_1, \ldots, V_t$ where for every $v \in V(G)$ and every $i \in \{1, \ldots, t\}$, either v is adjacent to every vertex in V_i or it is adjacent to none of them. Note that each part V_i of G is either a clique or an independent set.*

Neighborhood diversity is a stronger parameter than vertex cover, in the sense that bounded vertex cover graphs are a subclass of bounded neighborhood diversity graphs. It is known that an optimal neighborhood diversity decomposition of a graph G can be computed in time $O(n^3)$. See [17] for more details.

Lampis [17] showed that: (i) for every graph G we have $\mathbf{nd}(G) \leq 2^{vc(G)} + vc(G)$, where $vc(G)$ is the vertex cover number of G; $cw(G) \leq \mathbf{nd}(G) + 1$, where $cw(G)$ is the clique-width of G; (iii) there exist graphs of constant treewidth and unbounded neighborhood diversity and vice versa; (iv) an optimal neighborhood diversity decomposition of a graph G can be computed in polynomial time.

Lemma 7. *The* WELL-COVERED GRAPH *problem is* FPT *when parameterized by the neighborhood diversity.*

Proof. Given a graph G, we first obtain a neighborhood partition of G with minimum width using the polynomial-time algorithm of Lampis [17]. Let $t := \mathbf{nd}(G)$ and let $V_1, \ldots, V_t$ be the partition of $V(G)$. As we can observe, for any pair u, v of non-adjacent vertices belonging to the same part V_i, if u is in a maximal independent set S then v also belongs to S, otherwise S cannot be maximum. On the other hand, if $N[u] = N[v]$ then for any maximal independent set S_u such that $u \in S_u$ there exists another maximal independent set S_v such that $S_v = S_u \setminus \{u\} \cup \{v\}$. Hence, we can contract each partition V_i which is an independent set into a single vertex v_i with weight $\tau(v_i) = |S_i|$, and contract each partition V_i which is a clique into a single vertex v_i with weight $\tau(v_i) = 1$, in order to obtain a graph G_t with $|V(G_t)| = t$, where the weight of a vertex v_i of G_t means that any maximal independent set of G uses either none or exactly $\tau(v)$

vertices of V_i. At this point, we just need to analyze whether all maximal independent sets of G_t have the same weight (sum of the weights of its vertices), which can be done in $2^t \cdot n^{O(1)}$ time.   □

Corollary 5. *The* WELL-COVERED GRAPH *problem is* FPT *when parameterized by the vertex cover number* $n - \alpha$.

Acknowledgement. We would like to thank the anonymous reviewers for their thorough, pertinent, and very helpful remarks.

References

1. Bodlaender, H.L., Downey, R.G., Fellows, M.R., Hermelin, D.: On problems without polynomial kernels. J. Comput. System Sci. **75**(8), 423–434 (2009)
2. Brandstädt, A.: Partitions of graphs into one or two independent sets and cliques. Discrete Math. **152**(1–3), 47–54 (1996)
3. Chvátal, V., Slater, P.J.: A note on well-covered graphs. Ann. Discret. Math. **55**, 179–181 (1993)
4. Cygan, M., Fomin, F.V., Kowalik, L., Lokshtanov, D., Marx, D., Pilipczuk, M., Pilipczuk, M., Saurabh, S.: Parameterized Algorithms. Springer, Cham (2015)
5. Dell, H.: A simple proof that AND-compression of NP-complete problems is hard. Electronic Colloquium on Computational Complexity (ECCC), 21: 75 (2014)
6. Diestel, R.: Graph Theory. Graduate Texts in Mathematics, vol. 173, 4th edn. Springer, Heidelberg (2012)
7. Downey, R.G., Fellows, M.R.: Fundamentals of Parameterized Complexity. Texts in Computer Science. Springer, Heidelbreg (2013)
8. Drucker, A.: New limits to classical and quantum instance compression. SIAM J. Comput. **44**(5), 1443–1479 (2015)
9. Edmonds, J.: Paths, trees and flowers. Can. J. Math. **17**, 449–467 (1965)
10. Favaron, O.: Very well-covered graphs. Discrete Math. **42**, 177–187 (1982)
11. Feder, T., Hell, P., Klein, S., Motwani, R.: List partitions. SIAM J. Discrete Math. **16**(3), 449–478 (2003)
12. Feder, T., Hell, P., Klein, S., Nogueira, L.T., Protti, F.: List matrix partitions of chordal graphs. Theoret. Comput. Sci. **349**(1), 52–66 (2005)
13. Flum, J., Grohe, M.: Parameterized Complexity Theory. Springer, Heidelberg (2006)
14. Golzari, R.J., Zaare-Nahandi, R.: Unmixed r-partite graphs. CoRR, abs/1511.00228 (2015)
15. Haghighi, H.: A generalization of Villarreal's result for unmixed tripartite graphs. Bull. Iran. Math. Soc. **40**(6), 1505–1514 (2014)
16. Kolay, S., Panolan, F., Raman, V., Saurabh, S.: Parameterized algorithms on perfect graphs for deletion to (r, ℓ)-graphs. In: Proceedings of MFCS 2016, vol. 58, LIPIcs, pp. 75: 1–75: 13 (2016)
17. Lampis, M.: Algorithmic meta-theorems for restrictions of treewidth. Algorithmica **64**(1), 19–37 (2012)
18. Lesk, M., Plummer, M.D., Pulleyblank, W.R.: Equi-matchable graphs. In: Graph Theory and Combinatorics. Academic Press, pp. 239–254 (1984)
19. Niedermeier, R.: Invitation to Fixed-Parameter Algorithms, vol. 31. Oxford University Press, Oxford (2006)

20. Plummer, M.D.: Some covering concepts in graphs. J. Comb. Theory **8**(1), 91–98 (1970)
21. Ravindra, G.: Well-covered graphs. J. Comb. Inform. Syst. Sci. **2**(1), 20–21 (1977)
22. Sankaranarayana, R.S., Stewart, L.K.: Complexity results for well-covered graphs. Networks **22**(3), 247–262 (1992)
23. Stockmeyer, L.: Planar 3-colorability is polynomial complete. ACM SIGACT News **5**(3), 19–25 (1973)
24. Tankus, D., Tarsi, M.: Well-covered claw-free graphs. J. Comb.Theory, Ser. B **66**(2), 293–302 (1996)
25. Topp, J., Volkmann, L.: Well covered and well dominated block graphs and unicyclic graphs. Mathematica Pannonica **1**(2), 55–66 (1990)
26. Villarreal, R.H.: Unmixed bipartite graphs. Revista Colombiana de Matemáticas **41**(2), 393–395 (2007)

Algorithmic Analysis for Ridesharing of Personal Vehicles

Qian-Ping Gu[1(✉)], Jiajian Leo Liang[1], and Guochuan Zhang[2]

[1] School of Computing Science, Simon Fraser University, Burnaby, Canada
{qgu,jjl24}@sfu.ca
[2] College of Computer Science and Technology,
Zhejiang University, Hangzhou, China
zgc@zju.edu.cn

Abstract. The ridesharing problem is to share personal vehicles by individuals (participants) with similar itineraries. A trip in the ridesharing problem is a participant and his/her itinerary. To realize a trip is to deliver the participant to his/her destination by a vehicle satisfying the itinerary requirement. We study two optimization problems in ridesharing: minimize the number of vehicles and minimize the total travel distance of vehicles to realize all trips. The minimization problems are complex and NP-hard because of many parameters. We simplify the problems by considering only the source, destination, vehicle capacity, detour distance and preferred path parameters. We prove that the simplified minimization problems are still NP-hard while a further simplified variant is polynomial time solvable. These suggest a boundary between the NP-hard and polynomial time solvable cases.

Keywords: Ridesharing problem · Algorithmic analysis · Optimization algorithms

1 Introduction

The ridesharing problem is the shared use of personal vehicles by individuals (participants) who have similar itineraries. When a vehicle is selected to serve any participant, the owner of the vehicle is called a *driver* and a participant other than a driver is called a *passenger*. Ridgesharing can save the total cost of all drivers and passengers, reduce traffic congestion, conserve fuel, and reduce air pollution [6,13,14]. Despite the advantages of ridesharing, according to [8], the share of personal vehicles has decreased by almost 10 % in the past 30 years. The average occupancy rate of personal vehicles is 1.6 persons per vehicle mile based on reports published in 2011 [10,15]. Currently, ridesharing coordination is not fully regulated and organized in the industry. A major obstacle prevents ridesharing from being widely adopted is the lack of efficient and convenient methods to arrange vehicles for drivers and passengers. Other hurdles include privacy, safety, social discomfort, and pricing. Some of these issues can be addressed by introducing reputation building system, profiling, or preferences [10]. With today's

© Springer International Publishing AG 2016
T.-H.H. Chan et al. (Eds.): COCOA 2016, LNCS 10043, pp. 438–452, 2016.
DOI: 10.1007/978-3-319-48749-6_32

technology in GPS and smartphone, Internet-enabled mobile devices should be able to play an important role in popularizing ridesharing. There are new companies trying to reduce the gap between convenient transportation and ridesharing [2,12], such as Uber and Lyft. Readers may refer to recent surveys [2,8] about ridesharing in general. In the two surveys, methods for general ridesharing problems are reviewed along with some approaches for encouraging participation in ridesharing.

The ridesharing problem is similar to the dial-a-ride problem (DARP) [7]. In DARP, a set of drivers serves a set of passengers. One difference between the ridesharing problem and DARP is that, a driver in the ridesharing problem may only provide service to passengers of similar itineraries to that of the driver, while a driver in DARP may provide service for a wider range of passengers. The ridesharing problem is a generalization of the travel salesman problem and NP-hard in general. Most previous studies on the ridesharing problem focus on developing heuristics or solving some simplified variants of the problem, such as single passenger at a time, single pick-up of a driver's trip, not including pricing, or single objective function. Usually, the ridesharing problem is formulated as an IP (or MIP) problem and solved using some heuristics or meta-heuristics. In [4], Baldacci et al. propose both an exact method and a heuristic to solve the car pooling problem based on two integer programming formulations. Herbawi and Weber [11] give an IP formulation of the dynamic ridesharing problem where the objective function contains four optimization goals at the same time. Each of the components in the objective function is associated with a parameter that controls which optimization goal has more emphasis. They propose a generational genetic algorithm to solve such an IP formulation of the ridesharing problem [11]. In [1], Agatz et al. develop optimization-based approaches for solving dynamic ridesharing in a practical environment where drivers and ride requests from passengers dynamically enter and leave the system. They build a simulation environment based on the travel demand model data from the Atlanta Regional Commission, and use it to compare different methods. The ridesharing problem they study is simplified, a driver can serve only one passenger. In a recent paper [12], Huang et al. propose a method for large scale real-time ridesharing. They compare their method with some general approaches for the ridesharing problem, such as branch and bound algorithm and mixed integer programming approach. Their comparison is based on a large scale taxi data set made by Shanghai taxis. For a literature review of the ridesharing problem, readers are referred to [2,8].

A trip in ridesharing is a participant (driver or passenger) with an itinerary specified by many parameters such as source and destination in a road network, departure/arrival time, preferred path of a driver, distance/time detour limit a driver can tolerate, vehicle capacity, price, and so on. To realize a trip is to arrange a driver to deliver the participant to his/her destination satisfying the itinerary requirement. Usually such an arrangement is realized by a central matching agency which finds a driver arrangement to realize all trips. The ridesharing problem can be *static* or *dynamic*. In the static ridesharing, the driver arrangement is computed for a given set of trips. In the dynamic ridesharing, each trip arrives online and a driver is arranged for an arrived trip without the knowledge of trips in the future. The static and dynamic ridesharings are closely related. One can view a dynamic ridesharing instance as a sequence of static ridesharing instances. It is important to understand the fundamentals of the static ridesharing problem since the solutions and methods for the static problem can be used for the dynamic ridesharing systems for a batch of trips arrived in a specific time window. In this paper, we study the static rideshairing problem.

In many previous studies, drivers and passengers are considered different: passengers are served by drivers but a driver is not served by another partici-pant [2,8]. We consider a more general ridesharing problem: every participant can be a driver or passenger; a solution for the problem is a subset of selected drivers who can serve all trips and an assignment of passengers to drivers. We study two important optimization problems in ridesharing: minimize the num-ber of drivers and minimize the total travel distance of the drivers for realizing all trips. The large number of parameters make the problems very complex and NP-hard. The minimization problems can be simplified by considering only some of the parameters but little is known on to what extent of the simplification the minimization problems become polynomial time solvable. We give an algorith-mic analysis for the simplified minimization problems and explore a boundary between the NP-hard and polynomial time solvable cases. We prove that the simplified minimization problems, where only the parameters of source, vehicle capacity and one of destination, distance detour limit and preferred path are considered, are still NP-hard while a further simplified variant of minimizing the number of vehicles becomes polynomial time solvable. These results suggest a boundary between the NP-hard and polynomial time solvable cases.

The rest of this paper is organized as follows. Section 2 gives the preliminaries of the paper. In Sect. 3, we prove the NP-hardness results. Section 4 shows a polynomial time solvable case. The final section concludes the paper.

2   Preliminaries

A road network is expressed by a graph G which consists of a set $V(G)$ of vertices (locations in the network) and a set $E(G)$ of edges, each edge is a set $\{u, v\}$ of two vertices (a road between u and v). G is weighted if each edge is assigned a weight (cost to use the road). When the weight of each edge in G is not specified, we assume the weight is one. A *path* in G is a sequence $e_1, \ldots, e_k$ of edges, where $e_i = \{v_{i-1}, v_i\} \in E(G)$ for $1 \leq i \leq k$ and no vertex is repeated in the sequence. The *length* of a path is the sum of the weights assigned to the edges of the path.

In the ridesharing problem, we assume that every participant can be a driver or passenger. A trip is a participant and his/her itinerary. In general, each trip has a source, a destination, an earliest departure time, a latest arrival time, a preferred path (e.g., a shortest path) to reach the destination, a limit on the detour distance/time from the preferred path to serve other participants and a price limit the participant can pay if served. We simplify the ridesharing problem by considering only the source, destination, vehicle capacity, distance detour limit and preferred path parameters. For the simplified problem, we denote a trip by an integer i specified by $(s_i, t_i, n_i, d_i, \mathcal{P}_i)$, where

- s_i is the source (start location) of i (a vertex in G),
- t_i is the destination of i (a vertex in G),
- n_i is the number of seats (capacity) of i available for passengers,
- d_i is the detour distance limit i can tolerate for offering ridesharing services, and
- $\mathcal{P}_i$ is a set of preferred paths of i from s_i to t_i in G.

In what follows, we use the ridesharing problem for the simplified ridesharing problem unless otherwise stated. We say trip i can serve i itself and can serve a trip $j \neq i$ if i and j can arrive at their destinations with j a passenger of i and the detour of i is at most d_i. A trip i can serve a set $\sigma(i)$ of trips if trip i can serve all trips of $\sigma(i)$ and the total detour of i is at most d_i. A trip i can serve at most $n_i + 1$ trips (including the driver) at any specific time point but $|\sigma(i)|$ may be greater than $n_i + 1$ if i can serve some passengers after the delivery of some other passengers (known as *re-take passengers* in previous studies). If re-take passengers is not allowed then $|\sigma(i)| \leq n_i + 1$. The ridesharing problem is that given an instance (G, R), G is a graph and $R = \{1, \ldots, l\}$ is a set of trips, find a set $S \subseteq R$ of drivers and a mapping $\sigma : S \to 2^R$ such that (1) for each $i \in S$, i can serve $\sigma(i)$, (2) for each pair $i, j \in S$ with $i \neq j$, $\sigma(i) \cap \sigma(j) = \emptyset$, and (3) $\cup_{i \in S} \sigma(i) = R$. We call (S, σ) a solution for the ridesharing instance. We consider the problem of minimizing $|S|$ (the number of drivers) and the problem of minimizing the total travel distance of the drivers in S.

3   NP-hardness Results

The minimization problems can be further simplified, assuming some of parameters $s_i, t_i, n_i, d_i, \mathcal{P}_i$ satisfying certain conditions specified below:

(C1) Unique destination: all trips have the same destination, that is, $t_i = D$ for every $i \in R$.
(C2) Zero detour: each trip can only serve others on his/her preferred path, that is, $d_i = 0$ for every $i \in R$.
(C3) Fixed path: $\mathcal{P}_i$ has a unique preferred path P_i.

When any of the conditions is satisfied, the problems are simplified by dropping the corresponding parameter. We prove that the minimization problems are still NP-hard even if any two of the conditions are satisfied.

3.1   NP-hardness Result for Non-zero Detour

We prove that the minimization problems are NP-hard when conditions (C1) and (C3) are satisfied but the non-zero detour is allowed. The proof is a reduction from the 3-partition problem to the minimization problems. The decision problem of 3-partition is that given a set $A = \{a_1, \ldots, a_{3k}\}$ of $3k$ positive integers, where $k \geq 2$, $\sum_{i=1}^{3k} a_i = kM$ and $M/4 < a_i < M/2$, whether A can be partitioned into k disjoint subsets $A_1, \ldots, A_k$, each subset has three elements of A and the sum of integers in each subset is M. The decision problem of 3-partition is NP-complete even M is upper bounded by a polynomial in k [9].

Given an instance $A = \{a_1, \ldots, a_{3k}\}$ of the 3-partition problem, we construct an instance (G, R_A) of the ridesharing problem as follows (also see Fig. 1(a)):

- G is a weighted graph with $V(G) = \{I, D, v_1, \ldots, v_{4k}\}$ and $E(G)$ having edge $\{D, I\}$ of weight kM, edges $\{v_i, I\}$ of weight a_i, $1 \leq i \leq 3k$, and edges $\{v_i, I\}$ of weight kM, $3k + 1 \leq i \leq 4k$.
- $R_A = \{1, 2, \ldots, 4k\}$ has $4k$ trips.
  Each trip i, $1 \leq i \leq 4k$, has source $s_i = v_i$ and destination $t_i = D$.
  Each trip i, $1 \leq i \leq 3k$, has $n_i = 0$ (i can only serve i itself) and $d_i \geq 0$.
  Each trip i, $3k + 1 \leq i \leq 4k$, has $n_i = 3$ (i can serve up to three passengers at the same time) and $d_i = 2M$.
  Each trip i, $1 \leq i \leq 4k$, has a unique preferred path between v_i and D in G.

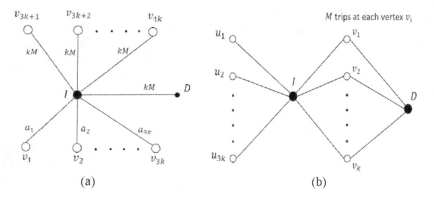

Fig. 1. Ridesharing instances based on a given 3-partition problem instance: (a) (C1) and (C3) are satisfied; (b) (C1) and (C2) are satisfied.

Lemma 1. *Any solution for the instance (G, R_A) has every trip i with $3k + 1 \leq i \leq 4k$ as a driver and total travel distance at least $2k(k+1)M$.*

Proof. In the instance (G, R_A), any trip i for $3k + 1 \leq i \leq 4k$ can not be served by any trip $j \neq i$ because for $3k + 1 \leq j \leq 4k$, the detour in such a service by j is $2kM > 2M$ for $k > 1$ and the detour limit of j is at most $2M$, and for $1 \leq j \leq 3k$, $n_j = 0$. Therefore, any solution for the instance must have every trip i with $3k + 1 \leq i \leq 4k$ as a driver (at least k drivers).

Let S be the set of k drivers i with $3k + 1 \leq i \leq 4k$. The total travel distance of the drivers in S is at least $2kkM$. For each trip j with $1 \leq j \leq 3k$, the total travel distance of drivers in S and trip j is at least $2kkM + 2a_j$ if j is served by a driver in S, otherwise is at least $2kkM + a_j + kM$. Since $a_j < kM$, the minimum total travel distance to realize all trips is to have every j, $1 \leq j \leq 3k$, as a passenger and the distance is $2kkM + \sum_{1 \leq j \leq 3k} 2a_j = 2kkM + 2kM = 2k(k+1)M$. □

Theorem 1. *Minimizing the number of drivers is NP-hard when conditions (C1) and (C3) are satisfied but condition (C2) is not.*

Proof. To get the theorem, we prove that an instance $A = \{a_1, \ldots, a_{3k}\}$ of the 3-partition problem has a solution if and only if the ridesharing problem instance (G, R_A) has a solution of k drivers.

Assume that the 3-partition instance has a solution $A_1, \ldots, A_k$, the sum of elements in each A_j is M. We assign each trip i with $3k + 1 \leq i \leq 4k$ to serve one set A_j for $j = i - 3k$. Driver i can serve the three passengers corresponding to A_j with the total detour distance $2M$ because $n_i = 3$, A_j has three elements and the sum of elements in A_j is M. Hence, we have a solution of k drivers for (G, R_A).

Assume that (G, R_A) has a solution of k drivers. By Lemma 1, every trip i with $3k + 1 \leq i \leq 4k$ must be a driver in the solution and each trip j for $1 \leq j \leq 3k$ can not be a driver in the solution. By the detour limit $d_i = 2M$ and $n_i = 3$, each driver i can serve at most 3 passengers. From this and there are $3k$ passengers, each driver i must serve exactly 3 passengers in the solution. Assume that some driver i has a detour smaller than $2M$. Then from the fact that the sum of elements in A is kM, some driver i' must have a detour greater than $2M$, a contradiction. So the detour of each driver i is exactly $2M$. For each driver i, $3k + 1 \leq i \leq 4k$, let A_j, $j = i - 3k$, be the subset of the three integers of A corresponding to the three passengers served by i. Then $A_1, \ldots, A_k$ is a solution for the 3-partition problem instance.

The size of (G, R_A) is linear in k. It takes a linear time to convert a solution of (G, R_A) to a solution of the 3-partition instance and vice versa. □

Theorem 2. *Minimizing the total travel distance of drivers is NP-hard when conditions (C1) and (C3) are satisfied but condition (C2) is not.*

The proof of the theorem is similar to that for Theorem 1 (details omitted).

3.2   NP-hardness Result for No Fixed Preferred Path

When conditions (C1) and (C2) are satisfied but each trip may have multiple preferred paths, we prove that the minimization problems are NP-hard. Again, the proof is a reduction from the 3-partition problem. Given an instance $A = \{a_1, \ldots, a_{3k}\}$ of the 3-partition problem, we construct an instance (G, R_A) of the ridesharing problem as follows (also see Fig. 1(b)):

- G is a graph with $V(G) = \{I, D, u_1, \ldots, u_{3k}, v_1, \ldots, v_k\}$ and $E(G)$ having edges $\{u_i, I\}, 1 \leq i \leq 3k$, $\{I, v_i\}$ and $\{v_i, D\}, 1 \leq i \leq k$. Each edge has weight 1.

– $R_A = \{1, \ldots, 3k + kM\}$ trips.

Each trip $i, 1 \leq i \leq 3k$, has source $s_i = u_i$, destination $t_i = D$, $n_i = a_i$, $d_i = 0$ and k preferred paths $\{u_i, I\}, \{I, v_j\}, \{v_j, D\}$, $1 \leq j \leq k$.

Each trip $i, 3k + 1 \leq i \leq 3k + kM$, has source $s_i = v_j, j = \lceil (i - 3k)/M \rceil$, destination $t_i = D$, $n_i = 0$, $d_i = 0$ and a unique preferred path $\{v_j, D\}$.

Lemma 2. *Any solution for the instance (G, R_A) has every trip $i, 1 \leq i \leq 3k$, as a driver and total travel distance at least $9k$.*

Proof. Since condition (C2) is satisfied (detour is not allowed), every trip i, $1 \leq i \leq 3k$, must be a driver in any solution. A solution with exactly $3k$ drivers $i, 1 \leq i \leq 3k$, has total travel distance $9k$ and any solution with a trip $i, 3k + 1 \leq i \leq 3k + kM$, as a driver has total travel distance greater than $9k$. $\square$

Theorem 3. *Minimizing the number of drivers is NP-hard when conditions (C1) and (C2) are satisfied but condition (C3) is not.*

Proof. We prove the theorem by showing that an instance $A = \{a_1, \ldots, a_{3k}\}$ of the 3-partition problem has a solution if and only if the ridesharing problem instance (G, R_A) has a solution of $3k$ drivers.

Assume that the 3-partition instance has a solution $A_1, \ldots, A_k$. Then for each $A_j = \{a_{j_1}, a_{j_2}, a_{j_3}\}, 1 \leq j \leq k$, we assign the three trips j_1, j_2, j_3 as drivers to serve the $n_{j_1} + n_{j_2} + n_{j_3} = a_{j_1} + a_{j_2} + a_{j_3} = M$ trips with sources at vertex v_j. This gives a solution of $3k$ drivers for (G, R_A).

Assume that (G, R_A) has a solution of $3k$ drivers. By Lemma 2, each trip $i, 1 \leq i \leq 3k$, is a driver and each trip $j, 3k + 1 \leq j \leq 3k + kM$ is a passenger in the solution, total kM passengers. Since $\sum_{1 \leq i \leq 3k} a_i = kM$, each driver $i, 1 \leq i \leq 3k$ serves exactly a_i passengers. Since $a_i < M/2$ for every $1 \leq i \leq 3k$, at least three drivers are required to serve the M passengers with sources at each vertex $v_j, 1 \leq j \leq k$. Therefore, the solution of $3k$ drivers has exactly three drivers j_1, j_2, j_3 to serve the M passengers with sources at the vertex v_j, implying $a_{j_1} + a_{j_2} + a_{j_3} = M$. Let $A_j = \{a_{j_1}, a_{j_2}, a_{j_3}\}, 1 \leq j \leq k$, we get a solution for the 3-partition instance.

The size of (G, R_A) is polynomial in k. It takes a polynomial time to convert a solution of (G, R_A) to a solution of the 3-partition instance and vice versa. $\square$

Theorem 4. *Minimizing the total travel distance of drivers is NP-hard when conditions (C1) and (C2) are satisfied but condition (C3) is not.*

The proof of the theorem is similar to that for Theorem 3 (details omitted).

3.3   NP-hardness Result for Non-unique Destinations

We show that the minimization problems are NP-hard when conditions (C2) and (C3) are satisfied but trips may have distinct destinations. The proof is a reduction from the Interval Scheduling with Machine Availability Problem (ISMAP) [3], which is also called the k-Track Assignment Problem [5].

ISMAP is a machine scheduling problem: Let I be an instance of m machines and n jobs. Each machine i, $1 \leq i \leq m$, has an operational interval $[a_i, b_i)$, a_i is the start time and b_i is the end time. Each job j, $1 \leq j \leq n$, has a process interval $[p_j, q_j)$, p_j is the start time and q_j is the end time. We may simply call $[a_i, b_i)$ and $[p_j, q_j)$ an *interval* and call each of a_i, b_i, p_j, q_j an *end point* of an interval. Each job can only be processed by one machine and each machine can process at most one job at any time point. A *schedule* for I is an assignment of all jobs to machines such that every job is assigned to one machine, if job j is assigned to machine i then $[p_j, q_j) \subseteq [a_i, b_i)$, and if jobs j and j' are assigned to machine i then $[p_j, q_j) \cap [p_{j'}, q_{j'}) = \emptyset$. The decision version of ISMAP is that: Given an instance I, is there a schedule for I? The decision problem is NP-complete [3,5].

Given an ISMAP instance I, we first construct another ISMAP instance I' such that there is a schedule for I if and only if there is a schedule for I'. Then we construct an instance of the ridesharing problem such that there is an optimal solution for the ridesharing problem if and only if there is a schedule for I'.

The construction of I' is as follows: We assume that $b_i \leq b_{i+1}$ for $1 \leq i \leq m - 1$ and $a_i \geq 0$ for $1 \leq i \leq m$. For each machine i in I, we extend the operational interval of i to $[i - m - 1, b_i + i)$ and include i as a machine in I'. Every job j in I is included in I'. For each machine $i \in I'$, we create a new job x_i with process interval $[p_{x_i}, q_{x_i}) = [i - m - 1, a_i)$ and a new job y_i with process interval $[p_{y_i}, q_{y_i}) = [b_i, b_i + i)$. Then each of jobs x_i and y_i can be processed only by machine i in any schedule for I'.

Lemma 3. *There is a schedule for I if and only if there is a schedule for I'.*

Proof. Assume that there is a schedule S for I. Then in addition to S, assigning jobs x_i and y_i to machine i, $1 \leq i \leq m$, gives a schedule for I'. Assume that there is a schedule S' for I'. Then removing jobs x_i and y_i, $1 \leq i \leq m$, from S' gives a schedule for I. The size of I' is linear in $|I|$. A schedule of I can be computed from a schedule of I' in linear time and vice versa. $\square$

Given an instance I' of m machines and $n + 2m$ jobs, we construct an instance (G, R_I) for the ridesharing problem as follows:

- G is a graph with $V(G) = \{u \mid u$ is an end point of an interval in $I'\}$ and $E(G) = \{\{u, v\} \mid u < v$ and no $w \in V(G)$ with $u < w < v\}$. Notice that G is a path.
- R_I is the set of trips defined below.
  For each machine i, $1 \leq i \leq m$, of interval $[i - m - 1, b_i + i)$, three trips are created,
  (1) trip i (corresponding to machine i) with source $s_i = i - m - 1$, destination $t_i = b_i + i$, $n_i = 1$, $d_i = 0$, and the unique preferred path between $i - m - 1$ and $b_i + i$ in G;
  (2) trip $m + i$ (corresponding to job x_i of interval $[i - m - 1, a_i)$) with source $s_{m+i} = i - m - 1$, destination $t_{m+i} = a_i$, $n_{m+i} = 0$, $d_{m+i} = 0$, and the unique preferred path between $i - m - 1$ and a_i in G; and

(3) trip $2m + i$ (corresponding to job y_i of interval $[b_i, b_i + i)$) with source $s_{2m+i} = b_i$, destination $t_{2m+i} = b_i + i$, $n_{2m+i} = 0$, $d_{2m+i} = 0$, and the unique preferred path between b_i and $b_i + i$ in G.

For each job j of interval $[p_j, q_j)$, $1 \leq j \leq n$, a trip $i = 3m + j$ is created with source $s_i = p_j$, destination $t_i = q_j$, $n_i = 0$, $d_i = 0$, and the unique preferred path between p_j and q_j in G.

Lemma 4. *Every solution for the instance (G, R_I) has every trip i, $1 \leq i \leq m$, as a driver.*

Proof. From the zero-detour condition (C2), a trip i can serve a trip $j \neq i$ if the interval of j is a subset of the interval of i in I' and $n_i > 0$. Since the interval of any trip i, $1 \leq i \leq m$, is not a subset of the interval of any trip other than i, trip i can not be served by any trip other than i. Further, any trip i, $m + 1 \leq i \leq 3m + n$, can not serve any trip other than i because $n_i = 0$. So any solution for the instance $(G.R_I)$ must include every trip i, $1 \leq i \leq m$, as a driver. $\qquad\square$

Theorem 5. *Minimizing the number of drivers is NP-hard when conditions (C2) and (C3) are satisfied but condition (C1) is not.*

Proof. Given an ISMAP instance I, we construct an instance I' as shown above. By Lemma 3, I has a schedule if and only if I' has a schedule. We prove that I' has a schedule if and only if instance (G, R_I) has a solution of m drivers.

Assume that I' has a schedule. Then for every job assigned to machine i, $1 \leq i \leq m$, the interval of the job is a subset of the interval of i and thus trip i can serve the trip corresponding to the assigned job. Further, there is no overlap between the intervals of any two jobs assigned to machine i. Therefore, trip i can serve all trips corresponding to the jobs assigned it. Thus, trips i, $1 \leq i \leq m$, and the assignment of jobs to every trip i give a solution of m drivers for (G, R_I).

Let (S, σ) be a solution of m drivers for (G, R_I). By Lemma 4, S has all trips i, $1 \leq i \leq m$, as the drivers and every trip j, $m + 1 \leq j \leq 3m + n$, is served by a driver. A schedule for I' can be obtained by assigning each job corresponding to a passenger in $\sigma(i)$ to machine i.

The size of (G, R_I) is linear in $|I'|$ and $|I|$. A schedule for I' and that for I can be computed from a solution of m drivers in linear time and vice versa. $\qquad\square$

Theorem 6. *Minimizing the total travel distance of drivers is NP-hard when conditions (C2) and (C3) are satisfied but condition (C1) is not.*

The proof of the theorem is similar to that for Theorem 5 (details omitted).

4   A Polynomial Time Solvable Case

We give a polynomial time exact algorithm for the problem of minimizing the number of drivers when all conditions (C1), (C2) and (C3) are satisfied, and the preferred paths of all trips R lie on a single path of the road network G

(i.e., the graph induced by the preferred paths of all trips is a path in G). For this minimization problem, a problem instance can be expressed by a set $R = \{1, \ldots, l\}$ of l trips on a graph G with $V(G) = \{0, 1, \ldots, l\}$ and $E(G) = \{\{i, i+1\} \mid 0 \leq i \leq l - 1\}$. Each trip i has $s_i \in \{1, \ldots, l\}$, destination $t_i = 0$, $n_i \geq 0$, $d_i = 0$ and the unique path P_i between s_i and 0. We can further assume without loss of generality that $s_i < s_{i+1}, 1 \leq i \leq l - 1$ (i.e., for each $i \in R$, $s_i = i$).

Algorithm RFP (Ridesharing-For-Path)
Input: Graph G and $R = \{1, .., l\}$ with $0 < s_1 < s_2 < ... < s_l$.
Output: a solution (S, σ) to the instance (G, R) such that $|S|$ is minimized.
begin
    $S := \{1\}$; $\sigma(1) := \{1\}$;
    for $i = 2$ **to** l **do**
        $S := S \cup \{i\}$; $\sigma(i) := \{i\}$; compute free(i) and S_f; $k := $Find-Target$(i)$;
        while $k \neq 0$ **do** /* serve $\sigma(k)$ by drivers in S_f and remove k from S */
            for each $j \in S_f$, $k < j < i$ **do**
                move free(j) trips from $\sigma(k)$ to $\sigma(j)$ and update free(j);
            end for
            move the remaining trips of $\sigma(k)$ to $\sigma(i)$ and update free(i);
            remove k from S and update S_f;
            if free$(i) \geq 1$ **and** $|S| \geq 2$ **then** $k := $Find-Target$(i)$;
        end while
    end for
end.
Procedure Find-Target(i)
begin
    for each $j \in S \setminus \{i\}$ **do** gap$(i, j) := |\sigma(j)| - \sum_{a \in S_f, j < a < i}$ free(a);
    Let $k = \arg_{j \in S \setminus \{i\}} \min\{gap(i, j)\}$;
    if free$(i) > $ gap(i, k) **then** return k **else** return 0;
end.

Fig. 2. Algorithm for minimizing the number of drivers in ridesharing.

4.1 Algorithm

Given an instance (G, R) and a (partial) solution (S, σ) for R, we define the following for each driver in S:

- free$(i) = n_i - |\sigma(i)| + 1$ is the number of additional trips i can serve.
- $S_f := \{i \mid i \in S$ and free$(i) > 0\}$ is the set of drivers in S who can serve additional trips.

Algorithm RFP (Fig. 2) processes every trip i of R, from $i = 1$ to l. In processing i, the algorithm finds a minimum number of drivers for the trips from 1 to i. At any execution point of Algorithm RFP, whenever a trip is assigned to be a passenger, it remains as a passenger throughout the algorithm. On the other hand, an assigned driver can be changed to a passenger when a new trip is processed. Procedure Find-Target computes a driver $k \in S$ to be removed from S and an integer gap. To remove a driver k from S, the trips of $\sigma(k)$ will be served by drivers j in S_f for $k < j \leq i$, first by those other than i and then by i. The value $gap(i, k)$ is the minimum number of seats required from i to remove a driver from S and k is the driver removed from S by $gap(i, k)$ seats from i.

4.2   Analysis of Algorithm

Obviously Algorithm RFP finds a solution in polynomial time. We prove that the algorithm finds a minimum solution. For $1 \leq j < i$, let $R(j, i) = \{k \mid j \leq k \leq i\}$. For a (partial) solution (S, σ) for R, let $S(j, i) = S \cap R(j, i)$ be the set of drivers in $R(j, i)$.

Lemma 5. *Let (S, σ) be the solution found by Algorithm RFP processed $R(1, i)$ only and (S^*, σ^*) be any solution for R. If there is a driver k in S that serves a passenger y, then for every driver j in $S(y + 1, k - 1)$, $|\sigma(j)| \geq |\sigma^*(j)|$.*

Proof. Since y is served by k in S, $\text{free}(j) = 0$ for $y < j < k$ by the algorithm when y is put into $\sigma(k)$. From this, $|\sigma(j)| = n_j + 1 \geq |\sigma^*(j)|$.   □

Lemma 6. *Let (S, σ) be the solution found by Algorithm RFP processed $R(1, i)$ only. If there is an optimal solution (S_1^*, σ_1^*) for R with $S_1^*(1, i) \subseteq S$, then an optimal solution (S^*, σ^*) for R can be constructed such that $S^* = S_1^*$ and for every $j \in S^*(1, i)$, $\sigma(j) \subseteq \sigma^*(j)$ (**inclusion property**).*

Proof sketch. Let (S, σ) be the solution found by Algorithm RFP processed $R(1, i)$ only and (S_1^*, σ_1^*) be an optimal solution for R with $S_1^*(1, i) \subseteq S$. We check the inclusion property for every element of $S_1^*(1, i)$ in the decreasing order. If the inclusion property does not hold on some driver $k \in S_1^*(1, i)$, we modify σ_1^* to make the inclusion property hold for every j with $k \leq j \leq i$. When the inclusion property is checked for every element of $S_1^*(1, i)$, we get the lemma.

Let $Y_j = \sigma(j) \setminus \sigma_1^*(j)$ for $j \in S_1^*(1, i)$. Assume that $Y_k \neq \emptyset$ for some $k \in S_1^*(1, i)$ and $Y_j = \emptyset$ (inclusion property holds) for every $j \in S_1^*(k + 1, i)$. Every $y_0 \in Y_k$ is served by a driver $j_0 \neq k$ in solution S_1^*. If $\sigma_1^*(k) \subset \sigma(k)$ then we move y_0 from $\sigma_1^*(j_0)$ to $\sigma_1^*(k)$. After the move, we have the following **progress:**

– The size of Y_k is reduced by one, the size of Y_j for every $j \in S_1^*(1, i)$ with $j \neq k$ is non-increasing, and the inclusion property holds for every $j \in S_1^*(k + 1, i)$.

Suppose that $\sigma_1^*(k) \not\subset \sigma(k)$. Let u be a passenger in $\sigma_1^*(k) \setminus \sigma(k)$. If $u < j_0$ then we move y_0 from $\sigma_1^*(j_0)$ to $\sigma_1^*(k)$ and move u from $\sigma_1^*(k)$ to $\sigma_1^*(j_0)$ (Fig. 3(a)). After this, only $\sigma_1^*(k)$ and $\sigma_1^*(j_0)$ are changed. The size of Y_k is reduced by one. Further, the size of Y_{j_0} is not increased and if $j_0 \in S_1^*(k+1, i)$ then the inclusion property holds for every $j \in S_1^*(k + 1, i)$ because $y_0 \notin \sigma(j_0)$. Therefore, we get a progress.

Assume that $u > j_0$. Then $y_0 < j_0 < u < k$ and $y_0 \in \sigma(k)$. By Lemma 5, $|\sigma(j_0)| \geq |\sigma_1^*(j_0)|$. From this and $y_0 \in \sigma_1^*(j_0)$ but $y_0 \notin \sigma(j_0)$, there is a $y_1 \in$

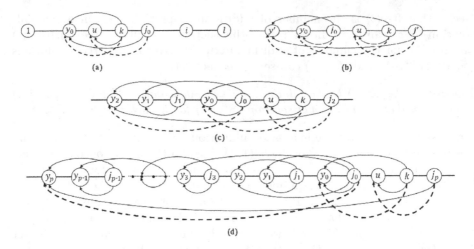

Fig. 3. Different modifications of (S_1^*, σ_1^*) (**move processes**) to have a **progress**. Trips in R are represented by vertices on a path. Each arc (u,v) from vertex u to vertex v denotes that driver u serves passenger v in some solution. The solid arcs above the path represent (S, σ), the solid arcs below the path represent (S_1^*, σ_1^*), and the dashed arcs below the path represent the modified (S_1^*, σ_1^*) after a move process.

$\sigma(j_0) \setminus \sigma_1^*(j_0)$ such that $y_1 \in \sigma_1^*(j_1)$ and $j_1 \neq j_0$. We try to move y_0 to $\sigma_1^*(k)$ to have a progress by the following **move process**:

– For $y_0 \in Y_k$, if there are $u \in \sigma_1^*(k) \setminus \sigma(k)$, j_0 and j' such that $y_0 \in \sigma_1^*(j_0)$, $y' \in \sigma(j_0)$, $y' \in \sigma_1^*(j')$ and $j' > u$, then we move y_0 from $\sigma_1^*(j_0)$ to $\sigma_1^*(k)$, move y' from $\sigma_1^*(j')$ to $\sigma_1^*(j_0)$ and move u from $\sigma_1^*(k)$ to $\sigma_1^*(j')$ (Fig. 3(b)).

If $j_1 > u$ then we apply the move process with $j_1 = j'$ and $y_1 = y'$. It can be shown that after the move process, we get a progress. Assume that $j_1 < u$ (**extend-case**). Similar to the existence of y_1, there is a passenger $y_2 \in \sigma(j_1) \setminus \sigma_1^*(j_1)$ such that $y_2 \in \sigma_1^*(j_2)$ and $j_2 \neq j_1$. Notice that $y_2 \neq y_1$ and $y_2 \neq y_0$ since $j_1 < u < k$ ($j_1 \neq j_0 \neq k$). If $j_2 > u$ then we apply the move process with $j_2 = j'$ and $y_2 = y'$ to get a progress (Fig. 3(c)). Otherwise ($j_2 < u$), we have the extend-case.

Assume that the extend case continues and we have a chain of distinct passengers $y_0, y_1, \ldots, y_{p-1}$ such that $y_c \in \sigma(j_{c-1})$, $y_c \in \sigma_1^*(j_c)$, and $j_{c-1} \neq j_c$ for $c = 1, \ldots, p-1$, and $y_0 \in \sigma(k)$, $y_0 \in \sigma_1^*(j_0)$, and $j_0, j_1, \ldots, j_{p-1} < u$. It can be shown that there is a $y_p \in \sigma(j_{p-1}) \setminus \sigma_1^*(j_{p-1})$ such that $y_p \in \sigma_1^*(j_p)$ and $j_p \neq j_c$ for every $c = 0, \ldots, p-1$. Since the elements in the chain are distinct, the length of the chain cannot exceed $u - 1$, and there is a j_p with $j_p > u$ for the chain $y_0, \ldots, y_p$. Then we apply the move process with $j_p = j'$ and $y_p = y'$ to get a progress (Fig. 3(d)).

From the above, we can modify (S_1^*, σ_1^*) such that S_1^* does not change and the size of Y_k is reduced by one. Repeat the above, we get $Y_k = \emptyset$ and have the lemma proved. $\qquad\square$

Recall that free(i) is the number of additional trips a driver i in a (partial) solution (S, σ) can serve. In the rest of this section, we will use free$_1(i)$, free$^*(i)$ and free$_1^*(i)$ for the numbers of additional trips a driver i in (partial) solutions $(S_1, \sigma_1), (S^*, \sigma^*)$ and (S_1^*, σ_1^*) can serve, respectively.

Lemma 7. *Let (S, σ) be the solution found by Algorithm RFP processed $R(1, i)$ only. Then there is an optimal solution (S^*, σ^*) for R such that $S^*(1, i) \subseteq S$.*

Proof sketch. We prove the lemma by induction on i. For $i = 1$, $S = \{1\}$ and for any optimal solution S^* for R, $S^*(1, 1) \subseteq S$. So the induction base is true. Assume that for $i - 1 \geq 1$, the lemma holds and we prove it for i. Let (S, σ) and (S_1, σ_1) be the solutions found by Algorithm RFP processed $R(1, i)$ and $R(1, i - 1)$ only, respectively. By the induction hypothesis, there is an optimal solution (S_1^*, σ_1^*) such that $S_1^*(1, i-1) \subseteq S_1$. If $S_1^*(1, i-1) \subseteq S$ then $S_1^*(1, i) \subseteq S$ because $i \in S$. Taking $S^* = S_1^*$, we get the lemma. So we assume that the set $W = \{w \mid w \in S_1^*(1, i - 1)$ and $w \notin S\}$ is not empty. We show that another optimal solution (S^*, σ^*) can be found by modifying S_1^* with every $w \in W$ removed from S_1^* such that $S^*(1, i) \subseteq S$.

We start the modification from a **trivial case** where $S \subseteq S_1^*(1, i) \cup \{i\}$. Let $S^* = (S_1^* \setminus W) \cup \{i\}$. Then $|S^*(1, i)| = |S_1^*(1, i)| - |W| + 1$ if $i \notin S_1^*$, otherwise $|S^*(1, i)| = |S_1^*(1, i)| - |W|$. Since S serves all trips in $R(1, i)$, we can move every trip in $\sigma_1(j), j \in W$ to $\sigma^*(k)$ for some $k \in S^*(1, i)$. Since $|W| \geq 1$, $|S^*| \leq |S_1^*|$ and S^* is optimal. Notice that in the trivial case, $i \notin S_1^*$ and $|W| = 1$, otherwise, $|S^*| < |S_1^*|$, a contradiction since S_1^* is optimal.

Assume that $S \nsubseteq S_1^*(1, i) \cup \{i\}$. The idea for finding (S^*, σ^*) is to modify S_1^* by replacing a driver $w \in W$ with a driver in $(S_1 \setminus S_1^*) \cup \{i\}$. Let w be an arbitrary element of W and $Z = S_1 \setminus S_1^*$. Notice that $S \cap Z \neq \emptyset$, otherwise $S \subseteq S_1^*(1, i) \cup \{i\}$. By Lemma 6, we assume that the inclusion property holds for every $j \in S_1^*(1, i - 1)$. The modification is divided into two cases.

Case 1. There is a $z \in Z$ with $z < w$. We further assume that for some $z \in Z$ with $z < w$, gap$(i, w) \leq$ gap(i, z). It can be shown that $|\sigma_1(z)| \geq |\sigma_1^*(w)| + \sum_{j \in S_1^*(z+1, w)}$ free$_1^*(j)$, implying that if we remove $|\sigma_1(z)|$ trips from $\sigma_1^*(j)$ for $j \in S_1^*$, $S_1^*(w + 1, l)$ can serve at least $|\sigma_1^*(w)|$ additional trips. We apply the following **update process** to modify (S_1^*, σ_1^*):

- Add z to S_1^*, define $\sigma_1^*(z) = \sigma_1(z)$ and remove the trips of $\sigma_1^*(z)$ from $\sigma_1^*(j)$ for $j \neq z$.

After the update, $S_1^*(w + 1, l)$ can serve at least $|\sigma_1^*(w)|$ additional trips. We move the trips of $\sigma_1^*(w)$ to $\sigma_1^*(j), j \in S_1^*(w + 1, l)$ and remove w from S_1^*. By this, the size of S_1^* is not changed, the inclusion property still holds for every $j \in S_1^*(1, i - 1)$, the size of W is reduced by one, and the size of Z is reduced by one.

Assume that for every $z \in Z$ with $z < w$, gap$(i, w) >$ gap(i, z). We apply the update process taking some $z \in S \cap Z$ ($S \cap Z \neq \emptyset$). After the update, it can be shown that $S_1^*(w + 1, l)$ can serve at least $n_z + 1 \geq |\sigma_1^*(w)|$ additional trips after the update. We move the trips of $\sigma_1^*(w)$ to $\sigma_1^*(j)$, $j \in S_1^*(w+1, l)$ and remove w from S_1^*. By this, the size of S_1^* is not changed, the inclusion property still holds for every $j \in S_1^*(1, i-1)$, the size of W is reduced by one, and the size of Z is reduced by one.

Case 2. For every $z \in Z$, $z > w$. Let z_0 be the minimum in Z. It can be shown that $S_1(w, z_0 - 1) = S_1^*(w, z_0 - 1)$ and $\text{free}_1^*(j) = \text{free}_1(j)$ for every $j \in S_1^*(w, z_0 - 1)$.

Assume that $\text{gap}(i, w) > \text{gap}(i, z_0)$. It can be shown that there is a $z \in S \cap Z$ with $w < z_0 < z$ and $\text{gap}(i, w) \le \text{gap}(i, z)$ ($n_z \ge n_w$). We apply the update process taking this z. After the update, $S_1^*(w+1, l)$ can serve at least $n_z + 1 \ge |\sigma_1^*(w)|$ additional trips. We move the trips of $\sigma_1^*(w)$ to $\sigma_1^*(j)$, $j \in S_1^*(w+1, l)$ and remove w from S_1^*. By this, the size of S_1^* is not changed, the inclusion property still holds for every $j \in S_1^*(1, i-1)$, the size of W is reduced by one, and the size of Z is reduced by one.

Assume that $\text{gap}(i, w) \le \text{gap}(i, z_0)$. We apply the update process with $z_0 = z$. It can be shown that after the update, $|\sigma_1^*(w)| = |\sigma_1(w)|$ and we can move the trips of $\sigma_1^*(w)$ to $\sigma_1^*(j)$, $j \in S_1^*(w+1, l)$, and remove w from S_1^*. By this, the size of S_1^* is not changed, the inclusion property still holds for every $j \in S_1^*(1, i-1)$, the size of W is reduced by one, and the size of Z is reduced by one.

Repeat the processes in Cases 1 and 2, we get an optimal solution S_1^* with either $W = \emptyset$ or $Z = \emptyset$. For $W = \emptyset$, $S^* = S_1^*$ is the solution we want to find. For $Z = \emptyset$, $S \subseteq S_1^*(1, i) \cup \{i\}$ and by the trivial case, we can get S^*.   □

Theorem 7. *Algorithm RFP finds a solution (S, σ) for input instance (G, R) with the minimum number of drivers in $O(l^2)$ time.*

Proof. First, observe that (S, σ) is a solution for R since every trip in R is processed one by one and assigned to be a driver initially. By Lemma 7, there is an optimal solution (S^*, σ^*) for R such that $S^* \subseteq S$. Assume that $S^* \subset S$. By Lemma 6, we can modify σ^* such that the inclusion property holds, that is, $\sigma(j) \subseteq \sigma^*(j)$ for every $j \in S \cap S^*$. Let z be a driver in $S \setminus S^*$. Then all trips of $\sigma(z)$ must be served by drivers in $S^*(z + 1, l)$. By the inclusion property, $\text{free}^*(j) \le \text{free}(j)$ for every $j \in S^*(1, l)$. This implies that $|\sigma(z)| \le \sum_{j \in S(z+1, l)} \text{free}(j)$, which is a contradiction to Algorithm RFP as $\sigma(z)$ should have been served by S. Therefore, it must be the case that $S^* = S$. It is not difficult to see that Algorithm RFP has running time $O(l^2)$.   □

Algorithm RFP works for a more general case: The preferred paths of trips lie on multiple vertex-disjoint paths (except at the destination) of the road network. Algorithm RFP finds an optimal solution for the trips on a same path and the union of the optimal solutions is an optimal solution for R.

5   Concluding Remarks

We proved that the problems of minimizing the number of vehicles and minimizing the total distance of vehicles for realizing all given trips in the ridesharing problem are still NP-hard even any two conditions of (C1), (C2) and (C3) are satisfied. We also showed that the problem of minimizing the number of vehicles is polynomial time solvable when all three conditions are satisfied, and the preferred paths of participants satisfy an additional condition. It is open whether the problem is NP-hard or polynomial time solvable when all three conditions are satisfied but each trip can have an arbitrary unique preferred path in the road network. It is interesting to develop efficient practical algorithms and approximation algorithms for the minimization problems.

Acknowledgment. The authors thank anonymous reviewers for their constructive comments. The work was partially supported by Canada NSERC Engage/Discovery Grants and China NSFC Grant 11531014.

References

1. Agatz, N., Erera, A., Savelsbergh, M., Wang, X.: Dynamic ride-sharing: a simulation study in metro Atlanta. Transp. Res. Part B **45**(9), 1450–1464 (2011)
2. Agatz, N., Erera, A., Savelsbergh, M., Wang, X.: Optimization for dynamic ridesharing: a review. Eur. J. Oper. Res. **223**, 295–303 (2012)
3. Antoon, K., Lenstra, J.K., Papadimitriou, C., Spieksma, F.: Interval scheduling: a survey. Naval Res. Logistics **54**(5), 530–543 (2007)
4. Baldacci, R., Maniezzo, V., Mingozzi, A.: An exact method for the car pooling problem based on lagrangean column generation. Oper. Res. **52**(3), 422–439 (2004)
5. Brucker, P., Nordmann, L.: The k-track assignment problem. Computing **54**, 97–122 (1994)
6. Chan, N.D., Shaheen, S.A.: Ridesharing in North America: past, present, and future. Transp. Rev. **32**(1), 93–112 (2012)
7. Cordeau, J.-F., Laporte, G.: The dial-a-ride problem: models and algorithms. Ann. Oper. Res. **153**(1), 29–46 (2007)
8. Furuhata, M., Dessouky, M., Ordóñez, F., Brunet, M., Wang, X., Koenig, S.: Ridesharing: the state-of-the-art and future directions. Transp. Res. Part B: Methodol. **57**, 28–46 (2013)
9. Garey, M.R., Johnson, D.S.: Computers and Intractability: A Guide to the Theory of NP-Completeness. W.H. Freeman and Company, New York (1979)
10. Ghoseiri, K., Haghani, A., Hamedi, M.: Real-time rideshare matching problem. Final report of UMD-2009-05. U.S. Department of Transportation (2011)
11. Herbawi, W., Weber, M.: The ridematching problem with time windows in dynamic ridesharing: a model and a genetic algorithm. In: Proceedings of ACM Genetic and Evolutionary Computation Conference (GECCO), pp. 1–8 (2012)
12. Huang, Y., Bastani, F., Jin, R., Wang, X.S.: Large scale real-time ridesharing with service guarantee on road networks. Proc. VLDB Endowment **7**(14), 2017–2028 (2014)
13. Kelley, K.: Casual carpooling enhanced. J. Public Transp. **10**(4), 119–130 (2007)
14. Morency, C.: The ambivalence of ridesharing. Transportation **34**(2), 239–253 (2007)
15. Santos, A., McGuckin, N., Nakamoto, H.Y., Gray, D., Liss, S.: Summary of travel trends: 2009 national household travel survey. Technical report, US Department of Transportation Federal Highway Administration (2011)

On the Complexity of Bounded Deletion Propagation

Dongjing Miao[1], Yingshu Li[1(✉)], Xianmin Liu[2], and Jianzhong Li[2]

[1] Department of Computer Science, Georgia State University,
Atlanta, GA 30303, USA
yili@gsu.edu

[2] Harbin Institute of Technology, Harbin, Heilongjiang 150001, China

Abstract. Deletion propagation problem is a class of view update problem in relational databases [1]. Given a source database D, a monotone relational algebraic query Q, the view V generated by the query $Q(D)$ and an update on view ΔV, deletion propagation is to find a side effect free update ΔD on database D such that $Q(D \backslash \Delta D) = V \backslash \Delta V$. In general, the database updated may be very distant from the original database. In this paper, we propose a new approach, bounded version deletion propagation problem (b-dp for short), where number of tuples deleted '$|\Delta D|$' is bounded by constant b, in which it aims to find the view side-effect free and bounded ΔD, then analyze its computational complexity. Our results show that in many cases both the data and combined complexity drop, even for functional dependency restricted version deletion propagation.

Keywords: Bounded deletion propagation · View update · Database · Complexity

1 Introduction

The ability to analyze the impact of view update on database is a central problem for data provenance and quality research. In the study of view update in database [2–6], as an important problem, propagation analysis [1] of view update has been studied for a long time. Propagation analysis mainly focuses on deciding whether there is a side-effect free deletion, and minimizing side-effect over source database or view which caused by the asymmetry between update on view and the source database.

The fundamental decision problem of propagation analysis is view side effect free deletion propagation, first defined in [1], which can be stated as follows: given a source database D, an monotone relational query Q, the view $V = Q(D)$, and update on view (a set of tuples) ΔV, the view side-effect free deletion propagation is to decide if there is a ΔD such that $Q(D \backslash \Delta D) = V \backslash \Delta V$, i.e., side effect free whenever such ΔD exists.

ⓒ Springer International Publishing AG 2016
T.-H.H. Chan et al. (Eds.): COCOA 2016, LNCS 10043, pp. 453–462, 2016.
DOI: 10.1007/978-3-319-48749-6_33

Example 1. Let's visit an example of the deletion propagation. Consider an file management database of a company including two relations, Dept(dept, user) records department each user belongs to, and Author(dept, file) records files that each group has the authority to access. There is also a view defined as a conjunctive query (Selection-Projection-Join) "show the file and users have authority to access it" as follow,

Dept :

dept	user
d1	u1
d2	u2
d2	u3

Author :

dept	file
d1	f1
d2	f2
d2	f3

$\pi_{user,file}(\text{Dept} \bowtie \text{Author})$:

user	file
u1	f1
u2	f2
u2	f3
u3	f2
u3	f3

now given a deletion on view V, '$\Delta V = (u1, f1)$', the task is to find a side effect free deletion. For this deletion, there does existing view side effect free update over source data D, actually three alternatives,

(*a*) delete '(d1, u1)' from Dept,
(*b*) delete '(d1, f1)' from Author,
(**c**) delete both '(d1, u1)' and '(d1, f1)' from both Dept and Author respectively.

any deletion will not bring side effect into the view. However, for another view deletion '$\Delta V = (u2, f2)$', there are several possible propagations as follows,

(*a*) delete '(d2, u2)' from Dept,
(*b*) delete '(d2, f2)' from Author,
(*c*) delete both '(d2, u2)' and '(d2, f2)' from both Dept and Author respectively.

Here, '(*a*)' will produce a side effect, '(u2, f3)' is also deleted from view besides ΔV; '(*b*)' will produce another side effect, '(u3, f2)' is also deleted from view besides ΔV; At last, '(*c*)' will produce more side effects, both '(u2, f3)' and '(u3, f2)' are deleted from view besides ΔV. It is easy to see that in this case, deleting ΔV has to result in deletion of some other tuples, therefore, we claim that there is no view side effect free deletion on source database D.   □

There have been some complexity results on the view side effect problem for insertion [7,8] and deletion propagation [1,8–13], moreover, on the data complexity of deletion propagation, Kimfield et al. [9] showed the dichotomy '*head domination*' for every conjunctive query without self joins, deletion propagation is either APX-hard or solvable (in polynomial time) by the unidimensional algorithm, they also showed the dichotomy '*functional head domination*' [10] for *fd*-restricted version, For multiple or group deletion [11], they especially showed the trichotomy for group deletion a more general case. On the combined complexity of deletion propagation, [8,13] showed the variety results for different combination of relational algebraic operators. At the same time, [12] studied the functional dependency restricted version deletion propagation problem and showed the tractable and intractable results on both data and combined complexity aspects. All the previous results showed that, for most cases, the deletion propagation is hard due to the huge searching space.

In this work, we introduce a bounded version of deletion propagation, where an additional bound on the number of tuple deleted from the source database is given. This new approach is interesting because it usually gives lower complexity bounds on formalisms. As we shown in this paper, for the bounded case all query classes become polynomial tractable on the data complexity and in many cases the combined complexity drops. One might argue that a way to overcome these problems is to consider minimizing side effect only. There are, however, instances for which this restriction is too strong and excludes potential updates. Moreover, the crucial difficulty is the high complexity. We can see that even the data complexity under conjunctive query is NP-hard.

Contributions. We introduce a natural version of deletion propagation where the number of tuple deleted from the source database is bounded. This new notion leads to more natural answers in certain scenarios, as in data provenance and quality management, preserves the nice properties of deletion propagation. Usually, the side effects are defined on source and view respectively. In this paper, we study the view side effect, investigate the *bounded view side-effect-free deletion propagation* problem, '*b*-dp' for short, mainly on the complexity aspect of this problem.

We give a detailed complexity analysis (of the combined complexity) for relational algebraic query classes and show that in many cases the complexity drops. For instance, for conjunctive queries, for the single deletion, the data complexity drops from NP-hard to PTime and the combined complexity drops from Σ_2^P-complete to Δ_2^P but not in NP$\cup$coNP, and for the group deletion, the combined complexity also drops to Δ_2^P but is in DP-hard.

2   New Approach and Overview of Results

Let's begin with some necessary definition. A *schema* is a finite sequence $\mathbf{R} = \langle R_1, \ldots, R_m \rangle$ of distinct relation symbols, where each R_i has an arity $r_i > 0$ and includes several attributes, denoted by $R_i = \{A_1, \ldots, A_{r_i}\}$. Each attribute A_i has a corresponding set $dom(A_i)$ which is the domain of values appearing in A_i. An database instance D (over $\mathbf{R}$) is a sequence $\langle R_1^D, \ldots, R_m^D \rangle$, such that each R_i^D is a finite set of tuples $\{t_1, \ldots, t_N\}$, each tuple t_k belongs to the set $dom(A_1) \times \cdots \times dom(A_{r_i})$. We use $R.A_i$ to indicate the attribute A_i of relation R, and also we denote R^D as R without loss of clarity.

Definition 1 (*bounded* view side − effect free deletion propagation, b-*dp*). *Given a database D, a positive integer b, a query Q, its view V and a set of tuples deleted ΔV, it is to decide whether there is a tuple set ΔD such that*

(1) $|\Delta D| \leq b$,
(2) $Q(D \backslash \Delta D) = V \backslash \Delta V$.

In this paper, we study the complexity of b-dp in two cases *single* deletion when $|\Delta V| = 1$, and *group* deletion when $|\Delta V| > 1$. The query Q is written by operations in relational algebra including S (selection), P (projection), J (join), U (union). For the sake of clarity, we list notations again as follows.

$D, \Delta D$	Source database, and its update
$V, \Delta V$	View, update on view
b, Q	Constant bound, relational algebraic query
S, P, J, U	Selection, Projection, Join, Union

We examine the impact of different combinations of these factors on both data and combined complexity of these problems. *Data complexity* is the complexity expressed in terms of the size of the database only, while *combined complexity* is the complexity expressed in terms of both the size of the database and the query expression [14] (Table 1).

Table 1. The data complexity of bounded, general, fd-restricted deletion propagation

Query	Single deletion			Group deletion		
	b-(fd-)dp	dp [1]	fd-dp [12]	b-(fd-)dp	dp [1]	fd-dp [12]
SJU	PTime	NP-hard	NP-hard	PTime	NP-hard	NP-hard
SPU		PTime			PTime	
SPJ		NP-hard			NP-hard	
SPJU						

Table 2. The combined complexity of bounded, general, fd-restricted deletion propagation

Query	Single deletion			Group deletion		
	b-(fd-)dp	dp [7,8]	fd-dp [12]	b-(fd-)dp	dp [7,8]	fd-dp [12]
SJU	PTime	NP-hard	NP-hard	PTime	NP-hard	NP-hard
SPU		PTime			PTime	
SPJ	NP-hard, coNP-hard, Δ_2^P	Σ_2^P	Σ_2^P-complete	DP-hard, Δ_2^P	Σ_2^P	Σ_2^P-complete
SPJU						

The complexity measure follows the work [1] where the complexity results of propagation problem were first established and the studies [3,15,16] where the complexity of view update problems was studied. We provide a complete picture of the complexity on these problems for views defined in various fragments of

SPJU queries, identifying all those cases that are intractable. Especially, when the input data is inconsistent and the deletion propagation is restricted by functional dependencies, the previous results showed it is even harder, for example, it belongs to Σ_2^P-complete for conjunctive query. However, our bounded method can also lower the complexity at presence of functional dependency constraints, concretely, the complete picture we map out as the follow table (Table 2).

3   Bounded Deletion Propagation

In the following, we show specifically the results in domination positions including both data and combined complexity of group and single deletion case.

Theorem 1. *b-dp is* PTime *for SPU and SJU query under group deletion on combined complexity.*

Proof

(a) This is easy that one can enumerate all the possible bounded deletions of the input database D, there are $O(|D|^b)$ candidates in all. Then, for each of them, compute the query result, and check if it equals to $V \backslash \Delta V$, if none of them can give the right answer, then we can claim that there is no valid bounded deletion for the input instance of b-dp.

(b) We should first briefly introduce some definitions. Given any SJ query Q with an equivalent standard form $\sigma_c(R_1 \times \cdots \times R_m)$, each R_i $(1 \leq i \leq m)$ is a table included in D, and any two tables in $R_1, \cdots, R_m$ may be the same table in the database, if Q includes self-join. For a tuple $t \in V$, we now define its *inverse* on D with respect to Q: for each i, let $t_{R_i}^-$ be the projection on R_i of t. We also know that an SJU query has the equivalent standard form $Q_1 \cup \cdots \cup Q_k$, where each Q_i is a standard SJ query. Then, the following algorithm works polynomially,

    (a) Given an instance $\langle D, Q, V, \Delta V, b \rangle$, enumerate each possible b-bounded deletion ΔD polynomially;

    (b) For each ΔD enumerated, check if $Q(D \backslash \Delta D) = V \backslash \Delta V$. Concretely, for each ΔD, we check if (1) for each tuple $t \in (V \backslash \Delta V)$, there is a SJ query Q_i involved in Q such that each R_i in $D \backslash \Delta D$ contains $t_{R_i}^-$, and (2) there is at least a tuple t in ΔV, in each SJ query Q_i involved in Q, there is at least a table R_i in $D \backslash \Delta D$ which does not contain the corresponding $t_{R_i}^-$;

    (c) If there is some ΔD satisfies the constraint $Q(D \backslash \Delta D) = V \backslash \Delta V$, then output $D \backslash \Delta D$; Otherwise, there is no side effect free deletion propagation.

Since there is no intermediate of join need to generate, it is a polynomial algorithm.   □

Compare with the result above, we next show SPJ query is even harder than NP on combined complexity unless P=NP.

Lemma 1. *b-dp for an* SPJ *query is* coNP-hard *under single deletion on combined complexity.*

Proof. We build a reduction from 3-DNF tautology problem. An instance of 3-DNF tautology problem includes a variable sets $X = \{x_1, ..., x_n\}$, and a 3-DNF boolean expression ϕ with m clauses $\{C_1, \ldots, C_m\}$, the task is to determine whether ϕ is satisfied by all assignments for X.

Base instance D. Let D has m relations $R_1, \cdots, R_m$, where R_i simulates clause C_i and variables in it. Concretely, for each clause C_i $(1 \le i < m)$, build a quintuple relation $R_i (A_1, A_2, A_3, A_4, A_5)$.

(a) Add 7 tuples into table R_i, whose values of A_1, A_2, A_3 refer to the 7 value assignments (0–1 value) of the 3 variables which make the corresponding clause unsatisfied, values of A_4 are the result of the corresponding assignments, i.e., seven '$-$'s, and the values of A_5 are the same sign '$\uparrow$'.

(b) Add b tuples into table R_i, respectively, they are $(\clubsuit_1, \clubsuit_1, \clubsuit_1, -, \downarrow_1)$, ...,
$(\clubsuit_b, \clubsuit_b, \clubsuit_b, -, \downarrow_b)$.

Here, the signs '$\uparrow$' and '$\downarrow_i$'s are used to identify join paths of variables from '$\clubsuit_i$'s.

Query Q. We first prepare all the join conditions used in query as follows.

(a) To identify join path of variables, build a join condition

$$con_{sign} := R_1.A_5 = \cdots = R_m.A_5,$$

so that the intermediate join paths of the query consists of just $b+1$ types, '$\uparrow$' and '$\downarrow_i$'s.

(b) For the occurrences of each variable x_i $(i \in [1, n])$ in clauses, say 'k_i', we build a join condition with conjunctive form such that

$$con_i := q_1 \wedge q_2 \wedge \cdots \wedge q_{k_i},$$

such that each q is also an equation $R_l.A_p = R_{l'}.A_{p'}$, if x_i occurs in the p-th and p'-th positions of clauses C_l and $C_{l'}$ where $p, p' \in \{1, 2, 3\}$ and $l, l' \in \{1, \cdots, m\}$.

(c) Let R is $R_1 \times \cdots \times R_m$, then we define the query Q such that,

$$Q := \pi_{R_1.A_4, ..., R_m.A_4} (\sigma_{con_1 \wedge \cdots \wedge con_n \wedge con_{sign}} R).$$

View V. Initially, one can verify that $V = \{(\underbrace{-, \ldots, -}_{m})\}$.

View single deletion ΔV. Let $\Delta V = V$.   □

Surprisingly, we next show that on combined complexity, for conjunctive query (SPJ), b-dp is also NP-hard,

Lemma 2. *b-dp for a* SJU *query is* NP-hard *under single deletion on combined complexity.*

Proof. We build a reduction form 3-SAT problem to b-dp for SJU query under group deletion. An instance of 3-SAT problem includes a set of variables $X = \{x_1, ..., x_n\}$ and a 3-CNF boolean expression ϕ with m clauses $\{C_1, \ldots, C_m\}$, the task is to decide whether there is an assignment τ for X such that ϕ is satisfied by τ.

Base instance D. Let D has m relations $R_1, \cdots, R_m$, where R_i simulates clause C_j and variables in it. Concretely, for each clause C_i ($1 \leq i \leq m$), build a quintuple relation $R_i (A_1, A_2, A_3, A_4, A_5)$.

(a) Add 7 tuples into table R_i, whose values of A_1, A_2, A_3 refer to the 7 value assignments (0–1 value) of the 3 variables which make the corresponding clause satisfied, values of A_4 are the result of the corresponding assignments, i.e., seven '+'s, and the values of A_5 are the same sign '↑'.
(b) Add b tuples into table R_i, respectively, they are $(\clubsuit_1, \clubsuit_1, \clubsuit_1, +, \downarrow)$, ..., $(\clubsuit_b, \clubsuit_b, \clubsuit_b, +, \downarrow)$.

Similar to the last lemma, the signs '↑' and '↓' are used to distinguish join paths from the query results.

Query Q. We first prepare all the join conditions used in query as follows.

(a) Similar to the last lemma, to identify join path of variables, build a join condition
$$con_{sign} := R_1.A_5 = \cdots = R_m.A_5,$$
so that the intermediate join paths of the query consists of just 2 types, '↑' and '↓.'

(b) For the occurrences of each variable x_i ($i \in [1, n]$) in clauses, say 'k_i', we build a join condition with conjunctive form such that
$$con_i := q_1 \wedge q_2 \wedge \cdots \wedge q_{k_i},$$
such that each q is also an equation $R_l.A_p = R_{l'}.A_{p'}$, if x_i occurs in the p-th and p'-th positions of clauses C_l and $C_{l'}$ where $p, p' \in \{1, 2, 3\}$ and $l, l' \in \{1, \cdots, m\}$.

(c) Let R is $R_1 \times \cdots \times R_m$, then we define the query $Q = Q_1 \times Q_2$ such that,
$$Q_1 := \pi_{R_1.A_4,...,R_m.A_4} (\sigma_{con_1 \wedge \cdots \wedge con_n \wedge con_{sign}} R),$$
$$Q_2 := \pi_{A_5} (R_1).$$

View V. Initially, one can verify that $V = \{(\underbrace{+, \ldots, +}_{m}, \uparrow), (\underbrace{+, \ldots, +}_{m}, \downarrow)\}$.

Single View deletion ΔV. Let $\Delta V = \{(\underbrace{+, \ldots, +}_{m}, \downarrow)\}$. □

Corollary 1. *b*-dp *is not in both* NP *and* coNP *unless* P=NP.

Theorem 2. *b-dp for an* SPJ *query is both* NP-hard *and* coNP-hard *under single deletion while* DP-hard *under group deletion on combined complexity.*

Proof. Due to Lemmas 1 and 2, the former can be proved obviously.

Now, we show why it is DP-hard under group deletion. We build a polynomial reduction from SAT-UNSAT problem which is a canonical DP-complete problem. Given two Boolean expressions ϕ, ϕ', both in conjunctive normal form with 3 literals per clause, the question is to decide if ϕ is satisfiable and ϕ' is not. We slightly modify and combine the two reductions of Lemmas 1 and 2 to show the reduction here.

To reduce the express ϕ, we use the reduction of Lemma 2 to build the base relation, the only modification is, instead of b ♣-like tuples, we now add only $\lfloor \frac{b}{2} \rfloor$ ones: $(♣_1, ♣_1, ♣_1, +, \downarrow_1), \cdots, (♣_{\lfloor \frac{b}{2} \rfloor}, ♣_{\lfloor \frac{b}{2} \rfloor}, ♣_{\lfloor \frac{b}{2} \rfloor}, +, \downarrow_{\lfloor \frac{b}{2} \rfloor})$.

To reduce the expression ϕ', we modify the reduction of Lemma 1 as follows, (1) change the first 7 tuples as the *true* assignments; (2) change all the value of A_4 into '+'; (3) similarly, instead of b ♣-like tuples, we now add only $\lceil \frac{b}{2} \rceil$ ones: $(♣_1, ♣_1, ♣_1, +, \downarrow), \cdots, (♣_{\lceil \frac{b}{2} \rceil}, ♣_{\lceil \frac{b}{2} \rceil}, ♣_{\lceil \frac{b}{2} \rceil}, +, \downarrow)$ (4) add a additional tuple $(\square, \square, \square, \square, \rightarrow)$ into each R_i.

Let Q^1 and Q^2 as the queries defined in Lemmas 1 and 2, we simply set the query as $Q = Q^1 \times Q^2$. Then one can verify that the initial view includes 4 tuples, $\{(+, \cdots, +, \uparrow), (+, \cdots, +, \downarrow)\} \times \{(-, \cdots, -), (\square, \cdots, \square)\}$.

At last, define the update $\Delta V = V \backslash \{(+, \cdots, +, \uparrow, \square, \cdots, \square)\}$.

One can verify that answer of SAT-UNSAT is *yes iff* there is a b-bounded side-effect-free deletion.   □

Lemma 3. *b-dp for an* SPJU *query belongs to* Δ_2^P *even under group deletion on combined complexity.*

Proof. There is a Δ_2^P algorithm for solving this problem.

(a) Given an instance $\langle D, Q, V, \Delta V, b \rangle$, enumerate each possible b-bounded deletion ΔD, actually $|D|^b$ in total;

(b) For each ΔD enumerated, query an NP oracle check if $Q(D \backslash \Delta D) = V \backslash \Delta V$;

(c) If there is some ΔD, the NP oracle returns '*yes*', it can be decided that there is a b-bounded deletion for instance $\langle D, Q, V, \Delta V, b \rangle$; If for all ΔD, the NP oracle return '*no*', then it can be decided that there is no solution for instance $\langle D, Q, V, \Delta V, b \rangle$.

4   Conclusion

We study the complexity of b-dp problem a bounded version of general deletion propagation, map out the complete picture of it. We find that, for most cases of relation algebraic query, bounded deletion decision problem is more easy than its general case, that is bound can lower the complexity of deletion propagation problem. We are currently finding the tractable condition and approximation algorithms for conjunctive query, besides the bounded version of deletion propagation. We also plan to study the cases at the present of other types of dependency constraints on database, such as independent dependencies in the future work.

References

1. Buneman, P., Khanna, S., Tan, W.C.: On propagation of deletions and annotations through views. In: Proceedings of the Twenty-First ACM SIGMOD-SIGACT-SIGART Symposium on Principles of Database Systems, PODS 2002, pp. 150–158. ACM, New York (2002)
2. Dayal, U., Bernstein, P.A.: On the correct translation of update operations on relational views. ACM Trans. Database Syst. **7**(3), 381–416 (1982)
3. Bancilhon, F., Spyratos, N.: Update semantics of relational views. ACM Trans. Database Syst. **6**(4), 557–575 (1981)
4. Cosmadakis, S.S., Papadimitriou, C.H.: Updates of relational views. J. ACM **31**(4), 742–760 (1984)
5. Bohannon, A., Pierce, B.C., Vaughan, J.A.: Relational lenses: a language for updatable views. In: Proceedings of the Twenty-Fifth ACM SIGMOD-SIGACT-SIGART Symposium on Principles of Database Systems, PODS 2006, pp. 338–347. ACM, New York (2006)
6. Keller, A.M.: Algorithms for translating view updates to database updates for views involving selections, projections, and joins. In: Proceedings of the Fourth ACM SIGACT-SIGMOD Symposium on Principles of Database Systems, PODS 1985, pp. 154–163. ACM, New York (1985)
7. Cong, G., Fan, W., Geerts, F., Li, J., Luo, J.: On the complexity of view update analysis and its application to annotation propagation. IEEE Trans. Knowl. Data Eng. **24**(3), 506–519 (2012)
8. Cong, G., Fan, W., Geerts, F.: Annotation propagation revisited for key preserving views. In: Proceedings of the 15th ACM International Conference on Information and Knowledge Management, CIKM 2006, pp. 632–641. ACM, New York (2006)
9. Kimelfeld, B., Vondrák, J., Williams, R.: Maximizing conjunctive views in deletion propagation. ACM Trans. Database Syst. **37**(4), 1–37 (2012)
10. Kimelfeld, B.: A dichotomy in the complexity of deletion propagation with functional dependencies. In: Proceedings of the 31st Symposium on Principles of Database Systems, PODS 2012, pp. 191–202. ACM, New York (2012)
11. Kimelfeld, B., Vondrák, J., Woodruff, D.P.: Multi-tuple deletion propagation: approximations and complexity. Proc. VLDB Endow. **6**(13), 1558–1569 (2013)
12. Miao, D., Liu, X., Li, J.: On the complexity of sampling query feedback restricted database repair of functional dependency violations. Theoret. Comput. Sci. **609**, 594–605 (2016)
13. Cong, G., Fan, W., Geerts, F., Li, J., Luo, J.: On the complexity of view update analysis and its application toannotation propagation. IEEE Trans. Knowl. Data Eng. **24**(3), 506–519 (2012)

14. Vardi, M.Y.: The complexity of relational query languages (extended abstract). In: Proceedings of the Fourteenth Annual ACM Symposium on Theory of Computing, STOC 1982, pp. 137–146. ACM, New York (1982)
15. Cosmadakis, S.S., Papadimitriou, C.H.: Updates of relational views. J. ACM **31**(4), 742–760 (1984)
16. Lechtenbörger, J., Vossen, G.: On the computation of relational view complements. ACM Trans. Database Syst. **28**(2), 175–208 (2003)

On Residual Approximation in Solution Extension Problems

Mathias Weller[1,2], Annie Chateau[1,2(✉)], Rodolphe Giroudeau[1],
Jean-Claude König[1], and Valentin Pollet[1]

[1] LIRMM - CNRS UMR 5506, Montpellier, France
{mathias.weller,annie.chateau,rgirou,konig,pollet}@lirmm.fr
[2] IBC, Montpellier, France

Abstract. The solution extension variant of a problem consists in, being given an instance and a partial solution, finding the best solution comprising the given partial solution. Many problems have been studied with a similar approach. For instance the Precoloring Extension problem, the clustered variant of the Travelling Salesman problem, or the General Routing Problem are in a way typical examples of solution extension variant problems. Motivated by practical applications of such variants, this work aims to explore different aspects around extension on classical optimization problems. We define *residue-approximations* as algorithms whose performance ratio on the non-prescribed part can be bounded, and corresponding complexity classes. Using residue-approximation, we classify several problems according to their *residue-approximability*.

1   Introduction

This paper presents a general study and exploration of the behavior of combinatorial problems when a partial solution is given, generally qualified as *extension problems*. We focus here on the complexity and approximation point of view, and especially on what changes, or not, when considering this partial solution. The motivation of this work is rooted in the search to improve practical applications of approximation algorithms. Indeed, it is sometimes possible to infer a limited part of an optimal solution in polynomial time, or to consider prerequisites of problems as a partial solution. Originating in the search of practical algorithms solving the SCAFFOLDING problem that represents a crucial step in the modern-day genome assembly process [21,22], it now appears that extension problems have wider impact. This may be the case, for instance, in application domains like scheduling, networks design, routing, resource assignment, etc. If inferring a partial solution is easy, and it is possible to prove that giving a partial solution does not alter the approximation ratio of a heuristic, this heuristic could be used to complete the solution. Depending on how much of a partial solution we can guess, this may lead to improvements of known heuristics, at no additional approximation cost. In the following, we refer to partial solutions as *residue*. As opposed to traditional optimization problems that, given an instance, aim at optimizing the (cost of the) solution, *extension problems* take an instance and a

© Springer International Publishing AG 2016
T.-H.H. Chan et al. (Eds.): COCOA 2016, LNCS 10043, pp. 463–476, 2016.
DOI: 10.1007/978-3-319-48749-6_34

Table 1. Synthetic view of complexity and approximation results.

Problem	Extension variant	Residue variant
VERTEX COVER	2-approx	2-approx.
FEEDBACK VERTEX SET	2-approx	2-approx.
STEINER TREE	3/2-approx	3/2-approx.
MAX LEAF	$\geq 1/3 \cdot OPT - 2/3 \cdot |P|$	Open
BIN PACKING	3/2-approx	no $\mathcal{APX}$ unless $\mathcal{P} = \mathcal{NP}$

partial solution and aim at optimizing the full solution (that extends the given partial solution) and *residue problems* take an instance and a partial solution and aim at optimizing the residue of the full solution.

Related Work. Several extension problems have already been studied. A series of papers dealt with PRECOLORING EXTENSION which, given a graph and a partial proper coloring of its vertices, asks to find a proper coloring extending the partial one. On proper interval graphs [17] and bipartite graphs [14], the problem is $\mathcal{NP}$-hard for ≥ 3 precolored vertices. On general interval graphs, PRECOLORING EXTENSION is polynomial-time solvable if each color is only used once, while $\mathcal{NP}$-hard if each color can be used at most twice [6]. A variant of SHORTEST CYCLE consists in finding a shortest simple cycle traversing k specified elements (vertices or edges). This variant is known to be $\mathcal{NP}$-hard but Björklund et al. [7] designed an $\mathcal{FPT}$-algorithm solving it in $\mathcal{O}(2^k poly(n))$ time. Previous work has been done on the TSP problem with prescribed edges and vertices, called GENERAL ROUTING PROBLEM (GRP) [15,18], including a 3/2-approximation using an adaptation of Christofides' algorithm [8]. Also, CLUSTERED TSP, where the set of vertices is partitioned into an ordered collection of to-be-visited clusters has been previously studied from the approximation point of view [1,2,10,13].

Our Contribution. One goal of the present work is to properly define underlying notions of extension and residue variants of classical problems, giving a general framework for approximation concerns. We provide a first collection of results and define two new complexity classes: $\mathcal{RAPX}$, which gathers problems allowing a constant ratio approximation on residue, and $\mathcal{FRAPX}$, in which this ratio is the same as for the underlying classical problem. In order to illustrate those notions and provide intuition on how they could meaningfully describe extension problems, we also give some results for a sample of classical problems, together with the description of residue-approximating algorithms when possible. Table 1 summarizes the main results obtained for the considered problems.

In Sect. 2, we give useful definition and examples, Sect. 3 lists a few examples of problems of $\mathcal{FRAPX}$. Thereafter, we develop two more involved examples of problems in $\mathcal{RAPX}$: Sect. 4 concerns the BIN PACKING problem, which we prove to not be residue-approximable despite it being in $\mathcal{APX}$. We show that the extension variant is, however, approximable with the same ratio as the original problem. Section 5 deals with the problem of finding a spanning tree with maximal number of leaves in a graph and its inverse, finding a spanning tree with a minimum number of internal vertices. We show that the extension version of this second problem is approximable with the same ratio as the original problem.

2   Definitions

Let Σ be a ground set. For a function $f : \Sigma \to \mathbb{Q}$, its *domain* $\mathrm{dom}(f)$ is the subset of Σ for which f is defined and its *image* $\mathrm{img}(f)$ is the set of rationals that are assigned by f. We write $f(x) = \bot$ if x is not in the domain of f. Given a set $\mathcal{S} \subseteq 2^{\Sigma}$, and a function $\omega : 2^{\Sigma} \to \mathbb{N}$ (all of which may be implicitly defined), a *minimization problem* or *maximization problem* asks to find an $S \in \mathcal{S}$ minimizing or maximizing $\omega(S)$, respectively. We call $\mathcal{S}$ the *solution space* of the problem and we call the elements of $\mathcal{S}$ *solutions* for it. Each $P \subseteq \Sigma$ is called *partial solution* for the problem and we define $\mathcal{S}^P := \{R \mid R \cap P = \emptyset \wedge R \cup P \in \mathcal{S}\}$. Let A be an optimization problem defined like the following:

> **Problem A**
> **Input:** $\mathcal{S}, \omega : 2^{\Sigma} \to \mathbb{N}$
> **Task:** Find a solution $S^* \in \mathcal{S}$ such that $\omega(S^*)$ is optimal.

Then, the *extension variant* A_{ext} of A is the following:

> **Problem A_{ext}**
> **Input:** $\mathcal{S}$, some $P \subseteq \Sigma$, and $\omega : 2^{\Sigma} \to \mathbb{N}$
> **Task:** Find a solution $S^* \in \mathcal{S}$ s.t. $P \subseteq S^*$ and $\omega(S^*)$ is optimal.

Consider for example the well-known MINIMUM VERTEX COVER (MVC) problem, where we are given a graph $G = (V, E)$ and the task is to find a smallest set $S^* \subseteq V$ such that each edge of G is incident with a vertex in S^*. The graph G implicitly defines the set $\mathcal{S}$ of all vertex sets that cover all edges of G and the function $\omega : 2^V \to \mathbb{N}$ as $\omega(Y) = |Y|$ for all $Y \subseteq V$. Then, the corresponding extension problem asks for a minimum-size vertex-cover $S^* \subseteq V$ of G that contains P.

   For algorithms finding an optimal solution, it often does not matter whether we optimize the weight of the overall solution S or the weight of the additional part $S \setminus P$, called the *residue*. However, in the context of approximation algorithms, this difference becomes significant as an algorithm producing an approximate residue may be more desirable than one producing an approximate solution containing the given partial solution. This motivates introducing the *residue variant* of optimization problems, asking to find a residue that optimizes a weight function ω_P (which might differ from ω and might depend on P).

> **Problem A_{res}**
> **Input:** $\mathcal{S}$, some $P \subseteq \Sigma$, and $\omega_P : 2^{\Sigma} \to \mathbb{Z}$
> **Task:** Find an $R \in \mathcal{S}^P$ such that $\omega_P(R)$ is optimal.

Since ω_P may differ from ω, this definition allows degenerated cases where an optimal solution for A_{res} does not correspond to an optimal solution for A. Therefore, we will focus on the special case of A_{res} for which $\omega(\emptyset) = 0$ and, for all $P \subseteq \Sigma$ and $R \in \mathcal{S}^P$, we have $\omega_P(R) = \omega(R \cup P) - \omega(P)$. In this case, we call A_{res} *canonical*.

In this work, we are interested in developing algorithms that produce good approximate residues. To this end, we define the class $\mathcal{RAPX}$ containing a problem A if and only if its canonical residue version A_{res} is ρ-approximable for some $\rho \in \mathbb{Q}$. Note that $\mathcal{RAPX} \subseteq \mathcal{APX}$ since A is a special case of A_{res} with $P = \emptyset$. Indeed, we show $\mathcal{RAPX} \subset \mathcal{APX}$ (unless $\mathcal{P} = \mathcal{NP}$) in Sect. 4.

Observation 1. *Let $\rho \in \mathbb{Q}$, let A be an optimization problem and let its canonical residue variant be ρ-approximable. Then, A is ρ-approximable.*

Further, we define the class $\mathcal{FRAPX} \subseteq \mathcal{RAPX}$ of problems for which the converse of Observation 1 holds too. That is, $\mathcal{FRAPX}$ contains the problems A such that, for all $\rho \in \mathbb{Q}$, its canonical residue variant A_{res} is ρ-approximable if and only if A is ρ-approximable.

Continuing the above example, the residue variant of MVC takes a graph $G = (V, E)$, a vertex set $P \subseteq V$ and a weight function ω_P and the task is to find a vertex set R such that $R \cup P$ is a vertex cover of G and $\omega_P(R)$ is minimized. In its canonical residue version, which we call MINIMUM RESIDUE VERTEX COVER (MVC$_{\text{res}}$), we have that $\omega_P(R) = \omega(R \cup P) - \omega(P) = |R \setminus P|$. Observe that we can trivially reduce MVC$_{\text{res}}$ back to MVC by simply removing the vertices of P from G, along with their incident edges. Now, the fact that MVC is 2-approximable [5] immediately implies that MVC$_{\text{res}}$ is 2-approximable. We call problems exhibiting such behavior "r-hereditary".

Definition 1 (r-hereditary). *Let A be an optimization problem and let A_{res} be a canonical residue variant of A. Let λ, τ be polynomial-time computable functions such that, for all instances $(\mathcal{S}, P, \omega_P)$ of A_{res},*

(a) $(\mathcal{S}', \omega') := \lambda(\mathcal{S}, P, \omega_P)$ is an instance of A,
(b) $\forall_X (X \subseteq \Sigma \setminus P \wedge X \cup P \in \mathcal{S}) \iff \exists_{Y \in \mathcal{S}'} \tau(Y) = X$, and
(c) $\forall_{Y \in \mathcal{S}'} \omega'(Y) = \omega_P(\tau(Y))$.

Then, A is called r-hereditary *by virtue of λ and τ.*

Noticeably, any r-hereditary problem in $\mathcal{APX}$ is in $\mathcal{FRAPX}$.

Proposition 1. *Let $\rho \in \mathbb{Q}$ and A be an r-hereditary, ρ-approximable optimization problem. Then, A_{res} is ρ-approximable.*

Proof. Without loss of generality, et A be a maximization problem and let λ and τ be the functions by virtue of which A is r-hereditary. Let $(\mathcal{S}, P, \omega_P)$ be an instance of A_{res} and let $(\mathcal{S}', \omega') := \lambda(\mathcal{S}, P, \omega_P)$, which, by Definition 1 (a), is an instance of A.

Consider an optimal residue $R \subseteq \Sigma \setminus P$ for A_{res}, that is, $R \cup P \in \mathcal{S}$ and $\omega_P(R) = OPT_{A_{\mathrm{res}}}(\mathcal{S}, P, \omega_P)$. By Definition 1(b) & (c), there is some $Y \in \mathcal{S}'$ with $\tau(Y) = R$ and $\omega'(Y) = \omega_P(R)$. Thus, $OPT_{A_{\mathrm{res}}}(\mathcal{S}, P, \omega_P) = \omega_P(R) = \omega'(Y) \leq OPT_A(\mathcal{S}', \omega')$. Since A is ρ-approximable, we can compute an $S' \in \mathcal{S}'$ with $\omega'(S') \geq \rho \cdot OPT_A(\mathcal{S}', \omega')$ in polynomial time. Note that, by Definition 1(b), $\tau(S')$ is a solution for $(\mathcal{S}, P, \omega_P)$ and, since

$$\omega_P(\tau(S')) \overset{\text{Def. 1(c)}}{=} \omega'(S') \geq \rho \cdot OPT_A(\mathcal{S}', \omega') \geq \rho \cdot OPT_{A_{\mathrm{res}}}(\mathcal{S}, P, \omega_P),$$

it is ρ-approximate. Finally, since λ and τ can be computed in polynomial time, so can $\tau(S')$. $\qquad\square$

3   Some $\mathcal{FRAPX}$ Problems

In this section, we propose three examples of r-hereditary problems in $\mathcal{APX}$, proving their containment in $\mathcal{FRAPX}$.

3.1   Feedback Vertex Set

FEEDBACK VERTEX SET (FVS)
Input: $G = (V, E)$ a graph, $\omega : V \rightarrow \{1\}$
Task: Find $S \subset V$ s.t. $G[V \setminus S]$ is a forest and $\omega(S)$ is minimum.

Proposition 2. FEEDBACK VERTEX SET $\in \mathcal{FRAPX}$.

Proof. Let (G, P, ω) be an instance of FEEDBACK VERTEX SET$_{\mathrm{res}}$. Let λ be the transformation such that $\lambda(G, P, \omega) = (G - P, \omega)$ is an instance of FVS. Let τ be the identity relation, that is, $\tau(S) = S$ for all solutions S of $(G - P, \omega)$. Then, for all sets X:

1. if X is disjoint from P and $X \cup P$ is a feedback vertex set of G, then $\tau(X) = X$ is a feedback vertex set of $G - P$ and
2. if there is some feedback vertex set Y of $G - P$ with $Y = \tau(X) = X$, then Y is disjoint from P and $Y \cup P$ is a feedback vertex set of G.

Finally, for each feedback vertex set Y of $G - P$, we have $\omega(\tau(Y)) = \omega(Y)$.

Thus, FVS is r-hereditary and, as (on undirected graphs) it is also 2-approximable [4], Proposition 1 implies the claim. $\qquad\square$

3.2   Satisfiability

For any formula φ and any (partial) assignment α of variables of φ, we let $\varphi \mid_\alpha$ denote the formula resulting from replacing the variables assigned by α by their assigned values.

WEIGHTED SAT (WSAT)
Input: a formula φ over the variable set V, some $\omega : V \to \mathbb{N}$
Task: Find a satisfying assignment $\alpha : V \to \{0,1\}$ for φ such that
$\sum_v \omega(v) \cdot \alpha(v)$ is optimal.

MIN/MAX SAT (MMSAT)
Input: a set φ of disjunctions over a variable set V, $\omega : \varphi \to \mathbb{N}$
Task: Find an assignment $\alpha : V \to \{0,1\}$ for φ such that
$\sum\{\omega(C) \mid \alpha$ satisfies $C \in \varphi\}$ is optimal.

Proposition 3. *Let $\mathcal{A}$ be a version of* WEIGHTED SAT *such that $\mathcal{A} \in \mathcal{APX}$ and for any instance (φ, ω) of $\mathcal{A}$ and any variable x of φ, also $(\varphi \mid_{x=0}, \omega)$ and $(\varphi \mid_{x=1}, \omega)$ are instances of $\mathcal{A}$. Then, $\mathcal{A} \in \mathcal{FRAPX}$.*

Proof. Let (φ, P, ω) be an instance of WEIGHTED SAT$_{\mathrm{res}}$. Let λ be the transformation such that $\lambda(\varphi, P, \omega) = (\psi, \omega)$ is an instance of $\mathcal{A}$ with $\psi = \varphi \mid_P$. Let τ be the identity relation, that is, $\tau(\alpha) = \alpha$ for all satisfying assignments α of ψ. Then, for all X:

1. if X is disjoint from P and $X \cup P$ is a satisfying assignment for φ, then $\tau(X) = X$ is a satisfying assignment for ψ and
2. if there is some satisfying assignment Y for ψ with $Y = \tau(X) = X$, then Y is disjoint from P and $Y \cup P$ is a satisfying assignment for φ.

Finally, for each satisfying assignment Y for ψ, we have $\omega(\tau(Y)) = \omega(Y)$. Thus, $\mathcal{A}$ is r-hereditary and Proposition 1 implies the claim.   □

For instance, the minimization variant where each clause in φ contains at most two variables and ω is uniform is 2-approximable [12] and Proposition 3 implies that this problem is in $\mathcal{FRAPX}$. A similar construction as in the proof of Proposition 3 in which λ also removes clauses containing a 1 from φ shows that MMSAT is r-hereditary, implying that its approximable variants [3,11] are in $\mathcal{FRAPX}$.

3.3   Steiner Spanning Subgraph

STEINER SPANNING SUBGRAPH (SSS)
Input: an edge-weighted graph $G = (V, E, \omega)$, terminals $T \subseteq V$
Task: Find $S \subseteq E$ such that $G[S]$ is connected, spanning T, and $\omega(S)$ is minimal.

Proposition 4. STEINER SPANNING SUBGRAPH $\in \mathcal{FRAPX}$.

Proof. Consider an instance (G, P, T, ω) of STEINER SPANNING SUBGRAPH$_{\mathrm{res}}$. To avoid ambiguities, suppose that there is no triangle (u, v, z) such that $uv \in P$ and $\omega(zu) = \omega(zv)$ (otherwise, delete zu without losing generality). Further, suppose that $\omega(a) = \infty$ for all $a \notin E(G)$ and suppose that no edge in P intersects T (otherwise, remove the intersection from T). Let λ be the transformation such that $\lambda(G, P, T, \omega) = (G', T', \omega')$ is an instance of SSS obtained as follows. (G', T', ω') is obtained from (G, P, T, ω) by contracting each connected component C of $G[P]$ into a vertex x_C, adding x_C as a new terminal, and setting $\omega'(ux_C) = \min_{v \in C} \omega(uv)$. For all solutions Y to (G', T', ω'), we define $\tau(Y)$ as the result of undoing this operation on Y, that is, replacing each $ux_C \in Y$ by uv with $v = \text{argmin}_{z \in C} \omega(uz)$ and likewise replacing each $x_C x_D$ by the lightest edge between C and D in G.

Then, for all X:

1. if X is disjoint from P and $X \cup P$ induces a connected spanning subgraph of G containing T, then $Y := \lambda(G[X \cup P], P, T, \omega)$ is a solution for (G', T', ω'), and $\tau(Y) = X$;
2. if there is a solution Y to (G', T', ω') with $\tau(Y) = X$, then Y is disjoint from P and $X \cup P$ is a solution to (G, P, T, ω).

Finally, for each solution Y to (G', T', ω'), we have $\omega(\tau(Y)) = \omega'(Y)$.

Thus, SSS is r-hereditary and, since it is also 3/2-approximable [19], STEINER SPANNING SUBGRAPH $\in \mathcal{FRAPX}$ by Proposition 1.   □

4   Non-symmetrical Approximability

In this section, we show that BIN PACKING is an example of a problem in $\mathcal{APX}$ whose canonical residue variant cannot be approximated within any constant factor, unless $\mathcal{P} = \mathcal{NP}$. It is defined as follows.

BIN PACKING (BP)
Input: a vector $[w]$ of rational weights
Task: Find $b : \mathbb{N} \to \mathbb{N}$ such that $\sum \{w_i \mid b(i) = j\} \leq 1$ for each $j \in \mathbb{N}$, minimizing $|\text{img}(b)|$.

Deciding whether the given weights fit into two bins or not is already strongly $\mathcal{NP}$-hard [9]. However, BP can be approximated to within a factor of 3/2 [20]. The canonical residue variant of BP is

BIN PACKING$_{\text{res}}$ (RBP)
Input: a vector $[w]$ of rational weights, an assignment $P : \mathbb{N} \to \mathbb{N}$
Task: Find some $r : \mathbb{N} \setminus \text{dom}(P) \to \mathbb{N}$ such that for each $j \in \mathbb{N}$, $\sum \{w_i \mid r(i) = j \vee P(i) = j\} \leq 1$, minimizing $|\text{img}(r) \setminus \text{img}(P)|$.

Now, consider the following reduction. Given an instance $\mathcal{I}$ of BP *i.e.* a vector $[w]$ of size k, we construct an instance $\mathcal{I}'$ of BIN PACKING$_{\text{res}}$ by dividing all weights by 2 from the vector $[w]$, and adding two new items to the vector $[w]$ with weights $w_k = w_{k+1} = 0.5$, and we set $P(k) = 0$ and $P(k+1) = 1$. Then, this BIN PACKING$_{\text{res}}$ instance has a solution r with $|\text{img}(r) \setminus \text{img}(P)| = 0$ if and only if the weights $[w]$ of the original instance fit into two bins. Thus, BIN PACKING$_{\text{res}}$ cannot be approximated to within any constant factor unless $\mathcal{P} = \mathcal{NP}$.

Observation 2. Bin Packing $\in \mathcal{APX} \setminus \mathcal{RAPX}$.

In the following, we show that the extension variant of BP can be approximated to within a factor of $\frac{3}{2}$.

> Bin Packing$_{\text{ext}}$ (EBP)
> **Input:** a vector $[w]$ of rational weights, an assignment $P : \mathbb{N} \to \mathbb{N}$
> **Task:** Find some $b : \mathbb{N} \to \mathbb{N}$ such that $P \subseteq b$ and, for each $j \in \mathbb{N}$,
> $\sum \{w_i \mid b(i) = j\} \leq 1$, minimizing $|\text{img}(b)|$.

Without loss of generality, we suppose that $[w]$ does not fit into a single bin. We call a solution b for an instance $([w], P)$ of EBP $\frac{2}{3}$-*filling* if at most one bin j is strictly less than $\frac{2}{3}$ full (that is, $\sum \{w_i \mid b(i) = j\} < 2/3$). In this case, we suppose that no item assigned to j by b fits into any other bin.

Lemma 1. *Let $([w], P)$ be an instance of EBP and let b be a $\frac{2}{3}$-filling solution. Then, b is $\frac{3}{2}$-approximate.*

Proof. For each $j \in \mathbb{N}$, let $h_j := \sum \{w_i \mid b(i) = j\}$ be the height of bin j with respect to b, and let $\kappa := |\text{img}(b)|$ be the number of bins used by b. Let κ^* be the optimal number of bins. Clearly, $\kappa^* \geq \sum_i w_i = \sum_j h_j$. Now, assuming every bin is at least $\frac{2}{3}$ full, we have

$$\kappa^* \geq \sum_j h_j \geq \kappa \cdot \min_j(h_j) \geq \frac{2}{3}\kappa \qquad \text{implying} \qquad \frac{3}{2}\kappa^* \geq \kappa.$$

Otherwise, there is a bin k that is less than $\frac{2}{3}$ full. Then, $h_k > \max_{j \neq k}(1 - h_j)$ since no weight from k can be assigned to another bin (b is $\frac{2}{3}$-filling). We still have $\kappa^* \geq \sum_j h_j$, which gives:

$$\kappa^* \geq h_k + \sum_{j \neq k} h_j > \max_{j \neq k}(1 - h_j) + (\kappa - 1)\min_{j \neq k}(h_j) > \frac{1}{3} + \frac{2}{3}(\kappa - 1)$$

Now we have $3\kappa^* > 2\kappa - 1$ giving $\frac{3}{2}\kappa^* \geq \kappa$. $\qquad\square$

Given an instance $([w], P)$ of EBP we call a weight w_i *heavy* if $w_i \in \text{dom}(P)$ or $w_i > \frac{1}{3}$. We call $([w], P)$ *heavy* if all weights of $[w]$ are heavy and we call a heavy instance *strict* if, for all bins $j \in \text{img}(P)$, we have $\sum \{w_i \mid P(i) = j\} \leq \frac{1}{3}$. Using the following rules, we prove that we can focus on strict instances.

Rule 1. *Let $([w], P)$ be an instance of EBP and let $w_i \leq \frac{1}{3}$ be a weight of $[w]$. Then, remove w_i from $[w]$.*

Rule 2. *Let $([w], P)$ be a heavy instance of EBP and let $j \in \text{img}(P)$ with $h_j := \sum \{w_i \mid P(i) = j\} > \frac{1}{3}$. Then, remove all w_i with $P(i) = j$ from $[w]$ and remove any heaviest weight $w_k \notin \text{dom}(P)$ with $w_k \leq 1 - h_j$ from $[w]$.*

Lemma 2. *Given a $\frac{3}{2}$-approximate solution b' for the result $([w'], P')$ of apply-ing Rule 1 or Rule 2 to $([w], P)$, a $\frac{3}{2}$-approximate solution for $([w], P)$ can be computed in polynomial time.*

Proof. Rule 1: Let w_k be the removed weight. If there is a bin $j \in \text{img}(b')$ with $\sum\{w_i \mid b'(i) = j\} \leq \frac{2}{3}$, then let $b := b' \cup \{(k \to j)\}$. In this case, b does not use more bins than b' and, hence, it is $\frac{3}{2}$-approximate for $([w], P)$. If all bins are $\frac{2}{3}$-filled by b', then b assigns w_k to a new bin. In this case, b is $\frac{2}{3}$-filling and, by Lemma 1, b is $\frac{3}{2}$-approximate.

Rule 2: Let P_j be the set of assignments to j in P. If there is no weight $w_k \notin \text{dom}(P)$ with $w_k \leq 1 - h_j$ in $[w]$, then no solution for $([w], P)$ assigns any additional weight to bin j. Thus, $b' \cup P_j$ is $\frac{3}{2}$-approximate for $([w], P)$. Otherwise, let w_k be heaviest among all weights outside $\text{dom}(P)$ with $w_k \leq 1 - h_j$. We show that there is an optimal solution b for $([w], P)$ that assigns w_k to bin j and, since b' is $\frac{3}{2}$-approximate for $([w'], P')$, we conclude that $b' \cup P_j \cup \{(k \to j)\}$ is $\frac{3}{2}$-approximate for $([w], P)$. Let b be an optimal solution for $([w], P)$ with $(k \to j) \notin b$ and let ℓ be the bin containing w_k. Further, since $([w], P)$ is heavy and j is $\frac{1}{3}$-filled by P, we know that b assigns at most one weight $w_{k'} \notin \text{dom}(P)$ to j. Also, $w_{k'} \leq w_k$ by maximality of w_k. Then, swapping $(k' \to j)$ and $(k \to \ell)$ for $(k' \to \ell)$ and $(k \to j)$ in b yields a solution using the same number of bins as b. □

As the result of exhaustive application of Rules 1 and 2 is strict, it remains only to show that EBF is $\frac{3}{2}$-approximable on strict instances.

Theorem 1. BIN PACKING$_{ext}$ *is $\frac{3}{2}$-approximable.*

Proof. Let $([w], P)$ be an instance of EBP and, using Lemma 2, suppose that $([w], P)$ is strict. We call the weights (bins) in $\text{dom}(P)$ ($\text{img}(P)$) *preassigned*. Let $[w']$ be the result of removing all preassigned weights from $[w]$, let n be the length of $[w']$ and let $p := |\text{img}(P)|$. Let b^* be an optimal solution for $([w], P)$ and let $\kappa^* := |\text{img}(b^*)|$. The graph $G = (V, E)$ with $V := \{1, \ldots, n\}$ and $E := \{ij \mid w'_i + w'_j \leq 1\}$ is called the *compatibility graph* of $[w']$. In G, we compute a maximum matching M and let $m := |M|$. Let $s := n - 2m$ be the number of vertices that M does not touch. Since $([w], P)$ is strict, no bin can hold more than two weights of $[w']$ and, by maximality of M, we have $\kappa^* \geq \max(p, m + s)$. Let b_g be an assignment resulting from exhaustive application of the following routine: take a heaviest, non-fixed, unmarked weight w'_i in $[w']$, mark it and, if possible, assign it to an arbitrary preassigned bin. Then,

Case 1: $\text{img}(b_g) = \text{img}(P)$. Let $[w'']$ be the vector of weights of $[w]$ that are not assigned by $b_g \cup P$ and note that its size is at most $n - p$. We build the compatibility graph G'' of $[w'']$ and compute a maximum matching M'' on G''. Let $m'' := |M''|$ and let s'' be the number of vertices that M'' does not touch. Extending b_g by assigning each pair of weights corresponding to an edge of M'' to a new bin and the s'' remaining weights to separate bins, we obtain a solution b for $([w], P)$ using $\kappa = p + m'' + s''$ bins. Let b_p be a solution for $([w], P)$ such that $b_p \setminus P$ uses at most $p/2$ preassigned bins. Then, b_p uses at least $m + s - \frac{p}{2}$ bins in total. By spreading the at most p weights that $b_p \setminus P$ assigns to preassigned bins, we obtain from b_p a solution for $([w], P)$ that puts at most one weight in each preassigned bin and, thus, uses at least as many bins as b. Hence, $|\text{img}(b)| = m'' + s'' + p \leq m + s - \frac{p}{2}$, implying $\kappa \leq p + m + s - \frac{p}{2} \leq \frac{3}{2} \max(p, m + s)$.

Case 2: $\text{img}(P) \subsetneq \text{img}(b_g)$. Let $[w'']$ be the vector of weights that are not assigned by $b_g \cup P$ and let w_m be minimal in $[w'']$. Since $([w], P)$ is strict, $w_m > \frac{2}{3}$. We show that the extension b of $b_g \cup P$ that assigns each weight in $[w'']$ to a separate bin is an optimal solution for $([w], P)$. To this end, assume that $|\text{img}(b)| - p > |\text{img}(b^*)| - p$. Then, there is a bin $j \in \text{img}(P)$ such that b^* assigns a weight $w_i \geq w_m$ of $[w]$ to j but b does not. Then, j cannot be outside $\text{img}(b_g)$ as otherwise, b_g would assign w_m, so b_g assigns a weight $w_k < w_m$ to j. But this contradicts the construction of b_g, as w_m would have been considered before w_k.

In both cases, we produce a $\frac{3}{2}$-approximate solution for $([w], P)$. $\square$

5   Spanning Tree with Many Leaves

Consider the problem of finding a spanning tree with a maximum number of leaves in a given graph. Its extension variant prescribes some edges.

MAX LEAF (ML)
Input: a graph $G = (V, E)$
Task: find $S \subseteq E$ such that (V, S) is a spanning tree of G, maximizing the number of leaves.

MAX LEAF$_{\text{ext}}$ (MLE)
Input: a graph $G = (V, E)$, an edge set $P \subseteq E$
Task: find $S \subseteq E$ such that $P \subseteq S$ and (V, S) is a spanning tree of G, maximizing the number of leaves.

We prove that solving MAX LEAF is equivalent to solving its extension variant MAX LEAF$_{\text{ext}}$. Further, we conclude that approximating the inverse problem, MIN INTERNAL SPANNING TREE is equivalent to approximating its extension variant MIN INTERNAL SPANNING TREE$_{\text{ext}}$. In the following, suppose that G has at least three vertices.

Transformation 1 *(see Fig. 1). Let $(G = (V, E), P)$ be an instance of MLE such that (V, P) is a forest. Let $P_M \subseteq P$ such that (V, P_M) is a matching and P_M is maximal under this condition. Then, construct an instance G' of ML as follows: For each $uv \in P_M$, add a vertex x_{uv} to V and the edges ux_{uv} and vx_{uv} to E. Finally, for all vertices w of degree at least two in $(V, P \setminus P_M)$, add a new vertex y_w to V and add the edge wy_w to E.*

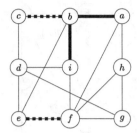

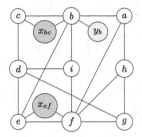

(a) Graph $G = (V, E)$ with P in bold. Maximal matching P_M consists in dashed edges.

(b) Graph $G' = (V', E')$. Two new vertices issued from P_M are gray, one new vertex from high degree vertex of P is lightgray.

Fig. 1. Transformation 1 turns the left instance of MLE into the right instance of ML.

Lemma 3. *Let $G' = (V', E')$ be the result of applying Transformation 1 to $(G = (V, E), P)$ and let $k \in \mathbb{N}$. there is a spanning tree with $\leq k$ non-leaves containing P in G if and only if there is a spanning tree with $\leq k$ non-leaves in G'.*

Proof. "$\Rightarrow$": Let (V, S) be a tree with ℓ leaves (k non-leaves) containing P. Clearly, for each w of degree at least two in $(V, P \setminus P_M)$, we can add wy_w to S, creating the leaf y_w. Further, for each $uv \in P_M$, at least one of u and v, say u, is not a leaf in (V, S). Then, we can add ux_{uv} to S, creating the leaf x_{uv}. Thus, the modified set induces a spanning tree for G' with $\ell + |V' \setminus V|$ leaves (k non-leaves).

"$\Leftarrow$": Let $T' := (V', S')$ be a spanning tree of G' with a maximum number of leaves. We show that there is a set S such that (a) $P \subseteq S$, (b) (V', S) has the same number of leaves as (V', S'), (c) all vertices in $V' \setminus V$ are leaves in (V', S), and (d) (V', S) is a spanning tree of G'. Then, $(V, S \cap E)$ is a spanning tree of G with at most $|V' \setminus V|$ leaves less than (V', S'). That is, $(V, S \cap E)$ has the same number of non-leaves as T'.

To construct S, let $uv \in P_M \setminus S'$. Since $x_{uv} \in V'$ has degree two in G', at least one of ux_{uv}, vx_{uv} is in S'. Without loss of generality, let $ux_{uv} \in S'$. If $vx_{uv} \in S'$, then we can swap vx_{uv} for uv in S', creating a new leaf x_{uv} and, thus, contradicting optimality of T'.

Hence, x_{uv} is a leaf and u is not a leaf in T'. If both u and v are non-leaves in T', then we swap the first edge on the unique u-v-path in T' for uv in S'. If v is a leaf in T' and z is its unique neighbor in T', then we swap vz for uv in S'. Next, let $uv \in P \setminus (P_M \cup S')$. Since $uv \notin P_M$, one of u, v, say u, is a non-leaf of (V, P). Then, since u is neighbor of some leaf y_u in T', we know that u is also a non-leaf in T'. If v is a non-leaf in T', then we swap uv for any edge of $S' \setminus P$ occurring in the unique u-v-path in T' (note that such an edge exists since otherwise P contains a cycle). If v is a leaf in T' then v is a leaf in (V, P), as otherwise, v has a neighbor y_v in G' that is not covered by T'. In this case, let $vz \in S' \setminus P$ and we swap uv for vz in S'. Note that each swap introduces an edge of P into S' and removes an edge of $S' \setminus P$ and, thus, the algorithm terminates after at

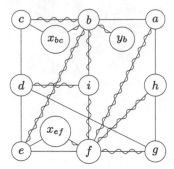

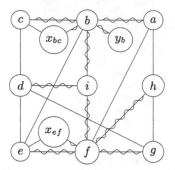

(a) Graph $G' = (V', E')$, with T' (waved edges) a spanning tree with a maximum number of leaves (=9).

(b) Construction of S' by swapping eb to ef and af to ab, yielding a spanning tree with maximum number of leaves in G.

Fig. 2. From a solution of ML (left) to a solution of MLE by successive swaps (right).

most $|P|$ swaps. We let S denote the result of an exhaustive application of the above swap-rules to S'. Figure 2 shows the application to the example of Fig. 1.

First, since the application is exhaustive, no edge of P remains outside S, implying (a). Second, since none of the swaps we do creates a cycle and $|S| = |S'|$ we know that (V', S) is still a spanning tree of G', implying (d). Third, (b) is implied by the fact that none of the swap operations turns a leaf into an internal vertex. Fourth, to show (c), assume there is some vertex in $V' \setminus V$ that is a non-leaf in (V', S). Clearly, it is not a y_u for any $u \in V$. Thus, let $uv \in P_M$ such that $x_{uv} \in V' \setminus V$ is a non-leaf in (V', S). Then, by (a), (u, v, x_{uv}) is a cycle in (V', S), contradicting (d).   □

We use the fact that the proof of Lemma 3 is constructive to transfer approximation algorithms for MAX LEAF to MAX LEAF$_{\text{ext}}$.

Corollary 1. *Every ρ-approximation for the problem* MAX LEAF *can be used to pseudo-approximate* MAX LEAF$_{ext}$.

Proof. Given an instance (G, P) of MAX LEAF$_{\text{ext}}$, use Transformation 1 to compute an instance G' of MAX LEAF and let $\Delta V := V' \setminus V$. Then, compute a ρ-approximate spanning tree (V', S') for G', and let f' denote its number of leaves. Then, morph S' into a set S as in the proof of Lemma 3, that is, all vertices of ΔV are leaves in (V', S) and (V', S) has at least f' leaves. Let f be the number of leaves in $(V, S \cap E)$.

Then, $f' \geq \rho \cdot OPT_{ML}(G')$ and $f \geq f' - |\Delta V|$ and, by Lemma 3, we have $OPT_{MLE}(G, P) = OPT_{ML}(G') - |\Delta V|$. Thus,

$$f \geq \rho \cdot OPT_{MLE}(G, P) - (1 - \rho)|\Delta V| \geq \rho \cdot OPT_{MLE}(G, P) - (1 - \rho)|P|$$

□

Corollary 2. *Every ρ-approximation for* MIN INTERNAL SPANNING TREE *can be used to ρ-approximate* MIN INTERNAL SPANNING TREE$_{ext}$.

Note that it exists a 5/3-approximation for the Maximum Internal Spanning Tree problem [16].

6 Conclusion

In this article, we tackle variants of several classical combinatorial optimization problems (VERTEX COVER, STEINER TREE, FEEDBACK VERTEX SET, ...) focused around extending a given partial solution. This allows us to define approximation with respect to the additional part of the solution, the residue. We propose the approximation classes $\mathcal{FRAPX}$ and $\mathcal{RAPX}$ capturing approximability of the residue. In this context, we design polynomial-time approximation algorithms and show approximation lower bounds. We hope to open new avenues of research in considering residue approximation, as many combinatorial problems may be examined under this point of view. In this way, our work can be understood as a factory for interesting open problems. Aside from considering more classical combinatorial problems, we are interested in researching the parameterized complexity of extension problems, parameterized (also) by the size of the imposed partial solution.

References

1. Anily, S., Bramel, J., Hertz, A.: A 5/3-approximation algorithm for the clustered traveling salesman tour and path problems. Oper. Res. Lett. **24**(1–2), 29–35 (1999)
2. Arkin, E.M., Hassin, R., Klein, L.: Restricted delivery problems on a network. Networks **29**(4), 205–216 (1997)
3. Avidor, A., Zwick, U.: Approximating MIN 2-SAT and MIN 3-SAT. Theory Comput. Syst. **38**(3), 329–345 (2005). ISSN 1433–0490
4. Bafna, V., Berman, P., Fujito, T.: A 2-approximation algorithm for the undirected feedback vertex set problem. SIAM J. Discrete Math. **12**(3), 289–297 (1999)
5. Bar-Yehuda, R., Even, S.: A linear-time approximation algorithm for the weighted vertex cover problem. J. Algorithms **2**(2), 198–203 (1981)
6. Biró, M., Hujter, M., Tuza, Z.: Precoloring extension. i. interval graphs. Discrete Math. **100**(1–3), 267–279 (1992)
7. Björklund, A., Husfeldt, T., Taslaman, N.: Shortest cycle through specified elements. In Proceedings of 23rd SODA, pp. 1747–1753 (2012)
8. Christofides, N.: Worst-case analysis of a new heuristic for the travelling salesman problem. TR 388, Graduate School of Industrial Administration. Carnegie Mellon University (1976)
9. Garey, M.R., Johnson, D.S.: Computers and Intractability: A Guide to the Theory of NP-Completeness. W. H. Freeman and Company, New York (1979)
10. Gendreau, M., Laporte, G., Hertz, A.: An approximation algorithm for the traveling salesman problem with backhauls. Oper. Res. **45**(4), 639–641 (1997)
11. Goemans, M.X., Williamson, D.P.: New 3/4-approximation algorithms for the maximum satisfiability problem. SIAM J. Discrete Math. **7**(4), 656–666 (1994)
12. Gusfield, D., Pitt, L.: A bounded approximation for the minimum cost 2-SAT problem. Algorithmica **8**(2), 103–117 (1992)

13. Guttmann-Beck, N., Hassin, R., Khuller, S., Raghavachari, B.: Approximation algorithms with bounded performance guarantees for the clustered traveling salesman problem. Algorithmica **28**(4), 422–437 (2000)
14. Hujter, M., Tuza, Z.: Precoloring extension. ii. graphs classes related to bipartite graphs. Acta Math. Univ. Comenian. (N.S.) **62**(1), 1–11 (1993)
15. Jansen, K.: An approximation algorithm for the general routing problem. Inf. Process. Lett. **41**(6), 333–339 (1992)
16. Knauer, M., Spoerhase, J.: Better approximation algorithms for the maximum internal spanning tree problem. Algorithmica **71**(4), 797–811 (2015)
17. Marx, D.: Precoloring extension on unit interval graphs. Discrete Appl. Math. **154**(6), 995–1002 (2006)
18. Orloff, C.S.: A fundamental problem in vehicle routing. Networks **4**(1), 35–64 (1974)
19. Robins, G., Zelikovsky, A.: Improved steiner tree approximation in graphs. In: Proceedings of 11th SODA, pp. 770–779 (2000)
20. Simchi-Levi, D.: New worst-case results for the bin packing problem. Naval Res. Logist. **41**, 579–585 (1994)
21. Weller, M., Chateau, A., Giroudeau, R.: Exact approaches for scaffolding. BMC Bioinform. **16**(Suppl. 14), S2 (2015)
22. Weller, M., Chateau, A., Giroudeau, R.: On the complexity of scaffolding problems: from cliques to sparse graphs. In: Lu, Z., Kim, D., Wu, W., Li, W., Du, D.-Z. (eds.) COCOA 2015. LNCS, vol. 9486, pp. 409–423. Springer, Heidelberg (2015). doi:10. 1007/978-3-319-26626-8_30

On the Parameterized Parallel Complexity and the Vertex Cover Problem

Faisal N. Abu-Khzam[1,4], Shouwei Li[2(✉)], Christine Markarian[3],
Friedhelm Meyer auf der Heide[2], and Pavel Podlipyan[2]

[1] Department of Computer Science and Mathematics,
Lebanese American University, Beirut, Lebanon
[2] Heinz Nixdorf Institute & Department of Computer Science,
Paderborn University, Fürstenallee 11, 33102 Paderborn, Germany
sli@mail.uni-paderborn.de
[3] Department of Mathematical Sciences, Haigazian University, Beirut, Lebanon
[4] School of Engineering and Information Technology,
Charles Darwin University, Darwin, Australia

Abstract. Efficiently parallelizable parameterized problems have been classified as being either in the class FPP (*fixed-parameter parallelizable*) or the class PNC (*parameterized analog of NC*), which contains FPP as a subclass. In this paper, we propose a more restrictive class of parallelizable parameterized problems called *fixed-parameter parallel-tractable* (FPPT). For a problem to be in FPPT, it should possess an efficient parallel algorithm not only from a theoretical standpoint but in practice as well. The primary distinction between FPPT and FPP is the parallel processor utilization, which is bounded by a polynomial function in the case of FPPT. We initiate the study of FPPT with the well-known k-vertex cover problem. In particular, we present a parallel algorithm that outperforms the best known parallel algorithm for this problem: using $\mathcal{O}(m)$ instead of $\mathcal{O}(n^2)$ parallel processors, the running time improves from $4\log n + \mathcal{O}(k^k)$ to $\mathcal{O}(k \cdot \log^3 n)$, where m is the number of edges, n is the number of vertices of the input graph, and k is an upper bound of the size of the sought vertex cover. We also note that a few P-complete problems fall into FPPT including the monotone circuit value problem (MCV) when the underlying graphs are bounded by a constant Euler genus.

1 Introduction

The area of *Parameterized Complexity* has witnessed tremendous growth in the last two decades and has become a central research area in theoretical computer science. A problem is *fixed-parameter tractable* (FPT) if it has an algorithm that runs in time $\mathcal{O}(f(k) \cdot n^{\mathcal{O}(1)})$, where n is the problem size and k is the input

This work was partially supported by the German Research Foundation (DFG) within the Collaborative Research Center "On-The-Fly Computing" (SFB 901) and the International Graduate School "Dynamic Intelligent Systems".

ⓒ Springer International Publishing AG 2016
T-H.H. Chan et al. (Eds.): COCOA 2016, LNCS 10043, pp. 477–488, 2016.
DOI: 10.1007/978-3-319-48749-6_35

parameter that is independent of n, for an arbitrary computable function f. A typical, well-known, example is the k-vertex cover problem which is solvable in time $\mathcal{O}(kn + 1.274^k)$, where n is the number of vertices of the input graph and k is an upper bound of the size of the sought vertex cover [6]. Numerous other NP-complete problems also fall into the class FPT.

The study of parameterized complexity has been extended to parallel computing and this is broadly known as *parameterized parallel complexity*. The first systematic work on parameterized parallel complexity appeared in [5], where the authors introduced two classes to characterize the efficiently parallelizable parameterized problems, according to the degree of efficiency required, known as *parameterized analog of NC* (PNC) and *fixed-parameter parallelizable* (FPP), respectively. The class PNC contains all parameterized problems that have a parallel deterministic algorithm running in time $\mathcal{O}(f(k) \cdot (\log n)^{h(k)})$ using $\mathcal{O}(g(k) \cdot n^\beta)$ parallel processors, where n is the input size, k is the parameter, f, g, and h are arbitrary computable functions, and β is a constant independent of n and k. The class FPP contains all parameterized problems that have a parallel deterministic algorithm running in time $\mathcal{O}(f(k) \cdot (\log n)^\alpha)$ using $\mathcal{O}(g(k) \cdot n^\beta)$ parallel processors, where n is the input size, k is the parameter, f and g are arbitrary computable functions, and α, β are constants independent of n and k. They also proposed a FPP algorithm that solves the k-vertex cover problem and proved that all problems involving *MS* (definable in monadic second-order logic with quantifications over vertex and edge sets) or *EMS properties* (that involve counting or summing evaluations over sets definable in monadic second-order logic) are in FPP when restricted to graphs of bounded *treewidth*.

Our Contribution. In this paper, we propose a new class of efficiently parallelizable parameterized problems called *fixed-parameter parallel-tractable* (FPPT). FPPT has several advantages over PNC and FPP. It eliminates the heavy exponent that depends on the parameter k in the logarithm of PNC. Moreover, it gets rid of the arbitrary function g, which does not seem to lead to a parallel algorithm that is efficient from a practical point of view of processor utilization. Thus, a problem in FPPT should possess an efficient parallel algorithm, not only from a theoretical standpoint but also in practice. We initiate the study of FPPT with the well-known k-vertex cover problem for which we propose a parallel algorithm and show that it belongs to FPPT. We also note that the classical monotone circuit value problem (MCV) belongs to FPPT when the underlying graph is bounded to a constant Euler genus.

Our proposed algorithm for the k-vertex cover problem outperforms the best known parallel algorithm for this problem, which appeared in [5]. The number of parallel processors is $\mathcal{O}(m)$ instead of $\mathcal{O}(n^2)$ and the running time improves from $4 \log n + \mathcal{O}(k^k)$ to $\mathcal{O}(k \cdot \log^3 n)$, where m is the number of edges, n is the number of vertices of the input graph, and k is an upper bound of the size of the sought vertex cover. Our algorithm employs kernelization using the *Crown Reduction* method of [2]. As for the MCV problem, we have recently presented a parallel algorithm when the Euler genus of the underlying graph is bounded by the parameter k [3]. The algorithm runs in time $\mathcal{O}((k + 1) \cdot \log^3 n)$

using $\mathcal{O}(n^c)$ parallel processors, where $\mathcal{O}(n^c)$ is the best processor boundary for parallel matrix multiplication [19].

2   Preliminaries

Given an undirected graph $G = (V, E)$, the *k-vertex cover* problem asks whether there is a set of vertices $V' \subseteq V$ of size at most k (k is assumed to be fixed), such that for every edge $(u, v) \in E$, at least one of its endpoints, u or v is in V'. In other words, the complement of V' is an independent set (i.e., a set that induces an edge-less subgraph).

A *matching* M is a subset of E such that no two edges in M share a common vertex. A vertex that is incident to an element of M is said to be matched under M. A matching is maximal if it is not contained in a larger matching. Such a matching is *maximum* if no matching of larger cardinality exists. If all vertices are matched under a matching, then it is a *perfect matching*. The problem of finding a maximum matching or a perfect matching, even in planar graphs, has received considerable attention in the field of parallel computation.

Merging a subset V' of the vertices consists of replacing all the vertices in V' with a single vertex w such that $N(w) = \bigcup_{v \in V'} N(v) \backslash V'$.

A *crown* in a graph G is an ordered pair (I, H) of subsets of V that satisfies the following criteria:

- $I \neq \emptyset$ is an independent set of G,
- $H = N(I)$, and
- there exists a matching M on the edges connecting I and H such that all elements of H are matched under M.

H is called the *head* of the crown. The *width* of the crown is $|H|$. A *straight crown* is a crown (I, H) that satisfies the condition $|I| = |H|$. A *flared crown* is a crown (I, H) that satisfies condition $|I| > |H|$. These notions are depicted in Fig. 1.

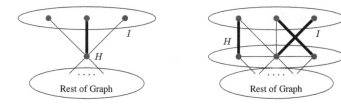

Fig. 1. Sample crowns (Bold edges denote a matching)

The following theorem was proved in the early work on crown decomposition.

Theorem 1 ([1,7]). *If G is a graph with a crown (I, H), then there is a vertex cover of G of minimum size that contains all the vertices in H and none of the vertices in I.*

Kernelization is a polynomial-time transformation that reduces an arbitrary instance (I, k) of a parameterized problem to an equivalent instance (I', k'), such that $k' \leq k$ and $|I'|$ is bounded by some function of k. The resulting instance is often referred to as a problem kernel of the instance. It is kernelization that distinguishes an arbitrary problem from one that is FPT. More specifically, a problem is FPT if and only if it has a kernelization algorithm [11].

3   FPPT Kernelization by Crown Reduction

The classes PNC and FPP introduced by Cesati and Ianni in [5] fail to capture some important aspects mainly caused by the "arbitrary" computable functions h and g bounding the running time and the processor utilization. For example, the exponent in the logarithm of PNC heavily depends on the parameter k. It grows at a rapid rate thus making the running time very close to a linear function even for not too large values of k. Moreover, according to its definition, g is allowed to be a super-polynomial function of the parameter k, which makes the processor utilization unreasonable even for relatively small values of k. Since the number of available processors is a strong physical constraint when employing a parallel algorithm, membership in PNC and FPP can sometimes be relevant mostly for theoretical purposes.

Motivated by the above, we refine the classes PNC and FPP and propose a new class called *fixed-parameter parallel-tractable* (FPPT). The main purpose is to have a better characterization of solvable fixed-parameter parallelizable problems at least from a practical standpoint. We formally define FPPT as follows:

Definition 1 (FPPT). The class of *fixed-parameter parallel-tractable* (FPPT) contains all parameterized problems that have a parallel deterministic algorithm whose running time is in $\mathcal{O}(f(k) \cdot (\log n)^{\alpha})$ using $\mathcal{O}(n^{\beta})$ parallel processors, where n is the size of input, k is the parameter, f is an arbitrary computable function, and α, β are constants independent of n and k.

The Lemma below shows the relation between FPT and PNC, given by Cesati and Ianni.

Lemma 1 ([5]). *PNC is a subset of FPT.*

It is then easy to observe that FPP $\subseteq$ PNC. Hence, we conclude that:

$$FPPT \subseteq FPP \subseteq PNC \subseteq FPT.$$

In the following, we present an FPPT kernelization algorithm for the k-vertex cover problem, which is based on a crown decomposition of the input graph in [2].

Algorithm 1. Parallel algorithm for finding a crown.

---

procedure PARALLELCROWN(G)

    Step 1: Find a maximal matching M_1 in parallel and identify the set of all unmatched vertices as the set O of outsiders.

    Step 2: Find a maximum auxiliary matching M_2 of the edges between O and $N(O)$ in parallel.

    Step 3: If every vertex in $N(O)$ is matched by M_2, then $H = N(O)$ and $I = O$ form a straight crown, and we are done.

    Step 4: Let I_0 be the set of vertices in O that are unmatched by M_2.

    Step 5: Repeat Steps 5a and 5b until $n = N$ so that $I_{N-1} = I_N$.

        5a. Let $H_n = N(I_n)$.

        5b. Let $I_{n+1} = I_n \cup N_{M_2}(H_n)$.

    Step 6: $I = I_N$ and $H = H_N$ form a flared crown.

end procedure

It has been shown in [7] that, if both matchings M_1 and M_2 are of a size less than or equal to k, then G has at most $3k$ vertices that are not in the crown. So we only analyze the running time and processor utilization here. Apparently, the most expensive part of this procedure are Steps 1 and 2.

Israeli and Shiloach presented a parallel algorithm for maximal matching. The performance of their algorithm is given by the following lemma:

Lemma 2 ([16]). *A maximal matching in general graphs can be found in time $\mathcal{O}(\log^3 n)$ using $\mathcal{O}(m)$ parallel processors, where m is the number of edges and n is the number of vertices of the input graph.*

So we only need to show that the maximum matching M_2 in Step 2 can be constructed (efficiently) in parallel, which will be discussed in Sect. 4 where we prove that the construction of such a maximum matching is in FPPT. Therefore we obtain the following.

Theorem 2. *Vertex cover kernelization via crown reduction is in FPPT.*

Consequently, and since crown reduction guarantees a $3k$-kernel, we can solve the reduced instance of vertex cover in $\mathcal{O}(f(k))$-time. This finishes the proof of our main result.

Theorem 3. *The k-vertex cover problem is in FPPT.*

4   Parameterized Maximum Matching

In this section, let us consider a parameterized version of the maximum matching problem stated below, as a subroutine of the parallel crown reduction procedure. Note that, in our definition of the parameterized maximum matching problem, we ask the question whether the cardinality of the maximum matching is equal to or less than parameter k. (This is to be contrasted with the usual parameterization of a maximization problem where one would ask for a solution of size k or more).

Input:	A graph $G = (V, E)$ and a positive integer k.
Problem:	Is there a maximum matching M of size at most k? If yes, how to construct one?

To the best of our knowledge, this is the first time that the maximum matching problem is studied with an input parameter given, especially in a way that upper-bounds the size of the sought matching. In classical computational complexity, the maximum matching problem has the same complexity as the perfect matching problem, since the former can be easily reduced to the latter in logarithmic space. Suppose we want to check whether there is a matching with k edges in G. If such a matching existed, $2k$ vertices would have been matched and $n - 2k$ would have been "free." We add $n - 2k$ new vertices to G and create edges between these new vertices and all old vertices in G in order to obtain a new graph G'. Thus, G' has a perfect matching if and only if G has a matching with exactly k edges. Conversely, perfect matching parameterized by the number of matching edges is trivially FPPT by a simple reduction to our version of parameterized maximum matching (if $n \neq 2k$ then it is a no instance). So the two problems have the same parameterized parallel complexity with respect to the parameter k. However, when the parameter is the number of perfect matchings, perfect matching falls in the class FPPT, more precisely in NC in this case [4], while it is not known yet whether maximum matching has an NC algorithm when the number of maximum matchings is bounded by a polynomial in n. This problem is open at this stage. The related studies on perfect matching, including NC algorithms for some special structures of graphs or parameters, such as: dense graphs [9], regular bipartite graphs [18], strongly chordal graphs [10], claw-free graphs [8], bipartite graphs with polynomially bounded number of perfect matchings [15], general graphs with polynomially bounded number of perfect matchings [4], bipartite planar and small genus graphs [20].

Suppose G is a graph with minimum vertex cover bounded by a constant k. It is not hard to see that the maximum matching of G must be bounded to k as well (since the edges in a maximum matching are pairwise disjoint, at least one end of each edge should be included in the minimum vertex cover). Therefore, in order to cover all edges in a maximum matching, the minimum vertex cover must be greater than or equal to the size of a maximum matching.

Our algorithm is based on the augmenting path approach used in the classical *Blossoms* algorithm. We show that, with careful analysis, the parameterized maximum matching problem can be solved in time $\mathcal{O}(f(k) \cdot \log^3 n)$ using $\mathcal{O}(m)$ parallel processors, where m is the number of edges and n is the number of vertices of the input graph. The algorithm is summarized in Algorithm 2.

Algorithm 2. Parallel algorithm for parameterized maximum matching

---

1: **procedure** FINDMAXIMUMMATCHING(G,k)   ▷ Graph G and parameter k
2:     Step 1: Find a maximal matching M_1 in parallel. If $|M_1| > k$, return false.
3:     Step 2: Construct a new graph by merging the unmatched vertices into a new
         vertex S.
4:     Step 3: Construct an alternating BFS tree rooted at S in parallel.
5:     Step 4: Either find an augmenting path with respect to M_1, or construct a new
         graph with a maximum matching of cardinality $k - 1$.
6:     Step 5: Repeat Step 1 to Step 4 at most k rounds.
7: **end procedure**

It is well known that the size of any maximal matching is at least half the size of a maximum matching. Therefore, we note that the maximal matching produced in Step 1 is not less than $k/2$.

In Step 2, we make a transformation as follows. We merge all unmatched vertices into a single vertex S so that all edges connected to the unmatched vertices become incident to S, thus making all augmenting paths in the graph G transfer into an odd alternating cycle in the new graph.

Next, we construct an alternating BFS tree with the following properties:

- S is the root (and layer 0);
- All vertices adjacent to S are in layer 1;
- All edges from odd layers to even layers are matching edges (elements of the matching M_1 constructed in Step 1).

Note that we also add the unmatched edges into the alternating BFS tree in Fig. 2 to help in the analysis. This will be made clear in the sequel.

The parallel alternating BFS tree can be constructed in time $\mathcal{O}(\log^2 n)$ using $\mathcal{O}(n^3)$ parallel processors [13]. However, the graph for which the BFS is constructed consists of at most $2k + 1$ vertices because all unmatched vertices are merged into S. Thus, the running time will be $\mathcal{O}(\log^2 k)$ with $\mathcal{O}(k^3)$ parallel processors. Consequently, we observe the following:

- All vertices in the tree are matched except for S. All edges between layer 0 and layer 1 are unmatched, otherwise S will be matched.
- If the nodes in layer 1 have no descendants (i.e., neighbors in a layer of a higher index), then they must be matched in this layer (e.g., b and c in Fig. 2). Otherwise, these nodes would be unmatched. We call such an edge type B edge. The unmatched edges in the same layer are also type B edges.
- The unmatched edges from odd layers to descendant layers are of type A (e.g., (d, f) and (a, k) in Fig. 2).
- The unmatched edges from even layers to the adjacent odd layer are of type C (e.g., (h, j) in Fig. 2).
- The depth of the tree is at most $2k$ because the size of the maximum matching is bounded by k.

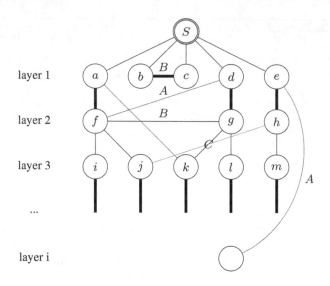

Fig. 2. Alternating BFS tree. (Bold edges denote matching edges)

Lemma 3. *If M_1 is not a maximum matching, then every alternating odd cycle must pass through at least one type B edge.*

Proof. Suppose there is an augmenting path $p = e_1, e_2, \ldots, e_h$ that consists of type A, type C, and matching edges. According to the observation above, the initial vertex of p must be S and e_1 must be an unmatched edge between layer 0 and layer 1 (e.g., (S, e) in Fig. 2). For p to be an augmenting path, e_2 must be a matched edge. Since it is not allowed to take type B edges, we have to go further down (e.g., (e, h) in Fig. 2). Then, e_3 must be a type C edge (e.g., (h, j) in Fig. 2). Because any type C edge starts from an even layer towards an adjacent odd layer, it is impossible to construct an augmenting path.   □

Since each augmenting path transforms into an alternating odd cycle and increases the matching by 1, we only need to check if there are type B edges in the alternating BFS tree. If yes, we proceed by checking whether the alternating odd cycle comes from a valid augmenting path. Otherwise, we either get a maximum matching with cardinality at most k or reduce the graph to a new graph with the maximum matching of size bounded by $k - 1$. Now, we analyze this procedure by considering the following two cases:

Case 1: Suppose an alternating odd cycle resulted from an augmenting path is $(s - a - b - s)$ and the edge (a, b) is matched. a, b connect to different vertices u_1 and u_2 in S. It is easy to find an augmenting path $u_1 - a - b - u_2$. We can argue the same way if the edge (a, b) is an unmatched type B edge. Refer to Fig. 3 for an illustration.

Case 2: Suppose that an alternating odd cycle transformed from an augmenting path is $(u - a - b - u)$ and the edge (a, b) is matched. Also a and b connect to the

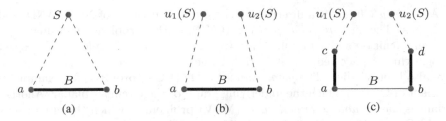

Fig. 3. An alternating odd cycle passes through type B edges. (a) An alternating odd cycle; (b) (a,b) is a matched type B edge; (c) (a,b) is an unmatched type B edge.

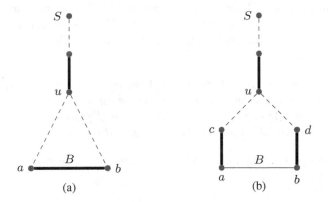

Fig. 4. Blossom structure.

same vertices u and there is a alternating path from S to u. We do not know if there exists an augmenting path, neither do we know how to find one, if there is any. For this case, the alternating odd cycle is called a blossom structure. Refer to Fig. 4 for an illustration. Since a blossom contains at least one matched edge and three nodes, we can shrink the blossom to a single vertex and the new graph obtained will have a maximum matching of cardinality at most $k - 1$. Since the size of the maximum matching is bounded by k, at most k rounds are needed to construct a maximum matching.

Thus, the running time will be $\mathcal{O}(k \cdot \log^3 n)$ with $\mathcal{O}(m)$ parallel processors, where m is the number of edges, n is the number of vertices of the input graph, and k is an upper bound of the size of the sought maximum matching. Given all of the above, we can now state Theorem 4.

Theorem 4. *Maximum matching parameterized by an upper bound on the matching size, is in FPPT.*

5   MCV with Bounded Genus Is in FPPT

In this section, we note that the problem known as the monotone circuit value (MCV) problem with bounded genus is also in FPPT.

A *Boolean Circuit* is a directed acyclic graph consisting of NOT, AND, and OR gates. The *circuit value problem* (CVP) is the problem of evaluating a boolean circuit on a given input. CVP was shown to be P-complete with respect to logarithmic space reductions in [17]. Some other restricted variants of CVP have also been studied. The *planar circuit value* (PCV) problem, for example, is a variant of CVP in which the underlying graph of the circuit is planar. Another variant is the *monotone circuit value* (MCV) problem in which the circuit only has AND and OR gates. Both the PCV and MCV problem were shown to be P-complete in [14]. Another variant in which the circuit is simultaneously planar and monotone is known as the PMCV problem. PMCV problem was shown to be in NC, and its first NC algorithm was given in [23]. Recently, we proposed a parallel algorithm for a variant of the general MCV problem in which the underlying graph bounded by a constant Euler genus k in [3]. The main result can be stated as Theorem 5.

Theorem 5. *Given a general monotone boolean circuit with n gates and an underlying graph bounded by a constant Euler genus k, the circuit can be evaluated in time $\mathcal{O}((k+1) \cdot \log^3 n)$ using $\mathcal{O}(n^c)$ parallel processors, where $\mathcal{O}(n^c)$ is the best processor boundary for parallel matrix multiplication.*

Hence, we deduce that the monotone circuit value problem with bounded Euler genus k is in FPPT.

6 Concluding Remarks and Future Work

In this paper, we initiated the study of a new class of efficiently parallelizable parameterized problems called fixed-parameter parallel-tractable (FPPT). A problem in FPPT, unlike one in PNC or FPP, must be solved by a polynomial number of processors independent of the parameter k. Hence it should possess an efficient parallel algorithm not only from a theoretical standpoint but also in practice.

We reconsidered the k-vertex cover problem and noted that it falls in FPPT due to the algorithm of Cesati and Ianni in [5]. Our improved algorithm is based on obtaining a quadratic kernel. We showed that obtaining a linear kernel for the problem is also in FPPT by proving that constructing a corresponding crown decomposition is in FPPT as well. We believe this will lead to proving other problems also, where a kernel is obtained via a crown reduction, are in FPPT.

Furthermore, we conclude that FPPT is somehow orthogonal to the NP-complete and P-complete classes in the sense that some, but not all, problems from each of these classes fall in FPPT. We further raise some open questions.

- At this stage, the first obvious question is which other problems belong to FPPT.
- Numerous NP-complete problems fall into the class FPT, but clearly not all of them do. It was shown that there is a $W[*]$ hierarchy in the NP class where

FPT $= W[0]$. With the introduction of parameterization into the field of parallel computing, some NP-complete and P-complete problems were shown to have a fixed-parameter parallel algorithm by fixing one (or more) parameter(s). These include the vertex cover problem, the graph genus problem [12], and the monotone circuit value problem, which all fall into FPPT. Now the question is what happens if we restrict our attention to P-complete problems: could it be that the P class also has a hierarchy, say $Z[*]$, analogous to $W[*]$ in the class NP, where FPPT $= Z[0]$? In fact, we believe such a hierarchy should exist because the following is true. The MCV problem and the NAND circuit value problem [22] are both P-complete problems. The first is in FPPT while the second is not, taking the graph genus as a parameter. Another conceivable example is the lexicographically first maximal subgraph problem (LFMIS). Miyano showed that the latter is P-complete even for bipartite graphs with bounded degree at most 3 in [21]. It would be interesting, and probably not too difficult, to obtain hardness results showing a problem cannot belong to FPPT unless some new parameterized parallel-complexity hierarchy collapses.

– Furthermore, there has been an interest in the so-called "gradually intractable problems" for the class NP. The question here is whether "gradually *unparallelizable* problems" can be analogously investigated in the class of P-complete. We suggest involving one or more parameters to characterize the "gradually" procedure. This would be helpful to understand the intrinsic difficulty of P-complete problems and to answer the question of whether P $=$ NC, which is one of the main motivations behind this paper.

Acknowledgments. We wish to thank the anonymous referees for their valuable comments to improve the quality and presentation of this paper.

References

1. Abu-Khzam, F.N., Collins, R.L., Fellows, M.R., Langston, M.A., Suters, W.H., Symons, C.T.: Kernelization algorithms for the vertex cover problem,: theory and experiments. In: Arge, L., Italiano, G.F., Sedgewick, R. (eds.) Proceedings of the Sixth Workshop on Algorithm Engineering and Experiments and the First Workshop on Analytic Algorithmics and Combinatorics, New Orleans, LA, USA, 10 January 2004, pp. 62–69. SIAM (2004)
2. Abu-Khzam, F.N., Fellows, M.R., Langston, M.A., Suters, W.H.: Crown structures for vertex cover kernelization. Theory Comput. Syst. **41**(3), 411–430 (2007)
3. Abu-Khzam, F.N., Li, S., Markarian, C., Meyer auf der Heide, F., Podlipyan, P.: The monotone circuit value problem with bounded genus is in NC. In: Dinh, T.N., Thai, M.T. (eds.) COCOON 2016. LNCS, vol. 9797, pp. 92–102. Springer, Heidelberg (2016). doi:10.1007/978-3-319-42634-1_8
4. Agrawal, M., Hoang, T.M., Thierauf, T.: The polynomially bounded perfect matching problem is in $\mathbf{NC}^2$. In: Thomas, W., Weil, P. (eds.) STACS 2007. LNCS, vol. 4393, pp. 489–499. Springer, Heidelberg (2007)
5. Cesati, M., Di Ianni, M.: Parameterized parallel complexity. In: Pritchard, D., Reeve, J.S. (eds.) Euro-Par 1998. LNCS, vol. 1470, pp. 892–896. Springer, Heidelberg (1998)

6. Chen, J., Kanj, I.A., Xia, G.: Improved upper bounds for vertex cover. Theor. Comput. Sci. **411**(40), 3736–3756 (2010)
7. Chor, B., Fellows, M., Juedes, D.W.: Linear kernels in linear time, or how to save k colors in $O(n^2)$ steps. In: Hromkovič, J., Nagl, M., Westfechtel, B. (eds.) WG 2004. LNCS, vol. 3353, pp. 257–269. Springer, Heidelberg (2004)
8. Chrobak, M., Naor, J., Novick, M.B.: Using bounded degree spanning trees in the design of efficient algorithms on claw-free graphs. In: Dehne, F., Sack, J.-R., Santoro, N. (eds.) WADS 1989. LNCS, vol. 382, pp. 147–162. Springer, Heidelberg (1989)
9. Dahlhaus, E., Hajnal, P., Karpinski, M.: On the parallel complexity of hamiltonian cycle and matching problem on dense graphs. J. Algorithms **15**(3), 367–384 (1993)
10. Dahlhaus, E., Karpinski, M.: The matching problem for strongly chordal graphs is in NC. Inst. für Informatik, Univ. (1986)
11. Downey, R.G., Fellows, M.R.: Fundamentals of Parameterized Complexity, vol. 4. Springer, Heidelberg (2013)
12. Elberfeld, M., Kawarabayashi, K.-I.: Embedding and canonizing graphs of bounded genus in logspace. In: Proceedings of the 46th Annual ACM Symposium on Theory of Computing, pp. 383–392. ACM (2014)
13. Ghosh, R.K., Bhattacharjee, G.: Parallel breadth-first search algorithms for trees and graphs. Int. J. Comput. Math. **15**(1–4), 255–268 (1984)
14. Goldschlager, L.M.: The monotone and planar circuit value problems are log space complete for P. SIGACT News **9**(2), 25–29 (1977)
15. Grigoriev, D.Y., Karpinski, M.: The matching problem for bipartite graphs with polynomially bounded permanents is in NC. In: 28th Annual Symposium on Foundations of Computer Science, 1987, pp. 166–172. IEEE (1987)
16. Israeli, A., Shiloach, Y.: An improved parallel algorithm for maximal matching. Inf. Process. Lett. **22**(2), 57–60 (1986)
17. Ladner, R.E.: The circuit value problem is log space complete for P. SIGACT News **7**(1), 18–20 (1975)
18. Lev, G.F., Pippenger, N., Valiant, L.G.: A fast parallel algorithm for routing in permutation networks. IEEE Trans. Comput. **100**(2), 93–100 (1981)
19. Li, K., Pan, V.Y.: Parallel matrix multiplication on a linear array with a reconfigurable pipelined bus system. IEEE Trans. Comput. **50**(5), 519–525 (2001)
20. Mahajan, M., Varadarajan, K.R.: A new NC-algorithm for finding a perfect matching in bipartite planar and small genus graphs. In: Proceedings of the Thirty-Second Annual ACM Symposium on Theory of Computing, pp. 351–357. ACM (2000)
21. Miyano, S.: The lexicographically first maximal subgraph problems: P-completeness and NC algorithms. Math. Syst. Theory **22**(1), 47–73 (1989)
22. Post, E.L.: The Two-Valued Iterative Systems of Mathematical Logic. (AM-5), vol. 5. Princeton University Press, Princeton (2016)
23. Yang, H.: An NC algorithm for the general planar monotone circuit value problem. In: Proceedings of the Third IEEE Symposium on Parallel and Distributed Processing, 1991, pp. 196–203, December 1991

A Linear Potential Function for Pairing Heaps

John Iacono and Mark Yagnatinsky[✉]

NYU Tandon School of Engineering, Brooklyn, USA
myagna01@students.poly.edu

Abstract. We present the first potential function for pairing heaps with linear range. This implies that the runtime of a short sequence of operations is faster than previously known. It is also simpler than the only other potential function known to give constant amortized time for insertion.

1 Introduction

The pairing heap is a data structure for implementing priority queues introduced in the mid-1980s [FSST86]. The inventors conjectured that pairing heaps and Fibonacci heaps [FT84] have the same amortized complexity for all operations. While this was eventually disproved [Fre99], pairing heaps are, unlike Fibonacci heaps, both easy to implement and also fast in practice [LST14]. Thus, they are widely used, while their asymptotic performance is still not fully understood. Here, we show that a short sequence of operations on a pairing heap takes less time than was previously known.

1.1 Motivation: Why a Linear Potential Function Is Useful

We begin with a brief review of potential functions. Given a sequence of operations $o_1, o_2, \ldots, o_m$ executed on a particular data structure, and an integer $i \geq 0$, a potential function is a function Φ that maps i to a real number Φ_i. The real number Φ_i is called the *potential* after the ith operation. Given Φ, we define the amortized time a_i of an operation o_i as the actual time t_i of the operation, plus the change in potential $\Phi_i - \Phi_{i-1}$.

Observe that since $a_k = t_k + \Phi_k - \Phi_{k-1}$, we have $t_k = a_k - \Phi_k + \Phi_{k-1}$. Thus, the total time taken for the subsequence of consecutive operations from i to j is $\sum_{k=i}^{j} t_k = \sum_{k=i}^{j}(a_k - \Phi_k + \Phi_{k-1}) = \Phi_{i-1} - \Phi_j + \sum_{k=i}^{j} a_k$. From this formula, we can derive the motivation for our result: namely, that a potential function with a small range is useful. (When we speak of the range of a potential function, we mean its maximum value as function of n minus its minimum value. Often the minimum is zero, and thus the range is simply the maximum value.) To see this, suppose you have a data structure with n elements, and you perform a sequence of k operations on it that don't change the size. If each operation takes $O(\log n)$ amortized time, then the total actual time is bounded by $O(k \log n)$ plus the

---

See [IY16] for the full version of this paper.

© Springer International Publishing AG 2016
T.-H.H. Chan et al. (Eds.): COCOA 2016, LNCS 10043, pp. 489–504, 2016.
DOI: 10.1007/978-3-319-48749-6_36

loss of potential. Thus, if the range of the potential function is $O(n \log n)$, then the total time is $O(k \log n + n \log n)$, but if the range of the potential function is linear, this is improved to $O(k \log n + n)$, which is asymptotically better whenever k is $o(n)$. Thus, a reduced range of a potential function improves the time bounds for short sequences that don't start from an empty data structure.

1.2 Pairing Heaps

A *pairing heap* [FSST86] is a heap-ordered general rooted ordered tree. That is, each node has zero or more children, which are listed from left to right, and a child's key value is always larger than its parent's. The basic operation on a pairing heap is the *pairing* operation, which combines two pairing heaps into one by attaching the root with the larger key value to the other root as its leftmost child. For the purposes of implementation, pairing heaps are stored as a binary tree using the leftmost-child, right-sibling correspondence. That is, a node's left child in the binary tree corresponds to its leftmost child in the general tree, and its right child in the binary tree corresponds to its right sibling in the general tree. In order to support decrease-key, there is also a parent pointer which points to the node's parent in the binary representation.

Priority queue operations are implemented in a pairing heap as follows: **make-heap()**: return null; **get-min(H)**: return H.val; **insert(H, x)**: create a new node and pair it with the existing root (if present); **decrease-key(p, y)**: let n be the node p points to, set the value of n's key to y, and if n is not the root, detach n from its parent and pair it with the root; **delete-min(H)**: first, we remove the root. The rest of delete-min proceeds in two stages, known as the first pairing pass and the second pairing pass. In the first pass, we pair the children of the old root in groups of two. In the second pass, we incrementally pair the remaining trees from right to left. Finally, we return the new root. See Fig. 1 for an example of a delete-min executing on a pairing heap. (Readers familiar with splay trees may notice that in the binary view, a delete-min resembles a splay operation.)

All pairing heap operations take constant actual time, except delete-min, which takes time linear in the number of children of the root. Pairing heaps naturally support one more operation in constant time: merge. This takes two independent heaps and pairs them. Unfortunately, this takes amortized linear time using our potential function.

History. Pairing heaps were originally inspired by splay trees [ST85]. Like splay trees, they are a self-adjusting data structure: the nodes of the heap don't store any information aside from the key value and whatever pointers are needed to traverse the structure. This is in contrast to, say, Fibonacci heaps [FT84], which store at each node an approximation of that node's subtree size. Fibonacci heaps support delete-min in logarithmic amortized time, and all the other heap operations in constant amortized time. However, they are complicated to implement, somewhat bulky, and therefore slow in practice [SV86]. Pairing heaps were introduced as a simpler alternative to Fibonacci heaps, and it was conjectured that

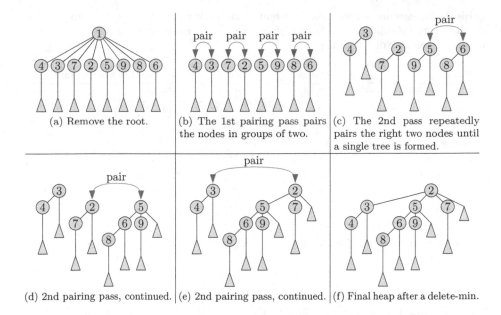

(a) Remove the root.

(b) The 1st pairing pass pairs the nodes in groups of two.

(c) The 2nd pass repeatedly pairs the right two nodes until a single tree is formed.

(d) 2nd pairing pass, continued. | (e) 2nd pairing pass, continued. | (f) Final heap after a delete-min.

Fig. 1. Delete-min on a heap where the root has eight children.

they have the same amortized complexity for all operations, although [FSST86] showed only an amortized logarithmic bound for insert, decrease-key, and delete-min. The conjecture was eventually disproved when it was shown that if insert and delete-min both take $O(\log n)$ amortized time, an adversary can force decrease-key to take $\Omega(\log \log n)$ amortized time [Fre99].

Present. Nevertheless, pairing heaps are fast in practice. For instance, the authors of [LST14] benchmarked a variety of priority queue data structures. They also tried to estimate difficulty of implementation, by counting lines of code, and pairing heaps were essentially tied for first place by that metric, losing to binary heaps by only two lines. Despite (or rather because of) their simplicity, pairing heaps had the best performance among over a dozen heap variants in two out of the six benchmarks.

1.3   Previous Work and Our Result

In [Col00, Theorem 3], Cole develops a linear potential function for splay trees (that is, the potential function ranges from zero to $O(n)$), improving on the potential function used in the original analysis of splay trees, which had a range of $O(n \log n)$ [ST85]. As explained above, this allows applying amortized analysis over shorter operation sequences.

There are several variants of pairing heaps such as [IO14] and [Elm09a], and one of them also has a potential function that is $o(n \log n)$ [IO14]. The main

Table 1. Various heaps and amortized bounds on their running times. Top: analyses of pairing heaps. Middle: close relatives of pairing heaps. Bottom: more distant relatives of pairing heaps. Note: $\lg = \log_2$.

Result	Range	Insert	Decrease-key	Delete-min
Pairing heap [FSST86]	$\Theta(n \lg n)$	$O(\lg n)$	$O(\lg n)$	$O(\lg n)$
Pairing heap [Pet05]	$O(n \cdot 4^{\sqrt{\lg \lg n}})$	$O(4^{\sqrt{\lg \lg n}})$	$O(4^{\sqrt{\lg \lg n}})$	$O(\lg n)$
Pairing heap [Iac00]	$O(n \lg n)$	$O(1)$	$O(\lg n)$	$O(\lg n)$
Pairing heap [This paper]	$\Theta(n)$	$O(1)$	$O(\lg n)$	$O(\lg n)$
Stasko/Vitter [SV86]	$O(n \lg n)$	$O(1)$	$O(\lg n)$	$O(\lg n)$
Elmasry [Elm09a, Elm09b]	$O(n \lg n)$	$O(1)$	$O(\lg \lg n)$	$O(\lg n)$
Sort heap [IO14]	$\Theta(n \lg \lg n)$	$O(\lg \lg n)$	$O(\lg \lg n)$	$O(\lg n \lg \lg n)$
Binomial heap [Vui78]	$\Theta(\lg n)$	$O(1)$	$O(\lg n)$	$O(\lg n)$
Fibonacci heap [FT84]	$\Theta(n)$	$O(1)$	$O(1)$	$O(\lg n)$
Rank-pairing heap [HST09]	$\Omega(n)$	$O(1)$	$O(1)$	$O(\lg n)$

theme in all the variants is to create a heap with provably fast decrease-key, while maintaining as much of the simplicity of pairing heaps as possible.

Our Result. We present a potential function for pairing heaps that is much simpler than the one found for splay trees in [Col00] and also simpler than the only previously known potential function for pairing heaps that is $o(n \log n)$ [Pet05]. Further, it is simpler than the only other potential function known to give constant amortized time for insertion [Iac00], and perhaps more importantly, it is the first potential function for pairing heaps whose range is $O(n)$, which allows the use of amortized analysis to bound the run times of shorter operation sequences than before. In the case of pairing heaps, this bound on the potential function range is asymptotically the best possible, since the worst-case time for delete-min is linear, and thus we need to store at least a linear potential to pay for it.

Previous Work. In Table 1, we list the amortized operation costs and ranges of several potential functions. Each row of the table corresponds to a single analysis of a specific heap variant. The table is divided into three parts. The top part is devoted to analyses of pairing heaps. The middle is for variants of pairing heaps, and the bottom is for heaps that are sufficiently different from pairing heaps that calling them a variant seems inaccurate. For an extended discussion of this table, see [Yag16] or [IY16].

2   The Potential Function

Our potential function is the sum of three components. The first is the node potential, which will give a value to each node. (The total node potential is the sum of the values for individual nodes.) The second is the edge potential:

each edge will have a potential of either 0 or -7. (The total edge potential is likewise the sum of the values of individual edges.) The third we shall call the size potential. We begin with explaining the concept of the sticky size, since we will need it to define all three components. The size n of a heap is how many elements it currently stores. The sticky size N is initially 1. After every heap update, the sticky size is updated as follows: if $n \geq 2N$, then N is doubled, and if $n \leq N/2$, then N is halved. The sticky size is the only aspect of the potential function which is not computable simply from the current state of the heap but is based on the history of operations.

The size potential is simply $900|N - n|$.

The node potential is slightly more complicated. Given a node x, let $|x|$ be the size of its subtree (in the binary view). Its left child in the binary view is x_L and its right child is x_R. The node potential ϕ_x of x depends on $|x_L|$ and $|x_R|$. Note that $|x| = |x_L| + |x_R| + 1$. Let $\lg = \log_2$. There are three cases. If $|x_L| > \lg N$ and $|x_R| > \lg N$, then x is a *large* node and $\phi_x = 400 + 100 \lg |x|$. If $|x_L| \leq \lg N < |x_R|$, then x is a *mixed* node and $\phi_x = 400 + 100 \frac{|x_L|}{\lg N} \lg |x|$. (The case where $|x_R| \leq \lg N < |x_L|$ is symmetric.) Finally, if $|x_L| \leq \lg N$ and $|x_R| \leq \lg N$, then x is a *small* node and $\phi_x = 0$.

If the right child of a large node (in the binary view) is also large, then the edge potential of the edge connecting them is -7. All other edges have zero edge potential.

We define the actual cost of an operation to be the number of pairings performed plus one. Through most of the analysis, it will seem that the node potential is a hundred times larger than what is needed. Near the end, we will see that we use the excess to pay for mixed-large pairings during a delete-min.

3   The Analysis

In the proofs below, we assume for convenience that the heap has at least four elements, so we can say things like "the root of the heap is always mixed, since it has no siblings and $n - 1$ descendants in the general representation," which assumes that $n - 1 > \lg N$, which may not be true for heaps with less than four elements. If we need stronger assumptions on the size, we will call those out explicitly.

Lemma 1. *The potential of a pairing heap is $O(N) = O(n)$.*

Proof. The size potential is linear by definition. The edge potential is slightly negative: most edges have potential zero, the exception being those edges that connect two large nodes. If there are L large nodes, the total edge potential may be as low as $-7(L - 1)$. But the node potential is at least $400L$, so the edge potential can never make the heap potential negative.

We now turn to the node potential. The small nodes have potential zero. Observe that the lowest common ancestor (in the binary representation) of two large nodes must itself be large. This immediately implies that if the left subtree (in the binary view) of a large node contains any large nodes, then this subtree contains a unique large node which is the common ancestor of all large nodes in this subtree. (And likewise for the right subtree.) Thus, it makes sense to speak of the tree induced by the large nodes, which is the binary tree obtained by taking the original pairing heap and performing the following two operations: first, erase all small nodes (they have no mixed or large descendants). After this, all mixed nodes have only one child, so they form paths. Second, contract these paths to edges. Now only large nodes remain, and ancestor-descendant relations are preserved. (One thing that is not preserved is the root: in the original tree, the root is a mixed node.) If a large node has less than two children in this shrunken tree, call it a *leaf-ish* large node, and otherwise call it an *internal* large node. In the original tree, a leaf-ish large node x has a left and a right subtree (in the binary representation), at least one of which has no large nodes in it: call that a *barren subtree* of x; see Fig. 2.

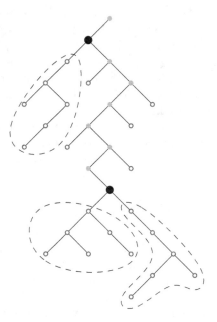

Let x and y be two leaf-ish large nodes, and observe that their barren subtrees are disjoint, even if x is an ancestor of y. We say that a leaf-ish large node *owns* the nodes in its barren subtree. Observe that, in the binary view, every leaf-ish large node has at least $\lg N$ descendants which are not owned by any other leaf-ish large node, which implies that there are at most $n/\lg N$ leaf-ish large nodes. There are more leaf-ish large nodes than there are internal large nodes (since a binary tree always has more leaf-ish nodes than internal nodes, as can be shown by induction), so there are at most $2n/\lg N$ large nodes in total, each of which has a potential of at most $100\lg n + 400 = 100\lg N + O(1)$, so the total potential of all large nodes is at most $200n + o(n)$.

This leaves the mixed nodes. Every mixed node has a *heavy subtree* with more than $\lg N$ nodes in the binary view, and a possibly empty *light subtree* with at most $\lg N$ nodes. Since the light subtree contains no mixed nodes, every node in the heap is in the light subtree of at most one mixed node; that is, the light subtrees of different nodes are disjoint. If x is a mixed node and L_x is

Fig. 2. Barren subtrees are marked by dashed bubbles; small nodes are marked with hollow red circles, large nodes are marked with bold black disks, and mixed nodes are green. (Color figure online)

the size of its light subtree, the node potential of x is $\phi_x = 400 + 100\frac{L_x}{\lg N}\lg|x| \le 400 + 100\frac{L_x}{\lg N}\lg n \le 400 + 100\frac{L_x}{\lg N}\lg 2N = 400 + 100\frac{L_x}{\lg N} + 100L_x \le 400 + 100\frac{L_x}{\lg 4} + 100L_x = 400 + 150L_x$.

Summing the potential over all mixed nodes, we have $\sum_x (400 + 150L_x) = 400n + 150 \sum_x L_x$. Since all the light subtrees are disjoint, $\sum_x L_x$ is at most the heap size: n. Thus, the combined potential of all mixed nodes is $550n$, and that of all nodes is $750n + o(n)$. □

We now analyze the five heap operations. All operations except delete-min take constant actual time. The get-min operation does not change the heap so its amortized time is also obviously constant. The make-heap operation creates a new heap with a potential of 900, due to the size potential. This leaves insertion, decrease-key, and deletion, which we handle in that order.

Lemma 2. *Insertion into a pairing heap takes $O(1)$ amortized time.*

Proof. The actual time is constant, so it suffices to bound the change in potential. We first bound the potential assuming N stays constant during the execution of the operation. Note that we need not worry about edge potentials here, since inserting a new node can not disturb any existing edges, and creates only one new edge, and the new node is never large, since if it becomes the root it will have no siblings in the general view, and if it does not become the root then it will have no children in the general view. (In the binary view this corresponds to having no right child or no left child.)

There are only two nodes whose node potentials change, the new node and the old root. If the old root has a larger key value than the new node, then the new node becomes the root. They both become mixed nodes with a potential of 400. If the new node is bigger than the old root, then the old root still has a potential of 400, and the new node becomes a mixed node, because it has no children in the general view, and thus also has a potential of 400. Thus, if N does not change, the amortized cost is constant.

However, N could increase to the next power of two. If it does, some mixed nodes may become small and some large nodes may become mixed or even small. These are decreases, so we can ignore them. Also, all mixed nodes that remain mixed will have their potential decrease. What we have to worry about is the edge potential. However, since there are only $O(n/\log n)$ large nodes, the total edge potential can only increase by an amount that is $o(n)$. Meanwhile, if N increases, its new value is the new heap size n, while the old value was $n/2$, so we release $450n$ units of size potential. The increase in potential if N doubles is thus $O(n/\log n) - \Theta(n)$, which is negative for large enough n. How large must n be for this to hold? There were previously at most $2n\lg(n/2)$ large nodes, and thus at most that many edges with negative edge potential. Thus, we need $7 \cdot 2n\lg(n/2) < 450n$. Solving for n yields $232 < 225\lg n$. Thus, the asymptotic statement actually holds for all $n > 3$. Since a heap that small contains no large nodes at all, the statement holds unconditionally. □

The following observation will be useful when we analyze decrease-key.

Lemma 3. *A node's potential is monotone nondecreasing in the size of both of its subtrees (in the binary view) if n is fixed.*

Proof. We must show that increasing the size of the left or right sub-tree of a node never causes its potential to drop. This follows immediately from several simple observations. As long as a node is small, its potential is identically zero and thus monotone. Since the potential is always non-negative, transitioning from small to non-small can only increase the potential. Observe that the formulas for mixed nodes and large nodes can be combined, as follows: $400 + 100 \min \left(1, \frac{\min(|x_L|, |x_R|)}{\lg N}\right) \times \lg(|x_L| + |x_R| + 1)$. This formula is monotone in $|x_L|$ and $|x_R|$, by inspection.   □

Lemma 4. *Decrease-key in a pairing heap takes $O(\log n)$ amortized time.*

Proof. The actual time is constant, so it suffices to bound the potential change. There are three types of nodes that can change potential: the old root, the decreased node, and the decreased node's ancestors in the binary representation. The ancestors' potential can only go down, since their subtree size is now smaller and potential is a monotone function of subtree size. This leaves just two nodes that might change potential: the root and the node that had its key decreased. But the potential of a node is between zero and $400 + 100 \lg n$ at all times. This leaves the edge potential and the size potential. The size potential does not change. Only $O(1)$ edges were created and destroyed, so the edge potential change due to directly affected edges is negligible. However, it is possible that there is a large indirect effect: some large ancestors of the decreased node might transition from large to mixed, and if those nodes had an incident edge with negative potential, its potential is now zero. Fortunately for us, at most $\lg n + O(1)$ edges can undergo this transition. To see this, let x be the decreased node. The parent of x may transition from large to mixed as a result of losing x, but it can't transition from large to small, because losing x can only affect the size of one of its subtrees, not both. Likewise, x's grandparent may transition from large to mixed, as well as x's great-grandparent, and x's great-great-grandparent $=$ great^2-grandparent, and so on. However, x's great$^{\lg n}$-grandparent still has more than $\lg n$ descendants in both subtrees despite losing x, and so will not undergo this transition. Thus, the change in edge potential is $O(\log n)$.   □

Lemma 5. *Delete-min in a pairing heap takes $O(\log n)$ amortized time.*

Proof. We break the analysis into two parts. Delete-min changes n, which means it may change N, which may affect the size potential, and the node potential, and the edge potential. The first part of our analysis bounds the change in potential due to changing N, and the second part deals with the delete-min and associated pairings.

Changing N may change small nodes into mixed or even large, and likewise it may change mixed nodes into large. In the case of the edge potentials, this works in our favor, since the edge potentials can only go down when then the number of large nodes increase.

If the new value of N is n and the old value was $2n$, then we release $900n$ units of size potential, while in Lemma 1 we showed that the sum of the mixed and large potentials is at most $750n + o(n)$. In fact, we can redo the calculation of that lemma slightly more precisely now, by taking advantage of the fact that we now have $N = n$. There are at most $\frac{2n}{\lg n}$ large nodes, each with a potential of at most $400 + 100 \lg n$, so the total potential of all large nodes is $200n + \frac{800n}{\lg n}$. We then add the mixed nodes for a total of $750n + \frac{800n}{\lg n}$. Thus, we release enough size potential if $900 - 750 = 150 > \frac{800}{\lg n}$. If the heap contains at least 64 items, this inequality holds, since $150 \cdot 6 = 900 > 800$. Could it be that for small heaps, we do not release enough size potential to pay for changing N? If the heap has size 8 or less, then there are no large nodes, and we are then also guaranteed to release enough size potential. Since the size of the heap must be a power of two when N changes, this leaves the heap sizes of 16 and 32 in question. A heap of size 16 has at most one large node, so the relevant inequality is $900 \cdot 16 > 550 \cdot 16 + 400 + 100 \lg 16$. Finally, a heap of size 32 has at most three large nodes, so we need $350 \cdot 32 > 3(400 + 100 \lg 32)$. It turns out that both of these inequalities hold; we omit the arithmetic calculations.

It remains to analyze the cost of the pairings performed. The number of pairings performed in the second pass is either k or $k - 1$. We will show that the second pass increases the potential by $O(\log n)$ and that the first pass increases the potential by $O(\log n) - 2k$, and thus the amortized cost of delete-min is $O(\log n)$.

If the root has $c > 0$ children, then a delete-min performs $c - 1$ pairings and thus takes c units of actual time. (If the root has $c = 0$ children, we are doing delete-min on a heap of size 1, which is trivial.) The loss of the root causes a potential drop of 400. Notice first that when two nodes are paired, this does not affect the subtree sizes of any other nodes. There are several cases to consider, depending on the sizes of nodes that get paired, and also depending on whether it is the first or second pairing pass. To avoid confusion: whenever we use the notation $|x|$ to refer to the size of a node x, if x changed size as a result of the pairing we are analyzing, we mean its initial size. Finally, when we pair two nodes, the node that becomes the parent of the other is said to have won the pairing, while the other is said to have lost it.

We give a brief sketch of the proof in this paragraph. There are six cases to consider, depending on the sizes of the nodes involved in the pairing. If both nodes are large, we use the classic analysis; if at least one node is small, we take advantage of the fact that there are few such pairings; if both nodes are mixed, then the loser is small and 400 units of potential are released; finally, if there are many medium-large pairings, then they must release some edge potential.

We now establish some vocabulary we will use throughout the analysis. Every pairing performed during the delete-min will be between two adjacent siblings (in the general view) x and y, where x is left of y; see Fig. 3. (In the binary view, y is the right child of x.) We use x_L to denote x's left subtree (in the binary view), y_L for y's left subtree, and y_R for y's right subtree (which is the subtree containing the siblings right of y in the general view).

Finally, let k_{ML} denote the number of pairings performed in the first pass of a delete-min that involve one mixed node and one large node, and let $\Delta\phi_{ML}$ denote the increase in potential as a result of those pairings. We also define k_{LL}, $\Delta\phi_{MM}$ and so on, analogously.

Large-Large. We start with a large-large pairing. In this case, the potential function is the same as the classic potential of [FSST86], and the analysis is nearly identical.

First Pass. We show that $\Delta\phi_{LL} \le -393k_{LL}+O(\log n)$. We will use the fact that $ab \le \frac{1}{4}(a+b)^2$. (Proof: $ab \le \frac{1}{4}(a+b)^2 \iff 4ab \le (a+b)^2 = a^2+b^2+2ab \iff 0 \le a^2 + b^2 - 2ab = (a-b)^2$, and the square of a real number is never negative.)

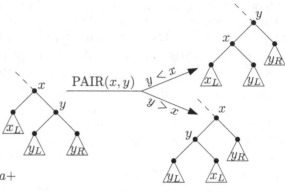

Fig. 3. A step of the first pairing pass (binary view).

We know that $|y_R| > 0$, for otherwise y could not be large (see Fig. 3). The initial potential of x is $400 + 100 \lg |x| = 400 + 100 \lg(|x_L| + 2 + |y_L| + |y_R|)$, and the initial potential of y is $400 + 100 \lg |y| = 400 + 100 \lg(|y_L| + 1 + |y_R|)$. The potential of x and y after the pairing depends on which of them won the pairing, but the sum of their potentials is the same in either case: $800 + 100 \lg(|x_L| + |y_L| + 1) + 100 \lg(|x_L| + |y_L| + 2 + |y_R|)$. The change P in node potential is $P = 100 \lg(|x_L|+|y_L|+1)-100 \lg(|y_L|+|y_R|+1) < 100 \lg(|x_L|+|y_L|+1)-100 \lg |y_R| = 100[\lg(|x_L| + |y_L| + 1) + \lg |y_R|] + 100[-\lg |y_R| - \lg |y_R|] \le 100 \lg \frac{1}{4}(|x_L| + |y_L| + 1 + |y_R|)^2 - 200 \lg |y_R| < 100 \lg \frac{1}{4}|x|^2 - 200 \lg |y_R| = 200 \lg |x| - 400 - 200 \lg |y_R|$.

We now sum the node potential change over all large-large pairings done in the first pass. Denote the nodes linked by large-large pairings during this pass as $x_1, \ldots, x_{2k}$, with x_i being left of x_{i+1}. As a notational convenience, let $L_i = 200 \lg |x_i|$. Also, let x_i' denote the right subtree of x_i. Note that for odd i, we have $x_i' = x_{i+1}$, but for even i, we don't, since there is no guarantee that all large nodes are adjacent to each other. If we also define $L_i' = 200 \lg |x_i'|$, then by the calculation above, the ith pairing raises the potential by at most $L_{2i-1} - 400 - L_{2i}'$. If all large pairings done in the first pass were adjacent, we'd have $L_{2i}' = L_{2i+1}$. Since they need not be, we have $L_{2i}' \ge L_{2i+1}$. Thus, we have a telescoping sum: $\le \sum_{i=1}^{k}(L_{2i-1} - 400 - L_{2i}') \le \sum_{i=1}^{k}(L_{2i-1} - 400 - L_{2i+1}) = -400k + L_1 - L_{2k+1} \le -400k + L_1 \le -400k + 200 \lg n$.

We should also consider the effects of edge potential. In the course of a pairing, five edges are destroyed and five new edges are created. Out of these five, three connect a node to its right child (in the binary view) and thus may have non-zero potential. In the case of a large-large pairing, if the edge that originally connected x to its parent p had negative potential, then the edge that connects the winner of the pairing to p also has negative potential. Likewise, if the edge that connects y to y_R had negative potential, then the edge that connects the winner to y_R does too. However, the edge that connected x to y had negative potential, while the edge that connects the loser to its new right child might not.

Thus, each pairing releases at least 400 units of node potential and costs at most 7 units of edge potential, and therefore $\Delta\phi_{LL} \leq -393k_{LL} + O(\log n)$.

Second Pass. The second pairing pass repeatedly pairs the two rightmost nodes. Therefore, one of them has no right siblings in the general representation, which means in the binary representation its right subtree has size zero, which implies that it is not a large node. Hence, in the second pairing pass there are no large-large pairings.

Mixed-Mixed. Since x and y are initially mixed, any incident edge has potential zero, so any edge potential change would be a decrease, and thus we can afford to neglect the edge potentials.

First Pass. We will show that $\Delta\phi_{MM} \leq -150k_{MM} + O(\log n)$. There are two cases: (1) x and y both have small left subtrees in the binary view, or (2) either x or y has a large left subtree in the binary view. We first handle the second case: at least one of the two nodes being paired has a large left subtree and hence a small right subtree. The left-heavy node can not be x, because y is contained in the right subtree of x, and y can not be mixed if the right subtree of x is small. Thus we conclude that y's right subtree is small. This can happen, but it can only happen once during the first pass of a delete-min, because in that case all right siblings of y in the general view will be small. An arbitrary pairing costs only $O(\log n)$ potential, so this case can not contribute more than $O(\log n)$ to the cost of the first pass.

We now turn to case (1), where x and y both have small left subtrees. The initial potential of x is $400 + 100\frac{|x_L|}{\lg N}\lg|x|$, and the initial potential of y is $400 + 100\frac{|y_L|}{\lg N}\lg|y|$. Observe that whichever node loses the pairing will have left and right subtrees with sizes $|x_L|$ and $|y_L|$. Thus, both its subtrees will be small, and so the loser becomes a small node with a potential of zero. There are two sub-cases to consider: (a) the winning node remains mixed, or (b) it becomes large. We first consider case (a). Since the winner remains mixed, its new potential is $400 + 100\frac{|x_L|+|y_L|+1}{\lg N}\lg|x|$. We will make use of the fact that $|y_L| + 1 < \lg N$, since the winner is mixed. The increase P in potential is:

$$P = 400 + 100\frac{|x_L|+|y_L|+1}{\lg N}\lg|x| - (400 + 100\frac{|y_L|}{\lg N}\lg|y|) - (400 + 100\frac{|x_L|}{\lg N}\lg|x|) =$$

$$-400 + 100\frac{|y_L|+1}{\lg N}\lg(|x_L| + 1 + |y|) - 100\frac{|y_L|}{\lg N}\lg|y| \leq -400 + 100\frac{|y_L|+1}{\lg N}\lg(\lg N +$$

$$1 + |y|) - 100\frac{|y_L|}{\lg N}\lg|y| \leq -400 + 100\frac{|y_L|+1}{\lg N}\lg(|y_R| + |y|) - 100\frac{|y_L|}{\lg N}\lg|y| <$$

$$-400 + 100\frac{|y_L|+1}{\lg N}\lg(|y|+|y|) - 100\frac{|y_L|}{\lg N}\lg|y| = -400 + 100\frac{|y_L|+1}{\lg N} + 100\frac{1}{\lg N}\lg|y| <$$

$$-400 + 100 + 100\frac{1}{\lg N}\lg n = -300 + 100\frac{1}{\lg N}\lg n < -300 + 100\frac{1}{\lg N}\lg 2N =$$

$$-200 + \frac{100}{\lg N} \leq -200 + \frac{100}{\lg 4} = -150. \text{ Thus, if the node that wins the pairing}$$

remains mixed, then at least 150 units of potential are released.

We now turn to case (b) where the node that wins the pairing becomes large. This can only happen if $|x_L| + |y_L| + 1 > \lg N$. In that case, the increase in potential is $P = 400 + 100\lg|x| - 100(4 + \frac{|y_L|}{\lg N}\lg|y|) - 100(4 + \frac{|x_L|}{\lg N}\lg|x|) =$

$-400+100 \lg |x| - 100 \frac{|y_L|}{\lg N} \lg |y| - 100 \frac{|x_L|}{\lg N} \lg |x| < -400+100 \lg |x| - 100 \frac{|y_L|}{\lg N} \lg |y| -$
$100 \frac{|x_L|}{\lg N} \lg |y| = -400 + 100 \lg |x| - 100 \frac{|x_L|+|y_L|}{\lg N} \lg |y| \leq -400 + 100 \lg |x| -$
$100 \lg |y| = -400+100 \lg(|x_L|+1+|y|) - 100 \lg |y| \leq -400+100 \lg 2|y| - 100 \lg |y| =$
$-400 + 100(\lg |y| + 1) - 100 \lg |y| = -300$, and thus at least 300 units of potential
are released.

Second Pass. In the second pass, we are guaranteed that $|y_R| = 0$. The initial potential of y is thus 400. The initial potential of x depends on which of its subtrees is small. But in fact we know that x is right-heavy, or else y could not be mixed. Thus the initial potential of x is $400 + 100 \frac{|x_L|}{\lg N} \lg |x| = 400 + 100 \frac{|x_L|}{\lg N} \lg(|x_L| + 2 + |y_L|)$. Whichever node wins the pairing will have a final potential of 400 (because it will have no right siblings). Whichever node loses the pairing will have a final potential of $400 + 100 \frac{|x_L|}{\lg N} \lg(|x_L| + 1 + |y_L|)$. Thus, there is no potential gain.

Mixed-Small and Large-Small. These types of pairings can only happen once during the first pass. To see this, observe that, in the general view, all right siblings of a small node are small. Therefore we have $k_{MS}+k_{LS} < 2$. An arbitrary pairing costs only $O(\log n)$ potential, so $\Delta\phi_{MS} + \Delta\phi_{LS}$ is $O(\log n)$. The winner of such a pairing is no longer small, so this type of pairing can happen only once during the second pass as well.

Small-Small. We show that the number k_{SS} of small-small pairings performed in both passes is $O(\log n)$, and that the potential increase $\Delta\phi_{SS}$ caused by said pairings is also $O(\log n)$. In one pass, there are fewer than $\lg N$ small-pairings, because a small node has few right siblings. By the same logic as for mixed-mixed, the loser of a small-small pairing remains small. The winner may remain small, in which case the pairing costs no potential. Or the winner may become mixed. However, that can only happen once per pass. For if the winner is mixed, that means that the combined subtree sizes of the two nodes exceeded $\lg N$, which means none of the left siblings were small. Thus, in the first pass, only the first small-small pairing may have the winner become mixed. Thus, $\Delta\phi_{SS}$ is $O(\log n)$.

Mixed-Large. We show that the heap potential can increase by at most $O(\log n)$ as a result of all mixed-large pairings performed. We begin with the second pass this time, to get the easy case out of the way first. The mixed-large pairings of the second pass cause no potential increase at all.

Second Pass. In the second pass, we are guaranteed that $|y_R| = 0$. The initial potential of y is thus 400. The initial potential of x is $400 + 100 \lg |x| = 400 + 100 \lg(|x_L|+2+|y_L|)$. Whichever node wins the pairing will have a final potential of 400 (because it will have no right siblings). Whichever node loses the pairing will have a final potential of $400 + 100 \lg(|x_L| + 1 + |y_L|)$. Thus, there is no increase in node potential. At first it seems that the edge potential could increase, for even though $|y_L| \geq \lg N$ (and likewise for x_L), there is no guarantee that y_L (or x_L) is a large node. On the other hand, there is also no guarantee that the parent of x in the binary view is not a large node. Thus, it is possible that we lose an edge with negative potential and have nothing to replace it with. However, this can only happen once, because the loser of such a pairing becomes a large node and will play the role of y_L in the next pairing, and once any pairing has a large y_L, all subsequent ones will too.

First Pass. We have three cases to consider, depending on which of $|x_L|$, $|y_L|$, or $|y_R|$ is small. The case where $|y_R| \leq \lg N$ is easy to dispense with, as it can only happen once during a delete-min. Observe that in the other two cases, the edge potential can't increase, because the winner of the pairing is large. If $|y_L| \leq \lg N$, the initial potential of x is $400 + 100 \lg|x|$, and the initial potential of y is $400 + 100 \frac{|y_L|}{\lg N} \lg|y|$. After the pairing, the loser is a mixed node with potential $400 + 100 \frac{|y_L|}{\lg N} \lg(|x_L| + 1 + |y_L|)$. The winner is large with a potential of $400 + 100 \lg|x|$. Thus, the increase P in potential is $P = 100 \frac{|y_L|}{\lg N} \lg(|x_L| + 1 + |y_L|) - 100 \frac{|y_L|}{\lg N} \lg|y| = 100 \frac{|y_L|}{\lg N}[\lg(|x_L| + 1 + |y_L|) - \lg(|y_L| + 1 + |y_R|)] \leq 100[\lg(|x_L| + 1 + |y_L|) - \lg(|y_L| + 1 + |y_R|)] < 100[\lg|x| - \lg(|y_L| + 1 + |y_R|)] < 100[\lg|x| - \lg|y_R|]$. Observe that when we sum over all such mixed-large pairings, we get a telescoping sum of the same form as the one that arose in the analysis of large-large pairings, and thus the combined potential increase for all such pairings is $O(\log n)$.

We now turn to the last case, where $|x_L| \leq \lg N$. The initial potential of x is now $400 + 100 \frac{|x_L|}{\lg N} \lg|x|$, and that of y is $400 + 100 \lg|y|$. The potential of the winner of the pairing is $400 + 100 \lg|x|$, and the potential of the loser is $400 + 100 \frac{|x_L|}{\lg N} \lg(|x_L| + |y_L| + 1)$, so the increase P in potential is $P = 100 \lg|x| + 100 \frac{|x_L|}{\lg N} \lg(|x_L| + |y_L| + 1) - 100 \lg|y| - 100 \frac{|x_L|}{\lg N} \lg|x| < 100 \lg|x| - 100 \lg|y| < 100 \lg|x| - 100 \lg|y_R|$. Summing over all such mixed-large pairings, this sum again telescopes in the same way as in the large-large case. Thus, all mixed-large pairings combined cost only $O(\log n)$ potential. However, unlike small-small, the actual cost can be far greater, and unlike large-large, we may not release enough node potential to pay for it. What saves us is the edge potentials.

Call a mixed-large pairing *normal* if $|y_R| > \lg N$. Observe that during the first pass, at most one mixed-large pairing can be abnormal, because if $|y_R| \leq \lg N$, all its right siblings (in the general view) are small nodes. We show that three consecutive normal mixed-large pairings release at least 7 units of potential: enough to pay for those three pairings, with 4 units left over. First, observe that the winner of a normal mixed-large pairing must be large. Thus, for every three consecutive normal mixed-large pairings, at least two large siblings become adjacent that were not adjacent before, and thus some edge has its potential go from 0 to -7. (We must of course be careful that this is not offset by some other edge nearby undergoing the opposite transition, but indeed, we are safe here, because the winner of the pairing is large.) Thus, even though we don't release enough node potential to pay for all mixed large pairings, there is hope that if there are so many of them that they tend to be consecutive, the edge potentials can pay for them instead, while if there are not so many that they tend to be consecutive, perhaps they do not dominate the cost of the first pass. We will soon see that this is indeed the case.

Bringing it All Together. We can now calculate the total amortized runtime of delete-min. The actual work done in the second pass is the same as that of the first pass, and the second pass causes at most a logarithmic increase in potential. Thus, we must show that $\Delta\phi \leq -2k + O(\log n)$, where k is the number of pairings done by the first pass and $\Delta\phi$ is the change in potential due to the first pass. We have $k = k_{LL} + k_{MM} + k_{SS} + k_{ML} + k_{MS} + k_{LS} \leq k_{LL} + k_{MM} + \lg N + k_{ML} + 1 = k_{LL} + k_{MM} + k_{ML} + O(\log n)$. The increase in potential from the first pass is $\Delta\phi = \Delta\phi_{LL} + \Delta\phi_{MM} + \Delta\phi_{SS} + \Delta\phi_{ML} + \Delta\phi_{MS} + \Delta\phi_{LS} \leq O(\log n) - 393 k_{LL} - 150 k_{MM}$. Summing the actual work with the node potential change, we obtain $O(\log n) - 392 k_{LL} - 149 k_{MM} + k_{ML} \leq O(\log n) - 149(k_{LL} + k_{MM}) + k_{ML}$. Thus, the question is whether most of those terms cancel, leaving us with a $O(\log n) - k$ amortized cost. There are two cases to consider, depending on how large k_{ML} is. If at most $\frac{74}{75}$ of all pairings done are mixed-large (that is, $k_{ML} < \frac{74}{75}k$, or rather $k_{ML} < \frac{74}{75}(k_{LL} + k_{MM} + k_{ML})$, or equivalently $k_{ML} < 74 k_{LL} + 74 k_{MM}$), then the amortized cost of the first pass is at most $O(\log n) - 149(k_{LL} + k_{MM}) + k_{ML} \leq O(\log n) - 149(k_{LL} + k_{MM}) + 74(k_{LL} + k_{MM}) = O(\log n) - 74(k_{LL} + k_{MM}) + k_{LL} + k_{MM} \leq O(\log n) - k_{ML} + k_{LL} + k_{MM} \leq O(\log n) - k$. Since the second pass only increases the potential by $O(\log n)$ and its actual cost is k, the cost for the whole delete-min is $O(\log n) - k + k = O(\log n)$.

That leaves the case where more than $\frac{74}{75}$ of parings are mixed-large ones. In fact, we will use the weaker assumption that at least 2021 of the pairings in the first pass are mixed-large. We divide the pairings into groups of three: the first three pairings, the second three, and so on. If a group consists of only mixed-large pairings, then it releases 7 units of potential, for a total amortized cost of -4. There are $k/3$ groups (give or take divisibility by three), and all but $k/21$ of those groups consist entirely of mixed-large pairings. These $k/3 - k/21 = 6k/21 = 2k/7$ groups release $8k/7$ units of spare potential. The remaining $k/21$ groups require $k/7$ units of potential to pay for them, leaving k units of spare potential, which we use to pay for the second pairing pass.   □

4   Final Words

It would be interesting to extend this potential function to so that merging two pairing heaps is fast. The amortized time using this potential function is, in the worst case, linear, because the two heaps might have sizes that are adjacent powers of 2 (e.g., 1024 and 2048), and thus the size potential of the new heap is linear in the combined size of the old heap, whereas the size potentials of the old heaps are both zero.

It would also be interesting to see whether a similar potential function can be made to work for splay trees; the one presented here does not. In particular, a splay where every double-rotation is a zig-zag does not release enough potential if the node x being accessed is large and all nodes on the path to x are mixed, so the amortized cost of the splay would be super-logarithmic.

References

[Col00] Cole, R.: On the dynamic finger conjecture for splay trees. Part II: the proof. SIAM J. Comput. **30**(1), 44–85 (2000)

[Elm09a] Elmasry, A.: Pairing heaps with $O(\log \log n)$ decrease cost. In: Proceedings of the Twentieth Annual ACM-SIAM Symposium on Discrete Algorithms, SODA, pp. 471–476, January 2009

[Elm09b] Elmasry, A.: Pairing heaps with costless meld, April 2009. http://arxiv.org/abs/0903.4130. In: Part I of the Proceedings of the 18th Annual European Symposium Algorithms, ESA, pp. 183–193, September 2010

[Fre99] Fredman, M.L.: On the efficiency of pairing heaps and related data structures. J. ACM **46**(4), 473–501 (1999)

[FSST86] Fredman, M.L., Sedgewick, R., Sleator, D.D., Tarjan, R.E.: The pairing heap: a new form of self-adjusting heap. Algorithmica **1**(1), 111–129 (1986)

[FT84] Fredman, M.L., Tarjan, R.E.: Fibonacci heaps and their uses in improved network optimization algorithms. In: Proceedings of the 25th IEEE Symposium on the Foundations of Computer Science, FOCS, pp. 338–346, October 1984. J. ACM **34**(3), 596–615 (1987)

[HST09] Haeupler, B., Sen, S., Tarjan, R.E.: Rank-pairing heaps. In: Proceedings of the 17th Annual European Symposium on Algorithms, ESA, pp. 659–670, September 2009. SIAM J. Comput. **40**(6), pp. 1463–1485 (2011)

[Iac00] Iacono, J.: Improved upper bounds for pairing heaps. In: Halldórsson, M.M. (ed.) SWAT 2000. LNCS, vol. 1851, pp. 32–45. Springer, Heidelberg (2000). arXiv.org/abs/1110.4428 (2011)

[IY16] Iacono, J., Yagnatinsky, M.: A Linear Potential Function for Pairing Heaps. arXiv.org/abs/1606.06389 (2016)

[IO14] Why some heaps support constant-amortized-time decrease-key operations, others do not. First version: Iacono, J.: arXiv.org/abs/1302.6641, February 2013. Later: Iacono, J., Özkan, Ö.: 41st International Colloquium on Automata, Languages, and Programming, ICALP, pp. 637–649, July 2014

[LST14] Larkin, D.H., Sen, S., Tarjan, R.E.: A back-to-basics empirical study of priority queues. In: 2014 Proceedings of the 16th Workshop on Algorithm Engineering, Experiments, ALENEX, pp. 61–72, January 2014. arXiv.org/abs/1403.0252, March 2014

[Pet05] Pettie, S.: Towards a final analysis of pairing heaps. In: 46th Annual IEEE Symposium on Foundations of Computer Science, FOCS, pp. 174–183, October 2005

[ST85] Sleator, D.D., Tarjan, R.E.: Self-adjusting binary search trees. J. ACM **32**(3), 652–686 (1985)

[SV86] Stasko, J.T., Vitter, J.S.: Pairing heaps: experiments and analysis. Commun. ACM **30**(3), 234–249 (1987)

[Vui78] Vuillemin, J.: A data structure for manipulating priority queues. Commun. ACM **21**(4), 309–315 (1978)

[Yag16] Yagnatinsky, M.: Thinnest V-shapes, biggest angles, and lowest potentials. Ph.D. thesis, NYU Tandon School of Engineering, May 2016. http://cse.poly.edu/~myag/thesis.pdf

Amortized Efficiency of Ranking and Unranking Left-Child Sequences in Lexicographic Order

Kung-Jui Pai[1]([✉]), Ro-Yu Wu[2], Jou-Ming Chang[3], and Shun-Chieh Chang[4,5]

[1] Department of Industrial Engineering and Management,
Ming Chi University of Technology, New Taipei City, Taiwan
poter@mail.mcut.edu.tw
[2] Department of Industrial Management,
Lunghwa University of Science and Technology, Taoyuan, Taiwan
[3] Institute of Information and Decision Sciences,
National Taipei College of Business, Taipei, Taiwan
[4] CyberTrust Technology Institute, Institute for Information Industry,
Taipei, Taiwan
[5] International Sino College, Siam University, Bangkok, Thailand

Abstract. A new type of sequences called left-child sequences (LC-sequences for short) was recently introduced by Wu et al. [19] for representing binary trees. In particular, they pointed out that such sequences have a natural interpretation from the view point of data structure and gave a characterization of them. Based on such a characterization, there is an algorithm to generate all LC-sequences of binary trees with n internal nodes in lexicographic order. In this paper, we extend our study to the ranking and unranking problems. By integrating a measure called "left distances" introduced by Mäkinen [8] to represent binary trees, we develop efficient ranking and unranking algorithms for LC-sequences in lexicographic order. With a help of aggregate analysis, we show that both ranking and unranking algorithms can be run in amortized cost of $\mathcal{O}(n)$ time and space.

Keywords: Binary trees · Left-child sequences · Lexicographic order · Ranking algorithms · Unranking algorithms · Amortized cost

1 Introduction

Exhaustively generating a class of combinatorial objects is an important research topic. This is due to many applications in computer science such as combinatorial group testing, algorithm performance analyzing, and counterexample searching. As usual, combinatorial objects are encoded by using integer sequences so that all sequences are generated in a particular order, such as the lexicographic order

This research was partially supported by MOST grants 104-2221-E-131-004 (Kung-Jui Pai), 104-2221-E-262-005 (Ro-Yu Wu) and 104-2221-E-141-002-MY3 (Jou-Ming Chang) from the Ministry of Science and Technology, Taiwan.

T.-H.H. Chan et al. (Eds.): COCOA 2016, LNCS 10043, pp. 505–518, 2016.
DOI: 10.1007/978-3-319-48749-6_37

[3, 13, 27, 28] or a Gray-code order [12, 14]. For algorithmic efficiency, generations in lexicographic order are demanded to run in constant amortized time (CAT for short) [1, 15]. By contrast, for Gray-code order, each generation is demanded to take a constant time (i.e., the so-called *loopless algorithms* defined by Ehrlich [2] and see [6, 17–19, 24, 26] for more references).

Given a specific order of objects, a *ranking algorithm* is a function that determines the rank of a given object in the generated list, and an *unranking algorithm* is one that produces the sequence of object corresponding to a given rank. Efficient ranking and unranking are important and useful for storing and retrieving elements in a class of combinatorial objects. Moreover, these techniques can be extended to the uses of database indexing, data compression and data encryption [4, 10].

Since binary trees are one of the most fundamental data structures in computer science, many types of integer sequences have been introduced to encode binary trees [5, 9]. In addition, there exist many algorithms for generating binary tree sequences, see e.g. [7, 11, 12, 17–19, 23]. Also, for ranking and unranking binary tree sequences and their generalizations, we refer to [3, 16, 20–22, 25]. Recently, Wu et al. [19] proposed a loopless algorithm associated with tree rotations for simultaneously generating four types of binary tree sequences. In particular, they suggested two new types of sequences called *left-child sequences* (LC-sequences for short) and their mirror images called *right-child sequences* (RC-sequences for short). It is the practice to implement binary trees by using structure pointer representation so that nodes in a tree are allocated by structures and children of nodes are accessed through pointers. Accordingly, Wu et al. [19] claimed that LC- and RC-sequences were inspired by such a natural structure representation of binary trees. From then on no further research is devoted to the study of LC- and RC-sequences. In this paper, we are interested to make an exploration on this topic. By integrating a measure called left-distances sequences (LD-sequences for short) introduced by Mäkinen [8] to represent binary trees, we develop efficient algorithms for ranking and unranking LC-sequences of binary trees in lexicographic order. By symmetry, a similar way can develop algorithms for RC-sequences.

The rest of this paper is organized as follows. Section 2 formally gives the definitions of LC-sequences and LD-sequences and shows the correspondence between them. Section 3 provides a lexicographic generation of LC-sequences and introduces a coding trees structure for representing such a generated list. According to the structure properties of coding tree, Sects. 4 and 5 respectively present our ranking and unranking algorithms and show the correctness. Finally, concluding remarks are given in the last section.

2 Preliminaries

2.1 Left-Distance Sequences and Left-Child Sequences

In the early stage, Mäkinen [8] used a measure called *left distance* to represent a binary tree. For a binary tree T, a *left-distance sequence* (LD-sequence for short) of T, denoted by $d(T) = (d_1, d_2, \ldots, d_n)$, is an integer sequence such that the term d_i for each node $i \in T$ is recursively defined as follows:

$$d_i = \begin{cases} 0 & \text{if } i \text{ is the root of } T; \\ d_{p(i)} & \text{if } i \text{ is a left child;} \\ d_{p(i)} + 1 & \text{if } i \text{ is a right child,} \end{cases}$$

where $p(i)$ stands for the parent of node i in T. For instance, the LD-sequence of the binary tree shown in Fig. 1 is $d(T) = (0, 0, 1, 2, 3, 3, 4, 3, 1)$. It is easy to observe that every node lying on the left arm of T has left distance 0. Moreover, for a node i, every node lying on the left arm of R_i has left distance $d_i + 1$. Mäkinen [8] also characterized an integer sequence $(d_1, d_2, \ldots, d_n)$ to be an LD-sequence if and only if the following conditions are fulfilled: $d_1 = 0$ and $0 \leqslant d_i \leqslant d_{i-1} + 1$ for $2 \leqslant i \leqslant n$.

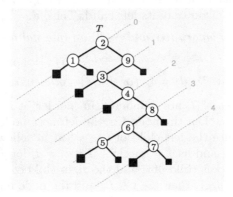

Fig. 1. A binary tree T with $d(T) = (0, 0, 1, 2, 3, 3, 4, 3, 1)$ and $c(T) = (0, 1, 0, 0, 0, 5, 0, 6, 3)$.

Wu et al. [19] recently introduced a new type of sequences called *left-child sequence* (LC-sequence for short) to represent binary trees. For a binary T with n internal nodes labeled by $1, 2, \ldots, n$ in inorder, the LC-sequence of T, denoted by $c(T) = (c_1, c_2, \ldots, c_n)$, is an integer sequence so that the term c_i, $1 \leqslant i \leqslant n$, is defined as follows:

$$c_i = \begin{cases} 0 & \text{if the left child of } i \text{ is a leaf;} \\ j & \text{if } j \text{ is the left child of } i \text{ in } T. \end{cases}$$

For instance, the LC-sequence of the binary tree T shown in Fig. 1 is $c(T) = (0, 1, 0, 0, 0, 5, 0, 6, 3)$. Wu et al. [19] also characterized LC-sequences as follows.

Theorem 1. *Let $c = (c_1, c_2, \ldots, c_n)$ be an integer sequence. Then, c is the LC-sequence of a binary tree T with n internal nodes if and only if the following conditions are fulfilled for all $i \in \{1, 2, \ldots, n\}$: (i) $0 \leqslant c_i < i$ and (ii) $c_j = 0$ or $c_j > c_i$ for all $c_i < j < i$.*

2.2 Sequences Transformation

Given a binary tree T, we now provide a linear time transformation between $c(T)$ and $d(T)$. Before this, we need some auxiliary properties.

Lemma 1. *Let T be a binary tree and $i \in T$ an internal node. Then*

$$
d_i = \begin{cases}
0 & \text{if } i = 1; \\
d_{i-1} + 1 & \text{if } i \geqslant 2 \text{ and } c_i = 0; \\
d_{c_i} & \text{otherwise.}
\end{cases}
$$

Proof. We may imagine that T is the right subtree of a dummy node numbered by 0 and let $d_0 = -1$. Suppose that the node i occurs in the left arm of a subtree R_j. If $c_i = 0$, then the left child of i must be a leaf. In this case, since all nodes of T are numbered in inorder, we have $i = j + 1$. Since every node lying on the left arm of R_j has left distance $d_j + 1$, it implies that $d_i = d_j + 1 = d_{i-1} + 1$. In particular, if $i = 1$, then $d_i = 0$. Also, for a node i with $c_i \neq 0$, it is clear that d_i is the same as the left distance of its left child. Thus, $d_i = d_{c_i}$. □

Lemma 2. *Let T be a binary tree and $i \in T$ an internal node. Then*

$$
c_i = \begin{cases}
0 & \text{if } i = 1 \text{ or } d_i > d_{i-1}; \\
\max\{j \in T \colon d_j = d_i \text{ for } j < i\} & \text{otherwise.}
\end{cases}
$$

Proof. Since all nodes of T are numbered in inorder, it is clear that $c_1 = 0$. For $i > 1$, if $d_i > d_{i-1}$, then the node i is the leftmost node in the left arm of R_{i-1}. In this case, since the left child of i is a leaf, it follows that $c_i = 0$. We now consider $d_i \leqslant d_{i-1}$ and let $t = \max\{j \in T \colon d_j = d_i \text{ for } j < i\}$. If $d_i = d_{i-1}$, then $i - 1$ must be the left child of i, and the right child of $i - 1$ is a leaf. Thus, $c_i = i - 1 = t$. If $d_i < d_{i-1}$, then $i - 1 \in L_i$ and the node numbered by t is the left child of i, and thus $c_i = t$. □

Based on Lemmas 1 and 2, linear time transformations between LC-sequences and LD-sequences are shown in Fig. 2. Therefore, we have the following theorem.

Function LC-to-LD$(c_1, c_2, \ldots, c_n)$	**Function** LD-to-LC$(d_1, d_2, \ldots, d_n)$
begin	**begin**
$\quad d_1 \leftarrow 0;$	$\quad c_1 \leftarrow 0;\ t(0) = 1;$
$\quad$ **for** $i \leftarrow 2$ **to** n **do**	$\quad$ **for** $i \leftarrow 2$ **to** n **do**
$\quad\quad$ **if** $c_i = 0$ **then** $d_i \leftarrow d_{i-1} + 1;$	$\quad\quad$ **if** $d_i > d_{i-1}$ **then** $c_i \leftarrow 0;$
$\quad\quad$ **else** $d_i \leftarrow d_{c_i};$	$\quad\quad$ **else** $c_i \leftarrow t(d_i);$
	$\quad\quad t(d_i) \leftarrow i;$
$\quad$ **return** $d_1, d_2, \ldots, d_n;$	$\quad$ **return** $c_1, c_2, \ldots, c_n;$

Fig. 2. Linear time transformations between LC- and LD-sequences.

Theorem 2. *Transformations between an LC-sequence and an LD-sequence of length n can be done in $\mathcal{O}(n)$ time.*

3   Lexicographic Generation and Coding Tree Structure

It is well-known that the number of binary trees with n internal nodes is totally $\frac{1}{n+1}\binom{2n}{n}$. There is a systematic way to depict all binary tree sequences by use of coding trees [19]. A coding tree $\mathbb{T}_n$ is a rooted tree consisting of n levels such that every node is associated with a label and the full labels along a path from the root to a leaf in $\mathbb{T}_n$ represent the sequence of a binary tree with n internal nodes. Figure 3 demonstrates the coding tree $\mathbb{T}_4$ with respect to LC-sequences and LD-sequences, where each node x in the ith level of the tree is labeled by c_i/d_i. For instance, labels in the left arm (respectively, right arm) of $\mathbb{T}_4$ represent the right-skewed tree with LC-sequence $(0,0,0,0)$ and LD-sequence $(0,1,2,3)$ (respectively, left-skewed tree with LC-sequence $(0,1,2,3)$ and LD-sequence $(0,0,0,0)$), where we omit the drawing of leaves in the corresponding tree. In this paper, we consider a specific coding tree $\mathbb{T}_n$ in which all LC-sequences of binary trees are emerged from left to right in lexicographic order (e.g., see $\mathbb{T}_4$ in Fig. 3). Although the list of LC-sequences produces a lexicographic order, it should be noted that the corresponding list of LD-sequences does not produce a reverse lexicographic order.

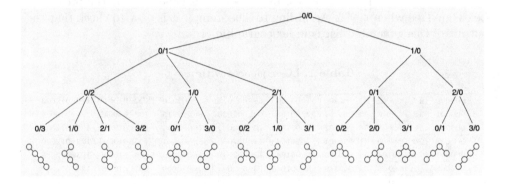

Fig. 3. A coding tree $\mathbb{T}_4$ for representing LC-sequences and LD-sequences.

3.1   Generating LC-Equences in Lexicographic Order

According to the characterization of LC-sequences described in Theorem 1, we now propose an algorithm to generate all LC-sequences of binary trees with n internal nodes in lexicographic order. Figure 4 shows the algorithm, where the outer loop specifies the range of condition (i) of Theorem 1, while the testing by **if** $\cdots$ **then** statement in the inner loop is the condition (ii) of Theorem 1. Initially, we set $c_1 = 0$, and then perform a procedure call Lex-Gen-Tree(2) to start the generation.

Theorem 3. Lex-Gen-Tree *can correctly generate all LC-sequences of binary trees with n internal nodes in lexicographic order.*

```
Procedure Lex-Gen-Tree(i)
  begin
    if i = n + 1 then  Print(c₁, c₂, ..., cₙ);
    else
        for cᵢ ← 0 to i − 1 do      // The range of condition (i) in Theorem 1.
            flag ← 1;  j ← cᵢ + 1;
            while flag = 1 and j < i do
                if cⱼ ≠ 0 and cⱼ ≤ cᵢ then      // The testing of the condition (ii) in Theorem 1.
                    flag ← 0;
                j ← j + 1;
            if flag = 1 then  call Lex-Gen-Tree(i + 1);
```

Fig. 4. An algorithm for generating LC-sequences in lexicographic order.

Proof. The correctness of the algorithm directly follows from Theorem 1. For each $i \in \{2, 3, \ldots, n\}$, since all valid $c_i \in \{0, 1, \ldots, i - 1\}$ in this procedure are emerged in increasing order, it implies that all output sequences are present in lexicographic order. □

For example, Table 1 shows the list of LC-equences generated by the procedure Lex-Gen-Tree when $n = 6$. According to Theorem 3, it is easy to check that the output of this generated list is in lexicographic order.

Table 1. LC-sequences with $n = 6$.

0: 000000	17: 000042	34: 000300	51: 001005	68: 002015	85: 002310	102: 010043	119: 012003
1: 000001	18: 000043	35: 000301	52: 001030	69: 002030	86: 002315	103: 010045	120: 012004
2: 000002	19: 000045	36: 000302	53: 001035	70: 002031	87: 002340	104: 010200	121: 012005
3: 000003	20: 000100	37: 000304	54: 001040	71: 002035	88: 002341	105: 010204	122: 012030
4: 000004	21: 000104	38: 000305	55: 001043	72: 002040	89: 002345	106: 010205	123: 012035
5: 000005	22: 000105	39: 000310	56: 001045	73: 002041	90: 010000	107: 010240	124: 012040
6: 000010	23: 000140	40: 000315	57: 001300	74: 002043	91: 010002	108: 010245	125: 012043
7: 000015	24: 000145	41: 000320	58: 001304	75: 002045	92: 010003	109: 010300	126: 012045
8: 000020	25: 000200	42: 000321	59: 001305	76: 002100	93: 010004	110: 010302	127: 012300
9: 000021	26: 000201	43: 000325	60: 001340	77: 002104	94: 010005	111: 010304	128: 012304
10: 000025	27: 000204	44: 000340	61: 001345	78: 002105	95: 010020	112: 010305	129: 012305
11: 000030	28: 000205	45: 000341	62: 002000	79: 002140	96: 010025	113: 010320	130: 012340
12: 000031	29: 000210	46: 000342	63: 002001	80: 002145	97: 010030	114: 010325	131: 012345
13: 000032	30: 000215	47: 000345	64: 002003	81: 002300	98: 010032	115: 010340	
14: 000035	31: 000240	48: 001000	65: 002004	82: 002301	99: 010035	116: 010342	
15: 000040	32: 000241	49: 001003	66: 002005	83: 002304	100: 010040	117: 010345	
16: 000041	33: 000245	50: 001004	67: 002010	84: 002305	101: 010042	118: 012000	

3.2   Structure Properties of Coding Trees

Since LD-sequences can be used to assist the design of ranking and unranking algorithms, we first observe some structure properties of coding trees from the view point of LD-sequences. Throughout this paper, we also use the following notation. For a node x in $\mathbb{T}_n$, if x is not the root, then the parent of x is denoted by $p(x)$. Also, the labels of left child and left distance of x are denoted by $\hat{c}(x)$ and $\hat{d}(x)$, respectively. The following property can be easily obtained from the coding tree structure.

Lemma 3. *If x is a non-leaf node with $\hat{d}(x) = k$ in the ith level of $\mathbb{T}_n$, then the sons of x have left distance labels $k + 1, 0, 1, \ldots, k$ from left to right in the $(i + 1)$th level of $\mathbb{T}_n$.*

Let $A_{i,k}$ denote the number of leaves in the subtree rooted at a node with left distance k in the ith level of $\mathbb{T}_n$. Obviously, $A_{n,k} = 1$ for $0 \leqslant k \leqslant n - 1$ and $A_{n-1,k} = k + 2$ for $0 \leqslant k \leqslant n - 2$. In general, we have

$$A_{i,k} = \sum_{j=0}^{k+1} A_{i+1,j} \tag{1}$$

where $1 \leqslant i \leqslant n-1$ and $0 \leqslant k \leqslant i-1$. For a concise expression and the efficiency of computation, we define the following formula:

$$B_{i,k} = \sum_{j=0}^{k} A_{i,j} \tag{2}$$

where $1 \leqslant i \leqslant n - 1$ and $0 \leqslant k \leqslant i - 1$. For example, Table 2 shows the results of $B_{i,k}$ for $n = 6$. In fact, the table built from Eqs. (1) and (2) is called the *Catalan's triangle table*, and thus it can be reformulated as the following closed form.

Proposition 1. *Let $n \geqslant 2$ be an integer. For $1 \leqslant i \leqslant n$ and $0 \leqslant k \leqslant i - 1$, we have*

$$B_{i,k} = \frac{k+1}{m} \binom{2m+k}{m-1} \tag{3}$$

where $m = n - i + 1$.

According to Eq. (3), the following corollaries can easily be obtained and are useful for designing ranking and unranking algorithms.

Corollary 1. *Let $n \geqslant 2$ be integer. For $2 \leqslant i \leqslant n$ and $0 \leqslant k \leqslant i - 1$, we have*

$$B_{i-1,k} = B_{i,k} \cdot \frac{(2m+k+1)(2m+k+2)}{(m+1)(m+k+2)} \tag{4}$$

where $m = n - i + 1$.

Corollary 2. *Let $n \geqslant 2$ be integer. For $2 \leqslant i \leqslant n$ and $1 \leqslant k \leqslant i - 1$, we have*

$$B_{i,k-1} = B_{i,k} \cdot \frac{k(m+k+1)}{(k+1)(2m+k)} \tag{5}$$

where $m = n - i + 1$.

Table 2. The Catalan's triangle table for $n = 6$

$B_{i,k}$	k					
	0	1	2	3	4	5
i 1	132					
2	42	132				
3	14	42	90			
4	5	14	28	48		
5	2	5	9	14	20	
6	1	2	3	4	5	6

Corollary 3. *Let $n \geqslant 2$ be integer. For $1 \leqslant i \leqslant n - 1$ and $0 \leqslant k \leqslant i - 1$, we have*

$$B_{i+1,k+1} = B_{i,k} \cdot \frac{m(k+2)}{(k+1)(2m+k)} \tag{6}$$

where $m = n - i + 1$.

4   A Ranking Algorithm

In this section, we develop an algorithm to deal with the ranking problem. From the structure of coding trees (e.g., see Fig. 3), we know that there is a one-to-one correspondence between binary trees with n internal nodes and the leaves of $\mathbb{T}_n$. Thus, ranking a designated tree T is equivalent to counting the number of leaves before the leaf corresponding to T in the arrangement of $\mathbb{T}_n$.

Let T be a binary tree with $c(T) = (c_1, c_2, \ldots, c_n)$ and let $(x_1, x_2, \ldots, x_n)$ be the corresponding path in $\mathbb{T}_n$ such that $\hat{c}(x_i) = c_i$ for $1 \leqslant i \leqslant n$. For each $i \in \{1, 2, \ldots, n\}$, we let $R(i)$ be the number of leaves turned up before x_i when we travel $\mathbb{T}_n$ in preorder (i.e., visit recursively the root and then the subtrees of $\mathbb{T}_n$ from left to right). For instance, if we consider the tree T with $c(T) = (0, 1, 0, 3)$ in Fig. 3, we have $R(1) = 0$, $R(2) = 9$, $R(3) = 9$ and $R(4) = 11$. As a result, the goal of our ranking algorithm is to compute $R(n)$. Because we do not really build the coding tree, introducing LD-sequence instead of traveling $\mathbb{T}_n$ can help us to compute $R(n)$.

Lemma 4. *For an LD-sequence $(d_1, d_2, \ldots, d_n)$, let $(x_1, x_2, \ldots, x_n)$ be the corresponding path in $\mathbb{T}_n$ such that $\hat{d}(x_i) = d_i$ for $i \in \{1, 2, \ldots, n\}$. Then the following relations holds for $2 \leqslant i \leqslant n$:*

$$R(i) = \begin{cases} R(i-1) & \text{if } d_{i-1} < d_i; \\ R(i-1) + (B_{i,d_{i-1}+1} - B_{i,d_{i-1}}) & \text{if } d_{i-1} \geqslant d_i = 0; \\ R(i-1) + (B_{i,d_{i-1}+1} - B_{i,d_{i-1}}) + B_{i,d_i-1} & \text{if } d_{i-1} \geqslant d_i \neq 0; \end{cases}$$

```
Function Ranking(c₁, c₂, ..., cₙ)
  begin
    (d₁, d₂, ..., dₙ) ← LC-to-LD(c₁, c₂, ..., cₙ);
    R ← 0;
    for i ← 2 to n do
      if dᵢ ⩽ dᵢ₋₁ then      // Is there an incremental change?
        R ← R + (B_{i+1,dᵢ₋₁+1} − B_{i+1,dᵢ₋₁});
        if dᵢ ≠ 0 then  R ← R + B_{i+1,dᵢ−1};
    return R;
```

Fig. 5. A ranking algorithm.

Proof. By Lemma 3, we may assume that y_j, $j = 1, 2, \ldots, d_{i-1}+2$, are the sons of x_{i-1} with left distance labels $d_{i-1}+1, 0, 1, \ldots, d_{i-1}$, respectively. Recall that the term $B_{i,k}$ defined in Eq. (2) is the total number of leaves in the subtrees rooted at nodes with labels from 0 to k, respectively, in the ith level of $\mathbb{T}_n$. Let $L(y_j)$ denote the number of leaves in the subtree rooted at y_j for $j = 1, 2, \ldots, d_{i-1}+2$. For $d_{i-1} < d_i$, since $\hat{d}(y_1) = d_{i-1} + 1$, we have $x_i = y_1$. This indicates that x_i is followed closely by x_{i-1} in the preorder traversal of $\mathbb{T}_n$, and thus $R(i) = R(i-1)$. For $d_{i-1} \geqslant d_i$, if $d_i = 0$, we have $x_i = y_2$. In this case, $R(i) - R(i-1)$ is equal to the number of leaves in the subtree rooted at y_1, and thus it can be expressed by the term

$$L(y_1) = A_{i,d_{i-1}+1} = B_{i,d_{i-1}+1} - B_{i,d_{i-1}}.$$

On the other hand, if $d_i \neq 0$, then x_i is the $(d_i + 2)$th son of x_{i-1} in $\mathbb{T}_n$. In this case, not only the term $L(y_1)$ but also the sum of leaves in the subtrees rooted at y_j for $j = 2, \ldots, d_i + 1$ needs to be counted in $R(i)$, and the latter can be expressed by the term

$$\sum_{j=2}^{d_i+1} L(y_j) = \sum_{k=0}^{d_i-1} A_{i,k} = B_{i,d_i-1}.$$

This proves the lemma.   □

For convenience of discussion, the nonzero term of $R(i) - R(i-1)$ in the statement of Lemma 4 is called the *incremental change* of $R(i-1)$. Based on Lemma 4, we provide a ranking algorithm as shown in Fig. 5.

Example 1. Let T be a tree with left-child sequence $c(T) = (0, 1, 0, 3, 0, 4)$. To perform Ranking$(0, 1, 0, 3, 0, 4)$, we first call the function LC-to-LD$(0, 1, 0, 3, 0, 4)$ to obtain the corresponding LD-sequence $d(T) = (0, 0, 1, 1, 2, 1)$. Initially, we set $R = 0$. When $i = 2$, since $d_2 = d_1 = 0$, we update
$\quad R = R + (B_{2,1} - B_{2,0}) = 0 + (132 - 42) = 90.$
When $i = 3$, since $d_3 = 1 > d_2 = 0$, there is no incremental change. When $i = 4$, since $d_4 = d_3 = 1 \neq 0$, we update

$$R = R + (B_{4,2} - B_{4,1}) + B_{4,0} = 90 + (28 - 14) + 5 = 109.$$
When $i = 5$, since $d_5 = 2 > d_4 = 1$, there is no incremental change. When $i = 6$, since $d_6 = 1 < d_5 = 2$ and $d_6 \neq 0$, we update
$$R = R + (B_{6,3} - B_{6,2}) + B_{6,0} = 109 + (4 - 3) + 1 = 111.$$
Finally, the algorithm outputs $R = 111$.

Obviously, we need to extract some elements of Catalan's triangle table in the above ranking algorithm. However, building Catalan's triangle table in advance requires $\Omega(n^2)$ time and space, and so does the ranking algorithm. In what follows, we will improve the complexity and show that every element of Catalan's triangle table can be accessed in constant time without really building the table. Conceptually, for calculating the rank R of an LC-sequence, we imagine that Catalan's triangle table is a chess board and only elements along a certain path where a chess B moves on the chess board are accumulated in R. Technically, we do not employ Eq. (2) to extract elements in the chess board because it requires $\mathcal{O}(n^2)$ time for each extraction. Alternatively, the movement of B can be done by using Eqs. (4)–(6).

Figure 6 shows the details of the algorithm. First of all, we convert the input to an LD-sequence $(d_1, d_2, \ldots, d_n)$ and put B on the initial location by using Eq. (4) (Lines 2 to 5). After this step, we have $i = 1$, $k = 0$ and $B = B_{1,0}$. For each $i = 1, 2, \ldots, n - 1$, we first update B by using Eq. (6) and move B to the right-down cell (Lines 7 to 8). According to Lemma 4, the next location where B moves can be determined by the relation between d_i and d_{i+1} as follows. If $d_i \leqslant d_{i-1}$, we compute the left cell of B, say B', using Eq. (5). Then, we update R by adding the incremental change $B - B'$, and move B to the left cell (Lines 10 to 12). In addition, if $d_i \neq 0$, an extra incremental change needs to be considered. In this case, we search for such an incremental change by moving B' to the left cell step by step until the desired cell is reached (Lines 13 to 16). After this search, R is updated and we have $k = d_i$ (Lines 17 to 18). On the other hand, if $d_{i-1} < d_i$, then no incremental change needs to be considered.

Theorem 4. *Given the LC-sequence of a binary tree T with n internal nodes, determining the rank of T in lexicographic order can be done in $\mathcal{O}(n)$ time and space.*

Proof. By Lemma 4, the function Ranking() can determine the rank of T in lexicographic order. Since it is the hard core of Refined-Ranking(), the correctness of the algorithm directly follows. Obviously, the space requirement is $2n + \mathcal{O}(1)$, which is due to the conversion of LC-to-LD(). We give the time complexity on amortized analysis as follows. Note that each of Eqs. (4)–(6) can be computed in constant time. To acquire the overall running time, we use aggregate method to determine an upper bound on the number of moves aroused by B (and its substitute B') on the chess board. For the initial stage, it is clear that B moves $n - 1$ steps from bottom to up. For each $i = 1, 2, \ldots, n - 1$, there is a right-down move for B. If $d_i \leqslant d_{i-1}$, it needs at least one left move for B. Moreover, more left moves are required for the searching of an extra incremental change. Let m_i be such a number of additional left moves aroused by B'. Thus, the total number

---

Function Refined-Ranking($c_1, c_2, \ldots, c_n$)

1 **begin**
2 $\quad (d_1, d_2, \ldots, d_n) \leftarrow$ LC-to-LD$(c_1, c_2, \ldots, c_n)$;
3 $\quad R \leftarrow 0; \quad B \leftarrow 1; \quad k \leftarrow 0; \qquad$ // Initialization.
4 $\quad$**for** $i \leftarrow n$ **downto** 2 **do** $\qquad$ // Compute $B_{1,0}$ to store in B.
5 $\quad\quad B \leftarrow B \cdot \frac{(2m+k+1)(2m+k+2)}{(m+1)(m+k+2)}$, where $m = n - i + 1$; $\quad$ // Ref. Eq. (4)
6 $\quad$**while** $i < n$ **do**
7 $\quad\quad B \leftarrow B \cdot \frac{m(k+2)}{(k+1)(2m+k)}$, where $m = n - i + 1$; $\qquad$ // Ref. Eq. (6)
8 $\quad\quad i \leftarrow i + 1; \quad k \leftarrow k + 1; \qquad$ // Move B to the right-down cell.
9 $\quad\quad$**if** $d_i \leqslant d_{i-1}$ **then** $\qquad$ // Is there an incremental change?
10 $\quad\quad\quad B' \leftarrow B \cdot \frac{k(m+k+1)}{(k+1)(2m+k)}$, where $m = n - i + 1$; $\quad$ // Ref. Eq. (5)
11 $\quad\quad\quad R \leftarrow R + (B - B')$;
12 $\quad\quad\quad k \leftarrow k - 1; \quad B \leftarrow B'; \qquad$ // Move B to the left cell.
13 $\quad\quad\quad$**while** $k \geqslant d_i$ **do** $\qquad$ // Search for an extra incremental change.
14 $\quad\quad\quad\quad$**if** $k = d_i$ **then** $B \leftarrow B'$;
15 $\quad\quad\quad\quad B' \leftarrow B' \cdot \frac{k(m+k+1)}{(k+1)(2m+k)}$, where $m = n - i + 1$; $\quad$ // Ref. Eq. (5)
16 $\quad\quad\quad\quad k \leftarrow k - 1; \qquad$ // Move the substitute B' to the left cell.
17 $\quad\quad\quad$**if** $d_i \neq 0$ **then** $R \leftarrow R + B'$;
18 $\quad\quad\quad k \leftarrow d_i$;
19 $\quad$**return** R;

Fig. 6. A refined version of ranking algorithm.

of movements of B and B' is at most $3(n-1) + \sum_{i=1}^{n-1} m_i$. By Lemma 3, we know that $m_i \leqslant d_{i-1} + 2$, and thus each m_i is bounded above by n moves. This shows that the algorithm can be run in $\mathcal{O}(n^2)$ time. Although this analysis is correct, the complexity obtained by considering the worst-case cost is not tight. It is fortunate that the sum of m_i for all $i = 1, 2, \ldots, n-1$ is caused by the left moves of B', and it is bounded above by the sum of the width of chess board and the number of right-down moves for B. Therefore, we have $\sum_{i=1}^{n-1} m_i \leqslant 2n - 1$. This leads to an algorithm with running cost in $\mathcal{O}(n)$ time. $\qquad\square$

5   An Unranking Algorithm

In this section, we provide a reverse function Unranking() that converts a positive integer N to an LC-sequence $(c_1, c_2, \ldots, c_n)$. As before, the corresponding LD-sequence has been chosen purposely to assist the conversion. The basic ideal of design inside the function is to decompose N into a sequence of incremental changes. From these incremental changes, the corresponding LD-sequence can be determined.

Let n and N be two parameters of the function. We begin to put B on the initial location by using Eq. (4) and set $d_i = 0$ for all $i = 1, 2, \ldots, n$ (Lines 2 to

---

Function Unranking$(n; N)$

1 **begin**
2 $\quad d_1 = 0; \quad B \leftarrow 1; \quad k \leftarrow 0; \quad$ // Initialization.
3 $\quad$ **for** $i \leftarrow n$ **downto** 2 **do** $\quad$ // Compute $B_{1,0}$ to store in B.
4 $\quad\quad B \leftarrow B \cdot \frac{(2m+k+1)(2m+k+2)}{(m+1)(m+k+2)}$, where $m = n - i + 1;$ $\quad$ // Ref. Eq. (4)
5 $\quad\quad d_i = 0;$
6 $\quad$ **while** $N > 0$ **do**
7 $\quad\quad B \leftarrow B \cdot \frac{m(k+2)}{(k+1)(2m+k)}$, where $m = n - i + 1;$ $\quad\quad$ // Ref. Eq. (6)
8 $\quad\quad i \leftarrow i + 1; \quad k \leftarrow k + 1;$ $\quad$ // Move B to the right-down cell.
9 $\quad\quad B' \leftarrow B \cdot \frac{k(m+k+1)}{(k+1)(2m+k)}$, where $m = n - i + 1;$ $\quad\quad$ // Ref. Eq. (5)
10 $\quad\quad$ **if** $N \geqslant B - B'$ **then** $\quad$ // Is there an incremental change?
11 $\quad\quad\quad N \leftarrow N - (B - B');$
12 $\quad\quad\quad k \leftarrow k - 1; \quad B \leftarrow B';$ $\quad$ // Move B to the left cell.
13 $\quad\quad\quad$ **while** $k > 0$ **do** $\quad$ // Search for an extra incremental change.
14 $\quad\quad\quad\quad B' \leftarrow B' \cdot \frac{k(m+k+1)}{(k+1)(2m+k)}$, where $m = n - i + 1;$ $\quad$ // Ref. Eq. (5)
15 $\quad\quad\quad\quad$ **if** $B' \leqslant N$ **then break**;
16 $\quad\quad\quad\quad k \leftarrow k - 1; \quad B \leftarrow B';$ $\quad$ // Move B to the left cell.
17 $\quad\quad\quad$ **if** $k \neq 0$ **then** $N \leftarrow N - B';$
18 $\quad\quad\quad d_i = k;$
19 $\quad\quad$ **else** $d_i \leftarrow d_{i-1} + 1;$
20 $\quad (c_1, c_2, \ldots, c_n) \leftarrow$ LD-to-LC$(d_1, d_2, \ldots, d_n);$
21 $\quad$ **return** $c_1, c_2, \ldots, c_n;$

Fig. 7. An unranking algorithm.

5). After this step, we have $i = 1$, $k = 0$ and $B = B_{1,0}$. Then, we carry out the following loop till N has been decreased to 0 or $i = n$. For each round in the loop, we move B to the right-down cell by using Eq. (6), and then compute the left cell of B, say B', using Eq. (5) (Lines 7 to 9). Note that the variable i is increased by one for each round of the loop. Now, we can test the condition $N \geqslant B - B'$ to determine if there exist incremental changes needed to be subtracted from N in the current status. If there is no such an incremental change, then d_i is directly set as d_{i-1} plus one. Otherwise, we update N by subtracting $B - B'$ from N, and move B to the left cell (Lines 11 to 12). In addition, by repeatedly looking ahead to extract the left cell B' of B until $B' \leqslant N$, we can further search for an additional incremental change. After this search, N is updated, and we have $d_i = k$ (Lines 13 to 18). Once the main loop has terminated, we convert $(d_1, d_2, \ldots, d_n)$ to the corresponding LC-sequence for the output. The detail of the unranking function is shown in Fig. 7.

Theorem 5. *Given a positive integer N, determining a binary tree with LC-sequence $(c_1, c_2, \ldots, c_n)$ such that its rank is N in lexicographic order can be done in $\mathcal{O}(n)$ time and space.*

Proof. Since Unranking() is the reverse procedure of Refined-Ranking(), the correctness of the algorithm directly follows from Lemma 4 again. Clearly, the space requirement is $\mathcal{O}(n)$. For algorithmic efficiency, we test the condition $N > 0$ instead of $i < n$ in the main loop (Line 8) to determine if the decomposition of N is finished. For each round in the loop, it is clear that the sequence of movements of B and it substitute B' is the same as that in Theorem 5. As a result, an argument similar above that uses aggregation can analyze the time complexity of the algorithm, and thus the theorem follows.   □

6   Concluding Remarks

In this paper, according to the characterization presented in [19], we show that it is easy to design an algorithm to generate all LC-sequences of binary trees with n internal nodes in lexicographic order. Moreover, based on such an ordering, we propose efficient ranking and unranking algorithms with time complexity and space requirement of $\mathcal{O}(n)$.

Although the algorithm proposed in [19] is a loopless algorithm, the generation of each LC-sequence can be obtained from its predecessor by changing possibly two digits. As a future work, an open challenging problem is to generate all LC-sequences of binary trees in Gray-code order. Furthermore, for such a Gray-code order, designing efficient ranking and unranking algorithms are especially interested. To the best of our knowledge, so far no such algorithms exist for generating, ranking and unranking LC-sequences in Gray-code order.

References

1. Effler, S., Ruskey, F.: A CAT algorithm for generating permutations with a fixed number of inversions. Inf. Process. Lett. **86**, 107–112 (2003)
2. Ehrlich, G.: Loopless algorithms for generating permutations, combinations, and other combinatorial configurations. J. ACM **20**, 500–513 (1973)
3. Er, M.C.: Lexicographic listing and ranking t-ary trees. Comput. J. **30**, 569–572 (1987)
4. Golynski, A.: Optimal lower bounds for rank and select indexes. Theor. Comput. Sci. **387**, 348–359 (2007)
5. Knuth, D.E.: The Art of Computer Programming: Fascicle 4A - Generating All Trees, vol. 4. Addison-Wesley, Boston (2005)
6. Korsh, J.F., LaFollette, P.: A loopless Gray code for rooted trees. ACM Trans. Algorithms **2**, 135–152 (2006)
7. Lucas, J.M., van Baronaigien, R., Ruskey, F.: On rotations and the generation. J. Algorithms **15**, 343–366 (1993)
8. Mäkinen, E.: Left distance binary tree representations. BIT **27**, 163–169 (1987)
9. Mäkinen, E.: A survey on binary tree codings. Comput. J. **34**, 438–443 (1991)
10. Mäkinen, V., Navarro, G.: Rank and select revisited and extended. Theor. Comput. Sci. **387**, 332–347 (2007)
11. Pallo, J.: Enumerating, ranking and unranking binary trees. Comput. J. **29**, 171–175 (1986)

12. Proskurowski, A., Ruskey, F.: Binary tree Gray codes. J. Algorithms **6**, 225–238 (1985)
13. Ruskey, F.: Generating t-ary trees lexicographically. SIAM J. Comput. **7**, 424–439 (1978)
14. Savage, C.D.: A survey of combinatorial Gray codes. SIAM Rev. **39**, 605–629 (1997)
15. Sawada, J.: Generating bracelets in constant amortized time. SIAM Comput. **31**, 259–268 (2001)
16. Trojanowaki, A.E.: Ranking and listing algorithms for k-ary trees. SIAM J. Comput. **7**, 492–509 (1978)
17. Vajnovszki, V.: On the loopless generation of binary tree sequences. Inf. Process. Lett. **68**, 113–117 (1998)
18. van Baronaigien, D.R.: A loopless algorithm for generating binary tree sequences. Inf. Process. Lett. **39**, 189–194 (1991)
19. Wu, R.-Y., Chang, J.-M., Chan, H.-C., Pai, K.-J.: A loopless algorithm for generating multiple binary tree sequences simultaneously. Theor. Comput. Sci. **556**, 25–33 (2014)
20. Wu, R.-Y., Chang, J.-M., Chang, C.-H.: Ranking and unranking of non-regular trees with a prescribed branching sequence. Math. Comput. Model **53**, 1331–1335 (2011)
21. Wu, R.-Y., Chang, J.-M., Chen, A.-H., Ko, M.-T.: Ranking and unranking of non-regular trees in Gray-code order. IEICE Trans. Fund. **E96-A**, 1059–1065 (2013)
22. Wu, R.-Y., Chang, J.-M., Chen, A.-H., Liu, C.-L.: Ranking and unranking t-ary trees in a Gray-code order. Comput. J. **56**, 1388–1395 (2013)
23. Wu, R.-Y., Chang, J.-M., Wang, Y.-L.: A linear time algorithm for binary tree sequences transformation using left-arm and right-arm rotations. Theor. Comput. Sci. **355**, 303–314 (2006)
24. Wu, R.-Y., Chang, J.-M., Wang, Y.-L.: Loopless generation of non-regular trees with a prescribed branching sequence. Comput. J. **53**, 661–666 (2010)
25. Wu, R.-Y., Chang, J.-M., Wang, Y.-L.: Ranking and unranking of t-ary trees using RD-sequences. IEICE Trans. Inf. Syst. **E94-D**, 226–232 (2011)
26. Xiang, L., Ushijima, K., Tang, C.: Efficient loopless generation of Gray codes for k-ary trees. Inf. Process. Lett. **76**, 169–174 (2000)
27. Zaks, S.: Lexicographic generation of ordered trees. Theor. Comput. Sci. **10**, 63–82 (1980)
28. Zaks, S., Richards, D.: Generating trees and other combinatorial objects lexicographically. SIAM J. Comput. **8**, 73–81 (1979)

Combinatorial Optimization

Optimal Speed Scaling with a Solar Cell

(Extended Abstract)

Neal Barcelo[1], Peter Kling[2]($\boxtimes$), Michael Nugent[1], and Kirk Pruhs[1]

[1] Department of Computer Science, University of Pittsburgh, Pittsburgh, USA
[2] Universität Hamburg, Hamburg, Germany
peter.kling@uni-hamburg.de

Abstract. We consider the setting of a sensor that consists of a speed-scalable processor, a battery, and a solar cell that harvests energy from its environment at a time-invariant recharge rate. The processor must process a collection of jobs of various sizes. Jobs arrive at different times and have different deadlines. The objective is to minimize the *recharge rate*, which is the rate at which the device has to harvest energy in order to feasibly schedule all jobs. The main result is a polynomial-time combinatorial algorithm for processors with a natural set of discrete speed/power pairs.

1 Introduction

Most of the algorithmic literature on scheduling devices to manage energy assume the objective of minimizing the total energy usage. This is an appropriate objective if the amount of available energy is bounded, say by the capacity of a battery. However, many devices (most notably sensors in hazardous environments) contain energy harvesting technologies. Solar cells are probably the most common example, but some sensors also harvest energy from ambient vibrations [9,10] or electromagnetic radiation [11] (e.g., from communication technologies such as television transmitters). To get a rough feeling for the involved scales (see also [11]), note that batteries can store on the order of a joule of energy per cubic millimeter, while solar cells provide several hundred microwatt per square millimeter in bright sunlight, and both vibrations and ambient radiation technologies provide on the order of nanowatt per cubic millimeter. Compared to non-harvesting technologies, the algorithmic challenge is to cope with a more dynamic setting, where the difference between *total available* and *total used* energy is non-monotonic (cf. Fig. 1).

See [5] for the full version.

P. Kling—Supported by fellowships of the Postdoc-Programme of the German Academic Exchange Service (DAAD) and the Pacific Institute of Mathematical Sciences (PIMS). Work done while at the University of Pittsburgh.

K. Pruhs—Supported, in part, by NSF grants CCF-1115575, CNS-1253218, CCF-1421508, CCF-1535755, and an IBM Faculty Award.

© Springer International Publishing AG 2016
T.-H.H. Chan et al. (Eds.): COCOA 2016, LNCS 10043, pp. 521–535, 2016.
DOI: 10.1007/978-3-319-48749-6_38

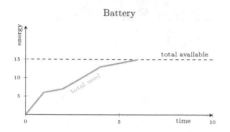
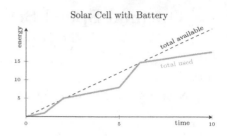

Fig. 1. The total available (dashed) and total used (solid) energy for a battery and for a solar cell with a battery. The solid line cannot cross the dashed line; when both lines meet, the battery is depleted. Depletion is permanent for a battery and temporarily for a solar cell with battery. Note that the difference between total available and total used energy is non-monotonic for solar cells.

Problem and Model in a Nutshell. The goal of this research is to use an algorithmic lens to investigate how the addition of energy harvesting technologies affects the complexity of scheduling such devices. As a test case, we consider the first (and most investigated) problem on energy-aware scheduling due to [12]. There, the authors assumed that (a) the processor is speed-scalable; (b) the power used is the square of the speed; and (c) each job has a certain size, an earliest (release-) time at which it can be run, and a deadline by which it must be finished. Their objective was to minimize the total energy used by the processor when finishing all jobs. We modify these assumptions as follows:

(a) The device has a speed-scalable processor with a *finite* number of speeds $s_1 < \cdots < s_k$, each associated with a power consumption rate $P_1 < \cdots < P_k$.
(b) The device harvests energy at a time-invariant recharge rate $R > 0$ (like a solar-cell in bright sunlight).
(c) The device has a battery (initially empty) to store harvested energy. To concentrate on the energy harvesting aspect, we assume that the battery's capacity isn't a limiting factor.

The objective becomes to find the minimal necessary recharge rate to finish all jobs between their release time and deadline.

As is the case with the (discrete) variant of [12], our solar cell problem can be written as a linear program. Thus, in principle it is solvable in polynomial time by standard mathematical programming methods (e.g., the Ellipsoid method). However, [12] showed that the total energy minimization problem is algorithmically much easier than linear programming by giving a simple, combinatorial greedy algorithm. In the same spirit, we study whether the solar cell version allows for a similarly simple, purely combinatorial algorithm.

Results in a Nutshell. Our main result is a polynomial-time combinatorial algorithm for well-separated processor speeds. Well-separation is a technical but natural requirement to ease the analysis. It ensures that the speed/power cover a good efficiency spectrum, as explained below. Let $\Delta_i := \frac{P_i - P_{i-1}}{s_i - s_{i-1}}$. The speeds are *well-separated* if there is a constant $c > 1$ such that $\Delta_{i+1} = c \cdot \Delta_i$ for all i. To understand this condition, note that there is a strong convex relationship between the speed and power in CMOS-based processors [6], typically modelled as Power $=$ Speed$^\alpha$ for some constant $\alpha > 1$ [12]. Thus, lower speeds give significantly better energy efficiency. A chip designer aims to choose discrete speeds (from the continuous range of options) that are well-separated in terms of performance and energy efficiency. A natural choice is to grow speeds exponentially (i.e., $s_{i+1} = c' \cdot s_i$ for a suitable $c' > 1$). With $P_i = s_i^\alpha$, we get that speeds are well-separated with the constant $c := c'^{\alpha-1}$.

Our algorithm can be viewed as a homotopic optimization algorithm that maintains an energy optimal schedule while the recharge rate is continuously decreased. Similar approaches for other speed scaling problems have been used in [1,2,7,8]. The resulting combinatorial algorithm exposes interesting structural properties and relations to be maintained while decreasing the recharge rate and adapting the schedule, not unlike (but much more complex than) the homotopic algorithm from [2]. While this allows us to prove a polynomial runtime for our algorithm, the actual bound is quite high and only barely superior to bounds derived by generic convex program solvers. We believe that this runtime is merely an artifact of our hierarchical analysis approach, which aims at simplifying the (already quite involved) analysis. However, this might also indicate that other, non-homotopical approaches might be more suitable to tackle this scheduling variant.

Context and Related Results. The only other theoretical work (we are aware of) on this solar cell problem is by [3]. They considered arbitrary (continuous) speeds $s \in \mathbb{R}_{\geq 0}$ and power consumption s^α (where $\alpha > 1$ is a constant). They showed that the offline problem can be expressed as a convex program. Thus, using the well-known KKT conditions one can efficiently recognize optimal solutions, and standard methods (e.g., the Ellipsoid Method) efficiently solve this problem to any desired accuracy. [3] also proved that the schedule that optimizes the total energy usage is a 2-approximation for the objective of recharge rate. Finally, they showed that the online algorithm BKP, which is known to be O (1)-competitive for total energy usage [4], is also O (1)-competitive with respect to the recharge rate. So, intuitively, the take-away from [3] was that schedules that naturally arise when minimizing energy usage are O (1) approximations with respect to the recharge rate. In particular, [3] left as an open question whether there is a simple, combinatorial algorithm for the solar cell problem.

Outline. Both our algorithm design and analysis are quite involved and require significant understanding of the relation between the recharge rate and optimal schedules. Thus, we start with an informal overview in the next section. The formal model description and definitions can be found in Sects. 3 and 4. The actual algorithm description is given in Sect. 5. Due to space restrictions, most proofs are left for the full version.

2   Approach and Overview

In the following, we state the central optimality conditions and give a simple illustrating example. Afterward, we explain how to improve upon a given schedule via suitable transformations guided by these optimality conditions. Finally, we explain how our algorithm realizes these transformations in polynomial time.

Optimality Conditions. As the first step in our algorithm design, we consider the natural linear program for our problem and translate the complementary slackness conditions (which characterize optimal solutions) into structural optimality conditions. This results in Theorem 1, which states[1] that optimal solutions can be characterized as follows:

(a) *Feasibility:* All jobs are fully processed between their release times and deadlines and the battery is never depleted.

(b) *Local Energy Optimality:* The job portions scheduled within each *depletion interval* (time between two moments when the battery is depleted) are scheduled in an energy optimal way.

(c) *Speed Level Relation (SLR):* Consider job j that runs in two depletion intervals I and I'. Let the average speed of j in I lie between discrete speeds s_{a-1} and s_a. Similarly, let it lie between s_{b-1} and s_b in I'. The SLR states that the difference $b - a$ is independent of the job j. In other words, jobs jump roughly the same amount of discrete speed levels between depletion intervals[2].

(d) *Split Depletion Point (SDP):* There is a depletion point (time when the battery is depleted) $\tau > 0$ such that no job with deadline $> \tau$ is run before τ.

As above for the SLR, we often consider the *average speed* of a scheduled job during a depletion interval. Note that one can easily derive an actual, discrete schedule from these average speeds: If a job j runs at average speed $s \in [s_{a-1}, s_a)$ during a time interval I of length $|I|$, we can interpolate the average speed with discrete speeds by scheduling j first for $\frac{s_a - s}{s_a - s_{a-1}} \cdot |I|$ time units at speed s_{a-1} and for $\frac{s - s_{a-1}}{s_a - s_{a-1}} \cdot |I|$ time units at speed s_a. Using that speeds are well-separated[3], it follows easily that this is an optimal discrete way to achieve average speed s.

A Simple Example. To build intuition, consider a simple example. The processor has two discrete speeds $s_1 = 1$ and $s_2 = 2$ with power consumption rates $P_1 = 1$ and $P_2 = 4$, respectively. Job j is released at time 0 with deadline 4 and work 3. Job j' is released at time 1 with deadline 2 and work 2. The *energy*

---

[1] Statements slightly simplified; Sect. 3 gives the full formal conditions.

[2] Figure 3 gives an example where the SLR can be observed: The orange and light-blue jobs run both in depletion interval I_3 and I_4. The orange job's average speed "jumps" one discrete speed level from I_3 to I_4 (from below s_2 to above s_2). Thus, the light-blue job must also jump one discrete speed level (from below s_3 to above s_3).

[3] In fact, $\Delta_{i+1} > \Delta_i$ is already sufficient. Also note that starting with the lower speed is essential: otherwise the battery's energy level might become negative.

optimal schedule runs job j' at speed 2 during the time interval $[1, 2]$ and job j at speed 1 during the time intervals $[0, 1]$ and $[2, 4]$. It needs recharge rate $R = 2.5$. There is a depletion point $\tau = 2$ and two depletion intervals $I_1 = [0, \tau)$ and $I_2 = [\tau, \infty)$. See the left side of Fig. 2 for an illustration. While this schedule fulfills the first three optimality conditions for *rate optimality*, the SDP condition is violated (j is run both before and after τ). Thus, while it is energy optimal it is not recharge rate optimal.

Consider what happens if we decrease the recharge rate R by an infinitesimal small amount ε (i.e., decrease the slope of the dotted line in the left part of Fig. 2). This results in a negative energy in the battery at time τ (the solid line in Fig. 2 "spikes" through the dotted line at $\tau = 2$). This is not allowed, so we have to decrease the energy used before τ. To do so, we move some work from a job that is processed on both sides of τ from I_1 to I_2 (the violation of the SDP guarantees the existence of such a job). Continuing to do so allows us to decrease the recharge rate until the SDP holds (see the right side of Fig. 2). Thus, the resulting schedule is recharge rate optimal (i.e., needs a solar cell of minimal recharge rate). Also note that this schedule is no longer energy optimal (the total amount of used energy increased).

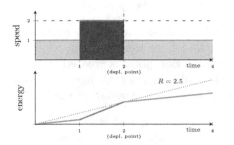 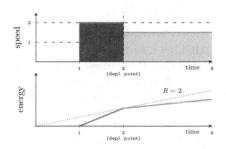

Fig. 2. The energy optimal (left) and recharge rate optimal (right) schedules. Job speeds are plotted as average speeds (i.e., the green job running at average speed $3/2$ in the depletion interval $[2, 4)$ on the right runs in the actual, discrete schedule at speed 1 during $[2, 3)$ and at speed 2 during $[3, 4)$). (Color figure online)

Algorithmic Intuition. Our algorithm extends on the schedule transformation we saw in the simple example above. We start with an energy optimal schedule S and a trivial bound on the recharge rate R such that the first three optimality conditions hold. We then lower R while maintaining a schedule satisfying these first three conditions until, additionally, the SDP holds. Lowering the recharge rate R means we have to move work out of each depletion interval (or we get a negative energy in the battery). Since we want to maintain the first three optimality conditions, we cannot move work arbitrarily. To capture all constraints while moving work we employ a *distribution muligraph* G_D. Its vertices are the depletion intervals, and there is a directed edge for each way in which work can be transferred between depletion intervals. See Fig. 3 for an illustration.

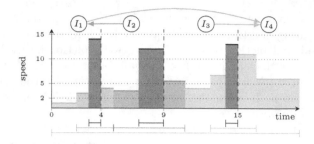

Fig. 3. Four discrete speeds ($s_1 = 2, s_2 = 5, s_3 = 10, s_4 = 15$), seven jobs (release/dead-lines indicated by the colored bars). Four depletion points $\tau \in \{0, 4, 9, 15\}$ form four depletion intervals $I_1 = [0, 4), I_2 = [4, 9), I_3 = [9, 15), I_4 = [0, \infty)$. A subgraph of the distribution graph G_D is shown above. (Color figure online)

The heart of our algorithm is to find a suitable *transfer path* for each depletion interval: a path over which work can be transferred to the rightmost depletion interval (possibly via multiple jobs). Given such transfer paths, we can move work out of every depletion interval. While this allows us to make progress, there are three types of events that can occur and must be handled:

- *Edge Removal Event:* It is no longer possible to transfer work on a particular edge because there is no more work left on the job we were moving.
- *Depletion Point Appearance Event:* A new depletion point is created.
- *Speed Level Event:* Further transfer of work would cause a job's average speed in a depletion interval to cross a discrete speed (possibly violating the SLR).

In these cases, our algorithm attempts to find a different collection of transfer paths. If this is not possible, the algorithm tries to update G_D as follows:

- *Depletion Point Removal Update:* Find a depletion point that can be removed. Removing the constraint that the battery is depleted at this point may allow for new ways to transfer work.
- *Cut Update:* Because of the SLR, jobs have to jump the same amount of discrete speed levels between two depletion intervals. Thus, all jobs must cross the next discrete speed at the same time. A cut update basically signals that all involved jobs reached a suitable discrete speed and can now cross the discrete speed level. See Sect. 4 and Definition 1 for details.

As an example, consider what happens when moving work of the light blue job from I_3 to I_4 in Fig. 3. After a while, its average speed in I_3 reaches the discrete speed s_2 (a speed level event). The SLR forbids to further decrease this job's speed (it would jump two discrete speed levels, while the orange job jumps only one). Instead, we start to move work of the orange job from I_3 to I_4 until it hits the discrete speed s_1. All jobs processed before and after depletion point $\tau = 15$ are now at suitable discrete speeds and we can allow both the orange and light blue job to further decrease their speeds (a cut update).

Our correctness proof shows that, if none of these updates is possible, the SDP holds. A technical difficulty is that events might influence each other, resulting in complex dependencies (which we ignored in the above example). A lot of the complexity of our algorithm/analysis stems from an urge to avoid these dependencies wherever possible. However, it seems likely that a more careful study of these dependencies would yield a significant simplification and improvement.

Events and Updates in Polynomial Time. Our description above assumes that we move work continuously and stop at the corresponding events. To implement this in our algorithm, we have to calculate the next event for the current collection of transfer paths and then compute the correct amount of work to move between all involved depletion intervals. While the involved calculations follow from a simple linear equation system, the main difficulty is to show that the number of events remains polynomial. To ensure this, our algorithm design facilitates the following hierarchy of invariants:

- *Cut Invariant:* Cut updates are at the top of the hierarchy. Intuitively, this invariant states that job-speeds (or speed levels) tend to increase toward the right (since, as a net effect, work is generally moved to the right). This is, for example, used to prove that there is only a polynomial number of cut updates.
- *Depletion Point Removal Invariant:* Depletion point updates are at the second level of the hierarchy. This invariant states that once a depletion point is removed it will not be added again (until the next cut update).
- *Speed Level Invariant:* Speed level updates are also at the second level of the hierarchy. This invariant states that this event creates a time interval to which no work is added (until the next cut update).
- *Edge Removal Invariant:* Edge removal events are at the bottom of the hierarchy. This invariant states that once work of a job was transferred to an earlier depletion interval ("to the left"), it will not be transferred to a later one ("to the right") until the next cut, depletion point removal, or speed level update.

This hierarchy provides a monotone progress measure, but complicates the algorithm/analysis quite a bit. In particular, we have to deal with two aspects: (a) G together with all transfer paths may be exponentially large. To handle this, we search for suitable transfer paths on a subgraph H of G, containing only the best transfers to move work between any given pair of depletion intervals. (b) We have to define how to select these collections of transfer paths. On a high level, the algorithm prefers transfers that move work right to transfers that move work left. Between transfers moving work right it prefers shorter transfers, while between transfers moving work left it prefers longer transfers (cf. Definition 8).

3   Structural Optimality via Primal-Dual Analysis

We model our problem as a linear program and use complementary slackness conditions to derive structural properties that are sufficient for optimality. These structural properties are used in both the design and the analysis of the algorithm.

3.1   Model

We consider the problem of scheduling a set of n jobs $J := \{1, 2, \ldots, n\}$ on a single processor that features k different speeds $0 < s_1 < s_2 < \cdots < s_k$ and that is equipped with a solar-powered battery. The battery is attached to a solar cell and recharges at a rate of $R \geq 0$. The power consumption when running at speed s_i is $P_i > 0$. That is, while running at speed s_i work is processed at a rate of s_i and the battery is drained at a rate of P_i. When the processor is idling (not processing any job) we say it runs at speed $s_0 := 0$ and power $P_0 := 0$.

Each job $j \in J$ comes with a *release time* r_j, a *deadline* d_j, and a processing volume (or work) p_j. For each time t, a schedule S must decide which job to process and at what speed. Preemption is allowed, so that a job may be suspended at any point in time and resumed later on. We model a schedule S by two functions $S(t)$ (*speed*) and $J(t)$ (*scheduling policy*) that map a time $t \in \mathbb{R}$ to a speed index $S(t) \in \{0, 1, \ldots, k\}$ and a job $J(t) \in J$. We say a job j is *active* at time t if $t \in [r_j, d_j)$. Jobs can only be processed when they are active. Thus, a *feasible schedule* must ensure that $J^{-1}(j) \subseteq [r_j, d_j)$ holds for all jobs j. Moreover, a feasible schedule must finish all jobs and must ensure that the energy level of the battery never falls below zero. More formally, we require $\int_{J^{-1}(j)} s_{S(t)} \mathrm{d}t \geq p_j$ for all jobs j and $\int_0^{t_0} P_{S(t)} \mathrm{d}t \leq R \cdot t_0$ for all times t_0. Our objective is to find a feasible schedule that requires the minimum recharge rate.

3.2   Linear Programming Formulation

For the following linear programming formulation, we discretize time into equal length time slots t. Without loss of generality, we assume that their length is such that there is a feasible schedule for the optimal recharge rate R that processes at most one job using at most one discrete speed in each single time slot[4]. Our linear program uses indicator variables x_{jit} that state whether a given job j is processed at a speed s_i during time slot t. Note that not only does this imply a possible huge number of variables but it is also not trivial to compute the length of the time slots. Nevertheless, this will not influence the running time of our algorithm, since we merely use the linear program to extract sufficient structural properties of optimal solutions. Our analysis will also use the fact that we can always further subdivide the given time slots into even smaller slots without changing the optimal schedule. By rescaling the problem parameters, we can assume that the (final) time slots are of unit length.

With the variables x_{jit} as defined above and a variable R for the recharge rate, the integer linear program (ILP) shown in Fig. 4a corresponds to our scheduling problem. The first set of constraints ensure that each job is finished during its release-deadline interval, while the second set of constraints ensures that the

---

[4] The existence of such a schedule follows from standard speed scaling arguments. To see this, note that any schedule can be transformed to use earliest deadline first and interpolate an average speed in a depletion interval by at most one speed change between two discrete speeds. Thus, the number of job changes and speed changes is finite (depending on n) and we merely have to choose the time slots suitably small.

battery's energy level does not fall below zero. The final set of constraints ensures that the processor runs at a constant speed and processes at most one job in each time slot.

$$\min \quad R$$

$$\text{s.t.} \quad \sum_{t\in[r_j,d_j)}\sum_i x_{jit}\cdot s_i \geq p_j \qquad \forall j$$

$$R\cdot t - \sum_{t'\leq t}\sum_{j\in J}\sum_{i=1}^{k} x_{jit'}\cdot P_i \geq 0 \qquad \forall t$$

$$\sum_{j\in J}\sum_{i=1}^{k} x_{jit} \leq 1 \qquad \forall t$$

$$x_{jit} \in \{0,1\} \qquad \forall j,i,t$$

$$\max \quad \sum_{j\in J}\alpha_j\cdot p_j - \sum_t \gamma_t$$

$$\text{s.t.} \quad \alpha_j\cdot s_i - \sum_{t'\geq t}\beta_{t'}\cdot P_i - \gamma_t \leq 0 \qquad \forall j,i,t$$

$$\sum_t \beta_t\cdot t \leq 1$$

$$\alpha_j,\beta_t,\gamma_t \geq 0 \qquad \forall j,t$$

(a) ILP for our scheduling problem.   (b) Dual program for the ILP's relaxation.

Fig. 4. ILP formulations used in our analysis.

Structural Properties for Optimality. The complementary slackness constraints for the programs shown in Fig. 4 give us necessary and sufficient properties for the optimality of a pair of feasible primal and dual solutions. Although these conditions are only necessary and sufficient for optimal solutions of the ILP's *relaxation*, our choice of the time slots ensures that there is an integral optimal solution to the relaxation. Based on these complementary slackness constraints, we derive some purely combinatorial structural properties (not based on the linear programming formulation) that will guarantee optimality. To this end, we will consider *speed levels* of jobs in depletion intervals – essentially the discrete speed a job reached in a specific depletion interval – and how they change at depletion points. In the following, if we speak of a speed s between two discrete speeds (e.g., $s_2 < s < s_3$) we implicitly assume s to refer to the average speed in the considered time interval.

Definition 1 (Speed Level Relation). *A schedule S and a recharge rate R obey the* Speed Level Relation *(SLR) if there exist natural numbers $\mathcal{L}(j,\ell) \in \mathbb{N}$ (speed levels) such that*

(a) *job j processed at speed $s_{j,\ell} \in (s_{i-1},s_i)$ in depletion interval $I_\ell \Rightarrow \mathcal{L}(j,\ell) = i$*
(b) *job j processed at speed $s_{j,\ell} = s_i$ in depletion interval $I_\ell \Rightarrow \mathcal{L}(j,\ell) \in \{i,i+1\}$*
(c) *jobs j,j' both active in depletion intervals I_{ℓ_1} and I_{ℓ_2} with $\ell_1 < \ell_2 \Rightarrow \mathcal{L}(j,\ell_2) - \mathcal{L}(j,\ell_1) = \mathcal{L}(j',\ell_2) - \mathcal{L}(j',\ell_1) \in \mathbb{N}_0$ (in particular, the speed levels of a job are non-decreasing)*
(d) *job j processed in $I_l \Rightarrow \mathcal{L}(j,l) \geq \mathcal{L}(j',l)$ for all j' active in $I_{l,j} = I_l \cap [r_j,d_j)$*

Intuitively, the SLR states that jobs jump the same number of discrete speeds between depletion intervals (cf. Sect. 2), that speed levels are non-decreasing, and that the currently processed job is one of maximum speed level among active jobs. Using this definition, we are ready to characterize optimal schedules in terms of the following combinatorial properties. Note that for (b) of the following theorem, one can simply use a YDS schedule (cf. [12]) for the workload assigned to the corresponding depletion interval.

Theorem 1. *Consider a schedule S and a recharge rate R. The following properties are sufficient[5] for S and R to be optimal:*

(a) S is feasible.
(b) The work in each depletion interval is scheduled energy optimal.
(c) The SLR holds.
(d) There is a split depletion point: a depletion point $\tau_k > 0$ such that no job with deadline greater than τ_k is processed before τ_k.

For the proof we derive the complementary slackness conditions and interpret them in a suitable way. Details are left for the full version.

4   Notation

Given a schedule S, we need a few additional notions to describe and analyze our algorithm.

Structuring the Input. Let us start by formally defining depletion points and depletion intervals. As noted earlier, depletion points represent time points where our algorithm maintains a battery level of zero and partition the time horizon into depletion intervals. Note that these definitions depend on the current state of the algorithm.

Definition 2 (Depletion Point). *Let $E_S(t)$ be the energy remaining at time t in schedule S. Then τ_i is a depletion point if $E_S(\tau_i) = 0$ (and the algorithm has labeled it as such). L is the number of depletion points, $\tau_0 := 0$, and $\tau_{L+1} := \infty$.*

Definition 3 (Depletion Interval). *For $\ell > 0$, the ℓ-th depletion interval is $I_\ell := [\tau_{\ell-1}, \tau_\ell)$. We also define $s_{j,\ell}$ as the (average) speed of job j during I_ℓ.*

To simplify the discussion, we sometimes identify a depletion interval I_ℓ with its index ℓ. While moving work between depletion intervals, our algorithm uses the jobs' *speed levels* together with the SLR as a guide:

Definition 4 (Speed Level). *For all j, ℓ with $I_\ell \cap [r_j, d_j) \neq \emptyset$, the speed level $\mathcal{L}(j, \ell)$ of j in I_ℓ is such that if j is processed in I_ℓ, then $s_{j,\ell} \in [s_{\mathcal{L}(j,\ell)-1}, s_{\mathcal{L}(j,\ell)}]$.*

---

[5] If we restrict ourselves to *normalized* (earliest deadline first, only one speed change per job in a depletion interval) schedules, they are in fact also necessary.

Note that this definition should be understood as a variable of our algorithm. In particular, it is not unique if the job runs at a discrete speed s_{i-1}. In these cases, $\mathcal{L}(j, \ell)$ can be either $i - 1$ or i (and the algorithm can set $\mathcal{L}(j, \ell)$ as it wishes). The algorithm initializes the speed level for every depletion interval where j is active based on the initial YDS schedule and assigns speed levels maintaining the SLR throughout its execution.

Next, we give a slightly weaker version of the well-known EDF (Earliest Deadline First) scheduling policy. The idea is to maintain EDF w.r.t. depletion intervals but to allow deviations within depletion intervals. For example, we avoid schedules with depletion intervals $I_1, \ldots, I_4$ where job j_1 is scheduled in I_1 and I_3 and j_2 in I_2 and I_4. This will ensure that the collection of transfer paths will be laminar, which is useful throughout the analysis.

Definition 5 (Weak EDF, Informal[6]). *Schedule S is* weak EDF *if there is a schedule that is EDF in which each job is run in the same depletion intervals as in S.*

Next, we consider to what extent a schedule adheres to the optimality conditions (Theorem 1). We distinguish between schedules that (essentially) adhere to the first two optimality conditions and schedules that also have the third optimality condition (SLR).

Definition 6 (Nice and Perfect). *Schedule S is* nice *if it is feasible, obeys YDS between depletion points, and satisfies weak EDF. If, additionally, S fulfills the SLR, we call it* perfect.

Distributing Workload. We now define ϵ-transfers, the building block for our algorithm. Intuitively, they formalize possible ways to move work around between depletion intervals. Our definition ensures that moving work over ϵ-transfers maintains niceness throughout the algorithm's execution. Moreover, we also ensure that ϵ-transfers only affect the schedule's speed profile at their sources/targets.

Definition 7 (ϵ-transfer). *The sequence $(\ell_a, j_a)_{a=0}^s$ is called an ϵ-transfer if we can, simultaneously for all a, move some non-zero workload of j_a from ℓ_{a-1} to ℓ_a while maintaining niceness and without changing any job speeds in $\ell_1, \ldots, \ell_{s-1}$. The pair (ℓ_0, j_0) (resp., (ℓ_s, j_s)) is the* source *and* source job *(resp.,* destination *and* destination job*) of the ϵ-transfer. The ϵ-transfer is* active *if it also maintains perfectness.*

Each edge drawn in Fig. 3 is a (trivial) ϵ-transfer.

Next we define the priority of an ϵ-transfer. Our algorithm compares ϵ-transfers based on source and destination. Once the source and destination have been fixed, the priority is used to determine which ϵ-transfer is used to transfer work. As mentioned in Sect. 2, the basic idea is to: prefer transfers that move work right to transfers that move work left, between transfers moving work right prefer shorter transfers, and between transfers moving work left prefer longer transfers.

---

[6] The formal definition is left for the full version.

Definition 8 (Transfer Priority). *Let* $T_1 = (\ell_a^1, j_a^1)_{a=0}^{s_1}$ *and* $T_2 = (\ell_a^2, j_a^2)_{a=0}^{s_2}$ *be two different ϵ-transfers with* $\ell_0^1 = \ell_0^2$ *and* $\ell_{s_1}^1 = \ell_{s_2}^2$. *Let* $a_1^* = \arg\min_a\{\ell_a^1 \neq \ell_a^2\}$ *and* $a_2^* = \arg\min_a\{j_a^1 \neq j_a^2\}$. *We say that T_1 is* higher priority *than T_2 if*

(a) $\ell_{a_1^*}^2 < \ell_{a_1^*-1}^1 < \ell_{a_1^*}^1$, *or*
(b) $\ell_{a_1^*}^1 < \ell_{a_1^*}^2$, *and either* $\ell_{a_1^*}^2 < \ell_{a_1^*-1}^1$ *or* $\ell_{a_1^*-1}^1 < \ell_{a_1^*}^1$, *or*
(c) a_1^* *does not exist and the deadline for* $j_{a_2^*}^1$ *is earlier than the deadline for* $j_{a_2^*}^2$.

Finally, we can define our multigraph of legal ϵ-transfers.

Definition 9 (Distribution Graph). *The* distribution graph *is a multigraph* $G_D = (V_D, E_D)$. V_D *is the set of depletion points and for every active ϵ-transfer* $(\ell_a, j_a)_{a=0}^s$, *there is a corresponding edge with source ℓ_0 and destination ℓ_s.*

5   Algorithm Description

This section provides a formal description of the algorithm. From a high level, the algorithm can be broken into two pieces: (a) choosing which ϵ-transfers to move work along (in order to lower the recharge rate), and (b) handling events that cause any structural changes. We start in Sect. 5.1 by describing the structural changes our algorithm keeps track of and by giving a short explanation of each event. Section 5.2 describes our algorithm. Due to space constraints, the full correctness and runtime proofs are left for the full version.

5.1   Keeping Track of Structural Changes

How much work is moved along each single ϵ-transfer depends inherently on the structure of the current schedule. Thus, intuitively, an event is any structural change to the distribution graph or the corresponding schedule while we are moving work. At any such event, our algorithm has to update the current schedule and distribution graph. The following are the basic structural changes our algorithm keeps track of:

– **Depletion Point Appearance:** For some job j, the remaining energy $E_S(d_j)$ at j's deadline becomes zero and the rate of change of energy at d_j is strictly negative. If we were not to add this depletion point, the amount of energy available at d_j would become negative, violating the schedule's feasibility. We can easily calculate when this happens by examining the rate of change of R as well as the rate of change of $s_{j,\ell}$ for all jobs j that run in the depletion interval I_ℓ containing d_j.
– **Edge Removal:** An edge removal occurs when, for some job j, the workload of j processed in a depletion interval I_ℓ becomes zero. In other words, all of j's work has been moved out of I_ℓ. Similar to before, we can easily keep track of the time when this occurs for any job j processed in a given depletion interval, since all involved quantities change linearly.

– **Edge Inactive:** An edge inactive event occurs when for some job j its speed $s_{j,\ell}$ in a depletion interval I_ℓ becomes equal to some discrete speed s_i. Once more, we keep track of when this happens for each job processed in a given depletion interval.

Handling of Critical Intervals. Note that by moving work along ϵ-transfers between two events e_1 and e_2, our algorithm causes (a) the speed of exactly one YDS critical interval in each depletion interval to decrease and (b) the speed of some YDS critical intervals to increase. For a single critical interval, these speed changes are monotone over time (between two events). However, critical intervals might merge or separate during this process (e.g., when the speed of a decreasing interval becomes equal to a neighbouring interval). In other words, the critical intervals of a given depletion interval might be different at events e_1 and e_2. On first glance, this might seem problematic, as a critical interval merge/separation could cause a change in the rate of change of the critical interval's speed, perhaps with the result that the algorithm stops for events spuriously, or misses events it should have stopped for. However, since only neighbouring critical intervals can merge and separate, this can be easily handled: In any depletion interval, there are at most O (n) critical intervals at event e_1. Since only neighbouring critical intervals can merge/separate when going from e_1 to e_2, for each critical interval changing speed there are at most O (n^2) possible candidate critical intervals that can be part of event e_2. We just compute the next event caused by each of these candidates, and whether or not each candidate event can feasibly occur. Then, the next event to be handled by our algorithm is simply the minimum of all feasible candidates. This is an inefficient way to handle critical interval changes, but it significantly simplifies the algorithm description. We leave the description of a more efficient way to handle "critical interval events" for the full version.

Handling Events. When we have identified the next event, we must update the distribution graph and recalculate the rates at which we move work along the ϵ-transfers. Given the definition of the distribution graph, updating the graph is fairly straightforward. However, after updating the graph there might no longer be a path from every depletion interval to the far right depletion interval. This can be seen as a cut in the distribution graph. In these cases, to make progress, we either have to remove a depletion point or adapt the jobs' speed levels; If both of these fixes are not possible, our algorithm has found an optimal solution. A detailed description of this can be found in the full version.

5.2   Main Algorithm

Now that we have a description of each event type, we can formalize the main algorithm. A formal description of the algorithm can be found in Listing 1.1. We give an informal description of its subroutines CALCULATERATES, UPDATE-GRAPH, and PATHFINDING below.

UPDATEGRAPH(T, G_D, S): This subroutine takes an event type T, the distribution graph G_D and the current schedule S and performs the required structural changes. It suffices to describe how to build the graph from scratch given

```
1   Set R to be recharge rate that ensures YDS schedule, S, is feasible
2   Let G_D = (V_D, E_D) be the corresponding Distribution Graph.
3   G'_D = PATHFINDING(G_D)
4   (Δ, δ_{j,ℓ}, T):= CALCULATERATES(G'_D, S)
5   while True:
6       for each job j and depletion interval ℓ:
7           set s_{j,ℓ} = s_{j,ℓ} + Δ · δ_{j,ℓ}
8       set R = R − Δ
9       UPDATEGRAPH(T, G_D, S)
10      if ∃ fixable cut:
11          fix cut with either a depletion point removal or SLR procedure
12      else :   exit
13      G'_D = PATHFINDING(G_D)
14      (Δ, δ_{j,ℓ}, T):= CALCULATERATES(G'_D, S)
```

Listing 1.1. The algorithm for computing minimum recharge rate schedule.

```
1   Let S = {v_L}, where v_L is rightmost vertex
2   while exists an edge e = (v_1, v_2) with v_1, v_2 ∈ S and e is the highest priority such edge:
3       add v_1 to S
```

Listing 1.2. The PATHFINDING subroutine.

a schedule (computing a schedule simply involves computing a YDS schedule between each depletion point). Now the question becomes: Given two depletion points, how do we choose the ϵ-transfer between these two? While perhaps daunting at first, this can be achieved via a depth-first search from the source depletion interval. Whenever the algorithm runs into a depletion interval it has previously visited in the search, it chooses the higher priority ϵ-transfer of the two as defined by the priority relation.

PATHFINDING(G_D): We define PATHFINDING(G_D) in Listing 1.2. Note the details of determining the highest priority edge are omitted but the implementation is rather straightforward. The priority relation for choosing edges is: First choose the shortest right going edge, and otherwise choose the longest left going edge. While this priority relation itself is rather straightforward, it requires a non-trivial amount of work to show that it yields suitable monotonicity properties to bound the runtime.

CALCULATERATES(G'_D, S): This subroutine takes as input the set of paths from the distribution graph G'_D and the current schedule S. It returns for each job j and each depletion interval ℓ, the rate $\delta_{j,\ell}$ at which $s_{j,\ell}$ should change, T, the next event type, and Δ the amount the recharge rate should be decreased. It is straightforward to see the set of paths chosen by the algorithm G'_D can be viewed as a tree with the root being the rightmost depletion interval. Assuming R is decreasing at a rate of 1, and working our way from the leaves to the root, we can calculate $\delta_{j,\ell}$ such that the rate of change of energy at all depletion points remains 0. With these rates, we can use the previously discussed methods to find both T and Δ.

References

1. Angelopoulos, S., Lucarelli, G., Nguyen, K.T.: Primal-dual and dual-fitting analysis of online scheduling algorithms for generalized flow time problems. In: Bansal, N., Finocchi, I. (eds.) ESA 2015. LNCS, vol. 9294, pp. 35–46. Springer, Heidelberg (2015)
2. Antoniadis, A., Barcelo, N., Consuegra, M.E., Kling, P., Nugent, M., Pruhs, K., Scquizzato, M.: Efficient computation of optimal energy and fractional weighted flow trade-off schedules. In: Symposium on Theoretical Aspects of Computer Science, pp. 63–74 (2014)
3. Bansal, N., Chan, H.L., Pruhs, K.: Speed scaling with a solar cell. Theor. Comput. Sci. **410**(45), 4580–4587 (2009)
4. Bansal, N., Kimbrel, T., Pruhs, K.: Speed scaling to manage energy and temperature. J. ACM **54**(1), 3:1–3:39 (2007)
5. Barcelo, N., Kling, P., Nugent, M., Pruhs, K.: Optimal speed scaling with a solar cell. arXiv:1609.02668 [cs.DS]
6. Brooks, D.M., Bose, P., Schuster, S.E., Jacobson, H., Kudva, P.N., Buyukto-sunoglu, A., Wellman, J.D., Zyuban, V., Gupta, M., Cook, P.W.: Power-aware microarchitecture: design and modeling challenges for next-generation microprocessors. IEEE Micro **20**(6), 26–44 (2000)
7. Cole, D., Letsios, D., Nugent, M., Pruhs, K.: Optimal energy trade-off schedules. In: International Green Computing Conference, pp. 1–10 (2012)
8. Pruhs, K., Uthaisombut, P., Woeginger, G.J.: Getting the best response for your ERG. ACM Trans. Algorithms **4**(3), 38:1–38:17 (2008)
9. Remick, K., Quinn, D.D., McFarland, D.M., Bergman, L., Vakakis, A.: High-frequency vibration energy harvesting from impulsive excitation utilizing intentional dynamic instability caused by strong nonlinearity. J. Sound Vib. **370**, 259–279 (2016)
10. Stephen, N.G.: On energy harvesting from ambient vibration. J. Sound Vib. **293**(1–2), 409–425 (2006)
11. Vullers, R., van Schaijk, R., Doms, I., Hoof, C.V., Mertens, R.: Micropower energy harvesting. Solid-State Electron. **53**(7), 684–693 (2009)
12. Yao, F.F., Demers, A.J., Shenker, S.: A scheduling model for reduced CPU energy. In: Foundations of Computer Science, pp. 374–382 (1995)

An Approximation Algorithm for the k-Median Problem with Uniform Penalties via Pseudo-Solutions

Chenchen Wu[1], Donglei Du[2], and Dachuan Xu[3]([✉])

[1] College of Science, Tianjin University of Technology,
NO. 399, Binshui West Street, Xiqing District, Tianjin, China
wu_chenchen_tjut@163.com
[2] Faculty of Business Administration, University of New Brunswick,
Fredericton, NB E3B 5A3, Canada
ddu@unb.ca
[3] Department of Information and Operations Research,
College of Applied Sciences, Beijing University of Technology,
100 Pingleyuan, Chaoyang District, Beijing 100124, People's Republic of China
xudc@bjut.edu.cn

Abstract. We present a $(1 + \sqrt{3} + \epsilon)$-approximation algorithm for the k-median problem with uniform penalties in polynomial time, extending the recent result by Li and Svensson for the classical k-median problem without penalties. One important difference of this work from that of Li and Svensson is a new definition of sparse instance to exploit the combinatorial structure of our problem.

Keywords: Approximation algorithm · k-median problem · Penalty · Pseudo-solution

1 Introduction

The k-median problem is a classical NP-hard problem in combinatorial optimization, with vast applications in computer science, management science, and operations research etc. Formally, given a facility set and a client set, we want to open at most k facilities and connect each client to exactly one opened facility such that the total connection cost is minimized. In real practice, some clients are too far from the facilities, who may choose not to serve these clients by paying some penalty cost. This consideration leads naturally to the k-median problem with uniform penalties, the main focus of this work.

1.1 Literature Review

Extensive research exists on approximation algorithms for the (metric) k-median problem; namely the version where the connection cost is non-negative, symmetric, and satisfying triangle inequalities. Jain et al. [8] show that no approximation algorithm exists with approximation ratio smaller than $1 + 2/e$ for this

© Springer International Publishing AG 2016
T.-H.H. Chan et al. (Eds.): COCOA 2016, LNCS 10043, pp. 536–546, 2016.
DOI: 10.1007/978-3-319-48749-6_39

problem, unless NP$\subseteq DTIME(n^{O(loglogn)})$. Charikar et al. [2] propose a 20/3-approximation algorithm based on LP-rounding technique. Arya et al. [1] give a $(3 + \epsilon)$-approximation algorithm based on local search technique, where $\epsilon > 0$ is an arbitrary constant.

In contrast to the 'usual' definition of approximation ratio, there is another performance measure of relevance to the k-median problem; namely, the bi-criteria (α, β), where an algorithm for the the k-median problem finds a solution with cost no more than α times the optimum, while using at most βk facilities. Lin and Vitter [12] give a polynomial-time algorithm for the metric k-median problem with bi-criteria of $(2(1 + \epsilon), (1 + 1/\epsilon)k)$ for any $\epsilon > 0$.

The k-median problems is also closely related to the facility location problem (FLP), also NP-hard. In this problem, we are given a facility set and a client set. Each facility has an open cost and each client pays a connection cost to be served by an open facility. The objective is to open some facilities F and connect each client to an open facility such that the total cost $C(F) := C_f(F) + C_c(F)$, including the facility opening cost $C_f(F)$ and the connection cost $C_c(F)$, is minimized.

The first constant approximation ratio for the (metric) FLP is proposed by Shmoys et al. [14] based on LP-rounding technique. They obtain two approximation ratios, one of 4 with deterministic rounding and the other of 3.16 with randomized rounding. The lower bound of the approximation ratio for the FLP is 1.463 unless NP$\subseteq DTIME(n^{O(loglogn)})$ (c.f. [5]). Sviridenko indicates that the hardness result can be strengthened to depend only on the assumption that P$\neq$NP (cited as personal communication in [4]).

Due to the close relationship between the k-median problem and the FLP, techniques designed for one usually also lead to insights for the other. For example, Jain and Vazirani [9] propose a 3-approximation algorithm for the FLP based on a dual-ascent procedure, which satisfies a nice property of Lagrange multiplier preservation (LMP). An algorithm has LMP property if $\alpha C_f(F) + C_c(F) \leq \alpha C(F^*)$ where F is the solution obtained by the algorithm and F^* is the optimal solution. They show that any ρ-approximation algorithm with LMP property for the FLP yields a 2ρ-approximation algorithm for the corresponding k-median problem. With this observation, they obtain a 6-approximation algorithm for the k-median problem from the afore-mentioned 3-approximation for the FLP. Following this line of research, Jain et al. [7] improve the approximation ratio for the FLP to 2 with LMP property, and hence also leading to improved ratio for the k-median problem. Li and Svensson [10] show that any ρ-approximation algorithm with LMP property for the FLP yields a $\left(\frac{1+\sqrt{3}+\epsilon}{2}\right) \rho$-approximation algorithm for the corresponding k-median problem which improves the corresponding result of [9]. One important contribution of the work in [10] is the introduction of the concept of *pseudo-solution*, where slightly more than k facilities can be opened. Their algorithm first constructs a pseudo-solution, which is then converted to a feasible solution without suffering too much on the approximation ratio.

Charikar et al. [3] introduce the k-median problem with (linear) penalties and propose a primal-dual based 4-approximation algorithm. This ratio is improved to $3 + \varepsilon$ by Hajiaghayi et al. [6] using local search approach. Charikar et al. [3] also introduce penalties into FLP. There are two versions of the problem studied in the literature, the linear penalty FLP where the penalty cost is linear with respect to the penalty client set, and the submodular penalty FLP where the penalty cost is a submodular function with respect to penalty client set. The currently best approximation ratios for the linear and submodular versions are 1.5148 and 2, respectively ([11]). Wang et al. [13] introduce the k-facility location problem with (linear) penalties and propose a local search $2 + 1/p + \sqrt{3 + 2/p + 1/p^2} + \varepsilon$-approximation algorithm, where $p \in \mathbb{Z}_+$ is a parameter of the algorithm.

1.2   Preliminaries

Formally, in one instance $\mathcal{I} = (\mathcal{F}, \mathcal{C}, d, p, k)$ of the k-median problem with (linear) penalties, we are given a set of facility $\mathcal{F}$ and a set of client $\mathcal{C}$. Each pair of nodes $i, j \in \mathcal{F} \cup \mathcal{C}$ has a connection cost $d(i, j)$ which is assumed to be metric. Each client $j \in \mathcal{C}$ has a penalty cost p_j. In this paper, we consider the case where all penalty costs are uniform, that is, p_j's for all client $j \in D$ are equal. This version is called the k-median problem with uniform penalties. The goal is to open at most k facilities in $\mathcal{F}$, choose some clients to be penalized, and connect each non-penalized client to exactly one open facility such that the total cost including the connection cost and the penalty cost is minimized.

For convenience, we denote $d(i, S)$ as the minimum connection cost from $i \in \mathcal{F} \cup \mathcal{C}$ to $S \subseteq \mathcal{F} \cup \mathcal{C}$, i.e., $d(i, S) = \min_{i' \in S} d(i, i')$. The k-median problem with uniform penalties can be formulated as an integer program. Below, the $0-1$ variable y_i denotes whether the facility i is opened or not; the $0-1$ variable x_{ij} denotes whether the client j is connected to facility i or not; and the $0-1$ variable z_j denotes whether the client j is penalized or not.

$$\max \sum_{i \in \mathcal{F}, j \in \mathcal{C}} d(i, j) x_{ij} + \sum_{j \in \mathcal{C}} p_j z_j$$

$$\text{s. t.} \sum_{i \in \mathcal{F}} x_{ij} + z_j \geq 1, \qquad \forall j \in \mathcal{C},$$

$$x_{ij} \leq y_i, \qquad \forall i \in \mathcal{F}, j \in \mathcal{C}, \qquad (1)$$

$$\sum_{i \in \mathcal{F}} y_i \leq k,$$

$$x_{ij}, y_i \in \{0, 1\}, \qquad \forall i \in \mathcal{F}, j \in \mathcal{C}.$$

In this paper, we propose an $(1 + \sqrt{3} + \epsilon)$-approximation algorithm for the k-median problem with uniform penalties, where ϵ is a given positive constant. Following the framework of [10], we define a *spare instance* as follows. A facility

$i \in \mathcal{F}$ in the instance $\mathcal{I} = (\mathcal{F}, \mathcal{C}, d, p, k)$ is A-*sparse* if

$$\sum_{j \in C_{\mathcal{I}}(i, \frac{1}{3}d(i, F^*))} \min \left\{ p_j, \frac{2}{3}d(i, F^*) \right\} \leq A,$$

where $C_{\mathcal{I}}(i, d)$ is the client set $\{j \in \mathcal{C} : d(i, j) < d\}$ for a facility i and real number d, and F^* is the optimal solution for the instance $\mathcal{I}$; otherwise, we call this facility i A-*dense*. An instance $\mathcal{I} = (\mathcal{F}, \mathcal{C}, d, p, k)$ is A-sparse if all facilities in $\mathcal{F}$ are all A-sparse.

In the remaining of the paper, we present a three-step algorithm and the corresponding analysis in Sect. 2. In Sect. 3, we discuss future directions of research.

2   Our Algorithm and Analysis

The main algorithm consists of three high-level steps.

Step 1 Convert the input instance into a $\frac{C_c(F^*)}{t}$-sparse instance for a given constant t. Note that $C_c(F^*)$ is the optimal cost of the k-median problem with uniform penalties.

Step 2 Construct a pseudo-solution from a fractional solution, a.k.a. bi-point solution, for the sparse instance. In the pseudo-solution, the number of opened facilities is $k + c$, where c is a constant.

Step 3 Construct a feasible integer solution from the pseudo-solution.

The following parameters can be used to adjust the algorithm.

– ϵ is a arbitrary constant.
– $\delta \in (0, 1/8)$ is a scalar.
– c is the constant that is the additive number of the open facilities in the pseudo-solution.
– $t > \max\{\frac{24c(1+\sqrt{3})}{\epsilon\delta}, \frac{6c}{\delta}\}$.

2.1   Step 1: Obtain a Sparse Instance

For an instance $\mathcal{I}$ and a constant t, the following process generates a $\frac{C_c(F^*)}{t}$-sparse instance.

Algorithm 1

Step 1.1 Initially, $l := 1$ and $\mathcal{F}' := \mathcal{F}$.

Step 2.2 Consider all possible series of facility-pairs with l facility-pairs as $(i_1, \bar{i}_1), (i_2, \bar{i}_2), \cdots, (i_l, \bar{i}_l)$. Set $q_{i_u} := d(i_u, \bar{i}_u)$; Update $\mathcal{F}' := \mathcal{F}' \setminus \cup_{u=1}^{l} F_{\mathcal{I}}(i_u, q_{i_u})$ to obtain a new instance $\mathcal{I}' = (\mathcal{F}', \mathcal{C}, d, p, k)$, where $F_{\mathcal{I}}(i, d) = \{i' \in \mathcal{F} : d(i, i') < d\}$ for a given facility i and real number d which is an open set.

Step 2.3 If $l < t$, update $l := l + 1$ and goto Step 2; otherwise, output all facility-pairs.

In fact, we use the cost of the optimal solution F^*, i.e., $C_c(F^*)$, which can be obtained by guessing the value with step length $1 + \epsilon$. Note that we can guess the cost of $C_c(F^*)$ in polynomial time although Algorithm 1 outputs $2^{O(t)}$ series of facility-pairs. We will show that one of the above series corresponds to a $\frac{C_c(F^*)}{t}$-sparse instance. Let us consider the following conceptional algorithm.

Algorithm 2

Step 1 Initially, set $l := 0$ and $\mathcal{F}' = \mathcal{F}$.
Step 2 Find a $\frac{C_c(F^*)}{t}$-dense facility i in F'. Update $l := l + 1$ and set $i_l = i$.
Find the closest facility in the optimal solution of $\mathcal{I}$ for i_l, and denote the facility as $\bar{i}_l$. Then, we obtain a facility-pair $(i_l, \bar{i}_l)$.
Step 3 Update $\mathcal{F}' := \mathcal{F} \backslash F_{\mathcal{I}}(i_l, d(i_l, \bar{i}_l))$.
Step 4 Repeat Steps 2 and 3 until there is no dense facility in $\mathcal{F}'$.

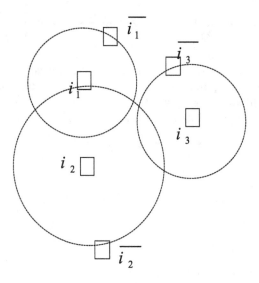

Fig. 1. Construct a sparse instance

In Algorithm 2, we obtain a series of facility-pairs $(i_1, \bar{i}_1)$, $(i_2, \bar{i}_2), \cdots , (i_l, \bar{i}_l)$ with length l (c.f. Fig. 1). Obviously, the instance $(\mathcal{F}', \mathcal{C}, d, p, k)$ given by the above procedure is sparse since we deleted all dense facilities. Moreover, all facilities in the optimal solution of $\mathcal{I}$ is still in $\mathcal{F}'$, because the set $F(i, d)$ is an open set. Thus, the optimal solution of $\mathcal{I}$ is the same as the optimal solution of $(\mathcal{F}', \mathcal{C}, d, p, k)$. Now we just need to show that the length l of the series of the facility-pair is less than t; that is, the series $(i_1, \bar{i}_1)$, $(i_2, \bar{i}_2), \cdots , (i_l, \bar{i}_l)$ is one of the outputs for Algorithm 1.

Lemma 1. *For the given constant t, we have $l \leq t$.*

Sketch of Proof: We proceed in two steps: (i) the clients in the balls of $C_{\mathcal{I}}(i_s, d(i_s, \bar{i}_s))$ are disjoint; and (ii) the optimal cost is no less than the total cost of the penalty cost or the connection cost of the clients in these balls.

Theorem 1. *By Algorithm 1, we can obtain a $\frac{C_c(F^*)}{t}$-sparse instance $\mathcal{I}'$ for an instance $\mathcal{I}$. And also, the optimal solution of the new instance $\mathcal{I}'$ is the same as that of $\mathcal{I}$.*

2.2   Step 2: Obtain a Pseudo-Solution

In this section, we show how to obtain a pseudo-solution from a fractional solution for the sparse instance. The linear programming relaxation of the k-median problem with uniform penalties (1) is as follows.

$$\max \sum_{i \in \mathcal{F}, j \in \mathcal{C}} d(i,j) x_{ij} + \sum_{j \in \mathcal{C}} p_j z_j$$

$$\text{s. t.} \sum_{i \in \mathcal{F}} x_{ij} + z_j \geq 1, \qquad \forall j \in \mathcal{C},$$

$$x_{ij} \leq y_i, \qquad \forall i \in \mathcal{F}, j \in \mathcal{C}, \qquad (2)$$

$$\sum_{i \in \mathcal{F}} y_i \leq k,$$

$$x_{ij}, y_i \geq 0, \qquad \forall i \in \mathcal{F}, j \in \mathcal{C}.$$

Any solution of program (2) is a fractional feasible solution. For each two opened facility sets F_1 and F_2 with $|F_1| \leq k < |F_2|$, we construct a fractional feasible solution, so-called *bi-point solution*, for the k-median problem with uniform penalties. That is, let $a, b > 0$ be two real numbers such that $a + b = 1$ and $a|F_1| + b|F_2| = k$. A bi-point solution is constructed as:

- $x_{ij} = aI_{j \in S_1(i)} + bI_{j \in S_1(i)}$,
- $y_i = aI_{i \in F_1} + bI_{i \in F_2}$,
- $z_j = aI_{p_j < d(j, F_1)} + bI_{p_j < d(j, F_2)}$,

where $I_{(.)}$ is an indicator function, $S_1(i)$ is the client set, each of which is connected to i in the solution F_1, and $S_2(i)$ is the client set, each of which is connected to i in the solution F_2. We view the penalty as the connection cost; that is, $C_c(F) = \sum_{j \in \mathcal{C}} \min\{p_j, d(j, F)\}$. Therefore, the cost of a bi-point solution for the k-median problem with uniform penalties is $aC_c(F_1) + bC_c(F_2)$.

It is easy to obtain the following theorem by combining the factor-revealing LP in [7].

Theorem 2. *There is a 2-approximation algorithm with LMP property for the facility location problem with uniform penalties.*

This is the only place in the entire proof which uses the assumption that the penalty costs are all equal.

Applying the algorithm by Jain et al. [7] to the facility location problem with uniform penalties to obtain two opened sets F_1 and F_2, where $|F_1| \leq k < |F_2|$, then, Theorem 2 implies that the cost of the bi-point solution $aC_c(F_1) + bC_c(F_2)$ is no more than 2 times the optimal cost for the k-median problem with uniform penalties.

The following algorithm outputs a pseudo-solution with constant additive facilities from the bi-point solution. We present the algorithms based on three cases depending on the parameter a; that is,

$$a \in \left(0, \frac{\sqrt{3} - 1}{4} \right],$$

$$a \in \left(\frac{\sqrt{3} - 1}{4}, \frac{2}{1 + \sqrt{3}} \right],$$

$$a \in \left(\frac{2}{1 + \sqrt{3}}, 1 \right].$$

Algorithm 3

Case 2.1 If $a \in \left(0, \frac{\sqrt{3}-1}{4} \right]$, output the solution using the algorithm in Li and Svensson [10].

Case 2.2 If $a \in \left(\frac{\sqrt{3}-1}{4}, \frac{2}{1+\sqrt{3}} \right]$, output the cheaper solution among F_1 and the one generated by the algorithm in Li and Svensson [10] in this case.

Case 2.3 If $a \in \left(\frac{2}{1+\sqrt{3}}, 1 \right]$, output F_1 since $|F_1| \leq k$.

Li and Svensson [10] show that

Lemma 2. *There are algorithms to obtain a pseudo-solution for the k-median problem with a factor of $\frac{\sqrt{3}+1+\epsilon}{2}$ for $a \in \left(0, \frac{\sqrt{3}-1}{4} \right]$ and $a \in \left(\frac{\sqrt{3}-1}{4}, \frac{2}{1+\sqrt{3}} \right]$, respectively. Moreover, the additive number of the pseudo-solution for both cases are a constant.*

For our problem, we can obtain the same lemma as follows.

Lemma 3. *The cost of the solution for both $a \in \left(0, \frac{\sqrt{3}-1}{4} \right]$ and $a \in \left(\frac{\sqrt{3}-1}{4}, \frac{2}{1+\sqrt{3}} \right]$ is no more than $\frac{\sqrt{3}+1+\epsilon/2}{2}(aC_c(F_1) + bC_c(F_2))$ for any $\epsilon > 0$.*

Sketch of the Proof: It is easy to prove the first case. For the second, if the client is penalized in the pseudo-solution, the factor will be 1. If the client is connected to an opened facility, we show that the upper bound of connection cost also holds due to linear penalty cost.

With respect to the case $a \in \left(\frac{2}{1+\sqrt{3}}, 1\right]$, we have that F_1 is the feasible solution with cost

$$C_c(F_1) \leq \frac{1}{a} a C_c(F_1) \leq \frac{1}{a} (a C_c(F_1) + b C_c(F_2))$$

$$\leq \frac{\sqrt{3}+1}{2} (a C_c(F_1) + b C_c(F_2)).$$

Assume that the additive constant number of the opened facility in Algorithm 3 is c. Lemma 3 and Theorem 2 together imply that

Theorem 3. *The cost $C_c(T)$ of pseudo-solution T by Algorithm 3 is no more than $(1 + \sqrt{3} + \frac{\epsilon}{2}) C_c(F^*)$.*

2.3 Step 3: Obtain a Feasible Solution

In Sect. 2.2, the solution obtained, denoted as T, is infeasible since there are $k+c$ opened facilities for the $\frac{C_c(F^*)}{t}$-sparse instance $(\mathcal{F}, \mathcal{C}, d, p, k)$. We now convert T into a feasible solution where exactly k facilities are opened.

Algorithm 4. *Set $T' = T$ as the initial solution. Let δ be a constant such that $\frac{1+3\delta}{1-3\delta} \leq 1 + \sqrt{3}$, $t > \frac{6c(2+\sqrt{3})}{\epsilon\delta}$, and $B = 6\frac{C_c(F^*)+C_c(T)}{\delta t}$.*

Step 3.1 If $|T'| > k$, find a facility $i \in T'$ such that $C_c(T' \setminus \{i\}) \leq C_c(T') + B$.
Step 3.2 Set $T' = T' \setminus \{i\}$.
Step 3.3 Repeat Steps 3.1 and 3.2 until $|T'| \leq k$ or no such facility exists.
Step 3.4 If $|T'| \leq k$, output $F = T'$ as the feasible solution and stop; otherwise, go to Step 3.5.
Step 3.5 Choose any possible $D \subseteq T'$ and $V \subseteq \mathcal{F}$ such that $|D| + |V| = k$ and $|V| < t$. For each $i \in D$, find

$$\bar{s}_i := \arg \min_{s_i \in F_{\mathcal{I}}(i,\delta Q_i)} \sum_{j \in C_{\mathcal{I}}(i,\frac{1}{3}Q_i)} \min\{d(s_i, j), d(j, D), p_j\},$$

where $Q_i = d(i, T' \setminus \{i\})$ is the connection cost from i to the closest facility in T' except itself. Set $F_{D,V} = V \cup \{\bar{s}_i : i \in D\}$.
Step 3.6 Output $F = \arg\min_{D,V} C_c(F_{D,V})$.

In Algorithm 4, the number of combinations of D and V is $C_n^t C_n^{k-t}$, implying the running time of Algorithm 4 is polynomial in terms of the inputs.

We give one choice of D and V obtained from Algorithm 4. The cost of this choice serves as an upper bound. A facility i is determined if $q_i \leq \delta Q_i$. Note that $q_i = d(i, F^*)$, $Q_i = d(i, T' \setminus \{i\})$, and δ is a constant used in Algorithm 4; otherwise, the facility is undetermined.

- Set $D_0 := \{i \in T' : q_i \leq \delta Q_i\}$ which contains all determined facilities in T'.
- For each $i \in D_0$, denote $f_i^* := \arg\min_{i' \in F^*} d(i, i')$ as the closest facility in the optimal solution from i.

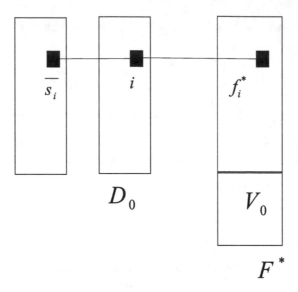

Fig. 2. Construct a feasible solution

- Set $V_0 := F^* \setminus \{f_i^* : i \in D_0\}$.

Subsequently, we show that D_0 and V_0 (c.f. Fig. 2) is one choice among all the $C_n^t C_n^{k-t}$ combinations in Algorithm 4.

Lemma 4. $|D_0| + |V_0| = k$ and $|V_0| < t$.

Sketch of Proof: We first prove that $f_i^* \neq f_{\bar{i}}^*$ for each pair $i \neq \bar{i} \in D_0$, implying $|D_0| + |V_0| = k$.

Next, let U_0 be the set including all undetermined facilities in T'. Note that $|D_0| = |T'| - |U_0|$ and $|V_0| = k - |D_0|$ together imply that

$$|V_0| = k - |T'| + |U_0| \leq |U_0|,$$

where the last inequality holds since T' is a pseudo-solution. It suffices to prove that $|U_0| < t$, which can be done by contradiction. Assume that $|U_0| \geq t$, we can find a facility i in T' such that $C_c(T' \setminus \{i\}) \leq C_c(T') + B$, which is a contradiction.

Finally, the cost of F_{D_0,V_0} is an upper bound for the cost obtained by the algorithm.

Lemma 5.
$$C_c(F_{D_0,V_0}) \leq \frac{1 + 3\delta}{1 - 3\delta} C_c(F^*).$$

Sketch of Proof: We consider the penalty cost or connection cost for the client depending on whether the client is penalized or not. If the client is penalized in the optimal solution, the cost of the client is always no more than the penalty cost.

If the client is not penalized in the optimal solution, we consider whether the client is in some ball $C_{\mathcal{I}}(i, Q_i/3)$ or not. If it is in some ball, we show that it is either connected to $\bar{s}_i$, or V_0, or penalized. Thus, the cost for these clients is no more than that in the optimal solution. In the case that a client does not belong to any ball, the cost will be no more than the that in the optimal solution if j is connected to V_0 in the optimal solution. If the client is connected to some f_i^* in the optimal solution, we estimate the cost from the client to $\bar{s}_i$.

Now, we are ready to estimate the cost of the feasible solution by Algorithms 1, 3, and 4.

Theorem 4. *The cost of the solution by Algorithm 4 is no more than*

$$\min \left\{ C_c(T) + cB, \frac{1 + 3\delta}{1 - 3\delta} C_c(F^*) \right\}.$$

Proof. If the output is obtained by Step 3.4, the cost will be no more than $C_c(T) + cB$ since there are at most c additive facilities. Thus, the result holds, combining with Lemma 5.

Theorem 5. *There is a $(1 + \sqrt{3} + \epsilon)$-approximation algorithm for the k-median problem with uniform penalties.*

Proof. On the one hand, the cost $C_c(T)$ of the pseudo-solution is no more than $(1 + \sqrt{3} + \epsilon)C_c(F^*)$ by Theorem 3. By the value of $B = 6\frac{C_c(F^*) + C_c(T)}{\delta t}$, we have

$$C_c(T) + cB = C_c(T) + c \cdot 6\frac{C_c(F^*) + C_c(T)}{t\delta}$$

$$\leq (1 + \sqrt{3} + \epsilon)C_c(F^*) + c \cdot 6\frac{C_c(F^*) + (1 + \sqrt{3} + \frac{\epsilon}{2})C_c(F^*)}{t\delta}$$

$$\leq (1 + \sqrt{3} + \frac{\epsilon}{2})C_c(F^*) + c \cdot \frac{12(1 + \sqrt{3} + \frac{\epsilon}{2})}{t\delta}C_c(F^*). \tag{3}$$

Since $t > \frac{24c(1+\sqrt{3})}{\epsilon\delta}$, we have

$$C_c(T) + cB \leq (1 + \sqrt{3} + \epsilon)C_c(F^*).$$

On the other hand, we can choose an appropriate $\delta \in (0, \frac{1}{8})$ such that $\frac{1+3\delta}{1-3\delta} \leq 1 + \sqrt{3}$, which is achievable because $\frac{1+3\delta}{1-3\delta}$ is increasing with δ. Therefore, the approximation ratio of the algorithm is $1 + \sqrt{3} + \epsilon$ for any given $\epsilon > 0$.

3   Conclusion Remarks

The work of [10] inspires us in applying the concept of pseudo-solution for a sparse instance to design our algorithm. We exploit the combinatorial structures of the sparse instance by enumerating. Combining with the algorithm of Jain et al. [7], we propose a $(1 + \sqrt{3} + \epsilon)$-approximation algorithm for the k-median problem with uniform penalties.

There are also several directions for future research.

- We believe that the idea in this paper can be extended into the generalized linear penalty, even submodular penalty version of the k-median problem.
- It will be interesting to develop an algorithm for the k-facility location problem, which adds opening cost to the k-median problem. The currently best approximation ratio is $3 + \epsilon$ for any $\epsilon > 0$ via local search scheme. The ideas in this paper may improve the approximation ratio.

Acknowledgements. The first author was supported by Natural Science Foundation of China (No. 11501412). The second author was supported by Natural Sciences and Engineering Research Council of Canada (NSERC) grant 06446 and by Natural Science Foundation of China (No. 10971187). The third author was supported by Natural Science Foundation of China (No. 11531014) and Collaborative Innovation Center on Beijing Society-Building and Social Governance.

References

1. Arya, V., Garg, N., Khandekar, R., Meyerson, A., Munagala, K., Pandit, V.: Local search heuristics for k-median and facility location problems. SIAM J. Comput. **33**, 544–562 (2004)
2. Charikar, M., Guha, S., Tardos, E., Shmoys, D.B.: A constant-factor approximation algorithm for the k-median problem. J. Comput. Sýst. Sci. **65**, 129–149 (2002)
3. Charikar M., Khuller S., Mount, D.M., Narasimhan, G.: Algorithms for facility location problems with outliers. In: Proceedings of SODA, pp. 642–651 (2001)
4. Chudak, F.A., Shmoys, D.B.: Improved approximation algorithms for the uncapacitated facility location problem. SIAM J. Comput. **33**, 1–25 (2003)
5. Guha, S., Khuller, S.: Greedy strikes back: improved facility location algorithms. In: Proceedings of SODA, pp. 649–657 (1998)
6. Hajiaghayi, M., Khandekar, R., Kortsarz, G.: Local search algorithms for the red-blue median problem. Algorithmica **63**, 795–814 (2012)
7. Jain, K., Mahdian, M., Markakis, E., Saberi, A., Vazirani, V.V.: Greedy facility location algorithms analyzed using dual fitting with factor-revealing LP. J. ACM **50**, 795–824 (2003)
8. Jain, K., Mahdian, M., Saberi, A.: A new greedy approach for facility location problems. In: Proceedings of STOC, pp. 731–740 (2002)
9. Jain, K., Vazirani, V.V.: Approximation algorithms for metric facility location and k-median problems using the primal-dual schema and lagrangian relaxation. J. ACM **48**, 274–296 (2001)
10. Li, S., Svensson, O.: Approximating k-median via pseudo-approximation. In: Proceedings of STOC, pp. 901–910 (2013)
11. Li, Y., Du, D., Xiu, N., Xu, D.: Improved approximation algorithms for the facility location problems with linear/submodular penalties. Algorithmica **73**, 460–482 (2015)
12. Lin, J.H., Vitter, J.S.: Approximation algorithms for geometric median problems. Inf. Process. Lett. **44**, 245–249 (1992)
13. Wang, Y., Xu, D., Du, D., Wu, C.: Local search algorithms for k-median and k-facility location problems with linear penalties. In: Proceedings of COCOA, pp. 60–71 (2015)
14. Shmoys, D.B., Tardos, E., Aardal, K.I.: Approximation algorithms for facility location problems. In: Proceedings of STOC, pp. 265–274 (1997)

On-Line Pattern Matching on Uncertain Sequences and Applications

Carl Barton[1], Chang Liu[2], and Solon P. Pissis[2(✉)]

[1] The Blizard Institute, Barts and The London School of Medicine and Dentistry,
Queen Mary University of London, London, UK
c.barton@qmul.ac.uk
[2] Department of Informatics, King's College London, London, UK
{chang.2.liu,solon.pissis}@kcl.ac.uk

Abstract. We study the fundamental problem of pattern matching in the case where the string data is *weighted*: for every position of the string and every letter of the alphabet a probability of occurrence for this letter at this position is given. Sequences of this type are commonly used to represent uncertain data. They are of particular interest in computational molecular biology as they can represent different kind of ambiguities in DNA sequences: distributions of SNPs in genomes populations; position frequency matrices of DNA binding profiles; or even sequencing-related uncertainties. A weighted string may thus represent many different strings, each with probability of occurrence equal to the product of probabilities of its letters at subsequent positions. In this article, we present new average-case results on pattern matching on weighted strings and show how they are applied effectively in several biological contexts. A free open-source implementation of our algorithms is made available.

1 Introduction

Uncertain sequences are common in many applications: (i) data measurements such as imprecise sensor measurements; (ii) flexible modelling of DNA sequences such as DNA binding profiles; (iii) when observations are private and thus sequences of observations may have artificial uncertainty introduced deliberately. For example, in computational molecular biology an uncertain sequence can be used to incorporate SNP distributions from a population of genomes into a reference sequence. This process can be realised by a IUPAC-encoded sequence [3,13] or by directly incorporating the results of SNP studies such as [11,19]. An uncertain sequence can also be used as a flexible model of DNA sequences such as DNA binding profiles, and is known as *position frequency matrix* [18].

As *pattern matching* is a core computational task in many real-world applications, we focus here on designing efficient on-line algorithms for pattern matching on uncertain sequences. On-line pattern matching algorithms process the text position-by-position, in the order that it is fed to the algorithm, without having the entire text at hand. Hence this type of algorithms is useful when one wishes

© Springer International Publishing AG 2016
T.-H.H. Chan et al. (Eds.): COCOA 2016, LNCS 10043, pp. 547–562, 2016.
DOI: 10.1007/978-3-319-48749-6_40

to query for one or a few patterns in potentially many texts without having to pre-compute and store an index over the texts.

We start with a few definitions to explain our results. An *alphabet* Σ is a finite non-empty set of size σ, whose elements are called *letters*. A *string* on an alphabet Σ is a finite, possibly empty, sequence of elements of Σ. The zero-letter sequence is called the *empty string*, and is denoted by ε. The *length* of a string x is defined as the length of the sequence associated with the string x, and is denoted by $|x|$. We denote by $x[i]$, for all $0 \leq i < |x|$, the letter at index i of x. Each index i, for all $0 \leq i < |x|$, is a position in x when $x \neq \varepsilon$. The i-th letter of x is the letter at position $i - 1$ in x. We refer to any string $x \in \Sigma^q$ as a *q-gram*.

The *concatenation* of two strings x and y is the string of the letters of x followed by the letters of y; it is denoted by xy. A string x is a *factor* of a string y if there exist two strings u and v, such that $y = uxv$. Consider the strings x, y, u, and v, such that $y = uxv$, if $u = \varepsilon$ then x is a *prefix* of y, if $v = \varepsilon$ then x is a *suffix* of y. Let x be a non-empty string and y be a string, we say that there exists an *occurrence* of x in y, or more simply, that x *occurs in* y, when x is a factor of y. Every occurrence of x can be characterised by a position in y; thus we say that x occurs at the *starting position* i in y when $y[i \,..\, i + |x| - 1] = x$.

A *weighted string* x of length n on an alphabet Σ is a finite sequence of n sets. Every $x[i]$, for all $0 \leq i < n$, is a set of ordered pairs $(s_j, \pi_i(s_j))$, where $s_j \in \Sigma$ and $\pi_i(s_j)$ is the probability of having letter s_j at position i. Formally, $x[i] = \{(s_j, \pi_i(s_j)) | s_j \neq s_\ell \text{ for } j \neq \ell, \text{ and } \sum_j \pi_i(s_j) = 1\}$. A letter s_j *occurs* at position i of a weighted string x if and only if the *occurrence probability* of letter s_j at position i, $\pi_i(s_j)$, is greater than 0. A string u of length m is a *factor* of a weighted string if and only if it occurs at starting position i with *cumulative occurrence probability* $\prod_{j=0}^{m-1} \pi_{i+j}(u[j]) > 0$. Given a *cumulative weight threshold* $1/z \in (0, 1]$, we say that factor u is *valid*, or equivalently that factor u has a valid occurrence, if it occurs at starting position i and $\prod_{j=0}^{m-1} \pi_{i+j}(u[j]) \geq 1/z$. Similarly, we say that letter s_j at position i is valid if $\pi_i(s_j) \geq 1/z$. For succinctness of presentation, if $\pi_i(s_j) = 1$ the set of pairs is denoted only by the letter s_j; otherwise it is denoted by $[(s_{j_1}, \pi_i(s_{j_1})), \ldots, (s_{j_k}, \pi_i(s_{j_k}))]$.

Suppose we are given a cumulative weight threshold $1/z$. Given a (weighted) string u and a weighted string v, both of length m, we say that u and v *match*, denoted by $u =_z v$, if there exists a (valid) factor of u of length m that is also a valid factor of v of length m. Given a string u and a weighted string v, both of length m, and a non-negative integer $k < m$, we say that u and v *match with k-mismatches*, denoted by $u =_{z,k} v$, if when at most k letters in u were replaced to create a new string u' then $u' =_z v$. We consider the following three problems.

WEIGHTEDTEXTMATCHING (WTM)
Input: a string x of length m, a weighted string y of length $n > m$, and a cumulative weight threshold $1/z \in (0, 1]$
Output: all positions i of y such that $x =_z y[i \,..\, i + m - 1]$

> GENERALWEIGHTEDPATTERNMATCHING (GWPM)
> **Input:** a weighted string x of length m, a weighted string y of length $n > m$, and a cumulative weight threshold $1/z \in (0, 1]$
> **Output:** all positions i of y such that $x =_z y[i .. i + m - 1]$

> APPROXWEIGHTEDTEXTMATCHING (AWTM)
> **Input:** a string x of length m, a weighted string y of length $n > m$, a non-negative integer $k < m$, and a cumulative weight threshold $1/z \in (0, 1]$
> **Output:** all positions i of y such that $x =_{z,k} y[i .. i + m - 1]$

Our computational model. We assume word-RAM model with word size $w = \Omega(\log(nz))$. We consider the log-probability model of representations of weighted strings in which probability operations can be realised exactly in $\mathcal{O}(1)$ time. We assume that $\sigma = \mathcal{O}(1)$ since the most commonly studied alphabet is $\{A, C, G, T\}$. In this case a weighted string of length n has a representation of size $\mathcal{O}(n)$. A position on a weighted string is viewed as a non-empty subset of the alphabet such that each letter of this subset has probability of occurrence greater than 0. For the analysis, we assume all possible non-empty subsets of the alphabet are independent and identically distributed random variables uniformly distributed.

Related results. Problem WEIGHTEDTEXTMATCHING can be solved in time $\mathcal{O}(n \log z)$ [14]. Moreover, in [14], the authors showed that their solution to WEIGHTEDTEXTMATCHING can be applied to the well-known *profile-matching* problem [17]. Problem GENERALWEIGHTEDPATTERNMATCHING can be solved in time $\mathcal{O}(zn)$ [4]. Problem APPROXWEIGHTEDTEXTMATCHING can be solved in time $\mathcal{O}(n\sqrt{m \log m})$ using FFTs [2]. All these results are worst-case complexities. Problem WEIGHTEDTEXTMATCHING can be solved in average-case search time $o(n)$ for *weight ratio* $\frac{z}{m} < \min\{\frac{1}{\log z}, \frac{\log \sigma}{\log z(\log m + \log \log \sigma)}\}$ [5].

Our contribution. We provide efficient on-line algorithms for solving these problems and provide their *average-case* analysis, obtaining the following results. Note that preprocessing resources below denote worst-case complexities, $0 < c < 1/2$ is an absolute constant, $v = \frac{2^\sigma - 1}{2^\sigma - 1}$, $d = 1 + (1 - c) \log_v(1 - c) + c \log_v c$, and $a = 4\sqrt{c(1-c)}$.

Problem	Preprocessing space	Preprocessing time	Search time	Conditions
WTM	$\mathcal{O}(m)$	$\mathcal{O}(m)$	$\mathcal{O}(\frac{nz \log m}{m})$	
GWPM	$\mathcal{O}(zm)$	$\mathcal{O}(zm)$	$\mathcal{O}(\frac{nz \log m}{m})$	
AWTM	$\mathcal{O}(\sigma^q)$	$\mathcal{O}(mq\sigma^q)$	$\mathcal{O}(\frac{nz(\log m + k)}{m})$	$q \geq \frac{3 \log_v m - \log_v a}{d}$, $\frac{k}{m} \leq c - \frac{2cq}{m}$

We also provide extensive experimental results, using both real and synthetic data: (i) we show that our implementations outperform state-of-the-art approaches by more than one order of magnitude; (ii) furthermore, we demonstrate the suitability of the proposed algorithms in a number of real biological contexts.

2   Tools for Standard and Weighted Strings

Suffix trees are used as computational tools; for an introduction see [9]. The *suffix tree* of a non-empty standard string y is denoted by $\mathcal{T}(y)$.

Fact 1 [10]. *Given a non-empty string y of length n, $\mathcal{T}(y)$ can be constructed in time and space $\mathcal{O}(n)$. Checking whether a string x of length m occurs in y can be performed in time $\mathcal{O}(m)$ using $\mathcal{T}(y)$.*

We next define some primitive operations on weighted strings. Suppose we are given a cumulative weight threshold $1/z$. Let u be a string of length m and v be a weighted string of length m. We define operation VER1(u, v, z): it returns *true* if $u =_z v$ and *false* otherwise. Let u be a weighted string of length m and v be a weighted string of length m. We define operation VER2(u, v, z): it returns *true* if $u =_z v$ and *false* otherwise. For a weighted string v, by $\mathcal{H}(v)$ we denote a string obtained from v by choosing at each position the heaviest letter, that is, the letter with the maximum probability (breaking ties arbitrarily). We call $\mathcal{H}(v)$ the *heavy string* of v. Let u be a string of length m and v be a weighted string of length m. Given a non-negative integer $k < m$, we can check whether $u =_{z,k} v$ using Function VER3. An implementation of VER3 is provided below. Intuitively, we replace at most k letters of u with the heaviest letter in v based on how much these letter replacements contribute to the cumulative probability.

Function *VER3(u, v, z, k)*

   $v' \leftarrow \mathcal{H}(v)$;

   $A \leftarrow$ EMPTYLIST();

   foreach i *such that* $u[i] \neq v'[i]$ **do**

      if $\pi_i(u[i]) = 0$ **then**

         $u[i] \leftarrow v'[i]$;

         $k \leftarrow k - 1$;

         if $k < 0$ **then**

            return *false*;

      else

         $\alpha \leftarrow \pi_i(v'[i])/\pi_i(u[i])$;

         $A \leftarrow$ INSERT($< i, \alpha >$);

   Find the kth largest element in A with respect to α;

   Add the k largest elements of A with respect to α to set A_k;

   foreach $< i, \alpha > \in A_k$ **do**

      $u[i] \leftarrow v'[i]$;

   if $\prod_{j=0}^{m-1} \pi_j(u[j]) \geq 1/z$ **then**

      return *true*;

   return *false*;

Lemma 1. *VER1, VER2, and VER3 can be implemented to work in time $\mathcal{O}(m)$, $\mathcal{O}(mz)$, and $\mathcal{O}(m)$, respectively.*

Proof. For VER1 we can check whether $u =_z v$ in time $\mathcal{O}(m)$ by checking $\prod_{j=0}^{m-1} \pi_j(u[j]) \geq 1/z$. For VER2 we can check whether $u =_z v$ in time $\mathcal{O}(mz)$ using the algorithm of [4].

Let us denote by $E = \{e_1, \ldots, e_{|E|}\}$, $|E| \leq k$, the set of positions of the input string u, which we replace in VER3 by the heaviest letter of v. We denote the resulting string by u'. Towards contradiction, assume we can guess a set $F = \{f_1, \ldots, f_{|F|}\}$, $|F| \leq k$, of positions over u resulting in another string u'' such that $P_{u''} = \prod_{j=0}^{m-1} \pi_j(u''[j]) > P_{u'} = \prod_{j=0}^{m-1} \pi_j(u'[j]) \geq P_u = \prod_{j=0}^{m-1} \pi_j(u[j])$. It must hold that

$$\frac{P_{u''}}{P_u} = \frac{\pi_{f_1}(u''[f_1]) \ldots \pi_{f_{|F|}}(u''[f_{|F|}])}{\pi_{f_1}(u[f_1]) \ldots \pi_{f_{|F|}}(u[f_{|F|}])} > \frac{P_{u'}}{P_u} = \frac{\pi_{e_1}(u'[e_1]) \ldots \pi_{e_{|E|}}(u'[e_{|E|}])}{\pi_{e_1}(u[e_1]) \ldots \pi_{e_{|E|}}(u[e_{|E|}])}.$$

In case $E = F$, this implies that there exist letters heavier than the corresponding heaviest letters, a contradiction. In case $E \neq F$, given that there exists no letter heavier than the heaviest letter at each position, this implies that there exists an element $< i, \alpha >$, $i \in F$, in list A which is the rth largest, $r \leq k$, with respect to α, and it is larger than the rth element picked by VER3, a contradiction. All operations in VER3 can be trivially done in time $\mathcal{O}(m)$ except for finding the kth largest element in a list of size $\mathcal{O}(m)$, which can be done in time $\mathcal{O}(m)$ using the *introselect* algorithm [16]. □

We say that u is a *(right-)maximal factor* of a weighted string x at position i if u is a valid factor of x starting at position i and no string $u' = u\alpha$, for $\alpha \in \Sigma$, is a valid factor of x at this position.

Fact 2 [1]. *A weighted string has at most z different maximal factors starting at a given position.*

3 Algorithms

Let us start with a few auxiliary definitions. An *indeterminate string* x of length m on an alphabet Σ is a finite sequence of m sets, such that $x[i] \subseteq \Sigma$, $x[i] \neq \emptyset$, for all $0 \leq i < m$. Naturally, we refer to any indeterminate string of length q as *indeterminate q-gram*. We say that two indeterminate strings x and y *match*, denoted by $x \approx y$, if $|x| = |y|$ and for each $i = 0, \ldots, |x|-1$, we have $x[i] \cap y[i] \neq \emptyset$. Intuitively, we view a weighted string as an indeterminate string in order to conduct the average-case analysis of the algorithms according to our model.

3.1 Weighted Text Matching

In this section we present a remarkably simple and efficient algorithm to solve problem WEIGHTEDTEXTMATCHING. We start by providing a lemma on the probability that a random indeterminate q-gram and a standard q-gram match.

Lemma 2. *Let u be a standard q-gram and v be a uniformly random indeterminate q-gram. The probability that $u \approx v$ is no more than $(\frac{2^{\sigma-1}}{2^\sigma-1})^q$, which tends to $\frac{1}{2^q}$ as σ increases.*

Proof. There are less than 2^σ non-empty subsets of an alphabet of size σ. Let $a \in \Sigma$, then $2^{\sigma-1}$ of these subsets include a. Clearly then the probability that two positions of u and v match is no more than $\frac{2^{\sigma-1}}{2^\sigma-1}$. Therefore the probability of them matching at every position is no more than $(\frac{2^{\sigma-1}}{2^\sigma-1})^q$. $\square$

Algorithm $WTM(x, m, y, n, z, q)$

    Construct $T(x)$;

    $i \leftarrow 0$;

    while $i < n - m + 1$ **do**

        $j \leftarrow i + m - q$;

        Let $\mathcal{A}$ denote the set of all valid q-grams starting at position j in y;

        foreach $s \in \mathcal{A}$ **do**

            Check if s occurs in x using $T(x)$;

        if *no* $s \in \mathcal{A}$ *occurs in* x **or** $\mathcal{A} = \emptyset$ **then**

            $i \leftarrow j + 1$;

        else

            if $VER1(x, y[i..i+m-1], z) = true$ **then**

                output i;

            $i \leftarrow i + 1$;

Theorem 3. *Algorithm WTM solves problem* WEIGHTEDTEXTMATCHING *in average-case search time $\mathcal{O}(\frac{nz \log m}{m})$ if we set $q \geq 3 \log_{\frac{2^\sigma-1}{2^{\sigma-1}}} m$, which tends to $3 \log_2 m$ as σ increases. The worst-case preprocessing time and space is $\mathcal{O}(m)$.*

Proof. By Fact 1 the time and space required for constructing $T(x)$ is $\mathcal{O}(m)$. We consider a sliding window of size m of y and read q-grams backwards from the end of this window and check if they occur anywhere within x. By Fact 1 this check can be done in time $\mathcal{O}(q)$ per q-gram. If a q-gram occurs anywhere in x then we verify the entire window, otherwise we shift the window $m - q + 1$ positions to the right. Clearly none of the skipped positions can be the starting position of any occurrence of x as if this was the case, the q-gram must occur in x; so the algorithm is correct. Verifying (all starting positions of) the window takes time $\mathcal{O}(m^2)$ by Lemma 1 and the probability that a q-gram matches within a pattern of length $m > q$ is no more than $m(\frac{2^{\sigma-1}}{2^\sigma-1})^q$ by Lemma 2. We note that reading the q-grams takes time $\mathcal{O}(zq)$ per position by Fact 2, so to achieve the claimed runtime we must pick a value for q such that the expected cost per window is $\mathcal{O}(zq)$. This is achieved when $\mathcal{O}(m^3(\frac{2^{\sigma-1}}{2^\sigma-1})^q) = \mathcal{O}(zq)$. It is always true when $q \geq 3\log_{\frac{2^{\sigma-1}}{2^{\sigma-1}}} m$, which tends to $3\log_2 m$ as σ increases. There are $\mathcal{O}(\frac{n}{m})$ non-overlapping windows of length m and this proves the theorem. $\square$

3.2   General Weighted Pattern Matching

In this section we present an algorithm, denoted by GWPM, to solve problem GENERALWEIGHTEDPATTERNMATCHING. Algorithm GWPM largely follows algorithm WTM.

Lemma 3. *Let u and v be uniformly random indeterminate q-grams. The probability that $u \approx v$ is no more than $(1-(1-(\frac{2^{\sigma-1}}{2^{\sigma}-1})^2)^{\sigma})^q$, which tends to $(1-(\frac{3}{4})^{\sigma})^q$ as σ increases.*

Proof. There are less than 2^{σ} non-empty subsets of an alphabet of size σ. Let $a \in \Sigma$, then $2^{\sigma-1}$ of these subsets include a. Clearly then the probability that a occurs at both positions is no more than $(\frac{2^{\sigma-1}}{2^{\sigma}-1})^2$. It then follows that the probability that the two sets have a non-empty intersection is no more than $1-(1-(\frac{2^{\sigma-1}}{2^{\sigma}-1})^2)^{\sigma}$. Therefore the probability that all positions have a non-empty intersection is no more than $(1-(1-(\frac{2^{\sigma-1}}{2^{\sigma}-1})^2)^{\sigma})^q$. $\square$

Algorithm *GWPM*$(x, m, y, n, z, q, \Sigma)$

   Let $\mathcal{F} = \{x_1, \ldots, x_f\}$ denote the set of all valid factors of length m of x;

   if $\mathcal{F} = \emptyset$ **then**

      return;

   Construct string $X = x_1\$ \ldots \$x_f$, where $\$ \notin \Sigma$;

   Construct $\mathcal{T}(X)$;

   $i \leftarrow 0$;

   while $i < n - m + 1$ **do**

      $j \leftarrow i + m - q$;

      Let $\mathcal{A}$ denote the set of all valid q-grams starting at position j in y;

      foreach $s \in \mathcal{A}$ **do**

         Check if s occurs in X using $\mathcal{T}(X)$;

      if *no $s \in \mathcal{A}$ occurs in X* **or** $\mathcal{A} = \emptyset$ **then**

         $i \leftarrow j + 1$;

      else

         if *VER2*$(x, y[i \ldots i + m - 1], z) = true$ **then**

            output i;

         $i \leftarrow i + 1$;

Theorem 4. *Algorithm GWPM solves problem* GENERALWEIGHTEDPATTERN-MATCHING *in average-case search time $\mathcal{O}(\frac{nz\log m}{m})$ if we set $q \geq 3\log_u m$, where*

$u = \frac{(2^\sigma-1)^{2\sigma}}{(2^\sigma-1)^{2\sigma}-((2^\sigma-1)^2-2^{2\sigma-2})^\sigma}$, *which tends to* $3\log_{1/(1-(\frac{3}{4})^\sigma)} m$ *as* σ *increases. The worst-case preprocessing time and space is* $\mathcal{O}(zm)$.

Proof. Let $x_1, x_2, \ldots, x_f$ denote all f valid factors of length m of x. We construct string $X = x_1\$x_2\$\ldots\$x_f$, where $\$ \notin \Sigma$. By Facts 1 and 2 the time and space required for constructing $\mathcal{T}(X)$ is $\mathcal{O}(zm)$. Plugging Lemmas 1 and 3 to the proof of Theorem 3 yields the result. This is achieved when $\mathcal{O}(m^3 z(1-(1-(\frac{2^\sigma-1}{2^\sigma-1})^2)^\sigma)^q) = \mathcal{O}(zq)$. It is always true when $q \geq 3\log_u m$, where $u = \frac{(2^\sigma-1)^{2\sigma}}{(2^\sigma-1)^{2\sigma}-((2^\sigma-1)^2-2^{2\sigma-2})^\sigma}$, which tends to $3\log_{1/(1-(\frac{3}{4})^\sigma)} m$ as σ increases. $\square$

3.3 Approximate Weighted Text Matching

In this section we present an algorithm to solve problem APPROXWEIGHTED-TEXTMATCHING. The algorithm, denoted by AWTM, is split into two distinct stages: preprocessing the pattern x and searching the weighted text y.

Preprocessing. We build a q-gram index in a similar way as that proposed by Chang and Marr in [8]. Intuitively, we wish to determine the minimum possible Hamming distance (mismatches) between every q-gram on Σ and any q-gram of x. An index like this allows us to lower bound the Hamming distance between a window of y and x without computing the Hamming distance between them. To build this index, we generate every string of length q on Σ, and find the minimum Hamming distance between it and all factors of length q of x. This information can easily be stored by generating a numerical representation of the q-gram and storing the minimum Hamming distance in an array at this location. If we know the numerical representation, we can then look up any entry in constant time. This index has size $\mathcal{O}(\sigma^q)$ and can be trivially constructed in worst-case time $\mathcal{O}(mq\sigma^q)$ and space $\mathcal{O}(\sigma^q)$.

Lemma 4. *Let* u *be a standard* q-gram *and* v *be a uniformly random indeterminate* q-gram. *The probability that* u *and* v *match with* cq-mismatches *is at most* $(\frac{2^\sigma-1}{2^\sigma-1})^q(1-c)^{-(1-c)q}c^{-cq}\frac{1}{4\sqrt{c(c-1)}}$, *where* $0 < c < 1/2$.

Proof. Without loss of generality assume that cq is an integer. By Lemma 2, the probability the q-grams match with *exactly* i mismatches is

$$\binom{q}{i}\left(\frac{2^{\sigma-1}}{2^\sigma-1}\right)^q.$$

Therefore, by Lemma 3.8.2 in [15] on the sum of binomial coefficients, the probability that u and v match with cq-mismatches is

$$\sum_{i=0}^{cq}\binom{q}{i}\left(\frac{2^{\sigma-1}}{2^\sigma-1}\right)^q = \left(\frac{2^{\sigma-1}}{2^\sigma-1}\right)^q\sum_{i=0}^{cq}\binom{q}{i} \leq \left(\frac{2^{\sigma-1}}{2^\sigma-1}\right)^q(1-c)^{-(1-c)q}c^{-cq}\frac{1}{4\sqrt{c(c-1)}}.$$

This decreases exponentially in q when $(1-c)^{-(1-c)}c^{-c} > 0$ which holds for $0 < c < 1/2$. It tends to $2^{-q}(1-c)^{-(1-c)q}c^{-cq}\frac{1}{4\sqrt{c(c-1)}}$ as σ increases. $\square$

Searching. We wish to read backwards enough indeterminate q-grams from a window of size m such that the probability that we must verify the window is small and the amount we can shift the window by is sufficiently large. By Lemma 4, we know that the probability of a random indeterminate q-gram occurring in a string of length m with cq-mismatches is no more than $m\left(\frac{2^{\sigma-1}}{2^{\sigma}-1}\right)^q(1-c)^{-(1-c)q}c^{-cq}\frac{1}{4\sqrt{c(1-c)}}$. For the rest of the discussion let $a = 4\sqrt{c(1-c)}$.

In the case when we read $k/(cq)$ indeterminate q-grams, we know that with probability at most $(k/(cq))m\left(\frac{2^{\sigma-1}}{2^{\sigma}-1}\right)^q(1-c)^{-(1-c)q}c^{-cq}1/a$ we have found at most k mismatches. This does not permit us to discard the window if all q-grams occur with at most cq mismatches. To fix this, we instead read $\ell = 1 + k/(cq)$ q-grams. If any indeterminate q-gram occurs with less than cq mismatches, we will need to verify the window; but if they all occur with at least cq mismatches, we must exceed the threshold k and can shift the window. When shifting the window we have the case that we shift after verifying the window and the case that the mismatches exceed k so we do not verify the window. If we have verified the window, we can shift past the last position we checked for an occurrence: we can shift by m positions. If we have not verified the window, as we read a fixed number of indeterminate q-grams, we know the minimum-length shift we can make is one position past this point. The length of this shift is at least $m - (q + k/c)$ positions. This means we will have at most $\frac{n}{m-(q+k/c)} = \mathcal{O}(\frac{n}{m})$ windows. The previous statement is only true assuming $m > q + k/c$, as then the denominator is positive. From there we see that we also have the condition that $q + k/c$ can be at most ϵm, where $\epsilon < 1$, so the denominator will be $\mathcal{O}(m)$. This puts the condition on c, that is, $c > \frac{k}{\epsilon m - q}$. Therefore, for each window, we verify with probability at most $(1+k/(cq))m\left(\frac{2^{\sigma-1}}{2^{\sigma}-1}\right)^q(1-c)^{-(1-c)q}c^{-cq}1/a$. So the probability that a verification is triggered is

$$(1+k/(cq))m\left(\frac{2^{\sigma-1}}{2^{\sigma}-1}\right)^q(1-c)^{-(1-c)q}c^{-cq}1/a.$$

By Lemma 1 verification takes time $\mathcal{O}(m^2)$; per window the expected cost is

$$\mathcal{O}(m^2)(1+k/(cq))m\left(\frac{2^{\sigma-1}}{2^{\sigma}-1}\right)^q(1-c)^{-(1-c)q}c^{-cq}1/a =$$
$$\mathcal{O}\left(\frac{(q+k)m^3\left(\frac{2^{\sigma-1}}{2^{\sigma}-1}\right)^q(1-c)^{-(1-c)q}c^{-cq}}{qa}\right).$$

We wish to ensure that the probability of verifying a window is small enough that the average work done is no more than the work we must do if we skip a window without verification. When we do not verify a window, we read $\ell = 1 + k/(cq)$ indeterminate q-grams and shift the window. This means that we process $q + k/c = \mathcal{O}(q + k)$ positions. So a sufficient *condition* is the following:

$$\frac{(q + k)m^3\left(\frac{2^{\sigma-1}}{2^\sigma-1}\right)^q(1 - c)^{-(1-c)q}c^{-cq}}{qa} = \mathcal{O}(q + k).$$

For sufficiently large m, this holds if we set $q \geq \frac{3\log_v m - \log_v a - \log_v q}{1+(1-c)\log_v(1-c)+c\log_v c}$, where $v = \frac{2^{\sigma-1}}{2^\sigma-1}$, which tends to $\frac{3\log_2 m - \log_2 a - \log_2 q}{1+(1-c)\log_2(1-c)+c\log_2 c}$ as σ increases.

Algorithm $AWTM(x, m, y, n, z, k, q, \ell, \Sigma)$

$D[0 .. |\Sigma|^q - 1] \leftarrow 0;$

foreach $s \in \Sigma^q$ **do**

> Compute the *minimal* Hamming distance e between s and any factor of x, and set $D[s] \leftarrow e;$

$i \leftarrow 0;$

while $i < n - m + 1$ **do**

> $d \leftarrow 0;$
>
> **foreach** $t \in [1, \ell]$ **do**
>
>> $j \leftarrow i + m - t \times q;$
>>
>> Let $\mathcal{A}$ denote the set of all valid q-grams starting at position j in $y;$
>>
>> **if** $\mathcal{A} = \emptyset$ **then**
>>
>>> **break;**
>>
>> **else**
>>
>>> $d_{\min} \leftarrow q;$
>>>
>>> **foreach** $s \in \mathcal{A}$ **do**
>>>
>>>> $d_{\min} \leftarrow \min\{d_{\min}, D[s]\};$
>>>
>>> $d \leftarrow d + d_{\min};$
>
> **if** $d > k$ **or** $\mathcal{A} = \emptyset$ **then**
>
>> $i \leftarrow j + 1;$
>
> **else**
>
>> **if** $VER3(x, y[i .. i + m - 1], z, k) = true$ **then**
>>
>>> **output** $i;$
>>
>> $i \leftarrow i + 1;$

For this analysis to hold we must be able to read the required number of indeterminate q-grams. Note that the above probability is the probability that at least one of q-grams matches with less than cq mismatches. To ensure we have enough unread random q-grams for the above analysis to hold, the window must be of size $m \geq 2q + 2k/c$. Consider the case where $2q + 2k/c > m \geq 2q + k/c$. If we have verified a window then we have enough new random q-grams, and if we have just shifted, then we know that all the q-grams we previously read matched with at least cq mismatches and we have at least one new q-gram. The probability that one of these matches with less than cq mismatches is less than the one used above so the analysis holds in both cases. Note that this technique can work for any ratio which satisfies $k/m \leq c - \frac{2cq}{m}$.

Now recall that in fact we are processing a *weighted* text, not an indeterminate one. By the aforementioned analysis, we can choose a suitable value for c and q, to obtain the following result, noting also that it takes time $\mathcal{O}(z(q+k))$, by Fact 2, to obtain the *valid* q-grams of $\mathcal{O}(q+k)$ positions of the weighted text.

Theorem 5. *Algorithm* *AWTM* *solves* *problem* ApproxWeightedText-Matching *in average-case search time* $\mathcal{O}\left(\frac{nz(\log m + k)}{m}\right)$ *if* $k/m \leq c - \frac{2cq}{m}$ *and we set* $q \geq \frac{3\log_v m - \log_v a}{1 + (1-c)\log_v(1-c) + c\log_v c}$, *where* $v = \frac{2^\sigma - 1}{2^\sigma - 1}$, $a = 4\sqrt{c(1-c)}$ *and* $0 < c < 1/2$, *which tends to* $\frac{3\log_2 m - \log_2 a}{1 + (1-c)\log_2(1-c) + c\log_2 c}$ *as* σ *increases. The worst-case preprocessing time and space is* $\mathcal{O}(mq\sigma^q)$ *and* $\mathcal{O}(\sigma^q)$, *respectively.*

4   Experimental Results

Algorithms WTM and GWPM were implemented as a program to perform exact weighted string matching, and algorithm AWTM was implemented as a program to perform approximate weighted string matching. The programs were implemented in the C++ programming language and developed under the GNU/Linux operating system. The input parameters for WTM and GWPM are a pattern (weighted string for GWPM), a weighted text, and a cumulative weight threshold. The output of this program is the starting positions of all valid occurrences. The input parameters for AWTM are a pattern, a weighted text, a cumulative weight threshold, and an integer k for mismatches. The output of this program is the starting positions of all valid occurrences. These implementations are distributed under the GNU General Public License (GPL). The implementation for WTM and GWPM is available at https://github.com/YagaoLiu/GWSM, and the implementation for AWTM is available at https://github.com/YagaoLiu/HDwtm. The experiments were conducted on a Desktop PC using one core of Intel Core i5-4690 CPU at 3.50 GHz under GNU/Linux. All programs were compiled with g++ version 4.8.4 at optimisation level 3 (-O3).

WTM vs.Stateoftheart. We first compared the time performance of WTM against two other state-of-the-art algorithms: the worst-case $\mathcal{O}(n \log z)$-time algorithm of [14], denoted by KPR; and the average-case $o(n)$-time algorithm of [5], denoted by BLP. For this experiment we used synthetic DNA data generated by a randomised script. The length of the input weighted texts was 32MB. Four texts with four different uncertain positions percentages, denoted by δ, were used: 2.5 %, 5 %, 7.5 %, and 10 %. The length of the input patterns ranged between 16 and 256. The cumulative weight threshold $1/z$ was set according to δ to make sure that all patterns could potentially have valid occurrences. Specifically we set: $z = 256$ when $\delta = 2.5$ %; $z = 16,384$ when $\delta = 5$ %; $z = 1,048,576$ when $\delta = 7.5$ %; and $z = 67,108,864$ when $\delta = 10$ %. The length of q-grams was

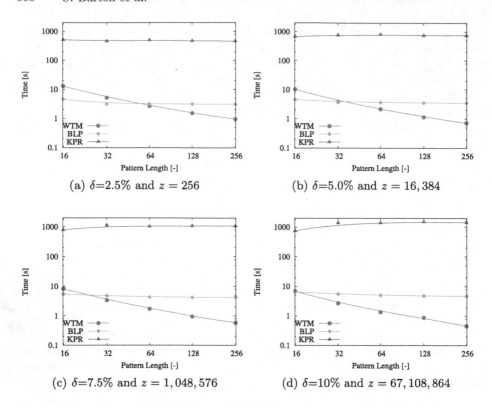

(a) δ=2.5% and $z = 256$

(b) δ=5.0% and $z = 16,384$

(c) δ=7.5% and $z = 1,048,576$

(d) δ=10% and $z = 67,108,864$

Fig. 1. Elapsed time of KPR, BLP, and WTM for on-line pattern matching using synthetic weighted texts of length 32MB over the DNA alphabet ($\sigma = 4$)

set to $q = 8$. The results plotted in Fig. 1 show that WTM was between one and three orders of magnitude faster than the state-of-the-art approaches with these datasets. In addition, our theoretical findings in Theorem 3 are confirmed: for increasing m and constant n and z, the elapsed time of WTM decreases.

Application I. DnaA proteins are the universal initiators of replication from the chromosomal replication origin (oriC) in bacteria. The DnaA protein recognises and binds specifically to the IUPAC-encoded sequence TTWTNCACA, named DnaA box, which is present in all studied bacterial chromosomal replication origins [7]. We transformed sequence TTWTNCACA into the weighted sequence TT[(A,0.5),(T,0.5)]T[(A,0.25),(T,0.25), (G,0.25),(T,0.25)]CACA; and then we searched for this sequence in a dozen of bacterial genomes obtain from the NCBI genome database. The length of q-gram was set to $q = 4$ and the cumulative weight threshold to $1/z = 1/8$ to ensure all factors of length 9 are valid. We compared the time performance of algorithm GWPM against the worst-case $\mathcal{O}(nz^2 \log z)$-time algorithm of [6], denoted by WPT, for this assignment. The bacteria used, the number of occurrences found, and the elapsed times are shown in Table 1. The results show that GWPM is one order of magnitude faster than WPT.

Table 1. Elapsed time of GWPM and WPT for searching for the DnaA box TTWTNCACA in 12 bacterial genomes

Bacteria	DNA length	Number of occurrences	Elapsed time (s) GWPM	Elapsed time (s) WPT
Bacillus subtilis	4,215,606	321	1.19	9.83
Escherichia coli	4,641,652	165	1.27	10.45
Haemophilus influenzae	1,830,138	177	0.56	4.83
Helicobacter pylori	1,667,867	172	0.48	3.96
Mycobacterium tuberculosis	4,411,532	21	1.14	9.54
Proteus mirabilis	4,063,606	294	1.16	9.23
Pseudomonas aeruginosa	6,264,404	66	1.60	14.66
Pseudomonas putida	6,181,873	141	1.62	13.55
Salmonella enterica	4,857,432	190	1.31	12.03
Staphylococcus aureus	2,821,361	202	0.80	7.75
Streptomyces lividans	8,345,283	18	2.08	18.42
Yersinia pestis	4,653,728	190	1.29	10.72

Application II. Mutations in 14 known genes on human chromosome 21 have been identified as the causes of monogenic disorders. These include one form of Alzheimer's disease, amyotrophic lateral sclerosis, autoimmune polyglandular disease, homocystinuria, and progressive myoclonus epilepsy; in addition, a locus for predisposition to leukaemia has been mapped to chromosome 21 [12]. To this end, we evaluated the time performance of AWTM for pattern matching in a genomes population. Pattern matching is useful for evaluating whether SNPs occur in experimentally derived transcription factor binding sites [11, 18]. As input text we used the human chromosome 21 augmented with a set of genomic variants obtained from the 1000 Genomes Project. The SNPs present in the population were incorporated to transform the chromosome sequence into a weighted text. The length of human chromosome 21 is $48,129,895$ base pairs. The input patterns were selected randomly from the text; their length ranged between 16 and 256. In this real scenario, δ was found to be 0.7 %; we therefore set the cumulative weight threshold to the constant value of $1/z = 1/1,024$. For a pattern of length m, the maximum allowed number of mismatches k was set to $2.5\% \times m$, $5\% \times m$, $7.5\% \times m$, and $10\% \times m$. The length of q-grams was set to $q = \log_2 m$ and the number of q-grams read backwards was set to $\ell = \lfloor \frac{k}{0.2 \times q} + 1 \rfloor$. The exact values for q, k, and ℓ are presented in Table 2. The results plotted in Fig. 2 demonstrate the effectiveness of AWTM: all pattern occurrences can be found within a few seconds, even for error rates of 10 %. In addition, our theoretical findings in Theorem 5 are confirmed: for increasing m and constant n, k, and z, the preprocessing time of AWTM increases but the search time decreases.

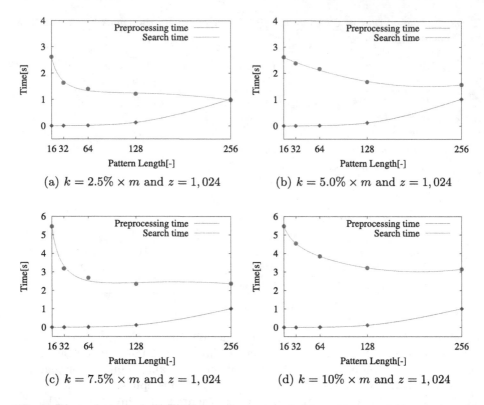

Fig. 2. Elapsed time of AWTM for on-line approximate pattern matching in human chromosome 21 augmented with SNPs from the 1000 Genomes Project

Table 2. Pattern length m, q-grams length q, number k of maximum allowed mismatches, and number ℓ of q-grams read backwards for different error rates

m	q	2.5%		5.0%		7.5%		10%	
		k	ℓ	k	ℓ	k	ℓ	k	ℓ
16	4	1	2	1	2	2	3	2	3
32	5	1	2	2	3	3	4	4	5
64	6	2	2	4	4	5	5	7	6
128	7	4	3	7	6	10	8	13	10
256	8	7	5	13	9	20	13	26	17

5   Final Remarks

In this article, we provided efficient on-line average-case algorithms for solving various weighted pattern matching problems. We also provided extensive experimental results, using both real and synthetic data, showing that our implementations outperform state-of-the-art approaches by more than one order of magnitude. Furthermore, we demonstrated the suitability of the proposed algorithms in a number of real biological contexts. We would like to stress though that the applicability of these algorithms is not exclusive to molecular biology.

Our immediate target is to investigate other ways to measure approximation in weighted pattern matching. Another direction is to study pattern matching on the following generalised notion of weighted strings. A *generalised weighted string* x of length n on an alphabet Σ is a finite sequence of n sets. Every $x[i]$, for all $0 \le i < n$, is a set of ordered pairs $(s_j, \pi_i(s_j))$, where s_j is a possibly empty string on Σ and $\pi_i(s_j)$ is the probability of having string s_j at position i.

References

1. Amir, A., Chencinski, E., Iliopoulos, C.S., Kopelowitz, T., Zhang, H.: Property matching and weighted matching. Theor. Comput. Sci. **395**(2–3), 298–310 (2008)
2. Amir, A., Iliopoulos, C., Kapah, O., Porat, E.: Approximate matching in weighted sequences. In: Lewenstein, M., Valiente, G. (eds.) CPM 2006. LNCS, vol. 4009, pp. 365–376. Springer, Heidelberg (2006). doi:10.1007/11780441_33
3. Barton, C., Iliopoulos, C.S., Pissis, S.P.: Optimal computation of all tandem repeats in a weighted sequence. Algorithms Mol. Biol. **9**(21), 1–12 (2014)
4. Barton, C., Kociumaka, T., Pissis, S.P., Radoszewski, J.: Efficient index for weighted sequences. In: CPM 2016, LIPIcs, vol. 54, pp. 4: 1–4: 13. Schloss Dagstuhl-Leibniz-Zentrum fuer Informatik (2016)
5. Barton, C., Liu, C., Pissis, S.P.: Fast average-case pattern matching on weighted sequences. CoRR abs/1512.01085 (2015). (submitted to IPL)
6. Barton, C., Pissis, S.P.: Linear-time computation of prefix table for weighted strings. In: Manea, F., Nowotka, D. (eds.) WORDS 2015. LNCS, vol. 9304, pp. 73–84. Springer, Heidelberg (2015). doi:10.1007/978-3-319-23660-5_7
7. Caspi, R., Helinski, D.R., Pacek, M., Konieczny, I.: Interactions of DnaA proteins from distantly related bacteria with the replication origin of the broad host range plasmid RK2. J. Biol. Chem. **275**(24), 18454–18461 (2000)
8. Chang, W.I., Marr, T.G.: Approximate string matching and local similarity. In: Crochemore, M., Gusfield, D. (eds.) CPM 1994. LNCS, vol. 807, pp. 259–273. Springer, Heidelberg (1994). doi:10.1007/3-540-58094-8_23
9. Crochemore, M., Hancart, C., Lecroq, T.: Algorithms on Strings. Cambridge University Press, New York (2007)
10. Farach, M.: Optimal suffix tree construction with large alphabets. In: FOCS 1997, pp. 137–143. IEEE Computer Society (1997)
11. Guo, Y., Jamison, D.C.: The distribution of SNPs in human gene regulatory regions. BMC Genom. **6**(1), 1–11 (2005)
12. Hattori, M., et al.: The DNA sequence of human chromosome 21. Nature **405**, 311–319 (2000)
13. Huang, L., Popic, V., Batzoglou, S.: Short read alignment with populations of genomes. Bioinformatics **29**(13), 361–370 (2013)
14. Kociumaka, T., Pissis, S.P., Radoszewski, J.: Pattern matching and consensus problems on weighted sequences and profiles. In: ISAAC 2016, LIPIcs. Schloss Dagstuhl-Leibniz-Zentrum fuer Informatik (2016)
15. Lovász, L., Pelikán, J., Vesztergombi, K.: Discrete Mathematics: Elementary and Beyond. Springer, New York (2003)

16. Musser, D.R.: Introspective sorting and selection algorithms. Softw. Pract. Exp. **27**(8), 983–993 (1997)
17. Pizzi, C., Ukkonen, E.: Fast profile matching algorithms - a survey. Theor. Comput. Sci. **395**(2–3), 137–157 (2008)
18. Sandelin, A., Alkema, W., Engström, P., Wasserman, W.W., Lenhard, B.: JASPAR: an open-access database for eukaryotic transcription factor binding profiles. Nucleic Acids Res. **32**(1), D91–D94 (2004)
19. Varela, M.A., Amos, W.: Heterogeneous distribution of SNPs in the human genome: microsatellites as predictors of nucleotide diversity and divergence. Genomics **95**(3), 151–159 (2010)

Scheduling with Interjob Communication on Parallel Processors

Jürgen König, Alexander Mäcker, Friedhelm Meyer auf der Heide, and Sören Riechers[✉]

Heinz Nixdorf Institute and Computer Science Department,
Paderborn University, Paderborn, Germany
{juergen.koenig,alexander.maecker,fmadh,soeren.riechers}@upb.de

Abstract. Consider a scheduling problem in which a set of jobs with interjob communication, canonically represented by a weighted tree, needs to be scheduled on m parallel processors interconnected by a shared communication channel. In each time step, we may allow any processed job to use a certain capacity of the channel in order to satisfy (parts of) its communication demands to adjacent jobs processed in parallel. The goal is to find a schedule with minimum length in which communication demands of all jobs are satisfied.

We show that this problem is NP-hard in the strong sense even if the number of processors and the maximum degree of the underlying tree is constant. Consequently, we design and analyze simple approximation algorithms with asymptotic approximation ratio $2 - 1/2(m-1)$ in case of paths and a ratio of $5/2$ in case of arbitrary trees.

1 Introduction

In "On-The-Fly Computing" [1,9], one main idea is that future software-based IT services are automatically composed from base services traded on global markets. Thereby, the functionality of a service is provided by the interaction of smaller pieces of software resulting in the exchange of data during the execution. This strengthens the necessity of taking into account communication when designing scheduling algorithms which enable efficient execution of such software. It might even shift the focus from processing times to planning communication, particularly if the exchange of data rather than actual computations becomes the major bottleneck in a system.

These observations lead to a new scheduling problem that we study in this paper. We are given a dependency graph describing a service composed of jobs (base services) by identifying nodes with jobs and using weighted edges to model the required interjob communication of jobs. These edge weights can, for instance, be thought of as communication volume in bytes. Also, we are given a system comprised of m parallel, identical processors connected by a shared

This work was partially supported by the German Research Foundation (DFG) within the Collaborative Research Centre "On-The-Fly Computing" (SFB 901).

© Springer International Publishing AG 2016
T.-H.H. Chan et al. (Eds.): COCOA 2016, LNCS 10043, pp. 563–577, 2016.
DOI: 10.1007/978-3-319-48749-6_41

communication channel (e.g. a data bus) enabling communication between the processors and hence between jobs processed in parallel. Given that the available communication channel constitutes a scarce resource with bounded capacity (e.g. available data rate in bytes per second), a fundamental question arising in this setting is: How to assign jobs to processors and share the channel among them in order to minimize the time at which all jobs with their related communication demands are done and hence, to minimize the time until the service is completed.

We model this scheduling problem as a novel bin packing variant and propose and analyze efficient, practical approximation algorithms. In the following, we give a formal description of the studied problem and review relevant related work. In Sect. 2, we study the computational hardness showing that the considered problem is (for the most relevant cases) NP-hard in the strong sense even for trees with constant degree and a constant number of machines. Consequently, we then focus on approximation algorithms and start by considering the most basic case for trees of degree 2 and show that a NEXTFIT strategy achieves an asymptotic approximation ratio of $2 - 1/2(m-1)$ in Sect. 3. For trees with arbitrary degree we show an asymptotic $5/2$-approximation algorithm in Sect. 4.

1.1 Model

We consider the following scheduling problem called SIC. Given a service described by a communication tree $G = (V, E)$ on a set V of jobs together with a weight function $w : E \rightarrow]0, C]$, each edge $\{u, v\} \in E$ represents the communication requirement between jobs u and v. Additionally, we are given a set of m identical, parallel processors connected by a shared communication channel with capacity C. Each processor can process at most one job per (discrete) time step while a job can be processed in several (not necessarily contiguous) time steps. Two jobs can communicate only when they are executed in parallel. Hence, in any time step t, at most m jobs can be processed and additionally, a schedule has to define how much capacity of the communication channel is allocated to pairs of jobs processed in t. Thereby the channel may not be overused, i.e. a capacity of at most C may be allocated to jobs per time step. As soon as for a pair of jobs with strictly positive communication demand the accumulated share of the channel it was assigned over time is at least its requirement $w(e)$, we call this edge to be completed. The objective is to find a schedule that minimizes the time where the last edge is completed (also called makespan).

Formally, the problem is defined as a bin packing variant which is equivalent to this scheduling problem. In the bin packing formulation, each edge $e \in E$ corresponds to an item e with size $s_e := w(e)$. The goal is to pack all items into as few bins with capacity C as possible while allowing items to be arbitrarily split into parts and subject to the following constraints:

1. *Capacity Constraint:* Each bin may contain (parts of) items of overall size of at most C,
2. *Edge Constraint:* Each bin may contain (parts of) items incident to at most m nodes in the underlying tree representation.

In the rest of the paper, we assume without loss of generality that $C = 1$. Observe that in terms of the original scheduling formulation each part of an item corresponds to one time step in which the corresponding jobs are scheduled and its size represents the channel capacity assigned to this pair of jobs in this time step. The edge constraint respresents the fact that only m machines are available while the capacity constraint represents the available channel capacity. The number of bins then coincides with the number of time steps required to finish the service.

Since in the next section we show that SIC is NP-hard in general, even with constant degree $d \geq 2$ and constant number of processors $m \geq 4$, we focus on designing approximation algorithms. A polynomial time algorithm A is called to have an (absolute) approximation ratio of α if, on any instance I, it holds $\frac{A(I)}{\text{OPT}(I)} \leq \alpha$, where $A(I)$ and $\text{OPT}(I)$ denote the number of bins used by algorithm A on I and by an optimal solution for I, respectively. A has an asymptotic ratio of α if $R^\infty \leq \alpha$, where $R^\infty := \lim_{k \to \infty} \sup_I \left\{ \frac{A(I)}{\text{OPT}(I)} : \text{OPT}(I) = k \right\}$.

1.2   Related Work

Bin packing has a long history in computer science research and a huge body of literature on several variants of bin packing problems emerged in the past.

In its most basic variant, n items of sizes $0 < s_i \leq 1$ need to be placed into as few bins of capacity 1 as possible. This problem is easily seen to be NP-hard by a straightforward reduction from the classical PARTITION problem, which also directly gives an inapproximability for an approximation ratio below $\frac{3}{2}$ unless $P = NP$. This bound is actually achieved by the well-known FIRSTFITDE-CREASING strategy, which first sorts the items in decreasing order by their sizes and then places the current item to be packed into the first bin it fits into. In [5], Dósa et al. also prove that FIRSTFITDECREASING uses at most $\frac{11}{9}\text{OPT} + \frac{6}{9}$ bins and that this bound is tight. When considering asymptotic approximation algorithms, even (fully) polynomial time approximation schemes (A(F)PTAS) are known [4,10]. Such algorithms provide solutions with asymptotic approximation ratio of $(1 + \varepsilon)$, for any $\varepsilon > 0$, and run in time polynomial in the input (and $\frac{1}{\varepsilon}$ in case of an AFPTAS).

While there are dozens of variants of this basic problem, the variant closest related to SIC is bin packing with cardinality constraints and splittable items introduced in [3]. In this problem items of arbitrary size need to be packed into as few unit size bins as possible. In contrast to classical bin packing, items may be split into parts which can be placed in different bins but any bin may contain at most m (parts of) items. Note that this constraint is similar to the edge constraint in SIC, though there is no underlying dependency graph. In [3], Chung et al. prove NP-hardness and give an algorithm with asymptotic approximation factor of $\frac{3}{2}$ for the case $m = 2$. Later, Epstein and van Stee [8] gave an NP-hardness result for $m > 2$ and they show a simple NEXTFIT strategy to achieve an exact (absolute) approximation factor of $2 - \frac{1}{m}$. In [6], an efficient polynomial time approximation scheme is given for the case $m = o(n)$, which essentially solves the problem as it is NP-hard in the strong sense and for $m = \Theta(n)$ the problem cannot be approximated better than $\frac{3}{2}$ unless $P = NP$.

In [2], Brinkmann et al. consider a different variant of scheduling with shared resources. Here, they focus on the distribution of the resource assuming the assignment of jobs to processors and their order on each processor is already fixed. The remaining challenge is to distribute the resource among the processors such that it is used efficiently and the makespan of the resulting schedule is minimized. They prove that their problem is NP-hard if the number of processors is part of the input and give a simple greedy approximation algorithm for a fixed number of processors m, which has an absolute approximation factor of $2 - \frac{1}{m}$. Using a dynamic program, they also give an exact polynomial time algorithm for $m = 2$.

2   Complexity

First, note that our problem is trivial to solve for constant $m = 2$. In this case, each item has to be packed alone, hence packing all items into distinct bins is optimal.

For larger values of m, as a first observation and a direct corollary from NP-hardness of cardinality constrained bin packing with splittable items [3], which, for a cardinality constraint set to $m-1$, is equivalent to SIC when the dependency graph forms a star, we have the following proposition.

Proposition 1. *The* SIC *problem is strongly NP-hard for constant* $m \geq 3$ *processors and the degree* d *of the communication tree being part of the input.*

Despite this hardness result, it is interesting to study the question whether the complexity changes when the degree d of the dependency graph is fixed. We will see that the problem is in P for $m = 3$ and, more interestingly and suprisingly, remains NP-hard for any constant degree d and constant $m \geq 4$.

2.1   Exact Algorithm for $m = 3$ Processors

In this section, we use a similar representation for packings as Epstein and van Stee [7] used when they introduced a PTAS for cardinality constrained bin packing with splittable items. Here, a packing is represented by a graph where nodes correspond to items and edges correspond to bins. For a bin containing two item parts, there is an edge between the two items. If a bin contains only one item, there is a loop on that item. The following lemma can be adapted from [7].

Lemma 1 (Lemma 4 [7]). *If the graph of a packing contains a cycle, it is possible to modify the packing such that this cycle is removed without increasing the number of bins.*

From Lemma 5 in [7], we can also directly derive the following lemma.

```
 1: for h := height of G = (V, E) to 1 do
 2:     for all nodes v with depth h − 1 do
 3:         Let E_v = {e_1, . . . , e_k} be the edges to the children of v
 4:         Let e_v be the edge to the parent of v
 5:         b = |E_v|; r = s_v
 6:         for all graphs (E_v, B) with degree ≤ 2, without cycles of length > 1 do
 7:             s'_v = s_v
 8:             Greedily fill items induced by E_v into bins induced by B
 9:             Fill bins with only one item with parts of e_v; reduce s'_v accordingly
10:             if |B| ≤ b and s'_v ≤ r then
11:                 Store current packing and set b = |B| and r = s'_v
12:         Pack bins according to stored packing
13:         Remove all children of v; reduce s_v to r
```

Fig. 1. Finding an optimal solution for $m = 3$ and constant degree d

Lemma 2. *For constant $m = 3$ and any star dependency graph, there exists an optimal packing where each item is split into at most 2 parts.*

Taken together, these two lemmas imply that the packing of any star subgraph can be represented by a graph of degree 2, i.e. a collection of paths.

Theorem 1. *The* SIC *problem can be solved in polynomial time for constant $d \geq 2$ and $m = 3$.*

Proof Sketch. Consider the algorithm in Fig. 1. It repeatedly packs subtrees rooted at the second lowest level of the tree optimally. This is done by generating all possible graph representations without cycles where each item is split at most twice (i.e. the graph representation is a collection of paths, possibly containing loops). Due to Lemmas 1 and 2, we know that one of the graphs generated in Line 6 represents an optimal packing for (the star subgraph) E_v. From [7], we also know that we can fill the bins greedily using the graph representation.

As induction hypothesis, assume that the items in $E \setminus \{E_v \cup \{e_v\}\}$ together with the remainder of e_v (not packed in Line 12) are packed optimally. Now, note that in any solution for the full graph, items from E_v can only be packed together with e_v or other items from E_v. Thus, in a given optimal solution of the full graph, we can partition the bins into two sets.

1. B_1^*: Bins containing no items from E_v.
2. B_2^*: Bins containing items from E_v (hence no items from $E \setminus \{E_v \cup \{e_v\}\}$).

As our algorithm finds the optimal solution for the bin set E_v packing as much as possible of e_v additionally, it must have found a packing with bins B, where $|B| \leq |B_2^*|$. If $|B| < |B_2^*|$, we are done, as in the induction hypothesis, the additional bin can always be used to fully pack e_v. Otherwise, we have $|B| = |B_2^*|$, and as out of all solutions fully packing E_v, our algorithm found the solution packing the highest amount of e_v possible, it follows that it packed at least as much of e_v as the optimal solution did in the bins B_2^*. Hence, the remainder to be packed from e_v in the packing from the induction hypothesis is less or equal than the remainder to be packed in the optimum. Thus, the number of bins used in the induction hypothesis is at most $|B_1^*|$. Inductively, it follows that the full solution is optimal.

Note that the number of graphs in Line 6 is definitely below 2^{d^2}. Also, the generation of such graphs happens at most n times, yielding polynomial runtime.   □

2.2   NP-Hardness for $m > 3$ Processors

Theorem 2. *The* SIC *problem is strongly NP-hard for* $d = 2$ *and constant* $m \geq 6$.

Proof. We start with the 3-PARTITION problem with a restricted size of the elements which is defined as follows. Given a multiset $A = \{a_1, \ldots, a_n\}$ of $n = 3k$ elements, a bound B with $B/4 < a_i < B/2 \; \forall \; i \in \{1, \ldots, n\}$ and $\sum_{a \in A} a = kB$, is there a partition into k sets $A_1, \ldots, A_k$ such that $|A_i| = 3$ and $\sum_{a \in A_i} a = B$ for all $i \in \{1, \ldots, k\}$?

Let our SIC instance consist of one path with $\ell + 1 = 3k \cdot (3m - 2) + 2k \cdot (m - 5) + 1$ nodes. We denote the edge between node i and $i + 1$ by e_i, yielding $E = \{e_1, \ldots, e_\ell\}$ with sizes $\{s_1, \ldots, s_\ell\}$. Now, let $s_{(i-1) \cdot (3m-2)+1} = \left(\frac{1}{2} + \frac{a_i}{2B}\right)\left(1 - \frac{m-5}{5m}\right) < \frac{3}{4}$ for all $i \in \{1, \ldots, 3k\}$, called *medium items*. Also, let $s_{(i-1) \cdot (3m-2)+m} = s_{(i-1) \cdot (3m-2)+\lfloor \frac{3}{2}m \rfloor} = s_{(i-1) \cdot (3m-2)+2m} = 1 - \frac{m-2}{5m} \; \forall \; i \in \{1, \ldots, 3k\}$, called *large items*. All other edges are assigned a size of $s_i = \frac{1}{5m}$, called *small items*. For a visualization, see Fig. 2. Now, is there a packing of size at most $11k$?

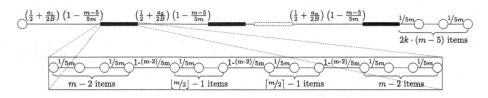

Fig. 2. Corresponding SIC instance for 3-PARTITION with input $\{a_1, \ldots, a_n\}$

In case that the 3-PARTITION instance is a YES-instance, we need to show that there is a corresponding packing with at most $11k$ bins for our SIC instance. Given a set A_i, the 3 (medium) items S_i derived from it can always be packed into two bins together with $m - 5$ of the last $2k \cdot (m - 5)$ items, i.e. items $e_{3k \cdot (3m-2)+(2i-2)(m-5)+1}, \ldots, e_{3k \cdot (3m-2)+(2i-1)(m-5)}$ for the first bin, and items $e_{3k \cdot (3m-2)+(2i-1)(m-5)+1}, \ldots, e_{3k \cdot (3m-2)+2i(m-5)}$ for the second bin. That is, because the 3 items from S_i only use 4 of the allowed incident nodes in both bins, and the $m - 5$ small items use $m - 4$ of the allowed incident nodes. Also, we have $\sum_{s \in S_i} s_j + 2 \cdot \frac{m-5}{5m} = (\frac{3}{2} + \sum_{a \in A_i} \frac{a}{2B})(1 - \frac{m-5}{5m}) + 2 \cdot \frac{m-5}{5m} = 2$ giving a valid packing if we split one of the medium items accordingly. Thus, we can pack all item sets S_i into two bins each, together with the last $2k \cdot (m-5)$ items, leading to $2k$ bins.

Observe that the $3m - 3$ items in each block filling the gaps between two medium items (i.e. the rectangles in Fig. 2) can be put into three bins, that is, the first $m - 1$ items of each block (i.e., $m - 2$ small items and one large item) can be put into one bin, as the sum of their sizes is exactly 1 by construction. The same holds for the second and third item set of $m - 1$ items, respectively. More formally, items $e_{(i-1) \cdot (3m-2)+2}, \ldots, e_{(i-1) \cdot (3m-2)+m}$ for each $i \in \{1, \ldots, 3k\}$ as well as item sets $\{e_{(i-1) \cdot (3m-2)+m+1}, \ldots, e_{(i-1) \cdot (3m-2)+3m-2}\}$ and $\{e_{(i-1) \cdot (3m-2)+2m}, \ldots, e_{(i-1) \cdot (3m-2)+3m-2}\}$ for each $i \in \{1, \ldots, 3k\}$ can be put into one bin each. This gives another $3 \cdot 3k$ bins and all items are packed.

On the other hand, we show that if there is a packing with size at most $11k$, we show that the respective 3-PARTITION instance is a YES-Instance. In order to do so, we show the following properties in the given order:

(1) The capacity of each bin must be fully utilized.
(2) At least k medium or large items need to be split.
(3) In order to pack the last $2k \cdot (m-5)$ items, $4k$ additional separate components of the dependency graph have to be packed together with them.
(4) Our $11k$ bins contain exactly $9k \cdot (m - 1) + 2k \cdot (m - 3)$ item parts.
(5) The last $2k \cdot (m - 5)$ items are packed into $2k$ bins containing exactly $m - 5$ of these items and two medium item parts each.
(6) Exactly k of the $3k$ medium items are split, and they are split into exactly two parts each.
(7) The corresponding 3-PARTITION is a YES-instance.

(1) The capacity must be fully utilized in any bin, because we have $\sum_{i=1}^{l} s_i = \sum_{i=1}^{3k} (\frac{1}{2} + \frac{a_i}{2B})(1 - \frac{m-5}{5m}) + 2k(m-5) \cdot \frac{1}{5m} + 3 \cdot 3k \cdot ((m-2) \cdot \frac{1}{5m} + (1 - \frac{m-2}{5m})) = (\frac{3k}{2} + \frac{kB}{2B})(1 - \frac{m-5}{5m}) + 2k \cdot \frac{m-5}{5m} + 3 \cdot 3k = 11k$.

(2) Now, we show that at least k medium or large items need to be split. There are $3 \cdot 3k$ large items as well as $3k$ medium items. If less than k of these $12k$ items were split, there would be at least one bin fully containing two of these items. This is a contradiction, as all the item sizes are greater than $\frac{1}{2}$. Hence, there are at least $l + k$ item parts.

(3) We now concentrate on the $2k \cdot (m - 5)$ last items, i.e. on the items $e_{3k \cdot (3m-2)+1}, \ldots, e_{3k \cdot (3m-2)+2k \cdot (m-5)}$. Note that these items cannot be packed with a medium or a large item without using two components of the dependency graph. However, if using two components, they contain at most $m-2$ items, which implies (as we always use the full capacity by (1)), that each bin containing one of these items also contains at least two medium or large items. As by construction, there are always at least $m - 2$ edges between a medium and a medium or large item, and at least $\lfloor \frac{m}{2} \rfloor - 1$ edges between two large items, there are only two possibilities how to obtain this:

(a) At most $m - 5$ of the considered small items are combined with at least two further components of the dependency graph, which contain exactly one medium or large item each.

(b) At most $\left\lceil \frac{m-6}{2} \right\rceil \leq \frac{m-5}{2}$ of the considered small items are combined with only one further component of the dependency graph, which contains exactly two large items.

Taken together, this implies that in order to pack all $2k \cdot (m - 5)$ items, there have to be taken at least $2k \cdot 2 = 4k$ additional separate components from the dependency graph.

(4) It follows that in our packing, the $11k$ bins can contain at most $11k \cdot (m - 1) - 4k = 9k \cdot (m - 1) + 2k \cdot (m - 3)$ item parts by (3). However, as we showed earlier in (2), the overall number of item parts is at least $l + k = 3k \cdot (3m - 2) + 2k \cdot (m - 5) + k = 9k \cdot (m - 1) + 2k \cdot (m - 3)$. Thus, both properties are tight and in all bins not containing any of the $2k \cdot (m - 5)$ last items, exactly $m - 1$ items must be packed.

Now, this can only be done by using one (complete) large item and $m - 2$ adjacent small items, as one medium item together with $m - 2$ adjacent small items does not use the full capacity.

(5) Considering the last $2k \cdot (m - 5)$ items again, note that all bins fulfilling (3) (a) now need to contain *exactly* $m - 5$ of these small items as well as all bins fulfilling (b) now need to contain *exactly* $\frac{m-5}{2}$ of these small items. However, this implies that (b) never happens. Otherwise, if there were $2j$ bins fulfilling (b) (giving $2k - j$ bins fulfilling (a)), each of them would make at least one large item incomplete. As this leads to exactly $2k - j$ bins containing $m - 3$ items and $2j$ bins containing $m - 2$ items, there must be exactly $9k - j$ bins containing $m - 1$ items. However, there are only at most $9k - 2j$ (complete) large items left, yielding $j = 0$.

(6) We now know that there are exactly $9k$ bins containing one (complete) large and $m - 2$ small items each. Hence, there are exactly $2k$ bins where each bin contains $m - 5$ of the last $2k \cdot (m - 5)$ items and two parts of medium items. We also know that only medium items are split, thus exactly k of the medium items are split into exactly two parts each.

W.l.o.g., let $\tilde{E} = \{e_1, \ldots, e_k\}$ be the (medium) items that are split. We observe that no two items from $\tilde{E}$ can be packed into the same bin. Otherwise, at least one of them uses a capacity of at least $\frac{1}{2}$ in that bin, hence a capacity of at most $\frac{1}{4}$ in the other bin, where the remaining part of the item is packed. Then, the capacity is not fully utilized in that other bin, as the other medium item (which has to exist by (5) and (6)) uses less than a capacity of $\frac{3}{4}$.

Now, for each medium item split into two parts, we know that exactly one complete other medium item is packed together with each part of it. For each split item, we build a set containing itself and the two medium items packed together with it. There are m such sets $S_1, \ldots, S_m$ with 3 elements each. Now, the sum of the sizes of these three items is $2 - 2 \cdot \frac{m-5}{5m}$ (as both bins additionally contain $m - 5$ small items), implying $\sum_{s \in S_i} \left(\frac{1}{2} + \frac{s}{2B} \right) = 2$ which yields $\sum_{s \in S_i} s = B$. This is a 3-PARTITION. $\qquad\square$

Corollary 1. *The* SIC *problem is strongly NP-hard for* $d = 2$ *and* $m = 4$.

Proof Sketch. We use the same reduction as in the proof of Theorem 2, but with medium item sizes $\frac{1}{2} + \frac{a_i}{2B}$ and without adding the last $2k(m - 5)$ auxiliary items. This implies removing step (3) and instead using the fact that any bin containing a medium item part can only be packed up to the full capacity using two separate components of the dependency graph. As the medium items alone have an overall capacity of $2k$, it follows that at least $2k$ additional separate components need to be used, yielding exactly $9k \cdot (m - 1) + 2k \cdot (m - 2)$ item parts to be packed in step (4). However, this directly implies that at least $9k$ bins have to contain the full number of $m - 1$ items, which (by (1)) is only possible using one large and $m - 2$ adjacent small items. It follows that the remaining $2k$ bins contain the $3k$ medium items. Hence, the remaining part of the proof remains the same.   □

Corollary 2. *The* SIC *problem is strongly NP-hard for* $d = 2$ *and* $m = 5$.

Proof Sketch. We modify the reduction of Corollary 1 by adding one additional small item adjacent to each medium item and reducing the medium item sizes by factor $(1 - \frac{1}{5 \cdot 5})$. Now, three medium items are always packed together with two of these adjacent items into two bins. There are k remaining small items adjacent to medium items. In order for these to be packed, we add k small gadgets to the end of the path (instead of the $2k \cdot (m - 5)$ small items from the proof of Theorem 2). These small gadgets consist of one *very large* item of size $1 - \frac{2}{5 \cdot 5}$ (in contrast to large items, which now have a size of $1 - \frac{3}{5 \cdot 5}$) and one small item. Two adjacent small gadgets are always separated by the usual rectangular gadgets from Fig. 2. With a similar argument as (3) from the proof of Theorem 2, we now ask how to pack the small items adjacent to medium items and conclude that they either have to be packed together with medium items or (for the remaining k items) they have to be packed together with the newly introduced small gadgets. This concludes the proof sketch for $m = 5$.

Corollary 3. *The* SIC *problem is strongly NP-hard for constant* $d \geq 2$ *and constant* $m \geq 4$.

Proof Sketch. In order to achieve a higher degree than 2, we add a similar gadget to the (rectangular) gadgets with the large (and small) items from the original reduction in Theorem 2. This new gadget is visualized in Fig. 3. To separate this gadget from the rest of the instance, we first add another rectangular gadget as used in the proof of Theorem 2 at the end of the path. The new part of the gadget (behind the rectangular gadget) consists of a star graph with many small and few large items to achieve the necessary degree. We fill it up with small items in a path towards the rectangular gadget. This is necessary to achieve an overall number of items divisible by $(m - 1)$ (i.e. an overall number of small items divisible by $(m - 2)$). Now, the items from the newly introduced gadget have to be packed with one large item and $m - 2$ small items into one bin each similar to the packing for the rectangular gadgets in Theorem 2. The remaining reduction is analogous.   □

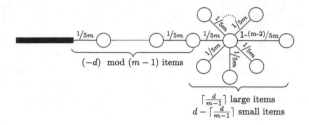

Fig. 3. Gadget to add for degree $d \geq 2$

3 Communication Trees of Degree Two

In this section, we present a simple greedy algorithm for instances with a dependency graph of degree two, i.e. we focus on paths. It is a straightforward adaption of the classical NEXTFIT algorithm. The formal description of the algorithm A_1 is given in Fig. 4. It dispatches the items in order of the path starting at one of the nodes with degree one. If an item fits into the current bin without violating neither the capacity nor edge constraint, it is placed in the current bin. If it does not fit into the current bin (since at least one of the constraints would be violated), as much of the item is placed in the current bin as possible and a new bin is opened in which the (remaining part of the) item is placed. That is, A_1 will always (except for the last bin) fully pack a bin or pack the maximum number of (parts of) items allowed in a bin.

```
1:  Index items in order of path, E := {e_0, e_1, . . .}
2:  B := new empty bin
3:  for all items e_i ∈ E do
4:      if e_i does not fit into B w.r.t. edge constraint then
5:          B := new empty bin
6:      if e_i fits into B then
7:          Pack e_i into B
8:      else
9:          Pack as much of e_i into B as possible
10:         B := new empty bin
11:         Pack remaining part of e_i into B
```

Fig. 4. Algorithm A_1 for instances of degree two.

We next analyze the approximation ratio of A_1 and show that any solution is asymptotically by a factor less than two worse than the optimum.

Theorem 3. *Algorithm A_1 has an asymptotic approximation ratio of at most $2 - \frac{1}{2(m-1)}$ and runs in time $O(n)$.*

Proof. For the sake of analysis, we partition the set of bins used by algorithm A_1 into the set B_1 containing those bins with a tight capacity constraint and B_2 containing those bins which are not full but have a tight edge constraint. We also define the following three sets of items based on how they are packed in the solution given by algorithm A_1. E_1 contains those items for which at least one part of the item is packed in a bin from B_1. In E_2 there are all items for which at least one part of the item is packed in a bin from B_2, and finally $E_R := E \setminus (E_1 \cup E_2)$ contains the remaining items. Note that in E_R there can only be items placed in the last bin that was opened, and that $E_1 \cap E_2$ need not be empty.

We now give a bound on the maximum number of bins used by A_1. We have

$$A_1 = |B_1| + |B_2| + \left\lceil \frac{|E_R|}{m-1} \right\rceil$$

since each bin either belongs to B_1 or B_2 or is the last bin and $E_R \neq \emptyset$. Additionally, since any bin belonging to B_1 only contains items from E_1, we have $|B_1| \leq \sum_{e \in E_1} s_e$. Any bin belonging to B_2 contains $m - 1$ (parts of) items from E_2 and each item from E_2 is packed in at most one bin from B_2 and hence, $|B_2| = \frac{|E_2|}{m-1}$. Therefore, we have

$$A_1 = |B_1| + |B_2| + \left\lceil \frac{|E_R|}{m-1} \right\rceil \leq \sum_{e \in E_1} s_e + \frac{|E_2|}{m-1} + \left\lceil \frac{|E_R|}{m-1} \right\rceil.$$

Next, we lower bound the number of bins used by OPT. We can state that:

1. $\text{OPT} \geq \left\lceil \sum_{e \in E} s_e \right\rceil$ and
2. $\text{OPT} \geq \frac{|E|}{m-1}$,

where the inequalities directly follow from the capacity and edge constraint, respectively. Denoting the set of items belonging to E_1 and E_2 by E', we have $|E| = |E_1| + |E_2| + |E_R| - |E'|$. Together with the claim that $|E'| \leq |E_1|/2$, we can conclude

$$A_1 \leq \sum_{e \in E_1} s_e + \frac{|E_2| + |E_R|}{m-1} + 1$$

$$\leq \frac{|E_1|}{2(m-1)} + \frac{|E_2| + |E_R|}{m-1} + \frac{2(m-1)-1}{2(m-1)} \sum_{e \in E} s_e + 1$$

$$\leq \frac{|E_1| + |E_2| + |E_R| - |E'|}{m-1} + \frac{2 \cdot (m-1) - 1}{2(m-1)} \sum_{e \in E} s_e + 1$$

$$\leq \left(1 + \frac{2 \cdot (m-1) - 1}{2(m-1)} \right) \text{OPT} + 1,$$

where we used the fact that $|E_1| \geq \sum_{e \in E_1} s_e$ and the aforementioned bound on A_1 in the first two estimations, the claimed bound on $|E'|$ in the second last and the bounds on OPT in the last inequality.

Hence, it remains to prove the claim. Recall that E' contains those items that are in E_1 and in E_2, i.e. any item $e \in E'$ is partly packed in a bin which is full and partly packed in a bin which is not full but has a tight edge constraint. By the definition of A_1, such an item $e \in E'$ fulfills the condition that it is first *partly* packed in a bin B' belonging to B_1 and then a bin belonging to B_2 is opened to pack the remaining part of item e. Note that consequently B' contains a different item belonging to $E_1 \setminus E'$. Hence, to any item $e \in E'$ we can associate a different item $\bar{e} \in E_1 \setminus E'$, proving the claim and concluding the proof of an asymptotic approximation ratio of $2 - \frac{1}{2(m-1)}$.

The runtime is $O(n)$ as we essentially traverse the path once.   □

We now show that our analysis of the approximation factor of A_1 is (almost) tight by giving an instance on which it obtains an approximation factor of at least $2 - \frac{2}{m}$.

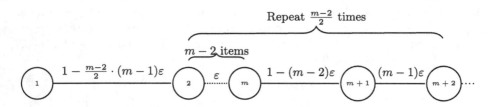

Fig. 5. Hard instance for algorithm A_1

Intuitively, this instance (cf. Fig. 5) exploits the two different optimization goals, i.e. using the full capacity of a bin and using the full number of allowed adjacent nodes. The optimal algorithm chooses to skip certain edges in order to always use the full capacity. In contrast, the greedy algorithm always uses the full capacity (packing only two items) in every second bin and the full number of allowed adjacent nodes (using only low capacity) in the other bins. This gives an approximation factor of almost 2.

Theorem 4. *The approximation factor of Theorem 3 is almost tight since there is an instance with approximation factor of at least $2 - \frac{2}{m}$. For $m \to \infty$ the upper as well as the lower bound converge to 2.*

Proof. Consider an instance as given in Fig. 5 and let $\varepsilon > 0$ be sufficiently small. The instance consists of a path of items of the following form: The leftmost item has a size of $1 - \frac{m-2}{2} \cdot (m-1)\varepsilon$ and is followed by a subpath P of $m - 2$ items with size ε, an item of size $1 - (m-2)\varepsilon$ and one of size $(m-1)\varepsilon$. This subpath is repeated several times such that we obtain $\frac{m-2}{2}$ copies of P. In the following, we refer to the nodes of any copy of P by the indices used in P.

On this instance, an optimal solution can pack the items of the subpath P from node 2 to m together with the item between node m and $m + 1$ into one bin. As there are $\frac{m-2}{2}$ copies of P, $\frac{m-2}{2}$ bins are needed for the respective items. Additionally, the item between node 1 and 2 can be packed together with the items between nodes $m+1$ and $m+2$ of all copies of P. Hence, $\text{OPT} \leq \frac{m-2}{2} + 1 = \frac{m}{2}$.

In contrast, A_1 processes the path from left to right. Hence, it packs all items of the subpath from node 1 to node m in one bin. Afterward, the items of the subpath from m to $m+2$ are packed in a second bin, leading to an unpacked part of the item between $m+1$ and $m+2$. This part is packed into a further bin together with the items between 2 and m of the next copy of P. As this repeats $\frac{m-2}{2}$ times, A_1 needs a total of $m-1$ bins. Note that this result would not change if our algorithm processed the path from right to left instead.

Taken the upper and lower bound together yields $\frac{A_1}{\text{OPT}} \geq \frac{2(m-1)}{m}$.   □

4 Communication Trees with Arbitrary Degree

In this section, we propose an algorithm A_2, which provides $\frac{5}{2}$-approximate solutions for instances described by any tree with degree at most d. Note that by splitting the tree at each node and losing a constant factor of 2, the problem can be reduced to cardinality constrained bin packing with splittable items. As there exists an EPTAS for this problem [6], we can get a $2 \cdot (1 + \varepsilon)$-approximation by using it. However, to achieve an approximation factor of 2.5 for our problem, we need to set the ε from [6] to at most $\frac{1}{36}$, and the constant in the runtime is around $\left(\frac{1}{\varepsilon^8}\right)^{\frac{1}{\varepsilon^2}}$, which is a very large number, rendering the usefulness rather questionable in practice. In the literature, the best algorithm of cardinality constrained bin packing with splittable items using a rather simple approach is NEXTFIT with a tight approximation factor of $2 - \frac{1}{m}$, which would only yield a $2 \cdot (2 - \frac{1}{m})$-approximation. The advantage of our algorithm is its simplicity and low runtime.

Roughly speaking, our algorithm consists of two steps: In a first step (cf. Fig. 6), A_2' computes a preliminary (generally infeasible) packing, which ignores the capacity constraints of bins. Then in a second step, we fix the preliminary packing by repacking parts of items that are packed in bins violating the capacity constraint. In the following, we assume that the communication tree is rooted. If this is not the case, we can root it at any node. Also, we use T_v to denote the nodes in the subtree rooted at node v.

Intuitively, A_2' processes nodes from the leaves to the root of the tree. That is, the algorithm finds a node v with maximal depth such that the tree rooted in v has size at least m. Having found such a node, it packs sets of $m-1$ items into one bin, which implies an efficient utilization of the edge constraint. It proceeds this way until it reaches the root of the tree.

Lemma 3. A_2' produces a preliminary packing with $A_2' \leq \frac{1.5|E|}{m-1} + 2$. It runs in time $O(n \log n)$.

Proof. Let $E_1 := E \setminus E_2$. We first show that $|E_1| + (m-1) \geq |E_2|$. Whenever items are added to E_2 in Line 11, a bin is packed with $m-1$ items not belonging

---

1: $E_2 := \emptyset$
2: **for** $h :=$ height of G **to** 1 **do**
3:    **for all** nodes v with depth h **do**
4:       **if** $|T_v| \geq m$ **then**
5:          Let $C_v = \{v_1, \ldots, v_k\}$ be the children of v such that $|T_{v_1}| \leq \ldots \leq |T_{v_k}|$
6:          $b := 1$
7:          **for** $i := 1$ **to** k **do**
8:             **if** $|T_{v_b}| + \ldots + |T_{v_i}| \geq m - 1$ **then**
9:                Let G' be the subgraph induced by $\{v\} \cup T_{v_b} \cup \ldots \cup T_{v_i}$
10:                Pack $m - 1$ items from G' incident to at most m nodes into new bin
11:                Put all remaining items from G' into E_2
12:                Remove G' (except for v if $i \neq k$) from G
13:                $b = i + 1$
14: Add any remaining items in G to E_2
15: Greedily pack the items from E_2 into at most $\frac{|E_2|}{m/2}$ bins

---

Fig. 6. A'_2 constructing a preliminary packing

to E_2 in Line 10. Additionally, in this case at most $m - 1$ items are added to E_2 because of the following reasoning. Assume to the contrary that more than $m-1$ items are added. This could only happen if $|T_{v_b}| + \ldots + |T_{v_i}| \geq 2m - 1$ holds in Line 8 at some point during the execution of the algorithm, since G' contains at most $|T_{v_b}| + \ldots + |T_{v_i}|$ items and $m - 1$ items are packed in Line 10. In this case $|T_{v_i}| \geq m$ needs to hold. However, then there would have been an earlier iteration of the for-loop in Line 2 in which $|T_{v_i}|$ was removed or $|T_{v_i}|$ became smaller than m, which is a contradiction. Consequently, $|E_1| \geq |E_2|$ before the execution of Line 14 and then at most $m - 1$ items may be added to E_2, yielding $|E_1| + (m - 1) \geq |E_2|$.

Also, it is always possible to pack the items from E_2 in $\lceil \frac{|E_2|}{m/2} \rceil$ bins in Line 14 since $\frac{m}{2}$ items can be incident to at most m nodes and thus, the edge constraint is met.

Therefore, we obtain

$$A'_2 \leq \left\lceil \frac{|E_1|}{m-1} \right\rceil + \left\lceil \frac{|E_2|}{m/2} \right\rceil \leq \frac{|E_1|}{m-1} + \frac{1.5|E_2|}{m} + \frac{0.5|E_2|}{m} + 2$$

$$\leq \frac{|E_1|}{m-1} + \frac{1.5|E_2|}{m} + \frac{0.5(|E_1| + (m - 1))}{m-1} + 2 \leq \frac{1.5|E|}{m-1} + 2.5 \ .$$

Concerning the runtime of A'_2 one can see that it can be implemented such that it runs in $O(n \log n)$ time. In a preprocessing step we can root the tree (if necessary) and compute the values $|T_{v_i}|$ and depth of all nodes by applying a depth-first search, which takes $O(n)$ time. Then, we essentially visit each node twice (once in the two upper level loops and once in the loop in Line 7) and the overall runtime is dominated by sorting nodes in Line 5. Hence, A'_2 has a runtime of $O(n \log n)$ (as each node is only sorted once). $\qquad \square$

Given a solution of A_2', we can simply transform it into a feasible packing by reallocating (parts of) items into new bins such that no capacity constraint is violated. To this end, algorithm A_2 considers each overfull bin B and greedily takes (parts of) items of overall size 1 out of B and places them in new bins until B's capacity constraint is met.

Theorem 5. A_2 *has an asymptotic approximation ratio of* $\frac{5}{2}$. *It has a runtime of* $O(n \log n)$.

Proof. By Lemma 3 we know that $A_2' \leq \frac{1.5|E|}{m-1} + 2.5$. Also we know that for each bin B filled with items of some size s in the solution of A_2', algorithm A_2 opens at most $\lceil s \rceil - 1 < s$ additional bins. Then we have $A_2 \leq \frac{1.5|E|}{m-1} + 2.5 + \sum s_i$. Since $\text{OPT} \geq \max \left\{ \frac{|E|}{m-1}, \sum s_i \right\}$ we obtain an approximation ratio of $\frac{\frac{1.5|E|}{m-1}}{\frac{|E|}{m-1}} + \frac{2.5}{\text{OPT}} + \frac{\sum s_i}{\sum s_i} \leq 2.5 \cdot \left(1 + \frac{1}{\text{OPT}}\right)$ for A_2, proving the theorem. $\square$

References

1. Collaborative Research Centre 901 - On-The-Fly-Computing. http://sfb901. uni-paderborn.de
2. Brinkmann, A., Kling, P., auf der Heide, F.M., Nagel, L., Riechers, S., Süß, T.: Scheduling shared continuous resources on many-cores. In: Proceedings of the 26th ACM Symposium on Parallelism in Algorithms and Architectures (SPAA 2014), pp. 128–137. ACM (2014)
3. Chung, F., Graham, R., Mao, J., Varghese, G.: Parallelism versus memory allocation in pipelined router forwarding engines. Theory Comput. Syst. **39**(6), 829–849 (2006)
4. de la Vega, W.F., Lueker, G.S.: Bin packing can be solved within 1+epsilon in linear time. Combinatorica **1**(4), 349–355 (1981)
5. Dósa, G., Li, R., Han, X., Tuza, Z.: Tight absolute bound for first fit decreasing bin-packing: FFD(l) ≤ 11/9 OPT(L) + 6/9. Theoret. Comput. Sci. **510**, 13–61 (2013)
6. Epstein, L., Levin, A., van Stee, R.: Approximation schemes for packing splittable items with cardinality constraints. Algorithmica **62**(1–2), 102–129 (2012)
7. Epstein, L., van Stee, R.: Approximation schemes for packing splittable items with cardinality constraints. In: Kaklamanis, C., Skutella, M. (eds.) WAOA 2007. LNCS, vol. 4927, pp. 232–245. Springer, Heidelberg (2008)
8. Epstein, L., van Stee, R.: Improved results for a memory allocation problem. Theory Comput. Syst. **48**(1), 79–92 (2011)
9. Happe, M., auf der Heide, F.M., Kling, P., Platzner, M., Plessl, C.: On-the-fly computing: a novel paradigm for individualized IT services. In: Proceedings of the 16th IEEE International Symposium on Object/Component/Service-Oriented Real-Time Distributed Computing (ISORC 2013), pp. 1–10. IEEE Computer Society (2013)
10. Karmarkar, N., Karp, R.M.: An efficient approximation scheme for the one-dimensional bin-packing problem. In: Proceedings of the 23rd Annual Symposium on Foundations of Computer Science (FOCS 1982), pp. 312–320. IEEE Computer Society (1982)

Cost-Efficient Scheduling on Machines from the Cloud

Alexander Mäcker$^{(\boxtimes)}$, Manuel Malatyali, Friedhelm Meyer auf der Heide, and Sören Riechers

Heinz Nixdorf Institute & Computer Science Department,
Paderborn University, Paderborn, Germany
{alexander.maecker,manuel.malatyali,fmadh,soeren.riechers}@upb.de

Abstract. We consider a scheduling problem where machines need to be rented from the cloud in order to process jobs. There are two types of machines available which can be rented for machine-type dependent prices and for arbitrary durations. However, a machine-type dependent setup time is required before a machine is available for processing. Jobs arrive online over time, have machine-type dependent sizes and have individual deadlines. The objective is to rent machines and schedule jobs so as to meet all deadlines while minimizing the rental cost.

Since we observe the slack of jobs to have a fundamental influence on the competitiveness, we study the model when instances are parameterized by their (minimum) slack. An instance is called to have a slack of β if, for all jobs, the difference between the job's release time and the latest point in time at which it needs to be started is at least β. While for $\beta < s$ no finite competitiveness is possible, our main result is an $O(c/\varepsilon + 1/\varepsilon^3)$-competitive online algorithm for $\beta = (1 + \varepsilon)s$ with $1/s \leq \varepsilon \leq 1$, where s and c denotes the largest setup time and the cost ratio of the machine-types, respectively. It is complemented by a lower bound of $\Omega(c/\varepsilon)$.

1 Introduction

Cloud computing provides a new concept for provisioning computing power, usually in the form of virtual machines (VMs), that offers users potential benefits but also poses new (algorithmic) challenges to be addressed. On the positive side, main advantages for users of moving their business to the cloud are manifold: Possessing huge computing infrastructures as well as asset and maintenance cost associated with it are no longer required and hence, costs and risks of large investments are significantly reduced. Instead, users are only charged to the extent they actually use computing power and it can be scaled up and down depending on current demands. In practice, two ways of renting machines from the cloud are typically available – on-demand-instances without long-term commitment, and reserved instances. In the former case, machines can be rented

This work was partially supported by the German Research Foundation (DFG) within the Collaborative Research Centre "On-The-Fly Computing" (SFB 901).

© Springer International Publishing AG 2016
T.-H.H. Chan et al. (Eds.): COCOA 2016, LNCS 10043, pp. 578–592, 2016.
DOI: 10.1007/978-3-319-48749-6_42

for any duration and the charging is by the hour (Amazon EC2 [2]) or by the minute (Google Cloud [9]); whereas in the latter case, specific (long-term) leases of various lengths are available. We focus on the former case and note that the latter might better be captured within the leasing framework introduced in [15].

Despite these potential benefits, the cloud also bears new challenges in terms of renting and utilizing resources in a (cost-)efficient way. Typically, the cloud offers diverse VM types differing in the provided resources and further characteristics. One might think of different machine types and each machine either being a high-CPU instance, which especially suits the requirements for compute-intensive jobs, or a high-memory instance for I/O-intensive jobs. Another example is the distinction of CPU and GPU instances. Certain computations can be accelerated by the use of GPUs and in [11], for instance, it is observed that, for certain workloads, one can significantly improve performance when not only using classical CPU but additionally GPU instances in Amazon EC2. Hence, the performance of a job can strongly depend on the VM type on which it is executed and one should account for this in order to be cost-efficient while guaranteeing a good performance, for example, in terms of user-defined due dates or deadlines implicitly given by a desired quality of service level. Consequently, introducing the option for a scheduling algorithm to choose between machine types seems to be reasonable and necessary.

Also, despite the fact that computing power can be scaled according to current demands, one needs to take into account that this scaling is not instantaneous and it initially could take some time for a VM to be ready for processing workload. A recent study [12] shows such setup times to typically be in the range of several minutes for common cloud providers and hence, to be non-ignorable for the overall performance [13].

1.1 Problem Description

The problem CLOUDSCHEDULING is extracted from the preceding observations.

Machine Environment. There are two types $\tau \in \{A, B\}$ of machines, which can be rented in order to process workload: A machine M of type τ can be opened at any time a_M and, after a setup taking some non-negligible s_τ time units was performed, can be closed at any time b_M with $b_M \geq a_M + s_\tau$, for values $a_M, b_M, s_\tau \in \mathbb{R}_{\geq 0}$. It can process workload during the interval $[a_M + s_\tau, b_M]$ and the rental cost incurred by a machine M is determined by the duration for which it is open and is formally given by $c_\tau \cdot (b_M - a_M)$. Note that a machine cannot process any workload nor can be closed during the setup. However, the user is still charged for the duration of the setup. We refer to the cost $c_\tau \cdot s_\tau$ incurred by a machine during its setup as *setup cost* and to the cost $c_\tau \cdot (b_M - a_M - s_\tau)$ incurred during the remaining time it is open as *processing cost*. Without loss of generality, we assume that $s_B \geq s_A$, $c_A = 1$ and define $c := c_B$. We restrict ourselves to the case $c \geq 1$ (which is consistent with practical observations where c is usually in the range of 1 to a few hundred [2,12]). However, one can obtain similar results for $c < 1$ by an analogous reasoning.

Job Characteristics. The workload of an instance is represented by a set J of n jobs, where each job j is characterized by a release time $r_j \in \mathbb{R}_{\geq 0}$, a deadline $d_j \in \mathbb{R}_{>0}$, and sizes $p_{j,\tau} \in \mathbb{R}_{>0}$ for all $\tau \in \{A, B\}$, describing the processing time of job j when assigned to a machine of type τ. Observe that therefore machines of the same type are considered to be identical while those of different types are unrelated. Throughout the paper, we assume by using a suitable time scale that $p_{j,\tau} \geq 1$ for all $j \in J$ and all $\tau \in \{A, B\}$. The jobs arrive online over time at their release times at which the scheduler gets to know the deadline and sizes of a job. Each job needs to be completely processed by *one* machine before its respective deadline, i.e., to any job j we have to assign a *processing interval* I_j of length $p_{j,\tau}$ that is contained in j's *time window* $[r_j, d_j] \supseteq I_j$ on a machine M of type τ that is open during the entire interval $I_j \subseteq [a_M + s_\tau, b_M]$. A machine M of type τ is called *exclusive* machine for a job j if it only processes j and $b_M - a_M = s_\tau + p_{j,\tau}$. We define the *slack* of a job j as the amount by which j is shiftable in its window: $\sigma_{j,\tau} := d_j - r_j - p_{j,\tau}$ for $\tau \in \{A, B\}$.

Objective Function. The objective of CLOUDSCHEDULING is to rent machines and compute a feasible schedule that minimizes the rental cost, i.e., $\sum_M c_{\tau(M)} \cdot (b_M - a_M)$, where $\tau(M)$ denotes the type of machine M.

We analyze the quality of our algorithms in terms of their competitiveness and assume that problem instances are parameterized by their *minimum slack*. An instance is said to have a minimum slack of β if $\max_\tau \{\sigma_{j,\tau}\} \geq \beta$, for all $j \in J$. Then, for a given β, an algorithm is called ρ-*competitive* if, on any instance with minimum slack β, the rental cost is at most by a factor ρ larger than the cost of an optimal offline algorithm.

In the following, by OPT we denote an optimal schedule as well as the cost it incurs. Throughout the paper (in all upper and lower bounds), we assume that OPT is not too small and in particular, OPT $= \Omega(c \cdot r_{max})$, where $r_{max} = \max_{j \in J} r_j$. This bound is true, for example, if the optimal solution has to maintain at least one open machine (of the more expensive machine type) during (a constant fraction of) the considered time horizon. This assumption seems to be reasonable, particularly for large-scale systems where, at any time, the decision to make is rather concerned with dozens of machines than whether a single machine is rented at all. Similar assumptions are made in the literature [4], where it is argued (for identically priced machines) that the case where OPT $= o(r_{max})$ is of minor interest as workload and costs are very small.

1.2   Related Work

Cloud scheduling has recently attracted the interest of theoretical researchers. Azar et al. [4] consider a scheduling problem where jobs arrive online over time and need to be processed by identical machines rented from the cloud. While machines are paid in a pay-as-you-go manner, a fixed setup time T_s is required before the respective machine is available for processing. In this setting, Azar et al. consider a bicriterial optimization problem where the rental cost is to be minimized while guaranteeing a reasonable maximum delay. An online algorithm that is $(1 + \varepsilon, O(1/\varepsilon))$-competitive regarding the cost and the maximum delay, respectively, is provided. A different model for cloud scheduling was considered by Saha [17]. She considers jobs that arrive over time and need to be finished before their respective deadlines. To process jobs, identical machines are available and need to be rented from the cloud. The goal is to minimize the rental cost where a machine that is rented for t time units incurs cost of $\lceil t/D \rceil$ for some fixed D. The problem is considered as an offline as well as an online problem and algorithms that guarantee solutions incurring costs of $O(\alpha)$OPT (where α is the approximation factor of the algorithm for machine minimization in use, see below) and $O(\log(p_{max}/p_{min}))$OPT, respectively, are designed.

A different, but closely related problem is that of *machine minimization*. In this problem, n jobs with release times and deadlines are considered and the objective is to finish each job before its deadline while minimizing the number of machines used. This problem has been studied in online and offline settings for the general and different special cases. The first result is due to [16] where an offline algorithm with approximation factor of $O(\log n/\log \log n)$ is given. This was later improved to $O(\sqrt{\log n/\log \log n})$ by Chuzhoy et al. [6]. Better bounds have been achieved for special cases; if all jobs have a common release date or equal processing times, constant approximation factors are achieved [19]. In the online case, a lower bound of $\Omega(n + \log(p_{max}/p_{min}))$ and an algorithm matching this bound is given in [17]. For jobs of equal size, an optimal e-competitive algorithm is presented in [7].

A further area of research that studies rental/leasing problems from an algorithmic perspective and which recently gained attention is that of *resource leasing*. Its focus is on infrastructure problems and while classically these problems are modeled such that resources are bought and then available all the time, in their leasing counterparts resources are assumed to be rented only for certain durations. In contrast to our model, in the leasing framework resources can not be rented for arbitrary durations. Instead, there is a given number K of different leases with individual costs and durations for which a resource can be leased. The leasing model was introduced by Meyerson [15] and problems like FACIL-ITYLEASING or SETCOVERLEASING have been studied in [1,3,10] afterward.

A last problem worthmentioning here is the problem of *scheduling with calibrations*[5,8]. Although it does not consider the aspect of minimizing resources and the number of machines is fixed, it is closely related to machine minimization and shares aspects with our model. There is given a set of m (identical) machines and a set of jobs with release times, deadlines and sizes. After a machine is calibrated at a time t, it is able to process workload in the interval $[t, t+T]$, for some fixed T, and the goal is to minimize the number of calibrations. For sufficiently large m, the problem is similar to a problem we need to solve in Sect. 3.1.

1.3 Our Results

We study algorithms for CLOUDSCHEDULING where jobs need to be scheduled on machines rented from the cloud such that the rental costs are minimized subject to quality of service constraints. Particularly, the problem introduces the possibility for a scheduler to choose between different machine types being heterogeneous in terms of prices and processing capabilities. It also captures the fact that due to times for preparing machines and acquiring resources, available computing power does not scale instantaneously. Our results show the competitiveness to heavily depend on the minimum slack β jobs exhibit. While for $\beta < s_B$ no finite competitiveness is possible, a trivial rule achieves an optimal competitiveness of $\Theta(cs_A + s_B)$ for $\beta = (1+\varepsilon)s_B$, $0 \leq \varepsilon < 1/s_B$. For $1/s_B \leq \varepsilon \leq 1$, we present an algorithm which is up to an $O(1/\varepsilon^2)$-factor optimal. Its competitiveness is $O(c/\varepsilon + 1/\varepsilon^3)$, while a general lower bound is proven to be $\Omega(c/\varepsilon)$.

2   Preliminaries

We begin our study by showing that setup times are hard to cope with for any online algorithm when the minimum slack is below a natural threshold. This is the case because an online algorithm needs to hold machines ready at any time for arriving jobs with small deadlines.

Proposition 1. *If $\beta < s_B$, no online algorithm has a finite competitiveness.*

Proof. Suppose there is an online algorithm with competitiveness k, where k is some constant. We observe that there needs to be a time t at which the algorithm has no machine of type B open. This is true because otherwise, the adversary can release only one job with $r_1 = 0$, $p_{1,B} = 1$, $d_1 = s_B + p_{1,B} + \beta$ and $p_{1,A} > s_B - s_A + p_{1,B} + \beta$. Then, the cost for the optimal algorithm is only $c \cdot (s_B + 1)$, whereas the online algorithm has cost of at least $c \cdot t$, which is a contradiction to the competitiveness as it grows in t and hence, without bound.

Therefore, let t be a time where the online algorithm has no machine of type B open. The adversary releases one job j with $r_j = t$, $d_j = r_j + p_{j,B} + \beta$ and an arbitrary size $p_{j,B}$ and $p_{j,A} > p_{j,B} + \beta$. To finish this job without violating its deadline, it needs to be started not after $t + \beta$ on a machine of type B. However, this is not possible since there is no machine of type B available at time t and the setup of a type B machine needs time $s_B > \beta$. Hence, any online algorithm must rent a machine of type B all the time, which is a contradiction.   □

Due to this impossibility result, we restrict our studies to cases where the minimum slack is $\beta = (1+\varepsilon)s_B$ for some $\varepsilon \geq 0$. It will turn out that for very small values of ε, $0 \leq \varepsilon < 1/s_B$, there is a high lower bound on the competitiveness and essentially, we cannot do better than processing each job on its own machine. However, for larger values of ε, the situation clearly improves and leaves room for designing non-trivial online algorithms. The next lemma is proven similarly to Proposition 1. The adversary releases a whole bunch of jobs as soon as an online algorithm has no machine available forcing it to use machines of the unfavorable type. For the few proofs (mostly fairly straightforward) that are omitted throughout the paper, the interested reader is refered to the full version [14].

Lemma 1. *If $\beta = (1+\varepsilon)s_B$, $1/s_B \leq \varepsilon \leq 1$, there is a lower bound on the competitiveness of $\Omega(c/\varepsilon)$. For $0 \leq \varepsilon < 1/s_B$ a lower bound is $\Omega(cs_A + s_B)$.*

2.1   Simple Heuristics

As a first step of studying algorithms in our model we discuss some natural heuristics. One of the doubtlessly most naive rules simply decides on the machine type to process a job on based on the cost it incurs on this type. The algorithm A_1 assigns each job to its own machine and chooses this machine to be of type A if $s_A + p_{j,A} \leq c(s_B + p_{j,B})$ and $\sigma_{j,A} \geq s_A$, or if $\sigma_{j,B} < s_B$. Otherwise, it chooses the machine to be of type B. A simple calculation yields the following proposition.

Proposition 2. *If* $\beta = (1+\varepsilon)s_B$ *and* $0 \leq \varepsilon < 1/s_B$, A_1 *is* $\Theta(cs_A + s_B)$-*competitive. For* $1/s_B \leq \varepsilon \leq 1$ *its competitiveness is* $\Theta(c/\varepsilon + s_B)$.

While this trivial rule is optimal for $0 \leq \varepsilon < 1/s_B$, for larger values of ε the dependence of the competitiveness on the setup time is undesired as it can be quite high and in particular, it is sensitive to the time scale (and recall that we chose a time scale such that the smallest processing time of any job is at least 1). Therefore, the rest of the paper is devoted to finding an algorithm with a competitiveness being independent of s_B and narrowing the gap to the lower bound of $\Omega(c/\varepsilon)$ for $1/s_B \leq \varepsilon \leq 1$.

One shortcoming of A_1 is the fact that jobs are never processed together on a common machine. A simple idea to fix this and to batch jobs is to extend A_1 by an ANYFIT rule as known from bin packing problems (e.g. see [18]). Such an algorithm dispatches the jobs one by one and only opens a new machine if the job to be assigned cannot be processed on any already open machine. Then, the job is assigned to some machine it fits into. It turns out that the competitiveness of this approach still depends on the setup time s_B, which we show by proving a slightly more general statement. Consider the class GREEDYFIT consisting of all deterministic heuristics ALG fitting into the following framework:

1. At each time t at which the jobs $J_t = \{j \in J : r_j = t\}$ arrive, ALG processes them in *any* order $j_1, j_2, \dots$
   1.1 If j_i cannot *reasonably* be processed on any open machine, ALG opens *some* machine.
   1.2 ALG assigns j_i to *some* open machine.
2. ALG *may* close a machine as soon as it is about to idle.

The terms *some, any* and *may* above should be understood as to be defined by the concrete algorithm. A job is said to be *reasonably* processable on a machine M if it does not violate any deadline and the processing cost it incurs does not exceed the cost for setting up and processing the job on a new machine.

Unfortunately, one can still easily trick such an algorithm to open a machine being unfavorable (though reasonable) for all upcoming jobs to process.

Proposition 3. *If* $\beta \geq s_B$, *any* GREEDYFIT *algorithm is* $\Omega(c/\varepsilon + s_B)$-*competitive.*

3   An $O(c/\varepsilon + 1/\varepsilon^3)$-Competitive Algorithm

Due to the discussed observations, it seems that decisions on the type of a machine to open as well as finding a good assignment of jobs requires more information than given by a single job. The rough outline of our approach is as follows. In a first step, we identify a variant of CLOUDSCHEDULING that we can solve by providing an Integer Linear Program (ILP). This formulation is heavily based on some structural lemma that we prove next before turning to the actual ILP. In a second step, we describe how to use the ILP solutions, giving infeasible subschedules, to come up with a competitive algorithm. Throughout the description we make the assumption that ε is known to the algorithm in advance, which, however, can easily be dropped by maintaining a guess on ε.

We now show a fundamental lemma that provides a way of suitably batching the processing of jobs and structuring the rental intervals. We define intervals of the form $[is_\tau, (i+1)s_\tau)$, for $i \in \mathbb{N}_0$ and $\tau \in \{A, B\}$, as the i-th τ-interval. Intuitively, we can partition the considered instance into subinstances each consisting only of jobs released during one B-interval and the times during which machines are open are aligned with these τ-intervals.

Lemma 2. *By losing a constant factor, we may assume that* OPT *fulfills the following properties:*

1. *each job j that is processed on a machine of type τ and fulfills $p_{j,\tau} \geq s_\tau$ is assigned to an exclusive machine,*
2. *each remaining machine M of type τ is open for exactly five τ-intervals, i.e., $[a_M, b_M) = [is_\tau, (i+5)s_\tau)$ for some $i \in \mathbb{N}_0$, and*
3. *if it is opened at $a_M = (i-1)s_\tau$, the beginning of the $(i-1)$-th τ-interval, it only processes jobs released during the i-th τ-interval.*

Proof. We prove the lemma by describing how to modify OPT to establish the three properties while increasing its cost only by a constant factor. Consider a fixed type $\tau \in \{A, B\}$ and let $J_\tau \subseteq J$ be the set of jobs processed on machines of type τ in OPT. Let J_M be the set of jobs processed by a machine M.

Property 1. Any job $j \in J_\tau$ with $p_{j,\tau} \geq s_\tau$ is moved to a new exclusive machine of type τ if not yet scheduled on an exclusive machine in OPT. This increases the cost due to an additional setup and the resulting idle time by $s_\tau + p_{j,\tau} \leq 2p_{j,\tau}$. Applying this modification to all jobs j with $p_{j,\tau} \geq s_\tau$ therefore increases the cost of OPT by a factor of at most three and establishes the desired property. From now on, we will assume for simplicity that all jobs j scheduled on a machine of type τ fulfill $p_{j,\tau} < s_\tau$.

Property 2. Next, we establish the property that each remaining machine is open for exactly four τ-intervals; when establishing the third property, this is extended to five intervals as claimed in the lemma. Consider a fixed machine M. Partition the time during which M is open into intervals of length s_τ by defining $I_k := [a_M + ks_\tau, a_M + (k+1)s_\tau)$ for $k \in \{1, \ldots, \lceil (b_M - a_M)/s_\tau - 1 \rceil\}$. Let t_j be the starting time of job $j \in J_M$ and let $T_k = \{j \in J_M : t_j \in I_k\}$ partition the jobs of J_M with respect to the interval during which they are started. We replace machine M by $\lceil (b_M - a_M)/s_\tau - 1 \rceil$ machines $M_1, M_2, \ldots$ such that M_i processes the jobs from T_i keeping the jobs' starting times as given by OPT and setting $a_{M_i} := \min_{j \in T_i} t_j - s_\tau$ and $b_{M_i} := \max_{j \in T_i} (t_j + p_{j,\tau})$. Observe that the cost originally incurred by M is $c_\tau (b_M - a_M)$ while those incurred by the replacing machines $M_1, M_2, \ldots$ are at most $c_\tau (b_M - a_M) + c_\tau (\lceil (b_M - a_M)/s_\tau - 1 \rceil) s_\tau \leq 2c_\tau (b_M - a_M)$ where the additional term stems from the additional setups. Since it holds that $p_{j,\tau} < s_\tau$ for all $j \in J_M$, we conclude that $b_{M_i} - a_{M_i} \leq 3s_\tau$ for all M_i. For each M_i, we can now simply decrease a_{M_i} and increase b_{M_i} such that M_i is open for exactly four τ-intervals. This maintains the feasibility and increases the cost incurred by each machine M_i from at least $c_\tau s_\tau$ to at most $4c_\tau s_\tau$. Applying these modifications to all machines M, establishes the claimed property while increasing the overall cost by a factor of at most eight.

Property 3. It remains to prove the third property. Again consider a fixed machine M. Let $N_{i,M} := \{j \in J_M : r_j \in [is_\tau, (i+1)s_\tau)\}$, $i \in \mathbb{N}_0$, be the (sub-)set of jobs released during the i-th τ-interval and processed by M. Furthermore, let $\#(M) := |\{i : N_{i,M} \neq \emptyset\}|$ be the number of τ-intervals from which M processes jobs. If $\#(M) \leq 3$, we replace M by $\#(M)$ machines M_i each processing only jobs from $N_{i,M}$. To define the schedule for M_i, let $N_{i,M} = \{j_{i_1}, j_{i_2}, \ldots\}$ such that $t_{i_1} < t_{i_2} < \ldots$ holds. We reset the starting times of the jobs in $N_{i,M}$ by setting $t_{i_1} := r_{i_1}$ and $t_{i_k} := \max\{r_{i_k}, t_{i_{k-1}} + p_{i_{k-1},\tau}\}$, for $k > 1$, and set $a_M := (i-1)s_\tau$ and $b_M := a_M + 5s_\tau$. This gives a feasible schedule for $N_{i,M}$ since no starting time is increased and according to Property 2 we have $\sum_{j \in N_{i,M}} p_{j,\tau} \leq 3s_\tau$ and hence, $\max_{j \in N_{i,M}} r_j + \sum_{j \in N_{i,M}} p_{j,\tau} < (i+4)s_\tau = b_M$. Also, the replacing machines fulfill the three properties and the cost is increased by a factor less than four.

For the complementary case where $\#(M) > 3$, we argue as follows. Observe that due to the already established property that M is open for exactly four τ-intervals, $N_{i,M} = \emptyset$ for all $i \geq a_M/s_\tau + 4$. Note that $a_M/s_\tau \in \mathbb{N}_0$ due to Property 2. Also, for each $i \in \{a_M/s_\tau - 1, a_M/s_\tau, a_M/s_\tau + 2, a_M/s_\tau + 3\}$, we can move the jobs from $N_{i,M}$ to new machines by an argument analogous to the one given for the case $\#(M) \leq 3$ above. Hence, we have established the property that M only processes jobs $j \in \bigcup_{i=0}^{a_M/s_\tau - 2} N_{i,M} \cup N_{a_M/s_\tau + 1, M}$ and applying the modification to all machines M from the set $\mathcal{M}$ of machines fulfilling $\#(M) > 3$, increases the cost by a constant factor.

In order to finally establish the third property, our last step proves how we can reassign all those jobs j with $j \in \bigcup_{i=0}^{a_M/s_\tau - 2} N_{i,M}$ for some $M \in \mathcal{M}$. Let $J' = \bigcup_{M \in \mathcal{M}} \bigcup_{i=0}^{a_M/s_\tau - 2} N_{i,M}$ be the set containing these jobs and partition them according to their release times by defining $N_i = (\bigcup_{M \in \mathcal{M}} N_{i,M}) \cap J'$. Note that for any $j \in N_i$ it holds $d_j \geq a_M + s_\tau \geq (i+3)s_\tau$, for all $i \in \mathbb{N}_0$. Let $w_i := \sum_{j \in N_i} p_{j,\tau}$. We can assign all jobs from N_i to new machines fulfilling the three properties as follows: We open $\lceil w_i/s_\tau \rceil$ new machines at time $(i-1)s_\tau$ and keep them open for exactly five τ-intervals. Due to the fact that for all jobs $j \in N_i$ it holds $r_j \leq (i+1)s_\tau$ and $d_j \geq (i+3)s_\tau$, we can accommodate a workload of at least s_τ in the interval $[(i+1)s_\tau, (i+3)s_\tau]$ on each machine by assigning jobs from N_i to it in any order. By these modifications the cost increase due to cases where $w_i \geq s_\tau$ is given by an additive of at most $c_\tau \sum_{i:w_i \geq s_\tau} \lceil w_i/s_\tau \rceil 5s_\tau$, which is $O(\text{OPT})$ since $\text{OPT} \geq c_\tau \sum_{i:w_i \geq s_\tau} w_i$. The overall increase in the cost due to the cases where $w_i < s_\tau$ is given by an additive of $c_\tau \sum_{i:w_i < s_\tau} 5s_\tau = O(\text{OPT})$ due to our assumption $\text{OPT} = \Omega(c \cdot r_{max})$. □

By Lemma 2, we can partition any instance into subinstances such that the i-th subinstance consists of those jobs released during the interval $[is_B, (i+1)s_B)$ and solve them separately. Hence, we assume without loss of generality that the entire instance only consists of jobs released during the interval $[0, s_B)$.

3.1   Tentative Subschedules

Before we turn to describing our algorithm for CLOUDSCHEDULING, we first identify and describe a special problem variant that will be helpful when designing our algorithm. In this variant, we assume setup costs instead of setup times, i.e., setups do not take any time but still incur the respective cost. Under this relaxed assumption, the goal is to compute a schedule for a whole batch of (already arrived) jobs. Precisely, for a given time t, we consider a subset $J' \subseteq J$ of jobs such that $r_j \leq t$ for all $j \in J'$ and our goal is to compute a schedule for J' that minimizes the rental cost and in which each job is finished at least s_B time units before its deadline (has *earliness* at least s_B). We do not require the starting times of jobs to be at least t and hence, a resulting schedule may be infeasible for CLOUDSCHEDULING if realized as computed. Therefore, to emphasize its character of not being final, we call such a schedule a *tentative schedule*.

We now divide the jobs into three sets according to their sizes. Let $J_1 := \{j \in J : p_{j,A} \geq s_A \wedge p_{j,B} \geq s_B\}$ contain those jobs for which the processing cost dominates the setup cost on both machine types. The set $J_2 := \{j \in J : p_{j,A} < s_A \wedge p_{j,B} < s_B\}$ contains those jobs for which the processing cost does not dominate the setup cost on either of the two machine types. Finally, $J_3 := J \backslash (J_1 \cup J_2)$ and we write $J_3 = J_{3,1} \cup J_{3,2} = \{j \in J_3 : p_{j,A} \geq s_A \wedge p_{j,B} < s_B\} \cup \{j \in J_3 : p_{j,A} < s_A \wedge p_{j,B} \geq s_B\}$. By Lemma 2, we directly have the following lemmas.

Lemma 3. *For an optimal schedule for jobs from J_2, we may assume each machine of type $\tau \in \{A, B\}$ to be open for exactly five τ-intervals. If a job $j \in J_2$ is processed on a machine of type τ and $r_j \in [is_\tau, (i+1)s_\tau)$, $i \in \mathbb{N}_0$, its processing interval is completely contained in the interval $[is_\tau, (i+4)s_\tau]$.*

Lemma 4. *For an optimal schedule for jobs from $J_{3,1}$, we may assume each machine of type B to be open for exactly five B-intervals. If a job $j \in J_{3,1}$ is processed on a machine of type A, it is processed on an exclusive machine and if j is processed on a machine of type B and $r_j \in [is_B, (i+1)s_B)$, $i \in \mathbb{N}_0$, its processing interval is completely contained in the interval $[is_B, (i+4)s_B]$.*

Analogous statements hold for jobs from $J_{3,2}$.

By Lemmas 3 and 4, we can now formulate the problem (which is NP-hard by the NP-hardness of classical BINPACKING) as an ILP (cf. Fig. 1) and compute $O(1)$-approximate tentative schedules. We use $\mathcal{I}_\tau(j)$ to denote all possible processing intervals of job j on a machine of type $\tau \in \{A, B\}$. Intuitively,

$$\min \quad 5s_A \sum_i z_i + 5cs_B \cdot z_B + \sum_{\substack{x(I,j):I \in \mathcal{I}_A(j) \\ j \in J_1 \cup J_{3,1}}} x(I,j)p_{j,A} + c \sum_{\substack{x(I,j):I \in \mathcal{I}_B(j) \\ j \in J_1 \cup J_{3,2}}} x(I,j)p_{j,B}$$

$$\text{s.t.} \quad \sum_{\substack{j \in J_2 \cup J_{3,2}: \\ r_j \in [(i-1)s_A, is_A)}} \sum_{I \in \mathcal{I}_A(j):t \in I} x(I,j) \le z_i \qquad \forall t \in L_i, \forall i \in \{1, \dots, {}^{s_B}/{s_A}\} \quad (1)$$

$$\sum_{j \in J_2 \cup J_{3,1}} \sum_{I \in \mathcal{I}_B(j):t \in I} x(I,j) \le z_B \qquad \forall t \in L_B \quad (2)$$

$$\sum_{I \in \mathcal{I}_A(j) \cup \mathcal{I}_B(j)} x(I,j) = 1 \qquad \forall j \in J \quad (3)$$

$$x(I,j) \in \{0,1\} \qquad \forall j \in J, \forall I \in \mathcal{I}_A(j) \cup \mathcal{I}_B(j) \quad (4)$$

Fig. 1. Integer linear program for variant with setup cost.

$\mathcal{I}_A(j) \cup \mathcal{I}_B(j)$ describes all possible ways how job j can be scheduled. Note that $\mathcal{I}_A(j)$ and $\mathcal{I}_B(j)$ are built under the assumptions from Lemmas 2, 3 and 4. For each $I \in \mathcal{I}_A(j) \cup \mathcal{I}_B(j)$, the indicator variable $x(I,j)$ states if job j is processed in interval I. We use L_B to denote all left endpoints of intervals in $\bigcup_{j \in J_2 \cup J_{3,1}} \mathcal{I}_B(j)$ and L_i to denote all left endpoints of intervals in $\bigcup_{j \in J_2 \cup J_{3,2}:r_j \in [(i-1)s_A, is_A)} \mathcal{I}_A(j)$. Additionally, we use a variable z_B to denote the number of (non-exclusive) machines of type B that we rent. The variable z_i describes the number of machines of type A that we rent and that process jobs released during the $(i-1)$-th A-interval. For simplicity we assume that s_B is an integer multiple of s_A.

We are now asked to minimize the cost for machines of type B plus those for machines of type A, taking into account that each machine of type τ is either open for a duration of exactly $5s_\tau$ time units (first two summands of the objective function) or is an exclusive machine (last two summands). The constraints of type (1) and type (2) ensure that, at any point in time, the number of jobs processed on (non-exclusive) machines does not exceed the number of open machines. Additionally, constraints of type (3) and type (4) ensure that each job is completely processed by exactly one machine in a contiguous interval.

Observe that in general this ILP has an infinite number of variables since each $\mathcal{I}_\tau(j)$ contains all possible processing intervals of j on a machine of type τ. Yet, there is an efficient way (adapted from [6], cf. full version [14]) to reduce the number of variables that need to be considered to $O(|J|^2)$ such that afterward a solution only being by a constant factor larger than the optimal one of the original formulation exists. Note that we solve the resulting ILP optimally and this may take non-polynomial time.

3.2 The BatchedDispatch Algorithm

In this section, we describe and analyze our competitive algorithm, which is essentially based on several observations concerning how to restrict and then turn tentative schedules, as described in the previous section, into feasible solutions.

We first show the following lemma, which relates the cost of an optimal schedule to the cost of one where jobs are finished early. For simplicity we use $\Delta := \frac{1}{2}\varepsilon s_B$.

Lemma 5. *There is a schedule with all jobs having earliness at least s_B, cost $O((^c/_\varepsilon + ^1/_{\varepsilon^2})\text{OPT})$ and no machine processes any two jobs j, j' with $r_j \in [i\Delta, (i+1)\Delta)$ and $r_{j'} \in [i'\Delta, (i'+1)\Delta)$ with $i \neq i'$ and $i, i' \in \mathbb{N}_0$.*
   Furthermore, the cost for jobs $j \in \bar{J} := \{j : \exists \tau \in \{A, B\}\ r_j + p_{j,\tau} > d_j - s_B\}$ is $O((^c/_\varepsilon + ^1/_{\varepsilon^2})\text{OPT})$. The cost for jobs $j \in J \backslash \bar{J}$ is $O(^1/_{\varepsilon^2}\text{OPT})$.

We are now ready to describe our approach. The formal description is given in Fig. 2. It relies on Δ and its relation to the slack of a job (*after* decreasing the d_j in Step 1.1) and uses the following partition of J into the sets:

$$J'_1 = \{j : \sigma_{j,B} \geq 2\Delta \wedge (\sigma_{j,A} \geq 2\Delta \vee p_{j,A} \leq \Delta)\},$$
$$J'_2 = \{j : \sigma_{j,B} \geq 2\Delta \wedge (\sigma_{j,A} < 2\Delta \wedge p_{j,A} > \Delta)\},$$
$$J'_3 = \{j : \sigma_{j,B} < 2\Delta \wedge \sigma_{j,A} \geq 2\Delta\}.$$

Note that $\{j : \sigma_{j,B} < 2\Delta \wedge \sigma_{j,A} < 2\Delta\} = \emptyset$, since $\max\{\sigma_{j,A}, \sigma_{j,B}\} \geq (1 + \varepsilon)s_B - s_B = \varepsilon s_B = 2\Delta$. Before proving the correctness and bounds on the cost, we shortly describe the high level ideas skipping technical details, which will become clear during the analysis. The algorithm proceeds in phases, where each phase is devoted to scheduling jobs released during an interval of length Δ. In a given phase, we first decrease the deadlines of jobs by s_B (cf. Step1) to ensure that later on (cf. Step 2.2) tentative schedules meeting these modified deadlines can be extended by the necessary setups without violating any original deadline. We also precautionary open machines for jobs that are required to be started early. Then at the end of the phase (cf. Step 2), we compute tentative schedules with additional restrictions on starting times and machines to use (cf. Step 2.1), which are carefully defined depending on the characteristics of jobs concerning their slacks. This approach ensures that we can turn solutions into feasible schedules while guaranteeing that costs are not increased too much. The feasibility is crucial since tentative schedules are not only delayed due to the added setups but also because of computing the schedules only at the end of a phase.

In the next lemmas, the analysis of the algorithm is carried out, leading to our main result in Theorem 1. Recall that due to Lemma 5 it is sufficient to show that the schedules for a single phase are feasible and the costs are increased by a factor of $O(1)$ and $O(^1/_\varepsilon)$, respectively, in comparison to the tentative schedules if they were computed without the additional restrictions of Step 2.1.

Lemma 6. *For jobs from J'_1, BATCHEDDISPATCH produces a feasible schedule with rental cost $O(^c/_\varepsilon + ^1/_{\varepsilon^2})\text{OPT}$.*

Proof. Fix an arbitrary phase i and define $N_i := \{j \in J'_1 : r_j \in [(i-1)\Delta, i\Delta)\}$ to be the set of jobs released during the i-th phase. Let S be the tentative schedule for N_i and let t_j denote the starting time of job $j \in N_i$ in S. Let S' and t'_j be defined analogously for the respective schedule produced by BATCHEDDISPATCH.

BatchedDispatch(ε)

Let $C = (s_B \geq s_A + \Delta \wedge \frac{s_A}{c} > \Delta)$.

In phase $i \in \mathbb{N}$ process all jobs j with $r_j \in [(i-1)\Delta, i\Delta)$:

1. Upon arrival of a job j at time t
   1.1 Set $d_j = d_j - s_B$.
   1.2 Classify j to belong to the set J_1', J_2' or J_3'.
       If C holds, also define $J_{2,1}' := J_2'$ and $J_{2,2}' := \emptyset$.
       Otherwise, $J_{2,1}' := \{j \in J_2' : c p_{j,B} > s_A\}$ and $J_{2,2}' := J_2' \setminus J_{2,1}'$.
   1.3 If C does not hold, open a machine of type A at t for each $j \in J_{2,1}'$.
2. At time $i\Delta$
   2.1 Compute tentative schedules for the job sets with these additional restrictions:
       – Starting times of jobs from J_1', $J_{2,2}'$ and J_3' are at least $i\Delta$,
       – starting times of jobs from $J_{2,1}'$ on machines of type B are at least $i\Delta$ and
       – jobs from $J_{2,2}'$ (J_3') only use machines of type B (type A).
   2.2 Realize the tentative schedules by increasing the starting times of jobs from
       – J_1' by s_τ on machines of type τ,
       – $J_{2,1}'$ by s_B on machines of type B and by $s_A + \Delta$ or s_A depending on whether C holds or not on machines of type A,
       – $J_{2,2}'$ by s_B and
       – J_3' by s_A,
       and doing the respective setups at time $i\Delta$. Use machines from Step 1.3 for $J_{2,1}'$ if necessary and otherwise close them after finishing the setups.

Fig. 2. Description of BatchedDispatch(ε) algorithm for CloudScheduling.

We first reason about the feasibility of S'. First of all, the tentative schedule S provides a solution with $t_j \geq i\Delta$ and $t_j + p_{j,\tau} \leq d_j$ for all $j \in N_1$ processed on a machine of type τ, i.e., no modified deadline (as given after Step 1 is executed) is violated and no job is started before time $i\Delta$. Since the starting times are increased in S' to $t_j' = t_j + s_\tau$, each job j processed on a machine of type τ is not started before $i\Delta + s_\tau$ and is finished by $t_j' + p_{j,\tau} \leq t_j + s_\tau + p_{j,\tau} \leq d_j + s_\tau$. Therefore, by starting the setups at $i\Delta$, S' is a feasible solution.

Second, we have to prove the bound on the cost of S'. Let S^* be the tentative schedule for N_i if computed without the restrictions on the starting times. Since the cost of S' are not larger than those of S, we need to show that the cost of S only increase by a constant factor in comparison to S^*, which according to Lemma 5 has cost $O((c/\varepsilon + 1/\varepsilon^2)\mathrm{OPT})$. Let $N_i^\tau \subseteq N_i$ be the set of jobs processed on machines of type τ in S^*, for $\tau \in \{A, B\}$. Consider an arbitrary machine M of type τ in S^* and let $J_M = \{j_1, j_2, \ldots\}$ denote the jobs processed on M such that $t_{j_1}^* < t_{j_2}^* < \ldots$ holds. Let $k \in \mathbb{N}$ such that $t_{j_k}^* + p_{j_k,\tau} < i\Delta$ and $t_{j_{k+1}}^* + p_{j_{k+1},\tau} \geq i\Delta$. Now, we can leave all jobs $j \in \{j_{k+2}, j_{k+3}, \ldots\}$ unaffected (i.e. $t_j = t_j^*$) and we can process j_{k+1} on an exclusive machine started at time $t_{j_{k+1}} = t_{j_{k+1}}^* = i\Delta > r_{j_{k+1}}$. This is feasible since j_{k+1} is finished at time $i\Delta + p_{j_{k+1},\tau} \leq d_{j_{k+1}}$. The last inequality either holds since $p_{j_{k+1}} \leq \Delta$ and $d_{j_{k+1}} \geq 2\Delta + r_{j_{k+1}}$ because $\sigma_{j_{k+1},B} \geq 2\Delta$, or since $\sigma_{j_{k+1},\tau} = d_{j_{k+1}} - r_{j_{k+1}} - p_{j_{k+1},\tau} \geq 2\Delta$. All remaining jobs $j \in \{j_1, \ldots, j_k\}$ can be moved to a new machine M' of type τ and $t_j = t_j^* + \Delta$. Each job $j \in \{j_1, \ldots, j_k\}$ is then finished at time $t_j + p_{j,\tau} = t_j^* + p_{j,\tau} + \Delta \leq d_j$.

The last inequality either holds since $d_j \geq 2\Delta + r_j \geq (i+1)\Delta \geq t_j^* + p_{j,\tau} + \Delta$, or since $t_j^* < r_j + \Delta$ and $d_j - r_j - p_{j,\tau} \geq 2\Delta$. Also, $t_{j_q}^* + \Delta + p_{j_q,\tau} \leq t_{j_{q+1}}^* + \Delta$ for all $q \in \{1, \ldots, k-1\}$. Hence, $t_j \geq i\Delta$ for all $j \in J_M$ and the cost at most triples in comparison to S^* since the two additional machines need not run longer than M. Applying the argument to all machines M gives $S \leq 3S^* = O((c/\varepsilon + 1/\varepsilon^2)\text{OPT})$, which concludes the proof. $\square$

Lemma 7. *For jobs from J_2' and J_3', BATCHEDDISPATCH produces a feasible schedule with rental cost $O(c/\varepsilon + 1/\varepsilon^3)\text{OPT}$.*

Proof. We first consider the set J_2' and argue about $J_{2,1}'$ and $J_{2,2}'$ separately.

Scheduling jobs from $J_{2,1}'$. Fix an arbitrary phase i and let $N_i := \{j \in J_{2,1}' : r_j \in [(i-1)\Delta, i\Delta)\}$. Let S be the tentative schedule for N_i and let t_j denote the starting time of job $j \in N_i$ in S. Let S' and t_j' be defined analogously for the respective schedule produced by BATCHEDDISPATCH. Let $N_i^\tau \subseteq N_i$ be the set of jobs processed on machines of type τ in S, for $\tau \in \{A, B\}$. Recall that we compute a tentative schedule for jobs from $J_{2,1}'$ at time $i\Delta$ with the additional constraint $t_j \geq i\Delta$ for all $i \in N_i^B$, that is, the starting times of all jobs processed on machines of type B are at least $i\Delta$. Similar to the previous lemma, with respect to machines of type B, S' is feasible and fulfills the bound on the cost. For machines of type A we can argue as follows. In case C holds, we have $t_j' = t_j + \Delta + s_A \geq i\Delta + s_A$ for all $j \in N_i^A$. Hence, we are able to realize the respective schedule by starting the respective setup processes at time $i\Delta$. Also, since $\Delta + s_A \leq s_B$ by the fact that C holds, $t_j' + p_{j,A} \leq t_j + \Delta + s_A + p_{j,A} \leq d_j + \Delta + s_A \leq d_j + s_B$, i.e., no original deadline is violated. Hence, S' is feasible and the bound on the cost is satisfied by a reasoning as in Lemma 6.

In case C does not hold, recall that at each arrival of a job $j \in J_{2,1}'$, we open a new machine of type A at time r_j. This ensures that a machine is definitely available for the respective job $j \in N_i^A$ at time $t_j + s_A$. Therefore, we obtain a feasible schedule and it only remains to prove the bound on the cost. Note that the additional setup cost in S' compared to S is upper bounded by $|J_{2,1}'| \cdot s_A$. Because $s_A < cp_{j,B}$ holds and S can process at most three jobs on a machine of type A (since $\sigma_{j,A} = d_j - r_j - p_{j,A} < 2\Delta$ implying $d_j < 2\Delta + p_{j,A} + r_j$ and $p_{j,A} > \Delta$ by the definition of J_2'), $|J_{2,1}'| \cdot s_A \leq 3S$ holds. Therefore, BATCHEDDISPATCH produces a feasible schedule for jobs from $J_{2,1}'$ incurring cost of $O(c/\varepsilon + 1/\varepsilon^2)\text{OPT}$.

Scheduling jobs from $J_{2,2}'$. Fix an arbitrary phase i and let $N_i := \{j \in J_{2,2}' : r_j \in [(i-1)\Delta, i\Delta)\}$. For these jobs we claim that the tentative schedule S has cost $O(c/\varepsilon + 1/\varepsilon^3)\text{OPT}$. Then, the same reasoning as in Lemma 6 proves the desired result. Hence, we only have to show the claim. To this end, we show how to shift any workload processed on a machine of type A in a tentative schedule S^* corresponding to S without the additional restriction on machine types to use to machines of type B. Recall that we only need to consider the case that C does not hold since otherwise $J_{2,2}' = \emptyset$. Let $N_i^A \subseteq N_i$ be the jobs processed on machines of type A in S^*. Note that $N_i^A \cap \{j : \exists \tau \in \{A, B\} \ r_j + p_{j,\tau} > d_j - s_B\} = \emptyset$ and thus according to Lemma 5, the cost for jobs from N_i^A is $O(1/\varepsilon^2)\text{OPT}$. We show that only using machines of type B increases the cost by $O(1/\varepsilon)$ and distinguish two cases depending on whether $\frac{s_A}{c} \leq \Delta$ holds.

First, we show the claim if $\frac{s_A}{c} \leq \Delta$ holds. On the one hand, any machine of type A in S^* can process at most three jobs j with $j \in N_i^A$ (since $\sigma_{j,A} = d_j - r_j - p_{j,A} < 2\Delta$ implying $d_j < 2\Delta + p_{j,A} + r_j$ and $p_{j,A} > \Delta$ by the definition of J_2'), leading to (setup) cost of $\Omega(|N_i^A|s_A)$. On the other hand, we can schedule all jobs from N_i^A on $O(\frac{|N_i|s_A}{\Delta c})$ machines of type B each open for $O(s_B)$ time units and thus, with cost of $O(\frac{|N_i^A|s_A}{\Delta c}cs_B)$. To see why this is true, observe that for all jobs $j \in N_i^A$ it holds $[r_j, d_j] \supseteq [i\Delta, (i+1)\Delta] =: I_i$ since $r_j \leq i\Delta$ and $\sigma_{j,A} \geq 2\Delta$. Because all jobs $j \in N_i^A$ fulfill $p_{j,B} \leq s_A/c$ by definition, we can accommodate $\lfloor c\Delta/s_A \rfloor \geq 1$ many jobs from N_i^A in I_i. Hence, there is a schedule with cost $O(1/\varepsilon^3)\text{OPT}$ only using machines of type B, proving the claim.

Finally, it needs to be proven that the claim also holds for the case $\frac{s_A}{c} > \Delta$ and hence, $s_B < s_A + \Delta$. Let $M_1, \ldots, M_m$ denote the machines of type A used by S^* and let $\kappa_1, \ldots, \kappa_m$ be the durations for which they are open. Since $p_{j,B} \leq p_{j,A}$ for all $j \in J_{2,2}'$, we can replace each machine $M \in \{M_1, \ldots, M_m\}$ by a machine M' of type B using the same schedule on M' as on M. We increase the cost by a factor of at most $\frac{\sum_{i=1}^m c(s_B + \kappa_i)}{\sum_{i=1}^m (s_A + \kappa_i)} \leq \frac{cs_B}{s_A} + c \leq \frac{c(\Delta + s_A)}{s_A} + c \leq \frac{c\Delta}{s_A} + 2c = O(s_A/\Delta) = O(1/\varepsilon)$, where the second last bound holds due to $s_A/c > \Delta$. Thus, we have shown a schedule fulfilling the desired properties and cost $O(1/\varepsilon^3)\text{OPT}$ to exist and since S fulfills the same bound on the cost, this proves the claim.

Since all jobs from J_3' can be assumed to be scheduled on machines of type A without any loss in the cost, we obtain the same result for J_3' as we have in Lemma 6. □

Theorem 1. *For* $\beta = (1 + \varepsilon)s_B$, $1/s_B \leq \varepsilon \leq 1$, BatchedDispatch *is* $O(c/\varepsilon + 1/\varepsilon^3)$*-competitive.*

4 Conclusion

We presented a competitive algorithm for CloudScheduling where jobs with deadlines need to be scheduled on machines rented from the cloud so as to minimize the rental cost. We parameterized instances by their minimum slack β and showed different results depending thereon. Alternatively, we could examine the problem without this instance parameter and instead understand β as a parameter of the algorithm describing the desired maximum tardiness of jobs.

Although the considered setting with $k = 2$ types of machines seems to be restrictive, it turned out to be challenging and closing the gap between our lower and upper bound remains open. Also, it is in line with other research (e.g. [4,17]) and in this regard it is a first step toward models for scheduling machines from the cloud addressing the heterogeneity of machines. For future work, however, it would be interesting to study even more general models for arbitrary $k > 1$.

References

1. Abshoff, S., Markarian, C., Meyer auf der Heide, F.: Randomized online algorithms for set cover leasing problems. In: Zhang, Z., Wu, L., Xu, W., Du, D.-Z. (eds.) COCOA 2014. LNCS, vol. 8881, pp. 25–34. Springer, Heidelberg (2014). doi:10.1007/978-3-319-12691-3_3

2. Amazon EC2. https://aws.amazon.com/ec2/
3. Anthony, B.M., Gupta, A.: Infrastructure leasing problems. In: Fischetti, M., Williamson, D.P. (eds.) IPCO 2007. LNCS, vol. 4513, pp. 424–438. Springer, Heidelberg (2007). doi:10.1007/978-3-540-72792-7_32
4. Azar, Y., Ben-Aroya, N., Devanur, N.-R., Jain, N.: Cloud scheduling with setup cost. In: Proceedings of the 25th ACM Symposium on Parallelism in Algorithms and Architectures (SPAA 2013), pp. 298–304. ACM (2013)
5. Bender, M.A., Bunde, D.P., Leung, V.J., McCauley, S., Phillips, C.A.: Efficient scheduling to minimize calibrations. In: Proceedings of the 25th ACM Symposium on Parallelism in Algorithms and Architectures (SPAA 2013), pp. 280–287. ACM (2013)
6. Chuzhoy, J., Guha, S., Khanna, S., Naor, J.: Machine minimization for scheduling jobs with interval constraints. In: Proceedings of the 45th Symposium on Foundations of Computer Science (FOCS 2004), pp. 81–90. IEEE (2004)
7. Devanur, N.R., Makarychev K., Panigrahi, D., Yaroslavtsev, G.: Online Algorithms for Machine Minimization. CoRR (2014). 1403.0486
8. Fineman, T.J., Sheridan, B.: Scheduling non-unit jobs to minimize calibrations. In: Proceedings of the 27th ACM Symposium on Parallelism in Algorithms and Architectures (SPAA 2015), pp. 161–170. ACM (2015)
9. Google Cloud. https://cloud.google.com/
10. Kling, P., Meyer auf der Heide, F., Pietrzyk, P.: An algorithm for online facility leasing. In: Even, G., Halldórsson, M.M. (eds.) SIROCCO 2012. LNCS, vol. 7355, pp. 61–72. Springer, Heidelberg (2012). doi:10.1007/978-3-642-31104-8_6
11. Lee, G., Chun, B.-G., Katz, R.H.: Heterogeneity-aware resource allocation and scheduling in the cloud. In: Proceedings of the 3rd USENIX Workshop on Hot Topics in Cloud Computing (HotCloud 2011). USENIX (2001)
12. Mao, M., Humphrey, M.: A performance study on the VM startup time in the cloud. In: Proceedings of the 2012 IEEE 5th International Conference on Cloud Computing (ICCC 2012), pp. 423–430. IEEE (2012)
13. Mao, M., Li, J., Humphrey, M.: Cloud auto-scaling with deadline and budget constraints. In: Proceedings of the 2010 11th IEEE/ACM International Conference on Grid Computing (GRID 2010), pp. 41–48. IEEE (2010)
14. Mäcker, A., Malatyali, M., Meyer auf der Heide, F., Riechers, S.: Cost-Efficient Scheduling on Machines from the Cloud. CoRR (2016). 1609.01184
15. Meyerson, A.: The parking permit problem. In: Proceedings of the 46th Annual Symposium on Foundations of Computer Science (FOCS 2005), pp. 274–282. IEEE (2005)
16. Raghavan, P., Thompson, C.D.: Randomized rounding: a technique for provably good algorithms and algorithmic proofs. Combinatorica 7(4), 365–374 (1987)
17. Saha, B.: Renting a cloud. In: Proceedings of the Annual Conference on Foundations of Software Technology and Theoretical Computer Science (FSTTCS 2013), pp. 437–448. LIPIcs (2013)
18. Sgall, J.: Online bin packing: old algorithms and new results. In: Beckmann, A., Csuhaj-Varjú, E., Meer, K. (eds.) CiE 2014. LNCS, vol. 8493, pp. 362–372. Springer, Heidelberg (2014). doi:10.1007/978-3-319-08019-2_38
19. Yu, G., Zhang, G.: Scheduling with a minimum number of machines. Oper. Res. Lett. 37(2), 97–101 (2009)

Strategic Online Facility Location

Maximilian Drees, Björn Feldkord$^{(\boxtimes)}$, and Alexander Skopalik

Heinz Nixdorf Institute, Paderborn University, Paderborn, Germany
{maxdrees,bjoernf,skopalik}@mail.upb.de

Abstract. In this paper we consider a strategic variant of the online facility location problem. Given is a graph in which each node serves two roles: it is a strategic client stating requests as well as a potential location for a facility. In each time step one client states a request which induces private costs equal to the distance to the closest facility. Before serving, the clients may collectively decide to open new facilities, sharing the corresponding price. Instead of optimizing the global costs, each client acts selfishly. The prices of new facilities vary between nodes and also change over time, but are always bounded by some fixed value α. Both the requests as well as the facility prices are given by an online sequence and are not known in advance.

We characterize the optimal strategies of the clients and analyze their overall performance in comparison to a centralized offline solution. If all players optimize their own competitiveness, the global performance of the system is $\mathcal{O}(\sqrt{\alpha} \cdot \alpha)$ times worse than the offline optimum. A restriction to a natural subclass of strategies improves this result to $\mathcal{O}(\alpha)$. We also show that for fixed facility costs, we can find strategies such that this bound further improves to $\mathcal{O}(\sqrt{\alpha})$.

Keywords: Online algorithms · Competitive analysis · Facility location · Algorithmic game theory

1 Introduction

In this paper, we consider selfish clients who each have to access a common essential IT service at irregular points in time. The clients can be regarded as intermediaries, offering services to customers while outsourcing some of the work to external service providers. As a result, a client does not know in advance when he needs to access an external service or how often he needs to do this over some period of time. The clients are arranged in an overlay network, with some of them having direct access to the service. Those without have to use the overlay network to access the closest instance of the service of another client. The resulting delay in time can be considered as additional individual costs for the client accessing the service. Depending on how often the service is needed and by whom, it can be profitable to establish an additional service instance

This work was partially supported by the German Research Foundation (DFG) within the Collaborative Research Centre "On-The-Fly Computing" (SFB 901).

T.-H.H. Chan et al. (Eds.): COCOA 2016, LNCS 10043, pp. 593–607, 2016.
DOI: 10.1007/978-3-319-48749-6_43

at one of the clients. This creates further (non-recurrent) costs but improves the accessibility of all nearby clients, which in turn share the establishing costs among each other. When receiving their own service instance, different clients are supplied by different service providers and as a result, the prices for establishing a new service instance can differ between the clients. In addition, these prices may change over time. Similar to accessing the service, the clients do not know how these prices will develop in the future.

We analyze this problem in a new setting that combines concepts from both online algorithms and algorithmic game theory. Online algorithms deal with problems in which the input is only partially known at the start of the computation and further information arrives over time. Our model is based on the classic online facility location problem, with a facility representing an instance of the essential service. The overlay network between the clients induces an undirected graph, with each node representing a client as well as a possible position for a facility. At each point in time, one of the clients has to serve a request (we say a request occurs for that client), which increases his individual costs by the shortest distance to the closest facility. The clients do not know the stream of requests in advance. Before a request is served, all clients decide whether to open up new facilities and where. In existing models, these decisions are made by a centralized online algorithm which determines the actions of all clients. For such an algorithm, the goal is to minimize the global costs of all clients combined. In this paper, however, the clients are selfishly trying to minimize their individual costs without any interest in how their actions affect the other clients or the system as a whole. The interaction of such selfish players is the topic of algorithmic game theory. When deciding on a new facility on a certain node, the clients collect sealed bids, i.e. no one knows the bids of the other. If their total sum exceeds the current price of opening a new facility on the node, then that facility is created and the individual cost of each player grows by his own bid. This process iterates over all possible facility locations in an arbitrary order. The prices of the new facilities are independent between the nodes and can also change over time. We require the prices only to be within the interval $[1, \alpha]$ for a constant α at all times. Just like with the requests, the clients do not know in advance how theses prices will develop.

Each client uses his own individual online algorithm which determines his bids. We analyze the competitiveness of such an algorithm in comparison to an optimal offline algorithm that knows all requests and prices in advance. This value also depends on the online algorithms of the other clients. We are interested in equilibria in which each client uses the best-possible online algorithm, assuming the decisions of the other clients are fixed.

Related Work. The Online Facility Location Problem has been introduced by Meyerson [8]. He gave an $\mathcal{O}(1)$-competitive algorithm for inputs arriving in a random order and a $\mathcal{O}(\log n)$-competitive algorithm for an adversarial order of inputs, where n is the total number of clients.

For general instances, Fotakis [4] showed that the best possible competitive ratio that is achievable for the problem is $\Theta\left(\frac{\log n}{\log \log n}\right)$. The lower bound does not carry over to our model since we allow the opening of facilities only on nodes in a graph where we have uniform distances.

The problem was also considered in a leasing variant, where facilities are not bought for an infinite amount of time but only for a fixed time span. There are several options on how much to pay for a certain amount of time in this problem. Kling et al. [7] developed the first algorithm for this problem with a competitive ratio independent of the input length.

Furthermore, Facility Location was also considered under game theoretical aspects [5]. A fixed number of facilities was assigned to positions determined by a mechanism that made the decision on the basis of the agents reporting their positions strategically. In contrast, in our model agents invest money into facilities directly as their strategic options.

An online problem that is in some ways similar to the Facility Location Problem is the Page Replication Problem, where clients need access to a page that can be copied to different nodes in the network. However, the costs of the copies are not fixed but proportional to the distance between an existing copy of the page and the new copy.

For general graphs it is shown in [2] that the competitive ratio for this problem is $\Theta(\log n)$. However, it was shown in [1] that for graph classes like trees and rings there are algorithms that achieve a constant competitive ratio.

In our model, prices are not fixed but fluctuate over time. The prices are also determined by an adversary in the online setting. Price fluctuations in a rent or buy mechanic have already been considered by Bienkowski [3], who showed for the Ski-Rental Problem that no online algorithm can do better than $\Theta(\sqrt{s})$ where s is the fixed ratio between renting and buying costs, even if prices only may change by an amount of 1 per time step.

Immorlica et al. [6] introduced the concept of dueling algorithms where two algorithms try to outperform each other on a search problem. The payoffs of the algorithms are like in a zero-sum game, meaning that benefits of one player directly translate to losses of another player. In our model, all players benefit from resources which they acquire collectively. In this sense, players cannot directly harm others without passing on beneficial resources themselves.

Our Contribution. We propose a variant of facility location where each client minimizes its local competitive ratio. We show that optimal strategies are $\mathcal{O}(\sqrt{\alpha})$-competitive and achieve overall costs that are at most $\mathcal{O}(\sqrt{\alpha} \cdot \alpha)$ times worse than the costs of an optimal offline solution.

We can improve the result by a factor of $\sqrt{\alpha}$ if we restrict the strategies to algorithms that buy a facility as soon as the accumulated costs have reached the costs of the facility on the node of the respective player. We further show a possible improvement of $\sqrt{\alpha}$ if the costs for facilities are fixed to α and that even under this condition, no centralized algorithm can achieve a constant competitiveness independent of α.

2 Preliminaries

An instance I of the Strategic Online Facility Location Problem is given by a graph $G = (V, E)$, a node $f_0 \in V$ and a constant $\alpha > 1$. Let $n = |V|$. Each node in the graph is a player as well as a possible location for a facility. We will mostly use the letter v when referring to nodes in their function as players and f in their function as facility locations. Initially, there is only one facility in G and it is located at f_0. We denote by $d(x, y)$ the shortest path distance between nodes x and y.

Requests arrive as an online sequence $\phi = v^1, v^2, \ldots, v^T$ with $v^t \in V$ for all time steps t. So in each time step, exactly one player v^t is chosen and has to serve a request. Let $F^t \subseteq V$ be the set of open facilities in step t and $F^0 = \{f_0\}$. Then the connection costs of player v^t induced by the request in step t is $\min_{f \in F^t} d(v^t, f)$.

Before serving each request, the players may collectively decide to open some new facilities. The price for this may change over time and can differ between facility locations. The prices $p^t = (p_f^t)_{f \in V}$ arrive as an online sequence $\pi = p^1, p^2, \ldots, p^T$, with $p_f^t \in [1, \alpha]$ being the price for opening a new facility on $f \in V$ in round t.

The decision to open a facility is made as follows. In each time step t, a bidding phase takes place after the current request v^t and the prices p^t have been revealed, but before the request has to be served. The bidding phase iterates over the unopened facilities in an arbitrary order. For each facility f, each player v submits a sealed bid $b_{v,f}^t$. If the sum of all bids for a location exceeds the price p_f^t, a facility is opened at that location and each player is charged his bid. The decision whether a facility is opened is made before the bidding on the next facility takes place so that players can adapt their behavior on the nodes accordingly. Thus, the total cost of the requested node v^t in time step t is $c_{v^t}^t = \min_{f \in F^t} d(v^t, f) + \sum_{f \in F^t \setminus F^{t-1}} b_{v^t, f}^t$ and for all remaining nodes v it is $c_v^t = \sum_{f \in F^t \setminus F^{t-1}} b_{v,f}^t$.

Each player v invokes an (online) algorithm $\mathcal{A}_v$ that determines his bids in each time step. The cumulative cost of a player for a given online sequence (ϕ, π) is denoted by $c_v := \sum_{t=1}^{T} c_v^t$ and defined as the sum of costs for serving his requests, plus the sum of his accepted bids. In lieu of standard game theoretic models, we assume the player to exhibit a very pessimistic and cautious behavior. Each player assumes the online sequence and even the strategies of all other players to be adversarial. Hence, he seeks to minimize his worst-case performance loss of the chosen online algorithm in comparison to the optimal solution in hindsight. For online algorithms, this is well known as the competitive ratio. Let β_{-v} denote the bids of the players except player v for all time steps and all facilities. Then the *local competitive ratio* of player v is $\rho_v := \sup_{I, \phi, \pi, \beta_{-v}} \left\{ \frac{c_v}{c_v^*} \right\}$ where c_v^* are the costs of the optimal offline solution to the optimization goal of player v.

If every player v chooses an algorithm A_v, we call $\mathcal{A} = (\mathcal{A}_v)_{v \in V}$ a strategy profile. We are interested in strategy profiles where each player chooses an algorithm with a minimal local competitive ratio, such that each player perceives his behavior as optimal. However, we allow the players to deviate from the best competitive ratio by a constant factor[1]. Hence, we define an equilibrium as a strategy profile in which each player chooses an online algorithm with an asymptotically optimal competitive ratio.

Denote by S_A the set of equilibria. We call the sum of all players' cost $C(A) = \sum_{v \in V} c_v$ the social costs of A. We seek to quantify the loss of efficiency due to the lack of coordination by the users and the uncertainty about the request sequence. We write C_{Opt} for the costs of the optimal centralized offline solution and call the worst-case cost ratio between an equilibrium and the optimal offline solution the online price of anarchy $\text{PoA} := \max_{I,\phi,\pi} \max_{A \in S_A} \left\{ \frac{C(A)}{C_{\text{Opt}}} \right\}$ and the cost ratio with respect to the best equilibrium the online price of stability $\text{PoS} := \max_{I,\phi,\pi} \min_{A \in S_A} \left\{ \frac{C(A)}{C_{\text{Opt}}} \right\}$.

3 Dynamic Facility Costs

We start to analyze our model by establishing a lower bound on the local competitive ratio.

Theorem 1. *The local competitive ratio is $\Omega(\sqrt{\alpha})$ for any strategy A_v.*

Proof. Consider a graph with two nodes v and f_0 which are connected by an edge (recall that f_0 is the initially opened facility). We construct an online sequence such that the price on v is $\sqrt{\alpha}$ in the first time step and α afterwards. There is a request on v in the first time step. If node v decides to bid at least $\sqrt{\alpha}$ on this first time step, the sequence stops and v has costs of at least $\sqrt{\alpha}$ for buying the facility. The optimal solution does not bid anything and has distance costs of 1, so we get a cost ratio of $\sqrt{\alpha}$.

If the node v bids less than $\sqrt{\alpha}$, the sequence continues with at least α more requests on v. Node v has a cost of at least $\alpha + 1$, whereas the optimal solution has a cost of $\sqrt{\alpha}$. The resulting cost ratio is at least $\alpha+1/\sqrt{\alpha} > \sqrt{\alpha}$. □

We now know that $\mathcal{O}(\sqrt{\alpha})$-competitive algorithms are strategies that are chosen in an equilibrium. In a later section we state an explicit algorithm which is $\mathcal{O}(\sqrt{\alpha})$-competitive, so we know that such an algorithm exists. Intuitively, such an algorithm should increase his bids when his connection costs increase and the bids for more distant facilities should be smaller. We denote $B(t, d, d_A)$ to be the bid of an algorithm A_v of node v at time step t for a facility at distance d if its currently closest facility is located at distance d_A. We define a class of algorithms by giving bounds on $B(t, d, d_A)$ such that an algorithm can choose a bid within these bounds at any time step and is $\mathcal{O}(\sqrt{\alpha})$-competitive as a result. Note that we assume that each algorithm has a value B_α for which it holds that if the costs accumulated in total by the player are at least B_α, it will bid α on its own node and thereby terminate the bidding process for itself.

---

[1] This may be justified since an optimal algorithm might be hard to determine for the players.

Definition 1. *The bids of an algorithm $\mathcal{A}_v$ are called bounded if the bids satisfy*

$$B(t, d, d_A) \leq \mathcal{O}(\sqrt{\alpha}) \cdot C_{Opt}(t+1) - C_{\mathcal{A}_v}(t) - d \tag{1}$$

$$B(t, d, d_A) \geq \begin{cases} \frac{B_\alpha + \alpha}{\mathcal{O}(\sqrt{\alpha})} - C_{Opt}(t), & d = 0 \\ \frac{C_{\mathcal{A}_v}(t) + \alpha}{\mathcal{O}(\sqrt{\alpha})} - C_{Opt}(t), & d > 0, \frac{d_A}{\mathcal{O}(\sqrt{\alpha})} - d \leq 0 \\ \Omega(\sqrt{\alpha}) + (B_\alpha - C_{\mathcal{A}_v}(t)) \left(\frac{1}{\mathcal{O}(\sqrt{\alpha})} - \frac{d}{d_A} \right), & d > 0, \frac{d_A}{\mathcal{O}(\sqrt{\alpha})} - d > 0 \end{cases}$$
$$\tag{2}$$

where $C_{\mathcal{A}_v}(t)$ and $C_{Opt}(t)$ are the costs gathered by $\mathcal{A}_v$ and its corresponding offline optimum, assuming that the online sequence ends after time step $t-1$, $d_A := \min_{f \in F^t} d(v, f)$, d is the distance to the potential facility that $\mathcal{A}_v$ is bidding on, and $\sqrt{\alpha} \leq B_\alpha \leq \sqrt{\alpha} \cdot \alpha$ are the accumulated costs which cause $\mathcal{A}_v$ to buy a facility for costs α on v.

Note that $C_{Opt}(t+1)$ can be calculated by an algorithm $\mathcal{A}_v$ in step t since the request is announced at the beginning of step t, and $C_{Opt}(t+1)$ represents the cost before serving the request at step $t+1$.

Theorem 2. *Any bounded algorithm is locally $\mathcal{O}(\sqrt{\alpha})$-competitive.*

Proof. Let $v \in V$ and $\mathcal{A}_v$ be a bounded algorithm. We may assume w.l.o.g. that there is a request on v in every time step for the benefit of easier notation. First, we describe the costs of $\mathcal{A}_v$ using the upper bound on the bids from the definition. Let t^* be the time step where $\mathcal{A}_v$ last had to pay for a new facility. Let B_{t^*} be the bid paid in the step t^* and $C_{\mathcal{A}_v}(t)$, as in the definition. Then the costs of the algorithm are

$$C_{\mathcal{A}_v} = C_{\mathcal{A}_v}(t^*) + B_{t^*} + \sum_{t=t^*}^{T} \min_{f \in F^t} d(v, f)$$
$$\leq \mathcal{O}(\sqrt{\alpha}) \cdot C_{Opt}(t^*+1) + \sum_{t=t^*+1}^{T} \min_{f \in F^t} d(v, f).$$

If the optimal offline algorithm does not open any facility which is closer than the facility $\mathcal{A}_v$ opened in step t^* via bidding, it holds that

$$C_{Opt} \geq C_{Opt}(t^*+1) + (T - t^*) \cdot \min_{f \in F^{t^*}} d(v, f) \geq \frac{1}{\mathcal{O}(\sqrt{\alpha})} \cdot C_{\mathcal{A}_v}.$$

Otherwise, we consider the last point in time t' where the optimal algorithm bids on a node at distance d_{Opt}, which is closer to v than the facility $\mathcal{A}_v$ gets in t^*. In order for the facility at node v to be opened, Opt must pay at least the bid $\mathcal{A}_v$ made at the same time. In case $d_{Opt} = 0$ we have

$$C_{Opt} \geq C_{Opt}(t') + B(t', 0, d_A) \geq \frac{B_\alpha + \alpha}{\mathcal{O}(\sqrt{\alpha})} \geq \frac{1}{\mathcal{O}(\sqrt{\alpha})} \cdot C_{\mathcal{A}_v}.$$

In case $d_{Opt} > 0$ and $\frac{d_A}{\mathcal{O}(\sqrt{\alpha})} - d \leq 0$ with $d_A := \min_{f \in F^{t'}} d(v, f)$ we get

$$
\begin{aligned}
C_{Opt} &\geq C_{Opt}(t') + B(t', d_{Opt}, d_A) + (T - t')d_{Opt} \\
&\geq \frac{C_{\mathcal{A}_v}(t') + \alpha}{\mathcal{O}(\sqrt{\alpha})} + \frac{1}{\mathcal{O}(\sqrt{\alpha})}(T - t') \min_{f \in F^{t'}} d(v, f) \\
&\geq \frac{1}{\mathcal{O}(\sqrt{\alpha})} \cdot C_{\mathcal{A}_v}.
\end{aligned}
$$

Finally, if $d_{Opt} > 0$ and $\frac{d_A}{\mathcal{O}(\sqrt{\alpha})} - d > 0$ it holds

$$
\begin{aligned}
C_{Opt} &\geq C_{Opt}(t') + B(t', d_{Opt}, d_A) + (T - t')d_{Opt} \\
&\geq C_{Opt}(t') + \Omega(\sqrt{\alpha}) + (B_\alpha - C_{\mathcal{A}_v}(t'))\left(\frac{1}{\mathcal{O}(\sqrt{\alpha})} - \frac{d_{Opt}}{d_A}\right) + (T - t')d_{Opt} \\
&\geq C_{Opt}(t') + \frac{\alpha + B_\alpha - C_{\mathcal{A}_v}(t')}{\mathcal{O}(\sqrt{\alpha})}.
\end{aligned}
$$

Suppose we have $C_{\mathcal{A}_v}(t') \leq \mathcal{O}(\sqrt{\alpha}) \cdot C_{Opt}(t')$, then

$$
\begin{aligned}
C_{Opt} &\geq C_{Opt}(t') + \frac{\alpha + B_\alpha - C_{\mathcal{A}_v}(t')}{\mathcal{O}(\sqrt{\alpha})} \\
&\geq \frac{B_\alpha + \alpha}{\mathcal{O}(\sqrt{\alpha})} \\
&\geq \frac{1}{\mathcal{O}(\sqrt{\alpha})} \cdot C_{\mathcal{A}_v}
\end{aligned}
$$

which we show by an inductive argument: Consider the part of the sequence which ends at time t'. We bound the costs of both $\mathcal{A}_v$ and Opt until that point using the same arguments as above on the reduced sequence.

This concludes the proof since we have shown $C_{\mathcal{A}_v} \leq \mathcal{O}(\sqrt{\alpha}) \cdot C_{Opt}$ for all possible cases. $\qquad\square$

The above result implies that there may be many algorithms for the players which are asymptotically optimal and hence this adds an additional factor of unpredictability for the players to the game.

Note that a bounded algorithm may choose not to bid α on its own location, even though it has already accumulated a cost larger than α but less than $\sqrt{\alpha}\alpha$. We will later show that such algorithms may have undesired properties in contrast to some more natural algorithms.

Definition 2. *An algorithm $\mathcal{A}_v$ is called* frugal *if he bids α for a facility at v if his accumulated costs would otherwise exceed α in the current time step.*

Frugal algorithms contain the class of natural algorithms $\mathcal{A}_v$, which always bid their accumulated costs on v. For our analysis of this class later in this section however, the property stated in the definition is sufficient.

We can make a similar statement about these algorithms as for general $\mathcal{O}(\sqrt{\alpha})$-competitive algorithms by setting $B_\alpha = \alpha$ in the bounds given above.

Corollary 1. *Any frugal algorithm* $\mathcal{A}_v$ *is* $\mathcal{O}(\sqrt{\alpha})$-*competitive if for the bids it holds*

$$B(t, d, d_A) \leq \mathcal{O}(\sqrt{\alpha}) \cdot C_{Opt}(t+1) - C_{\mathcal{A}_v}(t) - d \tag{3}$$

$$B(t, \dot{d}, d_A) \geq \begin{cases} \frac{\alpha}{\mathcal{O}(\sqrt{\alpha})} - C_{Opt}(t), & d = 0 \\ \frac{C_{\mathcal{A}_v}(t) + \alpha}{\mathcal{O}(\sqrt{\alpha})} - C_{Opt}(t), & d > 0, \frac{d_A}{\mathcal{O}(\sqrt{\alpha})} - d \leq 0 \\ \Omega(\sqrt{\alpha}) + (\alpha - C_{\mathcal{A}_v}(t)) \left(\frac{1}{\mathcal{O}(\sqrt{\alpha})} - \frac{d}{d_A} \right), & d > 0, \frac{d_A}{\mathcal{O}(\sqrt{\alpha})} - d > 0. \end{cases} \tag{4}$$

We characterized the optimal strategies for each node, assuming they seek to minimize their competitive ratio and have given a set of algorithms which achieve an optimal competitive ratio. In the next section we develop an algorithm based on the ideas from above which also performs well against a centralized offline solution, given that all players use it. We now study the impact on the overall performance if every node implements some $\mathcal{O}(\sqrt{\alpha})$-competitive algorithm without any further restrictions.

Theorem 3. *If every player uses an* $\mathcal{O}(\sqrt{\alpha})$-*competitive algorithm, the online price of anarchy is* $\Theta(\sqrt{\alpha} \cdot \alpha)$.

Proof. Since every player buys a facility as soon as its accumulated costs would exceed B_α, the costs of every player with at least one request in the online sequence will never exceed $\sqrt{\alpha} \cdot \alpha + \alpha$, and the costs of the other players are 0. In the optimal solution every player with at least one request has costs of at least 1, hence the cost ratio is at most $(\sqrt{\alpha} + 1)\alpha$.

For the proof of the lower bound, we construct a graph where there are n nodes $v_1, \ldots, v_n$ connected to a node x in a star-like formation and the initial facility f_0 is connected to x such that the distance $d(v_i, f_0)$ is $\alpha/4 \cdot \sqrt{\alpha}$. We construct the online sequence such that every possible facility has a cost of α in every time step and there is one request on each of the nodes $v_1, \ldots, v_n$.

We will choose n large enough such that the optimal solution is to open a facility on x in the first time step. The costs of that solution are therefore at most $\alpha + n$.

We determine the strategies of the nodes $v_1, \ldots, v_n$. The cost of the first request was $\alpha/4 \cdot \sqrt{\alpha}$ and since the price of all facilities is α, the costs of the local optimum of v_1 are α after its first request. Hence it is a viable strategy to not make any bids, since v_1 can still serve one request and buy a facility on his node while having costs less than $\sqrt{\alpha} \cdot \alpha$.

Since bidding 0 on all nodes in all following time steps is a viable strategy for v_1, it becomes a viable strategy for all other players as well since they observe the same prices for the facilities. As a result the players never open any facility and the costs of the resulting strategy profile are $\alpha/4 \cdot \sqrt{\alpha} \cdot n$. By choosing n to be at least α, we get the resulting cost ratio. $\square$

The bound for the price of anarchy seems to be unreasonably high since a competitive ratio of α could simply be reached by buying a facility on every node that is requested at least once. The higher factor stems from the fact that the algorithms compare themselves to their local optimum solution and not to the global optimum. If we impose the natural restriction of frugality on the algorithms, we gain a factor of $\sqrt{\alpha}$ in efficiency.

Theorem 4. *If every player uses a frugal $\mathcal{O}(\sqrt{\alpha})$-competitive algorithm, then the online price of anarchy is $\Theta(\alpha)$.*

Proof. We argue analogously to the previous theorem. The costs never exceed 2α for nodes which have at least 1 request and in the optimal solution these costs are at least 1. From this, the upper bound follows directly as before.

For the lower bound, we adapt the construction from the proof of Theorem 3. Again, we have nodes $v_1, \ldots, v_n$ connected to a node x with distance 1 and f_0 connected to x such that for the distances it now holds $d(v_i, f_0) = \alpha/4$.

We use the same request sequence as before and get that the costs of every local optimum is $\alpha/4$. Hence a player can still serve a request without his accumulated cost exceeding α and his competitive ratio becoming worse than $\mathcal{O}(\sqrt{\alpha})$. From this it follows that we can assign strategies such that no player makes any bid.

Hence the costs of the corresponding strategy profile are $n \cdot \Theta(\alpha)$, while the costs of the optimal solution are at most $\alpha + n$. This results in the lower bound for the Price of Anarchy by choosing $n > \alpha$. □

Note that the lower bound in Theorem 4 does not make use of price fluctuations. If we make explicit use of the price changes in our online sequence, we can show that no centralized, deterministic online algorithm and therefore no strategy profile can achieve an asymptotically better competitive ratio.

Theorem 5. *No deterministic online algorithm is better than $\Omega(\alpha)$-competitive.*

Proof. We construct a graph with the initial facility f_0 in the center and α straight lines connecting from f_0 to endpoints $v_1, \ldots, v_\alpha$ such that $d(f_0, v_i) = \alpha$. In the first time step there is a request on a neighbor of f_0 and the prices for all facilities on $v_1, \ldots, v_\alpha$ are 1. For all future time steps, the costs for the facilities are α.

If the algorithm buys all the facilities on $v_1, \ldots, v_\alpha$ then its costs are $\alpha + 1$ and the optimal costs are 1 if the sequence stops after this time step. If the algorithm does not buy a facility on v_i, there is a request on v_i and the costs are $\alpha + 1$, while the optimum can reach costs of 2 by buying a facility on v_i in the first step. □

4   Fixed Facility Costs

We now consider a fixed price of α for opening a facility. Observe that the lower bound for the local competitive ratio of Theorem 1 still holds. We can simulate any price p through a bid of $\alpha - p$ by another player. We have seen that there exist many strategies where the bid in a certain round is not predetermined so

---

Algorithm 1. A competitive algorithm for fixed facility costs.

$n_v \leftarrow 0$;   // number of requests on v
$c_v \leftarrow 0$;   // costs for requests on v
for *each round* $t \in \{1, \ldots, T\}$ **do**
   if *request for* v **then**
      $n_v \leftarrow n_v + 1$;
      $f^* \leftarrow$ closest open facility;
      for *each* $f \in V$ **do**
         $\mu_v(f) \leftarrow \mu_v(f) + d(v, f^*) - d(v, f)$; // savings if facility on f

   for *each* $f \in V$ **do**
      bid $b_v^t(f) \leftarrow \sqrt{\alpha} \cdot \mu_v(f)$;   // bidding phase
      if *facility is opened at* f **then**
         for *each* $z \in V$ **do**
            $\mu_v(z) \leftarrow \mu_v(z) - n_v \cdot (d(v, f^*) - d(v, f))$; // update counters

   if *request for* v **then**
      $c_v \leftarrow c_v +$ distance to closest open facility;   // update cost
   if $c_v \geq \sqrt{\alpha}$ **then**
      $\mu_v(v) \leftarrow \sqrt{\alpha}$;   // open facility on v at next request
      $\mu_v(f) \leftarrow 0$ for all $f \neq v$;

---

that it is reasonable to still assume that the prices show an adversarial behavior. Hence the players will still choose the same strategies as before.

Proposition 1. *The local competitive ratio is* $\Omega(\sqrt{\alpha})$ *even for fixed facility costs.*

Note that the lower bounds shown for the online price of anarchy with both the general and the frugal strategies still hold since they do not make use of the price changes. However, we can show that the online price of stability is an improvement from the price of anarchy in this model by a factor of at least $\Theta(\sqrt{\alpha})$.

We show that for fixed facility prices a simple algorithm has optimal local competitiveness and is not worse than a factor of $\mathcal{O}(\sqrt{\alpha})$ in comparison to the global optimum if every player implements that algorithm. In Algorithm 1 a node v stores a counter $\mu_v(f)$ for every node f, which indicates how much costs could have been avoided if a facility had existed on f. The counter is therefore increased by $d(v, f^*) - d(v, f)$ each time there is a request on v, where f^* is the currently closest facility to v. Node v will issue a bid of $b_v^t(f) = \sqrt{\alpha} \cdot \mu_v(f)$ in each time step t. The counters are updated accordingly when a new facility is opened. As an additional rule, node v bids α on itself as soon as its costs for requests exceed $\sqrt{\alpha}$ and sets all counters to 0.

Theorem 6. *Algorithm 1 has a local competitive ratio of $\mathcal{O}(\sqrt{\alpha})$.*

Proof. We consider a node v executing the Algorithm 1. Suppose the closest facility at the time of the first request of v is located on node f_1 and then new facilities are opened on $f_2, \ldots, f_k$ where each of the facilities is closer to v.

Let R_i be the number of requests of v before a facility is opened on f_{i+1}. The costs for requests are therefore $\sum_{i=1}^{k-1} R_i \cdot d(v, f_i)$. The costs for the bids are at most

$$\sum_{i=1}^{k-1} \sqrt{\alpha} \cdot \left(\sum_{j=1}^{i} R_j \right) \cdot (d(v, f_i) - d(v, f_{i+1})) = \sqrt{\alpha} \cdot \sum_{i=1}^{k-1} R_i \cdot (d(v, f_i) - d(v, f_k)).$$

For the optimal solution we may assume that a new facility is opened on f_i without v having to bid anything right after this copy is established in the online version. If the offline optimum never bids enough to contribute in the opening of a facility, the costs for requests are the same as for the online algorithm and therefore the cost ratio is at most $\sqrt{\alpha} + 1$.

If the optimal solution contributes to a facility on a node f' before the online algorithm has opened a facility at the same distance or closer to v, the costs for this facility must be at least the bid v gave at that time. The bid of v is $\sqrt{\alpha} \cdot \mu_v(f') \geq \sqrt{\alpha}$. Hence the cost of the optimal solution is at least $\sqrt{\alpha}$. The costs of Algorithm 1 are at most $3 \cdot \alpha$ since a facility is bought at v if the costs for requests exceed $\sqrt{\alpha}$.   □

We will now evaluate the profile $\mathcal{A}^*$ where every player uses Algorithm 1.

Theorem 7. *The online price of stability for fixed facility costs is $\mathcal{O}(\sqrt{\alpha})$.*

Proof. For the analysis we assume that whenever the online algorithm opens a new facility, the optimum gets the same facility for 0 costs in the same time step. Obviously, this does not increase the costs of the optimal solution.

It is obvious that the costs of the optimum for the new instance are not higher than the costs for the original instance. Furthermore, we show that the optimal algorithm will never open an additional facility. To see this, let $\nu(v)$ be the sum of all bids on v at any point in time. We have $\nu(v) = \sqrt{\alpha} \cdot \sum_{i \in R} [d(y_i', r_i) - d(r_i, v)]$, where R is the index set of requests so far and y_i' is the currently closest facility to node r_i. Furthermore, we define $\tau(v) := \sum_{i \in R} [d(y_i, r_i) - d(r_i, v)]$ where y_i is the facility which was used to serve the request on node r_i. Hence $\tau(v)$ denotes the costs which could have been saved in total if a facility would have existed on v from the beginning. A facility at v only lowers the cost of the offline algorithm if $\tau(v) > \alpha$ at some point in time.

We are only interested in requests where $d(y_i', r_i) - d(r_i, v) \geq 1$, otherwise both algorithms already have a facility at node r_i. Assume $d(y_i, r_i) - d(r_i, v) \geq \sqrt{\alpha}$, then r_i bids at least α on v. It follows that $\tau(v) < \nu(v) \leq \alpha$ and therefore no additional facility would pay off for the optimum.

We observe that when comparing the online algorithm to the optimum of the modified instance, both algorithms have the same costs for requests since the optimum does not open any additional facilities. In the proof of Theorem 6 we have seen that every node only pays a factor of at most $\mathcal{O}(\sqrt{\alpha})$ more for bids than for requests.

Hence, for the total costs of the strategy profile it holds that the costs are $\mathcal{O}(\sqrt{\alpha})$ times the costs for requests, while the optimal solution pays at least the costs for the requests.   □

One can easily see that the analysis for this profile is tight, using an instance with only one active node.

Proposition 2. *For strategy profile* $\mathcal{A}^*$ *it holds* $\max_{I,\phi,\pi} \left\{ \frac{C(\mathcal{A}^*)}{C_{Opt}} \right\} = \Omega(\sqrt{\alpha})$.

Proof. We construct a graph where a node v is the neighbor of the node f_0 which has opened a facility initially. Our sequence consists of $\sqrt{\alpha}$ requests of v. Since the cost of each request is exactly 1, this causes v to bid α on itself before the last request and therefore its costs are $\alpha + \sqrt{\alpha} - 1$ while the costs of the social optimum are $\sqrt{\alpha}$.   □

We have seen that the self-interested behavior of nodes and the online nature of the problem lead to an efficiency loss which grows in α. We note that the dependency on α cannot be dropped even if players would fully cooperate. To that end, we show that a centralized, deterministic online algorithm cannot achieve a competitive ratio independent of α.

Theorem 8. *No deterministic (centralized) online algorithm can achieve a constant competitive ratio independent of* α.

Proof. Consider a full binary tree where the distance from the root to every leaf is α and the distances from one level to the next start with $\alpha/2$ and decrease exponentially such that the distance between a level i and a level $i + 1$ node is $\alpha/2^i$. We assume that the initial facility is located at a distance more than α from any node within the tree.

The input sequence is a sequence of nodes $v_1, \ldots, v_{\log \alpha}$ such that v_1 is the root of the tree and v_i is on level $i - 1$ so that the nodes form a simple path from the root to the leaf $v_{\log \alpha}$.

Consider buying only one facility on $v_{\log \alpha}$. Then the costs are at most $\alpha + \sum_{i=0}^{\log \alpha} \frac{\alpha}{2^i} \le 3\alpha$, which is an upper bound for the costs of the optimal solution.

Now consider a deterministic online algorithm with a constant competitive ratio. The online algorithm can only buy a constant number of facilities c since its cost will be at least $c \cdot \alpha$. Beyond level $\log c$, the online algorithm cannot cover all paths down to the leafs, therefore it makes no sense to buy facilities on higher levels in advance of the sequence.

If the distance between two facilities on the chosen path is more than k (after level $\log c$), then the costs are at least $k \cdot \alpha/2^l$ for the nodes in between the levels l and $l + k$. It follows that the distance between facilities cannot be larger than a constant and hence the facility on the highest level built by the online algorithm can be assumed to be on a constant level l^*. This however implies the costs $k \cdot \alpha / 2^{l^*}$ for later requests, where k is $\log \alpha - l^*$. We have that the costs of the online algorithm are $\Omega(k \cdot \alpha)$ and k is not constant, which implies that an algorithm with a constant competitiveness cannot exist.   □

5   Local Facilities

We will now consider a restricted version of the problem where in addition to the fixed facility costs, players can only open a facility on their own node for the cost of α. Therefore the private costs for a player $v \in V$ can be described as

$$c(v) = |F^T \cap \{v\}| \cdot \alpha + \sum_{t=1}^{T} |\{v\} \cap \{v^t\}| \cdot \left[\min_{f \in F^t} d(v, f) \right].$$

It is easy to see that the problem a single node has to solve boils down to the classic ski-rental problem where the price for renting can decrease over time. Therefore the following holds:

Proposition 3. *No local strategy can be better than $\left(2 - \frac{1}{\alpha}\right)$-competitive and there exists a unique strategy for each player which is $\left(2 - \frac{1}{\alpha}\right)$-competitive.*

Proposition 3 implies that there is exactly one equilibrium in our model if we force the players to play a strategy with a tight competitive ratio. We observe that due to the limited options the player has in contrast to the model which included the bidding mechanism on all nodes, players can achieve a much better competitive ratio since the corresponding optimal offline player also has less options, which makes it easier to compete against. Unfortunately, the online price of anarchy is high despite the fact that the competitive ratio each player achieves individually can be bounded by a constant.

Theorem 9. *The online price of anarchy for local facilities with fixed facility costs is $\Theta(\alpha)$.*

Proof. For the lower bound consider a graph consisting of nodes $v_1, \ldots, v_n$ that are directly connected to a node x that is connected by a line of length $\alpha - 2$ to f_0, which is the node with the initial facility. Each of the nodes $v_1, \ldots, v_n$ receives exactly one request in an arbitrary order.

The private costs of each node are at most $\alpha - 1$. Hence no node will decide to open a facility. The optimal solution opens a facility on x and has a cost of at most $\alpha + n$. With growing n, the ratio between the social costs of the competitive strategy profile and the social optimum therefore tends to α.

To see that the online price of anarchy is also in $\mathcal{O}(\alpha)$ we simply observe that the costs for nodes which have a request are at least 1 in the social optimum, since they either open a facility which has costs of α or the distance to the next facility is at least 1. In the competitive strategy profile, the costs of such a node are at most 2α.   □

We observe that although the price of anarchy is not worse than in the model with the bidding mechanism, better solutions like in the previous chapter are not available despite the fixed facility costs. It is also noteworthy that every strategy profile consisting of arbitrary constant-competitive strategies still cannot achieve a better ratio than $\Omega(\alpha)$.

Note that the online price of anarchy is the same as for the frugal strategies in the general case, which leads us to the conclusion that the in worst case, the bidding mechanism for all nodes does not improve the global performance of the system despite its possibilities of cooperation for the players. As shown in the previous section, there are however strategies in the setting with fixed facility costs where the bidding mechanism allows for an improvement.

For the sake of completeness we finally investigate the effects of adding changing prices back to the model. Intuitively it should be clear that the results for the social costs of the possible strategy profiles are the same or worse as for the general model since the nodes face the same adversity while having fewer strategic options.

Theorem 10. *The online price of anarchy for local facilities with adversarial facility costs is* $\Theta(\sqrt{\alpha} \cdot \alpha)$.

Proof. The lower bound can be constructed exactly as the lower bound in the proof of Theroem 3.

The upper bound follows from the simple observation that no algorithm may accumulate more costs than $\mathcal{O}(\sqrt{\alpha}) \cdot \alpha$, since the costs of the local optimum are at most α.   □

Theorem 11. *The online price of stability for local facilities with adversarial facility costs is* $\Theta(\alpha)$.

Proof. The lower bound follows directly from the lower bound on the price of stability of the general case in Theorem 5.

A possible strategy for every node is to bid $\sqrt{\alpha}$ times the accumulated costs by serving requests. Therefore its costs never exceed $\mathcal{O}(\alpha)$ and are at most $\mathcal{O}(\sqrt{\alpha})$ times the costs of the local optimum. The latter can be seen by observing that the local optimum either pays the same costs for requests as the algorithm or at least $\sqrt{\alpha}$ for the facility.   □

6   Final Remarks and Future Work

In our work, we have focused on the competitive ratio as the optimization goal of the players. Due to the online nature of the problem, some kind of comparison to the offline optimum seems to be reasonable. Alternative cost measures might be interesting, but not all are suitable for our model. For example, if the players seek to optimize their worst-case absolute costs, each of them would be forced to buy a facility on their own node in the first time step. The resulting costs would be at most α, while any request sequence may feature an arbitrary number of requests for any given player. Until a facility is opened, each request for the player at the same location results in costs larger than 0. If, on the other hand, the players try to minimize their regret, i.e. the difference between their actual and the optimal online costs, then again always buying a facility in the first step optimizes the worst case, as it yields a regret of at most α. If the facility is bought during some later round and no more requests occur afterwards, then the regret increases only as long as its price is α in that round. If the total cost accumulated by requests is at most α at that point in time the regret is exactly α. Otherwise, for accumulated costs larger than α, it becomes even larger.

Our model implements a simple rent-or-buy mechanism for which it might be interesting to be extended to a leasing mechanism as used in [7], for example. In addition, the ideas of this paper could be extended to other online and possibly game-theoretical problems besides facility location to form our understanding of how a lack of coordination colludes with the online nature found in a lot of problems.

References

1. Albers, S., Koga, H.: New on-line algorithms for the page replication problem. J. Algorithms **27**(1), 75–96 (1998)
2. Awerbuch, B., Bartal, Y., Fiat, A.: Competitive distributed file allocation. In: Proceedings of the 25th Annual ACM Symposium on Theory of Computing (STOC), pp. 164–173. ACM (1993)
3. Bienkowski, M.: Price fluctuations: to buy or to rent. In: Bampis, E., Jansen, K. (eds.) WAOA 2009. LNCS, vol. 5893, pp. 25–36. Springer, Heidelberg (2010). doi:10. 1007/978-3-642-12450-1_3
4. Fotakis, D.: On the competitive ratio for online facility location. Algorithmica **50**(1), 1–57 (2008)
5. Fotakis, D., Tzamos, C.: On the power of deterministic mechanisms for facility location games. ACM Trans. Econ. Comput. **2**(4), 15:1–15:37 (2014)
6. Immorlica, N., Kalai, A.T., Lucier, B., Moitra, A., Postlewaite, A., Tennenholtz, M.: Dueling algorithms. In: Proceedings of the 43rd Annual ACM Symposium on Theory of Computing (STOC), pp. 215–224. ACM (2011)
7. Kling, P., auf der Heide, F.M., Pietrzyk, P.: An algorithm for online facility leasing. In: Even, G., Halldórsson, M.M. (eds.) SIROCCO 2012. LNCS, vol. 7355, pp. 61–72. Springer, Heidelberg (2012). doi:10.1007/978-3-642-31104-8_6
8. Meyerson, A.: Online facility location. In: Proceedings of the 42nd IEEE Symposium on Foundations of Computer Science (FOCS), pp. 426–431 (2001)

An Efficient PTAS for Parallel Machine Scheduling with Capacity Constraints

Lin Chen[1]([✉]), Klaus Jansen[2], Wenchang Luo[3], and Guochuan Zhang[4]

[1] MTA SZTAKI, Hungarian Academy of Science, Budapest, Hungary
chenlin198662@gmail.com
[2] Department of Computer Science, Christian-Albrechts-University Kiel, Kiel, Germany
kj@informatik.uni-kiel.de
[3] Faculty of Science, Ningbo University, Ningbo 315211, China
luowenchang@163.com
[4] College of Computer Science, Zhejiang University, Hangzhou 310027, China
zgc@zju.edu.cn

Abstract. In this paper, we consider the classical scheduling problem on parallel machines with capacity constraints. We are given m identical machines, where each machine k can process up to c_k jobs. The goal is to assign the $n \leq \sum_{k=1}^{m} c_k$ independent jobs on the machines subject to the capacity constraints such that the makespan is minimized. This problem is a generalization of the c-partition problem where $c_k = c$ for each machine. The c-partition problem is strongly NP-hard for $c \geq 3$ and the best known approximation algorithm of which has a performance ratio of 4/3 due to Babel et al. [2]. For the general problem where machines could have different capacities, the best known result is a 1.5-approximation algorithm with running time $O(n \log n + m^2 n)$ [14]. In this paper, we improve the previous result substantially by establishing an efficient polynomial time approximation scheme (EPTAS). The key idea is to establish a non-standard *ILP* (Integer Linear Programming) formulation for the scheduling problem, where a set of crucial constraints (called proportional constraints) is introduced. Such constraints, along with a greedy rounding technique, allow us to derive an integer solution from a relaxed fractional one without violating constraints.

Keywords: Scheduling · Approximation algorithms · Capacity constraints · Integer programming

1 Introduction

We consider a scheduling problem with machine capacity constraints (SMCC). There are m identical machines, where each machine k can process at most c_k jobs. The goal is to schedule $n \leq \sum_{k=1}^{m} c_k$ independent jobs on these machines such that the makespan is minimized subject to the capacity constraints.

G. Zhang—Research supported in part by NSFC (11271325).

T.-H.H. Chan et al. (Eds.): COCOA 2016, LNCS 10043, pp. 608–623, 2016.
DOI: 10.1007/978-3-319-48749-6_44

The SMCC problem has many real-world applications, such as distributing students in a university [20], crew scheduling in airline industries [18]. It is also of practical interests in the optimization of assembly lines for printed circuit boards (PCB) [17]. The cardinality constraint is actually a very natural and important consideration in combinatorial optimization and has received considerable much attention. We refer readers to a nice survey of this topic [4].

The classical scheduling problem without capacity constraints is one of the well-studied NP-hard problems in combinatorial optimization and computer science. For this problem, Hochbaum and Shmoys [8] give a polynomial-time approximation scheme (PTAS), which, given any instance I, outputs a schedule of length $(1+\epsilon)OPT(I)$ in time bounded by some polynomial in the length of the input |I|. An efficient polynomial time approximation scheme (EPTAS) which runs in $f(1/\epsilon)poly(|I|)$ is later given by Alon et al. [1] where the function f is doubly exponential in $1/\epsilon$. Such a result is further improved recently by Jansen and others [11,12], where an EPTAS of a running time that almost matches the lower bound [5] is presented. As a consequence, an EPTAS is also expected for scheduling with capacity constraints.

Previous Work. The special case of the SMCC problem where machines have a uniform capacity, i.e., $c_k = c$, is also know as the c-partition problem. Kellerer and Woeginger [16] first analyze the performance of LPT for $c = 3$ and show a worst-case ratio of $4/3 - 1/(3m)$. This result is improved later by Kellerer and Kotov [15], where a 7/6-approximation algorithm is given. For an arbitrary c, a 4/3-approximation algorithm is provided in [2].

As for the SMCC problem where there could be different capacities for machines, Dell'Amico et al. [6] analyzes its lower bounds and gives heuristic algorithms. Woeginger [19] devises an FPTAS when the number of machines is fixed. Later Zhang et al. [20] present a 3-approximation algorithm by applying the iterative rounding method proposed by Jain [9]. Subsequently, a 2-approximation algorithm is designed by Barna and Aravind [3] to solve the problem in a more general setting, i.e., the unrelated machine scheduling problem with capacity constraints. Recently, Kellerer and Kotov [14] give a 1.5-approximation algorithm.

Our Contribution. In this paper we consider the SMCC problem and provide a non-standard ILP (Integer Linear Programming) formulation for it. In this ILP we introduce a set of special constraints which we call as *proportional constraints*. An interesting fact is that, such constraints are unnecessary for the ILP formulation, but if we relax the integer variables, then they provide additional properties of the fractional solution. With a greedy rounding technique, we are able to round such a fractional solution into an integral one.

With the above-mentioned ILP formulation, we are able to prove the following theorem.

Theorem 1. *There is an EPTAS for the problem of scheduling n jobs on m machines with capacity constraints that outputs a $(1+\epsilon)$-approximation solution in almost linear time $f(1/\epsilon) + O(n \log n \log \log n)$, where*

$$f(1/\epsilon) = 2^{O(1/\epsilon^2 \log^2 (1/\epsilon) \log \log(1/\epsilon))}.$$

Briefly speaking, the algorithm is carried out in three steps.

First, we design an algorithm called Best-Fit schedule for the SMCC problem. The Best-Fit schedule returns a $(2 + \epsilon)$-approximation solution running in $O(n \log n (\log \log n + \log(1/\epsilon)))$ time. It is implemented first to provide an initial value T in $[OPT, 3OPT]$ (by choosing $\epsilon = 1$). Using this value we scale the processing times of jobs in the standard way. Notice that we aim to derive an EPTAS whose running time is nearly linear in n, yet all the previously known constant ratio approximation algorithms [3, 14, 20] run in $\Omega(n^2)$ time in the worst case, and could not be used to compute such an initial value.

Second, we introduce the proportional constraints and establish an ILP for the SMCC problem based on the scaling of jobs. Using the Best-Fit schedule in step 1 as a subroutine and also the special structure of the fractional solution based on the proportional constraints, we are able to relax the majority of integer variables in the $MILP$. The relaxation of the $MILP$ contains only constant number of integer variables, and could be solved efficiently using Kannan's algorithm [13]. Indeed, an EPTAS is already achievable now, except that the running time is still large (which is approximately $2^{2^{O(1/\epsilon)}} poly(n)$).

The third step is to modify the (relaxed) $MILP$ we derive in the second step so that we can solve it in a more efficient way. The key idea is to reduce the number of fractional variables and integer variables at the cost of a slight increase in the makespan.

2   Preliminaries

As described in Zhang et al. [20], we have the following standard integer linear program (ILP) formulation for the SMCC problem:

$$\text{minimize} \quad t \tag{1}$$

$$\text{subject to} \quad \sum_{j=1}^{n} p_j x_{jk} \leq t, \qquad\qquad 1 \leq k \leq m \tag{2}$$

$$\sum_{j=1}^{n} x_{jk} \leq c_k, \qquad\qquad 1 \leq k \leq m \tag{3}$$

$$\sum_{k=1}^{m} x_{jk} = 1, \qquad\qquad 1 \leq j \leq n \tag{4}$$

$$x_{jk} \in \{0, 1\}, \qquad\qquad 1 \leq j \leq n, 1 \leq k \leq m \tag{5}$$

Here p_j is the processing time of job j, and $x_{jk} = 1$ indicates that job j is on machine k, otherwise $x_{jk} = 0$.

If $c_k > n$ for some k, we just change it to n since no schedule could assign more than n jobs on this machine. For simplicity we use $\hat{n}$ to denote the original number of jobs and let $n = \sum_{j=1}^{m} c_j$ by adding $n - \hat{n}$ dummy jobs with processing time of 0. We replace the inequalities (3) by $\sum_{j=1}^{n} x_{jk} = c_k$ for $k = 1, 2, \ldots, m$. Note that dummy jobs do not contribute to the makespan, but each of them takes up a machine capacity of 1. Since $c_k \leq \hat{n}$ and $m \leq \hat{n}$, it can be easily seen that $\hat{n} \leq n \leq \hat{n}^2$.

3   Best-Fit Schedule

In this section we provide our Best-Fit Schedule that returns a $(2 + \epsilon)$-approximate solution in $O(\hat{n} \log n (\log \log n + \log(1/\epsilon)))$ time.

Throughout this section, we sort the m machines such that $1 \leq c_1 \leq c_2 \leq \cdots \leq c_m$, and the n $(= \sum_{j=1}^{m} c_j)$ jobs that $0 \leq p_1 \leq p_2 \cdots \leq p_n$. For simplicity we call a job as a zero job if its processing time is zero, and call it as a nonzero job otherwise.

The basic idea of this algorithm is simple: each time we pick one machine, say machine 1, and try to assign it exactly c_1 consecutive jobs in the job sequence (recall that they are in nondecreasing order of processing times) such that the load of this machine (the total processing time of jobs on it) could be slightly larger than the optimal makespan but not much, then we close this machine, delete scheduled jobs from the job list and consider the next machines in the same way. Formally speaking, we have the following lemma:

Lemma 2. *If there is a feasible solution for the following linear system LP_r:*

$$\sum_{j=1}^{n} p_j x_{jk} \leq t_k \qquad 1 \leq k \leq m$$

$$\sum_{j=1}^{n} x_{jk} = c_k \qquad 1 \leq k \leq m$$

$$\sum_{k=1}^{m} x_{jk} = 1 \qquad 1 \leq j \leq n$$

$$0 \leq x_{jk} \leq 1 \qquad 1 \leq j \leq n, 1 \leq k \leq m$$

then an integer solution satisfying:

$$\sum_{j=1}^{n} p_j x_{jk} \leq t_k + p_{max} \qquad 1 \leq k \leq m$$

$$\sum_{j=1}^{n} x_{jk} = c_k \qquad 1 \leq k \leq m$$

$$\sum_{k=1}^{m} x_{jk} = 1 \qquad 1 \leq j \leq n$$

$$x_{jk} \in \{0, 1\} \qquad 1 \leq j \leq n, 1 \leq k \leq m$$

could be obtained in $O(\hat{n} \log n)$ time, where $p_{max} = \max_j \{p_j\} = p_n$.

Proof. We prove the lemma by induction on m. When $m = 1$, this result is trivially true. Suppose this lemma holds for $m = m'$. Consider the problem when $m = m' + 1$.

Let x^*_{jk} be the feasible solution for LP_r. For machine 1, we know that $\sum_{j=1}^{c_1} p_j \leq t_1$, because if the total processing time of the smallest c_1 jobs is greater than t_1, then there would be no feasible solution for LP_r.

If $\sum_{j=n-c_1+1}^{n} p_j > t_1$, then there exists an index τ such that $\sum_{j=\tau-1}^{\tau+c_1-2} p_j \leq t_1$ and $\sum_{j=\tau}^{\tau+c_1-1} p_j > t_1$, we assign jobs from τ to $\tau+c_1-1$ on machine 1. Otherwise we have $\sum_{j=n-c_1+1}^{n} p_j \leq t_1$, and we put the biggest c_1 jobs on machine 1, in this case let $\tau = n - c_1 + 1$.

Notice that we could not determine the value of τ directly through binary search. The summation of c_1 nonzero jobs costs $\Omega(c_1)$ time, yet we might end up by putting c_1 zero jobs on it. Then in the worst case we use at least $\Omega(\sum_k c_k)$ time, which could be $\Omega(\hat{n}^2)$.

To avoid the above problem, we determine τ in the following way. Suppose i is the least index such that $p_i > 0$, and κ is the integer such that $2^{\kappa-1} < c_1 \leq 2^{\kappa}$. We then test one by one the sets of jobs $ST_l = \{p_i, p_{i+1}, \cdots, p_{i+2^{l-1}-1}\}$ for $l = 1, 2, \cdots, \kappa$, to see if their total processing times sum greater than t_1.

Let $L(ST_l) = \sum_{j=i}^{i+2^{l-1}-1} p_j$, if there exists some $l_0 \leq \kappa$ such that $L(ST_{l_0-1}) \leq t_1 < L(ST_{l_0})$, then $\tau \leq i + 2^{l_0-1} - c_1 + 1$, and we then apply binary search on τ. For the binary search, we need to test at most $\lceil \log n \rceil$ values of τ, and for each value, the summation from p_τ to $p_{\tau+c_1-1}$ takes at most $O(2^{l_0-1})$ computations since there are at most 2^{l_0-1} nonzero jobs as $\tau \leq i + 2^{l_0-1} - c_1 + 1$. Meanwhile the computation for $L(ST_1)$ to $L(ST_{l_0})$ takes at most $O(2^{l_0})$ time. Thus in all, we put at least 2^{l_0-2} nonzero jobs on machine 1 in $O(2^{l_0} \log n)$ time.

Otherwise, $L(ST_\kappa) \leq t_1$, then $\tau + c_1 - 1 \geq i + 2^{\kappa-1} - 1$. Again we apply binary search on τ and sum the processing times of c_1 jobs for each value of τ, this consumes $O(c_1 \log n)$ time, and the computation for $L(ST_1)$ to $L(ST_\kappa)$ takes $O(2^{\kappa+1})$ time. Meanwhile as $\tau + c_1 - 1 \geq i + 2^{\kappa-1} - 1$ we would put at least $2^{\kappa-1}$ nonzero jobs on machine 1.

In all, the above procedure determines τ in at most $O(4\phi \log n)$ time if it puts ϕ nonzero jobs on machine 1.

Let $S_1 = \{j | \text{job } j \text{ is put on machine 1}\}$, and $\hat{c}_1$ be the number of nonzero jobs on machine 1, we will show that the following linear system LP'_r:

$$\sum_{j \notin S_1} p_j x_{jk} \leq t_k \qquad 2 \leq k \leq m' + 1$$

$$\sum_{j \notin S_1} x_{jk} = c_k \qquad 2 \leq k \leq m' + 1$$

$$\sum_{k=2}^{m'+1} x_{jk} = 1 \qquad j \notin S_1$$

$$0 \leq x_{jk} \leq 1 \qquad j \notin S_1, \quad 2 \leq k \leq m' + 1$$

still admits a feasible solution, then according to the induction hypothesis, an integer solution satisfying

$$\sum_{j \notin S_1} p_j x_{jk} \le t_k + p_{max} \qquad 2 \le k \le m' + 1$$

$$\sum_{j \notin S_1} x_{jk} = c_j \qquad 2 \le k \le m' + 1$$

$$\sum_{k=2}^{m'+1} x_{jk} = 1 \qquad j \notin S_1$$

$$x_{jk} \in \{0, 1\} \qquad j \notin S_1, \quad 2 \le k \le m' + 1$$

could be constructed in $O((\hat{n} - \hat{c}_1) \log n)$ time. Since at most $O(\hat{c}_1 \log n)$ time is needed to determine jobs that are put on machine 1, in all $O(\hat{n} \log n)$ time is needed to construct the integer solution for all machines and Lemma 2 follows immediately.

The following part of this proof is devoted to show that LP'_τ admits a feasible solution, indeed, we will construct such a solution from x^*_{jk}.

Notice that we have $\sum_{j=\tau}^{\tau+c_1-1} p_j \ge \sum_{j=1}^{n} x^*_{j1} p_j$ (this inequality holds strictly if $\tau < n - c_1 + 1$). Rewrite it as

$$\sum_{j=\tau}^{\tau+c_1-1} \alpha_j p_j \ge \sum_{j=1}^{\tau-1} x^*_{j1} p_j + \sum_{j=\tau+c_1}^{n} x^*_{j1} p_j,$$

where $\alpha_j = 1 - x^*_{j1}$ for $j = \tau, \cdots, \tau + c_1 - 1$. For simplicity, let $\alpha_l = \sum_{j=1}^{\tau-1} x^*_{j1}$ and $\alpha_u = \sum_{j=\tau+c_1}^{n} x^*_{j1}$, define p_l and p_u as $\alpha_l p_l = \sum_{j=1}^{\tau-1} x^*_{j1} p_j$ and $\alpha_u p_u = \sum_{j=\tau+c_1}^{n} x^*_{j1} p_j$ ($p_l = p_{\tau-1}$ if $\alpha_l = 0$, and $p_u = p_{\tau+c_1}$ if $\alpha_u = 0$). Thus,

$$\sum_{j=\tau}^{\tau+c_1-1} \alpha_j p_j \ge \alpha_l p_l + \alpha_u p_u,$$

$$\sum_{j=\tau}^{\tau+c_1-1} \alpha_j = \alpha_l + \alpha_u.$$

Now it can be easily seen that in order to construct a feasible solution for LP'_τ, we only need to exchange $\alpha_l p_l$ and $\alpha_u p_u$ that are scheduled on machine 1 in the solution x^*_{jk} with $\alpha_j p_j$ ($\tau \le j \le \tau + c_1 - 1$) that are distributed among other machines. More precisely, if we can find ξ_j, η_j such that for each j we have $\tau \le j \le \tau + c_1 - 1$, $\xi_j + \eta_j = \alpha_j$, $\xi_j p_u + \eta_j p_l \le \alpha_j p_j$, and moreover, $\sum_{j=\tau}^{\tau+c_1-1} \xi_j = \alpha_u$, $\sum_{j=\tau}^{\tau+c_1-1} \eta_j = \alpha_l$, then starting from the solution x^*_{jk}, we exchange jobs like this: if $\beta_j p_j$ (here $\tau \le j \le \tau + c_1 - 1$ and $\beta_j > 0$) is scheduled on some machine, say k where $k \ge 2$, then we move $\beta_j p_j$ to machine 1 and put $\frac{\beta_j}{\alpha_j}(\xi_j p_u + \eta_j p_l)$ on machine k instead, or more precisely, put $\frac{\beta_j}{\alpha_j}(\xi_j \frac{\sum_{j=\tau+c_1}^{n} x^*_{j1} p_j}{\sum_{j=\tau+c_1}^{n} x^*_{j1}} + \eta_j \frac{\sum_{j=1}^{\tau-1} x^*_{j1} p_j}{\sum_{j=1}^{\tau-1} x^*_{j1}})$ on k. By doing so, the completion time of machine k doesn't increase while the total fractional number of jobs on it remains the same, and thus a fractional solution for LP'_r is constructed.

Next we prove that such desired ξ_j and η_j do exist by constructing them. If $\alpha_u = 0$, then setting $\eta_j = \alpha_j$ and $\xi_j = 0$ is enough. Else if $\alpha_l = 0$ and $\alpha_u = \sum_{j=\tau+c_1}^{n} x^*_{j1} \neq 0$, then $\tau < n - c_1 + 1$, it yields a contradiction with $\sum_{j=\tau}^{\tau+c_1-1} \alpha_j p_j > \alpha_l p_l + \alpha_u p_u$. Therefore, in the following we assume that $\alpha_u \neq 0$ and $\alpha_l \neq 0$, this implies that $\sum_{j=\tau}^{\tau+c_1-1} \alpha_j p_j > \alpha_l p_l + \alpha_u p_u$. We distinguish three cases and determine ξ_j and η_j $(\tau \leq j \leq \tau + c_1 - 1)$ recursively as follows.

Case 1. If $\alpha_j \frac{p_j - p_l}{p_u - p_l} \leq \alpha_u - \sum_{w=\tau}^{j-1} \xi_w$ and $\alpha_j \frac{p_u - p_j}{p_u - p_l} \leq \alpha_l - \sum_{w=\tau}^{j-1} \eta_w$, then let

$$\xi_j = \alpha_j \frac{p_j - p_l}{p_u - p_l} \text{ and } \eta_j = \alpha_j \frac{p_u - p_j}{p_u - p_l}.$$

Case 2. If Case 1 holds for any $j \leq j_0 - 1$, while $\alpha_{j_0} \frac{p_{j_0} - p_l}{p_u - p_l} > \alpha_u - \sum_{w=\tau}^{j_0-1} \xi_w$ (j_0 is the first index that satisfying $\alpha_{j_0} \frac{p_{j_0} - p_l}{p_u - p_l} > \alpha_u - \sum_{w=\tau}^{j_0-1} \xi_w$), then let

$$\xi_{j_0} = \alpha_u - \sum_{w=\tau}^{j_0-1} \xi_w, \ \eta_{j_0} = \alpha_{j_0} - \xi_{j_0}.$$

For $j_0 < j \leq \tau + c_1 - 1$, let $\xi_j = 0$ and $\eta_j = \alpha_j$.

Note that in both Case 1 and Case 2, we have $\xi_j + \eta_j = \alpha_j$, $\xi_j p_u + \eta_j p_l = \alpha_j p_j$, for $j \leq j_0 - 1$, and $\xi_j + \eta_j = \alpha_j$, $\xi_j p_u + \eta_j p_l \leq \alpha_j p_j$, for $j \geq j_0$.

Case 3. Case 1 holds for any $j \leq j_0 - 1$, while $\alpha_{j_0} \frac{p_u - p_{j_0}}{p_u - p_l} > \alpha_l - \sum_{w=\tau}^{j_0-1} \eta_w$. We show that this case is impossible.

From $\sum_{j=\tau}^{\tau+c_1-1} \alpha_j p_j > \alpha_l p_l + \alpha_u p_u$ we have $(\alpha_l - \sum_{w=\tau}^{j_0-1} \eta_w) p_l + (\alpha_u - \sum_{w=\tau}^{j_0-1} \xi_w) p_u < \sum_{j=j_0}^{\tau+c_1-1} \alpha_w p_w$. With $\alpha_{j_0} \frac{p_u - p_{j_0}}{p_u - p_l} > \alpha_l - \sum_{w=\tau}^{j_0-1} \eta_w$ we can deduce $(\alpha_l - \sum_{w=\tau}^{j_0-1} \eta_w) p_l + [\alpha_{j_0} - (\alpha_l - \sum_{w=\tau}^{j_0-1} \eta_w)] p_u > \alpha_{j_0} p_{j_0}$. Furthermore, with $\alpha_l = \sum_{i=\tau}^{\tau+c_1-1} \alpha_j - \alpha_u$ and $(\alpha_l - \sum_{w=\tau}^{j_0-1} \eta_w) p_l + [\alpha_{j_0} - (\alpha_l - \sum_{w=\tau}^{j_0-1} \eta_w)] p_u > \alpha_{j_0} p_{j_0}$, we can derive $(\alpha_l - \sum_{w=\tau}^{j_0-1} \eta_w) p_l + (\alpha_u - \sum_{w=\tau}^{j_0-1} \xi_w) p_u > \sum_{w=j_0}^{\tau+c_1-1} \alpha_w p_w$, which is a contradiction (note that $p_u \geq p_j$ for any $\tau \leq j \leq \tau + c_1 - 1$).

Thus Case 3 never happens, which means the ξ_j and η_j defined above satisfy what we expect, i.e., for each j $(\tau \leq j \leq \tau + c_1 - 1)$, $\xi_j + \eta_j = \alpha_j$, $\xi_j \alpha_u + \eta_j \alpha_l \leq \alpha_j p_j$, and $\sum_{j=\tau}^{\tau+c_1-1} \xi_j = \alpha_u$, $\sum_{j=\tau}^{\tau+c_1-1} \eta_j = \alpha_l$. $\quad\square$

Given the above lemma, a simple algorithm BFS could be derived. We choose $t_1 = t_2 = \cdots = t_m = t$. Then through binary search we find the minimum $t = t'$ such that the application of the algorithm in the above lemma do not fail.

Notice that as long as $t \geq OPT$ the linear system LP_r admits a feasible solution, thus $t' \leq OPT$ and our Best-Fit schedule returns a feasible solution with makepan no more than $OPT + p_{max}$, which is of 2-approximation and the overall running time is $O(n \log n \log(np_{max}))$.

Specifically, if we choose some constant $1/\delta$, and scale the numbers between p_{max} and np_{max} into $S = \{p_{max}, (1+\delta)p_{max}, (1+\delta)^2 p_{max}, \cdots, (1+\delta)^\omega p_{max}\}$ with $\omega = \lceil \log_{1+\delta} n \rceil \leq 1/\delta \log n$, then we can apply binary search on the numbers in S instead. A schedule of length at most $OPT(2+\delta)$ could be determined in $O(\hat{n} \log n \log(1/\delta \cdot \log n))$ time.

4   An Efficient Polynomial Time Approximation Scheme

The algorithm of the above section could provide us with some integer T satisfying $OPT \leq T \leq 3OPT$ in $O(\hat{n} \log n \log \log n)$ (by setting $c_0 = 1$) time, and given $\epsilon > 0$, we would formulate an $MILP$ in this section that for any fixed t satisfying $T/3 \leq t \leq T$, it either determines there is no feasible schedule with makespan no more than t, or produces a solution with makespan no more than $t(1 + \epsilon)$, in which all the split jobs have tiny processing times.

We now scale the processing time of jobs using T in the standard way. For simplicity we assume $\mu = 1/\epsilon$ to be an integer greater than 10, and let $\lambda = \lceil \log_{1+\epsilon} \mu \rceil$. Let $SV_b = \{T\epsilon(1 + \epsilon), T\epsilon(1 + \epsilon)^2, \cdots, T\epsilon(1 + \epsilon)^\lambda\}$ and $SV_s = \{T\epsilon^2, T\epsilon^2(1 + \epsilon), \cdots, T\epsilon^2(1 + \epsilon)^\lambda\}$ be the sets of scaled processing times. For each job whose processing time is larger than $T\epsilon$, we round it up to be its nearest value in SV_b, and call it as a big job. The remaining jobs are then called small jobs.

We define containers. For each ϕ such that $1 \leq \phi \leq \mu$, we define a ϕ-container which contains exactly ϕ jobs, and is denoted by a $(\lambda+2)$-tuple $(\nu_0, \nu_1, \cdots, \nu_\lambda, \phi)$, where $\nu_i \leq \phi$ $(1 \leq i \leq \lambda)$ is the number of big jobs with rounded processing time $T\epsilon(1+\epsilon)^i$ for $1 \leq i \leq \lambda$ in this container. Except for these big jobs, there are also $\phi - \sum_{i=1}^{\lambda} \nu_i$ small jobs in this container, we compute their average processing time (their total processing time divided by $\phi - \sum_{i=1}^{\lambda} \nu_i$) and then round this value up to be its nearest value in SV_s. We use ν_0 to represent such a scaled average processing time. Specifically, if there is no small job in a container, then $\nu_0 = T\epsilon^2$.

The reason for the introduction of containers is that the optimum solution could always be seen as a schedule for m containers and the remaining small jobs. To see why, consider each machine with capacity ϕ where $1 \leq \phi \leq \mu - 1$. Obviously all jobs on this machine could be represented by a ϕ-container. Meanwhile consider machine k with $c_k \geq \mu$. Obviously the largest μ jobs on this machine could be represented through a μ-container. Since there are at most μ big jobs on each machine, the remaining $(c_k - \mu)$ jobs on this machine are all small jobs.

Moreover, for each container $(\nu_0, \nu_1, \cdots, \nu_\lambda, \phi)$ where $1 \leq \phi \leq \mu$, we define $\sum_{\kappa=1}^{\lambda} \nu_\kappa T\epsilon(1 + \epsilon)^\kappa + \nu_0(\phi - \sum_{\kappa=1}^{\lambda} \nu_\kappa)$ as the load of this container. Suppose t_k is the load of machine k in the optimum solution (of the unrounded instance), then if we use the load of the container on this machine to substitute the total processing time of corresponding jobs, the new load of machine k becomes $t'_k \leq t_k(1 + \epsilon) + \mu T\epsilon^2 \leq OPT(1 + 4\epsilon)$.

We classify small jobs into two parts. A small job which is contained in one of the containers in the optimum solution is called an inner small job, otherwise it is an outer small job. Our above analysis implies the following observation:

Observation. There exists a feasible schedule of makespan at most $OPT(1+4\epsilon)$ such that

1. For $1 \leq \phi \leq \mu - 1$, there is one ϕ-container on each machine of capacity ϕ.
2. There is one μ-container on each machine whose capacity is at least μ.

3. Each big job is in one of the containers.
4. Each small job is either inside some container as an inner small job, or is scheduled as an outer small job outside containers.

We denote the above solution as Sol^*. Before establishing the $MILP$, we have the following estimation on the number of different containers.

Lemma 3. *The number of all different containers is bounded by* $2^{O(\mu \log \log \mu)}$.

We sort ϕ-containers in non-increasing order of their loads and denote by $v_i^\phi = (\nu_0^\phi(i), \nu_1^\phi(i), \cdots, \nu_\lambda^\phi(i), \phi)$ the i-th one. Specifically, for μ-containers, we also denote the i-th μ-container as $v_i = (\nu_0(i), \nu_1(i), \cdots, \nu_\lambda(i), \mu)$ (i.e., $v_i^\mu = v_i$, $\nu_\kappa^\mu(i) = \nu_\kappa(i)$) for simplicity and use Γ_i to denote its load.

To simplify our notation, we use GM_ϕ $(1 \le \phi \le \mu - 1)$ to denote the set of machines with capacity exactly ϕ, and GM_μ the set of machines with capacity greater than or equal to μ. We further renumber the machines so that GM_μ is made up of machine 1 to machine $m' = |GM_\mu|$, and $\mu \le c_1 \le c_2 \le \cdots \le c_{m'}$. All the containers could then be classified similarly into μ groups with GC_ϕ $(1 \le \phi \le \mu)$ being the set of all ϕ-containers. Meanwhile, containers could also be classified into $(\lambda + 1)$ groups according to the first coordinate of their tuples, i.e., we define $GS_l = \{v_i^\phi | \nu_0^\phi(i) = T\epsilon^2(1+\epsilon)^l, 1 \le \phi \le \mu, 1 \le i \le |GC_\phi|\}$ for $0 \le l \le \lambda$.

We first set up an ILP for the SMCC problem, in which the following integer variables are used:

z_i^ϕ: an integer variable denoting the number of the i-th ϕ-container (i.e., v_i^ϕ) that are used.

y_{ik}: a $0-1$ integer variable for the μ-container v_i. If $y_{ik} = 1$, then one container of v_i is on machine k, otherwise $y_{ik} = 0$.

x_{ijk}: a $0-1$ integer variable for outer small jobs. If $y_{ik} = 1$, job j is on machine k and is not inside the container v_i, then $x_{ijk} = 1$. Otherwise $x_{ijk} = 0$.

w_{jl}: a $0-1$ integer variable for inner small jobs. If job j is an outer small job, then $w_{jl} = 0$ for any $0 \le l \le \lambda$. Otherwise job j is an inner small job, and is contained in some container. If this container is in group GS_κ for some $0 \le \kappa \le \lambda$, then $w_{j\kappa} = 1$, $w_{jl} = 0$ for $l \ne \kappa$.

Let N_κ be the number of big jobs with rounded processing time $T\epsilon(1+\epsilon)^\kappa$ for $1 \le \kappa \le \lambda$, n' be the number of small jobs, and we sequence them as $p_1 \le p_2 \le \cdots \le p_{n'}$. Given any fixed t $(T/3 \le t \le T)$, we formulate an $ILP(t)$ as follows.

$$(1) \quad \sum_{i=1}^{|GC_\phi|} z_i^\phi = |GM_\phi|, \quad 1 \le \phi \le \mu$$

$$(2) \quad \sum_{k=1}^{m'} y_{ik} = z_i^\mu, \quad 1 \le i \le |GC_\mu|$$

$$(3) \quad \sum_{i=1}^{|GC_\mu|} y_{ik} = 1, \quad 1 \le k \le m'$$

$$(4) \quad \sum_{\phi=1}^{\mu} \sum_{i=1}^{|GC_\phi|} \nu_\kappa^\phi(i) z_i^\phi = N_\kappa, \quad 1 \le \kappa \le \lambda$$

$$(5) \quad \sum_{j=1}^{n'} x_{ijk} = (c_k - \mu) y_{ik}, \quad 1 \le i \le |GC_\mu|, 1 \le k \le m'$$

$$(6) \quad \sum_{j=1}^{n'} p_j x_{ijk} \le (t - \Gamma_i) y_{ik}, \quad 1 \le i \le |GC_\mu|, 1 \le k \le m'$$

$$(7) \quad \sum_{i=1}^{|GC_\mu|} \sum_{k=1}^{m'} x_{ijk} + \sum_{l=0}^{\lambda} w_{jl} = 1, \quad 1 \le j \le n'$$

$$(8) \quad \sum_{j=1}^{n'} w_{jl} = \sum_{\phi=1}^{\mu} \sum_{i:v_i^\phi \in GS_l \cap GC_\phi} (\phi - \sum_{\kappa=1}^{\lambda} \nu_\kappa^\phi(i)) z_i^\phi, \quad 0 \le l \le \lambda$$

$$(9) \quad \sum_{j=1}^{n'} p_j w_{jl} \le T\epsilon^2 (1+\epsilon)^l \sum_{\phi=1}^{\mu} \sum_{i:v_i^\phi \in GS_l \cap GC_\phi} (\phi - \sum_{\kappa=1}^{\lambda} \nu_\kappa^\phi(i)) z_i^\phi, \quad 0 \le l \le \lambda$$

$$(10) \quad z_i^\phi = 0 \quad \text{if} \quad \sum_{\kappa=1}^{\lambda} \nu_\kappa^\phi(i) T\epsilon(1+\epsilon)^\kappa + \nu_0^\phi(i)(\phi - \sum_{\kappa=1}^{\lambda} \nu_\kappa) > t$$

$$(11) \quad z_i^\phi \in \mathbb{N}, y_{ik}, x_{ijk}, w_{jl} \in \{0, 1\}.$$

We explain these constraints. Constraint (1) shows that for $1 \le \phi \le \mu$, the number of ϕ-containers used equals to the number of machines in group GM_ϕ. Constraints (2) and (3) ensure that there is exactly one μ-container on each machine of group GM_μ, and the total number of container v_i that are used is z_i^μ. Constraint (4) shows that each big job is contained in one of the containers. As to constraints (5) and (6), notice that y_{ik} is either 0 or 1, thus for each k there could be exactly one i such that $y_{ik} = 1$, while $y_{i'k} = 0$ for $i' \ne i$. Then from constraint (5), we know that $x_{i'jk} = 0$. So, constraints (5) and (6) actually show that for each machine k of group GM_μ, the total number of jobs scheduled won't exceed c_k, and the total processing time won't exceed t. Constraint (7) implies that each small job is scheduled, either as an inner small job or as an outer one. Constraints (8) and (9) are for inner small jobs. Both sides of the constraint (8) equals to the total number of inner jobs we use. The left side of constraint (9) is the exact total processing time of all the inner small jobs, while the right side is an upper bound of it (notice that we have round up the average processing time of inner small jobs in each container). Constraint (11) ensures that every container we use is feasible in the sense that its load won't exceed t.

It can be easily seen that once we set $t = t_0 \ge OPT(1 + 4\epsilon)$, then Sol^* satisfies all the above constraints and is thus a feasible solution for $ILP(t_0)$. Thus, the smallest t such that $ILP(t)$ admits a feasible solution is no greater than $OPT(1 + 4\epsilon)$.

Here the constraints (5) and (6) are the so-called *proportional constraints* we mention before.

We remark that if y_{ik} are restricted to be $0-1$ integer variables, then there are actually other formulations which can be much simpler. However, in order to solve the programming, we need to relax most of the integer variables so that there are only a constant number of integer variables, and these proportional constraints become crucial for the relaxation, as we will show in Lemma 5.

Consider $ILP(t)$. By replacing the constraints $x_{ijk}, w_{jl} \in \{0, 1\}$ with $0 \leq x_{ijk} \leq 1$ and $0 \leq w_{jl} \leq 1$, we derive a mixed integer linear programming and denote it as $MILP(t)$. Obviously any feasible solution of $ILP(t)$ is a feasible solution for $MILP(t)$, thus the minimum integer t such that $MILP(t)$ admits a feasible solution is also no greater than $OPT(1+4\epsilon)$. Furthermore, the following lemma follows from Lemma 2.

Lemma 4. *Given a feasible solution of $MILP(t)$ with $t = t^*$, a feasible schedule for the SMCC problem with makespan no more than $t^* + T\epsilon$ could be constructed in $O(\hat{n} \log n)$ time.*

With the above lemma, we only need to find the minimum t such that $MILP(t)$ admits a feasible solution. Through binary search, we need to determine whether $MILP(t)$ admits a feasible solution for each fixed t. However, we could not solve the $MILP(t)$ directly in polynomial time since the number of integer variables is $\Omega(m)$. We further replace the constraint $y_{ik} \in \{0, 1\}$ in $MILP(t)$ to be $0 \leq y_{ik} \leq 1$, and call such a relaxed $MILP(t)$ as $MILP_r(t)$. We have the following.

Lemma 5. *Given a feasible solution of $MILP_r(t)$ for $t = t^*$, a feasible solution of $MILP(t)$ with $t = t^*$ could be constructed.*

Proof. The proof is a constructive one. Note that there are $m' = |GM_\mu|$ machines with capacity no less than μ and they are renumbered as machine 1 to machine m'. All the containers we mention afterwards refer to μ-containers. All the containers used in the fractional solution are sorted in non-increasing order of their load. In the following we will construct a feasible solution such that the h-th largest container is on machine h through iteratively moving (fractions of) containers. To achieve this, it suffices to move container 1 onto machine 1 whereas all the other containers could be moved in the same way.

To move container 1, we will alter x_{ijk}^* and y_{ik}^* while keeping other variables (e.g., $z_i^{\phi*}$) intact. Specifically, we alter x_{ijk}^* and y_{ik}^* only in two ways.

One is to move jobs from some machine to another. We either move outer small jobs as indicated by x_{ijk}^* or move containers.

The other way is to re-assign small jobs to containers. Notice that y_{ik} could be fractional, thus there might be several fractions of containers on the same machine. For simplicity, if $x_{ijk}^* > 0$, then we say x_{ijk}^* fraction of job j is assigned to container i. By re-assigning, we could change x_{ijk}^* to $x_{ijk}^* - \delta$ ($\delta \leq x_{ijk}^*$) while let $x_{i'jk}^*$ become $x_{i'jk} + \delta$ for some i', which means now a fraction δ of job j on machine k is assigned to a new container i'.

We need to ensure that constraints (1) to (11) are never violated during our rounding procedure. Notice that if we only carry out the above mentioned two methods to alter variables, then $z_i^{\phi*}, w_{jl}^*$ won't change. Moreover, $\sum_{i=1}^{|GC_\mu|} \sum_{k=1}^{m'}$ x_{ijk}^* and $\sum_{k=1}^{m'} y_{ik}^*$ as a whole won't change either, so to ensure that constraints (1) to (11) are never violated, we only need to ensure that constraints (3), (5) and (6) are never violated.

Now we will describe our rounding procedure. We focus on machine 1. Suppose after applying the rounding procedure for some times we arrive at some feasible solution x_{ijk}^h and y_{ik}^h for $MILP_r$. If among the containers used the one with the largest load is on machine 1, then y_{i1}^h are already $0 - 1$ variables and we close machine 1 and fix y_{i1}^h and x_{ij1}^h forever. Then we could simply renumber machines so the remaining machines are machines 1 to $m' - 1$, and again focus on the first machine.

Otherwise we suppose without loss of generality that $z_1^h > 0$ and $0 \le y_{11}^h < 1$, then there exists a machine w with $y_{1w}^h > 0$. If $y_{11}^h + y_{1w}^h \le 1$, we would try to move the fraction y_{1w}^h of container 1 to machine 1, otherwise it's enough by moving only $1 - y_{11}^h$ fraction out of y_{1w}^h of container 1. Define $\eta = \min\{y_{1w}^h, 1 - y_{11}^h\}$. To ensure (3), we exchange η fraction of container 1 on machine w with the same fraction of other containers on machine 1, i.e., we let $y_{11}' = y_{11}^h + \eta$, $y_{i1}' = y_{i1}^h(1 - \alpha)$ for $i > 1$ where $\alpha = \frac{\eta}{\sum_{i>1} y_{i1}^h} = \frac{\eta}{1 - y_{11}^h}$, and meanwhile $y_{1w}' = y_{1w}^h - \eta$, $y_{iw}' = y_{iw}^h + \alpha y_{i1}^h$. It can be easily verified that by doing so (4) is satisfied on both machine 1 and w. If we could then alter x_{ijk}^h so that (5) and (6) are satisfied, then either y_{11}^h becomes 1, or we have successfully moved the fraction of container 1 on another machine to machine 1, and thus carry on such a procedure at most m' times y_{11}^h would become 1.

The above mentioned altering from y_{i1}^h to y_{i1}' would actually be accomplished in several steps, and in each step we only consider one of those y_{i1}^h ($i > 1$). Assume without loss of generality that $y_{21} > 0$, then let

$$y_{11}^{h+1} = y_{11}^h + \alpha y_{21}^h, y_{21}^{h+1} = y_{21}^h(1 - \alpha),$$

$$y_{1w}^{h+1} = y_{1w}^h - \alpha y_{21}^h, y_{2w}^{h+1} = y_{2w}^h + \alpha y_{21}^h.$$

All the other y_{ik}^h are kept intact, then (4) is satisfied, and we then need to change x_{ijk}^h on the two machines so that (5) and (6) are also satisfied.

Consider the following simple altering of x_{1j1}^h, x_{2j1}^h, x_{1jw}^h and x_{2jw}^h.

$$x_{1j1}^{h+1} = \alpha x_{2j1}^h + x_{1j1}^h, x_{2j1}^{h+1} = (1 - \alpha)x_{2j1}^h,$$

$$x_{1jw}^{h+1} = x_{1jw}^h(1 - \alpha y_{21}^h/y_{1w}^h), x_{2jw}^{h+1} = x_{2jw}^h + \alpha y_{21}^h/y_{1w}^h x_{1jw}^h.$$

By altering them this way we don't actually move jobs between machine 1 and w, but rather re-assign outer small jobs on the two machines, so the total number of jobs along with total processing times on the two machines remain unchanged. Also notice that $x_{2j1}^{h+1}/x_{2j1}^h = y_{21}^{h+1}/y_{21}^h$ and $x_{1jw}^{h+1}/x_{1jw}^h = y_{1w}^{h+1}/y_{1w}^h$, so it follows directly that

$$(12) \sum_{j=1}^{n'} x_{2j1}^{h+1} = c_1 y_{21}^{h+1}, \qquad \sum_{j=1}^{n'} x_{1j1}^{h+1} = c_1 y_{11}^{h+1}$$

$$(13) \sum_{j=1}^{n'} p_j x_{2j1}^{h+1} \le (t^* - \Gamma_2) y_{21}^{h+1}, \qquad \sum_{j=1}^{n'} x_{1jw}^{h+1} = c_w y_{1w}^{h+1}$$

$$(14) \sum_{j=1}^{n'} x_{2jw}^{h+1} = c_w y_{2w}^{h+1}, \qquad \sum_{j=1}^{n'} p_j x_{1jw}^{h+1} \le (t^* - \Gamma_1) y_{1w}^{h+1}$$

We could also conclude that $\sum_{j=1}^{n'} p_j x_{2jw}^{h+1} \le (t^* - \Gamma_2) y_{2w}^{h+1}$, since $\sum_{j=1}^{n'} p_j x_{2jw}^{h+1} + \sum_{j=1}^{n'} p_j x_{1jw}^{h+1} = \sum_{j=1}^{n'} p_j x_{2jw}^{h} + \sum_{j=1}^{n'} p_j x_{1jw}^{h}$, thus

$$\sum_{j=1}^{n'} p_j x_{2jw}^{h+1} = \sum_{j=1}^{n'} p_j x_{2jw}^{h} + \sum_{j=1}^{n'} p_j x_{1jw}^{h} \alpha y_{21}^{h} / y_{1w}^{h}$$
$$\le (t^* - \Gamma_2) y_{2w}^{h} + (t^* - \Gamma_1) y_{1w}^{h} \alpha y_{21}^{h} / y_{1w}^{h}$$
$$\le (t^* - \Gamma_2) y_{2w}^{h+1}$$

The last inequality follows from $t^* - \Gamma_1 \le t^* - \Gamma_2$.

So if $\sum_{j=1}^{n'} p_j x_{1j1}^{h+1} \le (t^* - \Gamma_1) y_{11}^{h+1}$, then we have successfully derived a new feasible solution of $MILP_r(t^*)$.

Otherwise $\sum_{j=1}^{n'} p_j x_{1j1}^{h+1} > (t^* - \Gamma_1) y_{11}^{h+1}$, and we would have to exchange outer small jobs on machine 1 with w, and meanwhile re-assign them.

Let y_{ik}^{h+1} be defined as before. First we define again $x_{2j1}^{h+1} = (1-\alpha) x_{2j1}^{h}$, then for each outer small job j on machine 1, we need to reduce a fraction of αx_{2j1}^{h} (either to move it out or to re-assign it) We put $\theta \alpha x_{2j1}^{h}$ onto machine w, where $\theta \in [0,1]$ is a parameter that to be fixed later. Then we re-assign the remaining fraction $(1-\theta)\alpha x_{2j1}^{h}$ for each j to container 1.

Since in all $\sum_{j=1}^{n'} \theta \alpha x_{2j1}^{h}$ jobs are moved to machine w, we then need to move the same number of jobs from machine w back to machine 1. Among the $\sum_{j=1}^{n'} x_{1jw}^{h}$ outer small jobs on machine w, suppose we move $\varphi \sum_{j=1}^{n'} x_{1jw}^{h} = \sum_{j=1}^{n'} \theta \alpha x_{2j1}^{h}$, it follows that $\varphi = \frac{c_1}{c_w} \frac{\eta}{y_{1w}^{h}} \frac{y_{21}^{h}}{1-y_{11}^{h}} \theta \in [0,1]$.

Meanwhile we set $x_{1jw}^{h+1} = x_{1jw}^{h}(1 - \frac{y_{21}^{h}}{1-y_{11}^{h}} \frac{\eta}{y_{1w}})$, then adding up the fraction of φx_{1jw}^{h} that is moved out, there is still a fraction of $(1 - \frac{c_1}{c_w}\theta) \frac{y_{21}^{h}}{1-y_{11}^{h}} \frac{\eta}{y_{1w}^{h}} x_{1jw}^{h}$ left for each j, and we re-assign them to container 2.

The above discussion implies that we set x_{ijk}^{h+1} as follows

$$x_{1j1}^{h+1} = (\alpha - \theta\alpha) x_{2j1}^{h} + x_{1j1}^{h} + \varphi x_{1jw}^{h},$$

$$x_{2j1}^{h+1} = (1-\alpha) x_{2j1}^{h},$$

$$x_{1jw}^{h+1} = x_{1jw}^{h}(1 - \alpha y_{21}^{h} / y_{1w}^{h}),$$

$$x_{2jw}^{h+1} = x_{2jw}^{h} + \theta\alpha x_{2j1}^{h} + \left(\frac{y_{21}^{h}}{1-y_{11}^{h}} \frac{\eta}{y_{1w}} - \varphi\right) x_{1jw}^{h}.$$

The same argument shows that inequalities (12) to (14) are also satisfied.

We then set $\sum_{j=1}^{n'} p_j x_{1j1}^{h+1} = (t^* - \Gamma_1) y_{11}^{h+1}$. This is possible since

$$\sum_{j=1}^{n'} p_j x_{1j1}^{h+1} = \sum_{j=1}^{n'} (\alpha - \theta\alpha) p_j x_{2j1}^{h} + \sum_{j=1}^{n'} p_j x_{1j1}^{h} + \frac{c_1}{c_w} \frac{y_{21}^{h}}{1 - y_{11}^{h}} \frac{\eta}{y_{1w}^{h}} \theta \sum_{j=1}^{n'} p_j x_{1jw}^{h}$$

is linear in variable θ. And if $\theta = 0$, then we actually do not exchange outer small jobs between machine 1 and w, and thus $\sum_{j=1}^{n'} p_j x_{1j1}^{h+1} > (t^* - \Gamma_1) y_{11}^{h+1}$. If $\theta = 1$, simple computation shows that $\sum_{j=1}^{n'} p_j x_{1j1}^{h+1} \leq (t^* - \Gamma_1)(y_{11}^{h} + \frac{c_1}{c_w} \alpha y_{21}^{h}) \leq (t^* - \Gamma_1) y_{11}^{h+1}$. Thus there exists some $\theta \in [0,1]$, such that $\sum_{j=1}^{n'} p_j x_{1j1}^{h+1} = (t^* - \Gamma_1) y_{11}^{h+1}$.

Now we only need to prove for such a θ, $\sum_{j=1}^{n'} p_j x_{2jw}^{h+1} \leq (t^* - \Gamma_2) y_{2w}^{h+1}$. Consider machine 1.

$$\Gamma_1(y_{11}^{h+1} - y_{11}^{h}) + \Gamma_2(y_{21}^{h+1} - y_{21}^{h}) + \sum_{j=1}^{n'} (x_{1j1}^{h+1} + x_{2j1}^{h+1} - x_{1j1}^{h} - x_{2j1}^{h}) p_j$$

$$= t^* y_{11}^{h+1} - \Gamma_1 y_{11}^{h} - \sum_{j=1}^{n'} p_j x_{1j1}^{h} - \alpha(\Gamma_2 y_{21}^{h} + \sum_{j=1}^{n'} x_{2j1}^{h} p_j)$$

$$\geq t^* y_{11}^{h+1} - \Gamma_1 y_{11}^{h} - (t^* - \Gamma_1) y_{11}^{h} - \alpha[\Gamma_2 y_{21}^{h} + (t^* - \Gamma_2) y_{21}^{h}] = 0$$

The total processing time of jobs on machine 1 doesn't decrease, then the total processing time jobs on machine w doesn't increase.

$$\Gamma_2 y_{2w}^{h+1} + \sum_{j=1}^{n'} p_j x_{2jw}^{h+1}$$

$$\leq \Gamma_2 y_{2w}^{h} + \sum_{j=1}^{n'} p_j x_{2jw}^{h} + \Gamma_1 y_{1w}^{h} + \sum_{j=1}^{n'} p_j x_{1jw}^{h} - \Gamma_1 y_{1w}^{h+1} - \sum_{j=1}^{n'} p_j x_{1jw}^{h+1}$$

$$= \Gamma_2 y_{2w}^{h} + \sum_{j=1}^{n'} p_j x_{2jw}^{h} + \alpha y_{21}^{h} \Gamma_1 + \alpha y_{21}^{h} \frac{1}{y_{1w}^{h}} \sum_{j=1}^{n'} p_j x_{1jw}^{h} \leq t^* y_{2w}^{h+1}$$

Then it follows $\sum_{j=1}^{n'} p_j x_{2jw}^{h+1} \leq (t^* - \Gamma_2) y_{2w}^{h+1}$, and we complete our proof. $\square$

With the above lemma, given a feasible solution of $MILP_r(t^*)$, a feasible solution of $MILP(t^*)$ could be constructed in the following way. The integer variables z_i^ϕ of $MILP(t^*)$ are taking the same value of that in $MILP_r(t^*)$. For each $1 \leq \phi \leq \mu - 1$, we put (arbitrarily) a ϕ-container on a machine with capacity ϕ. Then we schedule μ-containers greedily as Lemma 5 implies, i.e., we put the μ-container with a larger load onto a machine (in GM_μ) with a smaller capacity, as we have mentioned in the proof. Such a schedule actually determines the value of y_{ik} in $MILP(t^*)$ in $O(m)$ time. By fixing the value of all the integer variables in $MILP(t^*)$ in this way, it becomes a linear programming and Lemma 5 implies that it admits a feasible solution.

Now it suffices to solve $MILP_r(t)$, which only contains a constant number of integer variables (which is $2^{O(\mu \log \log \mu)}$), and thus could be solved in polynomial time via Kannan's algorithm [13]. However, such a running time is far from linear. To achieve the desired running time, we need to further modify $MILP_r(t)$ by reducing the number of its fractional variables. The reader may refer to the appendix for a full proof of the following lemma.

Lemma 6. *There exists an algorithm that solves $MILP_r(t)$ in $f(1/\epsilon) + O(n \log n \log \log n)$ time, where $f(1/\epsilon) = 2^{O(1/\epsilon^2 \log^2(1/\epsilon) \log \log(1/\epsilon))}$.*

Combining Lemma 6 with Lemma 4, Theorem 1 is proved.

References

1. Alon, N., Azar, Y., Woeginger, G.J., Yadid, T.: Approximation schemes for scheduling on parallel machines. J. Sched. **1**, 55–66 (1998)
2. Babel, L., Kellerer, H., Kotov, V.: The k-partitioning problem. Math. Methods Oper. Res. **47**, 59–82 (1998)
3. Barna, S., Aravind, S.: A new approximation technique for resource-allocation problems. In: Proceedings of the 1st Annual Symposium on Innovations in Computer Science, pp. 342–357 (2010)
4. Bruglieri, M., Ehrgott, M., Hamacher, H.W., Maffioli, F.: An annotated bibliography of combinatorial optimization problems with fixed cardinality constraints. Discret. Appl. Math. **154**(9), 1344–1357 (2006)
5. Chen, L., Jansen, K., Zhang, G.C.: On the optimality of approximation schemes for the classical scheduling problem. In: Proceedings of SODA 2014, pp. 657–668 (2014)
6. Dell'Amico, M., Lori, M., Martello, S., Monaci, M.: Lower bounds and heuristic algorithms for the k_i-partitioning problem. Eur. J. Oper. Res. **171**, 725–742 (2004)
7. Eisenbrand, F., Shmonin, G.: Caratheodory bounds for integer cones. Oper. Res. Lett. **34**, 564–568 (2006)
8. Hochbaum, D.S., Shmoys, D.B.: Using dual approximation algorithms for scheduling problems: theoretical and practical results. J. ACM **34**, 144–162 (1987)
9. Jain, K.: A factor 2 approximation algorithm for the generalized Steiner network problem. Combinatorica **21**, 39–60 (2001)
10. Jansen, K.: An EPTAS for scheduling jobs on uniform processors: using an MILP relaxation with a constant number of integral variables. SIAM J. Discret. Math. **24**, 457–485 (2010)
11. Jansen, K., Robenek, C.: Scheduling jobs on identical and uniform processors revisited. In: Solis-Oba, R., Persiano, G. (eds.) WAOA 2011. LNCS, vol. 7164, pp. 109–122. Springer, Heidelberg (2012). doi:10.1007/978-3-642-29116-6_10
12. Jansen, K., Klein, K., Verschae, J.: Closing the gap for makespan scheduling via sparsification techniques. In: Proceedings of ICALP, pp. 72:1–72:13 (2016)
13. Kannan, R.: Minkowski's convex body theorem and integer programming. Math. Oper. Res. **12**, 415–440 (1987)
14. Kellerer, H., Kotov, V.: A 3/2-approximation algorithm for k_i-partitioning. Oper. Res. Lett. **39**, 359–362 (2011)
15. Kellerer, H., Kotov, V.: A 7/6-approximation algorithm for 3-partitioning and its application to multiprocessor scheduling. Inf. Syst. Oper. Res. **37**, 48–56 (1999)

16. Kellerer, H., Woeginger, G.: A tight bound for 3-partitioning. Discret. Appl. Math. **45**, 249–259 (1993)
17. Magazine, M.J., Ball, M.O.: Sequencing of insertions in printed circuit board assembly. Oper. Res. **36**, 192–201 (1988)
18. Rushmeier, R.A., Hoffman, K.L., Padberg, M.: Recent advances in exact optimization of airline scheduling problems. Technical report, George Mason University (1995)
19. Woeginger, G.: A comment on scheduling two parallel machines with capacity constraints. Discret. Optim. **2**, 269–272 (2005)
20. Zhang, C., Wang, G., Liu, X., Liu, J.: Approximating scheduling machines with capacity constraints. In: Deng, X., Hopcroft, J.E., Xue, J. (eds.) FAW 2009. LNCS, vol. 5598, pp. 283–292. Springer, Heidelberg (2009). doi:10.1007/978-3-642-02270-8_29

A Pseudo-Polynomial Time Algorithm
for Solving the Knapsack Problem
in Polynomial Space

Noriyuki Fujimoto[✉]

Osaka Prefecture University, Sakai, Osaka 599-8531, Japan
fujimoto@cs.osakafu-u.ac.jp

Abstract. It is well-known that the knapsack problem is NP-complete and can be solved in pseudo-polynomial time if we use dynamic programming. However, such dynamic programming approach requires pseudo-polynomial space that makes the resultant algorithm impractical since the computer memory is exhausted before the running time gets too long to tolerate. To remedy this, recently, Lokshtanov and Nederlof presented the first algorithm to solve the knapsack problem in pseudo-polynomial time and polynomial space. This paper presents another algorithm for solving the knapsack problem in pseudo-polynomial time and polynomial space.

Keywords: Combinatorial optimization · Exact algorithm · Space efficiency · Generating function

1 Introduction

The knapsack problem is to decide whether there exists a subset of a given set of n items with weight and value such that the sum of weight in the subset is at most a given positive integer W and the sum of value in the subset is at least a given positive integer V. The knapsack problem is $\mathcal{NP}$-complete [7]. Hence, it is unlikely that the knapsack problem can be solved in polynomial time. On the other hand, it is well-known that the knapsack problem can be solved in pseudo-polynomial time $O(nW)$ if we use dynamic programming [8]. However, such dynamic programming approach requires pseudo-polynomial space $O(n+W)$ [8] that makes the resultant algorithm impractical with an increase of W since the computer memory is exhausted before the running time gets too long to tolerate.

To remedy this, recently, Lokshtanov and Nederlof presented the first algorithm LN [11,12] to solve the knapsack problem in pseudo-polynomial time and polynomial space. The algorithm LN runs in $\tilde{O}(n^4 \log^2(VW)VW)$ time and $O(n^2 \log(VW))$ space where $\tilde{O}(\cdot)$ suppresses factors that are poly-logarithmic in the input size. The algorithm LN reduces the knapsack problem to calculating one of the coefficients of the corresponding generating function of two-variables. To compute the coefficient, the algorithm LN utilizes the Discrete Fourier Transform (DFT).

© Springer International Publishing AG 2016
T.-H.H. Chan et al. (Eds.): COCOA 2016, LNCS 10043, pp. 624–638, 2016.
DOI: 10.1007/978-3-319-48749-6_45

This paper presents another algorithm for solving the knapsack problem in pseudo-polynomial time and polynomial space. The proposed algorithm runs in $\tilde{O}((n + \log((\sum_{j=1}^{n} w_j)(\sum_{j=1}^{n} v_j)))(\sum_{j=1}^{n} w_j)(\sum_{j=1}^{n} v_j)(n + \log \sum_{j=1}^{n}(w_j + v_j)))$ time and $O((n + \log \sum_{j=1}^{n}(w_j + v_j))(\log(\sum_{j=1}^{n} w_j)(\sum_{j=1}^{n} v_j)))$ space where w_j and v_j are respectively the weight and the value of the item j. The proposed algorithm also reduces the knapsack problem to calculating one of the coefficients of the corresponding generating function in the factorized form. However, the proposed algorithm never uses DFT. Instead, the proposed algorithm calculates the coefficient by gradually reducing the number of terms in the function while preserving the coefficient without expanding the function. That is, we present a novel technique to compute any coefficient of a complex generating function only by dynamically transforming the generating function.

The remainder of this paper is organized as follows. First, Sect. 2 gives some basic facts on generating functions and complex numbers. Next, Sect. 3 describes the proposed algorithm in detail. The proposed algorithm requires multiple-precision arithmetic. Therefore, Sect. 4 theoretically analyzes the required precision to show the time complexity and the space complexity of the proposed algorithm. Finally, Sect. 5 gives concluding remarks and indicates some future works.

2   Preliminaries

This section briefly reviews the basic facts required to understand this paper. See [4,9] and so on for details.

2.1   Generating Functions

Let $G(z) = \sum_{j=0}^{\infty} c_j z^j$ be a complex generating function. Then, the following expressions hold:

$$(G(z) + G(-z))/2 = \sum_{j=0}^{\infty} c_{2j} z^{2j} \tag{1}$$

$$(G(z) - G(-z))/2 = \sum_{j=0}^{\infty} c_{2j+1} z^{2j+1} \tag{2}$$

$$G(z)/(1 - z) = \sum_{j=0}^{\infty} \left(\left(\sum_{k=0}^{j} c_k \right) z^j \right) \tag{3}$$

Note that the result of each expression is again a complex generating function.

2.2   Complex Numbers

Let $z = x + yi = re^{i\theta} = r(\cos\theta + i\sin\theta)$ be a complex number where i is the imaginary unit. We denote the real part and the imaginary part of z respectively by $Re(z)$ and $Im(z)$. The absolute value of $|z|$ is defined as $r = \sqrt{x^2 + y^2}$. It follows that $|z| \geq 0$. Let v and w be complex numbers. Let m, p, and q be positive integers. Then, the following expressions hold:

$$|v \pm w| \leq |v| + |w| \tag{4}$$

$$|vw| = |v||w| \tag{5}$$

$$|v^m| = |v|^m \tag{6}$$

$$|v/w| = |v|/|w| \tag{7}$$

$$|v/2| = |v|/2 \tag{8}$$

$$z = re^{i\theta} \Rightarrow z^m = r^m e^{im\theta} \tag{9}$$

$$z^{\frac{q}{p}} = (z^{\frac{1}{p}})^q \tag{10}$$

$$z = re^{i\theta} \text{ with } -\pi < \theta \leq \pi \Rightarrow \sqrt{z} = \sqrt{r}e^{i\theta/2} \tag{11}$$

3   The Proposed Algorithm

For a problem instance $n, w_1, w_2, \cdots, w_n, v_1, v_2, \cdots, v_n, W, V$ of the knapsack problem, consider a polynomial function $P(x, y) = \prod_{j=1}^{n}(1 + x^{w_j}y^{v_j})$. Let $W_{total} = \sum_{j=1}^{n} w_j$ and $V_{total} = \sum_{j=1}^{n} v_j$. Then, the polynomial $P(x, y)$ can be regarded as a polynomial $\sum_{j=0}^{W_{total}} c_j x^j$ of x where each c_j is a polynomial $\sum_{k=0}^{V_{total}} d_{j,k} y^k$ of y. The coefficients $\{d_{j,k}\}$ imply that there exist $d_{j,k}$ subsets with weight j and value k. Therefore, the problem instance is a yes instance if and only if $\sum_{j=0}^{W} \sum_{k=V}^{V_{total}} d_{j,k} \geq 1$. Hence, we can solve the knapsack problem by computing $\sum_{j=0}^{W} \sum_{k=V}^{V_{total}} d_{j,k}$.

The polynomial $P(x, y)$ is a generating function of two-variables. Therefore, from Eq. (3), $P(x, y)/(1-x) = \sum_{j=0}^{W_{total}} (\sum_{k \leq j} c_k)x^j + \sum_{j=W_{total}+1}^{\infty} cx^j$ where c is some polynomial of y (In fact, $c = \sum_{k=0}^{W_{total}} c_k$). Let $Q(y)$ be the coefficient of the term x^W. Then, $Q(y) = \sum_{j=0}^{W} c_j = \sum_{j=0}^{W} \sum_{k=0}^{V_{total}} d_{j,k} y^k = \sum_{k=0}^{V_{total}} (\sum_{j=0}^{W} d_{j,k})y^k$. The polynomial $Q(y)$ is a generating function of one-variable. Hence, from Eq. (3), $Q(y)/(1-y) = \sum_{k=0}^{V_{total}} \sum_{h \leq k} (\sum_{j=0}^{W} d_{j,h})y^k + \sum_{k=V_{total}+1}^{\infty} dy^k$ where d is some constant. Let R_{V-1} (resp. $R_{V_{total}}$) be the coefficient of the term y^{V-1} (resp. $y^{V_{total}}$). Then, we have $R_{V_{total}} - R_{V-1} = \sum_{h \leq V_{total}} (\sum_{j=0}^{W} d_{j,h}) - \sum_{h \leq V-1} (\sum_{j=0}^{W} d_{j,h}) = \sum_{h=V}^{V_{total}} \sum_{j=0}^{W} d_{j,h} = \sum_{j=0}^{W} \sum_{k=V}^{V_{total}} d_{j,k}$. Thus, if we can compute arbitrary coefficients of a generating function in the above form, we can solve the knapsack problem.

In Algorithm 1, we present an algorithm to compute any coefficient of $P(x, y)/(1-x)$ and $Q(y)/(1-y)$. For a given generating function f_0 in the above form, Algorithm 1 recursively defines a sequence $\langle f_1, f_2, \cdots, f_D \rangle$ of generating functions in the above form and finally returns the coefficient computed by f_D. Note that Algorithm 1 never expands f_0 (i.e., $P(x, y)/(1-x)$ and $Q(y)/(1-y)$) and that the representation of the generating functions defined in Algorithm 1 is brief.

The idea behind Algorithm 1 is as follows. A given generating function $f_0(z)$ is of the form $\sum_{j=0}^{d_0} c_j z^j + \sum_{j=d_0+1}^{\infty} cz^j$. We want to extract t_0th order coefficient of f_0. Assume t_0 is even. Let $f_1(z) = (f_0(\sqrt{z}) + f_0(-\sqrt{z}))/2$, $d_1 = \lfloor d_0/2 \rfloor$, and $t_1 = \lfloor t_0/2 \rfloor$. Then, from Eq. (1), the defined $f_1(z)$ is of the form $\sum_{j=0}^{d_1} c_{2j} z^j + \sum_{j=d_1+1}^{\infty} cz^j$ and t_1th order coefficient of f_1 is equal to t_0th order coefficient of f_0. Also in the case that t_0 is odd, a similar argument holds if let $f_1(z) = (\sqrt{z} f_0(\sqrt{z}) - \sqrt{z} f_0(-\sqrt{z}))/2$, $d_1 = \lceil d_0/2 \rceil$, and $t_1 = \lceil t_0/2 \rceil$ because the defined $f_1(z)$ is of the form $\sum_{j=0}^{d_1} c_{2j+1} z^j + \sum_{j=d_1+1}^{\infty} cz^j$ from Eq. (2). This process can be repeated until d_D is reduced to 1 for some positive integer D. At that time, the defined $f_D(z)$ is a generating function of the form $c_0' + c_1' z + \sum_{j=2}^{\infty} cz^j$ and $t_D = 1$ (because $t_k \leq d_k$ and t_k is never zero for any k). Let $g(z) = f_D(z) - ((f_D(\sqrt{z}) + f_D(-\sqrt{z}))/2 - f_D(0))z$. Then, $g(z)$ is equivalent to $c_0' + c_1' z$. Hence, we can extract c_1' by computing $(g(z') - g(0))/z' = (g(z') - f_k(0))/z'$ for any complex number z' except $1 + 0i$. Note that we cannot use $1 + 0i$ as z' because $f_0(z)$ has a denominator $(1 - z)$. Algorithm 1 selected $0 + i$ as z' because complexity analysis in Sect. 4 requires $|z'| = 1$ and also because the division can be easily computed $((a + bi)/i = (ai - b)/(-1) = b - ai)$.

In Algorithm 2, we present an algorithm to decide, using Algorithm 1 as a subroutine, whether a given knapsack problem instance is a yes instance or a no instance.

Algorithms 1 and 2 are theoretical algorithms in the sense that it correctly works if every computation is performed with infinite precision. Algorithm 1 computes square roots of complex numbers (e.g., $\sqrt{-i}$) as shown in Lemma 7 in Sect. 4. These numbers cannot be represented exactly in binary notation with finite precision. Therefore, we must do calculations with modestly large precision. We analytically show in Sect. 4 that the required precision is at most $n + O(\log \sum_{j=1}^{n}(w_j + v_j))$ bits. The analysis imposes some constraints to Algorithm 1 and Algorithm 2 in order to guarantee that truncation errors are within acceptable range. Hence, we need to slightly modify Algorithms 1 and 2 to work correctly with finite precision. We present a finite precision version (a practical version) of Algorithm 2 in Algorithm 3. As for the modification to Algorithm 1, we describe it in Sect. 4. The modification of both Algorithms 1 and 2 depends on our current analysis. Better analysis may lead another finite precision version.

Algorithm 3 is slightly different from the idea described in the first two paragraphs in this section. The difference is the use of division-by-two and the factors $(1 - i^{1/(2W_{total})})$ and $(1 - i^{1/(2V_{total})})$ in the computations. These are required to keep the absolute value of the operands of each multiplication and each addition at most 1, which is required by our current analysis. The detail is described in Sect. 4.

Note that Algorithm 1 can be used to extract any coefficient of a given polynomial $f_k(z) = \sum_{j=0}^{d_k} c_j z^j$. In this case, Algorithm 1 can be slightly simplified by replacing line 3 to 4 with "return $f_k(1) - f_k(0)$" because it follows that $c = 0$ for a polynomial.

Algorithm 1. Calculate a coefficient of a given complex generating function (an infinite precision version)

Input: a complex generating function f_k such that $f_k(z) = \sum_{j=0}^{d_k} c_j z^j + \sum_{j=d_k+1}^{\infty} c z^j$ where each coefficient is a complex constant or a complex generating function of a variable other than z, target order $t_k (> 0)$

Output: the t_kth order coefficient of f_k

1: **function** COEFF(d_k, f_k, t_k)   ▷ Symbol $\triangleq$ represents definition.
2:     **if** $d_k = 1$ **then**   ▷ Note that d_k is not always 1 even if t_k is 1.
3:         $h(z) \triangleq f_k(z) - ((f_k(\sqrt{z}) + f_k(-\sqrt{z}))/2 - f_k(0))z$
4:             **return** $(h(\mathrm{i}) - f_k(0))/\mathrm{i}$
5:     **end if**
6:     **if** t_k is even **then**
7:         $t_{k+1} \leftarrow \lfloor t_k/2 \rfloor$
8:         $d_{k+1} \leftarrow \lfloor d_k/2 \rfloor$
9:         $f_{k+1}(z) \triangleq (f_k(\sqrt{z}) + f_k(-\sqrt{z}))/2$
10:    **else**
11:        $t_{k+1} \leftarrow \lceil t_k/2 \rceil$
12:        $d_{k+1} \leftarrow \lceil d_k/2 \rceil$
13:        $f_{k+1}(z) \triangleq (\sqrt{z} f_k(\sqrt{z}) - \sqrt{z} f_k(-\sqrt{z}))/2$
14:    **end if**
15:    **return** COEFF($d_{k+1}, f_{k+1}, t_{k+1}$)
16: **end function**

Algorithm 2. Solve the decision counterpart of the knapsack problem (an infinite precision version)

Input: positive integers $n, w_1, w_2, \cdots, w_n, v_1, v_2, \cdots, v_n, W, V$ such that $n \geq 1$, $\forall j (w_j \leq W, v_j \leq V)$, $W \leq \sum_{j=1}^{n} w_j$ and $V \leq \sum_{j=1}^{n} v_j$

Output: $\begin{cases} \text{yes} & \text{(if the given problem instance is a yes instance)} \\ \text{no} & \text{(otherwise)} \end{cases}$

1: **function** KNAPSACKDECISION($n, w_1, w_2, \cdots, w_n, v_1, v_2, \cdots, v_n, W, V$)
2:     $W_{total} \leftarrow \sum_{j=1}^{n} w_j$; $V_{total} \leftarrow \sum_{j=1}^{n} v_j$
3:     $f_0(x, y) \triangleq (\prod_{j=1}^{n}(1 + x^{w_j} y^{v_j}))/(1 - x)$
4:     $f(y) \triangleq$ COEFF(W_{total}, f_0 as a generating function of variable x, W)
5:     $g_0(y) \triangleq f(y)/(1 - y)$
6:     $c_1 \leftarrow$ COEFF($V_{total}, g_0, V_{total}$); $c_2 \leftarrow$ COEFF($V_{total}, g_0, V - 1$)
7:     $c \leftarrow c_1 - c_2$
8:     **if** $Re(c) \geq 1$ **then return** yes
9:         **else return** no
10:    **end if**
11: **end function**

Algorithm 3. Solve the decision counterpart of the knapsack problem (a finite precision version)

Input: positive integers $n, w_1, w_2, \cdots, w_n, v_1, v_2, \cdots, v_n, W, V$ such that $n \geq 1$, $\forall j (w_j \leq W, v_j \leq V)$, $W \leq \sum_{j=1}^n w_j$ and $V \leq \sum_{j=1}^n v_j$

Output: $\begin{cases} \text{yes} & \text{(if the given problem instance is a yes instance)} \\ \text{no} & \text{(otherwise)} \end{cases}$

1: **function** KNAPSACKDECISION($n, w_1, w_2, \cdots, w_n, v_1, v_2, \cdots, v_n, W, V$)
2:     $W_{total} \leftarrow \sum_{j=1}^n w_j$; $V_{total} \leftarrow \sum_{j=1}^n v_j$
3:     Set arithmetic precision of fixed-point data to 2 bits for the integral part and ℓ bits for the fractional part where $\ell = n + O(\log \sum_{j=1}^n (w_j + v_j))$ bits.
4:     Set rounding mode to "round toward zero" except the computations of $(1 - x)$ and $(1 - y)$ during evaluation of instances of f_0 and g_0, which require "round away from zero".
5:     $f_0(x, y) \triangleq (\prod_{j=1}^n ((1 + x^{w_j} y^{v_j})/2))((1 - i^{1/(2W_{total})})/(1 - x))$
6:     $f(y) \triangleq \text{COEFF}(W_{total}, f_0$ as a generating function of variable $x, W)$
7:     $g_0(y) \triangleq f(y)((1 - i^{1/(2V_{total})})/(1 - y))$
8:     $c_1 \leftarrow \text{COEFF}(V_{total}, g_0, V_{total})$; $c_2 \leftarrow \text{COEFF}(V_{total}, g_0, V - 1)$
9:     $c \leftarrow (c_1 - c_2)/(1 - i^{1/(2W_{total})})/(1 - i^{1/(2V_{total})})$
10:    **if** $Re(c) < 2^{-n-1}$ **then return** no   ▷ if $2^n Re(c)$ is approximately zero
11:    **else return** yes   ▷ if $2^n Re(c)$ is approximately at least one
12:    **end if**
13: **end function**

---

4   Complexity Analysis

4.1   An Overview of the Analysis

Algorithm 3 computes the number of yes instances with finite precision. The computed number includes accumulated error. If the accumulated error is small enough, then the answer of Algorithm 3 is correct. In the following, we analyze the accumulation in bottom-up style from basic operations to higher level computations in order to see how large the precision needs to be. For simplicity, we shall be content with safe estimates of the precision, instead of finding the best possible bounds. Therefore, the required precision obtained by our current analysis may be improved.

4.2   Error Accumulation by Basic Operations

Lemma 1 [10]. *Let v' and w' be computed approximate values of complex numbers v and w respectively. Then, we have $|(v' + w') - (v + w)| \leq |v' - v| + |w' - w|$.*

Proof. From Eq. (4), $|(v' + w') - (v + w)| = |(v' - v) + (w' - w)| \leq |v' - v| + |w' - w|$ □

This lemma implies that the absolute error of the sum of two approximate values is at most the sum of the absolute errors of the two. Similar lemma follows for the difference of two approximate values.

Lemma 2. *Let v' and w' be computed approximate values of complex numbers v and w respectively. Then, we have $|(v' - w') - (v - w)| \leq |v' - v| + |w' - w|$.*

Lemma 3 [10]. *Let v' and w' be computed approximate values of complex numbers v and w respectively. Then, we have $|v| \leq 1, |w| \leq 1, |v'| \leq |v|, |w'| \leq |w| \Rightarrow |v'w' - vw| \leq |v' - v| + |w' - w|$.*

Proof. From Eqs. (4) and (5), $|v'w' - vw| = |(v' - v)w' + v(w' - w)| \leq |(v' - v)w'| + |v(w' - w)| = |v' - v||w'| + |v||w' - w|$ □

Note that $|v'| \leq |v|$ and $|w'| \leq |w|$ are guaranteed if we round towards zero instead of rounding-off when we compute approximate values [10]. This lemma implies that the absolute error of the product of two approximate values is at most the sum of the absolute errors of the two if the absolute values of the two approximate values and the corresponding true values are at most 1.

Lemma 4. *Let v' and w' be computed approximate values of complex numbers v and w respectively. Then, we have*

$$|v| \leq 1, |v'| \leq |v|, |w'| \geq |w| \Rightarrow |\frac{v'}{w'} - \frac{v}{w}| \leq \frac{|v' - v|}{|w|} + \frac{|w' - w|}{|w|^2}$$

Proof. From Eqs. (4), (5), and (7),

$$\begin{aligned}
|\frac{v'}{w'} - \frac{v}{w}| &= |\frac{v'w - vw'}{w'w}| = |\frac{(v' - v)w - (w' - w)v}{w'w}| \\
&\leq |\frac{(v' - v)w}{w'w}| + |\frac{(w' - w)v}{w'w}| = \frac{|v' - v||w|}{|w'||w|} + \frac{|w' - w||v|}{|w'||w|} \\
&= \frac{|v' - v|}{|w'|} + \frac{|w' - w||v|}{|w'||w|} \leq \frac{|v' - v|}{|w|} + \frac{|w' - w|}{|w|^2}
\end{aligned}$$

□

Note that $|w'| \geq |w|$ is guaranteed if we round away from zero instead of rounding-off. This lemma means that the absolute error of a division is at most the sum, of the absolute errors of the denominator and the numerator, divided by the square of the true value of the denominator if we compute the denominator by rounding away from zero and the numerator by rounding towards zero.

Lemma 5. *Let z' be a computed approximate value of a complex number z. Then, $|z| \leq 1, |z'| \leq |z| \Rightarrow |(z')^m - z^m| \leq m|z' - z|$ for any positive integer m.*

Proof. For $m = 1$, clearly the lemma follows. Assume that $|(z')^k - z^k| \leq k|z' - z|$ follows for some $k(\geq 1)$. Then, from Lemma 3, $|(z')^{k+1} - z^{k+1}| = |(z')^k z' - z^k z| \leq |(z')^k - z^k| + |z' - z| \leq k|z' - z| + |z' - z| = (k + 1)|z' - z|$. Therefore, the lemma follows by induction.   □

This lemma implies that the absolute error of a given approximate value to the power m is at most m times the absolute error of the given value.

4.3   Upper Bounds on Absolute Values of Complex Operands

The analysis in Sect. 4.2 assumes that the absolute values of complex operands of basic operations are at most 1. This section proves that the absolute values of the corresponding complex operands appeared in the computation of Algorithm 3 are at most 1.

In the following, let D be the number of functions dynamically defined by Algorithm 1 for given d_0. That is, we assume Algorithm 1 dynamically defines functions $f_1, f_2, \cdots, f_D$ for given d_0.

Lemma 6. *It follows that $D = \Theta(\log d_0)$.*

Proof. Algorithm 1 calls itself for roughly halved d_k until $d_k = 1$. Hence, we can estimate the depth (i.e., the maximum k) of recurrence by estimating a sequence $\langle d_0, d_1, \cdots \rangle$. It follows that $d_{k+1} \leq \max(\lfloor d_k/2 \rfloor, \lceil d_k/2 \rceil) = \lceil d_k/2 \rceil \leq d_k/2 + 1/2 = (d_k + 1)/2$ for any k. Therefore, $d_j \leq (d_0 + 2^j - 1)/2^j$ for any $j \geq 1$. Since d_j is an integer, it follows that $j \geq \log_2 d_0 \Rightarrow d_j \leq 1$. This implies that $D \leq \lceil \log d_0 \rceil$. Similarly, we have $D > \log d_0 - 1$ from $d_{k+1} \geq \min(\lfloor d_k/2 \rfloor, \lceil d_k/2 \rceil)$. $\square$

Lemma 7. *Let z' be any argument of f_0 called by Algorithm 1. Note that z' is a computed approximate value and there exist 2^D instances of f_0. Let z be the true value of z'. Then, $z \in \{0\} \cup \{\pm i^{\frac{1+4j}{2^D}} | j \in \{0, 1, 2, \cdots, 2^{D-1} - 1\}\} \cup \{\pm i^{\frac{1+8j}{2^{D+1}}} | j \in \{0, 1, 2, \cdots, 2^{D-1} - 1\}\} \cup \{\pm i^{\frac{5+8j}{2^{D+1}}} | j \in \{0, 1, 2, \cdots, 2^{D-1} - 1\}\}$.*

Proof. From line 3 to 4 in Algorithm 1, the true value of the argument of f_D is in $\{0, i, \pm \sqrt{i}\}$.

When the true value of the argument of f_D is zero, the true value of the argument of f_0 is zero because the true value of the argument of f_k is zero for any k.

When the true value of the argument of f_D is i, we prove by induction the proposition that the true value of the argument of f_0 is in $\{\pm i^{\frac{1+4j}{2^D}} | j \in \{0, 1, 2, \cdots, 2^{D-1} - 1\}\}$. When $D = 1$, the true value of the argument of f_0 is in $\{\pm \sqrt{i}\}$. Hence the proposition follows for $D = 1$. Assume the proposition follows for $D = k$ where k is some positive integer. Then, the true value of the argument of f_0 for $D = k$ is in $S = \{\pm i^{\frac{1+4j}{2^k}} | j \in \{0, 1, 2, \cdots, 2^{k-1} - 1\}\}$. Thus, the true value of the argument of f_0 for $D = k + 1$ is in

$$\{\pm \sqrt{s} | s \in S\}$$

$$= \{\pm \sqrt{+i^{\frac{1+4j}{2^k}}} | j \in \{0, 1, 2, \cdots, 2^{k-1} - 1\}\}$$

$$\cup \{\pm \sqrt{-i^{\frac{1+4j}{2^k}}} | j \in \{0, 1, 2, \cdots, 2^{k-1} - 1\}\}$$

$$= \{\pm i^{\frac{1+4j}{2^{k+1}}} | j \in \{0, 1, 2, \cdots, 2^{k-1} - 1\}\} \cup \{\pm i^{\frac{1+4j}{2^{k+1}}} i | j \in \{0, 1, 2, \cdots, 2^{k-1} - 1\}\}$$

$$= \{\pm i^{\frac{1+4j}{2^{k+1}}} | j \in \{0, 1, 2, \cdots, 2^{k-1} - 1\}\}$$

$$\cup \{\pm i^{\frac{1+4(j+2^{k-1})}{2^{k+1}}} | j \in \{0, 1, 2, \cdots, 2^{k-1} - 1\}\}$$

$$= \{\pm i^{\frac{1+4j}{2^{k+1}}} | j \in \{0, 1, 2, \cdots, 2^{k+1-1} - 1\}\}$$

Hence the proposition follows by induction.

Similarly, we can prove the similar propositions in the case of the true value of the argument of f_D is in $\{\pm\sqrt{i}\}$.

Therefore, the lemma follows.   □

Lemma 8. *Let z' be any argument of f_0 called by Algorithm 1. Let z be the true value of z'. Then, $|1 - z| \geq |1 - i^{1/2^{D+1}}|$.*

Proof. From Lemma 7,

$$
|1 - z| = \begin{cases}
1 \\
|1 \mp i^{\frac{1+4j}{2^D}}| \ (j \in \{0, 1, 2, \cdots, 2^{D-1} - 1\}) \\
|1 \mp i^{\frac{1+8j}{2^{D+1}}}| \ (j \in \{0, 1, 2, \cdots, 2^{D-1} - 1\}) \\
|1 \mp i^{\frac{5+8j}{2^{D+1}}}| \ (j \in \{0, 1, 2, \cdots, 2^{D-1} - 1\})
\end{cases}
$$

Since, for any positive integer p, $i^{1/2^p}$ is the result of p times application of a square root operation to i, from Eqs. (9), (10), and (11), it follows that, for any positive integer q, $|1 \mp i^{\frac{q}{2^p}}| = 1 \mp (i^{\frac{1}{2^p}})^q| = (1 \mp \cos\frac{q\pi}{2^p})^2 + (\mp\sin\frac{q\pi}{2^p})^2 = 2(1 \mp \cos\frac{q\pi}{2^p})$. Hence the lemma follows.   □

Lemma 9. *Let x and y be complex numbers. Let $f_k(x, y)$ $(k \in \{1, 2, \cdots\})$ be the function dynamically defined by Algorithm 1 for the function $f_0(x, y)$ defined by Algorithm 3. Then, $|x| \leq 1, |y| \leq 1 \Rightarrow |f_k(x, y)| \leq 1$ for any non-negative integer k.*

Proof. From Eqs. (4) to (7),

$$
|f_0(x, y)| = |(\prod_{j=1}^{n} \frac{1 + x^{w_j} y^{v_j}}{2})((1 - i^{1/(2\sum_{j=1}^{n} w_j)})/(1 - x))|
$$

$$
= (\prod_{j=1}^{n} \frac{|1 + x^{w_j} y^{v_j}|}{|2|})|1 - i^{1/(2\sum_{j=1}^{n} w_j)}|/|1 - x|
$$

$$
\leq (\prod_{j=1}^{n} \frac{|1| + |x^{w_j} y^{v_j}|}{2})|1 - i^{1/(2\sum_{j=1}^{n} w_j)}|/|1 - x|
$$

$$
= (\prod_{j=1}^{n} \frac{1 + |x|^{w_j}|y|^{v_j}}{2})|1 - i^{1/(2\sum_{j=1}^{n} w_j)}|/|1 - x|
$$

From Lemmas 6 and 8,

$$
|1 - x| \geq |1 - i^{1/2^{D+1}}|
$$

$$
\geq |1 - i^{1/2^{1+\log\sum_{j=1}^{n} w_j}}|
$$

$$
= |1 - i^{1/(2\sum_{j=1}^{n} w_j)}|
$$

Hence the lemma follows for $k = 0$.

Assume the lemma follows for $k = j(\geq 0)$.

- in the case that $f_{j+1}(x,y) = (f_j(\sqrt{x},y) + f_j(-\sqrt{x},y))/2$
  From Eqs. (4) and (8),

$$|f_{j+1}(x,y)| = |\frac{f_j(\sqrt{x},y) + f_j(-\sqrt{x},y)}{2}|$$
$$= \frac{|f_j(\sqrt{x},y) + f_j(-\sqrt{x},y|)}{2}$$
$$\leq \frac{|f_j(\sqrt{x},y)| + |f_j(-\sqrt{x},y)|}{2}$$

- in the case that $f_{j+1}(x,y) = (\sqrt{x}f_j(\sqrt{x},y) - \sqrt{x}f_j(-\sqrt{x},y))/2$
  From Eqs. (4), (5), and (8),

$$|f_{j+1}(x,y)| = |\frac{\sqrt{x}f_j(\sqrt{x},y) - \sqrt{x}f_j(-\sqrt{x},y)}{2}|$$
$$= \frac{|\sqrt{x}f_j(\sqrt{x},y) - \sqrt{x}f_j(-\sqrt{x},y)|}{2}$$
$$\leq \frac{|\sqrt{x}f_j(\sqrt{x},y)| + |\sqrt{x}f_j(-\sqrt{x},y)|}{2}$$
$$= \frac{|\sqrt{x}||f_j(\sqrt{x},y)| + |\sqrt{x}||f_j(-\sqrt{x},y)|}{2}$$

Since, from Eq. (11), $|x| \leq 1 \Rightarrow |\sqrt{x}| \leq 1, |-\sqrt{x}| \leq 1$, the lemma follows by induction. $\square$

A similar lemma follows also for the function $g_k(y)$. The proof is essentially the same as that of Lemma 9.

Lemma 10. *Let y be a complex number. Let $g_k(y)$ ($k \in \{1,2,\cdots\}$) be the function dynamically defined by Algorithm 1 for the function $g_0(y)$ defined by Algorithm 3. Then, $|y| \leq 1 \Rightarrow |g_k(y)| \leq 1$ for any non-negative integer k.*

Lemma 11. *Let z' be any argument of f_k called by Algorithm 1. Let z be the true value of z'. Then, $z = 0$ or $|z| = 1$.*

Proof. From line 3 to 4 in Algorithm 1, if $k = D$ then $z \in \{0, i, \pm\sqrt{i}\}$. In the case of $z = 0$ for $k = D$, the lemma follows since $z' = z = 0$ also for any other k. Therefore, in the following, we consider the case of $z \neq 0$ for $k = D$. In this case, $|z| = 1$ for $k = D$. Since $f_{k+1}(z)$ is defined in terms of $f_k(\sqrt{z})$ and $f_k(-\sqrt{z})$ for any other k, it follows that $|z| = 1$ also for any other k. Hence, the lemma follows. $\square$

This lemma implies that when $f_k(x,y)$ and $g_k(y)$ are evaluated in Algorithm 3, it follows that $|x| \leq 1, |y| \leq 1$ for any k.

4.4 Upper Bounds on Error of Complex Operands

From Lemmas 9 to 11, all complex numbers which appear in the computation of Algorithm 3 (including intermediate results) have the absolute value at most two. Therefore, in the following, we assume that any complex number is implemented as rectangular form whose real part and imaginary part are respectively implemented as a fixed point number such that integer part is of length 2 bits and fractional part is of length ℓ bits. Let δ_ℓ be the maximum truncation error of such complex numbers. We have $\delta_\ell = |2^{-\ell} + 2^{-\ell}i| = 2^{1/2-\ell}$ [10].

Let ω_r be the complex 2^r-th power root of unity. That is, $\omega_1 = -1, \omega_2 = i, \omega_3 = (1+i)/\sqrt{2}, \cdots$. Let $M(m)$ be the time complexity of a multiplication of two integers of length m bits. Then, the following lemma follows.

Lemma 12. *Let k be a given positive integer. Let ω be ω_k. Then, for a given non-negative integer j, an approximate value $(\omega^j)'$ of ω^j such that $|(\omega^j)' - \omega^j| < (2k-1)\delta_\ell$ can be computed in $O(kM(\ell))$ time and $O(k\ell)$ space with preprocessing in $O(kM(\ell))$ time and $O(k\ell)$ space.*

Proof. If $\omega_r = x_r + y_r i$, then $\omega_{r+1} = x_{r+1} + y_{r+1}i$ where $x_{r+1} = \sqrt{\frac{1+x_r}{2}}$ and $y_{r+1} = \frac{y_r}{2x_{r+1}}$ [13]. Each square root operation for a real number of length m bits can be done in $O(M(m))$ [1]. As preprocessing, we compute $\omega_1, \omega_2, \cdots, \omega_k$ in $O(kM(\ell))$ time and keep these values in $O(k\ell)$ space. Once the ω_r have been calculated, we can compute any power ω^j by noting that $\omega^j = \omega_1^{j_{k-1}} \cdots \omega_{k-1}^{j_1} \omega_k^{j_0}$ if $j = (j_{k-1} \cdots j_1 j_0)_2$ [10]. Since we can perform the computation with sufficient precision to make $|\omega_r' - \omega_r| < \delta_\ell$, it follows that $|(\omega^j)' - \omega^j| < (2k-1)\delta_\ell$ for all j, because the error is due to at most k approximations and $(k-1)$ truncations [10] (See Lemma 3). Hence the lemma follows.

Lemma 13. *Let z' be any argument of f_k called by Algorithm 1. Let z be the true value of z'. Then, $|z' - z| \le (2D - 2k + 5)\delta_\ell$.*

Proof. From line 3 to 4 in Algorithm 1, if $k = D$ then $z \in \{0, i, \pm\sqrt{i}\}$. When $z = 0$ for $k = D$, the lemma follows because $z' = z = 0$ for any k. When $z = i$ (resp. $\pm\sqrt{i}$) for $k = D$, from the proof of Lemma 7, it follows that $z = \pm\omega_{2+D-k}^j$ (resp. $\pm\omega_{3+D-k}^j$) for any other k and some positive integer j. Therefore, from Lemma 12, $|z' - z| < (2(3 + D - k) - 1)\delta_\ell$. Hence the lemma follows. □

Let $\langle f_1, f_2, \cdots, f_{D_f} \rangle$ be a sequence of complex functions dynamically defined by Algorithm 1 for $f_0(x, y)$ defined in Algorithm 3. Then, based on the one-variable complex function returned by Algorithm 1, Algorithm 3 defines $g_0(y)$ as

$$g_0(y) \triangleq \frac{f_{D_f}(i, y) - (\frac{f_{D_f}(\sqrt{i}, y) + f_{D_f}(-\sqrt{i}, y)}{2} - f_{D_f}(0, y))i - f_{D_f}(0, y)}{i}$$

$$((1 - i^{1/(2\sum_{j=1}^n v_j)})/(1 - y)) \tag{12}$$

Let $\langle g_{1,1}, g_{1,2}, \cdots, g_{1,D_1} \rangle$ and $\langle g_{2,1}, g_{2,2}, \cdots, g_{2,D_2} \rangle$ be two sequences of functions dynamically defined by Algorithm 3, using Algorithm 1 twice, to compute two complex numbers c_1 and c_2 respectively.

Lemma 14. *When $f_0(x, y)$ is evaluated in Algorithm 3, $|f_0(x', y') - f_0(x, y)| \leq (\frac{2D_f + 5}{2} \sum_{j=1}^{n}(w_j + v_j) + \frac{5}{2}n + (2D_f + 6)/|1 - i^{1/(2\sum_{j=1}^{n} w_j)}|^2)\delta_\ell$.*

Proof. Since the result of the power operation is truncated, from Lemma 5 and Lemma 13, $|(x')^{w_j} - x^{w_j}| \leq w_j(2D_f + 5)\delta_\ell + \delta_\ell$ and $|(y')^{v_j} - y^{v_j}| \leq v_j(2D_f + 5)\delta_\ell + \delta_\ell$. Therefore, from Lemma 3, $|(x')^{w_j}(y')^{v_j} - x^{w_j}y^{v_j}| \leq (w_j + v_j)(2D_f + 5)\delta_\ell + 3\delta_\ell$ because the result of the multiplication is truncated. Since the result of the division-by-two is truncated and an addition of fixed point numbers never yields truncation, $|(1 + (x')^{w_j}(y')^{v_j})/2 - (1 + x^{w_j}y^{v_j})/2| \leq ((w_j + v_j)(2D_f + 5)\delta_\ell + 3\delta_\ell)/2 + \delta_\ell$. On the other hand, we may assume that $|(i')^{1/(2\sum_{j=1}^{n} w_j)} - i^{1/(2\sum_{j=1}^{n} w_j)}| \leq \delta_\ell$ because $i^{1/(2\sum_{j=1}^{n} w_j)}$ is the $(1 + \log\sum_{j=1}^{n} w_j)$th root of unity and we can compute the kth root of unity within the precision for any non-negative integer k [10] as shown in the proof of Lemma 12. Since $|1 - x| \geq |1 - i^{1/(2\sum_{j=1}^{n} w_j)}|$ (as shown in the proof of Lemma 9), $|1 - i^{1/(2\sum_{j=1}^{n} w_j)}| \leq 1$, and a division is truncated, it follows that $|(1 - (i')^{1/(2\sum_{j=1}^{n} w_j)})/(1 - x') - (1 - i^{1/(2\sum_{j=1}^{n} w_j)})/(1 - x)| \leq ((1 + 2D_f + 5)\delta_\ell)/|1 - i^{1/(2\sum_{j=1}^{n} w_j)}|^2 + \delta_\ell$ from Lemma 4. Since the result of each multiplication is truncated, $|(\prod_{j=1}^{n}((1 + (x')^{w_j}(y')^{v_j})/2))((1 - (i')^{1/(2\sum_{j=1}^{n} w_j)})/(1 - x)) - (\prod_{j=1}^{n}((1 + x^{w_j}y^{v_j})/2))((1 - i^{1/(2\sum_{j=1}^{n} w_j)})/(1 - x))| \leq \sum_{j=1}^{n}(((w_j + v_j)(2D_f + 5)\delta_\ell + 3\delta_\ell)/2 + \delta_\ell) + ((2D_f + 6)\delta_\ell)/|1 - i^{1/(2\sum_{j=1}^{n} w_j)}|^2 + \delta_\ell + n\delta_\ell$ from Lemma 3. $\square$

Let $\langle U_{f,0}, U_{f,1}, \cdots, U_{f,D}\rangle$ be a sequence of real numbers defined as follows.

$$U_{f,0} = (\frac{2D_f + 5}{2}\sum_{j=1}^{n}(w_j + v_j) + \frac{5}{2}n + (2D_f + 6)/|1 - i^{1/(2\sum_{j=1}^{n} w_j)}|^2)\delta_\ell$$

$$U_{f,k} = U_{f,k-1} + (2D_f - 2k + 9)\delta_\ell$$

Then, the following lemma follows.

Lemma 15. *When $f_k(x, y)$ is evaluated in Algorithm 3, $|f_k(x', y') - f_k(x, y)| \leq U_{f,k}$.*

Proof. From Lemma 14, the lemma follows for $k = 0$.

Assume the lemma follows for $k = j(\geq 0)$. Then, we have $|f_j(\sqrt{x'}, y') - f_j(\sqrt{x}, y)| \leq U_{f,j}$ and $|f_j(-\sqrt{x'}, y') - f_j(-\sqrt{x}, y)| \leq U_{f,j}$.

- in the case that $f_{j+1}(x, y) = (f_j(\sqrt{x}, y) + f_j(-\sqrt{x}, y))/2$
  $|f_{j+1}(x', y') - f_{j+1}(x, y)| \leq 2U_{f,j}/2 + \delta_\ell$.
- in the case that $f_{j+1}(x, y) = (\sqrt{x}, yf_j(\sqrt{x}, y) - \sqrt{x}f_j(-\sqrt{x}, y))/2$
  From Lemma 13, $|\pm\sqrt{x'} - \pm\sqrt{x}| \leq (2D - 2j + 5)\delta_\ell$. Therefore, $|f_{j+1}(x', y') - f_{j+1}(x, y)| \leq 2(U_{f,j} + (2D - 2j + 5)\delta_\ell + \delta_\ell)/2 + \delta_\ell$.

Hence the lemma follows by induction. $\square$

It follows that $D_f \geq 1 \Rightarrow U_{f,D_f} = U_{f,0} + \sum_{k=1}^{D_f}(2D_f - 2k + 9)\delta_\ell = (\frac{2D_f + 5}{2}\sum_{j=1}^{n}(w_j + v_j) + \frac{5}{2}n + (2D_f + 6)/|1 - i^{1/(2\sum_{j=1}^{n} w_j)}|^2 + D_f^2 + 8D_f)\delta_\ell$. This equation follows also for $D_f = 0$. Hence, we have the following corollary.

Corollary 1. *It follows that* $U_{f,D_f} = (\frac{2D_f+5}{2}\sum_{j=1}^{n}(w_j + v_j) + \frac{5}{2}n + (2D_f + 6)/|1 - i^{1/(2\sum_{j=1}^{n} w_j)}|^2 + D_f^2 + 8D_f)\delta_\ell$.

Similar propositions follow also for the sequences $\langle g_0, g_{1,1}, g_{1,2}, \cdots, g_{1,D_1}\rangle$ and $\langle g_0, g_{2,1}, g_{2,2}, \cdots, g_{2,D_2}\rangle$. The proofs are essentially the same as that for $\langle f_0, f_1, \cdots, f_{D_f}\rangle$.

From Eq. (12), we have the following lemmata.

Lemma 16. *When $g_0(y)$ is evaluated in Algorithm 3, $|g_0(y') - g_0(y)| \leq 4U_{D_f} + 3\delta_\ell + \frac{(2D_1+6)\delta_\ell}{|1-i^{1/(2\sum_{j=1}^{n} v_j)}|^2}$ for $\langle g_0, g_{1,1}, g_{1,2}, \cdots, g_{1,D_1}\rangle$.*

Lemma 17. *When $g_0(y)$ is evaluated in Algorithm 3, $|g_0(y') - g_0(y)| \leq 4U_{D_f} + 3\delta_\ell + \frac{(2D_2+6)\delta_\ell}{|1-i^{1/(2\sum_{j=1}^{n} v_j)}|^2}$ for $\langle g_0, g_{1,1}, g_{1,2}, \cdots, g_{1,D_2}\rangle$.*

Let $U_{1,D_1} = 4U_{f,D_f} + 3\delta_\ell + \frac{(2D_1+6)\delta_\ell}{|1-i^{1/(2\sum_{j=1}^{n} v_j)}|^2} + (D_1^2 + 8D_1)\delta_\ell$ and $U_{2,D_2} = 4U_{f,D_f} + 3\delta_\ell + \frac{(2D_2+6)\delta_\ell}{|1-i^{1/(2\sum_{j=1}^{n} v_j)}|^2} + (D_2^2 + 8D_2)\delta_\ell$. Then, the following lemmata follow.

Lemma 18. *When $g_{1,D_1}(y)$ is evaluated in Algorithm 3, $|g_{1,D_1}(y') - g_{1,D_1}(y)| \leq U_{1,D_1}$.*

Lemma 19. *When $g_{2,D_2}(y)$ is evaluated in Algorithm 3, $|g_{2,D_2}(y') - g_{2,D_2}(y)| \leq U_{2,D_2}$.*

4.5   Required Precision

Theorem 1. *Algorithm 3 correctly works if $\ell = n + O(\log \sum_{j=1}^{n}(w_j + v_j))$.*

Proof. In line 9 of Algorithm 3, the computed c is the number of solutions of a given knapsack instance devided by 2^n. Since the value of c includes accumulated truncation errors, Algorithm 3 judges a given knapsack instance is a no instance iff the computed number of solutions (i.e., c times 2^n) is less than $1/2$. Hence, Algorithm 3 works correctly if the truncation errors accumlated in the computed c is less than 2^{-n-1}. Thus we obtain $|c' - c| < 2^{-n-1}$ as a sufficient condition for Algorithm 3 to work correctly. On the other hand, from Lemmas 18 and 19, we have $|g_{1,D_1}(y') - g_{1,D_1}(y)| \leq U_{1,D_1}$ and $|g_{2,D_2}(y') - g_{2,D_2}(y)| \leq U_{2,D_2}$. Since it follows that

$$|c' - c| \leq \frac{\frac{(U_{1,D_1}+U_{2,D_2})+\delta_\ell}{|1-i^{1/(2\sum_{j=1}^{n} w_j)}|^2} + \delta_\ell + \delta_\ell}{|1 - i^{1/(2\sum_{j=1}^{n} v_j)}|^2} + \delta_\ell$$

and $\log(x + y) = \log x + \log(1 + y/x)$, from Lemma 6, we have $\ell = n + O(\log \sum_{j=1}^{n}(w_j + v_j)) - O(\log|1 - i^{1/(2\sum_{j=1}^{n} w_j)}|) - O(\log|1 - i^{1/(2\sum_{j=1}^{n} v_j)}|)$. Since $O(-\log|1 - i^{1/(2x)}|) = \Theta(\log x)$, the theorem follows.   □

4.6   Complexity of the Proposed Algorithm

Theorem 2. *Algorithm 3 runs in* $\tilde{O}((n + \log((\sum_{j=1}^{n} w_j)(\sum_{j=1}^{n} v_j)))(\sum_{j=1}^{n} w_j)$
$(\sum_{j=1}^{n} v_j)(n + \log \sum_{j=1}^{n}(w_j + v_j)))$ *time and* $O((n + \log \sum_{j=1}^{n}(w_j + v_j))$
$(\log(\sum_{j=1}^{n} w_j)(\sum_{j=1}^{n} v_j)))$ *space if* $\ell = n + O(\log \sum_{j=1}^{n}(w_j + v_j))$.

Proof. The complex numbers $(1 - \mathrm{i}^{1/(2\sum_{j=1}^{n} w_j)})$ and $(1 - \mathrm{i}^{1/(2\sum_{j=1}^{n} v_j)})$ are enough to be computed only once. We compute them using the method described in the proof of Lemma 12. The complexity of them is $O((\log \sum_{j=1}^{n} w_j + \log \sum_{j=1}^{n} v_j)M(\ell))$ time and $O((\log \sum_{j=1}^{n} w_j + \log \sum_{j=1}^{n} v_j)\ell)$ space.

The complexity of computing c_1 is greater than that of computing c_2 because $D_1 \geq D_2$. The dominant part of computing c_1 is computing instances of $f_0, f_1, \cdots, f_{D_f}$. To keep the accumulated truncation errors within acceptable range, each instance of $f_0(x, y)$ generates the value of its argument x and y using the methods described in the proofs of Lemmas 7 and 12 before starting its computation. Similarly, each instance of $f_{k+1}(x, y)$ $(k \geq 0)$ generates the value $\sqrt{x}$ if necessary and the value y before starting its computation. This requires $O((D_f + D_1)M(\ell))$ time each and $O(\ell \max(D_1, D_f))$ space as a whole. (The required preprocessing is already done when we compute $(1 - \mathrm{i}^{1/(2\sum_{j=1}^{n} w_j)})$ and $(1 - \mathrm{i}^{1/(2\sum_{j=1}^{n} v_j)})$.) Computing an instance of f_0 requires $2n$ power operations, $(n + 1)$ additions, n divisions-by-two, $2n$ multiplications, and one general division. Computing an instance of f_{k+1} $(k \geq 0)$ requires one negation, at most two multiplications, one addition, and one division-by-two. The number of instances of f_k $(k \geq 0)$ is $2^{D_1+D_f-k}$.

Therefore, from Lemma 6, Algorithm 3 performs $O(n(\sum_{j=1}^{n} w_j)(\sum_{j=1}^{n} v_j))$ power operations, $O(n(\sum_{j=1}^{n} w_j)(\sum_{j=1}^{n} v_j))$ additions, $O(n(\sum_{j=1}^{n} w_j)$ $(\sum_{j=1}^{n} v_j))$ divisions-by-two, $O(n(\sum_{j=1}^{n} w_j)(\sum_{j=1}^{n} v_j))$ multiplications, $O((\sum_{j=1}^{n} w_j)(\sum_{j=1}^{n} v_j))$ negations, and $O((\sum_{j=1}^{n} w_j)(\sum_{j=1}^{n} v_j))$ general divisions for complex numbers such that mantissas of real part and imaginary part are of length $O(\ell)$ bits.

Each negation, addition, multiplication, and division-by-two for complex numbers can be implemented respectively with $O(1)$ negations, $O(1)$ additions, $O(1)$ multiplications followed by $O(1)$ additions, and $O(1)$ divisions-by-two for real numbers. For a complex number z, an integer d, and real numbers x and y, it follows that $z^d = e^{d\mathrm{Log}\, z}$, $\mathrm{Log}\, z = \ln|z| + \mathrm{i}Arg(z)$, and $e^{x+\mathrm{i}y} = e^x \cos y + \mathrm{i}e^x \sin y$ where $Arg(z)$ is the principal value of argument of z which can be computed with $\arctan \frac{Im(z)}{Re(z)}$. Therefore, raising a complex number to an integer power can be implemented with $O(1)$ elementary functions, $O(1)$ multiplications, $O(1)$ additions, and a square root operation for real numbers.

Each negation, addition, multiplication, division-by-two, power operation, general division, and each of the elementary functions for real numbers of length ℓ bits respectively can be done in $O(1)$, $O(\ell)$, $O(M(\ell))$, $O(\ell)$, $O(M(\ell)\log \ell)$ [10], $O(M(\ell))$ [2], and $O(M(\ell)\log \ell)$ [3], time. The best known upper bound of $M(m)$ is $O(m \log m \cdot 2^{O(\log^* m)})$ [5,6] where $\log^* m \leq 6$ even if m is astronomically large.

Hence the theorem follows.   □

Note that the proof of Theorem 2 requires that Algorithm 3 should be added the preprocessing to compute $\omega_1, \omega_2, \cdots, \omega_{1+\max(\log\sum_{j=1}^n w_j, \log\sum_{j=1}^n v_j)}$ and also that each of functions $f_k(x, y)$ and $g_k(y)$ defined in Algorithms 1 and 3 should be modified to compute its arguments x and y by itself (rather than to be passed by f_{k+1} or g_{k+1}) from the pre-computed values $\omega_1, \omega_2, \cdots$.

5   Conclusions and Future Works

We have presented an algorithm for the knapsack problem which runs in pseudo-polynomial time and polynomial space.

Our analysis of the required precision may be not tight. Therefore, one of future works is to estimate the complexity of the presented algorithm more precisely by giving more precise analysis. In this paper, we have tackled the knapsack problem. However, our algorithmic technique to represent a combinatorial optimization problem as a two-variable generating function and to compute an arbitrary coefficient of the complex generating function can be probably applied to other problems. Hence, another future work is to apply our technique to other various combinatorial optimization problems.

References

1. Brent, R.P.: Multiple-precision zero-finding methods and the complexity of elementary function evaluation. In: Analytic Computational Complexity, pp. 151–176. Academic Press (1975)
2. Brent, R.P.: The complexity of multiple-precision arithmetic. In: Complexity of Computational Problem Solving. University of Queensland Press (1976)
3. Brent, R.P.: Fast multiple-precision evaluation of elementary functions. J. ACM **23**, 242–251 (1976)
4. Cooke, R.L.: Classical Algebra: Its Nature, Origins, and Uses. Wiley, Hoboken (2008)
5. De, A., Kurur, P.P., Saha, C., Saptharishi, R.: Fast integer multiplication using modular arithmetic. In: ACM Symposium on Theory of Computation (STOC), pp. 499–506 (2008)
6. Fürer, M.: Faster integer multiplication. In: ACM Symposium on Theory of computing (STOC), pp. 57–66 (2007)
7. Garey, M.R., Johnson, D.S.: Computers and Intractability: a Guide to the Theory of NP-Completeness. W H Freeman & Co. (Sd), San Francisco (1979)
8. Kellerer, H., Pferschy, U., Pisinger, D.: Knapsack Problems. Springer, Heidelberg (2004)
9. Knuth, D.E.: Concrete Mathematics. Addison-Wesley Professional, Boston (1994)
10. Knuth, D.E.: Seminumerical Algorithms. The Art of Computer Programming, vol. 2. Addison-Wesley Professional, Boston (1997)
11. Lokshtanov, D., Nederlof, J.: Saving space by algebraization. In: ACM Symposium on Theory of Computing (STOC), pp. 321–330 (2010)
12. Nederlof, J.: Exact algorithms and time/space tradeoffs. In: Kao, M.-Y. (ed.) Encyclopedia of Algorithms, pp. 1–5. Springer, Heidelberg (2015)
13. Tate, S.R.: Stable computation of the complex roots of unity. IEEE Trans. Sig. Process. **43**(7), 1709–1711 (1995)

Game Theory

An Incentive Mechanism for Selfish Bin Covering

Weian Li, Qizhi Fang, and Wenjing Liu[✉]

School of Mathematical Sciences, Ocean University of China, Qingdao, China
opaass@163.com, qfang@ouc.edu.cn, liuwj_123@126.com

Abstract. In this paper, we consider the selfish bin covering problems, which can be viewed as the bin covering problems in game theoretic settings. Our main contribution is an incentive mechanism with better price of anarchy. Under this mechanism, for any instance with a Nash equilibrium (NE), we show that price of anarchy is 2/3. For the cases that the NE does not exist, we propose a concept of modified NE, named M-NE, which can be obtained in finite steps from any initial state. We further show that for M-NE, the price of anarchy is 1/2 and the price of stability is 1.

Keywords: Selfish bin covering · Incentive mechanism · Nash equilibrium · Price of anarchy (PoA) · Price of stability (PoS)

1 Introduction

Consider an item set $L = \{a_1, a_2, \cdots, a_n\}$, each item has its own size $s(a_i) \in (0,1]$. Assume that there are sufficiently many bins with unit capacity 1. A bin is called covered if the total size of items assigned to it is no less than 1. The bin covering problem is to assign the items into bins in order to maximize the number of covered bins. It was shown that the bin covering problem is NP-hard, and there exist no approximation algorithm with approximation ratio better than 1/2, unless $P = NP$ [2,3]. Although Assmann et al. [3] showed that the simple next-fit algorithm for bin covering problems has the worst case performance ratio of exactly 1/2, quite a few approximation algorithms with better asymptotic approximate ratios were investigated, such as the algorithms with asymptotic ratio 2/3, 3/4 [2,3,6] and AFPTAS in [7,11].

There are many interpretations of the simple bin covering model in the real world. Take a company with n employees as an instance. Let $s(a_i)$ denote the ith employee's ability, the company has many projects to be completed and each employee can only participate one project in the meantime. To complete each project needs at least 1 ability in total and will bring one unit benefit to the company. The company wishes to give an assignment of employees to maximize the number of completed projects. Note that there is a premise that every employee should obey the assignment without complaint. However, the

The work is partially supported by National Natural Science Foundation of China (NSFC) (NO. 11271341 and 11501316).

T.-H.H. Chan et al. (Eds.): COCOA 2016, LNCS 10043, pp. 641–654, 2016.
DOI: 10.1007/978-3-319-48749-6_46

premise may be difficult to realize in the real world since employees are usually selfish, that is, they prefer taking actions to pursue their own interests rather than the overall interests. This kind of problem in which players are considered to be self-interested is called selfish bin covering problem.

The selfish bin covering problem (SBC) was first studied by Cao and Yang [5]. A critical question of SBC is how to design a mechanism, i.e., a payoff scheme for the actions of players (i.e., employees), to lead the selfish players to achieve the desired social welfare (in SBC, the social welfare is the number of covered bins). The Price of Anarchy (PoA) and Price of Stability (PoS) [12,14] are usually used as the measure of the quality of the equilibria under a mechanism. PoA is defined as the ratio between the social welfare of the worst equilibrium and the social optimum, and PoS is defined formally as the ratio between the social welfare of the best equilibrium and the social optimum. Given an instance I of SBC, denote $\text{OPT}(I)$ the maximum number of covered bins and $\text{NE}(I_{\mathcal{M}})$ the number of covered bins in the state of NE under a given mechanism $\mathcal{M}$. For a class of SBC problems under a given mechanism $\mathcal{M}$, the corresponding PoA and PoS are defined as

$$\text{PoA(SBC)} = \inf_{I \in \text{SBC}} \min_{\text{NE}} \frac{\text{NE}(I_{\mathcal{M}})}{\text{OPT}(I)}$$

$$\text{PoS(SBC)} = \inf_{I \in \text{SBC}} \max_{\text{NE}} \frac{\text{NE}(I_{\mathcal{M}})}{\text{OPT}(I)}.$$

The bin covering problem, which can be considered as a dual version of the classical bin packing problem, is also called the dual bin packing problem. Except for Cao and Yang's paper, there are few results on the mechanism design of SBC, compared with that of selfish bin packing problems(SBP). The mechanism of SBP with proportional rule was first introduced by Bilò [4] in 2006. He proved that the bin packing game converges to a pure NE in a finite sequence of selfish improving steps (however, the number of steps may be exponential), starting from any feasible packing and provided non-tight bounds on PoA[1], i.e., $8/5 \leq PoA \leq 5/3$. In [10], Epstein and Kleiman gave nearly tight bounds for the PoA of NE, i.e., $1.6416 \leq PoA \leq 1.6428$. More investigation on various mechanisms were shown in [1,8,9,13,15]. Unfortunately, the mechanism design techniques of SBP offer little to that of SBC.

Cao and Yang [5] analyzed the selfish bin covering problem in the proportional rule. Under the mechanism, there always exists an NE where all the items are assigned in one bin, which makes the PoA bad enough (PoA=0). To get rid of it, they defined three types of fireable Nash equilibrium, denoted by FNE(I), FNE(II) and FNE(III), the PoAs and PoSs of which are 1/2 and 1, respectively. In this paper, we design a new mechanism to prevent this situation and improve the PoA. The payoff scheme, i.e., the mechanism, is designed based on the idea that gives redundant items the motivation to move from covered bins into uncovered bins. Under this mechanism, when an NE exists, we show that the PoA is 2/3. To overcome the weakness of the non-existence of NE, we further propose

---

[1] The PoA of selfish bin packing is defined contrary to that of SBC.

a modified Nash equilibrium (M-NE), and show that PoA is $1/2$ and PoS is 1. Our work provides a new idea for mechanism design for SBC.

This paper is organized as follows. In Sect. 1, we introduce the background and related work on bin covering problem and the selfish bin packing problem. In Sect. 2, an incentive mechanism called No-Redundance mechanism for SBC is proposed. In Sect. 3, we show for the instances which admit an NE, the PoA of No-Redundance mechanism is $2/3$. In Sect. 4, we propose a modified Nash equilibrium which can be obtained in finite steps. Under the No-Redundance mechanism, the PoA and PoS of the modified Nash equilibrium is $1/2$ and 1, respectively. Further discussion is given in Sect. 5.

2   Incentive Mechanism

Denote the item set as $L=\{a_1, a_2, \cdots, a_n\}$. For $i = 1, \cdots, n$, the size of item a_i is denoted by $s(a_i)$ and we assume $0 < s(a_i) \leq 1$. There are sufficiently many bins, each of which has the same capacity 1. Let $\pi = \{B_1, B_2, \cdots, B_m\}$ be a partition of L, i.e., $\cup_{i=1}^{m} B_i = L$ and $B_i \cap B_j = \emptyset$ for all $1 \leq i \neq j \leq m$. B_i is regarded as a bin or the set of items assigned in this bin. Denote by $B_{j,\pi}$ the member in π which item j is assigned in and $s(B_i)$ the total size of items in B_i, i.e., $s(B_i) = \sum_{a_j \in B_i} s(a_j)$. In particular, $s(L) = \sum_{a_i \in L} s(a_i)$.

A bin B is covered if the total size of items in B is greater than or equal to 1, i.e., $\sum_{a_i \in B} s(a_i) \geq 1$. A covered bin B is minimally covered if the removal of any of its member causes it uncovered, that is, let $a_{min}(B) = \min\{a_j : j \in B\}$ and $1 \leq s(B) < 1 + a_{min}(B)$. In particular, a bin B is exactly covered if $s(B) = 1$. A partition π is reasonable if there is at most one uncovered bin. Further, π is rational if it is reasonable and all the covered bins are minimally covered. Denote by $p(\pi)$ the social welfare of the partition π, i.e., the number of covered bins in π.

We will formulate an incentive mechanism for SBC under which redundance can be avoided spontaneously. The payoff scheme of the No-Redundance (NR) mechanism are formulated as follows.

No-Redundance Mechanism

Payoff scheme for a covered bin

Denote $B = \{a_1, \ldots, a_m\}$ as a covered bin, sort the items in non-increasing order of their sizes. Without loss of generality, assume that $s(a_1) \geq \cdots \geq s(a_m)$. There must be a positive integer k, $1 \leq k < m$, such that $\sum_{i=1}^{k} s(a_i) < 1$ and $\sum_{i=1}^{k+1} s(a_i) \geq 1$. $a_{k+2}, \cdots, a_m$ are called redundant items.

If $s(a_1) > 1/2$, the payoff function is

$$p(a_i) = \begin{cases} s(a_1), & i = 1, \\ (1 - s(a_1)) \times \frac{s(a_i)}{\sum_{i=2}^{k+1} s(a_i)}, & i = 2, 3, \ldots, k+1, \\ -\varepsilon, & k+1 < i \leq m. \end{cases} \quad (1)$$

If $s(a_1) \leq 1/2$, the payoff function is

$$p(a_i) = \begin{cases} \frac{s(a_i)}{\sum_{i=1}^{k+1} s(a_i)}, & i = 1, 2, \ldots, k+1, \\ -\varepsilon, & k+1 < i \leq m. \end{cases} \tag{2}$$

(Where $\varepsilon > 0$ and can be considered as the penalty of redundance. If there are two or more items of the same size, we assume that items of smaller index have higher priority.)

Payoff scheme for an uncovered bin

The payoff function $p(a_i)$ is defined by proportion rule, that is,

$$p(a_i) = -\frac{s(a_i)}{s(B)} \times \varepsilon_1, \tag{3}$$

(where $\varepsilon_1 > 0$ and can be considered as a penalty of unproductive occupation. Assume that $\varepsilon > \varepsilon_1$.)

Remark. Under the NR mechanism, the migration of items have the following properties.

(1) Items whose size is larger than $1/2$ have an absolute advantage on smaller items—that is, any migration of smaller ones will not influence their benefit, which is different from the proportional rule.
(2) Smaller redundant items would rather migrate to an uncovered bin or even open an empty bin than stay in a covered bin.

3   Nash Equilibrium

In this section, we discuss existence of NE and PoA under the NR mechanism.

Lemma 1. *In an NE, there exists at most one uncovered bin.*

Proof. Suppose there exist two or more uncovered bins in an NE. Denote two of them as B_0, B_0'. Without loss of generality, let $s(B_0) \geq s(B_0')$. Then any item in B_0' will be better off by moving into B_0, which is a contradiction.   □

Lemma 2. *In an NE, every covered bin is minimal covered.*

Proof. Suppose there exists a covered but not minimal covered bin in an NE. Denote the smallest item in this bin as a_{min} and the payoff of a_{min} is $-\varepsilon$. It is straightforward that a_{min} will benefit by moving into an empty bin, which is a contradiction.   □

Based on the above two lemmas, we have the following conclusion.

Theorem 1. *For a given instance of SBC, an NE must be a rational partition.*

3.1   The Existence of an NE

Before discussing the PoA of SBC under the NR mechanism, we need to verify whether an NE exists. In this subsection, we prove that to decide whether an NE exists is NP-hard.

Theorem 2. *Under the NR mechanism, to decide whether an NE exists is NP-hard.*

Proof. We show this by reduction from the partition problem: given a set of integers: $e_1, e_2, \ldots, e_n$, is there a subset $S \subseteq \{e_1, e_2, \ldots, e_n\}$ such that $\sum_{e_i \in S} e_i = \sum_{e_i \notin S} e_i$?

For any instance of the partition problem: $e_1, e_2, \ldots, e_n$, we construct an instance of the selfish bin covering problem as follows. Consider an item set $L = \{a_1, a_2, \ldots, a_{n+2}\}$, which satisfies $s(a_1) = 2e_1/(\sum_{i=1}^{n} 2e_i + 1), s(a_2) = 2e_2/(\sum_{i=1}^{n} 2e_i + 1), \ldots, s(a_n) = 2e_n/(\sum_{i=1}^{n} 2e_i + 1), and s(a_{n+1}) = s(a_{n+2}) = (\sum_{i=1}^{n} e_i + 1)/(\sum_{i=1}^{n} 2e_i + 1)$. There are sufficiently many bins, each of which has the capacity of one unit. It is not difficult to find that the sizes of a_{n+1}, a_{n+2} are greater than $1/2$ and the total size of all the items is exactly 2. Then we only need to prove that the answer of the partition problem is "yes" if and only if there exists an NE for the item set under the NR mechanism.

The necessity is trivial. If the answer of the Partition problem is "Yes", i.e., there exists a subset $S \subseteq \{e_1, e_2, \ldots, e_n\}$ such that $\sum_{e_i \in S} e_i = \sum_{e_i \notin S} e_i$. Let $B_0 = \{a_i | e_i \in S\} \cup \{a_{n+1}\}$ and $B_1 = L \setminus B_0$. It is not difficult to check the partition $\{B_0, B_1\}$ is an NE.

Now we need to show the sufficiency—that is, if there exists an NE, prove that the answer to the partition problem is "yes".

First, we claim that a_{n+1} and a_{n+2} are not in the same bin in an NE. Suppose they are in the same bin, this bin only consists of these two items and the other ones are packed in another uncovered bin, according to Lemmas 1 and 2. Then the item a_{n+2} will move to the uncovered bin, which is a contradiction with NE.

Considering that the total size of all the items is 2 and Lemmas 1 and 2, the NE must be composed of two bins—that is, the NE can be expressed as $\{B_0, B_1\}$, where $s(B_0) \geq s(B_1)$. We claim that both B_0 and B_1 are exactly covered. Suppose the claim is invalid, then B_0 is a minimal covered but not exactly covered bin and B_1 is an uncovered bin. By the first claim, a_{n+1} and a_{n+2} must be in different bins. Without loss of generality, suppose $a_{n+1} \in B_0, a_{n+2} \in B_1$. a_{n+2} will be strictly better off by move to B_0, which contradicts NE.

Above all, an NE is composed of two exactly covered bins B_0, B_1, where $a_{n+1} \in B_0$ and $a_{n+2} \in B_1$. Let $S = \{e_i | a_i \in B_0 \setminus \{a_{n+1}\}\}$ and the answer to the Partition problem is "Yes".   □

In some special cases it is polynomial-time to decide whether an NE exists.

Theorem 3. *For a given item set L, if each of them has a size larger than $1/2$, to decide whether an NE exists can be done in polynomial time.*

Proof. We discuss the question in two cases.

Case 1. The number of items is even. Denote $L = \{a_1, a_2, \ldots, a_{2n}\}$ in non-increasing order of their sizes. Let $B_1 = \{a_1, a_{2n}\}$, $B_2 = \{a_2, a_{2n-1}\}$, ..., $B_n = \{a_n, a_{n+1}\}$, and it is straightforward that the partition $\{B_1, \ldots, B_n\}$ is an NE.

Case 2. The number of items is odd. Denote $L = \{a_1, a_2, \ldots, a_{2n+1}\}$ in non-increasing order of their sizes and we will show that there exists no NE. Suppose an NE exists. By Lemmas 1 and 2, in the NE there must be an unique uncovered bin which only contains one item, and n minimal covered bins, each of which contains exactly two items. Denote the uncovered bin as B_0. There are three possibilities: **Case 2.1.** $B_0 = \{a_1\}$; **Case 2.2.** $B_0 = \{a_{2n+1}\}$; **Case 2.3.** $B_0 = \{a_k\}$, $2 \leq k \leq 2n$.

In case 2.1, a_1 will benefit by moving to any covered bin, which contradicts the NE.

In case 2.2, a_{2n} is in the same covered bin with one larger item and will benefit by moving to B_0, which contradicts the NE.

In case 2.3, denote by B_1 the covered bin which contains a_{k+1}. a_k will be better off by moving to B_1, which contradicts the NE.

So there exists no NE in case 2.

Above all, we can decide whether an NE exists by counting the number of items which can certainly be done in polynomial time.   □

3.2   The Tight Lower Bound for an NE

In the case which admits an NE, we will show that the PoA of the NR mechanism is $2/3$.

Lemma 3. *Under the NR mechanism, for an item set* $L = \{a_1, \ldots, a_n\}$ *which admits an NE,* $|NE(L)| \geq 2/3 \cdot (OPT(L) - 1)$, *where* $OPT(L)$ *is the social optimum of* L.

Proof. Sort all the items in non-increasing order of their sizes, and suppose that $s(a_1) \geq s(a_2) \geq \cdots \geq s(a_n)$, without loss of generality. Denote $|NE(L)|$ as $|NE|$, without confusion. Our discussion is started in two cases.

Case 1. $s(a_1) \leq 1/2$. Then the size of each item is less than or equal to $1/2$. By Lemma 2, in an NE each covered bin is also minimal covered, so the size of each covered bin is less than $3/2$. As there is at most one uncovered bin in an NE, it implies

$$3/2 \cdot |NE| + 1 \geq s(L). \tag{4}$$

Since $s(L) \geq OPT(L)$, we get

$$|NE| \geq 2/3 \cdot (OPT(L) - 1). \tag{5}$$

Hence the conclusion holds in this case.

Case 2. $s(a_1) > 1/2$. First, we divide all the items into two categories, according to their sizes.

$$(1) \quad s(x_1) \geq s(x_2) \geq \cdots \geq s(x_p) > 1/2 \quad (X - subset)$$
$$(2) \; 1/2 \geq s(y_1) \geq s(y_2) \geq \cdots \geq s(y_r) > 0 \; (Y - subset)$$

Clearly, $p + r = n$. And we analyze the lower bound for the NE in two cases.

Case 2.1. $p \leq |NE|$. We characterize the NE as Proposition 1.

Proposition 1. *In an NE, when $p \leq |NE|$, there exists no covered bin which contains two X-subset items.*

Proof. Suppose that there exists a covered bin which contains two X-subset items. By Lemma 2, the bin only consists of these two items, denoted as x_i, x_j, and assume $s(x_i) \geq s(x_j)$. Under the NR mechanism, $p(x_i) = s(x_i)$ and $p(x_j) = 1 - s(x_i) < s(x_j)$. Since $p \leq |NE|$, there must exist a covered bin which only consists of Y-subset items, denoted as B_0. x_j will be better off by moving to B_0, which contradicts the NE. □

By Lemma 2 and Proposition 1, each covered bin is minimal covered and contains at most one X-subset item, which implies the size of each covered bin does not exceed $3/2$. Thus the conclusion also holds in case 2.1.

Case 2.2. $p > |NE|$. We characterize the NE as Proposition 2.

Proposition 2. *Under the NR mechanism, when $p > |NE|$, an NE has the following structure, see Figure 1.*

(1) $x_1, x_2, \ldots, x_{|NE|}$ must be in different covered bins, denoted as $B_1, \ldots, B_{|NE|}$ respectively;

(2) $x_{|NE|+1}$ and $x_{|NE|}$, $x_{|NE|+2}$ and $x_{|NE|-1}, \ldots$, x_p and $x_{2|NE|+1-p}$ must be in the same (covered) bins, respectively;

(3) The size of the unique uncovered bin (if there exists, denoted as B_0) is less than $1 - s(x_{|NE|+1})$.

Proof. (1) Suppose there exists an item x_i $(1 \leq i \leq |NE|)$ in the uncovered bin. Then must be a covered bin which does not contain any of $x_1, x_2, \ldots, x_{|NE|}$. x_i will benefit by moving to the bin, which contradicts the NE. Thus $x_1, x_2, \ldots, x_{|NE|}$ must be in covered bins.

Suppose x_i and x_j $(1 \leq i < j \leq |NE|)$ are in a same bin. Then the bin is covered and $s(x_i) \geq s(x_j)$. The payoff of x_j is $p(x_j) = 1 - s(x_i) < s(x_j)$. On the other hand, there must exist a covered bin that contains none of $x_1, x_2, \ldots, x_{|NE|}$. Then x_j will be better off by moving to the bin, which contradicts the NE.

Thus (1) holds.

(2) First prove that $x_{|NE|}$ and $x_{|NE|+1}$ are in the same bin. Since $p > |NE|$, the item $x_{|NE|+1}$ must be in the same bin with some x_i, where $i \in \{1, 2, \ldots, |NE|\}$, and have a payoff $1 - s(x_i)$. Then $x_i = x_{|NE|}$, otherwise $x_{|NE|+1}$ will migrate for more benefit, which contradicts the NE.

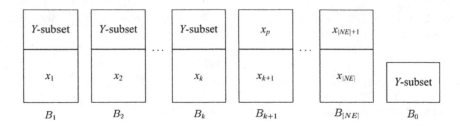

Fig. 1. The structure of NE

The other pairs can be proved similarly.

(3) If it does not hold, $x_{|NE|+1}$ will move to the uncovered bin for more benefit, which contradicts the NE.   □

Based on the structure, call the set of all the largest and the smallest items in B_i $(i = 1, 2, \ldots, |NE|)$ type-B and the set of the remaining items type-A. $k = 2|NE| - p$. Consider the optimal covering and define

$k_A :=$ the number of covered bins in the optimal covering with only type-A items,

$k_{AB} :=$ the number of covered bins in the optimal covering with exactly one type-B item,

$k_{BB} :=$ the number of covered bins in the optimal covering with more than one type-B items.

Clearly,

$$\text{OPT}(L) = k_A + k_{AB} + k_{BB}. \tag{6}$$

Since each covered bin in an NE is minimally covered, we have

$$\sum_{a_i \in \text{type-A}} s(a_i) + \sum_{i=1}^{k} s(x_i) < k + 1/2. \tag{7}$$

By (7), we obtain an upper bound of $k_A + k_{AB}$, that is

$$k_A + k_{AB} \le \sum_{a_i \in \text{type-A}} s(a_i) + \sum_{i=1}^{k_{AB}} s(x_i) < k + 1/2 + \sum_{i=1}^{k_{AB}} s(x_i) - \sum_{i=1}^{k} s(x_i) \tag{8}$$

To make the upper bound precisely, we first prove the following claim.

Proposition 3. $k_{AB} \le |NE|$.

Proof. Suppose $k_{AB} > |NE|$, then it holds that

$k_{AB} \le$ the total size of the k_{AB} covered bins in the optimal covering
　　　　with exactly one type-B item

$\le \sum_{a_i \in \text{type-A}} s(a_i) +$ the total size of the largest k_{AB} type-B items

$$= [\sum_{a_i \in \text{type-A}} s(a_i) + \sum_{i=1}^{|NE|+1} s(x_i)] + \text{the total size of the largest } (k_{AB}$$

$$-(|NE|+1)) \text{ type-B items excluding } x_1, \ldots, x_{|NE|+1}$$

$$< |NE| + 1 + (k_{AB} - (|NE|+1)) = k_{AB}. \tag{9}$$

It is a contradiction. Thus $k_{AB} \leq |NE|$.   □

Now we discuss the upper bound of k_{BB}. Considering that the number of type-B is $2|NE|$, the number of the remaining type-B items in covered bins which contain more than one type-B item is $2|NE| - k_{AB}$. Therefore

$$k_{BB} \leq (2|NE| - k_{AB})/2. \tag{10}$$

The conclusion is proved in three cases.

Case 2.2.1. $k_{AB} < k$. By (8) and considering that $s(x_i) > 1/2$, we have

$$k_A + k_{AB} \leq k + 1/2 - \sum_{i=k_{AB}+1}^{k} s(x_i) < (k + k_{AB})/2 + 1/2$$

$$\leq (|NE| + k_{AB})/2 + 1/2. \tag{11}$$

Thus, we obtain

$$\text{OPT}(L) \leq (|NE| + k_{AB})/2 + (2|NE| - k_{AB})/2 + 1/2 = 3|NE|/2 + 1/2. \tag{12}$$

Case 2.2.2. $k_{AB} = k$. By (8) and considering that $s(x_i) < 1$, we have

$$k_A + k_{AB} = k_A + k < k + 1/2, \tag{13}$$

which implies that $k_A < 1/2$.

Since k_A is an integer, we have $k_A = 0$, thus

$$\text{OPT}(L) = k_{AB} + k_{BB} \leq (2|NE| + k_{AB})/2 \leq 3|NE|/2. \tag{14}$$

Case 2.2.3. $k < k_{AB} \leq |NE|$. By (8) and considering that $s(x_i) > 1/2$, we have

$$k_A + k_{AB} < k + 1/2 + \sum_{i=k+1}^{k_{AB}} s(x_i)$$

$$< k + 1/2 + k_{AB} - k = k_{AB} + 1/2, \tag{15}$$

which implies that $k_A < 1/2$.

Similar to Case 2.2.2, we have

$$\text{OPT}(L) = k_{AB} + k_{BB} \leq (2|NE| + k_{AB})/2 \leq 3|NE|/2. \tag{16}$$

Above all, $|NE(L)| \geq 2/3 \cdot (\text{OPT}(L) - 1)$.   □

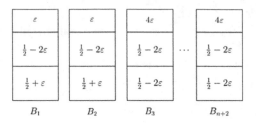

Fig. 2. OPT$(L) = n + 2$

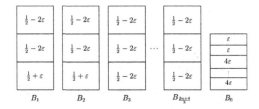

Fig. 3. $|\text{NE}(L)| = \frac{2n+4}{3}$

Considering that there exists no NE in some cases, we focus on those instances which admit an NE. Denote the set of instances of SBC which admit an NE as $\mathcal{A}$ and we can define the PoA of NE confined on $\mathcal{A}$ as

$$\text{PoA}(\mathcal{A}) = \inf_{L \in \mathcal{A}} \min_{\text{NE}} \frac{|\text{NE}(L)|}{\text{OPT}(L)}. \tag{17}$$

Theorem 4. *Under the NR mechanism, PoA$(\mathcal{A}) = 2/3$.*

Proof. By Lemma 3, we have PoA$(\mathcal{A}) \geq 2/3$.

For any n and a positive number ε which is small enough, construct an instance as follows. There are $3n + 6$ items in total: two items with the size $1/2 + \varepsilon$, $2n + 2$ items with the size $1/2 - 2\varepsilon$, n items with the size 4ε and two items with the size ε. There exists an optimal partition, where all the bins are exactly covered, see Fig. 2. OPT$(L) = n + 2$.

Let $\varepsilon = 1/(9n)$. It is not hard to check that the partition in Fig. 3 is an NE under the NR mechanism, thus we have

$$|\text{NE}(L)| = \frac{2n + 4}{3}. \tag{18}$$

Since

$$\lim_{n \to \infty} \frac{|\text{NE}(L)|}{\text{OPT}(L)} = \frac{2}{3}, \tag{19}$$

we have the theorem. □

4 Modified Nash Equilibrium

The No-Redundance mechanism energizes and mobilizes all redundant items, however, by assigning larger items priorities. An item will always egoistically try to migrate so long as there exists a bin consisting of smaller items since smaller ones will be first kicked out. The flexibility of larger items causes the situation in which the migration is difficult to terminate and there may not be an NE.

Based on the idea that the migration of items should be constrained and the size of the resulted bin should maintain small enough to make the remaining items cover more bins, we induce a concept of modified Nash equilibrium (M-NE) under the NR mechanism. For simplicity, we denote the set of redundant items in covered bin B as $R(B)$.

Definition 1. *A partition $\pi = (B_1, B_2, \ldots, B_m)$ is called a modified Nash equilibrium if and only if there exist no item $a_j \in L$ and a bin $B_i (1 \leq i \leq m, B_i \neq B_{j,\pi})$ such that:*

(1) a_j will strictly benefit by moving to B_i;
(2) $s(B_i') < \min\{\delta(B_{j,\pi}), \delta(B_i)\}$,

where $B_i' = B_i \cup \{a_j\}$, if $s(B_i) + s(a_j) < 1$; and B_i' is the minimally covered bin after redundant items migrate from $B_i \cup \{a_j\}$, i.e. $B_i' = B_i \cup \{a_j\} \setminus R(B_i \cup \{a_j\})$, if $s(B_i) + s(a_j) \geq 1$. For any bin B,

$$\delta(B) = \begin{cases} s(B), & \text{if } B \text{ is covered,} \\ +\infty, & \text{otherwise.} \end{cases}$$

Theorem 5. *Under the NR mechanism, an M-NE must be a rational partition, i.e. there exists at most one uncovered bin and each covered bin is minimally covered.*

Proof. The proof is similar to Theorem 1. □

In the setting of M-NE, the migration will continue until there exists no item and a bin satisfying the two conditions in Definition 1. Will the constrained best response dynamic process eventually terminate? Or can an M-NE always be obtained from any initial partition in finite steps?

To prove the existence of M-NE, we first define a vector valued potential function. Let $P(\cdot)$ be the potential function, $\pi = (B_1, B_2, \ldots, B_m)$ is a reasonable partition, and B_m is the unique uncovered bin, then

$$P(\pi) = (s(B_1'), s(B_2'), \ldots, s(B_{m-1}')), \tag{20}$$

where $(B_1', \ldots, B_{m-1}')$ is a rearrangement of $(B_1, B_2, \ldots, B_{m-1})$ in non-decreasing order of their sizes.

For two vectors $v = (v_1, \ldots, v_s), w = (w_1, \ldots, w_t)$, we say that v is lexicographically smaller than w, which is denoted by $v \prec w$, iff there exists an integer $j_0, 1 \leq j_0 \leq \min\{s, t\}$, such that $v_{j_0} < w_{j_0}$ and $v_j = w_j, \forall j < j_0$.

Lemma 4. *For any reasonable partition π which is not an M-NE, there exists a reasonable partition π', such that $P(\pi') \prec P(\pi)$ and $p(\pi') \geq p(\pi)$.*

Proof. Denote the unique uncovered bin (if there exists) as B_m. We can design a constrained best response (CBR) algorithm as follows.

Since π is not an M-NE, there exists an item a_j and a bin B_i satisfying the two conditions in Definition 1. The fact that a_j will strictly benefit by moving to B_i implies that $a_j \notin R(B_i \cup \{a_j\})$. Then the migration of a_j can be classified into three cases.

Case 1. Both $B_{j,\pi}$ and B_i are covered. We can construct a new reasonable partition π' as follows.

1. Move a_j to B_i.
2. Take out redundant items in $B_i \cup \{a_j\}$ and obtain a minimally covered bin B_i'.
3. Run FFD[2] on items in $B_{j,\pi} \setminus \{a_j\} \cup R(B_i \cup \{a_j\}) \cup B_m$ and we get a reasonable partition π'.

Since $s(B_i') < s(B_i)$, we have $s(a_j) < s(R(B_i \cup \{a_j\}))$. Thus $s(B_{j,\pi} \setminus \{a_j\} \cup R(B_i \cup \{a_j\})) > s(B_{j,\pi}) \geq 1$.

Case 2. $B_{j,\pi}$ is covered and B_i is the unique uncovered bin. We construct π' in two cases.

Case 2.1. $s(B_i) + s(a_j) \geq 1$.

1. Move a_j to B_i.
2. If $R(B_i \cup \{a_j\}) \neq \emptyset$, take out redundant items in $B_i \cup \{a_j\}$ and obtain a minimally covered bin B_i'.
3. Run FFD on items in $B_{j,\pi} \setminus \{a_j\} \cup R(B_i \cup \{a_j\})$ and we get a reasonable partition π'.

Case 2.2. $s(B_i) + s(a_j) < 1$. It implies that $a_j \in R(B_{j,\pi})$. Move a_j to B_i. B_i' is uncovered and $B_{j,\pi} \setminus \{a_j\}$ is still covered. Thus we get a reasonable partition π'.

Case 3. $B_{j,\pi}$ is the unique uncovered bin and B_i is covered.

1. Move a_j to B_i.
2. Take out redundant items in $B_i \cup \{a_j\}$ and obtain a minimally covered bin B_i'.
3. Run FFD on items in $B_{j,\pi} \setminus \{a_j\} \cup R(B_i \cup \{a_j\})$ and we get a reasonable partition π'.

It is not hard to check that $P(\pi') \prec P(\pi)$ and $p(\pi') \geq p(\pi)$. $\square$

---

[2] First Fit Decreasing, which is described as follows. First sort the items in non-increasing order of their sizes, then puts the items one by one in this order to a bin until the bin is covered. Open a new bin and repeat the above actions until all the items are assigned.

Theorem 6. *M-NE always exists and can be obtained from any partition in finite steps without decreasing the social welfare.*

Proof. For any partition, first unify all the uncovered bins and obtain a reasonable partition without decreasing the social welfare. Denote the reasonable partition as π^0 and the unique uncovered bin (if there exists) as B_m.

If π^0 is not an M-NE, run CBR in Lemma 4 on π^0. And we get a reasonable partition π^1 such that $P(\pi^1) \prec P(\pi^0)$ and $p(\pi^1) \geq p(\pi^0)$.

The migration of items will continue until an M-NE arrives. Denote the partition after the t-th migration be π^t. If π^t is not an M-NE, run CBR on π^t. We will obtain a reasonable partition π^{t+1} such that $P(\pi^{t+1}) \prec P(\pi^t)$ and $p(\pi^{t+1}) \geq p(\pi^t)$, and so on.

Since the number of reasonable partitions is finite for a given instance of SBC, we get immediately that M-NE exists and can be obtained in finite steps without decreasing the social welfare. $\square$

For a given instance I of SBC, denote by $NE(I)$ the set of NEs of I and $M{-}NE(I)$ the set of M-NEs of I. It is straightforward that $NE(I) \subseteq M{-}NE(I)$.

Theorem 7. *(1)$PoA^{M-NE} = 1/2$; (2)$PoS^{M-NE} = 1$.*

Proof. (1) Theorem 5 guarantees that $PoA^{M-NE} \geq 1/2$. We construct an instance as follows. There are $6n$ items in total: $2n$ large items with $s(a_1) = \cdots = s(a_{2n}) = 1 - 1/(2n)$, $4n$ small items with $s(a_{2n+1}) = \cdots = s(a_{6n}) = 1/(4n)$. It is easy to check that the partition that each bin contains either two large items or $4n$ small items is an M-NE. And the partition that each bin contains one large item and two small items is optimal. Since $(n+1)/(2n) \to 1/2$ as $n \to \infty$, we complete the proof.

(2) It is followed directly by Theorem 6. $\square$

5   Conclusion and Further Discussion

In this paper, we mainly discussed mechanism design for the selfish bin covering problem. We formulated an incentive mechanism called as no-redundance mechanism under which the worst NE has good properties. Considering that there exist instances of SBC which does not admit NE, we proposed a modified Nash equilibrium which always exists under the NR mechanism, and showed that PoA and PoS of M-NE is 1/2 and 1, respectively.

There are several directions for the future research, such as, to study what kind of instances of SBC admit an NE under the NR mechanism, to design a new mechanism under which there always exists a pure NE and the PoA is good enough.

References

1. Adar, R., Epstein, L.: Selfish bin packing with cardinality constraints. Theoret. Comput. Sci. **495**, 66–80 (2013)
2. Assmann, S.B.: Problems in discrete applied mathematics. Ph.D. thesis, Department of Mathematics, MIT, Cambridge, MA (1983)
3. Assmann, S.F., Johnson, D.S., Kleitman, D.J., Leung, J.Y.-T.: On a dual version of the one-dimensional bin packing problem. J. Algorithms **5**(4), 502–525 (1984)
4. Bilò, V.: On the packing of selfish items. In: Proceedings 20th IEEE International Parallel & Distributed Processing Symposium, p. 9. IEEE (2006)
5. Cao, Z., Yang, X.: Selfish bin covering. Theoret. Comput. Sci. **412**(50), 7049–7058 (2011)
6. Csirik, J., Frenk, J.B.G., Labbé, M., Zhang, S.: Two simple algorithms for bin covering. Acta Cybernetica **14**(1), 13–25 (1999)
7. Csirik, J., Johnson, D.S., Kenyon, C.: Better approximation algorithms for bin covering. In: Proceedings of the Twelfth Annual ACM-SIAM Symposium on Discrete Algorithms, pp. 557–566. Society for Industrial and Applied Mathematics (2001)
8. Coffman, E.G., Csirik, J., Leung, J.Y.-T.: Variants of classical one-dimensional bin packing. In: Handbook of Approximation Algorithms and Metaheuristics, p. 13 (2007). Chap. 33
9. Dósa, G., Epstein, L.: Generalized selfish bin packing. arXiv preprint arXiv:1202.4080 (2012)
10. Epstein, L., Kleiman, E.: Selfish bin packing. Algorithmica **60**(2), 368–394 (2011)
11. Jansen, K., Solis-Oba, R.: An asymptotic fully polynomial time approximation scheme for bin covering. Theoret. Comput. Sci. **306**(1), 543–551 (2003)
12. Koutsoupias, E., Papadimitriou, C.: Worst-case equilibria. In: Meinel, C., Tison, S. (eds.) STACS 1999. LNCS, vol. 1563, pp. 404–413. Springer, Heidelberg (1999)
13. Ma, R., Dósa, G., Han, X., Ting, H.-F., Ye, D., Zhang, Y.: A note on a selfish bin packing problem. J. Global Optim. **56**(4), 1457–1462 (2013)
14. Papadimitriou, C.: Algorithms, games, and the internet. In: Proceedings of the Thirty-Third Annual ACM Symposium on Theory of Computing, 749–753. ACM (2001)
15. Yu, G., Zhang, G.: Bin packing of selfish items. In: Papadimitriou, C., Zhang, S. (eds.) WINE 2008. LNCS, vol. 5385, pp. 446–453. Springer, Heidelberg (2008)

Congestion Games with Mixed Objectives

Matthias Feldotto[(✉)], Lennart Leder, and Alexander Skopalik

Heinz Nixdorf Institute & Department of Computer Science,
Paderborn University, Paderborn, Germany
{feldi,lleder,skopalik}@mail.upb.de

Abstract. We study a new class of games which generalizes congestion games and its bottleneck variant. We introduce *congestion games with mixed objectives* to model network scenarios in which players seek to optimize for latency and bandwidths alike. We characterize the existence of pure Nash equilibria (PNE) and the convergence of improvement dynamics. For games that do not possess PNE we give bounds on the approximation ratio of approximate pure Nash equilibria.

Keywords: Congestion games · Bottleneck congestion games · Pure nash equilibrium · Existence · Convergence · Complexity · Approximation

1 Introduction

Resource allocation problems in large-scale scenarios such as networks often cannot be solved as a single optimization problem. The size of the problem, the distributed nature of information, or control preclude a centralized approach. As a consequence, decisions are delegated to local actors or players. This gives rise to strategic behavior as these players often have economic interests. Game theory has studied the effect of such strategic interaction in various models of resource allocation and scheduling. One of the most prominent ones is the class of atomic congestion games [20], in which players allocate sets of resources. The cost of a resource depends on the number of players allocating it. The cost of a player is the sum of the costs of her allocated resources. The appeal of this model stems not only from its applicability to prominent problems like scheduling, routing and load balancing, but also from desirable game theoretic properties. Congestion games always possess pure Nash equilibria and the natural improvement dynamics converge to a pure Nash equilibrium since these games are potential games. In fact, the class of congestion games is isomorphic to the class of potential games [19], which shows their expressiveness. When modeling network routing with congestion games and most of its variants like weighted [11] or player-specific congestion games [18] one faces deficiency due to the nature of

---

This work was partially supported by the German Research Foundation (DFG) within the Collaborative Research Centre "On-The-Fly Computing" (SFB 901) and by the EU within FET project MULTIPLEX under contract no. 317532.

T.-H.H. Chan et al. (Eds.): COCOA 2016, LNCS 10043, pp. 655–669, 2016.
DOI: 10.1007/978-3-319-48749-6_47

the players' cost functions. As a player's cost is determined by the sum of the resource costs, congestion games are not well suited to model effects like bandwidth allocation, as here the cost of a player is determined only by the bottleneck resource. Hence, Banner and Orda [4] introduced bottleneck congestion games where the cost of a player is the maximum cost of her chosen resources. Again, due to the nature of the cost functions, this class of games and most of its variants [14,16] only model the bottleneck effects and are unable to describe latency effects. It is not difficult to envision scenarios in which both effects, latency and bandwidth, are relevant to decision makers - especially in today's IT infrastructures where we find techniques with shared resources, for example in the context of cloud computing or in software-defined networking. Many users with lots of different applications and therefore different objectives interact in one network and compete for the same resources. Consider, for example, on the one hand media streaming and on the other hand video gaming. In one application, bandwidth is the most important property, in the other it is latency.

Therefore, we study a game theoretic model in which players may have heterogeneous objectives. We introduce the model of *congestion games with mixed objectives*. In this model, resources have two types of costs, latency cost and bottleneck cost, where the latter corresponds to the inverse of bandwidth. The players' costs may depend on both types of cost, where we allow different players to have different preferences regarding the two cost types.

Our Contribution. We show that pure Nash equilibria exist and can be computed in polynomial time in singleton games and in some matroid games. However, we show that the matroid property alone is not sufficient for the existence. Additionally, it is necessary that either the players are only interested in latency or bottleneck cost, or that the cost functions have a monotone dependence. For the latter case, we show convergence of best-response dynamics while the remaining cases are only weakly acyclic. For matroid games that do not satisfy one of the additional properties, we show that pure equilibria might not exist and it is even NP-hard to decide whether one exists. To overcome these non-existence results, we consider approximate pure Nash equilibria. For several classes, we can show that there exist β-approximate pure Nash equilibria where β depends on the size of the largest strategy.

Related Work. Milchtaich [18] studies the concept of player-specific congestion games and shows that in the singleton case these games always admit pure Nash equilibria. Ackermann et al. [1] generalize these results to matroid strategy spaces and show that the result also holds for weighted congestion games. Furthermore, they point out that in a natural sense the matroid property is maximal for the guaranteed existence of pure Nash equilibria in player-specific and weighted congestion games. Moreover, Milchtaich [18] examines congestion games in which players are both weighted and have player-specific cost functions. By constructing a game with three players, he shows that these games do not necessarily possess pure Nash equilibria, even in the case of singleton strategies. Mavronicalas et al. [17] study a special case of these games in which cost functions are not entirely player-specific. Instead, the player-specific resource costs are derived by combining the general resource cost function and a player-specific constant via a specified operation (e.g. addition or multiplication). They show that this restriction is sufficient to guarantee the existence of pure Nash equilibria in games with three players. Dunkel and Schulz [9] show that the decision problem of whether a weighted network congestion game possesses a pure Nash equilibrium is NP-hard. The equivalent result is achieved for player-specific congestion games by Ackermann and Skopalik [3].

Banner and Orda [4] study the applicability of game-theoretic concepts in network routing scenarios. In particular, they derive bounds on the price of anarchy in network bottleneck congestion games with restricted cost functions and show that a pure Nash equilibrium which is socially optimal always exists. Cole et al. [8] further investigate the non-atomic case, they especially consider the impacts of variable traffic rates. In contrast, Harks et al. [16] concentrate on the atomic case and study the lexicographical improvement property, which guarantees the existence of pure Nash equilibria through a potential function argument. They show that bottleneck congestion games fulfill this property and, hence, they are potential games. Harks et al. [15] consider the complexity of computing pure Nash equilibria and strong equilibria in bottleneck congestion games. Moreover, they show this property in matroid bottleneck congestion games.

Chien and Sinclair [7] study the convergence towards approximate pure Nash equilibria in symmetric congestion games. Skopalik and Vöcking [21] show inapproximability in asymmetric congestion games, which is complemented by approximation algorithms for linear and polynomial delay functions [5,10], even for weighted games [6]. Hansknecht et al. [13] use the concept of approximate potential functions to examine of approximate pure Nash equilibria in weighted congestion games under different restrictions on the cost functions.

Preliminaries. A *congestion game with mixed objectives* is defined by a tuple $\Gamma = (N, R, (\Sigma_i)_{i \in N}, (\alpha_i)_{i \in N}, (\ell_r)_{r \in R}, (e_r)_{r \in R})$, where $N = \{1, \ldots, n\}$ denotes the set of players and R denotes the set of resources. For each player i let $\Sigma_i \subseteq 2^R$ denote the strategy space of player i and $\alpha_i \in [0, 1]$ the preference value of player i. For each resource r let $\ell_r : N \to \mathbb{R}$ denote the non-decreasing latency cost function associated to resource r, and let $e_r : N \to \mathbb{R}$ denote the non-decreasing bottleneck cost function associated to resource r. For a state $S = (S_1, \ldots, S_n) \in \Sigma_1 \times \ldots \times \Sigma_n$, we define for each resource $r \in R$ by $n_r(S) = |\{i \in N \mid r \in S_i\}|$ the congestion of r. The latency cost of r in state S is given by $\ell_r(S) = \ell_r(n_r(S))$, and the bottleneck cost by $e_r(S) = e_r(n_r(S))$. The total cost of player i in state S depends on α_i and is defined as $c_i(S) = \alpha_i \cdot \sum_{r \in S_i} \ell_r(S) + (1 - \alpha_i) \cdot \max_{r \in S_i} e_r(S)$.

For a state $S = (S_1, \ldots, S_i, \ldots, S_n)$, we denote by (S_i', S_{-i}) the state that is reached if player i plays strategy S_i' while all other strategies remain unchanged. A state $S = (S_1, \ldots, S_n)$ is called a *pure Nash equilibrium (PNE)* if for all $i \in N$ and all $S_i' \in \Sigma_i$ it holds that $c_i(S) \leq c_i(S_i', S_{-i})$ and a *β-approximate pure Nash equilibrium* for a $\beta \geq 1$, if for all $i \in N$ and all $S_i' \in \Sigma_i$ it holds that $c_i(S) \leq \beta \cdot c_i(S_i', S_{-i})$.

A *singleton congestion game with mixed objectives* is a congestion game with mixed objectives Γ with the additional restriction that all strategies are singletons, i.e., for all $i \in N$ and all $S_i \in \Sigma_i$ we have that $|S_i| = 1$. A *matroid congestion game with mixed objectives* is a congestion game with mixed objectives in which the strategy spaces of all players form the bases of a matroid on the set of resources. We say that the cost functions of a congestion game with mixed objectives have a *monotone dependence* if there is a monotone non-decreasing function $f : \mathbb{R} \to \mathbb{R}$, such that $e_r(x) = f(\ell_r(x))$ for all $r \in R$. We call the players α-*uniform* if there is an $\alpha \in [0,1]$ such that $\alpha_i = \alpha$ for all players $i \in N$. We say the players have *pure preferences* if $\alpha_i \in \{0,1\}$ for all players $i \in N$.

2 Existence of Pure Nash Equilibria

Congestion games with mixed objectives are more expressive than standard or bottleneck congestion games. Consequently, the existence of pure Nash equilibria is guaranteed only for special cases. Unlike, e.g., player-specific congestion games, the matroid property is not sufficient for the existence of PNE. We show that we have the existence of PNE in singleton games or for matroid games with players that have pure preferences or cost functions that have a monotone dependence.

Theorem 1. *A congestion game with mixed objectives Γ contains a pure Nash equilibrium if Γ is a*

1. *singleton congestion game, or*
2. *matroid congestion game and the players have pure preferences, or*
3. *matroid congestion game and the cost functions have a monotone dependence.*

A pure Nash equilibrium can be computed in polynomial time.

Proof. We prove the theorem by reducing the existence problem of a pure Nash equilibrium in a congestion game with mixed objectives to the existence problem of a PNE in a congestion game with player-specific cost functions. The existence of PNE is guaranteed in singleton [18] and matroid [1] player-specific congestion games and polynomial time complexity immediately follows [1,2].

   We will utilize the following lemma which states that an optimal basis with respect to sum costs is also optimal w.r.t. maximum costs.

Lemma 1. *Let M be a matroid, and let $B = \{b_1, \ldots, b_m\}$ be a basis of M which minimizes the sum of the element costs. Then for any other basis $B' = \{b'_1, \ldots, b'_m\}$ it holds that $\max_{1 \le i \le m} b_i \le \max_{1 \le i \le m} b'_i$.*

Proof (Lemma). Let $B = \{b_1, \ldots, b_m\}$ be an optimal basis, and assume by contradiction that there is a different basis $B' = \{b'_1, \ldots, b'_m\}$ with $b'_m < b_m$ (w.l.o.g. assume that b_m and b'_m are the most expensive resources in B and B', respectively). Since B and B' are both bases and $b_m \notin B'$, there is an element $b'_i \in B'$ such that $B'' = B \setminus \{b_m\} \cup \{b'_i\}$ is a basis of M. By assumption we have $b'_i \le b'_m < b_m$, which implies that B'' has a smaller total cost than B. Therefore, B cannot be optimal, which gives a contradiction.   □

We now proceed to prove Theorem 1 and consider the three different cases:

1. The cost of player i in a state S with $S_i = \{r\}$ is $c_i(S) = \alpha_i \cdot \ell_r(S) + (1 - \alpha_i) \cdot e_r(S)$. By defining the player-specific cost functions $c_r^i(x) = \alpha_i \cdot \ell_r(x) + (1 - \alpha_i) \cdot e_r(x)$ for every $i \in N$ and $r \in R$, we obtain an equivalent singleton player-specific congestion game.

2. Using Lemma 1, we can treat all the players who strive to minimize their bottleneck costs as if they were striving to minimize the sum of the bottleneck costs of their resources. Hence, we can construct a player-specific congestion game in which the player-specific cost functions correspond to the latency functions for those players with preference value 1, and to the bottleneck cost functions for those players with preference value 0.

3. A player i who allocates the resources $\{r_1, \dots, r_k\}$, where w.l.o.g. r_k is the most expensive one, in a state S, incurs a total cost of $c_i(S) = \alpha_i \cdot \sum_{j=1}^{k} \ell_{r_j}(S) + (1 - \alpha_i) \cdot e_{r_k}(S)$. Due to Lemma 1, we know that $\ell_{r_k}(S)$ is minimized if $\sum_{j=1}^{k} \ell_{r_j}(S)$ is minimized. The monotonicity of f with $e_r(x) = f(\ell_r(x))$ implies that e_{r_k} is also minimized. Observe that monotonicity also ensures that r_k is the bottleneck resource. Hence, a PNE of a congestion game with cost functions ℓ_r for every $r \in R$ is a PNE of Γ. $\qquad\square$

We show that the matroid property is not sufficient for the existence of PNE. Even for linear cost functions and uniform players, there are games without PNE.

Theorem 2. *There is a matroid congestion game with mixed objectives Γ with linear cost functions and α-uniform players which does not possess a pure Nash equilibrium.*

Proof. We construct a two-player game with linear cost functions and $\alpha_i = 0.5$ for both players. The set of resources is $R = \{r_1, \dots, r_7\}$. The strategies for player 1 are $\Sigma_1 = \{\{r_i, r_j, r_k\} \mid i, j, k \in \{1, \dots, 6\}\}$ and the strategies for player 2 are $\Sigma_2 = \{\{r_i, r_j\} \mid i, j \in \{4, \dots, 7\}\}$. The latency and bottleneck cost functions for the first three resources are $\ell_{r_1}(x) = \ell_{r_2}(x) = \ell_{r_3}(x) = 0$ and $e_{r_1}(x) = e_{r_2}(x) = e_{r_3}(x) = 200 \cdot x$, respectively. For resource r_4 and r_5 the cost functions are $\ell_{r_4}(x) = \ell_{r_5}(x) = 20 \cdot x$ and $e_{r_4}(x) = e_{r_5}(x) = 50 \cdot x$. For resource r_6 the cost functions are $\ell_{r_6}(x) = 8 \cdot x$ and $e_{r_6}(x) = 80 \cdot x$. For resource r_7 the cost functions are $\ell_{r_7}(x) = 0$ and $e_{r_7}(x) = 160 \cdot x$.

We note that for player 1 only the strategies $S_{1,1} := \{r_1, r_2, r_3\}$ and $S_{1,2} := \{r_4, r_5, r_6\}$ can be best-response strategies in any state, since $\{r_1, r_2, r_3\}$ strictly dominates all remaining strategies. Hence, with respect to the existence of pure Nash equilibria, we can restrict player 1 to these two strategies. In the analogous way, we can restrict player 2 to the strategies $S_{2,1} := \{r_4, r_5\}$ and $S_{2,2} := \{r_6, r_7\}$. This yields a game with only four states and, as we can easily verify, a best-response improvement step sequence starting from any of these states runs in cycles:

$$\begin{pmatrix} 100 & 45 \\ S_{1,1} & S_{2,1} \end{pmatrix} \xrightarrow{1} \begin{pmatrix} 94 & 90 \\ S_{1,2} & S_{2,1} \end{pmatrix} \xrightarrow{2} \begin{pmatrix} 108 & 88 \\ S_{1,2} & S_{2,2} \end{pmatrix} \xrightarrow{1} \begin{pmatrix} 100 & 84 \\ S_{1,1} & S_{2,2} \end{pmatrix} \xrightarrow{2} \begin{pmatrix} 100 & 45 \\ S_{1,1} & S_{2,1} \end{pmatrix}$$

The numbers above the strategies give the costs of the respective player in the described state, and the numbers on the arrows indicate which player changes her strategy from one state to the next one. $\square$

Note that the nature of the existence proofs of Theorem 1 implies that a PNE can be computed in polynomial time. However, if existence is not guaranteed, the decision problem whether a matroid game has a pure Nash equilibrium is NP-hard:

Theorem 3. *It is* NP-*hard to decide whether a matroid congestion game with mixed objectives possesses at least one pure Nash equilibrium even if the players are α-uniform.*

Proof. We reduce from Independent Set (IS), which is known to be NP-complete [12]. Let the graph $G = (V, E)$ and $k \in \mathbb{N}$ be an instance of IS. We construct a matroid congestion game Γ that has a pure Nash equilibrium if and only if G has an independent set of size at least k.

We begin by describing the structure of Γ. The game contains the following groups of players:

- For each node $v \in V$, there is one player who can allocate all edges adjacent to v, but has a profitable deviation to a special strategy if and only if a player of a neighboring node allocates one of the adjacent edges.
- There are two players who play a game that is equivalent to the game in the proof of Theorem 2 if an additional player allocates a certain resource r_7, but possesses a PNE otherwise.
- Finally, there is one connection player for whom it is profitable to allocate r_7 if and only if at least $n - k + 1$ of the node players deviate from their "edge strategy".

Clearly, if we can achieve this dynamic, the existence of a PNE in Γ is equivalent to the existence of an independent set in G. Let $V = \{v_1, \ldots, v_n\}$, and for every $v_i \in V$ we denote by $E_{v_i} = \{e \in E \mid v_i \in e\}$ the set of edges adjacent to v_i, and by $d(v_i) = |E_{v_i}|$ the degree of v_i in G. Let $d = \max_{v \in V} d(v)$ be the maximum degree in G. We can assume that $d \geq 2$.

We now give a formal definition of $\Gamma = (N, R, (\Sigma_i)_{i \in N}, (\alpha_i)_{i \in N}, (\ell_r)_{r \in R}, (e_r)_{r \in R})$. The set of players is $N = \{v_1, \ldots, v_n, c, 1, 2\}$ and the set of resources is $R = \{r_e \mid e \in E\} \cup \{r_i^j \mid i \in \{1, \ldots, n\}, j \in \{1, \ldots, d(v_i) - 1\}\} \cup \{r_c, r_1, \ldots, r_7\}$. The strategies for the vertex players v_i are all subsets of size $d(v_i)$ from a set consisting of the adjacent edge resources, some alternative resources, and resource r_c:

$$\Sigma_{v_i} = \left\{ \{X \mid X \subseteq \left(\{r_e \mid e \in E_{v_i}\} \cup \{q_i^1, \ldots, q_i^{d(v_i)-1}, r_c\} \right) \text{ and } |X| = d(v_i) \right\}.$$

Our choice of cost functions will ensure that in an equilibrium every vertex player v_i either allocates all resources r_e that belong to her adjacent edges $e \in E_{v_i}$ or the resources q_i and r_c. The two strategies of the connection player are $\Sigma_c = \{\{r_c\}, \{r_7\}\}$. Finally, there are the players 1 and 2 with strategies $\Sigma_1 = \{\{r_i, r_j, r_k\} \mid i, j, k \in \{1, \ldots, 6\}\}$ and, $\Sigma_2 = \{\{r_i, r_j\} \mid i, j \in \{4, \ldots, 7\}\}$, respectively.

The cost functions for the edge resources are $\ell_{r_e}(x) = 1000 \cdot x$ and $e_{r_e}(x) = 0$ for all $e \in E$, for the alternative resources $\ell_{q_i^j}(x) = 0$ and $e_{q_i^j}(x) = 1000 \cdot d(v_i) + 1$ for all $1 \le i \le n$ and $1 \le j \le d$, and for the connection resource $\ell_{r_c}(x) = 0$ and $e_{r_c}(x) = 0$ for $x \le n - k + 1$, $e_{r_c}(x) = 1000$ for $x > n - k + 1$. The cost functions of the resources $r_1, \ldots, r_7$ are $\ell_{r_1}(x) = \ell_{r_2}(x) = \ell_{r_3}(x) = 0, e_{r_1}(x) = e_{r_2}(x) = e_{r_3}(x) = 200 \cdot x, \ell_{r_4}(x) = \ell_{r_5}(x) = 20 \cdot x, \ e_{r_4}(x) = e_{r_5}(x) = 50 \cdot x, \ell_{r_6}(x) = 8 \cdot x, \ e_{r_6}(x) = 80 \cdot x, \ \ell_{r_7}(x) = 0, \ e_{r_7}(x) = 80 \cdot x$. We choose the value $\alpha_i = 0.5$ for all players.

It remains to show that this game has a pure Nash equilibrium if and only if G has an independent set of size k. If there is an independent set, we can construct an equilibrium as follows: Each node player that corresponds to a node in the independent set chooses the strategy that contains all her edge resources. Each remaining node player chooses a strategy that contains only her q-resources and the resource r_c. The connection player chooses resource r_c. Player 1 chooses $\{r_1, r_2, r_3\}$ and player 2 chooses $\{r_6, r_7\}$. It is easy to verify that this is indeed a pure Nash equilibrium.

If there is no independent set of size at least k, we argue that in an equilibrium there are more than $n - k$ of the node players on resource r_c. Observe that for a node player that allocates at least one of the q-resources it is the best response to allocate the remaining q-resources and the resource r_c for no additional cost. Furthermore, if two node players allocate the same edge resource, their best response is to choose the q-resources and r_c. Hence, the best response of the connection player is $\{r_7\}$ and players 1 and 2 will play the subgame defined in Theorem 2 that does not have a pure Nash equilibrium.   □

In Theorem 1 we characterized restrictions on the preference values and cost functions that guarantee the existence of PNE in congestion games with mixed objectives, when combined with the matroid property of strategy spaces. The following theorem shows that the matroid property is necessary even if we impose the additional constraint that bottleneck and latency cost functions are identical and linear.

Theorem 4. *There exists a congestion game with mixed objectives with linear cost functions $\ell_r = e_r$ for all resources $r \in R$, and either*

1. *pure preferences, or*
2. *α-uniform players*

which does not possess a pure Nash equilibrium.

Proof. We show the correctness of the statement by constructing two games which fulfill the preconditions stated in the two cases of the theorem and which do not have a pure Nash equilibrium.

1. We define the game with two players with $\alpha_1 = 0$ and $\alpha_2 = 1$. The game has six resources $R = \{r_1, r_2, \ldots, r_6\}$. Each player has two strategies, thus $\Sigma_1 = \{\{r_1\}, \{r_2, r_3, r_4, r_5\}\}$ and $\Sigma_2 = \{\{r_2, r_3, r_4\}, \{r_5, r_6\}\}$. The latency and bottleneck costs are given by $\ell_{r_1}(x) = 6 \cdot x, \ \ell_{r_2}(x) = \ell_{r_3}(x) = \ell_{r_4}(x) = 2 \cdot x$,

$\ell_{r_5}(x) = 4 \cdot x$, $\ell_{r_6}(x) = 3 \cdot x$ with $e_r(x) = \ell_r(x)$ for all $r \in R$ We utilize the fact that bottleneck players prefer to allocate many cheap resources, while players who are interested in latency are more willing to share a single expensive resource.

Let $S_{1,1}$ and $S_{1,2}$ denote the strategies of player 1, and $S_{2,1}$ and $S_{2,2}$ the strategies of player 2. Then we have the following cycle of improvement steps which visits all four states:

$$\begin{pmatrix} 6 & 6 \\ S_{1,1} & S_{2,1} \end{pmatrix} \xrightarrow{1} \begin{pmatrix} 4 & 12 \\ S_{1,2} & S_{2,1} \end{pmatrix} \xrightarrow{2} \begin{pmatrix} 8 & 11 \\ S_{1,2} & S_{2,2} \end{pmatrix} \xrightarrow{1} \begin{pmatrix} 6 & 7 \\ S_{1,1} & S_{2,2} \end{pmatrix} \xrightarrow{2} \begin{pmatrix} 6 & 6 \\ S_{1,1} & S_{2,1} \end{pmatrix}$$

The numbers above the strategies give the cost of the respective player in the associated state, and the numbers on the arrows indicate which player has to change her strategy in order to get to the next state. As we see, every change in strategy decreases the cost of the player performing it. Hence, none of the four states is a pure Nash equilibrium.

2. We prove the theorem for the example $\alpha = 0.5$. However, it is easily generalizable to arbitrary values between 0 and 1. The idea is to construct a game with two players in which one player always allocates an expensive resource. In addition this, both players allocate two resources and share exactly one of these resources. In detail we have the resources $R = \{r_1, r_2, \ldots, r_5\}$ and the strategy sets $\Sigma_1 = \{\{r_1, r_2, r_4\}, \{r_1, r_3, r_5\}\}$ and $\Sigma_2 = \{\{r_2, r_5\}, \{r_3, r_4\}\}$. The latency and bottleneck costs are given by $\ell_{r_1}(x) = 32 \cdot x$, $\ell_{r_2}(x) = \ell_{r_3}(x) = 14 \cdot x$ and $\ell_{r_4}(x) = \ell_{r_5}(x) = 12 \cdot x + 8$ with $e_r(x) = \ell_r(x)$ for all $r \in R$. Depending on which resource is shared, the players allocate either two resources with medium costs or one cheap and one expensive resource, where the sum of the two resource costs is slightly smaller in the latter case.

   The second player prefers the first alternative, since she has to pay an additional price for her most expensive resource. On the other hand, the first player always allocates an expensive resource, hence she incurs no additional costs when allocating a cheap and an expensive resource.

   The strategy spaces are constructed in such a way that in every state the players share exactly one of the resources $\{r_2, \ldots, r_5\}$. Let S_1 denote a state in which r_2 or r_3 is shared, and S_2 a state in which r_4 or r_5 is shared. Then the players incur the following costs: $c_1(S_1) = 56, c_2(S_1) = 38, c_1(S_2) = 55, c_2(S_2) = 39$.

   As we see, player 1 prefers the state S_2, while player 2 prefers S_1. Since both players always have the possibility to deviate to the other state, there is no state in which none of the players can improve her costs, and hence Γ possesses no pure Nash equilibrium.   □

3   Convergence

In this section we investigate in which games convergence of best-response improvement sequences to a pure Nash equilibrium can be guaranteed.

Perhaps surprisingly, there are singleton games in which best-response improvement sequences may run in cycles. This is even true for games with pure preferences.

Theorem 5. *There are singleton congestion games with mixed objectives and pure preferences in which best-response improvement sequences may run in cycles.*

Proof. We prove the theorem by constructing a singleton game with three players and showing that there exists a cyclic best-response improvement sequence in certain states. The game consists of three resources $R = \{r_1, r_2, r_3\}$ and the players have the strategy sets $\Sigma_1 = \{\{r_1\}, \{r_2\}\}$, $\Sigma_2 = \{\{r_1\}, \{r_3\}\}$ and $\Sigma_3 = \{\{r_2\}, \{r_3\}\}$. Two players prefer the latency costs $\alpha_1 = \alpha_2 = 1$, one the bottleneck costs $\alpha_3 = 0$. The latency and bottleneck costs are given by $\ell_{r_1} = (2, 5)$, $\ell_{r_2} = (3, 4)$, $\ell_{r_3} = (1, 6)$ and $e_{r_2} = (1, 4)$, $e_{r_3} = (2, 3)$. The first number in the cost functions gives the cost if the resource is used by one player, the second number gives the cost if two players use it. The bottleneck cost function of r_1 is irrelevant since it is only used by players 1 and 2.

In this game, the following cycling best-response improvement sequence can occur (set braces omitted for better readability):

$$\begin{pmatrix} 5 & 5 & 1 \\ r_1 & r_1 & r_2 \end{pmatrix} \xrightarrow{1} \begin{pmatrix} 4 & 2 & 4 \\ r_2 & r_1 & r_2 \end{pmatrix} \xrightarrow{2} \begin{pmatrix} 4 & 1 & 4 \\ r_2 & r_3 & r_2 \end{pmatrix} \xrightarrow{3} \begin{pmatrix} 3 & 6 & 3 \\ r_2 & r_3 & r_3 \end{pmatrix}$$

$$\xrightarrow{1} \begin{pmatrix} 2 & 6 & 3 \\ r_1 & r_3 & r_3 \end{pmatrix} \xrightarrow{2} \begin{pmatrix} 5 & 5 & 2 \\ r_1 & r_1 & r_3 \end{pmatrix} \xrightarrow{3} \begin{pmatrix} 5 & 5 & 1 \\ r_1 & r_1 & r_2 \end{pmatrix}$$

The numbers on the arrows indicate which player has to change her strategy in order to reach the next state. The variable r_i denotes the resource which is used by the corresponding player and the number on top gives the cost value for this player. We can verify that each change in strategy is beneficial for the player performing it (since every player has only two strategies, every improving strategy is a best-response strategy).

Hence, we have a cycle of six states that are visited during this best-response improvement sequence. The pure Nash equilibria (r_1, r_3, r_2) and (r_2, r_1, r_3) are never reached. □

Note that due to our reduction in the proof of Theorem 1, we know that there exists a sequence that leads to an equilibrium [1].

Corollary 1

1. *Singleton congestion games with mixed objectives are weakly acyclic.*
2. *Matroid congestion games with mixed objectives that have pure preferences are weakly acyclic.*

We now turn to matroid games with a monotone dependence and show that they converge quickly to a PNE if the players perform lazy best-response moves.

That is, if players perform a best response move, they choose the best-response strategy that has as many resources in common with the previous strategy as possible.

Theorem 6. *Let Γ be a matroid congestion game with mixed objectives with cost functions that have a monotone dependence. Then any sequence of lazy best-response improvement steps starting from an arbitrary state in Γ converges to a pure Nash equilibrium after a polynomial number of steps.*

Proof. The proof idea is based on the proof by Ackermann et al. [2] which shows that matroid congestion games guarantee polynomial convergence to PNE.

We consider an increasingly ordered enumeration of all latency values that can occur in Γ (an enumeration of the values $\{\ell_r(x) \mid r \in R, x \in N\}$). Let $\ell'_r(x)$ denote the position of the respective cost value in the enumeration.

We define the following variant of Rosenthal's potential function: $\Phi(S) = \sum_{r \in R} \sum_{i=1}^{n_r(S)} \ell'_r(i)$. If $n = |N|$ denotes the number of players, and $m = |R|$ the number of resources in Γ, then there are at most $n \cdot m$ different cost values in the game. Hence, the value of Φ is upper bounded by $n^2 \cdot m^2$. Thus, it suffices to show that every lazy best-response improvement step decreases the value of Φ by at least 1.

If a player replaces a single resource r by another resource r' in a lazy best-response with $\alpha_i \cdot \ell_{r'}(S') + (1-\alpha_i) \cdot e_{r'}(S') < \alpha_i \cdot \ell_r(S) + (1-\alpha_i) \cdot e_r(S)$, then due to the monotone dependence, we have $\ell_{r'}(S') < \ell_r(S)$. Hence $\ell_{r'}(S')$ must occur before $\ell_r(S)$ in the increasingly ordered enumeration of the cost values and we have $\ell'_{r'}(S') < \ell'_r(S)$. Thus, every sequence of lazy best-response improvement steps in Γ terminates after a polynomial number of steps. □

We remark that the only reason to restrict the players to *lazy* instead of arbitrary best-response strategies is that the players may have a preference value of exactly 0. If a player's cost is determined solely by her most expensive resource, she might be playing a best-response strategy by replacing her most expensive resource by a cheaper one and additionally replace another resource by a more expensive one. This additional exchange does not necessarily increase her costs, but it could lead to an increase in the value of the potential function. However, if all players have preference values different from 0, the theorem holds for arbitrary best-response improvement steps.

4   Approximate Pure Nash Equilibria

As PNE do not exist in general, we study the existence of approximate equilibria. However, in general we cannot achieve an approximation factor better than 3.

Theorem 7. *There is a congestion game with mixed objectives, in which all cost functions are linear, that does not contain a β-approximate pure Nash equilibrium for any $\beta < 3$.*

Proof. We show the theorem by constructing a game with 10 players, in which in every state there is at least one player who can improve her costs by a factor of 3. A formal definition of the game is given by:

$$\Gamma = (N, R, (\Sigma_i)_{i \in N}, (\alpha_i)_{i \in N}, (\ell_r)_{r \in R}, (e_r)_{r \in R}) \text{with}$$

$$N = \{1, \ldots, 10\}, \ R = \{r_1^i, r_2^i \mid i \in N \setminus \{5, 6\}\} \cup$$
$$\left\{ r_{1,1}^{i,j}, r_{1,2}^{i,j}, r_{2,1}^{i,j}, r_{2,2}^{i,j} \mid i < j \text{ and } (i, j \in \{1, 2, 3, 4\} \text{ or } i, j \in \{7, 8, 9, 10\}) \right\},$$

$$\Sigma_i = \Big\{ \{r_1^i\} \cup \{r_{1,1}^{j,k}, r_{2,1}^{j,k} \mid j = i \text{ or } k = i\},$$
$$\{r_2^i\} \cup \{r_{1,2}^{j,k}, r_{2,2}^{j,k} \mid j = i \text{ or } k = i\} \Big\} \text{ for all } i \in N \setminus \{5, 6\}$$

$$\Sigma_5 = \Big\{ \{r_1^1, \ldots, r_1^4, r_1^7, \ldots, r_1^{10}, r_{1,2}^{i,j}, r_{1,2}^{k,l} \mid i, j \in \{1, \ldots, 4\}, \ k, l \in \{7, \ldots, 10\}\} \Big\}$$
$$\cup \Big\{ \{r_2^1, \ldots, r_2^4, r_2^7, \ldots, r_2^{10}, r_{1,1}^{i,j}, r_{1,1}^{k,l} \mid i, j \in \{1, \ldots, 4\}, \ k, l \in \{7, \ldots, 10\}\} \Big\}$$

$$\Sigma_6 = \Big\{ \{r_1^1, \ldots, r_1^4, r_2^7, \ldots, r_2^{10}, r_{2,2}^{i,j}, r_{2,1}^{k,l} \mid i, j \in \{1, \ldots, 4\}, \ k, l \in \{7, \ldots, 10\}\} \Big\}$$
$$\cup \Big\{ \{r_2^1, \ldots, r_2^4, r_1^7, \ldots, r_1^{10}, r_{2,1}^{i,j}, r_{2,2}^{k,l} \mid i, j \in \{1, \ldots, 4\}, \ k, l \in \{7, \ldots, 10\}\} \Big\}$$

$\alpha_i = 1$ for all $i \in N \setminus \{5, 6\}$ and $\alpha_i = 0$ for $i \in \{5, 6\}$,
$\ell_r(x) = x$ and $e_r(x) = 0$ for all $r \in \{r_j^i \mid i \in N \setminus \{5, 6\}, j \in \{1, 2\}\}$,
$\ell_r(x) = 0$ and $e_r(x) = x$ for all $r \in \left\{ r_{k,l}^{i,j} \mid i, j \in N \setminus \{5, 6\}, \ k, l \in \{1, 2\} \right\}$

The game contains three different groups of players and two different groups of resources. Players 1 to 4 and 7 to 10 each have two personal resources r_1^i and r_2^i among which they have to choose. These are the only resources on which they incur costs.

In addition to these resources, there are two times two resources corresponding to each set consisting of two players from the same group (either 1 to 4 or 7 to 10). The two resources represent the two strategies which are available to these players, and exist distinctly for both players 5 and 6. If player i plays her first strategy, then she also allocates all resources that correspond to sets in which i is contained and represent the first strategy.

In every state, the players 5 and 6 allocate one of the personal resources r_j^i for each player $i \in \{1, \ldots, 4\}$ and $i \in \{7, \ldots, 10\}$, where the j is the same for all players among a group. Additionally, for both groups they have to allocate a resource that corresponds to one pair of players and represents the j that they are not using (e. g., if player 5 allocates resource r_1^i for all players $i \in \{1, \ldots, 4\}$, she must also allocate a resource that corresponds to a pair of players from $\{1, \ldots, 4\}$ and represents strategy 2, and an analogous resource for the players in $\{7, \ldots, 10\}$).

These two additional resources are the ones on which players 5 and 6 incur costs. Since they are interested in their bottleneck costs, their costs are equal to the more expensive of the two.

We can analyze this game by considering an arbitrary state. The strategy sets of players 5 and 6 are constructed in such a way that in every state they allocate the same personal resources of all players in $\{1, \ldots, 4\}$ or $\{7, \ldots, 10\}$.

W. l. o. g. assume that they both allocate the resources $r_1^1, \ldots, r_1^4$. This implies that the players 1 to 4 have a cost of 3 if they use the first resource, and a cost of 1 if they allocate their second resource. Hence, this state cannot be a β-approximate PNE with $\beta < 3$ if any of them play their first strategy.

If, on the other hand, all these players play their second strategy, players 5 and 6 incur a cost of 3 on their resources $r_{1,2}^{i,j}$ and $r_{2,2}^{i,j}$, respectively, independent of which i and j they are using, since all players in $\{1, \ldots, 4\}$ allocate the resources corresponding to strategy 2. Both player 5 and 6 could decrease the cost of the resource corresponding to the first group to 1 by switching the strategy. However, we have to take the other group into account as well.

If player 5 switches her strategy, she allocates the resource $r_{1,1}^{k,l}$ for freely selectable k and l in $\{7, \ldots, 10\}$, which yields a cost of 1 if at least two players from $\{7, \ldots, 10\}$ play the second strategy. Analogously, player 6 incurs a cost of 1 on the resource corresponding to the second group if at least two players from this group play their first strategy. One of the two cases must hold, which implies that either player 5 or player 6 can improve her cost from 3 to 1. Hence, in every state there is at least one player who can improve her cost by a factor of 3, and there exists no β-approximate pure Nash equilibrium for any $\beta < 3$. $\square$

On the positive side we can show small approximation factors for small strategy sets. Besides matroid games, we can show approximation factors which are independent of the structure of the strategy sets, but depend on either α-uniform players or on equal cost functions:

Theorem 8. *Let Γ be a congestion game with mixed objectives. Let $d = \max_{i \in N, S_i \in \Sigma_i} |S_i|$ be the maximal number of resources a player can allocate.*

1. *If Γ is a matroid congestion game, then Γ contains a d-approximate pure Nash equilibrium.*
2. *If the players are α-uniform, then Γ contains a d-approximate pure Nash equilibrium.*
3. *If $e_r = \ell_r$ for all resources $r \in R$, then Γ contains a $\sqrt{d}$-approximate pure Nash equilibrium.*
4. *If the players are α-uniform and $e_r = \ell_r$ for all resources $r \in R$, then Γ contains a β-approximate pure Nash equilibrium for $\beta = \frac{d}{\alpha \cdot (d-1) + 1}$.*

Proof

1. The proof relies on the fact that PNE always exist in player-specific matroid congestion games [1]. We define a player-specific congestion game Γ' with the following cost functions: $c_r^i(S) = \alpha_i \cdot \ell_r(S) + (1 - \alpha_i) \cdot e_r(S)$.

   We show that every PNE in Γ' is a d-approximate pure Nash equilibrium in Γ. Since PNE always exist in matroid player-specific congestion games, the claim follows. We denote by $c_i(S)$ the costs of player i in state S in Γ, and by $c_i^p(S)$ the costs in Γ'. Let S_i be a best-response strategy w. r. t. S_{-i} in Γ'. Then we get for all strategies $S_i' \in \Sigma_i$:

$$c_i(S_i', S_{-i}) = \alpha_i \cdot \sum_{r \in S_i'} \ell_r(S_i', S_{-i}) + (1 - \alpha_i) \cdot \max_{r \in S_i'} e_r(S_i', S_{-i})$$

$$\geq \alpha_i \cdot \sum_{r \in S_i'} \ell_r(S_i', S_{-i}) + (1 - \alpha_i) \cdot \frac{1}{d} \cdot \sum_{r \in S_i'} e_r(S_i', S_{-i})$$

$$\geq \frac{1}{d} \cdot c_i^p(S_i', S_{-i}) \geq \frac{1}{d} \cdot c_i^p(S_i, S_{-i}) \geq \frac{1}{d} \cdot c_i(S_i, S_{-i})$$

2. We show that the function $\Phi(S) = \sum_{r \in R} \sum_{i=1}^{n_r(S)} (\alpha \cdot \ell_r(i) + (1 - \alpha) \cdot e_r(i))$ is a d-approximate potential function, i.e., its value decreases in every d-improvement step. Consider a state S and a player i who improves her costs by a factor of more than d by deviating to the strategy S_i':

$$\Phi(S_i', S_{-i}) - \Phi(S)$$

$$= \sum_{r \in R} \sum_{i=1}^{n_r(S')} (\alpha \cdot \ell_r(i) + (1 - \alpha) \cdot e_r(i)) - \sum_{r \in R} \sum_{i=1}^{n_r(S)} (\alpha \cdot \ell_r(i) + (1 - \alpha) \cdot e_r(i))$$

$$\leq \alpha \cdot \sum_{r \in S_i'} \ell_r(S') + (1 - \alpha) \cdot \max_{r \in S_i'} e_r(S') + (1 - \alpha) \cdot \sum_{r \in S_i'} e_r(S') - (1 - \alpha) \cdot \max_{r \in S_i'} e_r(S')$$

$$- \left(\alpha \cdot \sum_{r \in S_i} \ell_r(S) + (1 - \alpha) \cdot \max_{r \in S_i} e_r(S) \right)$$

$$\leq c_i(S') - c_i(S) + \sum_{r \in S_i'} (1 - \alpha) \cdot e_r(S') - (1 - \alpha) \cdot \max_{r \in S_i'} e_r(S')$$

$$\leq c_i(S') - c_i(S) + (d - 1) \cdot c_i(S') < c_i(S') - d \cdot c_i(S') + (d - 1) \cdot c_i(S') = 0$$

3. We show that $\Phi(S) = \sum_{r \in R} \sum_{i=1}^{n_r(S)} \ell_r(i)^2$ is a $\sqrt{d}$-approximate potential function. Consider a state S that is minimizing Φ and player i who deviates to the strategy S_i'. Note that $\Phi(S) - \Phi(S_i', S_{-i}) = \sum_{r \in S} \ell_r(S)^2 - \sum_{r \in S'} \ell_r(S')^2$. Hence, $\sum_{r \in S} \ell_r(S)^2 \leq \sum_{r \in S'} \ell_r(S')^2$.

$$c_i(S_i', S_{-i}) = \sum_{r \in S_i'} \alpha_i \cdot \ell_r(S') + (1 - \alpha_i) \max_{r \in S_i'} \ell_r(S')$$

$$\geq \alpha_i \cdot \left(\sum_{r \in S_i'} \ell_r(S')^2 \right)^{\frac{1}{2}} + (1 - \alpha_i) \cdot \left(\max_{r \in S_i'} \ell_r(S')^2 \right)^{\frac{1}{2}}$$

$$\geq \alpha_i \cdot \left(\frac{1}{\sqrt{d}} \cdot \sum_{r \in S_i} \ell_r(S) \right) + (1 - \alpha_i) \cdot \left(\frac{1}{d} \cdot \left(\sum_{r \in S_i} \ell_r(S)^2 \right) \right)^{\frac{1}{2}}$$

$$\geq \frac{\alpha_i}{\sqrt{d}} \cdot \sum_{r \in S_i} \ell_r(S) + \frac{1 - \alpha_i}{\sqrt{d}} \cdot \max_{r \in S_i} \ell_r(S) = \frac{1}{\sqrt{d}} \cdot c_i(S).$$

4. We argue that $\Phi(S) = \sum_{r \in R} \sum_{i=1}^{n_r(S)} \ell_r(i)$ is an approximate potential function. If the latency and bottleneck cost functions are identical for each resource, we use the fact that the cost of the bottleneck resource is at least as high as the average latency cost, i.e., $\max_{r \in S_i} e_r(S) \geq \frac{1}{|S_i|} \sum_{r \in S_i} \ell_r(S) \geq \frac{1}{d} \sum_{r \in S_i} \ell_r(S)$, which implies $c_i(S) = \alpha \cdot \sum_{r \in S_i} \ell_r(S) + (1-\alpha) \cdot \max_{r \in S_i} \ell_r(S) \geq (\alpha + \frac{1-\alpha}{d}) \sum_{r \in S_i} \ell_r(S)$.

Consider a state S and a player i who improves her costs by a factor of more than $\beta = \frac{d}{\alpha \cdot (d-1)+1}$ by deviating to the strategy S_i'.

$$\Phi(S_i', S_{-i}) - \Phi(S) \leq c_i(S') - c_i(S) + \sum_{r \in S_i'}(1-\alpha) \cdot \ell_r(S') - (1-\alpha) \cdot \max_{r \in S_i'} \ell_r(S')$$

$$\leq c_i(S') - c_i(S) + (1-\alpha) \cdot \left(1 - \frac{1}{d}\right) \cdot \sum_{r \in S_i'} \ell_r(S')$$

$$\leq c_i(S') - c_i(S) + (1-\alpha) \cdot \left(1 - \frac{1}{d}\right) \cdot \frac{c_i(S')}{\alpha + \frac{1-\alpha}{d}}$$

$$= \left(1 + \frac{(1-\alpha) \cdot \frac{d-1}{d}}{\alpha + \frac{1-\alpha}{d}}\right) c_i(S') - c_i(S)$$

$$= \left(1 + \frac{d-1}{\alpha \cdot (d-1)+1} - 1 + \frac{1}{\alpha \cdot (d-1)+1}\right) c_i(S') - c_i(S)$$

$$= \frac{d}{\alpha \cdot (d-1)+1} c_i(S') - c_i(S) < 0$$

$\square$

We remark that the β given in the fourth case of Theorem 8 is bounded from above by $\frac{1}{\alpha}$. However, if α is close to 0, the bound of $\sqrt{d}$ derived for general games with $\ell_r = e_r$ for all resources, but without restrictions on the preference values, may give a better approximation guarantee.

5   Conclusions

We studied a new class of games in which players seek to minimize the sum of latency costs, the maximum of bottleneck costs, or a combination thereof. As a promising avenue for future work it would be interesting to consider other types of cost aggregation. This would be useful in scenarios with heterogeneous players with different interests in which the resources represent not only links in a network but also servers, routers, or any network functions in general.

References

1. Ackermann, H., Röglin, H., Vöcking, B.: Pure nash equilibria in player-specific and weighted congestion games. Theoret. Comput. Sci. **410**(17), 1552–1563 (2009)
2. Ackermann, H., Röglin, H., Vöcking, B.: On the impact of combinatorial structure on congestion games. J. ACM **55**(6), 25:1–25:22 (2008)

3. Ackermann, H., Skopalik, A.: Complexity of pure nash equilibria in player-specific network congestion games. Internet Math. **5**(4), 323–342 (2008)
4. Banner, R., Orda, A.: Bottleneck routing games in communication networks. IEEE J. Sel. Areas Commun. **25**(6), 1173–1179 (2007)
5. Caragiannis, I., Fanelli, A., Gravin, N., Skopalik, A.: Efficient computation of approximate pure nash equilibria in congestion games. In: IEEE 52nd Annual Symposium on Foundations of Computer Science (FOCS), pp. 532–541 (2011)
6. Caragiannis, I., Fanelli, A., Gravin, N., Skopalik, A.: Approximate pure nash equilibria in weighted congestion games: existence, efficient computation, and structure. ACM Trans. Econ. Comput. **3**(1), 2:1–2:32 (2015)
7. Chien, S., Sinclair, A.: Convergence to approximate nash equilibria in congestion games. Games Econ. Behav. **71**(2), 315–327 (2011)
8. Cole, R., Dodis, Y., Roughgarden, T.: Bottleneck links, variable demand, and the tragedy of the commons. Networks **60**(3), 194–203 (2012)
9. Dunkel, J., Schulz, A.S.: On the complexity of pure-strategy nash equilibria in congestion and local-effect games. Math. Oper. Res. **33**(4), 851–868 (2008)
10. Feldotto, M., Gairing, M., Skopalik, A.: Bounding the potential function in congestion games and approximate pure nash equilibria. In: Liu, T.-Y., Qi, Q., Ye, Y. (eds.) WINE 2014. LNCS, vol. 8877, pp. 30–43. Springer, Heidelberg (2014). doi:10.1007/978-3-319-13129-0_3
11. Fotakis, D., Kontogiannis, S., Spirakis, P.: Selfish unsplittable flows. Theoret. Comput. Sci. **348**(2), 226–239 (2005)
12. Garey, M.R., Johnson, D.S.: Computers and Intractability: A Guide to the Theory of NP-Completeness. W. H. Freeman, New York (1979)
13. Hansknecht, C., Klimm, M., Skopalik, A.: Approximate pure nash equilibria in weighted congestion games. In: Jansen, K., Rolim, J.D.P., Devanur, N.R., Moore, C. (eds.) Approximation, Randomization, and Combinatorial Optimization, Algorithms and Techniques (APPROX/RANDOM 2014). Leibniz International Proceedings in Informatics (LIPIcs), vol. 28, pp. 242–257. Schloss Dagstuhl-Leibniz-Zentrum fuer Informatik, Dagstuhl, Germany (2014)
14. Harks, T., Hoefer, M., Schewior, K., Skopalik, A.: Routing games with progressive filling. IEEE/ACM Trans. Netw. **24**(4), 2553–2562 (2016)
15. Harks, T., Hoefer, M., Klimm, M., Skopalik, A.: Computing pure nash and strong equilibria in bottleneck congestion games. Math. Program. **141**(1), 193–215 (2013)
16. Harks, T., Klimm, M., Möhring, R.H.: Strong nash equilibria in games with the lexicographical improvement property. In: Leonardi, S. (ed.) WINE 2009. LNCS, vol. 5929, pp. 463–470. Springer, Heidelberg (2009). doi:10.1007/978-3-642-10841-9_43
17. Mavronicolas, M., Milchtaich, I., Monien, B., Tiemann, K.: Congestion games with player-specific constants. In: Kučera, L., Kučera, A. (eds.) MFCS 2007. LNCS, vol. 4708, pp. 633–644. Springer, Heidelberg (2007). doi:10.1007/978-3-540-74456-6_56
18. Milchtaich, I.: Congestion games with player-specific payoff functions. Games Econ. Behav. **13**(1), 111–124 (1996)
19. Monderer, D., Shapley, L.S.: Potential games. Games Econ. Behav. **14**(1), 124–143 (1996)
20. Rosenthal, R.W.: A class of games possessing pure-strategy nash equilibria. Int. J. Game Theor. **2**(1), 65–67 (1973)
21. Skopalik, A., Vöcking, B.: Inapproximability of pure nash equilibria. In: Proceedings of the Fortieth Annual ACM Symposium on Theory of Computing, STOC 2008, pp. 355–364. ACM, New York (2008)

An Optimal Strategy for Static Black-Peg Mastermind with Two Pegs

Gerold Jäger[✉]

Department of Mathematics and Mathematical Statistics,
University of Umeå, 90187 Umeå, Sweden
gerold.jaeger@math.umu.se

Abstract. Mastermind is a famous two-player game which has attracted much attention in literature within the last years. In this work we investigate the static (also called non-adaptive) variant of Mastermind. The principal rule is that the codemaker has to choose a secret consisting of p pegs and c colors for each peg and the codebreaker may give a number of guesses at once, where for each guess he receives information from the codemaker. Using this information he has a final guess for the correct secret. The aim of the game is to minimize the number of guesses. Whereas Goddard has investigated the static version of original Mastermind in 2003, we do such an investigation of its black-peg variant, where the received information consists only of a number of black pegs which corresponds to the number of pegs matching in the corresponding question and the secret. As main result we present a strategy for this game for $p = 2$ pegs and arbitrarily many colors $c \geq 3$ colors and prove its feasibility and optimality. Furthermore, by computer search we found optimal strategies for 9 other pairs (p, c).

Keywords: Game theory · Logic game · Strategy · Static Mastermind

1 Introduction

Mastermind is a board game invented by Meirowitz in 1970 with interesting applications in cryptography [7] and bioinformatics [8]. Mastermind is played by two players, the *codemaker* and the *codebreaker*. Whereas the codemaker chooses a secret code consisting of 4 pegs and 6 possible colors for each peg, the codebreaker does not know this code and has to give several guesses, also called questions, until the correct secret has been found. After each guess the codemaker gives an answer how good the guess is. This answer consists of black and white pegs, where each black peg corresponds to a peg of the codebreaker's guess which is correct in position and color, and each white peg corresponds to another peg which is correct only in color. Clearly, the codebreaker wants to minimize the number of questions needed to find the secret. For original Mastermind with 4 pegs and 6 colors an optimal strategy needs $5625/1296 \approx 4.34$ questions in the average case [18], and 5 questions in the worst case [17]. Within the last years

© Springer International Publishing AG 2016
T.-H.H. Chan et al. (Eds.): COCOA 2016, LNCS 10043, pp. 670–682, 2016.
DOI: 10.1007/978-3-319-48749-6_48

much research has been done for Mastermind and its variants. Stuckman and Zhang showed that the decision problem corresponding to Generalized Mastermind with p pegs and c colors, i.e., the problem, whether a secret exists which satisfies a given set of questions and answers, is $\mathcal{NP}$-complete [19]. Further theoretical results about the hardness of Mastermind can be found in [5,20]. An analysis of optimal strategies has been presented for original Mastermind [3,10,12], and for several of its variants, e.g., for the black-peg variant, where the white pegs are ignored in the answers [11,13], the AB game, where the colors in all question and in the secret must be different [2,15], and Constant-Size Memory Mastermind, where the codebreaker may only store a constant number of questions and answers during the game [6,14,16].

In this work we investigate a further variant of Mastermind, called *Static Mastermind* (or *Non-Adaptive Mastermind*), where the codebreaker has to give all questions at the beginning of the game, then receives all codemaker's answers, and finally, has to give the correct answer. For fixed $p, c \in \mathbb{N}$, define $s(p, c)$ as the worst case number of questions of an optimal strategy of (p, c)-Static Mastermind. Chvátal showed that for fixed c it holds that $s(p, c) = \mathcal{O}(p/\log p)$ [4]. Goddard developed an optimal $\lceil 2c/3 \rceil$-strategy for two pegs, an optimal $(c-1)$-strategy for three pegs and an optimal $(c-1)$-strategy for four pegs, where c is sufficiently large [9]. An application of Static Mastermind for string and vector databases was given in [1].

Consider the black-peg variant of this game, i.e., we only allow to give a number of black pegs as an answer, where each black peg stands for one peg which is correct in position and color. We call this variant Static Black-Peg Mastermind. To the best of our knowledge, this variant has not been investigated in literature.

As main result we present a strategy for the case of two pegs and arbitrarily many colors $c \geq 3$ and prove that this strategy is feasible and optimal.

This work is organized as follows. In Sect. 2 we give some basic notations and explanations. In Sect. 3 we present a $\lceil (4c-1)/3 \rceil$-strategy for the case of two pegs and arbitrarily many colors $c \geq 3$, in Sect. 4 we prove its feasibility and in Sect. 5 its optimality. In Sect. 6 based on computer search we provide optimal strategies for 9 other pairs (p, c) and feasible strategies for 11 other pairs (p, c). Finally, we give some suggestions for future work.

2 Preliminaries

Let p denote the number of pegs and c the number of colors. If p and c are fixed, we call the game (p, c)-Static Black-Peg Mastermind. The possible answers are written as 0B, 1B, 2B, ..., pB. Each strategy for Static Black-Peg Mastermind with r questions starts with $r - 1$ questions which the codebreaker has to ask at the beginning of the game. We call these questions *main questions*. After the codebreaker has received all $r - 1$ answers to these main questions, he has to ask the *final question*, which has to be correct and thus receive the answer pB. We call each such strategy a *feasible r-strategy*, and only *feasible strategy*, if we do not focus on the number of questions. If after answering the $r - 1$ main questions at least two secrets are possible, then we call the corresponding strategy an *infeasible r-strategy* or only *infeasible strategy*. The strategy is called *optimal*, if there is no feasible r'-strategy with $r' < r$. For fixed $p, c \in \mathbb{N}$, define $sb(p, c)$ as the worst case number of questions of a an optimal strategy of (p, c)-Static Black-Peg Mastermind. Obviously, regarding optimal strategies it makes no sense to ask the same question twice. Thus, we can assume that the questions within a strategy are pairwise disjoint. In the following let the pegs be numbered by $1, 2 \ldots, p$ and the colors by $1, 2, \ldots, c$.

The following observation enables us to speed up a computer based brute force search.

Observation 1. *For Static Black-Peg Mastermind (and also for original Black-peg Mastermind) it holds that for a given question the codemaker can answer each peg independently by $0B$ or $1B$ and receive the overall answer by adding all answers $1B$. E.g., if we consider p pegs and c colors, there would be an equivalent game containing pc colors, where each peg has c separate colors.*

It follows:

Proposition 1. *Each strategy of Static Mastermind can be assumed to start with $(1, 1, \ldots, 1)$ as first main question.*

We implemented a computer program in the programming language C++ which checks for given values of p, c and $r \in \mathbb{N}$, whether there exists a feasible r-strategy for (p, c)-Static Black-Peg Mastermind or not. This program is a naive brute force search algorithm, which checks all possible strategies, i.e., all combinations of questions, for being a feasible strategy or not. Note that by Proposition 1 we can fix the first question in this computer program and thus significantly reduce the search space. The program also allows to check whether a given strategy is feasible or not, and thus it can verify the theoretical result of this work for $p = 2$. All experiments were done on a standard desktop in a Unix-based system. The source code of this computer program is available online at [21].

Remark 1. Note that the cases $c = 1$ and $p = 1$ are trivial:

(a) For $c = 1$, without any main questions the correct secret can be identified in one final question as $(1, 1, \ldots, 1)$. Thus, it holds that $sb(p, 1) = 1$ for all p.
(b) For $p = 1$, after $c - 1$ main questions the correct secret can be identified in the final question. Thus, it holds that $sb(1, c) = c$ for all c.

3   Two Pegs Strategy

In this section and the following sections let $p = 2$. We introduce a $\lceil (4c - 1)/3 \rceil$-strategy for each c, where we distinguish between the cases $c \equiv 0 \bmod 3$,

$c \equiv 1 \bmod 3$, $c \equiv 2 \bmod 3$. For the number $k := \lceil (4c-1)/3 \rceil - 1$ of main questions it holds that

$$
k = \begin{cases}
\dfrac{4}{3} \cdot c - 1 = 4 \cdot \dfrac{c}{3} - 1 & \equiv 3 \bmod 4 & \text{for } c \equiv 0 \bmod 3 \\[2ex]
\dfrac{4}{3} \cdot c - \dfrac{4}{3} = 4 \cdot \dfrac{c-1}{3} & \equiv 0 \bmod 4 & \text{for } c \equiv 1 \bmod 3 \\[2ex]
\dfrac{4}{3} \cdot c - \dfrac{2}{3} = 4 \cdot \dfrac{c-2}{3} + 2 & \equiv 2 \bmod 4 & \text{for } c \equiv 2 \bmod 3
\end{cases}
$$

Strategy 1 ($\lceil (4c-1)/3 \rceil$-strategy for $p = 2$ and arbitrary $c \equiv 0 \bmod 3$).

1. Divide the k main questions into two blocks of questions, the first $(k-1)/2$ questions and the second $(k+1)/2$ questions.
2. The first peg contains the colors $1, 2, \ldots, (k-1)/2$ in the first block and the colors $(k+1)/2, (k+1)/2, (k+1)/2+1, (k+1)/2+1, \ldots, 3(k+1)/4-1, 3(k+1)/4-1(=c-1)$ in the second block.
3. The second peg contains the colors $(k+1)/2+1, (k+1)/2+1, (k+1)/2+1, (k+1)/2+2, (k+1)/2+2, \ldots, 3(k+1)/4-1, 3(k+1)/4-1$ in the first block and the colors $1, 2, \ldots, (k+1)/2$ in the second block (note that the first number $(k+1)/2+1$ occurs three times, not only twice).
4. The last color $k = 3(k+1)/4 = c$ is not used in the questions at all.
5. Finally, the secret has to be asked as final question Q_{k+1}.

Strategy 2 ($\lceil (4c-1)/3 \rceil$-strategy for $p = 2$ and arbitrary $c \equiv 1 \bmod 3$).

1. Divide the k main questions into two blocks of questions, the first $k/2$ questions and the second $k/2$ questions.
2. The first peg contains the colors $1, 2, \ldots, k/2$ in the first block and the colors $k/2+1, k/2+1, k/2+2, k/2+2 \ldots, 3k/4, 3k/4(=c-1)$ in the second block.
3. The second peg is received from the first peg by switching the role of the two blocks.
4. The last color $k = 3k/4 + 1 = c$ is not used in the questions at all.
5. Finally, the secret has to be asked as final question Q_{k+1}.

Strategy 3 ($\lceil (4c-1)/3 \rceil$-strategy for $p = 2$ and arbitrary $c \equiv 2 \bmod 3$, $c \neq 2$).

1. Divide the k main questions into two blocks of questions, the first $k/2$ questions and the second $k/2$ questions.
2. The first peg contains the colors $1, 2, \ldots, k/2$ in the first block and the colors $k/2+1, k/2+1, k/2+1, k/2+2, k/2+2 \ldots, 3(k-2)/4+1, 3(k-2)/4+1(=c-1)$ in the second block (note that the first number $k/2+1$ occurs three times, not only twice).
3. The second peg is received from the first peg by switching the role of the two blocks.
4. The last color $k = 3(k-2)/4 + 2 = c$ is not used in the questions at all.
5. Finally, the secret has to be asked as final question Q_{k+1}.

As examples, the main questions of Strategy 1 for $p = 2$ and $c = 9$ with $k = 11$ questions, Strategy 2 for $p = 2$ and $c = 10$ with $k = 12$ questions, and Strategy 3 for $p = 2$ and $c = 11$ with $k = 14$ questions are listed in Table 1a–c, respectively.

Table 1. Examples for Strategies 1, 2 and 3 with $p = 2$.

Peg	1	2
Q_1	1	7
Q_2	2	7
Q_3	3	7
Q_4	4	8
Q_5	5	8
Q_6	6	1
Q_7	6	2
Q_8	7	3
Q_9	7	4
Q_{10}	8	5
Q_{11}	8	6

(a) c=9.

Peg	1	2
Q_1	1	7
Q_2	2	7
Q_3	3	8
Q_4	4	8
Q_5	5	9
Q_6	6	9
Q_7	7	1
Q_8	7	2
Q_9	8	3
Q_{10}	8	4
Q_{11}	9	5
Q_{12}	9	6

(b) c=10.

Peg	1	2
Q_1	1	8
Q_2	2	8
Q_3	3	8
Q_4	4	9
Q_5	5	9
Q_6	6	10
Q_7	7	10
Q_8	8	1
Q_9	8	2
Q_{10}	8	3
Q_{11}	9	4
Q_{12}	9	5
Q_{13}	10	6
Q_{14}	10	7

(c) c=11.

Remark 2. Strategy 3 for the trivial case $p = 2$ and $c = 2$ containing 2 questions (the main question $(1, 1)$ and the final question) is not feasible, as the possible secrets $(1, 2)$ and $(2, 1)$ are indistinguishable. By Propostion 1, there is also no other feasible strategy with only two questions. However, the strategy with the two main questions $(1, 1)$ and $(1, 2)$ is feasible.

For proving the following result we need to define the notion of *neighboring main questions*.

Definition 1. *For a given strategy for two pegs, two main questions are called* neighboring main questions, *if they have one color in common in exactly one peg. We say that they* overlap *in this peg.*

E.g., in Table 1a the main question 1, which is $(1, 7)$, and the main question 3, which is $(3, 7)$, are neighboring main questions as well as the main question 4, which is $(4, 8)$, and the main question 5, which is $(5, 8)$.

Observation 2. *For the three Strategies 1, 2 and 3 it holds:*

1. *The last color c does not occur in any main question, but all other colors occur in both pegs of the main questions.*
2. *There are exactly two blocks of questions.*
3. *Each main question contains one color which occurs only once in the corresponding peg of the main questions, and another color which occurs twice or three times in the corresponding peg.*
   E.g., consider the main question $(6, 2)$ in Table 1a. Then the first color 6 occurs twice, namely in the main questions $(6, 1)$ and $(6, 2)$, whereas the second color 2 occurs only once, namely in the main question $(6, 2)$.
4. *Each main question has one or two neighboring main questions.*
   E.g., the main question $(6, 2)$ has one neighboring main question, which is $(6, 1)$.
5. *The case that one color occurs more than twice, namely three times, appears only in the second peg of Strategy 1 and in the first and second peg of Strategy 3.*

4   Feasibility of Two Pegs Strategy

Theorem 1. *Strategies 1, 2 and 3 are feasible $\lceil (4c - 1)/3 \rceil$-strategies for $p = 2$ and for the corresponding $c \geq 3$.*

Proof. Consider the Strategies 1, 2 and 3. We use Observation 2 and illustrate the argumentation by examples from Table 1a. We have to show that each possible secret (a, b) can be uniquely determined by the combination of answers. We distinguish the following cases:

1. One answer is 2B.
   Clearly, the possible secret is uniquely determined by this answer.
2. All answers are 0B.
   By Observation 2.1, the color c occurs twice in the possible secret, i.e., the only possible secret is (c, c).
   E.g., in Table 1a, this possible secret is $(9, 9)$.
3. One answer is 1B and the other answers are 0B.
   Let the main question corresponding to the answer 1B be (a, b).
   If a occurs only once in the first peg of the main questions, then by Observation 2.3, b occurs at least twice in the second peg of the main questions. As there is only one answer 1B, a is correct in the first peg, and for the second peg only the color c remains. I.e., the only possible secret is (a, c).
   E.g., for the possible secret $(4, 9)$ the main question $(4, 8)$ receives the answer 1B, and all other answers are 0B.
   If a occurs at least twice in the first peg, analogously the only possible secret is (c, b).
   E.g., for the possible secret $(9, 3)$ the main question $(7, 3)$ receives the answer 1B, and all other answers are 0B.

4. Two answers are 1B and the two corresponding main questions are not neighboring main questions. The other answers are 0B.

Let the two main questions corresponding to the answers 1B be (a, b) and (d, e)[1], where $b \neq e$ and w.l.o.g., $a < d$. Then a occurs only once in the first peg of the main questions and e occurs only once in the second peg of the main questions, while by Observation 2.3, b occurs at least twice in the second peg of the the main questions and d at least twice in the first peg of the main questions. As there are only two answers 1B, a is correct in the first peg, and e in the second peg. I.e., the only possible secret is (a, e).

E.g., for the possible secret $(3, 2)$ the main questions $(3, 7)$ and $(6, 2)$ receive the answer 1B, and all other answers are 0B.

5. Two answers are 1B and the two corresponding main questions are neighboring main questions. The other answers are 0B.

W.l.o.g., the two main questions overlap in the second peg. Let the two main questions corresponding to the answers 1B be (a, b) and (d, b). By Observation 2.3, a and d occur only once in the first peg of the main questions, and b is correct in the second peg. For the first peg only the color c remains. I.e., the only possible secret is (c, b).

E.g., for the possible secret $(9, 8)$ the main questions $(4, 8)$ and $(5, 8)$ receive the answer 1B, and all other answers are 0B.

6. Three answers are 1B and the three corresponding main questions are pairwise neighboring main questions. The other answers are 0B.

W.l.o.g., the three pairwise neighboring main questions overlap in the second peg. Let the three main questions corresponding to the answers 1B be (a, b), (d, b) and (e, b). Analogously to case 5, the only possible secret is (c, b).

E.g., for the possible secret $(9, 7)$ the main questions $(1, 7)$, $(2, 7)$ and $(3, 7)$ receive the answer 1B, and all other answers are 0B.

7. Three answers are 1B and the three corresponding main questions are not pairwise neighboring main questions. The other answers are 0B.

Then two of the three main questions are neighboring main questions. W.l.o.g., they overlap in the second peg. Let the three main questions corresponding to the answers 1B be (a, b), (d, b) and (e, f). Analogously to case 5, b is correct in the second peg. As also the main question (e, f) receives the answer 1B, the only possible secret is (e, b).

E.g., for the possible secret $(3, 8)$ the main questions $(3, 7)$, $(4, 8)$ and $(5, 8)$ receive the answer 1B, and all other answers are 0B.

8. Four answers are 1B and two pairs of corresponding main questions are neighboring main questions. The other answers are 0B.

One pair of neighboring main questions overlaps in the first peg, and the other pair in the second peg. Let the four main questions corresponding to the answers 1B be (a, b), (a, d), (e, f) and (g, f). By Observation 2.3, a is correct in the first peg and f in the second peg. Thus, the only possible secret is (a, f).

---

[1] Note that the parameter c is reserved for the number of colors. So we use the parameters $a, b, d, e, \ldots$ here.

E.g., for the possible secret $(6, 8)$ the main questions $(4, 8)$, $(5, 8)$, $(6, 1)$ and $(6, 2)$ receive the answer 1B, and all other answers are 0B.

9. Four answers are 1B and three corresponding main questions are pairwise neighboring main questions. The other answers are 0B.

W.l.o.g., let the three pairwise neighboring main questions overlap in the second peg. Let the four main questions corresponding to the answers 1B be (a, b), (d, b), (e, b) and (f, g). Analogously to case 5, b is correct in the second peg. As also the main question (f, g) receives the answer 1B, the only possible secret is (f, b).

E.g., for the possible secret $(4, 7)$ the main questions $(1, 7)$, $(2, 7)$, $(3, 7)$ and $(4, 8)$ receive the answer 1B and all other answers are 0B.

10. Five answers are 1B, and the other answers are 0B.

Then three main questions are pairwise neighboring main questions. W.l.o.g, they overlap in the second peg. Let the five main questions corresponding to the answers 1B be (a, b) (d, b), (e, b), (f, g) and (f, h). Analogously to case 5, b is correct in the second peg and f in the first peg. Thus, the only possible secret is (f, b).

E.g., for the possible secret $(6, 7)$ the main questions $(1, 7)$, $(2, 7)$, $(3, 7)$, $(6, 1)$ and $(6, 2)$ receive the answer 1B, and all other answers are 0B.

11. Six answers are 1B, and the other answers are 0B.

This case can only hold for Strategy 3. Then there are three main questions which are pairwise neighboring main questions and three further main questions which are also pairwise neighboring main questions. Let the six main questions corresponding to the answers 1B be (a, b) (d, b), (e, b), (f, g), (f, h) and (f, i). Analogously to case 5, b is correct in the second peg and f in the first peg. Thus, the only possible secret is (f, b).

E.g., for the possible secret $(8, 8)$ (here in Table 1c!) the main questions $(1, 8)$, $(2, 8)$, $(3, 8)$, $(8, 1)$, $(8, 2)$ and $(8, 3)$ receive the answer 1B, and all other answers are 0B.   □

5   Optimality of Two Pegs Strategy

Theorem 2. *Strategies 1, 2 and 3 are optimal for $p = 2$ and for the corresponding $c \geq 3$, i.e., $sb(2, c) = \lceil (4c - 1)/3 \rceil$ for arbitrary c.*

For proving this main theorem we need one definition and two lemmas.

Definition 2. *Let a strategy for Static Black-Peg Mastermind with two pegs be given. A main question of this strategy is called $(1, 1)$-question, if the first color of the question occurs only once in the first peg of all main questions and the second color of the question occurs only once in the second peg of all main questions.*

Lemma 1. *For Static Black-Peg Mastermind with two pegs it holds for each feasible strategy that there exists at most one color which does not occur in the first peg of all main questions of the strategy, and at most one color which does not occur in the second peg of all main questions.*

Proof. Assume that for a feasible strategy there are two colors a and b which do not occur in the first peg of the main questions. Consider an arbitrary color x. Then the possible secrets (a, x) and (b, x) receive the same combination of answers in the main questions. Thus, they are indistinguishable, and the secret cannot be found after the main questions in all cases. So we have a contradiction, and the strategy is not feasible. Thus, there exists at most one color which does not occur in the first peg of the main questions.

Analogously, it can be shown that there exists at most one color which does not occur in the second peg of the main questions.   □

Lemma 2. *For Static Black-Peg Mastermind with two pegs it holds:*

(a) *Each feasible strategy contains at most one $(1, 1)$-question in the main questions.*
(b) *If a feasible strategy contains exactly one $(1, 1)$-question in the main questions, then in at least one of the two pegs of the main questions, all c colors must occur.*

Proof. **(a)** Assume that a feasible strategy contains two different $(1, 1)$-questions in the main questions, say (a, b) and (d, e), where $a \neq d$ and $b \neq e$. Then the possible secrets (a, e) and (d, b) receive the answer 1B for each of the questions (a, b) and (d, e), and 0B for all other questions of the main questions, i.e., the same combination of answers. Thus, these possible secrets (a, e) and (d, b) are indistinguishable, and the secret cannot be found after the main questions in all cases. So we have a contradiction, and the strategy is not feasible. Therefore, the assertion follows.

(b) Let a feasible strategy contain exactly one $(1, 1)$-question in the main questions, say (a, b). Assume that there is a color d which does not occur in the first peg of the main questions and a color e which does not occur in the second peg of the main questions. Then the possible secrets (a, e) and (d, b) receive the answer 1B for the question (a, b), and 0B for all other questions of the main questions, i.e., the same combination of answers. Thus, these possible secrets (a, e) and (d, b) are indistinguishable, and the secret cannot be found after the main questions in all cases. So we have a contradiction, and the strategy is not feasible. Therefore, the assertion follows.   □

Proof of Theorem 2. The case $c = 1$ is trivial, as $\lceil (4 \cdot 1 - 1)/3 \rceil = 1$. The case $c = 2$ has been considered in Remark 2 and is also clear, as $\lceil (4 \cdot 2 - 1)/3 \rceil = 3$. Thus, in the following let $c > 2$. Assume that there are better strategies than the Strategies 1, 2 and 3.

By Lemma 1, in each of the two pegs we have at least $c - 1$ colors. To find a better strategy, there should be as many colors as possible which occur only once in the corresponding peg. However, by Lemma 2(a) there is a restriction that at most one $(1, 1)$-question may occur in the main questions. Clearly, all other of the $c - 1$ or c colors should occur twice, and if some main questions remain, also three or more times. We distinguish the following cases:

1. $c \equiv 0 \mod 3$:

   Assume now that there is a better feasible strategy than Strategy 1. It is sufficient to assume that there is a strategy which is better by one question, i.e., which needs $\frac{4}{3} \cdot c - 2 = \lceil (4c - 1)/3 \rceil - 2$ main questions. Furthermore, assume that at most $\frac{2}{3} \cdot c - 1$ colors occur only once in one peg. Then at least $c - 1 - (\frac{2}{3} \cdot c - 1)$ colors occur twice or more times. Then the cardinality of main questions is at least

   $$\frac{2}{3} \cdot c - 1 + 2 \cdot \left(c - 1 - \left(\frac{2}{3} \cdot c - 1 \right) \right) = \frac{4}{3} \cdot c - 1$$

   I.e., there is one main question too much. Thus, at least $\frac{2}{3} \cdot c$ colors occur only once. Because of

   $$2 \cdot \frac{2}{3} \cdot c = \frac{4}{3} \cdot c - 2 + 2$$

   there are two $(1, 1)$-questions in the main questions. By Lemma 2(a), the received strategy is not feasible. Thus, there is no better feasible strategy than Strategy 1.

2. $c \equiv 1 \mod 3$:

   Assume now that there is a better feasible strategy than Strategy 2. It is sufficient to assume that there is a strategy which is better by one question, i.e., which needs $\frac{4}{3} \cdot c - \frac{2}{3} = \lceil (4c - 1)/3 \rceil - 2$ main questions. Furthermore, assume that at most $\frac{2}{3} \cdot c - \frac{2}{3}$ colors occur only once in one peg. Then at least $c - 1 - (\frac{2}{3} \cdot c - \frac{2}{3})$ colors occur twice or more times. Then the cardinality of main questions is at least

   $$\frac{2}{3} \cdot c - \frac{2}{3} + 2 \cdot \left(c - 1 - \left(\frac{2}{3} \cdot c - \frac{2}{3} \right) \right) = \frac{4}{3} \cdot c - \frac{4}{3}$$

   I.e., there is one main question too much. Thus, at least $\frac{2}{3} \cdot c + \frac{1}{3}$ colors occur only once. Because of

   $$2 \cdot \left(\frac{2}{3} \cdot c + \frac{1}{3} \right) = \frac{4}{3} \cdot c - \frac{7}{3} + 3$$

   there are three $(1, 1)$-questions in the main questions. By Lemma 2(a), the received strategy is not feasible. Thus, there is no better feasible strategy than Strategy 2.

3. $c \equiv 2 \mod 3$:

   Assume now that there is a better feasible strategy than Strategy 3. It is sufficient to assume that there is a strategy which is better by one question, i.e., which needs $\frac{4}{3} \cdot c - \frac{5}{3} = \lceil (4c - 1)/3 \rceil - 2$ main questions. We consider two sub-cases:

(i) One color is not used in any main question.

Assume that at most $\frac{2}{3} \cdot c - \frac{4}{3}$ colors occur only once in one peg. Then at least $c - 1 - \left(\frac{2}{3} \cdot c - \frac{4}{3}\right)$ colors occur twice or more times. Then the cardinality of main questions is at least

$$\frac{2}{3} \cdot c - \frac{4}{3} + 2 \cdot \left(c - 1 - \left(\frac{2}{3} \cdot c - \frac{4}{3}\right)\right) = \frac{4}{3} \cdot c - \frac{2}{3}$$

I.e., there is one main question too much. Thus, at least $\frac{2}{3} \cdot c - \frac{1}{3}$ colors occur only once. Because of

$$2 \cdot \left(\frac{2}{3} \cdot c - \frac{1}{3}\right) = \frac{4}{3} \cdot c - \frac{5}{3} + 1$$

there is one $(1,1)$-question in the main questions. By assumption and by Lemma 2(b), the received strategy is not feasible.

(ii) In one peg of the main questions all colors are used.

W.l.o.g., let this peg be the first peg. Assume that at most $\frac{2}{3} \cdot c - \frac{1}{3}$ colors occur only once in the first peg. Then at least $c - \left(\frac{2}{3} \cdot c - \frac{1}{3}\right)$ colors occur twice or more times in the first peg. Then the cardinality of main questions is at least

$$\frac{2}{3} \cdot c - \frac{1}{3} + 2 \cdot \left(c - \left(\frac{2}{3} \cdot c - \frac{1}{3}\right)\right) = \frac{4}{3} \cdot c + \frac{1}{3}$$

I.e., there are two main questions too much. Thus, at least $\frac{2}{3} \cdot c + \frac{2}{3}$ colors occur only once in the first peg.

Analogously to (i) it can be shown that at least $\frac{2}{3} \cdot c - \frac{1}{3}$ colors occur only once in the second peg.

Because of

$$\left(\frac{2}{3} \cdot c + \frac{2}{3}\right) + \left(\frac{2}{3} \cdot c - \frac{1}{3}\right) = \frac{4}{3} \cdot c - \frac{5}{3} + 2$$

there are two $(1,1)$-questions in the main questions. By Lemma 2(a), the received strategy is not feasible. Thus, there is no better feasible strategy than Strategy 3. □

6   Summary of Results and Future Work

As main result of this work we have introduced a $\lceil(4c-1)/3\rceil$-strategy for Static Black-Peg Mastermind for the case of $p = 2$ pegs and arbitrarily many colors $c \geq 3$ and proved its feasibility and optimality.

By doing additional tests with our computer program we received optimal strategies for the following 9 pairs (p, c): $(3, 2)$, $(3, 3)$, $(3, 3)$, $(4, 2)$, $(4, 3)$, $(5, 2)$, $(5, 3)$, $(6, 2)$ and $(7, 2)$, and give upper bounds for 11 other pairs (p, c).

Table 2. Summary of results for values $sb(p,c)$.

c \ p	1	2	3	4	5	6	7	8	9
1	1	1	1	1	1	1	1	1	1
2	2	3	4	5	5	6	7	≤ 8	≤ 9
3	3	4	5	6	6	≤ 7	≤ 8	≤ 9	≤ 10
4	4	5	7						
5	5	7	≤ 8						
6	6	8	≤ 10						
7	7	9	≤ 12						
8	8	11	≤ 13						
9	9	12	≤ 15						

We summarize our results in Table 2, where we give for the relevant pairs of values (p, c) either exact values $sb(p, c)$ or upper bounds.

Our suggestion for future work is to find optimal strategies for more pairs (p, c), where p is a small constant (other than $p = 2$) and c is arbitrary, or c is a small constant and p is arbitrary.

Another research direction would be to create a more effective program for finding optimal strategies, which is able to find optimal strategies for significantly more pairs (p, c).

References

1. Asuncion, A.U., Goodrich, M.T.: Nonadaptive Mastermind algorithms for string and vector databases, with case studies. IEEE Trans. Knowl. Data Eng. **25**(1), 131–144 (2013)
2. Chen, S.T., Lin, S.S.: Optimal algorithms for $2 \times n$ AB games - a graph-partition approach. J. Inf. Sci. Eng. **20**(1), 105–126 (2004)
3. Chen, S.T., Lin, S.S., Huang, L.T.: Optimal algorithms for $2 \times n$ Mastermind games - a graph-partition approach. Comput. J. **47**(5), 602–611 (2004)
4. Chvátal, V.: Mastermind. Combinatorica **3**, 325–329 (1983)
5. Doerr, B., Spöhel, R., Thomas, H., Winzen, C.: Playing Mastermind with many colors. In: Proceedings of 24th Annual ACM-SIAM Symposium on Discrete Algorithms (SODA 2013), SIAM, pp. 695–704 (2013)
6. Doerr, B., Winzen, C.: Playing Mastermind with constant-size memory. Theory Comput. Syst. **55**(4), 658–684 (2014)
7. Focardi, R., Luccio, F.L.: Guessing bank PINs by winning a Mastermind game. Theory Comput. Syst. **50**(1), 52–71 (2012)
8. Gagneur, J., Elze, M.C., Tresch, A.: Selective phenotyping, entropy reduction and the Mastermind game. BMC Bioinform. (BMCBI) **12**, 406 (2011)
9. Goddard, W.: Static Mastermind. J. Combin. Math. Combin. Comput. **47**, 225–236 (2003)
10. Goddard, W.: Mastermind revisited. J. Combin. Math. Combin. Comput. **51**, 215–220 (2004)

11. Goodrich, M.T.: On the algorithmic complexity of the Mastermind game with black-peg results. Inf. Process. Lett. **109**(13), 675–678 (2009)
12. Jäger, G., Peczarski, M.: The number of pessimistic guesses in Generalized Mastermind. Inf. Process. Lett. **109**(12), 635–641 (2009)
13. Jäger, G., Peczarski, M.: The number of pessimistic guesses in Generalized black-peg Mastermind. Inf. Process. Lett. **111**(19), 933–940 (2011)
14. Jäger, G., Peczarski, M.: Playing several variants of Mastermind with constant-size memory is not harder than with unbounded memory. In: Kratochvíl, J., Miller, M., Froncek, D. (eds.) IWOCA 2014. LNCS, vol. 8986, pp. 188–199. Springer, Heidelberg (2015). doi:10.1007/978-3-319-19315-1_17
15. Jäger, G., Peczarski, M.: The worst case number of questions in generalized AB game with and without white-peg answers. Discret. Appl. Math. **184**, 20–31 (2015)
16. Jäger, G., Peczarski, M.: Bounding memory for Mastermind might not make it harder. Theoret. Comput. Sci. **596**, 55–66 (2015)
17. Knuth, D.E.: The computer as Mastermind. J. Recr. Math. **9**, 1–6 (1976)
18. Koyama, K., Lai, T.W.: An optimal Mastermind strategy. J. Recr. Math. **25**(4), 251–256 (1993)
19. Stuckman, J., Zhang, G.Q.: Mastermind is NP-complete. INFOCOMP J. Comput. Sci. **5**, 25–28 (2006)
20. Viglietta, G.: Hardness of Mastermind. In: Kranakis, E., Krizanc, D., Luccio, F. (eds.) FUN 2012. LNCS, vol. 7288, pp. 368–378. Springer, Heidelberg (2012)
21. Source code of the computer program of this article. http://snovit.math.umu.se/~gerold/source_code_static_mastermind.tar.gz

Miscellaneous

The Incentive Ratio in Exchange Economies

Ido Polak[(✉)]

Nanyang Technological University, Singapore, Singapore
S120059@e.ntu.edu.sg

Abstract. The incentive ratio measures the utility gains from strategic behaviour. Without any restrictions on the setup, ratios for linear, Leontief and Cobb–Douglas exchange markets are unbounded, showing that manipulating the equilibrium is a worthwhile endeavour, even if it is computationally challenging. Such unbounded improvements can be achieved even if agents only misreport their utility functions. This provides a sharp contrast with previous results from Fisher markets. When the Cobb–Douglas setup is more restrictive, the maximum utility gain is bounded by the number of commodities. By means of an example, we show that it is possible to exceed a known upper bound for Fisher markets in exchange economies.

Keywords: Incentive ratio · Competitive equilibrium · Equilibrium manipulation · Utility function · Exchange economy

1 Introduction

General equilibrium theory and (noncooperative) game theory are among the most succesful and well-studied areas in economic theory. The former seeks to explain the existence of equilibria in multiple markets at the same time. The latter serves as the primary tool for predicting, analysing and describing the behaviour of rational agents' actions both in and out of equilibrium.

Here we try to combine the two approaches for exchange economies. Specifically, we ask how much any individual agent can gain from strategically misreporting his utility function. The results presented here suggest that, contrary to previous findings in Fisher markets, the gains from strategic behaviour may be significant, even allowing an agent to improve his equilibrium utility without bound. If we impose (common) restrictions, the utility gain in Cobb–Douglas markets is bounded by the number of commodities, but it may exceed the upper bound from Fisher markets, which can be shown by means of an example. The results obtained show a sharp contrast with the findings in the Fisher market setup [6,7]: there, incentive ratios are bounded by the small constants 2, 2 and $e^{1/e} \approx 1.44$ for linear, Leontief and Cobb–Douglas markets, respectively.

1.1 Related Work

In a Fisher market [3] agents possess an amount of money rather than a bundle of commodities as in exchange economies. The idea that an agent may act

© Springer International Publishing AG 2016
T.-H.H. Chan et al. (Eds.): COCOA 2016, LNCS 10043, pp. 685–692, 2016.
DOI: 10.1007/978-3-319-48749-6_49

strategically in such a market by misreporting his utility function in order to get a better equilibrium bundle, compared to the scenario where everyone is truthful, was already considered in [1] for the case of linear utility functions. The incentive ratio was first coined in [7], where the strategic variable of interest is the (Leontief) utility function of a player and bidding the true budget is a dominant strategy. In [5,6], a slightly more sophisticated version of the incentive ratio is presented, in which players may also strategise on their endowments. The "exchange market game" is introduced in [11], and agents have linear utility functions. They may lie about their utility function to manipulate the outcome of the exchange process. It is consequently shown that a symmetric strategy profile is a Nash equilibrium if and only if it is conflict-free. In [4], price of anarchy bounds are computed for linear, Leontief and Cobb–Douglas markets in Fisher markets. The strategic variable of interest is the utility function. It primarily differs from the analysis presented here in that it focuses on welfare of all agents rather than measuring the benefits of strategic behaviour for one specific agent. It is known that the Walrasian mechanism is susceptible to manipulation via endowments: via withholding endowments and recovering it fully [13], recovering part of it [15,16] and even destroying part of one's initial endowment [2]. However, the aforementioned studies only show that manipulation of the equilibrium is possible, but do not quantify it.

The rest of this paper is organised as follows. Section 2 discusses the necessary machinery, definitions and introduces some notation. Section 3 presents the results for incentive ratios in Linear, Leontief and Cobb–Douglas exchange economies. The latter receives most attention. Finally, Sect. 4 concludes and provides some directions for future research.

2   Preliminaries

We use the following notation. Suppose $x, y \in \mathbb{R}^n$. Then $x \cdot y = \sum_{k=1}^{n} x_k y_k$ denotes the dot product of x and y. $x \leq y$ means $x_k \leq y_k$ for $k = 1, \ldots, n$. For a vector $u = (u_1, \ldots, u_n)$, by u_{-i} we mean the vector $(u_1, \ldots, u_{i-1}, u_{i+1}, \ldots, u_n)$ (i.e. all entries except the i-th). We write (u_i, u_{-i}) for $(u_1, \ldots, u_{i-1}, u_i, u_{i+1}, \ldots, u_n)$. For positive integer n, we use $[n]$ as shorthand notation for the set $\{1, \ldots, n\}$. I_m is the $m \times m$ identity matrix. The transpose of a matrix M is denoted by M^T, its determinant by $|M|$ and its adjugate by $Adj(M)$. If $f : \mathbb{R}^m \to \mathbb{R}^n$, then $Df(x)$ represents the Jacobian matrix of f at x.

We start with the definition of an exchange economy.

Definition 1 (Exchange Economy). *An (exchange) economy is a tuple $\xi = ((u_i)_{i=1}^n, (e_i)_{i=1}^n)$, where $u_i : \mathbb{R}_+^m \to \mathbb{R}$ is the utility function of agent $i \in [n]$ and $e_i \in \mathbb{R}_+^m$ is a vector where e_{ij} indicates how much agent $i \in [n]$ possesses of commodity $j \in [m]$.*

In an economy, agents obtain a bundle $x_i \in \mathbb{R}_+^m$ by trading commodities given a price vector $p \in \mathbb{R}^m$. If p is such a price vector, then every agent solves the following consumer problem ($\mathcal{CP}$).

Definition 2 (Demand).

$$\begin{aligned}
maximize \quad & u_i(x_i) \\
subject\ to \quad & p \cdot x_i \leq p \cdot e_i \\
& x_i \geq 0
\end{aligned} \qquad (\mathcal{CP})$$

We call the set of solutions to problem ($\mathcal{CP}$) the demand of agent i (at prices p).

We can write $x_i(p, p \cdot e_i)$ to show explicitly that demand depends on endowments and prices. Since prices are in turn determined by endowments and utility functions, we may also write $x_i(u_i, u_{-i}, e)$ or, when it is understood that u_{-i} and e are fixed, simply as $x_i(u_i)$. We also need the notion of Walrasian or competitive equilibrium.

Definition 3 (Competitive/Walrasian Equilibrium). *A competitive equilibrium is a pair $(p, x) \in \mathbb{R}^m \times (\mathbb{R}_+^m)^n$ such that:*

1. *For all $j \in [m]$, $\sum_{i=1}^{n} x_{ij} = \sum_{i=1}^{n} e_{ij}$ i.e. markets clear.*
2. *For all $i \in [n]$, x_i is a solution to ($\mathcal{CP}$), i.e. x_i is the best bundle among the bundles that he can afford.*

2.1   Incentive Ratio

Every agent is characterized by two parameters, his endowment e_i and his utility function u_i. Generally, different endowments and different utility functions will lead to different equilibria. What if an agent purposely misreports his utility function, thereby trying to get a better equilibrium allocation?

The incentive ratio is a concept introduced in [7]. It attempts to measure the (maximum) benefits of manipulating the equilibrium by strategically misreporting personal parameters. Formally, we define it as follows (adapted for exchange economies, the original definition was for Fisher markets, see also [5–7]):

Definition 4 (Incentive Ratio). *The incentive ratio of agent i in a market M (e.g. linear, Cobb–Douglas or Leontief), denoted ζ_i^M, is defined as:*

$$\zeta_i^M = \max_{u_{-i} \in U_{-i}, e_{-i} \in (\mathbb{R}_+^m)^{n-1}} \max_{u_i' \in U_i} \frac{\max_{x' \in \mathcal{E}(u_i')} u_i(x_i'(u_i', u_{-i}, e))}{\min_{x \in \mathcal{E}(u_i)} u_i(x_i(u_i, u_{-i}, e))} .$$

The incentive ratio of market M is subsequently defined as $\zeta^M = \max_{i \in [n]} \zeta_i^M$.

Remark 1. In this definition:

– Variables with a prime ($'$) refer to the scenario in which agent i misreports his parameters (and all other agents report truthfully). That is, he reports u_i' and as a result, obtains a bundle $x_i'(u_i', u_{-i}, e)$. Notice that this bundle is evaluated by the true utility function.

- Given that player i reports $\tilde{u}_i$ (i.e. truthful or not) as his utility function (the other players report u_{-i}), we denote by $\mathcal{E}(\tilde{u}_i)$ the set of equilibrium allocations, that is, $\mathcal{E}(\tilde{u}_i) = \{x \in (\mathbb{R}_+^m)^n | \exists p \in \mathbb{R}_+^m \ (p, x)$ is a Walras equilibrium$\}$. Under some (mild) assumptions this set is nonempty, but it could contain multiple equilibrium allocations.
- U_i contains the admissible strategies/utility functions for player i, including the one that agent he chooses when he misreports his utility function. We denote $U_{-i} = \prod_{k \neq i} U_k$. We will only consider the case where all U_k's are equal, thus $U_{-i} = U^{n-1}$.
- From the preceding arguments, we may restrict attention to agent i, thus we may rewrite the incentive ratio for the market M as

$$\zeta^M = \max_{u_{-i} \in U^{n-1}, e_{-i} \in (\mathbb{R}_+^m)^{n-1}} \max_{u'_i \in U} \frac{\max_{x' \in \mathcal{E}(u'_i)} u_i(x'_i(u'_i, u_{-i}, e))}{\min_{x \in \mathcal{E}(u_i)} u_i(x_i(u_i, u_{-i}, e))}.$$

We will, without loss of generality, restrict ourselves to scenarios where $e_i \in [0, 1]^m$ for all $i \in [n]$ and $\sum_{i=1}^n e_{ij} = 1$ for all $j \in [m]$. The following Leontief example shows that the consequences from the nonuniqueness of equilibrium can be significant.

Example 1. $e_1 = (1 - \epsilon, \epsilon)$, $e_2 = (\epsilon, 1 - \epsilon)$, $u_2(x_2) = \min\{x_{21}, x_{22}\}$, $\epsilon > 0$, small

Truthful	**Nontruthful**
$u_1(x_1) = \min\{x_{11}, x_{12}\}$	$u'_1(x'_1) = \min\{x'_{11}, x'_{12}\}$
then	then
$p = (\delta, 1)$	$p' = (1, 1)$
$x_1 = ((\epsilon + \delta - \delta\epsilon)/(1 + \delta), (\epsilon + \delta - \delta\epsilon)/(1 + \delta))$	$x'_1 = (1/2, 1/2)$
$x_2 = (1 - \epsilon + \delta\epsilon)/(1 + \delta), (1 - \epsilon + \delta\epsilon)/(1 + \delta))$	$x'_2 = (1/2, 1/2)$

We have $u_1(x'_1)/u_1(x_1) = (1 + \delta)/(2(\epsilon + \delta - \delta\epsilon))$. Letting δ, ϵ tend to 0, the incentive ratio tends to ∞.

In [6,7], the following markets are considered, with arbitrarily many agents and commodities: Linear, i.e. $U = \{u(x) = \alpha \cdot x \mid \alpha \in \mathbb{R}_+^m\}$; Leontief, i.e. $U = \{u(x) = \min_{j \in [m]}\{x_j/\alpha_j\} \mid \alpha \in \mathbb{R}_{++}^m\}$; Cobb–Douglas, i.e. $U = \{u(x) = \prod_{j=1}^m x_j^{\alpha_j} \mid 0 \leq \alpha_j \leq 1$ for all $j \in [m], \sum_{j=1}^m \alpha_j = 1\}$. The tight bounds for Fisher markets are 2, 2 and $e^{1/e}$ for linear, Leontief and Cobb–Douglas respectively.

3 Results

The incentive ratio is a first step to quantifying the possible gains of misreporting in exchange economies. From Example 1, even destroying part of one's initial endowment (in the case e_i is the strategic variable of interest), can let the incentive ratio tend to infinity.

Without any further restrictions on the setup presented in [6,7], also incentive ratios in linear and Cobb–Douglas exchange economies are unbounded. We treat here linear markets; see [12] for details on the Cobb–Douglas case.

Proposition 1. *The incentive ratio for linear exchange economies equals* $+\infty$.

Proof. $e_1 = (\epsilon, 1 - \epsilon)$, $e_2 = (1 - \epsilon, \epsilon)$, $u_2(x_2) = x_{21} + \frac{\delta}{1-\epsilon} x_{22}$, $\delta, \epsilon > 0$, small

Truthful	Nontruthful
$u_1(x_1) = x_{11}$	$u_1'(x_1') = x_{11}' + \dfrac{\delta}{1 - \epsilon} x_{12}'$
$p = (1, \delta/(1 - \epsilon))$	$p' = (1, \delta/(1 - \epsilon))$
$x_1 = (\delta + \epsilon, 0), x_2 = (1 - \delta - \epsilon, 1)$	$x_1' = (1, 0), x_2' = (0, 1)$

We have $u_1(x_1')/u_1(x_1) = 1/(\delta + \epsilon)$. Letting both δ, ϵ tend to 0, the incentive ratio tends to ∞.   □

Henceforth we focus on Cobb–Douglas markets and make the following assumption to ensure all equilibrium prices are positive; this is rather standard in algorithmic game theory.[1]

Assumption 1

- (Positivity of endowments). Every agent possesses a strictly positive amount of every commodity: $\forall i \in [n], \forall j \in [m]\ \ e_{ij} > 0$.
- (Strong competitiveness (see e.g. [4])). Every commodity is demanded by at least one agent: $\forall j \in [m]\ \exists i \in [n]\ \alpha_{ij} > 0$ and this remains true when agent i reports α_i'.

This entails that the demand of agent $i \in [n]$ for commodity $j \in [m]$ is given by $x_{ij}(p, p \cdot e_i) = \alpha_{ij} p \cdot e_i / p_j$ and the economy excess demand function $z(p) := \sum_{i=1}^{n}(x_i(p, p \cdot e_i) - e_i) = \sum_{i=1}^{n} x_i(p, p \cdot e_i) - 1$ has the gross substitute property, which implies that the equilibrium price is unique (see e.g. [10]). For markets with two commodities we have (for a proof see [12]), as in Fisher markets [6]:

Proposition 2. *Consider a Cobb–Douglas market, $n \geq 2$, $m = 2$. The incentive ratio is $e^{1/e}$ and this bound is tight.*

The following example shows we can exceed the $e^{1/e}$ bound.

Example 2 (Incentive ratio $> e^{1/e}$). Suppose the market is as follows:

$$
\begin{cases}
e_1 = (.99, .01, .01), e_2 = (.01, .99, .99) \\
u_1(x_1) = x_{11}^{.2} x_{12}^{.3} x_{13}^{.5} \\
u_2(x_2) = x_{21}^{.4} x_{12}^{.6} \\
u_1'(x_1') = x_{11}'^{.85} x_{12}'^{.1} x_{13}'^{.05}
\end{cases}
\text{then}
\begin{cases}
p = (.398, .597, .201) \\
x_1 \approx (.202, .202, 1) \\
p' = (.4045, .1344, .0201) \\
x_1' \approx (.845, .299, 1),
\end{cases}
$$

Therefore the incentive ratio is $u_1(x_1')/u_1(x_1) \approx 1.50$.

The remainder of this section is devoted to the proof that the incentive ratio for Cobb–Douglas markets is bounded. The following lemma is crucial. We only provide a sketch the proof here due to space constraints; see [12] for complete proofs.

---

[1] Alternatively, we could assume the existence of a nonmanipulating agent who possesses at least a little bit of all commodities and who desires every commodity.

Lemma 1

(i) $p_j(\alpha_i)$ *reaches its maximum when* $\alpha_{ij} = 1$ *and* $\alpha_{ik} = 0$ *for all* $k \neq j$. *I.e. for any chosen normalisation,* p_j *is maximal when* $\alpha_{ij} = 1$.

(ii) $\alpha'_{ij}/p'_j(\alpha'_i)$ *reaches its maximum when* $\alpha'_{ij} = 1$ *and* $\alpha'_{ik} = 0$ *for all* $k \neq j$. *I.e. for any chosen normalisation,* $\alpha'_{ij}/p'_j(\alpha'_i)$ *is maximal when* $\alpha'_{ij} = 1$.

(iii) *Let* $A = (\alpha_{ji})_{1 \leq i \leq n, 1 \leq j \leq m}$ *and similarly* A' *when agent i reports* α'_i, $E = (e_{ji})_{1 \leq i \leq n, 1 \leq j \leq m}$. *Then the first row of* $Adj(EA^T - I_m)$ $(Adj(E(A')^T - I_m))$ *contains the equilibrium price vector* p (p') *(upto a nonzero constant). Moreover,* $p \cdot e_i = p' \cdot e_i$.

Proof (sketch). For the first two points, w.l.o.g. we focus on commodity m.

(i) This follows from the implicit function theorem applied to $[D_{\hat{p}}\hat{z}(p(\alpha); \alpha)]^{-1}$, a matrix in which all entries are negative, where $p(\alpha)$ is the equilibrium price when players report strategies according to $\alpha = (\alpha_1, \ldots, \alpha_n)$, $\hat{p}$ and $\hat{z}$ are vectors with the first $m - 1$ entries from p and $z(p(\alpha))$.

(ii) The extreme value theorem assures a maximum is attained. Using necessary conditions for a maximum (see [9]) and homogeneity of degree 0 we get the proof.

(iii) This uses an argument along the lines of [14] and the fact that an equilibrium price vector p satisfies (see [8]) $p^T(EA^T - I_m) = 0$. The budget of agent i, $p \cdot e_i$, can be written as a matrix determinant that does not change following the increase of α_{ij} to α'_{ij} and a decrease of equal magnitude of α_{ik} to α'_{ik}, $1 \leq j \neq k \leq m$. □

Theorem 1. *The incentive ratio for Cobb–Douglas markets is at most* m.

Proof

$$\frac{u_i(x'_i)}{u_i(x_i)} = \prod_{j=1}^{m} \left(\frac{\alpha'_{ij}}{\alpha_{ij}} \frac{p' \cdot e_i}{p \cdot e_i} \frac{p_j}{p'_j} \right)^{\alpha_{ij}} \leq \prod_{j=1}^{m} \left(\frac{1}{\alpha_{ij}} \frac{p' \cdot e_i}{p \cdot e_i} \max_{\alpha'_i} \frac{\alpha'_{ij}}{p'_j} \max_{\alpha_i} p_j \right)^{\alpha_{ij}}$$

$$\leq \prod_{j=1}^{m} \left(\frac{1}{\alpha_{ij}} \right)^{\alpha_{ij}} \leq m,$$

where the second inequality follows from the lemma above and the last inequality from the weighted AM–GM inequality.   □

We summarize the results presented here in Table 1.

Table 1. Upper bounds on the incentive ratio in $n \times m$ exchange economies.

Market	Fisher [6,7]	Exchange
Leontief	2	∞
Linear	2	∞
Cobb–Douglas (Without Assumption 1, $n = 2$, $m = 2$)	$e^{1/e}$	∞
Cobb–Douglas (With Assumption 1, any n, $m = 2$)	$e^{1/e}$	$e^{1/e}$
Cobb–Douglas (With Assumption 1, any n, $m \geq 3$)	$e^{1/e}$	m

4   Conclusion

Results for incentive ratios in Fisher markets were encouraging: the maximum gains from strategic behaviour were bounded by small constants and therefore equilibrium mechanisms could be expected to work rather well, meaning that the profits from (computationally challenging) strategic behaviour were small relative to the costs, and thus, not worthwhile on most occassions. However, the results here indicate that in the more general setup of exchange economies, results are diametrically different and without further restrictions, all ratios are unbounded. The Cobb–Douglas case demonstrates that, when equilibrium prices (and hence allocations and utilities) are unique, the incentive ratio is bounded by the number of commodities. Therefore it may be argued that, unlike in Fisher markets, gains from strategic behaviour can be significant and manipulation could be worthwhile.

For Cobb–Douglas markets, the case $m = 2$ demonstrates that the bound m is unlikely to be tight. This, and the question of the incentive ratio when agents are allowed to misreport their endowments in linear and Cobb–Douglas markets, are left as directions for future research. Allowing groups of agents to misreport their utility function and extending the results from Cobb–Douglas markets to, for example, any other market satisfying weak gross substitution, are other interesting possibilites.

Acknowledgments. Research funding through the Nanyang Technological University PhD research scholarship is gratefully acknowledged. I would like to thank Xiaohui Bei and Satoru Takahashi for discussions on the topic and many useful suggestions and comments on earlier versions of this manuscript. I also thank anonymous referees for helpful remarks. All remaining errors are of course my own.

References

1. Adsul, B., Babu, C.S., Garg, J., Mehta, R., Sohoni, M.: Nash equilibria in fisher market. In: Kontogiannis, S., Koutsoupias, E., Spirakis, P.G. (eds.) SAGT 2010, pp. 30–41. Springer, Heidelberg (2010)
2. Aumann, R., Peleg, B.: A note on Gale's example. J. Math. Econ. **1**(2), 209–211 (1974)
3. Brainard, W.C., Scarf, H.: How to compute equilibrium prices in 1891. In: Cowles Foundation Discussion Papers 1272, Cowles Foundation forResearch in Economics, Yale University (2000)
4. Brânzei, S., Chen, Y., Deng, X., Filos-Ratsikas, A., Frederiksen, S., Zhang, J.: The fisher market game: equilibrium and welfare. In: Twenty-Eighth AAAI Conference on Artificial Intelligence (2014)
5. Chen, N., Deng, X., Tang, B., Zhang, H.: Incentives for strategic behavior in fisher market games. In: AAAI Conference on Artificial Intelligence (2016)
6. Chen, N., Deng, X., Zhang, H., Zhang, J.: Incentive ratios of fisher markets. In: Czumaj, A., Mehlhorn, K., Pitts, A., Wattenhofer, R. (eds.) Automata, Languages, and Programming. LNCS, pp. 464–475. Springer, Heidelberg (2012)

7. Chen, N., Deng, X., Zhang, J.: How profitable are strategic behaviors in a market? In: Demetrescu, C., Halldórsson, M.M. (eds.) Algorithms- ESA 2011. LNCS, pp. 106–118. Springer, Heidelberg (2011)
8. Curtis Eaves, B.: Finite solution of pure trade markets with Cobb-Douglas utilities. In: Manne, A.S. (ed.) Economic Equilibrium: Model Formulationand Solution, Part II. Solution Methods. Mathematical Programming Studies, vol. 23, pp. 226–239. Springer, Heidelberg (1985)
9. de la Fuente, A.: Mathematical Methods and Models for Economists. Cambridge University Press, Cambridge (2000). Cambridge Books Online
10. Mas-Colell, A., Whinston, M., Green, J.: Microeconomic Theory. Oxford University Press, Oxford (1995)
11. Mehta, R., Sohoni, M.: Exchange markets: strategy meets supply-awareness. In: Chen, Y., Immorlica, N. (eds.) WINE 2013. LNCS, vol. 8289, pp. 361–362. Springer, Heidelberg (2013)
12. Polak, I.: The Incentive Ratio in Exchange Economies. arXiv:1609.02423 [cs.GT] (2016)
13. Postlewaite, A.: Manipulation via endowments. Rev. Econ. Stud. **46**(2), 255–262 (1979)
14. Press, W.H., Dyson, F.J.: Iterated prisoners dilemma contains strategies that dominate any evolutionary opponent. Proc. Natl. Acad. Sci. **109**(26), 10409–10413 (2012)
15. Thomson, W.: Monotonic Allocation Mechanisms. Working Paper No. 116, University of Rochester (1987)
16. Yi, G.: Manipulation via withholding: a generalization. Rev. Econ. Stud. **58**(4), 817–820 (1991)

w-Centroids and Least (w, l)-Central Subtrees in Weighted Trees

Erfang Shan[1] and Liying Kang[2(✉)]

[1] School of Management, Shanghai University, Shanghai 200444, P.R. China
efshan@i.shu.edu.cn
[2] Department of Mathematics, Shanghai University, Shanghai 200444, P.R. China
lykang@shu.edu.cn

Abstract. Let T be a weighted tree with a positive number $w(v)$ associated with each of vertices and a positive number $l(e)$ associated with each of its edges. In this paper we show that each least (w, l)-central subtree of a weighted tree either contains a vertex of the w-centroid or is adjacent to a vertex of the w-centroid. Also, we show that any two least (w, l)-central subtrees of a weighted tree either have a nonempty intersection or are adjacent.

Keywords: Tree · w-centroid · (w, l)-central subtree · Least (w, l)-central subtree

1 Introduction

The "central part" of a graph has many important applications in the location of industrial plants, warehouses, distribution centers, and public service facilities in transportation networks, as well as the location of various facilities in telecommunication networks. Much research has been devoted to the topic on central part of a tree [1,3,4].

The central subtrees, and least central subtrees in trees were introduced in [4]. Nieminen and Peltola [4] described the general properties of a least central subtree of a tree, and gave some connections between the least central subtree and the center/centroid of a tree. Moreover, they proved that the intersection of two least central subtrees is nonempty. Hamina and Peltola [2] further studied the relationship between the least central subtrees and the center and the centroid of a tree, and proved that every least central subtree of a tree contains the center and at least one vertex of the centroid of the tree.

Let T be a weighted tree with a positive number $w(v)$ associated with each of vertices and a positive number $l(e)$ associated with each of its edges. The w-centroid for weighted trees was defined in [3]. For a subtree S of T, we define the *weight* of S by $w(S) = \sum_{v \in V(S)} w(v)$. For $v \in V(T)$, denote by $T - v$ the graph obtained from T by removing v, and let $T_{v,1}, T_{v,2}, \ldots, T_{v,d(v)}$ be all the

---

Research was partially supported by NSFC (grant numbers 11571222, 11471210).

T.-H.H. Chan et al. (Eds.): COCOA 2016, LNCS 10043, pp. 693–701, 2016.
DOI: 10.1007/978-3-319-48749-6_50

subtrees of $T - v$, where $d(v)$ is the degree of v in T. We define the *branch weight* of the vertex v to be $B(v) = \max_{1 \le i \le d(v)}\{w(T_{v,i})\}$. The *$w$-centroid $C_R(T)$* of a tree T is the set of all vertices v for which the largest component of $T - v$ has as small weight as possible, i.e., $B(v) = \min_{u \in V(T)}\{B(u)\}$. In particular, if every vertex of T has weight 1, then the w-centroid is the centroid in trees.

The aim of this paper is to generalize the concept of the least central subtree in trees to weighted trees. We introduce the (w, l)-central subtree of weighted trees, and we give the connection between least (w, l)-central subtrees and the w-centroid of a weighted tree. Furthermore, we find the differences between the two versions of unweighted trees and weighted trees.

The *median graph G_{LW}* for a weighted tree T is defined as a weighted graph with vertex set $V(G_{LW})$ and edge set $E(G_{LW})$, where

$$V(G_{LW}) = \{S \mid S \text{ is a subtree of a weighted tree } T\},$$
$$E(G_{LW}) = \{S_i S_j \mid S_i \subset S_j \text{ and } |S_j| = |S_i| + 1, \text{ or } S_j \subset S_i \text{ and } |S_i| = |S_j| + 1\},$$

and the length (weight) of each edge $S_i S_j$ in G_{LW} is defined as

$$l(S_i S_j) = \begin{cases} w(S_j) - w(S_i) & \text{if } S_i \subset S_j; \\ w(S_i) - w(S_j) & \text{if } S_j \subset S_i. \end{cases}$$

Let $P(S_1, S_2)$ be a path joining vertices S_1 and S_2 in the graph G_{LW}, the *length $l(P(S_1, S_2))$* of path $P(S_1, S_2)$ be the sum of the weights of edges in $P(S_1, S_2)$. The *distance $d_L(S_1, S_2)$* between vertices S_1 and S_2 in G_{LW} is the length of a shortest S_1-S_2 path in G_{LW}. Let $e_L(S) = \max\{d_L(S, S') \mid S' \text{ is a sub-tree of } T\}$ be the *L-eccentricity* of the subtree S. Clearly, $e_L(S)$ is the eccentricity of the vertex S_1 in G_{LW}. A subtree S is a *(w, l)-central subtree* of a weighted tree T if it has the minimum eccentricity $e_L(S)$ in G_{LW}. A (w, l)-central sub-tree with the minimum vertex weight is called a *least (w, l)-central subtree* of a weighted tree T. When $w(v) = l(e) = 1$ for each vertex v and edge e in T, the (w, l)-central subtree and the least (w, l)-central subtree are the ordinary central subtree and least central subtree of T defined in [2,4], respectively.

In this paper we show the connection between least (w, l)-central subtrees and the w-centroid of a weighted tree. Formally, we prove that each least (w, l)-central subtree of a weighted tree either contains a vertex of the w-centroid or is adjacent to a vertex of the w-centroid. Also, we prove that any two least (w, l)-central subtrees of a weighted tree either have a nonempty intersection or are adjacent.

2   Notation and Preliminaries

In this section let us introduce some notation and terminology.

The *vertex set* of a graph G is referred to as $V(G)$ and its *edge set* as $E(G)$. The number of vertices of G is its *order*, written as $|G|$. If $U \subseteq V(G)$, $G[U]$ is the subgraph of G induced by U and we write $G - U$ for $G[V(G) - U]$. In other words, $G - U$ is obtained from G by deleting all the vertices in $U \cap V(G)$ and their incident edges. If $U = \{v\}$ is a singleton, we write $G - v$ rather than $G - \{v\}$.

For two subtrees S_1 and S_2 of a weighted tree T, the *distance* $d_T(S_1, S_2)$ between S_1 and S_2 in T is the length of the shortest weighted path joining two vertices of S_1 and S_2 in T. In particular, if $S_1 = \{x\}$ and $S_2 = \{y\}$, write $d_T(x, y)$ instead of $d_T(\{x\}, \{y\})$. For subtrees S_1 and S_2 of a tree T, the *meet* $S_1 \wedge S_2$ of subtrees S_1 and S_2 is defined as the subtree induced by the intersection of the vertex sets of S_1 and S_2 whenever the intersection is nonempty, and the *join* $S_1 \vee S_2$ is the least subtree of T containing the subtrees S_1 and S_2. In other words, $S_1 \vee S_2$ is the subtree induced by the union of vertices of S_1 and S_2 whenever the intersection of the vertex sets of S_1 and S_2 is nonempty. In the case of non-intersection subtrees, $S_1 \vee S_2$ is the subtree induced by the union of vertices of S_1 and S_2 together with the vertices of the path from S_1 to S_2.

3   Least (w, l)-Central Subtrees and w-Centroids

In this section we describe the connection between (w, l)-least central subtrees and the w-centroid of a weighted tree. For this purpose, we first give the following lemmas.

Lemma 1. *Let S_1 and S_2 be two subtrees of a weighted tree T. If $V(S_1) \subset V(S_2)$, then $d_L(S_1, S_2) = w(S_2) - w(S_1)$.*

Lemma 2. *Let S_1 and S_2 be two subtrees of a weighted tree T. If $V(S_1) \cap V(S_2) \neq \emptyset$, then $d_L(S_1, S_2) = w(S_1) + w(S_2) - 2w(S_1 \wedge S_2)$.*

If $V(S_1) \cap V(S_2) = \emptyset$, then S_1 and S_2 is connected in T by a path from S_1 to S_2. Let us denote $P_T(S_1, S_2)$ the path in T from S_1 to S_2, and $P_{in}(S_1, S_2)$ the path induced by the internal vertices of path $P_T(S_1, S_2)$.

Lemma 3. *Let S_1 and S_2 be two subtrees of a weighted tree T. If $V(S_1) \cap V(S_2) = \emptyset$, then $d_L(S_1, S_2) = w(S_1) + w(S_2) + 2w(P_{in}(S_1, S_2))$.*

Proof. Let $P_{in}(S_1, S_2) = v_1 v_2 \ldots v_k$ and let $P(S_1, S_2)$ be a shortest S_1-S_2 path in G_{LW}. To obtain S_2 from S_1, each vertex in S_2 needs to be added to S_1, each vertex in S_1 needs to be removed from S_1, and each vertex in $P_{in}(S_1, S_2)$ needs to be added to S_1 and then removed. For each vertex $v \in V(S_2)$, if S_{i_j} is the first subtree containing v among all subtrees in $P(S_1, S_2)$, i.e., $v \notin V(S_{i_{j-1}}), v \in V(S_{i_j})$, then $l(S_{i_{j-1}} S_{i_j}) = w(v)$. For each vertex $v \in V(S_1)$, if S_{i_k} is the last subtree containing v among all subtrees in $P(S_1, S_2)$, i.e., $v \in V(S_{i_k}), v \notin V(S_{i_{k+1}})$, then $l(S_{i_k} S_{i_{k+1}}) = w(v)$. For each vertex v in $P_{in}(S_1, S_2)$, if S_{i_l} is the first subtree containing v among all subtrees in $P(S_1, S_2)$, i.e., $v \in V(S_{i_l}), v \notin V(S_{i_{l-1}})$, then $l(S_{i_{l-1}} S_{i_l}) = w(v)$. Assume that S_{i_m} is the last subtree containing v among all subtrees in $P(S_1, S_2)$, i.e., $v \in V(S_{i_m}), v \notin V(S_{i_{m+1}})$, then $l(S_{i_m} S_{i_{m+1}}) = w(v)$. So $l(P(S_1, S_2)) \geq w(S_1) + w(S_2) + 2w(P_{in}(S_1, S_2))$. Hence,

$$d_L(S_1, S_2) = l(P(S_1, S_2)) \geq w(S_1) + w(S_2) + 2w(P_{in}(S_1, S_2)). \tag{1}$$

On the other hand, among all S_1-S_2 paths that pass through the vertex $S_1 \vee S_2$ in G_{LW}, let $P^*(S_1, S_2)$ be a shortest one. Then $l(P^*(S_1, S_2)) = d_L(S_1, S_1 \vee S_2) + d_L(S_2, S_1 \vee S_2)$. By Lemma 1, $d_L(S_1, S_1 \vee S_2) = w(S_1 \vee S_2) - w(S_1) = w(S_2) + w(P_{in}(S_1, S_2))$ and $d_L(S_2, S_1 \vee S_2) = w(S_1 \vee S_2) - w(S_2) = w(S_1) + w(P_{in}(S_1, S_2))$. So $d_L(S_1, S_2) \leq l(P^*(S_1, S_2)) = w(S_1) + w(S_2) + 2w(P_{in}(S_1, S_2))$. Combining this with (1), we obtain $d_L(S_1, S_2) = w(S_1) + w(S_2) + 2w(P_{in}(S_1, S_2))$. $\qquad\square$

By the above lemmas and using the similar approach in [2], we obtain the following key properties of least (w, l)-central subtrees.

Lemma 4. *Let C_L be a least (w, l)-central subtree of a weighted tree T and v be a vertex adjacent to C_L, let C_v be the subtree of T induced by $V(C_L) \cup \{v\}$. If subtree S_v^* satisfies $e_L(C_v) = d_L(C_v, S_v^*)$, then $S_v^* \neq T$ and*

(1) if $V(C_L) \cap V(S_v^*) \neq \emptyset$, then $v \notin V(S_v^*)$,
(2) if $V(C_L) \cap V(S_v^*) = \emptyset$, then v does not lie in the path joining C_L and S_v^* in T.

Proof. First by Lemma 1, we observe that $d_L(C_v, T) = w(T) - w(C_L) - w(v) < w(T) - w(C_L) \leq e_L(C_L)$. So $S_v^* \neq T$. We distinguish two cases depending on whether or not $V(C_L)$ and $V(S_v^*)$ meet.

Case 1. $V(C_L) \cap V(S_v^*) \neq \emptyset$. If $v \in V(S_v^*)$, then, by Lemma 2, we have

$$
\begin{aligned}
e_L(C_v) &= d_L(C_v, S_v^*) \\
&= w(C_v) + w(S_v^*) - 2w(C_v \wedge S_v^*) \\
&= w(C_L) + w(v) + w(S_v^*) - 2(w(C_L \wedge S_v^*) + w(v)) \\
&= d_L(C_L, S_v^*) - w(v) \\
&\leq e_L(C_L) - w(v) < e_L(C_L),
\end{aligned}
$$

which contradicts the fact that C_L is the least (w, l)-central subtree of T. Thus $v \notin V(S_v^*)$.

Case 2. $V(C_L) \cap V(S_v^*) = \emptyset$. If $v \in V(S_v^*)$, then $w(P_{in}(C_L, S_v^*)) = 0$. By Lemmas 2 and 3, we have

$$
\begin{aligned}
e_L(C_v) &= d_L(C_v, S_v^*) \\
&= w(C_v) + w(S_v^*) - 2w(C_v \wedge S_v^*) \\
&= w(C_L) + w(v) + w(S_v^*) - 2w(v) \\
&= w(C_L) + w(S_v^*) + w(P_{in}(C_L, S_v^*)) - w(v) \\
&= d_L(C_L, S_v^*) - w(v) \\
&\leq e_L(C_L) - w(v) < e_L(C_L).
\end{aligned}
$$

This contradiction implies that $v \notin V(S_v^*)$. If $v \notin V(S_v^*)$, and v lies in the C_L-S_v^* path in T. Hence, $w(P_{in}(C_L, S_v^*)) = w(P_{in}(C_v, S_v^*)) + w(v)$. By Lemma 3,

$$
\begin{aligned}
e_L(C_v) &= d_L(C_v, S_v^*) \\
&= w(C_v) + w(S_v^*) + 2w(P_{in}(C_v, S_v^*)) \\
&= w(C_L) + w(v) + w(S_v^*) + 2(w(P_{in}(C_L, S_v^*)) - w(v)) \\
&= w(C_L) + w(S_v^*) + 2w(P_{in}(C_L, S_v^*)) - w(v) \\
&= d_L(C_L, S_v^*) - w(v) \\
&\leq e_L(C_L) - w(v) < e_L(C_L),
\end{aligned}
$$

a contradiction. Thus v does not lie in the C_L-S_v^* path in T. $\qquad\square$

Let S_L be the component of $T - v$ containing C_L, and let $S_v = T \setminus V(S_L)$.

Lemma 5. Let C_v, S_v^* be defined as in Lemma 4. Then $d_L(C_v, S_v^* \vee S_v) = d_L(C_v, S_v^*) + w(S_v) - w(P_{in}(S_v^*, S_v)) - 2w(v)$.

Proof. By Lemma 4, $V(S_v^*) \subseteq V(S_L)$. So there exists a vertex $x \in V(S_v^*)$ such that $d_T(v, x) = d_T(v, S_v^*) = d_T(S_v, S_v^*)$. We consider the following two cases depending on whether C_L meets S_v^* in T.

Case 1. $V(C_L) \cap V(S_v^*) \neq \emptyset$. By Lemma 4, $v \notin V(S_v^*)$, and so $x \in V(C_L)$. For the subtree $S_v^* \vee S_v$, we have

$$
\begin{aligned}
w(S_v^* \vee S_v) &= w(S_v^*) + w(S_v) + w(P_{in}(S_v^*, S_v)), \\
w((S_v^* \vee S_v) \wedge C_v)) &= w(S_v^* \wedge C_v) + w(P_{in}(S_v^*, S_v)) + w(v).
\end{aligned}
$$

Then, by Lemma 2, we have

$$
\begin{aligned}
d_L(C_v, &S_v^* \vee S_v) \\
&= w(C_v) + w(S_v^* \vee S_v) - 2w((S_v^* \vee S_v) \wedge C_v)) \\
&= w(C_v) + w(S_v^*) + w(S_v) + w(P_{in}(S_v^*, S_v)) \\
&\quad - 2(w(S_v^* \wedge C_v) + w(P_{in}(S_v^*, S_v)) + w(v)) \\
&= d_L(C_v, S_v^*) + w(S_v) - w(P_{in}(S_v^*, S_v)) - 2w(v).
\end{aligned}
$$

Case 2. $V(C_L) \cap V(S_v^*) = \emptyset$. By Lemma 4, v does not lie in the C_L-S_v^* path in tree T. Then $x \notin V(C_L)$ and so $x \neq v$. Hence,

$$
\begin{aligned}
w(S_v^* \vee S_v) &= w(S_v^*) + w(S_v) + w(P_{in}(S_v^*, S_v)) \\
&= w(S_v^*) + w(S_v) + w(P_{in}(v, x)).
\end{aligned}
$$

Let $v' \in V(C_v)$ be the vertex satisfying $d_T(v, v') = d_T(S_v, S_v^*) - d_T(C_v, S_v^*)$. Note that $(S_v^* \vee S_v) \wedge C_v$ is v'-v path $P_T(v', v)$ in T. So, by Lemma 3, we have

$$
w((S_v^* \vee S_v) \wedge C_v) = w(P_T(v', v)) = w(P_{in}(x, v)) - w(P_{in}(x, v')) + w(v).
$$

Hence, by Lemmas 2–3, we have

$$
\begin{aligned}
d_L(C_v, S_v^* \vee S_v) &= w(C_v) + w(S_v^* \vee S_v) - 2w((S_v^* \vee S_v) \wedge C_v)) \\
&= w(C_v) + w(S_v^*) + w(S_v) + w(P_{in}(S_v^*, S_v)) - 2w(P_T(v', v)) \\
&= w(C_v) + w(S_v^*) + w(S_v) + w(P_{in}(S_v^*, S_v)) \\
&\quad - 2(w(P_{in}(x, v)) - w(P_{in}(x, v'))) + w(v)) \\
&= w(C_v) + w(S_v^*) + 2w(P_{in}(x, v')) + w(S_v) + w(P_{in}(S_v^*, S_v)) \\
&\quad - 2w(P_{in}(x, v)) - 2w(v) \\
&= d_L(C_v, S_v^*) + w(S_v) - w(P_{in}(S_v^*, S_v)) - 2w(v),
\end{aligned}
$$

as desired.   □

Lemma 6. *Let* C_v, S_v^* *be defined as in Lemma 4. Then* $w(S_v) \leq w(P_{in}(S_v^*, S_v)) + 2w(v)$.

Proof. By Lemma 5, we obtain

$$
\begin{aligned}
e_L(C_v) &\geq d_L(C_v, S_v^* \vee S_v) \\
&= d_L(C_v, S_v^*) + w(S_v) - w(P_{in}(S_v^*, S_v)) - 2w(v) \\
&= e_L(C_v) + w(S_v) - w(P_{in}(S_v^*, S_v)) - 2w(v).
\end{aligned}
$$

Therefore, $w(S_v) \leq w(P_{in}(S_v^*, S_v)) + 2w(v)$.   □

The following lemmas provide essential properties of the w-centroid in a weighted tree.

Lemma 7 [3]. *For a vertex* v *of a weighted tree* T, $v \in C_R(T)$ *if and only if* $B(v) \leq w(T)/2$.

Lemma 8 [1,3]. *For a weighted tree* T, *the* w-centroid $C_R(T)$ *in* T *consists of either one vertex or two adjacent vertices. In the latter case, we have* $B(v) = w(T)/2$ *for each* $v \in C_R(T)$.

By the above lemmas, we are now ready to prove the main results in this paper.

Theorem 1. *For a weighted tree* T, *if* w-centroid $C_R(T)$ *consists of two adjacent vertices, then any least* (w, l)-central subtree of T contains at least a vertex of $C_R(T)$.

Proof. Suppose to the contrary that there does exist a least (w, l)-central subtree C_L of T such that $V(C_L) \cap C_R(T) = \emptyset$. Let $C_R(T) = \{u, u_1\}$. By Lemma 8, u_1 and u are adjacent in T. For notational simplicity, we identify $C_R(T)$ with the subtree induced by $C_R(T)$ in T. Let $u_1 \in C_R(T)$ be the vertex such that $d_T(u_1, C_L) = d_T(C_R(T), C_L)$. Let v be the neighbor of C_L such that v lies in the path from u to C_L. Then the u-C_L path passes through u_1, and so $v \neq u$. Let S_L, S_R be the subtrees of $T - v, T - u$ containing C_L respectively, and let

$S_u = T \setminus V(S_R)$. Clearly, $V(S_v^*) \subseteq V(S_L)$ by Lemma 4 (S_v^* is defined as in Lemma 4). As noted already in Lemma 5, there exists a vertex $x \in V(S_v^*)$ such that $d_T(v, x) = d_T(v, S_v^*) = d_T(S_v, S_v^*)$ (S_v is defined as in Lemma 5).

Note that $V(S_L) \cup \{v\} \subseteq V(S_R)$. By Lemma 7, $w(S_R) \leq w(T)/2$. Then

$$w(S_L \cup \{v\}) \leq w(S_R) \leq \frac{w(T)}{2}. \tag{2}$$

Hence,

$$w(S_v) = w(T \setminus (S_L \cup \{v\})) + w(v) \geq \frac{w(T)}{2} + w(v). \tag{3}$$

By Lemma 6, we have

$$\begin{aligned} w(S_v) &\leq w(P_{in}(S_v^*, S_v)) + 2w(v) \\ &\leq w(S_L \cup \{v\}) - w(v) - w(x) + 2w(v) \\ &= w(S_L \cup \{v\}) - w(x) + w(v). \end{aligned}$$

Then

$$\begin{aligned} w(S_L \cup \{v\}) &\geq w(S_v) - w(v) + w(x) \\ &\geq \frac{w(T)}{2} + w(x) \qquad \text{(by (3))} \\ &> \frac{w(T)}{2}. \end{aligned}$$

This is a contradiction to (2). □

Theorem 2. *For a weighted tree T, if the w-centroid $C_R(T)$ consists of a single vertex, then any least (w, l)-central subtree of T either contains $C_R(T)$ or is adjacent to $C_R(T)$.*

Proof. Let $C_R(T) = \{u\}$. It is sufficient to prove that for any least (w, l)-central subtree C_L of T, if it is not adjacent to $C_R(T)$ then it contains the vertex in $C_R(T)$. Suppose to the contrary that there does exist a least (w, l)-central subtree C_L of T such that $V(C_L) \cap C_R(T) = \emptyset$ and C_L is not adjacent to $C_R(T)$. Let v be the neighbor of C_L such that v lies on the path from C_L to $C_R(T)$. Since C_L is not adjacent to $C_R(T)$, $u \neq v$. A similar argument, as described in Theorem 1, yields a proof of this result. □

Theorem 3. *Any two least (w, l)-central subtrees of a weighted tree T either have a nonempty intersection or are adjacent.*

Proof. For the case when the w-centroid $C_R(T)$ of T is not a single vertex, we know that the set $C_R(T)$ consists of two adjacent vertices, so the result follows directly from Theorem 1. We next may assume that the set $C_R(T)$ consists of a single vertex, say v. Let C_{L_1} and C_{L_2} be two least (w, l)-central subtrees of T.

If either C_{L_1} or C_{L_2} contains the centroid v, then C_{L_1} and C_{L_2} are adjacent by Theorem 2. So we may assume that neither C_{L_1} nor C_{L_2} contain v.

The following proof is by contradiction. Suppose that $V(C_{L_1}) \cap V(C_{L_2}) = \emptyset$. By Theorem 2, both C_{L_1} and C_{L_2} are adjacent to v. Let $C_v = C_{L_1} \cup \{v\}$ and S_v^* be the subtree such that $e_L(C_v) = d_L(C_v, S_v^*)$. By Lemma 4, $S_v^* \neq T$, and

$$d_L(C_v, S_v^*) \geq e_L(C_{L_1}). \qquad (4)$$

Suppose that $V(S_v^*) \cap V(C_v) \neq \emptyset$. By Lemma 4, $v \notin V(S_v^*)$. Then $V(S_v^*) \cap V(C_{L_2}) = \emptyset$. Since $V(C_{L_1}) \cap V(C_{L_2}) = \emptyset$ and v is adjacent to both C_{L_1} and C_{L_2}, the path from C_{L_2} to S_v^* in T goes through the centroid v. Then $w(P_{in}(C_{L_2}, S_v^*)) \geq w(v)$, so, by Lemmas 2–3,

$$
\begin{aligned}
e_L(C_{L_2}) &\geq d_L(C_{L_2}, S_v^*) \\
&= w(C_{L_2}) + w(S_v^*) + 2w(P_{in}(C_{L_2}, S_v^*)) \\
&\geq w(C_{L_2}) + w(S_v^*) + 2w(v) \\
&\geq w(C_{L_1}) + w(v) + w(S_v^*) - 2w(C_v \wedge S_v^*) + w(v) \\
&= w(C_v) + w(S_v^*) - 2w(C_v \wedge S_v^*) + w(v) \\
&= d_L(C_v, S_v^*) + w(v) \\
&> e_L(C_{L_1}). \qquad \text{(by (4))}
\end{aligned}
$$

This contradicts the fact that $e(C_{L_1}) = e(C_{L_2})$.

Suppose that $V(S_v^*) \cap V(C_{L_1}) = \emptyset$. By Lemma 4, v is not on the C_{L_1}-S_v^* path in T. Since $V(C_{L_1}) \cap V(C_{L_2}) = \emptyset$ and v is adjacent to both C_{L_1} and C_{L_2}, the path from C_{L_2} to S_v^* in T goes through the vertex v. Then $w(P_{in}(C_{L_2}, S_v^*)) \geq w(v) + w(P_{in}(C_{L_1}, S_v^*))$, so, by Lemmas 3–4,

$$
\begin{aligned}
e_L(C_{L_2}) &\geq d_L(C_{L_2}, S_v^*) \\
&= w(C_{L_2}) + w(S_v^*) + 2w(P_{in}(C_{L_2}, S_v^*)) \\
&\geq w(C_{L_2}) + w(S_v^*) + 2w(P_{in}(C_{L_1}, S_v^*)) + 2w(v) \\
&= w(C_v) + w(S_v^*) + 2w(P_{in}(C_v, S_v^*)) + w(v) \\
&= d_L(C_v, S_v^*) + w(v) \\
&> e_L(C_{L_1}). \qquad \text{(by (4))}
\end{aligned}
$$

We obtain a contradiction again. Therefore, $V(C_{L_1}) \cap V(C_{L_2}) \neq \emptyset$, the assertion follows. $\qquad \square$

4   Conclusion

In this paper we describe the connection between least (w, l)-central subtrees and the w-centroid of a weighted tree. It is an interesting question to determine the structures of the trees with unique least (w, l)-central subtree. Another question of interest is how to find a least (w, l)-central subtree of a weighted tree?

References

1. Bielak, H., Pańczyk, M.: A self-stabilizing algorithm for finding weighted centroid in trees. Ann. UMCS Inform. AI XII. **2**, 27–37 (2012)
2. Hamina, M., Peltola, M.: Least central subtrees, center, and centroid of a tree. Networks **57**, 328–332 (2011)
3. Kariv, O., Hakimi, S.L.: An algorithm approach to network location problems. II: the p-medians. SIAM J. Appl. Math. **37**, 539–560 (1979)
4. Nieminen, J., Peltola, M.: The subtree center of a tree. Networks **34**, 272–278 (1999)

Solving Dynamic Vehicle Routing Problem with Soft Time Window by iLNS and hPSO

Xiaohan He[1], Xiaoli Zeng[1], Liang Song[1], Hejiao Huang[1,2(✉)],
and Hongwei Du[1,2]

[1] Harbin Institute of Technology Shenzhen Graduate School, Shenzhen, China
hjhuang@hitsz.edu.cn
[2] Shenzhen Key Laboratory of Internet Information Collaboration, Shenzhen, China

Abstract. Vehicle Routing Problem (VRP) is widely studied under the real logistics environments. For the reason that customers' demands could appear dynamically and need to be served within fuzzy time windows, the dynamic vehicle routing problem with soft time windows (DVRPSTW) is studied in this paper. We use the improved large neighborhood search algorithm(iLNS) and the hybrid Particle Swarm Optimization(hPSO) to solve the problem. The performance of both algorithms comparing with benchmarks shows that our methods can solve DVRPSTW efficiently with more customers.

Keywords: DVRPSTW · iLNS · hPSO

1 Introduction

The vehicle routing problem(VRP) has been well studied since it was proposed by Dantzig and Ramser [1]. VRP is a combinatorial optimization problem which is NP hard, and hence cannot be exactly solved in polynomial time. In real logistics environments, customers' demands appear dynamically and need to be served in a fuzzy time windows. As far as we know, the Dynamic Vehicle Routing Problem with Soft Time Window (DVRPSTW) has not been studied before. There are many related works on dynamic VRP [2–6] and VRP with soft time windows [7–12], however, none of them considered the environments in which customers may allow vehicle to come earlier before the releasing time or later after the deadline. So, we use the natural idea of dividing the working day into multiple time slices, in which customers' demands are collected, and then design the improved large neighborhood search(iLNS) and the hybrid particle swarm optimization(hPSO) to generate the optimized routes in each time slice.

The remaining parts of this paper are organized as follows. Section 2 describes DVRPSTW and gives the mathematical formulation of the studied problem. Section 3 proposes the iLNS algorithm and the hPSO algorithm. Finally, computational results are described in Sect. 4.

© Springer International Publishing AG 2016
T.-H.H. Chan et al. (Eds.): COCOA 2016, LNCS 10043, pp. 702–709, 2016.
DOI: 10.1007/978-3-319-48749-6_51

2 Problem Description and Mathematical Formulation

The dynamic vehicle routing problem with soft time window has one depot, K vehicles to serve N customers among which M customers are the those not serviced in yesterday and left to today. The route for each vehicle is denoted as p_i $(1 \leq i \leq K)$. $V = \{v_i | i \in (0, 1, \cdots n)\}$ is the vertex set containing n vertices and one depot. Each vehicle starts from the depot to service customers with the speed F and goes back to depot when capacity Q is met. $E = \{(v_i, v_j) | v_i, v_j \in V, i \neq j\}$ is the set of edges, with the distance d_{ij} $(i, j = 0, 1, \cdots n)$ for each edge. A customer's demand is denoted as a quantity variable C_i $(i = 1, 2, \cdots n)$ and a location variable (x_i, y_i) $(i = 1, 2, \cdots n, j = 1, 2, \cdots n)$. Each vertex is associated with a time window $[a_i, b_i]$ $(i = 0, 1, \cdots, n)$, where a_i is releasing time and b_i is the deadline. The arrival time and the leaving time of each vehicle at vertex i are s_i and l_i, respectively. The service duration for vertex i is st_i. We take punitive measures denoted by a vector $[\alpha, \beta, pv_i]$ where α is punishment coefficient for the arriving earlier than the time window and β is the punishment coefficient for the arriving later than the time window. pv_i is the largest punishment interval so that a customer cannot be served if the vehicle comes out of pv_i, and T is the longest service time in one day.

The mathematical formulation of DVRPSTW can be described as follows:

$$x_{ij}^k = \begin{cases} 1 & \text{if vehicle k drives from customer i to j} \\ 0 & \text{else} \end{cases} \tag{1}$$

$$y_i^k = \begin{cases} 1 & \text{if vehicle k goes througth customer i} \\ 0 & \text{else} \end{cases} \tag{2}$$

All vehicles must satisfy the following constraints:

$$\sum_{i=0}^{n} c_i y_i^k \leq Q \tag{3}$$

$$\sum_{j=0}^{n} \sum_{k=1}^{k} x_{ij}^k = 1 \qquad i \in (1, 2, \cdots, n) \tag{4}$$

$$\sum_{i=0}^{n} \sum_{k=1}^{k} x_{ij}^k = 1 \qquad j \in (1, 2, \cdots, n) \tag{5}$$

$$\sum_{j=1}^{n} x_{0j}^k \leq 1 \qquad k \in (1, 2, \cdots, K) \tag{6}$$

$$\sum_{i=1}^{n} x_{i0}^k \leq 1 \qquad k \in (1, 2, \cdots, K) \tag{7}$$

$$\sum_{i=0}^{n} x_{ir}^k - \sum_{j=0}^{n} x_{rj}^k = 0 \qquad r \in (1, 2, \cdots, n) \, k \in (1, 2, \cdots, K) \tag{8}$$

$$s_i + st_i = l_i \qquad i \in (1, 2, \cdots, n) \tag{9}$$

$$a_i - pv_i \le s_i \le b_i + pv_i \qquad i \in (1, 2, \cdots, n) \tag{10}$$

$$\sum_{i=0}^{n} \sum_{j=0}^{n} \frac{x_{ij}^l \cdot d_{ij}}{F} + \sum_{i=1}^{n} St_i \le K \cdot T \tag{11}$$

The expression (3) is the capacity constraint. For each customer, the expression (4) guarantees that there is only one vehicle leaves out it and expression (5) guarantees that only one vehicle arrives at it. Expressions (6) and (7) together ensure that all the vehicles must start from and end at the depot. The expression (8) ensures that each route is continuous. Expression (9) means that the leaving time of a vehicle is equal to the service duration plus the arriving time. Expression (10) indicates the arriving time should be inside the soft time window. The expression (11) requests all the customers only can be served within the prescribed time, otherwise they will serve in the next day.

The objective function of DVRPSTW is as follows, which minimizes the driving distance, penalty value and the number of abandon customers.

$$\min \left(\alpha \cdot \sum_{i=0}^{n} \sum_{j=0}^{n} \sum_{l=1}^{k} d_{ij} x_{ij}^l + \beta \cdot \sum_{i=l}^{n} \text{penalty}(x) + \delta \cdot \text{count}(\text{Abandon}) \right) \tag{12}$$

3   The iLNS and hPSO Algorithms

The solving framework for DVRPSTW contains two main components, scheduler and solver. Scheduler mainly receives and collects new customers' demands, and then sends them to solver in every time slice. Solver containing iLNS and hPSO will deal with customers' demands and generate the route for each vehicle. Algorithm 1 describs the scheduler in detail.

Algorithm 1. Scheduler Procedure

1: Initialize the position of vehicle and get non-served demands
2: Initialize the time slices and the capacity of vehicle
3: Get part solution by the optimized algorithm
4: Write the solution to database
5: Increase the time slice
6: Update the current position and the remaining capacity of vehicle
7: **while** the time slice does not reach to the end **do**
8:     Get rid of the served vertex from the solution.
9:     Insert the customer vertex in the new time slice
10:     Get part Solution by the optimized algorithm
11:     Write the solution to database
12:     Update the current position and the remaining capacity of vehicle
13:     Increase the time slice
14: **end while**

If a vehicle violates the time window, a penalty value related with customers' satisfaction degree will be given. As shown in expression (13) (EET and ELT are the boundaries of the penalty interval, and $y_1 = Penalty(EET)$), if the vehicle arrives earlier than a_i and later than EET, there is no loss for customer but the costed time increases for the logistics company. When a customer is served between b_i and ELT, the unsatisfied degree of customer grows exponentially. The relationship between penalty value and time window is described as below:

$$Penalty\,(x) = \begin{cases} e^{x-b_i} - 1 & (pv_i = ELT - b_i, b_i \leq x \leq ELT) \\ -\frac{y_1}{pv_i}\,(x - a_i) & (pv_i = a_i - EET, EET \leq x \leq a_i) \end{cases} \tag{13}$$

3.1   Improved Large Neighborhood Search

Large neighborhood search mainly uses the insert-remove heuristic to optimize result. In this paper, we will take an activity scheduling based greedy algorithm shown in Algorithm 3, which is more suitable for VRPSTW. Algorithm 2 describes the iLNS algorithm.

Algorithm 2. improved Large Neighborhood Search(iLNS)

1: Get initial solution $Pi0$ by activity scheduling based greedy
2: Initialize the current solution pool $Pigc = \phi$ and previous solution pool $Pigp = \phi$;
3: **while** stopping criteria is not satisfied **do**
4:      Fill up the current solution pool by the initial solution
5: **end while**
6: **while** <= iteration Li **do**
7:      Transfer the solution into $Pigc$ to $Pigp$ and $Pigc = \phi$.
8:      bestResult = findBestResult($Pigp$);
9:      **while** <= iteration Ni **do**
10:         Random select a solution Pi from $Pigp$;
11:         Remove customer vertices from Pi by removing rules;
12:         Re-insert the removed vertices into Pi:
13:         Check whether the result of Pi is improved. If the count of the unimproved
     of Pi more than Mi, break this loop.
14:         **if** Pi is better than the worst solution in $Pigc$ **then**
15:             insert Pi into $Pigc$
16:         **end if**
17:     **end while**
18: **end while**

3.2   Hybrid Particle Swarm Optimization

Particle swarm optimization [13,14] mainly simulates social behaviors like bird swarm. Every particle in swarm is a candidate solution which searches in the solution space in the way of approaching to the best solution of its own history and the best solution of the whole swarm. The pseudo code of hPSO is shown in Algorithm 4. The structure of particle is crucial to the hPSO. We take a $2 \times D$ dimensional vectors as the structure of particle and D is the number of customer vertices.

Algorithm 3. Activity scheduling based greedy strategy

1: Initialize solution = null
2: **while** The static vertex set is not null **do**:
3:     Sort all the vertices according to the releasing time in ascending order
4:     Scan the array from the first vertex and put the vertex into the route if the subsequent vertices satisfy the time window constraint
5:     Delete the selected vertices from the static vertex set
6:     put the route into solution
7: **end while**

Algorithm 4. Hybrid PSO Optimization Algorithm

1: Initialize the number of swarm and particles
2: Generate the initial solution by Grasp()
3: Calculate the fitness of each particle
4: Get optimal solution of each particle
5: Get optimal solution of the whole particles
6: **while** stopping criteria is not satisfied **do**
7:     Update the velocity and position of each particle by formulation (18) and (19)
8:     Search in adjacent space of each particle by PathRelink();
9:     Update optimal solution of each particle
10:     Update optimal solution of the whole paricle
11:     Update the inertia weight w by formulation (20).
12: **end while**
13: return the best solution

Generating the Initial Solution. A good initial solution for PSO can accelerate convergence velocity, nevertheless many researchers simply generate random initial solution. In this paper, we will use greedy randomized adaptive search procedure (GRASP) which was first proposed by Feo and Resende [15]. This method needs to build a restricted candidate table and select candidate vertices from the table, then insert it into routes by greedy strategy.

Optimizing the Result. This method called path relinking strategy is mainly to optimize the solution further by exploring in adjacent routes. The current solution will be regard as start solution and the optimal solution will be as target solution. And the start solution and target solution may exchange the role of each other. When exploring the space between start solution and target solution, a better solution may be found, then the start position or target solution may change as the result of target solution must be better than the start solution. Due to the solution is numerical arrays, we explore between the starting solution and ending solution by exchanging some vertices. In this way, the inversion number corresponding to the numerical array will change. The difference between two inversion number will be as judgment standard for optimal process.

Table 1. Computational results of the type 1 problem

Instance	hLNS				hPSO						iLNS					
	Td	Ra	Cpu(s)	Nv	Td	Re(%)	Ra	Re(%)	Cpu(s)	Nv	Td	Re(%)	Ra	Re(%)	Cpu(s)	Nv
C101	1995.13	0.38	19.21	10	3004.06	50.57	0.20	-48.28	20.98	10	3078.77	54.31	0.13	-65.08	19.95	10
C102	1424.97	0.28	15.90	9	2092.93	46.88	0.35	25.58	25.24	8	2103.46	47.61	0.17	-40.89	28.70	10
C103	1105.09	0.41	18.72	8	1784.10	61.44	0.46	11.87	30.79	8	1425.08	28.96	0.26	-36.99	29.93	8
C104	1206.91	0.17	21.87	10	1710.74	41.75	0.12	-33.31	27.33	10	1383.20	14.61	0.10	-44.46	21.79	10
C105	1526.53	0.50	13.95	6	2279.10	49.30	0.58	17.36	29.32	8	2341.77	53.40	0.15	-69.65	23.14	10
C106	2681.94	0.19	24.70	12	2365.81	-11.79	0.32	64.14	29.41	8	2923.01	8.99	0.12	-35.27	23.41	11
C107	2369.38	0.15	20.06	13	2521.17	6.41	0.28	89.33	30.28	10	2478.40	4.60	0.11	-23.43	20.38	11
C108	1921.06	0.17	16.47	11	2129.95	10.87	0.33	91.66	24.59	8	2094.16	9.01	0.17	-2.35	33.82	11
C109	1495.13	0.16	19.46	9	2115.29	41.48	0.38	138.82	29.67	9	1839.08	23.00	0.15	-7.65	26.63	10
Average	1747.35	0.27	18.93	9.78	2222.57	27.20	0.33	24.87	27.51	8.78	2185.21	25.06	0.15	-43.74	25.31	10.11
R101	1247.33	0.16	44.00	9	1632.85	30.91	0.12	-24.10	20.01	10	2641.31	111.76	0.04	-74.29	34.24	14
R102	1189.29	0.23	42.13	8	1623.17	36.48	0.14	-36.98	19.70	8	2406.52	102.35	0.07	-71.25	31.63	13
R103	1236.85	0.09	51.63	10	1652.60	33.61	0.09	4.65	24.19	10	2106.59	70.32	0.03	-65.01	37.26	12
R104	846.71	0.74	19.07	4	1725.07	103.74	0.23	-68.42	22.75	9	2396.49	183.03	0.14	-80.85	28.61	13
R105	1355.62	0.38	29.61	8	1216.63	-10.25	0.33	-13.77	22.36	7	2144.21	58.17	0.22	-42.77	30.03	9
R106	798.21	0.45	29.29	6	1787.37	123.92	0.12	-73.44	19.67	10	1764.60	121.07	0.20	-56.41	27.87	8
R107	1045.70	0.17	28.37	8	1566.02	49.76	0.10	-39.36	19.72	9	1356.98	29.77	0.12	-29.49	45.69	8
R108	1199.30	0.43	36.57	7	1536.44	28.11	0.22	-48.31	23.02	9	2680.47	123.50	0.10	-76.84	57.35	12
R109	1356.25	0.21	32.09	9	1249.33	-7.88	0.31	49.70	18.99	6	2226.83	64.19	0.14	-30.56	33.19	9
R110	1215.08	0.21	34.71	8	1521.62	25.23	0.19	-12.89	23.46	7	1888.17	55.40	0.12	-42.24	37.21	8
R111	1064.27	0.15	34.12	8	1590.41	49.44	0.11	-26.95	23.77	9	1499.77	40.92	0.07	-50.92	51.17	9
R112	1593.91	0.06	39.52	11	1797.67	12.78	0.05	-14.83	23.88	10	2514.03	57.73	0.10	60.88	32.26	11
Average	1179.04	0.27	35.09	8.00	1574.93	33.58	0.17	-38.32	21.79	8.67	2135.50	81.12	0.11	-58.65	37.21	10.50
RC101	1381.15	0.53	38.17	7	1625.40	17.68	0.26	-51.61	19.22	10	3526.89	155.36	0.04	-92.15	20.99	15
RC102	1680.18	0.22	46.82	10	2074.40	23.46	0.17	-22.02	21.12	10	2975.89	77.12	0.11	-48.40	57.02	12
RC103	1769.04	0.09	38.95	11	2352.54	32.98	0.05	-47.59	19.90	12	2575.86	45.61	0.04	-50.67	34.47	12
RC104	1504.52	0.12	43.94	10	1994.19	32.55	0.09	-25.87	19.85	10	1986.10	32.01	0.04	-62.08	25.03	11
RC105	1856.86	0.21	53.19	10	1815.28	-2.24	0.24	13.69	19.19	9	2851.83	53.58	0.20	-6.93	31.73	11
RC106	1656.42	0.16	56.33	10	2235.91	34.98	0.07	-57.37	23.65	11	3432.99	107.25	0.05	-67.94	37.61	13
RC107	1594.86	0.15	54.29	10	2013.12	26.23	0.11	-27.00	19.80	10	3035.12	90.31	0.06	-59.96	39.86	12
RC108	1650.39	0.09	69.69	11	2457.16	48.88	0.02	-77.42	21.55	12	3086.30	87.00	0.02	-77.32	41.28	14
Average	1636.68	0.20	50.17	9.88	2071.00	26.54	0.12	-36.21	20.53	10.50	2933.87	79.26	0.07	-63.50	36.00	12.50

4   Computional Results

This paper proposes iLNS and hPSO working on Solomon's benchmark in http://web.cba.neu.edu/~msolomon/problems.htm. In order to evaluate the performance of our algorithms, we compare these two methods with a hybrid large neighborhood search algorithm in [5], and we call it hLNS short for hybrid Large Neighborhood Search. There are two differences between hLNS and iLNS. One is the method to create initial solution. The hLNS uses Clarke-Wright algorithm [5] while our solution takes the activity scheduling based greedy algorithm. Another one is the removing strategy. We abandon the complicated removing strategy about time window and select the simple random removing strategy considering complicated constraints. Table 1 shows the comparison between hPSO, iLNS and hLNS on type 1 benchmark. The hLNS in [5] works on DVRP with hard time windows, and the results of hPSO and iLNS are about DVRPSTW. The dynamic degree is set to 0.5 and penalty interval is set to 20. For description convenience, the column (1) is called C1 and the column (2) is called C2 and so on. In the computational results of parameter, Td denotes travel distance, Ra denotes ratio of abandoned customers, time of CPU and Nv denotes number of vehicles. The Re in C7 is short for relative error of Td which equals to (C6-C2)/C2*100 %, Re in C9 is relative error of Ra in C8 equals to (C8-C3)/C3*100 %, Re in C13 is relative error of Td between iLNS and hLNS equaling to (C12-C2)/C2*100 %, Re in C15 equals to (C14-C3)/C3*100 %. In order to appropriately evaluate the algorithms, we calculate the average of one type instance, like in row 12, the average row is the average of 9 instances of type C1. The Re of Td is positive and the Re of Ra is negative, meaning that the solution of these two methods can serve more customers and travel longer distance for all the type instances. The reason is that the problem of DVRPSTW allows vehicles to come earlier or later which is more relaxed than DVRPTW for vehicles. The iLNS and hPSO need more computing time than hLNS, meanwhile the vehicles of these two methods for serving are more than hLNS, since more customers will be served.

Acknowledgment. This work was financially supported by National Natural Science Foundation of China with Grant No.11371004 and No. 61672195, Shenzhen Science and Technology Plan with Grant No. JCYJ20160318094336513 and No. JCYJ20160318094101317, and Shenzhen Overseas High Level Talent Innovation and Entrepreneurship Special Fund with Grant No. KQCX20150326141251370.

References

1. Ramser, D.G.B.: Greedy randomized adaptive search procedures. J. Glob. Optim. **6**, 109–133 (1995)
2. Larsen, A., Madsen, O., Solomon, M.: Partially dynamic vehicle routing-models and algorithms. J. Oper. Res. Soc. **53**, 637–646 (2002)
3. Hashimoto, H., Ibaraki, T., Imahori, S., Yagiura, M.: The vehicle routing problem with flexible time windows and traveling times. Discrete Appl. Math. **154**, 2271–2290 (2006)

4. Li, J., Mirchandani, P., Borenstein, D.: Real-time vehicle rerouting problems with timewindows. Eur. J. Oper. Res. **194**, 711–727 (2009)
5. Hong, L.: An improved LNS algorithm for real-time vehicle routing problem with time windows. Comput. Oper. Res. **39**, 151–163 (2012)
6. Yang, D., et al.: A hybrid large neighborhood search for dynamic vehicle routing problem with time deadline. In: The 9th Annual International Conference on Combinatorial Optimization and Applications (2015)
7. Calvete, H., Galé, C., Sánchez-Valverde, B., Oliveros, M.: Vehicle routing problems with soft time windows: an optimization based approach. Monografias del Seminario Matematico Garcia de Galdeano **31**, 295–304 (2004)
8. Figliozzi, M.: An iterative route construction and improvement algorithm for the vehicle routing problem with soft time windows. Transp. Res. Part C **18**, 668–679 (2010)
9. Qureshi, A., Taniguchi, E., Yamada, T.: Exact solution for the vehicle routing problem with semi soft time windows and its application. Proc.Soc. Behav. Sci. **2**, 5931–5943 (2010)
10. Fan, X., Li, N., Zhang, B., Liu, Z.: Research on vehicle routing problem with soft time windows based on tabu search algorithm. In: IEEE International Conference on Industrial Engineering and Engineering Management (2011)
11. Xu, J., Yan, F., Li, S.: Vehicle routing optimization with soft time windows in a fuzzy random environment. Transp. Res. Part E **47**, 1075–1091 (2011)
12. Iqbal, S., Rahman, M.: Vehicle routing problems with soft time windows. In: 7th International Conference on Electrical and Computer Engineering (2012)
13. Marinakis, Y., Marinaki, M.: A hybrid genetic-particle swarm optimization algorithm for the vehicle routing problem. Expert Syst. Appl. **37**, 1446–1455 (2010)
14. Khouadjia, M., Sarasola, B., Alba, E., Jourdan, L., Talbi, E.: A comparative study between dynamic adapted PSO and VNS for the vehicle routing problem with dynamic requests. Appl. Soft Comput. **12**, 1426–1439 (2012)
15. Feo, T.: Greedy randomized adaptive search procedures. J. Glob. Optim. **6**, 109–133 (1995)

Convex Independence in Permutation Graphs

Wing-Kai Hon[1], Ton Kloks[1], Fu-Hong Liu[1(✉)], and Hsiang-Hsuan Liu[1,2]

[1] National Tsing Hua University, Hsinchu, Taiwan
{wkhon,fhliu,hhliu}@cs.nthu.edu.tw, klokston@gmail.com
[2] University of Liverpool, Liverpool, UK
hhliu@liverpool.ac.uk

Abstract. A set C of vertices of a graph is P_3-convex if every vertex outside C has at most one neighbor in C. The convex hull $\sigma(A)$ of a set A is the smallest P_3-convex set that contains A. A set M is convexly independent if for every vertex $x \in M$, $x \notin \sigma(M - x)$. We show that the maximal number of vertices that a convexly independent set in a permutation graph can have, can be computed in polynomial time. (Due to space limit, the missing proofs are presented in the full paper. Please see https://drive.google.com/file/d/0B1Ilu0-p1dDsSkpsZFZsR1Y4Uk0/ view or http://arxiv.org/abs/1609.02657).

1 Introduction

Popular models for the spread of disease and of opinion are graph convexities. The P_3-convexity is one such convexity, and it is defined as follows.

Definition 1. *A set S of vertices in a graph G is P_3-convex if every vertex outside S has at most one neighbor in S.*

The P_3-convexity will be the only convexity studied in this paper, so from now on we use the term convex, instead of P_3-convex. For a set A of vertices we let $\sigma(A)$ denote its <u>convex hull</u>, that is, the smallest convex set that contains A.[1]

For a set of points A in $\mathbb{R}^d$ and a point x in its Euclidean convex hull, there exists a set $F \subseteq A$ of at most $d + 1$ points such that $x \in \sigma(F)$, i.e., x is in the Euclidean convex hull of F. This is Carathéodory's theorem. For convexities in graphs one defines the Carathéodory number as the smallest number k such that, for any set A of vertices, and any vertex $x \in \sigma(A)$, there exists a set $F \subseteq A$ with $|F| \leqslant k$ and $x \in \sigma(F)$. For a set S, let

$$\partial(S) = \sigma(S) \setminus \bigcup_{x \in S} \sigma(S - x). \tag{1}$$

A set is irredundant if $\partial(S) \neq \varnothing$. Duchet showed that the Carathéodory number is the maximal cardinality of an irredundant set.

---

[1] In his classic paper, Duchet defines a graph convexity as a collection of 'convex' subsets of a (finite) set V that contains $\varnothing$ and V, and that is closed under intersections, and that, furthermore, has the property that each convex subset induces a connected subgraph. This last condition is, here, omitted.

© Springer International Publishing AG 2016
T.-H.H. Chan et al. (Eds.): COCOA 2016, LNCS 10043, pp. 710–717, 2016.
DOI: 10.1007/978-3-319-48749-6_52

Definition 2. *A set* S *is convexly independent if*

$$\text{for all } x \in S, \quad x \notin \sigma(S - x). \tag{2}$$

Notice that, if a set is convexly independent then so is every subset of it (since σ is a closure operator).

It appears that there is no universal notation for the maximal cardinality of a convexly independent set.[2] In this paper we denote it by $\beta_c(G)$. Every irredundant set is convexly independent, thus the convex-independence number $\beta_c(G)$ is an upperbound for the Carathéodory number. For example, for paths P_n with n vertices, and for cycles C_n with n vertices, we have equality;

$$\beta_c(P_n) = 2 \cdot \left\lfloor \frac{n}{3} \right\rfloor + (n \bmod 3) \quad \text{and} \quad \beta_c(C_n) = \beta_c(P_{n-1}). \tag{3}$$

Other examples, for which the Carathéodory number equals the convex independence number, are <u>leafy trees</u>, which are trees with at most one vertex of degree two. It is easy to <u>check, that</u>

$$\text{T is a leafy tree} \quad \Rightarrow \quad \beta_c(T) = \text{the number of leaves in T.} \tag{4}$$

Examples for which the Carathéodory number is strictly less than the convex-independence number are disconnected graphs. If S is an irredundant set then $\sigma(S)$ is necessarily connected. However, $\beta_c(G)$ is the sum of the convexly independence numbers of G's components. Notice also that $\beta_c(P_6) = 4$, but there exists a maximum convexly independent set S for which $\sigma(S) = S$ and is disconnected, and, thence, redundant.

A set S is a 2-packing if it is an independent set in G^2, that is, no two vertices of S are adjacent or have a common neighbor. Every 2-packing S is convexly independent, as $\sigma(S) = S$. For splitgraphs with minimal degree at least two, a maximal convex-independent set is a 2-packing, unless it has only two vertices. It follows that computing the convexly independence number is NP-complete for splitgraphs (it is Karp's SET PACKING, problem 4). For biconnected chordal graphs (including the splitgraphs mentioned above), every vertex is in the convex hull of any set of two vertices at distance at most two. Thus, the Carathéodory number for those is two.

Ramos et al. show that computing the convexly independence number remains NP-complete for bipartite graphs, and they show that it is polynomial for trees and for threshold graphs.

The intersection graph of a collection of straight line segments, with endpoints on two parallel (horizontal) lines, is called a permutation graph. Dushnik and Miller characterize them as the comparability graphs for which the complement is a comparability graph as well. In this paper we show that the convexly independence number of permutation graphs is computable in polynomial time.

This seems a good time to do a warm-up; let's have a close look at convex independence in cographs (Fig. 1).

---

[2] Ramos et al. call it the 'rank' of the graph, but this word has been used for so many different concepts that it has lost all meaning.

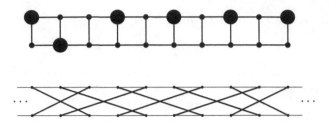

Fig. 1. The heavy dots specify an irredundant set S with $\sigma(S) = V$. The second figure shows the intersection model, i.e., the 'permutation diagram,' for ladders. This example shows that the Carathéodory number for biconnected permutation graphs is unbounded.

2 Convex Independence in Cographs

Definition 3. *A graph is a cograph if it has no induced* P_4, *the path with 4 vertices.*

Cographs are characterized by the property that every induced subgraph is disconnected or else, its complement is disconnected. In other words, cographs allow a complete decomposition by joins and unions. It follows that cographs are permutation graphs, as also this class is closed under joins and unions.

Ramos et al. analyze the convex-independence number for threshold graphs. Threshold graphs are the graphs without induced P_4, C_4 and $2K_2$, hence, threshold graphs are properly contained in the class of cographs. In the following theorem we extend their results.

Theorem 1. *There exists a linear-time algorithm to compute the convex-independence number of cographs.*

Proof. Let G be a cograph. First assume that G is a union of two smaller cographs, G_1 and G_2. In that case, the convex-independence number of G is the sum of $\beta_c(G_1)$ and $\beta_c(G_2)$, that is,

$$G = G_1 \oplus G_2 \quad \Rightarrow \quad \beta_c(G) = \beta_c(G_1) + \beta_c(G_2). \tag{5}$$

Now, assume that G is a join of two smaller cographs G_1 and G_2. In that case, every vertex of G_1 is adjacent to every vertex of G_2. Let S be a convex-independent set. If S has at least one vertex in G_1 and at least one vertex in G_2, then $|S| = 2$, since G[S] cannot have an induced P_3 or K_3.

Consider a convex-independent set $S \subseteq V(G_1)$. Assume that $|S| > 1$ and that $|V(G_2)| \geqslant 2$. Then, any two vertices of $S \cap V(G_1)$ generate $V(G_2) \subseteq \sigma(S)$, and, in turn, $V(G)$ is in their convex hull. This implies that S cannot have any other vertices, that is,

$$|V(G_2)| \geqslant 2 \quad \Rightarrow \quad |S| \leqslant 2.$$

Next, assume

$$\boxed{S \subseteq V(G_1) \quad \text{and} \quad |V(G_2)| = 1.}$$

Say u is in the singleton $V(G_2)$, that is, u is a universal vertex. Let $C_1, \ldots, C_t$ be the components of G_1. We claim that

$$|S \cap C_i| \leqslant \min\{2, |C_i|\}.$$

To see that, assume that $|C_i| \geqslant 2$. Then, since $G[C_i]$ is a connected cograph, $G[C_i]$ is the join of two cographs, say with vertex sets A and B. If S has three vertices in A, then each of them is in the convex hull of the other two, since $B \cup \{u\}$ is contained in their common neighborhood, and this set contains at least two vertices.

Assume S has vertices in at least two different components of G_1. Assume furthermore that one component C_i has at least two vertices of S, say p and q. Let ζ be a vertex of S in another component. Then $u \in \sigma(\{p, \zeta\})$, because $[p, u, \zeta]$ is an induced P_3.

The induced subgraph $G[C_i]$ is a join of two smaller cographs, say with vertex sets A and B. If p and q are both in A, then $q \in \sigma(S-q)$, since p and u generate $B \subset \sigma(S)$, and $B \cup \{u\}$ contains two neighbors of q. If $p \in A$ and $q \in B$, then q has two neighbors in $\sigma(S-q)$, namely p and u. Thus, again, $q \in \sigma(S-q)$.

In fine, either each component of G_1 contains one vertex of S, or else $|S| \leqslant 2$. □

3 Monadic Second-Order Logic

In this section we show that the maximal cardinality of a convex-independent set is computable in linear time for graphs of bounded treewidth or rankwidth. To do that, we show that there is a formulation of the problem in monadic second-order logic. The claim then follows from Courcelle's theorem.

By definition, a set of vertices $W \subseteq V$ is convex if

$$\forall_{x \in V} \ x \notin W \quad \Rightarrow \quad |N(x) \cap W| \leqslant 1. \tag{6}$$

Let $S \subseteq V$. To formulate that a set $W = \sigma(S)$ we formulate that (i) $S \subseteq W$, (ii) W satisfies (6), and (iii) For all W' for which the previous two conditions hold, $W \subseteq W'$. Finally, a set S is convexly independent if

$$\forall_{x \in V} \ x \in S \quad \Rightarrow \quad x \notin \sigma(S-x). \tag{7}$$

Actually, to show that $x \notin \sigma(S-x)$ it is sufficient to formulate that (for every vertex $x \in S$) there is a set W_x such that

$$W_x \text{ is convex} \quad \text{and} \quad S \setminus \{x\} \subseteq W_x \quad \text{and} \quad x \notin W_x. \tag{8}$$

The formulas (6)—(8) show that convex independence can be formulated in monadic second-order logic (without quantification over subsets of edges). By Courcelle's theorem we obtain the following.

Theorem 2. *There exists a linear-time algorithm to compute the convex-independence number for graphs of bounded treewidth or rankwidth.*

3.1   Trees

Let T be a tree with n vertices and maximal degree Δ. Ramos et al. present an involved algorithm, that runs in $O(n \log \Delta)$ time, to compute a convexly independent set. By Theorem 2, there exists a linear-time algorithm that accomplishes this. We propose a different algorithm.

Theorem 3. *There exists a linear-time algorithm that computes the convexly independence number of trees.*

Proof. Let T be a tree. Decompose T into a minimal number of maximal, vertex-disjoint, leafy trees, $F_1, \ldots, F_s$, and a collection of paths. The endpoints of the paths are separate leaves of the trees F_i or pendant vertices. By Equation (4), each leafy tree F_i has a maximum convexly independent set consisting of its leaves. The convexly independence numbers of the connecting paths are given by Equation (3). Notice that each path has a maximum, convexly independent set that contains the two endpoints. $\qquad\square$

4   The Convex-Independence Number of Permutation Graphs

In the following discussion, let G be a permutation graph with a fixed permutation diagram. We refer to S as a generic convex-independent set in G.

Definition 4. *Let S be a convex-independent set in G. Let $x \in V \setminus S$. A 2-path connecting x to S is a sequence of vertices*

$$\Delta = [s_1, s_2, x_1, \ldots, x] \tag{9}$$

in which every vertex has two neighbors that appear earlier in the sequence, or else it is in S.

Lemma 1

$$x \in \sigma(S) \quad \Leftrightarrow \quad \text{there is a 2-path connecting } x \text{ to S.} \tag{10}$$

Proof. Following Duchet, let $I(x, y)$ be the 'interval function' of the P_3-convexity, that is, for two vertices x and y, $I(x, y)$ is $\{x, y\}$ plus the set of vertices that are adjacent to both x and y.[3] For a set S, we let

$$I^0(S) = S \quad \text{and} \quad I^{k+1}(S) = I(I^k(S) \times I^k(S)). \tag{11}$$

Then,

$$\sigma(S) = \bigcup_{k \in \mathbb{N} \cup \{0\}} I^k(S). \tag{12}$$

In other words, a vertex is in $I^{k+1}(S)$ if it is in $I^k(S)$, or else it has two neighbors in $I^k(S)$. This is expressed by the existence of a 2-path (Fig. 2). $\qquad\square$

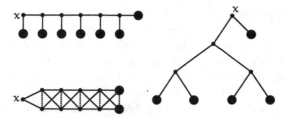

Fig. 2. The figure shows three examples of 2-paths from a vertex x to a set S. The heavy dots represent vertices of S. Notice, however, that the binary tree is not a permutation graph (since it has an asteroidal triple). The second example is a simple path in which each vertex, except x, is replaced by a twin. Permutation graphs are closed under creating twins, so, since paths are permutation graphs, this second example is so also.

Lemma 2. *Each component of* G[S]*, i.e., the subgraph induced by* S*, is a single vertex or an edge. In a permutation diagram for* G*, there is a linear left-to-right ordering of the components of* G[S]*.*

Proof. Since S is convexly independent, G[S] cannot contain K_3 or P_3. Thus each component of G[S] is an edge or a vertex.

Fix a permutation diagram for G. The line segments that correspond to the vertices of S form a permutation diagram for G[S]. Each component of G[S] is a connected part of the diagram, and the left-to-right ordering of the connected parts in the diagram yields a total ordering of the components of G[S].   □

Definition 5. *The last component of* S *is the rightmost component in the linear ordering as specified in Lemma 2.*

We say that a vertex $\notin$ S is to the right of the last component if its line segment appears to the right of the last component, i.e., the endpoints of the line segment, on the top line and bottom line of the diagram, appear to the right of the endpoints of the last component.

Definition 6. *The <u>border</u> of* σ(S) *is the set of the two rightmost endpoints, on the top line and bottom line of the permutation diagram, that are endpoints of line segments corresponding to vertices of* σ(S)*.*

We say that a line segment is to the left of the border if both its endpoints are left of the appertaining endpoints that constitute the border.

Lemma 3. *(i) If the elements of the border of* σ(S) *are the endpoints of a single line segment, then this is the line segment of a vertex in* S*. (ii) For every vertex in* σ(S) \ S *there exist two vertex-disjoint paths to* S *with all vertices in* σ(S)*. (iii) Let the line segment of a vertex* x *be to the right of the last component of* S*. Then* x ∈ σ(S) *if and only if* x*'s line segment is to the left of the border.*

---

[3] Duchet proved that, for interval convexities, the Carathéodory number is the smallest integer k ∈ ℕ such that every (k + 1)-set is redundant. Thus, the fixed-parameter Carathéodory number is polynomial.

Our algorithm performs a dynamic programming on feasible last components of S and the border of $\sigma(S)$. By Lemma 3, the last element of S and the border supply sufficient information to decide whether a 'new last component,' to the right of the previous last component, has a vertex in $\sigma(S)$ or not.

Let $S^* = S \cup X$, where X is either a single vertex or an edge, and assume that X has no vertex $x \in X \cap \sigma(S^* - x)$. To guarantee that S^* is a convex-independent set (with last component X), we need to check that, for any $s \in S$, $s \notin \sigma(S^* - s)$. The figure below shows that the last component of S and the border of $\sigma(S)$ do not convey sufficient information to guarantee that S^* is convexly independent.

In the following, let L be the last component of S, let X be a feasible, 'new,' last component of $G[S^*]$, to the right of L, where $S^* = S \cup X$. (Of course, both L and X are either single vertices or edges, and no vertex of L is adjacent to any vertex of X.) The feasibility of X is defined so that:

$$\forall_{x \in X} \; x \notin \sigma(S^* - x). \tag{13}$$

In our final theorem, below, we prove that it is sufficient to maintain a constant amount of information, to enable the algorithm to check the convexly independence of S^*. First we define a partial 2-path. We consider two cases, namely where $|X| = 1$ and where $|X| = 2$. To define it, we 'simulate' X by an auxiliary vertex, or an auxiliary true twin, that we place immediately to the right of L.

Definition 7. *When $|X| = 1$, add one vertex s' with a line segment whose endpoints are immediately to the right of the rightmost endpoint of L on the top line and the rightmost endpoint of L on the bottom line. When $|X| = 2$, then replace the vertex s' above by a true twin s_1' and s_2'. Let $X' = \{s'\}$ when $|X| = 1$ and $X' = \{s_1', s_2'\}$ when $|X| = 2$. Finally, let $S' = S \cup X'$. A partial 2-path from $u \in S$ to $S^* - u$ is a 2-path from u to $S' - u$, from which S' is removed.*

Theorem 4. *There exists a polynomial-time algorithm to compute a convexly independent set of maximal cardinality in permutation graphs.*

Proof. Consider a vertex $u \in S$ for which $u \in \sigma(S^* - u)$. We may assume that $u \notin L$, because L is available to the algorithm and so, it is easy to check the condition for elements of L. Then there is a 2-path $\Delta = [s_1, s_2, \ldots, u]$ from u to $S^* - u$. If no vertex of X is in this 2-path, then $u \in \sigma(S - u)$, which contradicts our assumption that S is convexly independent. We may assume that at least one of s_1 and s_2 is an element of X.

Since Δ contains two vertex-disjoint paths from u to $S^* - u$, at least one of these paths must contain some vertex of $N(L)$. Partition the vertices of $N(L)$ in two parts. One part contains those vertices that have their endpoint on the top line to the right of L, and the other part contains those vertices that have their endpoint on the bottom line to the right of L. We claim that both parts are totally ordered by set-inclusion of their neighborhoods in the component of $G - N[L]$ that contains X. To see that, consider two elements a and b of $N(L)$. Say that a and b both have an endpoint on the top line, to the right of L. If that endpoint of a is to the left of the endpoint of b, then every neighbor of a in the component of $G - N[L]$ that contains X is also a neighbor of b.

We store subsets with two vertices, y_1 and y_2 in $N(L)$, for which there is a partial 2-path from some vertex $u \in S$ to y_1 and y_2. It is sufficient to store only those two vertices y_1 and y_2 that they have a maximal neighborhood. In other words, we choose y_1 and y_2 such that their endpoints on the top line and bottom line are furthest to the right, or, if they are both in the same part, the two that have a maximal neighborhood.

One other possibility is, that a 2-path from u to $\sigma(S^* - u)$ has only one vertex $y \in N(L)$ on a path from u to X. Of those 2-paths, We also store the element y, with a largest neighborhood in the component of $G - N[L]$ that contains X.

To check if S^* is convexly independent it is now sufficient to check if one of the partial paths to y_1 and y_2, or to the single element y, extend to X.   □

References

1. Baker, K., Fishburn, P., Roberts, F.: Partial orders of dimension two. Networks **2**, 11–28 (1971)
2. Barbosa, R., Coelho, E., Dourado, M., Rautenbach, D., Szwarcfiter, J.: On the Carathéodory number for the convexity of paths of order three. SIAM J. Discrete Math. **27**, 929–939 (2012)
3. Bisztriczky, T., Tóth, G.: Convexly independent sets. Combinatorica **10**, 195–202 (1990)
4. Borie, R., Parker, R., Tovey, C.: Automatic generation of linear-time algorithms from predicate calculus descriptions of problems on recursively constructed graph families. Algorithmica **7**, 555–581 (1992)
5. Courcelle, B.: The monadic second-order logic of graphs. I. Recognizable sets of finite graphs. Inf. Comput. **85**, 12–75 (1990)
6. Danzer, L., Grünbaum, B., Klee, V.: Helly's theorem and its relatives. In: Klee, V., (ed.) Convexity, Proceedings of the 7th Symposium in Pure Mathematics held at the University of Washington, Seattle, Washington June 13–15, 1961, pp. 101–180. American Mathematical Society (1963)
7. Dreyer, P., Roberts, F.: Irreversible k-threshold processes: graph-theoretical threshold models of the spread of disease and of opinion. Discrete Appl. Math. **157**, 1615–1627 (2009)
8. Duchet, P.: Convex sets in graphs II: minimal path convexity. J. Comb. Theory, Ser. B **44**, 307–316 (1988)
9. Dushnik, B., Miller, E.: Partially ordered sets. Am. J. Math. **63**, 600–610 (1941)
10. Edelman, P., Jamison, R.: The theory of convex geometries. Geom. Dedicata. **19**, 247–270 (1985)
11. Franklin, S.: Some results on order convexity. Am. Math. Mon. **69**, 357–359 (1962)
12. Karp, R.: Reducibility among combinatorial problems. In: Miller, R., Thatcher, J. (eds.) Complexity of Computer Computations, pp. 85–103. Plenum, New York (1972)
13. Kloks, T., Wang, Y.: Advances in Graph Algorithms. Manuscript on ViXra: 1409.0165 (2014)
14. Ramos, I., dos Santos, V., Szwarcfiter, J.: Complexity aspects of the computation of the rank of a graph. Discrete Math. Theoret. Comput. Sci. **16**, 73–86 (2014)

The Connected p-Center Problem
on Cactus Graphs

Chunsong Bai[1], Liying Kang[2], and Erfang Shan[3(✉)]

[1] Fuyang Normal College, Fuyang 236041, P.R. China
csbai@fync.edu.cn
[2] Department of Mathematics, Shanghai University, Shanghai 200444, P.R. China
lykang@shu.edu.cn
[3] School of Management, Shanghai University, Shanghai 200444, P.R. China
efshan@i.shu.edu.cn

Abstract. In this paper, we study a variant of the p-center problem on cactus graphs in which the p-center is asked to be connected, and this problem is called the *connected p-center problem*. For the connected p-center problem on cactus graphs, we propose an dynamic programming algorithm and show that the time complexity is $O(n^2 p^2)$, where n is number of vertices.

Keywords: Location problem · Connected p-center problem · Cactus graph · Dynamic programming

1 Introduction

This paper concerns the connected p-center location problem on cactus graphs. Given a simple graph $G = (V, E)$ with n vertices and m edges, a classical p-center problem on a graph $G = (V, E)$ is to determine a p-vertex set V_p in G such that the maximum distance between V_p and V is minimized.

The p-center problem on an arbitrary graph has been known to be NP-hard [3,4]. Olariu [5] presented an $O(n)$ time algorithm for the 1-center problem on interval graphs. Tamir [6] showed that the weighted and un-weighted p-center problems on networks can be solved in $O(n^p m^p \log^2 n)$ time and $O(n^{p-1} m^p \log^3 n)$ time, respectively. Frederickson [2] showed how to solve this problem for trees in optimal linear time using parametric search.

The connected p-center problem is proposed by Yen and Chen [7]. They showed that the CpC problem is NP-hard even when the underlying graph is a bipartite graph or a split graph, and gave an $O(n)$ time algorithm to solve the problem on tree graphs. In [8], Yen proved that the CpC problem on block graph is NP-hard even when (1) $w(v) = 1$, for all $v \in V$, and $l(e) \in \{1, 2\}$, for all $e \in E$, and (2) $w(v) \in \{1, 2\}$, for all $v \in V$, and $l(e) = 1$, for all $e \in E$, respectively.

Research was partially supported by NSFC (grant numbers 11571222, 11471210).

T.-H.H. Chan et al. (Eds.): COCOA 2016, LNCS 10043, pp. 718–725, 2016.
DOI: 10.1007/978-3-319-48749-6_53

2   Notations and Basic Properties

Let $G = (V, E)$ be a simple cactus graph, where each vertex $v \in V$ is associated with a unit weight $w(v) = 1$ and each edge $e \in E$ is associated with a length $l(e) > 0$. Denote by $P[u, v]$ the shortest path in G from u to v, $u, v \in V$.

In order to facilitate the overview of the proposed algorithms for the center problems in cactus networks, we start with the well-known tree structure of a cactus network [1]. The vertex set V is partitioned into three different subsets: C-vertices, G-vertices and hinges.

It is easy to see that a cactus consists of blocks, which are either a cycle or a graft. Thus, we can use a tree T_G to represent the skeleton over G, where each element in T_G represents a block or a hinge of G.

To make the tree T_G ready for use as intended, we convert it into a *rooted tree* as follows: We pick an arbitrary block, e.g., B_0, as the "root" of T_G. For each block B in T_G, we define the *level* $Lev(B)$ of B to be the number of edges on $P[B, B_0]$. Denote by $L = \max_{B \in T_G}\{Lev(B)\}$. If it exists, the *father* of a block B is always a hinge h, called its *companion hinge*. For simplicity, we pick an arbitrary vertex $h_0 \in B_0$ as the *virtual* companion hinge of B_0. Denote by B_h the block B whose companion hinge is h.

For each block B_h in T_G, denote by G_h the sub-cactus of G induced by the vertices of B_h and all sub-cacti hanging from B_h. Specially, $G = G_{h_0}$. For each hinge h of G_{h_0}, denote by g_h the vertex of $G_{h_0} \setminus G_h$ which is the farthest to h. Denote by $g(h) = d(h, g_h)$.

Let $\delta_{G_h}(V_k)$ be the *maximum weighted distance from a k-vertex set V_k to a sub-cactus G_h*, that is,

$$\delta_{G_h}(V_k) = \max_{u \in V(G_h)} \{w(u)d(u, V_k)\},$$

where $d(u, V_k) = \min_{v \in V_k} d(u, v)$.

The Connected p-Center (CpC) Problem: Given a connected graph $G = (V, E)$ and a positive integer $p \geq 2$, identify a p-vertex set $V_p \subseteq V$ such that $\delta_G(V_p)$ is minimized under the restriction that the subgraph induced by V_p is connected. V_p is called a *connected p-center* of G.

For each graft B_h, we define a problem $P(G_h, v, k)$: Given a vertex v of B_h and a positive integer $k \leq p$, identify a connected k-vertex set $V(G_h, v, k)$ of G_h such that $\delta_{G_h}(V(G_h, v, k))$ is minimized, under the restriction that v is the closest vertex to h in $V(G_h, v, k) \cap V(B_h)$. $V(G_h, v, k)$ is called a *v-restricted connected k-center of G_h*.

For each cycle B_h with s indexed vertices $v_1 = h, v_2, \ldots, v_s$, we define a problem $P(G_h, \{v_i, v_j\}, k)$ $(P^{co}(G_h, \{v_i, v_j\}, k))$: Given two vertices $v_i, v_j \in V(B_h)$ with $i \leq j$, and a positive integer $k \leq p$, identify a connected k-vertex set $V(G_h, \{v_i, v_j\}, k)$ $(V^{co}(G_h, \{v_i, v_j\}, k))$ of G_h such that $\delta_{G_h}(V(G_h, \{v_i, v_j\}, k))$ $(\delta_{G_h}(V^{co}(G_h, \{v_i, v_j\}, k)))$ is minimized, under the restriction that $V(G_h, \{v_i, v_j\}, k) \cap V(B_h)$ contains only the vertices of the path from v_i to v_j on B_h in clockwise (counter-clockwise) direction. $V(G_h, \{v_i, v_j\}, k)$

$(V^{co}(G_h, \{v_i, v_j\}, k))$ is called a $\{v_i, v_j\}$-restricted clockwise (counter-clockwise) connected k-center of G_h.

For all sub-cacti G_h, denote by $\mathcal{V}_1$ (resp. $\mathcal{V}_2$) the set of $V(G_h, v, p)$ (resp. $V(G_h, \{v_i, v_j\}, p)$ and $V^{co}(G_h, \{v_i, v_j\}, p)$).

Lemma 1. *There exists a connected p-center of G_{h_0} in $\mathcal{V}_1 \cup \mathcal{V}_2$.*

Proof. Let V_p be a connected p-center of G_{h_0}. We assume that $v \in V_p$ is the closest vertex to h_0, and B_h is the block that contains v. We distinguish the following two cases.

Case 1. B_h is a graft of G_{h_0}, assume that $V(G_h, v, p)$ is a v-restricted connected p-center of G_h. It is easy to see that:

$$\delta_{G_{h_0}}(V_p) = \max\{\delta_{G_h}(V_p), d(v, h) + g(h)\}$$
$$\geq \max\{\delta_{G_h}(V(G_h, v, p)), d(v, h) + g(h)\}$$
$$= \delta_{G_{h_0}}(V(G_h, v, p)),$$

which implies $V(G_h, v, p)$ is also an optimal solution to CpC problem.

Case 2. B_h is a cycle of G_{h_0}. W.l.o.g., we only consider the case $V_p \cap V(B_h)$ contains only the vertices of the path from v_i to v_j on B_h in clockwise direction, where $i \leq j$ (the other case can be handled similarly). Assume that $V(G_h, \{v_i, v_j\}, p)$ is a $\{v_i, v_j\}$-restricted connected p-center of G_h. By the similar discussion in Case 1, we have:

$$\delta_{G_{h_0}}(V_p) = \max\{\delta_{G_h}(V_p), \max\{d(v_i, h), d(v_j, h)\} + g(h)\}$$
$$\geq \max\{\delta_{G_h}(V(G_h, \{v_i, v_j\}, p)), \max\{d(v_i, h), d(v_j, h)\} + g(h)\}$$
$$= \delta_{G_{h_0}}(V(G_h, \{v_i, v_j\}, p)),$$

which implies $V(G_h, \{v_i, v_j\}, p)$ is also an optimal solution to CpC problem. $\square$

Based on Lemma 1, we are going to devise an algorithm to identify all restricted connected p-centers in $\mathcal{V}_1 \cup \mathcal{V}_2$.

3   Algorithm for the CpC Problem on Cactus Graphs

3.1   Procedure GRAFT(B, h)

Given a graft $T = B_h$. Root T at the vertex h. Let $leaf(T)$ be all leaves of T. For each vertex v of T, we define the *level* $lev(v)$ of v to be the number of edges on $P[h, v]$ and $L'_m = \max_{v \in V(T)} lev(v)$. If $v \neq h$, then by removing the last edge of $P[h, v]$, we obtain two subtrees of T. Let T_v be the subtree that contains v, and let $T^c_v = T \setminus T_v$. Similarly, we let G_v be the subgraph of G_h induced by the vertices of T_v and the sub-cacti hanging from T_v, and $G^c_v = G_h \setminus G_v$.

For each vertex v in T, let $E(v)$ be the edges of T_v which are adjacent to v. Denote by $s(v) = |E(v)|$. We define an arbitrary order among the edges of $E(v)$, and we denote the lth edge in $E(v)$ by $e(v, l)$. If v_l is the other endpoint of the $e(v, l)$, then we say that v_l is the lth *son* of v, and v is the father $fa(v_l)$ of v_l. Denote by $son(v)$ be the sons of v. Denote by $T_{e(v,l)}$ the maximal connected subgraph of T_v which contains v but does not contain any edge $e(v, j)$ for $j > l$. In particular, $T_{e(v,0)} = v$ and $T_{e(v,s(v))} = T_v$. Similarly, we define $G_{e(v,l)}$ to be the subgraph of G_v induced by the vertices of $T_{e(v,l)}$ and all sub-cacti hanging from $T_{e(v,l)}$.

Let $e(v, l)$ be an arbitrary edge of T. Let $S(e(v, l), k)$ be a connected k-vertex set of G_v which contains v but does not contain any vertex $v_j \in son(v)$ for $j > l$. Then we define a partial distance-value of $S(e(v, l), k)$ over $G_{e(v,l)}$.

Definition 1. *Let $e(v, l)$ be any arbitrary edge of T. For each positive integer k, $1 \le k \le \min\{p, |G_{e(v,l)}|\}$, we define:*

$$R^*(e(v, l), k) = \min_{S(e(v,l),k) \subseteq G_{e(v,l)}} \delta_{G_{e(v,l)}}(S(e(v, l), k)).$$

The corresponding set to $R^*(e(v, l), k)$ is denoted by $S^*(e(v, l), k)$.

Next, for the vertices in G_v^c, we define the value $R^*(G_v^c, v)$ as follow:

$$R^*(G_v^c, v) = \delta_{G_v^c}(v),$$

which is in fact the distance-value of the 1-center v over G_v^c.

Once we obtain the values $R^*(e(v, s(v)), k)$ and $R^*(G_v^c, v)$, the distance-value of $V(G_h, v, k)$ can be computed as:

$$\delta_{G_h}(V(G_h, v, k)) = \max\{R^*(e(v, s(v)), k), \; R^*(G_v^c, v)\}. \tag{1}$$

According to our assumption, when the block B_h to be processed, we can assign for each v in $leaf(T)$ the following values. For each vertex v of degree 0 in G, we assign:

$$R^*(e(v, 0), 1) = 0$$

and

$$S^*(e(v, 0), 1) \leftarrow \{v\}.$$

For each vertex v which is the companion hinge of some block B_v, if B_v is a graft, we assign:

$$R^*(e(v, 0), k) = \delta_{G_v}(V(G_v, v, k))$$

and

$$S^*(e(v, 0), k) \leftarrow V(G_v, v, k).$$

Otherwise, we assign:

$$R^*(e(v, 0), k) = \delta_{G_v}(V(G_v, \{v, v\}, k))$$

and

$$S^*(e(v, 0), k) \leftarrow V(G_v, \{v, v\}, k).$$

The Computation of $R^*(e(v,l),k)$ and $S^*(e(v,l),k)$. We assume that, when the jth stage begins, the value $R^*(e(v,s(v)),k)$ has been computed for each vertex $v \in T$ of level $lev(v) \geq L'_m - j + 1$. During the jth stage, we search through all vertices of level $L'_m - j$. For each such a vertex v, we compute all values $R^*(e(v,s(v)),k)$ and go on the next vertex of level $L'_m - j$.

Let v be a vertex of level $L'_m - j$. We start by assigning:

$$R^*(e(v,0),1) = \max_{u \in son(v)} \{d(v,u) + R^*(e(u,0),1)\}.$$

Assume that we already known the values $R^*(e(v,l'),k)$ for all $l' < l$, and we now compute the value $R^*(e(v,l),k)$ as follows:

$$R^*(e(v,l),k) = \min \{ \min_{0 \leq k' \leq k-1} \max\{R^*(e(v,l-1),k'), R^*(e(v_l,s(v_l)),k-k')\},$$
$$\max\{R^*(e(v,l-1),k), \ell(v,v_l) + R^*(e(v_l,s(v_l)),1)\}\}. \quad (2)$$

On the right-hand side of (2), the first term corresponds to $v_l \in S^*(e(v,l),k)$, and the second term corresponds to $v_l \notin S^*(e(v,l),k)$.

If $v_l \in S^*(e(v,l),k)$, assign:

$$S^*(e(v,l),k) \leftarrow S^*(e(v,l-1),k'') \cup S^*(e(v_l,s(v_l)),k-k''), \quad (3)$$

where k'' is the number such that the first term of the right-hand side of (3) is minimized. Otherwise, assign:

$$S^*(e(v,l),k) \leftarrow S^*(e(v,l-1),k).$$

We can compute all values $R^*(e(v,l),k)$ by passing through all edges in T. Note that there are at most $|T|p$ values $R^*(e(v,l),k)$ must be computed, each of those computations involved the finding of a minimum over at most $2k$ terms. Thus, the total time complexity is $O(|T|p^2)$.

The Computation of $R^*(G_v^c,v)$. The value $R^*(G_v^c,v)$ can be computed by using the distances matrix of T and the values $R^*(e(u,0),1)$, $u \in leaf(T)$, that is:

$$R^*(G_v^c,v) = \max_{u \in leaf(T)} \{d(v,u) + R^*(e(u,0),1)\}.$$

It is easy to see that the total time is $O(|T|^2)$ to compute the values $R^*(G_v^c,v)$.

3.2   The Procedure CYCLE(C,h)

Let C be the cycle B_h with s clockwise indexed vertices $v_1 = h, v_2, \ldots, v_s$. For any pair $v_i, v_j \in V(C)$ ($i \leq j$), denote by C_{v_i,v_j} (C_{v_i,v_j}^{co}) the subgraph induced by the vertices of the path from v_i to v_j in clockwise direction (in counter-clockwise direction), and G_{v_i,v_j} (G_{v_i,v_j}^{co}) the subgraph induced by C_{v_i,v_j} (C_{v_i,v_j}^{co}) and the sub-cacti hanging from it. Let $G_{v_i,v_j}^c = G_h \setminus G_{v_i,v_j}$.

The Computation of $V(G_h, \{v_i, v_j\}, k)$. Let $V(\{v_i, v_j\}, k)$ be a connected k-vertex set of G_{v_i, v_j} that contains the vertices v_i and v_j. Then we define the partial distance-value of $V(\{v_i, v_j\}, k)$ over G_{v_i, v_j} as follow:

$$R_1^*(\{v_i, v_j\}, k) = \min_{V(\{v_i, v_j\}, k) \subseteq G_{v_i, v_j}} \delta_{G_{v_i, v_j}}(V(\{v_i, v_j\}, k)).$$

Let $e_{m(j,i)}$ be the edge that contains the midpoint of the path from v_j to v_i in clockwise direction. Particularly, if the midpoint happens to be a vertex, then it coincides with $v_{m(j,i)}$. By deleting the edge $e_{m(j,i)}$ from G_{v_i, v_j}^c, we obtain two subgraphs $G_{v_i, v_j}^{c,1}$ and $G_{v_i, v_j}^{c,2}$, which contain $v_{m(j,i)}$ and $v_{m(j,i)+1}$, respectively. Now we define the following values:

$$R_2^*(\{v_i, v_j\}, v_j) = \delta_{G_{v_i, v_j}^{c,1}}(\{v_j\})$$

and

$$R_3^*(\{v_i, v_j\}, v_i) = \delta_{G_{v_i, v_j}^{c,2}}(\{v_i\})$$

to represent the partial distance-values of v_j and v_i, respectively.

Once we obtain all values defined above, the distance-value of $V(G_h, \{v_i, v_j\}, k)$ can be computed as:

$$\delta_{G_h}(V(G_h, \{v_i, v_j\}, k)) = \max\{R_1^*(\{v_i, v_j\}, k), R_2^*(\{v_i, v_j\}, v_j), R_3^*(\{v_i, v_j\}, v_i)\}.$$

Note that the values $R_1^*(\{v_i, v_j\}, k)$ can be computed by applying the procedure GRAFT(B, h), and the total time is $O(|C|^2 p^2)$.

Given an edge $e_m = (v_m, v_{m+1})$ in C. Let $\mathcal{P}(e_m) = \{\{v_{l_1}, v_{r_1}\}, \{v_{l_2}, v_{r_2}\}, \ldots, \{v_{l_t}, v_{r_t}\}\}$ be all vertex pairs of C with their middle edges are e_m, where $l_1 \geq l_2 \geq \ldots \geq l_t$. Let $\mathcal{V} = \{v_{l_1}, v_{l_2}, \ldots, v_{l_t}\}$.

Because of the recursion

$$R_2^*(\{v_{r_k}, v_{l_k}\}) = \max\{R_2^*(\{v_{r_{k-1}}, v_{l_{k-1}}\}, v_{l_{k-1}}) + d(v_{l_k} + v_{l_{k-1}}),$$
$$\max_{l_{k-1} > j' \geq l_k} \{d(v_{l_k}, v_{j'}) + R_1^*(\{v_{j'}, v_{j'}\}, 1)\}\}, \qquad (4)$$

we can calculate all values $R_2^*(\{v_{r_k}, v_{l_k}\}, v_{l_k})$ for $l_1 \leq l_k \leq l_t$ by passing through all vertices in $\mathcal{V}$ and cost $O(|C|)$ time for comparing and adding operations. Thus, all values can be computed in $O(|C|^2)$ time since there $O(|C|^2)$ values must be computed and $O(|C|)$ edges in C.

The Computation of $V^{co}(G_h, \{v_i, v_j\}, k)$. Let $V^{co}(\{v_i, v_j\}, k)$ be a connected k-vertex set of G_{v_i, v_j}^{co} that contains the vertices v_i and v_j, let $e_{m(i,j)}$ be the edge that contains the midpoint of the path from v_i to v_j in clockwise direction. Particularly, if the midpoint happens to be a vertex, then it coincides with $v_{m(i,j)}$.

Next we define the partial value:

$$R_4^*(\{v_i, v_j\}, k) = \min_{V^{co}(\{v_i, v_j\}, k) \subseteq G_{v_i, v_j}^{co}} \delta_{G_{v_i, v_j}^{co}}(V^{co}(\{v_i, v_j\}, k)),$$

as well as the values $R_5^*(\{v_i, v_j\}, v_i)$ and $R_6^*(\{v_i, v_j\}, v_j)$, similar to $R_2^*(\{v_i, v_j\}, v_j)$ and $R_3^*(\{v_i, v_j\}, v_i)$, respectively. Therefore, we can compute the distance-value of $V^{co}(G_h, \{v_i, v_j\}, k)$ as:

$$\delta_{G_h}(V^{co}(G_h, \{v_i, v_j\}, k)) = \max\{R_4^*(\{v_i, v_j\}, k),\ R_5^*(\{v_i, v_j\}, v_i),\ R_6^*(\{v_i, v_j\}, v_j)\}.$$

It is easy to see that all values $R_4^*(\{v_i, v_j\}, k)$, $R_5^*(\{v_i, v_j\}, v_i)$, $R_6^*(\{v_i, v_j\}, v_j)$ can be computed similarly as above, and the time complexity is $O(|C|^2 p^2)$.

3.3   Algorithm for the CpC Problem

By Lemma 1, we can now identify a connected p-center V_p^* from $\mathcal{V}_1 \cup \mathcal{V}_2$. The distance-value of V_p^* can be computed by the following relation:

$$\delta(V_p^*) = \min \Big\{ \min_{V(G_h, v, p) \in \mathcal{V}_1} \{\max\{\delta_{G_h}(V(G_h, v, p)), d(v, h) + g(h)\}\},$$
$$\min_{V(G_h, \{v_i, v_j\}, p) \in \mathcal{V}_2} \max\{\delta_{G_h}(V(G_h, \{v_i, v_j\}, p)), \max\{d(v_j, h), d(v_i, h)\} + g(h)\},$$
$$\min_{V^{co}(G_h, \{v_i, v_j\}, p) \in \mathcal{V}_2} \max\{\delta_{G_h}(V^{co}(G_h, \{v_i, v_j\}, p)), g(h)\}\Big\}. \tag{5}$$

We can now formulate the algorithm for the CpC problem.

Algorithm 1. Connected_p-Center_on_Cactus_Graphs.

Input: A cactus graph $G(h_0)$, the corresponding skeleton $T_S(B_0)$ and its
          maximal level L_m.
  Output: A connected p-center V_p^* and its distance-value.
2 **for** $i = 1; i <= L_m; i++$ **do**
3   **for** *each block B of level $2L_m - 2i + 1$* **do**
4   **if** *B is a graft* **then**
5   Let h be the companion hinge of B, let L_m' be the maximal
                 level of B;
6   **for** $j = 1; j <= L_m'; j++$ **do**
7   Call **GRAFT**(B, h) to compute the values $V(G_h, v, k)$ for
                    each vertex v of level j and $1 \le k \le \min\{p, |G_v|\}$;
8   **end**
9   **end**
10   **if** *B is a cycle* **then**
11   Let h be the companion hinge of C;
12   Call **CYCLE**(B, h) to compute the values $V(G_h, \{v_i, v_j\}, k)$
                 and $V^{co}(G_h, \{v_i, v_j\}, k)$ for all pair $v_i, v_j \in V(C)$ $(i \le j)$ and
                 all possible numbers k;
13   **end**
14   **end**
15 **end**
16 **return** Identify a connected p-center V_p^* by using the Eq. (5).

As a preprocessing for Algorithm 1, we first compute the distance-matrix of the given cactus. Then we find a skeleton of the given cactus and compute $g(h)$ for each companion hinge h in the skeleton. This preprocessing requires $O(n^2)$ steps. Then we can find a p-center from $\mathcal{V}_1 \cup \mathcal{V}_2$ by using the binary search method.

Theorem 1. *The CpC problem on a cactus graph of n vertices can be solved in $O(n^2 p^2)$ time.*

4   Conclusions

In this paper we consider the connected p-center on graphs. We devise a dynamic programming algorithm of the complexity $O(n^2 p^2)$ for the problem on cactus graphs. In the future, it is very meaningful to extend our algorithm to other classes of graphs, such as interval graphs, circular-arc graphs and planar graphs, etc.

References

1. Burkard, R.E., Krarup, J.: A linear algorithm for the pos/neg-weighted 1-median problem on cactus. Computing **60**, 498–509 (1998)
2. Frederickson, G.: Parametric search and locating supply centers in trees. In: Dehne, F., Sack, J.-R., Santoro, N. (eds.) WADS 1991. LNCS, vol. 519, pp. 299–319. Springer, Heidelberg (1991)
3. Garey, M.R., Johnso, D.S.: Computers and Intractability: A Guide to the Theory of NP-Completeness. Bell Laboratories, Murray Hill, Freeman & Co., New York (1978)
4. Kariv, O., Hakimi, S.L.: An algorithmic approach to network location problems, part II: p-medians. SIAM J. Appl. Math. **37**, 539–560 (1979)
5. Olariu, S.: A simple linear-time algorithm for computing the center of an interval graph. Int. J. Comput. Math. **24**, 121–128 (1990)
6. Tamir, A.: Improved complexity bounds for center location problems on networks by using dynamic data structures. SIAM J. Discrete Math. **1**, 377–396 (1988)
7. Yen, W.C.-K., Chen, C.-T.: The p-center problem with connectivity constraint. Appl. Math. Sci. **1**, 1311–1324 (2007)
8. Yen, W.C.-K.: The connceted p-center problem on block graphs with forbidden vertices. Theor. Comput. Sci. **426–427**, 13–24 (2012)

Comparison of Quadratic Convex Reformulations to Solve the Quadratic Assignment Problem

Sourour Elloumi[1] and Amélie Lambert[2(✉)]

[1] ENSTA ParisTech (and CEDRIC-Cnam) 828,
Boulevard des Maréchaux, F-91762 Palaiseau cedex, France
[2] CEDRIC-Cnam, 292 rue Saint-Martin, F-75141 Paris cedex 03, France
amelie.lambert@cnam.fr

Abstract. We consider the (QAP) that consists in minimizing a quadratic function subject to assignment constraints where the variables are binary. In this paper, we build two families of equivalent quadratic convex formulations of (QAP). The continuous relaxation of each equivalent formulation is then a convex problem and can be used within a B&B. In this work, we focus on finding the "best" equivalent formulation within each family, and we prove that it can be computed using semidefinite programming. Finally, we get two convex formulations of (QAP) that differ from their sizes and from the tightness of their continuous relaxation bound. We present computational experiments that prove the practical usefulness of using quadratic convex formulation to solve instances of (QAP) of medium sizes.

Keywords: Quadratic assignment problem · Convex quadratic programming · Semidefinite programming · Experiments

1 Introduction

We consider the general version of the Quadratic Assignment Problem (QAP) where we are given a four-dimensional array $Q = (q_{ijkl})$ of coefficients. The problem can be formulated as:

$$
(QAP) \begin{cases}
\min f(x) = \displaystyle\sum_{(i,j,k,l)\in \mathcal{I}^4} q_{ijkl} x_{ij} x_{kl} & \\[2mm]
\displaystyle\sum_{i=1}^{n} x_{ij} = 1 & j \in \mathcal{I} & (1) \\[2mm]
\displaystyle\sum_{j=1}^{n} x_{ij} = 1 & i \in \mathcal{I} & (2) \\[2mm]
x_{ij} \in \{0,1\} & (i,j) \in \mathcal{I}^2 & (3)
\end{cases}
$$

where $\mathcal{I}$ is the set of n facilities or locations and the decision variable x_{ij} corresponds to facility i being assigned to location j, and q_{ijkl} is the cost incurred by

© Springer International Publishing AG 2016
T.-H.H. Chan et al. (Eds.): COCOA 2016, LNCS 10043, pp. 726–734, 2016.
DOI: 10.1007/978-3-319-48749-6_54

assigning facility i to location j and facility k to location l. Sahni and Gonzalez [11] have shown that the (QAP) is $\mathcal{NP}$-hard. The (QAP) remains one of the hardest optimization problems. In the more general case, when the Hessian matrix is fully dense, no exact algorithm can solve problems of size $n > 35$. Several algorithms were introduced for solving this problem. The majority of them are based on B&B technics. To compute polynomial bounds, many authors have proposed linearizations of the quadratic objective function by introducing additional variables. Semidefinite programming relaxations for the (QAP) were also considered by several authors. In terms of quality, the bounds obtained by semidefinite programming and convex programming are competitive with the best existing lower bounds for the (QAP). We refer the reader to the recent survey [6] for more references.

We introduced in [2,3] the Mixed Integer Quadratic Convex Reformulation (MIQCR) approach. This method handles general mixed-integer quadratic problems (QP) by designing the tightest possible equivalent program to (QP) with a convex objective function, within a convex reformulation scheme. In this paper, we present two algorithms to solve (QAP) based on the reformulation of the original problem into an equivalent quadratic problem whose continuous relaxation is convex. These algorithms are specializations of method MIQCR to (QAP).

In Sect. 2, we introduce a new family of equivalent formulations to (QAP) where we stay in the same space of variables than in the original (QAP). In Sect. 3, we introduce another family of equivalent formulations to (QAP) in an extended space of variables. In Sect. 4, we present computational results on instances of (QAP) to show the effectiveness of the approaches. Section 5 draws a conclusion.

Notation: We denote by $v(P)$ the optimal value of (P), by $\mathcal{S}_n$ the set of $n \times n$ symmetric matrices, by $\mathcal{S}_n^+$ the set of positive semidefinite matrices of $\mathcal{S}_n$ and $M \succeq 0$ means that $M \in \mathcal{S}_n^+$, by $\langle A, B \rangle$ the scalar product of matrices $A \in \mathcal{S}_n$ and $B \in \mathcal{S}_n$, i.e. $\langle A, B \rangle = tr(AB) = \sum_{i,j} a_{ij} b_{ij}$.

2   A Compact Equivalent Quadratic Convex Formulation

We consider the problem of reformulating (QAP) into an equivalent quadratic 0–1 program. This reformulation consists in a perturbation of the Hessian matrix of the objective function $f(x)$ such that the new objective function is convex. The continuous relaxation of the new quadratic 0–1 program is thus a polynomial problem, and it is easy to develop a B&B based on this relaxation. To build this reformulation, we add to $f(x)$ a combination of three sets of functions where each of these functions vanishes on the feasible solution set. More precisely, we consider: (i) equalities $(x_{ij}^2 - x_{ij}) = 0$ for all $(i,j) \in \mathcal{I}^2$, that are satisfied for any $0 - 1$ variable. Using the matrix notation, each identity $(x_{ij}^2 - x_{ij}) = 0$ can be rewritten $\langle Q_{ij}^1, xx^T \rangle - x_{ij} = 0$, where Q_{ij}^1 is the matrix associated to the equality (i, j) which all entries equal to 0 except the diagonal entry corresponding to variable x_{ij} which is equal to 1. (ii) Equalities $x_{ij}x_{il} = 0$ for all $(i, j, l) \in \mathcal{I}^3$: $j < l$ coming from the fact that a facility cannot be affected to more than one location. Using the matrix notation, each identity $x_{ij}x_{il} = 0$ can be rewritten $\langle Q^2_{ijl}, xx^T \rangle = 0$, where Q^2_{ijl} is the matrix associated to the equality (i, j, l) which all entries equal to 0 except the entry corresponding to the row of variable x_{ij} and the column of variable x_{il} which is equal to 1. (iii) Equalities $x_{ij}x_{kj} = 0$ for all $(i, j, k) \in \mathcal{I}^3$: $i < k$ coming from the fact that a location cannot be affected to more than one facility. Using the matrix notation, each identity $x_{ij}x_{kj} = 0$ can be rewritten $\langle Q^3_{ijk}, xx^T \rangle = 0$, where Q^3_{ijk} is the matrix associated to the equality (i, j, k) which all entries equal to 0 except the entry corresponding to the row of variable x_{ij} and the column of variable x_{kj} which is equal to 1.

We now introduce real scalar parameters: $\beta_{ij} \ \forall \in (i, j) \in \mathcal{I}^2$, $\lambda_{ijl} \ \forall (i, j, l) \in \mathcal{I}^3$: $j < l$, and $\lambda'_{ijk} \ \forall (i, j, k) \in \mathcal{I}^3$: $i < k$. We define a new objective function as follows:

$$f_{\beta,\lambda,\lambda'}(x) = f(x) + \sum_{\substack{(i,j) \in \mathcal{I}^2}} \beta_{ij}(x^2_{ij} - x_{ij}) + \sum_{\substack{(i,j,l) \in \mathcal{I}^3 \\ j < l}} \lambda_{ijl} x_{ij} x_{il} + \sum_{\substack{(i,j,k) \in \mathcal{I}^3 \\ i < k}} \lambda'_{ijk} x_{ij} x_{kj}$$

It is clear that if x is feasible for (QAP) (i.e. if x satisfies Constraints (1)–(3)), then $f(x) = f_{\beta,\lambda,\lambda'}(x)$. We then obtain the following equivalent problem:

$$(QAP_{\beta,\lambda,\lambda'}) \left\{ \max_{(1)(2)(3)} \ f_{\beta,\lambda,\lambda'}(x) \right.$$

where the objective function can be rewritten as follows: $f_{\beta,\lambda,\lambda'}(x) = \langle Q_{\beta,\lambda,\lambda'}, xx^T \rangle - $

$\beta^T x$ with $Q_{\beta,\lambda,\lambda'} = Q + \sum_{\substack{(i,j) \in \mathcal{I}^2}} \beta_{ij} Q^1_{ij} + \sum_{\substack{(i,j,l) \in \mathcal{I}^3 \\ j < l}} \lambda_{ijl} Q^2_{ijl} + \sum_{\substack{(i,j,k) \in \mathcal{I}^3 \\ i < k}} \lambda'_{ijk} Q^3_{ijk}$

Varying parameters $(\beta, \lambda, \lambda')$ gives rise to $(QAP_{\beta,\lambda,\lambda'})$ as a family of equivalent problems to (QAP). To use these equivalent problems into a B&B, we are interested by parameters $(\beta, \lambda, \lambda')$ such that $f_{\beta,\lambda,\lambda'}(x)$ is a convex function. Moreover, in order to have a good behavior of the B&B, we will focus on those that give the tightest continuous relaxation bound.

Denoting $(\overline{QAP}_{\beta,\lambda,\lambda'})$ the continuous relaxation of $(QAP_{\beta,\lambda,\lambda'})$, this amounts to solve the following problem

$$(P) \left\{ \max_{Q_{\beta,\lambda,\lambda'} \succeq 0} \ \{v(\overline{QAP}_{\beta,\lambda,\lambda'})\} \right.$$

Here, we prove that an optimal solution to (P) can be deduced from a dual solution of the following program (SDP_{R2}) introduced in [12] which is also a semidefinite relaxation of (QAP):

$$(SDP_{R2}) \begin{cases} \min f(X,x) = \langle Q, X \rangle \\ (1)(2) \\ X_{ijij} - x_{ij} = \langle Q^1_{ij}, X \rangle - x_{ij} = 0 & (i,j) \in \mathcal{I}^2 & (4) \\ X_{ijil} = \langle Q^2_{ijl}, X \rangle = 0 & (i,j,l) \in \mathcal{I}^3,\ j < l & (5) \\ X_{ijkj} = \langle Q^3_{ijk}, X \rangle = 0 & (i,j,k) \in \mathcal{I}^3,\ i < k & (6) \\ \begin{pmatrix} 1 & x \\ x^T & X \end{pmatrix} \succeq 0, x \in \mathbb{R}^{n^2} \quad X \in \mathcal{S}_{n^2} & & (7) \end{cases}$$

Theorem 1. *The optimal value of* (P) *is equal to the optimal value of* (SDP_{R2}).

Proof. Let us firstly prove that $v(P) \le v(SDP_{R2})$. Let $(\bar{\beta}, \bar{\lambda}, \bar{\lambda}')$ be any feasible solution to (P). To prove that $v(P) \le v(SDP_{R2})$ we prove that from any feasible solution $(\bar{X}, \bar{x})$ to (SDP_{R2}), we can build a feasible solution x to $(\overline{QAP}_{\bar{\beta}, \bar{\lambda}, \bar{\lambda}'})$ with a lower objective value, i.e., satisfying $f_{\bar{\beta}, \bar{\lambda}, \bar{\lambda}'}(x) \le f(\bar{X}, \bar{x})$. Take $x = \bar{x}$, this solution is feasible to $(\overline{QAP}_{\bar{\beta}, \bar{\lambda}, \bar{\lambda}'})$. Indeed, Constraints (1) and (2) are obviously satisfied, and by Constraints (4) and (7), we have $0 \le x \le 1$. We now prove that $f_{\bar{\beta}, \bar{\lambda}, \bar{\lambda}'}(x) \le f(\bar{X}, \bar{x})$ or equivalently that $\Delta = f_{\bar{\beta}, \bar{\lambda}, \bar{\lambda}'}(x) - f(\bar{X}, \bar{x}) \le 0$. We have:

$$\Delta = \langle Q_{\bar{\beta}, \bar{\lambda}, \bar{\lambda}'}, \bar{x}\bar{x}^T \rangle - \bar{\beta}^T \bar{x} - \langle Q, \bar{X} \rangle = \langle Q_{\bar{\beta}, \bar{\lambda}, \bar{\lambda}'}, \bar{x}\bar{x}^T \rangle - \bar{\beta}^T \bar{x} - \langle Q, \bar{X} \rangle - \sum_{(i,j) \in \mathcal{I}^2} \bar{\beta}_{ij}(\langle Q^1_{ij}, \bar{X} \rangle - \bar{x}_{ij})$$

$$- \sum_{\substack{(i,j,l) \in \mathcal{I}^3 \\ j < l}} \bar{\lambda}_{ijl} \langle Q^2_{ijl}, \bar{X} \rangle - \sum_{\substack{(i,j,k) \in \mathcal{I}^3 \\ i < k}} \bar{\lambda}'_{ijk} \langle Q^3_{ijk}, \bar{X} \rangle = \langle Q_{\bar{\beta}, \bar{\lambda}, \bar{\lambda}'}, \bar{x}\bar{x}^T - \bar{X} \rangle - \bar{\beta}^T \bar{x} + \bar{\beta}^T \bar{x} \le 0$$

Let us secondly prove that $v(P) \ge v(SDP_{R2})$ or equivalently $v(P) \ge v(DSDP_{R2})$ where $(DSDP_{R2})$ is the dual of (SDP_{R2}):

$$(DSDP_{R2}) \begin{cases} \max g(\beta, \lambda, \lambda', \rho, \rho') = -\sum_{i \in \mathcal{I}} (\rho_i + \rho'_i) \\ Q + \sum_{(i,j) \in \mathcal{I}^2} \beta_{ij} Q^1_{ij} + \sum_{\substack{(i,j,l) \in \mathcal{I}^3 \\ j < l}} \lambda_{ijl} Q^2_{ijl} + \sum_{\substack{(i,j,k) \in \mathcal{I}^3 \\ i < k}} \lambda'_{ijk} Q^3_{ijk} \succeq 0 & (8) \\ \beta + \sum_{i \in \mathcal{I}} (\rho_i + \rho'_i) \ge 0 & (9) \\ \beta_{ij} \in \mathbb{R},\ \lambda_{ijl} \in \mathbb{R},\ \lambda'_{ijk} \in \mathbb{R},\ \rho_i \in \mathbb{R},\ \rho'_i \in \mathbb{R} \end{cases}$$

where $\beta_{ij} \in \mathbb{R}$ are the dual variables associated with constraints (4), $\lambda_{ijl} \in \mathbb{R}$, and $\lambda'_{ijk} \in \mathbb{R}$ are the dual variables associated with constraints (5) and (6), respectively, and $\rho_i \in \mathbb{R}$, and $\rho'_i \in \mathbb{R}$ are the dual variables associated with constraints (1) and (2), respectively. Let $(\bar{\beta}, \bar{\lambda}, \bar{\lambda}', \bar{\rho}, \bar{\rho}')$ be a feasible solution to $(DSDP_{R2})$, then, we build the following positive semidefinite matrix:

$$Q_{\bar{\beta}, \bar{\lambda}, \bar{\lambda}'} = Q + \sum_{(i,j) \in \mathcal{I}^2} \bar{\beta}_{ij} Q^1_{ij} + \sum_{\substack{(i,j,l) \in \mathcal{I}^3 \\ j < l}} \bar{\lambda}_{ijl} Q^2_{ijl} + \sum_{\substack{(i,j,k) \in \mathcal{I}^3 \\ i < k}} \bar{\lambda}'_{ijk} Q^3_{ijk}$$

By Constraint (8), we have $Q_{\bar{\beta}, \bar{\lambda}, \bar{\lambda}'} \succeq 0$, and therefore $(\bar{\beta}, \bar{\lambda}, \bar{\lambda}')$ form a feasible solution to (P). The objective value of this solution is equal to $v(\overline{QAP}_{\bar{\beta}, \bar{\lambda}, \bar{\lambda}'})$.

We now prove that $v(\overline{QAP}_{\bar{\beta},\bar{\lambda},\bar{\lambda}'}) \geq v(DSDP_{R2})$. For this, we prove that for any feasible solution $\bar{x}$ to $(\overline{QAP}_{\bar{\beta},\bar{\lambda},\bar{\lambda}'})$, the associated objective value is not smaller than $g(\bar{\beta}, \bar{\lambda}, \bar{\lambda}', \bar{\rho}, \bar{\rho}')$. Denote by Δ the difference between the objective values:

$$\Delta = \langle Q_{\bar{\beta},\bar{\lambda},\bar{\lambda}'}, \bar{x}\bar{x}^T \rangle + \bar{\beta}^T\bar{x} + \sum_{i \in \mathcal{I}}(\bar{\rho}_i + \bar{\rho}'_i) \geq \bar{\beta}^T\bar{x} + \sum_{i \in \mathcal{I}}(\bar{\rho}_i + \bar{\rho}'_i) \geq (\bar{\beta} + \sum_{i \in \mathcal{I}}(\bar{\rho}_i + \bar{\rho}'_i))^T\bar{x} \geq 0 \qquad \square$$

Algorithm 1.

---

step 1: Solve the semidefinite program (SDP_{R2})

step 2: Deduce an optimal solution $(\beta^*, \lambda^*, \lambda'^*)$: β^* is the vector of optimal dual variables associated with constraints (4), λ^* is the vector of optimal dual variables associated with constraints (5), λ'^* is the vector of optimal dual variables associated with constraints (6).

step 3: Solve the program $(QAP_{\beta^*,\lambda^*,\lambda'^*})$ by a MIQP solver.

(Its continuous relaxation is a convex program with an optimal value equal to the optimal value of (SDP_{R2}))

From Theorem 1 we can deduce Algorithm 1 to solve (QAP). An interesting remark concerning this equivalent convex formulation is that we do not add any variable and constraint to build it. Moreover, the Step 1 of Algorithm 1 requires the solution of a semidefinite problem with a reasonable size (i.e. $\mathcal{O}(n^4)$ variables, and $\mathcal{O}(n^3)$ constraints). We will see in the experiments of Sect. 4 that solving (SDP_{R2}) with an appropriate algorithm based on a bundle method, as in [4], is practicable. Furthermore, the reasonable size (n^2) of $(QAP_{\beta^*,\lambda^*,\lambda'^*})$, that will be solved at each node of the B&B is clearly an advantage. However, by remaining in the same space of variables, we limit the sharpness of the bound at the root node of the B&B and consequently, this method cannot handle instances of (QAP) of large sizes.

3   An Extended Quadratic Convex Equivalent Formulation

In this section, we describe another family of quadratic convex reformulations of (QAP) that is built in an extended space of variables. The basic idea is the same as for the algorithm described in Sect. 2, but in this family of reformulations, we will use additional variables and constraints in order to strengthen the continuous relaxation value of the equivalent formulation. In other words, we will use a tighter semidefinite relaxation of (QAP) to build an equivalent convex formulation. We first introduce new variables y that will satisfy:

$$y_{ijkl} = x_{ij}x_{kl} \quad \forall(i, j, k, l) \in I^4 : i \neq k \text{ and } j \neq l \qquad (C1)$$

Then, as in Sect. 2, we consider functions that vanish on the feasible solution set defined by the original (QAP) (i.e. Constraints (1)–(3)) and by $(C1)$. More formally, in addition to the combination of the three sets of functions introduced in Sect. 2, we consider the following equalities: $x_{ij}x_{kl} - y_{ijkl} = 0$ for all $(i, j, k, l) \in \mathcal{I}^4 : i \neq k$ and $j \neq l$. Using the matrix notation, these equalities can be rewritten as $\langle Q^4_{ijkl}, xx^T \rangle - y_{ijkl} = 0$, where Q^4_{ijkl} is the matrix which all entries are equal to 0 except the entry corresponding to the row of variable x_{ij} and the column of variable x_{kl} which is equal to 1. Then, we introduce non-negative real parameters ϕ_{ijkl} and we consider the following new quadratic function:

$$f_{\beta,\lambda,\lambda',\phi}(x,y) = f_{\beta,\lambda,\lambda'}(x) - \sum_{\substack{(i,j,k,l)\,\in\,\mathcal{I}^4 \\ i \neq k;\, j \neq l}} \phi_{ijkl}(\langle Q^4_{ijkl}, \bar{x}\bar{x}^T \rangle - y_{ijkl}) = \langle Q_{\beta,\lambda,\lambda',\phi}, xx^T \rangle - \beta^T x + \phi^T y$$

where $\quad Q_{\beta,\lambda,\lambda',\phi} = Q_{\beta,\lambda,\lambda'} - \displaystyle\sum_{\substack{(i,j,k,l)\,\in\,\mathcal{I}^4 \\ i \neq k;\, j \neq l}} \phi_{ijkl} Q^4_{ijkl}.$

Finally, we have to enforce Condition $(C1)$ using linear constraints. As $\phi_{ijkl} \geq 0$, and since additional variables y_{ijkl} only appear in the objective function, it is easy to see that in an optimal solution y_{ijkl} will be set to its smallest possible value. Hence, problem (QAP) is equivalently stated as:

$$(QAP_{\beta,\lambda,\lambda',\phi}) \begin{cases} \min\ f_{\beta,\lambda,\lambda',\phi}(x,y) = \langle Q_{\beta,\lambda,\lambda',\phi}, xx^T \rangle - \beta^T x + \phi^T y \\ (1)(2)(3) \\ y_{ijkl} \geq x_{ij} + x_{kl} - 1 & (10) \\ y_{ijkl} \geq 0 & (11) \end{cases}$$

Here again, we build an infinite family of equivalent problems to (QAP), and we now state the problem to find the best semidefinite matrix $Q_{\beta,\lambda,\lambda',\phi}$ in terms of continuous relaxation bound:

$$(P') \left\{ \max_{Q_{\beta,\lambda,\lambda',\phi} \succeq 0} \{ v(\overline{QAP}_{\beta,\lambda,\lambda',\phi}) \} \right.$$

Following the same reasoning steps as in Sect. 2, we claim that this best matrix can be deduced from the dual of the following semidefinite relaxation of (QAP) denoted by (SDP_{R4}) and introduced for (QAP) in [10]:

$$(SDP_{R4}) \begin{cases} \min f(X,x) = \langle Q, X \rangle \\ (1)(2)(4)(5)(6)(7) \\ -X_{ijkl} = \langle -Q^4_{ijkl}, X \rangle \leq 0 & (i,j,k,l) \in \mathcal{I}^4,\ k \neq i,\ j \neq l & (12) \end{cases}$$

Theorem 2. *The optimal value of (P') is equal to the optimal value of (SDP_{R4}).*

Proof. A proof can be deduced from the proof of Theorem 1 and the fact that constraints: $X_{ijkl} - 1 - x_{ij} - x_{kl} \geq 0$ are redundant in (SDP_{R4}) (this can be deduced from [1]). $\qquad\square$

From Theorem 2 we can deduce Algorithm 2 to solve (QAP). In this quadratic convex reformulation we use a semidefinite relaxation that is well known to be very sharp. Hence, we will start Step 3 of our algorithm with a good continuous relaxation bound, but it is obtained by solving a quite large semi-definite problem. We will see in our experiments that solving (SDP_{R4}) with an appropriate algorithm based on a bundle method [4], is still practicable.

Algorithm 2.

step 1: Solve the semidefinite program (SDP_{R4})

step 2: Deduce an optimal solution $(\beta^*, \lambda^*, \lambda'^*, \phi^*)$: $\beta^*, \lambda^*, \lambda'^*$ are deduced as in Algorithm 1, ϕ^* is the vector of optimal dual variables associated with constraint (12).

step 3: Solve the program $(QAP_{\beta^*, \lambda^*, \lambda^{*\prime}, \phi^*})$ by a MIQP solver.

(Its continuous relaxation is a convex program with an optimal value equal to the optimal value of (SDP_{R4}))

4   Computational Results

In this section, we present experiments on the Nugent instances from QAPLIB [7]. In these instances, the distance matrix contains Manhattan distances of rectangular grids. We compare our algorithms, Algorithms 1 and 2, with the solver Cplex12.6.2.

Experimental Environment and Parameters: Our experiments were carried out on a server with 2 CPU Intel Xeon each of them having 12 cores and 2 threads of 2.5 GHz and $4 * 16$ GB of RAM using a Linux operating system. For solving (SDP_{R2}) and (SDP_{R4}) we used CSDP [5] together with the Conic Bundle [8], as done in [4]. We used the C interface of the solver Cplex 12.6.2 [9]

Table 1. Computational results for Algorithm 1, Algorithm 2, and Cplex12.6.2 for Nugent instances from QAPLIB (time limit of 3 hours)

Name	Algorithm 1 (R2)			Algorithm 2 (R4)			Cplex12.6.2		
	Gap	Time	Nodes	Gap	Time	Nodes	Gap	Time	Nodes
nug03	0.0	1	0	0.0	0	0	100	2	0
nug06	5.5	4	0	0.9	1	0	100	3	0
nug12	9.3	33	8900	3.5	69	2674	100	1273	2989838
nug14	6.7	87	64725	1.7	213	2164	100	(49.0 %)	1377491
nug15	8.4	153	306446	2.4	380	7126	100	(58.2 %)	1074602
nug16a	6.2	309	668740	1.8	658	8172	100	(71.3 %)	745267
nug16b	12.9	1847	13305362	4.2	779	92580	100	(56.6 %)	870424
nug17	7.5	1119	4308236	2.5	1281	37483	100	(75.6 %)	643233
nug18	7.8	6436	26403844	3.0	3499	196915	100	(76.9 %)	470523
nug20	9.1	(5.0 %)	29662294	3.6	(2.6 %)	326868	100	(86.8 %)	249492
Average	7.3	1110 (9)	5007361	2.4	764 (9)	38568	100	426 (3)	996613

Name: **nug_n**, where **n** is the number of facilities and of locations. *Gap:* $\left|\dfrac{Opt - Cont}{Opt}\right| * 100$, where *Cont* is the optimal value of the continuous relaxation, and *Opt* the optimal solution value of the instance. *Time* (Total CPU in seconds - limited to 3 hours): *Ref + Cplex* where *Ref* is the CPU time for *Step 1* and *Cplex* the CPU time for *Step 3*. We present the final gap (*g%*), $g = \left|\dfrac{Opt - b}{Opt}\right| * 100$ where b is the best bound obtained within the time limit. *Nodes:* Number of nodes visited by the B&B.

for solving $(\overline{QAP}_{\beta^*, \lambda^*, \lambda'^*})$ and $(\overline{QAP}_{\beta^*, \lambda^*, \lambda'^*, \phi^*})$ and for the direct submission of the problems to Cplex. We set parameters as follows: *Step 1:* Parameters axtol, aytol of CSDP [5] are set to 10^{-5}. The *Conic Bundle* [8] stops if the Euclidean norm of the latest aggregate subgradient is smaller than 10^{-1}, or if the precision of 10^{-5} is satisfied. *Step 3 or direct Cplex:* the relative mipgap is 10^{-5} and the absolute gap is 0.999. objdiff is set to 0.999, and varsel to 4. We use the multi-threading version of Cplex12.6.2 with up to 48 threads.

In Table 1, we compare the solution of the 10 Nugent instances with up to $n = 20$ by Algorithm 1, Algorithm 2, and Cplex12.6.2. We observe that Algorithms 1 and 2 outperform Cplex12.6.2 in terms of initial gap and Total CPU time. Then, we notice that the initial gap obtained with Algorithm 2 is always smaller (factor of 3 in average) than the gap obtained by Algorithm 1. As a consequence, the number of nodes visited during the B&B is significantly smaller with Algorithm 2 in comparison with Algorithm 1 (factor 128) and with Cplex12.6.2 (factor 29). Finally, Cplex12.6.2 is unable to solve instances with $n > 12$ while Algorithms 1 and 2 solve instances with up to $n = 18$ within the 3 hours of CPU time. A last remark concerns the CPU time for *Step 1*, for Algorithm 1 it is about 3 times faster than for Algorithm 2. This is due to the number of Constraints of (SDP_{R4}). However, this longer time spent is profitable when we consider the total computational time.

5   Conclusion

Our two algorithms differ by the sizes of their reformulations and by the tightness of their continuous relaxation bounds. While the first is compact since it remains in the same space of variables as in the initial (QAP), it is limited by its initial gap when we solve the reformulation. The second works in an extended space of variables and in spite of a larger size, it allows us to solve the considered instances faster. We report computational results and we show that our methods are efficient for solving instances of medium sizes.

References

1. Billionnet, A., Elloumi, S.: Best reduction of the quadratic semi-assignment problem. DAMATH: Discret. Appl. Math. Comb. Oper. Res. Comput. Sci. **109**, 197–213 (2001)
2. Billionnet, A., Elloumi, S., Lambert, A.: Extending the QCR method to the case of general mixed integer program. Math. Program. **131**(1), 381–401 (2012)
3. Billionnet, A., Elloumi, S., Lambert, A.: Exact quadratic convex reformulations of mixed-integer quadratically constrained problems. Math. Program. **158**(1), 235–266 (2016)
4. Billionnet, A., Elloumi, S., Lambert, A., Wiegele, A.: Using a conic bundle method to accelerate both phases of a quadratic convex reformulation. Inf. J. Comput. (2016, to appear)
5. Borchers, B.: CSDP, AC library for semidefinite programming. Optim. Methods Softw. **11**(1), 613–623 (1999)

6. Burkard, R.E.: Quadratic Assignment Problems, pp. 2741–2814. Springer, New York (2013)
7. Burkard, R.E., Karisch, S., Rendl, F.: QAPLIB - a quadratic assignment problem library. J. Glob. Optim. **10**, 391–403 (1997)
8. Helmberg, C.: Conic Bundle v0.3.10 (2011)
9. IBM-ILOG: IBM ILOG CPLEX 12.6 Reference Manual (2014). http://www-01.ibm.com/support/knowledgecenter/SSSA5P_12.6.0/ilog.odms.studio.help/Optimization_Studio/topics/COS_home.html
10. Roupin, F.: Semidefinite relaxations of the quadratic assignment problem in a lagrangian framework. Int. J. Math. Oper. Res. **1**, 144–162 (2009)
11. Sahni, S., Gonzalez, T.: P-complete approximation problems. J. Assoc. Comput. Mach. **23**, 555–565 (1976)
12. Zhao, Q., Karisch, S.E., Rendl, F., Wolkowicz, H.: Semidefinite relaxations for the quadratic assignment problem. J. Combin. Optim. **2**, 71–109 (1998)

Using Unified Model Checking to Verify Heaps

Xu Lu, Zhenhua Duan$^{(\boxtimes)}$, and Cong Tian$^{(\boxtimes)}$

ICTT and ISN Lab, Xidian University, Xi'an 710071, P.R. China
{zhhduan,ctian}@mail.xidian.edu.cn

Abstract. This paper addresses the problem of verifying heap evolution properties of pointer programs. To this end, a new unified model checking approach with MSVL (Modeling, Simulation and Verification Language) and PPTL^{SL} is presented. The former is an executable subset of PTL (Projection Temporal Logic) while the latter is an extension of PPTL (Propositional Projection Temporal Logic) with separation logic. MSVL is used to model pointer programs, and PPTL^{SL} to specify heap evolution properties. In addition, we implement a prototype in order to demonstrate our approach.

Keywords: Heap verification · Model checking · MSVL · PPTL · Separation logic

1 Introduction

Pointers are indispensable in real-world programs or applications. Reasoning about pointer programs is quite challenging since pointer usage is often complex and flexible. Potential bugs encountered in pointer programs such as null pointer dereference, memory leaks, or shape destruction are due to the nature of pointers. The problem is more serious for concurrent programs since we need to consider all possible execution sequences of processes. Alias analysis, as the name implies, is a point-to analysis which naively checks whether pointers can be aliased. Shape analysis is another form of pointer analysis that attempts to discover the possible shapes of heap structures. It aims to prove that these structures are not misused or corrupted.

Reynolds [1] proposes a famous Hoare-style logic known as separation logic which has received much attention. For the last decade, many works extend separation logic to do automated assertion checking [2] and shape analysis [3] in real-world applications. PTL (Pointer Assertion Logic) is a notation for expressing assertions about the heap structures of imperative languages. PALE (PAL Engine) is a complete implementation of PAL that encodes both programs and partial assertions as formulas in monadic second-order logic. However, loop invariants have to be manually provided so that it is not fully automatic.

This research is supported by the National Natural Science Foundation of China Grant Nos. 61133001, 61322202, 61420106004, and 91418201.

T.-H.H. Chan et al. (Eds.): COCOA 2016, LNCS 10043, pp. 735–743, 2016.
DOI: 10.1007/978-3-319-48749-6_55

In this paper we intend to apply the model checking framework to verify heap evolution properties. As the property-specification language we use a variant of temporal logic namely PPTL^{SL} [4]. PPTL^{SL} is a two-dimensional (spatial and temporal) logic by extending PPTL (Propositional Projection Temporal Logic) [5] with a decidable fragment of separation logic. For the program part, our method makes use of a temporal logic programming language, which is an executable subset of PTL (Projection Temporal Logic), called MSVL (Modeling, Simulation and Verification Language) [6], to model heap programs. PPTL^{SL} can be translated (preserving satisfiability) into a strict subset of PTL. Specifications and models lie in the same logic framework, hence the name unified model checking. The previous unified model checking approach [7] cannot verify heap evolution properties. We extend the approach in [7] by replacing PPTL with the more expressive specification language PPTL^{SL} such that heap evolution properties can be verified, and also the corresponding model checking approach is developed in this paper.

The work in [8] studies the problem of establishing temporal properties, including liveness properties of Java programs with evolving heaps. A specification language Evolution Temporal Logic (ETL) is defined which is a first-order linear temporal logic with transitive closure. ETL mainly focuses on describing behaviors of large granularity heap objects and high-level threads. Navigation Temporal Logic (NTL) [9] extends LTL with pointer assertions on single-reference structures including primitives for the birth and death of entities. The abstracted model checking algorithm for NTL is a non-trivial extension of the tableau based algorithm for LTL, which can be applied for both sequential and concurrent pointer programs. The major disadvantage of the approach of [9] is that it can only verify programs manipulating singly-linked lists. In [10], Rieger presents an abstraction and verification framework for pointer programs operating on unbounded heaps. In his work, two abstraction techniques are introduced, one is for singly-linked structures and the other employs context-free hyperedge replacement graph grammars to model more general heap structures. A two-dimensional (time and space) logic named maLTL is developed in [11] by combining temporal logic LTL and CTL, which is suitable to deal with pointers and heap management in the context of C programs. Since both dimensions of maLTL are realized by temporal logics that makes the difference between the two dimensions unclear.

2   Projection Temporal Logic

Let Var be a countable set of typed variables consisting of static and dynamic variables, and $Prop$ a countable set of propositions. $\mathbb{B}$ represents the boolean domain $\{true, false\}$, and $\mathbb{D}$ denotes the data domain. The terms e and formulas Q of PTL are given by the following grammar:

$$e ::= x \mid \bigcirc e \mid \ominus e \mid fun(e_1,\ldots,e_n)$$
$$Q ::= q \mid e_1 = e_2 \mid Pred(e_1,\ldots,e_n) \mid \neg Q \mid Q_1 \vee Q_2 \mid \exists y \colon Q \mid \bigcirc Q \mid Q^* \mid (Q_1,\ldots,Q_m)\,prj\,Q$$

where $q \in Prop$ is a proposition, $x \in Var$ a variable, and fun a function of arity n and $Pred$ is a predicate of arity n.

A state s is defined to be a pair (I_v, I_p) of state interpretations I_v and I_p, $I_v : Var \rightarrow \mathbb{D} \cup \{nil\}, I_p : Prop \rightarrow \mathbb{B}$. An interval $\sigma = \langle s_0, s_1, \ldots \rangle$ is a nonempty sequence of states, finite or infinite. The length of σ, denoted by $|\sigma|$, is ω if σ is infinite, otherwise it is the number of states minus one. To have a uniform notation for both finite and infinite intervals, we will use extended integers as indices. That is, we consider the set N_0 of non-negative integers, and define $N_\omega = N_0 \cup \{\omega\}$, and extend the comparison operators, $=, <, \leq$, to N_ω by considering $\omega = \omega$, and for all $i \in N_0$, $i < \omega$. Moreover, we define $\preceq$ as $\leq -\{(\omega, \omega)\}$. With such a notation, $\sigma_{(i..j)}(0 \leq i \preceq j \leq |\sigma|)$ denotes the subinterval $\langle s_i, \ldots, s_j \rangle$ and $\sigma^{(k)}(0 \leq k \preceq |\sigma|)$ denotes the suffix interval $\langle s_k, \ldots, s_{|\sigma|} \rangle$ of σ. The concatenation of σ with another interval σ' is denoted by $\sigma \cdot \sigma'$. Further, let $\sigma = \langle s_k, \ldots, s_{|\sigma|} \rangle$ be an interval and $r_1, \ldots, r_h$ be integers ($h \geq 1$) such that $0 \leq r_1 \leq r_2 \leq \cdots \leq r_h \preceq |\sigma|$. The projection of σ onto $r_1, \ldots, r_h$ is the interval, $\sigma \downarrow (r_1, \ldots, r_h) = \langle s_{t_1}, \ldots, s_{t_l} \rangle$, where $t_1, \ldots, t_l$ is obtained from $r_1, \ldots, r_h$ by deleting all duplicates.

An interpretation for a PTL formula is a triple $\mathcal{I} = (\sigma, k, j)$ where $\sigma = \langle s_0, s_1, \ldots \rangle$ is an interval, k a non-negative integer and j an integer or ω such that $k \preceq j \leq |\sigma|$. We write $(\sigma, k, j) \models Q$ to mean that a formula Q is interpreted over a sub-interval $\sigma_{(k..j)}$ of σ with the current state being s_k. The notation $s_k = (I_v^k, I_p^k)$ indexed by k represents the k-th state of an interval σ. The semantics of terms and PTL formulas are defined by:

$$\mathcal{I}[x] = \begin{cases} I_v^k[x] = I_v^i[x] & x \text{ is static,} \\ I_v^k[x] & \text{otherwise.} \end{cases} \qquad \mathcal{I}[fun(e_1, \ldots, e_n)] = \begin{cases} \mathcal{I}[fun](\mathcal{I}[e_1], \ldots, \mathcal{I}[e_n]) & \text{if } i < k, \\ nil & \text{otherwise.} \end{cases}$$

$$\mathcal{I}[\bigcirc e] = \begin{cases} (\sigma, i, k+1, j)[e] & \text{if } k < j, \\ nil & \text{otherwise.} \end{cases} \qquad \mathcal{I}[\ominus e] = \begin{cases} (\sigma, i, k-1, j)[e] & \text{if } i < k, \\ nil & \text{otherwise.} \end{cases}$$

$\mathcal{I} \models q$ iff $I_p^k(q) = true$. $\mathcal{I} \models e_1 = e_2$ iff $\mathcal{I}(e_1) = \mathcal{I}(e_2)$.

$\mathcal{I} \models Pred(e_1, \ldots, e_n)$ iff $Pred(\mathcal{I}[e_1], \ldots, \mathcal{I}[e_n]) = true$ and $\mathcal{I}[e_i] \neq nil$, for all i.

$\mathcal{I} \models \neg Q$ iff $\mathcal{I} \not\models Q$. $\mathcal{I} \models \exists y : Q$ iff $\exists \sigma'$ such that $\sigma'_{(k..j)} \stackrel{x}{=} \sigma_{(k..j)}$ and $(\sigma', k, j) \models Q$.

$\mathcal{I} \models Q_1 \vee Q_2$ iff $\mathcal{I} \models Q_1$ or $\mathcal{I} \models Q_2$. $\mathcal{I} \models \bigcirc Q$ iff $k < j$ and $(\sigma, k+1, j) \models Q$.

$\mathcal{I} \models (Q_1, \ldots, Q_m) prj\, Q$ iff $\exists k = r_0 \leq r_1 \leq \cdots \leq r_m \preceq j$ such that $(\sigma, r_0, r_1) \models Q_1$, $(\sigma, r_{l-1}, r_l) \models Q_l (1 < l \leq m), (\sigma', 0, 0, |\sigma'|) \models Q$ for one of the σ' : (a) $r_m < j$ and $\sigma' = \sigma \downarrow (r_0, \ldots, r_m) \cdot \sigma_{(r_m+1..j)}$ (b) $r_m = j$ and $\sigma' = \sigma \downarrow (r_0, \ldots, r_h)$ for $0 \leq h \leq m$.

$\mathcal{I} \models Q^*$ iff $\exists r_0, \ldots, r_n \in N_\omega$ such that $k = r_0 \leq r_1 \leq \cdots \leq r_{n-1} \preceq r_n = j(n \geq 0)$ and $(\sigma, r_0, r_1) \models Q$ and for all $1 < l \leq n, (\sigma, r_{l-1}, r_l) \models Q$; or $\exists k = r_0 \leq r_1 \leq r_2 \leq \cdots$ such that $\lim_{i \to \infty} r_i = \omega$ and $(\sigma, r_0, r_1) \models Q$ and for $l > 1, (\sigma, r_{l-1}, r_l) \models Q$.

A formula Q is satisfied over an interval σ, written $\sigma \models Q$, if $(\sigma, 0, |\sigma|) \models Q$ holds. Also we have the following derived formulas:

$$\varepsilon \stackrel{def}{=} \neg \bigcirc true \qquad more \stackrel{def}{=} \neg\varepsilon \qquad Q_1; Q_2 \stackrel{def}{=} (Q_1, Q_2)\, prj\, \varepsilon \qquad Q^+ \stackrel{def}{=} Q; Q^*$$

$$\Diamond Q \stackrel{def}{=} true; Q \qquad \Box Q \stackrel{def}{=} \neg\Diamond\neg Q \qquad len(n) \stackrel{def}{=} \bigcirc len(n-1) \qquad skip \stackrel{def}{=} len(1)$$

where ε (or $len(0)$) denotes an interval with zero length, ";" and "$+(*)$" are used to describe sequential and loop properties respectively.

3   Modeling, Simulation and Verification Language

MSVL is an executable temporal logic with framing technique which is recently extended with function calls [12] and groups of types such as basic data types, pointer types and struct types [13]. Thus it is capable of modeling pointer programs. The arithmetic and boolean expressions of MSVL can be defined as:

$$e ::= n \mid x \mid \bigcirc x \mid \ominus x \mid e_1 + e_2 \mid e_1 - e_2 \mid e_1 \times e_2 \mid e_1/e_2 \qquad b ::= \neg b \mid b_1 \vee b_2 \mid e_1 = e_2 \mid e_1 < e_2$$

Some useful elementary statements of MSVL can be inductively defined as follows. A convenient way to execute MSVL programs is to transform them into their equivalent normal forms (Definition 1).

$$\textbf{\textit{empty}} \stackrel{def}{=} \varepsilon \quad \textbf{\textit{x}} <== \textbf{\textit{e}} \stackrel{def}{=} x = e \wedge p_x \quad \textbf{\textit{Q}}_1 \, \textbf{\textit{and}} \, \textbf{\textit{Q}}_2 \stackrel{def}{=} Q_1 \wedge Q_2 \quad \textbf{\textit{Q}}_1 \, \textbf{\textit{or}} \, \textbf{\textit{Q}}_2 \stackrel{def}{=} Q_1 \vee Q_2$$

$$\textbf{\textit{x}} := \textbf{\textit{e}} \stackrel{def}{=} \bigcirc x = e \wedge \bigcirc p_x \wedge skip \quad \textbf{\textit{if}} \, \textbf{\textit{b}} \, \textbf{\textit{then}} \, \textbf{\textit{Q}}_1 \, \textbf{\textit{else}} \, \textbf{\textit{Q}}_2 \stackrel{def}{=} (b \rightarrow Q_1) \wedge (\neg b \rightarrow Q_2)$$

$$\textbf{\textit{while}} \, \textbf{\textit{b}} \, \textbf{\textit{do}} \, \textbf{\textit{Q}} \stackrel{def}{=} (Q \wedge b)^* \wedge \Box(empty \rightarrow \neg b) \quad \textbf{\textit{lbf}}(\textbf{\textit{x}}) \stackrel{def}{=} \neg af(x) \rightarrow \exists b : (\ominus x = b \wedge x = b)$$

$$\textbf{\textit{frame}}(\textbf{\textit{x}}) \stackrel{def}{=} \Box(more \rightarrow \bigcirc lbf(x)) \quad \textbf{\textit{Q}}_1 \| \textbf{\textit{Q}}_2 \stackrel{def}{=} (Q_1 \wedge (Q_2; true)) \vee (Q_2 \wedge (Q_1; true))$$

$$\textbf{\textit{await}}(\textbf{\textit{b}}) \stackrel{def}{=} frame(x_1, \ldots, x_n) \wedge \Box(empty \leftrightarrow b) \text{ where } x_i \in \{x \mid x \text{ appears in } b\}$$

Definition 1 (Normal Form of MSVL). An MSVL program Q is in normal form if $Q \stackrel{def}{=} \bigvee_{i=1}^{n'} Q_{e_i} \wedge \varepsilon \vee \bigvee_{j=1}^{n} Q_{c_j} \wedge \bigcirc Q_{f_j}$, where Q_{f_j} is a general MSVL program, whereas Q_{e_i} and Q_{c_j} are true or all are state formulas of the form: $(x_1 = e_1) \wedge \cdots \wedge (x_m = e_m) \wedge \dot{q}_{x_1} \wedge \cdots \wedge \dot{q}_{x_m}$. $\dot{q}$ denotes either q or $\neg q$.

The normal form divides the formula into two parts: the present part and the future part. A key conclusion is that any MSVL program can be reduced to its normal form [5,6]. Therefore, we can use an incremental way to execute MSVL programs based on normal form.

Theorem 1. Any MSVL program Q can be reduced to its normal form.

Using normal form of MSVL as a basis, a graph can be constructed, namely Normal Form Graph (NFG) [5,6], by recursively progressing the future part of a normal form, which explicitly illustrates the state space of an MSVL program.

3.1   Examples of MSVL Programs Manipulating Pointers

Producer-consumer program. Consider the producer-consumer problem encoded in MSVL below. The producer process and the consumer process share a buffer which is realized as a global singly-linked list. The producer repeatedly generates new items by allocating new memory heap cells, and adds them to the tail x of the buffer, whereas the consumer removes items from the head y of the buffer and disposes them.

The parallel operator "||" in MSVL considers the true concurrency semantics of programs. In order to simulate the interleaving semantics of the two concurrent processes, the *await* statement is employed to force a process to sleep when the waiting condition is false, otherwise the process will continue to execute.

```
struct Node { Node *nxt };
frame(PC, x, y, t, r) and (
   int PC<==0 and Node *x<==NULL, *y<==NULL, *t<==NULL, *r<==NULL and empty;
   y:=(Node*)malloc(sizeof(Node)) and x := (Node*)malloc(sizeof(Node));
   y->nxt:=x and (PC:=0 or PC:=1);
   //Producer                              //Consumer
   while(true) {                           while(true) {
     await(PC=0);                            if(x!=y) then {
     t:=(Node*)malloc(sizeof(Node));           await(PC=1);
     x->nxt:=t and (PC:=0 or PC:=1); ||        r:=y and (PC:=0 or PC:=1);
     await(PC=0);                              await(PC=1);
     x:=x->nxt and (PC:=0 or PC:=1)            y:=y->nxt and (PC:=0 or PC:=1);
   }                                           await(PC=1);
                                               next(free(r)) and (PC:=0 or PC:=1)
                                             } else {
                                               await(PC=1);(PC:=0 or PC:=1)
                                             }
                                           }
)
```

4   The Two-Dimensional Logic PPTL^{SL}

Previously, we integrate a decidable fragment of separation logic (referred to as SL) with PPTL to obtain a two-dimensional logic. The logic, referred to as PPTL^{SL} [4], allows us to express heap evolution properties. We assume a countable set $PVar$ of variables with pointer type, and a finite set Loc of memory locations. $PVal = Loc \cup \{null\}$ denotes the set of pointer values which are either locations or $null$. The syntax of PPTL^{SL} formulas P is defined by the grammar:

$$e ::= null \mid l \mid x \qquad \phi ::= e_1 = e_2 \mid e_0 \mapsto \{e_1, \ldots, e_n\} \mid \neg\phi \mid \phi_1 \vee \phi_2 \mid \phi_1 \# \phi_2 \mid \exists x : \phi$$
$$P ::= \phi \mid \neg P \mid P_1 \vee P_2 \mid \bigcirc P \mid (P_1, \ldots, P_m) \, prj \, P \mid P^*$$

$l \in Loc$, $x \in PVar$, ϕ represents SL formulas, and P PPTL^{SL} formulas. Formula $e_0 \mapsto \{e_1, \ldots, e_n\}$ denotes that e_0 points to $e_1, \ldots, e_n$, where e_0 represents an address in the heap and $e_1, \ldots, e_n$ the consecutive values held in that address.

The formula $\phi_1 \# \phi_2$ specifies properties holding respectively for disjoint portions of the current heap, one makes ϕ_1 true and the other makes ϕ_2 true. The temporal operators as well as their semantics are taken from PTL.

We refer to a pair (I_s, I_h) as a memory state, $I_s : PVar \rightharpoonup PVal, I_h : Loc \rightharpoonup \bigcup_{i=1}^n PVal^i$, where I_s represents a stack and I_h a heap. I_s serves as valuations of pointer variables and I_h as valuations of heap cells. We write $dom(f)$ to denote the domain of mapping f. Given two mappings f_1 and f_2, the notation $f_1 \perp f_2$ means that f_1 and f_2 have disjoint domains. Moreover, we use $f_1 \cdot f_2$ to denote the union of f_1 and f_2. The semantics of SL formulas is given by:

$$(I_s, I_h)[null] = null \qquad (I_s, I_h)[l] = l \qquad (I_s, I_h)[x] = I_h(x)$$

$$\boldsymbol{I_s, I_h} \models_{\boldsymbol{SL}} \boldsymbol{e_1 = e_2} \text{ iff } (I_s, I_h)[e_1] = (I_s, I_h)[e_2]. \quad \boldsymbol{I_s, I_h} \models_{\boldsymbol{SL}} \neg\phi \text{ iff } I_s, I_h \not\models_{SL} \phi.$$

$$\boldsymbol{I_s, I_h} \models_{\boldsymbol{SL}} \boldsymbol{e_0 \mapsto \{e_1, \dots, e_n\}} \text{ iff } dom(I_h) = \{(I_s, I_h)[e_0]\} \text{ and } I_h((I_s, I_h)[e_0]) = ((I_s, I_h)[e_1], \dots, (I_s, I_h)[e_n]). \quad \boldsymbol{I_s, I_h} \models_{\boldsymbol{SL}} \boldsymbol{\phi_1 \vee \phi_2} \text{ iff } I_s, I_h \models_{SL} \phi_1 \text{ or } I_s, I_h \models_{SL} \phi_2.$$

$$\boldsymbol{I_s, I_h} \models_{\boldsymbol{SL}} \boldsymbol{\phi_1 \# \phi_2} \text{ iff } \exists I_{h_1}, I_{h_2} : I_{h_1} \perp I_{h_2} \text{ and } I_h = I_{h_1} \cdot I_{h_2} \text{ and } I_s, I_{h_1} \models_{SL} \phi_1$$
and $I_s, I_{h_2} \models_{SL} \phi_2. \quad \boldsymbol{I_s, I_h} \models_{\boldsymbol{SL}} \boldsymbol{\exists x : \phi} \text{ iff } \exists v \in PVal \text{ such that } I_s[x \rightarrow v], I_h \models_{SL} \phi.$

The semantics of P is similar to that of Q since the only difference is their state formulas (formulas without temporal operators). Therefore, we only give the interpretation of state formulas, i.e., $\mathcal{I} \models \phi$ iff $I_s^k, I_h^k \models_{SL} \phi$. We abusively use the notation $\mathcal{I} \models P$, and in this case P is interpreted over an interval of memory states.

SL can describe various heap structures, we present the following derived formulas expressed in SL which are related to singly-linked lists.

$$e \mapsto \{-_1, \dots, -_n\} \overset{\text{def}}{=} \exists x_1, \dots, x_n : e \mapsto \{x_1, \dots, x_n\} \# true$$

$$e \hookrightarrow \{-, \dots, -\} \overset{\text{def}}{=} e \mapsto \{-, \dots, -\} \# true \quad alloc(e, n) \overset{\text{def}}{=} e \hookrightarrow \{-_1, \dots, -_n\}$$

$$alloc(e) \overset{\text{def}}{=} \bigvee_{i=1}^n alloc(e, i) \quad emp \overset{\text{def}}{=} \neg\exists x : alloc(x) \quad \sharp e \geq n \overset{\text{def}}{=} \#_{i=1}^n (\exists x_i : x_i \hookrightarrow \{e\})$$

$$e_1 \overset{\circlearrowleft}{\longrightarrow} e_2 \overset{\text{def}}{=} alloc(e_1, 1) \wedge (e_2 \neq e_1 \rightarrow \neg alloc(e_2, 1) \wedge \sharp e_1 = 0) \wedge (\forall x : x \neq e_2 \rightarrow$$
$$(\sharp x = 1 \rightarrow alloc(x, 1))) \wedge (\forall x : x \neq null \rightarrow \sharp x \leq 1) \wedge (\forall x : alloc(x) \rightarrow alloc(x, 1))$$

$$ls(e_1, e_2) \overset{\text{def}}{=} e_1 \overset{\circlearrowleft}{\longrightarrow} e_2 \wedge \neg(e_1 \overset{\circlearrowleft}{\longrightarrow} e_2 \# \neg emp)$$

$$e_1 \rightarrow^+ e_2 \overset{\text{def}}{=} ls(e_1, e_2) \# true \quad e_1 \rightarrow^* e_2 \overset{\text{def}}{=} e_1 = e_2 \vee e_1 \rightarrow^+ e_2$$

Here, $ls(e_1, e_2)$ describes a list segment starting from the location e_1 to e_2, $e_1 \rightarrow^+ e_2$ and $e_1 \rightarrow^* e_2$ mean that e_2 is reachable from e_1 via certain pointer links. emp denotes an empty heap, and $alloc(e)$ indicates the address e is allocated in the current heap. Formulas describing other heap structures can also be derived.

4.1 Specify Heap Evolution Properties

For the producer-consumer program with a shared list, some interesting heap evolution properties can be specified by PPTL^{SL}:

(1) Absence of memory leaks, i.e., any item can be reached by certain variable during the execution of the program: $\Box(\forall z : alloc(z) \rightarrow (x \rightarrow^* z \lor y \rightarrow^* z \lor t \rightarrow^* z \lor r \rightarrow^* z))$.
(2) The tail of the list is never deleted nor disconnected from the head: $\bigcirc^2 \Box(alloc(x) \land y \rightarrow^* x)$.
(3) Shape integrity of the buffer, i.e., the shape of the buffer is repeatedly formed as a linked list within every five time units: $\bigcirc^2((len(5) \land \Diamond ls(y, null))^+)$.

In [4], we have proved that PPTL^{SL} is decidable by an equisatisfiable translation. The key idea is that all the formulas expressing heaps in SL are reduced into the first-order theory. Since the model of first-order set has no heap ingredient, we use extra variables to simulate it. Let H denote a vector of n tuples of pointer variables, i.e., $H = ((H_{1,1}, \ldots, H_{1,m_1}), \ldots, (H_{n,1}, \ldots, H_{n,m_n}))$. The translations $f(\phi, H)$ and $F(P, H)$ generate an equisatisfiable state formula ϕ' to ϕ and an equisatisfiable temporal formula P' to P respectively. Both ϕ' and P' belong to PTL. The detailed definitions of f and F can be found in [4]. Analogous to MSVL, by using a very similar approach, we can construct a graph structure (also called NFG) that explicitly characterizes the model for any reduced PPTL^{SL} formula. The detailed proofs are given in [4], here we only give a brief summary.

Theorem 2. PPTL^{SL} is decidable, and its complexity is the same as PPTL, i.e., non-elementary.

5   Model Checking with MSVL and PPTL^{SL}

Our model checking approach is similar to the traditional automata-based model checking except that ours is based on NFG. The starting point is a pointer program modeled by an MSVL program M, and a PPTL^{SL} formula P that formalizes the desired heap evolution property on M.

In general, the model checking procedure in this framework first creates the NFG G_M of an MSVL program and the NFG $G_{\neg P}$ of the negation of the input PPTL^{SL} formula, then constructs the product G of the two NFGs. The nodes (edges) of G are conjunctions of nodes (edges) in G_M and $G_{\neg P}$. If there exist a valid path in G, a counterexample is found, otherwise M satisfies P.

We have developed a unified model checking tool (prototype) based on the approach presented in this paper. As shown Fig. 1, the tool structure consists of three essential modules: MSV, PPTL^{SL} solver and unified model checker. An MSVL program M is fed into MSV, and a property P is given to the PPTL^{SL} solver. MSV constructs the NFG of M and PPTL^{SL} solver builds the NFG of $\neg P$. The Model checker does not try to build the complete production of the two NFGs in practice. Instead, it works in "on the fly" manner and tries to find one valid path as early as possible. The SMT solver Z3 is called when checking whether an edge in the product NFG is satisfied or not. We have successfully verified the producer-consumer program with respect to the corresponding heap evolution properties mentioned in Sect. 4.

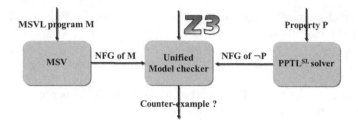

Fig. 1. Tool architecture.

6   Conclusion

In this paper, we propose a unified model checking approach with MSVL and PPTL^{SL}. We can apply this approach to verify heap evolution properties of pointer programs, including both safety and liveness properties on heap structures. Since PPTL^{SL} is able to be reduced to a subset of PTL so that programs and properties both belong to the same logic framework which makes the verification more convenient. We have developed a model checking tool that exploits the SMT solver Z3 as the verification engine. In the future, more case studies on various heap structures in addition to singly-linked structures will be carried out.

References

1. Reynolds, J.C.: Separation logic: a logic for shared mutable data structures. In: 17th Annual IEEE Symposium on Logic in Computer Science, Computer Society, pp. 55–74. IEEE, Washington (2002)
2. Berdine, J., Calcagno, C., O'Hearn, P.W.: Symbolic execution with separation logic. In: Yi, K. (ed.) APLAS 2005. LNCS, vol. 3780, pp. 52–68. Springer, Heidelberg (2005)
3. Calcagno, C., Distefano, D., O'Hearn, P.W., Yang, H.: Compositional shape analysis by means of bi-abduction. J. ACM (JACM) **58**(6), 26 (2011)
4. Lu, X., Duan, Z., Tian, C.: Extending PPTL for verifying heap evolution properties. arXiv preprint arXiv:1507.08426 (2015)
5. Duan, Z.: An extended interval temporal logic and a framing technique for temporal logic programming. Ph.D. thesis, University of Newcastle upon Tyne (1996)
6. Duan, Z., Yang, X., Koutny, M.: Framed temporal logic programming. Sci. Comput. Program. **70**(1), 31–61 (2008)
7. Duan, Z., Tian, C.: A unified model checking approach with projection temporal logic. In: Liu, S., Maibaum, T., Araki, K. (eds.) ICFEM 2008. LNCS, vol. 5256, pp. 167–186. Springer, Heidelberg (2008). doi:10.1007/978-3-540-88194-0_12
8. Yahav, E., Reps, T.W., Sagiv, S., Wilhelm, R.: Verifying temporal heap properties specified via evolution logic. Logic J. IGPL **14**(5), 755–783 (2006)
9. Distefano, D., Katoen, J.-P., Rensink, A.: Safety and liveness in concurrent pointer programs. In: Boer, F.S., Bonsangue, M.M., Graf, S., Roever, W.-P. (eds.) FMCO 2005. LNCS, vol. 4111, pp. 280–312. Springer, Heidelberg (2006). doi:10.1007/11804192_14

10. Rieger, S.: Verification of pointer programs. Ph.D. thesis, RWTH Aachen University (2009)
11. del Mar Gallardo, M., Merino, P., Sanán, D.: Model checking dynamic memory allocation in operating systems. J. Autom. Reasoning **42**(2–4), 229–264 (2009)
12. Zhang, N., Duan, Z., Tian, C.: Extending MSVL with function calls. In: Merz, S., Pang, J. (eds.) ICFEM 2014. LNCS, vol. 8829, pp. 446–458. Springer, Heidelberg (2014). doi:10.1007/978-3-319-11737-9_29
13. Wang, X., Duan, Z., Zhao, L.: Formalizing and implementing types in MSVL. In: Liu, S., Duan, Z. (eds.) SOFL+MSVL 2013. LNCS, vol. 8332, pp. 60–73. Springer, Heidelberg (2014)

A Filtering Heuristic for the Computation of Minimum-Volume Enclosing Ellipsoids

Linus Källberg$^{(\boxtimes)}$ and Thomas Larsson

Mälardalen University, Västerås, Sweden
{linus.kallberg,thomas.larsson}@mdh.se

Abstract. We study heuristics to accelerate existing state-of-the-art algorithms for the minimum-volume enclosing ellipsoid problem. We propose a new filtering heuristic that can significantly reduce the number of distance computations performed in algorithms derived from Khachiyan's first-order algorithm. Our experiments indicate that in high dimensions, the filtering heuristic is more effective than the elimination heuristic proposed by Harman and Pronzato. In lower dimensions, the elimination heuristic is superior.

Keywords: Minimum-volume enclosing ellipsoids · Löwner ellipsoids · Approximation algorithms · Heuristics · Filtering · Pruning

1   Introduction

The problem of computing the minimum-volume enclosing ellipsoid (MVEE), also known as the Löwner ellipsoid, of a given set of points recurs in a variety of application areas such as computational geometry [5,20], computer graphics [4], robotics [15], statistics [2,18], and optimal experimental design [17]. For example, MVEEs can provide satisfactory approximations of complex geometric shapes, which makes them useful in, e.g., collision detection [14]. Indeed, it was shown by John [9] that the MVEE of any d-dimensional set, if scaled about its center by $1/d$, is fully enclosed by the convex hull of the set.

Given a set of points $\mathcal{P} := \{p_1, p_2, \ldots, p_n\} \subset \mathbb{R}^d$, an enclosing ellipsoid $\mathcal{E}_{Q,c}$ of $\mathcal{P}$ is specified by a $d \times d$ symmetric positive-definite matrix Q and a center point $c \in \mathbb{R}^d$ satisfying

$$(p_i - c)^\top Q(p_i - c) \leq 1 \quad \text{for } i = 1, 2, \ldots, n. \tag{1}$$

The orthonormal axes of $\mathcal{E}_{Q,c}$ are given by the eigenvectors of Q, and their lengths are given by the square roots of the reciprocals of the corresponding eigenvalues. Provided that the affine hull of $\mathcal{P}$ is $\mathbb{R}^d$, $\mathcal{E}_{Q,c}$ has a positive volume given by

$$\text{vol}\,\mathcal{E}_{Q,c} = \eta \det Q^{-1/2}, \tag{2}$$

where η is the volume of the d-dimensional unit ball. For ease of discussion, we will assume this is the case throughout the rest of this paper. The enclosing

© Springer International Publishing AG 2016
T.-H.H. Chan et al. (Eds.): COCOA 2016, LNCS 10043, pp. 744–753, 2016.
DOI: 10.1007/978-3-319-48749-6_56

ellipsoid of $\mathcal{P}$ with minimum volume, which we denote by MVEE($\mathcal{P}$), is thus given by the Q and c that minimize (2) subject to the n conditions (1) and the condition that Q is symmetric positive-definite.

In low dimensions, the MVEE problem can be solved exactly by locating the set of at most $d(d+3)/2$ *support points*, which are points p_i that satisfy (1) with equality [5,20]. In applications dealing with higher dimensions, on the other hand, it is more practical to settle for approximate solutions. Given a tolerance parameter $\epsilon > 0$, a $(1 + \epsilon)$-approximation of MVEE($\mathcal{P}$) is an enclosing ellipsoid $\mathcal{E}_\epsilon$ of $\mathcal{P}$ such that $\mathrm{vol}\,\mathcal{E}_\epsilon \leq (1 + \epsilon)\,\mathrm{vol}\,\mathrm{MVEE}(\mathcal{P})$. The problem of efficiently computing $(1+\epsilon)$-approximate MVEEs has received considerable attention, and several algorithms have been proposed in the literature [12,13,16,18,19].

In this paper, we study acceleration techniques that can be applied to some of these algorithms. Specifically, we focus on their application to the first-order algorithm by Khachiyan [12], as well as some improved variants derived from it [13,19]. Khachiyan's algorithm can be viewed as the Franke–Wolfe algorithm adapted for a dual formulation of the MVEE problem. It computes a $(1 + \epsilon)$-approximation of the MVEE in $O(d([(1 + \epsilon)^{2/(d+1)} - 1]^{-1} + \log d + \log\log n))$ iterations, or $O(d(d/\epsilon + \log\log n))$ iterations if $\epsilon \in (0, 1]$. The cost of each iteration is $O(nd)$, which is due to a search for the farthest point in $\mathcal{P}$ from the center of the current candidate ellipsoid, measured in the ellipsoidal norm. We describe a new filtering heuristic to reduce the number of distance computations performed during this search. This is achieved by caching distance information across iterations and using conservative bounds on the new distances. As this search constitutes the main bottleneck of the algorithm, this can improve the performance substantially, especially for large problems with $n \gg d$.

2   Related Work

Prior to this work, a few acceleration heuristics have been developed for the MVEE problem. Galkovskyi et al. [7] propose a *domination* heuristic for the general class of LP-type optimization problems, and present a specialization for the MVEE problem. The specialized heuristic is mainly applicable to MVEE algorithms that generate candidate support sets as part of their operation. The key idea is to compute the convex hull of each such candidate support set, and then eliminate any input points that fall in its interior, as these points clearly cannot be on the boundary of the MVEE. The main drawback of this approach is that computing these convex hulls becomes prohibitively slow even in moderately high dimensions. Nevertheless, they report large performance gains for the two-dimensional case.

Harman and Pronzato [8] present a heuristic for the D-optimal design problem, which is equivalent to a dual formulation of the MVEE problem. Based on a conservative bound on the distance from the center of the current ellipsoid to the boundary of the MVEE, their heuristic can identify and remove points from the input that cannot be support points. The bound itself takes $O(nd)$ time to compute, or $O(1)$ time if the distance to the farthest input point is already given.

Thus, the heuristic has a negligible overhead in any dimension. We will use this heuristic for comparison in our experimental evaluation in Sect. 5.

Similar elimination heuristics have been investigated for the problem of computing the minimum enclosing ball (MEB) of a set of points [1,11]. Moreover, we have previously developed [10] a filtering heuristic for the MEB problem that, similarly to the heuristic proposed here, reduces the number of distance computations by reusing distances computed in earlier iterations of the MEB algorithm. Being specifically designed for the MEB problem, however, this heuristic is not immediately applicable to the computation of MVEEs.

3   Khachiyan's Algorithm

We begin with a brief review Khachiyan's algorithm and its variants. For more details, we refer to the original publications [12,13,19]. The algorithm is commonly described for the special case where $\mathcal{P} := \{p_1, p_2, \ldots, p_n\}$ is centrally symmetric, i.e., where $\mathcal{P} = -\mathcal{P}$, in which case the MVEE is always centered at the origin. The general case can be handled by replacing $\mathcal{P}$ by the lifted set $\hat{\mathcal{P}} := \{\pm\hat{p}_i : i = 1, 2, \ldots, n\} \subset \mathbb{R}^{d+1}$, where $\hat{p}_i := (p_i^\top, 1)^\top$, which is centrally symmetric in $d + 1$ dimensions. A $(1 + \epsilon)$-approximation of MVEE($\mathcal{P}$) can then be recovered from the intersection of the computed $(1 + \epsilon)$-approximation of MVEE($\hat{\mathcal{P}}$) and the hyperplane $\mathcal{H} := \{(x^\top, 1)^\top : x \in \mathbb{R}^d\} \subset \mathbb{R}^{d+1}$. We will use this lifting in the subsequent presentation.

The algorithm iterates through a sequence of feasible solutions to the following dual of the MVEE problem:

$$\text{maximize}_u \ \log \det V(u)$$

$$\text{subject to } \sum_{i=1}^{n} u_i = 1,$$

$$u_i \geq 0 \quad \text{for } i = 1, 2, \ldots, n,$$

where $u := (u_1, u_2, \ldots, u_n)$ is the decision variable and

$$V(u) := \sum_{i=1}^{n} u_i \hat{p}_i \hat{p}_i^\top.$$

This problem is also known as the D-optimal design problem [17]. Each feasible solution u^k gives a trial ellipsoid $\mathcal{E}_{M_k,0} \subset \mathbb{R}^{d+1}$ by

$$M_k := \frac{1}{d+1} V(u^k)^{-1}. \tag{3}$$

Khachiyan's original algorithm starts out with the initial solution $u = u^1$, where $u_i^1 := 1/n$ for $i = 1, 2, \ldots, n$. Then in each iteration k, the farthest point $\hat{p}_j$ in the ellipsoidal norm $\sqrt{x^\top M_k x}$ is located, and its squared distance κ is recorded, i.e.,

$$j := \arg\max_{i=1}^{n} \hat{p}_i^\top M_k \hat{p}_i, \qquad \kappa := \hat{p}_j^\top M_k \hat{p}_j.$$

Note that due to weak duality, $\mathcal{E}_{M_k,0}$ given by (3) is an under-approximation of $\text{MVEE}(\hat{\mathcal{P}})$, i.e., $\text{vol}\,\mathcal{E}_{M_k,0} \leq \text{vol}\,\text{MVEE}(\hat{\mathcal{P}})$. Thus, $\hat{p}_j$ necessarily lies outside of $\mathcal{E}_{M_k,0}$ or on its boundary, which means that $\kappa \geq 1$. If $\kappa \leq (1+\epsilon)^{2/(d+1)}$ holds, then the algorithm terminates. The current ellipsoid can then be scaled about its center by $\sqrt{\kappa}$ to provide the sought $(1+\epsilon)$-approximation of $\text{MVEE}(\hat{\mathcal{P}})$. Otherwise, the algorithm proceeds with the updated solution

$$u_i^{k+1} := \begin{cases} (1-\beta)u_i^k + \beta & i = j, \\ (1-\beta)u_i^k & i \neq j, \end{cases} \tag{4}$$

where $\beta := (\kappa - 1)/((d+1)\kappa - 1)$.

The matrices $V(u^{k+1})$ and $V(u^{k+1})^{-1}$ can be computed in $O(d^2)$ time as

$$V(u^{k+1}) = (1-\beta)V(u^k) + \beta\hat{p}_j\hat{p}_j^\top, \quad V(u^{k+1})^{-1} = \frac{1}{1-\beta}V(u^k)^{-1} - \gamma\hat{q}_j\hat{q}_j^\top, \tag{5}$$

where $\gamma := [(1-\beta)^2/\beta + (1-\beta)(d+1)\kappa]^{-1}$ and $\hat{q}_j := V(u^k)^{-1}\hat{p}_j$. Similarly, if denoting $g_i^k := \hat{p}_i^\top M_k\hat{p}_i$, the next distance $g_i^{k+1} := \hat{p}_i^\top M_{k+1}\hat{p}_i$ is given by

$$g_i^{k+1} = \frac{1}{1-\beta}g_i^k - \gamma(\hat{q}_j^\top\hat{p}_i)^2/(d+1). \tag{6}$$

By caching each distance g_i^k, the new distance g_i^{k+1} can thus be computed in $O(d)$ time as opposed to $O(d^2)$ time. The total cost of each iteration becomes $O(d^2 + nd)$, which is $O(nd)$ since we assume that the affine hull of $\mathcal{P}$ is $\mathbb{R}^d$.

Kumar and Yıldırım [13] present a more sophisticated procedure to compute an initial solution u^1. It uses the volume approximation method by Betke and Henk [3] to find $\min(2d, n)$ affinely independent points in $\mathcal{P}$ that define a reasonably large initial ellipsoid. The multipliers u_i^1 corresponding to these points are set to $1/\min(2d, n)$ and the remaining multipliers are set to zero. This initialization reduces the number of iterations to $O(d([(1+\epsilon)^{2/(d+1)} - 1]^{-1} + \log d))$, or $O(d^2/\epsilon)$ if $\epsilon \in (0,1]$.

As a byproduct, their modified algorithm also computes an ϵ-*coreset* $\mathcal{X}$, which is a subset of $\mathcal{P}$ with the property that it serves as a good approximation of the whole set. Specifically, this means that there exists a $(1+\epsilon)$-approximation $\mathcal{E}_\epsilon$ of $\text{MVEE}(\mathcal{X})$ such that $\mathcal{E}_\epsilon \supset \mathcal{P}$. Because $\text{vol}\,\text{MVEE}(\mathcal{X}) \leq \text{vol}\,\text{MVEE}(\mathcal{P})$, such an ellipsoid is at the same time a $(1+\epsilon)$-approximation of $\text{MVEE}(\mathcal{P})$. The returned coreset contains exactly the points p_i such that $u_i > 0$ in the final solution, and satisfies $|\mathcal{X}| \in O(d^2/\epsilon)$ for $\epsilon \in (0,1]$.

Todd and Yıldırım [19] further introduce *away steps*. Apart from finding $j_+ := j$ and $\kappa_+ := \kappa$ in each iteration, their modification also locates $j_- := \arg\min_{1 \leq i \leq n: u_i^k > 0} \hat{p}_i^\top M_k\hat{p}_i$ and $\kappa_- := \hat{p}_{j_-}^\top M_k\hat{p}_{j_-}$, i.e., the closest point $\hat{p}_{j_-}$ to the center of the current ellipsoid among points $\hat{p}_i$ with a positive weight u_i^k (i.e., coreset points). If $\kappa_+ - 1 \geq 1 - \kappa_-$ holds, i.e., the point $\hat{p}_{j_-}$ is not farther from the ellipsoid boundary than $\hat{p}_{j_+}$, the solution is updated as in the original algorithm. Otherwise, it is updated using (4) with the substitutions $j = j_-$ and

$$\beta = \max\left(\frac{\kappa_- - 1}{(d+1)\kappa_- - 1}, \frac{u_{j_-}}{u_{j_-} - 1}\right). \tag{7}$$

If this results in $u_{j_-}^{k+1} = 0$, the point p_{j_-} is dropped from the coreset. Their modified algorithm retains the same asymptotic bounds as the Kumar–Yıldırım algorithm on the number of iterations and the size of the coreset, but typically requires fewer iterations and computes a smaller coreset in practice. It also uses a stronger termination criterion that helps reducing the coreset size.

4   Approach

In iteration $k+1$, the farthest point $\hat{p}_j$ given by $j := \arg\max_{i=1}^n \hat{p}_i^\top M_{k+1}\hat{p}_i$ can be found in $O(nd)$ time by applying (6) for each of the n points and recording the largest distance. The goal of our heuristic is to reduce the number of applications of (6) from n to a much smaller value, while still finding the true farthest point. Using the matrix M_k from the previous iteration, define the matrix $H := M_k^{1/2} M_{k+1}^{-1} M_k^{1/2}$ and denote its eigenvalues by $\lambda_1 \le \lambda_2 \le \cdots \le \lambda_{d+1}$. The following then holds for the farthest point $\hat{p}_j$:

$$1 \le \hat{p}_j^\top M_{k+1}\hat{p}_j = \hat{p}_j^\top M_k^{1/2} H^{-1} M_k^{1/2}\hat{p}_j \le \frac{1}{\lambda_1}\hat{p}_j^\top M_k\hat{p}_j,$$

where the last step follows from $1/\lambda_1$ being the largest eigenvalue of H^{-1}. Since this implies $\hat{p}_j^\top M_k\hat{p}_j \ge \lambda_1$, any point $\hat{p}_i$ satisfying

$$\hat{p}_i^\top M_k\hat{p}_i < \lambda_1 \tag{8}$$

cannot be the farthest point $\hat{p}_j$. Figure 1 provides a geometric interpretation of this. Condition (8) can be evaluated in constant time if the distance $\hat{p}_i^\top M_k\hat{p}_i$ was cached from iteration k. When it is satisfied, the computation of $\hat{p}_i^\top M_{k+1}\hat{p}_i$ can be skipped, or filtered out, which saves a $O(d)$ cost when using (6). However, because this means that the exact distance is never computed in iteration $k+1$, the same condition cannot be used again in iteration $k+2$. We return shortly to how this problem can be solved.

Let $\mathrm{eig}(\cdot)$ denote the set of distinct eigenvalues of a matrix. Letting $V_k := V(u^k)$ and $V_{k+1} := V(u^{k+1})$, and using (3) and (5), we have

$$
\begin{aligned}
\mathrm{eig}(H) &= \mathrm{eig}(M_k^{1/2} M_{k+1}^{-1} M_k^{1/2}) \\
&= \mathrm{eig}(V_k^{-1/2} V_{k+1} V_k^{-1/2}) \\
&= \mathrm{eig}(V_k^{-1/2}((1-\beta)V_k + \beta\hat{p}_j\hat{p}_j^\top)V_k^{-1/2}) \\
&= \mathrm{eig}((1-\beta)I + \beta(V_k^{-1/2}\hat{p}_j)(V_k^{-1/2}\hat{p}_j)^\top) \\
&= \{1-\beta, 1-\beta + \beta(V_k^{-1/2}\hat{p}_j)^\top(V_k^{-1/2}\hat{p}_j)\} \\
&= \{1-\beta, 1-\beta + \beta(d+1)\kappa\},
\end{aligned} \tag{9}
$$

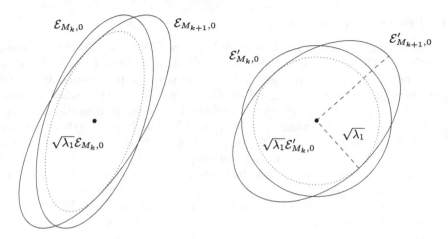

Fig. 1. Geometric interpretation of condition (8). Left: the trial ellipsoids $\mathcal{E}_{M_k,0}$ and $\mathcal{E}_{M_{k+1},0}$. Note that condition (8) implicitly defines the ellipsoid $\sqrt{\lambda_1}\mathcal{E}_{M_k,0}$ (shown dotted), which is $\mathcal{E}_{M_k,0}$ scaled about its center by $\sqrt{\lambda_1}$. Clearly, only points outside of $\sqrt{\lambda_1}\mathcal{E}_{M_k,0}$ make candidates to be the farthest point outside of $\mathcal{E}_{M_{k+1},0}$. Right: The same ellipsoids under the linear transformation that turns $\mathcal{E}_{M_k,0}$ into the unit ball. Here it is clear that $\sqrt{\lambda_1}$ is simply the length of the shortest axis of the transformed $\mathcal{E}_{M_{k+1},0}$. Therefore, $\sqrt{\lambda_1}\mathcal{E}_{M_k,0}$ is $\mathcal{E}_{M_k,0}$ shrunk just enough to fit inside $\mathcal{E}_{M_{k+1},0}$.

where I is the identity matrix. The expansion of $\mathrm{eig}(\cdot)$ above follows from the result of [6]. Note that the eigenvalue $1 - \beta$ has algebraic multiplicity d, but is included only once in the set $\mathrm{eig}(H)$. Thus, we have $\lambda_1 = \min\mathrm{eig}(H) = \min(1 - \beta, 1 - \beta + \beta(d + 1)\kappa)$. In Khachiyan's original algorithm, it always holds that $\beta > 0$, which gives $\min\mathrm{eig}(H) = 1 - \beta$. In Todd and Yıldırım's modified algorithm, however, $\beta < 0$ after an away step (see (7)), which gives $\lambda_1 = \min\mathrm{eig}(H) = 1 - \beta + \beta(d + 1)\kappa_-$. Note that in this case, j and κ in (9) should be replaced by j_- and κ_-, respectively.

By replacing the left-hand side of (8) by an upper bound $h_i^k \geq \hat{p}_i^\top M_k \hat{p}_i$, we get the more conservative condition

$$h_i^k < \lambda_1. \tag{10}$$

Note that this condition implies (8), and can therefore be used instead as the filtering condition. If condition (10) fails, the corresponding upper bound h_i^{k+1} for the next iteration is simply set to the exact value of $\hat{p}_i^\top M_{k+1} \hat{p}_i$. Otherwise, h_i^{k+1} is set to h_i^k/λ_1, which gives an upper bound on $\hat{p}_i^\top M_{k+1} \hat{p}_i$ because

$$\hat{p}_i^\top M_{k+1} \hat{p}_i \leq \frac{1}{\lambda_1} \hat{p}_i^\top M_k \hat{p}_i \leq h_i^k/\lambda_1.$$

Thus, regardless of whether the filtering condition succeeds or not, each h_i^{k+1} can be computed in constant time.

4.1 Practical Considerations

We now address a few important practical considerations for the proposed heuristic. Firstly, note that if the distance $\hat{p}_i^\top M_k \hat{p}_i$ is filtered out in iteration k, and the filtering condition then fails for the point $\hat{p}_i$ in iteration $k + 1$, we cannot use (6) to compute $\hat{p}_i^\top M_{k+1} \hat{p}_i$ in $O(d)$ time because the cached distance g_i^k is not available. Our solution to this problem is to store, alongside each cached distance g_i, the iteration number when g_i was last updated. Furthermore, the values of β, γ, and $\hat{q}_j$ from the $d + 1$ most recent iterations are saved in a cyclic buffer of size $O(d^2)$. When a filtering attempt fails, but no more than $d + 1$ iterations have passed since g_i was last updated, these saved values are used to compute the distance using repeated applications of (6). This takes $O(md)$ time, where m is the number of iterations since g_i was updated. In effect, this "reverts" all filtering performed for point $\hat{p}_i$ in the last m iterations. If more than $d + 1$ iterations have passed, i.e., $m > d + 1$, it is more efficient to compute the distance as $\hat{p}_i^\top M_{k+1} \hat{p}_i$ in $O(d^2)$ time. In difficult cases where the filtering heuristic is ineffective, this measure ensures that the time complexity of the algorithm remains unaffected.

Secondly, we note that if the search for the farthest point is implemented as a sequential loop, then the right-hand side of the filtering condition (10) can be replaced by $\lambda_1 \kappa'$ whenever a new largest distance $\kappa' > 1$ is encountered during the search. This is clearly beneficial because it progressively improves the filtering effectiveness throughout the loop.

Finally, we point out that when a coreset is maintained, as is done in our implementation used in Sect. 5, additional optimizations are possible. We begin each iteration by scanning the coreset $\mathcal{X}$ for the farthest point. This gives a reasonably large candidate value $\kappa_\mathcal{X}$ for κ. Then the filtering condition $h_i^k < \lambda_1 \kappa_\mathcal{X}$ is evaluated for $i = 1, 2, \ldots, n$, while at the same time setting each h_i^{k+1} to h_i^k / λ_1. No exact distances are computed at this stage. Instead, the values of i for which the filtering test fails are collected in a list. Then a separate loop computes the exact distances for the corresponding points $\hat{p}_i$ and updates the corresponding upper bounds h_i^{k+1}. As the main overhead of the heuristic comes from computing the h_i^{k+1} and testing the filtering condition n times in each iteration, it is beneficial to do this in a tight dedicated loop in this way. Although this means that the right-hand side of the filtering condition cannot be continuously enlarged during the farthest point search, we found empirically that the benefits of using a dedicated filtering loop outweighed the benefits of the above-mentioned optimization.

5 Experiments

For our experimental evaluation, we implemented the Todd–Yıldırım algorithm, since it is the most efficient variant in practice. The algorithm was implemented in C++ under Visual Studio 2012, and was run on a laptop PC with a 2.7 GHz CPU and 16 GB of main memory. The filtering heuristic could be turned on and off in the code using preprocessor directives. For comparison, we also implemented the elimination heuristic proposed by Harman and Pronzato [8], which identifies and removes points from the input that cannot be support points of the MVEE. For input we generated random point sets from a d-variate normal distribution, a uniform distribution confined to a d-dimensional hypercube, and a uniform distribution confined to a d-dimensional ball. In all runs, we used double-precision floating-point arithmetic, with $\epsilon = 10^{-3}$ as the target approximation quality.

Our computational results are shown in Table 1. The elimination heuristic tends to be more effective than the filtering heuristic for low-dimensional problems. This also leads to superior speedups in these dimensions. Indeed, the speedup factors differ by almost an order of magnitude in a few cases. The elimination heuristic is thus clearly the one to recommend for such input cases. As the dimension is increased, however, the relative difference in effectiveness decreases and eventually filtering becomes more effective. This can be observed appreciably from $d = 25$ on the first two distributions. On the uniform distribution in a ball, filtering is consistently more effective. We remark that this type of input is more challenging for the algorithm itself—which requires an order of magnitude more iterations than on the normal distribution—as well as for the heuristics. This is because a large portion of the input points tend to be close to the boundary of the ball containing the points, particularly in the higher dimensions. However, our filtering method appears to handle this type of input better than the elimination method.

The elimination heuristic still exhibits the largest performance improvements even in some cases where it is less effective than filtering in reducing distance computations. This is explained by the larger overhead of the filtering heuristic: in each iteration, all of the n upper bounds h_i^{k+1} need to be computed and the filtering condition tested for each point, resulting in a $O(n)$ overhead per iteration. With the elimination heuristic, on the other hand, it only takes $O(n')$ time per iteration to test the elimination condition for the n' points that have not yet been eliminated, which might be significantly fewer than the points in the original set. However, as the dimension increases, the $O(n)$ overhead of the filtering heuristic becomes increasingly negligible compared to the $O(d)$ saving of each skipped distance computation. Furthermore, we remark that the heuristic is applicable also for other types of input objects, such as ellipsoids, and might pay off even more in those cases due to the higher costs involved in computing the distances. While this is true for the elimination heuristic as well, its advantage of having a smaller overhead can be expected to be negligible in those scenarios. For example, Yıldırım [21] proposes a procedure that takes $O(d^C)$ arithmetic operations per input ellipsoid, where C is a constant larger than 3.

It is therefore an interesting direction for future work to evaluate both heuristics for MVEE computations with, e.g., sets of ellipsoids as input. It would also be interesting to evaluate a combination of the heuristics. Judging from Table 1, the two appear to complement each other: the elimination heuristic has the largest effect in low dimensions, whereas the filtering heuristic is more effective in the higher dimensions.

Table 1. Each row displays average results from ten point sets randomly generated with the shown dimensions d and n. Execution times are shown for the baseline version of the algorithm, with no heuristics enabled. Speedup factors and effectiveness are shown for filtering and elimination, respectively. The effectiveness of each heuristic is measured as the percentage of the distance computations that were avoided, compared to the baseline. In all runs, $\epsilon = 10^{-3}$ was used.

Distrib.	d	n	Iters	$|\mathcal{X}|$	Time (sec)	Speedup		Effect. (%)	
						Filt.	Elim.	Filt.	Elim.
Normal	5	10000	266.9	14.7	0.016	3.2	9.7	91.2	95.5
	5	100000	559.0	16.5	0.333	3.9	25.5	94.1	98.1
	5	1000000	665.8	17.6	4.002	3.6	31.1	96.0	99.2
	10	10000	883.0	44.2	0.078	3.6	7.1	88.3	90.3
	10	100000	879.5	45.6	0.790	3.8	8.9	89.9	93.2
	10	1000000	1514.0	49.0	13.858	4.5	15.2	93.2	96.5
	25	1000	1173.4	131.2	0.025	1.5	1.6	58.6	50.1
	25	10000	2471.1	175.6	0.442	2.9	3.1	78.5	75.5
	25	100000	3212.3	193.8	5.910	3.9	4.5	84.2	84.6
	50	1000	2234.2	328.1	0.103	1.2	1.2	35.8	23.9
	50	10000	4495.9	469.5	1.653	2.2	1.9	63.5	53.3
	50	100000	6738.5	563.9	24.999	3.3	2.6	74.6	69.7
	100	1000	3313.3	664.3	0.371	1.0	1.0	12.8	5.5
	100	10000	8737.8	1222.7	7.023	1.6	1.4	46.9	31.4
	100	100000	13688.4	1578.9	105.726	2.6	1.8	62.6	50.7
	200	1000	3582.1	972.1	1.279	1.0	1.0	0.7	0.1
	200	10000	15798.1	2866.5	28.811	1.2	1.1	28.3	12.9
Uniform (cube)	5	10000	770.3	15.7	0.046	3.3	10.6	91.3	93.8
	5	100000	1248.6	15.6	0.735	4.0	25.8	94.1	97.1
	5	1000000	939.0	15.1	5.608	2.8	19.3	93.1	97.3
	10	10000	2021.4	51.0	0.177	3.7	6.5	88.6	88.2
	10	100000	1681.9	49.0	1.498	3.7	7.4	89.5	90.5
	10	1000000	2446.0	52.1	22.288	3.9	9.6	91.6	93.8
	25	1000	2076.6	168.2	0.044	1.4	1.4	54.5	40.6
	25	10000	4035.5	207.6	0.722	2.6	2.4	75.3	66.4
	25	100000	6247.8	235.9	11.464	3.9	3.6	83.7	80.1
	50	1000	3298.8	408.6	0.155	1.2	1.1	29.8	15.5
	50	10000	7364.9	586.4	2.716	2.0	1.6	60.6	43.9
	50	100000	11691.7	709.0	43.271	3.1	2.1	72.6	61.3
	100	1000	4145.3	782.1	0.462	1.0	1.0	7.3	2.0
	100	10000	13663.3	1579.4	11.033	1.5	1.2	42.4	21.6
	100	100000	23255.3	2063.8	179.835	2.4	1.5	59.6	40.5
	200	1000	3640.1	995.0	1.296	1.0	1.0	0.1	0.0
	200	10000	23472.2	3759.5	43.229	1.1	1.0	22.4	6.2
Uniform (ball)	5	1000	2837.0	22.0	0.018	2.0	2.9	85.6	75.0
	5	10000	5414.8	30.3	0.323	2.9	3.6	90.3	80.3
	5	100000	6550.0	112.6	3.871	3.3	4.1	93.0	82.9
	5	1000000	4027.0	320.8	23.928	2.1	2.1	92.0	80.9
	10	1000	5476.2	63.6	0.054	1.8	1.7	73.8	53.1
	10	10000	16524.3	79.3	1.467	2.7	2.1	83.9	63.3
	10	100000	26764.0	170.6	24.110	3.3	2.1	87.9	68.6

References

1. Ahipaşaoğlu, S.D., Yıldırım, E.A.: Identification and elimination of interior points for the minimum enclosing ball problem. SIAM J. Optim. **19**(3), 1392–1396 (2008)
2. Barnett, V.: The ordering of multivariate data. J. Roy. Stat. Soc. Ser. A (Gen.) **139**(3), 318–355 (1976)
3. Betke, U., Henk, M.: Approximating the volume of convex bodies. Discret. Comput. Geom. **10**(1), 15–21 (1993)
4. Bouville, C.: Bounding ellipsoids for ray-fractal intersection. In: Proceedings of SIGGRAPH 1985, pp. 45–52. ACM (1985)
5. Chazelle, B., Matoušek, J.: On linear-time deterministic algorithms for optimization problems in fixed dimension. J. Algorithms **21**(3), 579–597 (1996)
6. Ding, J., Zhou, A.: Eigenvalues of rank-one updated matrices with some applications. Appl. Math. Lett. **20**(12), 1223–1226 (2007)
7. Galkovskyi, T., Gärtner, B., Rublev, B.: The domination heuristic for LP-type problems. In: Proceedings of the Meeting on Algorithm Engineering and Experiments, pp. 74–84. Society for Industrial and Applied Mathematics (2009)
8. Harman, R., Pronzato, L.: Improvements on removing nonoptimal support points in D-optimum design algorithms. Stat. Probab. Lett. **77**(1), 90–94 (2007)
9. John, F.: Extremum problems with inequalities as subsidiary conditions. In: Studies and Essays, Presented to R. Courant on His 60th Birthday, pp. 187–204. Wiley Interscience (1984)
10. Källberg, L., Larsson, T.: Faster approximation of minimum enclosing balls by distance filtering and GPU parallelization. J. Graph. Tools **17**(3), 67–84 (2013)
11. Källberg, L., Larsson, T.: Improved pruning of large data sets for the minimum enclosing ball problem. Graph. Models **76**(6), 609–619 (2014)
12. Khachiyan, L.G.: Rounding of polytopes in the real number model of computation. Math. Oper. Res. **21**(2), 307–320 (1996)
13. Kumar, P., Yildirim, E.A.: Minimum-volume enclosing ellipsoids and core sets. J. Optim. Theor. Appl. **126**(1), 1–21 (2005)
14. Liu, S., Wang, C.C.L., Hui, K.-C., Jin, X., Zhao, H.: Ellipsoid-tree construction for solid objects. In: Proceedings of the 2007 ACM Symposium on Solid and Physical Modeling, pp. 303–308 (2007)
15. Rimon, E., Boyd, S.P.: Obstacle collision detection using best ellipsoid fit. J. Intell. Rob. Syst. **18**(2), 105–126 (1997)
16. Sun, P., Freund, R.M.: Computation of minimum-volume covering ellipsoids. Oper. Res. **52**(5), 690–706 (2004)
17. Titterington, D.M.: Optimal design: some geometrical aspects of D-optimality. Biometrika **62**(2), 313–320 (1975)
18. Titterington, D.M.: Estimation of correlation coefficients by ellipsoidal trimming. Appl. Stat. **27**(3), 227–234 (1978)
19. Todd, M.J., Yıldırım, E.A.: On Khachiyan's algorithm for the computation of minimum-volume enclosing ellipsoids. Discret. Appl. Math. **155**(13), 1731–1744 (2007)
20. Welzl, E.: Smallest enclosing disks (balls and ellipsoids). In: Maurer, H. (ed.) New Results and New Trends in Computer Science. LNCS, vol. 555, pp. 359–370. Springer, Heidelberg (1991)
21. Yıldırım, E.A.: On the minimum volume covering ellipsoid of ellipsoids. SIAM J. Optim. **17**(3), 621–641 (2006)

Relaxations of Discrete Sets
with Semicontinuous Variables

Gustavo Angulo$^{(\boxtimes)}$

Department of Industrial and Systems Engineering,
Pontificia Universidad Católica de Chile,
Vicuña Mackenna 4860, Macul, Santiago, Chile
gangulo@ing.puc.cl

Abstract. A variable is said semicontinuous if its domain is given by
the union of two disjoint nonempty closed intervals. As such, they can
be regarded as relaxations of binary or integrality constraints appearing
in combinatorial optimization problems. For knapsacks and a class of
single-node flow sets, we consider relaxations involving unbounded semicontinuous variables. We analyze the complexity of linear optimization
over such relaxations and provide descriptions of their convex hulls in
terms of linear inequalities and extended formulations.

Keywords: Semicontinuous variable · Semicontinuous knapsack set ·
Semicontinuous single-node flow set · Indicator variable

1 Introduction

In this work we consider sets of the form $S := \{x \in \mathbb{R}^n : x \in P, x_i \in [0, p_i] \cup [l_i, \infty) \; \forall i \in N\}$, where $n \geq 2$, $N := \{1, \ldots, n\}$, $P \subseteq \mathbb{R}_+^n$ is a nonempty polyhedron, and $0 \leq p_i \leq l_i$ for all $i \in N$. If $p_i < l_i$, we say that variable x_i is semicontinuous, and continuous otherwise. Semicontinuity is a natural relaxation of binary and integrality constraints found in combinatorial optimization problems. For instance, a binary variable belongs to the set $\{0\} \cup [1, \infty)$, while a nonnegative integer variable is contained in $[0, k] \cup [k + 1, \infty)$ for any $k \in \mathbb{Z}_+$. They also provide a means of modeling some simple disjunctions without including additional binary variables as commonly done in textbook formulations. Recent work using semicontinuous variables include scheduling in health care [3] and lot-sizing with minimum order quantity [7], among others. We are interested in studying conv(S), the convex hull of S, in terms of linear descriptions for two classes of sets: knapsacks and single-node flow sets with variable upper bounds and indicator variables. This extends the works of [4,5] and [1] as follows:

- For knapsacks with unbounded semicontinuous variables having positive and negative signs, in Sect. 2 we show present a decomposition property and provide a description of its convex hull via an extended formulation of polynomial size. Moreover, when all variables have positive signs, we provide complete linear descriptions in the original space.

© Springer International Publishing AG 2016
T.-H.H. Chan et al. (Eds.): COCOA 2016, LNCS 10043, pp. 754–762, 2016.
DOI: 10.1007/978-3-319-48749-6_57

– For single-node flow sets with variable upper bounds and indicator variables, in Sect. 3 we propose to treat the indicator variables as semicontinuous. We show that a decomposition property also holds and provide a complete linear description without additional variables.

For $i \in N$, let e^i denote the i-th unit vector in $\mathbb{R}^n$. The polyhedrality of conv(S) for any polyhedron $P \subseteq \mathbb{R}^n_+$ follows from [1, Proposition 23] and its proof.

Proposition 1. If $S \neq \emptyset$, then conv(S) is a nonempty pointed polyhedron and its recession cone is equal to that of P.

2 Knapsack Sets

Given a real $b > 0$ and a partition (N^+, N^-) of N, let $S_K \subseteq \mathbb{R}^n$ be defined by

$$\sum_{i \in N^+} x_i - \sum_{i \in N^-} x_i \geq b \tag{1}$$

$$x_i \in [0, p_i] \cup [l_i, \infty) \ \forall i \in N. \tag{2}$$

This set is general enough to be regarded as a relaxation of more complex sets. Note that the coefficients 1 and -1 in (1) are without loss of generality since we can redefine p and l in order to bring general coefficients to the above form. For ease of exposition, we assume that $p < l$. Replacing (2) with

$$x_i \geq 0 \ \forall i \in N, \tag{3}$$

we obtain the linear relaxation P_K given by (1) and (3).

2.1 Basic Structure and Complexity

Proposition 2. conv(S_K) is full-dimensional.

Let $N^+_*(x) := \{i \in N^+ : x_i \geq l_i > 0\}$ and $N^-_*(x) := \{i \in N^- : x_i \geq l_i > 0\}$.

Proposition 3. If x is a vertex of conv(S_K), $|N^+_*(x)| \leq 1$ and $|N^-_*(x)| \leq 1$.

Proof. Let $x \in S_K$ and suppose there exist $i, j \in N^+_*(x)$ with $i \neq j$, the case $i, j \in N^-_*(x)$ being analogous. Consider the points $x' := x - x_j e^j + x_j e^i$ and $x'' := x - x_i e^i + x_i e^j$. We have $x', x'' \in S_K$ and $x' \neq x''$. Setting $\lambda = \frac{x_i}{x_i + x_j} \in (0,1)$ we obtain $\lambda x' + (1 - \lambda)x'' = \lambda(x - x_j e^j + x_j e^i) + (1 - \lambda)(x - x_i e^i + x_i e^j) = x + (\lambda x_j - (1 - \lambda)x_i)e^i + (-\lambda x_j + (1 - \lambda)x_i)e^j = x + 0e^i + 0e^j = x.$ □

Let (Opt$_K$) denote the problem $\min\{c^\top x : x \in S_K\}$ for some $c \in \mathbb{R}^n$. Recall that by Proposition 1, conv(S_K) is a polyhedron with the same recession cone as P_K, which is given by $C_K := \{r \in \mathbb{R}^n_+ : \sum_{i \in N^+} r_i - \sum_{j \in N^-} r_j \geq 0\}$.

Proposition 4. *Assume S_K is nonempty. Then (Opt_K) is unbounded if and only if either there exists $i \in N^+$ with $c_i < 0$ or there exist $i \in N^+$ and $j \in N^-$ such that $c_i + c_j < 0$.*

Proof. Since (Opt_K) is equivalent to $\min\{c^\top x : x \in \mathrm{conv}(S_K)\}$ and $S_K \neq \emptyset$, (Opt_K) is unbounded if and only if there exists $r \in C_K$ such that $c^\top r < 0$.

If there exists $i \in N^+$ with $c_i < 0$, then taking $r = e^i$, we have $r \in C_K$ and $c^\top r = c_i < 0$. If there exist $i \in N^+$ and $j \in N^-$ such that $c_i + c_j < 0$, then taking $r = e^i + e^j$, we have $r \in C_K$ and $c^\top r = c_i + c_j < 0$.

Now, suppose that $c_i \geq 0$ for all $i \in N^+$ and $c_i + c_j \geq 0$ for all $i \in N^+$ and $j \in N^-$. Let $c^+ := \min_{i \in N^+}\{c_i\}$ and $c^- := \min_{j \in N^-}\{c_j\}$. Note that $c^+ \geq 0$ and $c^+ + c^- \geq 0$. Let $r \in C_K$. Then $c^\top r = \sum_{i \in N^+} c_i r_i + \sum_{j \in N^-} c_j r_j \geq \sum_{i \in N^+} c^+ r_i + \sum_{j \in N^-} c^- r_j \geq c^+ \sum_{i \in N^+} r_i - c^+ \sum_{j \in N^-} r_j \geq 0$. Therefore, (Opt_K) cannot be unbounded. $\square$

In view of Propositions 3 and 4, let us define $P_K^0 := \{x \in S_K : x_k \in [0, p_k]\ \forall k \in N\}$, $P_K^i := \{x \in S_K : x_i \geq l_i,\ x_k \in [0, p_k]\ \forall k \neq i\}$ for $i \in N$, and finally $P_K^{ij} := \{x \in S_K : x_i \geq l_i,\ x_j \geq l_j,\ x_k \in [0, p_k]\ \forall k \neq i, j\}$ for $i \in N^+$, $j \in N^-$.

Let $\mathcal{P}_K$ denote the collection comprising the $\mathcal{O}(n^2)$ sets above defined. Note that by Proposition 3, the vertices of $\mathrm{conv}(S_K)$ are contained within the vertices of the sets in $\mathcal{P}_K$. We obtain the following result.

Proposition 5. *Solving (Opt_K) can be done by solving $\mathcal{O}(n^2)$ LPs.*

2.2   The Case $N = N^+$

$l_i \leq b$ *for all* $i \in N$

Proposition 6. *If $l_i \leq b$ for all $i \in N$, then $\mathrm{conv}(S_K) = P_K$.*

Proof. The vertices of (1) and (3) are of the form be^i for $i \in N$, and since $b \geq l_i$ for $i \in N$, all such points belong to S_K. $\square$

$l_i \geq b$ *for all* $i \in N$

Proposition 7. *If $\sum_{i \in N} p_i < b$, then $\mathrm{conv}(S_K)$ is given by (3) and*

$$\sum_{i \in N} \frac{x_i}{l_i} \geq 1. \tag{4}$$

Proof. Let $x \in S_K$ be a vertex of $\mathrm{conv}(S_K)$. Since $\sum_{i \in N} x_i \geq b$ and $\sum_{i \in N} p_i < b$, by Proposition 3, there exists a unique $i \in N$ such that $x_i \geq l_i$ and $x_j \in [0, p_j]$ for all $j \neq i$. Moreover, since $l_i \geq b$, then $x_i = l_i$ must hold for x to be a vertex. Now, if there exists $j \neq i$ with $x_j > 0$, then consider $x' = x - x_j e^j$ and $x'' = x + l_j e^j$, which belong to S. Taking $\lambda = \frac{l_j}{x_j + l_j} \in (0, 1)$, we obtain $x = \lambda x' + (1 - \lambda)x''$, a contradiction. Therefore $x_j = 0$ for $j \neq i$ and thus $x = l_i e^i$. In particular, all the vertices of $\mathrm{conv}(S_K)$ must be of this form, and the convex hull of all such points is precisely the simplex given by (3) and (4). $\square$

We now analyze the case $\sum_{i \in N} p_i \geq b$.

Definition 1. *A subset $R \subseteq N$ is a reverse cover if $b_R := b - \sum_{i \in R} p_i > 0$.*

For each reverse cover R, consider the inequality

$$\sum_{i \in R} \frac{x_i}{l_i} + \left(1 - \sum_{i \in R} \frac{p_i}{l_i}\right) \sum_{i \notin R} \frac{x_i}{b_R} \geq 1. \tag{5}$$

Proposition 8. *For each reverse cover R, (5) is valid for $\mathrm{conv}(S_K)$.*

Proposition 9. *When $\sum_{i \in N} p_i = b$, (5) is facet-defining if $R = N \setminus \{j\}$ for some $j \in N$ such that $p_j > 0$. When $\sum_{i \in N} p_i > b$, (5) is facet-defining if either $R = \emptyset$ and $l_i = b$ for all i having $p_i = 0$, or if $p_i > 0$ for all $i \notin R$.*

Proof. Assume $\sum_{i \in N} p_i = b$. Let $R = N \setminus \{j\}$ for $j \in N$ with $p_j > 0$. Consider the points $x^i := l_i e^i$ for $i \neq j$ and $x^j := p$, which belong to S_K. Note that $b_R = p_j$ and (5) takes the form $\sum_{k \neq j} \frac{x_k}{l_k} + \left(1 - \sum_{k \neq j} \frac{p_k}{l_k}\right) \frac{x_j}{p_j} \geq 1$. Then it is easy to see that the n points above defined satisfy (5) at equality. Finally, they form a linearly independent set, and therefore (5) defines a facet.

Now assume $\sum_{i \in N} p_i > b$ and $l_i = b$ if $p_i = 0$. Let $R = \emptyset$ and $\bar{p} := \sum_{i \in N} p_i$. For $0 < \epsilon < 1$, define the point $\bar{x} := \frac{b}{\bar{p}}(1 - \epsilon)p$. Then $\bar{x}$ satisfies $\bar{x}_i = 0$ if $p_i = 0$, $0 < \bar{x}_i < p_i$ if $p_i > 0$, and $\sum_{i \in N} \bar{x}_i = b - \epsilon b$. Consider the points $x^i := \bar{x} + \epsilon b e^i$ if $p_i > 0$ and $x^i := l_i e^i = b e^i$ if $p_i = 0$, all of them satisfying (1). For ϵ sufficiently small, they also satisfy (2) and thus belong to S_K. Note that $b_R = b$ and (5) takes the form $\sum_{k \in N} x_k \geq b$. These n points satisfy (5) at equality. Finally, they form a linearly independent set, and therefore (5) defines a facet.

Assume again $\sum_{i \in N} p_i > b$, and let R be such that $p_i > 0$ for all $i \notin R$. Let $\delta := \frac{b_R}{\sum_{i \notin R} p_i} \in (0, 1)$ and define the point $\bar{x}$ with $\bar{x}_i = p_i$ for $i \in R$ and $\bar{x}_i = (1 - \epsilon)\delta p_i$ for $i \notin R$, where $0 < \epsilon < 1$. Note that $0 < \bar{x}_i < p_i$ if $i \notin R$. Also, $\sum_{i \notin R} \bar{x}_i = (1 - \epsilon)b_R$ and thus $\sum_{i \in N} \bar{x}_i = b - \epsilon b_R$. Finally, consider the points $x^i := l_i e^i$ for $i \in R$ and $x^i := \bar{x} + \epsilon b_R e^i$ for $i \notin R$, which satisfy (1). For ϵ sufficiently small, they also satisfy (2) and thus belong to S_K. For $i \in R$, it is clear that x^i satisfies (5) at equality. For $i \notin R$, we have

$$\sum_{k \in R} \frac{x_k^i}{l_k} + \left(1 - \sum_{k \in R} \frac{p_k}{l_k}\right) \sum_{k \notin R} \frac{x_k^i}{b_R} = \sum_{k \in R} \frac{\bar{x}_k}{l_k} + \left(1 - \sum_{k \in R} \frac{p_k}{l_k}\right) \sum_{k \notin R} \frac{\bar{x}_k}{b_R} + \left(1 - \sum_{k \in R} \frac{p_k}{l_k}\right) \frac{\epsilon b_R}{b_R}$$

$$= \sum_{k \in R} \frac{p_k}{l_k} + \left(1 - \sum_{k \in R} \frac{p_k}{l_k}\right)(1 - \epsilon) + \left(1 - \sum_{k \in R} \frac{p_k}{l_k}\right)\epsilon = 1$$

Thus, these points satisfy (5) at equality as well. Finally, the n points form a linearly independent set, and therefore (5) defines a facet. □

Theorem 1. *If $\sum_{i \in N} p_i \geq b$, then $\mathrm{conv}(S_K)$ is given by (3) and (5).*

Proof. Let $c \in \mathbb{R}^n \setminus \{0\}$ and consider the problem $\min\{c^\top x : x \in S_K\}$. We assume that $c \geq 0$ since otherwise the problem is unbounded. Following the technique introduced in [6], we will show that there exists an inequality among (3) and (5) that is satisfied at equality by all optimal solutions to $\min\{c^\top x : x \in S_K\}$. Let $x^* \in S_K$ be any optimal solution.

If $c_j = 0$ for some $j \in N$, then there exists some i with $c_i > 0$ and therefore, by optimality, x^* satisfies $x_i \geq 0$ at equality. Thus we may assume $c > 0$. This and $l_i \geq b$ for $i \in N$, together with the optimality of x^*, imply $x^* \leq l$ and $|\{i \in N : x_i^* = l_i\}| \leq 1$.

Suppose that all the entries of c are equal to a constant $\bar{c}$. Since $\sum_{i \in N} p_i \geq b$, by optimality, we have that x^* satisfies (5) with $R = \emptyset$ at equality. If c is nonconstant, let $(N_1, \ldots, N_q)$ be a partition of N such that the entries of c indexed by N_l are equal to a constant $\bar{c}_l$, and $\bar{c}_l < \bar{c}_{l+1}$ for all $1 \leq l < q$. If $\sum_{i \in N_1} p_i \geq b$, then (5) with $R = \emptyset$ is satisfied by x^* at equality. Otherwise, let $1 \leq r < q$ be such that $\sum_{l=1}^r \sum_{i \in N_l} p_i < b$ and $\sum_{l=1}^{r+1} \sum_{i \in N_l} p_i \geq b$, and set $R := \cup_{l=1}^r N_l$. We claim that x^* satisfies (5) for R at equality.

If $x_j^* = l_j$ for some $j \in R$, then (5) is tight. Thus assume $x_i^* \leq p_i$ for all $i \in R$. Now, suppose $x_j^* = l_j$ for some $j \notin R$. Consider a vector $y \in \mathbb{R}_+^n$ where $y_i = p_i$ for $i \in R$, the entries indexed by N_{r+1} are set so that $0 \leq y_i \leq p_i$ and $\sum_{i \in N_{r+1}} y_i = b_R$, and the remaining entries are set to zero. Clearly, $y \in S_K$. Moreover, since $c_j \geq \bar{c}_{r+1}$, we have $c^\top y = \sum_{l=1}^k \sum_{i \in N_l} \bar{c}_l y_i = \sum_{l=1}^r \sum_{i \in N_l} \bar{c}_l p_i + \bar{c}_{r+1} b_R < c_j(b - b_R) + c_j b_R = c_j b \leq c^\top x^*$, a contradiction with the optimality of x^*. Thus, we must have $x_i^* \leq p_i$ for all $i \in N$. Moreover, it is easy to see that x^* must have the form of y, and therefore (5) is tight.   □

2.3 The General Case

Theorem 2. $conv(S_K) = conv\left(P_K^0 \cup (\cup_{i \in N} P_K^i) \cup (\cup_{i \in N^+, \, j \in N^-} P_K^{ij})\right)$.

Proof. The reverse inclusion is easy since each set in $\mathcal{P}_K$ is contained in S_K. For the forward inclusion, by Proposition 3, we already know that the vertices of $conv(S_K)$ are contained in the sets in $\mathcal{P}_K$. It remains to show that the recession cone of $conv(S_K)$, given by C_K, is contained in that of the set on right-hand side. Let $r \in C_K \setminus \{0\}$. Define $r^+ := \sum_{i \in N^+} r_i$ and $r^- := \sum_{j \in N^-} r_j$. Note that $r^+ > 0$.

If $r^- = 0$, then we have $r = \sum_{i \in N^+} \frac{r_i}{r^+}(r^+ e^i)$ with $r^+ e^i$ in the recession cone of P_K^i. If $r^- > 0$, then we have $r = \sum_{i \in N^+} \sum_{j \in N^-} \frac{r_i r_j}{r^+ + r^-}(r^+ e^i + r^- e^j)$ with $r^+ e^i + r^- e^j$ in the recession cone of P_K^{ij}.   □

3   Single-Node Flow Sets

Let N represent a set of nodes and let $b > 0$ be the required total flow from nodes in N into another node 0. For $i \in N$, let y_i be the flow from node i to node 0, and x_i be the flow into node i. Let l_i and h_i be the lower bounds on flows x_i

and y_i whenever these variables are positive. Let z_i and w_i be semicontinuous indicator variables satisfying constraints $x_i > 0 \Rightarrow z_i \geq 1$, $y_i > 0 \Rightarrow w_i \geq 1$, $x_i = 0 \Rightarrow z_i = 0$ if $l_i > 0$, and $y_i = 0 \Rightarrow w_i = 0$ if $h_i > 0$. Define $S_F \subseteq \mathbb{R}^{4n}$ by

$$\sum_{i \in N} y_i \geq b \tag{6}$$

$$y_i \leq x_i \ \forall i \in N \tag{7}$$

$$(x_i, z_i) \in \{(0,0)\} \cup [l_i, \infty) \times [1, \infty) \ \forall i \in N \tag{8}$$

$$(y_i, w_i) \in \{(0,0)\} \cup [h_i, \infty) \times [1, \infty) \ \forall i \in N. \tag{9}$$

Note that if we minimize linear function $c^\top x + d^\top y + p^\top z + q^\top w$ over S_F with $p, q > 0$, then any optimal solution satisfies $z, w \in \{0,1\}^n$.

3.1 Basic Structure and Complexity

Proposition 10. $\mathrm{conv}(S_F)$ *is full-dimensional.*

Proposition 11. *Let (x, y, z, w) be a vertex of $\mathrm{conv}(S_F)$. Then there exists i such that $x_i > 0$ and $x_j = 0$ for all $j \neq i$.*

Proof. Let $(x, y, z, w) \in \mathrm{conv}(S_F)$ be such that $x_i > 0$ and $y_i = 0$. Noting that $z_i \geq 1$, we have that (x, y, z, w) is the middle point between $(x + x_i e^i, y, z + z_i e^i, w)$ and $(x - x_i e^i, y, z - z_i e^i, w)$, which are different points contained in S_F.

Suppose now that $x_i > 0$ and $x_j > 0$. By the above observation, we may assume $y_i > 0$ and $y_j > 0$. In particular $z_i, z_j, w_i, w_j \geq 1$. Let $\lambda = \frac{y_i}{y_j}$ and note that $x_i \geq \lambda y_j$ and $\lambda x_j \geq y_i$. Then

$$\left(x + \frac{1}{\lambda} x_i e^i - x_j e^j, y + y_j e^i - y_j e^j, z + \frac{1}{\lambda} z_i e^i - z_j e^j, w + \frac{1}{\lambda} w_i e^i - w_j e^j \right),$$

$$(x - x_i e^i + \lambda x_j e^j, y - y_i e^i + y_i e^j, z - z_i e^i + \lambda z_j e^j, w - w_i e^i + \lambda w_j w^j)$$

are distinct points contained in S_F and their convex combination with multipliers λ and $1 - \lambda$ yields (x, y, z, w). $\qquad \square$

In a fashion similar to that of Sect. 2, $\mathrm{conv}(S_F)$ can be decomposed into n pieces which are enough for linear optimization and provide an extended formulation of polynomial size. More precisely, for $i \in N$, let

$$P_F^i := \{(x, y, z, w) \in S_F : \ x_j = y_j = z_j = w_j = 0 \ \forall j \neq i\}$$

$$= \{(x, y, z, w) : \ x_i \geq l_i, \ y_i \geq \max\{h_i, b\}, \ x_i \geq y_i, \ x_j = y_j = z_j = w_j = 0 \ \forall j \neq i\}.$$

Proposition 12. *Optimizing a linear function over S_F can be done by solving n linear programming problems. Moreover, $\mathrm{conv}(S_F) = \mathrm{conv}(\cup_{i \in N} P_F^i)$.*

3.2 Polyhedral Description

Let $L := \{i \in N : \max\{h_i, b\} < l_i\}$. Let $\mathcal{N}$ be the collection of partitions (N_x, N_y, N_z, N_w) of N where $N_x \subseteq L$. Consider the inequality

$$\sum_{i \in N_x} \frac{x_i}{l_i} + \sum_{i \in N_y} \frac{y_i}{\max\{h_i, b\}} + \sum_{i \in N_z} z_i + \sum_{i \in N_w} w_i \geq 1. \tag{10}$$

Proposition 13. *Inequality (10) is valid for conv(S_F).*

Proposition 14. *Inequality (10) is facet-defining for conv(S_F).*

Proof. Let $\phi^i := (\max\{l_i, h_i, b\}e^i, \max\{h_i, b\}e^i, e^i, e^i)$, $i \in N$. Consider

$$\phi^i, \ \phi^i + (0, \epsilon e^i, 0, 0), \ \phi^i + (0, 0, \epsilon e^i, 0), \ \phi^i + (0, 0, 0, \epsilon e^i) \quad i \in N_x,$$

$$\phi^i, \ \phi^i + (\epsilon e^i, 0, 0, 0), \ \phi^i + (0, 0, \epsilon e^i, 0), \ \phi^i + (0, 0, 0, \epsilon e^i) \quad i \in N_y,$$

$$\phi^i, \ \phi^i + (\epsilon e^i, 0, 0, 0), \ \phi^i + (0, \epsilon e^i, 0, 0), \ \phi^i + (0, 0, 0, \epsilon e^i) \quad i \in N_z,$$

$$\phi^i, \ \phi^i + (\epsilon e^i, 0, 0, 0), \ \phi^i + (0, \epsilon e^i, 0, 0), \ \phi^i + (0, 0, \epsilon e^i, 0) \quad i \in N_w.$$

The $4n$ points above belong to S_F, satisfy (10) at equality, and are affinely independent. Therefore, (10) defines a facet. $\qquad\square$

Theorem 3. *The set conv(S_F) is given by (10) and the trivial inequalities $y \geq 0$, $z \geq 0$, $w \geq 0$, and $x - y \geq 0$.*

Proof. Let $\mathcal{N}^j$ be the collection of partitions (N_x, N_y, N_z, N_w) of $\{1, \ldots, j\}$ with $N_x \subseteq \{i \in L : i \leq j\}$. By induction on $1 \leq j \leq n$, we will show that $Q^j := \mathrm{conv}(\cup_{i \leq j} P_F^i)$ is given by (10) restricted to $\mathcal{N}^j$, the trivial inequalities, and $x_i = z_i = w_i = 0$ for $i > j$. Then the result will follow from the case $j = n$.

For any $1 \leq j \leq n$, P_F^j is given by

$$x_j \geq l_i, \quad y_j \geq \max\{h_j, b\}, \quad z_j \geq 1, \quad w_j \geq 1, \quad x_j - y_j \geq 0$$
$$x_i = y_i = z_i = w_i = 0 \quad i \neq j.$$

Note that the first inequality is redundant if and only if $\max\{h_j, b\} \geq l_j$, while all remaining inequalities are irredundant. Also, for $j = 1$, the first four inequalities are obtained from the last row of equalities together with (10) for $\mathcal{N}^1$. Thus $Q^1 = P_F^1$ has the desired form.

Assume the result holds for $1 \leq j < n$. For $j+1$, we have $\mathrm{conv}(\cup_{i \leq j+1} P_F^i) = \mathrm{conv}(Q^j \cup P_F^{j+1})$. By the inductive hypothesis, we have that Q^j is given by (10) for $\mathcal{N}^j$, the trivial inequalities, and $x_i = z_i = w_i = 0$ for $i > j$. On the other hand, P_F^{j+1} is given by the above system. By disjunctive programming, Q^{j+1} admits an extended formulation with additional variables $\lambda \in [0, 1]$, $(x', y', z', w') \in (1 - \lambda)Q^j$, and $(x'', y'', z'', w'') \in \lambda P_F^{j+1}$. The system has the form

$$(x, y, z, w) = (x', y', z', w') + (x'', y'', z'', w'')$$

$$\sum_{i \in N'_x} \frac{x'_i}{l_i} + \sum_{i \in N'_y} \frac{y'_i}{\max\{h_i, b\}} + \sum_{i \in N'_z} z'_i + \sum_{i \in N'_w} w'_i \geq 1 - \lambda \quad (N'_x, N'_y, N'_z, N'_w) \in \mathcal{N}^j$$

$$x'_i - y'_i \geq 0, \quad y'_i, z'_i, w'_i \geq 0 \qquad i \leq j$$

$$x'_i = y'_i = z'_i = w'_i = 0 \qquad i > j$$

$$x''_{j+1} \geq l_{j+1}\lambda, \quad y''_{j+1} \geq \max\{h_{j+1}, b\}\lambda$$

$$z''_{j+1} \geq \lambda, \quad w''_{j+1} \geq \lambda, \quad x''_{j+1} - y''_{j+1} \geq 0$$

$$x''_i = y''_i = z''_i = w''_i = 0 \qquad i \neq j+1$$

$$0 \leq \lambda \leq 1.$$

From the above system, we have

$$(x_i, y_i, z_i, w_i) = \begin{cases} (x'_i, y'_i, z'_i, w'_i) & i \leq j \\ (x''_{j+1}, y''_{j+1}, z''_{j+1}, w''_{j+1}) & i = j+1 \\ (0, 0, 0, 0) & i > j+1. \end{cases}$$

Replacing and projecting out (x', y', z', w') and (x'', y'', z'', w''), we obtain

$$\sum_{i \in N'_x} \frac{x_i}{l_i} + \sum_{i \in N'_y} \frac{y_i}{\max\{h_i, b\}} + \sum_{i \in N'_z} z_i + \sum_{i \in N'_w} w_i \geq 1 - \lambda \quad (N'_x, N'_y, N'_z, N'_w) \in \mathcal{N}^j$$

$$x_i - y_i \geq 0, \quad y_i, z_i, w_i \geq 0 \qquad i \leq j$$

$$x_{j+1} \geq l_{j+1}\lambda, \quad y_{j+1} \geq \max\{h_{j+1}, b\}\lambda$$

$$z_{j+1} \geq \lambda, \quad w_{j+1} \geq \lambda, \quad x_{j+1} - y_{j+1} \geq 0$$

$$x_i = y_i = z_i = w_i = 0 \qquad i > j+1$$

$$0 \leq \lambda \leq 1.$$

Finally, projecting out λ, we arrive at

$$\sum_{i \in N_x} \frac{x_i}{l_i} + \sum_{i \in N_y} \frac{y_i}{\max\{h_i, b\}} + \sum_{i \in N_z} z_i + \sum_{i \in N_w} w_i \geq 1 \quad (N_x, N_y, N_z, N_w) \in \mathcal{N}^{j+1}$$

$$x_i - y_i \geq 0, \quad y_i, z_i, w_i \geq 0 \qquad i \leq j+1$$

$$x_i = y_i = z_i = w_i = 0 \qquad i > j+1.$$

$\square$

Proposition 15. *Inequalities (10) can be separated in $\mathcal{O}(n)$ time.*

Proof. A point (x^*, y^*, z^*, w^*) satisfies all inequalities (10) if and only if we have $\sum_{i \in L} \min\left\{ \frac{x_i^*}{l_i}, \frac{y_i^*}{\max\{h_i, b\}}, z_i^*, w_i^* \right\} + \sum_{i \in N \setminus L} \min\left\{ \frac{y_i^*}{\max\{h_i, b\}}, z_i^*, w_i^* \right\} \geq 1.$ $\square$

References

1. Angulo, G., Ahmed, S., Dey, S.S.: Semi-continuous network flow problems. Math. Program. **145**(1), 565–599 (2014)
2. Balas, E.: Disjunctive programming. In: Johnson, E.L., Hammer, P.L., Korte, B.H. (eds.) Annals of Discrete Mathematics, Discrete Optimization II, vol. 5, pp. 3–51. Elsevier, Amsterdam (1979)
3. Bard, J.F., Shu, Z., Leykum, L.: A network-based approach for monthly scheduling of residents in primary care clinics. Oper. Res. Health Care **3**(4), 200–214 (2014)
4. de Farias Jr., I.R.: Semi-continuous cuts for mixed-integer programming. In: Bienstock, D., Nemhauser, G.L. (eds.) IPCO 2004. LNCS, vol. 3064, pp. 163–177. Springer, Heidelberg (2004)
5. de Farias, I.R., Zhao, M.: A polyhedral study of the semi-continuous knapsack problem. Math. Program. **142**(1), 169–203 (2013)
6. Lovász, L.: Graph theory and integer programming. In: Johnson, E.L., Hammer, P.L., Korte, B.H. (eds.) Annals of Discrete Mathematics, Discrete Optimization I, vol. 4, pp. 141–158. Elsevier, Amsterdam (1979)
7. Park, Y.W., Klabjan, D.: Lot sizing with minimum order quantity. Discrete Appl. Math. **181**, 235–254 (2015)

Unfolding the Core Structure of the Reciprocal Graph of a Massive Online Social Network

Braulio Dumba$^{(\boxtimes)}$ and Zhi-Li Zhang

University of Minnesota, Twin Cities, MN, USA
{braulio,zhzhang}@cs.umn.edu

Abstract. Google+ (G+ in short) is a *directed* online social network where nodes have either *reciprocal* (bidirectional) edges or *parasocial* (one-way) edges. As reciprocal edges represent strong social ties, we study the core structure of the subgraph formed by them, referred to as the *reciprocal network* of G+. We develop an effective three-step procedure to *hierarchically* extract and unfold the *core* structure of this reciprocal network. This procedure builds up and generalizes ideas from the existing k-shell decomposition and clique percolation approaches, and produces higher-level representations of the core structure of the G+ reciprocal network. Our analysis shows that there are seven subgraphs ("communities") comprising of dense clusters of cliques lying at the center of the core structure of the G+ reciprocal network, through which other communities of cliques are richly connected. Together they form the core to which "peripheral" sparse subgraphs are attached.

Keywords: Reciprocal network · Google+ · Network core · Reciprocity

1 Introduction

Many online social networks (OSNs) such as Twitter, Google+, Flickr contain both *reciprocal* edges, i.e., edges that have already been linked back, and *parasocial* edges, i.e., edges that have not been or is not linked back [1], and thus directed in nature. Reciprocity is defined as the ratio of the number of reciprocal edges to the total number of edges in the network, and has been widely studied in the literature in various contexts, see, e.g., [1–6]. It is believed to reciprocity plays an important role in the structural properties, formation and evolution of online social networks. Empirical studies have shown that many OSNs exhibit a nontrivial amount of reciprocity: Twitter is estimated to have a reciprocity value of 0.22 [7], Google+ 0.32 [8] and Flickr 0.62 [9].

Reciprocal edges represent the most stable type of connections or relations in directed OSNs: for example, in Twitter it represents users are mutually "following" each other, and in Google+ it represents two users are in each other's circles. Hence, reciprocal edges reflect strong ties between nodes or users [10–12]. Most existing studies have focused on reciprocity (a single-valued aggregate metric) to characterize massive *directed* OSNs, which we believe is inadequate.

© Springer International Publishing AG 2016
T.-H.H. Chan et al. (Eds.): COCOA 2016, LNCS 10043, pp. 763–771, 2016.
DOI: 10.1007/978-3-319-48749-6_58

Instead, we consider the *reciprocal graph* (or *reciprocal network*) of a directed OSN – namely, the *bidirectional* subgraph formed by the reciprocal edges among users in a directed OSN. In a sense, this reciprocal network can be viewed as the stable "skeleton" network of the directed OSN that holds it together. We are interested in analyzing and uncovering the *core* structural properties of the reciprocal network of a directed OSN, as they could reveal the possible organizing principles shaping the observed network topology of an OSN [2].

Using Google+ (thereafter referred to as *G+* in short) as a case study, in this paper we perform a comprehensive empirical analysis of the "core structure" of the reciprocal network of G+. Based on a massive G+ dataset (see Sect. 2 for a brief overview of G+ and a description of the dataset), we find that out of more than 74 million nodes and ≈ 1.4 billion edges in (a snapshot of) the directed G+ OSN, more than two-third of the nodes are part of G+'s *reciprocal* network and more than a third of the edges are reciprocal edges (with a reciprocity value of roughly 0.34). This reciprocal network contains a *giant connected subgraph* with more than 40 million nodes and close to 400 million edges (see Sect. 2 for more details). Existence of this massive (giant connected) reciprocal (sub)graph in G+ raises many interesting and challenging questions. How is this reciprocal network formed? Does it contain a "core" network structure? If yes, what does this structure look like?

In an attempt to address these questions, we develop an effective three-step procedure to *hierarchically* extract and unfold the *core* structure of G+'s reciprocal network, building up and generalizing ideas from the existing k-shell decomposition [13] and clique percolation [14] approaches. We first applied a modified version of the k-shell decomposition method to prune nodes and edges of sparse subgraphs that are likely to lie at the peripherals of the G+ reciprocal network (see Sect. 3). We then performed a form of clique percolation to generate a new *directed* (hyper)graphs where vertices are maximal cliques containing the nodes in the dense "core" graph generated in the previous step, and there exists a directed edge from clique C_i to clique C_j if half of the nodes in C_i are contained in C_j (see Sect. 4). We found that this (hyper)graph of cliques comprises of 2000+ connected components (CCs), which represent the core "communities" of the G+ reciprocal network. Finally, we introduced three metrics to study the relations among these CCs in the underlying G+ reciprocal network: the number of nodes shared by two CCs, the number of nodes that are neighbors in the two CCs, and the number of edges connecting these neighboring nodes (see Sect. 5). These metrics produce a set of new (hyber)graphs that succinctly summarize the (high-level) structural relations among the core "community" structures and provide a "big picture" view of the core structure of the G+ reciprocal network and how it is formed. In particular, we found that there are seven CCs that lie at the center of this core structure through which the other CCs are most richly connected. In Sect. 6, we conclude the paper with a brief discussion of the future work.

We summarize the major contributions of our paper as follows. To the best our knowledge, our paper is the first study on the core structure of a "reciprocal network" extracted from a massive *directed* social graph. While this paper focuses on G+, we believe that our approach is applicable to other directed OSNs: (i) we develop an effective three-step procedure to *hierarchically* extract and unfold the *core* structure of a reciprocal network arising from a directed OSN; (ii) We apply our method to the reciprocal network of the massive Google+ social network, and unfold its core structure. In particular, we find that there are seven subgraphs ("communities") comprising of dense clusters of cliques that lie at the center of the core structure of the G+ reciprocal networks, through which other communities of cliques are richly connected; together they form the core to which other nodes and edges that are part of sparse subgraphs on the peripherals of the network are attached.

2   G+ Dataset and Overview of the Reciprocal Network

In this section, we briefly describe key features of the Google+ service, a summary of our dataset and our methodology to extract the reciprocal network.

Platform Description: On June 2011 Google launched its own social networking service called Google+ (G+). Previous works in the literature [5,6] labelled G+ as a hybrid online social network because its features are similar to both Facebook and Twitter. Similar to Twitter (and different from Facebook) the relationships in G+ are unidirectional. Similar to Facebook, each user has a stream, where any activity performed by the user appears (like the Facebook wall). For more information about the features of G+ the reader is referred to [15,16].

Dataset and Reciprocal Network: We obtained our dataset from an earlier study on G+ [6]. The dataset is a directed graph (denoted as Γ) of the social links of the users[1] in $G+$, collected from August 24th, 2012 to September 10th, 2012. It consists of 74,419,981 nodes and 1,396,943,404 edges. We use Breadth-First-Search (BFS) to extract the *largest weakly connected component* (LWCC) of Γ. We label the extracted LWCC as subgraph Ω (66,237,724 nodes and 1,291,890,737 edges). Since the users Ω form the most important component of the G+ network [6], we extract the subgraph G composed of nodes with at

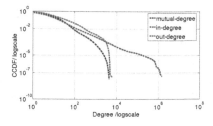

Fig. 1. Degree distributions for subgraph H

---

[1] In this paper we use the terms "user" and "node" interchangeable.

least one reciprocal edge from Ω. Afterwards, we use BFS to extract $G's$ *largest connected component* (LCC); we label this new subgraph as H. In this paper, we consider this subgraph H as the "reciprocal network" of G+[2]. It consists of 40,403,216 nodes and 395,677,038 edges, with a density of 4.85×10^{-7}, slightly larger than the density of Ω (2.95×10^{-7}).

Figure 1 shows the complementary cumulative distribution function (CCDF) of the degrees of nodes in H. We can see that these curves have approximately the shape of a power law distribution. The CCDF of a power law distribution is given by $Cx^{-\alpha}$ and $x, \alpha, C > 0$. By using the tool in [17,18], we estimate the exponent α that best models each of our distributions. We obtain $\alpha = 2.72$ for mutual degree, $\alpha = 2.41$ for out-degree and $\alpha = 2.03$ for in-degree distributions. The observed power-law trend in the distributions implies that a small fraction of users have a disproportionately large number of connections, while most users have a small number of connections – *this is characteristics of many social networks.*

3 Extracting the Core Graph of the Reciprocal Network

Given its massive size (with more than 40 million nodes and nearly 400M edges), we apply a modified version of the *k-shell decomposition* method [13] to prune nodes and edges of *sparse* subgraphs that are likely to lie at the "periphery" of the G+ reciprocal network. K-shell decomposition is a classical graph decomposition technique which has been used as an analysis and visualization tool to extract and study the "core" structure of complex networks, such as that of the Internet AS graph [13]. The classical k-shell decomposition method works as follow: (a) first, remove all nodes in the network with degree 1 (and their respective edges) – these nodes are assigned to the 1-shell; (b) more generally, at step $k = 2, \ldots,$ remove all nodes in the remaining network with degree k or less (and their respective edges) – these nodes are assigned to the k-shell; and (c) the process

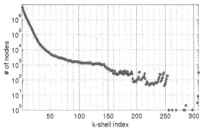

(a) Number of nodes per k-shell

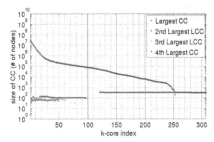

(b) 1st, 2nd, 3rd and 4th k-core LCCs

Fig. 2. k-shell (a) and k-core (b) as k varies from 1 to 308

stops when all nodes are removed at the last step – the highest shell index is labelled k_{max}. The network can be viewed as the union of all k_{max} shells, and for each k, we define the k-core as the union of all shells with indices larger or equal to k.

We apply the k-shell decomposition method to the G+ reciprocal network. We find that the $k_{max} = 308$, and the k_{max}-core is a clique of size 298 nodes (the maximum clique in the G+ reciprocal network). Figure 2(a) shows the number of nodes belonging to the k-shell as k varies from 1 to 308: we see that 99 % of the nodes in our network fall in the lower k-shells (from $k = 1$ to 100). This is not surprising, as the majority of the nodes in our network have degree less than 100. Figure 2(b) shows the size of the largest as well as those of the 2nd, 3rd and 4th largest connected components in the k-core, as k varies from 1 to 308. We note that at step $k = 121$, a small subgraph containing the maximum clique (of size 298) breaks off from the largest connected component which dissolves after $k = 253$, whereas this subgraph containing the maximum clique persists after $k = 252$ and becomes the largest component, and at $k_{max} = 308$, we are left with the maximum clique plus 10 additional nodes that are connected to the maximum clique. Closer inspection of nodes in the maximum clique reveals that its users belong to a single institution in Taiwan, forming a close-knit community where each user follows everyone else. We see that directly applying the standard k-shell decomposition to the G+ reciprocal network produces a clique of size 298, which we believe is unlikely to be the "core" of the G+ reciprocal network.

In order to extract a meaningful "core" of the G+ reciprocal network, we therefore modify the standard k-shell decomposition method to stop the process earlier using the following criterion: we terminate the process at k_C when the largest connected component breaks apart in two or more pieces where each contains a dense subgraph (e.g., a clique of size $q \gg k_C$, here we use a threshold of $q = 200$). Applying this criterion, we terminate the k-shell decomposition at $k_C = 120$, which yields the k_C-core graph with $k_C = 120$: this core graph G_{120} has 51,189 nodes and 7,133,227 edges, with an average degree of 139.4 and a density of approximately 0.0054, which is much smaller than that of the reciprocal network H as a whole.

4   Clique Percolation Analysis and Core Clique Graph

In this section, given the dense core subgraph G_{120}, we want to extract the minimal set of the largest maximal cliques that cover every node in G_{120} and use these cliques to build a new (hyber)graph that provides a higher-level representation of the core structure of the G+ reciprocal network. To achieve this, we implement an algorithm to extract a maximal clique containing a given node in a network. The algorithm is a variation of the popular Bron-Kerbosh algorithm [19]. Hence, we name it Simplified Bron-Kerbosh (SBK) and it is described in Algorithm 1 – the parameter t is used to set an upper bound on the size of the recursion tree. Then, we develop a procedure to extract the minimal set of the largest maximal cliques that cover every node in a given graph (Algorithm 2). The resulting set of cliques returned from this procedure is always guaranteed to contain at least an unique node per clique. We apply this procedure to subgraph G_{120}. We obtain 37,005 maximal cliques with an average clique size of 22.26 nodes.

Algorithm 1. Simplified Bron-Kerbosh (SBK)

1: u : pivot vertex
2: R : currently growing maximal clique
3: $P := N[u]$: set of neighbors of vertex u
4: t : size of the largest clique allowed
5: **SBK**(R, P, u, t)
6: **if** $P := 0$ or $R := t$ **then**
7:     Report R as a maximal clique
8: **else**
9:     Let u_{new} be the vertex with highest number of neighbors in P
10:     $R_{new} := R \cup \{u_{new}\}$
11:     $P_{new} := P \cap N[u_{new}]$
12:     **SBK**$(R_{new}, P_{new}, u_{new}, t)$

Using the extracted 37,005 maximal cliques, we generate a new *directed* (hyber)graph, where the vertices are (unique) cliques of various sizes, and there exists a *directed* edge from clique C_i to clique C_j if more than half of the nodes in C_i are contained in C_j, i.e., $C_i \rightarrow C_j$ if $(|C_i| \cap |C_j|)/|C_i| \geq \theta = 0.5$. We vary the parameter θ from 0.5 to 0.7, and find that it does not fundamentally alter the connectivity structure of the (hyper)graph of cliques thus generated. We remark that the maximal clique containing each node v can be viewed as the most stable structure that node v is part of. The directed (hyper)graph of cliques captures the relations among these stable structures each node is part of: intuitively, each directed edge in a sense reflects the attraction (or gravitational pull) that one clique (a constellation of nodes) has over the other. Hence this (hyber)graph of cliques provides us with a higher-level representation of the dense core graph of the G+ reciprocal network – how the most stable structures are related to each other. This procedure can be viewed as a form of clique percolation [14].

Algorithm 2. Extract Minimal Set of Maximal Cliques from a Graph

---

1: **procedure** EMC$(G(V, E))$
2:     construct a set W and $W := V$
3:     construct a ordered list S of the nodes in V based on their degree (decreasing order)
4:     select the first item in S, vertex i, as the pivot
5:     apply the SBK algorithm using i as the pivot vertex
6:     add the reported maximal clique c_i containing i to the clique set $C_{total} = [c_n, c_m, ..]$
7:     remove the nodes in c_i from W : $W_j = W_i - c_i$
8:     select the next item in S, vertex j, as the next pivot vertex such that $j \notin C_{total}$ and repeat steps(5), (6) and (7) until $W = \emptyset$

We find that this (hyper)graph of cliques comprises of 2,328 connected components (CCs). The largest component has 4,411 cliques, 5,697 nodes and 799,076 edges, while the smallest has 1 clique, 21 nodes and 210 edges respectively. We regard these connected components (CCs) as forming the *core communities* of the core graph of the G+ reciprocal graph: each CC is composed of either one single clique (such a CC shares few than half of its members with other cliques or CCs), or two or more cliques (stable structures) (where one clique shares at least half of its member with another clique in the same CC, thus forming a closely knit community).

5   Core Community Structure Analysis

In this section, we investigate the relationship between the connected components (CCs) in our (hyper)graph of cliques constructed in the previous section (Sect. 4), in particular the 65 largest CCs. First, regarding these CCs as the core community structures (a dense cluster of cliques) of the G+ reciprocal network, we define three metrics to study the relations among these CCs in the underlying G+ reciprocal network: (i) *Shared nodes*: the number of nodes that CC_i and CC_j have in common; (ii) *Shared neighbors*: the number of nodes in CC_i that have at least one edge to a node in CC_j; (iii) *Cross-edges*: the number of cross edges between CC_i and CC_j. These metrics produce a set of new (hyber)graphs that succinctly summarize the (high-level) structural relations among the core community structures. They provide a big picture view of the core graph of the G+ reciprocal network and yield insights as to how it is formed.

Figure 3(a) shows the (hyber)graph of the relationship between the components based on the number of shared nodes. We observe that there are seven CCs that lie at the center of this (hyber)graph through which the other CCs are most richly connected. For the remaining two metrics, we observe that every CC_i has at least one cross-edge and consequently one neighboring node with every other CC_j; thus the CC graph generated based on cross-edge or shared neighbors forms a complete graph – a clique. Hence, we focus our analysis on the strongest relationship between the CCs: for every CC_i, we extract the CC_j that has the largest number of cross-edges with CC_i; likewise, for the neighboring nodes. Figure 3 also shows the (hyber)graph of the relationship between the CCs based on their number of "cross-edges" and "shared neighbors": a node represents a CC and a directed edge $CC_i \rightarrow CC_j$ implies that CC_i has the largest number of cross edges (Fig. 3(b)) or neighboring nodes (Fig. 3(c)) to nodes in CC_j. These figures show that most CCs have the largest number of cross edges and shared neighbors with the same seven CCs identified in Fig. 3(a). Based on these results, we conclude that there are seven subgraphs (core communities) comprising of dense clusters of cliques that lie at the center of the core graph of the G+ reciprocal network, through which other communities of cliques are richly connected. The 2,328 connected components (CCs) in the clique (hyper)graph form the core graph of the G+ reciprocal network, to which other nodes and edges that are part of sparse subgraphs on the peripherals of the network are attached.

We note in particular that in the periphery of our (hyber)graphs, we find a small CC composed with 35 of the largest cliques in the G+ reciprocal network. The average, minimum and maximum sizes of the cliques in this CC are 237, 109

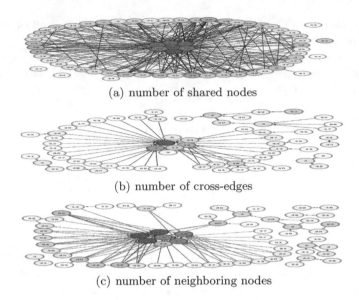

(a) number of shared nodes

(b) number of cross-edges

(c) number of neighboring nodes

Fig. 3. (Hyper)Graphs for the core communities of the reciprocal network of G+ (Color figure online)

and 298 – the latter is the maximum clique of the G+ reciprocal network. This CC is highlighted by a "red circle" in the (hyper)graphs in Fig. 3. It shows this CC lies more at the outer ring of G+s dense core structure. As mentioned earlier in Sect. 3, the 298 users in this maximum clique of the G+ reciprocal network belong to a single institution in Taiwan where every user follows every other. The users in this clique also form close relations with many other users, forming 34 other cliques. Together, these 35 cliques form a close-knit community. However, we see that this community in fact does not lie at the very "center" – instead lies more at the outer ring – of the core graph of the G+ reciprocal network. Hence, we see that simply applying the conventional k-shell decomposition method to the G+ reciprocal network would yield the maximum clique in the G+ reciprocal network, but not its *core* structure. In contrast, the seven CCs mentioned above more likely lie at the "center" of the core graph of the G+ reciprocal network.

6   Conclusion

In this paper we have developed an effective three-step procedure to *hierarchically* extract and unfold the *core* structure of the reciprocal network of Google+, building up and generalizing ideas from the existing k-shell decomposition and clique percolation approaches. As part of ongoing and future work, we will develop a more rigorous characterization of the core graph of the G+ reciprocal network based on the (modified) k-shell decomposition, and provide a more in-depth analysis of the (hyber)graph structures of the clique core graph and the (high-level) structural relations among the core "community" structures. We also plan to apply our method to other directed OSNs such as Twitter.

Acknowledgments. This research was supported in part by DoD ARO MURI Award W911NF-12-1-0385, DTRA grant HDTRA1- 14-1-0040 and NSF grant CNS-1411636. We thank the authors of [6] for the datasets.

References

1. Gong, N.Z., Xu, W.: Reciprocal versus parasocial relationships in online social networks. Soc. Netw. Anal. Min. **4**(1), 184–197 (2014)
2. Garlaschelli, D., Loffredo, M.I.: Patterns of link reciprocity in directed networks. Phys. Rev. Lett. **93**, 268–701 (2004)
3. Jiang, B., Zhang, Z.-L., Towsley, D.: Reciprocity in social networks with capacity constraints. In: KDD 2015, pp. 457–466. ACM (2015)
4. Hai, P.H., Shin, H.: Effective clustering of dense and concentrated online communities. In: Asia-Pacific Web Conference (APWEB) 2010, pp. 133–139. IEEE (2010)
5. Gong, N.Z., Xu, W., Huang, L., Mittal, P., Stefanov, E., Sekar, V., Song, D.: Evolution of the social-attribute networks: measurements, modeling, and implications using Google+. In: IMC 2015, pp. 131–144. ACM (2015)
6. Gonzalez, R., Cuevas, R., Motamedi, R., Rejaie, R., Cuevas, A.: Google+ or Google–? Dissecting the evolution of the new OSN in its first year. In: WWW 2013, pp. 483–494. ACM (2013)
7. Kwak, H., Lee, C., Park, H., Moon, S.: What is Twitter, a social network or a news media? In: WWW 2010, pp. 591–600. ACM (2010)
8. Magno, G., Comarela, G., Saez-Trumper, D., Cha, M., Almeida, V.: New kid on the block: exploring the Google+ social graph. In: IMC 2012, pp. 159–170. ACM (2012)
9. Mislove, A., Marcon, M., Gummadi, K.P., Druschel, P., Bhattacharjee, B.: Measurement and analysis of online social networks. In: IMC 2007, pp. 29–42. ACM (2007)
10. Wolfe, A.: Social network analysis: methods and applications. Am. Ethnol. **24**(1), 219–220 (1997)
11. Jamali, M., Haffari, G., Ester, M.: Modeling the temporal dynamics of social rating networks using bidirectional effects of social relations and rating patterns. In: WWW 2011, pp. 527–536. ACM (2011)
12. Li, Y., Zhang, Z.-L., Bao, J.: Mutual or unrequited love: identifying stable clusters in social networks with uni- and bi-directional links. In: Bonato, A., Janssen, J. (eds.) WAW 2012. LNCS, vol. 7323, pp. 113–125. Springer, Heidelberg (2012). doi:10.1007/978-3-642-30541-2_9
13. Carmi, S., Havlin, S., Kirkpatrick, S., Shavitt, Y., Shir, E.: A model of Internet topology using k-shell decomposition. PNAS **104**, 11150–11154 (2007)
14. Palla, G., Dernyi, I., Farkas, I., Vicsek, T.: Uncovering the overlapping community structure of complex networks in nature and society. Nature **435**(7043), 814–818 (2005)
15. Google+ Platform. http://www.google.com/intl/en/+/learnmore/
16. Google+. http://en.wikipedia.org/wiki/Google+
17. Clauset, A., Shalizi, C.R., Newman, M.E.J.: Power-law distributions in empirical data. SIAM Rev. **51**, 661–703 (2009)
18. Fitting Power Law Distribution. http://tuvalu.santafe.edu/~aaronc/powerlaws/
19. Cazals, F., Karande, C.: A note on the problem of reporting maximal cliques. Theoret. Comput. Sci. **407**(1), 564–568 (2008)

Tackling Common Due Window Problem
with a Two-Layered Approach

Abhishek Awasthi[1(✉)], Jörg Lässig[1], Thomas Weise[2], and Oliver Kramer[3]

[1] Department of Computer Science, University of Applied Sciences Zittau/Görlitz,
Görlitz, Germany
{abhishek.awasthi,joerg.laessig}@hszg.de
[2] School of Computer Science and Technology,
University of Science and Technology of China, Hefei, Anhui, China
tweise@ustc.edu.cn
[3] Department of Computing Science, Carl von Ossietzky University of Oldenburg,
Oldenburg, Germany
oliver.kramer@uni-oldenburg.de

Abstract. This work presents a polynomial algorithm to optimize any given job sequence for the Common Due-Window (CDW) Problem. The CDW problem comprises of scheduling and sequencing a set of jobs against a due-window to minimize the total weighted earliness/tardiness penalty. This due-window is defined by the *left* and *right* common due-dates. Jobs that finish before (after) the left (right) due-date are termed as early (tardy) jobs. We present an exact polynomial algorithm for optimally scheduling a given fixed job sequence for a single machine with the runtime complexity of $O(n)$, where n is the number of jobs. The linear algorithm and a heuristic based on the V-shaped property are then incorporated with a modified Simulated Annealing (SA) algorithm to obtain the optimal/near-optimal solutions. We carry out computational experiments to demonstrate the utility of our approach over the benchmark instances and previous work on this problem.

1 Introduction

The Common Due-Window (CDW) scheduling problem consists of finding an optimal sequence of jobs along with the optimal completion times of the jobs such that the total weighted earliness and tardiness penalties is minimized. Each job possesses a distinct processing time and asymmetric penalties per unit time in case the job finishes its processing before or after the due-window. The jobs which are completed between or at the due-window are called straddle jobs and do not incur any penalty. Similar to the Common Due-Date (CDD) problem, the CDW also occurs in mass production industry adopting the Just-in-Time philosophy. The problem can be formulated as follows. Let n be the total number of jobs and d_l, d_r be the left and right due-dates, respectively, of the common due-window. We also define the processing time of any job as P_i and C_i as the completion time of the jobs for all $i = 1, 2, \ldots, n$. Any job which finishes its processing

© Springer International Publishing AG 2016
T.-H.H. Chan et al. (Eds.): COCOA 2016, LNCS 10043, pp. 772–781, 2016.
DOI: 10.1007/978-3-319-48749-6_59

before or after the due-date incurs an earliness or tardiness penalty, represented as g_i or h_i, respectively. Mathematically, we can write $g_i = \max\{0, d_l - C_i\}$, and $h_i = \max\{0, C_i - d_r\}$, for all i. Corresponding to the earliness/tardiness values, we also have the associated penalties which are expressed as α_i and β_i, respectively. With this formulation, the objective of the problem is to schedule the jobs against the due-window to minimize the total weighted penalty incurred by the earliness and tardiness of all the jobs, as shown in Eq. (1).

$$\min \sum_{i=1}^{n} \{\alpha_i \cdot g_i + \beta_i \cdot h_i\} \,. \tag{1}$$

CDW is an extension of CDD with the presence of a common due-window instead of a common due-date. However, several important similar properties hold for both problems. In 1994, Krämer and Lee studied due-window scheduling for the parallel machine case and presented useful properties for the CDW [7], explained later in the paper. Krämer and Lee also showed that the CDW with unit weight case is also NP-complete and provided a dynamic programming algorithm for the two machine case [7]. In 2005, Biskup and Feldmann dealt with the general case of the CDW problem and approached it with three different meta-heuristic algorithms, namely, Evolutionary Strategy, Simulated Annealing and Threshold Accepting. They also validated their approaches on 250 benchmark instances up to 200 jobs [2]. Wan studied the common due-window problem with controllable processing times with constant earliness/tardiness penalties and distinct compression costs, and discussed some properties of the optimal solution along with a polynomial algorithm for the solving the problem in 2007 [10]. In 2010, Yeung et al. formulated a supply chain scheduling control problem involving single supplier and manufacturer and multiple retailers. They formulated the problem as a two machine CDW and presented a pseudo-polynomial algorithm to solve the problem optimally [12]. Ji et al. study the due-window assignment problem, where each job has a job-dependent due-window and an additional penalty is associated with the due-window position [5]. In 2013, Janiak et al. presented a survey paper on the common due-window assignment scheduling problem and discussed more than 30 different variations of the problem [3]. In 2015, Janiak et al. also presented an extensive survey on scheduling against due-window and provided computational complexities for several variants of the problem [4].

In this work, we consider the single machine general case for the CDW problem with asymmetric penalties for the general case, which is an NP-hard problem [2]. Our solution approach is divided in two layers, (i) finding the optimal/near-optimal job sequence, and, (ii) optimizing any given processing sequence with an efficient polynomial algorithm. We present a linear exact algorithm to optimize a given job sequence on a single machine. Henceforth, we present a simple heuristic algorithm based on a V-shaped property, to evolve and improve any job sequence. Our algorithm and the heuristic are then combined with the Simulated Annealing algorithm to find the optimal/best job sequence. We test the effectiveness of our approach on the benchmark instances provided in [2] with the help of experimental analyses.

2   The Exact Algorithm for a Given Job Sequence

In this section, we present and explain our exact algorithm for a given job sequence of CDW, using useful and important properties proved in previous works in the literature.

Property 1. There exists an optimal schedule without any machine idle time where either the first job starts at time $t = 0$ or one of the jobs finishes at d_l or d_r, i.e., $C_i = d_l$ or $C_i = d_r$, for some i [7, 11].

Property 2. In any optimal schedule, the early jobs are sequenced in non-increasing order of the ratio P_i/α_i and the tardy jobs are sequenced in non-decreasing order of the ratio P_i/β_i [2].

One important observation that can be inferred from the objective function and the above properties is that moving the job sequence to the right gives us a V-shaped trend for the objective function values, depending on where the due-window falls on the schedule.

Property 3. If in the optimal schedule of any given job sequence of CDW problem $J_E = \{1, 2, \ldots, r-1, r\}$ be the set of early jobs that are processed in that order and $r = \arg\max(C_i \leq d_l)$ for $i = 1, 2, \ldots, n$ and $J_T = \{k, k+1 \ldots, n\}$ be the set of tardy jobs such that $k = \arg\min(C_i > d_r)$ for $i = 1, 2, \ldots, n$. Then we have, $\sum_{i \in J_T} \beta_i < \sum_{i \in J_E} \alpha_i$ for minimum possible values of k and r [1].

This is an important property provided in [1] which basically proves that the optimal schedule of any given job sequence can be obtained by right shifting the jobs, as long as $\sum_{i \in J_T} \beta_i < \sum_{i \in J_E} \alpha_i$ holds. Using the above properties, we now present our exact polynomial algorithm for the CDW problem to optimize any given job sequence. As mentioned above in Property 1, we know that the optimal schedule of the CDW has no idle time of the machine between C_1 and C_n. Hence, for our algorithm, the jobs are initialized with the first job starting at time $t = 0$ and are shifted to the right by minimum deviation of the completion times from the right and the left due-dates. This way, every shift ensures that one of the jobs finishes at one of the due-dates (Property 1) and we do not skip over the optimal position of the due-dates. Once Property 3 is satisfied, we have our optimal schedule and no more shifting is required. This approach of right-shifting-the-jobs is implemented in [1]. However, it requires the update of the completion times C_i of all the jobs and thus accounts for a runtime complexity of $O(n)$ for each shift. We present a much faster approach where the calculation of the completion times of all the jobs needs to be done only once, throughout the algorithm.

The idea behind our approach lies in the fact that the calculation of the objective function or checking Property 3, only requires the relative deviation of the completion times of all the jobs with the left and right due-dates. Hence, shifting all the jobs and due-dates together with the same amount does not affect the objective function value, as well as the set of early and tardy jobs. For our algorithm, we initialize the completion times of the jobs such that $C_1 = P_1$ and the subsequent jobs are followed without any machine idle time. The C_i values remain fix for the whole algorithm. Thereafter, we find the optimal position of a movable due-window (d'_l, d'_r) of the same length as of the original due-window. This optimal position is calculated using Property 3, by shifting the movable due-window from extreme right to left as long as the sum of the tardiness penalties is less than the sum of the earliness penalties. If the optimal position of this new due-window lies to the left of the original due-window $i.e.$, $d'_l < d_l$ or $d'_r < d_r$, (note that both the inequalities will be satisfied simultaneously, since the due-windows are of same lengths, $d'_l < d_l \Rightarrow d'_r < d_r$), then we take d'_l and d'_r for calculating the final earliness/tardiness of the jobs. However, if the position of this movable due-window lies to the right of the original due-window, $i.e.$, $d'_l > d_l$ or $d'_r > d_r$, then we retain the original due-dates for calculating the final earliness/tardiness of the jobs. The reason for the above statements can be proved by considering the two cases separately. If the optimal position of the movable due-window is such that $d'_l < d_l$, then it means that Property 3 is satisfied for some value k but the original due-date falls at some job index i where $i > k$. Thus, we need to shift all the jobs to the right such that jobs $k, k+1, \ldots, n$ are tardy. Instead, we can just take the d'_l and d'_r as the due-dates to calculate the final earliness/tardiness of the jobs to obtain the objective function value, because of the fact that the earliness/tardiness are relative deviations with the due-dates. However, if the optimal position of the movable due-window is such that $d'_l > d_l$, then it means that Property 3 is satisfied for some value k but the original due-date falls at some job index i where $i < k$. In this case, we are actually required to practically shift the jobs to the left, which can not be done as the schedule of the jobs is already starting at time $t = 0$. Hence, for this case we need to take the original due-window (d_l, d_r) to calculate the final earliness/tardiness of the jobs. The case $d'_l = d_l$ is obvious.

Our algorithm first assigns the right due-date (d'_r) of the movable due-window as C_n and $d'_l = d'_r - d_r + d_l$. This ensures that the right due-date falls at the completion time of the last job and the left due-date is placed such that $C_i \leq d'_l < C_{i+1}$, where $i < n$. In the next steps, this movable due-window is shifted to the left as long as we obtain the optimal position using Property 3. We now formulate some parameters which are essential for the understanding of the exact algorithm. We define η_l, φ_l, η_r, and φ_r as

$$\eta_l = \arg\max(C_i - d'_l < 0), \quad \varphi_l = \arg\max(C_i - d'_l \leq 0),$$
$$\eta_r = \arg\max(C_i - d'_r < 0), \quad \varphi_r = \arg\max(C_i - d'_r \leq 0), \quad i = 1, 2, \ldots, n. \quad (2)$$

In the above equation, η_l depicts the last job which finishes strictly before the left due-date d'_l, while φ_l is the last job which finishes at or before d'_l. η_r and φ_r can be understood on the same lines, with respect to the right due-date d'_r. Hence, one can write $\delta_l = d'_l - C_{\eta_l}$ and $\delta_r = d'_r - C_{\eta_r}$. Clearly, $\min\{\delta_l, \delta_r\}$ is the minimum possible left shift of the due-window required such that either one of the left/right due-dates falls at the completion time of a job. Any left shift of the due-window is basically made depending on the position of the left and the right due-dates.

Algorithm 1. Linear algorithm for any CDW job sequence.

---

1 $C_i \leftarrow \sum_{k=1}^{i} P_k \ \forall \ i = 1, 2, \ldots, n$
2 $d'_r \leftarrow C_n$
3 $d'_l \leftarrow d'_r - d_r + d_l$
4 Compute $\eta_l, \varphi_l, \eta_r, \varphi_r$
5 $\delta_l \leftarrow d'_l - C_{\eta_l}$
6 $\delta_r \leftarrow d'_r - C_{\eta_r}$
7 $pe \leftarrow \sum_{i=1}^{\varphi_l} \alpha_i$
8 $pl \leftarrow 0$
9 while $pl < pe$ **do**
10 $\quad LeftDueDate \leftarrow d'_l$
11 $\quad RightDueDate \leftarrow d'_r$
12 $\quad \delta \leftarrow \min\{\delta_l, \delta_r\}$
13 $\quad d'_l \leftarrow d'_l - \delta$
14 $\quad d'_r \leftarrow d'_r - \delta$
15 $\quad$ **if** $(\eta_l < \varphi_l)$ **then**
16 $\quad\quad pe \leftarrow pe - \alpha_{\varphi_l}$
17 $\quad\quad \varphi_l \leftarrow \varphi_l - 1$
18 $\quad$ **if** $(\eta_r < \varphi_r)$ **then**
19 $\quad\quad pl \leftarrow pl + \beta_{\varphi_r}$
20 $\quad\quad \varphi_r \leftarrow \varphi_r - 1$
21 $\quad$ **if** $\delta_l < \delta_r$ **then**
22 $\quad\quad \eta_l \leftarrow \eta_l - 1$
23 $\quad$ **else if** $\delta_r < \delta_l$ **then**
24 $\quad\quad \eta_r \leftarrow \eta_r - 1$
25 $\quad$ **else if** $\delta_r = \delta_l$ **then**
26 $\quad\quad \eta_l \leftarrow \eta_l - 1$
27 $\quad\quad \eta_r \leftarrow \eta_r - 1$
28 $\quad$ **if** $\eta_l > 0$ **then**
29 $\quad\quad \delta_l \leftarrow d'_l - C_{\eta_l}$
30 $\quad\quad \delta_r \leftarrow d'_r - C_{\eta_r}$
31 $\quad$ **else**
32 $\quad\quad$ **break**
33 if $LeftDueDate \leq d_l$ **then**
34 $\quad d_l \leftarrow LeftDueDate$
35 $\quad d_r \leftarrow RightDueDate$
36 Compute $g_i, h_i, \forall i$
37 return $Sol \leftarrow \sum_{i=1}^{n} (g_i \cdot \alpha_i + h_i \cdot \beta_i)$

---

3   Proof of Optimality and Runtime Complexity

In this section, we present the optimality of Algorithm 1 with respect to the objective function value of Eq. (1), for any given job sequence.

Theorem 1. *Algorithm 1 is optimal for any given job sequence of the CDW problem, with respect to the objective function value, with a runtime complexity of $O(n)$.*

Proof. We first schedule the given job sequence such that the processing of the first job starts at time $t = 0$ and the remaining jobs are processed without any machine idle time, maintaining the sequence of the jobs. We then place the movable due-window in a way such that $d'_r = C_n$ for some n and $d'_l = d'_r - d_r + d_l$, maintaining the length of the due-window same as the original. From this point on, we refer to this *movable* due-window as the due-window itself, unless mentioned otherwise. Meanwhile, we also calculate $\eta_l, \varphi_l, \eta_r$ and φ_r to keep track of the jobs that are closest to the due-dates. One can interpret the η and φ values to indicate the position of the due-window. At each step of the iterative left shift of δ where $\delta = \min\{\delta_l, \delta_r\}$, where $\delta_l = d'_l - C_{\eta_l}$ and $\delta_r = d'_r - C_{\eta_r}$. Apparently, any left shift of δ ensures that we do not skip over any job while checking for Property 3. However, depending on the position of the due-window (or the η_l, φ_l and η_r, φ_r values), we can have several different cases that can occur during the left shift.

During the course of any left shift, if $\eta_l < \varphi_l$ then we are certain that d'_l falls at the completion time of some job. Hence, $\eta_l < \varphi_l$ necessarily implies that $\varphi_l = \arg\max(C_i - d'_l \leq 0) = u$ for some job u, such that $C_u = d'_l$ and $\eta_l = u - 1$. Moreover, at any instance of the algorithm, we never make a left shift that is greater than δ_l. Hence, the value of pe will be reduced by α_{φ_l}, since for any shift which is greater than zero, job u will fall after the left due-date. Likewise, whatever the left shift, the value of φ_l will also be reduced by 1, for the same reason that u will no longer fall at d'_l, after the shift of the due-window. Also note that the only possible case such that $\eta_l = \varphi_l$, is the one when d'_l falls in between the completion time of two jobs, after the left shift, say $u - 1$ and u. Since, $\delta = \min\{\delta_l, \delta_r\}$, the maximum possible shift will lead to $d'_l = C_u$. However, in this case we do not need to update pe and φ_l as per their definitions. The value of η_l will indeed get reduced by 1, if the shift is made by δ_l, *i.e.*, the left due-date falls at the completion time of a job after the left shift of the due-window. Hence, to check if η_l needs to be updated or not, we only need to check if the shift if made by δ_l, as implemented in Algorithm 1. The same procedure is adopted for the case when $\eta_r < \varphi_r$. After every left shift, we update the values for $\eta_l, \varphi_l, \eta_r, pe$ and pl depending on the position of the due-window and the amount of left shift. Henceforth, we update the values of δ_l and δ_r with the new values of η_l and η_r, respectively. Note that we need to check for a special case where the earliness penalty of the first job is higher than the sum of the tardiness penalties of the jobs which are completed after the right due-date. In this case, the optimal schedule will occur when the left due-date falls at C_1. It is for this case, that we need to check that $\eta_l > 0$ before making a left shift, as depicted in Algorithm 1 line 28. For the next iteration of the *while* loop in line 9 of Algorithm 1, we update the positions of the left and right due-date by $\delta = \min\{\delta_l, \delta_r\}$ and repeat the same procedure as long as $pl < pe$, according to Property 3. We then check if the optimal position of this *movable* due-window lies to the left or the right of the original due-window as explained in Sect. 2.

As far as the runtime complexity is concerned, it can easily observed that the complexity of the Algorithm 1 is $O(n)$, since all the initialization steps and the calculation of the parameters all require $O(n)$ runtime. As for the iterative *while* loop, all the computations inside the loop are of $O(1)$ as we update the values of any parameter by simple one step computation. However, this iterative left shift can take $2 \cdot n$ steps in the worst case, however, it does not affect the complexity of Algorithm 1. Hence, the complexity of Algorithm 1 is $O(n)$.   □

4   Computational Results

We now present the results of computational experiments for the problem discussed in this work. The experiments are carried out on the CDW benchmark instances provided by Biskup and Feldmann in [2]. All the computations are carried out using MATLAB on a 1.73 GHz PC with 2 GB RAM. We implement a modified Simulated Annealing algorithm to generate the job sequences, while each job sequence is optimized with Algorithm 1 and further improved by arranging the early jobs in non-increasing order of P_i/α_i, while the tardy jobs are arranged in the non-decreasing order of P_i/β_i as per Property 2. Our experimental analyses suggest that an ensemble size of 10 and the maximum number of iterations of $150 \cdot n$ work best for the provided instances in general. The initial temperature for SA is kept as twice the standard deviation of the energy at infinite temperature: $\theta_{E_{T=\infty}} = \sqrt{\langle E^2 \rangle_{T=\infty} - \langle E \rangle^2_{T=\infty}}$. This quantity is estimated by randomly sampling the configuration space [9]. An exponential schedule for cooling is adopted with a cooling rate of 0.9999. We implement two acceptance criteria: the Metropolis acceptance probability, $\min\{1, \exp((-\Delta E)/T)\}$ [9] and a constant acceptance probability of $c \cdot 10^{-2}$, where c is a constant less than 10. A solution is accepted with this constant probability if it is rejected by the Metropolis criterion. This concept of a constant probability is useful when the SA is run for many iterations and the metropolis acceptance probability is almost zero, since the temperature would become infinitesimally small. Apart from this, we also incorporate elitism in our SA. Elitism has been successfully adopted in evolutionary algorithms for several complex optimization problems [6]. Theoretical studies have been made analysing speed-ups in parallel evolutionary algorithms combinatorial optimization problems in [8]. As for the perturbation rule, we first randomly select a certain number of jobs in any job sequence and permute them randomly to create a new sequence. The number of jobs selected for this permutation is taken as $3 + \lfloor \sqrt{n/10} \rfloor$, where n is the number of jobs. In addition to this, we also incorporate swapping of one job each from the set J'_E and J'_T with each other. This swapping is specially useful for the instances with large due-window size, as random perturbation might cause little to no effect on the job sequence if all the jobs that are perturbed belong to the straddling set of jobs J_S. The reason for this is that changing the sequence of straddling jobs does not change the objective function value. All our computation results are averaged over 100 different replications of the SA algorithm, for each instance. We present our results for the CDW where the due-window size for any instance is calculated using the values of $h1$ and $h2$. A due-window has a left (d_l) and right (d_r) due-date, where $d_l = \lfloor h1 \cdot \sum_{i=1}^{n} P_i \rfloor$ and $d_r = \lfloor h2 \cdot \sum_{i=1}^{n} P_i \rfloor$, as described in [2].

Table 1. Average runtime in seconds and percentage gap of our solutions with the benchmark results of [2], averaged over 10 different instances for each job size.

Jobs	h1–h2	0.1–0.2	0.1–0.3	0.2–0.5	0.3–0.4	0.3–0.5	Average
10	%$\overline{\Delta}$	0.000	0.000	0.000	0.000	0.000	**0.000**
	Time	0.000	0.000	0.000	0.000	0.001	**0.000**
20	%$\overline{\Delta}$	−0.018	0.003	−0.010	−0.004	0.000	**−0.006**
	Time	0.014	0.030	0.020	0.020	0.021	**0.021**
50	%$\overline{\Delta}$	−0.037	0.020	0.219	−0.124	−0.021	**0.011**
	Time	0.189	0.191	0.138	0.133	0.103	**0.151**
100	%$\overline{\Delta}$	−0.029	−0.016	−0.079	−0.100	−0.214	**−0.088**
	Time	0.680	0.804	0.897	0.846	0.885	**0.822**
200	%$\overline{\Delta}$	−0.062	0.001	0.073	−0.109	−0.054	**−0.030**
	Time	3.675	4.044	4.392	4.310	4.298	**4.144**

Let F_i^{REF} represent the benchmark results provided in [2] and F_i be the solution value obtained in this work for any instance i, then the percentage gap %Δ is defined as $(F_i - F_i^{\mathrm{REF}}) \cdot 100/F_i^{\mathrm{REF}}$. Table 1 shows the average values of percentage gap (%$\overline{\Delta}$) and the average runtime required by our algorithm. The percentage gaps and the runtimes are averaged over 10 different benchmark instances for each due-window, for 100 different replications of SA. For the first 50 instances with 10 jobs we obtain the optimal solution for all the instances. For higher instances with 20, 50, 100 and 200 jobs, we achieve better results on average with the average percentage gap of −0.006, 0.011, −0.088 and −0.030 %, respectively. The negative values of the percentage gap in Table 1 indicate that we obtain better results than the benchmark results of [2]. The runtime for all the results is the time after which the solutions mentioned in Table 1 are obtained on average after 10 different replication. Biskup and Feldmann do not provide the runtime for their algorithms, hence we cannot compare our results on the basis of the runtime. However, the quality of our results is certainly superior than the benchmark results, owing to our approach. The average runtime over 10 different replications of our algorithm for job size of 20 and above is only 0.021, 0.151, 0.822 and 4.144 s. The previous approach for this problem involved an $O(n^2)$ algorithm to optimize any job sequence [1]. However, the runtimes with that approach for 10, 20 and 50 jobs on the same machine were 0.173, 0.465 and 6.028 s, respectively. Moreover, we also obtain better results than the previous approach mentioned in [1]. We also present the graphical representation of the average percentage error obtained by our algorithm. The negative value indicates that we obtain better results than the benchmark solutions, in Fig. 1. Evidently, the worst possible solution values by our approach are for the case when the due-window size is as big as the 30 % of the total length of the schedule, *i.e.*, with the due-window restriction factor of 0.2 to 0.5.

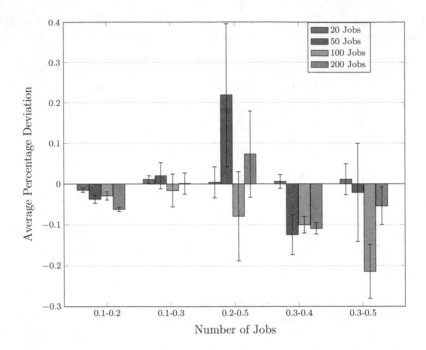

Fig. 1. Comparative average percentage deviation of CDW along with its standard deviation.

The reason behind it is the fact that the perturbation causes the least change in the processing sequence of the jobs as several jobs belong to the straddling set. In such a case, swapping of jobs from J'_E to J'_T is highly useful. However, it must be noted that the percentage error values presented in Fig. 1 are averaged over 100 different replications of SA and the worst value for average percentage deviation over all the instances is only 0.2 %. Regardless, over all the replications of SA, in the average case we still obtain solutions better than the best known values as is clear from Table 1. Furthermore, the robustness of our approach is highlighted by the small standard deviation values over all the instances, as shown in Fig. 1. Our algorithm consistently obtains good quality solutions with the worst possible standard deviation of just 0.177 %. As far as the solution quality is concerned, not only our approach is robust over all the instances, we also improve several benchmark results provided in [2].

5   Conclusion

In this paper we present a novel $O(n)$ algorithm for a the general case of the CDW to optimize a given job sequence, proving its runtime complexity and its optimality with respect to the solution value. Additionally, we also incorporate the V-shaped property to locally improve a job sequence. We apply our algorithms to the benchmark instances provided by Biskup and Feldmann [2] and obtain better results for 137 instances out of 150 benchmark instances for the job size of 50 and higher. Our algorithm works in the same manner for the case when the processing times of all the jobs are equal or of unit time length. Our approach for that problem would be same except for the fact that on Algorithm 1 each shift will be equal to the length of the processing time. The remaining procedure of improvement heuristics and the SA can be utilized as described in this work. Finally, we would also like to assert that our linear algorithm is well suited for the CDW problem with assignable due-window of a given size, incorporating penalty for due-window location. The only addition to our algorithm would be to balance out the effect of the due-window assignment and the objective function value, which is monotonic for constant penalty associated with the due-window location.

References

1. Awasthi, A., Lässig, J., Kramer, O., Weise, O.: Common due-window problem: polynomial algorithms for a given processing sequence. In: IEEE Symposium on Computational Intelligence in Production and Logistics Systems (IEEE SSCI-CIPLS), pp. 32–39 (2014)
2. Biskup, D., Feldmann, M.: On scheduling around large restrictive common due windows. Eur. J. Oper. Res. **162**(3), 740–761 (2005)
3. Janiak, A., Janiak, W., Kovalyov, M., Kozan, E., Pesch, E.: Parallel machine scheduling and common due window assignment with job independent earliness and tardiness costs. Inf. Sci. **224**, 109–117 (2013)
4. Janiak, A., Janiak, W., Krysiak, T., Kwiatkowski, T.: A survey on scheduling problems with due windows. Eur. J. Oper. Res. **242**(2), 347–357 (2015)
5. Ji, M., Chen, K., Ge, J., Cheng, T.: Group scheduling and job-dependent due window assignment based on a common flow allowance. Comput. Ind. Eng. **68**, 35–41 (2014)
6. Kim, J.: Genetic algorithm stopping criteria for optimization of construction resource scheduling problems. Constr. Manag. Econ. **31**(1), 3–19 (2013)
7. Krämer, F., Lee, C.: Due window scheduling for parallel machines. Math. Comput. Model. **20**(2), 69–89 (1994)
8. Lässig, J., Sudholt, D.: General upper bounds on the runtime of parallel evolutionary algorithms. Evol. Comput. **22**(3), 405–437 (2014)
9. Salamon, P., Sibani, P., Frost, R.: Facts, Conjectures, and Improvements for Simulated Annealing. Society for Industrial and Applied Mathematics, Philadelphia (2002)
10. Wan, Guohua: Single machine common due window scheduling with controllable job processing times. In: Dress, Andreas, Xu, Yinfeng, Zhu, Binhai (eds.) COCOA 2007. LNCS, vol. 4616, pp. 279–290. Springer, Heidelberg (2007). doi:10.1007/978-3-540-73556-4_30
11. Yeung, W., Oguz, C., Cheng, T.: Single-machine scheduling with a common due window. Comput. Oper. Res. **28**(2), 157–175 (2001)
12. Yeung, W., Choi, T., Cheng, T.: Optimal scheduling of a single-supplier single-manufacturer supply chain with common due windows. IEEE Trans. Autom. Control **55**(12), 2767–2777 (2010)

A Polynomial Time Solution for Permutation Scaffold Filling

Nan Liu[1], Peng Zou[2], and Binhai Zhu[2(✉)]

[1] School of Computer Science and Technology, Shandong Jianzhu University,
Jinan, Shandong, China
belovedmilk@126.com
[2] Gianforte School of Computing, Montana State University,
Bozeman, MT 59717-3880, USA
peng.zou@msu.montana.edu, bhz@montana.edu

Abstract. Recently, the scaffold filling problem has attracted a lot of attention due to its potential applications in processing incomplete genomic data. However, almost all the current research assumes that a scaffold is given as an incomplete sequence (i.e., missing genes can be inserted anywhere in the incomplete sequence). This differs significantly from most of the real genomic dataset, where a scaffold is given as a sequence of contigs. We show in this paper that when a scaffold is given as a sequence of contigs, and when the genome contains no duplication of genes, the corresponding scaffold filling problem, with the objective being maximizing the number of adjacencies between the filled scaffold and a complete reference genome, is polynomially solvable.

1 Introduction

Since the first human genome was sequenced in 2001, the cost of sequencing a genome has been reduced greatly, with the current cost being only around $1k. On the other hand, the cost to finish these genomes completely has not been decreased as much compared with a decade ago [3]. This results in a lot of *draft* genomes, i.e., genomes not completely finished. Nonetheless, to analyze these genomic data, many tools do need complete genomes as input. For instance, to compute the reversal distance between two genomes, two complete genomes must be given as input. Hence, there is a need to transform a draft genome into a complete genome.

To make the result biologically meaningful, Munoz *et al.* first proposed the *scaffold filling* problem (initially on multichromosomal genomes with no gene repetitions) as follows [7]. Given a complete (permutation) genome R and an incomplete scaffold S (composed of a list of contigs), fill the missing genes in $R - S$ into S to obtain S' such that the genomic distance (or DCJ distance [11]) between R and S' is minimized. It was shown that this problem can be solved in polynomial time. In [5], Jiang *et al.* studied the case for singleton genomes without gene repetitions (i.e., permutations), using the simplest *breakpoint* distance

© Springer International Publishing AG 2016
T.-H.H. Chan et al. (Eds.): COCOA 2016, LNCS 10043, pp. 782–789, 2016.
DOI: 10.1007/978-3-319-48749-6_60

as the similarity measure. It was shown that this problem is solvable in polynomial time; in fact, even for the two-sided case when both the input scaffolds, being a reference to each other, are incomplete.

Most of the related research focuses on the case when the genomes and scaffolds contain gene repetitions, though they are not directly related to the research in this paper. When there are gene duplications, using number of (string) adjacencies, the problem becomes NP-hard. (Formally, the problem is to fill an incomplete sequence scaffold I into I', with respect to a complete reference genome G, such that the missing letters in $G - I$ are inserted back to I to have I' and the number of common adjacencies between G and I' is maximized.) Hence, most of these research focuses on designing approximation algorithms and interested readers are referred to [12] for a recent survey. Using the number of common adjacencies as a parameter k, it was shown that this problem is also fixed-parameter tractable (FPT) — but this only handles that case when G and I' are not quite similar so it is only theoretically meaningful [2].

Among the existing results, almost all papers covered genomes with gene duplications and genomic scaffolds as incomplete sequences, with the exception for [5,7] — as we have just mentioned. In [7], the genomes in consideration have no gene duplication but are multichromosomal and the scaffolds are sequences of contigs (after some preprocessing). In the first part of [5], the genomes in consideration have no duplication but the scaffolds are incomplete sequences (which means missing genes can be inserted anywhere in a scaffold). It is noted that in practical genomic datasets, a scaffold is usually given as a sequence of contigs, where a contig is usually computed through mature software tools like Celera Assembler [1], hence should not be altered arbitrarily. Therefore, in [5], a question was raised to solve the case when a scaffold is given as a sequence of contigs.

In this paper, we consider this *permutation scaffold filling* (PSF) problem which was left as an open problem in [5]. (In [5], a scaffold is just an incomplete sequence and a missing gene can be inserted anywhere.) A very simple example can be used to illustrate the difference between the two versions: $R = \langle 1, 2, 3, 4, 5 \rangle$, $S = \boxed{1,4,2,3}$, and $X = \{5\}$. If S is a contig which cannot be altered, then inserting 5 at the two ends of S cannot generate any adjacency. If 5 can be inserted anywhere in S, then one adjacency can be formed.

Now let us come back to the PSF problem. Assume henceforth that all the genomes in consideration do not have gene duplications. The reference genome R is a complete permutation. The scaffold S is given as a sequence of contigs. We need to insert the missing genes in $R - S$ into S to have S^+ such that the number of adjacencies between R and S^+ is maximized. (As for two complete permutations the number of breakpoints and adjacencies in either one sum to their length minus one, i.e., a fixed number, the objective function is equivalent to minimizing the number of breakpoints between R and S^+).

The paper is organized as follows. In Sect. 2, we give the preliminaries. In Sect. 3, we present the algorithm and the corresponding proofs. We conclude the paper in Sect. 4.

2   Preliminaries

Throughout this paper we focus only on singleton genomes with no gene dupli-
cation/repetitions (i.e., each is a permutation). But the results can be easily
adapted to multichromosomal genomes. At first, we review some necessary definitions on *breakpoints* and *adjacencies* in permutations. These ideas were orig-
inated in 1920s [8,9], though the more formal definitions were formulated in
1980s [10].

We assume that all genes and genomes are unsigned, and it is straightforward
to generalize the result to signed genomes. Given a gene set $\Sigma = \{1, \ldots, n\}$, a
string P is called *permutation* if each element in Σ appears exactly once in
P. We use $c(P)$ to denote the set of elements in permutation P. Given Σ, an
incomplete permutation P' is one where $c(P') \subset \Sigma$.

Given $\Sigma = \{1, \ldots, n\}$, and two (complete) permutations P_1, P_2 on Σ, a 2-
substring $ab \in P_1$ forms an *adjacency* if $ab \in P_2$ or $ba \in P_2$; otherwise ab is called
a *breakpoint*. (In this context, ab and ba are considered the same.) Let P_1 and P_2
be a reference to each other. Let $a(P_i), b(P_i), i = 1, 2$, be the set of adjacencies
and breakpoints in P_i respectively. It is a well-known fact that $a(P_1) = a(P_2)$,
$|b(P_1)| = |b(P_2)|$; moreover, $|a(P_1)| + |b(P_1)| = |a(P_2)| + |b(P_2)| = n - 1$. For
example, when $P_1 = \langle 1, 2, 3, 4, 5, 6 \rangle$ and $P_2 = \langle 3, 2, 4, 5, 1, 6 \rangle$, we have $a(P_1) =$
$\{23, 45\}$, $b(P_1) = \{12, 34, 56\}$, $a(P_2) = \{32, 45\}$, $b(P_2) = \{24, 51, 16\}$. Clearly,
$|a(P_1)| + |b(P_1)| = |a(P_2)| + |b(P_2)| = 5$. (Note that when P_1 or P_2 contains
duplicated letters, this does not hold anymore).

We define a *contig* as a string over a gene set Σ whose contents should not
be altered. A *scaffold* S is simply a sequence of contigs $\langle C_1, \ldots, C_m \rangle$. We define
$c(S) = c(C_1) \cup \cdots \cup c(C_m)$. Given a set of genes/letters X, let $S + X$ be the set
of all possible resulting permutations after filling all the elements in X into S
such that the contigs in S are not altered. In this case, the elements in X can
only be inserted before or after C_i's.

Now, we define the problem on permutation scaffold filling formally. As R is
a complete reference permutation, WLOG, we assume from now on that $\Sigma =$
$\{1, 2, \ldots, n\}$ and $R = [n]$ is the identity permutation.

Definition 1. *Permutation Scaffold Filling.*

Input: *A complete reference permutation $R = [n]$ and a scaffold $S =$
$\langle C_1, C_2, \ldots, C_m \rangle$ where R and the contig C_i's are over a gene set Σ, a set
$X = c(R) - c(S) \neq \emptyset$.*
Question: *Find $S^+ \in S + X$ such that $|a(S^+)|$, with respect to R, is maximized.*

3   A Polynomial Time Algorithm for Permutation Scaffold Filling

Before presenting our algorithm, we make the following definitions.

Let $\alpha(C_i), \beta(C_i)$ be the first and last letter of C_i respectively. (For conve-
nience, we use α_i, β_i from now on; and, α_i could be the same as β_i when $|C_i| = 1$).

Then $\langle \beta_i, \alpha_{i+1} \rangle$ (or simply $\beta_i \alpha_{i+1}$) constitutes a *slot* where missing genes can inserted between β_i and α_{i+1}. We write $\beta_0 = -\infty$ and $\alpha_{m+1} = +\infty$, where $\langle -\infty, \alpha_1 \rangle$ and $\langle \beta_m, +\infty \rangle$ are the leftmost and rightmost (open) slot respectively.

Define a type-1 (resp. type-2) substring s of length $\ell \geq 1$, over X, as one which can be inserted into the slot $\langle \beta_i, \alpha_{i+1} \rangle$, for some i, to increase the total number of adjacencies by $\ell + 1$ (resp. ℓ).

Note that if $\beta_i \alpha_{i+1}$ is already an adjacency with respect to R, then for a general R (with gene repetitions) it is possible that s is inserted in the slot to generate $|s| + 1$ adjacencies (while destroying the adjacency $\beta_i \alpha_{i+1}$). When R is a permutation, we will show that it is still possible that in an optimal solution an existing adjacency $\beta_i \alpha_{i+1}$ could be broken.

Similarly, define a type-3 substring s of length $\ell \geq 1$, over X, as one which can be inserted in the slot $\langle \beta_i, \alpha_{i+1} \rangle$, for some i, to increase the number of adjacencies by $\ell - 1$. Note that a type-3 substring can only form adjacencies internally, hence it does not matter where we insert s — provided that it does not destroy any existing adjacency. Due to the order of processing type-1, 2 and 3 substrings, a potential type-2 substring t could become type-3 since the corresponding slot for it might not be available by the time t is inserted.

We show an example as follows:

$$R = \langle 1, 2, 3, 4, 5, 6, 7, 8, 9, 10, 11, 12, 13 \rangle,$$

$$S = \langle \boxed{1,3}, \boxed{5,6,12}, \boxed{2}, \boxed{10,9,13} \rangle.$$

We have $\alpha_1 = 1, \beta_1 = 3, \alpha_2 = 5, \beta_2 = 12, \alpha_3 = 2, \beta_3 = 2, \alpha_4 = 10, \beta_4 = 13$. Then, $X = \{4, 7, 8, 11\}$ are missing from S; moreover, 4 is type-1, 11 is type-2, and $\langle 7, 8 \rangle$ is type-3.

We now proceed with the following lemma.

Lemma 1. *Let s be a type-1 substring to be inserted into S. There is an optimal solution which does not break s.*

Proof. Suppose that if s is inserted into the slot $\langle \beta_i, \alpha_{i+1} \rangle$ then $|s| + 1$ adjacencies can be obtained. WLOG, suppose that in some solution s is broken into s_1 and s_2; moreover, s_1 and s_2 are inserted at two different slots at least one of which is not $\langle \beta_i, \alpha_{i+1} \rangle$. Then, WLOG, inserting $s_1 t$ into $\langle \beta_i, \alpha_{i+1} \rangle$ and inserting s_2 into another slot in S can obtain at most $(|s_1| + |t|) + (|s_2| - 1) = |s| + |t| - 1$ adjacencies. Then, swap the inserted contents t in the slot $\langle \beta_i, \alpha_{i+1} \rangle$, which is after s_1, with s_2 (or its reversal) would give us at least $(|s| + 1) + (|t| - 1) = |s| + |t|$ adjacencies. This is because that t must be a type-2 or type-3 substring in the slot $\langle \beta_i, \alpha_{i+1} \rangle$, and s_2 must also be a type-2 or type-3 substring if not inserted in the same slot after s_1. The reason is that in R and any valid solution S' each gene appears exactly once. $\square$

This proof of lemma has a strong implication. If we color all letters in R that are also in X *red*, and focus on all *maximal* red substrings in R. Then we have the following corollary.

Corollary 1. *There is an optimal solution for PSF where all letters inserted into S are maximal (red) substrings (or their reversals) in R, or, all type-i substrings, $i = 1, 2, 3$, are maximal (red) substrings in R (or their reversals).*

Proof. The above lemma shows the case for type-1 substrings (and the proof works for type-2 and type-3 substrings). Obviously, an optimal solution can be obtained by processing type-i substrings in the order of $i = 1$, $i = 2$ and $i = 3$. □

We continue with more properties.

Lemma 2. *Let s be a type-1 substring for the slot $\langle \beta_i, \alpha_{i+1} \rangle$. Then, there is an optimal solution which inserts s into the slot $\langle \beta_i, \alpha_{i+1} \rangle$; moreover, no more gene is inserted into the slot after s is inserted.*

Proof. WLOG, suppose that, in some solution S', s has been inserted into the slot $\langle \beta_i, \alpha_{i+1} \rangle$ and later some substring t is inserted in the same slot. By Lemma 1, t could only be inserted right after β_i or right before α_{i+1}. Apparently, at this point t is either type-2 or type-3. Then we could move t out of the slot $\langle \beta_i, \alpha_{i+1} \rangle$, say inserting it at the leftmost or the rightmost open slot in S'. We hence obtain a new solution satisfying the conditions in this lemma, without decreasing the number of adjacencies. □

The above lemmas basically show that if s is a type-1 substring then we should insert it at the corresponding slot and lock that slot so that no more gene can be inserted there. The above lemma does not hold for the special case when $s = \emptyset$ and $\beta_i \alpha_{i+1}$ is already an adjacency. The reason is that we could insert two type-2 substrings, one after β_i and the other before α_{i+1}. This case needs to be specially handled. We show a simple example as follows. $R = \langle 1, 2, 3, 4, 5, 6, 7, 8 \rangle$, $S = \langle \boxed{8,6,2}, \boxed{3,7} \rangle$, and $X = \{1, 4, 5\}$. The optimal solution is to have a $S' = \langle \boxed{8,6,2}, 1, 5, 4, \boxed{3,7} \rangle$, resulting in 3 adjacencies. If the adjacency $\langle 2, 3 \rangle$ must be kept, we could only obtain a total of two adjacencies.

After all the type-1 substrings have been inserted and the corresponding slots locked, we will process all type-2 substrings. We have the following lemma.

Lemma 3. *After all type-1 substrings have been inserted and the corresponding slots locked, no new type-1 substring can formed by inserting type-2 substrings.*

Proof. WLOG, assume to the contrary that t_1 and t_2 are both type-2 and $t_1 t_2$ could be inserted into slot $\alpha_i \beta_{i+1}$ to obtain a type-1 substring. Then $t_1 t_2$ would have been a type-1 substring and would have been processed by Lemmas 1 and 2. □

With Lemma 3, we try to optimally insert the type-2 substrings as follows. First, assume that all the type-1 maximal *red* substrings in R have been inserted into S and the corresponding slots are locked. We then build a bipartite graph B. For each (unlocked) slot $\langle \beta_i, \alpha_{i+1} \rangle$, we create two black vertices $\beta_i \circ$ and $\circ \alpha_{i+1}$, representing the spots right after β_i and before α_{i+1} respectively. (For the leftmost and rightmost open slot, we would have only one black vertex.) Each remaining maximal *red* substring in R forms a blue vertex. There is an edge $\langle u, v \rangle$, where u, v are black and blue respectively, if the maximal red substring v in R can be inserted at the position u to obtain $|v|$ adjacencies. An example of the bipartite graph B is shown in Fig. 1, where we use letters to denote genes.

Lemma 4. *The maximum vertex degree of the bipartite graph B is four.*

Proof. It is easy to see that (1) B is bipartite, and (2) in each connected component of B the vertex degree is at most four. To see why each connected component in B has a maximum degree 4, note that a contig could have only one gene. Then, if we have two such singleton contigs $\boxed{i}$ and $\boxed{j}$, the substring $\langle i+1, i+2, \ldots, j-1 \rangle$ (or its reversal) could be inserted at 4 possible locations: $\circ i$, $i \circ$, $\circ j$ and $j \circ$. $\square$

With the above lemma, it is easy to obtain and insert the type-2 substrings: we simply compute the maximum matching of B, which can be done in $O(n^{2.5})$ time [4], and all the corresponding maximal red substrings inserted according to the matching are type-2. The remaining maximal red substrings are all type-3, and we could put them at the two open slots in an arbitrary order, as long as no existing adjacency is destroyed.

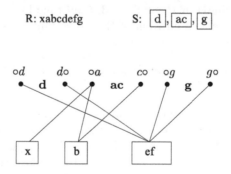

Fig. 1. A bipartite graph B for the type-2 substrings and the possible slots.

We now show the detailed algorithm. For convenience, when we say that a letter x is appended at the beginning (resp. end) of a contig C_i, we mean $C_i \leftarrow x \circ C_i$ (resp. $C_i \leftarrow C_i \circ x$).

Algorithm 1: *PSF(R, S)*

Input: $R = [n]$, $S = \langle C_1, \ldots, C_m \rangle$, $X = c(R) - c(S) \neq \emptyset$

1   In R, color all letters in X red. Identify all maximal red substrings in R.

2   Insert all (type-1) maximal red substrings at the right slots in between C_i and C_{i+1}. Lock these slots so that no more genes can be inserted.

3   For all slots $\langle \beta_i, \alpha_{i+1} \rangle$, if $\beta_i \alpha_{i+1} = <j, j+1>$ (resp. $\beta_i \alpha_{i+1} = <j+1, j>$) is already an adjacency and both $j-1, j+2 \in X$ then append $j-1$ at the end of C_i and append $j+2$ at the beginning of C_{i+1} (resp. append $j+2$ at the end of C_i and append $j-1$ at the beginning of C_{i+1}), update X and the maximal red substrings accordingly; if at most one of $j-1, j+2$ is in X then lock the slot $\langle \beta_i, \alpha_{i+1} \rangle$ so that no gene can be inserted.

4   Build a bipartite graph B for all of the remaining maximal red strings and the possible (unlocked) slots where they can be located and be type-2. For each connected component of B, use the standard method to compute the maximum matching and insert these (type-2) maximal red substrings accordingly.

5   Put all the remaining (type-3) maximal red substrings in the leftmost or rightmost slot, provided that no existing adjacency is destroyed.

The running time analysis can be done as follows. Step 1 takes $O(n^2)$ time. Without using any advanced data structure, we could scan the genes in R and S, color all the genes in R but not in S red, and then identify the correspondence between α_i, β_i in S, for $i = 1, \ldots, m$, with their copies in R. This can be done in $O(n^2)$ time. Then, naturally we could identify all maximal red substrings (in fact, all type-1 substrings) in R. Steps 2, 3, 5 all take linear time. Step 4 takes $O(n^{2.5})$ time due to the computation of the maximum matching [4]. Hence, we have the following theorem.

Theorem 1. *The Permutation Scaffold Filling problem can be solved in $O(n^{2.5})$ time.*

We comment that our algorithm in fact solves the dual version of Permutation Scaffold Filling, where the objective function is to minimize the number of breakpoints between R and S^+. Also, the bipartite graph we have used has a maximum degree of 4, so it might be possible to improve the running time for computing the maximum matching — hence improving the overall running time of the algorithm.

4   Concluding Remarks

In this paper, we show that the permutation scaffold filling problem is polynomially solvable. This answers partially an open question in [5]. We are currently investigating the corresponding NP-complete problem where the genomes and scaffolds could contain gene duplications. The current best approximation algorithm only has a factor of 2 [6].

Acknowledgments. This research is partially supported by NSF of China under project 61628207, by the Foundation for Outstanding Young Scientists in Shandong Province (project no. BS2014DX017), by the Doctoral Foundation of Shandong Jianzhu University, and by the China Scholarship Council. We also thank anonymous reviewers for several useful comments.

References

1. http://wgs-assembler.sourceforge.net/
2. Bulteau, L., Carrieri, A.P., Dondi, R.: Fixed-parameter algorithms for scaffold filling. Theoret. Comput. Sci. **568**, 72–83 (2015)
3. Chain, P.S., Grafham, D.V., Fulton, R.S., et al.: Genome project standards in a new era of sequencing. Science **326**, 236–237 (2009)
4. Hopcroft, J., Karp, R.: An $n^{5/2}$ algorithm for maximum matchings in bipartite graphs. SIAM J. Comput. **2**(4), 225–231 (1973)
5. Jiang, H., Zheng, C., Sankoff, D., Zhu, B.: Scaffold filling under the breakpoint, related distances. IEEE/ACM Trans. Comput. Biol. Bioinform. **9**(4), 1220–1229 (2012)
6. Jiang, H., Fan, C., Yang, B., Zhong, F., Zhu, D., Zhu, B.: Genomic scaffold filling revisited. In: Proceedings of the 27th Annual Symposium on Combinatorial Pattern Matching (CPM 2016), Tel Aviv, Israel, 27–29 June 2016, pp. 15:1–15:13 (2016)
7. Muñoz, A., Zheng, C., Zhu, Q., Albert, V., Rounsley, S., Sankoff, D.: Scaffold filling, contig fusion and gene order comparison. BMC Bioinform. **11**, 304 (2010)
8. Sturtevant, A.: A crossover reducer in Drosophila melanogaster due to inversion of a section of the third chromosome. Biol. Zent. Bl. **46**, 697–702 (1926)
9. Sturtevant, A., Dobzhansky, T.: Inversions in the third chromosome of wild races of drosophila pseudoobscura, and their use in the study of the history of the species. Proc. Nat. Acad. Sci. USA **22**, 448–450 (1936)
10. Watterson, G., Ewens, W., Hall, T., Morgan, A.: The chromosome inversion problem. J. Theor. Biol. **99**, 1–7 (1982)
11. Yancopoulos, S., Attie, O., Friedberg, R.: Efficient sorting of genomic permutations by translocation, inversion and block interchange. Bioinformatics **21**, 3340–3346 (2005)
12. Zhu, B.: Genomic scaffold filling: a progress report. In: Zhu, D., Bereg, S. (eds.) FAW 2016. LNCS, vol. 9711, pp. 8–16. Springer, Heidelberg (2016). doi:10.1007/978-3-319-39817-4_2

Author Index

Abu-Khzam, Faisal N.   477
Águeda, Raquel   241
Alves, Sancrey Rodrigues   423
Angulo, Gustavo   754
Aurora, Pawan   187
Awasthi, Abhishek   772

Bae, Sung Eun   326
Bai, Chunsong   718
Barcelo, Neal   521
Barton, Carl   547
Bazgan, Cristina   92
Bereg, Sergey   365
Bhore, Sujoy   254
Birmelé, Étienne   216
Brandes, Ulrik   175

Cai, Zhipeng   230
Chakraborty, Dibyayan   254
Chang, Jou-Ming   505
Chang, Shun-Chieh   505
Chateau, Annie   294, 463
Chen, Guihai   33
Chen, Lin   608
Chen, Xujin   128
Cheng, Dun-Wei   107
Cohen, Nathann   241

Dabrowski, Konrad K.   423
Dallard, Clément   294
Dang, Yaru   33
Das, Sandip   254
de Montgolfier, Fabien   216
der Heide, Friedhelm Meyer auf   563, 578
Diao, Zhuo   128
Ding, Wei   408
Dondi, Riccardo   113
dos Santos Souza, Uéverton   423
Dragan, Feodor F.   62
Drees, Maximilian   593
Du, Donglei   536
Du, Hongwei   702
Duan, Zhenhua   735
Dumba, Braulio   763

Elloumi, Sourour   726
Eto, Hiroshi   270

Fang, Qizhi   641
Faria, Luerbio   423
Feldkord, Björn   593
Feldotto, Matthias   655
Feng, Cuiying   350
Foucaud, Florent   92
Fujimoto, Noriyuki   624
Fujita, Shinya   241
Fukunaga, Takuro   77

Gao, Xiaofeng   33
Geraerts, Roland   311
Ghosh, Shamik   202
Giroudeau, Rodolphe   294, 463
Gregor, Petr   144
Gu, Qian-Ping   438

Hanawa, Yosuke   18
He, Xiaohan   702
Higashikawa, Yuya   18
Hillebrand, Arne   311
Holm, Eugenia   175
Hon, Wing-Kai   710
Hoogeveen, Han   311
Hsieh, Sun-Yuan   107
Hu, Xiaodong   128
Huang, Hejiao   702

Iacono, John   489
Ito, Takehiro   270

Jäger, Gerold   670
Jansen, Klaus   608
Jena, Monalisa   187

Källberg, Linus   744
Kamiyama, Naoyuki   18
Kang, Liying   693, 718
Karrenbauer, Andreas   175
Katoh, Naoki   18

Klein, Sulamita   423
Kling, Peter   521
Kloks, Ton   710
König, Jean-Claude   463
König, Jürgen   563
Kramer, Oliver   772

Lambert, Amélie   726
Larsson, Thomas   744
Lässig, Jörg   772
Leder, Lennart   655
Legay, Sylvain   241
Leitert, Arne   62
Li, Jianzhong   230, 453
Li, Shouwei   477
Li, Weian   641
Li, Yingshu   453
Liang, Jiajian Leo   438
Lin, Chin-Fu   285
Liu, Chang   547
Liu, Fu-Hong   710
Liu, Hsiang-Hsuan   710
Liu, Nan   782
Liu, Wenjing   641
Liu, Xianmin   230, 453
Liu, Zhilong   270
Lu, Xu   735
Luo, Wenchang   608

Mäcker, Alexander   563, 578
Maehara, Takanori   77
Malatyali, Manuel   578
Manoussakis, Yannis   241
Markarian, Christine   477
Matsui, Yasuko   241
Meyer auf der Heide, Friedhelm   477
Miao, Dongjing   230, 453
Miyano, Eiji   270
Montero, Leandro   241

Naserasr, Reza   241
Nugent, Michael   521

Otachi, Yota   241

Pai, Kung-Jui   505
Peng, Jigen   340

Peng, Sheng-Lung   285
Pisantechakool, Photchchara   3
Pissis, Solon P.   547
Planche, Léo   216
Podlipyan, Pavel   477
Polak, Ido   685
Pollet, Valentin   463
Pruhs, Kirk   521

Qiu, Ke   408

Rahman, Md. Saidur   393
Raman, Rajiv   187
Riechers, Sören   563, 578

Sakuma, Tadashi   241
Sau, Ignasi   423
Sen Gupta, Raibatak   202
Sen, M.K.   202
Sen, Sagnik   254
Shan, Erfang   693, 718
Shinn, Tong-Wook   326
Sikora, Florian   92, 113
Skopalik, Alexander   593, 655
Škrekovski, Riste   144
Song, Liang   702
Stephen, Tamon   378
Sultana, Shaheena   393

Takaoka, Tadao   326
Takizawa, Atsushi   18
Tan, Xuehou   3
Tang, Zhongzheng   128
Tian, Cong   735
Tian, Huamei   350
Tuza, Zsolt   241

van den Akker, Marjan   311
Vukašinović, Vida   144

Wang, Xinglong   33
Weise, Thomas   772
Weller, Mathias   294, 463
Whitesides, Sue   350
Wu, Chenchen   49, 536
Wu, Kui   350
Wu, Ro-Yu   505

Xu, Dachuan   49, 536
Xu, Renyu   241
Xu, Yi   340
Xu, Yinfeng   340
Xue, Yuan   159

Yagnatinsky, Mark   489
Yang, Boting   159
Ye, Tai-Ling   107
You, Xiaotian   33

Zasenko, Olga   378
Zeng, Xiaoli   702
Zhang, Guochuan   438, 608
Zhang, Peng   49
Zhang, Xinghe   49
Zhang, Zhi-Li   763
Zhong, Farong   159
Zhu, Binhai   782
Zilles, Sandra   159
Zou, Peng   782

Printed in the United States
By Bookmasters